GEOLOGICAL SURVEY OF CANADA
COMMISSION GÉOLOGIQUE DU CANADA

CATALOGUE OF TYPES AND FIGURED SPECIMENS OF FOSSIL PLANTS IN THE GEOLOGICAL SURVEY OF CANADA COLLECTIONS

Volume II

by
Thomas E. Bolton

1994

Don

Available in Canada through authorized
bookstore agents and other bookstores

or by mail from

Canada Communication Group - Publishing
Ottawa, Canada K1A 0S9

and from

Geological Survey of Canada offices:

601 Booth Street
Ottawa, Canada K1A 0E8

3303-33rd Street N.W.,
Calgary, Alberta T2L 2A7

A deposit copy of this publication is also available for reference
in public libraries across Canada

Cat. No. M41-4/2-1994E
ISBN 0-660-15404-8

Price subject to change without notice

CONTENTS

Page	
v	Introduction
1	Megaplants
1	Precambrian
5	Cambrian
7	Ordovician
10	Silurian
13	Devonian
16	Carboniferous-Permian
31	Triassic
33	Jurassic
33	Cretaceous
36	Tertiary-Recent
42	Palynomorphs
42	Pollen-Spore
42	Silurian
43	Devonian
109	Carboniferous-Permian
173	Triassic
178	Jurassic
189	Cretaceous
262	Tertiary-Holocene
297	Dinoflagellata
357	Acritarcha
357	Precambrian
362	Paleozoic
362	Cambrian
370	Ordovician
385	Silurian
390	Devonian
396	Carboniferous
397	Mesozoic
398	Chitinozoa
414	Bacteria
417	Diatoms
419	Plant Incertae Sedis

INTRODUCTION

The first comprehensive "Catalogue of types and figured specimens of fossil plants in the Geological Survey of Canada collections" was published by W.A. Bell in 1962. All primary and secondary megaplant and microplant types described to 1960 were included in that volume. This was followed by a "Catalogue of types and figured specimens of fossils plants in the Geological Survey of Canada collections (megaplant supplement 1963 - 67)" released by Bell in 1969. The present Volume II includes all megaplants and microplants added to the National Type Collection of Invertebrate and Plant Fossils between 1967 and 1991. Combined within this volume is data compiled under contract by W.A. Bell; an unedited manuscript draft was submitted in 1968, devoted to palynomorphs.

Volume II follows the same concept of types as adhered to by Bell and in the "Catalogue of type invertebrate fossils of the Geological Survey of Canada" compiled by T.E. Bolton (Volume I:1962, II:1965, III:1966, IV:1968, V:1974, VI:1977, VII:1982, VIII:1992), namely primary type categories Holotype, Paratype, Syntype, Lectotype, Neotype and secondary Hypotype and Figured specimen. All types are cited as objectively as possible. Two series of GSC type numbers have been followed: specimens described up to 1969 were recorded within a separate Plant catalogue series, after which plants and invertebrates were recorded within one catalogue.

The biostratigraphical application of these plants is most important, so that within each category of plant remains they are arranged alphabetically within the various geological periods wherever feasible. Listing within a general Pollen-Spore grouping or more specifically under Dinoflagellata-Diatoms depended upon the classification followed within the original publication, since exact assignment of many of the genera described is still fluctuating.

Collections of invertebrate, vertebrate and plant fossils constitute significant national resources; **all type fossil specimens should be deposited in an established repository**. *The tendency for individuals and organizations to deposit invertebrate and plant types with the Geological Survey of Canada has continued. Such donations make the specimens readily accessible under proper storage and curatorial supervision.*

INTRODUCTION

En 1962, W.A. Bell publiait la première version complète du document intitulé Catalogue of types and figured specimens of fossil plants in the Geological Survey of Canada collections, lequel réunit tous les types primaires et secondaires de plantes macroscopiques et microscopiques décrites jusqu'en 1960. En 1969, le même auteur publiait le Catalogue of types and figured specimens of fossil plants in the Geological Survey of Canada collections (megaplant supplement 1963-67). Le présent ouvrage, le volume II, comprend toutes les plantes macroscopiques et microscopiques qui ont été ajoutées à la Collection nationale de spécimens types d'invertébrés et de végétaux fossiles entre 1967 et 1991. On y trouve aussi des données compilées à forfait par Bell; un manuscrit inédit portant sur les palynomorphes a été remis en 1968.

Dans le volume II, on adopte la même classification des types que dans les documents de Bell et la compilation de T.E. Bolton intitulée Catalogue of invertebrate fossils of the Geological Survey of Canada (volumes I:1962, II:1965; III:1966; IV:1968; V:1974; VI:1977; VII:1982 et VIII:1992); il y a donc la catégorie primaire, qui regroupe les holotypes, les paratypes, les syntypes, les lectotypes et les néotypes, ainsi que la catégorie secondaire, qui comprend les hypotypes et les spécimens illustrés. On cite tous les types aussi objectivement que possible. On utilise deux séries de numéros de types de la CGC : les spécimens décrits en 1969 et avant sont inscrits dans une série distincte de catalogues des végétaux; les plantes et les invertébrés décrits après 1969 sont regroupés dans le même catalogue.

L'utilisation la plus importante de ces plantes se fait en biostratigraphie; par conséquent, dans la mesure du possible, pour chaque catégorie de restes de plantes, les spécimens sont classifiés alphabétiquement par époque géologique. Les inscriptions au sein d'un groupement général de pollens et de spores ou, plus précisément, de dinoflagellés et de diatomées, dépendent de la classification adoptée dans la publication d'origine, puisque l'attribution exacte de bon nombre des genres décrits est encore en voie de modification.

Les collections de fossiles d'invertébrés, de vertébrés et de végétaux sont des ressources nationales importantes; **tous les spécimens types de fossiles devraient donc être conservés dans un lieu établi, destiné à cette fin**. *La tendance des particuliers et des organismes de déposer les spécimens types d'invertébrés et de plantes à la Commission géologique du Canada se maintient. Ces dons facilitent l'accès à ces spécimens, en plus de permettre de les entreposer et de les conserver dans des conditions convenables.*

MEGAPLANTS

Precambrian

Archaeoellipsoides grandis Horodyski and Donaldson [Algae Incertae Sedis]
Holotype 57988; paratypes 57988-57990 [thin sections]
Horodyski, R.J. and Donaldson, J.A., 1980, Precambrian Research, vol. 11, p. 154, figs. 16A-H.
Horodyski, R.J. and Donaldson, J.A., 1983, J. Paleontol., vol. 57, no. 2, p. 274, figs. 5W (57988), X (57989).
Dismal Lakes Group, Precambrian, north shore unnamed lake 8 km west of northwest bay of Dismal Lakes, lat. 67°31.6′N, long. 117°46.4′W, District of Mackenzie.

Archaeotrichion sp. [Algae Incertae Sedis]
Fig. spec. 57903b
Hofmann, H.J. and Aitken, J.D., 1979, Can. J. Earth Sci., vol. 16, no. 1, p. 155, fig. 9D.
Little Dal Group, Precambrian, headwaters Gayna River, lat. 64°48′N, long. 129°47′W, Mackenzie Mountains, District of Mackenzie.

Beltina danai Walcott [Algae Incertae Sedis]
Hypotypes 57900, 57901
Hofmann, H.J. and Aitken, J.D., 1979, Can. J. Earth Sci., vol. 16, no. 1, p. 162, figs. 17A, B.
Little Dal Group, Precambrian, headwaters Gayna River, lat. 64°48′N, long. 129°47′W, Mackenzie Mountains, District of Mackenzie.

Beltina danai Walcott, 1899 [Algae Incertae Sedis]
Hypotypes 77205-77208
Hofmann, H.J., 1985, Palaeontology, vol. 28, pt. 2, p. 344, pl. 39, figs. 5-8.
Little Dal Group, Precambrian, lat. 64°48′N, long. 129°52.5′W, and lat. 64°37′N, long. 128°50′W (77206, 77208), Mackenzie Mountains, District of Mackenzie.

Biocatenoides? sp. [Algae Incertae Sedis]
Fig. spec. 57987
Horodyski, R.J. and Donaldson, J.A., 1980, J. Paleontol., vol. 57, no. 2, p. 275, fig. 5Y.
Dismal Lakes Group, Precambrian, north shore unnamed lake 8 km west of northwest bay of Dismal Lakes, lat. 67°31.6′N, long. 117°46.4′W, District of Mackenzie.

Biocatenoides? sp. [Algae Incertae Sedis]
Fig. specs. 57989
Horodyski, R.J. and Donaldson, J.A., 1980, J. Paleontol., vol. 11, p. 156, figs. 13L, M.
Dismal Lakes Group, Precambrian, north shore unnamed lake 8 km west of northwest bay of Dismal Lakes, lat. 67°31.6′N, long. 117°46.4′W, District of Mackenzie.

Chuaria circularis Walcott, 1899 [Algae Incertae Sedis]
Hypotypes 77177, 77179, 77180, 77185, 77189-77191, 77200, 77203
Hofmann, H.J., 1985, Palaeontology, vol. 28, pt. 2, p. 342, pl. 35, figs. 2, 4, 5-8; pl. 36, figs. 6, 12; pl. 37, fig. 1; pl. 38, fig. 4; pl. 39, fig. 2.
Little Dal Group, Precambrian, lat. 64°48′N, long. 129°52.5′W, and lat. 63°50′20″N, long. 127°33′55″W (77200), Mackenzie Mountains, District of Mackenzie.

Cluster of compressed spheroids [Algae Incertae Sedis]
Fig. spec. 66657 (thin section)
Horodyski, R.J. and Donaldson, J.A., 1983, J. Paleontol, vol. 57, no. 2, p. 275, fig. 6M.
Dismal Lakes Group, Precambrian, west of northwest bay of Dismal Lakes, lat. 67°26′N, long. 118°17.9′W, District of Mackenzie.

Cluster of spherical and elliptical envelopes [Algae Incertae Sedis]
Fig. spec. 66656 (thin section)
Horodyski, R.J. and Donaldson, J.A., 1983, J. Paleontol, vol. 57, no. 2, p. 275, fig. 6E.
Dismal Lakes Group, Precambrian, north shore unnamed lake 8 km west of northwest bay of Dismal Lakes, lat. 67°31.7′N, long. 117°44.8′W, District of Mackenzie.

Daltaenia mackenziensis Hofmann [Algae Incertae Sedis]
Holotype and paratypes 77203
Hofmann, H.J., 1985, Palaeontology, vol. 28, pt. 2, p. 346, pl. 39, figs. 1-3; text-fig. 6.
Little Dal Group, Precambrian, lat. 64°48′N, long. 129°52.5′W, Mackenzie Mountains, District of Mackenzie.

Elliptical envelope [Algae Incertae Sedis]
Fig. spec. 66656 (thin section)
Horodyski, R.J. and Donaldson, J.A., 1983, J. Paleontol., vol. 57, no. 2, p. 275, fig. 6F.
Dismal Lakes Group, Precambrian, north shore unnamed lake 8 km west of northwest bay of Dismal Lakes, lat. 67°31.7′N, long. 117°44.8′W, District of Mackenzie.

Eoentophysalis dismallakesensis Horodyski and Donaldson [Algae Incertae Sedis]
Holotype 57987; paratypes 57987-57990 (thin sections)
Horodyski, R.J. and Donaldson, J.A., 1980, Precambrian Research, vol. 11, p. 146, figs. 10A-D, 11A-F.
Horodyski, R.J. and Donaldson, J.A., 1983, J. Paleontol., vol. 57, no. 2, p. 274, figs. 5S (57987), T, V (57899), U (57988).
Dismal Lakes Group, Precambrian, north shore unnamed lake 8 km west of northwest bay of Dismal Lakes, lat. 67°31.6′N, long. 117°46.4′W, District of Mackenzie.

Eoentophysalis dismallakesensis(?) Horodyski and Donaldson [Algae Incertae Sedis]
Hypotype 57989

Horodyski, R.J. and Donaldson, J.A., 1980, Precambrian Research, vol. 11, p. 146, fig. 11G.
Dismal Lakes Group, Precambrian, north shore unnamed lake 8 km west of northwest bay of Dismal Lakes, lat. 67°31.6′N, long. 117°46.4′W, District of Mackenzie.

cf. *Eoentophysalis dismallakesensis* Horodyski and Donaldson [Algae Incertae Sedis]
Hypotype 66656 (thin section)
Horodyski, R.J. and Donaldson, J.A., 1983, J. Paleontol., vol. 57, no. 2, p. 275, fig. 6D.
Dismal Lakes Group, Precambrian, north shore unnamed lake 8 km west of northwest bay of Dismal Lakes, lat. 67°31.7′N, long. 117°44.8′W, District of Mackenzie.

Eomicrocoleus crassus Horodyski and Donaldson [Algae Incertae Sedis]
Holotype 57990
Horodyski, R.J. and Donaldson, J.A., 1980, Precambrian Research, vol. 11, p. 154, figs. 15A, B.
Dismal Lakes Group, Precambrian, north shore unnamed lake 8 km west of northwest bay of Dismal Lakes, lat. 67°31.6′N, long. 117°46.4′W, District of Mackenzie.

Filament compression [Algae Incertae Sedis]
Fig. spec. 61576
Hofmann, H.J., Hill, J. and King, A.F., 1979, Geol. Surv. Can., Paper 79-1B, p. 87, text-fig. 12.3.
Snows Pond Formation, Precambrian, east side of road 2.5 km southwest of North Harbour, lat. 47°08′N, long. 53°41′24″W, northeastern Newfoundland.

Filamentous microfossil [Algae Incertae Sedis]
Fig. spec. 57987
Horodyski, R.J. and Donaldson, J.A., 1980, J. Paleontol., vol. 57, no. 2, p. 275, fig. 5Z.
Dismal Lakes Group, Precambrian, north shore unnamed lake 8 km west of northwest bay of Dismal Lakes, lat. 67°31.6′N, long. 117°46.4′W, District of Mackenzie.

Filiconstrictosus sp. [Algae Incertae Sedis]
Fig. spec. 57990
Horodyski, R.J. and Donaldson, J.A., 1980, Precambrian Research, vol. 11, p. 152, fig. 13I.
Dismal Lakes Group, Precambrian, north shore unnamed lake 8 km. west of northwest bay of Dismal Lakes, lat. 67°31.6′N, long. 117°46.4′W, District of Mackenzie.

Grypania spiralis Walter, Oehler and Oehler, 1976 [Algae Incertae Sedis]
Hypotype 77204
Hofmann, H.J., 1985, Palaeontology, vol. 28, pt. 2, p. 349, pl. 39, fig. 4.
Little Dal Group, Precambrian, lat. 64°48′N, long. 129°52.5′W, Mackenzie Mountains, District of Mackenzie.

Incertae Sedis Sphaeromorphs [Algae Incertae Sedis]
Fig. specs. 64157, 64159
Horodyski, R.J., Donaldson, J.A. and Kerans, C., 1980, Can. J. Earth Sci., vol. 17, no. 9, p. 1170, figs. 5a, b.
Dismal Lakes Group, Precambrian, about 6 km northeast of northernmost point on Dismal Lakes, lat. 67°34′N, long. 117°26′20″W, District of Mackenzie.

Laminated, spherical envelop [Algae Incertae Sedis]
Fig. spec. 66656 (thin section)
Horodyski, R.J. and Donaldson, J.A., 1983, J. Paleontol., vol. 57, no. 2, p. 275, fig. 6C.
Dismal Lakes Group, Precambrian, north shore unnamed lake 8 km west of northwest bay of Dismal Lakes, lat. 67°31.7′N, long. 117°44.8′W, District of Mackenzie.

Longfengshania stipitata Du, 1982 [Algae Incertae Sedis]
Hypotype 77200
Hofmann, H.J., 1985, Palaeontology, vol. 28, pt. 2, p. 343, pl. 38, fig. 4.
Little Dal Group, Precambrian, lat. 63°50′20″N, long. 127°33′55″W, Mackenzie Mountains, District of Mackenzie.

'Molar-tooth' structure
Fig. spec. 77190
Hofmann, H.J., 1985, Paleontology, vol. 28, pt. 2, p. 337, pl 37, figs. 1, 3.
Little Dal Group, Precambrian, lat. 64°48′N, long. 129°52.5′W, District of Mackenzie.

Morania? *antiqua* Fenton and Fenton [Algae Incertae Sedis]
Hypotypes 57897, 57898, 57900, 57901
Hofmann, H.J. and Aitken, J.D., 1979, Can. J. Earth Sci., vol. 16, no. 1, p. 160, figs. 13I, J, 17A, B.
Little Dal Group, Precambrian, lat. 64°47′N, long. 129°40′W; lat. 64°37′N, long. 128°41′W (57898); and headwaters Gayna River, lat. 64°48′N, long. 129°47′W (57900, 57901), Mackenzie Mountains, District of Mackenzie.

Myxococcoides grandis Horodyski and Donaldson [Algae Incertae Sedis]
Holotype 57988; paratypes 57988-57990 (thin sections)
Horodyski, R.J. and Donaldson, J.A., 1980, Precambrian Research, vol. 11, p. 142, figs. 7A-N.
Horodyski, R.J. and Donaldson, J.A., 1983, J. Paleontol., vol. 57, no. 2, p. 274, figs. 5J-N (57987), O, P (57988).
Dismal Lakes Group, Precambrian, north shore unnamed lake 8 km west of northwest bay of Dismal Lakes, lat. 67°31.6′N, long. 117°46.4′W, District of Mackenzie.

Myxococcoides? sp. [Algae Incertae Sedis]
Fig. specs. 57987, 57989
Horodyski, R.J. and Donaldson, J.A., 1983, J. Paleontol., p. 274, figs. 5Q, R.
Dismal Lakes Group, Precambrian, north shore unnamed lake 8 km west of northwest bay of Dismal Lakes, lat. 67°31.6′N, long. 117°46.4′W, District of Mackenzie.

Myxococcoides? sp. [Algae Incertae Sedis]
Fig. spec. 57988
Horodyski, R.J. and Donaldson, J.A., 1980, Precambrian Research, vol. 11, p. 142, fig. 70.

Dismal Lakes Group, Precambrian, north shore unnamed lake 8 km west of northwest bay of Dismal Lakes, lat. 67°31.6′N, long. 117°46.4′W, District of Mackenzie.

Oscillatoriopsis curta Horodyski and Donaldson [Algae Incertae Sedis]
Holotype 57988; paratypes 57987, 57988, 57990 (thin sections)
Horodyski, R.J. and Donaldson, J.A., 1980, Precambrian Research, vol. 11, p. 149, figs. 13A-G.
Horodyski, R.J. and Donaldson, J.A., 1983, J. Paleontol., vol. 57, no. 2, p. 275, fig. 5AA (57988).
Dismal Lakes Group, Precambrian, north shore unnamed lake 8 km west of northwest bay of Dismal Lakes, lat. 67°31.6′N, long. 117°46.4′W, District of Mackenzie.

Oscillatoriopsis robusta Horodyski and Donaldson [Algae Incertae Sedis]
Holotype 57989
Horodyski, R.J. and Donaldson, J.A., 1980, Precambrian Research, vol. 11, p. 149, fig. 13H.
Dismal Lakes Group, Precambrian, north shore unnamed lake 8 km west of northwest bay of Dismal Lakes, lat. 67°31.6′N, long. 117°46.4′W, District of Mackenzie.

Oscillatoriopsis sp. [Algae Incertae Sedis]
Fig. spec. 57988
Horodyski, R.J. and Donaldson, J.A., 1980, Precambrian Research, vol. 11, p. 152, fig. 13J.
Dismal Lakes Group, Precambrian, north shore unnamed lake 8 km west of northwest bay of Dismal Lakes, lat. 67°31.6′N, long. 117°46.4′W, District of Mackenzie.

Proterozoic microfossils [Algae Incertae Sedis]
Hypotypes 66652-66654 (thin sections)
Horodyski, R.J. and Donaldson, J.A., 1983, J. Paleontol., vol. 57, no. 2, p. 275.
Dismal Lakes Group, Precambrian, north shore unnamed lake 8 km west of northwest bay of Dismal Lakes, lat. 67°31.6′N, long. 117°46.4′W, District of Mackenzie.

Sheath-like filament [Algae Incertae Sedis]
Fig. spec. 66655 (thin section)
Horodyski, R.J. and Donaldson, J.A., 1983, J. Paleontol., vol. 57, no. 2, p. 275, fig. 6J.
Dismal Lakes Group, Precambrian, north shore unnamed lake 8 km west of northwest bay of Dismal Lakes, lat. 67°31.7′N, long. 117°44.8′W, District of Mackenzie.

Siphonophycus beltensis Horodyski [Algae Incertae Sedis]
Hypotype 64159
Horodyski, R.J., Donaldson, J.A. and Kerans, C., 1980, Can. J. Earth Sci., vol. 17, no. 9, p. 1168, fig. 3a.
Dismal Lakes Group, Precambrian, about 6 km northeast of northernmost point on Dismal Lakes, lat. 67°34′N, long. 117°26′20″W, District of Mackenzie.

Siphonophycus crassiusculum Horodyski [Algae Incertae Sedis]
Hypotypes 64157, 64158
Horodyski, R.J., Donaldson, J.A. and Kerans, C., 1980, Can. J. Earth Sci., vol. 17, no. 9, p. 1170, figs. 3b-e.
Dismal Lakes Group, Precambrian, about 6 km northeast of northernmost point on Dismal Lakes, lat. 67°34′N, long. 117°26′20″W, District of Mackenzie.

Siphonophycus spp. [Algae Incertae Sedis]
Fig. specs. 57896b, 57903b
Hofmann, H.J. and Aitken, J.D., 1979, Can. J. Earth Sci., vol. 16, no. 1, p. 156, figs. 9E-H.
Little Dal Group, Precambrian, headwaters Gayna River, lat. 64°48′N, long. 129°47′W, Mackenzie Mountains, District of Mackenzie.

Siphonophycus spp. and *Archaeotrichion* spp. [Algae Incertae Sedis]
Fig. spec. 64158
Horodyski, R.J., Donaldson, J.A. and Kerans, C., 1980, Can. J. Earth Sci., vol. 17, no. 9, p. 1170, fig. 3f.
Dismal Lakes Group, Precambrian, about 6 km northeast of northernmost point on Dismal Lakes, lat. 67°34′N, long. 117°26′20″W, District of Mackenzie.

Sphaerophycus medium Horodyski and Donaldson [Algae Incertae Sedis]
Holotype 57989
Horodyski, R.J. and Donaldson, J.A., 1980, Precambrian Research, vol. 11, p. 140, fig. 5J.
Horodyski, R.J. and Donaldson, J.A., 1983, J. Paleontol., vol. 57, no. 2, p. 274, fig. 5C (57989).
Dismal Lakes Group, Precambrian, north shore unnamed lake 8 km west of northwest bay of Dismal Lakes, lat. 67°31.6′N, long. 117°46.4′W, District of Mackenzie.

Shaerophycus medium(?) Horodyski and Donaldson [Algae Incertae Sedis]
Hypotypes 57987-57989 (thin sections)
Horodyski, R.J. and Donaldson, J.A., 1980, Precambrian Research, vol. 11, p. 140, figs. 5K-P.
Horodyski, R.J. and Donaldson, J.A., 1983, J. Paleontol., vol. 57, no. 2, p. 274, figs. 5D-F, H (57987), G. (57988).
Dismal Lakes Group, Precambrian, north shore unnamed lake 8 km west of northwest bay of Dismal Lakes, lat. 67°31.6′N, long. 117°46.4′W, District of Mackenzie.

cf. *Sphaerophycus medium* Horodyski and Donaldson [Algae Incertae Sedis]
Fig. specs. 66656 (thin sections)
Horodyski, R.J. and Donaldson, J.A., 1983, J. Paleontol., vol. 57, no. 2, p. 275, figs. 6A, B.
Dismal Lakes Group, Precambrian, north shore unnamed lake 8 km west of northwest bay of Dismal Lakes, lat. 67°31.7′N, long. 117°44.8′W, District of Mackenzie.

Shaerophycus parvum Schopf [Algae Incertae Sedis]
Hypotypes 57987, 57988 (thin sections)
Horodyski, R.J. and Donaldson, J.A., 1980, Precambrian Research, vol. 11, p. 140, figs. 5A-D.
Horodyski, R.J. and Donaldson, J.A., 1983, J. Paleontol., vol. 57, no. 2, p. 274, figs. 5A (57988), B (57987).
Dismal Lakes Group, Precambrian, north shore unnamed lake 8 km west of northwest bay of Dismal Lakes, lat. 67°31.6′N, long. 117°46.4′W, District of Mackenzie.

Shaerophycus parvum(?) Schopf [Algae Incertae Sedis]
Hypotypes 57987, 57989 (thin sections)
Horodyski, R.J. and Donaldson, J.A., 1980, Precambrian Research, vol. 11, p. 140, figs. 5E-I.
Dismal Lakes Group, Precambrian, north shore unnamed lake 8 km west of northwest bay of Dismal Lakes, lat. 67°31.6′N, long. 117°46.4′W, District of Mackenzie.

Spherical envelope
Fig. specs. 66656, 66657 (thin sections)
Horodyski, R.J. and Donaldson, J.A., 1983, J. Paleontol., vol. 57, no. 2, p. 275, fig. 6G, K.
Dismal Lakes Group, Precambrian, north shore unnamed lake 8 km west of northwest bay of Dismal Lakes, lat. 67°31.7′N, long. 117°44.8′W, and west of northwest bay, lat. 67°26.1′N, long. 118°17.9′W, District of Mackenzie.

Shaerophycus? sp. [Algae Incertae Sedis]
Fig. spec. 57989
Horodyski, R.J. and Donaldson, J.A., 1980, Precambrian Research, vol. 11, p. 141, fig. 5Q.
Horodyski, R.J. and Donaldson, J.A., 1983, J. Paleontol., vol. 57, no. 2, p. 274, fig. 5I.
Dismal Lakes Group, Precambrian, north shore unnamed lake 8 km west of northwest bay of Dismal Lakes, lat. 67°31.6′N, long. 117°46.4′W, District of Mackenzie.

Spindle-shaped microfossils [Algae Incertae Sedis]
Fig. spec. 57990
Horodyski, R.J. and Donaldson, J.A., 1980, Precambrian Research, vol. 11, p. 157, fig. 13N.
Dismal Lakes Group, Precambrian, north shore unnamed lake 8 km west of northwest bay of Dismal Lakes, lat. 67°31.6′N, long. 117°46.4′W, District of Mackenzie.

Taeniatum spp. [Algae Incertae Sedis]
Fig. specs. 57896b
Hofmann, H.J. and Aitken, J.D., 1979, Can. J. Earth Sci., vol. 16, no. 1, p. 156, figs. 9A-C.
Little Dal Group, Precambrian, headwaters Gayna River, lat. 64°48′N, long. 129°47′W, Mackenzie Mountains, District of Mackenzie.

Taeniatum sp. [Algae Incertae Sedis]
Fig. specs. 61571-1 to 6, 61572, 61574-1 to 4, 61575-1, 2, 61580-1, 2, 61581, 61585, 61591 - 1 to 3, 61592, 61599, 61601, 61603, 61604-1 to 3.
Hofmann, H.J., Hill, J. and King, A.F., 1979, Geol. Surv. Can., Paper 79-1B, pl. 12.1, figs. 1-22, 24-29.
Precambrian, Random Formation, White Lake coastal exposure 750 m east-northeast of Smith Point, lat. 48°11′20″N, long. 53°50′18″W; Musgravetown Group, 1.5 km north of Melrose, lat. 48°29′30″N, long. 53°04′58″W (61572); Snows Pond Formation, 4 km southwest of Halls Town, 0.4 km northeast of The Pond that Feeds the Brook, lat. 47°31′12″N, long. 53°22′W (61574-1 to 4) and 4.8 km southwest of Halls Town, 1.5 km east of The Pond that Feeds the Brook, lat. 47°30′30″N, long. 53°22′17″W (61575-1, 2); Halls Town Formation, 2 km southwest of Harricott, lat. 47°09′30″N, long. 53°33′24″W (61580-1, 2, 61581); Cappahayden Formation, Burnt Pond, 1.2 km north of Cappahayden at end of road, lat. 46°51′45″N, long. 52°56′23″W (61585); Fermeuse Formation, west side Gripple Cove, 3.5 km southwest of Cape Race, lat. 46°38′30″N, long. 53°06′36″W (61591-1 to 3); Connecting Point Group, 0.5 km south of Cutwater Head, Canning's Cove, lat. 48°26′45″N, long. 53°50′40″W (61599), and 2 km west of Southern Bay Station on north side of Cabot Highway, Lethbridge, lat. 48°22′N, long. 53°42′W (61601); Drook Formation, road-cut on east side of Highway 5, ca. 1 km east-northeast of bridge, Peter's River, lat. 46°46′N, long. 53°35′45″W (61603); and Mall Bay Formation, ca. 220 m north of lighthouse, La Haye Point, lat. 46°54′22″N, long. 53°36′51″W (61604-1 to 3), southeastern Newfoundland.

Tawunia dalensis Hofmann [Algae Incertae Sedis]
Holotype 57893; paratypes 57890-57892, 57894-57897
Hofmann, H.J. and Aitken, J.D., 1979, Can. J. Earth Sci., vol. 16, no. 1, p. 158, figs. 13A-I.
Little Dal Group, Precambrian, headwaters Gayna River, lat. 64°48′N, long. 129°47′W; lat. 64°47′N, long. 129°40′W (57890, 57892, 57894, 57897); lat. 64°37′N, long. 128°41′W (57891); and lat. 64°58′N, long. 130°36′W (57895), Mackenzie Mountains, District of Mackenzie.

Tawunia dalensis Hofmann [Algae Incertae Sedis]
Hypotypes 77176-77178, 77183, 77184, 77186-77188, 77190,a-77199
Hofmann, H.J., 1985, Palaeontology, vol. 28, pt. 2, p. 334, pl. 35, figs. 1-35; pl. 36, figs. 1-5, 7-11; pl. 37, figs. 1, 2, 4-7; pl. 38, figs. 1-3; text-fig. 4.
Little Dal Group, Precambrian, lat. 64°48′N, long. 129°52.5′W, lat. 64°06′25″N, long. 128°27′30″W (77192), and lat. 63°50′20″N, long. 127°33′55″W (77193, 77198, 77199), Mackenzie Mountains, District of Mackenzie.

Thin filaments [Algae Incertae Sedis]
Fig. specs. 57989, 57990
Horodyski, R.J. and Donaldson, J.A., 1980, Precambrian Research, vol. 11, p. 157, figs. 13K, O.
Dismal Lakes Group, Precambrian, north shore unnamed lake 8 km west of northwest bay of Dismal Lakes, lat. 67°31.6′N, long. 117°46.4′W, District of Mackenzie.

Tyrasotaenia sp. [Algae Incertae Sedis]
Fig. specs. 77182, 77209-77211
Hofmann, H.J., 1985, Palaeontology, vol. 28, pt. 2, p. 349, pl. 35, fig. 10; pl. 39, figs. 9-11.
Little Dal Group, Precambrian, lat. 64°48′N, long. 129°52.5′W, and lat. 64°37′N, long. 128°50′W (77209), Mackenzie Mountains, District of Mackenzie.

Tyrasotaenia? sp. [Algae Incertae Sedis]
Fig. specs. 77179-77181, 77184, 77189, 77190,b.
Hofmann, H.J., 1985, Palaeontology, vol. 28, pt. 2, p. 349, pl. 35, figs. 4, 8, 9; pl. 36, figs. 5, 12; pl. 37, figs. 1, 8.
Little Dal Group, Precambrian, lat. 64°48′N, long. 129°52.5′W, Mackenzie Mountains, District of Mackenzie.

Vendotaenia? sp. [Algae]
Fig. spec. 83050
Narbonne, G.S. and Hofmann, H.J., 1987, Palaeontology, vol. 30, pt. 4, p. 671, text-fig. 10g.
Siltstone unit 2, Precambrian, south of Corn Creek, lat. 64°31′45″N, long. 132°54′45″, eastern Wernecke Mountains, Yukon.

3.5 µm, 1.5 µu, 2 µm wide filaments [Algae Incertae Sedis]
Fig. specs. 66656, 66657 (thin sections)
Horodyski, R.J. and Donaldson, J.A., 1983, J. Paleontol., vol. 57, no. 2, p. 275, figs. 6H, I, L.
Dismal Lakes Group, Precambrian, north shore unnamed lake 8 km west of northwest bay of Dismal Lakes, lat. 67°31.7′N, long. 117°44.8′W, and west of northwest bay, lat. 67°26.1′N, long. 118°17.9′W, District of Mackenzie.

Cambrian

Archeomyces dimakeloides Allison [Fungus]
Holotype 81907
Allison, C.W. and Awrimak, S.M., 1989, Precambrian Res., vol. 43, no. 4, p. 291, figs. 12.1-12.4.
Tindir Group, earliest Cambrian or latest Precambrian, headwaters of Tindir Creek between lat. 65°15′N and 65°17′N, long. 141°56′W and 141°57′W, Yukon.

Batinevia bajankolica Korde, 1973 [Algae]
Hypotype 90283
Voronova, L.G. et al., 1987, Acad. Nauk SSSR, Trans. Palaeontol. Instit., vol. 224, p. 60, pl. 27, fig. 5.
Sekwi Formation, Lower Cambrian, Ekwi River, Mackenzie Mountains, lat. 63°28 1/2′-29′N, long. 129°11 1/2′-13′W, District of Mackenzie.

Batinevia ramosa Korde, 1966 [Algae]
Hypotype 90281
Voronova, L.G. et al., 1987, Acad. Nauk SSSR, Trans. Palaeontol. Instit., vol. 224, p. 59, pl. 27, fig. 3.
Sekwi Formation, Lower Cambrian, Ekwi River, Mackenzie Mountains, lat. 63°28 1/2′-29′N, long. 129°11 1/2′-13′W, District of Mackenzie.

Bija canadensis Voronova et Drosdova [Algae]
Holotype 90285
Voronova, L.G. et al., 1987, Acad. Nauk SSSR, Trans. Palaeontol. Instit., vol. 224, p. 61, pl. 28, fig. 1.
Sekwi Formation, Lower Cambrian, Ekwi River, Mackenzie Mountains, lat. 63°28 1/2′-29′N, long. 129°11 1/2′-13′W, District of Mackenzie.

Botomaella aequalis Voronova et Drosdova [Algae Incertae Sedis]
Holotype 90293
Voronova, L.G. et al., 1987, Acad. Nauk SSSR, Trans. Palaeontol. Instit., vol. 224, p. 66, pl. 30, figs. 1, 4.
Sekwi Formation, Lower Cambrian, Ekwi River, Mackenzie Mountains, lat. 63°28 1/2′-29′N, long. 129°11 1/2′-13′W, District of Mackenzie.

Botomaella dubia Voronova et Drosdova [Algae Incertae Sedis]
Holotype 90295
Voronova, L.G. et al., 1987, Acad. Nauk SSSR, Trans. Palaeontol. Instit., vol. 224, p. 66, pl. 30, fig. 2; pl. 31, fig. 1.
Sekwi Formation, Lower Cambrian, Ekwi River, Mackenzie Mountains, lat. 63°28 1/2′-29′N, long. 129°11 1/2′-13′W, District of Mackenzie.

Botomaella crassa Voronova et Drosdova [Algae Incertae Sedis]
Holotype 90302
Voronova, L.G. et al., 1987, Acad. Nauk SSSR, Trans. Palaeontol. Instit., vol. 224, p. 67, pl. 32, fig. 4.
Sekwi Formation, Lower Cambrian, Ekwi River, Mackenzie Mountains, lat. 63°28 1/2′-29′N, long. 129°11 1/2′-13′W, District of Mackenzie.

Botomaella mitis Voronova, 1969 [Algae Incertae Sedis]
Hypotype 90299
Voronova, L.G. et al., 1987, Acad. Nauk SSSR, Trans. Palaeontol. Instit., vol. 224, p. 67, pl. 32, fig. 1.
Sekwi Formation, Lower Cambrian, Ekwi River, Mackenzie Mountains, lat. 63°28 1/2′-29′N, long. 129°11 1/2′-13′W, District of Mackenzie.

Canadiophycus fibrosus Voronova et Drosdova [Algae]
Holotype 90287
Voronova, L.G. et al., 1987, Acad. Nauk SSSR, Trans. Palaeontol. Instit., vol. 224, p. 65, pl. 28, fig. 3; pl. 29, fig. 4.
Sekwi Formation, Lower Cambrian, Ekwi River, Mackenzie Mountains, lat. 63°28 1/2′-29′N, long. 129°11 1/2′-13′W, District of Mackenzie.

Eophycomyces herkoides Allison [Fungus]
Holotype 81906
Allison, C.W. and Awrimak, S.M., 1989, Precambrian Res., vol. 43, no. 4, p. 290, figs. 12.5-12.7.
Tindir Group, earliest Cambrian or latest Precambrian, headwaters of Tindir Creek between lat. 65°15′N and 65°17′N, long. 141°56′W and 141°57′W, Yukon.

Epiphyton furcatum Korde, 1961 [Algae]
Hypotype 90282
Voronova, L.G. et al., 1987, Acad. Nauk SSSR, Trans. Palaeontol. Instit., vol. 224, p. 63, pl. 27, fig. 4.
Sekwi Formation, Lower Cambrian, lat. 63°26 1/3′-2/3′N, long. 129°26′-27′W, Mackenzie Mountains, District of Mackenzie.

Epiphyton racemosum Korde, 1961 [Algae]
Hypotype 90288
Voronova, L.G. et al., 1987, Acad. Nauk SSSR, Trans. Palaeontol. Instit., vol. 224, p. 62, pl. 28, fig. 4.

Sekwi Formation, Lower Cambrian, Ekwi River, Mackenzie Mountains, lat. 63°28 1/2′-29′N, long. 129°11 1/2′-13′W, District of Mackenzie.

Epiphyton spissum Korde, 1961 [Algae]
Hypotype 90289
Voronova, L.G. et al., 1987, Acad. Nauk SSSR, Trans. Palaeontol. Instit., vol. 224, p. 62, pl. 28, fig. 5.
Sekwi Formation, Lower Cambrian, Ekwi River, Mackenzie Mountains, lat. 63°28 1/2′-29′N, long. 129°11 1/2′-13′W, District of Mackenzie.

Epiphyton sp. [Algae]
Fig. specs. 25400, 25401
Handfield, R.C., 1971, Geol. Surv. Can., Bull. 201, pl. 1, figs. 2, 3.
Lower Cambrian, lat. 61°39′N, long. 128°11′W, Yukon; Atan Group, Lower Cambrian, lat. 59°16′N, long. 129°11′W, Good Hope Lake, British Columbia.

Girvanella problematica Nicholson et Etheridge, 1878 [Algae]
Hypotype 90291
Voronova, L.G. et al., 1987, Acad. Nauk SSSR, Trans. Palaeontol. Instit., vol. 224, p. 68, pl. 29, fig. 2.
Sekwi Formation, Lower Cambrian, Ekwi River, Mackenzie Mountains, lat. 63°28 1/2′-29′N, long. 129°11 1/2′-13′W, District of Mackenzie.

Girvanella sibirica Maslov, 1956 [Algae]
Hypotype 90296
Voronova, L.G. et al., 1987, Acad. Nauk SSSR, Trans. Palaeontol. Instit., vol. 224, p. 68, pl. 31, fig. 2.
Sekwi Formation, Lower Cambrian, Ekwi River, Mackenzie Mountains, lat. 63°28 1/2′-29′N, long. 129°11 1/2′-13′W, District of Mackenzie.

Globulus gregalis Voronova et Drosdova [Algae]
Holotype 90286
Voronova, L.G. et al., 1987, Acad. Nauk SSSR, Trans. Palaeontol. Instit., vol. 224, p. 70, pl. 28, fig. 2.
Sekwi Formation, Lower Cambrian, lat. 63°26 1/3′-2/3′N, long. 129°26′-27′W, Mackenzie Mountains, District of Mackenzie.

Gordonophyton parvulum Voronova et Drosdova, 1983 [Algae]
Hypotype 90297
Voronova, L.G. et al., 1987, Acad. Nauk SSSR, Trans. Palaeontol. Instit., vol. 224, p. 63, pl. 31, fig. 3.
Sekwi Formation, Lower Cambrian, Ekwi River, Mackenzie Mountains, lat. 63°28 1/2′-29′N, long. 129°11 1/2′-13′W, District of Mackenzie.

Hedstromia flabellata (Voronova) Voronova et Drosdova [Algae]
Hypotype 90287
Voronova, L.G. et al., 1987, Acad. Nauk SSSR, Trans. Palaeontol. Instit., vol. 224, p. 64, pl. 29, fig. 1.
Sekwi Formation, Lower Cambrian, lat. 63°26 1/3′-2/3′N, long. 129°26′-27′W, Mackenzie Mountains, District of Mackenzie.

Hedstromia series Voronova et Drosdova [Algae]
Holotype 90292
Voronova, L.G. et al., 1987, Acad. Nauk SSSR, Trans. Palaeontol. Instit., vol. 224, p. 65, pl. 29, fig. 3.
Sekwi Formation, Lower Cambrian, lat. 63°26 1/3′-2/3′N, long. 129°26′-27′W, Mackenzie Mountains, District of Mackenzie.

Kordephyton crinitum (Korde) Radugin et Stepanova, 1964 [Algae]
Hypotype 90300
Voronova, L.G. et al., 1987, Acad. Nauk SSSR, Trans. Palaeontol. Instit., vol. 224, p. 64, pl. 32, fig. 2.
Sekwi Formation, Lower Cambrian, Ekwi River, Mackenzie Mountains, lat. 63°28 1/2′-29′N, long. 129°11 1/2′-13′W, District of Mackenzie.

Mackenziephycus medullaris Voronova et Drosdova [Algae]
Holotype 90284
Voronova, L.G. et al., 1987, Acad. Nauk SSSR, Trans. Palaeontol. Instit., vol. 224, p. 71, pl. 27, fig. 7.
Sekwi Formation, Lower Cambrian, Ekwi River, Mackenzie Mountains, lat. 63°28 1/2′-29′N, long. 129°11 1/2′-13′W, District of Mackenzie.

Obruchevella delicata Reitlinger, 1948 [Algae]
Hypotype 90276
Voronova, L.G. et al., 1987, Acad. Nauk SSSR, Trans. Palaeontol. Instit., vol. 224, p. 58, pl. 26, fig. 3.
Sekwi Formation, Lower Cambrian, Ekwi River, Mackenzie Mountains, lat. 63°28 1/2′-29′N, long. 129°11 1/2′-13′W, District of Mackenzie.

Proaulopora extincta Korde, 1973 [Algae]
Hypotype 90301
Voronova, L.G. et al., 1987, Acad. Nauk SSSR, Trans. Palaeontol. Instit., vol. 224, p. 69, pl. 32, fig. 3.
Sekwi Formation, Lower Cambrian, Ekwi River, Mackenzie Mountains, lat. 63°28 1/2′-29′N, long. 129°11 1/2′-13′W, District of Mackenzie.

Proaulopora glabra Krasnopeeva, 1937 [Algae]
Hypotype 90294
Voronova, L.G. et al., 1987, Acad. Nauk SSSR, Trans. Palaeontol. Instit., vol. 224, p. 69, pl. 30, fig. 3.
Sekwi Formation, Lower Cambrian, Ekwi River, Mackenzie Mountains, lat. 63°28 1/2′-29′N, long. 129°11 1/2′-13′W, District of Mackenzie.

Razumovskia hispida Korde, 1973 [Algae]
Hypotype 90298
Voronova, L.G. et al., 1987, Acad. Nauk SSSR, Trans. Palaeontol. Instit., vol. 224, p. 68, pl. 31, fig. 4.
Sekwi Formation, Lower Cambrian, lat. 63°26 1/3′-2/3′N, long. 129°26′-27′W, Mackenzie Mountains, District of Mackenzie.

Renalcis chabakoviformis Voronova, 1973 [Algae]
Hypotype 90279
Voronova, L.G. et al., 1987, Acad. Nauk SSSR, Trans. Palaeontol. Instit., vol. 224, p. 57, pl. 26, fig. 6.

Sekwi Formation, Lower Cambrian, Ekwi River, Mackenzie Mountains, lat. 63°28 1/2′-29′N, long. 129°11 1/2′-13′W, District of Mackenzie.

Renalcis granosus Vologdin, 1932 [Algae]
Hypotype 90274
Voronova, L.G. et al., 1987, Acad. Nauk SSSR, Trans. Palaeontol. Instit., vol. 224, p. 56, pl. 26, fig. 1.
Sekwi Formation, Lower Cambrian, Ekwi River, Mackenzie Mountains, lat. 63°28 1/2′-29′N, long. 129°11 1/2′-13′W, District of Mackenzie.

Renalcis nodularis Voronova et Drosdova [Algae]
Holotype 90275
Voronova, L.G. et al., 1987, Acad. Nauk SSSR, Trans. Palaeontol. Instit., vol. 224, p. 58, pl. 26, fig. 2.
Sekwi Formation, Lower Cambrian, Ekwi River, Mackenzie Mountains, lat. 63°28 1/2′-29′N, long. 129°11 1/2′-13′W, District of Mackenzie.

Renalcis novus Voronova, 1976 [Algae]
Hypotype 90278
Voronova, L.G. et al., 1987, Acad. Nauk SSSR, Trans. Palaeontol. Instit., vol. 224, p. 57, pl. 26, fig. 5.
Sekwi Formation, Lower Cambrian, Ekwi River, Mackenzie Mountains, lat. 63°28 1/2′-29′N, long. 129°11 1/2′-13′W, District of Mackenzie.

Renalcis sp. [Algae]
Fig. spec. 25402
Handfield, R.C., 1971, Geol. Surv. Can., Bull. 201, pl. 1, fig. 4.
Sekwi Formation, Lower Cambrian, lat. 63°31′N, long. 129°34′W, Northwest Territories.

?*Siphonophycus* sp. [Algae]
Fig. spec. 75749
Baudet, D., Aitken, J.D. and Vanguestaine, M., 1989, Can. J. Earth Sci., vol. 26, no. 1, p. 143, pl. 2, fig. 17.
Vampire Formation, Early Cambrian, June Lake, lat. 63°29′30″N, long. 128°40′15″-41′30″W, Sekwi Mountain, Northwest Territories.

Solenopora compacta (Billings) Rothplentz (1908) [Algae]
Hypotype 90280
Voronova, L.G. et al., 1987, Acad. Nauk SSSR, Trans. Palaeontol. Instit., vol. 224, p. 61, pl. 27, figs. 1, 6.
Sekwi Formation, Lower Cambrian, Ekwi River, Mackenzie Mountains, lat. 63°28 1/2′-29′N, long. 129°11 1/2′-13′W, District of Mackenzie.

Tasmanites sp. [Algae Incertae Sedis]
Fig. specs. 75750, 75751
Baudet, D., Aitken, J.D. and Vanguestaine, M., 1989, Can. J. Earth Sci., vol. 26, no. 1, p. 144, pl. 2, figs. 18, 19.
Vampire Formation, Early Cambrian, June Lake, lat. 63°29′30″N, long. 128°40′15″-41′30″W, Sekwi Mountain, Northwest Territories.

Ordovician

Algal and fungal microborings
Fig. specs. 78250, 78251, 78239
Fillion, D. and Pickerill, R.K., 1990, Palaeontographics Canadiana No. 7, p. 63, pl. 18, figs. 2-10.
Dominion Formation, Bell Island Group, Lower Ordovician, Gull Island South Head, Bell Island, Conception Bay, eastern Newfoundland.

Algal borings
Fig. specs. 67754, 67715
Elias, R.J., 1982, Can. J. Earth Sci., vol. 19, no. 8, figs. 4h, i.
Stony Mountain Formation, Upper Ordovician, northwest pit in City of Winnipeg quarry, S.E. 1/4, S.W. 1/4, sec. 14, tp. 13N, rge. 2, E. Prin. mer., Stony Mountain, Manitoba.

Amia hexagona Fry
Holotype 13246; paratypes 13245, 13247-13249 (Plant catalogue)
Fry, W.L., 1983, Rev. Palaeobot. Palynol., vol. 39, p. 331, pl. VIII, figs. 1, 2, 4.
Cat Head Member, Red River Formation, Late Ordovician, between McBeth Point and Cat Head, Lake Winnipeg, Manitoba.

Chondrites (Bythotrephis) cuneatus Whiteaves
=*Winnipegia cuneata*, Fry, W.L., 1983, Rev. Palaeobot. Palynol., vol. 39, p. 320, pl. III, fig. 3 (holotype, invertebrate cat. no. 4383).

Chondrites cupressinus Whiteaves
=*Dowlingia cupressina*, Fry, W.L., 1983, Rev. Palaeobot. Palynol., vol. 39, p. 327, pl. IX, fig. 1 (holotype, invertebrate cat. no. 6851).

Chondrites gracillimus Whiteaves
=*Dowlingia gracillima*, Fry, W.L., 1983, Rev. Palaeobot. Palynol., vol. 39, p. 327, pl. IX, fig. 3 (holotype, invertebrate cat. no. 6852).

Chondrites patulus Whiteaves
=*Manitobia patula*, Fry, W.L., 1983, Rev. Palaeobot. Palynol., vol. 39, p. 323, pl. III, fig. 1 (paratype, invertebrate cat. no. 4386); pl. IV, figs. 1 (paratype, invertebrate cat. no. 4385), 3 (holotype, invertebrate cat. no. 4384); pl. VI, fig. 6 (hypotype, invertebrate cat. no. 1894g).

Dimorphosiphon rectangulare Høeg [Algae]
Hypotype 45549
Guilbault, J.P. and Mamet, B.L., 1976, Can. J. Earth Sci., vol. 13, no. 5, p. 644, pl. 3, fig. 3.
Ouareau Formation, Middle Ordovician, Ouareau River near Crabtree village, Joliette county, Québec.

Dimorphosiphon sp. [Algae]
Fig. specs. 85355
Elias, R.J., Nowlan, G.S. and Bolton, T.B., 1988, New Mexico Bur. Mines Mineral Res., Mem. 44, p. 343, pl. 1, fig. 3.
Fort Garry Member, Red River Formation, Upper Ordovician, Mowat Farm Quarry (Mulder Bros. Pit No. 12), about 20 km north of Winnipeg, sec. 27, tp. 13N, rge. 3, E. Prin. mer., Manitoba.

Dimorphosiphonoides hofmanni Guilbault and Mamet [Algae]
Holotype 45552; paratypes 45550, 45551, 45553-45555
Guilbault, J.P. and Mamet, B.L., 1976, Can. J. Earth Sci., vol. 13, no. 5, p. 646, pl. 4, figs. 1-6.
Lowville Formation, Middle Ordovician, Ouareau River near Crabtree village, Joliette county, and Saint-Vincent-de-Paul quarry, lat. 45°36′45″N, long. 75°39′30″W, Laval (45550), Québec; Stewart quarry, Rockland (45553-45555), Ontario.

Dimorphosiphonoides lespérancei Guilbault and Mamet [Algae]
Holotype 45546; paratypes 45545, 45547, 45548, 45556, 45557
Guilbault, J.P. and Mamet, B.L., 1976, Can. J. Earth Sci., vol. 13, no. 5, p. 645, pl. 3, figs. 1, 2, 4, 5; pl. 4, figs. 7, 8.
Lowville Formation, Middle Ordovician, Stewart quarry, Rockland, Ontario, and Saint-Vincent-de-Paul quarry, lat. 45°36′45″N, long. 75°39′30″W, Laval (45548, 45556, 45557), Québec.

Dimorphosiphonoides sp. [Algae]
Fig. specs. 45558, 45559
Guilbault, J.P. and Mamet, B.L., 1976, Can. J. Earth Sci., vol. 13, no. 5, p. 646, pl. 4, figs. 9, 10.
Lowville Formation, Middle Ordovician, Stewart quarry, Rockland, Ontario, and Ouareau River near Crabtree village, Joliette county, Québec.

Dowlingia penna Fry
Holotype 13259 (Plant catalogue)
Fry, W.L., 1983, Rev. Palaeobot. Palynol., vol. 39, p. 329, pl. VII, fig. 2.
Cat Head Member, Red River Formation, Late Ordovician, between McBeth Point and Cat Head, Lake Winnipeg, Manitoba.

Enfieldia lata Fry
Holotype 13230 (Plant catalogue)
Fry, W.L., 1983, Rev. Palaeobot. Palynol., vol. 39, p. 336, pl. IX, fig. 2.
Cat Head Member, Red River Formation, Late Ordovician, between McBeth Point and Cat Head, Lake Winnipeg, Manitoba.

Fisherites occidentalis (Salter) [Algae]
Hypotypes 90717-90720
Fisher, D.C. and Nitecki, M.W., 1982, Paleontological Soc., Mem. 13, p. 15, pl. 1, fig. 5; text-figs. 10A-D, 11A-C.
Middle Ordovician, Paquette Rapids, Ottawa River, Ontario-Québec.

Garwoodia gregaria (Nicholson) [Algae]
Hypotypes 45537, 45538
Guilbault, J.P. and Mamet, B.L., 1976, Can. J. Earth Sci., vol. 13, no. 5, p. 642, pl. 2, figs. 2, 3.
Leray Formation, Middle Ordovician, Ouareau River near Crabtree village, Joliette county, Québec.

Garwoodia sp. cf. *G. primitiva* Johnson [Algae]
Fig. spec. 45536
Guilbault, J.P. and Mamet, B.L., 1976, Can. J. Earth Sci., vol. 13, no. 5, p. 642, pl. 2, fig. 1.
Leray Formation, Middle Ordovician, Ouareau River near Crabtree village, Joliette county, Québec.

Gloecapsomorpha prisca Zalessky [Algae Incertae Sedis]
Hypotypes 26259-26262
McGregor, D.C. and Cramer, F.H., 1971, Geol. Surv. Canada, Bull. 202, p. 2, pl. 1, figs. 1-4.
Cat Head Member, Red River Formation, Ordovician, between McBeth Point and Cat Head, Lake Winnipeg, Manitoba.

Gloecapsomorpha prisca Zalessky [Algae Incertae Sedis]
Hypotypes 26263, 26264
McGregor, D.C. and Cramer, F.H., 1971, Geol. Surv. Canada, Bull. 202, p. 2, pl. 1, figs. 5, 6.
Ordovician, Estonia.

Hedstroemia bifilosa Rothpletz [Algae]
Hypotype 45539
Guilbault, J.P. and Mamet, B.L., 1976, Can. J. Earth Sci., vol. 13, no. 5, p. 644, pl. 2, fig. 4.
Black River Group, Middle Ordovician, Saint-Dominique, lat. 45°34′28″N, long. 75°52′15″W, Québec.

Hedstroemia halimedoidea Rothpletz [Algae]
Hypotypes 45540-45544
Guilbault, J.P. and Mamet, B.L., 1976, Can. J. Earth Sci., vol. 13, no. 5, p. 644, pl. 2, figs. 5-9.
Middle Ordovician, Lowville Formation, Ouareau River near Crabtree village, Joliette county, Québec, and Stewart quarry, Rockland, Ontario (45544); Black River Group, Saint-Dominique, lat. 45°34′28″N, long. 75°52′15″W (45541, 45542), Québec.

Inmostia reticulata Fry
Holotype 13238 (Plant catalogue)
Fry, W.L., 1983, Rev. Palaeobot. Palynol., vol. 39, p. 333, pl. VIII, fig. 5.
Cat Head Member, Red River Formation, Late Ordovician, between McBeth Point and Cat Head, Lake Winnipeg, Manitoba.

Kazakhstaniella(?) sp. [Algae]
Fig. spec. 53553
Bolton, T.E., 1980, Geol. Surv. Can., Paper 80-1C, pl. 2.5, fig. 4.
White Head Formation, Upper Ordovician, Grande Coupe, Percé, Gaspé Peninsula, Québec.

Kinwowia articulata Fry
Holotype 13256; paratype 13257,a (Plant catalogue)
Fry, W.L., 1983, Rev. Palaeobot. Palynol., vol. 39, p. 325, pl. VII, figs. 1, 3.
Cat Head Member, Red River Formation, Late Ordovician, between McBeth Point and Cat Head, Lake Winnipeg, Manitoba.

Leiosphaeridia spp. [Algae Incertae Sedis]
Fig. specs. 26273-26275
McGregor, D.C. and Cramer, F.H., 1971, Geol. Surv. Canada, Bull. 202, p. 3, pl. 1, figs. 15-18.
Cat Head Member, Red River Formation, Ordovician, between McBeth Point and Cat Head, Lake Winnipeg, Manitoba.

Leiosphaeridia spp. or *Protoleiosphaeridium* spp. [Algae Incertae Sedis]
Fig. specs. 26265-26272
McGregor, D.C. and Cramer, F.H., 1971, Geol. Surv. Canada, Bull. 202, p. 3, pl. 1, figs. 7-14.
Cat Head Member, Red River Formation, Ordovician, between McBeth Point and Cat Head, Lake Winnipeg, Manitoba.

Lepidocoleus sp. [Algae Incertae Sedis]
Fig. spec. 21988
Sinclair, G.W. et al., 1969, in Hewitt, D.F., Ontario Dept. Mines, Geol. Guide Book 3, photo 9, fig. 25.
Middle Ordovician, Lakefield quarry, Ontario.

Lowvillia grandis Guilbault and Mamet [Algae]
Holotype 45567
Guilbault, J.P. and Mamet, B.L., 1976, Can. J. Earth Sci., vol. 13, no. 5, p. 648, pl. 6, fig. 1.
Lowville Formation, Middle Ordovician, Ouareau River near Crabtree village, Joliette county, Québec.

Lowvillia multipora Guilbault and Mamet [Algae]
Holotype 45562; paratypes 45561, 45563-45565, 45581
Guilbault, J.P. and Mamet, B.L., 1976, Can. J. Earth Sci., vol. 13, no. 5, p. 650, pl. 5, figs. 2, 4-7; pl. 7, fig. 5.
Leray and Lowville formations, Middle Ordovician, Ouareau River near Crabtree village, Joliette county, Québec.

Lowvillia raripora Guilbault and Mamet [Algae]
Holotype 45578; paratypes 45560, 45568-45577, 45579, 45580
Guilbault, J.P. and Mamet, B.L., 1976, Can. J. Earth Sci., vol. 13, no. 5, p. 650, pl. 5, fig. 1; pl. 6, figs. 2-7; pl. 7, figs. 1-4, 6-8.
Middle Ordovician, Black River Group, Saint-Dominique, lat. 45°34′28″N, long. 75°52′15″W (45578), Québec; Leray and Lowville formations, Ouareau River near Crabtree village, Joliette county, Québec; Lowville Formation, Fallowfield quarry, 3/4 mile northeast of Fallowfield village, lat. 45°16′36″N, long. 75°48′53″W (45576), Ontario.

Lowvillia sp. [Algae]
Fig. spec. 45566
Guilbault, J.P. and Mamet, B.L., 1976, Can. J. Earth Sci., vol. 13, no. 5, p. 652, pl. 5, fig. 3.
Lowville Formation, Middle Ordovician, Ouareau River near Crabtree village, Joliette county, Québec.

Manitobia patula (Whiteaves)
Hypotypes 13232-13237, 13239-13242 (Plant catalogue)
Fry, W.L., 1983, Rev. Palaeobot. Palynol., vol. 39, p. 323, pl. IV, fig. 2; pl. V, figs. 1, 2; pl. VI, figs. 1-5, 7; pl. VII, fig. 5; pl. VIII, fig. 3.
Cat Head Member, Red River Formation, Late Ordovician, between McBeth Point and Cat Head, Lake Winnipeg, Manitoba.

Ortonella sequalis (Høeg) [Algae]
Hypotypes 45529-45534
Guilbault, J.P. and Mamet, B.L., 1976, Can. J. Earth Sci., vol. 13, no. 5, p. 640, pl. 1, figs. 1-4, 6, 7.
Middle Ordovician, Lowville Formation, Ouareau River near Crabtree village, Joliette county, and Black River Group, Saint-Dominique, lat. 45°34′28″N, long. 75°52′15″W (45530, 45533), Québec.

Ortonella sp. [Algae]
Fig. spec. 45535
Guilbault, J.P. and Mamet, B.L., 1976, Can. J. Earth Sci., vol. 13, no. 5, p. 640, pl. 1, fig. 5.
Black River Group, Middle Ordovician, Saint-Dominique, lat. 45°34′28″N, long. 75°52′15″W, Québec.

Pasceolus globosus Billings [Algae]
=*Cyclocrinites globosus*, Nitecki, M.H., 1970, Field Mus. Nat. Hist., Fieldiania: Geol., vol. 21, p. 81, figs. 32, 37 (hypotype 9333; holotype 1376a, syntypes 1376, b-e.).

Pasceolus halli Billings [Algae]
=*Cyclocrinites halli*, Nitecki, M.H., 1970, Field Mus. Nat. Hist., Fieldiania: Geol., vol. 21, p. 75, figs. 19, 20, 30, 31 (holotype 2227).

Rauserina notata (Antropov) [Algae]
Hypotypes 53549-53552
Bolton, T.E., 1980, Geol. Surv. Can., Paper 80-1C, pl. 2.4, figs. 5-7; pl. 2.5, fig. 2.
White Head Formation, Upper Ordovician, Priest Road (53549) and Grande Coupe, Percé, Gaspé Peninsula, Québec.

Receptaculites occidentalis Salter [Algae]
Norford, B.S. et al., 1970, Geol. Surv. Can., Econ. Geol. Rept. 1, pl. 4, fig. 4 (syntype 1125 p).

Receptaculites sp. [Algae]
Fig. spec. 42905
Bolton, T.E., 1977, Geol. Surv. Can., Bull. 269, p. 27, pl. 1, fig. 2.

Bad Cache Rapids Formation, Middle Ordovician, south of Igloolik settlement, lat. 69°20′N, long. 81°47′W, Foxe Basin, District of Franklin.

Solenopora sp. [Algae]
Fig. spec. 67014
Bolton, T.E., 1981, IUGS Subcommission on Silurian Stratigraphy, Ordovician-Silurian Boundary Working Group. Field Meeting, Anticosti-Gaspé, Québec, vol. II, p. 42, pl. I, figs. 1, 2.
Member 4 (bioherm), Ellis Bay Formation, Upper Ordovician, east bank Vauréal River second bend above bridge, Anticosti Island, Québec.

Tasmanites spp. [Algae Incertae Sedis]
Fig. specs. 26276-26278
McGregor, D.C. and Cramer, F.H., 1971, Geol. Surv. Can., Bull. 202, p. 3, pl. 1, figs. 19, 20, 33.
Cat Head Member, Red River Formation, Ordovician, between McBeth Point and Cat Head, Lake Winnipeg, Manitoba.

Unknown plant part
Fig. spec. 13258 (Plant catalogue)
Fry, W.L., 1983, Rev. Palaeobot. Palynol., vol. 39, p. 338, pl. III, fig. 2.
Cat Head Member, Red River Formation, Late Ordovician, between McBeth Point and Cat Head, Lake Winnipeg, Manitoba.

Westonia flexuosa Fry
Hypotype 13244 (Plant catalogue)
Fry, W.L., 1983, Rev. Palaeobot. Palynol., vol. 39, p. 338, pl. IX, fig. 4.
Cat Head Member, Red River Formation, Late Ordovician, between McBeth Point and Cat Head, Lake Winnipeg, Manitoba.

Wetheredella tumulus Copper [Algae[
Holotype 45371
Copper, P., 1976, Lethaia, vol. 9, p. 277, figs. 2A-C.
Ellis Bay Formation, Upper Ordovician, along Vaureal-Salmon River road, 0.3 km south of Petits Jardin River, Anticosti Island, Québec.

Whiteavesia obscura Fry
Holotype 13243 (Plant catalogue)
Fry, W.L., 1983, Rev. Palaeobot. Palynol., vol. 39, p. 335, pl. VIII, fig. 6.
Cat Head Member, Red River Formation, Late Ordovician, between McBeth Point and Cat Head, Lake Winnipeg, Manitoba.

Winnipegia cuneata (Whiteaves)
Hypotypes 1893a (Invertebrate catalogue), 13250-13254 (Plant catalogue)
Fry, W.L., 1983, Rev. Palaeobot. Palynol., vol. 39, p. 320, pl. I, figs. 1, 2 (same specimen), 3; pl. II, figs. 1, 2; pl. 7, fig. 4 (1893a).
Cat Head Member, Red River Formation, Late Ordovician, between McBeth Point and Cat Head, Lake Winnipeg, Manitoba.

Silurian

Aphralysia aff. *A. ferreoli* Mamet and Roux, 1975 [Algae]
Hypotypes 48758, 48759
Héroux, Y., Hubert, C., Mamet, B. and Roux, A., 1977, Can. J. Earth Sci., vol. 14, no. 12, p. 2880, pl. 4, figs. 3,4.
Sayabec Formation, Upper-Middle Silurian, 3.6 km east-northeast of La Rédemption, lat. 48°27′20″-48°27′30″N, long. 67°50′20″-67°49′52″W, Lac Matapédia region, Québec.

Bevocastria conglobata Garwood, 1931 [Algae]
Hypotypes 48817-48819
Héroux, Y., Hubert, C., Mamet, B. and Roux, A., 1977, Can. J. Earth Sci., vol. 14, no. 12, p. 2896, pl. 10, figs. 5-7.
Sayabec Formation, Upper-Middle Silurian, on Sainte-Jeanne-d'Arc highway, 2.6 km northwest of La Rédemption, lat. 48°27′41″N, long. 67°54′20″W and 3.6 km east-northeast of La Rédemption, lat. 48°27′20″-48°27′30″N, long. 67°50′20″-67°49′52″W (48819), Lac Matapédia region, Québec.

Cyclocrinites intermedius (Billings) [Algae]
=*Cyclocrinites gregarius*, Nitecki, M.H., 1970, Field Mus. Nat. Hist., Fieldiana: Geol., vol. 21, p. 86, fig. 40 (hypotype 8131).

Dimorphosiphon cf. *D. diadromum* Gnilovskaja, 1972 [Algae]
Hypotypes 48797-48801
Héroux, Y., Hubert, C., Mamet, B. and Roux, A., 1977, Can. J. Earth Sci., vol. 14, no. 12, p. 2889, pl. 8, figs. 2-6.
Sayabec Formation, Upper-Middle Silurian, 7.6 km south of Saint-Moise, lat. 48°29′-48°29′28″N, long. 67°51′12″-67°49′55″W and 7 km south-southeast of Saint-Moise, lat. 48°29′40″N, long. 67°49′23″W (48800), Lac Matapédia region, Québec.

Dimorphosiphon magnum Gnilovskaja, 1972 [Algae]
Hypotypes 48794, 48795
Héroux, Y., Hubert, C., Mamet, B. and Roux, A., 1977, Can. J. Earth Sci., vol. 14, no. 12, p. 2889, pl. 7, figs. 13-14.
Sayabec Formation, Upper-Middle Silurian, 3.6 km east-northeast of La Rédemption, lat. 48°27′20″-48°27′30″N, long. 67°50′20″-67°49′52″W, Lac Matapédia region, Québec.

Garwoodia gregaria (Nicholson), 1888 *emend.* Wood, 1941 [Algae]
Hypotype 48796

Héroux, Y., Hubert, C., Mamet, B. and Roux, A., 1977, Can. J. Earth Sci., vol. 14, no. 12, p. 2889, pl. 8, fig. 1.
Sayabec Formation, Upper-Middle Silurian, 7 km south-southeast of Saint-Moise, lat. 48°29′40″N, long. 67°49′23″W, Lac Matapédia region, Québec.

Girvanella problematica Nicholson and Etheridge, 1878, *emend.* Wood, 1957 [Algae]
Hypotypes 48740-48744
Héroux, Y., Hubert, C., Mamet, B. and Roux, A., 1977, Can. J. Earth Sci., vol. 14, no. 12, p. 2870, pl. 1, figs. 2-6.
Sayabec Formation, Upper Silurian, 8 km south of Saint-Moise, lat. 48°29′03″N, long. 67°50′52″W; 6.8 km southeast of Saint-Moise, lat. 48°31′11″N, long. 67°46′45″W (48741); and 5 km east of Sayabec, lat. 48°33′53″N, long. 67°37′30″W (48743), Lac Matapédia region, Québec.

Girvanella wetherdii Chapman, 1908 [Algae]
Hypotype 48739
Héroux, Y., Hubert, C., Mamet, B. and Roux, A., 1977, Can. J. Earth Sci., vol. 14, no. 12, p. 2869, pl. 1, fig. 1.
Sayabec Formation, Upper-Middle Silurian, 3.6 km east-northeast of La Rédemption, lat. 48°27′20″-48°27′30″N, long. 67°50′20″-67°49′52″W, Lac Matapédia region, Québec.

Halysis cf. *H. moniliformis* Høeg, 1932 [Algae]
Hypotypes 48826, 48827
Héroux, Y., Hubert, C., Mamet, B. and Roux, A., 1977, Can. J. Earth Sci., vol. 14, no. 12, p. 2897, pl. 11, figs. 7, 8.
Sayabec Formation, Upper-Middle Silurian, 3.6 km east-northeast of La Rédemption, lat. 48°27′20″-48°27′30″N, long. 67°50′20″-67°49′52″W, Lac Matapédia region, Québec.

Hedstroemia bernierensis Mamet and Roux [Algae]
Holotype 48807; paratypes 48808-48811
Héroux, Y., Hubert, C., Mamet, B. and Roux, A., 1977, Can. J. Earth Sci., vol. 14, no. 12, p. 2894, pl. 9, figs. 4-8.
Sayabec Formation, Upper-Middle Silurian, 3.3 km west-northwest of La Rédemption, lat. 48°27′27″N, long. 67°55′12″W; 2.1 east-northeast of La Rédemption, lat. 48°27′13″N, long. 67°51′20″ (48808); and on Sainte-Jeanne-D'Arc highway, 2.6 km northwest of La Rédemption, lat. 48°27′41″N, long. 67°54′20″W (48809, 48811), Lac Matapédia region, Québec.

Hedstroemia halimedoidea Rothpletz, 1913 [Algae]
Hypotypes 48813-48816
Héroux, Y., Hubert, C., Mamet, B. and Roux, A., 1977, Can. J. Earth Sci., vol. 14, no. 12, p. 2894, pl. 10, figs. 1-4.
Sayabec Formation, Upper-Middle Silurian, intersection Rivière Saint-Pierre and Val Brillant highway, 2 km east of Sayabec, lat. 48°31′N, long. 67°39′42″W (48813); 7 km south-southeast of Saint-Moise, lat. 48°29′40″N, long. 67°49′23″W; and on Sainte-Jeanne-D'Arc highway, 2.6 km northwest of La Rédemption, lat. 48°27′41″N, long. 67°54′20″W (48815), Lac Matapédia region, Québec.

Ischadites canadensis Billings [Algae]
=*Ischadites koenigii*, Nitecki, M.H., 1972, Field Mus. Nat. Hist., Fieldiana: Geol., vol. 28, p. 55, fig. 27b (hypotype 2590).

Litanaia? *matapediensis* Mamet and Roux [Algae]
Holotype 48784; paratypes 48782, 48783, 48785, 48802, 48803
Héroux, Y., Hubert, C., Mamet, B. and Roux, A., 1977, Can. J. Earth Sci., vol. 14, no. 12, p. 2886, pl. 7, figs. 1-4; pl. 8, figs 7, 8.
Sayabec Formation, Upper-Middle Silurian 7.6 km south of Saint-Moise, lat. 48°29′-48°29′28″N, long. 67°51′12″-67°49′55″W and 3.6 km east-northweat of La Rédemption, lat. 48°27′20″-48°27′30″N, long. 67°50′20″-67°49′52″W (48802, 48803), Lac Matapédia region, Québec.

Ortonella aequalis (Hoeg) [Algae]
Hypotypes 48804-48806
Héroux, Y., Hubert, C., Mamet, B. and Roux, A., 1977, Can. J. Earth Sci., vol. 14, no. 12, p. 2890, pl. 9, figs. 1-3.
Sayabec Formation, Upper-Middle Silurian, quarry northeast of Sayabec highway, about 2 km northwest of Val Brillant, lat. 48°32′28″N, long. 67°34′22″W, Lac Matapédia region, Québec.

Ortenella sp. [Algae]
Fig. spec. 48812
Héroux, Y., Hubert, C., Mamet, B. and Roux, A., 1977, Can. J. Earth Sci., vol. 14, no. 12, p. 2892, pl. 9, fig. 9.
Sayabec Formation, Upper-Middle Silurian, 3.6 km east-northeast of La Rédemption, lat. 48°27′20″-48°27′30″N, long. 67°50′20″-67°49′52″W, Lac Matapédia region, Québec.

Pasceolus gregarius Billings [Algae]
=*Cyclocrinites gregarius*, Nitecki, M.H., 1970, Field Mus. Nat. Hist., Fieldiana: Geol., vol. 21, p. 86, figs. 38, 39 (holotype 2230; syntypes 2230a-e).

Pasceolus intermedius Billings [Algae]
=*Cyclocrinites gregarius*, Nitecki, M.H., 1970, Field Mus. Nat. Hist., Fieldiana: Geol., vol. 21, p. 86 (hypotypes 2338, a-d, f, g).
Nitecki, M.H., 1972, ibid., vol. 28, p. 38, fig. 14.

Petrophyton kiaeri Høeg, 1932 [Algae]
Hypotypes 48820-48822
Héroux, Y., Hubert, C., Mamet, B. and Roux, A., 1977, Can. J. Earth Sci., vol. 14, no. 12, p. 2898, pl. 11, figs. 1-3.

Sayabec Formation, Upper-Middle Silurian, 3.3 km northwest of La Rédemption, lat. 48°27′51″N, long. 67°54′47″W (48820) and 2.6 km northwest of La Rédemption, lat. 48°27′41″N, long. 67°54′20″W, Lac Matapédia region, Québec.

Petrophyton aff. *P. kiaeri* Høeg, 1932 [Algae]
Hypotype 48823
Héroux, Y., Hubert, C., Mamet, B. and Roux, A., 1977, Can. J. Earth Sci., vol. 14, no. 12, p. 2900, pl. 11, fig. 4.
Sayabec Formation, Upper Silurian, 5.6 km west-northwest of La Rédemption, lat. 48°27′41″N, long. 67°56′57″W, Lac Matapédia region, Québec.

Pseudochaetetes filiformis (Nicholson), 1888 [Algae]
Hypotypes 48824, 48825
Héroux, Y., Hubert, C., Mamet, B. and Roux, A., 1977, Can. J. Earth Sci., vol. 14, no. 12, p. 2900, pl. 11, figs. 5-6.
Sayabec Formation, Upper-Middle Silurian, 2.6 km northwest of La Rédemption on Sainte-Jeanne-d'Arc highway, lat. 48°27′41″N, long. 67°54′20″W and 7 km south-southeast of Saint-Moise, lat. 48°29′40″, long. 67°49′23″W, Lac Matapédia region, Québec.

Rhabdoporella pachyderma Rothpletz, 1913 [Algae]
Hypotypes 48761-48781
Héroux, Y., Hubert, C., Mamet, B. and Roux, A., 1977, Can. J. Earth Sci., vol. 14, no. 12, p. 2884, pl. 6, figs. 1-21.
Sayabec Formation, Upper-Middle Silurian, 4.3 km east-northeast of La Rédemption, lat. 48°27′45″N, long. 67°49′30″W; islet and shore of Lac des Iles, 5 km west of La Rédemption, lat. 48°26′38″N, long. 67°56′50″W (48762, 48764, 48774, 48778, 48781); 4.3 km ENE of La Rédemption, lat. 48°27′48″N, long. 67°49′37″W (48768, 48770, 48772, 48777, 48780); and 6.5 km west of La Rédemption, lat. 48°26′53″N, long. 67°57′52″W (48771, 48775), Lac Matapédia region, Québec.

Rhabdoporella stolleyi Rothpletz [Algae]
Hypotypes 48787-48793
Héroux, Y., Hubert, C., Mamet, B. and Roux, A., 1977, Can. J. Earth Sci., vol. 14, no. 12, p. 2883, pl. 7, figs. 6-12.
Sayabec Formation, Upper-Middle Silurian, 4.3 km east-northeast of La Rédemption, lat. 48°27′45″N, long. 67°49′30″W and 6.5 km west of La Rédemption, lat. 48°26′53″N, long. 67°57′52″W, Lac Matapédia region, Québec.

Rothpletzella gotlandica (Rothpletz) [Algae]
Hypotype 45372
Copper, P., 1976, Lethaea, vol. 9, figs. 3A-C.
Eke Marls, Upper Silurian, Kattlunds 1, Grötlingbo, Gotland, Sweden.

Sphaerocodicum gotlandicum Rothpletz, 1908 *emend.* Wood, 1948 [Algae]
Hypotypes 48739, 48748-48753
Héroux, Y., Hubert, C., Mamet, B. and Roux, A., 1977, Can. J. Earth Sci., vol. 14, no. 12, p. 2873, pl. 1, figs. 1, 10; pl. 2, figs. 1-5.
Sayabec Formation, Upper-Middle Silurian, 3.6 km east-northeast of La Rédemption, lat. 48°27′20″-48°27′30″N, long. 67°50′20″-67°49′52″W; lat. 48°27′21″N, long. 67°50′10″W (48748, 48750, 48752); 3.3 km east-northwest of La Rédemption, lat. 48°27′25″N, long. 67°50′37″W (48749); and 3.8 km northeast of La Rédemption on Sainte-Jeanne-d'Arc highway, lat. 48°28′N, long. 67°55′W (48753), Lac Matapédia region, Québec.

Sphaerocodium munthei Rothpletz, 1913 *emend.* Wood, 1948 [Algae]
Hypotypes 48745-48747
Héroux, Y., Hubert, C., Mamet, B. and Roux, A., 1977, Can. J. Earth Sci., vol. 14, no. 12, p. 2873, pl. 1, figs. 7-9.
Sayabec Formation, Upper Silurian, 3.6 km east-northeast of La Rédemption, lat. 48°27′20″-48°27′30″N, long. 67°50′20″-67°49′52″W, and creek south of Sayabec highway, 3.3 km northwest of Val Brillant, lat. 48°32′45″N, long. 67°35′15″W, Lac Matapédia region, Québec.

Tasmanites aff. *T. minutus* Eisenack
Hypotype 66967
Duffield, S.L. and Legault, J.A., 1981, IUGS Subcommision on Silurian Stratigraphy, Ordovician-Silurian Boundary Working Group. Field Meeting, Anticosti-Gaspé, Québec, vol. II, pl. I, fig. 6.
Member 6, Ellis Bay Formation, Lower Silurian, '9-mile' pool section, Salmon River, Anticosti Island, Québec.

Vermiporella cf. *V. fragilis* Stolley, 1893 [Algae]
Hypotype 48786
Héroux, Y., Hubert, C., Mamet, B. and Roux, A., 1977, Can. J. Earth Sci., vol. 14, no. 12, p. 2884, pl. 7, fig. 5.
Sayabec Formation, Upper-Middle Silurian, Sainte-Jeanne-d'Arc highway, 3.8 km northwest of La Rédemption, lat. 48°28′N, long. 67°55′W, Lac Matapédia region, Québec.

Wetheredella pachytheca Mamet and Roux [Algae]
Holotype 48830; paratypes 48828, 48829, 48831-48833
Héroux, Y., Hubert, C., Mamet, B. and Roux, A., 1977, Can. J. Earth Sci., vol. 14, no. 12, p. 2901, pl. 12, figs. 1-6.
Sayabec Formation, Upper-Middle Silurian, 2 km east-northeast of La Rédemption, lat. 48°27′03″N, long. 67°51′20″W; 7.600 km south of Saint-Moise, lat. 48°29′-48°29′28″N, long. 67°51′12″-67°49′55″W (48828); and 1.3 km northeast of La Rédemption, lat. 48°27′10″N, long. 67°52′W (48833), Lac Matapédia region, Québec.

Wetherdella silurica Wood, 1948 [Algae]
Hypotypes 48834-48838
Héroux, Y., Hubert, C., Mamet, B. and Roux, A., 1977, Can. J. Earth Sci., vol. 14, no. 12, p. 2901, pl. 12, figs. 7-11.
Sayabec Formation, Upper-Middle Silurian, 2 km east-northeast of La Rédemption, lat. 48°27′03″N, long. 67°51′20″W (48834);5 km east-northeast of La Rédemption, lat. 48°27′37″N, long. 67°49′10″W; 3.600 km east-northeast of La Rédemption, lat. 48°27′20-48°27′30″N, long. 67°50′20″-67°49′52″W (48837); and 5 km east-northeast of La Rédemption, lat. 48°27′36″N, long. 67°49′07″W (48838), Lac Matapédia region, Québec.

Devonian

Baragwanathia abitibiensis Hueber
Holotype 69288; paratypes 69289-69318
Hueber, F.M., 1983, Botanical J. Linnean Soc., vol. 86, p. 65, Figs. 1-49.
Sextant Formation, Lower Devonian, west bank Abitibi River, first outcrop north of spit opposite north end of large east bank exposure, below rapids about 300 yards north of north end of large east bank exposure (69297, 69298), and about 100 yards north of large outcrop at point near head of Sextant River (69299, 69300), northern Ontario.

Bathurstia denticulata Hueber
Holotype 29900; paratypes 29901, 29902, 30022
Hueber, F.M., 1972, Geol. Surv. Can. Paper 71-28, p. 9, pl. 1, fig. 4; pl. 3, figs. 1-5.
Bathurst Island Formation, Early Devonian, 4 miles east of head of Young Inlet, Bathurst Island, District of Franklin.

Cf.*Bathurstia* sp.
Fig. spec. 29903
Hueber, F.M., 1972, Geol. Surv. Can, Paper 71-28, p. 13, pl. 2, fig. 3.
Bathurst Island Formation, Early Devonian, 4 miles east of head of Young Inlet, Bathurst Island, District of Franklin.

Chaleuria cirrosa Andrews, Gensel and Forbes
Holotype 33258; paratypes 33259-33267, 34596, 34597
Andrews, H.N., Gensel, P.G and Forbes, W.H, 1974, Palaentology, vol. 17, pt. 2, p. 398, pls. 52, 53, 54, figs. 1-6, 55, figs. 1-4, 56, figs. 1-6, 57, figs. 1-7.
McGregor, D.C. and Camfield, M., 1976, Geol. Surv. Can., Bull. 263, pl. 8, figs. 1, 2, 4-6 [small and large spores associated] .
Gensel, P.G., 1980, Rev. Palaeobot. Palynol., vol. 30, nos. 1/2, pl. 4, figs. 4, 5 (spores 33267), 6, 7 (spores 33265).
Campbellton Formation, Middle Devonian, beach outcrop 3/4 mile west of Dalhousie junction, New Brunswick.

Chovanella burgessi Peck and Eyer [Charophyte]
Hypotype 96608
Loranger, D.M., 1965, Devonian microfauna from northeastern
Alberta Part 4-Charophyta-Foraminfera-Branchiopoda-Conodontophorida, Evelyn de Mille Books Ltd., Calgary, p. 5, photograph 1, figs. 8-10.
Elk Point group, Middle Devonian, Christina River Hardy, No. 1 well, l.s.d. 2, sec. 25, tp. 77, rge. 9, W.4th mer., Alberta.

Chovanella burgessi Peck and Eyer [Charophyte]
Hypotypes 25602-25607
Eyer, J.A., 1971, Geol. Surv. Can., Bull. 197, pl. 5, figs. 4-5; pl.6, figs. 1-4.
Cedared Formation, Middle Devonian, Lussier River, lat. 50°04′N, long. 115°31′W and Stanford Range, lat. 50°30′N, long. 115°54.3′W, British Columbia.

Chovanella burgessi Peck and Eyer [Charophyte]
Hypotypes 27632-27635
Rice, H.M.A., 1971, Geol. Surv. Can., Paper 70-60 (1970), p. 8, pl. 1, figs. 9-12.
Pine Point and Slave Point formations, Middle Devonian, Midland Pine Point No. 1, about lat. 60°47′N, long. 114°15′W and Falconbridge No. 2, about 3.5 miles south of mouth of Buffalo River, District of Mackenzie.

Chovanella burgessi Peck and Eyer and *Eochara wickendeni* Choquette [Charophyte]
Hypotypes 25601, 25609
Eyer, J.A., 1971, Geol. Surv. Can., Bull. 197, pl. 5, fig. 1; pl. 6, fig. 6.
Cedared Formation, Middle Devonian, Stanford Range, lat. 50°27′N, long. 115°46′W and Lussier River, lat. 50°04′N, long. 115°31′W, British Columbia.

Drephanophycus spinaeformis Göppert 1852
Hypotype 29905
Hueber, F.M., 1972, Geol. Surv. Can. Paper 71-28, p. 14. pl. 1, figs. 1, 3.
Bathurst Island Formation, Early Devonian, 4 miles east of head of Young Inlet, Bathurst Island, District of Franklin.

Eochara wickendeni Choquette [Charophyte]
Hypotype 25608
Eyer, J.A., 1971, Geol. Surv. Can., Bull. 197, pl. 6, fig. 5.
Cedared Formation, Middle Devonian, Stanford Range, lat. 50°27′N, long. 115°46′W, British Columbia.

Eochara wickendeni Choquette [Charophyte]
Hypotypes 27624-27631
Rice, H.M.A., 1971, Geol. Surv. Can., Paper 70-60 (1970), p. 8, pl. 1, figs. 1-8.

Slave Point and Pine Point formations, Middle Devonian, Silver Summit No. 1, Buffalo River area(?) and Midland Pine Point No. 1, about lat. 60°47'N, long. 114°15'W, District of Mackenzie.

?*Moellerina greeni* Ulrich
Hypotypes 27636-27638
Rice, H.M.A., 1971, Geol. Surv. Can., Paper 70-60 (1970), p. 11, pl. 1, figs. 13-15.
Hare Indian Formation, Middle Devonian, Ramparts Gorge No. 3, lower end of gorge, right bank Mackenzie River, District of Mackenzie.

New genus, *Zosterophyllophyte*
Fig. specs. 67806, 67809, 67816
Gensel, P.G., 1982, Third North American Paleontol. Convention, Proc., vol. 1, pl. 1, figs. 6, 9, 12.
Gaspé Sandstone Group, Lower Devonian, beach outcrop on Restigouche River approximently 1 km west of Dalhousie Junction, New Brunswick.
=*Oricilla bilinearis*, Gensel, P.G., 1982, Rev. Palaeobot. Palynol., vol. 37, nos. 3/4, p. 354, pl. I, figs. 1 (holotype 67806), 3, 6; pl. III, fig. 2 (hypotype 67809); pl. II, fig. 4; pl. III, fig. 1 (hypotype 67816).

Oocampsa catheta Andrews, Gensel and Kasper
Holotype 41696; paratypes 41697-41703
Andrews, H.N., Gensel, P.G. and Kasper, A.E., 1975, Can. J. Botany, vol. 53, no. 16, p. 1722, figs. 2-13.
Gensel, P.G., 1980, Rev. Palaeobot. Palynol., vol. 30, nos. 1/2, pl.4, figs. 1 (spores 41702), 2, 3 (spores 41703).
Campbellton Formation, Middle Devonian, slump blocks near high-tide south shore Restigouche River 1/2 mile west of Dalhousie Junction, New Brunswick.

Oricilla bilinearis Gensel
Paratype 67807; hypotypes 67808, 67810-67815, 67817-67822
Gensel, P.G., 1982, Rev. Palaeobot. Palynol., vol. 37, nos. 3/4, p. 354, pl. I, figs. 1, 2, 4-8; pl. II, figs. 2, 3, 5, 6; pl. III, figs. 3-8.
Gaspé Sandstone Group, Lower Devonian, beach outcrop on Restigouche River approximately 1 km west of Dalhousie Junction, New Brunswick.

Pertica dalhousii Doran, Gensel and Andrews
Holotype 54738; paratypes 54739, 54746; hypotypes 54740-54743, 54745
Doran, J.B., Gensel, P.G. and Andrews, H.N., 1978, Can. J. Botany, vol. 56, no. 24, p. 3054, figs. 1E-F, 2, 3, 7-13, 15D.
Lower-Middle Devonian, south shore Restigouche River about 1/2 mile west of railway crossing at Dalhousie Junction, New Brunswick.

Pertica varis Granoff, Gensel and Andrews
Holotype 43218; paratypes 43219-43221; hypotypes 43222-43230
Granoff, J.A., Gensel, P.G. and Andrews, H.N., 1976, Palaeontographica, Abt. B, vol. 155, nos 5-6, p. 124, pls. 1-4.
Gensel, P.G., 1980, Rev. Palaeobot. Palynol., vol. 30, nos. 1/2, pl. 3, figs. 8, 9 (spores 43229).
Battery Point Formation, Lower Devonian, 1/2 mile east of Grand Cap-aux-Os, north shore Gaspé Bay, Québec.

cf. *Pertica* sp.
Fig. specs. 54747-54750
Doran, J.B., Gensel, P.G. and Andrews, H.N., 1978, Can. J. Botany, vol. 56, no. 24, figs. 16-22.
Battery Point Formation, Lower Devonian, north shore Gaspé Bay about 1/2 mile east of Cap-aux-Os, Québec.

Psilophyton charientos Gensel
Holotype 57864; paratypes 57853, 57855, 57859; hypotypes 57854, 57856-57858, 57860-57862, 57864-57872
Gensel, P.G., 1979, Palaeontographica, Abt. B, vol. 168, no.4, p. 85, pl. 1, figs. 1-9; pl. 2, figs. 1-5, 7; pl. 3, fig. 1-8; pl. 4, fig. 1-5.
Gensel, P.G., 1982, Third North American Paleontol. Convention, Proc., vol. 1, pl. 1, figs. 3 (57862), 5 (57859).
Gensel, P.G. and White, A.R., 1983, Palynology, vol. 7, p. 223, pl. 1, figs. 9, 10 (spores 57872).
Lower Devonian, beach outcrop about 0.5 mile west of Dalhousie Junction, New Brunswick.

Psilophyton coniculum Trant and Gensel
Holotype 79920; paratypes 79921-79962
Trant, C.A. and Gensel, P.G., 1985, Amer. J. Botany, vol. 72, no. 8, p. 1270, figs. 1-6, 9-20, 21A-21F, 23A-23H, 25-30, 31A-E, 33A-E, 35A-G.
Lower Devonian, beach outcrop near Pin Sec Point, approximately 10 miles east of Campbellton, lat. 48°N, long. 66°25'W, New Brunswick.

Psilophlyton dawsonii Banks, Leclercq and Hueber
Paratypes 32719-32722
Banks, H.P., Leclercq, S. and Hueber, F.M., 1975, Palaeontogr. Americana, vol. 8, no. 48, p. 105, pl. 17, figs. 2, 3, 7-10; pl. 23, figs. 51-53.
Sextant Formation, Lower Devonian, west bank Abitibi River below Sextant Rapids and above Coral Rapids, lat. 50°13'N, long. 81°38'W, Ontario.

Psilophyton forbesi Andrews
Hypotypes 57863, 57873-57889
Gensel, P.G., 1979, Palaeontographica, Abt. B, vol. 168, no. 4, p. 89, pl. 5, figs. 1-6; pl. 6, figs. 1-7; pl. 7, figs. 1-8, 10, 11; pl. 8, figs. 1-8, 11.
Gensel, P.G., 1980, Rev. Palaeobot. Palynol., vol. 30, nos. 1/2, pl.3, figs. 4-7, 10(spores 57883).
Gensel, P.G and White, A.R., 1983, Palynology, vol. 7, p. 223, pl. 1, figs. 6, 8 (spores 57883).
Battery Point Formation, Lower Devonian, near Seal Rock on north shore Gaspé Bay, Québec.

Quisquilites widderensis Legault [Algae]
Holotype 30662; paratypes 30660, 30661, 30663, 30664
Legault, J.A., 1973, Geol. Surv. Can., Bull 221, p. 60, pl. 11, figs. 17-21.

Hamilton Formation, Middle Devonian, Widder and Arkona (30664) members, GSC Arkona No. 1 well, Rock Glen Park, lot 4, con. SB, Bosanquet tp., Lambton co., and Arkona Member (30663), GSC Ipperwash No. 1 well, lot 9, con. A, Bosanquet tp., Lambton co., Ontario.

Rebuchia capitanea Hueber
Holotype 29899
Hueber, F.M., 1972, Geol. Surv. Can., Paper 71-28, p. 5, pl. 1, figs. 2, 5; pl. 2., figs. 1, 2.
Bathurst Island Formation, Early Devonian, 4 miles east of head of Young Inlet, Bathurst Island, District of Franklin.

Renalia hueberi Gensel
Holotype 43178; paratypes 43179, 43180; hypotypes 43181-43200, 43340, 43341
Gensel, P.G., 1976, Rev. Palaeobot. Palynol., vol. 22, no. 1, p. 25, pls. 1-5.
Gensel, P.G., 1980, Rev. Palaeobot. Palynol., vol. 30, nos. 1/2, pl. 1, figs. 5, 6(spores).
Battery Point Formation, Lower Devonian, beach outcrop below gun emplacement at Peninsula Gaspé (Fort Peninsule) on north shore Gaspé Bay, Québec.

Sawdonia acanthoteca Gensel, Andrews and Forbes
Holotype 35422; paratype 35423; hypotypes 35424-35433
Gensel, P.G., Andrews, N.N. and Forbes, W.H., 1975, Botanical Gazette, vol. 136, no. 1, p. 56, figs. 1, 3-22.
Gensel, P.G., 1980, Rev. Palaeobot. Palynol., vol. 30, nos. 1/2, pl. 2, figs. 1 (35432), 2, 3 (35431) [spores].
Gaspé Sandstone, Middle Devonian, coastal section about 2 miles west of Dalhousie, long. 66°25′W, New Brunswick.

Sawdonia ornata (Dawson) Hueber 1971
Hypotypes 29897, 29898
Hueber, F.M., 1972, Geol. Surv. Can. Paper 71-28, p. 4, pl. 1, fig. 4.
Bathurst Island and Stuart Bay formations, Early Devonian, 4 miles east of head of Young Inlet, Bathurst Island, District of Franklin.

Spongiophyton minutissimum Kräusel, 1954
Hypotypes 93003-93020
Gensel, P.G., Chaloner, W.G. and Forbes, W.H., 1991, Paleontology, vol. 34, pt. 1, p. 154, pl. 1, figs. 1-5, 7; pl. 3, figs. 5, 6; text-figs. 2A-E; text-figs. 3A-C; text-figs. 5A, B, E.
Lower Devonian, outcrop along Beauvista Drive (93003, 93005, 93008), Atholville, and along Restigouche River near Dalhousie Junction (93006, 93007, 93011), northern New Brunswick; just north of Bordeaux Quarry along Route 132, Cross Point, Gaspé, Québec.

Spores of *Psilophyton princeps*
Hypotypes 72163-72165
Gensel, P.G. and White, A.R., 1983, Palynology, vol. 2, p. 223, pl. 1, figs. 1-4.
Lower Devonian, New Brunswick.

Tasmanites sp. [Algae Incertae sedis)
Fig. specs. 30685, 30686
Legault, J.A., 1973, Geol. Surv. Can., Bull. 221, p. 61, pl. 13, figs. 8, 10.
Rockport Quarry Member, Hamilton Formation, Middle Devonian, GSC Arkona No. 1 well, Rock Glen Park, lot 4, con. SB, Bosanquet p., Lambton co., Ontario.

Uraloporella variabilis Korde [Algae]
Hypotypes 38427-38429
Riding, R. and Jansa, L.F., 1974, Can. J. Earth Sci., vol. 11, no. 10, p. 1421, pl. 1, figs. 2-4, 6.
Swan Hills and Cairo formations, Middle and Upper Devonian, SOBC Snipe Lake well, depth 8920 feet, l.s.d. 10, sec. 21, tp. 10, rge. 18, W.5th mer., and Mount Simla (38429), Alberta.
=*Jansaella ridingi*, Mamet, B. and Roux, A., 1975, Can. J. Earth Sci., vol. 12, no. 8, p. 1482 (holotype 38429).

Zosterophylloid axis
Fig. spec. 29904
Hueber, F.M., 1972, Geol. Surv. Can., Paper 71-28. p. 13, pl. 2, figs. 4,5.
Bathurst Island Formation, Early Devonian, 4 miles east of head of Young Inlet, Bathurst Island, District of Franklin.

Zosterophyllum divaricatum Gensel
Paratypes 65485a, b, 65500; hypotypes 65481-65484, 65487-65496, 65498, 65499, 65502-65513
Gensel, P.G., 1982, Am. J. Botany, vol. 69, no. 5, p. 652, figs. 1, 2, 4-9, 11-24, 27-32A, B, D-G, 33, 34, 36, 37, 39-67.
Gaspé Sandstone Group, Lower Devonian, beach outcrop approximately 0.8 km from Dalhousie Junction, long. 66°25′W, New Brunswick.

Zosterophyllum sp. n.
Fig. specs. 65486, 65497, 65501
Gensel, P.G., 1982, Third North American Paleontol. Convention, Proc., vol. 1, pl. 1, figs. 7, 11, 13
Gaspé Sandstone Group, Lower Devonian, beach outcrop approximately 0.8 km from Dalhousie Junction, long. 66°25′W, New Brunswick.
=*Zosterophyllum divaricatum*, Gensel, P.G., 1982, Am. J. Botany, vol. 69, no. 5, p. 652, figs. 25, 26, 32C (holotype 65486 a, b), 3 (hypotype 65497), 35 (hypotype 65501).

Carboniferous-Permian

Albertaporella involuta Johnson [Algae]
Hypotype 29218
Petryk, A.A. and Mamet, B.L., 1972, Can. J. Earth Sci., vol. 9, no. 7, p. 777, pl. 3, fig. 6.
Shunda Formation, Lower Carboniferous, Sheep Creek, lat. 51°35′N, long. 115°29′W, Alberta.

Alethopteris serlii (Brongniart, 1834) Goeppert, 1836
Hypotype 77145
Zodrow, E.L. and Vasey, G.M., 1986, J. Paleontol., vol. 60, no. 2, p. 210, fig. 2.2.
Pictou Group, Pennsylvanian, Mabou Mines section near Finlay Point, Cape Breton Island, Nova Scotia.

Alethopteris cf. *valida* Boulay, 1876
Hypotype 77146
Zodrow, E.L. and Vasey, G.M., 1986, J. Paleontol., vol. 60, no. 2, p. 211, fig. 3.
Pictou Group, Pennsylvanian, footage 621-624 feet, Borehole 5, Coal Mine Point, Mabou Mines coalfield, Cape Breton Island, Nova Scotia.

Algue indéterminée
Fig. specs. 39816-39819
Mamet, B. and Rudloff, B., 1972, Rev. Micropaléont., vol. 15, no. 2, p. 92, pl. 10, figs. 14-17.
Carboniferous, Debolt Formation, Pacific Highway a-47-I(2), British Columbia; Opal Member, Mount Head Formation, lat. 50°34′N, long. 114°56′W, South Misty Range, Alberta.

Anatolipora af. *A. carbonica* Konishi [Algae]
Hypotype 29216
Petryk, A.A. and Mamet, B.L., 1972, Can. J. Earth Sci., vol. 9, no. 7, p. 776, pl. 3, fig. 7.
Shunda Formation, Lower Carboniferous, Sheep Creek, lat. 51°35′N, long. 115°29′W, Alberta.

Anatolipora sp. [Algae]
Fig. spec. 29217
Petryk, A.A. and Mamet, B.L., 1972, Can. J. Earth Sci., vol. 9, no. 7, p. 777, pl. 3, fig. 8.
Shunda Formation, Lower Carboniferous, Clearwater River, lat. 51°47′N, long. 116°4′W, Alberta.

Anthracoporella girtyi Mamet et Roux [Algae]
Holotype 60994; paratypes 60997, 61003
Mamet, B.L., Roux, A. and Nassichuk, W.W., 1987, Geol. Surv. Can., Bull. 342, p. 37, pl. 10, figs. 6, 9; pl. 11, fig. 6.
Nansen Formation, Pennsylvanian and Permian(61003), Girty Creek, Head of Hare Fiord, lat. 81°12′N, long. 82°6′W, Ellesmere Island, District of Franklin.

Anthracoporellopsis machaevii Maslov, 1956 [Algae]
Hypotypes 60995, 60996, 61001, 61002
Mamet, B.L., Roux, A. and Nassichuk, W.W., 1987, Geol. Surv. Can., Bull. 342, p. 23, pl. 10, figs. 7, 8; pl. 11, figs. 4,5.
Nansen Formation, Pennsylvanian, 20 km west of head of Otto Fiord, lat. 81°20′N, long. 85°30′W and Girty Creek, head of Hare Fiord, lat. 81°12′N, long. 82°6′W; Belcher Channel Formation, Permian, west side of head of Blind Fiord, Ellesmere Island, and 12.8 km south of Cape Ogle, lat. 76°56′30″N, long. 95°23′W, Grinnel Peninsula, Devon Island, District of Franklin.

Aoujgalia elliotti Mamet et Roux, 1977 [Algae]
Hypotypes 61206, 61207
Mamet, B.L., Roux, A. and Nassichuk, W.W., 1987, Geol. Surv. Can., Bull. 342, p. 50, pl. 25, figs. 3, 4.
Belcher Channel Formation, Permian, north of Trold Bay, lat. 77°56′N, long. 85°30′W, Ellesmere Island, District of Franklin.

Aoujgalia richi Mamet et Roux, 1978 [Algae]
Hypotypes 61191-61193
Mamet, B.L., Roux, A. and Nassichuk, W.W., 1987, Geol. Surv. Can., Bull. 342, p. 50, pl. 24, figs. 1-3.
Missippian or Pennsylvanian, Little Fish Creek, lat. 68°12′N, long. 136°28′W, Yukon.

Aoujgalia variabilis Termier and Termier [Algae]
Hypotypes 29282-29286
Petryk, A.A. and Mamet, B.L., 1972, Can. J. Earth Sci., vol. 9, no. 7, p. 791, pl. 9, figs. 1-5.
Opal Member, Mount Head Formation, Carboniferous, Elbow Lake, lat. 50°38′N, long. 115°W, Alberta.

Aoujgalia variabilis Termier and Termier [Algae]
Hypotypes 39796, 39797, 31351
Mamet, B. and Rudloff, B., 1972, Rev. Micropaléont., vol. 15, no. 2, p. 90, pl. 6, figs. 15-17.
Carnarvon and Loomis members, Mount Head Formation, Carboniferous, Cataract Creek, Alberta, and Connor Lakes (31351), British Columbia.

Aoujgalia variabilis Termier and Termier, 1950 [Algae]
Hypotypes 61204, 61205
Mamet, B.L., Roux, A. and Nassichuk, W.W., 1987, Geol. Surv. Can., Bull. 342, p. 49, pl. 25, figs. 1, 2.
Missippian or Pennsylvanian, Little Fish Creek, lat. 68°12′N, long. 136°28′W, Yukon.

Archaeolithophyllum sp. cf. *A. johnsoni* Rácz, 1964 [Algae]
Fig. specs. 61244, 61245
Mamet, B.L., Roux, A. and Nassichuk, W.W., 1987, Geol. Surv. Can., Bull. 342, p. 54, pl. 28, figs. 1, 2.
Otto Fiord Formation, Pennsylvanian, Wood Glacier, lat. 80°54′N, long. 84°11′W, Ellesmere Island, District of Franklin.

Archaeolithophyllum missouriensum Johnson, 1956
Hypotypes 61246, 61247, 61260, 61261
Mamet, B.L., Roux, A. and Nassichuk, W.W., 1987, Geol. Surv. Can., Bull. 342, p. 54, pl. 28, figs. 3, 4; pl. 29, figs. 1, 2.

Nansen Formation, Pennsylvanian, Girty Creek head of Hare Fiord, lat. 81°12′N, long. 82°6′W, Ellesmere Island, District of Franklin; Permian, Little Fish Creek, lat. 68°12′N, long. 136°28′W, Yukon(61260, 61261).

Archaeolithophyllum sp. [Algae]
Fig. spec. 49042
Jansa, L.F., Mamet, B. and Roux, A., 1978, Can. J. Earth Sci., vol. 15, no. 9, p. 1436, pl. 2, fig. 7.
Windsor Group, Lower Carboniferous, core depth 57 cm, lat. 50°44.1′N, long. 54°57.1′W, northeastern Newfoundland shelf.

Asphaltina cordillerensis Mamet [Algae]
Holotype 29294; paratypes 29295, 29296
Petryk, A.A. and Mamet, B.L., 1972, Can. J. Earth Sci., vol. 9, no. 7, p. 795, pl. 10, figs. 4, 5.
Carnarvon, Baril, and Loomis members, Mount Head Formation, Carboniferous, South Misty Range, lat. 50°34′N, long. 114°56′W, Boulder Creek, lat. 50°37′N, long. 114°52′W, and Picklejar Lakes, lat. 50°32′N, long. 114°46′W, Alberta.

Asphaltina cordillerensis Mamet *dans* Petryk et Mamet, 1972 [Algae]
Hypotypes 61258, 61259
Mamet, B.L., Roux, A. and Nassichuk, W.W., 1987, Geol. Surv. Can., Bull. 342, p. 57, pl. 28, figs. 15, 16.
Pennsylvanian, Nansen Formation, 20 km west of head of Otto Fiord, lat. 81°20′N, long. 85°30′W; Otto Fiord Formation, 2.5 km northeast of van Hauen Pass, Ellesmere Island, District of Franklin.

Asphaltinella horowitzi Mamet et Roux, 1978 [Algae]
Hypotypes 61255-61257, 61265, 61266, 61269, 61270
Mamet, B.L., Roux, A. and Nassichuk, W.W., 1987, Geol. Surv. Can., Bull. 342, p. 57, pl. 28, figs. 12-14; pl. 29, figs. 6, 7; pl. 30, figs. 3, 4.
Otto Fiord Formation, Pennsylvanian, east of Girty Creek, and "Helicopter section", head of Hare Fiord(61266), Ellesmere Island, District of Franklin.

Atractyliopsis cumberlandensis Rich [Algae]
Hypotype 39421
Mamet, B.L., 1976, Geol. Surv. Can., Bull. 255, pl. 45, fig. 1.
Mount Head Formation, Lower Carboniferous, lat. 50°24′N, long. 114°37′W, Highwood Valley, Alberta.

Atractyliopsis sp. [Algae]
Fig. specs. 39757-39759
Mamet, B. and Rudloff, B., 1972, Rev. Micropaléont., vol. 15, no. 2, p. 83, pl. 3, figs. 15-17.
Carnarvon Member, Mount Head Formation, Carboniferous, Cataract Creek, Alberta.

Beresella sp. cf. *B. ishimica* Kulik, 1964 [Algae]
Fig. specs. 61024, 61025
Mamet, B.L., Roux, A. and Nassichuk, W.W., 1987, Geol. Surv. Can., Bull. 342, p. 30, pl. 12, figs. 9, 10.
Otto Fiord Formation, Pennsylvanian, Wood Glacier, lat. 80°54′N, long. 84°11′W, Ellesmere Island, District of Franklin.

Beresella ex. gr. *B. polyramosa* Kulik boundstone
Hypotype 61283
Mamet, B, Nassichuk, W.W. and Roux, A., 1979, Bull. Cent. Rech. Explor.-Prod. Elf-Aquitaine, vol. 3, no.2, pl. 1, fig. 4.
Belcher Channel Formation, Permian, Blind Fiord, Ellesmere Island, District of Franklin.

Beresella ex. gr.*B. polyramosa* Kulik, 1964 [Algae]
Hypotype 61044-61057
Mamet, B.L., Roux, A. and Nassichuk, W.W., 1987, Geol. Surv. Can., Bull. 342, p. 28, pl. 13, figs. 6, 19.
Pennsylvanian, Otto Fiord Formation, Wood Glacier, lat. 80°54′N, long. 84°11′W; Nansen Formation, Girty Creek, head of Hare Fiord, lat. 81°12′N, long. 82°6′W(61047, 61054, 61056), Ellesmere Island, District of Franklin.

Beresella sp. [Algae]
Fig. specs. 39770a, b
Mamet, B. and Rudloff, B., 1972, Rev. Micropaléont., vol. 15, no. 2, p. 85, pl. 5, figs. 5, 6.
Canyon Fiord Formation, Carboniferous, Ellesmere Island, District of Franklin.

Beresellid-algal peckstone [Algae]
Fig. spec. 61287
Mamet, B., Nassichuk, W.W. and Roux, A., 1979, Bull. Cent. Rech. Explor.-Prod. Elf-Aquitaine, vol. 3, no. 2, pl. 2, fig. 3.
Nansen Formation, Upper Carboniferous, Girty Creek, Ellesmere Island, District of Franklin.

Beresellid boundstone [Algae]
Fig. spec. 61286
Mamet, B, Nassichuk, W.W. and Roux, A., 1979, Bull. Cent. Rech. Explor.-Prod. Elf-Aquitaine, vol. 3, no. 2, pl. 2, fig. 2.
Nansen Formation, Upper Carboniferous, Girty Creek, Ellesmere Island, District of Franklin.

Bevocostria conglobata Garwood, 1931 [Algae]
Hypotypes 60928-60931
Mamet, B.L., Roux, A. and Nassichuk, W.W., 1987, Geol. Surv. Can., Bull. 342, p. 13, pl. 4, figs. 3, 6.
Nansen Formation, Carboniferous, 20 km east of head of Otto Fiord, lat. 81°20′N, long. 85°30′W, and Permian, Girty Creek, head of Hare Fiord, lat. 81°12′N, long. 82°6′W(60929, 60930); Belcher Channel Formation, Permian, west side of head of Blind Fiord(60931), Ellesmere Island, District of Franklin.

Calcisphaera laevis Williamson [Algae]
Hypotype 39282
Mamet, B.L., 1976, Geol. Surv. Can., Bull. 255, pl. 14, fig. 11.
Shunda Formation, Lower Carboniferous, lat. 52°32′N, long. 116°51′30″W, Brazeau River, Alberta.

Calcisphaera laevis Williamson, 1881 [Algae]
Hypotype 65840
Mamet, B.L., 1986, Geol. Surv. Can., Bull. 353, p. 31, pl. 12, fig. 12.

Turner Valley Formation, Lower Carboniferous, East Fellers Creek, lat. 54°42′30″N, long. 120°54′W, British Columbia.

Calcisphaera pachysphaerica (Pronina) [Algae]
Hypotype 39714
Mamet, B.L., 1976, Geol. Surv. Can., Bull. 255, pl. 95, fig. 11.
Kayak Shale/Alapah Limestone contact, Lower Carboniferous, lat. 68°51′N, long. 138°58′W, Mount Sedgwick, Yukon.

Calcisphaera sp. [Algae]
Fig. spec. 39715
Mamet, B., 1976, Geol. Surv. Can., Bull. 255, pl. 95, fig. 12.
Kayak Shale/Alapah Limestone contact, Lower Carboniferous, lat. 68°51′N, long. 138°58′W, Mount Sedgwick, Yukon.

Clavaporella reinae Rácz, 1966 [Algae]
Hypotypes 61135-61152
Mamet, B.L., Roux, A. and Nassichuk, W.W., 1987, Geol. Surv. Can., Bull. 342, p. 44, pl. 19, figs. 1-11; pl. 20, figs. 1-7.
Nansen Formation, Pennsylvanian, Upper Carboniferous, 20km west of head of Otto Fiord, lat. 81°20′N, long. 85°30′W, Ellesmere Island, District of Franklin.

Columbiapora johnsoni Mamet [Algae]
Holotype 39847; paratypes 39846, 39848-39850
Mamet, B.L., 1974, Rev. Micropaléont., vol. 17, no. 1, p. 44, pl. 3, figs. 5-7, 9, 11, 12.
Pekisko Formation, Lower Carboniferous, lat. 54°22′N, long. 120°29′W, Belcourt Creek, British Columbia.

Cuneiphycus johnsoni Flügel, 1966 [Algae]
Hypotypes 61248-61253
Mamet, B.L., Roux, A. and Nassichuk, W.W., 1987, Geol. Surv. Can., Bull. 342, p. 55, pl. 28, figs. 5-10.
Belcher Channel Formation, Permian, north of Trold Bay, lat. 77°56′N, long. 85°30′W; Nansen Formation, Pennsylvanian, Girty Creek, head of Hare Fiord, lat. 81°12′N, long. 82°6′W(61251) and 20 km west of head of Otto Fiord, lat. 81°20′N, long. 85°30′W(61252), Ellesmere Island, District of Franklin; Pennsylvanian, Little Fish Creek, lat. 68°12′N, long. 136°28′W, Yukon(61253).

Cuneiphycus? sp. ou nouveau genre? [Algae]
Fig. spec. 61254
Mamet, B.L., Roux, A. and Nassichuk, W.W., 1987, Geol. Surv. Can., Bull. 342, p. 56, pl. 28, fig. 11.
Otto Fiord Formation, Pennsylvanian, east of Girty Creek, head of Hare Fiord, Ellesmere Island, District of Franklin.

Dasycladacée indéterminée [Algae]
Fig. specs. 39761, 39769
Mamet, B. and Rudloff, B., 1972, Rev. Micropaléont., vol. 15, no. 2, p. 84, pl. 4, figs. 16(39769), 17(39761).
Turner Valley Formation, Carboniferous, Folding Mountain, Alberta.

Diplopora belcherensis Mamet et Roux [Algae]
Holotype 61156; paratypes 61157, 61158
Mamet, B.L., Roux, A. and Nassichuk, W.W., 1987, Geol. Surv. Can., Bull. 342, p. 39, pl. 21, figs. 1-3.
Belcher Channel Formation, Permian, west side of end of Blind Fiord, Ellesmere Island, District of Franklin.

Diplopora sp. cf.*D.pusilla* Kochansky-Devidé and Herak, 1960
Fig. specs. 61154, 61155
Mamet, B.L., Roux, A. and Nassichuk, W.W., 1987, Geol. Surv. Can., Bull. 342, p. 38, pl. 20, figs. 9, 10.
Nansen Formation , Middle Carboniferous, Girty Creek, head of Hare Fiord, Ellesmere Island, District of Franklin.

Displosphaerina inaequalis (Derville) [Algae]
Hypotype 39426
Mamet, B.L., 1976, Geol. Surv. Can., Bull. 255, pl. 45, fig 6.
Mount Head Formation, Lower Carboniferous, lat. 50°19′N, long. 115°05′W, Connor Lakes, British Columbia.

Donezella lutugini Maslov boundstone
Hypotypes 61280-61282
Mamet, B., Nassichuk, W.W. and Roux, A., 1979, Bull. Cent. Rech. Explor.-Prod. Elf-Aquitaine, vol. 3, no.2, pl. 1, fig. 1-3.
Otto Fiord and Nansen formations, Upper Carboniferous, Wood Glacier and Girty Creek(61282), Ellesmere Island, District of Franklin.

Donezella lutugini Maslov, 1929, *emend.* [Algae]
Hypotypes 61072-61085
Mamet, B.L., Roux, A. and Nassichuk, W.W., 1987, Geol. Surv. Can., Bull. 342, p. 32, pl. 14, figs. 15-23; pl. 15, figs. 1-5.
Pennsylvanian, Tellevak Member, Hare Fiord Formation, Glacier Lake, Blue Mountain; Nansen Formation, northwest of Girty Creek, head of Hare Fiord, lat. 81°12′N, long. 82°6′W(61073, 61083, 61085); Otto Fiord Formation, 2.2 km east of Girty Creek(61075, 61076, 61078, 61079), "Helicopter section", head of Hare Fiord(61077), Wood Glacier, lat. 80°54′N, long. 84°11′W(61080), and Blue Mountain, lat. 80°42′N, long. 85°52′W(61084), Ellesmere Island, District of Franklin.

Donezellid boundstone [Algae]
Fig. spec. 61293
Mamet, B., Nassichuk, W.W. and Roux, A., 1979, Bull. Cent. Rech. Explor.-Prod. Elf-Aquitaine, vol. 3, no.2, pl. 3, fig. 1.
Otto Fiord Formation, Upper Carboniferous, 2.5 km northwest of van Hauen Pass, Ellesmere Island, District of Franklin.

Dvinella bifurcata Maslov and Kulik-*Tubiphytes obscurus* Maslov boundstone [Algae]
Hypotype 61294
Mamet, B., Nassichuk, W.W. and Roux, A., 1979, Bull. Cent. Rech. Explor.-Prod. Elf-Aquitaine, vol. 3, no.2, pl. 3, fig. 2.
Nansen Formation, Upper Carboniferous, Girty Creek, Ellesmere Island, District of Franklin.

Dvinella bifurcata Maslov et Kulik, 1956 [Algae]
Hypotypes 61004, 61016-61023, 61026-61038
Mamet, B.L., Roux, A. and Nassichuk, W.W., 1987, Geol. Surv. Can., Bull. 342, p. 26, pl. 11, fig. 7; pl. 12, figs. 1-8, 11-23.
Pennsylvanian, Nansen Formation, northeast of Girty Creek, head of Hare, lat. 81°12′N, long. 82°6′W, and 6.5 km north end of Oobloyah Bay(61034); Otto Fiord Formation, Wood Glacier, lat. 80°54′N, long. 84°11′W(61019, 61031), and van Hauen Pass(61036), Ellesmere Island; Nansen Formation, Permian, Griesbach Creek south of Bunde Fiord, lat. 80°28′10″N, long. 94°28′30″W, western Axel Heiberg Island(61027, 61037), and 20 km west of head of Otto Fiord, lat. 81°20′N, long. 85°30′W, Ellesmere Island(61018), District of Franklin.

Dvinella comata Khvorova, 1949 [Algae]
Hypotypes 61009-61015
Mamet, B.L., Roux, A. and Nassichuk, W.W., 1987, Geol. Surv. Can., Bull. 342, p. 26, pl. 11, figs. 12-18.
Pennsylvanian, Nansen Formation, near Girty Creek, head of Hare Fiord, lat. 81°12′N, long. 82°6′W, and 20 km west of head of Otto Fiord, lat. 81°20′N, long. 85°30′W(61012-61015); Otto Fiord Formation, "Helicopter section", head of Hare Fiord(61011), Ellesmere Island, District of Franklin; Arkhingel Province, USSR(61010).

Dvinella crassitheca Kulik, 1964 [Algae]
Hypotypes 61039, 61040
Mamet, B.L., Roux, A. and Nassichuk, W.W., 1987, Geol. Surv. Can., Bull. 342, p. 27, pl. 13, figs. 1, 2.
Nansen Formation, Pennsylvanian, near Girty Creek, head of Hare Fiord, lat. 81°12′N, long. 82°6′W, Ellesmere Island, District of Franklin.

Dvinella gracilis Kulik, 1964 [Algae]
Hypotypes 61041-61043
Mamet, B.L., Roux, A. and Nassichuk, W.W., 1987, Geol. Surv. Can., Bull. 342, p. 28, pl. 13, figs. 3-5.
Pennsylvanian, Canyon Fiord Formation, northwestern Melville Island between McCormick Inlet and Ibbett Bay(61041), and Tellevak member, Hare Fiord Formation, Wood Glacier, lat. 80°54′N, long. 84°11′W, Ellesmere Island, District of Franklin.

Dvinella boundstone [Algae]
Fig. specs. 61285, 61288
Mamet, B, Nassichuk, W.W. and Roux, A., 1979, Bull. Cent. Rech. Explor.-Prod. Elf-Aquitaine, vol. 3, no.2, pl. 2, figs. 1, 4.
Nansen Formation, Upper Carboniferous, Girty Creek, Ellesmere Island, District of Franklin.

"*Earlandia*" *consternatio* (Conkin) [Algae]
Hypotype 39275
Mamet, B.L., 1976, Geol. Surv. Can., Bull. 255, pl. 14, fig. 4.
Shunda Formation, Lower Carboniferous, lat. 53°14′30″N, long. 117°45′30″W, Folding Mountain, Alberta.

Ellesmerella permica (Pia), 1937 [Algae]
Hypotypes 60923-60925
Mamet, B.L., Roux, A. and Nassichuk, W.W., 1987, Geol. Surv. Can., Bull. 342, p. 15, pl. 3, figs. 9-11.
Belcher Channel Formation, Permian, north of Trold Bay, lat. 77°56′N, long. 85°30′W, Ellesmere Island, District of Franklin.

Eotuberitina grandis (Golubsov) [Algae]
Hypotype 39503
Mamet, B.L., 1976, Geol. Surv. Can., Bull. 255, pl. 60, fig. 14.
Shunda Formation, Lower Carboniferous, depth 12,033 feet, Sohio-Un. Prod. et al. Tupper Creek 6-70-B/93-P-9 well, lat. 55°33′30″N, long. 120°15′W, British Columbia.

Eotuberitina reitlingerae (Mikluko-Malkai) [Algae]
Hypotype 39425
Mamet, B.L., 1976, Geol. Surv. Can., Bull. 255, pl. 45, fig. 5.
Opal Member, Mount Head Formation, Lower Carboniferous, lat. 50°39′N, long. 115°04′W, Elpoca Creek, Alberta.

Eovolutina tujmasensis Lipina [Algae]
Hypotype 39277
Mamet, B.L., 1976, Geol. Surv. Can., Bull. 255, pl. 14, fig. 6.
Shunda Formation, Lower Carboniferous, lat. 53°14′30″N, long. 117°45′30″W, Folding Mountain, Alberta.

Eovolutina tujmasensis Lipina, 1955 [Algae]
Hypotype 65841
Mamet, B.L., 1986, Geol. Surv. Can., Bull. 353, p. 31, pl. 12, fig. 13.
Shunda Formation, Lower Carboniferous, Mount Becker, lat. 54°31′30″N, long. 120°39′W, British Columbia.

cf. "Epimastopora piae" Bilgütay, 1960 [Algae]
Hypotype 61153
Mamet, B.L., Roux, A. and Nassichuk, W.W., 1987, Geol. Surv. Can., Bull. 342, p. 37, pl. 20, fig. 8.
Belcher Channel Formation, Permian, west side of head of Blind Fiord, Ellesmere Island, District of Franklin.

Epimastopora symetrica (Johnson) grainstone [Algae]
Hypotype 61300
Mamet, B., Nassichuk, W.W. and Roux, A., 1979, Bull. Cent. Rech. Explor.-Prod. Elf-Aquitaine, vol. 3, no. 2, pl. 3, fig. 8.
Nansen Formation, Upper Carboniferous, Otto Fiord, Ellesmere Island, District of Franklin.

Epimastopora symmetrica (Johnson) 1951
Hypotypes 61115-61134
Mamet, B.L., Roux, A. and Nassichuk, W.W., 1986, Geol. Surv. Can., Bull. 342, p. 36, pl. 17, figs. 8-11; pl. 18, figs. 1-16.
Permian, Belcher Channel Formation, 12.8 km south of Ogle Cape, lat. 76°56′30″N, long. 95°23′W, Grinnel Peninsula, Devon Island, and north of Troll Bay, lat. 77°56′N, long. 85°30′W, Ellesmere Island (61126), District of Franklin; Little Fish Creek, lat. 68°12′N, long. 136°28′W, Yukon(61116-61118, 61120-61123, 61127, 61128, 61130-61134); Nansen Formation, 20 km west of head of Otto Fiord, lat. 81°W′N, long. 85°30′W(61119, 61124), Ellesmere Island, District of Franklin.

Epimastopora grainstone [Algae]
Fig. specs. 61298, 61299
Mamet, B., Nassichuk, W. and Roux, A., 1979, Bull. Cent. Rech. Explor.-Prod. Elf-Aquitaine, vol. 3, no. 2, pl. 3, figs. 6, 7.
Permian, Belcher Channel Formation, Troll Bay, Ellesmere Island, District of Franklin; Fish Creek, Yukon.

Epimastoporella hunzaensis (Zanin Buri) 1965 [Algae]
Hypotypes 61093, 61094
Mamet, B.L., Roux, A. and Nassichuk, W.W., 1987, Geol. Surv. Can., Bull. 342, p. 35, pl. 15, figs. 13, 14.
Nansen Formation, Pennsylvanian, Girty Creek, head of Hare Fiord, lat. 81°12′N, long. 82°6′W, and 20 km west of head of Otto Fiord, lat. 81°20′N, long. 85°30′W, Ellesmere Island, District of Franklin.

Epimastoporella iwaizakiensis (Endo) grainstone [Algae]
Hypotypes 61296, 61297
Mamet, B., Nassichuk, W.W. and Roux, A., 1979, Bull. Cent. Rech. Explor.-Prod. Elf-Aquitaine, vol. 3, no. 2, pl. 3, figs. 4, 5.
Nansen Formation, Upper Carboniferous, Otto Fiord, Ellesmere Island, District of Franklin.

Epimastoporella iwaizakiensis (Endo) 1953 [Algae]
Hypotypes 61098-61107
Mamet, B.L., Roux, A. and Nassichuk, W.W., 1987, Geol. Surv. Can., Bull. 342, p. 33, pl. 16, figs. 4-13.
Belcher Channel Formation, Permian, west side of head of Blind Fiord; Nansen Formation, Upper Carboniferous, 20 km west of head of Otto Fiord, lat. 81°20′N, long. 85°30′W (61104, 61106), Ellesmere Island, District of Franklin.

Epimastoporella japonica (Endo) 1951 [Algae]
Hypotypes 61086-61092, 61095-61097
Mamet, B.L., Roux, A. and Nassichuk, W.W., 1987, Geol. Surv. Can., Bull. 342, p. 34, pl. 15, figs. 6-12; pl. 16, figs. 1-3.
Belcher Channel Formation, Permian, west side of head of Blind Fiord, and Blind Fiord (61090); Nansen Formation, Permian and Upper Carboniferous (61091, 61097), 20 km west of head of Otto Fiord, lat. 81°20′N, long. 85°30′W (61087-61089, 61095); Antoinette Formation, Permian, east of Bayley Mountain, lat. 82°52′N, long. 78°30′W (61096), Ellesmere Island, District of Franklin.

Epistacheoides connorensis Mamet and Rudloff [Algae]
Holotype 39792; paratype 39791
Mamet, B. and Rudloff, B., 1972, Rev. Micropaléont., vol. 15, no. 2, p. 89, pl. 6, figs. 10, 11.
Etherington Formation and Opal Member, Mount Head Formation, Carboniferous, Connor Lakes, British Columbia, and King Creek, Alberta.

Epistacheoides connorensis Mamet and Rudloff, 1972 [Algae]
Hypotype 49049
Jansa, L.F., Mamet, B. and Roux, A., 1978, Can. J. Earth Sci., vol. 15, no. 9, p. 1436, pl. 2, fig. 14.
Windsor Group, Lower Carboniferous, core depth 28 cm, lat. 50°44.4′N, long. 55°09.9′W, northeastern Newfoundland shelf.

Epistacheoides connorensis Mamet et Rudloff, 1972 [Algae]
Hypotypes 61194-61203
Mamet, B.L., Roux, A. and Nassichuk, W.W., 1987, Geol. Surv. Can., Bull. 342, p. 48, pl. 24, figs. 4-13.
Pennsylvanian, Little Fish Creek, lat. 68°12′N, long. 136°28′W, Yukon; Otto Fiord Formation, "Helicopter section", head of Hare Fiord (61196); Nansen Formation, 20 km west of head of Otto Fiord, lat. 81°20′N, long. 85°30′W (61197, 61199-61203), and Girty Creek, head of Hare Fiord, lat. 81°12′N, long. 82°6′W (61198), Ellesmere Island, District of Franklin.

Epistacheoides nephroformis Petryk and Mamet [Algae]
Holotype 29268; paratypes 29269-29281
Petryk, A.A. and Mamet, B.L., 1972, Can. J. Earth Sci., vol. 9, no. 7, p. 789, pl. 8, figs. 1-10.
Carnarvon, Opal, and Loomis members, Mount Head Formation, and Etherington, Carboniferous, Elbow Lake, lat. 50°38′N, long. 115°′W, and King Creek, lat. 50°43′N, long. 115°6′W, Alberta.

Epistacheoides nephroformis Petryk and Mamet [Algae]
Hypotype 39790
Mamet, B. and Rudloff, B., 1972, Rev. Micropaléont., vol. 15, no. 2, p. 89, pl. 6, figs. 8.
Etherington Formation, Carboniferous, King Creek, Alberta.

Epistacheoides nephroformis Petryk and Mamet [Algae]
Hypotype 39379
Mamet, B.L., 1976, Geol. Surv. Can., Bull. 255, pl. 38, fig. 2.
Carnarvon Member, Mount Head Formation, Lower Carboniferous, lat. 50°04′N, long. 114°43′30″W, Ewin Creek, British Columbia.

Epistaecheoides aff. *E. nephroformis* Petryk and Mamet, 1972 [Algae]
Hypotype 49048
Jansa, L.F., Mamet, B. and Roux, A., 1978, Can. J. Earth Sci., vol. 15, no. 9, p. 1436, pl. 2, fig. 13.

Windsor Group, Lower Carboniferous, core depth 82 cm, lat. 50°44.4′N, long. 55°09.9′W, northeastern Newfoundland shelf.

"Eugonophyllum" johsoni Konishi and Wray, 1961 [Algae]
Hypotypes 60939-60952, 60976, 60977
Mamet, B.L., Roux, A. and Nassichuk, W.W., 1987, Geol. Surv. Can., Bull. 342, p. 17, pl. 5, figs. 7-12; pl. 6, figs. 1-8; pl. 8, figs. 6, 7.
Nansen Formation, Pennsylvanian and Permian(60941), 20 km west of head of Otto Fiord, lat. 81°20′N, long. 85°30′W, Ellesmere Island, District of Franklin.

Eusphenopteris sp.
Fig. spec. 77162
Zodrow, E.L. and Vasey, G.M., 1986, J. Palentol., vol. 60, no. 2, p. 219, fig. 12.
Pictou Group, Pennsylvanian, Mabou Mines section, near Finlay Point, Cape Breton Island, Nova Scotia.

Exvotarisella index (Ehrenberg, 1854 emend. von Möller, 1879) [Algae]
Hypotypes 49028, 49029
Jansa, L.F., Mamet, B. and Roux, A., 1978, Can. J. Earth Sci., vol. 15, no. 9, p. 1434, pl. 1, figs. 14, 15.
Windsor Group, Lower Carboniferous, core depths 130 and 148 cm, lat. 50°44′N, long. 54°54.1′W, northeastern Newfoundland shelf.

Fourstonella fusiformis (Brady), 1876 [Algae]
Hypotype 49035
Jansa, L.F., Mamet, B. and Roux, A., 1978, Can. J. Earth Sci., vol. 15, no. 9, p. 1434, pl. 1, fig. 21.
Windsor Group, Lower Carboniferous, core depth 95 cm, lat. 50°44′N, long. 54°54.1′W, northeastern Newfoundland shelf.

Fourstonella sp. [Algae]
Fig. specs. 39788, 39789
Mamet, B. and Rudloff, B., 1972, Rev. Micropaléont., vol. 15, no. 2, p. 88, pl. 6, figs. 1, 2.
Herbert River Formation, Carboniferous, Windsor, Nova Scotia.

Fourstonella sp. [Algae]
Fig. spec. 61181
Mamet, B.L., Roux, A. and Nassichuk, W.W., 1987, Geol. Surv. Can., Bull. 342, p. 45, pl. 23, fig. 3.
Nansen Formation, Pennsylvanian, Girty Creek, head of Hare Fiord, lat. 81°′12N, long. 82°6′W, Ellesmere Island, District of Franklin.

Garwoodia gregaria (Nicholson), 1888 *emend.* Wood, 1941 [Algae]
Hypotype 60932
Mamet, B.L., Roux, A. and Nassichuk, W.W., 1987, Geol. Surv. Can., Bull. 342, p. 13, pl. 4, fig. 7.
Nansen Formation, Permian, 20 km west of head of Otto Fiord, lat. 81°20′N, long. 85°30′W, Ellesmere Island, District of Franklin.

Girvanella ducii Wethered [Algae]
Hypotypes 29188-29196, 29301
Petryk, A.A. and Mamet, B.L., 1972, Can. J. Earth Sci., vol. 9, no. 7, p. 771, pl. 1, figs. 3-5, 7.
Opal, Manston, and Loomis members, Mount Head Formation, Lower Carboniferous, Elbow Lake, lat. 50°38′N, long. 115°′W, South Misty Range, lat. 50°34′N, long. 114°56′W, and Boulder Creek, lat. 50°37′N, long. 114°52′W, Alberta.

Girvanella ducii Wethered [Algae]
Hypotypes 39716, 39717
Mamet, B. and Rudloff, B., 1972, Rev. Micropaléont., vol. 15, no. 2, p. 79, pl. 1, figs. 1, 4.
Banff Formation and Opal Member, Mount Head Formation, Carboniferous, Hudson Bay Imperial Union Paddy a-49-B(1), British Columbia, and lat. 50°34′N, long. 114°56′W, South Misty Range, Alberta.

Girvanella staminea Garwood, 1931 [Algae]
Hypotype 49037
Jansa, L.F., Mamet, B. and Roux, A., 1978, Can. J. Earth Sci., vol. 15, no. 9, p. 1431, pl. 2, fig. 2.
Windsor Group, Lower Carboniferous, core depth 144 cm, lat. 50°44′N, long. 54°1′W, northeastern Newfoundland shelf.

Girvanella wetheredii Chapman [Algae]
Hypotypes 29185-29187, 29300
Petryk, A.A. and Mamet, B.L., 1972, Can. J. Earth Sci., vol. 9, no. 7, p. 771, pl. 1, figs. 1, 2, 6.
Banff Formation and Rundle Group, Lower Carboniferous, Mount Lorette, lat. 50°59′N, long. 115°8′W, Clearwater River, lat. 51°47′N, long. 116°4′W, and South Misty Range, lat. 50°34′N, long. 114°56′W, Alberta.

Girvanella wetheredii Chapman, 1908 [Algae]
Hypotype 49036
Jansa, L.F., Mamet, B. and Roux, A., 1978, Can. J. Earth Sci., vol. 15, no. 9, p. 1431, pl. 2, fig. 1.
Windsor Group, Lower Carboniferous, core depth 144 cm, lat. 50°44′N, long. 55°54.1′W, northeastern Newfoundland shelf.

Girvanella? sp. [Algae]
Fig. spec. 39718
Mamet, B. and Rudloff, B., 1972, Rev. Micropaléont., vol. 15, no. 2, p. 80, pl. 1, fig. 6.
Debolt Formation, Carboniferous, Pacific Highway a-69-13, British Columbia.

Gyroporella? *clavata* Chuvaskov, 1974 [Algae]
Hypotypes 61163-61168
Mamet, B.L., Roux, A. and Nassichuk, W.W., 1987, Geol. Surv. Can., Bull. 342, p. 42, pl. 21, figs. 8-13.
Nansen Formation, Permian, 20 km west of head of Otto Fiord, lat. 81°20′N, long. 85°30′W, Ellesmere Island, District of Franklin.

Gyroporella? *constricta* Kochansky-Devidé, 1970 [Algae]
Hypotypes 61162, 61169
Mamet, B.L., Roux, A. and Nassichuk, W.W., 1987, Geol. Surv. Can., Bull. 342, p. 41, pl. 21, figs. 7, 14.
Nansen Formation, Permian, 20 km west of head of Otto Fiord, lat. 81°21′N, long. 85°30′W, Ellesmere Island, District of Franklin.

Gyroporella nipponica Endo et Hashimoto, 1955 [Algae]
Hypotypes 61174-61178
Mamet, B.L., Roux, A. and Nassichuk, W.W., 1987, Geol. Surv. Can., Bull. 342, p. 40, pl. 22, figs. 3-7.
Nansen Formation, Upper Carboniferous, 20 km west of head of Otto Fiord, lat. 81°20′N, long. 85°30′W, Ellesmere Island, District of Franklin.

Gyroporella sp. [Algae]
Fig. specs. 61170, 61171
Mamet, B.L., Roux, A. and Nassichuk, W.W., 1987, Geol. Surv. Can., Bull. 342, p. 42, pl. 21, figs. 15, 16.
Nansen Formation, Permian, 20 km west of head of Otto Fiord, lat. 81°20′N, long. 85°30′W, Ellesmere Island, District of Franklin.

Halimedoides arctica Mamet and Roux [Algae]
Holotype 60935; hypotypes 60933, 60934, 60936-60938
Mamet, B.L., Roux, A. and Nassichuk, W.W., 1987, Geol. Surv. Can., Bull. 342, p. 14, pl. 5, figs. 1-6.
Antoinette Formation, Permian, east of Mount Bayley on Greely Fiord , lat. 82°52′N, long. 78°30′W, northeastern Ellesmere Island, District of Franklin.

Intextulella agglometera Petryk [Algae]
Holotype 29214; paratype 29215
Petryk, A.A. and Mamet, B.L., 1972, Can. J. Earth Sci., vol. 9, no. 7, p. 797, pl. 2, figs. 8, 9.
Baril Member, Mount Head and Livingstone formations, Lower Carboniferous, Boulder Creek, lat. 50°37′N, long. 114°52′W, and Middle Mountain area, lat. 50°44′N, long. 115°3′W, Alberta.

Issinella devonica Reitlinger, 1954 [Algae]
Hypotypes 49026, 49027
Jansa, L.F., Mamet, B. and Roux, A., 1978, Can. J. Earth Sci., vol. 15, no. 9, p. 1431, pl. 1, figs. 12, 13.
Windsor Group, Lower Carboniferous, core depths 44.3 and 71 cm, lat. 50°44.1′N, long. 54°57.1.′W and lat. 50°44′N, long. 54°54.1′W, northeastern Newfoundland shelf.

Issinella sp. [Algae]
Fig. specs. 39771-39773
Mamet, B. and Rudloff, B., 1972, Rev. Micropaléont., vol. 15, no. 2, p. 86, pl. 5, figs. 8-10.
Windsor Group, Carboniferous, Edwardville, Nova Scotia.

Ivanovia tenuissima Khvorova boundstone [Algae]
Hypotype 61284
Mamet, B, Nassichuk, W.W. and Roux, A., 1979, Bull. Cent. Rech. Explor.-Prod. Elf-Aquitaine, vol. 3, no. 2, pl. 1, fig. 5.
Hare Fiord Formation, Carboniferous, van Hauen Pass, Ellesmere Island, District of Franklin.

Ivanovia tenuissima Khvorova, 1946 [Algae]
Hypotypes 60960-60965, 60971-60975
Mamet, B.L., Roux, A. and Nassichuk, W.W., 1987, Geol. Surv. Can., Bull. 342, p. 19, pl. 7, figs. 1-6; pl. 8, figs. 1-5.
Carboniferous, Hare Fiord Formation, van Hauen Pass, and Nansen Formation, 20 km west of head of Otto Fiord, lat. 81°20′N, long. 85°30′W(60963, 60965), Ellesmere Island, District of Franklin.

Kamaena delicata Antropov, 1967 [Algae]
Hypotype 49030
Jansa, L.F., Mamet, B. and Roux, A., 1978, Can. J. Earth Sci., vol. 15, no. 9, p. 1432, pl. 1, fig. 16.
Windsor Group, Lower Carboniferous, core depth 112 cm, lat. 50°44′N, long. 54°54.1′W, northeastern Newfoundland shelf.

Kamaena itkillikensis Mamet and Rudloff, 1972 [Algae]
Hypotypes 65835, 65836
Mamet, B.L., 1986, Geol. Surv. Can., Bull. 353, p. 29, pl. 12, figs. 7, 8.
Shunda Formation, Lower Carboniferous, Hook Creek, lat. 54°48′N, long. 121°19′W and Mount Becker, lat. 54°31′30″N, long. 120°39′W, British Columbia.

Kamaena aff. *lahuseni* (von Möller) [Algae]
Hypotypes 39776, 39777
Mamet, B. and Rudloff, B., 1972, Rev. Micropaléont., vol. 15, no. 2, p. 87, pl. 5, figs. 13-14.
Debolt and Shunda formations, Carboniferous, Ducette and Mount Hanington, British Columbia.

Kamaena maclareni Mamet and Rudloff [Algae]
Holotype 39779
Mamet, B. and Rudloff, B., 1972, Rev. Micropaléont., vol. 15, no. 2, p. 87, pl. 5, fig. 16.
Pekisko Formation, Carboniferous, Clearwater River, Alberta.

Kamaena? *tenuisepta* Mamet and Rudloff [Algae]
Holotype 39774; paratype 39775
Mamet, B. and Rudloff, B., 1972, Rev. Micropaléont., vol. 15, no. 2, p. 86, pl. 5, figs. 11, 12.
Opal Member, Mount Head Formation, and Pekisko Formation, Carboniferous, Elpoca Creek and Clearwater River, Alberta.

Kamaena sp. [Algae]
Fig. specs. 29220-29222
Petryk, A.A. and Mamet, B.L., 1972, Can. J. Earth Sci., vol. 9, no. 7, p. 777, pl. 3, figs. 9-11.
Pekusko and Shunda formations, Lower Carboniferous, Mount Merlin and Fagan's Lake, lat. 51°36′N, long. 115°46′W, Alberta.

Kamaena sp. [Algae]
Fig. spec. 39778
Mamet, B. and Rudloff, B., 1972, Rev. Micropaléont., vol. 15, no. 2, p. 87, pl. 5, fig. 15.
Opal Member, Mount Head Formation, Carboniferous, Storm Creek, Alberta.

Kamaenella denbighi Mamet and Roux, 1974 [Algae]
Hypotypes 49015-49021
Jansa, L.F., Mamet, B. and Roux, A., 1978, Can. J. Earth Sci., vol. 15, no. 9, p. 1434, pl. 1, figs. 1-7.
Windsor Group, Lower Carboniferous, core depth 57 cm, lat. 50°44.1′N, long. 54°57.1′W, northeastern Newfoundland shelf.

Komia abundans Korde, 1951 [Algae]
Hypotypes 61231, 61233-61239
Mamet, B.L., Roux, A. and Nassichuk, W.W., 1987, Geol. Surv. Can., Bull. 342, p. 53, pl. 26, fig. 11; pl. 27, figs. 2-8.
Nansen Formation, Pennsylvanian, 20 km north of head of Otto Fiord, lat. 81°20′N, long. 85°30′W, and 6.5 km north of end of Oobloyah Bay(61233-61238, Ellesmere Island; Arthaber Creek south of Bunde Fiord, lat. 80°34′30″N, long. 95°34′55″W, Axel Heiberg Island(61239), District of Franklin.

Komia abundans Korde boundstone [Algae]
Hypotype 61290
Mamet, B, Nassichuk, W. and Roux, A., 1979, Bull. Cent. Rech. Explor.-Prod. Elf-Aquitaine, vol. 3, no. 2, pl. 2, fig. 6.
Nansen Formation, Upper Carboniferous, Ooblayah Bay, Ellesmere Island, District of Franklin.

Komia cf. *abundans* Korde [Algae]
Hypotype 39812
Mamet, B. and Rudloff, B., 1972, Rev. Micropaléont., vol. 15, no. 2, p. 91, pl. 9, fig. 6.
Nizi Formation, Carboniferous, McDame, British Columbia.

Koninckopora inflata (de Koninck) [Algae]
Hypotypes 29237-29239
Petryk, A.A. and Mamet, B.L., 1972, Can. J. Earth Sci., vol. 9, no. 7, p. 781, pl. 4, figs. 7, 8.
Loomis, Opal and Marston members, Mount Head Formation, Carboniferous, Elbow Lake, lat. 50°38′N, long. 115°W, Mount McDougall, lat. 50°55′N, long. 115°5′W, and Highwood Valley, lat. 50°24′N, long. 114°37′W, Alberta.

Koninckopora inflata (de Koninck) [Algae]
Hypotypes 39738-39745
Mamet, B. and Rudloff, B., 1972, Rev. Micropaléont., vol. 15, no. 2, p. 82, pl. 2, figs. 14-18; pl. 3, figs. 1-3.
Carboniferous, Alapah Formation, Sedgwick Mountain, Yukon(39738, 39740); Carnarvon and Opal(39742) members, Mount Head Formation, Highwood Pass and Grizzly Creek, Alberta; Windsor Group, Edwardsville, Nova Scotia(39743-39745).

Koninckopora inflata (de Koninck), 1842 [Algae]
Hypotype 49038
Jansa, L.F., Mamet, B. and Roux, A., 1978, Can. J. Earth Sci., vol. 15, no. 9, p. 1432, pl. 2, fig. 3.
Windsor Group, Lower Carboniferous, core depth 82 cm, lat. 50°44.4′N, long. 55°9.9′W, northeastern Newfoundland shelf.

Koninckopora minuta Weyer [Algae]
Hypotypes 39736, 39737, 39746-39748
Mamet, B. and Rudloff, B., 1972, Rev. Micropaléont., vol. 15, no. 5, p. 82, pl. 2, figs. 8, 13; pl. 3, fig. 3-5.
Carboniferous, Opal and Carnarvon members (39746), Mount Head Formation, Picklejar Lakes and Storm Creek, Alberta; Kayak Formation, Sedgwick Mountain, Yukon (39747).

Konickopora tenuiramiosa Wood [Algae]
Hypotypes 29224-29236, 29785
Petryk, A.A. and Mamet, B.L., 1972, Can. J. Earth Sci., vol. 9, no. 7, p. 781, pl. 4, figs. 1-6.
Loomis, Marston and Opal members, Mount Head Formation, Lower Carboniferous, Elbow Lake, lat. 50°38′N, long. 115°′W, Highwood Range, lat. 50°24′N, long. 114°37′W, Mount McDougall, lat. 50°55′N, long. 115°5′W, Elpoca Creek, lat. 50°39′N, long. 115°4′W, South Misty Range, lat. 50°34′N, long. 114°56′W, and King Creek, lat. 43°N, long. 115°6′W, Alberta.

Koninckopora tenuiramosa Wood [Algae]
Hypotype 39422
Mamet, B.L., 1976, Geol. Surv. Can., Bull. 255, pl. 45, fig. 2.
Opal Member, Mount Head Formation, Lower Carboniferous, lat. 50°43′N, long. 115°6′W, King Creek, Alberta.

Koninckopora tenuiramosa Wood [Algae]
Hypotype 39705
Mamet, B.L., 1976, Geol. Surv. Can., Bull. 255, pl. 95, fig. 2.
Kayak Shale/Alapah Limestone talus, Lower Carboniferous, lat. 68°51′N, long. 138°58′W, Mount Sedgwick, Yukon.

Koninckopora tenuiramosa Wood [Algae]
Hypotypes 39725-39735
Mamet, B. and Rudloff, B., 1972, Rev. Micropaléont., vol. 15, no. 2, p. 82, pl. 2, figs. 1-7, 9-12.
Opal Member, Mount Head Formation, Carboniferous, Storm Creek(39725, 39731), Grizzly Creek, and Elbow Lake(39733, 39735), Alberta; Kayak Formation, Carboniferous, Sedgwick Mountain, Alberta(39732, 39734).

Koninckopora tenuiramosa Wood, 1942 [Algae]
Hypotypes 49039-49041
Jansa, L.F., Mamet, B. and Roux, A., 1978, Can. J. Earth Sci., vol. 15, no. 9, p. 1432, pl. 2, figs. 4-6.
Windsor Group, Lower Carboniferous, core depths 28, 82 and 72 cm, lat. 50°44.4′N, long. 55°9.9′W, northeastern Newfoundland shelf.

Linopteris muensteri (Eichwald)
=*Reticulopteris muensteri*, Zodrow, E.L. and Vasey, G.M., 1986, J. Palentol., vol. 60, no. 2, p. 213, fig. 8.3 (plant catalogue hypotype 877).

Linopteris obliqua var. *bunburii* Bell
=*Linopteris bunburii*, Zodrow, E.L. and McCandlish, K., 1980, Upper Carboniferous Fossil Flora in the Collections of Nova Scotia Museum; with Special Reference to the Sydney Coalfield; Nova Scotia Mus., pl. 21, figs. 2(paratype 7052), 4(paratype 7052B); pl. 21, fig. 1(holotype 2762).

Litostroma sp.
Fig, specs. 49024, 49025 [Algae]
Jansa, L.F., Mamet, B. and Roux, A., 1978, Can. J. Earth Sci., vol. 15, no. 9, p. 1436, pl. 1, figs. 10, 11.
Windsor Group, Lower Carboniferous, core depths 127 and 71 cm, lat. 50°44′N, long. 54°54.1′W, northeastern Newfoundland shelf.

Macroporella sp. [Algae]
Fig. specs. 61172, 61173
Mamet, B.L., Roux, A. and Nassichuk, W.W., 1987, Geol. Surv. Can., Bull. 342, p. 38, pl. 22, figs. 1, 2.
Lower Permian, Little Fish Creek, lat. 68°12′N, long. 136°28′W, Yukon.

Mariopteris nervosa? (Brongniart, 1834) Zeiller, 1879
Hypotypes 77157, 77158
Zodrow, E.L. and Vasey, G.M., 1986, J. Paleontol., vol. 60, no. 2, p. 217, figs. 11.1, 11.2.
Pictou Group, Pennsylvanian, Mabou Mines section near Finlay Point, Cape Breton Island, Nova Scotia.

Mariopteris sp.
Fig. spec. 77156
Zodrow, E.L. and Vasey, G.M., 1986, J. Paleontol., vol. 60, no. 2, p. 217, fig. 10.
Pictou Group, Pennsylvanian, Mabou Mines section near Finlay Point, Cape Breton Island, Nova Scotia.

Microcodium [Algae]
Fig. specs. 67826-67843
Mamet, B.L. et Roux, A., 1982, Can. J. Earth Sci., vol. 19, no. 2, p. 358, pl. 1, figs. 1-6; pl. 2, figs. 1-12.
Canyon Fiord Formation, Upper Carboniferous, Raanes Peninsula between Blind and Trold Fiord, lat. 78°23′30″N, long. 85°23′W, Ellesmere Island, District of Franklin; Lower Permian, Little Fish (Cache) Creek, lat. 68°16′30″N, long. 136°22′30″W (67828-67838); Jungle Creek Formation, Lower Permian, Fishing Branch River, lat. 66°08′49″N, long. 139°48′W (67839), and headwaters Nation River, lat. 65°37′N, long. 140°399′W (67841-67845), Yukon.

Microcodium sp. [Algae]
Fig. specs. 76525-76527
Nassichuk, W.W., Davies, G.R. and Mamet, B.L., 1986, Geol. Surv. Can., Paper 86-1B, p. 468, figs. 50.2 a-c.
Emma Fiord Formation, Lower Carboniferous, Lyall River, Grinnell Peninsula, lat. 76°54′30″N, long. 95°15′W, Devon Island, District of Franklin.

Microcodium sp. [Algae]
Fig. specs. 60978-60993
Mamet, B.L., Roux, A. and Nassichuk, W.W., 1987, Geol. Surv. Can., Bull. 342, p. 20, pl. 9, figs. 1-11; pl. 10, figs. 1-5
Permian, Little Fish Creek, lat. 68°12′N, long. 136°28′W, Yukon; Canyon Fiord Formation, Pennsylvanian, Raanes Peninsula between Blind and Trold fiords, lat. 78°28′30″N, long. 85°23′W (60981-60984, 60986, 60986), Ellesmere Island, District of Franklin.

Nanopora cf. *fragilissima* (Maslov) [Algae]
Hypotypes 39749-39756
Mamet, B. and Rudloff, B., 1972, Rev. Micropaléont., vol. 15, no. 2, p. 82, pl. 3, figs. 7-14.
Windsor Group, Carboniferous, Antigonish Milk Plant, Nova Scotia.

Nanopora cf. *N. fragilissima* (Maslov), 1939 [Algae]
Hypotypes 49033, 49034
Jansa, L.F., Mamet, B. and Roux, A., 1978, Can. J. Earth Sci., vol. 15, no. 9, p. 1431, pl. 1, figs. 19, 20.
Windsor Group, Lower Carboniferous, core depth 44 cm, lat. 50°44′N, long. 54°54.1′W, northeastern Newfoundland shelf.

Nansenella multifurcata Mamet and Roux [Algae]
Holotype 60922; hypotypes 60915-60921
Mamet, B.L., Roux, A. and Nassichuk, W.W., 1987, Geol. Surv. Can., Bull. 342, p. 15, pl. 3, figs. 1-8.
Nansen Formation, Pennsylvanian, Girty Creek, head of Hare Fiord, lat. 81°12′W, long. 82°6′N, and Permian, 20 km west of head of Otto Fiord, lat. 81°20′N, long. 85°30′W (60920); Otto Fiord Formation, Pennsylvanian, Weber Glacier (60915, 60918), and Wood Glacier, lat. 80°54′N, long. 84′11″W (60916, 60917, 60919, 60921), Ellesmere Island, District of Franklin.

Neoanchicodium catenoides Endo *in* Endo and Kanuma, 1954 [Algae]
Hypotypes 60953-60959, 60966, 60967
Mamet, B.L., Roux, A. and Nassichuk, W.W., 1987, Geol. Surv. Can., Bull. 342 p. 18, pl 6, figs. 9-15; pl. 7, figs. 7, 8.
Permian, Nansen Formation, Girty Creek, head of Hare Fiord, lat. 81°12′N, long. 82°6′W, and 20 km west of head of Otto Fiord, lat. 81°20′, long. 85°30′W (60955, 60958), and Belcher Channel Formation, west side of end of Blind Fiord (60956), Ellesmere Island; Nansen Formation, Arthabes Creek, south of Bunde Fiord, lat. 80°34′30″N, long. 95°34′55″W, Axel Heiberg Island (60967), District of Franklin.

Neuropteris cf. *heterophylla* (Brongniart, 1831) Sternberg, 1833
Hypotype 77147
Zodrow, E.L. and Vasey, G.M., 1986, J. Paleontol., vol. 60, no. 2, p. 212, fig. 4.
Pictou Group, Pennsylvanian, roof of 3-Foot Seam(?), Mabou Point west of Coal Mine Point, Mabou Mines coalfield, Cape Breton Island, Nova Scotia.

Neuropteris tenuifolia (Schlotheim, 1820) Sternberg, 1833
Hypotypes 77148-77153
Zodrow, E.L. and Vasey, G.M., 1986, J. Paleontol., vol. 60, no. 2, p. 213, figs. 6.1-6.4, 6.8, 7.1, 7.2.
Pictou Group, Pennsylvanian, footages 600′4″-607′and 621′-624′(77150), Borehole 5, Coal Mine Point, Mabou Mines section near Finlay Point (77151, 77152) and unknown locality (77153), Cape Breton Island, Nova Scotia.

Neuropteris (Mixoneura) flexuosa (Sternberg) Crookall, 1959
Hypotypes 85304,a-85306,a
Zodrow, E.L. and McCandlish, K., 1980, Rev. Palaeobot. Palynol., vol. 30, p. 61, pl. I; text fig. 2.
Pennsylvanian, dump at Lingan coal mine, Cape Breton Island, Nova Scotia.

Nostocites vesiculosa Maslov, 1929 [Algae]
Hypotypes 61267, 61268
Mamet, B.L., Roux, A. and Nassichuk, W.W., 1987, Geol. Surv. Can., Bull. 342, p. 58, pl. 30, figs. 1, 2.
Nansen Formation, Upper Carboniferous, 20 km west of head of Otto Fiord, lat. 81°20′N, long. 85°30′W, and Otto Fiord Formation, Mississippian-Pennsylvanian, Blue Mountain, lat. 80°42′N, long. 85°52′W, south of Hare Fiord, Ellesmere Island, District of Franklin.

Odontopteris minor Brongniart
=*?Odontopteris* spp., Zodrow, E.L., 1985, Palaeontographica, Abt. B, vol. 196, nos. 4-6, p. 92, text-figs. 9 (fig. spec. 2549), 10 (fig. spec. 2550).

Odonotopteris subcuneata Bunbury
=*Neuropteris scheuchzeri*, Zodrow, E.L., 1985, Palaeontographica, Abt. B, vol. 196, nos. 4-6, p. 95, text-fig. 11 (hypotype 2536).

Orthriosiphon saskatchewanensis Johnson and Konishi [Algae]
Hypotypes 29201-29207, 29775-29777
Petryk, A.A. and Mamet, B.L., 1972, Can. J. Earth Sci., vol. 9, no. 7, p. 775, pl. 2, figs. 1-7; pl. 3, figs. 1-3.
Livingstone, Mount Head (Salter Member), and Shunda formations, Lower Carboniferous, Mount Lorette, lat. 50°59′N, long. 115°8′W, Middle Mountain area, lat. 50°44′N, long. 115°3′W, Elbow Lake, lat. 50°38′N, long. 115°W, Divide Creek, lat. 51°45′N, long. 115°53′W, Clearwater River, lat. 51°47′N, long. 116°4′W, and Forbidden Creek, lat. 51°48′N, long. 115°50′W, Alberta.

Orthriosiphon saskatchewanensis Johnson and Konishi [Algae]
Hypotypes 39291, 39292, 39310
Mamet, B.L., 1976, Geol. Surv. Can., Bull. 255, pl. 17, figs. 1, 2; pl, 21, fig. 1.
Turner Valley Formation, Lower Carboniferous, lat. 53°05′N, long. 118°02′W, Jasper Lake, and lat. 50°17′N, long. 114°32′W, Mount Burke, Alberta.

Orthriosiphon saskatchewanensis Johnson and Konishi, 1956 [Algae]
Hypotypes 65829, 65830
Mamet, B.L., 1986, Geol. Surv. Can., Bull. 353, p. 28, pl. 12, figs. 1, 2.
Shunda Formation, Lower Carboniferous, East Fellers Creek, lat. 54°42′30″N, long. 120°54′W, British Columbia.

Orthriosiphon aff. *saskatchewanensis* Johnson and Konishi [Algae]
Hypotypes 39763, 39764
Mamet, B. and Rudloff, B., 1972, Rev. Micropaléont., vol. 15, no. 2, p. 85, pl. 4, figs. 10, 11.
Shunda Formation, Carboniferous, Folding Mountain, Alberta.

Orthriosiphon sp. cf. *O. saskatchewanensis* Johnson and Konishi, 1956 [Algae]
Fig. spec. 60968
Mamet, B.L., Roux, A. and Nassichuk, W.W., 1987, Geol. Surv. Can., Bull. 342, p. 21, pl. 7, fig. 9.
Nansen Formation, Pennsylvanian, 20 km west of head of Otto Fiord, lat. 81°20′N, long. 85°30′W, Ellesmere Island, District of Franklin.

Orthriosiphonoides salterensis Petryk and Mamet [Algae]
Holotype 29208; paratypes 29209-29213, 29302, 29303
Petryk, A.A. and Mamet, B.L., 1972, Can. J. Earth Sci., vol. 9, no. 7, p. 776, pl. 2, figs. 10-14; pl. 3, figs. 4, 5.
Mamet, B. and Rudloff, B., 1972, Rev. Micropaléont., vol. 15, no. 2, p. 85, pl. 4, fig. 8 (29302).
Salter, Baril and Opal members, Mount Head Formation, Lower Carboniferous, Highwood Valley, lat. 50°24′N, long. 114°37′W, South Misty Range, lat. 50°34′N, long. 114°56′W, Plateau Mountain, lat. 50°12′N, long. 114°32′W, Elbow Lake, lat. 50°38′N, long. 115°W, King Creek, lat. 50°43′N, long. 115°6′W, and Grotto Creek, lat. 51°6′N, long. 115°19′W, Alberta.

Orthriosiphonoides salterensis Petryk and Mamet [Algae]
Hypotype 39762
Mamet, B. and Rudloff, B., 1972, Rev. Micropaléont, vol. 15, no. 2, p. 85, pl. 4, fig. 9.
Salter Member, Mount Head Formation, Carboniferous, Grotto Creek, Alberta.

Orthriosiphonoides sp. cf. *O. saltерersis* Petryk *in* Petryk and Mamet 1972, [Algae]
Fig. specs. 60969, 60970
Mamet, B.L., Roux, A. and Nassichuk, W.W., 1987, Geol. Surv. Can., Bull. 342, p. 22, pl. 7, figs. 10, 11.
Permian, Little Fish Creek, lat. 68°12′N, long. 136°28′W, Yukon; Nansen Formation, Pennsylvanian, 20 km west of Otto Fiord, lat. 81°20′N, long. 85°30′W, Ellesmere Island, District of Franklin.

Orthriosiphonoides tenuiramosa Mamet and Rudloff [Algae]
Holotype 39765; paratypes 39766-39768
Mamet, B. and Rudloff, B., 1972, Rev. Micropaléont., vol. 15, no. 2, p. 86, pl. 4, figs. 12-15.
Salter Member, Mount Head Formation, Carboniferous, Grotto Creek, Alberta.

Ortonella coloradoensis Johnson [Algae]
Hypotypes 29197-29199
Petryk, A.A. and Mamet, B.L., 1972, Can. J. Earth Sci., vol. 9, no. 7, p. 773, pl. 1, figs. 8-10.
Opal and Carnarvon members, Mount Head Formation, Lower Carboniferous, Mount McDougall, lat. 50°55′N, long. 115°5′W, and Boulder Creek, lat. 50°37′N, long. 114°52′W, Alberta.

Ortonella coloradoensis Johnson, 1945 [Algae]
Hypotypes 65832, 65833
Mamet, B.L., 1986, Geol. Surv. Can., Bull. 353, p. 29, pl. 12, figs. 4, 5.
Shunda Formation, Lower Carboniferous, Belcourt Creek, lat. 54°22′N, long. 120°29′30″W and East Fellers Creek, lat. 54°42′30″N, long. 120°54′W, British Columbia.

Ortonella aff. *coloradoensis* Johnson [Algae]
Hypotype 39724
Mamet, B. and Rudloff, B., 1972, Rev. Micropaléont, vol. 15, no. 2, p. 81, pl. 1, fig. 12.
Banff Formation, Carboniferous, Mount Hanington, British Columbia.

Ortonella sp. aff. *O. gracilis* Johnson, 1951 [Algae]
Fig. spec. 60927
Mamet, B.L., Roux, A. and Nassichuk, W.W., 1987, Geol. Surv. Can., Bull. 342, p. 13, pl. 4, fig. 2.
Nansen Formation, Permian, 20 km west of head of Otto Fiord, lat. 81°20′N, long. 85°30′W, Ellesmere Island, District of Franklin.

Ortonella kershopensis Garwood [Algae]
Hypotype 39721
Mamet, B. and Rudloff, B,. 1972, Rev. Micropaléont, vol. 15, no. 2, p. 81, pl. 1, fig. 9.
Carnarvon Member, Mount Head Formation, Carboniferous, Cataract Creek, Alberta.

Ortonella tyrrelensis Mamet and Rudloff [Algae]
Holotype 39719; paratype 39720
Mamet, B. and Rudloff, B., 1972, Rev. Micropaléont, vol. 15, no. 2, p. 81, pl. 1, figs. 7, 8.
Shunda Formation, Carboniferous, Mount Tyrrell, Alberta; Carboniferous, Monks Head, Nova Scotia.

Ortonella tyrrelensis Mamet and Rudloff, 1972 [Algae]
Hypotype 65831
Mamet, B.L., 1986, Geol. Surv. Can., Bull. 353, p. 29, pl. 12, fig. 3.
Shunda Formation, Lower Carboniferous, Belcourt Creek, lat. 54°22′N, long. 120°29′30″W, British Columbia.

Ortonella sp. [Algae]
Fig. specs. 39722, 39723
Mamet, B. and Rudloff, B., 1972, Rev. Micropaléont., vol. 15, no. 2, p. 81, pl. 1, figs. 10, 11.
Pekisko and Kayak formations, Carboniferous, Mount Hanington, British Columbia, and Sedgwick Mountain, Yukon.

"Osagia incrustata" Twenhofel, 1919 [Algae]
Hypotypes 60907-60914, 60926
Mamet, B.L., Roux, A. and Nassichuk, W.W., 1987, Geol. Surv. Can., Bull. 342, p. 12. pl. 1, figs. 7-10; pl. 2, figs. 1-4; pl. 4, fig. 1.
Otto Fiord Formation, Pennsylvanian, 2.5-3 km northeast of van Hauen Pass; Wood Glacier, lat. 80°54′N, long. 84°11′W (60912, 60912); van Hauen Pass (60914, 60926), Ellesmere Island, District of Franklin.

Palaeoberesella lahuseni (von Möller), 1879 [Algae]
Hypotype 49032
Jansa, L.F., Mamet, B. and Roux, A., 1978, Can. J. Earth Sci., vol. 15, no. 9, p. 1434, pl. 1, fig. 18.
Windsor Group, Lower Carboniferous, core depth 112 cm, lat. 50°44′N, long. 54°54.1′W, northeastern Newfoundland shelf.

Palaeoberesella lahuseni (von Möller, 1879) [Algae]
Hypotype 65837
Mamet, B.L., 1986, Geol. Surv. Can., Bull. 353, p. 30, pl. 12, fig. 9.
Pekisko Formation, Lower Carboniferous, Mount Becker, lat. 54°31′30″N, long. 120°39′W, British Columbia.

Palaeoberesella aff. *P. lahuseni* (von Möller), 1879 [Algae]
Hypotype 49031
Jansa, L.F., Mamet, B. and Roux, A., 1978, Can. J. Earth Sci., vol. 15, no. 9, p. 1432, pl. 1, fig. 17.
Windsor Group, Lower Carboniferous, core depth 135 cm, lat. 50°44′N, long. 54°54.1′W, northeastern Newfoundland shelf.

Palaeocancellus aff. *P. cancellatus* (Williamson) [Algae]
Hypotype 39505
Mamet, B.L., 1976, Geol. Surv. Can., Bull. 255, pl. 60, fig. 16.
Shunda Formation, Lower Carboniferous, depth 12,033 feet, Sohio-Un. Prod. et al. Tupper Creek C-70-B/93-P-9 well, lat. 55°33′30″N, long. 120°15′W, British Columbia.

Paracalligelloides sp. [Algae]
Fig. specs. 39427-39429
Mamet, B.L., 1976, Geol. Surv. Can., Bull. 255, pl. 45, figs. 7-9.
Opal Member, Mount Head Formation, Lower Carboniferous, lat. 50°39′N, long. 115°04′W, Elpoca Creek, Alberta.

Parachaetetes garwoodi Hume *in* Garwood [Algae]
Hypotype 39813
Mamet, B. and Rudloff, B., 1972, Rev. Micropaléont., vol. 15, no. 2, p. 92, pl. 10, fig. 5.
Banff Formation, Carboniferous, Mount Hanington, British Columbia.

Parachaetetes garwoodi (Hinde *in* Garwood, 1912) [Algae]
Hypotype 65834
Mamet, B.L., 1986, Geol. Surv. Can., Bull. 353, p. 30, pl. 12, fig. 6.
Shunda Formation, Lower Carboniferous, East Fellers Creek, lat. 54°42′30″N, long. 120°54′W, British Columbia.

Paraepimastopora kansasensis (Johnson) 1946 [Algae}
Hypotypes 61108-61111
Mamet, B.L., Roux, A. and Nassichuk, W.W., 1987, Geol. Surv. Can., Bull. 342, p. 35, pl. 17, figs, 1-4.
Nansen Formation, Pennsylvanian, 20 km west of head of Otto Fiord, lat. 81°20′N, long. 85°30′W, Ellesmere Island, District of Franklin; Permian, Little Fish Creek, lat. 68°12′N, long. 136°28′W, Yukon.

Paraepimastopora sp. [Algae]
Fig. specs. 61112-61114
Mamet, B.L., Roux, A. and Nassichuk, W.W., 1987, Geol. Surv. Can., Bull. 342, p. 36, pl. 17, figs. 5-7.
Canyon Fiord Formation, Pennsylvanian, Henrietta Nesmith Glacier, north of Lake Hazen, northern Ellesmere Island, District of Franklin.

Parathurammina of the group *P. dagmarae* Suleimanov - *P.* aff. *P. subvasta* (Bykova) [Algae]
Fig. spec. 39502
Mamet, B.L., 1976, Geol. Surv. Can., Bull. 255, pl. 60, fig. 13.
Shunda Formation, Lower Carboniferous, depth 12,033 feet, Sohio-Un. Prod. Tupper Creek 6-70-B/93-P-9 well, lat. 55°33′30″N, long. 120°15′W, British Columbia.

Parathurammina of the group *P. paracushmani* Reitlinger [Algae]
Fig. spec. 39504
Mamet, B.L., 1976, Geol. Surv. Can., Bull. 255, pl. 60, fig. 15.
Shunda Formation, Lower Carboniferous, depth 6855 feet, Texaco NFA Boundary Lake (1) 6-6-86-13W6 well, lat. 56°26′N, long. 120°01′W, British Columbia.

Parathurammina of the group *P. spinosa* (Williamson) [Algae]
Fig. spec. 39279
Mamet, B.L., 1976, Geol. Surv. Can., Bull. 255, pl. 14, fig. 8.
Shunda Formation, Lower Carboniferous, lat. 52°32′N, long. 116°51′30″W, Brazeau River, Alberta.

Parathurammina of the group *P. spinosa* (Williamson) [Algae]
Fig. specs. 39494-39501
Mamet, B.L., 1976, Geol. Surv. Can., Bull. 255, pl. 60, figs. 5-12.
Shunda Formation, Lower Carboniferous, depth 3846 feet, Imperial Sikanni Chief (1)b-92-D/94-I-4 well, lat. 58°05′N, long. 121°53′W; depth 12,033 feet, Sohio-Un. Prod. et al. Tupper Creek c-70-B93-P-9 well, lat. 55°33′30″N, long. 120°15′W (39495-39497, 39499, 39501); and depth 6855 feet, Texaco NFA Boundary Lake (1)6-6-86-13W6 well, lat. 56°26′N, long 120°0′W (39498) British Columbia.

Pecopteris herdii (Bell, 1938) Jongmans, 1962
Hypotype 77159
Zodrow, E.L. and Vasey, G.M., 1986, J. Paleontol., vol. 60, no. 2, p. 221, fig. 13.2.
Pictou Group, Pennsylvanian, footage 328-333 feet, Borehole 4, Finlay Point, Mabou Mines coalfield, Cape Breton Island, Nova Scotia.

Pecopteris cf. *miltonii - lamuriana* complex sensu Darrah, 1969
Hypotypes 77160, a, 77161
Zodrow, E.L. and Vasey, G.M., 1986, J. Paleontol., vol. 60, no.2, p. 221, figs. 13.4-13.6.
Pictou Group, Pennsylvanian, Mabou Mines section near Finlay Point, Cape Breton Island, Nova Scotia.

Pekiskopora macqueeni Mamet [Algae]
Holotype 39821; paratypes 39820, 39822-39841
Mamet, B.L., 1974, Rev. Micropaléont., vol. 17, no. 1, p. 42, pl. 1, figs. 1-11; pl. 2, figs. 1-11.
Pekisko Formation, Lower Carboniferous, lat. 54°23′N, long. 120°29′W, Belcourt Creek, British Columbia.

Pekiskopora sp. [Algae]
Fig. specs. 39842-39845
Mamet, B.L., 1974, Rev. Micropaléont., vol. 17, no. 1, p. 42, pl. 3, figs. 1-4.
Pekisko Formation, Lower Carboniferous, lat. 54°22′N, long. 120°29′W, Belcourt Creek, British Columbia.

Pekiskopora sp. [Algae]
Fig. specs. 49043-49046
Jansa, L.F., Mamet, B. and Roux, A., 1978, Can. J. Earth Sci., vol. 15, no. 9, p. 1432, pl. 2, figs. 8-11.
Windsor Group, Lower Carboniferous, core depths 36, 156, 95 and 144 cm, lat. 50°44′N, long. 54°54.1′W, northeastern Newfoundland shelf.

Pekiskopora sp. [Algae]
Fig. spec. 65839
Mamet, B.L., 1986, Geol. Surv. Can., Bull. 353, p. 30, pl. 12, fig. 11.
Banff Formation, Lower Carboniferous, Mount Hanington, lat. 54°10′30″N, long. 120°15′W, British Columbia.

Petschoria elegans Korde, 1951 [Algae]
Hypotypes 61219-61223
Mamet, B.L., Roux, A. and Nassichuk, W.W., 1987, Geol. Surv. Can., Bull. 342, p. 51, pl. 25, figs. 16, 17; pl. 26, figs. 1-3.
Pennsylvanian, Little Fish Creek, lat. 68°12″N, long. 136°28′W, Yukon.

Polymorphocodium lapparenti Derville [Algae]
Hypotypes 29290-29293
Petryk, A.A. and Mamet, B.L., 1972, Can. J. Earth Sci., vol. 9, no. 7, p. 783, pl. 10, figs. 1, 2.
Loomis and Opal members, Mount Head Formation, Carboniferous, Elbow Lake, lat. 50°38″N, long. 115°W, and Elpoca Creek, lat. 50°39′N, long. 115°4′W, Alberta.

Polysphaerinella bulla (Conil and Lys), 1968 [Algae]
Hypotype 49047
Jansa, L.F., Mamet, B. and Roux, A., 1978, Can. J. Earth. Sci., vol. 15, no. 9, p. 1436, pl. 2, fig. 12.

Windsor Group, Lower Carboniferous, core depth 45 cm, lat. 50°44.1′N, long. 54°57.1′W, northeastern Newfoundland shelf.

Pseudogyroporella mizziaformis Endo, 1959 [Algae]
Hypotypes 61159-61161
Mamet, B.L., Roux, A. and Nassichuk, W.W., 1987, Geol. Surv. Can., Bull. 342, p. 39, pl. 21, figs. 4-6.
Nansen Formation, Upper Carboniferous, 20 km west of head of Otto Fiord, lat 81°20′N, long. 85°30″W, Ellesmere Island; Permian, Griesbach Creek, south of Bunde Fiord, lat. 80°28′10″N, long. 94°28′30″W, western Axel Heiberg Island (61160), District of Franklin.

Pseudohedstroemia polyfurcata Mamet and Roux, 1978 [Algae]
Hypotype 65838
Mamet, B.L., 1986, Geol. Surv. Can., Bull. 353, p. 29, pl. 12, fig. 10.
Shunda Formation, Lower Carboniferous, Belcourt Creek, lat. 54°22′N, long. 120°29′30″W, British Columbia.

Pseudoissinella alaskensis Mamet and Rudloff [Algae]
Paratype 39780
Mamet, B. and Rudloff, B., 1972, Rev. Micropaléont., vol. 15, no. 2, p. 86, pl. 5, fig. 19.
Opal Member, Mount Head Formation, Carboniferous, Tunnel Mountain, Alberta.

Pseudokamaena armstrongi Mamet [Algae]
Hypotype 39781
Mamet, B. and Rudloff, B., 1972, Rev. Micropaléont., vol. 15, no. 2, p. 87, pl. 5, fig. 20.
Pekisko Formation, Carboniferous, Clearwater River, Alberta.

Pseudokamaena armstrongi Mamet *in* Petryk and Mamet, 1972 [Algae]
Hypotypes 61006, 61007
Mamet, B.L., Roux, A. and Nassichuk, W.W., 1987, Geol. Surv. Can., Bull. 342, p. 22, pl. 11, figs. 9, 10.
Otto Fiord Formation, Pennsylvanian, east of Girty Creek, head of Hare Fiord, Ellesmere Island, District of Franklin.

Pseudokamaena boulderensis Mamet [Algae]
Holotype 29223
Petryk, A.A. and Mamet, B.L., 1972, Can. J. Earth Sci., vol. 9, no. 7, p. 780, pl. 3, fig. 13.
Carnarvon Member, Mount Head Formation, Lower Carboniferous, Boulder Creek, lat. 50°37′N, long. 114°52′W, Alberta.

Pseudokamaena sp. [Algae]
Fig. spec. 61005
Mamet, B.L., Roux, A. and Nassichuk, W.W., 1987, Geol. Surv. Can., Bull. 342, p. 23, pl. 11, fig. 8.
Otto Fiord Formation, Pennsylvanian, east of Girty Creek, head of Hare Fiord, Ellesmere Island, District of Franklin.

Pseudokomia cansecoensis Rácz, 1964
Hypotype 61208
Mamet, B.L., Roux, A. and Nassichuk, W.W., 1987, Geol. Surv. Can., Bull. 342, p. 51, pl. 25, fig. 5.
Nansen Formation, Pennsylvanian, Oobloyah Bay, Ellesmere Island, District of Franklin.

Pseudostacheoides loomisi Petryk and Mamet [Algae]
Holotype 29289; paratype 39795 [not 29303]
Petryk, A.A. and Mamet, B.L., 1972, Can. J. Earth Sci., vol. 9, no. 7, p. 793, pl. 9, figs. 6-8
Mamet, B. and Rudloff, B., 1972, Rev. Micropaléont., vol. 15, no. 2, p. 89, pl. 6, fig. 14 (39795)
Loomis and Marston members, Mount Head Formation, Carboniferous, Picklejar Lakes, lat. 50°32′N, long. 114°46′W, and Boulder Creek, lat. 50°37′N, long. 114°52′W, Alberta.

Pseudostacheoides loomisi Petryk and Mamet [Algae]
Hypotypes 39793, 39794
Mamet, B. and Rudloff, B., 1972, Rev. Micropaléont., vol. 15, no. 2, p. 89, pl. 6, figs. 12, 13.
Opal Member, Mount Head Formation, Carboniferous, South Misty Range, Alberta.

"Radiosphaera" aff. "R." basilica (Reitlinger) [Algae]
Hypotype 39276
Mamet, B.L., 1976, Geol. Surv. Can., Bull. 255, pl. 14, fig. 5.
Shunda Formation, Lower Carboniferous, lat. 52°32′N, long. 116°51′30″W, Brazeau River, Alberta.

Radiosphaera denticulata Malakhova, 1975 [Algae]
Hypotypes 49022, 49023
Jansa, L.F., Mamet, B. and Roux, A., 1978, Can. J. Earth Sci., vol. 15, no. 9, p. 1436, pl. 1, figs. 8, 9.
Windsor Group, Lower Carboniferous, core depths 44.3 and 135 cm, lat. 50°44.1′N, long. 54°57.1′W. and lat. 50°44′N, long. 54°54.1′W, northeastern Newfoundland shelf.

"Radiosphaera" of the group "R" solo (Williamson) [Algae]
Fig. specs. 39557, 39558
Mamet, B.L., 1976, Geol. Surv. Can., Bull. 255, pl. 72, figs. 3, 4.
Debolt Formation, Lower Carboniferous, depth 7054 feet, Pacific Highway (2)a-47-I/94-B-16 well, lat. 55°57′N, long. 122°04′W, British Columbia.

"Radiosphaera" sp. (="Asterosphaera" sp.) [Algae]
Fig. spec. 39424
Mamet, B.L., 1976, Geol. Surv. Can., Bull. 255, pl. 45, fig. 4.
Carnarvon Member, Mount Head Formation, Lower Carboniferous, lat. 50°37′N, long. 114°52′W, Boulder Creek, Alberta.

Rectangulina tortuosa (Antropov), 1950 [Algae]
Hypotypes 60901-60904
Mamet, B.L., Roux, A. and Nassichuk, W.W., 1987, Geol. Surv. Can., Bull. 342, p. 6, pl. 1, figs. 1-4.

Pennsylvanian, Tellevak Member, Hare Fiord Formation, Wood Glacier, Krieger Mountain, lat. 80°54′N, long. 84°11′W, Otto Fiord Formation, 25.3 km northeast of van Hauen Pass (60902, 60904), and Nansen Formation, Girty Creek, head of Hare Fiord, lat. 81°12′N, long. 82°6′W (60903), Ellesmere Island, District of Franklin.

Rectangulina sp. [Algae]
Fig. spec. 39815
Mamet, B. and Rudloff, B., 1972, Rev. Micropaléont., vol. 15, no. 2, p. 80, pl. 10, fig. 13,
Windsor Group, Carboniferous, Antigonish Milk Plant, Nova Scotia.

Rectangulina sp. [Algae]
Fig. specs. 60905, 60906
Mamet, B.L., Roux, A. and Nassichuk, W.W., 1987, Geol. Surv. Can., Bull. 342, p. 7, pl. 1, figs. 5, 6.
Otto Fiord Formation, Pennsylvanian, 2.5 km northeast of van Hauen Pass, Ellesmere Island, District of Franklin.

Reticulopteris muensteri (Eichwald, 1840) Gothan, 1941
Hypotypes 77154, 77155
Zodrow, E.L. and Vasey, G.M., 1986, J. Paleontol., vol. 60, no. 2, p. 213, figs. 8.1, 8.4.
Pictou Group, Pennsylvanian, footage 654′2″-655′, Borehole 2, Coal Mine Point and footage 262′-264′4″, Finlay Point, Mabou Mines coalfield, Cape Breton Island, Nova Scotia.

Richella incrustata Mamet et Roux [Algae]
Holotype 61275; paratypes 61276-61279
Mamet, B.L., Roux, A. and Nassichuk, W.W., 1987, Geol. Surv. Can., Bull. 342, p. 58, pl. 30, figs. 9-13.
Nansen Formation, Pennsylvanian, 20 km west of head of Otto Fiord, lat 81°20′N, long. 85°30′W, Ellesmere Island, District of Franklin.

Sphenophyllum majus (Bronn, 1828) Bronn, 1835 sensu Remy, 1962
Hypotype 85291
Zodrow, E.L., 1989, Rev. Palaeobot. Palynol., vol. 58, p. 317, pl. 1, fig. 5; text-fig. 9.
Minto Formation, Pictou Group, Pennsylvanian, Newcastle Coal Mine, Minto, New Brunswick.

Sphaeroporella sp. [Algae]
Fig. spec. 29297
Petryk, A.A., and Mamet, B.L., 1972, Can. J. Earth Sci., vol. 9, no. 7, p. 795, pl. 10, fig. 7.
Opal Member, Mount Head Formation, Carboniferous, Elbow Lake, lat. 50°38″N, long. 115°W, Alberta.

Sphaeroporella sp. [Algae]
Fig. spec. 39814
Mamet, B. and Rudloff, B., 1972, Rev. Micropaléont., vol. 15, no. 2, p. 88, pl. 10, fig. 12.
Livingstone Formation, Carboniferous, Carrot Creek, Alberta.

?Sphenopteris sp.
Fig. spec. 77163
Zodrow, E.L. and Vasey, G.M., 1986, J. Paleontol., vol. 60, no. 2, p. 223, fig. 14.
Pictou Group, Pennsylvanian, Finlay Point, Mabou Mines section, Cape Breton Island, Nova Scotia.

Sphinctoporella sp. [Algae]
Fig. spec. 39760
Mamet, B. and Rudloff, B., 1972, Rev. Micropaléont., vol. 15, no. 2, p. 84, pl. 4, fig. 7.
Windsor Group, Carboniferous, Antigonish Milk Plant, Nova Scotia.

Stacheia marginulinoides Brady, 1876 [Algae]
Hypotypes 61179, 61180
Mamet, B.L., Roux, A. and Nassichuk, W.W., 1987, Geol. Surv. Can., Bull. 342, p. 45, pl. 23, figs. 1, 2.
Otto Fiord Formation, Pennsylvanian, immediately east of Girty Creek, head of Hare Fiord, and 2.5 km northeast of van Hauen Pass, Ellesmere Island, District of Franklin.

Stacheia sp. [Algae]
Fig. spec. 29266
Petryk, A.A. and Mamet, B.L., 1972, Can. J. Earth Sci., vol. 9, no. 7, p. 784, pl. 7, fig. 7.
Opal Member, Mount Head Formation, Carboniferous, Elpoca Creek, lat. 50°39′N, long. 115°4′W, Alberta.

Stacheoidella spissa (Petryk and Mamet) 1972 [Algae]
Hypotype 61190
Mamet, B.L., Roux, A. and Nassichuk, W.W., 1987, Geol. Surv. Can., Bull. 342, p. 47, pl. 23, fig. 12.
Otto Fiord Formation, Pennsylvanian, Wood Glacier, lat. 80°54′N, long. 84°11′W, Ellesmere Island, District of Franklin.

Stacheoides? meandriformis Mamet and Rudloff [Algae]
Holotype 39804; paratypes 39802, 39803, 39805
Mamet, B. and Rudloff, B., 1972, Rev. Micropaléont., vol. 15, no. 2, p. 90, pl. 7, figs. 9, 10, 12, 14.
Carboniferous, Mount Head Formation, Connor Lakes, British Columbia; Cache Creek Formation, Richardson Mountains, Yukon.

Stacheoides meandriformis Mamet et Rudloff, 1972 [Algae]
Hypotypes 61186-61189
Mamet, B., Roux, A. and Nassichuk, W.W., 1987, Geol. Surv. Can., Bull. 342, p. 46, pl. 23, figs. 8-11.
Nansen Formation, Lower Permian, Griesbach Creek, south of Bunde Fiord, lat. 80°28′10″N, long. 94°28′30″W, and Upper Carboniferous (61188), Arthaber Creek, lat. 80°34′30″N, long. 95°34′55″W, Axel Heiberg Island; Nansen Formation, Upper Carboniferous (61187), 6.5 km north of end of Oobloyah Bay, Ellesmere Island, District of Franklin; Upper Carboniferous, Little Fish Creek, Yukon (61189).

Stacheiodes? spissa Petryk and Mamet [Algae]
Holotype 29241; paratypes 29240, 29242-29247, 29250, 29780-29782, 29304

Petryk, A.A., and Mamet, B.L., 1972, Can. J. Earth Sci., vol. 9, no. 7, p. 785, pl. 5, figs. 1-7 [fig. 2 = holotype].
Carnarvon, Opal and Salter members, Mount Head Formation, and Livingstone Formation, Carboniferous, South Misty Range, lat. 50°34'N, long. 114°56'W, Mount Burke, lat. 50°17'N, long. 114°32'W, Elbow Lake, lat. 50°38'N, long. 115°W, Mount Aylmer, lat. 51°20'N, long. 115°28'W, and Mount Lorette, lat. 50°59'N, long. 115°8'W, Alberta.

Stacheiodes(?) spissa Petryk and Mamet [Algae]
Hypotypes 39808, 39809
Mamet, B. and Rudloff, B., 1972, Rev. Micropaléont., vol. 15, no. 2, p. 90, pl. 7, figs. 16, 17.
Etherington Formation and Salter Member, Mount Head Formation, Carboniferous, South Misty Range and Grotto Creek, Alberta.

Stacheiodes? cf. *S? spissa* Petryk and Mamet [Algae]
Hypotypes 29248, 29249, 29251
Petryk, A.A. and Mamet, B.L., 1972, Can. J. Earth Sci., vol. 9, no. 7, p. 785, pl. 5, fig. 8.
Loomis Member, Mount Head Formation, Carboniferous, Middle Mountain area, lat. 50°44'N, long. 115°3'W, Alberta.

Stacheoides tenuis Petryk and Mamet [Algae]
Holotype 29252; paratypes 29253-29264, 29783, 29784
Petryk, A.A. and Mamet, B.L., 1972, Can. J. Earth Sci., vol. 9, no. 7, p. 787, pl. 6, fig. 1-6; pl. 7, figs. 1-4, 5?, 6.
Opal, Loomis and Marston members, Mount Head Formation, Carboniferous, South Misty Range, lat. 50°34'N, long. 114°56'W, Elbow Lake, lat. 50°38'N, long. 115°W, Elpoca Creek, lat. 50°39'N, long. 115°4'W, King Creek, lat. 50°43'N, long. 115°6'W, Highwood Valley, lat. 50°24'N, long. 114°37'W, and Boulder Creek, lat. 50°37'N, long. 114°52'W, Alberta.

Stacheoides tenuis Petryk and Mamet [Algae]
Hypotypes 39798-39801
Mamet, B. and Rudloff, B., 1972, Rev. Micropaléont., vol. 15, no. 2, p. 90, pl. 7, figs. 4, 6-8.
Opal Member, Mount Head Formation, Carboniferous, Elpoca Creek and King Creek (39799), Alberta.

Stacheoides tenuis Petryk and Mamet, 1972 [Algae]
Hypotypes 61182-61185
Mamet, B.L., Roux, A. and Nassichuk, W.W., 1987, Geol. Surv. Can., Bull. 342, p. 46, pl. 23, figs. 4-7.
Pennsylvanian and Lower Permian (61183), Little Fish Creek, lat. 68°12'N, long. 136°28'W, Yukon.

Tetranguladinium conspicuum Yu Jingxian, Guo Zhengying and Mao Shaozhi, 1983
Hypotype 68822
Ricketts, B.D. and Sweet, A.R., 1986, Geol. Surv. Can., Paper 84-15, p. 23, pl. 4, fig. 5.
Pocaterra Member, Cadomin Formation, Cretaceous, Highwood Pass, lat. 50°35'N, long. 115 1'W, Alberta.

Tubiphytes obscurus Maslov, 1956 [Algae]
Hypotypes 61262-61264, 61271-61274
Mamet, B.L., Roux, A. and Nassichuk, W.W., 1987, Geol. Surv. Can., Bull. 342, p. 59, pl. 29, figs. 3-5; pl. 30, figs. 5-8.
Permian, Little Fish Creek, lat. 68°12'N, long. 136°28'W, Yukon, and Belcher Channel Formation, 12.8 km south of Cape Ogle, lat. 76°56'30"N, long. 95°23'W, Grinnell Peninsula, Devon Island (61272, 61274); Nansen Formation, Pennsylvanian (61263, 61264, 61273) and Permian (61271), Girty Creek, head of Hare Fiord, lat. 81°12'N, long. 82°6'W, Ellesmere Island, District of Franklin.

Trigonocarpus sp. attached to *Neuropteris (Mixoneura) flexuosa* (Sternberg) Crookall, 1959
Fig. specs. 85307, 85308
Zodrow, E.L. and McCandlish, K., 1980, Rev. Palaeobot. Palynol., vol. 30, p. 65, pls. II-IV; text-fig. 1.
Pennsylvanian, dump at Lingan coal mine, Cape Breton Island, Nova Scotia.

Turoholia sp. [Algae]
Fig. spec. 65842
Mamet, B.L., 1986, Geol. Surv. Can., Bull. 353, pl. 12, fig. 14.
Banff Formation, Lower Carboniferous, Mount Becker, lat. 54°31'30"N, long. 120°39'W, British Columbia.

undetermined calcispheres? (*Palaeocancellus?* sp.) [Algae]
Fig. specs. 39506-39508
Mamet, B.L., 1976, Geol. Surv. Can., Bull. 255, pl. 60, figs. 17-19.
Shunda Formation, Lower Carboniferous, depth 12,033 feet, Sohio-Un. Prod. et al. Tupper Creek C-70-B/93-P-9 well, lat. 55°33'30"N, long. 120°15'W, British Columbia.

Ungdarella peratrovichensis Mamet and Rudloff [Algae]
Paratypes 39810, 39811
Mamet, B. and Rudloff, B., 1972, Rev. Micropaléont., vol. 15, no. 2, p. 91, pl. 8, figs. 6, 7.
Nizi Formation, Carboniferous, McDame, British Columbia.

Ungdarella peratrovichensis Mamet and Rudloff, 1972 [Algae]
Hypotypes 61218, 61224-61227, 61243
Mamet, B.L., Roux, A. and Nassichuk, W.W., 1987, Geol. Surv. Can., Bull. 342, p. 52, pl. 25, fig. 15; pl. 26, figs. 4-7; pl. 27, fig. 12.
Pennsylvanian, Nansen Formation, 20 km west of head of Otto Fiord, lat. 81°20'N, long. 85°30'W, and Otto Fiord Formation, east of Girty Creek, head of Hare Fiord (61243), Ellesmere Island; Belcher Channel Formation, Permian, Grinnel Peninsula, 12.8 km south of Cape Ogle, lat. 76°56'30"N, long. 95°23'W, Devon Island (61227), District of Franklin.

Ungdarella uralica Maslov boundstone [Algae]
Hypotype 61291
Mamet, B. Nassichuk, W. and Roux, A., 1979, Bull. Cent. Rech. Explor.-Prod. Elf-Aquitaine, vol. 3, no. 2, pl. 2, fig. 7.
Nansen Formation, Upper Carboniferous, Girty Creek, Ellesmere Island, District of Franklin.

Ungdarella uralica Maslov, 1956 [Algae]
Hypotypes 61209-61217, 61228-61230, 61232, 61240-61242
Mamet, B.L., Roux, A. and Nassichuk, W.W., 1987, Geol. Surv. Can., Bull. 342, p. 52, pl. 25, figs. 6-14; pl. 26, figs. 8-10; pl, 27, figs. 1, 9-11.
Pennsylvanian, Otto Fiord Formation, immediately east of Girty Creek; Canyon Fiord Formation, Victoria and Albert Mountains, 22 km northeast of East Cape River delta, lat. 80°10′N, long. 81°W (61214); Nansen Formation, 6.5 km north of end of Oobloyah Bay (61215-61217), and Girty Creek, head of Hare Fiord, lat. 81°12′N, long. 82°6′W (61240, 61242), Ellesmere Island; Canyon Fiord Formation, Raglan Ridge, Melville Island (61228, 61229, 61241), District of Franklin.

Ungdarella sp. [Algae]
Fig. spec. 39680
Mamet, B., 1976, Geol. Surv. Can., Bull. 255, pl. 89, fig. 6.
Ettrain Formation, Upper Carboniferous, lat. 65°45′N, long. 139°03′30″W, Nahoni Range, Ogilvie Mountains, Yukon.

Ungdarella grainstone [Algae]
Fig. spec. 61292
Mamet, B., Nassichuk, W. and Roux, A., 1979, Bull. Cent. Rech. Explor.-Prod. Elf-Aquitaine, vol. 3, no. 2, pl. 2, fig. 8.
Nansen Formation, Carboniferous, Girty Creek, Ellesmere Island, District of Franklin.

Uraloporella variabilis Korde, 1950 [Algae]
Hypotypes 61058-61071
Mamet, B.L., Roux, A. and Nassichuk, W.W., 1987, Geol. Surv. Can., Bull. 342, p. 30, pl. 14, figs. 1-14.
Pennsylvanian, Nansen Formation, Girty Creek, head of Hare Fiord, lat. 81°12′N, long. 82°6′W; Otto Fiord Formation, Wood Glacier, lat. 80°54′N, long. 84°11′W (61060-61063, 61067, 61068); Belcher Channel Formation, Permian, west side of head of Blind Fiord (61065, 61071), Ellesmere Island, District of Franklin.

Uraloporella sp. [Algae]
Fig. specs. 43089-43091
Mamet, B. and Roux, A., 1975, Can. J. Earth Sci., vol. 12, no. 8, p. 1480, figs. 4-7.
Otto Fiord Formation, Carboniferous, Wood Glacier west, Ellesmere Island, District of Franklin.

Uraloporella - Dvinella - Beresella boundstone [Algae]
Fig. spec. 61289
Mamet, B., Nassichuk, W. and Roux, A., 1979, Bull. Cent. Rech. Explor.-Prod. Elf-Aquitaine, vol. 3, no. 2, pl. 2, fig. 5.
Belcher Channel Formation, Permian, Blind Fiord, Ellesmere Island, District of Franklin.

Uraloporellid boundstone
Fig. spec. 61295
Mamet, B., Nassichuk, W.W. and Roux, A., 1979, Bull. Cent. Rech. Explor.-Prod. Elf-Aquitaine, vol. 3, no. 2, pl. 3, fig. 3.
Carboniferous.

Vermiporella? sp. 1 [Algae]
Fig. specs. 60999, 61000
Mamet, B.L., Roux, A. and Nassichuk, W.W., 1987, Geol. Surv. Can., Bull. 342, p. 38, pl. 11, figs. 2, 3.
Nansen Formation, Pennsylvanian, 20 km west of head of Otto Fiord, lat. 81°20″N, long. 85°30′W; Belcher Channel Formation, Permian, Blind Fiord, Ellesmere Island, District of Franklin.

Vermiporella? sp. 2 [Algae]
Fig. specs. 60998, 61008
Mamet, B.L., Roux, A. and Nassichuk, W.W., 1987, Geol. Surv. Can., Bull. 342, p. 38, pl. 11, figs. 1, 11.
Nansen Formation, Permian, Arthaber Creek, south of Bunde Fiord, lat. 80°34′30″N, long. 95°34′55″W, eastern Axel Heiberg Island, District of Franklin.

Windsoporella radiata Mamet and Rudloff [Algae]
Holotype 32709; paratypes 32709a-g
Mamet, B. and Rudloff, B., 1972, Rev. Micropaléont., vol. 15, no. 2, p. 83, pl. 3, figs. 18-25.
Windsor Group, Carboniferous, Antigonish Milk Plant, Nova Scotia.

Yukonella bamberi Mamet and Rudloff [Algae]
Holotype 39782; paratypes 39783-39787
Mamet, B. and Rudloff, B., 1972, Rev. Micropaléont., vol. 15, no. 2, p. 87, pl. 5, figs. 24-29.
Kayak Formation, Carboniferous, Sedgwick Mountain, Yukon.

Triassic

Anomozamites sp.
Fig. spec. 90713
Ash, S.R. and Basinger, J.F., 1991, Geol. Surv. Can., Bull. 412, p. 111, pl. 5.2, fig. 10.
Fosheim Member, Heiberg Formation, Upper Triassic, 9 miles east of head Strand Fiord, Axel Heiberg Island, District of Franklin.

Anthrophyopsis sp. cf. *A. crassinervis*
Fig. spec. 96654
Ash, S.R. and Basinger, J.F., 1991, Geol. Surv. Can., Bull. 412, p. 117, text-fig. 5.4.

Fosheim Member, Heiberg Formation, Upper Triassic, west limb of Fosheim Anticline on south side of a major westwardly flowing stream at about lat. 79°37′N, long. 84°50′W, Ellesmere Island, District of Franklin.

Cladophlebis sp.
Fig. spec. 96639a
Ash, S.R. and Basinger, J.F., 1991, Geol. Surv. Can., Bull. 412, p. 111, pl. 5.1, fig. 15.
Fosheim Member, Heiberg Formation, Upper Triassic, west limb of Fosheim Anticline on south side of a major westwardly flowing stream at about lat. 79°37′N, long. 84°50′W, Ellesmere Island, District of Franklin.

Clathropteris meniscoides Brongniart
Hypotypes 96636-96638, 90708, 90709
Ash, S.R. and Basinger, J.F., 1991, Geol. Surv. Can., Bull. 412, p. 110, pl. 5.1, figs. 8-12.
Fosheim Member, Heiberg Formation, Upper Triassic, west limb of Fosheim Anticline on south side of a major westwardly flowing stream at about lat. 79°37′N, long. 84°50′W, Ellesmere Island; Wolf Fiord area, Axel Heiberg Island (90708, 90709), District of Franklin.

Camptopteris spiralis Nathorst
Hypotype 90706a
Ash, S.R. and Basinger, J.F., 1991, Geol. Surv. Can., Bull. 412, p. 110, pl. 5.1, fig. 6.
Fosheim or Romulus Member, Heiberg Formation, Upper Triassic, 15 miles east of head of Strand Fiord, Axel Heiberg Island, District of Franklin.

Cone A
Fig. spec. 96651
Ash, S.R. and Basinger, J.F., 1991, Geol. Surv. Can., Bull. 412, p. 115, pl. 5.3, fig. 4.
Fosheim Member, Heiberg Formation, Upper Triassic, west limb of Fosheim Anticline on south side of a major westwardly flowing stream at about lat. 79°37′N, long. 84°50′W, Ellesmere Island, District of Franklin.

Czekanowoskia sp.
Fig. spec. 96646
Ash, S.R. and Basinger, J.F., 1991, Geol. Surv. Can., Bull. 412, p. 113, pl. 5.2, fig. 12.
Fosheim Member, Heiberg Formation, Upper Triassic, about 19 km east of Mt. Lockwood in sea cliffs on south side of Greeley Fiord at about lat. 80°16′N, long. 84°52′W, Ellesmere Island, District of Franklin.

Equisetites grosphodon Harris
Hypotype 90705
Ash, S.R. and Basinger, J.F., 1991, Geol. Surv. Can., Bull. 412, p. 108, pl. 5.1, fig. 5.
Fosheim Member, Heiberg Formation, Upper Triassic, Wolf Fiord area, Axel Heiberg Island, District of Franklin.

Ginkgo siberica (Heer) Seward
Hypotypes 96640, 96641
Ash, .S.R. and Basinger, J.F., 1991, Geol. Surv. Can., Bull. 412, p. 112, pl. 5.2, figs. 1, 4.
Fosheim Member, Heiberg Formation, Upper Triassic, about 19 km east of Mt. Lockwood in sea cliffs on south side of Greely Fiord at about lat. 80°16′N, long. 84°52′W, and on west limb of Fosheim Anticline on south side of a major westwardly flowing stream at about lat. 79°37′N, long. 84°50′W, Ellesmere Island, District of Franklin.

Marattiopsis sp. cf. *M. asiatica* Kawasaki
Fig. spec. 96633
Ash, S.R. and Basinger, J.F., 1991, Geol. Surv. Can., Bull. 412, p. 108, pl. 5.1, fig. 4.
Fosheim Member, Heiberg Formation, Upper Triassic, southeast of Blue Mountains on south side of a major easterly flowing stream at about lat. 80°39′N, long. 85°37′W, Ellesmere Island, District of Franklin.

Mesembrioxylon sp. A
Fig. specs. 90716a-c
Ash, S.R. and Basinger, J.F., 1991, Geol. Surv. Can., Bull. 412, p. 115, pl. 5.3, figs. 11-16.
Fosheim Member, Heiberg Formation, Upper Triassic, south and southeast of Mount Nicolay, Cornwall Island, District of Franklin.

Neocalamites hoerensis (Schimper) Halle
Hypotypes 96630-96632
Ash, S.R. and Basinger, J.F., 1991, Geol. Surv. Can., Bull. 412, p. 108, pl. 5.1, figs. 1-3.
Fosheim Member, Heiberg Formation, Upper Triassic, west limb of Fosheim Anticline on south side of a major westwardly flowing stream at about lat. 79°37′N, long. 84°50′W, and about 19 km east of Mt. Lockwood in sea cliffs on south side of Greely Fiord at about lat. 80°16′N, long. 84°52′W (96631), Ellesmere Island, District of Franklin.

Phlebopteris angustiloba
Hypotype 90707
Ash, S.R. and Basinger, J.F., 1991, Geol. Surv. Can., Bull. 412, p. 109, pl. 5.1, fig. 7.
Fosheim or Romulus Member, Heiberg Formation, Upper Triassic, 15 miles east of head of Strand Fiord, Axel Heiberg Island, District of Franklin.

Pelourdea sp.
Fig. spec. 96650
Ash, S.R. and Basinger, J.F., 1991, Geol. Surv. Can., Bull. 412, p. 114, pl. 5.3, fig. 9.
Fosheim Member, Heiberg Formation, Upper Triassic, west limb of Fosheim Anticline on south side of a major westwardly flowing stream at about lat. 79°37′N, long. 84°50′W, Ellesmere Island, District of Franklin.

Phylloid algae [Algae]
Fig. specs. 83223-83225
Reid, R.P., 1986, Can. J. Earth Sci., vol. 23, no. 12, p. 2069, figs. 3a-c.
Avens Reef, Late Triassic, Lime Peak, southern Yukon.

Podozamites sp.
Fig. specs. 96648, 96649
Ash, S.R and Basinger, J.F., 1991, Geol. Surv. Can., Bull. 412, p. 114, pl. 5.3, figs. 3, 8.

Fosheim Member, Heiberg Formation, Upper Triassic, about 19 km east of Mt. Lockwood in sea cliffs on south side of Greely Fiord at about lat. 80°16′N, long. 84°52′W and west limb of Fosheim Anticline on south side of a major westwardly flowing stream at about lat. 79°37′N, long. 84°50′W, Ellesmere Island, District of Franklin.

Pterophyllum subaequale Hartz
Hypotypes 90714, 90715
Ash, S.R. and Basinger, J.F., 1991, Geol. Surv. Can., Bull. 412, p. 112, pl. 5.2, figs. 6, 7.
Fosheim Member, Heiberg Formation, Upper Triassic, east side Buchanan Lake near Savik Creek, Axel Heiberg Island, District of Franklin.

Ptilophyllum sp.
Fig. spec. 90712
Ash, S.R. and Basinger, J.F., 1991, Geol. Surv. Can., Bull. 412, p. 111, pl. 5.2, fig. 9.
Fosheim or Romulus Member, Heiberg Formation, Upper Triassic, 15 miles east of head of Strand Fiord, Axel Heiberg Island, District of Franklin.

Sphenobaiera spectabilis (Nathorst) Florin
Hypotypes 96642a, b-96645
Ash, S.R. and Basinger, J.F., 1991, Geol. Surv. Can., Bull. 412, p. 112, pl. 5.2, figs. 2, 3, 5, 8, 11.
Fosheim Member, Heiberg Formation, Upper Triassic, about 19 km east of Mt. Lockwood in sea cliffs on south side of Greely Fiord at about lat. 80°16′N, long. 84°52′W, Ellesmere Island, District of Franklin.

Swedenborgia major Harris
Hypotype 96647
Ash, S.R. and Basinger, J.F., 1991, Geol. Surv. Can., Bull. 412, p. 113, pl. 5.3, fig. 1.
Fosheim Member, Heiberg Formation, Upper Triassic, about 19 km east of Mt. Lockwood in sea cliffs on south side of Greely Fiord at about lat. 80°16′N, long. 84°52′W, Ellesmere Island, District of Franklin.

Taeniopteris tenuinerois Braun
Hypotypes 96652, 96653
Ash, S.R. and Basinger, J.F., 1991, Geol. Surv. Can., Bull. 412, p. 116, pl. 5.3, figs. 6, 7.
Fosheim Member, Heiberg Formation, Upper Triassic, west limb of Fosheim Anticline on south side of a major westwardly flowing stream at about lat. 79°37′N, long. 84°50′W, Ellesmere Island, District of Franklin.

Todites recurvatus Harris
Hypotypes 96634, 96635, 90711
Ash, S.R. and Basinger, J.F., 1991, Geol. Surv. Can., Bull. 412, p. 109, pl. 5.1, fig. 13, 14.
Fosheim Member, Heiberg Formation, Upper Triassic, west limb of Fosheim Anticline on south side of a major westwardly flowing stream at about lat. 79°37′N, long. 84°50′W, Ellesmere Island, District of Franklin.

Jurassic

Aclistochara mundula Peck [Charophyte]
Hypotypes 84821, 84822
Loranger, D.M., 1955, Proc. Geol. Assoc. Can., vol. 7, pt. 1, p. 54, pl. 9, figs. 21-24.
Shaunavon Formation, Upper Jurassic, depths 4230-4235 feet, Norcanols-Radville No. 1 well, l.s.d. 16, sec. 36, tp. 5, rge. 19, W.2nd mer., Saskatchewan.

Tasmanites newtonii Wall [Algae Incertae Sedis]
Hypotype 75511
Davies, E.H., 1985, Palynology, vol. 9, pl. 4, fig. 23.
Late Pliensbachian, Liassic, Brehna Road Section, Lusitanian Basin, west-central Portugal.

Cretaceous

Aclistochara cylindrica? Peck [Charophyte]
Hypotype 84759
Loranger, D.M., 1951, Bull. Amer. Assoc. Petrol. Geol., vol. 35, no. 11, p. 2354, pl. 2, fig. 5.
Loranger, D.M., 1954, Western Canada Sedimentary Basin, Ralph Leslie Rutherford Memorial Vol., Clark, L.M., (ed.), Amer. Assoc. Petrol. Geol., p. 285, pl. 2, fig. 5.
Blairmore formation, Lower Cretaceous, Lower Limestone Creek, tp. 35, rge.11, W.5th mer., Alberta.

Aclistochara mundula Peck [Charophyte]
Hypotypes 84760, 84761
Loranger, D.M., 1951, Bull. Amer. Assoc. Petrol Geol., vol. 35, no. 11, p. 2357, pl. 2, figs. 3, 4.
Loranger, D.M., 1954, Western Canada Sedimentary Basin, Ralph Leslie Rutherford Memorial Vol., Clark, L.M., (ed.), Amer. Assoc. Petrol. Geol., p. 285, pl. 2, figs. 3, 4.
Blairmore formation, Lower Cretaceous, depth 3455-3461 feet, Imperial Morinville No. 1 well, l.s.d. 9, sec. 30, tp. 56, rge. 24, W.4th mer., and depth 3947-3978 feet, Imperial Looma No. 1 well, l.s.d. 4, sec. 10, tp. 50, rge. 23, W.4th mer., Alberta.

Aclistochara symmetrica Loranger [Charophyte]
Holotype 84762
Loranger, D.M., 1951, Bull. Amer. Assoc. Petrol Geol., vol. 35, no. 11, pl. 2, figs. 1, 2.

Loranger, D.M., 1954, Western Canada Sedimentary Basin, Ralph Leslie Rutherford Memorial Vol., Clark, L.M., (ed.), Amer. Assoc. Petrol. Geol., p. 285, pl.2, figs. 1, 2.
Blairmore formation, Lower Cretaceous, depth 2876-2896 feet, Imperial Provost no. 2 well, l.s.d. 1, sec. 33, tp. 37, rge. 3, W.4th mer., Alberta.

Arctopitys fforinii Bose and Manum
Hypotypes 6702-30, 6702-31
Bose, M.N. and Manum, S.B., 1990, Norsk-Polarinstitutt Skr. Nr. 192, p. 35, pl. 2, fig. 11; pl. 3, fig. 3; text-figs. 13A, 13D.
Lower Cretaceous, Padloping Island, north side of Cumberland Peninsula, Baffin Island, District of Franklin.

Arctopitys ineffecta Bose and Manum
Holotype 6702-33; hypotype 6702-32
Bose, M.N. and Manum, S.B., 1990, Norsk-Polarinstitutt Skr. Nr. 192, p. 39, text-figs. 16B, C.
Lower Cretaceous, Padloping Island, north side of Cumberland Peninsula, Baffin Island, District of Franklin.

Botryococcus sp. [Algae]
Fig. spec. 68825
Ricketts, B.D. and Sweet, A.R., 1986, Geol. Surv. Can., Paper 84-15, p. 23, pl. 4, fig. 8.
Elk Formation, Cretaceous, Coal Creek, lat. 49°30′N, long. 114°59′W, British Columbia.

Campenia(?) sp. [Algae]
Fig. spec. 84755
Leckie, D.A. et al., 1990, J. Sed. Petrology, vol. 60, no. 1, fig. 7C.
Shaftesbury Formation, Lower Cretaceous, west bank Peace River near town of Peace River, Alberta.

Charaxis sp. [Charophyte]
Fig. spec. 64763
McLean, J.R. and Wall, J.H., 1982, Bull. Can. Petrol. Geol., vol. 29, no. 3, 1981, pl. 6, figs 6, 7.
Gladstone Formation, 34 km below top, Lower Cretaceous, Malcolm Creek, sec. 6, tp. 57, rge. 8, W.6th mer., Alberta.

Elatocladus sp. *A*
Fig. specs. 6702-41, 6702-42
Bose, M.N. and Manum, S.B., 1990, Norsk-Polarinstitutt Skr. Nr. 192, p. 58, text-figs. 24J, 24L.
Lower Cretaceous, Padloping Island, north side of Cumberland Peninsula, Baffin Island, District of Franklin.

Holkopitys hoegii Bose and Manum
Hypotypes 6702-37, 6702-38
Bose, M.N. and Manum, S.B., 1990, Norsk-Polarinstitutt Skr. Nr. 192, p. 49, pl. 6, fig. 5-7; text-figs. 19C, 20A.
Lower Cretaceous, Padloping Island, north side of Cumberland Peninsula, Baffin Island, District of Franklin.

cf. *Lancettopsis lanceolata* Mädler, 1963 [Algae]
Hypotypes 84756-84758
Leckie, D.A. et al., 1990, J. Sed. Petrology, vol. 60, no. 1, p. 114, figs. 7D-F.
Shaftesbury Formation, Lower Cretaceous, west bank Peace River near town of Peace River, Alberta.

Leiosphaeridia sp. [Algae]
Fig. specs. 84753, 84754
Leckie, D.A. et al., 1990, J. Sed. Petrology, vol. 60, no. 1, p. 114, figs. 7A, B.
Shaftesbury Formation, Lower Cretaceous, west bank Peace River near town of Peace River, Alberta.

Microcarpolithes sp.
Fig. spec. 64764
McLean, J.R. and Wall, J.H., 1982, Bull. Can. Petrol Geol., vol. 29, no. 3, 1981, pl. 6, figs. 8, 9.
Gladstone Formation, 16 to 18 m above base, Lower Cretaceous, Little Berland T.H. 70-02 well, N.W. 1/4 sec. 9, tp. 53, rge. 2, W.6th mer., Alberta.

Oswaldheeria artica Bose and Manum
Hypotypes 6702-35, 6702-36
Bose, M.N. and Manum, S.B., 1990, Norsk-Polarinstitutt Skr. Nr. 192, p. 44, text-figs. 18E, 19P.
Lower Cretaceous, Padloping Island, north side of Cumberland Peninsula, Baffin Island, District of Franklin.

Oswaldheeria hallei(Florin) comb. nov.
Hypotype 6702-34
Bose, M.N. and Manum, S.B., 1990, Norsk-Polarinstitutt Skr. Nr. 192, p. 41, text-figs. D, Q.
Lower Cretaceous, Padloping Island, north side of Cumberland Peninsula, Baffin Island, District of Franklin.

Pediastrum sp. [Algae]
Fig. spec. 89729
Jerzykiewicz, T. and Sweet, A.R., 1988, Sedimentary Geol., vol. 59, no. 1/2, p. 58, pl. III, fig. 2.
St. Mary River Formation, Upper Cretaceous, Crowsnest River, lat. 49°30′N, long. 114°3′W, Alberta.

Pesavis parva Kalgutkar and Sweet [Fungus]
Holotype 83330; paratypes 83328, 83329, 83331-83343
Kalgutkar, R.M. and Sweet, A.R., 1988, Geol. Surv. Can., Bull. 379, p. 123, pl. 6.1, figs. 1-16.
Upper Cretaceous, Tent Island Formation, Big Fish River near mouth of Little Fish River, lat. 68°31′30″N-68°33′N, long. 136°15′50″-136°17′W, Little Fish River, lat. 68°32′15″N-68°32′20N, long. 136°12-136°11′38″W (83334-83336), and Bonnet Plume Formation, Peel River, lat. 65°52′N, long. 135°12′W(83337-83339); Moose Channel Formation, Paleocene, Big Fish River, lat. 68°33′40″N, long. 136°13′45″W to lat. 68°32′21″N,

long. 136°10′5″W(83340-83342), Yukon; Summit Creek Formation, Paleocene, southeast end of Tertiary Hills, lat. 64°27′30″N, long. 125°33′45″W, District of Mackenzie (83343).

Praechara sp. cf. *P. symmetrica* Peck [Charophyte]
Fig. spec. 64672
McLean, J.R. and Wall, J.H., 1982, Bull. Can. Petrol. Geol., vol. 29, no. 3, 1981, pl. 6, figs 4, 5.
Gladstone Formation, 36 m above base, Lower Cretaceous, Malcom Creek, sec. 6, tp. 57, rge. 8, W.6th mer., Alberta.

Pseudotorellia kordylina Bose and Manum
Hypotypes 6702-39, 6702-40
Bose, M.N. and Manum, S.B., 1990, Norsk-Polarinstitutt Skr. Nr. 192, p. 51, text-figs. 21G, 21H.
Lower Cretaceous, Padloping Island, north side of Cumberland Peninsula, Baffin Island, District of Franklin.

Schizophacus rugulatus (Cookson and Dettmann) Pierce, 1976 [Algae]
Hypotype 68828
Ricketts, B.D. and Sweet, A.R., 1986, Geol. Surv. Can., Paper 84-15, p. 23, pl. 4, fig, 11.
Elk Formation, Cretaceous, Coal Creek, lat. 49°30′N, long. 114°59′W, British Columbia.

Schizophacus spriggii (Cookson and Dettmann) Pierce, 1976 [Algae]
Hypotype 68230
Ricketts, B.D. and Sweet, A.R., 1986, Geol. Surv. Can., Paper 84-15, p. 25, pl. 4, fig, 1.
Elk Formation, Cretaceous, Coal Creek, lat. 49°30′N, long. 114°59′W, British Columbia.

Schizosporis reticulatus Cookson and Dettmann, 1959 emend. Pierce, 1979 [Algae]
Hypotype 68827
Ricketts, B.D. and Sweet, A.R., 1986, Geol. Surv. Can., Paper 84-15, p. 23, pl. 4, fig, 10.
Elk Formation, Cretaceous, Coal Creek, lat. 49°30′N, long. 114°59′W, British Columbia.

Sciadopityoides microphylla (Heer) comb. nov.
Hypotypes 6702-28, 6702-29
Bose, M.N. and Manum, S.B., 1990, Norsk-Polarinstitutt Skr. Nr. 192, p. 24 , text-figs. 8H, 8I.
Lower Cretaceous, Padloping Island, north side of Cumberland Peninsula, Baffin Island, District of Franklin.

Sciadopitytes variabilis Bose
Holotype 6702.1; paratypes 6702, 6702.1-6702.5, 6702.8, 6702.9, 6702.13-23, 6702.26
Bose, M.N. , 1955, Norsk. Geol. Tidssk., vol. 35, p. 53-58, pl. 1, figs. 1, 2, 5-8; text-figs. 1A-N, 2A-F, 3A-D.
Lower Cretaceous, Padloping Island, north side of Cumberland Peninsula, Baffin Island, District of Franklin.
=*Sciadopityoides variabilis*, Bose, M.N. and Manum, S.B., 1990, Norsk-Polarinstitutt Skr. Nr. 192, p. 32, pl. 2, figs. 7, 8 (6702-18); text-figs. 12A (6702-8), B (6702-1), D (6702-9), E, F (6702-13), G-I (6702-22), J (6702-23).

Sciadopityoides variabilis (Bose) Sveshnikova
Hypotypes 6702.10-6702.12, 6702.27
Bose, M.N. and Manum, S.B., 1990, Norsk-Polarinstitutt Skr. Nr. 192, p. 32, text-figs. 12C, 12K.
Lower Cretaceous, Padloping Island, north side of Cumberland Peninsula, Baffin Island, District of Franklin.

Scidoptys-like pollen grain
Fig. spec. 6702.6
Bose, M.N. , 1955, Norsk. Geol. Tidssk., vol. 35, p. 53-58, pl. 1, fig. 4.
Lower Cretaceous, Padloping Island, north side of Cumberland Peninsula, Baffin Island, District of Franklin.

Sigmopollis sp. [Algae]
Fig. spec. 68824
Ricketts, B.D. and Sweet, A.R., 1986, Geol. Surv. Can., Paper 84-15, p. 23, pl. 4, fig. 7.
Pocaterra Member, Cadomin Formation, Cretaceous, Highwood Pass, lat. 50°35′N, long. 115°1′W, Alberta.

Stellatochara mundula (Peck) [Charophyte]
Hypotypes 64760, 64761
McLean, J.R. and Wall, J.H., 1982, Bull. Can. Petrol. Geol., vol. 29, no. 3, 1981, pl. 6, figs. 1-3.
Gladstone Formation, 36.2-36.6 m and 17 m below top, Lower Cretaceous, Ruby Creek, sec. 11, tp. 45, rge. 21, W.5th mer., and Malcolm Creek, sec. 6 , tp. 57, rge. 8, W.6th mer., Alberta.

Tetranguladinium conspicuum Yu Jingxian, Guo Zhengying and Mao Shaozhi, 1983
Hypotype 68822
Ricketts, B.D. and Sweet, A.R., 1986, Geol. Surv. Can., Paper 84-15, p. 23, pl. 4, fig. 5.
Pocaterra Member, Cadomin Formation, Cretaceous, Highwood Pass, lat. 50°35′N, long. 115°1′W, Alberta.

Trichopeltinites sp. [Fungus]
Fig. specs. 91328-91333
Sweet, A.R. and Kalgutkar, R.M., 1989, Geol. Surv. Can., Bull. 396, p. 224, pl. 9.1, figs. 1-6.
Late Cretaceous, Scollard Formation, depth 47.29-47.44 m, Transalta Utilities C-HV-83-07 borehole, near Highvale Coal Mine Pit 3 on south side of Wabamun Lake, sec. 35, tp. 2, rge. 5, W.5th mer., Alberta (91328, 91330); Frenchman Formation, Morgan Creek, east locality, l.s.d. 6, sec. 18, tp. 1, rge. 4, W.3rd mer., Saskatchewan.

Williamsonia recentior Dawson
="*Williamsonia*" *recentior*, Crane, P.R. and Dilcher, D.L., 1981, Annals Missouri Botanical Gardens, vol. 71, no. 2, p. 395, fig. 37 (holotype Plant catalogue 5105).

Tertiary-Recent

Abietoxylon koreanoides Roy and Hills
Holotype 27938; paratype 27936
Roy, S.K. and Hills, L.V., 1972, Can. J. Botany, vol. 50, no. 12, p. 2642, figs. 20-26.
Beaufort Formation, late Miocene-early Pliocene, north side Ballast Brook, about 8.5 miles south of M'Clure Strait, lat. 74°21'N, long. 123°6'W, northwestern Banks Island, District of Franklin.

Abietoxylon traumatiductus Roy and Hills
Holotype 27953
Roy, S.K. and Hills, L.V., 1972, Can. J. Botany, vol. 50, no. 12, p. 2643, figs. 27-32.
Beaufort Formation, late Miocene-early Pliocene, north side Ballast Brook, 9 miles south-southeast of M'Clure Strait, lat. 74°19'30"N, long. 123°5'W, northwestern Banks Island, District of Franklin.

Acer busamarum fingerrockense Wolfe et Tanai
Paratype 82966A, B
Wolfe, J.A. and Tanai, T., 1987, J. Fac. Sci. Hokkaida Univ., vol. 22, no. 1, p. 154, pl. 50, fig. 2.
Miocene, north bank Chilcotin River, 4 miles west of Lee's Store, Hanceville, British Columbia.

Acer medianum Knowlton
Hypotype 82972
Wolfe, J.A. and Tanai, T., 1987, J. Fac. Sci. Hokkaida Univ., vol. 22, no. 1, p. 111, 234,.
Miocene, north bank Chilcotin River, 4 miles west of Lee's Store, Hanceville, British Columbia.

Acer minutifolium Chaney
Hypotypes 82967, 82968
Wolfe, J.A. and Tanai, T., 1987, J. Fac. Sci. Hokkaida Univ., vol. 22, no. 1, p. 174, pl. 54, figs. 3, 4.
Miocene, north bank Chilcotin River, 4 miles west of Lee's Store, Hanceville, British Columbia.

Acer traini Wolfe and Tanai
Paratype 82971A, B
Wolfe, J.A. and Tanai, T., 1987, J. Fac. Sci. Hokkaida Univ., vol. 22, no. 1, p. 81, 244.
Miocene, north bank Chilcotin River, 4 miles west of Lee's Store, Hanceville, British Columbia.

Acer wehri Wolfe and Tanai
Paratype 82970A, B
Wolfe, J.A. and Tanai, T., 1987, J. Fac. Sci. Hokkaida Univ., vol. 22, no. 1, p. 128, pl. 4, fig. 10.
Chu Chua Formation, Middle Eocene, Joseph Creek about 50 yards southeast of bridge, about 1 mile east of lodge, British Columbia.

Acer sp. 3
Fig. spec. 82969
Wolfe, J.A. and Tanai, T., 1987, J. Fac. Sci. Hokkaida Univ., vol. 22, no. 1, p. 136, pl. 2, fig. 5.
Princeton Group, Middle Miocene, road cut 0.4 miles up from highway 3, 0.8 miles north of bridge over Whipsaw Creek, British Columbia.

Actinidia sp. (Actinidiaceae): seed
Fig spec. 99130
Matthews, Jr., J.V. and Ovenden, L.E., 1990, Arctic, vol. 43, no. 4, p. 375, fig. 3.1.
Mary Sacks gravel, Miocene, Duck Hawk Bluffs, southwestern Banks Island, District of Franklin.

Aldrovanda sp. (Droseraceae): seed
Fig. spec. 99137
Matthews, Jr., J.V. and Ovenden, L.E., 1990, Arctic, vol. 43, no. 4, p. 370, fig. 3.9.
Miocene,upper Ramparts site southside of Porcupine River, lat. 67°19.83'N, long. 141°19.8'W, Alaska.

Alnus cf. *incana* (Linnaeus) Moench
Hypotypes 85393, 85398, 85399
Matthews, Jr., J.V., 1987, Geol. Surv. Can., Paper 87-1A, p. 83, figs. 9.3h-9.3j.
Beaufort Formation, Miocene-Pliocene, approximately lat. 79°50'N, near western coast of Meighen Island, Ballast Brook, northwestern Banks Island, and Duck Hawk Bluffs, southwestern Banks Island, District of Franklin.

Alnus cf. *tertiaria* Dorofeev
Hypotypes 85394-85397
Matthews, Jr., J.V., 1987, Geol. Surv. Can., Paper 87-1A, p. 83, figs. 9.3d-9.3g.
Beaufort Formation, Miocene-Pliocene, approximately lat. 79°50'N, near western coast of Meighen Island, District of Franklin.

Alnus (Alnobetula) sp.
Fig. specs. 85390-85392, 85400
Matthews, Jr., J.V., 1987, Geol. Surv. Can., Paper 87-1A, p. 83, figs. 9.3a-9.3c, 9.3k.
Beaufort Formation, Miocene-Pliocene, approximately lat. 79°50'N, near western coast of Meighen Island, and Duck Hawk Bluffs, southwestern Banks Island (85399, 85400), District of Franklin.

Amentotaxus sp. cf. *campbelli* Florin
= *Amentotaxus gladifolia*, Jähnichen, H., 1990, Tertiary Research, vol. 12, no. 2, p. 70, text-figs. 1a (hypotype 6400), b (hypotype 6401).

Andromeda polifolia Linnaeus
Hypotype 60881
Blake, Jr., W. and Matthews, Jr., J.V., 1979, Geol Surv. Can., Paper 79-1A, fig. 28.6 D.
Pre-Holocene, 8 km east of head of Makinson Inlets' north arm, east-central Ellesmere Island, District of Franklin.

Aracispermum sp.
Fig. spec. 85366, 85367
Matthews, Jr., J.V., 1987, Geol. Surv. Can., Paper 87-1A, p. 75, pl. 9.1, figs. 3, 4.
Beaufort Formation, Miocene-Pliocene, Ballast Brook, northwestern Banks Island, District of Franklin.

Azolla schopfi Dijkstra
Hypotypes 31061, 31063, 31064, 31066, 31068, a, c, 31071a-c, e, h, 31074a
Sweet, A.R. and Chandrasekharam, A., 1973, Can. J. Botany, vol. 51, no. 8, p. 1493, pl. 2, figs. 12-16; pl. 3, figs. 18-23.
Scollard Member, Paskapoo Formation, Paleocene, Research Council of Alberta core hole 65-1, lat. 53°6.5′N, long. 113°55′W; North Saskatchewan River west of Genesse, lat. 53°23′N, long. 113°55′W (31066); Red Deer River at Huxley, lat. 51°56′N, long. 112°59′W (31068 a, c, 31074a), Alberta.

Betula spp. Bractlet
Fig. specs. 85372-85375, 85382
Matthews, Jr., J.V., 1987, Geol. Surv. Can., Paper 87-1A, p. 82, figs. 9.2a-9.2d, 9.2k.
Beaufort Formation, Miocene-Pliocene, approximately lat. 79°50′N, near western coast of Meighen Island, and Ballast Brook, northwestern Banks Island (85382), District of Franklin.

Betula spp. Nutlet
Fig. specs. 85378, 85379, 85381, 85383-85389
Matthews, Jr., J.V., 1987, Geol. Surv. Can., Paper 87-1A, p. 82, figs. 9.2g, 9.2h, 9.2j, 9.2l-9.2r.
Beaufort Formation, Miocene-Pliocene, approximately lat. 79°50′N, near western coast of Meighen Island, Ballast Brook, northwestern Banks Island(85383-85386), and Duck Hawk Bluffs, southwestern Banks Island(85387-853890), District of Franklin.

Betula spp. Winged nutlet (samara)
Fig.specs. 85376, 85377, 85380
Matthews, Jr., J.V., 1987, Geol. Surv. Can., Paper 87-1A, p. 82, figs. 9.2e, 9.2f, 9.2i.
Beaufort Formation, Miocene-Pliocene, approximately lat. 79°50′N, near western coast of Meighen Island, District of Franklin.

Bolboforma badenensis Szczechura [Algae Incertae Sedis)
Hypotypes 68804-68808
McNeil, D.H., 1988, Micropaleontology, vol. 34, no. 1, p. 90, pl. 1, figs. 1a-e - 5a, b.
Late Miocene, depths 1685-1705m and 1745-1765m (68806), Gulf North Issungnak L-86 well, lat. 70°5′33′N, long. 134°26′45″W, Beaufort Sea, Arctic Canada.

Boschniakia rossica (Cham. and Schlecht.) Fedtsch.: seed
Hypotype 99142
Matthews, Jr., J.V. and Ovenden, L.E., 1990, Arctic, vol. 43, no. 4, p. 384, fig. 4.7.
Pliocene, southeast of head of Strathcona Fiord., lat. 76°26.5′N, long. 81°53′W, Ellesmere Island, District of Franklin.

Callimothallus pertusus Dilcher 1965 [Fungus)
Hypotypes 75869-75872
Kalgutkar, R.M., 1985, Geol. Surv. Can., Paper 85-1B, p. 262, pl. 31.1, figs. 1-4.
Bonnet Plum Formation, Late Paleocene-Middle Eocene, lat. 65°54′N, long. 135°4′W, Peel River, Yukon.

Carex maritima Gunnerus
Hypotype 60880
Blake, Jr., W., and Matthews, Jr., J.V., 1979, Geol. Surv. Can., Paper 79-1A, fig 28.6C.
Pre-Holocene, 8 km east of head of Makinson Inlets' north arm, east-central Ellesmere Island, District of Franklin.

Cleome sp.
Fig. specs. 85361, 85362
Matthews, Jr., J.V., 1987, Geol. Surv. Can., Paper 87-1A, p. 83, pl. 9.3, figs. 1, 2.
Beaufort Formation, Miocene-Pliocene, Ballast Brook, northwestern Banks Island, District of Franklin.

Comptonia cf. *peregrina* (Linnaeus) Coulter
Hypotypes 85370, 85371
Matthews, Jr., J.V., 1987, Geol. Surv. Can., Paper 87-1A, p. 77, pl. 9.2, figs. 1, 2.
Beaufort Formation, Miocene-Pliocene, Ballast Brook, northwestern Banks Island, District of Franklin.

Comptonia sp. A
Fig. specs. 85368, 85369
Matthews, Jr., J.V., 1987, Geol. Surv. Can., Paper 87-1A, p. 82, pl. 9.2, figs. 3, 4.
Beaufort Formation, Miocene-Pliocene, approximately lat. 79°50′N, near western coast of Meighen Island, District of Franklin.

Cyperaceae type A: fruit?
Fig. spec. 99133
Matthews, Jr., J.V. and Ovenden,L.E., 1990, Arctic, vol. 43, no. 4, p.384, fig. 3.5.
Pliocene, near head of Makinson Inlet, Ellesmere Island, District of Franklin.

Decodon sp. (Lythraceae): seed
Fig. spec. 99143
Matthews, Jr., J.V. and Ovenden, L.E., 1990, Arctic, vol. 43, no. 4, fig. 4.8.
Mary Sacks gravel, Miocene, Duck Hawk Bluffs, southwestern Banks Island, District of Franklin.

Diervilla sp. (Caprifoliaceae): seed
Fig. spec. 99139
Matthews, Jr., J.V. and Ovenden, L.E., 1990, Arctic, vol. 43, no. 4, fig. 4.4.
Mary Sacks gravel, Miocene, Duck Hawk Bluffs, southwestern Banks Island, District of Franklin.

Ditrichites fylesi Kuc [Moss]
Holotype 34741; paratype 34742
Kuc, M., 1974, Can. J. Earth Sci., vol. 11, no. 3, p. 411, figs. 3A-F.
Allenby Formation, Eocene, 6 miles north of Princeton, British Columbia.

Elaeagnaceoxylon shepherdioides Roy and Hills
Holotype 27924
Roy, S.K. and Hills, L.V., 1972, Can. J. Botany, vol. 50, no. 12, p. 2644, figs. 33-41.
Beaufort Formation, late Miocene-early Pliocene, north side Ballast Brook, 9 miles south-southeast of M'Clure Strait, lat. 74°19'30'N, long. 123°5'W, northwestern Banks Island, District of Franklin.

Epipremnum cf. *crassum* Reid and Reid
Hypotypes 85358, 85359
Matthews, Jr., J.V., 1987, Geol. Surv. Can., Paper 87-1A, p. 74, pl. 9.1, figs. 1, 2.
Beaufort Formation, Miocene-Pliocene, Ballast Brook, northwestern Banks Island, District of Franklin.
=*Epipremnum crassum*, Matthews, Jr., J.V. and Ovenden, L.E., 1990, Arctic, vol. 43, no. 4, p. 377, fig. 3.2.

Juglans eocinerea Hills, Klovan and Sweet
Holotype 31886; paratypes 31883-31885, 31887-31904
Hills, L.V., Klovan, J.E. and Sweet, A.R., 1974, Can. J. Botany, Vol. 52, no. 1, p. 84, figs. 13-16, 19-22, 25-38, 41-55, 57-59, 79, 80, 82.
Beaufort Formation, Tertiary, west side of second gulley west of triangulation station on cliffs west of Sacks Harbour, lat. 70°57'N, long. 125°47'21"W, Banks Island, District of Franklin.

Laricioxylon occidentaloides Roy and Hills
Holotype 27954; paratypes 27940, 27944, 27945, 27948, 27972, 27978, 27981
Roy, S.K. and Hills, L.V., 1972, Can. J. Botany, vol. 50, no. 12, p. 2641, figs. 13-19.
Beaufort Formation, late Miocene-early Pliocene, north side Ballast Brook, 9 miles south-southeast of M'Clure Strait, lat. 74°19'30"N, long. 123°5'W, about 8.5 miles south of M'Clure Strait, lat. 74°21'N, long. 123°6'W(27940), and north side of small tributary on west side of M'Clure Brook, 7 miles south of M'Clure Strait, lat. 74°21'N, long. 123°7'30'W(27978), prominent gulley on north side of Ballast Brook, 9.75 miles south-southeast of M'Clure Strait, lat. 74°19'N, long. 123°4'W(27944, 27972, 27981), and north side of small tributary on west side of Ballast Brook, 7 miles south of M'Clure Strait, lat. 74°21'N, long. 123°8'W (27945),northwestern Banks Island, District of Franklin.

Leaf undetermined
Fig. spec. 32381
Kennedy, H., 1973, Fossils and geological features in the Ottawa Lowlands, pl. 3, fig. 11.
Pleistocene, Green Creek, southeast of Ottawa, Ontario.
=Leaf impression in a concretion, Wagner, F.J.E., 1984, Fossils of Ontario Part 2, Royal Ontario Mus. Life Sci. Misc. Publ., Fig. 33B.

Liriodendron sp.(Magnoliaceae): seed
Fig. spec. 99141
Matthews, Jr., J.V. and Ovenden, L.E., 1990, Arctic, vol. 43, no. 4, p. 375, fig. 4.6.
Mary Sacks gravel, Miocene, Duck Hawk Bluffs, southwestern Banks Island, District of Franklin.

Lycopus sp.
Fig. spec. 95888
Matthews, Jr., J.V., Ovenden, L.E. and Fyles, J.G., 1990, Canada's Missing Dimension, vol. 1, C.R. Harington, ed., Canadian Mus. Nature, fig. 6i.
Beaufort Formation, Pliocene, Devaney section, 14.5 km west of Mould Bay station, lat. 76°15'N, long. 119°55'W, Prince Patrick Island, District of Franklin.

cf. *Melolia* sp. [Fungus]
Fig. specs. 75880, 75881
Kalgutkar, R.M., 1985, Geol. Surv. Can., Paper 85-1B, p. 266, pl. 31.2, figs. 4, 5.
Bonnet Plume Formation, Late Paleocene-Middle Eocene, lat. 65°54'N, long. 135°4'W, Peel River, Yukon.

Menyanthes trifoliata Linnaeus
Hypotype 60878
Blake, Jr., W., and Matthews, Jr., J.V., 1979, Geol. Surv. Can., Paper 79-1A, fig. 28.6A.
Pre-Holocene, 8 km east of head of Makinson Inlets' north arm, east-central Ellesmere Island, District of Franklin.

Minuartria Rossil (R. Br.)House
Hypotype 85093
Dyke, A.S. and Matthews, Jr., J.V., 1987, Géographie Physique et Quaternaire, vol. XLl, no. 3, fig. 12d.
Quaternary, North Bluff west of Pasley River, west-central Boothia Peninsula, District of Franklin.

cf. *Mneme*
Fig. specs. 85364, 85365
Matthews, Jr., J.V., 1987, Geol. Surv. Can., Paper 87-1A, p. 83, pl. 9.3, figs. 4, 5.
Beaufort Formation, Miocene-Pliocene, Duck Hawk Bluffs, southwestern Banks Island, District of Franklin.

Muscites eocenicus Kuc [Moss)
Holotype 31230, a
Kuc, M., 1972, Can. J. Earth Sci., vol. 9, no. 5, p. 602, figs. 1a, b.
Allenby Formation, Eocene, 6 miles north of Princeton, approximately lat. 49°32'43"N, long. 120°25'5"W, British Columbia.

Muscites maycocki Kuc [Moss)
Holotype 34743
Kuc, M., 1974, Can. J. Earth Sci., vol. 11, no. 3, p. 411, figs. 4A-C.
Allenby Formation, Eocene, 6 miles north of Princeton, British Columbia.

Muscites ritchiei Kuc [Moss]
Holotype 34744
Kuc, M., 1974, Can. J. Earth Sci., vol. 11, no. 3, p. 411, figs. 4D-F.
Allenby Formation, Eocene, 6 miles north of Princeton, British Columbia.

Myrica cf. *eogale* Nickit
Hypotype 95886
Matthews, Jr., J.V., Ovenden, L.E. and Fyles, J.G., 1990, Canada's Missing Dimension, Vol. 1, C.R Harington, ed., Canadian Mus. Nature, fig. 6h.
Beaufort Formation, Pliocene, 150 m downstream, 12 km northwest of Green Bay, Prince Patrick Island, District of Franklin.

Nymphoides sp. (Gentianaceae): seed
Fig. spec. 99145
Matthews, Jr., J.V. and Ovenden, L.E., 1990, Arctic, vol. 43, no. 4, fig. 4.10.
Pliocene, Beaver peat site, near of head Strathcona Fiord, Ellesmere Island, District of Franklin.

Palaeohypnum jovet-asti Kuc [Moss]
Holotype 34745; paratype 34746
Kuc, M., 1974, Can. J. Earth Sci., vol. 11, no. 3, p. 414, figs. 5A, B, 6A-D.
Allenby Formation, Eocene, 6 miles north of Princeton, British Columbia.

Palaeohypnum steerei Kuc [Moss]
Holotype 34747
Kuc, M., 1974, Can. J. Earth Sci., vol. 11, no. 3, p. 414, figs. 7A, B.
Allenby Formation, Eocene, 6 miles north of Princeton, British Columbia.

cf. *Paliurus* sp. (Rhamnaceae)
Fig. specs. 95882, 95883
Matthews, Jr., J.V. and Ovenden, L.E., 1990, Arctic, vol. 43, no. 4, fig. 4.2, 4.3.
Pliocene, lower end of Henderson Slough, Porcupine River, lat. 66°52.5′N, long. 143°34.79′W, Alaska, and Beaver peat site, near head of Strathcona Fiord, Ellesmere Island, District of Franklin.
=*Paliurus* sp., Matthews, Jr., J.V., Ovenden, L.E. and Fyles, J.G., 1990, Canada's Missing Dimension, vol. 1, C.R. Harington, ed., Canadian Mus. Nature, p. 118, fig. 6b (95882).
=*Paliurus* type, ibid., fig. 6c (95883).

Paramicrothallites canadensis Kalgutkar [Fungus]
Holotype 75874; paratypes 75873, 75875
Kalgutkar, R.M., 1985, Geol. Surv. Can., Paper 85-1B, p. 264, pl. 31.1, figs. 5-7.
Bonnet Plume Formation, Late Paleocene-Middle Eocene, lat. 65°54′N, long. 135°4′W, Peel River, Yukon.

Pesavis cf. *P. parva* Kalgutkar and Sweet [Fungus]
Hypotype 83350
Kalgutkar, R.M. and Sweet, A.R., 1988, Geol. Surv. Can., Bull. 379, p. 120, pl. 6.2, fig. 3.
Aklak Member, Reindeer Formation, Paleocene, Aklak Creek, lat. 68°40′50″N, long. 136°20′50″W, Yukon.

Pesavis tagluensis Elsik and Jansonius 1974 [Fungus]
Hypotypes 83353-83359
Kalgutkar, R.M. and Sweet, A.R., 1988, Geol. Surv. Can., Bull. 379, p. 123, pl. 6.2, fig. 6-12.
Bonnet Plume Formation, Paleocene-Eocene, Peel River, lat. 65°54′N, long. 135°4′W and lat. 65°52′N, long. 135°12′W(83358), Yukon.

Pesavis cf. *P. tagluensis* Elsik and Jansonius 1974 [Fengus]
Hypotypes 83344-83348, 83352, 83352
Kalgutkar, R.M. and Sweet, A.R., 1988, Geol. Surv. Can., Bull. 379, p. 120, pl. 6.1, figs. 17-20; pl. 6.2, figs. 1, 2, 4, 5.
Paleocene, Summit Creek Formation, southeast end of Tertiary Hills, lat. 64°27′30N, long. 125°33′45″W, District of McKenzie; Aklak Member Reindeer Formation, Aklak Creek, lat. 68°40′25″N, long. 136°20′35″W(83351), and Bonnet Plume Formation, Paleocene-Eocene, Peel River, lat. 65°52′N, long. 135°12′W(83352), Yukon.

Phyllanthus (Phyllanthus) sp. (Euphorbiaceae): seed
Fig. spec. 99135
Matthews, Jr., J.V. and Ovenden, L.E., 1990, Arctic, vol. 43, no. 43, no. 4, p. 377, fig. 3.7.
Mary Sacks gravel, Miocene, Duck Hawk Bluffs, southwestern Banks Island, District of Franklin.

Picea banskii Hills and Ogilvie
Holotype 25202; paratypes 25203-25209
Hills, L.V. and Ogilvie, R.T., 1970, Can. J. Botany, vol. 48, no. 3, p. 460, figs. 11a-e, 12a-e, 13a-e.
Beaufort Formation, Late Tertiary, Ballast Brook, northwestern Banks Island, District of Franklin.

Picea banskii Hills and Ogilvie
Hypotypes 42545-42550
Hills, L.V. and Bustin, R.M., 1976, Geol. Surv. Can., Paper 76-1B, pl. 13.1, figs. a-f.
Beaufort Formation, Miocene, lat. 79°53′N, long. 88°15′W, west of Gibs Fiord, Axel Heiberg Island, District of Franklin.

Piceoxylon beaufortense Roy and Hills
Holotype 27976; paratype 27939
Roy, S.K. and Hills, L.V., 1972, Can. J. Botany, vol. 50, no. 12, p. 2639, figs. 7-12.
Beaufort Formation, late Miocene-early Pliocene, north side of north fork of Ballast Brook, 7 miles south and 5.75 miles east of mouth of Ballast Brook, lat. 74°20′30″N, long. 122°48′W, northwestern Banks Island, District of Franklin.

Pinus escalantensis Banks, Ortiz-Sotomayor and Hartman
Holotype 54016
Banks, H.P., Ortiz-Sotomayor, A. and Hartman, C.M, 1981, Botanical Gazette, vol. 142, no. 2, p. 291, figs. 1-13.
Hesquiat Formation, Lower to mid-Oligocene, Escalante Island off Escalante Point, western Vancouver Island, British Columbia.

Pinus subsect. *Cembrae* Engelm. (Pinaceae)
Fig. spec. 99132
Matthews, Jr., J.V. and Ovenden, L.E., 1990, Arctic, vol. 43, no. 4, p. 383, fig. 3.4.
Beaufort Formation, Pliocene, Prince Patrick Island, District of Franklin.

Pinus subsect. *contortae*(?) sp. (Pinaceae)
Fig. spec. 99138
Matthews, Jr., J.V. and Ovenden, L.E., 1990, Arctic, vol. 43, no. 4, p. 380, fig. 4.3.
Beaufort Formation, Pliocene, Prince Patrick Island, District of Franklin.

Pinus subsect. *Eustrobi* Engelm
Fig. spec. 99131
Matthews, Jr., J.V. and Ovenden, L.E., 1990, Arctic, vol. 43, no. 4, p. 383, fig. 3.3.
Pliocene, Beaver peat site, near head of Strathcona Sound, Ellesmere Island, District of Franklin.

Pinus trunculus Dawson
Hypotype 7946a
Penhallow, D.P., 1908, Geol. Surv. Can., Rept. Tert. Pl. B.C., Publ. 1013, p. 71, text-fig. 27.
Tertiary [Princeton Group], Tulameen River opposite Vermilion Cliff, British Columbia.

Pinus tulameenensis Penhallow
Holotype 7698, a [obverse and reverse]
Penhallow, D.P., 1908, Geol. Surv. Can., Rept. Tert, Pl. B.C., Publ. 1013, p. 73, text-fig. 34.
Tertiary, 1 mile above Princeton, Tulameen River, British Columbia.

Pinuxylon albicauloides Greguss, 1954
Hypotypes 27926, 27949, 27968
Roy, S.K. and Hills, L.V., 1972, Can. J. Botany, vol. 50, no. 12, p. 2638, figs. 1-6(27968).
Beaufort Formation, late Miocene-early Pliocene, north side of Ballast Brook, 9 miles south-southeast of M'Clure Strait, lat. 74°19'30"N, long. 123°5'W, and in prominent gully on north side of Ballast Brook, 9.75 miles south-southeast of M'Clure Strait, lat. 74°19'N, long. 123°4'W(27968), northwestern Banks Island, District of Franklin.

Plochmopeltinites masonii Cookson 1947 [Fungus]
Hypotypes 75876-75879
Kalgutkar, R.M., 1985, Geol. Surv. Can., Paper 85-1B, p. 264, pl. 31.1, fig. 8; pl. 31.2, figs. 1-3.
Bonnet Plume Formation, Late Paleocene-Middle Eocene, lat. 65°54'N, long. 135°4'W, Peel River, Yukon.

Polanisia sp.
Fig. spec. 85363
Matthews, Jr., J.V., 1987, Geol. Surv. Can., Paper 87-1A, p. 83, pl. 9.3, fig. 3.
Beaufort Formation, Miocene-Pliocene, Duck Hawk Bluffs, southwestern Banks Island, District of Franklin.

Prunus sp.
Fig. spec. 95887
Matthews, Jr., J.V., Ovenden, L.E. and Fyles, J.G., 1990, Canada's Missing Dimension, vol. 1, C.R. Harington, ed., Canadian Mus. Nature, fig. 6g.
Beaufort Formation, Pliocene, Devaney section, 14.5 km west of Mould Bay Station, lat. 76°15'N, long. 119°35'W, Prince Patrick Island, District of Franklin.

Rhynchospora capitellata (Michx.) Vahl.
Hypotype 95885
Matthews, Jr., J.V., Ovenden, L.E. and Fyles, J.G., 1990, Canada's Missing Dimension, vol. 1, C.R. Harington, ed., Canadian Mus. Nature, fig. 6f.
Beaufort Formation, Pliocene, Devaney section, 14.5 km west of Mould Bay Station, lat. 76°15'N, long. 119°35'W, Prince Patrick Island, District of Franklin.

Rumex sp.
Fig. spec. 95881
Matthews, Jr., J.V., Ovenden, L.E. and Fyles, J.G., 1990, Canada's Missing Dimension, vol. 1, C.R. Harington, ed., Canadian Mus. Nature, fig. 6a.
Beaufort Formation, Pliocene, north of Mould Bay Station, Prince Patrick Island, District of Franklin.

cf. *Sagisma* Nikitin (Alismataceae): fruit
Fig. spec. 99144
Matthews, Jr., J.V. and Ovenden, L.E., 1990, Arctic, vol. 43, no. 4, p. 377, fig. 4.9.
Mary Sacks gravel, Miocene, Duck Hawk Bluffs, southwestern Banks Island, District of Franklin.

Scheuchzeria sp. (Scheuchzeriaceae): seed
Fig. spec. 99134
Matthews, Jr., J.V. and Ovenden, L.E., 1990, Arctic, vol. 43, no. 4, p. 383, fig. 3.6.
Pliocene, head of Strathcona Fiord. Ellesmere Island, District of Franklin.

Scirpus microcarpus Presl. (Cyperaceae)
Hypotype 99136
Matthews, Jr., J.V. and Ovenden, L.E., 1990, Arctic, vol. 43, no. 4, fig. 3.8.
Pliocene, Beaver Peat site, near head of Strathcona Fiord, Ellesmere Island, District of Franklin.

"seed" of *Callitriche*
Fig. spec. 85095
Dyke, A.S and Matthews, Jr., J.V., 1987, Géographie Physique et Quaternaire, vol. XLI, no. 3, fig. 12h.
Quaternary, South Bluffs west side of Pasley River, west-central Boothia Peninsula, District of Franklin.

"seed" of *Hypericum* sp.
Fig. spec. 85094
Dyke, A.S and Matthews, Jr., J.V., 1987, Géographie Physique et Quaternaire, vol. XLI, no. 3, fig. 12g.
Quaternary, South Bluffs west side of Pasley River, west-central Boothia Peninsula, District of Franklin.

Selaginella ruprestus (L.)
Hypotype 85092
Dyke, A.S. and Matthews, Jr., J.V., 1987, Géographie Physique et Quaternaire, vol. XLI, no. 3, fig. 12f.
Quaternary, South Bluffs west side of Pasley River, west-central Boothia Peninsula, District of Franklin.

Specimens Nos. 1-VII [Moss]
Fig. specs. 34748-34754
Kuc, M., 1974, Can. J. Earth Sci., vol. 11, no. 3, p. 414, figs. 8A-I.
Allenby Formation, Eocene, 6 miles north of Princeton, British Columbia.

Unknown seed
Fig. spec. 85360
Matthews, Jr., J.V., Geol. Surv. Can., Paper 87-1A, p. 83, pl. 9.3, fig. 6.
Beaufort Formation, Miocene-Pliocene, Ballast Brook, northwestern Bank Island, District of Franklin.

Unknown seed
Fig. spec. 95884
Matthews, Jr., J.V., Ovenden, L.E. and Fyles, J.G., 1990, Canada's Missing Dimension, vol. 1, C.R. Harington, ed., Canadian Mus. Nature, figs. 6d, e.
Beaufort Formation, Pliocene, Devaney section 14.5 km west of Mould Bay Station, lat. 76°15′N, long. 119°55′W, Prince Patrick Island, District of Franklin.

Weigela sp. (Caprifoliaceae): seed
Fig. spec. 99140
Matthews, Jr., J.V. and Ovenden, L.E., 1990, Arctic, vol. 43, no. 4, fig. 4.5.
Mary Sacks gravel, Miocene, Duck Hawk Bluffs, southwestern Banks Island, District of Franklin.

PALYNOMORPHS

POLLEN-SPORE

Silurian

Ambitisporites avitus Hoffmeister, 1959
Hypotype 67780
McGregor, D.C., 1984, Academia Nacional de Ciencias, Misc. No. 69, pl. 3, fig. 13.
Tarabuco Formation, Upper Silurian, Tarabuco section, Chuquisaca Dept., northwest of Tarabuco, Cordillera Oriental, Bolivia.

Apiculiretusispora spicula Richardson and Lister
Hypotype 55503
McGregor, D.C. and Narbonne, G.M., 1978, Can. J. Earth Sci., vol. 15, no. 8, p. 1295, pl. 1, figs. 5, 6.
Member B 20 m above base, Read Bay Formation, Upper Silurian, south side Goodsir Creek, eastern Cornwallis Island, District of Franklin.

Apiculiretusispora sp.
Fig. spec. 55502
McGregor, D.C. and Narbonne, G.M., 1978, Can. J. Earth Sci., vol. 15, no. 8, p. 1295, pl. 1, fig. 4.
Member B 20 m above base, Read Bay Formation, Upper Silurian, south side Goodsir Creek, eastern Cornwallis Island, District of Franklin.

Cluster of smooth palynormorphs
Fig. spec. 55513
McGregor, D.C. and Narbonne, G.M., 1978, Can. J. Earth Sci., vol. 15, no. 8, p. 1296, pl. 1, fig. 18.
Member B 20 m above base, Read Bay Formation, Upper Silurian, south side Goodsir Creek, eastern Cornwallis Island, District of Franklin.

dyad
Fig. spec. 73067
Duffield, S.L., 1985, J. Paleontol., vol. 59, no. 4, p. 1007, fig. 3.3.
Jupiter Formation, member 3, Early Silurian, cliff north of mouth of Jupiter River, Anticosti Island, Québec.

Emphanisporites multicostatus Rodriguez, 1978
Hypotype 67186
McGregor, D.C., 1984, Academia Nacional de Ciencias, Misc. No. 69, p. 29, pl. 1, fig. 17.
Tarabuco Formation, Upper Silurian, Tarabuco section, Chuquisaca Dept., northwest of Tarabuco, Cordillera Oriental, Bolivia.

?*Emphanisporites protophanus* Richardson and Ioannides, 1973
Hypotype 55506
McGregor, D.C. and Narbonne, G.M., 1978, Can. J. Earth Sci., vol. 15, no. 8, p. 1295, pl. 1, fig. 9.
Member B 20 m above base, Read Bay Formation, Upper Silurian, south side Goodsir Creek, eastern Cornwallis Island, District of Franklin.

Hispanaediscus sp.
Fig. specs. 55515-55517
McGregor, D.C. and Narbonne, G.M., 1978, Can. J. Earth Sci., vol. 15, no. 8, p. 1296, pl. 1, figs. 20-22.
Member B 20 m above base, Read Bay Formation, Upper Silurian, south side Goodsir Creek, eastern Cornwallis Island, District of Franklin.

Nodospora burnhamensis Strother and Traverse 1979
Hypotypes 73065, 73066, 73068-73072
Duffield, S.L., 1985, J. Paleontol., vol. 59, no. 4, p. 1006, figs. 3.1, 3.2, 3.4-3.6, 3.8, 3.9.
Jupiter Formation, member 3, Early Silurian, cliff north of mouth of Jupiter River, Anticosti Island, Québec.

Parts of tracheid-like fragment
Fig. spec. 55512
McGregor, D.C. and Narbonne, G.M., 1978, Can. J. Earth Sci., vol. 15, no. 8, p. 1296, pl. 1, figs. 16, 17.
Member B 20 m above base, Read Bay Formation, Upper Silurian, south side Goodsir Creek, eastern Cornwallis Island, District of Franklin.

Punctatisporites sp.
Fig. spec. 55499
McGregor, D.C. and Narbonne, G.M., 1978, Can. J. Earth Sci., vol. 15, no. 8, p. 1295, pl. 1, fig. 1.
Member B 20 m above base, Read Bay Formation, Upper Silurian, south side Goodsir Creek, eastern Cornwallis Island, District of Franklin.

Retusotriletes chartulatus McGregor
Holotype 55507; hypotype 55508
McGregor, D.C. and Narbonne, G.M., 1978, Can. J. Earth Sci., vol. 15, no. 8, p. 1301, pl. 1, figs. 10-12.
Member B 20 m above base, Read Bay Formation, Upper Silurian, south side Goodsir Creek, eastern Cornwallis Island, District of Franklin.

Retusotriletes sp.
Fig. specs. 55500, 55501, 55504
McGregor, D.C. and Narbonne, G.M., 1978, Can. J. Earth Sci., vol. 15, no. 8, p. 1295, pl. 1, figs. 2, 3, 7.
Member B 20 m above base, Read Bay Formation, Upper Silurian, south side Goodsir Creek, eastern Cornwallis Island, District of Franklin.

rugose spore(?)
Fig. spec. 71119
Duffield, S.L., 1985, J. Paleontol., vol. 59, no. 4, p. 1007, fig. 3.7.
Jupiter Formation, member 3, Early Silurian, cliff north of mouth of Jupiter River, Anticosti Island, Québec.

"cf. *Synorisporites verrucatus*" of Richardson and Ioannides, 1973
Hypotype 55505
McGregor, D.C. and Narbonne, G.M., 1978, Can. J. Earth Sci., vol. 15, no. 8, p. 1295, pl. 1, fig. 8.
Member B 20 m above base, Read Bay Formation, Upper Silurian, south side Goodsir Creek, eastern Cornwallis Island, District of Franklin.

Synorisporites sp.
Fig. spec. 67776
McGregor, D.C., 1984, Academia Nacional de Ciencias, Misc. No. 69, pl. 3, fig. 9.
Santa Rosa Formation, Upper Silurian, Laurani section, Chuquisaca Dept., northwest of Tarabuco, Cordillera Oriental, Bolivia.

Tetrads
Fig. specs. 55509-55511, 55514, 55526
McGregor, D.C. and Narbonne, G.M., 1978, Can. J. Earth Sci., vol. 15, no. 8, p. 1295, pl. 1, figs. 13-15, 19; pl. 2, fig. 1.
Member B 20 m above base, Read Bay Formation, Upper Silurian, south side Goodsir Creek, eastern Cornwallis Island, District of Franklin.

Devonian

Acanthotriletes denticulatus Naumova, 1953
Hypotype 15403
McGregor, D.C. and Owens, B., 1966, Geol. Surv. Can., Paper 66-30, pl. 26, fig. 5.
Griper Bay Formation, Upper Devonian, stream valley west of Sargeant Point, 2 miles from shore, northeastern Bathurst Island, District of Franklin.

Acanthotriletes famensis Naumova
Hypotype 13582
Kerr, J.W., McGregor, D.C. and McLaren, D.J., 1965, Bull. Can. Petrol. Geol., vol. 13, no. 3, pl. 3, fig. 3.
Griper Bay Formation, Upper Devonian, stream valley west of Sargeant Point, 2 miles from shore, northeastern Bathurst Island, District of Franklin.

cf. *Acanthotriletes gratus* Naumova
Hypotype 13600
Kerr, J.W., McGregor, D.C. and McLaren, D.J., 1965, Bull. Can. Petrol. Geol., vol. 13, no. 3, pl. 3, fig. 18.
Griper Bay Formation, Upper Devonian, stream valley west of Sargeant Point, 2 miles from shore, northeastern Bathurst Island, District of Franklin.

Acanthotriletes hirsutus Chibrikova
Hypotype 13616
Kerr, J.W., McGregor, D.C. and McLaren, D.J., 1965, Bull. Can. Petrol. Geol., vol. 13, no. 3, pl. 4, fig. 8.
Griper Bay Formation, Upper Devonian, 3/4 mile from shore in stream entering Reindeer Bay, 4 miles south of largest river entering the bay, northeastern Bathurst Island, District of Franklin.

Acanthotriletes multisetus Luber var. *major* Richardson, 1965
Hypotype 15286
McGregor, D.C. and Owens, B., 1966, Geol. Surv. Can., Paper 66-30, pl. 9, fig. 1.
Weatherall Formation, Middle Devonian, south limb of Robertson Point anticline, 8 miles northeast of Beverley Inlet, Melville Island, District of Franklin.

Acanthotriletes multisetus (Luber) Potonié and Kremp, 1955, var. *multisetus*
Hypotype 67154
McGregor, D.C., 1984, Academia Nacional de Ciencias, Misc. No. 69, p. 16, pl. 3, fig. 7.
Icla Formation, Lower or Middle Devonian, Laurani section, Chuquisaca Dept., northwest of Tarabuco, Cordillera Oriental, Bolivia.

Acanthotriletes perpusillus Naumova
Hypotype 32813
McGregor, D.C., 1974, Can. J. Earth Sci., vol. 11, no. 1, pl. 1, figs. 32, 33.
Vendom Fiord Formation, Early Devonian, lat. 79°28′40″N, long. 80°02′W, 12 miles south of head of Cañon Fiord, central Ellesmere Island, District of Franklin.

cf. *Acanthotriletes rarisetosus* Kedo
Hypotype 25442
McGregor, D.C., 1970, Colloque Stratig. Carbonifère, Congr. Colloques Univi. Liège, vol. 55, pl. 21, fig. 5.
Bedford Formation, Upper Devonian, Courtright bore hole, Courtright village, Lambton County, Ontario.

Acanthotriletes serratus Naumova
Hypotype 16592
McGregor, D.C., 1967, Review Palaeobotany and Palynology, I, p. 180, pl. 1, fig. 13.
Hecla Bay Formation, Melville Island Group, Middle or Upper Devonian, Helena Island, District of Franklin.

cf. *Acanthotriletes uncatus* Naumova
Hypotype 13613
Kerr, J.W., McGregor, D.C. and McLaren, D.J., 1965, Bull. Can. Petrol. Geol., vol. 13, no. 3, pl. 4, fig. 5.
Griper Bay Formation, Upper Devonian, 3/4 mile from shore in stream entering Reindeer Bay, 4 miles south of largest river entering the bay, northeastern Bathurst Island, District of Franklin.

Acanthotriletes sp.
Fig. spec. 15173
McGregor, D.C. and Owens, B., 1966, Geol. Surv. Can., Paper 66-30, pl. 3, fig. 8.
Battery Point Formation, Lower Devonian, south side Gaspé Bay, shore section from Anse-à-Brillant to Bois Brulé, eastern Gaspé Peninsula, Québec.

Acanthotriletes sp.
Fig. spec. 67155
McGregor, D.C., 1984, Academia Nacional de Ciencias, Misc. No. 69, p. 16, pl. 1, fig. 4.
Santa Rosa Formation, Lower Devonian, Laurani section, Chuquisaca Dept., northwest of Tarabuco, Cordillera Oriental, Bolivia.

Acinosporites acanthomammillatus Richardson
Hypotype 41776
McGregor, D.C. and Camfield, M., 1976, Geol. Surv. Can., Bull. 263, p. 6, pl. 6, figs. 14, 15.
Williams Island Formation, Middle Devonian, 157.5 feet depth, ODM Jaab Lake No. 1 well, lat. 51°11′54″N, long. 82°56′W, Ontario.

Acinosporites acanthomammillatus Richardson 1965
Hypotype 49832
McGregor, D.C., 1979, Palynology, vol. 3, pl. 2, fig. 10.
Weatherall Formation, Middle Devonian, creek flowing south, 10 miles east of Weatherall Bay, Melville Island, District of Franklin.

Acinosporites acanthomammillatus Richardson
Hypotypes 66187-66189, 66272
McGregor, D.C. and Camfield, M., 1982, Geol. Surv. Can., Bull. 348, p. 11, pl. 1, figs. 1-6.
Weatherall and Hecla Bay (66272) formations, Middle Devonian, stream section 16.2 km east of east arm of Weatherall Bay, northeastern Melville Island, District of Franklin.

?*Acinosporites acanthomammillatus* Richardson, 1965
Hypotype 67156
McGregor, D.C., 1984, Academia Nacional de Ciencias, Misc. No. 69, p. 17, pl. 3, fig. 5.
Huamampampa Formation, Middle Devonian, Laurani section, Chuquisaca Dept., northwest of Tarabuco, Cordillera Oriental, Bolivia.

Acinosporites hirsutus (Brideaux and Radforth) n. comb.
Hypotypes 66190, 66191
McGregor, D.C. and Camfield, M., 1982, Geol. Surv. Can., Bull, 348, p. 11, pl. 1, figs, 7, 8, 12, 13.
Weatherall Formation, Middle Devonian, stream section 16.2 km east of east arm of Weatherall Bay, northeastern Melville Island, District of Franklin.

Acinosporites lindlarensis Riegel
Hypotypes 32010, 32011
McGregor, D.C., 1973, Palaeontographica, Abt. B, vol. 142, no. 1-3, p. 49, pl. 6, figs. 17-21.
Battery Point and Malbaie formations, Middle Devonian, 550 yards southeast of small brook south of Cape Rouge, on point, and 65 yards southeast of conglomerate of Pointe Jaune, south shore Gaspé Bay, Québec.

cf. *Acinosporites lindlarensis* Riegel
Hypotype 61688
McGregor, D.C., 1979, Geol. Surv. Can., Paper 79-1B, pl. 22.2, figs. 40, 41.
Daleje Shale, Lower Devonian, roadside outcrop mear Karlštejn, 1 km north of Budnany in the Hluboké valley, Czechoslovakia.

Acinosporites lindlarensis var. *lindlarensis* Riegel
Hypotype 41757
McGregor, D.C. and Camfield, M., 1976, Geol. Surv. Can., Bull. 263, p. 6, pl. 5, figs. 2, 3.
Williams Island Formation, Middle Devonian, 157.5-159.5 feet depth, ODM Jaab Lake No. 1 well, lat. 51°11′54″N, long. 82°56′W, Ontario.

Acinosporites lindlarensis Riegel var. *lindlarensis*
Hypotype 66192
McGregor, D.C. and Camfield, M., 1982, Geol. Surv. Can., Bull. 348, p. 13, pl. 1, figs. 9, 10.
Weatherall Formation, Middle Devonian, stream section 16.2 km east of east arm of Weatherall Bay, northeastern Melville Island, District of Franklin.

?*Acinosporites lindlarensis* Riegel, 1968 var. *lindlarensis*
Hypotype 67157
McGregor, D.C., 1984, Academia Nacional de Ciencias, Misc. No. 69, p. 17, pl. 2, fig. 21.
Huamampampa Formation, Middle Devonian, Laurani section, Chuquisaca Dept., northwest of Tarabuco, Cordillera Oriental, Bolivia.

Acinosporites lindlarensis Riegel var. *lindlarensis*
Hypotype 73289
Richardson, J.B. and McGregor, D.C., 1986, Geol. Surv. Can., Bull. 364, pl. 6, fig. 8.
Williams Island Formation, Middle Devonian, ODM Jaab Lake No. 1 well, lat. 51°11.9′N, long. 82°57′W, Moose River Basin, Ontario.

Acinosporites lindlarensis var. *minor* McGregor and Camfield
Holotype 41758
McGregor, D.C. and Camfield, M., 1976, Geol. Surv. Can., Bull. 263, p. 8, pl. 5, figs. 4, 5.
Williams Island Formation, Middle Devonian, 157.5-159.5 feet depth, ODM Jaab Lake No. 1 well, lat. 51°11′54″N, long. 82°56′W, Ontario.

Acinosporites macrospinosus Richardson 1965
Hypotype 48171
McGregor, D.C., 1977, Palaeontographica, Abt. B, vol. 163, p. 129, pl. 1, figs. 13, 14.
Battery Point Formation, Middle Devonian, Tar Point to Whale Head section, south shore Gaspé Bay, Québec.

Acinosporites macrospinosus Richardson 1965
Hypotype 49828
McGregor, D.C., 1979, Palynology, vol. 3, pl. 2, fig. 4.
Devonian, 4170-4180 feet, Dome et al. Weatherall 0-10 well, northeastern Melville Island, District of Franklin.

Acinosporites macrospinosus Richardson
Hypotypes 66193-66195
McGregor, D.C. and Camfield, M., 1982, Geol. Surv. Can., Bull. 348, p. 13, pl. 1, figs. 11, 14-16.
Weatherall Formation, Middle Devonian, stream section 16.2 km east of east arm of Weatherall Bay, northeastern Melville Island, District of Franklin.

Acinosporites sp.
Fig. specs. 15174, 15182, 15313
McGregor, D.C. and Owens, B., 1966, Geol. Surv. Can., Paper 66-30, pl. 3, figs. 9, 10; pl. 15, figs. 4, 5.
Battery Point Formation, Lower Devonian, south side Gaspé Bay, shore section from Anse-à-Brillant to Bois Brulé Brook, and north side Gaspé Bay, shore cliffs between Gaspé Peninsula and Laurencelle Road, eastern Gaspé Peninsula, Québec; Griper Bay Formation, Upper Devonian, south limb of Robertson Point Anticline, 8 miles northeast of Beverley Inlet, Melville Island, District of Franklin (15313).

?*Acinosporites* sp.
Fig. spec. 61694
McGregor, D.C., 1979, Geol. Surv. Can., Paper 79-1B, pl. 22.2, fig. 48.
Daleje Shale, Lower Devonian, roadside outcrop near Karlštejn, 1 km north of Budnany in the Hluboké Valley, Czechoslovakia.

Acinosporites sp.
Fig. specs. 67790, 67794
McGregor, D.C., 1984, Academia Nacional de Ciencias, Misc. No. 69, pl. 4, figs. 6, 10.
Icla and Huamampampa formations, Middle Devonian, Laurani section, Chuquisaca Dept., northwest of Tarabuco, Cordillera Oriental, Bolivia.

?*Aculeispores* sp.
Fig. spec. 15485
McGregor, D.C. and Owens, B., 1966, Geol. Surv. Can., Paper 66-30, pl. 29, fig. 14.
Kettle Point Formation, Upper Devonian, depth 295-299 feet, borehole, Courtright village, Lambton County, Ontario.

Amicosporites splendidus Cramer, 1967
Hypotype 67783
McGregor, D.C., 1984, Academia Nacional de Ciencias, Misc. No. 69, pl. 3, fig. 16.
Santa Rosa Formation, Lower Devonian, Laurani section, Chuquisaca Dept., northwest of Tarabuco, Cordillera Oriental, Bolivia.

cf. *Anapiculatisporites acerosus* (Naumova) Lanniger
Hypotype 61712
McGregor, D.C., 1979, Geol. Surv. Can., Paper 79-1B, pl. 22.2, fig. 67.
Daleje Shale, Lower Devonian, upper part of southeast slope of Císaská Rokle, southeast of Srbsko, Czechoslovakia.

Anapiculatisporites petilus Richardson 1965
Hypotype 49823
McGregor, D.C., 1979, Palynology, vol. 3, pl. 1, fig. 23.
Weatherall Formation, Middle Devonian, creek section 5.5 to 10 miles north of head of Beverley Inlet, Melville Island, District of Franklin.

Anapiculatisporites petilus Richardson
Hypotype 66196
McGregor, D.C. and Camfield, M., 1982, Geol. Surv. Can., Bull. 348, p. 14, pl. 2, figs. 1, 2.
Weatherall Formation, Middle Devonian, stream section 16.2 km east of arm of Weatherall Bay, northeastern Melville Island, District of Franklin.

Anapiculatisporites sp.
Fig. spec. 15159
McGregor, D.C. and Owens, B., 1966, Geol. Surv. Can., Paper 66-30, pl. 2, fig. 28.
Battery Point Formation, Lower Devonian, south side Gaspé Bay, shore section from Anse-á-Brillant to Bois Brulé Brook, eastern Gaspé Peninsula, Québec.

Anapiculatisporites sp.
Fig. spec. 32793
McGregor, D.C., 1974, Can. J. Earth Sci., vol. 11, no. 1, pl. 1, figs. 8, 10.
Unnamed Early Devonian, lat. 79°28′40″N, long. 80°02′W, 12 miles south of head of Cañon Fiord, central Ellesmere Island, District of Franklin.

Anapiculatisporites sp.
Fig. spec. 48165
McGregor, D.C., 1977, Palaeontographica, Abt. B., vol. 163, p. 129, pl. 1, figs. 1, 2.
Battery Point Formation, Lower Devonian, Tar Point to Whale Head section, south shore Gaspé Bay, Québec.

Anapiculatisporites sp.
Fig. spec. 67789
McGregor, D.C., 1984, Academia Nacional de Ciencias, Misc. No. 69, pl. 4, fig. 5.
Huamampampa Formation, Middle Devonian, Laurani section, Chuquisaca Dept., northwest of Tarabuco, Cordillera Oriental, Bolivia.

Anapiculatisporites sp. A, Streel in Becker et al., 1974
Fig. spec. 86087
McGregor, D.C. and McCutcheon, S.R., 1988, Can. J. Earth Sci., vol. 25, no. 9, p. 1352, pl. 1, fig. 14.
Carrow Formation, Piskahegon Group, Late Devonian, depth 72.85 m, drill hole M-6, lat. 45°34′43″N, long. 66°35′27″W, southwestern New Brunswick.

Anapiculatisporites spp. 1, 2
Fig. specs. 67159, 67160
McGregor, D.C., 1984, Academia Nacional de Ciencias, Misc. No. 69, p. 18, pl. 1, figs. 1, 7.

Santa Rosa Formation, Lower Devonian, Laurani section, Chuquisaca Dept., northwest of Tarabuco, Cordillera Oriental, Bolivia.

Anaplanisporites extremus McGregor and Camfield
Holotype 41755
McGregor, D.C. and Camfield, M., 1976, Geol. Surv. Can., Bull. 263, p. 8, pl. 4, figs. 25, 26.
Stooping River Formation, Lower Devonian, 639-641 feet depth, ODM Jaab Lake No. 1 well, lat. 51°11'54"N, long. 82°56'W, Ontario.

Anaplanisporites sp.
Fig. specs. 15107, 15113, 15149-15151, 15160, 15265
McGregor, D.C. and Owens, B. 1966, Geol. Surv. Can., Paper 66-30, pl. 1, figs. 5, 12, 13; pl. 2, figs. 12-15, 29; pl. 5, fig. 19.
Forillon Member, Cape Bon Ami (15107), Shiphead Member, Grand Grève (15113) and Battery Point formations, Lower Devonian, north side Forillon Peninsula, Cape Rosier Cove; Cape Gaspé shore cliff below lighthouse; southside Gaspé Bay, shore section from Anse-à-Brillant to Bois Brulé Brook, and from Bois Brulé Brook to Pointe Jaune (15265), eastern Gaspé Peninsula, Québec.

Anaplanisporites sp.
Fig. spec. 15404
McGregor, D.C. and Owens, B., 1966, Geol. Surv. Can., Paper 66-30, pl. 26, fig. 12.
Griper Bay Formation, Upper Devonian, stream valley west of Sargeant Point, 2 miles from shore, northeastern Bathurst Island, District of Franklin.

Ancyrospora acutispinosus Chi and Hills
Holotype 43771; paratypes 43772-43774
Chi, B.I. and Hills, L.V., 1977, Bull. Can. Petrol. Geol., vol. 24, no. 4, p. 765, pl. 16, figs. 5-13; pl. 17, figs. 1, 2.
Beverley Inlet Formation, Upper Devonian, Cape Terrace, west side Melville Island, District of Franklin.

Ancyrospora ampulla? Owens
Hypotype 66197
McGregor, D.C. and Camfield, M., 1982, Geol. Surv. Can., Bull. 348, p. 15, pl. 2, figs. 3, 4.
Hecla Bay Formation, Middle Devonian, stream section 16.2 km east of east arm of Weatherall Bay, northeastern Melville Island, District of Franklin.

Ancyrospora ancyrea (Eisenack) Richardson, 1962
Hypotype 15281
McGregor, D.C. and Owens, B., 1966, Geol. Surv. Can., Paper 66-30, pl. 8, fig. 1.
Battery Point Formation, Lower Devonian, south side Gaspé Bay, shore section from Bois Brulé Brook to Pointe Jaune, eastern Gaspé Peninsula, Québec.

Ancyrospora ancyrea (Eisenack) Richardson 1962
Hypotype 49840
McGregor, D.C., 1979, Palynology, vol. 3, pl. 3, fig. 8.
Battery Point Formation, Middle Devonian, south shore Gaspé Bay, 350 yards southeast of first small brook south of Cape Rouge, Gaspé, Québec.

Ancyrospora ancyrea (Eisenack) Richardson var. *ancyrea*
Hypotype 32043
McGregor, D.C., 1973, Palaeontographica, Abt. B., vol. 142, no. 1-3, p. 64, pl. 9, fig. 13.
Battery Point Formation, Middle Devonian, 100 yards south of most prominent point between Cape Rouge and Pointe Jaune, south shore Gaspé Bay, Québec.

Ancyrospora ancyrea (Eisenack) Richardson var. *ancyrea*
Hypotypes 66198, 66199
McGregor, D.C. and Camfield, M., 1982, Geol. Surv. Can., Bull. 348, p. 11, pl. 2, figs. 5, 6.
Helca Bay and Weatherall formations, Middle Devonian, stream section 16.2 km east of east arm of Weatherall Bay, northeastern Melville Island, District of Franklin.

Ancyrospora ancyrea (Eisenack) Richardson var. *brevispinosa* Richardson
Hypotype 66200
McGregor, D.C. and Camfield, M., 1982, Geol. Surv. Can., Bull. 348, p. 15, pl. 2, fig. 7.
Hecla Bay Formation, Middle Devonian, stream section 16.2 km east of east arm of Weatherall Bay, northeastern Melville Island, District of Franklin.

Ancyrospora angulata (Tiwari and Schaarschmidt) n. comb.
Hypotype 66201
McGregor, D.C. and Camfield, M., 1982, Geol. Surv. Can., Bull.
348, p. 16, pl. 2, fig. 8.
Weatherall Formation, Middle Devonian, stream section 16.2 km east of east arm of Weatherall Bay, northeastern Melville Island, District of Franklin.

Ancyrospora bacillaris Chi and Hills
Holotype 43775; paratypes 43776, 43777
Chi, B.I. and Hills, L.V., 1977, Bull. Can. Petrol. Geol., vol. 24, no. 4, p. 766, pl. 17, figs. 4-9.
Beverley Inlet Formation, Upper Devonian, Cape Terrace, west side Melville Island, District of Franklin.

Ancyrospora deltatus Chi and Hills
Holotype 43778; paratypes 43779, 43380
Chi, B.I. and Hills, L.V., 1977, Bull. Can. Petrol. Geol., vol. 24, no. 4, p. 769, pl. 17, figs. 10-12; pl. 18, figs. 1-3.
Beverley Inlet Formation, Upper Devonian, Cape Terrace, west side Melville Island, District of Franklin.

Ancyrospora eurypterota Riegel 1973
Hypotype 49826
McGregor, D.C., 1979, Palynology, vol. 3, pl. 2, figs. 1, 2.
Devonian, 4170-4180 feet, Dome et al. Weatherall 0-10 well, northeastern Melville Island, District of Franklin.

Ancyrospora eurypterota? Riegel
Hypotypes 66204, 66205
McGregor, D.C. and Camfield, M., 1982, Geol. Surv. Can., Bull, 348, p. 16, pl. 2, figs. 11, 12.

Weatherall Formation, Middle Devonian, stream section 16.2 km east of east of Weatherall Bay, northeastern Melville Island, District of Franklin.

Ancyrospora furcula Owens 1971
Hypotype 49847
McGregor, D.C., 1979, Palynology, vol. 3, pl. 4, fig. 9. Richardson, J.B. and McGregor, D.C., 1986, Geol. Surv. Can., Bull. 364, pl. 17, figs. 4, 5.
Griper Bay Formation, Upper Devonian, creek section 5.5 to 10 miles north of head of Beverley Inlet. Melville Island, District of Franklin.

Ancyrospora ?grandispinosa Richardson, 1960
Hypotype 15283
McGregor, D.C. and Owens, B., 1966, Geol. Surv. Can., Paper 66-30, pl. 8, fig. 3.
Malbaie Formation, Lower Devonian, south side Gaspé Bay, shore section between Point Jaune and Whale Head, eastern Gaspé Peninsula, Québec.

Ancyrospora cf. *A. grandispinosa* Richardson
Hypotype 15284
McGregor, D.C. and Owens, B., 1966, Geol. Surv. Can., Paper 66-30, pl. 8, fig. 4.
Malbaie Formation, Lower Devonian, south side Gaspé Bay, shore section between Pointe Jaune and Whale Head, eastern Gaspé Peninsula, Québec.

Ancyrospora kedoae (Riegel) Turnau
Hypotypes 66202, 66203
McGregor, D.C. and Camfield, M., 1982, Geol. Surv. Can., Bull. 348, p. 17, pl. 5, figs. 9, 10, 13.
Weatherall Formation, Middle Devonian, stream section 16.2 km east of east arm of Weatherall Bay, northeastern Melville Island, District of Franklin.

Ancyrospora langii (Taugourdeau-Lantz) Allen
Hypotype 73293
Richardson, J.B. and McGregor, D.C., 1986, Geol. Surv. Can., Bull. 364, pl. 16, figs. 2, 3.
Hecla Bay Formation, Middle Devonian, southeastern Melville Island, District of Franklin.

Ancyrospora loganii McGregor
Holotype 32042; hypotypes 32041, 15282
McGregor, D.C., 1973, Palaeontographica, Abt. B., vol. 142, no. 1-3, p. 65, pl. 9, figs. 6, 7, 10-12.
Battery Point Formation, Middle and Lower Devonian, 100 yards south of most prominent point between Cap Rouge and Pointe Jaune, and 1300 yards south of second stream south of Bois Brulé Brook, south shore Gaspé Bay, Québec.

Ancyrospora ?loganii McGregor
Hypotype 41772
McGregor, D.C. and Camfield, M., 1976, Geol. Surv. Can., Bull. 263, p. 9, pl. 6, fig. 7.
Williams Island Formation, Middle Devonian, 169.5-171.5 feet depth, ODM Jaab Lake No. 1 well, lat. 51°11′54″N, long. 82°56′W, Ontario.

Ancyrospora loganii McGregor
Hypotype 66206
McGregor, D.C. and Camfield, M., 1982, Geol. Surv. Can., Bull. 348, p. 17, pl. 3, fig. 1.
Cape De Bray Formation, Middle Devonian, stream section 16.2 km east of east arm of Weatherall Bay, northeastern Melville Island, District of Franklin.

Ancyrospora longispinosa Richardson
Hypotype 66207
McGregor, D.C. and Camfield, M., 1982, Geol. Surv. Can., Bull. 348, p. 17, pl. 3, fig. 2.
Hecla Bay Formation, Middle Devonian, stream section 16.2 km east of east arm of Weatherall Bay, northeastern Melville Island, District of Franklin.

?Ancyrospora magnifica Owens
Hypotype 30704
Owens, B. and Richardson, J.B., 1972, 7th Internat. Congr. Carboniferous Strat. Geol. Compte Rendu, vol. 1, pl. 2, fig. 7.
Griper Bay Formation, Upper Devonian?, Prince Patrick Island, District of Franklin.

Ancyrospora melvillensis Owens
Hypotype 73295
Richardson, J.B. and McGregor, D.C., 1986, Geol. Surv. Can., Bull. 364, pl. 16, figs. 6, 7.
Fram Formation, Upper Devonian, east side of inner part of southeast arm of Bird Fiord, southwestern Ellesmere Island, District of Franklin.

Ancyrospora multiacicularis Chi and Hills
Holotype 43781; paratypes 43782, 43783
Chi, B.I. and Hills, L.V., 1977, Bull. Can. Petrol. Geol., vol. 24, no. 4, p. 771, pl. 18, figs. 4-10.
Beverley Inlet Formation, Upper Devonian, Cape Terrace, west side of Melville Island, District of Franklin.

Ancyrospora sp. cf. *A. nettersheimensis* Riegel
Fig. spec. 66208
McGregor, D.C. and Camfield, M., 1982, Geol. Surv. Can., Bull. 348, p. 18, pl. 3, fig. 3.
Weatherall Formation, Middle Devonian, stream section 16.2 km east of east arm of Weatherall Bay, northeastern Melville Island, District of Franklin.

cf. Ancyrospora simplex Guennel, 1963
Hypotypes 15391, 15394
McGregor, D.C. and Owens, B., 1966, Geol. Surv. Can., Paper 66-30, pl. 24, figs. 8, 9, 12.
Escarpment Member, Hay River Formation, Upper Devonian, left bank Hay River at base of Louise Falls, Northwest Territories.

Ancyrospora sp.
Fig. specs. 15282, 15285, 15331-15335
McGregor, D.C. and Owens, B., 1966, Geol. Surv. Can., Paper 66-30, pl. 8, figs. 2, 5; pl. 19, figs. 1, 2; pl. 20, figs. 1-5.
Battery Point Formation, Lower Devonian, south side Gaspé Bay, shore section from Boise Brulé Brook to Pointe Jaune, eastern Gaspé Peninsula, Québec; Griper

Bay Formation, Upper Devonian, south limb of Robertson Point anticline, 8 miles northeast of Beverley Inlet, Melville Island, District of Franklin (15331-15335).

Ancyrospora sp.
Fig. spec. 86115
McGregor, D.C. and McCutcheon, S.R., 1988, Can. J. Earth Sci., vol. 25, no. 9, p. 1352, pl. 2, fig. 21.
Carrow Formation, Piskahegan Group, Late Devonian, depth 72.85 m, drill hole M-6, lat. 45°34′43″N, long. 66°35′27″W, southwestern New Brunswick.

Ancyrospora n. sp. 1, 2
Fig. specs. 29409, 29421
McGregor, D.C. and Uyeno, T.T., 1972, Geol. Surv. Can., Paper 71-13, pl. 4, figs. 1, 13.
Griper Bay Formation, Upper Devonian, Beverley Inlet, southeastern Melville Island, District of Franklin.

Aneurospora goensis Streel 1964
Hypotype 49838
McGregor, D.C., 1979, Palynology, vol. 3, pl. 3, fig. 6.
Richardson, J.B. and McGregor, D.C., 1986, Geol. Surv. Can., Bull. 364, pl. 16, fig. 4.
Weatherall Formation, Middle Devonian, 3100-3140 feet, Elfex et al. Tiritchik M-48 well, Banks Island, District of Franklin.

Aneurospora minuta McGregor
Holotype 15265; hypotype 32012
McGregor, D.C., 1973, Palaeontographica, Abt. B, vol. 142, no. 1-3, p. 50, pl. 6, figs. 22-25.
Battery Point Formation, Middle Devonian, 100 yards south of most prominent point between Cap Rouge and Pointe Jaune, south shore Gaspé Bay, Quebec.

Aneurospora sp.
Fig. spec. 67779
McGregor, D.C., 1984, Academia Nacional de Ciencias, Misc. No. 69, pl. 3, fig. 12.
Santa Rosa Formation, Lower Devonian, Laurani section, Chuquisaca Dept., northwest of Tarabuco, Cordillera Oriental, Bolivia.

Aneurospora spp. 1, 2, 4
Fig. specs. 67161-67163
McGregor, D.C., 1984, Academia Nacional de Ciencias, Misc. No. 69, p. 19, 20, pl. 1, figs. 11, 13, 23.
Santa Rosa Formation, Lower Devonian, Laurani section, Chuquisaca Dept., northwest of Tarabuco, Cordillera Oriental, Bolivia.

Apiculatasporites brevidenticulatus (Chibrikova) n. comb.
Hypotype 66209
McGregor, D.C. and Camfield, M., 1982, Geol. Surv. Can., Bull. 348, p. 18, pl. 3, figs. 4, 5.
Hecla Bay Formation, Middle Devonian, stream section 16.2 km east of east arm of Weatherall Bay, northeastern Melville Island, District of Franklin.

Apiculatasporites dilucidus (McGregor, 1960) McGregor
Hypotypes 13203, 13204
McGregor, D.C., 1964, Geol. Surv. Can., Bull. 109, p. 13, pl. 2, figs. 12-14.
Ghost River Formation (sensu stricto), Middle or Upper Devonian, Phantom Crag, 4 miles northeast of eastern end of Lake Minnewanka, Alberta.

Apiculatasporites microconus (Richardson) n. comb.
Hypotype 66215
McGregor, D.C. and Camfield, M., 1982, Geol. Surv. Can., Bull. 348, p. 19, pl. 3, fig. 10.
Weatherall Formation, Middle Devonian, stream section 16.2 km east of east arm of Weatherall Bay, northeastern Melville Island, District of Franklin.

Apiculatasporites microechinatus Owens
Hypotype 32823
McGregor, D.C., 1974, Can. J. Earth Sci., vol. 11, no. 1, pl. 2, fig. 46.
Vendom Fiord Formation, Early Devonian, lat. 79°28′40″N, long. 80°02′W, 12 miles of head of Cañon Fiord, central Ellesmere Island, District of Franklin.

Apiculatasporites perpusillus (Naumova) McGregor
Hypotypes 15151, 31957
McGregor, D.C., 1973, Palaeontographica, Abt. B., vol. 142, no. 103, p. 23, pl. 2, figs. 8, 10-12.
Battery Point Formation, Lower Devonian, 400 yards east of Rivière de l′Anse-à-Brillant, south shore Gaspé Bay, Québec.

Apiculatasporites perpusillus (Naumova) McGregor
Hypotypes 61717, 61741
McGregor, D.C., 1979, Geol. Surv. Can., Paper 79-1B, pl. 22.2, fig. 72; pl. 22.3, fig. 96.
Lower Devonian, Daleje Shale, upper part of southeast slope of Císaská rokle, southeast of Srbsko; Trebotov Limestone, northwest side of railway, north side of Prokopské Valley at Hluboepy near Prague, Czechoslovakia.

Apiculatasporites perpusillus (Naumova ex Chibrikova) McGregor
Hypotype 66210
McGregor, D.C. and Camfield, M., 1982, Geol. Surv. Can., Bull. 348, p. 19, pl. 3, figs. 6, 7.
Hecla Bay Formation, Middle Devonian, stream section 16.2 km east of east arm of Weatherall Bay, northeastern Melville Island, District of Franklin.

Apiculatasporites perpusillus (Naumova) McGregor, 1973
Hypotype 67771
McGregor, D.C., 1984, Academia Nacional de Ciencias, Misc. No. 69, p. 20, pl. 2, fig. 4.
Huamampampa Formation, Middle Devonian, Lausani section, Chuquisaca Dept. northwest of Tarabuco, Cordillera Oriental, Bolivia.

Apiculatasporites sp. cf. *A. perpusillus* (Naumova) McGregor
Fig. spec. 61667
McGregor, D.C., 1979, Geol. Surv. Can., Paper 79-1B, pl. 22.1, fig. 19.
Zlíckov Limestone, Lower Devonian, near Pekárek Mill west of village of Solopysky, Czechoslavakia.

Apiculatasporites sp.
Fig. specs. 15112, 15114, 15122, 15154, 15253, 15351
McGregor, D.C. and Owens, B., 1966, Geol. Surv. Can., Paper 66-30, pl. 1, figs. 11, 14, 22; pl. 2, fig. 23; pl. 5, fig. 6; pl. 22, fig. 12.
Shiphead Member, Grande Grève and Battery Point (15154, 15253) formations, Lower Devonian, Cape Gaspé shore cliff below lighthouse; shore section from Anse-à-Brillant to Bois Brulé Brook (15154), and from Bois Brulé Brook to Pointe Jaune (15253), eastern Gaspe Peninsula, Québec; Escarpment Member, Hay River Formation, Upper Devonian, left bank Hay River at base of Louise Falls, Northwest Territories (15351).

Apiculatasporites sp.
Fig. spec. 61718
McGregor, D.C., 1979, Geol. Surv. Can., Paper 79-1B, pl. 22.2, fig. 73,
Daleje Shale, Lower Devonian, upper part of southeastern slope of Císaská rokle, southeast of Srbsko, Czechoslovakia.

Apiculatasporites n. sp.
Fig. spec. 29375
McGregor, D.C. and Uyeno, T.T., 1972, Geol. Surv. Can., Paper 71-13, pl. 1, fig. 7.
Stuart Bay Formation, Lower Devonian, Young Inlet, northeastern Bathurst Island, District of Franklin.

Apiculatisporis microconus Richardson
Hyptotypes 31962, 15252
McGregor, D.C., 1973, Palaeontographica, Abt. B, vol. 142, no. 1-3, p. 23, pl. 2, figs. 16, 21, 22.
Malbaie Formation, Middle Devonian, 65 yards southeast of conglomerate of Pointe Jaune, south shore Gaspé Bay, Québec.

Apiculatisporis microconus Richardson
Hyptotype 41756
McGregor, D.C. and Camfield, M., 1976, Geol. Surv. Can., Bull. 263, p. 11, pl. 5, fig. 1.
Williams Island Formation, Middle Devonian, 169.5-171.5 feet depth, ODM Jaab Lake No. 1 well, lat. 51°11′54″N, long. 82°56′W, Ontario.

Apiculatisporis microconus Richardson 1965
Hyptotype 49822
McGregor, D.C., 1979, Palynology, vol. 3, pl. 1, fig. 22.
Sextant Formation, Lower Devonian, 147 feet, Argor A-ETA No. 2 well, Kiasko River, lat. 50°41′30″N, long. 80°48′W, Ontario.

Apiculatisporis microconus Richardson
Hyptotype 61707
McGregor, D.C., 1979, Geol. Surv. Can., Paper 79-1B, pl. 22.2, fig. 61.
Daleje Shale, Lower Devonian, upper part of southeast slope of Císaská rokle, southeast of Srbsko, Czechoslovakia.

Apiculatisporis microconus Richardson, 1975
Hyptotype 67164
McGregor, D.C., 1984, Academia Nacional de Ciencias, Misc. No. 69, p. 21, pl. 2, fig. 17,
Icla Formation, Middle Devonian, Laurani section, Chuquisaca Dept., northwest of Tarabuco, Cordillera Oriental, Bolivia.

Apiculatisporis sp.
Fig. specs 15155, 15169
McGregor, D.C. and Owens, B., 1966, Geol. Surv. Can., Paper 66-30, pl. 2, fig. 24; pl. 3, fig. 4.
Battery Point Formation, Lower Devonian, south side Gaspé Bay, shore section from Anse-à-Brillant to Bois Brulé Brook, eastern Gaspé Peninsula, Québec.
=*Dibolisporites quebecensis*, McGregor, D.C., 1979, Palynology, vol. 3, pl. 1, fig. 14 (15155).

Apiculatisporis n. sp.
Fig. spec. 29380
McGregor, D.C. and Uyeno, T.T., 1972, Geol. Surv. Can., Paper 71-13, pl. 1, fig. 12,
Weatherall Formation, Middle Devonian, Beverley Inlet, southeastern Melville Island, District of Franklin.

Apiculiretusispora arenorugosa McGregor
Holotype 31961; hypotype 15143
McGregor, D.C., 1973, Palaeontographica, Abt. B, vol. 142, no. 1-3, p. 27, pl. 2, figs. 15, 25.
Battery Point Formation, Lower Devonian, 100 yards east of first small brook north of Rivière de l'Anse-à-Brillant, and 420 yards north of Bois Brulé Brook, south shore Gaspé Bay, Québec.

Apiculiretusispora arenorugosa McGregor
Hypotype 41738
McGregor, D.C. and Camfield, M., 1976, Geol. Surv. Can., Bull. 263, p. 11, pl. 3, fig. 15.
Stooping River Formation, Lower Devonian, 960-962 feet depth, ODM Jaab Lake No. 1 well, lat 51°11′54″N, long. 82°56′W, Ontario.

Apiculiretusispora arenorugosa McGregor
Hypotype 61731
McGregor, D.C., 1979, Geol. Surv. Can., Paper 79-1B, pl. 22.3, fig. 86.
Trebotov Limestone, Lower Devoninan, northwest side of railway, north side of Prokopské Valley at Hluboepy near Prague, Czechoslovakia.

Apiculiretusispora brandtii Streel
Hypotypes 15142, 31967
McGregor, D.C., 1973, Palaeontographica, Abt. B, vol. 142, no. 1-3, p. 26, pl. 2, figs. 28-30.
Battery Point Formation, Lower Devonian, 100 yards east of first small brook south of, and 500 yards east of Rivière de l'Anse-à-Brillant, south shore Gaspé Bay, Québec.

Apiculiretusispora brandtii? Streel
Hypotype 61725
McGregor, D.C., 1979, Geol. Surv. Can., Paper 79-1B, pl. 22.3, fig. 80.
Daleje Shale, Lower Devonian, upper part of southeast slope of Císaská rokle, southeast of Srbsko, Czechoslovakia.

Apiculiretusispora densiconata Tiwari and Schaarschmidt
Hypotype 66214
McGregor, D.C. and Camfield, M., 1982, Geol. Surv. Can., Bull. 348, p. 19, pl. 3, fig. 9.
Weatherall Formation, Middle Devonian, stream section 16.2 km east of east arm of Weatherall Bay, northeastern Melville Island, District of Franklin.

Apiculiretusispora gaspiensis McGregor
Holotype 31970; hypotypes 31968, 31969
McGregor, D.C., 1973, Palaeontographica, Abt. B, vol. 142, no. 1-3, p. 28, pl. 3, figs. 1-4.
Battery Point Formation, Lower and Middle Devonian, 1450 yards south of second stream south of Bois Brulé Brook; 130 yards south of prominent point between Cap Rouge and Pointe Jaune; and 1/2 mile northeast of Pointe Jaune, south shore Gaspé Bay, Québec.

Apiculiretusispora gaspiensis McGregor
Hypotypes 41774, 33263, 33264
McGregor, D.C. and Camfield, M., 1976, Geol. Surv. Can., Bull. 263, p. 11, pl. 6, figs. 11, 12; pl. 8, figs. 4, 5.
Williams Island Formation, Middle Devonian, 157.5-159.5 feet depth, ODM Jaab Lake No. 1 well, lat. 51°11″54″5DN, long. 82°5′W, Ontario; Campbellton Formation, Middle Devonian, beach outcrop 3/4 mile west of Dalhousie Junction, New Brunswick.

Apiculiretusispora gaspiensis McGregor
Hypotype 61706
McGregor, D.C., 1979, Geol. Surv. Can., Paper 79-1B, pl. 22.2, fig. 60.
Daleje Shale, Lower Devonian, upper part of southeast slope of Císaská rokle, southeast of Srbsko, Czehoslovakia.

Apiculiretusispora gaspiensis McGregor 1965
Hypotype 96772
Arkhangelskaya, A.D., McGregor, D.C. and Richardson, J.B., 1990, Geol. Surv. Can., Paper 90-1D, pl. 1, fig. b.
Strathcona Fiord Formation, Middle Devonian, east of Bird Fiord, Ellesmere Island, District of Franklin.

Apiculiretusispora leberidos McGregor and Camfield
Holotype 66211; paratypes 66212, 66213
McGregor, D.C. and Camfield, M., 1982, Geol. Surv. Can., Bull. 348, p. 20, pl. 3, figs. 8, 11, 12.
Hecla Bay Formation, Middle Devonian, stream section 16.2 km east of east arm of Weatherall Bay, northeastern Melville Island, District of Franklin.

Apiculiretusispora minor McGregor
Holotype 15162; hypotype 31965
McGregor, D.C., 1973, Palaeontographica, Abt. B, vol. 142, no. 1-3, p. 27, pl. 2, figs. 19, 23, 24, 26, 27.
Battery Point Formation, Lower Devonian, 45 yards east of first point east of, and 380 yards northwest of first small brook south of Rivière de l'Anse-à-Brillant, south shore Gaspé Bay, Quebec.

Apiculiretusispora minor McGregor
Hypotype 41742
McGregor, D.C. and Camfield, M., 1976, Geol. Surv. Can., Bull. 263, p. 12, pl. 3, figs. 25-27.
Stooping River Formation, Lower Devonian, 935-937 feet depth, ODM Jaab Lake No. 1 well, lat. 51°11′54″N, long. 82°56′W, Ontario.

Apiculiretusispora ?plicata (Allen) Streel
Hypotype 29371
McGregor, D.C. and Uyeno. T.T., 1972, Geol. Surv. Can., Paper 71-13, pl. 1, fig. 3.
Stuart Bay Formation, Lower Devonian, Young Inlet, northeastern Bathurst Island, District of Franklin.

Apiculiretusispora plicata (Allen) Streel
Hypotypes 31959, 31960
McGregor, D.C., 1973, Palaeontographica, Abt. B, vol. 142, no. 1-3, pl. 25, pl. 2, figs. 13, 14.
Battery Point Formation, Lower Devonian, 600 yards northwest of first small brook south of, and 259 yards north of Rivière de l'Anse-à-Brillant, south shore Gaspé Bay, Québec.

?*Apiculiretusispora plicata* (Allen) Streel
Hypotype 32819
McGregor, D.C., 1974, Can. J. Earth Sci., vol. 11, no. 1, pl. 1, fig. 39.
Vendom Fiord Formation, Early Devonian, lat. 79°28′40″N, long. 80°02′W, 12 miles south of head of Cañon Fiord, central Ellesmere Island, District of Franklin.

Apiculiretusispora plicata (Allen) Streel
Hypotype 41718
McGregor, D.C. and Camfield, M., 1976, Geol. Surv. Can., Bull. 263, p. 12, pl. 1, figs. 29, 30.
Kenogami River Formation, Lower Devonian, 1127-1129 feet depth, ODM Jaab Lake No. 1 well, lat. 51°11′54″N, long. 82°56′W, Ontario.

Apiculiretusispora plicata (Allen) Streel
Hypotypes 61684, 61702
McGregor, D.C. 1979, Geol. Surv. Can., Paper 79-1B, pl. 22.1, fig. 37; pl. 22.2, fig. 56.
Daleje Shale, Lower Devonian, roadside outcrop near Karlštejn, 1 km north of Budnany in the Hluboké Valley, and upper part of southeast slope of Císaská rokle, southeast of Srbsko, Czechoslovakia.

Apiculiretusispora plicata (Allen) Streel, 1967
Hypotype 67165
McGregor, D.C., 1984, Academia Nacional de Ciencias, Misc. No. 69, p. 21, pl. 2, fig. 15.

Santa Rosa Formation, Lower Devonian, Laurani section, Chuquisaca Dept., northwest of Tarabuco, Cordillera Oriental, Bolivia.

Apiculiretusispora porcatus Chi and Hills
Holotype 43661; paratypes 43662, 43663
Chi, B.I. and Hills, L.V., 1972, Bull. Can. Petrol. Geol., vol. 24, no. 4, p. 695, pl. 1, figs. 4-6.
Imperial Formation, Upper Devonian, lat. 65°07′N, long. 127°51′W, Imperial River, District of Mackenzie; Hecla Bay Formation, Middle Devonian, Cape Terrace, west side Melville Island, District of Franklin.

Apiculiretusispora pygmaea McGregor
Holotype 319971
McGregor, D.C., 1973, Palaeontographica, Abt. B, vol. 142, no. 1-3, p. 29, pl. 3, figs. 5-7.
Battery Point Formation, Lower Devonian, 150 feet east of road to beach from end of Laurencelle Road, north shore Gaspé Bay, Québec.

Apiculiretusispora pygmaea McGregor
Hypotype 41741
McGregor, D.C. and Camfield, M., 1976, Geol. Surv. Can., Bull. 263, p. 12, pl. 3, figs. 18, 19.
Stooping River Formation, Lower Devonian, 915-917 feet depth, ODM Jaab Lake No. 1 well, lat. 51°11′54″N, long. 82°56′W, Ontario.

Apiculiretusispora rotundus Streel
Hypotype 61730
McGregor, D.C., 1979, Geol. Surv. Can., Paper 79-1B, pl. 22.3, fig. 85.
Trebotov Limestone, Lower Devonian, northwest side of railway, north side of Prokopské Valley at Hluboepy near Prague, Czechoslovakia.

Apiculiretusispora sp.
Fig. spec. 15200
McGregor, D.C. and Owens, B., 1966, Geol. Surv. Can., Paper 66-30, pl. 5, fig. 3.
Malbaie Formation, Lower Devonian, south side Gaspé Bay, shore section between Point Jaune and Whale Head, eastern Gaspé Peninsula, Québec.

?*Apiculiretusispora* sp.
Fig. spec. 27090
McGregor, D.C., Sanford, B.V. and Norris, A.W., 1970, Proc. Geol. Assoc. Can., vol. 22, pl. 2, fig. 11.
Kwataboahegan Formation, Middle Devonian, 533 feet depth, ODM Jaab Lake No. 1 well, lat. 51°11.9′N, long. 82°57′W, Moose River Basin, Ontario.

Apiculiretusispora sp.
Fig. spec. 32818
McGregor, D.C., 1974, Can. J. Earth Sci., vol. 11, no. 1, pl. 1, figs. 37, 38.
Vendom Fiord Formation, Early Devonian, lat. 79°28′40″N, long. 80°02′W, 12 miles south of head of Cañon Fiord, central Ellesmere Island, District of Franklin.

?*Apiculiretusispora* sp.
Fig. spec. 32814
McGregor, D.C., 1974, Can. J. Earth Sci., vol. 11, no. 1, pl. 1, fig. 34.
Vendom Fiord Formation, Early Devonian, lat. 79°28′40″N, long. 80°02′W, 12 miles south of head of Cañon Fiord, central Ellesmere Island, District of Franklin.

Apiculiretusispora sp.
Fig. specs. 61664, 61666, 61683, 61685
McGregor, D.C., 1979, Geol. Surv. Can., Paper 79-1B, pl. 22.1, figs. 16, 18, 35, 37.
Lower Devonian, Zlíckov Limestone, near Pekárek Mill west of village of Solopysky; Daleje Shale, roadside outcrop near Karlštejn, 1 km north of Budnany in the Hluboké Valley (61683, 61685), Czechoslovakia.

?*Apiculiretusispora* sp.
Fig. specs. 61651, 61686
McGregor, D.C., 1979, Geol. Surv. Can., Paper 79-1B, pl. 22.1, fig. 3; pl. 22.2, fig. 38.
Lower Devonian, Dvorce-Prokop Limestone, road cut on west side of Vitava River below Barrandov, southern Prague, north of old quarry "U.Kapliky"; Daleje Shale, roadside outcrop near Karlštejn, 1 km north of Budnany in the Hluboké Valley, Czechoslovakia.

Apiculiretusispora sp.
Fig. spec. 67166
McGregor, D.C., 1984, Academia Nacional de Ciencias, Misc. No. 69, p. 21, pl. 3, fig. 3.
Icla Formation, Middle Devonian, Laurani section, Chuquisaca Dept., northwest of Tarabuco, Cordillera Oriental, Bolivia.

Archaeoperisaccus ?completus Naum.
Hypotype 29416
McGregor, D.C. and Uyeno, T.T., 1972, Geol. Surv. Can., Paper 71-13, pl. 4, fig. 8.
Griper Bay Formation, Upper Devonian, Beverley Inlet, southeastern Melville Island, District of Franklin.

Archaeoperisaccus elongatus Naum.
Hypotype 29414
McGregor, D.C. and Uyeno, T.T., 1972, Geol. Surv. Can., Paper 71-13, pl. 4, fig. 6.
Griper Bay Formation, Upper Devonian, Beverley Inlet, southeastern Melville Island, District of Franklin.

Archaeoperisaccus mirus Naum.
Hypotype 29417
McGregor, D.C. and Uyeno, T.T., 1972, Geol. Surv. Can., Paper 71-13, pl. 4, fig. 9.
Griper Bay Formation, Upper Devonian, Beverley Inlet, southeastern Melville Island, District of Franklin.

Archaeoperisaccus opiparus Owens
Hypotype 73298
Richardson, J.B. and McGregor, D.C., 1986, Geol. Surv. Can., Bull. 364, pl. 17, fig. 3.
Weatherall Formation, Middle Devonian, southeastern Melville Island, District of Franklin.

Archaeoperisaccus oblongus Owens
Hypotype 30701
Owens, B. and Richardson, J.B., 1972, 7th Internat. Congr. Carboniferous Strat. Geol. Compte Rendu, vol. 1, pl. 2, fig. 4.
Griper Bay Formation, Upper Devonian, Melville Island, District of Franklin.

Archaeoperisaccus opiparus Owens
Hypotype 30703
Owens, B. and Richardson, J.B., 1972, 7th Internat. Congr. Carboniferous Strat. Geol. Compte Rendu, vol. 1, pl. 2, fig. 6.
Griper Bay Formation, Upper Devonian, Melville Island, District of Franklin.

Archaeoperisaccus ovalis Naum.
Hypotype 29413
McGregor D.C. and Uyeno. T.T., 1972, Geol. Surv. Can., Paper 71-13, pl. 4, fig. 5.
Griper Bay Formation, Upper Devonian, Beverley Inlet, southeastern Melville Island, District of Franklin.

Archaeoperisaccus ovalis Naumova
Hypotype 73296
Richardson, J.B. and McGregor D.C., 1986, Geol. Surv. Can., Bull. 364, pl. 17, fig. 1.
Beverley Inlet Formation, Upper Devonian, southeastern Melville Island, District of Franklin.

Archaeoperisaccus cf. *ovalis* Naumova *in* Ozolin'a
Fig. spec. 16596
McGregor, D.C., 1967, Review Palaeobotany and Palynology, I, p. 181, pl. 1, fig. 20.
Griper Bay Formation, Upper Devonian, Helena Island, District of Franklin.

Archaeoperisaccus scabratus Owens
Hypotype 30702
Owens, B. and Richardson, J.B., 1972, 7th Internat. Congr. Carboniferous Strat. Geol. Compte Rendu, vol. 1, pl. 2, fig. 5.
Griper Bay Formation, Upper Devonian, Melville Island, District of Franklin.

Archaeoperisaccus scabratus Owens 1971
Hypotype 49846
McGregor, D.C., 1979, Palynology, vol. 3, pl. 4, fig. 7.
Griper Bay Formation, Upper Devonian, creek section 5.5 to 10 miles north of head of Beverley Inlet, Melville Island, District of Franklin.

Archaeoperisaccus cf. *A. timanicus* Pashkevich, 1964
Hypotype 15327
McGregor, D.C. and Owens, B., 1966, Geol. Surv. Can., Paper 66-30, pl. 18, figs. 4, 5.
Griper Bay Formation, Upper Devonian, south limb of Robertson Point anticline, 8 miles northeast of Beverley Inlet, Melville Island, District of Franklin.

Archaeoperisaccus cf. *A. timanicus* Pashkevich
Hypotype 16595
McGregor, D.C., 1967, Review Palaeobotany and Palynology, I, p. 181, pl. 1, figs. 14, 15.
Griper Bay Formation, Upper Devonian, Helena Island, District of Franklin.

Archaeoperisaccus sp.
Fig. specs. 15398-15400
McGregor, D.C. and Owens, B., 1966, Geol. Surv. Can., Paper 66-30, pl. 25, figs. 6-8.
Escarpment Member, Hay River Formation, Upper Devonian, left bank Hay River at base of Louise Falls, Northwest Territories.

?*Archaeoperisaccus* sp.
Fig. specs. 15325, 15326, 15328, 15329
McGregor, D.C. and Owens, B., 1966, Geol. Surv. Can., Paper 66-30, pl. 18, figs. 1-3, 6, 7.
Griper Bay Formation, Upper Devonian, south limb of Robertson Point anticline, 8 miles northeast of Beverley Inlet, Melville Island, District of Franklin.

Archaeoperisaccus n. sp. 1, 2
Fig. specs. 29412, 29415
McGregor, D.C., and Uyeno, T.T., 1972, Geol. Surv. Can., Paper 71-13, pl. 4, figs. 4, 7.
Griper Bay Formation, Upper Devonian, Beverley Inlet, southeastern Melville Island, District of Franklin.

cf. *Archaeotriletes setigerus* Kedo, 1955
Fig. spec. 15176
McGregor, D.C. and Owens, B., 1966, Geol. Surv. Can., Paper 66-30, pl. 3, fig. 12.
Battery Point Formation, Lower Devonian, south side Gaspé Bay, shore section from Anse-à-Brillant to Bois Brulé Brook, eastern Gaspé Peninsula, Québec.

Archaeozonotriletes amplectus (Naumova) Kedo
Hypotype 25453
McGregor, D.C., 1970, Colloque Stratig. Carbonifére, Congr. Colloques Univ. Liége, vol. 55, pl. 21, fig. 15.
Bedford Formation, Upper Devonian, Courtright bore hole, Courtright village, Lambton County, Ontario.

Archaeozonotriletes antaxios Chibrikova, 1962
Hypotype 15323
McGregor, D.C. and Owens, B., 1966, Geol. Surv. Can., Paper 66-30, pl. 17, figs. 4, 5.
Griper Bay Formation, Upper Devonian, south limb of Robertson Point anticline, 8 miles northeast of Beverley Inlet, Melville Island, District of Franklin.

Archaeozonotriletes antaxios Chibr.
Hypotype 29406
McGregor, D.C. and Uyeno, T.T., 1972, Geol. Surv. Can., Paper 71-13, pl. 3, fig. 7.
Hecla Bay Formation, Upper Devonian, Beverley Inlet, southeastern Melville Island, District of Franklin.

?*Archaeozonotriletes arduus* Arkli.
Hypotype 29391
McGregor, D.C. and Uyeno, T.T., 1972, Geol. Surv. Can., Paper 71-13, pl. 2, fig. 4.
Weatherall Formation, Middle Devonian, Beverley Inlet, southeastern Melville Island, District of Franklin.

Archaeozonotriletes chulus (Cramer) Richardson and Lister var. *chulus* Richardson and Lister
Hypotype 27064
McGregor, D.C., Sanford, B.V. and Norris, A.W., 1970, Proc. Geol. Assoc. Can., vol. 22, pl. 1, fig. 5.
Kenogami River Formation, Lower Devonian, 1,153 feet depth, ODM Jaab Lake No. 1 well, lat. 51°11.9′N, long. 82°57′W, Moose River Basin, Ontario.

Archaeozonotriletes chulus (Cramer) Richardson and Lister var. *inframurinatus* Richardson and Lister
Hypotype 27065
McGregor, D.C., Sanford, B.V. and Norris, A.W., 1970, Proc. Geol. Assoc. Can., vol. 22, pl. 1, fig. 6.
Kenogami River Formation, Lower Devonian, 1,072 feet depth, ODM Jaab Lake No. 1 well, lat. 51°11.9′N, long. 82°57′W, Moose River Basin, Ontario.

?*Archaeozonotriletes divellomedium* Chibrikova
Hypotypes 32815, 32816
McGregor, D.C., 1974, Can. J. Earth Sci., vol. 11, no. 1, pl. 1, figs. 35, 40.
Vendom Fiord Formation, Early Devonian, lat. 79°28′40″N, long. 80°02′W, 12 miles south of head of Cañon Fiord, central Ellesmere Island, Disrtict of Franklin.

cf. *Archaeozonotriletes famenensis* Naumova
Hypotype 13599
Kerr, J.W., McGregor, D.C. and McLaren, D.J., 1965, Bull, Can. Petrol. Geol., vol. 13, no. 3, p. 425, pl. 3, fig. 17.
Griper Bay Formation, Upper Devonian, 3/4 mile from shore in stream valley entering Reindeer Bay, 4 miles south of largest river entering the bay, northeastern Bathurst Island, District of Franklin.
=*Archaeozonotriletes famenensis* McGregor, D.C., 1967, Review Palaeobotany and Palynology, I, pl. 1, fig. 23.

Archaeozonotriletes famenensis Naumova 1953
Hypotype 49851
McGregor, D.C., 1979, Palynology, vol. 3, pl. 4, fig. 17.
Richardson, J.B. and McGregor, D.C., 1986, Geol. Surv. Can., Bull. 364, pl. 18, fig. 4.
Griper Bay Formation, Upper Devonian, east bank, 3 miles north of mouth of stream entering Graham Moore Bay 7 miles due East of Schomberg Point, Bathurst Island, District of Franklin.

cf. *Archaeozonotriletes famenensis* Naumova, 1953
Hypotype 15438
McGregor, D.C. and Owens, B., 1966, Geol. Surv. Can., Paper 66-30, pl. 27, figs. 11, 12.
Griper Bay Formation, Upper Devonian, stream valley west of Sargeant Point, 2 miles from shore, northeastern Bathurst Island, District of Franklin.

?*Archaeozonotriletes gorodkensis* Kedo
Hypotype 13623
Kerr, J.W., McGregor, D.C. and McLaren, D.J., 1965, Bull. Can. Petrol. Geol., vol. 13, no. 3, p. 428, pl. 4, fig. 12.
Griper Bay Formation, Upper Devonian, 2.8 miles from north coast, in valley of stream entering the sea at Noel Point, northern Helena Island, District of Franklin.

Archaeozonotriletes gracilis Kedo
Hypotype 29418
McGregor, D.C. and Uyeno, T.T., 1972, Geol. Surv. Can., Paper 71-13, pl. 4, fig. 10.
Griper Bay Formation, Upper Devonian, Beverley Inlet, southeastern Melville Island, District of Franklin.

cf. *Archaeozonotriletes gracilis* Kedo, 1957
Hypotype 15441
McGregor, D.C. and Owens, B., 1966, Geol. Surv. Can., Paper 66-30, pl. 27, figs. 15, 16.
Griper Bay Formation, Upper Devonian, stream valley west of Sargeant Point, 2 miles from shore, northeastern Bathurst Island, District of Franklin.

cf. *Archaeozonotriletes intertextus* Nekryata and Sergeeva in Kedo et al. 1971
Hypotype 86119
McGregor, D.C. and McCutcheon, S.R., 1988, Can. J. Earth Sci., vol. 25, no. 9, p. 1352, pl. 3, figs. 4, 5.
Carrow Formation, Piskahegan Group, Late Devonian, depth 72.85 m, drill hole M-6, lat. 45°34′43″N, long.66°35′27″W, southwestern New Brunswick.

cf. *Archaeozonotriletes laetus* Sergeeva
Hypotype 73300
Richardson, J.B. and McGregor, D.C., 1986, Geol. Surv. Can., Bull. 364, pl. 17, fig. 7.
Pedrix Formation, Upper Devonian, Monkman Pass area, Hart Range, lat. 53°52′34″N, long. 120°1′8″W, east-central British Columbia.

Archaeozonotriletes literatus (Waltz) Naumova
Hypotypes 25450, 25491
McGregor, D.C., 1970, Colloque Stratig. Carbonifère, Congr. Colloques Univ. Liége, vol. 55, pl. 21, fig. 13; pl. 23, fig. 15.
Upper Devonian, Bedford Formation, Courtright borehole, Courtright village, Lambton County, Ontario, and Imperial Formation, lat. 66°27.5′N, long. 135°22′W, Yukon.
=*Knoxisporites literatus*, Richardson, J.B. and McGregor, D.C., 1986, Geol. Surv. Can., Bull. 364, pl. 21, fig. 5 (25450).

?*Archaeozonotriletes literatus* (Waltz) Naumova in litt. *in* Kedo, 1963
Hypotype 13596
McGregor, D.C. and Owens, B., 1966, Geol. Surv. Can., Paper 66-30, pl. 26, fig. 25.

Griper Bay Formation, Upper Devonian, stream valley west of Sargeant Point, 2 miles from shore, northeastern Bathurst Island, District of Franklin.

?*Archaeozonotriletes macriomplanus* Arkh.
Hypotype 29400
McGregor, D.C. and Uyeno, T.T., 1972, Geol. Surv. Can., Paper 71-13, pl. 3, fig. 1.
Weatherall Formation, Middle Devonian, Beverley Inlet, southeastern Melville Island, District of Franklin.

?*Archaeozonotriletes macromanifestus* Naumova, 1953
Hypotype 15390
McGregor, D.C. and Owens, B., 1966, Geol. Surv. Can., Paper 66-30, pl. 24, fig. 7.
Escarpment Member, Hay River Formation, left bank Hay River at base of Louise Falls, Northwest Territories.

Archaeozonotriletes microaculeatus Kedo, 1974
Hypotype 86101
McGregor, D.C. and McCutcheon, S.R., 1988, Can. J. Earth Sci., vol. 25, no. 9, p. 1353, pl. 2, fig. 5.
Carrow Formation, Piskahegan Group, Late Devonian, depth 72.85 m, drill hole M-6, lat. 45°34′43″N, long. 66°35′27″W, southwestern New Brunswick.

cf. *Archaeozonotriletes micromanifestus* Naumova, 1953
Hypotypes 15379-15381
McGregor, D.C. and Owens, B., 1966, Geol. Surv. Can., Paper 66-30, pl. 23, figs. 18, 20, 24.
Escarpment Member, Hay River Formation, left bank Hay River at base of Louise Falls, Northwest Territories.

?*Archaeozonotriletes micromanifestus* var. *famenensis* Naumova
Hypotype 13609
Kerr, J.W., McGregor, D.C. and McLaren, D.J., 1965, Bull. Can. Petrol. Geol., vol. 13, no. 3, p. 425, pl. 4, fig. 1.
Griper Bay Formation, Upper Devonian, 3/4 mile from shore in stream entering Reindeer Bay, 4 miles south of largest river entering the bay, northeastern Bathurst Island, District of Franklin.

Archaeozonotriletes micromanifestus var. *minor* Naumova, 1953
Hypotype 15388
McGregor, D.C. and Owens, B., 1966, Geol. Surv. Can., Paper 66-30, pl. 24, fig. 4.
Escarpment Member, Hay River Formation, left bank Hay River at base of Louise Falls, Northwest Territories.

?*Archaeozonotriletes notatus* Naumova
Hypotype 13598
Kerr, J.W., McGregor, D.C. and McLaren, D.J., 1965, Bull. Can. Petrol. Geol., vol. 13, no. 3, p. 425, pl. 3, fig. 16.
McGregor, D.C. and Owens, B., 1966, Geol. Surv. Can., Paper 66-30, pl. 26, fig. 37.
Griper Bay Formation, Upper Devonian, stream valley west of Sargeant Point, 2 miles from shore, northeastern Bathurst Island, District of Franklin.

Archaeozonotriletes optivus Chibrikova, 1959
Hypotype 15324
McGregor, D.C. and Owens, B., 1966, Geol. Surv. Can., Paper 66-30, pl. 17, fig. 6.
Griper Bay Formation, Upper Devonian, south limb of Robertson Point anticline, 8 miles northeast of Beverley Inlet, Melville Island, District of Franklin.

Archaeozonotriletes optivus Chibr.
Hypotype 29408
McGregor, D.C. and Uyeno, T.T., 1972, Geol. Surv. Can., Paper 71-13, pl. 3, fig. 9.
Hecla Bay Formation, Upper Devonian, Beverley Inlet, southeastern Melville Island, District of Franklin.

Archaeozonotriletes optivus var. *vorobjevensis* Chibrikova
Hypotype 15319
McGregor, D.C. and Owens, B., 1966, Geol. Surv. Can., Paper 66-30, pl. 16, figs. 5, 6.
Griper Bay Formation, Upper Devonian, south limb of Robertson Point anticline, 8 miles northeast of Beverley Inlet, Melville Island, District of Franklin.

Archaeozonotriletes cf. *A. optivus* var. *vorobjevensis* Chibrikova, 1962
Hypotype 15318
McGregor, D.C. and Owens, B., 1966, Geol. Surv. Can., Paper 66-30, pl. 16, figs. 3, 4.
Griper Bay Formation, Upper Devonian, south limb of Robertson Point anticline, 8 miles northeast of Beverley Inlet, Melville Island, District of Franklin.

?*Archaeozonotriletes plicatus* Naumova in litt. *in* Chibrikova, 1959
Hypotype 15378
McGregor, D.C. and Owens, B., 1966, Geol. Surv. Can., Paper 66-30, pl. 23, fig. 17.
Escarpment Member, Hay River Formation, left bank Hay River at base of Louise Falls, Northwest Territories.

cf. *Archaeozonotriletes semilucensis* Naumova
Hypotype 13597
Kerr, J.W., McGregor, D.C. and McLaren, D.J., 1965, Bull. Can. Petrol. Geol., vol. 13, no. 3, p. 425, pl. 3, fig. 15.
Griper Bay Formation, Upper Devonian, stream valley west of Sargeant Point, 2 miles from shore, northeastern Bathurst Island, District of Franklin.

?*Archaeozonotriletes subnotatus* Chibrikova, 1959
Hypotype 15448
McGregor, D.C. and Owens, B., 1966, Geol. Surv. Can., Paper 66-30, pl. 27, fig. 25.
Griper Bay Formation, Upper Devonian, stream valley west of Sargeant Point, 2 miles from shore, northeastern Bathurst Island, District of Franklin.

Archaeozonotriletes timanicus Naumova
Hypotypes 66216, 66217
McGregor, D.C. and Camfield, M., 1982, Geol. Surv. Can., Bull. 348, p. 20, pl. 3, figs. 13, 14, 15.
Richardson, J.B. and McGregor, D.C., 1986, Geol. Surv. Can., Bull. 364, pl. 15, figs. 4, 5 (66216).

Hecla Bay Formation, Middle Devonian, stream section 16.2 km east of east arm of Weatherall Bay, northeastern Melville Island, District of Franklin.

?*Archaeozonotriletes tortuosus* Chibrikova, 1962
Hypotype 15256
McGregor, D.C. and Owens, B., 1966, Geol. Surv. Can., Paper 66-30, pl. 5, fig. 9.
Battery Point Formation, Lower Devonian, south side Gaspé Bay, shore section from Bois Brulé Brook to Pointe Jaune, eastern Gaspé Peninsula, Québec.

Archaeozonotriletes variabilis Naumova, 1953
Hypotype 15425
McGregor, D.C. and Owens, B., 1966, Geol. Surv. Can., Paper 66-30, pl. 26, fig. 36.
Griper Bay Formation, Upper Devonian, stream valley west of Sargeant Point, 2 miles from shore, northeastern Bathurst Island, District of Franklin.

Archaeozonotriletes variabilis Naum.
Hypotype 29407
McGregor, D.C. and Uyeno, T.T., 1972, Geol. Surv. Can., Paper 71-13, pl. 3, fig. 8.
Griper Bay Formation, Upper Devonian, Beverley Inlet, southeastern Melville Island, District of Franklin.

Archaeozonotriletes variabilis (Naumova) Allen 1965
Hypotype 30694
Owens, B. and Richardson, J.B., 1972, 7th Internat. Congr. Carboniferous Strat. Geol. Compte Rendu, vol. 1, pl. 1, fig. 10.
Weatherall Formation, Middle Devonian, Melville Island, District of Franklin.

Archaeozonotriletes variabilis Naumova emend. Allen
Hypotypes 66220-66222
McGregor, D.C. and Camfield, M., 1982, Geol. Surv. Can., Bull. 348, p. 21, pl. 3, figs. 18-20.
Hecla Bay and Weatherall (66222) formations, Middle Devonian, stream section 16.2 km east of east arm of Weatherall Bay, northeastern Melville Island, District of Franklin.

?*Archaeozonotriletes variabilis* Naumova
Hypotype 13624
Kerr, J.W., McGregor, D.C. and McLaren, D.J., 1965, Bull. Can. Petrol. Geol., vol. 13, no. 3, p. 428, pl. 4, fig. 13.
McGregor, D.C., 1979, Palynology, vol. 3, pl. 4, fig. 10.
Richardson, J.B. and McGregor, D.C., 1986, Geol. Surv. Can., Bull. 364, pl. 14, fig. 10.
Griper Bay or Parry Islands Formation, Upper Devonian, 2.8 miles from north east, in valley of a stream entering the sea at Noel Point, northern Helena Island, District of Franklin.

cf. *Archaeozonotriletes variabilis* Naumova, 1953
Hypotype 67792
McGregor, D.C., 1984, Academia Nacional de Ciencias, Misc. No. 69, pl. 4, fig. 8.
Huamampampa Formation, Middle Devonian, Laurani section, Chuquisaca Dept., northwest of Tarabuco, Cordillera Oriental, Bolivia.

?*Archaeozonotriletes varius* Nadler
Hypotype 61693
McGregor, D.C., 1979, Geol. Surv. Can., Paper 79-1B, pl. 22.2, fig. 47.
Daleje Shale, Lower Devonian, roadside outcrop near Karlštejn, 1 km north of Budnany in the Hluboké Valley, Czechoslovakia.

Archaeozonotriletes sp.
Fig. spec. 15321
McGregor, D.C. and Owens, B., 1966, Geol. Surv. Can., Paper 66-30, pl. 17, fig. 2.
Griper Bay Formation, Upper Devonian, south limb of Robertson Point anticline, 8 miles northeast of Beverley Inlet, Melville Island, District of Franklin.

Archaeozonotriletes sp.
Fig. spec. 32014
McGregor, D.C., 1973, Palaeontographica, Abt. B., vol. 142, no. 1-3, p. 56, pl. 7, fig. 19.
Battery Point Formation, Lower Devonian, 480 yards northwest of first small brook south of Rivière de l'Anse-à-Brillant, south shore Gaspé Bay, Québec.

?*Archaeozonotriletes* sp.
Fig. spec. 15322
McGregor, D.C. and Owens, B., 1966, Geol. Surv. Can., Paper 66-30, pl. 17, fig. 3.
Griper Bay Formation, Upper Devonian, south limb of Robertson Point anticline, 8 miles northeast of Beverley Inlet, Melville Island, District of Franklin.

Archaeozonotriletes sp. (sensu Naumova)
Fig. spec. 16598
McGregor, D.C., 1967, Review Palaeobotany and Palynology, I, pl. 1, fig. 22.
Griper Bay Formation, Upper Devonian, northern Helena Island, District of Franklin.

Asterocalamotriletes sp.
Fig. spec. 15146
McGregor, D.C. and Owens, B., 1966, Geol. Surv. Can., Paper 66-30, pl. 3, fig. 12.
Battery Point Formation, Lower Devonian, south side Gaspé Bay, shore section from Anse-à-Brillant to Bois Brulé Brook, eastern Gaspé Peninsula, Québec.

Auroraspora aurora Richardson 1960
Hypotype 96774
Arkhangelskaya, A.D., McGregor, D.C. and Richardson, J.B., 1990, Geol. Surv. Can., Paper 90-1D, pl. 1, fig. d.
Strathcona Fiord Formation, Middle Devonian, east of Bird Fiord, Ellesmere Island, District of Franklin.

Auroraspora hyalina (Naumova) Streel
Hypotype 73304
Richardson, J.B. and McGregor, D.C., 1986, Geol. Surv. Can., Bull. 364, pl. 18, fig. 6.

Devonian, west side of St. George's Bay, north of MsIsaac Point, lat. 45°44′24.4″N, long. 61°53′28.46″W, Antigonish County, Nova Scotia.

Auroraspora macra Sullivan
Hypotype 73309
Richardson, J.B. and McGregor, D.C., 1986, Geol. Surv. Can., Bull. 364, pl. 19, fig. 8.
Kotcho Formation, Upper Devonian, Imperial Island River No. 1 well, lat. 60°9′29″N, long. 121°8′16″W, District of McKenzie.

Auroraspora macra Sullivan, 1968
Hypotype 86113
McGregor, D.C. and McCutcheon, S.R., 1988, Can. J. Earth Sci., vol. 25, no. 9, p. 1353, pl. 2, fig. 19.
Carrow Formation, Piskahegan Group, Late Devonian, depth 72.85 m, drill hole M-6, lat. 45°34′43″N, long. 66°35′27″W, northwestern New Brunswick.

Auroraspora macromanifestus (Hacquebard) Richardson, 1960
Hypotype 15301
McGregor, D.C. and Owens, B., 1966, Geol. Surv. Can., Paper 66-30, pl. 11, fig. 3.
Weatherall Formation, Middle Devonian, south limb of Robertson Point anticline, 8 miles northeast of Beverley Inlet, Melville Island, District of Franklin.

Auroraspora macromanifestus (Hacquebard) Richardson 1960
Hypotype 43748
Chi, B.I. and Hills, L.V., 1971, Bull. Can. Petrol. Geol., vol. 24, no. 4, p. 747, pl. 13, fig. 12.
Hecla Bay Formation, Middle Devonian, Weatherall Bay, East Arm, Melville Island, District of Franklin.

Auroraspora micromanifestus (Hacquebard) Richardson
Hypotype 66218
McGregor, D.C. and Camfield, M., 1982, Geol. Surv. Can., Bull. 348, p. 21, pl. 3, fig. 16.
Weatherall Formation, Middle Devonian, stream section 16.2 east of east arm of Weatherall Bay, northeastern Melville Island, District of Franklin.

Auroraspora minuta Richardson, 1965
Hypotype 15272
McGregor, D.C. and Owens, B., 1966, Geol. Surv. Can., Paper 66-30, pl. 6, fig. 8.
Malbaie Formation, Lower Devonian, south side Gaspé Bay, shore section between Pointe Jaune and Whale Head, eastern Gaspé Peninsula, Québec.

Auroraspora pseudocrista Ahmed, 1980
Hypotype 86117
McGregor, D.C. and McCutcheon, S.R., 1988, Can. J. Earth Sci., vol. 25, no. 9, p. 1353, pl. 2, fig. 24.
Carrow Formation, Piskahegan Group, Late Devonian, depth 72.85 m, drill hole M-6, lat. 45°34′43″N, long. 61°35′27″W, southwestern New Brunswick.

Auroraspora torquata Higgs, 1975
Hypotype 86124
McGregor, D.C. and McCutcheon, S.R., 1988, Can. J. Earth Sci., vol. 25, no. 9, p. 1353, pl. 3, fig. 13.
Carrow Formation, Piskahegan Group, Late Devonian, depth 72.85 m, drill hole M-6, lat. 45°34′43″N, long. 61°35′27″W, southwestern Ontario.

"Azonomonoletes" sp.
Fig. spec. 29384
McGregor, D.C. and Uyeno, T.T., 1972, Geol. Surv. Can., Paper 71-13, pl. 1, fig. 16.
Weatherall Formation, Middle Devonian, Beverley Inlet, southeastern Melville Island, District of Franklin.

cf. Baculatisporites fusticulus Sullivan, 1968
Hypotype 86079
McGregor, D.C. and McCutcheon, S.R., 1988, Can. J. Earth Sci., vol. 25, no. 9, p. 1353, pl. 1, fig. 3.
Carrow Formation, Piskahegan Group, Late Devonian, depth 72.85 m, drill hole M-6, lat. 45°34′43″N, long. 61°35′27″W, southwestern New Brunswick.

Baculatisporites semilucensis (Naumova) n. comb.
Hypotype 66219
McGregor, D.C. and Camfield, M., 1982, Geol. Surv. Can., Bull. 348, p. 21, pl. 3, fig. 17.
Weatherall Formation, Middle Devonian, stream section 16.2 km east of east arm of Weatherall Bay, northeastern Melville Island, District of Franklin.

Baculatisporites sp.
Fig. spec. 29382
McGregor, D.C. and Uyeno, T.T., Geol. Surv. Can., Paper 71-13, pl. 1, fig. 14.
Bird Fiord Formation, Middle Devonian, Young Inlet, northeastern Bathurst Island, District of Franklin.

Biharispsorites arcticus var. arcticus Chi and Hills
Holotype 43683; paratype 43684
Chi, B.I. and Hills, L.V., 1977, Bull. Can. Petrol. Geol., vol. 24, no. 4, p. 707, pl. 3, figs. 9-11.
Weatherall Formation, Middle Devonian, Cape Terrace, west side Melville Island, District of Franklin.

Biharispsorites arcticus var. productus Chi and Hills
Holotype 43687; paratype 43688
Chi, B.I. and Hills, L.V., 1977, Bull. Can. Petrol. Geol., vol. 24, no. 4, p. 709, pl. 4, figs. 1-3.
Weatherall Formation, Middle Devonian, Cape Terrace, west side Melville Island, District of Franklin.

Biharispsorites arcticus var. spinosus Chi and Hills
Holotype 43685; paratype 43686
Chi, B.I. and Hills, L.V., 1977, Bull. Can. Petrol. Geol., vol. 24, no. 4, p. 708, pl. 3, figs. 12-14.
Weatherall Formation, Middle Devonian, Cape Terrace, west side Melville Island, District of Franklin.

Brochotriletes ?foveolatus Naumova
Hypotype 27086
McGregor, D.C., Sanford, B.V. and Norris, A.W., 1970, Proc. Geol. Assoc. Can., vol. 22, pl. 2, figs. 5, 6.

Stooping River Formation, Lower Devonian, 617 feet depth, ODM Jaab Lake No. 1 well, lat. 51°11.9′N, long. 82°57′W, Moose River Basin, Ontario.

Brochotriletes ?foveolatus Naumova
Hypotype 31990
McGregor, D.C., 1973, Palaeontographica, Abt. B, vol. 142, no. 1-3, p. 39, pl. 4, figs. 20, 23.
Battery Point Formation, Lower Devonian, 1/2 mile north of Bois Brulé Brook, south shore Gaspé Bay, Québec.

Brochotriletes ?foveolatus Naumova
Hypotype 41764
McGregor, D.C. and Camfield, M., 1976, Geol. Surv. Can., Bull. 263, p. 12, pl. 5, fig. 10.
Kwataboahegan Formation, Lower Devonian, 493-495 feet depth, ODM Jaab Lake No. 1 well, lat. 51°11′54″N, long. 82°56′W, Ontario.

Brochotriletes cf. B. foveolatus Naumova, 1953
Hypotype 15127
McGregor, D.C. and Owens, B., 1966, Geol. Surv. Can., Paper 66-30, pl. 1, fig. 28.
Shiphead Member, Grand Grève Formation, Cape Gaspé shore cliff below lighthouse, eastern Gaspé Pininsula, Québec.

Brochotriletes hudsonii McGregor and Camfield
Holotype 41729
McGregor, D.C. and Camfield, M., 1976, Geol. Surv. Can., Bull. 263, p. 12, pl. 3, figs. 1, 2.
Stooping River Formation, Lower Devonian, 1031-1033 feet depth, ODM Jaab Lake No. 1 well, lat. 51°11′54″N, long. 82°56′W, Ontario.

cf. *Brochotriletes hudsonii* McGregor and Camfield, 1976
Hypotype 48166
McGregor, D.C., 1977, Palaeontographica, Abt. B., vol. 163, p. 130, pl. 1, figs. 3, 4.
Battery Point Formation, Lower Devonian, Tar Point to Whale Head section, south shore Gaspé Bay, Québec.

Brochotriletes robustus (Scott and Rouse) n. comb.
Hypotype 27088
McGregor, D.C., Sanford, B.V. and Norris, A.W., 1970, Proc. Geol. Assoc. Can., vol. 22, pl. 2, figs. 8, 9.
Kwataboahegan Formation, Lower Devonian, 533 feet depth, ODM Jaab Lake No. 1 well, lat. 51°11.9′N, long. 82°57′W, Moose River Basin, Ontario.

Brochotriletes robustus (Scott and Rouse) McGregor
Hypotype 31991
McGregor, D.C., 1973, Palaeontographica, Abt. B, vol. 142, no. 1-3, p. 40, pl. 5, figs. 1, 6.
Battery Point Formation, Lower Devonian, 600 yards northwest of first small brook south of Rivière de l'Anse-à-Brillant, south shore Gaspé Bay, Québec.

Brochotriletes robustus (Scott and Rouse) McGregor
Hypotype 41754
McGregor, D.C. and Camfield, M., 1976, Geol. Surv. Can., Bull. 263, p. 12, pl. 4, figs. 22-24.
Stooping River Formation, Lower Devonian, 599-600 feet depth, ODM Jaab Lake No. 1 well, lat. 51°11′54″N, long. 82°56′W, Ontario.

Brochotriletes sp.
Fig. specs. 15110, 15186, 15187
McGregor, D.C. and Owens, B., 1966, Geol. Surv. Can., Paper 66-30, pl. 1, figs. 8, 9; pl. 4, figs. 5, 8.
Lower Devonian, Forillon Member, Cape Bon Ami Formation, north side Forillon Peninsula, Cape Rosier Cove; Battery Point Formation, north side Gaspé Bay, shore cliffs between Gaspé Peninsula and Laurencelle Road; and south side Gaspé Bay, shore section from Anse-à-Brillant to Bois Brulé Brook, eastern Gaspé Peninsula, Québec.

Brochotriletes sp.
Fig. spec. 16588
McGregor, D.C., 1967, Review Palaeobotany and Palynology, I, p. 176, pl. 1, fig. 3.
Grand Grève Formation, Lower Devonian, Gaspé Bay, Québec.

Brochotriletes sp.
Fig. spec. 27074
McGregor, D.C., Sanford, B.V. and Norris, A.W., 1970, Proc. Geol. Assoc. Can., vol. 22, pl. 1, figs. 14, 15.
Stooping River Formation, Lower Devonian, 909 feet depth, ODM Jaab Lake No. 1 well, lat. 51°11.9′N, long. 82°57′W, Moose River Basin, Ontario.

Brochotriletes sp.
Fig. spec. 27074
McGregor, D.C., 1979, Geol. Surv. Can., Paper 79-1B, pl. 22.3, fig. 95.
Trebotov Limestone, Lower Devonian, northwest side of railway, north side of Prokopské Valley at Hluboepy near Prague, Czechoslovakia.

Brochotriletes sp.
Fig. specs. 67158, 67168
McGregor, D.C., 1984, Academia Nacional de Ciencias, Misc. No. 69, p. 23, pl. 1, figs. 20, 25.
Santa Rosa Formation, Lower Devonian, Laurani section, Chuquisaca Dept., northwest of Tarabuco, Cordillera Oriental, Bolivia.

"cf.*Brochotriletes* sp. A" Richardson and Ioannides, 1973
Fig. spec. 67770
McGregor, D.C., 1984, Academia Nacional de Ciencias, Misc. No. 69, p. 22, pl. 2, fig. 3.
Santa Rosa Formation, Lower Devonian, Laurani section, Chuquisaca Dept., northwest of Tarabuco, Cordillera Oriental, Bolivia.

Brochotriletes sp. A, B, C
Fig. specs. 31992-31994
McGregor, D.C., 1973, Palaeontographica, Abt. B, vol. 142, no. 1-3, pp. 40, 41, pl. 5, figs. 2-5, 7, 8.
Battery Point Formation, Lower Devonian, 40 feet east of road to beach from end of Laurencelle Road, opposite small offshore island; 550 yards east of small stream,

3/4 mile north of Bois Brulé Brook; and 500 yards east of Rivière de l'Anse-à-Brillant, south shore Gaspé Bay, Québec.

Brochotriletes sp.
Fig. spec. 48169
McGregor, D.C., 1977, Palaeontographica, Abt. B., vol. 163, p. 130, pl. 1, figs. 9, 10.
Battery Point Formation, Lower Devonian, Tar Point to Whale Head section, south shore Gaspé Bay, Québec.

Bullatisporites bullatus Allen
Hypotype 32822
McGregor, D.C., 1974, Can. J. Earth Sci. vol. 11, no. 1, pl. 2, figs. 44, 45.
Vendom Fiord Formation, Early Devonian, lat. 79°28′40″N, long. 80°02′W, 12 miles south of head of Cañon Fiord, central Ellesmere Island, District of Franklin.

Calamospora ateva (Naumova, 1953) n. comb.
Hypotype 13207
McGregor, D.C., 1964, Geol. Surv. Can., Bull. 109, p. 6, pl. 2, fig. 17.
Ghost River Formation, Middle-Upper Devonian, Phantom Crag, 4 miles northeast of eastern end of Lake Minnewanka, Alberta.

Calamospora atava (Naumova) McGregor
Hypotype 13611
Kerr, J.W., McGregor, D.C. and McLaren, D.J., 1965, Bull. Can. Petrol. Geol., vol. 13, no. 3, p. 425, pl. 4, fig. 3.
Griper Bay Formation, Upper Devonian, 3/4 miles from shore in stream valley entering Reindeer Bay, 4 miles south of largest river entering the bay, northeastern Bathurst Island, District of Franklin.

Calamospora atava (Naumova) McGregor, 1964
Hypotype 15343
McGregor, D.C. and Owens, B., 1966, Geol. Surv. Can., Paper 66-30, pl. 22, fig. 4.
Escarpment Member, Hay River Formation, left bank of Hay River at base of Louise Falls, Northwest Territories.

Calamospora atava (Naumova) McGregor
Hypotype 31940
McGregor, D.C., 1973, Palaeontographica, Abt. B, vol. 142, no. 1-3, p. 13, pl. 1, fig. 1.
Battery Point Formation, Lower Devonian, 105 yards east of first point east of Rivière de l'Anse-à-Brillant, south shore Gaspé Bay, Québec.

Calamospora atava (Naumova) McGregor
Hypotype 13611
McGregor, D.C. and Camfield, M., 1982, Geol. Surv. Can., Bull. 348, p. 22, pl. 3, fig. 21.
Hecla Bay Formation, Middle Devonian, stream section 16.2 km east of east arm of Weatherall Bay, northeastern Melville Island, District of Franklin.

Calamospora cf. *C. microrugosa* (Ibrahim) S.W. & B *in* Balme, 1962
Hypotype 15342
McGregor, D.C. and Owens, B., 1966, Geol. Surv. Can., Paper 66-30, pl. 22, fig. 3.
Escarpment Member, Hay River Formation, left bank Hay River at base of Louise Falls, Northwest Territories.

Calamospora pannucea Richardson
Hypotype 31941
McGregor, D.C., 1973, Palaeontographica, Abt. B, vol. 142, no. 1-3, p. 14, pl. 1, fig. 2.
Battery Point Formation, Lower Devonian, 550 yards east of small stream 3/4 mile north of Bois Brulé Brook, south shore Gaspé Bay, Québec.

Calamospora pannucea Richardson
Hypotype 61678
McGregor, D.C., 1979, Geol. Surv. Can., Paper 79-1B, pl. 22.1, fig. 30.
Daleje Shale, Lower Devonian, roadside outcrop near Karlštejn, 1 km north of Budnany in the Hluboké Valley, Czechoslovakia.

Calamospora sp.
Fig. spec. 31942
McGregor, D.C., 1973, Palaeontographica, Abt. B, vol. 142, no. 1-3, p. 15, pl. 1, fig. 3.
Battery Point Formation, Middle Devonian, 850 yards east of Pointe Sandy Beach, south shore Gaspé Bay, Québec.

Calamospora sp.
Fig. spec. 32790
McGregor, D.C., 1974, Can. J. Earth Sci., vol. 11, no. 1, pl. 1, fig. 5.
Unamed Early Devonian, lat. 79°28′40″N, long. 80°02′W, 12 miles south of head of Cañon Fiord, central Ellesmere Island, District of Franklin.

Calyptosporites proteus (Naumova) Allen 1965
Hypotype 49834
McGregor, D.C., 1979, Palynology, vol. 3, pl. 3, fig. 1.
Richardson, J.B. and McGregor, D.C., 1986, Geol. Surv. Can., Bull. 364, pl. 9, fig. 4.
Weatherall Formation, Middle Devonian, creek flowing south, 10 miles east of Weatherall Bay, Melville Island, District of Franklin.

Calyptosporites velatus (Eisenack) Richardson, 1962
Hypotypes 15280, 15299, 15302
McGregor, D.C. and Owens, B., 1966, Geol. Surv. Can., Paper 66-30, pl. 7, fig. 7; pl. 11, fig. 1; pl. 12, fig. 1.
Battery Point Formation, Lower Devonian, south side Gaspé Bay, shore section from Bois Brulé Brook to Pointe Jaune, eastern Gaspé Peninsula, Québec (15280); Weatherall Formation, Middle Devonian, south limb of Robertson Point Anticline, 8 miles northeast of Beverley Inlet, Melville Island, District of Franklin.

Calyptosporites velatus (Eisenack) Richardson 1960
Hypotype 30696
Owens, B. and Richardson, J.B., 1972, 7th Internat. Congr. Carboniferous Strat. Geol. Compte Rendu, vol. 1, pl. 1, fig. 12.
Weatherall Formation, Middle Devonian, Melville Island, District of Franklin.

Calyptosporites sp.
Fig. spec. 15305
McGregor, D.C. and Owens, B., 1966, Geol. Surv. Can., Paper 66-30, pl. 11, figs. 1, 2.
Weatherall Formation, Middle Devonian, south limb of Robertson Point anticline, 8 miles northeast of Beverley Inlet, Melville Island, District of Franklin.
=*Grandispora longa*, Richardson, J.B. and McGregor, D.C., 1986, Geol. Surv. Can., Bull. 364, pl. 11, figs. 7, 8.

Camarozonotriletes antiquus Kedo
Hypotype 66224
McGregor, D.C. and Camfield, M., 1982, Geol. Surv. Can., Bull. 348, p. 23, pl. 3, figs. 22, 23.
Weatherall Formation, Middle Devonian, stream section 16.2 km east of east arm of Weatherall Bay, northeastern Melville Island, District of Franklin.

Camarozonotriletes sp. cf. C. breviclus Ishchenko, 1958
Fig. spec. 15290
McGregor, D.C. and Owens, B., 1966, Geol. Surv. Can., Paper 66-30, pl. 9, fig. 5.
Weatherall Formation, Middle Devonian, south limb of Robertson Point anticline, 8 miles northeast of Beverley Inlet, Melville Island, District of Franklin.

Camarozonotriletes laevigatus McGregor and Camfield
Holotype 66225
McGregor, D.C. and Camfield, M., 1982, Geol. Surv. Can., Bull. 348, p. 23, pl. 4, figs. 1, 2.
Hecla Bay Formation, Middle Devonian, stream section 16.2 km east of east arm of Weatherall Bay, northeastern Melville Island, District of Franklin.

Camarozonotriletes mosolovicus Naum.
Hypotype 29381
McGregor, D.C. and Uyeno, T.T., 1972, Geol. Surv. Can., Paper 71-13, pl. 1, fig. 13.
Weatherall Formation, Middle Devonian, Beverley Inlet, southeastern Melville Island, District of Franklin.

Camarozonotriletes mosolovicus Naumova ex Kedo
Hypotypes 66226, 66227
McGregor, D.C. and Camfield, M., 1982, Geol. Surv. Can., Bull. 348, p. 23, pl. 4, figs. 3-6.
Hecla Bay and Weatherall formations, Middle Devonian, stream section 16.2 km east of east arm of Weatherall Bay, northeastern Melville Island, District of Franklin.

Camarozonotriletes? *notatus* (Owens) n. comb.
Hypotype 66231
McGregor, D.C. and Camfield, M., 1982, Geol. Surv. Can., Bull. 348, p. 23, pl. 4, fig. 14.
Hecla Bay Formation, Middle Devonian, stream section 16.2 km east of east arm of Weatherall Bay, northeastern Melville Island, District of Franklin.

Camarozonotriletes cf. *C. obtusus* Naumova
Hypotype 16591
McGregor, D.C., 1967, Review Palaeobotany and Palynology, I, pl. 1, fig. 7.
Hecla Bay Formation, Melville Island Group, Middle or Upper Devonian, Helena Island, District of Franklin.

Camarozonotriletes parvus Owens
Hypotype 30693
Owens, B. and Richardson, J.B., 1972, 7th Internat. Congr. Carboniferous Strat. Geol. Compte Rendu, vol. 1, pl. 1, fig. 9.
Weatherall Formation, Middle Devonian, Melville Island, District of Franklin.

Camarozonotriletes parvus Owens
Hypotypes 66228, 66229
McGregor, D.C. and Camfield, M., 1982, Geol. Surv. Can., Bull. 348, p. 24, pl. 4, figs. 7-11.
Hecla Bay Formation, Middle Devonian, stream section 16.2 km east of east arm of Weatherall Bay, northeastern Melville Island, District of Franklin.

Camarozonotriletes pusillus Naumova ex Chibrikova
Hypotype 66230
McGregor, D.C. and Camfield, M., 1982, Geol. Surv. Can., Bull. 348, p. 24, pl. 4, figs. 12, 13.
Hecla Bay Formation, Middle Devonian, stream section 16.2 km east of east arm of Weatherall Bay, northeastern Melville Island, District of Franklin.

Camarozonotriletes sextantii McGregor and Camfield
Holotype 41915; paratype 41750
McGregor, D.C. and Camfield, M., 1976, Geol. Surv. Can., Bull. 263, p. 12, pl. 4, figs. 13, 14, 16-18.
Stooping River Formation, Lower Devonian, 599-600 and 639-641 feet depth, ODM Jaab Lake No. 1 well, lat. 51°11′54″N, long. 82°56′W, Ontario.

Camarozonotriletes sextantii McGregor and Camfield 1976
Hypotypes 48167, 48168
McGregor, D.C., 1977, Palaeontographica, Abt. B., vol. 163, p. 130, pl. 1, figs. 5-8.
Battery Point Formation, Lower Devonian, Gaspé, south shore Gaspé Bay, and Litte Gaspe Cove to Pinouil section, north shore Gaspé Bay, Québec.

Camarozonotriletes sextantii McGregor and Camfield 1976
Hypotype 49818
McGregor, D.C., 1979, Palynology, vol. 3, pl. 1, fig. 11.
Sextant Formation, Lower Devonian, west bank Abitibi River, about 1/2 mile south of Sextant Rapids, Ontario.

Camarozonotriletes sextantii McGregor and Camfield
Hypotype 61708
McGregor, D.C., 1979, Geol. Surv. Can., Paper 79-1B, pl. 22.2, figs. 62, 63
Daleje Shale, Lower Devonian, upper part of southeast slope of Císařská rokle, southeast of Srbsko, Czechoslovakia.

Camarozonotriletes sp.
Fig. spec. 15350
McGregor, D.C. and Owens, B., 1966, Geol. Surv. Can., Paper 66-30, pl. 22, fig. 11.
Escarpment Member, Hay River Formation, Upper Devonian, left bank Hay River at base of Louise Falls, Northwest Territories.

Camarozonotriletes n. sp.
Fig. spec. 29389
McGregor, D.C. and Uyeno, T.T., 1972, Geol. Surv. Can., Paper 71-13, pl. 2, fig. 2.
Weatherall Formation, Middle Devonian, Beverley Inlet, southeastern Melville Island, District of Franklin.

Camptotriletes dubuis McGregor
Holotype 31995; hypotype 31996
McGregor, D.C., 1973, Palaeontographica, Abt. B, vol. 142, no. 1-3, p. 42, pl. 5, figs. 9-11, 13, 14.
Battery Point Formation, Lower Devonian, at tip of first point east of, and 200 yards northwest of first small brook south of Rivière de l'Anse-à-Brillant, south shore Gaspé Bay, Québec.

Camptotriletes sp.
Fig. spec. 15168
McGregor, D.C. and Owens, B., 1966, Geol. Surv. Can., Paper 66-30, pl. 3, fig. 3.
Battery Point Formation, Lower Devonian, south side Gaspé Bay, shore section from Anse-à-Brillant to Bois Brulé Brook, eastern Gaspé Peninsula, Québec.

Camptotriletes sp.
Fig. spec. 67169
McGregor, D.C., 1984, Academia Nacional de Ciencias, Misc. No. 69, p. 23, pl. 2, fig. 16.
Icla Formation, Lower or Middle Devonian, Laurani section, Chuquisaca Dept., northwest of Tarabuco, Cordillera Oriental, Bolivia.

?*Camptozonotriletes aliquantus* Allen
Hypotype 29377
McGregor, D.C. and Uyeno, T.T., 1972, Geol. Surv. Can., Paper 71-13, pl. 1, fig. 9.
Eids Formation, Middle Devonian, Young Inlet, northeastern Bathurst Island, District of Franklin.

Camptozonotriletes aliquantus Allen 1965
Hypotype 96773
Arkhangelskaya, A.D., McGregor, D.C. and Richardson, J.B., 1990, Geol. Surv. Can., Paper 90-1D, pl. 1, fig. C.
Strathcona Fiord Formation, Middle Devonian, east of Bird Fiord, Ellesmere Island, District of Franklin.

cf. *Camptozonotriletes aliquantus* Allen, 1965
Hypotype 15134
McGregor, D.C. and Owens, B., 1966, Geol. Surv. Can., Paper 66-30, pl. 1, fig. 35.
Shiphead Member, Grand Grève Formation, Lower Devonian, Cape Gaspé, shore cliff below lighthouse, eastern Gaspé Peninsula, Québec.

Camptozonotriletes caperatus McGregor
Holotype 15197; hypotype 32016
McGregor, D.C., 1973, Palaeontographica, Abt. B, vol. 142, no. 1-3, p. 52, pl. 7, figs. 1, 2.
Battery Point Formation, Lower Devonian, 250 yards north of, and at tip of first point east of Rivière de l'Anse-à-Brillant, south shore Gaspé Bay, Québec.

Camptozonotriletes caperatus McGregor
Hypotype 41733
McGregor, D.C. and Camfield, M., 1976, Geol. Surv. Can., Bull. 263, p. 14, pl. 3, fig. 7.
Stooping River Formation, Lower Devonian, 909-911 feet depth, ODM Jaab Lake No. 1 well, lat. 51°11'54"N, long. 82°56'W, Ontario.

Camptozonotriletes sp.
Fig. spec. 27087
McGregor, D.C., Sanford, B.V. and Norris, A.W., 1970, Proc. Geol. Assoc. Can., vol. 22, pl. 2, fig. 7.
Stooping River Formation, Lower Devonian, 617 feet depth, ODM Jaab Lake No. 1 well, lat. 51°11.9'N, long. 82°57'W, Moose River Basin, Ontario.

Camptozonotriletes sp.
Fig. spec. 15193
McGregor, D.C., 1973, Palaeontographica, Abt. B, vol. 142, no. 1-3, p. 53, pl. 7, figs. 3, 8.
Battery Point Formation, Lower Devonian, 1 mile east of Watering Creek, north shore Gaspé Bay, Québec.

Ceratosporites sp. cf. Acanthotriletes multisetus (Luber) Potonié and Kremp var. *multisetus*
Fig. specs. 66232, 66233,
McGregor, D.C. and Camfield, M., 1982, Geol. Surv. Can., Bull. 348, p. 24, pl. 4, figs. 15, 17.
Weatherall Formation, Middle Devonian, stream section 16.2 km east of east arm of Weatherall Bay, northeastern Melville Island, District of Franklin.

cf. *Chelinospora cassicula* Richardson and Lister, 1969
Hypotype 67170
McGregor, D.C., 1984, Academia Nacional de Ciencias, Misc. No. 69, p. 24, pl. 1, fig. 8.
Santa Rosa Formation, Lower Devonian, Laurani section, Chuquisaca Dept., northwest of Tarabuco, Cordillera Oriental, Bolivia.

Chelinospora concinna Allen 1965
Hypotype 49839
McGregor, D.C., 1979, Palynology, vol. 3, pl. 3, fig. 7.
Richardson, J.B. and McGregor, D.C., 1986, Geol. Surv. Can., Bull. 364, pl. 16, fig. 10.
Griper Bay Formation, Upper Devonian, creek section 5.5 to 10 miles north of head of Beverley Inlet, Melville Island, District of Franklin.

?*Chelinospora ligurata* Allen
Hypotype 29396
McGregor, D.C. and Uyeno, T.T., 1972, Geol. Surv. Can., Paper 71-13, pl. 2, fig. 9.
Weatherall Formation, Middle Devonian, Beverley Inlet, southeastern Melville Island, District of Franklin.

Chelinospora ligurata Allen 1965
Hypotype 49835
McGregor, D.C., 1979, Palynology, vol. 3, pl. 3, fig. 2.
Weatherall Formation, Middle Devonian, creek section 5.5 to 10 miles north of head of Beverley Inlet, Melville Island, District of Franklin.

Chelinospora ligurata Allen
Hypotype 73294
Richardson, J.B., and McGregor, D.C., 1986, Geol. Surv. Can., Bull. 364, pl. 16, fig. 5.
Fram Formation, Upper Devonian, east side of inner part of southeast arm of Bird Fiord, southwestern Ellesmere Island, District of Franklin.

Chelinospora ?vermiculata Chaloner and Streel
Hypotype 41704
McGregor, D.C. and Camfield, M., 1976, Geol. Surv. Can., Bull. 263, p. 14, pl. 1, figs. 1-3.
Kenogami River Formation, Lower Devonian, 1150-1151 feet depth, ODM Jaab Lake No. 1 well, lat. 51°11′54″N, long. 82°56′W, Ontario.

Chelinospora sp.
Fig. spec. 15128
McGregor, D.C. and Owens, B., 1966, Geol. Surv. Can., Paper 66-30, pl. 1, fig. 29.
Shiphead Member, Grand Grève Formation, Lower Devonian, Cape Gaspé, shore cliff below lighthouse, eastern Gaspé Peninsula, Québec.

Cirratriradiata variverrucatus McGregor
Holotype 32018
McGregor, D.C., 1973, Palaeontographica, Abt. B, vol. 142, no. 1-3, p. 53, pl. 7, figs. 6, 7, 9.
Battery Point Formation, Lower Devonian, 1 1/2 mile west of dyke at Petit Cap-aux-Os, north shore Gaspé Bay, Québec.

Cirratriradiates grandis Chi and Hills
Holotype 43745; paratype 43746
Chi, B.I. and Hills, L.V., 1977, Bull. Can. Petrol. Geol., vol. 24, no. 4, p. 743, pl. 11, figs. 7-10.
Weatherall Formation, Middle Devonian, Cape Terrace, west side Melville Island, District of Franklin.

Cirratriradiates sp.
Fig. spec. 27082
McGregor, D.C., Sanford, B.V. and Norris, A.W., 1970, Proc. Geol. Assoc. Can., vol. 22, pl. 1, fig. 30.
Stooping River Formation, Lower Devonian, 909 feet depth, ODM Jaab Lake No. 1 well, lat. 51°11.9′N, long. 82°57′W, Moose River Basin, Ontario.

Clivosispora verrucata McGregor
Hypotype 32017
McGregor, D.C., 1973, Palaeontographica, Abt. B, vol. 142, no. 1-3, p. 54, pl. 7, fig. 10.
Battery Point Formation, Lower Devonian, first point east of first small brook south of Rivière de l'Anse-à-Brillant, south shore Gaspé Bay, Québec.

Clivosispora verrucata var. *convoluta* McGregor and Camfield
Holotype 41725; paratypes 41724, 41726-41728
McGregor, D.C. and Camfield, M., 1976, Geol. Surv. Can., Bull. 263, p. 15, pl. 2, figs. 13-21.
Stooping River Formation, Lower Devonian, 91 feet, Puskwuche Point No. 1 well, lat. 51°46′12″N, long. 80°40′06″W; 885-886, 944-945 and 617-619 feet depths, ODM Jaab Lake No.1 well, lat. 51°11′54″N, long. 82°56′W, Ontario.

Clivosispora verrucata var. *convoluta* McGregor and Camfield 1976
Hypotype 48170
McGregor, D.C., 1977, Palaeontographica, Abt. B., vol. 163, p. 131, pl. 1, figs. 11, 12.
Battery Point Formation, Lower Devonian, Tar Point to Whale Head section, south shore Gaspé Bay, Québec.

Clivosispora verrucata McGregor var. *convoluta*
Hypotype 61719
McGregor, D.C., 1979, Geol. Surv. Can., Paper 79-1B, pl. 22.2, fig. 74.
Daleje Shale, Lower Devonian, upper part of southeast slope of Císařská rokle, southeast of Srbsko, Czechoslovakia.

Clivosiopora verrucata var. *verrucata* McGregor
Hypotypes 41736, 41737
McGregor, D.C. and Camfield, M., 1976, Geol. Surv. Can., Bull. 263, p. 15, pl. 3, figs. 11-14.
Stooping River Formation, Lower Devonian, 960-962 and 885-886 feet depths, ODM Jaab Lake No. 1 well, lat. 51°11′54″N, long. 82°56′W, Ontario.

Clivosispora sp.
Fig. spec. 27081
McGregor, D.C., Sanford, B.V. and Norris, A.W., 1970, Proc. Geol. Assoc. Can., vol. 22, pl. 1, figs. 28, 29.
Stooping River Formation, Lower Devonian, 909 feet depth, ODM Jaab Lake No. 1 well, lat. 51°11.9′N, long. 82°57′W, Moose River Basin, Ontario.
=*Clivosispora verrucata* McGregor var. *convoluta*, Richardson, J.B. and McGregor, D.C., 1986, Geol. Surv. Can., Bull. 364, pl. 5, figs. 10, 11.

cf. *Clivosispora* sp.
Fig. spec. 15179
McGregor, D.C. and Owens, B., 1966, Geol. Surv. Can., Paper 66-30, pl. 3, figs. 16, 17.
Battery Point Formation (lower part), Lower Devonian, south side Gaspé Bay, shore section from Anse-à-Brillant to Bois Brulé Brook, Gaspé, Québec.
=*Clivosispora verrucata*, McGregor, D.C., 1973, Palaeontographica, Abt. B., vol. 142, no. 1-3, p. 54, pl. 7, figs. 4, 5. (holotype).
=*Clivosispora verrucata* McGregor var. *verrucata*, McGregor, D.C., 1979, Palynology, vol. 3, pl. 1, fig. 8.

Contagisporites optivus (Chibrikova) Owens
Hypotype 30705
Owens, B. and Richardson, J.B., 1972, 7th Internat. Congr. Carboniferous Strat. Geol. Compte Rendu, vol. 1, pl. 2, fig. 8.
Griper Bay Formation, Upper Devonian, Melville Island, District of Franklin.

Contagisporites optivus (Chibrikova) Owens
Hypotype 49848
McGregor, D.C., 1979, Palynology, vol. 3, pl. 4, fig. 13.
Hecla Bay Formation, Middle Devonian, south limb of Robertson Point anticline, 8 miles northeast of Beverley Inlet, Melville Island, District of Franklin.
=*Contagisporites optivus* (Chibrikova) Owens var. *optivus*, Richardson, J.B., and McGregor, D.C., 1986, Geol. Surv. Can., Bull. 364, pl. 16, fig. 1.

Contagisporites optivus (Chibrikova) var. *optivus* Owens 1971
Hypotype 43769
Chi, B.I. and Hills, L.V., 1977, Bull. Can. Petrol. Geol., vol. 24, no. 4, p. 762, pl. 16, fig. 3.
Hecla Bay Formation, Middle Devonian, Cape Terrace, west side Melville Island, District of Franklin.

Contagisporites optivus var. *vorobjevensis* (Chibrikova) Owens 1971
Hypotype 43770
Chi, B.I. and Hills, L.V., 1977, Bull. Can. Petrol. Geol., vol. 24, no. 4, p. 763, pl. 16, fig. 4.
Weatherall Formation, Middle Devonian, Cape Terrace, west side Melville Island, District of Franklin.

Converrucosisporites sp.
Fig. specs. 13208, 13209
McGregor, D.C., 1964, Geol. Surv. Can., Bull. 109, p. 12, pl. 2, figs. 18, 23, 24.
Ghost River Formation (sensu stricto), Middle or Upper Devonian, Phantom Crag, 4 miles northeast of eastern end of Lake Minnewanka, Alberta.

Converrucosisporites sp.
Fig. spec. 15416
McGregor, D.C. and Owens, B., 1966, Geol. Surv. Can., Paper 66-30, pl. 26, figs. 19, 20.
Griper Bay Formation, Upper Devonian, stream valley west of Sargeant Point, 2 miles from shore, northeastern Bathurst Island, District of Franklin.

?*Converrucosisporites* sp.
Fig. spec. 15449
McGregor, D.C. and Owens, B., 1966, Geol. Surv. Can., Paper 66-30, pl. 27, fig. 26.
Griper Bay Formation, Upper Devonian, stream valley west of Sargeant Point, 2 miles from shore, northeastern Bathurst Island, District of Franklin.

Convolutispora crassata? (Naumova) n. comb.
Hypotypes 66234, 66235
McGregor, D.C. and Camfield, M., 1982, Geol. Surv. Can., Bull. 348, p. 25, pl. 4, figs. 18-20.
Weatherall Formation, Middle Devonian, stream section 16.2 km east of east arm of Weatherall Bay, northeastern Melville Island, District of Franklin.

Convolutispora crassata? (Naumova) n. comb., corroded, cf. *Geminospora verrucosa*
Hypotype 66406
McGregor, D.C. and Camfield, M., 1982, Geol. Surv. Can., Bull. 348, p. 25, pl. 18, fig. 23.
Hecla Bay Formation, Middle Devonian, stream section 16.2 km east of east arm of Weatherall Bay, northeastern Melville Island, District of Franklin.

Convolutispora paraverrucata McGregor
Holotype 13202; paratype 13201
McGregor, D.C., 1964, Geol. Surv. Can., Bull. 109, p. 17, pl. 2, figs. 9, 10, 11.
Ghost River Formation (sensu stricto), Upper Devonian, Phantom Crag, 4 miles northeast of eastern end of Lake Minnewanka, Alberta.

Convolutispora paraverrucata McGregor
Holotype 13202; paratype 13201
McGregor, D.C., 1964, Geol. Surv. Can., Bull. 109, p. 17, pl. 21, figs. 9-11.
Ghost River Formation (sensu stricto), Middle or Upper Devonian, Phantom Crag, 4 miles northeast of eastern end of Lake Minnewanka, Alberta.

Convolutispora porcata McGregor and Camfield
Holotype 66236
McGregor, D.C. and Camfield, M., 1982, Geol. Surv. Can., Bull. 348, p. 25, pl. 4, fig. 21.
Hecla Bay Formation, Middle Devonian, stream section 16.2 km east of east arm of Weatherall Bay, northeastern Melville Island, District of Franklin.

Convolutispora subtilis Owens
Hypotype 66237
McGregor, D.C. and Camfield, M., 1982, Geol. Surv. Can., Bull. 348, p. 26, pl. 4, figs. 22, 23.
Weatherall Formation, Middle Devonian, stream section 16.2 km east of east arm of Weatherall Bay, northeastern Melville Island, District of Franklin.

Convolutispora tegula Allen 1965
Hypotype 49836
McGregor, D.C., 1979, Palynology, vol. 3, pl. 3, fig. 3.
Weatherall Formation, Middle Devonian, creek section 5.5 to 10 miles north of head of Beverley Inlet, Melville Island, District of Franklin.

Convolutispora tegula Allen
Hypotypes 66238, 66239
McGregor, D.C. and Camfield, M., 1982, Geol. Surv. Can., Bull. 348, p. 26, pl. 4, figs. 24-27.
Hecla Bay Formation, Middle Devonian, stream section 16.2 km east of east arm of Weatherall Bay, northeastern Melville Island, District of Franklin.

Convolutispora
Fig. spec. 13436
Kerr, J.W., McGregor, D.C. and McLaren, D.J., 1965, Bull. Can. Petrol. Geol., vol. 11, no. 3, fig. 11.

Ghost River Formation, Upper Devonian, Phantom Crag, 4 miles northeast of eastern end of Lake Minnewanka, Alberta.

Convolutispora sp.
Fig. spec. 13436
McGregor, D.C., 1963, Bull. Can. Petrol. Geol., vol. 11, no. 3, fig. 11.
Ghost River Formation (sensu stricto), Middle or Upper Devonian, Phantom Crag, 4 miles northeast of eastern end of Lake Minnewanka, Alberta.

Convolutispora sp.
Fig. spec. 15178
McGregor, D.C. and Owens, B., 1966, Geol. Surv. Can., Paper 66-30, pl. 3, figs. 14, 15.
Battery Point Formation, Lower Devonian, south side Gaspé Bay, shore section from Anse-à-Brillant to Bois Brulé Brook, eastern Gaspé Peninsula, Québec.

Convolutispora sp.
Fig. specs. 15352, 15355, 15357, 15358, 15460, 15461, 15467
McGregor, D.C. and Owens, B., 1966, Geol. Surv. Can., Paper 66-30, pl. 22, figs. 13, 14, 17, 18; pl. 28, figs. 11, 12, 19.
Upper Devonian, Escarpment Member, Hay River Formation, left bank Hay River at base of Louise Falls, Northwest Territories;
Kettle Point Formation, depth 295-299 feet, Courtright borehole, Courtright village, Lambton County, Ontario (15460, 15461, 15467).

Convolutispora sp.
Fig. spec. 29398
McGregor, D.C. and Uyeno, T.T., 1972, Geol. Surv. Can., Paper 71-13, pl. 2, fig. 11.
Griper Bay Formation, Upper Devonian, Beverley Inlet, southeastern Melville Island, District of Franklin.

Convolutispora sp.
Fig. spec. 86078
McGregor, D.C. and McCutcheon, S.R., 1988, Can. J. Earth Sci., vol. 25, no. 9, p. 1353, pl. 1, fig. 2.
Carrow Formation, Piskahegan Group, Late Devonian, depth 72.85 m, drill hole M-6, lat. 45°34′47″N, long. 61°35′27″W, southwestern New Brunswick.

?*Convolutispora* sp.
Fig. spec. 15359
McGregor, D.C. and Owens, B., 1966, Geol. Surv. Can., Paper 66-30, pl. 22, fig. 23.
Escarpment Member, Hay River Formation, left bank Hay River at base of Louise Falls, Alberta.

?*Corbulispora subalveolaris* (Luber) Sullivan
Hypotype 25474
McGregor, D.C., 1970, Colloque Stratig. Carbonifère, Congr. Colloques Univ. Liége, vol. 55, pl. 22, fig. 17.
Sunbury Formation, Upper Devonian, Gibraltar 77 borehole, 6 miles south of Courtright village, Lambton County, Ontario.

Cornispora bicornata Nazarenko 1971
Hypotype 49852
McGregor, D.C., 1979, Palynology, vol. 3, pl. 4, fig. 18.
Griper Bay Formation, Upper Devonian, stream valley 3/4 mile from mouth, 4 miles south of largest river entering Reindeer Bay, eastern Bathurst Island, District of Franklin,
=*Cornispora varicornata*, Richardson, J.B. and McGregor, D.C., 1986, Geol. Surv. Can., Bull. 364, pl. 18, fig. 5.

?*Corystisporites collaris* Tiwari and Schaarschmidt
Hypotype 66243
McGregor, D.C. and Camfield, M., 1982, Geol. Surv. Can., Bull. 348, p. 26, pl. 4, figs. 30-32.
Weatherall Formation, Middle Devonian, stream section 16.2 km east of east arm of Weatherall Bay, northeastern Melville Island, District of Franklin.

?*Corystisporites multispinosus* Richardson, 1965
Hypotype 15275
McGregor, D.C. and Owens, B., 1966, Geol. Surv. Can., Paper 66-30, pl. 7, fig. 1.
Battery Point Formation, Lower Devonian, south side Gaspé Bay, shore section from Bois Brulé Brook to Pointe Jaune, eastern Gaspé Peninsula, Québec.
=*Acinosporites macrospinosus*, McGregor, 1977, Palaeontographica, Abt.B, vol. 163, p. 129, pl. 1, figs. 13, 14 (hypotype 48171 = 15275)
=*Corystisporites macrospinosus*, McGregor, D.C., 1979, Palynology, vol. 3, pl. 1, fig. 20.

Corystisporites multispinosus Richardson var. *multispinosus*
Hypotype 66244
McGregor, D.C. and Camfield, M., 1982, Geol. Surv. Can., Bull. 348, p. 27, pl. 5, figs. 1, 2.
Hecla Bay Formation, Middle Devonian, stream section 16.2 km east of east arm of Weatherall Bay, northeastern Melville Island, District of Franklin.

Corystisporites multispinosus Richardson var. *spinulosus* McGregor and Camfield
Holotype 66245; paratype 66246
McGregor, D.C. and Camfield, M., 1982, Geol. Surv. Can., Bull. 348, p. 27, pl. 5, figs. 3, 4.
Hecla Bay and Weatherall formations, Middle Devonian, stream section 16.2 km east of east arm of Weatherall Bay, northeastern Melville Island, District of Franklin.

Corystisporites serratus? (Naumova) n. comb.
Hypotypes 66240-66242
McGregor, D.C. and Camfield, M., 1982, Geol. Surv. Can., Bull. 348, p. 27, pl. 4, figs. 28, 29, 33, 34.
Weatherall and Hecla Bay (66242) formations, Middle Devonian, stream section 16.2 km east of east arm of Weatherall Bay, northeastern Melville Island, District of Franklin.

Corystisporites serratus (Naumova) McGregor and Camfield
Hypotype 73292
Richardson, J.B. and McGregor, D.C., 1986, Geol. Surv. Can., Bull. 364, pl. 15, fig. 6.

Strathcona Fiord Formation, Middle Devonian, east of Bird Fiord, southwestern Ellesmere Island, District of Franklin.

?*Corystisporites* sp.
Fig. spec. 15175
McGregor, D.C. and Owens, B., 1966, Geol. Surv. Can., Paper 66-30, pl. 3, fig. 11.
Battery Point Formation (lower part), Lower Devonian, south side Gaspé Bay, section from Anse-à-Brillant to Bois Brulé Brook, eastern Gaspé Peninsula, Québec.
=*Kraeuselisporites gaspesiensis*, McGregor, D.C., 1973, Palaeontographica, Abt. B, vol. 142, p. 55, pl. 7, fig. 11 (holotype).

cf. *Corystisporites* sp.
Fig. spec. 67791
McGregor, D.C., 1984, Academia Nacional de Ciencias, Misc. No. 69, pl. 4, fig. 7.
Huamampampa Formation, Middle Devonian, Laurani section, Chuquisaca Dept., northwest of Tarabuco, Cordillera Oriental, Bolivia.

Corystisporites sp. cf. Acanthotriletes horridus Hacquebard
Fig. spec. 66254
McGregor, D.C. and Camfield, M., 1982, Geol. Surv. Can., Bull. 348, p. 28, pl. 5, fig. 11.
Hecla Bay Formation, Middle Devonian, stream section 16.2 km east of east arm of Weatherall Bay, northeastern Melville Island, District of Franklin.

Craspedispora arctica McGregor and Camfield
Holotype 66247; paratype 66248
McGregor, D.C. and Camfield, M., 1982, Geol. Surv. Can., Bull. 348, p. 28, pl. 5, figs. 5-9.
Hecla Bay and Weatherall formations, Middle Devonian, stream section 16.2 km east of east arm of Weatherall Bay, northeastern Melville Island, District of Franklin.

Cristatisporites albus? (Arkhangelskaya) n. comb.
Hypotypes 66249-66253
McGregor, D.C. and Camfield, M., 1982, Geol. Surv. Can., Bull. 348, p. 29, pl. 5, figs. 10, 12-16.
Weatherall Formation, Middle Devonian, stream section 16.2 km east of east arm of Weatherall Bay, northeastern Melville Island, District of Franklin.

Cristatisporites sp.
Fig. spec. 66255
McGregor, D.C. and Camfield, M., 1982, Geol. Surv. Can., Bull. 348, p. 31, pl. 6, fig. 1.
Weatherall Formation, Middle Devonian, stream section 16.2 km east of east arm of Weatherall Bay, northeastern Melville Island, District of Franklin.

Cristatisporites? sp.
Fig. specs. 13211, 13212
McGregor, D.C., 1964, Geol. Surv. Can., Bull. 109, p. 17, pl. 2, figs. 20-22.
Ghost River Formation (sensu stricto), Middle or Upper Devonian, Phantom Crag, 4 miles northeast of eastern end of Lake Minnewanka, Alberta.

Cyclogranisporites amplus? McGregor
Hypotypes 66256, 66257
McGregor, D.C. and Camfield, M., 1982, Geol. Surv. Can., Bull. 348, p. 31, pl. 6, figs. 2, 3.
Weatherall and Hecla Bay formations, Middle Devonian, stream section 16.2 km east of east arm of Weatherall Bay, northeastern Melville Island, District of Franklin.

Cyclogranisporites sp.
Fig. specs. 15142-15144, 15153, 15171, 15252, 15261, 15402, 13585, 15455
McGregor, D.C. and Owens, B., 1966, Geol. Surv. Can., Paper 66-30, pl. 2, figs. 7, 17-19, 21; pl. 3, fig. 6; pl. 5, figs. 5, 15; pl, 26, figs. 3, 4; pl. 28, fig. 6.
Lower Devonian, Battery Point Formation, south side Gaspé Bay, shore section from Anse-à-Brillant to Bois Brulé Brook, Malbaie Formation, shore section between Pointe Jaune and Whale Head (15252), and Battery Point Formation, 1/2 mile southeast of Sandy Beach spit (15261), eastern Gaspé Peninsula, Québec; Upper Devonian, Griper Bay Formation, stream valley west of Sargeant Point, 2 miles from shore, northeastern Banks Island, District of Franklin (15402, 13585), and Kettle Point Formation, depth 295-299 feet, Courtright borehole, Courtright village, Lambton County, Ontario (15455).
=*Apiculiretusispora plicati* McGregor, D.C., 1979, Palynology, vol. 3, pl. 1, fig. 6 (15153).
Richardson, J.B. and McGregor, D.C., 1986, Geol. Surv. Can., Bull. 364, pl. 4, fig. 2 (15153).

Cyclogranisporites sp.
Fig. spec. 61701
McGregor, D.C., 1979, Geol. Surv. Can., Paper 79-1B, pl. 22.2, fig. 55.
Daleje Shale, Lower Devonian, upper part of southeast slope of Císařská rokle, south east of Srbsko, Czechoslovakia.

?*Cyclogranisporites* sp.
Fig. specs. 15108, 15111
McGregor, D.C. and Owens, B., 1966, Geol. Surv. Can., Paper 66-30, pl. 1, figs. 6, 10.
Forillon Member, Cape Bon Ami Formation, Lower Devonian, north side Forillon Peninsula, Cape Rosier Cove, eastern Gaspé Peninsula, Québec.

Cymbosporites? cf. *catillus* Allen of Richardson and Lister
Hypotype 27067
McGregor, D.C., Sanford, B.V. and Norris, A.W., 1970, Proc. Geol. Assoc. Can., vol. 22, pl. 1, figs. 8, 9.
Kenogami River Formation, Lower Devonian, 1,072 feet depth, ODM Jaab Lake No. 1 well, lat. 51°11.9′N, long. 82°57′W, Moose River Basin, Ontario.

Cymbosporites ?*cyathus* Allen
Hypotype 32826
McGregor, D.C., 1974, Can. J. Earth Sci., vol. 11, no. 1, pl. 2, fig. 50.

Vendom Fiord Formation, Early Devonian, lat. 79°28′40″N, long. 80°02′W, 12 miles south of head of Cañon Fiord, central Ellesmere Island, District of Franklin.

cf. *Cymbosporites echinatus* Richardson and Lister 1969
Hypotype 48172
McGregor, D.C., 1977, Palaeontographica, Abt. B., vol. 163, p. 131, pl. 1, figs. 15, 16.
Battery Point Formation, Lower Devonaian, Tar Point to Whale Head section, south shore of Gaspé Bay, Québec.

Cymbosporites echinatus? Richardson and Lister, 1969
Hypotype 67777
McGregor, D.C., 1984, Academia Nacional de Ciencias, Misc. No. 69, pl. 3, fig. 10.
Santa Rosa Formation, Lower Devonian, Laurani section, Chuquisaca Dept., northwest of Tarabuco, Cordillera Oriental, Bolivia.

Cymbosporites magnificus (McGregor 1960) n. comb
Hypotype 66258
McGregor, D.C. and Camfield, M., 1982, Geol. Surv. Can., Bull. 348, p. 32, pl. 6, figs. 4, 5.
Hecla Bay Formation, Middle Devonian, stream section 16.2 km east of east arm of Weatherall Bay, northeastern Melville Island, District of Franklin.

Cymbosporites? *opacus* McGregor and Camfield
Holotype 66261; paratypes 66259, 66260
McGregor, D.C. and Camfield, M., 1982, Geol. Surv. Can., Bull. 348, p. 32, pl. 6, figs. 6, 9, 10.
Hecla Bay Formation, Middle Devonian, stream section 16.2 km east of east arm of Weatherall Bay, northeastern Melville Island, District of Franklin.

Cymbosporites paulus McGregor and Camfield
Holotype 41722
McGregor, D.C. and Camfield, M., 1976, Geol. Surv. Can., Bull. 263, p. 15, pl. 2, figs. 7-9.
Stooping River Formation, Lower Devonian, 815-817 feet depth, ODM Jaab Lake No. 1 well, lat. 51°11′54″N, long. 82°56′W, Ontario.

Cymbosporites proteus McGregor and Camfield
Holotype 41710; paratypes 41708, 41709, 41711-41713
McGregor, D.C. and Camfield, M., 1976, Geol. Surv. Can., Bull. 263, p. 16, pl. 1, figs. 8-16, 21-23.
Lower Devonian, Stooping River and Kenogomi River formations, 909-911 (41710, 41713), 755-757 (41709), 1072-1073.6 (41708) feet depths, ODM Jaab Lake No. 1 well, lat. 51°11′54″N, long. 82°56′W,; Sextant Formation, 136.6-136.8 feet depth, Argor ETA Kiasko River No. 2 well, lat. 50°41′30″N, long. 80°48′W; and Kenogami River Formation, 170-170.6 feet depth, Puskwuche Point No. 1 well, lat. 51°46′12″N, long. 80°40′06″W, Ontario.

Cymbosporites? *senex* McGregor and Camfield
Holotype 41719; paratype 41720
McGregor, D.C. and Camfield, M., 1976, Geol. Surv. Can., Bull. 263, p. 16, pl. 2, figs. 1-4.
Stooping River and Kenogami River formations, Lower Devonian, 909-911 feet depth, ODM Jaab Lake No. 1 well, lat. 51°11′54″N, long. 82°56′W, and 1440 feet depth, Puskwuche Point No. 1 well, lat. 51°46′12″N, long. 80°40′06″W, Ontario.

Cymbosporites yorkensis McGregor
Holotype 32025
McGregor, D.C., 1973, Palaeontographica, Abt. B., vol. 142, no. 1-3, p. 58, pl. 7, figs. 20, 21.
York River Formation, Lower Devonaian, 400 yards southeast of Tar Point dyke, south shore of Gaspé Bay, Québec.

Cymbosporites cf. C. yorkensis McGregor
Hypotype 41717
McGregor, D.C. and Camfield, M., 1976, Geol. Surv. Can., Bull. 263, p. 17, pl. 1, figs. 26, 31, 32.
Kenogami River Formation, Lower Devonian, 1072-1073.6 feet depth, ODM Jaab Lake No. 1 well, lat. 51°11′54″N, long. 82°56′W, Ontario.

Cymbosporites sp.
Fig. spec. 66268
McGregor, D.C. and Camfield, M., 1982, Geol. Surv. Can., Bull. 348, p. 33, pl. 6, figs. 16, 17.
Hecla Bay Formation, Middle Devonian, stream section 16.2 km east of east arm of Weatherall Bay, northeastern Melville Island, District of Franklin.

Cymbosporites sp.
Fig. spec. 67781
McGregor, D.C., 1984, Academia Nacional de Ciencias, Misc. No. 69, pl. 3, fig. 14.
Santa Rosa Formation, Lower Devonian, Laurani section, Chuquisaca Dept., northwest of Tarabuco, Cordillera Oriental, Bolivia.

Cymbosporites sp. cf. Retusotriletes tschibrikovii Mikhailova
Fig. specs. 66262, 66263
McGregor, D.C. and Camfield, M., 1982, Geol. Surv. Can., Bull. 348, p. 33, pl. 6, figs. 7, 8, 11, 12.
Hecla Bay Formation, Middle Devonian, stream section 16.2 km east of east arm of Weatherall Bay, northeastern Melville Island, District of Franklin.

Deltoidospora (Leiotriletes) microdeltoidus (McGregor)
Hypotype 32810
McGregor, D.C., 1974, Can. J. Earth Sci., vol. 11, no. 1, pl. 1, fig. 28.
Vendom Fiord Formation, Early Devonian, lat. 79°28′40″N, long. 80°02′W, 12 miles south of head of Cañon Fiord, central Ellesmere Island, District of Franklin.

Deltoidospora pridddyi (Berry) McGregor
Hypotype 31945
McGregor, D.C., 1973, Palaeontographica, Abt. B, vol. 142, no. 1-3, p. 16, pl. 1, figs. 6, 7.
Battery Point Formation, Lower Devonian, 150 feet west of road to beach from end of Laurencelle Road, north shore Gaspé Bay, Québec.

Deltoidospora sp.
Fig. specs. 15136-15138, 15405
McGregor, D.C. and Owens, B., 1966, Geol. Surv. Can., Paper 66-30, pl. 2, figs. 1-3; pl. 26, fig. 7.
Battery Point Formation, Lower Devonian, south side Gaspé Bay, shore section from Anse-à-Brillant to Bois Brulé Brook, eastern Gaspé Peninsula, Québec; Griper Bay Formation, stream valley west of Sargeant Point, 2 miles from shore, northeastern Bathurst Island, District of Franklin (15405).

Deltoidospora sp.
Fig. spec. 61679
McGregor, D.C., 1979, Geol. Surv. Can., Paper 79-1B, pl. 22.1, fig. 31.
Daleje Shale, Lower Devonian, roadside outcrop near Karlštejn, 1 km north of Budnany in the Hluboké valley, Czechoslovakia.

Deltoidospora sp.
Fig. spec. 66269
McGregor, D.C. and Camfield, M., 1982, Geol. Surv. Can., Bull. 348, p. 33, pl. 6, fig. 15.
Hecla Bay Formation, Middle Devonian, stream section 16.2 km east of east arm of Weatherall Bay, northeastern Melville Island, District of Franklin.

Deltoidospora spp.
Fig. specs. 32786-32788
McGregor, D.C., 1974, Can. J. Earth Sci., vol. 11, no. 1, pl. 1, figs. 1-3.
Unnamed Early Devonian, lat. 79°28′40″N, long. 80°02′W, 12 miles south of head of Cañon Fiord, central Ellesmere Island, District of Franklin.

Densosporites concinnus (Owens) n.comb.
Hypotypes 66264-66267
McGregor, D.C. and Camfield, M., 1982, Geol. Surv. Can., Bull. 348, p. 34, pl. 6, figs. 13, 14, 18, 19.
Hecla Bay Formation, Middle Devonian, stream section 16.2 km east of east arm of Weatherall Bay, northeastern Melville Island, District of Franklin.

Densosporites concinnus (Owens) McGregor and Camfield
Hypotype 73291
Richardson, J.B. and McGregor, D.C., 1986, Geol. Surv. Can., Bull. 364, pl. 12, fig. 2.
Weatherall Formation, Middle Devonian, southeastern Melville Island, District of Franklin.

Densosporites devonicus Richardson, 1960
Hypotype 15317
McGregor, D.C. and Owens, B., 1966, Geol. Surv. Can., Paper 66-30, pl. 16, figs. 1, 2.
Griper Bay Formation, Upper Devonian, south limb of Robertson Point anticline, 8 miles northeast of Beverley Inlet, Melville Island, District of Franklin.

Densosporites devonicus Richardson 1960
Hypotype 49837
McGregor, D.C., 1979, Palynology, vol. 3, pl. 3, fig. 4.
Williams Island Formation, Middle Devonian, 147-149 feet depth, ODM Jaab Lake well, lat. 51°12′N, long. 83°57′W, Ontario.

Densosporites devonicus Richardson
Hypotypes 66270, 66271
McGregor, D.C. and Camfield, M., 1982, Geol. Surv. Can., Bull. 348, p. 34, pl. 6, figs. 20, 21.
Weatherall Formation, Middle Devonian, stream section 16.2 km east of east arm of Weatherall Bay, northeastern Melville Island, District of Franklin.

Densosporites inaequus (McGregor) n.comb.
Hypotypes 66273-66277
McGregor, D.C. and Camfield, M., 1982, Geol. Surv. Can., Bull. 348, p. 35, pl. 7, figs. 1-6.
Richardson, J.B. and McGregor, D.C., 1986, Geol. Surv. Can., Bull. 364, pl. 12, fig. 7 (66274).
Weatherall Formation, Middle Devonian, stream section 16.2 km east of east arm of Weatherall Bay, northeastern Melville Island, District of Franklin.

Densosporites orcadensis Richardson
Hypotype 41769
McGregor, D.C. and Camfield, M., 1976, Geol. Surv. Can., Bull. 263, p. 17, pl. 6, fig. 3.
Williams Island Formation, Middle Devonian, 157.5-159.5 feet depth, ODM Jaab Lake No. 1 well, lat. 51°11′54″N, long. 82°56′W, Ontario.

Densosporites weatherallensis McGregor and Camfield
Holotype 66278; paratype 66279
McGregor, D.C. and Camfield, M., 1982, Geol. Surv. Can., Bull. 348, p. 35, pl. 7, figs. 7-9, 14.
Weatherall Formation, Middle Devonian, stream section 16.2 km east of east arm of Weatherall Bay, northeastern Melville Island, District of Franklin.

?*Densosporites* sp.
Fig. spec. 27071
McGregor, D.C., Sanford, B.V. and Norris, A.W., 1970, Proc. Geol. Assoc. Can., vol. 22, pl. 1, fig. 17.
Stooping River Formation, Lower Devonian, 1027 feet depth, ODM Jaab Lake No. 1 well, lat. 51°11.9′N, long. 82°57′W, Moose River Basin, Ontario.

Detached bifurcate process of megaspore
Fig. spec. 86095
McGregor, D.C. and McCutcheon, S.R., 1988, Can. J. Earth Sci., vol. 25, no. 9, p. 1353, pl. 1, fig. 24.
Carrow Formation, Piskahegan Group, Late Devonian, depth 72.85 m, drill hole M-6, lat. 45°34′43″N, long. 66°35′27″W, southwestern New Brunswick.

Diaphanospora perplexa Balme and Hassell, 1962
Hypotype 15431
McGregor, D.C. and Owens, B., 1966, Geol. Surv. Can., Paper 66-30, pl. 27, fig. 3.
Griper Bay Formation, Upper Devonian, stream valley west of Sargeant Point, 2 miles from shore, northeastern Bathurst Island, District of Franklin.

Diaphanospora reticulata Guennel
Hypotype 73303
Richardson, J.B. and McGregor, D.C., 1986, Geol. Surv. Can., Bull. 364, pl. 17, fig. 11.
Long Rapids Formation, Upper Devonian, Abitibi River, Ontario.

Diaphanospora riciniata Balme and Hassell, 1962
Hypotype 15433
McGregor, D.C. and Owens, B., 1966, Geol. Surv. Can., Paper 66-30, pl. 27, fig. 5.
Griper Bay Formation, Upper Devonian, stream valley west of Sargeant Point, 2 miles from shore, northeastern Bathurst Island, District of Franklin.

cf.*Diaphanospora submirabilis* (Luber) Byosheva, 1985
Hypotype 86107
McGregor, D.C. and McCutcheon, S.R., 1988, Can. J. Earth Sci., vol. 25, no. 9, p. 1353, pl. 2, figs. 11.
Carrow Formation, Piskahegan Group, Late Devonian, depth 72.85 m, drill hole M-6, lat. 45°34′43″N, long. 66°35′27″W, southwestern New Brunswick.

Diaphanospora sp.
Fig. spec. 86093
McGregor, D.C. and McCutcheon, S.R., 1988, Can. J. Earth Sci., vol. 25, no. 9, p. 1353, pl. 1, fig. 22.
Carrow Formation, Piskahegan Group, Late Devonian, depth 72.85 m, drill hole M-6, lat. 45°34′43″N, long. 66°35′27″W, southwestern New Brunswick.

Diatomozonotriletes devonicus Naum.
Hypotype 29393
McGregor, D.C. and Uyeno, T.T., 1972, Geol. Surv. Can., Paper 71-13, pl. 2, fig. 6.
Hecla Bay Formation, Middle Devonian, Beverley Inlet, southeastern Melville Island, District of Franklin.

Diatomozonotriletes franklinii McGregor and Camfield
Holotype 66280; paratype 66281
McGregor, D.C. and Camfield, M., 1982, Geol. Surv. Can., Bull. 348, p. 36, pl. 7, figs. 10-13.
Weatherall and Hecla Bay formations, Middle Devonian, stream section 16.2 km east of east arm of Weatherall Bay, northeastern Melville Island, District of Franklin.

Diatomozonotriletes franklinii? McGregor and Camfield, 1982
Hypotype 67793
McGregor, D.C., 1984, Academia Nacional de Ciencias, Misc. No. 69, pl. 4, fig. 9.
Huamampampa Formation, Middle Devonian, Laurani section, Chuquisaca Dept., northwest of Tarabuco, Cordillera Oriental, Bolivia.

Diatomozonotriletes oligodontus? Chibrikova
Hypotypes 66282, 66283
McGregor, D.C. and Camfield, M., 1982, Geol. Surv. Can., Bull. 348, p. 36, pl. 7, figs. 15-17.
Weatherall and Hecla Bay formations, Middle Devonian, stream section 16.2 km east of east arm of Weatherall Bay, northeastern Melville Island, District of Franklin.

Dibolisporites abitibiensis McGregor and Camfield
Holotype 41743; paratype 41761
McGregor, D.C. and Camfield, M., 1976, Geol. Surv. Can., Bull. 263, p. 18, pl. 3, figs. 20, 21, 28, 29.
Sextant Formation, Lower Devonian, 155.6-155.8 and 132.2-132.4 feet, Argor ETA Kiasko River No. 2 well, lat. 50°41′30″N, long. 80°48′W, Ontario.

Dibolisporites antiquus? (Naumova ex Kedo) n. comb.
Hypotype 66284
McGregor, D.C. and Camfield, M., 1982, Geol. Surv. Can., Bull. 348, p. 37, pl. 8, fig. 1.
Weatherall Formation, Middle Devonian, stream section 16.2 km east of east arm of Weatherall Bay, northeastern Melville Island, District of Franklin.

Dibolisporites echinaceus (Eisenack) Richardson, 1965
Hypotypes 15257, 15266
McGregor, D.C. and Owens, B., 1966, Geol. Surv. Can., Paper 66-30, pl. 5, fig. 10; pl. 6, fig. 1.
Battery Point and Malbaie formations, Lower Devonian, south side Gaspé Bay, shore section from Bois Brulé Brook to Pointe Jaune and between Pointe Jaune and Whale Head, eastern Gaspé Peninsula, Québec.

Diboliporites echinaceus (Eisenack) Richardson
Hypotype 27095
McGregor, D.C., Sanford, B.V. and Norris, A.W., 1970, Proc. Geol. Assoc. Can., vol. 22, pl. 22, fig. 16.
Williams Island Formation, Middle Devonian, 157 feet depth, ODM Jaab Lake No. 1 well, lat. 51°11.9′N, long. 82°57′W, Moose River Basin, Ontario.

Dibolisporites echinaceus (Eisenack) Richardson
Hypotypes 31972, 31973
McGregor, D.C., 1973, Palaeontographica, Abt. B, vol. 142, no. 1-3, p. 29, pl. 8, figs. 8-10, 12, 13.
Malbaie Formation, Middle Devonian, 65 yards southeast of conglomerate of Pointe Jaune, south shore Gaspé Bay, Québec.

Dibolisporitess echinaceus (Eisenack) Richardson
Hypotype 41768
McGregor, D.C. and Camfield, M., 1976, Geol. Surv. Can., Bull. 263, p. 19, pl. 6, figs. 1, 2.
Williams Island Formation, Middle Devonian, 157.5-159.5 feet depth, ODM Jaab Lake No. 1 well, lat. 51°11′54″N, long. 82°56′W, Ontario.

Dibolisporites echinaceus (Eisenack) Richardson
Hypotype 49824
McGregor, D.C., 1979, Palynology, vol. 3, pl. 1, fig. 24.
Malbaie Formation, Middle Devonian, south shore Gaspé Bay, 65 yards southeast of Point Jaune, Gaspé, Québec.

Dibolisporites echinaceus (Eisenack) Richardson
Hypotype 66285
McGregor, D.C. and Camfield, M., 1982, Geol. Surv. Can., Bull. 348, p. 37, pl. 8, fig. 2.
Weatherall Formation, Middle Devonian, stream section 16.2 km east of east arm of Weatherall Bay, northeastern Melville Island, District of Franklin.

Dibolisporites echinaceus? (Eisenack) Richardson, 1965 sensu, McGregor, 1973
Hypotype 67172
McGregor, D.C., 1984, Academia Nacional de Ciencias, Misc. No. 69, p. 25, pl. 3, fig. 2.
Icla Formation, Lower or Middle Devonian, Laurani section, Chuquisaca Dept., northwest of Tarabuco, Cordillera Oriental, Bolivia.

Dibolisporites cf. *D. echinaceus* (Eisenack) Richardson, 1965
Fig. spec. 15258
McGregor, D.C. and Owens, B., 1966, Geol. Surv. Can., Paper 66-30, pl. 5, figs. 11, 12.
Battery Point Formation, Lower Devonian, south side Gaspé Bay, shore section from Bois Brulé Brook to Pointe Jaune, eastern Gaspé Peninsula, Québec.

Dibolisporitees echinaceus (Eisenack) Richardson, *echinaceus* type
Hypotype 61737
McGregor, D.C., 1979, Geol. Surv. Can., Paper 79-1B, pl. 22.3, fig. 92.
Trebotov Limestone, Lower Devonian, northwest side of railway, north side of Prokopské Valley at Hluboepy near Prague, Czechoslovakia.

?*Dibolisporites echinaceus* (Eisenack) Richardson, *echinaceus* type
Hypotype 61723
McGregor, D.C., 1979, Geol. Surv. Can., Paper 79-1B, pl. 22.3, fig. 78.
Daleje Shale, Lower Devonian, upper part of southeast slope of Císařská rokle, southeast of Srbsko, Czechoslovakia.

Dibolisporites echinaceus (Eisenack) Richardson, *gibberosus* type
Hypotypes 61687, 61738
McGregor, D.C., 1979, Geol. Surv. Can., Paper 79-1B, pl. 22.2, fig. 39; pl. 22.3, fig. 93.
Lower Devonian, Daleje Shale, roadside outcrop near Karlštejn, 1 km north of Budnany in the Hluboké Valley; Trebotov Limestone, northwest side of railway, north side of Prokopské Valley at Hluboepy near Prague, Czechoslovakia.

?*Dibolisporites* sp. cf. *D. echinaceus* (Eisenack) Richardson, *gibberosus* type
Fig. spec. 61663
McGregor, D.C., 1979, Geol. Surv. Can., Paper 79-1B, pl. 22.1, fig. 15.
Zlíchov Limestone, Lower Devonian, near Pekárek mill west village of Solopysky, Czechoslovakia.

cf. *Dibolisporites* sp.
Fig. spec. 15260
McGregor, D.C. and Owens, B., 1966, Geol. Surv. Can., Paper 66-30, pl. 5, fig. 14.
Malbaie Formation, Lower Devonian, south side Gaspé Bay, shore section between Point Jaune and Whale Head, eastern Gaspé Peninsula, Québec.

Dibolisporites eifeliensis (Lanninger) McGregor n.comb.
Hypotypes 31975-31977
McGregor, D.C., 1973, Palaeontographica, Abt. B, vol. 142, no. 1-3, p. 31, pl. 3, figs. 17-22, 26.
Battery Point Formation, Middle and Lower Devonian, 100 yards south of most prominent point between Cap Rouge and Pointe Jaune; 550 yards east of small stream 3/4 mile north of Bois Brulé Brook; and 1300 yards south of second stream south of Bois Brulé Brook, south shore Gaspé Bay, Québec.

Dibolisporites eifeliensis (Lanninger) McGregor
Hypotype 32824
McGregor, D.C., 1974, Can. J. Earth Sci., vol. 11, no. 1, pl. 2, figs. 47, 51.
Vendom Fiord Formation, Early Devonian, lat. 79°28′40″N, long. 80°02′W, 12 miles south of head of Cañon Fiord, central Ellesmere Island, District of Franklin.

Dibolisporites eifeliensis (Lanninger) McGregor
Hypotype 41753
McGregor, D.C. and Camfield, M., 1976, Geol. Surv. Can., Bull. 263, p. 19, pl. 4, fig. 21.
Stooping River Formation, Lower Devonian, 617-619 feet depth, ODM Jaab Lake No. 1 well, lat. 51°11′54″N, long. 82°56′W, Ontario.

Dibolisporites eifeliensis (Lanninger) McGregor 1973
Hypotype 49817
McGregor, D.C., 1979, Palynology, vol. 3, pl. 1, fig. 10.
Richardson, J.B. and McGregor, D.C., 1986, Geol. Surv. Can., Bull. 364, pl. 5, fig. 6.
Stooping River Formation, Lower Devonian, 599-600 feet depth, ODM Jaab Lake No. 1 well, lat. 51°12′N, long. 82°57′W, Ontario.

Dibolisporites farranginis McGregor and Camfield
Holotype 66287; paratype 66286
McGregor, D.C. and Camfield, M., 1982, Geol. Surv. Can., Bull. 348, p. 38, pl. 8, figs. 3-4.
Hecla Bay Formation, Middle Devonian, stream section 16.2 km east of east arm of Weatherall Bay, northeastern Melville Island, District of Franklin.

Dibolisporites cf. *gibberosus* var. *major* (Kedo) Richardson
Hypotype 32825
McGregor, D.C., 1974, Can. J. Earth Sci., vol. 11, no. 1, pl. 2, figs. 48, 49.
Vendom Fiord Formation, Early Devonian, lat. 79°28′40″N, long. 80°02′W, 12 miles south of head of Cañon Fiord, central Ellesmere Island, District of Franklin.

Dibolisporites quebecensis McGregor
Holotype 15155; hypotype 31974
McGregor, D.C., 1973, Palaeontographica, Abt. B, vol. 142, no. 1-4, p. 32, pl. 3, figs. 11, 14-16.
Battery Point Formation, Lower Devonian, 400 yards east of Rivière de l'Anse-à-Brillant, south shore Gaspé Bay, Québec.

Dibolisporites quebecensis McGregor
Hypotypes 32820, 32821
McGregor, D.C., 1974, Can. J. Earth Sci., vol. 11, no. 1, pl. 2, figs. 41-43.
Vendom Fiord Formation, Early Devonian, lat. 79°28′40″N, long. 80°02′W, 12 miles south of head of Cañon Fiord, central Ellesmere Island, District of Franklin.

?*Dibolisporites quebecensis* McGregor, 1973
Hypotype 67174
McGregor, D.C., 1984, Academia Nacional de Ciencias, Misc. No. 69, p. 25, pl. 2, fig. 19.
Huamampampa Formation, Lower Devonian, Laurani section, Chuquisaca Dept., northwest of Tarabuco, Cordillera Oriental, Bolivia.

Dibolisporites cf. *D. quebecensis* McGregor
Hypotype 41749
McGregor, D.C. and Camfield, M., 1976, Geol. Surv. Can., Bull. 263, p. 19, pl. 4, figs. 9, 10.
Stooping River Formation, Lower Devonian, 649-650 feet depth, ODM Jaab Lake No. 1 well, lat. 51°11′54″N, long. 82°56′W, Ontario.

Dibolisporites ?radiatus Tiwari and Schaarschimdt 1975
Hypotype 48173
McGregor, D.C., 1977, Palaeontographica, Abt. B, vol. 163, p. 131, pl. 1, figs. 17-19.
Battery Point Formation, Lower Devonian, Sandy Beach to Haldimaand section, south shore Gaspé Bay, Québec.

Dibolisporites radatus? Tiwari and Schaarschmidt 1975
Hypotypes 67786, 67787
McGregor, D.C., 1984, Academia Nacional de Ciencias, Misc. No. 69, pl. 4, figs. 2, 3.
Icla Formation, Lower Devonian, Laurani section, Chuquisaca Dept., northwest of Tarabuco, Cordillera Oriental, Bolivia.

Dibolisporites uncatus (Naumova) n. comb.
Hypotypes 66288-66290
McGregor, D.C. and Camfield, M., 1982, Geol. Surv. Can., Bull. 348, p. 38, pl. 8, figs. 5, 6, 11.
Weatherall and Hecla Bay (66290) formations, Middle Devonian, stream section 16.2 km east of east arm of Weatherall Bay, northeastern Melville Island, District of Franklin.

Dibolisporites variegatus McGregor
Holotype 31978
McGregor, D.C., 1973, Palaeontographica, Abt. B, vol. 142, no. 1-3, p. 33, pl. 4, figs. 1, 2.
Battery Point Formation, Lower Devonian, 500 yards north of southern extremity of cliff north of Rivière de l'Anse-à-Brillant, south shore Gaspé Bay, Québec.

Dibolisporites vegrandis McGregor and Camfield
Holotype 66293
McGregor, D.C. and Camfield, M., 1982, Geol. Surv. Can., Bull. 348, p. 39, pl. 8, figs. 9, 10.
Hecla Bay Formation, Middle Devonian, stream section 16.2 km east of east arm of Weatherall Bay, northeastern Melville Island, District of Franklin.

Dibolisporites wetteldorfensis Lanninger
Hypotype 15161
McGregor, D.C., 1973, Palaeontographica, Abt. B, vol. 142, no. 1-3, p. 31, pl. 3, figs. 23-25.
Battery Point Formation, Lower Devonian, 600 yards northwest of first small brook south of Rivière de l'Anse-à-Brillant, south shore Gaspé Bay, Québec.

Dibolisporites ?wetteldorfensis Lanninger
Hypotype 41746
McGregor, D.C. and Camfield, M., 1976, Geol. Surv. Can., Bull. 263, p. 20, pl. 4, figs. 4-6.
Stooping River Formation, Lower Devonian, 960-962 feet depth, ODM Jaab Lake No. 1 well, lat. 51°11′54″N, long. 82°56′W, Ontario.

?*Dibolisporites wetteldorfensis* Lanninger
Hypotype 61659
McGregor, D.C., 1979, Geol. Surv. Can., Paper 79-1B, pl. 22.1, fig. 11.
Dvorce-Prokop Limestone, Lower Devonian, road-cut on west side of Vitava River below Barrandov, southern Prague, north of old quarry "U Kapliky", Czechoslovakia.

Dibolisporites wetteldorfensis Lanninger
Hypotypes 61668, 61669
McGregor, D.C., 1979, Geol. Surv. Can., Paper 79-1B, pl. 22.1, figs. 20, 25.
Zlíchov Limestone, Lower Devonian, near Pekárek mill west of village of Solopysky, Czechoslovakia.

Dibolisporites sp.
Fig. specs. 15132, 15254
McGregor, D.C. and Owens, B., 1966, Geol. Surv. Can., Paper 66-30, pl. 1, fig. 33; pl. 5, fig. 7.
Lower Devonian, Shiphead Member, Grand Grève Formation, Cape Gaspé shore section below lighthouse, and Battery Point Formation, south side Gaspé Bay, shore section from Bois Brulé Brook to Pointe Jaune, eastern Gaspé Peninsula, Québec.

Dibolisporites sp.
Fig. specs. 61695, 61711, 61733
McGregor, D.C., 1979, Geol. Surv. Can., Paper 79-1B, pl. 22.2, figs. 49, 66; pl. 22.3, fig. 88.
Lower Devonian, Daleje Shale, roadside outcrop near Karlštejn, 1 km north of Budnany in the Hluboké Valley, and upper part of southeast slope of Císařská rokle, southeast of Srbsko (61711); Trebotov Limestone, northwest side of railway, north side of Prokopské Valley at Hluboepy near Prague (61673), Czechoslovakia.

Dibolisporites sp.
Fig. spec. 67175
McGregor, D.C., 1984, Academia Nacional de Ciencias, Misc. No. 69, p. 26, pl. 2, fig. 20.
Icla Formation, Lower Devonian, Laurani section, Chuquisaca Dept., northwest of Tarabuco, Cordillera Oriental, Bolivia.

?*Dibolisporites* sp.
Fig. specs. 61652, 61675
McGregor, D.C., 1979, Geol. Surv. Can., Paper 79-1B, pl. 22.1, figs. 4, 27.
Lower Devonian, Dvorce-Prokop Limestone, road-cut on west side of Vitava River below Barrandov, southern Prague, north of old quarry "U Kapliky"; Zlíchov Limestone, near Pekárek mill west of village of Solopysky, Czechoslovakia.

Dibolisporites n.sp.
Fig. spec. 29373
McGregor, D.C. and Uyeno, T.T., 1972, Geol. Surv. Can., Paper 71-13, pl. 1, fig. 5.
Stuart Bay Formation, Lower Devonian, Young Inlet, northeastern Bathurst Island, District of Franklin.

?*Dicrospora multifurcata* Winslow
Hypotype 25457
McGregor, D.C., 1970, Colloque Stratig. Carbonifère, Congr. Colloques Univ. Liège, vol. 55, pl. 21, figs. 20, 21.
Bedford Formation, Upper Devonian, Courtright borehole, Courtright village, Lambton County, Ontario.

Dictyotriletes canadensis McGregor
Holotype 15177; hypotype 32001
McGregor, D.C., 1973, Palaeontographica, Abt. B, vol. 142, no. 1-3, p. 44, pl. 5, figs. 21-23.
Battery Point Formation, Lower Devonian, 480 yards northwest of first small brook south of, and 150 yards north of Rivière de l'Anse-à-Brillant, south shore Gaspé Bay, Québec.

Dictyotriletes emsiensis (Allen) McGregor
Hypotypes 31998
McGregor, D.C., 1973, Palaeontographica, Abt. B, vol. 142, no. 1-3, p. 42, pl. 5, fig. 15.
Battery Point Formation, Lower Devonian, 260 yards of first small brook south of Rivière de l'Anse-à-Brillant, south shore Gaspé Bay, Québec.

Dictyotriletes emsiensis (Allen) McGregor
Hypotype 41734
McGregor, D.C. and Camfield, M., 1976, Geol. Surv. Can., Bull. 263, p. 20, pl. 3, figs. 8, 9.
Stooping River Formation, Lower Devonian, 925-927 feet depth, ODM Jaab Lake No. 1 well, lat. 51°11′54″N, long. 82°56′W, Ontario.

Dictyotriletes emsiensis (Allen) McGregor, 1973
Hypotypes 67173, 67176, 67177
McGregor, D.C., 1984, Academia Nacional de Ciencias, Misc. No. 69, p. 26, pl. 1, figs. 21, 22, 24.
Santa Rosa Formation, Lower Devonian, Laurani section, Chuquisaca Dept., northwest of Tarabuco, Cordillera Oriental, Bolivia.

Dictyotriletes emsiensis? (Allen) McGregor
Hypotype 61670
McGregor, D.C., 1979, Geol. Surv. Can., Paper 79-1B, pl. 22.1, fig. 21.
Zlíchov Limestone, Lower Devonian, near Pekárek mill west of village of Solopysky, Czechoslovakia.

Dictyotriletes cf.*D. emsiensis* (Allen) McGregor
Hypotype 41731
McGregor, D.C. and Camfield, M., 1976, Geol. Surv. Can., Bull. 263, p. 21, pl. 3, figs. 5, 6.
Stooping River Formation, Lower Devonian, 599-600 feet depth, ODM Jaab Lake No. 1 well, lat. 51°11′54″N, long. 82°56′W, Ontario.

Dictyotriletes favosus McGregor and Camfield
Holotype 41721
McGregor, D.C. and Camfield, M., 1976, Geol. Surv. Can., Bull. 263, p. 21, pl. 2, figs. 5, 6.
Kenogami River Formation, Lower Devonian, 140 feet depth, Puskwuche Point No. 1 well, lat. 51°46′12″N, long. 80°40′06″W, Ontario.

Dictyotriletes favosus? McGregor and Camfield
Hypotype 61676
McGregor, D.C., 1979, Geol. Surv. Can., Paper 79-1B, pl. 22.1, fig. 28.
Zlíchov Limestone, Lower Devonian, near Pekárek mill west of village of Solopysky, Czechoslovakia.

Dictyotriletes sp. cf. *D. favosus* McGregor and Camfield
Fig. spec. 61714
McGregor, D.C., 1979, Geol. Surv. Can., Paper 79-1B, pl. 22.2, fig. 69.
Daleje Shale, Lower Devonian, upper part of southeast slope of Císařská rokle, southeast of Srbsko, Czechoslovakia.

Dictyotriletes ?*gorgoneus* Cramer
Hypotype 31997
McGregor, D.C., 1973, Palaeontographica, Abt. B, vol. 142, no. 1-3, p. 43, pl. 5, figs. 12, 17.
Battery Point Formation, Lower Devonian, 1050 yards northwest of first small brook south of Rivière de l'Anse-à-Brillant, south shore Gaspé Bay, Québec.

Dictyotriletes gorgoneus? Cramer
Hypotypes 61716, 61739
McGregor, D.C., 1979, Geol. Surv. Can., Paper 79-1B, pl. 22.2, fig. 71; pl. 22.3, fig. 94.
Lower Devonian, Daleje Shale, upper part of southeast slope of Císařská rokle, southeast of Srbsko; Trebotov Limestone, northwest side of railway, north side of Prokopské Valley at Hluboepy near Prague, Czechoslovakia.

Dictyotriletes sp. cf. *D. gorgoneus* Cramer, 1967
Fig. spec. 67179
McGregor, D.C., 1984, Academia Nacional de Ciencias, Misc. No. 69, p. 27, pl. 1, fig. 6.
Santa Rosa Formation, Lower Devonian, Laurani section, Chuquisaca Dept., northwest of Tarabuco, Cordillera Oriental, Bolivia.

Dictyotriletes subgranifer McGregor
Holotype 32000; hypotype 31999
McGregor, D.C., 1973, Palaeontographica, Abt. B, vol. 142, no. 1-3, p. 43, pl. 5, figs. 16, 18-20.
Battery Point Formation, Lower Devonian, 400 and 500 yards east of Rivière de l'Anse-à-Brillant, south shore Gaspé Bay, Québec.

Dictyotriletes subgranifer McGregor
Hypotype 41752
McGregor, D.C. and Camfield, M., 1976, Geol. Surv. Can., Bull. 263, p. 21, pl. 4, fig. 19, 20.
Stooping River Formation, Lower Devonian, 609-610 feet depth, ODM Jaab Lake No. 1 well, lat. 51°11′54″N, long. 82°56′W, Ontario.

Dictyotriletes subgranifer? McGregor
Hypotype 61720
McGregor, D.C., 1979, Geol. Surv. Can., Paper 79-1B, pl. 22.3, fig. 75.
Daleje Shale, Lower Devonian, upper part of southeast slope of Císařská rokle, southeast of Srbsko, Czechoslovakia.

cf.*Dictyotriletes subgranifer* McGregor, 1973
Hypotype 67180
McGregor, D.C., 1984, Academia Nacional de Ciencias, Misc. No. 69, p. 27, pl. 2, fig. 2.
Santa Rosa Formation, Lower Devonian, Laurani section, Chuquisaca Dept., northwest of Tarabuco, Cordillera Oriental, Bolivia.

Dictyotriletes sp.
Fig. specs. 15177, 15181, 15183-15185, 15361
McGregor, D.C. and Owens, B., 1966, Geol. Surv. Can., Paper 66-30, pl. 3, fig. 13; pl. 4, figs. 1, 2, 4, 6, 9; pl. 22, fig. 26.
Battery Point Formation, Lower Devonian, south side Gaspé Bay, shore section from Anse-à-Brillant to Bois Brulé Brook, and north side Gaspé Bay, shore cliffs between Gaspé Peninsula and Laurencelle Road (15181), eastern Gaspé Peninsula, Québec; Escarpment Member, Hay River Formation, Upper Devonian, left bank Hay River at base of Louise Falls, Northwest Territories (15361).
=*Dictyotriletes favosus*, McGregor, D.C., 1979, Palynology, vol. 3, pl. 1, fig. 1 (15181).
Richardson, J.B. and McGregor, D.C., 1986, Geol. Surv. Can., Bull. 364, pl. 6, figs. 1, 2 (15181).

Dictyotriletes sp.
Fig. specs. 27069, 27072, 27073, 27089
McGregor, D.C., Sanford, B.V. and Norris, A.W., 1970, Proc. Geol. Assoc. Can., vol. 22, pl. 1, figs. 11-13, 18; pl. 2, fig. 10.
Stooping River Formation, Lower Devonian and Kwataboahegan Formation, Middle Devonian (27089), 1027, 960 (27072, 27073) and 533 (27089) feet depths, ODM Jaab Lake No. 1 well, lat. 51°11.9′N, long. 82°57′W, Moose River Basin, Ontario.

Dictyotriletes sp.
Fig. spec. 29372
McGregor, D.C. and Uyeno, T.T., 1972, Geol. Surv. Can., Paper 71-13, pl. 1, fig. 4.
Stuart Bay Bay Formation, Lower Devonian, Young Inlet, northeastern Bathurst Island, District of Franklin.

Dictyotriletes sp.
Fig. spec. 15181
McGregor, D.C., 1973, Palaeontographica, Abt. B, vol. 142, no. 1-3, p. 44, pl. 6, figs. 1, 2.
Battery Point Formation, Lower Devonian, 1 mile east of Watering Creek, north shore Gaspé Bay, Québec.

Dictyotriletes sp.
Fig. specs. 61653, 61665, 61690
McGregor, D.C., 1979, Geol. Surv. Can., Paper 79-1B, pl. 22.1, figs. 5, 17; pl. 22.2, fig. 44.
Lower Devonian, Dvorce-Prokop Limestone, road-cut on west side of Vitava River below Barrandov, southern Prague, north of old qyarry ″U Kapliky″; Zclíchov Limestone, near Pekárek Mill west of village of Solopysky (61665); and Daleje Shale, roadside outcrop near Karlštejn, 1 km north of Budnany in the Hluboké Valley (61690), Czechoslovakia.

Dictyotriletes sp.
Fig. spec. 67785
McGregor, D.C., 1984, Academia Nacional de Ciencias, Misc. No. 69, pl. 4, fig. 1.
Santa Rosa Formation, Lower Devonian, Laurani section, Chuquisaca Dept., northwest of Tarabuco, Cordillera Oriental, Bolivia.

Dictyotriletes sp.
Fig. spec. 86081
McGregor, D.C. and McCutcheon, S.R., 1988, Can. J. Earth Sci., vol. 25, no. 9, p. 1353, pl. 2, figs. 5, 6.
Carrow Formation, Piskahegan Group, Late Devonian, depth 72.85 m, drill hole M-6, lat. 45°34′43″N, long. 66°35′27″W, southwestern New Brunswick.

?*Dictyotriletes* sp.
Fig. spec. 15356
McGregor, D.C. and Owens, B., 1966, Geol. Surv. Can., Paper 66-30, pl. 22, figs. 19, 20.
Escarpment Member, Hay River Formation, left bank Hay River at base of Louise Falls, Northwest Territories.

?*Dictyotriletes* sp.
Fig. spec. 32829
McGregor, D.C., 1974, Can. J. Earth Sci., vol. 11, no. 1, pl. 2, fig. 54.
Vendom Fiord Formation, Early Devonian, lat. 79°28′40″N, long. 80°02′W, 12 miles south of head of Cañon Fiord, central Ellesmere Island, District of Franklin.

?*Dictyotriletes* sp.
Fig. spec. 61673
McGregor, D.C., 1979, Geol. Surv. Can., Paper 79-1B, pl. 22.1, fig. 24.
Zlíchov Limestone, Lower Devonian, near Pekárek mill west of village of Solopysky, Czechoslovakia.

Dictyotriletes sp. A?, Keegan, 1977
Fig. spec. 86083
McGregor, D.C. and McCutcheon, S.R., 1988, Can. J. Earth Sci., vol. 25, no. 9, p. 1353, pl. 1, fig. 9.

Carrow Formation, Piskahegan Group, Late Devonian, depth 72.85 m, drill hole M-6, lat. 45°34′43″N, long. 66°35′27″W, southwestern New Brunswick.

?*Dictyotriletes* sp. B of Richardson and Lister, 1969
Fig. spec. 32795
McGregor, D.C., 1974, Can. J. Earth Sci., vol. 11, no. 1, pl. 1, fig. 11.
Unnamed Early Devonian, lat. 79°28′40″N, long. 80°02′W, 12 miles south of head of Cañon Fiord, central Ellesmere Island, District of Franklin.

Dictyotriletes spp. 1, 2
Fig. specs. 67178, 67181
McGregor, D.C., 1984, Academia Nacional de Ciencias, Misc. No. 69, p. 27, 28, pl. 2, fig. 5; pl. 1, fig. 5.
Santa Rosa Formation, Lower Devonian, Laurani section, Chuquisaca Dept., northwest of Tarabuco, Cordillera Oriental, Bolivia.

Diducites mucronatus (Kedo) Van Veen, 1981
Hypotype 86103
McGregor, D.C. and McCutcheon, S.R., 1988, Can. J. Earth Sci., vol. 25, no. 9, p. 1353, pl. 2, fig. 7.
Carrow Formation, Piskahegan Group, Late Devonian, depth 72.85 m, drill hole M-6, lat. 45°34′43″N, long. 66°35′27″W, southwestern New Brunswick.

Diducites pliacabilis Van Veen, 1981
Hypotypes 86104, 86105
McGregor, D.C. and McCutcheon, S.R., 1988, Can. J. Earth Sci., vol. 25, no. 9, p. 1353, pl. 2, figs. 8, 9.
Carrow Formation, Piskahegan Group, Late Devonian, depth 72.85 m, drill hole M-6, lat. 45°34′43″N, long. 66°35′27″W, southwestern New Brunswick.

Diducites versabilis (Kedo) Van Veen, 1981
Hypotype 86102
McGregor, D.C. and McCutcheon, S.R., 1988, Can. J. Earth Sci., vol. 25, no. 9, p. 1353, pl. 2, fig. 6.
Carrow Formation, Piskahegan Group, Late Devonian, depth 72.85 m, drill hole M-6, lat. 45°34′43″N, long. 66°35′27″W, southwestern New Brunswick.

Discernisporites micromanifestus (Hacquebard) Sabry and Neves
Hypotype 73311
Richardson, J.B. and McGregor, D.C., 1986, Geol. Surv. Can., Bull. 364, pl. 20, fig. 7.
Devonian, west side of St George's Bay, north of McIsaac Point, lat. 45°44′24.4″N, long. 61°58′28.46″W, Antigonish County, Nova Scotia.

Discernisporites micromanifestus (Hacquebard) Sabry and Neves, 1971, sensu Clayton et al. 1977
Hypotypes 86099, 86100
McGregor, D.C. and McCutcheon, S.R., 1988, Can. J. Earth Sci., vol. 25, no. 9, p. 1353, pl. 2, figs. 3, 4.
Carrow Formation, Piskahegan Group, Late Devonian, depth 72.85 m, drill hole M-6, lat. 45°34′43″N, long. 66°35′27″W, southwestern New Brunswick.

Elenisporis biformis (Arkhangelskaya) Arkhangelskaya 1985
Hypotype 96771
Arkhangelskaya, A.D., McGregor, D.C. and Richardson, J.B., 1990, Geol. Surv. Can., Paper 90-1D, pl. 1, fig. a.
Strathcona Fiord Formation, Middle Devonian east of Bird Fiord, Ellesmere Island, District of Franklin.

Emphanisporites annulatus McGregor, 1961
Hypotype 15270
McGregor, D.C. and Owens, B., 1966, Geol. Surv. Can., Paper 66-30, pl. 6, fig. 6.
Battery Point Formation, Lower Devonian, north side Southwest Arm, halfway between Gaspé Bridge and Cape O'Hara, Gaspé, Québec.

Emphanisporites annulatus McGregor
Hypotype 27084
McGregor, D.C., Sanford, B.V. and Norris, A.W., 1970, Proc. Geol. Assoc. Can., vol. 22, pl. 2, fig. 3.
Stooping River Formation, Lower Devonian, 617 feet depth, ODM Jaab Lake No. 1 well, lat. 51°11.9′N, long. 82°57′W, Moose River Basin , Ontario.

Emphanisporites annulatus McGregor
Hypotype 32002
McGregor, D.C., 1973, Palaeontographica, Abt. B, vol. 142, no. 1-3, p. 45, pl. 6, figs. 3, 4.
Battery Point Formation, Lower Devonian, 1 mile east of Douglastown P.O., south shore Gaspé Bay, Québec.

Emphanisporites annulatus McGregor
Hypotype 41759
McGregor, D.C. and Camfield, M., 1976, Geol. Surv. Can., Bull. 263, p. 21, pl. 5, fig. 6.
Stooping River Formation, Lower Devonian, 571-573 feet depth, ODM Jaab Lake No. 1 well, lat. 51°11′54″N, long. 82°56′W, Ontario.

Emphanisporites annulatus McGregor, 1961
Hypotype 67182
McGregor, D.C., 1984, Academia Nacional de Ciencias, Misc. No. 69, p. 28, pl. 3, fig. 1.
Icla Formation, Lower or Middle Devonian, Laurani section, Chuquisaca Dept., northwest of Tarabuco, Cordillera Oriental, Bolivia.

Emphanisporites annulatus McGregor 1961
Hypotype 96775
Arkhangelskaya, A.D., McGregor, D.C. and Richardson, J.B., 1990, Geol. Surv. Can., Paper 90-1D, pl. 1, fig. e.
Strathcona Fiord Formation, Middle Devonian, east of Bird Fiord, Ellesmere Island, District of Franklin.

Emphanisporites annulatus? McGregor
Hypotype 66292
McGregor, D.C. and Camfield, M., 1982, Geol. Surv. Can., Bull. 348, p. 39, pl. 8, fig. 8.
Cape De Bray Formation, Middle Devonian, stream section 16.2 km east of east arm of Weatherall Bay, northeastern Melville Island, District of Franklin.

Emphanisporites decoratus Allen
Hypotype 27068
McGregor, D.C., Sanford, B.V. and Norris, A.W., 1970, Proc. Geol. Assoc. Can., vol. 22, pl. 1, fig. 10.
Kenogami River Formation, Lower Devonian, 1072 feet depth, ODM Jaab Lake No. 1 well, lat. 51°11.9′N, long. 82°57′W, Moose River Basin, Ontario.

Emphanisporites decoratus Allen
Hypotype 27076
McGregor, D.C., Sanford, B.V. and Norris, A.W., 1970, Proc. Geol. Assoc. Can., vol. 22, pl. 1, fig. 20.
Richardson, J.B. and McGregor, D.C., 1986, Geol. Surv. Can., Bull. 364, pl. 5, figs. 12, 13.
Stooping River Formation, Lower Devonian, 909 feet depth, ODM Jaab Lake No. 1 well, lat. 51°11.9′N, long. 82°57′W, Moose River Basin , Ontario.

Emphanisporites decoratus Allen
Hypotype 41723
McGregor, D.C. and Camfield, M., 1976, Geol. Surv. Can., Bull. 263, p. 21, pl. 2, fig. 10-12.
Stooping River Formation, Lower Devonian, 960-962 feet depth, ODM Jaab Lake No. 1 well, lat. 51°11′54″N, long. 82°56′W, Ontario.

Emphanisporites decoratus Allen
Hypotype 49816
McGregor, D.C., 1979, Palynology, vol. 3, pl. 1, fig. 9.
Stooping River Formation, Lower Devonian, 917 feet, Campbell Lake D.D.H. No. 1, lat. 50°18′30″N, long. 82°09′30″W, Ontario.

Emphanisporites decoratus Allen
Hypotype 61657
McGregor, D.C., 1979, Geol. Surv. Can., Paper 79-1B, pl. 22.1, fig. 9.
Dvorce-Prokop Limestone, Lower Devonian, road-cut on west side of Vetava River below Barrandov, southern Prague, north of old quarry "U Kapliky", Czechoslovakia.

Emphanisporites cf. *E. decoratus* Allen
Hypotype 27079
McGregor, D.C., Sanford, B.V. and Norris, A.W., 1970, Proc. Geol. Assoc. Can., vol. 22, pl. 1, fig. 25.
Stooping River Formation, Lower Devonian, 909 feet depth, ODM Jaab Lake No. 1 well, lat. 51°11.9′N, long. 82°57′W, Moose River Basin, Ontario.

Emphanisporites cf. *E. decoratus* Allen
Hypotype 32796
McGregor, D.C., 1974, Can. J. Earth Sci., vol. 11, no. 1, pl. 11, figs. 12, 16.
Unnamed Early Devonian, lat. 79°28′40″N, long. 80°02′W, 12 miles south of head of Cañon Fiord, central Ellesmere Island, District of Franklin.

Emphanisporites erraticus (Eisenack) McGregor, 1961
Hypotype 15195
McGregor, D.C. and Owens, B., 1966, Geol. Surv. Can., Paper 66-30, pl. 4, figs. 16, 17.
Battery Point Formation, Lower Devonian, south side Gaspé Bay, shore section from Anse-à-Brillant to Bois Brulé Brook, eastern Gaspé Peninsula, Québec.

Emphanisporites erraticus (Eisenack) McGregor
Hypotype 32003
McGregor, D.C., 1973, Palaeontographica, Abt. B, vol. 142, no. 1-3, p. 45, pl. 6, figs. 5, 6.
Battery Point Formation, Lower Devonian, 1 mile east of Douglastown P.O., south shore Gaspé Bay, Québec.

Emphanisporites foveolatus Schultz
Hypotype 61715
McGregor, D.C., 1979, Geol. Surv. Can., Paper 79-1B, pl. 22.2, fig. 70.
Daleje Shale, Lower Devonian, upper part of southeast slope of Císařská rokle, southeast of Srbsko, Czechoslovakia.

Emphanisporites macgregori Schultz
Hypotype 27085
McGregor, D.C., Sanford, B.V. and Norris, A.W., 1970, Proc. Geol. Assoc. Can., vol. 22, pl. 2. fig. 4.
Stooping River Formation, Lower Devonian, 617 feet depth, ODM Jaab Lake No. 1 well, lat. 51°11.9′N, long. 82°57′W, Moose River Basin, Ontario.

Emphanisporites micrornatus Richardson and Lister
Hypotype 27061
McGregor, D.C., Sanford, B.V. and Norris, A.W., 1970, Proc. Geol. Assoc. Can., vol. 22, pl. 1, fig. 2.
Kenogami River Formation, Lower Devonian, 1153 feet depth, ODM Jaab Lake No. 1 well, lat. 51°11.9′N, long. 82°57′W, Moose River Basin, Ontario.

Emphanisporites micrornatus Richardson and Lister
Hypotype 41714
McGregor, D.C. and Camfield, M., 1976, Geol. Surv. Can., Bull. 263, p. 22, pl. 1, figs. 17, 18.
Stooping River Formation, Lower Devonian, 725-727 feet depth, ODM Jaab Lake No. 1 well, lat. 51°11′54″N, long. 82°56′W, Ontario.

Emphanisporites micrornatus Richardson and Lister 1969
Hypotype 48174
McGregor, D.C., 1977, Palaeontographica, Abt. B, vol. 163, p. 131, pl. 2, figs. 1-3.
McGregor, D.C., 1979, Palynology, vol. 3, pl. 1, fig. 4.
York River Formation, Lower Devonian, south shore Gaspé Bay, north of Anse-à-Brillant River, Gaspé, Québec.

Emphanisporites micrornatus? Richardson and Lister
Hypotype 61654
McGregor, D.C., 1979, Geol. Surv. Can., Paper 79-1B, pl. 22.1, fig. 6.
Dvorce-Prokop Limestone, Lower Devonian, road-cut on west side of Vitava River below Barrandov, southern Prague, north of old quarry "U Kapliky", Czechoslovakia.

Emphanisporites cf. *E. micrornatus* Richardson and Lister
Hypotype 27062
McGregor, D.C., Sanford, B.V. and Norris, A.W., 1970, Proc. Geol. Assoc. Can., vol. 22, pl. 1, fig. 3.
Kenogami River Formation, Lower Devonian, 1153 feet depth, ODM Jaab Lake No. 11 well, lat. 51°11.9′N, long. 82°57′W, Moose River Basin, Ontario.

Emphanisporites cf.*micrornatus* of Richardson and Lister
Hypotype 27078
McGregor, D.C., Sanford, B.V. and Norris, A.W., 1970, Proc. Geol. Assoc. Can., vol. 22, pl. 1, fig. 23.
Stooping River Formation, Lower Devonian, 909 feet depth, ODM Jaab Lake No. 1 well, lat. 51°11.9′N, long. 82°57′W, Moose River Basin, Ontario.

Emphanisporites minutus Allen
Hypotype 32794
McGregor, D.C., 1974, Can. J. Earth Sci., vol, 11, no. 1, pl. 1, figs. 9, 14.
Unnamed Early Devonian, lat. 79°28′40″N, long. 80°02′W, 12 miles south of head of Cañon Fiord, central Ellesmere Island, District of Franklin.

Emphanisporites minutus Allen, 1965
Hypotype 67184
McGregor, D.C., 1984, Academia Nacional de Ciencias, Misc. No. 69, p. 29, pl. 2, fig. 1.
Santa Rosa Formation, Lower Devonian, Laurani section, Chuquisaca Dept., northwest of Tarabuco, Cordillera Oriental, Bolivia.

Emphanisporites neglectus Vigran, 1964
Hypotype 15190
McGregor, D.C. and Owens, B., 1966, Geol. Surv. Can., Paper 66-30, pl. 4, fig. 11.
Battery Point Formation, Lower Devonian, south side Gaspé Bay, shore section from Anse-à-Brillant to Bois Brulé Brook, eastern Gaspé Peninsula, Québec.

Emphanisporites ?neglectus Vigran
Hypotype 32004
McGregor, D.C., 1973, Palaeontographica, Abt. B, vol. 142, no. 1-3, p. 46, pl. 6, figs. 7, 8.
Battery Point Formation, Lower Devonian, 1050 yards northwest of first small brook south of Rivière Anse-à-Brillant, south shore Gspé Bay, Québec.

Emphanisporites novellus McGregor and Camfield
Holotype 41745
McGregor, D.C. and Camfield, M., 1976, Geol. Surv. Can., Bull. 263, p. 22, pl. 4, figs. 1-3.
Stooping River Formation, Lower Devonian, 909-911 feet depth, ODM Jaab Lake No. 1 well, lat. 51°11′54″N, long. 82°56′W, Ontario.

Emphanisporites novellus McGregor and Camfield, 1976
Hypotype 67183
McGregor, D.C., 1984, Academia Nacional de Ciencias, Misc. No. 69, p. 29, pl. 1, fig. 9.
Santa Rosa Formation, Lower Devonian, Laurani section, Chuquisaca Dept., northwest of Tarabuco, Cordillera Oriental, Bolivia.

Emphanisporites robustus McGregor, 1961
Hypotype 15269
McGregor, D.C. and Owens, B., 1966, Geol. Surv. Can., Paper 66-30, pl. 6, fig. 5.
Battery Point Formation, Lower Devonian, south side Gaspé Bay, shore section from Bois Brulé Brook to Pointe Jaune, eastern Gaspé Peninsula, Québec.

Emphanisporites rotatus McGregor, 1961
Hypotype 15191
McGregor, D.C. and Owens, B., 1966, Geol. Surv. Can., Paper 66-30, pl. 4, fig. 12.
Battery Point Formation, Lower Devonian, south side Gaspé Bay, shore section from Anse-à-Brillant to Bois Brulé Brook, eastern Gaspé Peninsula, Québec.

Emphanisporites rotatus McGregor
Hypotype 27075
McGregor, D.C., Sanford, B.V. and Norris, A.W., 1970, Proc. Geol. Assoc. Can., vol. 22, pl. 1, fig. 19.
Stooping River Formation, Lower Devonian, 909 feet depth, ODM Jaab Lake No. 1 well, lat. 51°11.9′N, long. 82°57′W, Moose River Basin, Ontario.

Emphanisporites rotatus McGregor
Lectotype 32005; hypotypes 32006, 32007
McGregor, D.C., 1973, Palaeontographica, Abt. B, vol. 142, no. 1-3, p. 46, pl. 6, figs. 9-13.
Battery Point Formation, Lower Devonian, near Douglastown and 45 yards east of first point east of Rivière de l'Anse-à-Brillant, south shore Gaspé Bay, Québec.

Emphanisporties rotatus McGregor
Hypotype 32828
McGregor, D.C., 1974, Can. J. Earth Sci., vol. 11, no. 1, pl. 2, fig. 53.
Vendom Fiord Formation, Early Devonian, lat. 79°28′40″N, long. 80°02′W, 12 miles south of head of Cañon Fiord, central Ellesmere Island, District of Franklin.

Emphanisporites rotatus McGregor
Hypotype 41739
McGregor, D.C. and Camfield, M., 1976, Geol. Surv. Can., Bull. 263, p. 22, pl. 3, fig. 16.
Stooping River Formation, Lower Devonian, 599-600 feet depth, ODM Jaab Lake No. 1 well, lat. 51°11′54″N, long. 82°56′W, Ontario.

Emphanisporites rotatus McGregor
Hypotypes 61655, 61656, 61672, 61710, 61736
McGregor, D.C., 1979, Geol. Surv. Can., Paper 79-1B, pl. 22.1, figs. 7, 8, 23; pl. 22.2, fig. 65; pl. 22.3, fig. 91.
Lower Devonian, Dvorce-Prokop Limestone, road-cut on west side of Vitava River below Barrandov, southern Prague, north of old quarry "U Kapliky"; Zlíchov Limestone, near Pekárek mill west of village of Solopysky (61672); Dalje Shale, upper part of southeast slope of Císařská rokle, southeast of Srbsko (61710);

Trebotov Limestone, Lower Devonian, northwest side of railway, north side of Prokopské Valley at Hluboepy near Prague (61736), Czechoslovakia.

Emphanisporites rotatus McGregor
Hypotype 66291
McGregor, D.C. and Camfield, M., 1982, Geol. Surv. Can., Bull. 348, p. 39, pl. 8, fig. 7.
Cape De Bray Formation, Middle Devonian, stream section 16.2 km east of east arm of Weatherall Bay, northeastern Melville Island, District of Franklin.

Emphanisporites rotatus McGregor, 1961
Hypotype 67185
McGregor, D.C., 1984, Academia Nacional de Ciencias, Misc. No. 69, p. 30, pl. 1, fig. 18.
Huamampampa Formation, Lower Devonian, Laurani section, Chuquisaca Dept., northwest of Tarabuco, Cordillera Oriental, Bolivia.

Emphanisporites rotatus McGregor, *robustus* type
Hypotypes 61692, 61735
McGregor, D.C., 1979, Geol. Surv. Can., Paper 79-1B, pl. 22.2, fig. 46; pl. 22.3, fig. 90.
Lower Devonian, Daleje Shale, roadside outcrop near Karlštejn, 1 km north of Budnany in the Hlubocké Valley; Trebotov Limestone, northwest side of railway, north side of Prokopské Valley at Hluboepy near Prague, Czechoslovakia.

Emphanisporites cf. *E. rotatus* McGregor, 1961
Fig. specs. 15105, 15126, 15130, 15453
McGregor, D.C. and Owens, B., 1966, Geol. Surv. Can., Paper 66-30, pl. 1, figs. 3, 27, 31; pl. 28, fig. 4.
Lower Devonian, Forillon Member, Cape Bon Ami Formation, north side Forillon Peninsula, Cape Rosier Cove (15105), and Shiphead Member, Grand Grève Formation, Cape Gaspé shore cliff section below lighthouse, eastern Gaspé Peninsula, Québec; Kettle Point Formation, Upper Devonian, depth 295-299 feet, Courtright borehole, Courtright village, Lambton County, Ontario (15453).

Emphanisporites schultzii McGregor
Hypotype 32008
McGregor, D.C., 1973, Palaeontographica, Abt. B, vol. 142, no. 1-3, p. 48, pl. 6, fig. 14.
Richardson, J.B. and McGregor, D.C., 1986, Geol. Surv. Can., Bull. 364, pl. 7, fig. 9.
Battery Point Formation, Lower Devonian, 480 yards northwest of fist small brook south of Rivière-à-Brillant, south shore Gaspé Bay, Québec.

Emphanisporites schultzii McGregor
Hypotype 41760
McGregor, D.C. and Camfield, M., 1976, Geol. Surv. Can., Bull. 263, p. 23, pl. 5, fig. 70.
Stooping River Formation, Lower Devonian, 599-600 feet depth, ODM Jaab Lake No. 1 well, lat. 51°11′54″N, long. 82°56′W, Ontario.

Emphanisporites schultzii McGregor 1973
Hypotype 96776
Arkhangelskaya, A.D., McGregor, D.C. and Richardson, J.B., 1990, Geol. Surv. Can., Paper 90-1D, pl. 1, fig. f.
Strathcona Fiord Formation, Middle Devonian east of Bird Fiord, Ellesmere Island, District of Franklin.

Emphanisporites sp.
Fig. spec. 15065
McGregor, D.C. and Owens, B., 1966, Geol. Surv. Can., Paper 66-30, pl. 4, fig. 10.
Battery Point Formation, Lower Devonian, south side Gaspé Bay, shore section from Anse-à-Brillant to Bois Brulé Brook, eastern Gaspé Peninsula, Québec.

Emphanisporites sp.
Fig. spec. 15065 [32008]
McGregor, D.C., 1970, Geol. Surv. Can., Econ. Geol. Rept. No. 1, 5th edit., pl. 31, fig. 2.
Battery Point Formation, Lower Devonian, 480 yards northwest of first small brook south of Rivière de l'Anse-à-Brillant, south shore Gaspé Bay, Québec.
=*Emphanisporites schultzii*, McGregor, D.C., 1973, Palaeontographica, Abt. B, vol. 142, no. 1-3, p. 48, pl. 6, fig. 14 (32008).
McGregor, D.C., 1979, Palynology, vol. 3, pl. 1, fig. 13.

Emphanisporites sp.
Fig. spec. 27080
McGregor, D.C., Sanford, B.V. and Norris, A.W., 1970, Proc. Geol. Assoc. Can., vol. 22, pl. 1, figs. 26, 27.
Stooping River Formation, Lower Devonian, 909 feet depth, ODM Jaab Lake No. 1 well, lat. 51°11.9′N, long. 82°57′W, Moose River Basin, Ontario.

Emphanisporites sp.
Fig. spec. 32798
McGregor, D.C., 1974, Can. J. Earth Sci., vol. 11, no. 1, pl. 1, fig. 15.
Unnamed Early Devonian, lat. 79°28′40″N, long. 80°02′W, 12 miles south of head of Cañon Fiord, central Ellesmere Island, District of Franklin.

?*Emphanisporites* sp.
Fig. spec. 15125
McGregor, D.C. and Owens, B., 1966, Geol. Surv. Can., Paper 66-30, pl. 1, fig. 26.
McGregor, D.C., 1967, Review Palaeobotany and Palynology, I, pl. 1, fig. 2.
Shiphead Member, Grand Grève Formation, Lower Devonian, Cape Gaspé, shore cliff below lighthouse, eastern Gaspé Peninsula, Québec.

Emphanisporites sp. B. Richardson and Lister 1969
Hypotype 41716
McGregor, D.C. and Camfield, M., 1976, Geol. Surv. Can., Bull. 263, p. 23, pl. 1, fig. 25, 28.
Stooping River Formation, Lower Devonian, 960-962 feet depth, ODM Jaab Lake No. 1 well, lat. 51°11′54″N, long. 82°56′W, Ontario.

Endoculevspora gradzinskii Turnau, 1975, variant B, Van der Zwan, 1980
Hypotypes 86111, 86112
McGregor, D.C. and McCutcheon, S.R., 1988, Can. J. Earth Sci., vol. 25, no. 9, p. 1353, pl. 2, figs. 17, 18.
Carrow Formation, Piskahegan Group, Late Devonian, depth 72.85 m, drill hole M-6, lat. 45°34′43″N, long. 66°35′27″W, southwestern New Brunswick.

Endoculeospora (?*Archaeozonotriletes subnotatus* Chibrikova)
Fig. spec. 13603
Kerr, J.W., McGregor, D.C. and McLaren, D.J., 1965, Bull. Can. Petrol. Geol., vol. 13, no. 3, pl. 3, fig. 21.
Griper Bay Formation, Upper Devonian, stream valley west of Sargeant Point, 2 miles from shore, northeastern Bathurst Island, District of Franklin.

Endosporites micromanifestus Hacquebard
McGregor, D.C. and McCutcheon, S.R., 1988, Can. J. Earth Sci., vol. 25, no. 9, p. 1353, pl. 3, fig. 14 (holotype 4284).

Endosporites sp.
Fig. spec. 13213
McGregor, D.C., 1964, Geol. Surv. Can., Bull. 109, p. 19, pl. 2, fig. 25.
Ghost River Formation (sensu stricto), Middle or Upper Devonian, Phantom Crag, 4 miles northeast of eastern end of Lake Minnewanka, Alberta.

Enigmophytospora simplex Vigran
Hypotypes 15164, 15167
McGregor, D.C., 1973, Palaeontographica, Abt. B, vol. 142, no. 1-3, p. 16, pl. 1, figs. 8, 9.
Battery Point Formation, Lower Devonian, 250 yards north of and 600 yards northwest of first small brook south of Rivière de l'Anse-à-Brillant, south shore Gaspé Bay, Québec.

Enigmophytospora simplex Vigran 1964
Hypotype 43658
Chi, B.I. and Hills, L.V., 1977, Bull. Can. Petrol. Geol., vol. 24, no. 4, p. 691, pl. 1, fig. 1.
Weatherall Formation, Middle Devonian, Cape Terrace, west side Melville Island, District of Franklin.

?*Foveosporites insculptus* Playford, 1962
Hypotype 15353
McGregor, D.C. and Owens, B., 1966, Geol. Surv. Can., Paper 66-30, pl. 22, fig. 15.
Escarpment Member, Hay River Formation, left bank Hay River at base of Louise Falls, Northwest Territories.

Foveosporites cf. *F. insculptus* Playford, 1962
Hypotype 15312
McGregor, D.C. and Owens, B., 1966, Geol. Surv. Can., Paper 66-30, pl. 15, fig. 3.
Griper Bay Formation, Upper Devonian, south limb of Robertson Point anticline, 8 miles northeast of Beverley Inlet, Melville Island, District of Franklin.

Foveosporites pertusus Vigran
Hypotype 13619
McGregor, D.C., 1967, Review Palaeobotany and Palynology, I, pl. 1, figs. 18, 19.
Griper Bay Formation, Melville Island Group, Upper Devonian, Helena Island, District of Franklin.

?*Foveosporites pertusus* Vigran, 1964
Hypotype 15429
McGregor, D.C. and Owens, B., 1966, Geol. Surv. Can., Paper 66-30, pl. 27, fig. 1.
Griper Bay Formation, Upper Devonian, stream valley west of Sargeant Point, 2 miles from shore, northeastern Bathurst Island, District of Franklin.

Geminospora antaxios (Chibrikova) Owens
Hypotype 30699
Owens, B. and Richardson, J.B., 1972, 7th Internat. Congr. Carboniferous Strat. Geol. Compte Rendu, vol. 1, pl. 2, fig. 2.
Griper Bay Formation, Upper Devonian, Melville Island, District of Franklin.

Geminospora? *bislimbata* (Chibrikova) n.comb.
Hypotypes 66294, 66295
McGregor, D.C. and Camfield, M., 1982, Geol. Surv. Can., Bull. 348, p. 40, pl. 8, figs. 12, 13.
Richardson, J.B. and McGregor, D.C., 1986, Geol. Surv. Can., Bull. 364, pl. 15, fig. 2 (66295).
Hecla Bay Formation, Middle Devonian, stream section 16.2 km east of east arm of Weatherall Bay, northeastern Melville Island, District of Franklin.

?*Geminospora lemurata* Balme, 1962
Hypotype 15383
McGregor, D.C. and Owens, B., 1966, Geol. Surv. Can., Paper 66-30, pl. 23, fig. 21.
Escarpment Member, Hay River Formation, Upper Devonian, left bank Hay River at base of Louise Falls, Northwest Territories.

Geminospora lemurata (Playford)
Hypotypes 90583-90585
Chilaley, S. and McGregor, D.C, 1988, Palaeontographica, Abt. B, vol. 210, nos. 4-6, p. 137, pl. 11, figs. 7-9.
Gneudna Formation, Upper Devonian, Carnarvon Basin, Western Australia.

Geminospora micromanifesta (Naumova) n.comb.var. *minor* Naumova
Hypotypes 66296-66298
McGregor, D.C. and Camfield, M., 1982, Geol. Surv. Can., Bull. 348, p. 40, pl. 8, figs. 14, 15, 19-22.
Hecla Bay Formation, Middle Devonian, stream section 16.2 km east of east arm of Weatherall Bay, northeastern Melville Island, District of Franklin.
=*Geminospora lemurata*, Richardson, J.B. and McGregor, D.C., 1986, Geol. Surv. Can., Bull. 364, pl. 14, figs. 8, 9 (66296).

Geminospora micropaxilla (Owens) n.comb.
Hypotypes 66299-66301
McGregor, D.C. and Camfield, M., 1982, Geol. Surv. Can., Bull. 348, p. 41, pl. 8, figs. 16-18; pl. 9, figs. 1-3.
Hecla Bay Formation, Middle Devonian, stream section 16.2 km east of east arm of Weatherall Bay, northeastern Melville Island, District of Franklin.

Geminospora punctata Owens
Hypotype 30700
Owens, B. and Richardson, J.B., 1972, 7th Internat. Congr. Carboniferous Strat. Geol. Compte Rendu, vol. 1, pl. 2, fig. 3.
Griper Bay Formation, Upper Devonian, Prince Patrick Island, District of Franklin.

Geminospora spinosa Allen
Hypotype 32015
McGregor, D.C., 1973, Palaeontographica, Abt. B, vol. 142, no. 1-3, p. 52, pl. 6, figs. 32, 33.
Battery Point Formation, Lower Devonian, 420 yards north of Bois Brulé Brook, south shore Gaspé Bay, Québec.

?*Geminospora* cf. *G. spinosa* Allen
Hypotype 32831
McGregor, D.C., 1974, Can. J. Earth Sci., vol. 11, no. 1, pl. 2, fig. 56.
Vendom Fiord Formation, Early Devonian, lat. 79°28′40″N, long. 80°02′W, 12 miles south of head of Cañon Fiord, central Ellesmere Island, District of Franklin.

Geminospora ?svalbardiae Allen
Hypotype 29402
McGregor, D.C. and Uyeno, T.T., 1972, Geol. Surv. Can., Paper 71-13, pl. 3, fig. 3.
Griper Bay Formation, Upper Devonian, Beverley Inlet, southeastern Melville Island, District of Franklin.

?*Geminospora svalbardiae* (Vigran) Allen
Hypotype 32830
McGregor, D.C., 1974, Can. J. Earth Sci., vol. 11, no. 1, pl. 2, fig. 55.
Vendom Fiord Formation, Early Devonian, lat. 79°28′40″N, long. 80°02′W, 12 miles south of head of Cañon Fiord, central Ellesmere Island, District of Franklin.

cf. *Geminospora svalbardiae* (Vigran) Allen, 1965
Hypotype 15262
McGregor, D.C. and Owens, B., 1966, Geol. Surv. Can., Paper 66-30, pl. 5, fig. 16.
Battery Point Formation, Lower Devonian, south side Gaspé Bay, shore section from Bois Brulé Brook to Pointe Jaune, eastern Gaspé Peninsula, Québec.

Geminospora tuberculata (Kedo) Allen var. *micrornata* McGregor and Camfield
Holotype 66302
McGregor, D.C. and Camfield, M., 1982, Geol. Surv. Can., Bull. 348, p. 41, pl. 9, figs. 4, 5.
Hecla Bay Formation, Middle Devonian, stream section 16.2 km east of east arm of Weatherall Bay, northeastern Melville Island, District of Franklin.

Geminospora tuberculata (Kedo) Allen var.*tuberculata*
Hypotypes 66303-66305
McGregor, D.C. and Camfield, M., 1982, Geol. Surv. Can., Bull. 348, p. 41, pl. 9, figs. 6, 7, 10-13.
Hecla Bay Formation, Middle Devonian, stream section 16.2 km east of east arm of Weatherall Bay, northeastern Melville Island, District of Franklin.

Geminospora venusta? (Naumova) n.comb.
Hypotype 66307
McGregor, D.C. and Camfield, M., 1982, Geol. Surv. Can., Bull. 348, p. 42, pl. 9, figs. 14, 15.
Hecla Bay Formation, Middle Devonian, stream section 16.2 km east of east arm of Weatherall Bay, northeastern Melville Island, District of Franklin.

Geminospora verrucosa Owens
Hypotype 66308
McGregor, D.C. and Camfield, M., 1982, Geol. Surv. Can., Bull. 348, p. 42, pl. 9, fig. 16.
Weatherall Formation, Middle Devonian, stream section 16.2 km east of east arm of Weatherall Bay, northeastern Melville Island, District of Franklin.

Geminospora sp.
Fig. spec. 66306
McGregor, D.C. and Camfield, M., 1982, Geol. Surv. Can., Bull. 348, p. 43, pl. 9, figs. 8, 9.
Hecla Bay Formation, Middle Devonian, stream section 16.2 km east of east arm of Weatherall Bay, northeastern Melville Island, District of Franklin.

?*Geminospora* sp.
Fig. specs. 15315, 15376, 15382, 15386
McGregor, D.C. and Owens, B., 1966, Geol. Surv. Can., Paper 66-30, pl. 15, figs. 7, 10; pl. 23, figs. 15, 19; pl. 24, fig. 1.
Upper Devonian, Griper Bay Formation, south limb of Robertson Point anticline, 8 miles northeast of Beverley Inlet, Melville Island, District of Franklin (15315); Escarpment Member, Hay River Formation, left bank Hay River at base of Louise Falls, Northwest Territories.

Geminospora n.sp. 1, 2
Fig. specs. 29401, 29403
McGregor, D.C. and Uyeno, T.T., 1972, Geol. Surv. Can., Paper 71-13, pl. 3, figs. 2, 4.
Griper Bay Formation, Upper Devonian, Beverley Inlet, southeastern Melville Island, District of Franklin.

Grandispora cassideus (Owens)
Hypotypes 43754-43757
Chi, B.I. and Hills, L.V., 1977, Bull. Can. Petrol. Geol., vol. 24, no. 4, p. 753, pl. 14, figs.7-10.
Weatherall Formation, Middle Devonian, Mould Bay, Prince Patrick Island, District of Franklin.

Gneudnospora divellomedium (Chibrokova) Balme 1988
Hypotype 96777
Arkhangelskaya, A.D., McGregor, D.C. and Richardson, J.B., 1990, Geol. Surv. Can., Paper 90-1D, pl. 1, fig. g.
Strathcona Fiord Formation, Middle Devonian east of Bird Fiord, Ellesmere Island, District of Franklin.

Grandispora biornata (Lanninger) McGregor
Hypotype 61728
McGregor, D.C., 1979, Geol. Surv. Can., Paper 79-1B, pl. 22.3, fig. 83.
Daleje Shale, Lower Devonian, upper part of southeast slope of Císařská rokle, southeast of Srbsko, Czechoslovakia.

Grandispora douglastownense McGregor
Holotype 32034; hypotypes 32032, 32033
McGregor, D.C., 1973, Palaeontographica, Abt. B, vol. 142, no. 1-3, p. 62, pl. 8, figs. 8, 9, 12-14.
Battery Point Formation, Middle Devonian, 500 yards southwest of Battery Point below 'Battery Park Hotel', north shore southwest arm of Gaspé Bay, Québec.

Grandispora douglastownense McGregor
Hypotype 41767
McGregor, D.C. and Camfield, M., 1978, Geol. Surv. Can., Bull. 263, p. 23, pl. 5, fig. 13.
Sextant Formation, Lower Devonian, 158.3-158.5 feet, Argor ETA Kiasko River No. 2 well, lat. 50°41′30″N, long. 80°48′W, Ontario.

Grandispora douglastownense McGregor 1973
Hypotype 49819
McGregor, D.C., 1979, Palynology, vol. 3, pl. 1, fig. 16.
Battery Point Devonian, Devonian, about 800 yards south of first outcrop on beach south of Douglastown, 200 feet north of old wharf by low wooded point, Gaspé, Québec.

Grandispora douglastownense McGregor
Hypotype 61743
McGregor, D.C., 1979, Geol. Surv. Can., Paper 79-1B, pl. 22.3, fig. 98.
Trebotov Limestone, Lower Devonian, northwest side of railway, north side of Prokopské Valley near Prague, Czechoslovakia.

Grandispora douglastownense McGregor
Hypotype 66309
McGregor, D.C. and Camfield, M., 1982, Geol. Surv. Can., Bull. 348, p. 43, pl. 10, fig. 1.
Weatherall Formation, Middle Devonian, stream section 16.2 km east of east arm of Weatherall Bay, northeastern Melville Island, District of Franklin.

Grandispora echinata Hacquebard
McGregor, D.C. and McCutcheon, S.R., 1988, Can. J. Earth Sci., vol. 25, no. 9, p. 1353, pl. 3, fig. 15 (holotype 4285).

Grandispora echinata Hacquebard
Hypotype 25496
McGregor, D.C., 1970, Colloque Stratig. Carbonifère, Congr. Colloques Univ. Liège, vol. 55, pl. 23, fig. 19.
Richardson, J.B. and McGregor, D.C., 1986, Geol. Surv. Can., Bull. 364, pl. 20, fig. 6.
Imperial Formation, Upper Devonian, lat. 66°27.5′N, long. 135°22′W, Yukon.

cf. *Grandispora echinata* Hacquebard, 1957
Hypotype 86097
McGregor, D.C. and McCutcheon, S.R., 1988, Can. J. Earth Sci., vol. 25, no. 9, p. 1353, pl. 2, fig. 1.
Carrow Formation, Piskahegan Group, Late Devonian, depth 72.85 m, drill hole M-6, lat. 45°34′43″N, long. 66°35′27″W, southwestern New Brunswick.

Grandispora eximia (Allen) n.comb.
Hypotypes 66310-66312
McGregor, D.C. and Camfield, M., 1982, Geol. Surv. Can., Bull. 348, p. 44, pl. 10, figs. 2, 6, 7.
Richardson, J.B. and McGregor, D.C., 1986, Geol. Surv. Can., Bull. 364, pl. 9, fig. 5 (66310).
Hecla Bay Formation, Middle Devonian, stream section 16.2 km east of east arm of Weatherall Bay, northeastern Melville Island, District of Franklin.

Grandispora sp. cf. *G. eximia* (Allen) n.comb.
Fig. specs. 66313, 66314
McGregor, D.C. and Camfield, M., 1982, Geol. Surv. Can., Bull. 348, p. 44, pl. 10, figs. 3, 4.
Hecla Bay Formation, Middle Devonian, stream section 16.2 km east of east arm of Weatherall Bay, northeastern Melville Island, District of Franklin.

Grandispora incognita (Kedo)
Hypotype 41773
McGregor, D.C. and Camfield, M., 1976, Geol. Surv. Can., Bull. 263, p. 23, pl. 6, figs. 9, 10.
Williams Island Formation, Middle Devonian, 169.5-171.5 feet depth, ODM Jaab Lake No. 1 well, lat. 51°11′54″N, long. 82°56′W, Ontario.

Grandispora inculta Allen
Hypotypes 66317, 66318
McGregor, D.C. and Camfield, M., 1982, Geol. Surv. Can., Bull. 348, p. 44, pl. 11, figs. 2, 3.
Richardson, J.B. and McGregor, D.C., 1986, Geol. Surv. Can., Bull. 364, pl. 13, fig. 5 (66317).
Weatherall and Hecla Bay formations, Middle Devonian, stream section 16.2 km east of east arm of Weatherall Bay, northeastern Melville Island, District of Franklin.

Grandispora longa Chi and Hills
Hypotypes 66315, 66316
McGregor, D.C. and Camfield, M., 1982, Geol. Surv. Can., Bull. 348, p. 45, pl. 10, fig. 5; pl. 11, fig. 1.
Weatherall and Hecla Bay formations, Middle Devonian, stream section 16.2 km east of east arm of Weatherall Bay, northeastern Melville Island, District of Franklin.

Grandispora longus Chi and Hills
Holotype 43751; paratypes 43752, 43753
Chi, B.I. and Hills, L.V., 1977, Bull. Can. Petrol. Geol., vol. 24, no. 4, p. 752, pl. 14, figs.1-6.
Weatherall Formation, Middle Devonian, Mould Bay, Prince Patrick Island (43751) and Cape Terrace, west side of Melville Island, District of Franklin.

Grandispora ?macrotuberculata (Arkhangelskaya)
Hypotypes 32028-32030
McGregor, D.C., 1973, Palaeontographica, Abt. B, vol. 142, no. 1-3, p. 59, pl. 8, figs. 1-5.
Battery Point Formation, Lower Devonian, 400 yards north of Bois Brulé Brook, south shore Gaspé Bay, Québec.

Grandispora ?macrotuberculata (Arkhangelskaya)
Hypotype 41765
McGregor, D.C. and Camfield, M., 1976, Geol. Surv. Can., Bull. 263, p. 23, pl. 5, fig. 12.
Stooping River Formation, Lower Devonian, 617-619 feet depth, ODM Jaab Lake No. 1 well, lat. 51°11′54″N, long. 82°56′W, Ontario.

Grandispora ?macrotuberculata (Arkhangelskaya) McGregor 1973
Hypotype 49821
McGregor, D.C., 1979, Palynology, vol. 3, pl. 1, fig. 21.
?Griper Bay Formation, Upper Devonian, 1950-1980 feet, Panarctic Hotspur J-20 well, Vanier Island, District of Franklin.

Grandispora ?macrotuberculata (Arkhangelskaya) McGregor
Hypotype 61727
McGregor, D.C., 1979, Geol. Surv. Can., Paper 79-1B, pl. 22.3, fig. 82.
Daleje Shale, Lower Devonian, upper part of southeast slope of Císařská rokle, southeast of Srbsko, Czechoslovakia.

Grandispora magnifica (Owens)
Hypotype 43758
Chi, B.I. and Hills, L.V., 1977, Bull. Can. Petrol. Geol., vol. 24, no. 4, p. 754, pl. 14, fig 11.
Beverley Inlet Formation, Upper Devonian, Robertson Anticline, southern Melville Island, District of Franklin.

Grandispora mammillata Owens
Hypotype 41766
McGregor, D.C. and Camfield, M., 1976, Geol. Surv. Can., Bull. 263, p. 23, pl. 5, fig. 11.
Kwataboahegan Formation, Middle Devonian, 461-463 feet depth, ODM Jaab Lake No. 1 well, lat. 51°11′54″N, long. 82°56′W, Ontario.

Grandispora mammillata Owens 1971
Hypotype 49833
McGregor, D.C., 1979, Palynology, vol. 3, pl. 2, fig. 12.
Weatherall Formation, Middle Devonian, creek flowing south, 10 miles east of Weatherall Bay, Melville Island, District of Franklin.

Grandispora mammillata Owens
Hypotypes 66319, 66320
McGregor, D.C. and Camfield, M., 1982, Geol. Surv. Can., Bull. 348, p. 45, pl. 11, figs. 4-6.
Richardson, J.B. and McGregor, D.C., 1986, Geol. Surv. Can., Bull. 364, pl. 12, fig. 8 (66320).
Weatherall and Hecla Bay formations, Middle Devonian, stream section 16.2 km east of east arm of Weatherall Bay, northeastern Melville Island, District of Franklin.

Grandispora maxima (Lanninger)
Hypotype 41780
McGregor, D.C. and Camfield, M., 1976, Geol. Surv. Can., Bull. 263, p. 24, pl. 7, figs. 6, 8.
Williams Island Formation, Middle Devonian, 157.5-159.5 feet depth, ODM Jaab Lake No. 1 well, lat. 51°11′54″N, long. 82°56′W, Ontario.

Grandispora megaformis (Richardson)
Hypotype 32031
McGregor, D.C., 1973, Palaeontographica, Abt. B, vol. 142, no. 1-3, p. 60, pl. 8, figs. 6, 7.
Malbaie Formation, Middle Devonian, 1 mile east of Douglastown P.O., south shore Gaspé Bay, Québec.

Grandispora megaformis (Richardson)
Hypotype 41777
McGregor, D.C. and Camfield, M., 1976, Geol. Surv. Can., Bull. 263, p. 24, pl. 7, figs. 1, 4.
Williams Island Formation, Middle Devonian, 157.5-159.5 feet depth, ODM Jaab Lake No. 1 well, lat. 51°11′54″N, long. 82°56′W, Ontario.

Grandispora megafornis (Richardson) McGregor
Hypotypes 66321, 66322
McGregor, D.C. and Camfield, M., 1982, Geol. Surv. Can., Bull. 348, p. 46, pl. 11, figs. 7-9.
Hecla Bay Formation, Middle Devonian, stream section 16.2 km east of east arm of Weatherall Bay, northeastern Melville Island, District of Franklin.

Grandispora microfoveolatus Chi and Hills
Holotype 43759; paratypes 43760-43762
Chi, B.I. and Hills, L.V., 1977, Bull. Can. Petrol. Geol., vol. 24, no. 4, p. 756, pl. 14, figs. 12-15; pl. 15, figs. 1-3.
Beverley Inlet Formation, Upper Devonian, Cape Terrace, west side Melville Island, District of Franklin.

Grandispora multiapicalis Chi and Hills
Holotype 43763; paratypes 43764-43766
Chi, B.I. and Hills, L.V., 1977, Bull. Can. Petrol. Geol., vol. 24, no. 4, p. 758, pl. 15, figs.4-13.
Weatherall Formatiion, Middle Devonian, Mould Bay, Prince Patrick Island, District of Franklin.

Grandispora ?naumovii (Kedo)
Hypotypes 32036, 32037 [not 32077]
McGregor, D.C., 1973, Palaeontographica, Abt. B, vol. 142, no. 1-3, p. 61, pl. 9, figs. 1-3.
Malbaie Formation, Lower Devonian, 65 yards southeast of conglomerate of Pointe Jaune, south shore Gaspé Bay, Québec.

Grandispora ?naumovii (Kedo)
Hypotype 41779
McGregor, D.C. and Camfield, M., 1976, Geol. Surv. Can., Bull. 263, p. 24, pl. 7, fig. 5.
Williams Island Formation, Middle Devonian, 157.5-159.5 feet depth, ODM Jaab Lake No. 1 well, lat. 51°11′54″N, long. 82°56′W, Ontario.

Grandispora Naumovii (Kedo) McGregor 1973
Hypotype 49830
McGregor, D.C., 1979, Palynology, vol. 3, pl. 2, fig. 7.
Malbaie Formation, Middle Devonian, south shore Gaspé Bay, about 550 yards southeast of Point Jaune, Gaspé, Québec.

Grandispora protea (Naumova) Moreau-Benoit
Hypotypes 66323, 66324
McGregor, D.C. and Camfield, M., 1982, Geol. Surv. Can., Bull. 348, p. 46, pl. 11, fig. 10; pl. 12, fig. 1.
Weatherall Formation, Middle Devonian, stream section 16.2 km east of east arm of Weatherall Bay, northeastern Melville Island, District of Franklin.

Grandispora tiwarii McGregor and Camfield
Hypotype 66325
McGregor, D.C. and Camfield, M., 1982, Geol. Surv. Can., Bull. 348, p. 47, pl. 12, figs. 2, 4.
Cape De Bay Formation, Middle Devonian, stream section 16.2 km east of east arm of Weatherall Bay, northeastern Melville Island, District of Franklin.

Grandispora tomentosa Taugourdeau-Lantz
Hypotype 41770
McGregor, D.C. and Camfield, M., 1976, Geol. Surv. Can., Bull. 263, p. 24, pl. 6, figs. 4, 5, 8.
Williams Island Formation, Middle Devonian, 157.5-159.5 feet depth, ODM Jaab Lake No. 1 well, lat. 51°11′54″N, long. 82°56′W, Ontario.

Grandispora cf.*G. tomentosa* Taugourdeau-Lantz
Hypotype 41778
McGregor, D.C. and Camfield, M., 1976, Geol. Surv. Can., Bull. 263, p. 24, pl. 7, figs. 2, 3.
Williams Island Formation, Middle Devonian, 157.5-159.5 feet depth, ODM Jaab Lake No. 1 well, lat. 51°11′54″N, long. 82°56′W, Ontario.

Grandispora uyenoi McGregor and Camfield
Holotype 66326
McGregor, D.C. and Camfield, M., 1982, Geol. Surv. Can., Bull. 348, p. 47, pl. 12, figs. 3, 7.
Weatherall Formation, Middle Devonian, stream section 16.2 km east of east arm of Weatherall Bay, northeastern Melville Island, District of Franklin.

Grandispora variospinosa McGregor and Camfield
Holotype 66328; paratype 66327
McGregor, D.C. and Camfield, M., 1982, Geol. Surv. Can., Bull. 348, p. 47, pl. 12, figs. 5, 6.
Hecla Bay and Weatherall formations, Middle Devonian, stream section 16.2 km east of east arm of Weatherall Bay, northeastern Melville Island, District of Franklin.

Grandispora velata (Richardson)
Hypotype 32035
McGregor, D.C., 1973, Palaeontographica, Abt. B, vol. 142, no. 1-3, p. 61, pl. 8, figs. 10, 11.
Battery Point Formation, Middle Devonian, 100 yards south of most prominent point between Cap Rouge and Pointe Jaune, south shore Gaspé Bay, Québec.

Grandispora velata (Richardson) McGregor
Hypotype 41775
McGregor, D.C. and Camfield, M., 1976, Geol. Surv. Can., Bull. 263, p. 24, pl. 6, fig. 13.
Williams Island Formation, Middle Devonian, 157.5-159.5 feet depth, ODM Jaab Lake No. 1 well, lat. 51°11′54″N, long. 82°56′W, Ontario.

Grandispora velata (Richardson) McGregor 1973
Hypotypes 43749, 43750
Chi, B.I. and Hills, L.V., 1977, Bull. Can. Petrol. Geol., vol. 24, no. 4, p. 750, pl. 13, figs.13, 14.
Hecla Bay and Weatherall formations, Middle Devonian, Cape Terrace, west side Melville Island, District of Franklin.

Grandispora velata (Eisenack) Playford
Hypotype 66329
McGregor, D.C. and Camfield, M., 1982, Geol. Surv. Can., Bull. 348, p. 48, pl. 12, fig. 8.
Weatherall Formation, Middle Devonian, stream section 16.2 km east of east arm of Weatherall Bay, northeastern Melville Island, District of Franklin.

Grandispora velata (Eisenack) Playford 1971
Hypotype 96781
Arkhangelskaya, A.D., McGregor, D.C. and Richardson, J.B., 1990, Geol. Surv. Can., Paper 90-1D, pl. 1, fig. k.
Strathcona Fiord Formation, Middle Devonian, east of Bird Fiord, Ellesmere Island, District of Franklin.

Grandispora sp.
Fig. specs. 15306-15308, 15447
McGregor, D.C. and Owens, B., 1966, Geol. Surv. Can., Paper 66-30, pl. 13, figs. 3-5; pl. 14, figs. 1, 2; pl. 27, figs. 24, 27.
Weatherall Formation, Middle Devonian, south limb of Robertson anticline, 8 miles northeast of Beverley Inlet, Melville Island; Griper Bay Formation, Upper Devonian, stream valley west of Sargeant Point, 2 miles from shore, northeast Bathurst Island (15447), District of Franklin.

Grandispora sp.
Fig. spec. 15447
McGregor, D.C. and Owens, B., 1966, Geol. Surv. Can., Paper 66-30, pl. 27, figs. 24, 27.
Griper Bay Formation, Upper Devonian, stream valley west of Sargeant Point, 2 miles from shore, northeastern Bathurst Island, District of Franklin.

Grandispora sp.
Fig. spec. 32040
McGregor, D.C., 1973, Palaeontographica, Abt. B, vol. 142, no. 1-3, p. 63, pl. 9, figs. 5, 8.
Battery Point Formation, Lower Devonian, 1/2 mile north of Bois Brulé Brook, south shore Gaspé Bay, Québec.

Grandispora sp.
Fig. spec. 61726
McGregor, D.C., 1979, Geol. Surv. Can., Paper 79-1B, pl. 22.3, fig. 81.
Daleje Shale, Lower Devonian, upper part of southeast slope of Císařská rokle, southeast of Srbsko, Czechoslovakia.

Grandispora sp.
Fig. spec. 67795
McGregor, D.C., 1984, Academia Nacional de Ciencias, Misc. No. 69, pl. 4, fig. 11.
Huamampampa Formation, Lower Devonian, Laurani section, Chuquisaca Dept., northwest of Tarabuco, Cordillera Oriental, Bolivia.

Grandispora sp.
Fig. specs. 86098, 86118
McGregor, D.C. and McCutcheon, S.R., 1988, Can. J. Earth Sci., vol. 25, no. 9, p. 1353, pl. 2, fig. 2; pl. 3, figs. 1-3.
Carrow Formation, Piskahegan Group, Late Devonian, depth 72.85 m, drill hole M-6, lat. 45°34′43″N, long. 66°35′27″W, southwestern New Brunswick.

?*Grandispora* sp.
Fig. spec. 15471
McGregor, D.C. and Owens, B., 1966, Geol. Surv. Can., Paper 66-30, pl. 28, fig. 23.
Kettle Point Formation, Upper Devonian, depth 295-299 feet, Courtright borehole, Courtright village, Lambton County, Ontario.

Grandispora n.sp.
Fig. spec. 29392
McGregor, D.C. and Uyeno, T.T., 1972, Geol. Surv. Can., Paper 71-13, pl. 2, fig. 5.
Weatherall Formation, Middle Devonian, Beverley Inlet, southeastern Melville Island, District of Franklin.

Grandispora sp. 1, 2, 3
Fig. specs. 66331, 66332, 66330
McGregor, D.C. and Camfield, M., 1982, Geol. Surv. Can., Bull. 348, p. 48, 49, pl. 12, figs. 9, 10; pl. 13, fig. 1.
Hecla Bay Formation, Middle Devonian, stream section 16.2 km east of east arm of Weatherall Bay, northeastern Melville Island, District of Franklin.

Granulatisporites muninensis Allen
Hypotype 32799
McGregor, D.C., 1974, Can. J. Earth Sci., vol. 11, no. 1, pl. 1, figs. 17, 18.
Unnamed Early Devonian, lat. 79°28′40″N, long. 80°02′W, 12 miles south of head of Cañon Fiord, central Ellesmere Island, District of Franklin.

Granulatisporites muninensis? Allen
Hypotypes 66334-66336
McGregor, D.C. and Camfield, M., 1982, Geol. Surv. Can., Bull. 348, p. 49, pl. 13, figs. 5-11.
Hecla Bay and Weatherall (66336) formations, Middle Devonian, stream section 16.2 km east of east arm of Weatherall Bay, northeastern Melville Island, District of Franklin.

Granulatisporites sp.
Fig. specs. 15118, 15120
McGregor, D.C. and Owens, B., 1966, Geol. Surv. Can., Paper 66-30, pl. 1, figs. 18, 20.
Shiphead Member, Grand Grève Formation, Lower Devonian, Cape Gaspé, shore cliff below lighthouse, eastern Gaspé Peninsula, Québec.

Granulatisporites sp.
Fig. spec. 66333
McGregor, D.C. and Camfield, M., 1982, Geol. Surv. Can., Bull. 348, p. 50, pl. 13, figs. 2-4.
Hecla Bay Formation, Middle Devonian, stream section 16.2 km east of east arm of Weatherall Bay, northeastern Melville Island, District of Franklin.

Granulatisporites sp.
Fig. spec. 67775
McGregor, D.C., 1984, Academia Nacional de Ciencias, Misc. No. 69, p. 31, pl. 3, fig. 4.
Huamampampa Formation, Lower Devonian, Laurani section, Chuquisaca Dept., northwest of Tarabuco, Cordillera Oriental, Bolivia.

?*Heliotriletes* sp.
Fig. spec. 86084
McGregor, D.C. and McCutcheon, S.R., 1988, Can. J. Earth Sci., vol. 25, no. 9, p. 1353, pl. 1, fig. 10.
Carrow Formation, Piskahegan Group, Late Devonian, depth 72.85 m, drill hole M-6, lat. 45°34′43″N, long. 66°35′27″W, southwestern New Brunswick.

?*Hymenozonotriletes albus* Arkhangelskaya
Hypotype 27093
McGregor, D.C., Sanford, B.V. and Norris, A.W., 1970, Proc. Geol. Assoc. Can., vol. 22, pl. 2, fig. 14.
Williams Island Formation, Middle Devonian, 157 feet depth, ODM Jaab Lake No. 1 well, lat. 51°11.9′N, long. 82°57′W, Moose River Basin, Ontario.

?*Hymenozonotriletes antiquus* Gao, 1978
Hypotype 67202
McGregor, D.C., 1984, Academia Nacional de Ciencias, Misc. No. 69, p. 31, pl. 1, fig. 19.
Santa Rosa Formation, Lower Devonian, Laurani section, Chuquisaca Dept., northwest of Tarabuco, Cordillera Oriental, Bolivia.

Hymenozonotriletes archaelepidophytus Kedo 1955
Hypotype 49843
McGregor, D.C., 1979, Palynology, vol. 3, pl. 4, fig. 1.
Griper Bay Formation, Upper Devonian, Rookery Creek, northwestern Cornwallis Island, District of Franklin.

=*Retispora archaelepidophyta*, Richardson, J.B. and McGregor, D.C., 1986, Geol. Surv. Can., Bull. 364, pl. 14, fig. 3.

?*Hymenozonotriletes asper* Chibrikova
Hypotypes 61689, 61721, 61722
McGregor, D.C., 1979, Geol. Surv. Can., Paper 79-1B, pl. 22.2, figs. 42, 43; pl. 22.3, figs. 76, 77.
Daleje Shale, Lower Devonian, roadside outcrop near Karlštejn, 1 km north Budnany in the Hluboké Valley (61689), and upper part of southeast slope of Císařská rokle, southeast of Srbsko, Czechoslovakia.

cf. *Hymenozonotriletes asperitus* Kedo, 1963
Hypotype 13590
McGregor, D.C. and Owens, B., 1966, Geol. Surv. Can., Paper 66-30, pl. 26, fig. 26.
Griper Bay Formation, Upper Devonian, stream valley west of Sargeant Bay, two miles from shore, northeastern Bathurst Island, District of Franklin.

Hymenozonotriletes breviradius Chibrikova, 1962
Hypotype 15296
McGregor, D.C. and Owens, B., 1966, Geol. Surv. Can., Paper 66-30, pl. 10, figs. 3, 4.
Weatherall Formation, Middle Devonian, south limb of Robertson Point anticline, 8 miles northeast of Beverley Inlet, Melville Island, District of Franklin.

cf.*Hymenozonotriletes beviradius* Chibrikova 1962
Hypotype 48175
McGregor, D.C., 1977, Palaeontographica, Abt. B, vol. 163, p. 132, pl. 2, figs. 4-6.
Battery Point Formation, Lower Devonian, Tar Point to Douglastown section, south shore Gaspé Bay, Québec.

Hymenozonotriletes ?celeber Chibr.
Hypotype 29405
McGregor, D.C. and Uyeno, T.T., 1972, Geol. Surv. Can., Paper 71-13, pl. 3, fig. 6.
Hecla Bay Formation, Middle-Upper Devonian, Beverley Inlet, southeastern Melville Island, District of Franklin.

Hymenozonotriletes celeber Chibrikova 1959
Hypotype 49845
McGregor, D.C., 1979, Palynology, vol. 3, pl. 4, fig. 6.
Richardson, J.B. and McGregor, D.C., 1986, Geol. Surv. Can., Bull. 364, pl. 15, fig. 7.
Hare Indian Formation, Middle Devonian, 640-730 feet, Banff Aquit GPD Oscar Creek H-71 well, lat. 65°30′20″N, long. 127°13′26″W, District of Mackenzie.

Hymenozonotriletes commutatus Naumova, 1953, sensu Higgs 1975
Hypotype 86106
McGregor, D.C. and McCutcheon, S.R., 1988, Can. J. Earth Sci., vol. 25, no. 9, p. 1353, pl. 2, fig. 9.
Carrow Formation, Piskahegan Group, Late Devonian, depth 72.85 m, drill hole M-6, lat. 45°34′43″N, long. 66°35′27″W, southwestern New Brunswick.

Hymenozonotriletes deliquescens Naumova var. *cinctus* Chibrikova
Hypotype 13629
Kerr, J.W., McGregor, D.C. and McLaren, D.J., 1965, Bull. Can. Petrol. Geol., vol. 13, no. 3, p. 425, pl. 4, fig. 21.
Griper Bay Formation, Upper Devonian, 3/4 mile from shore in stream valley entering Reindeer Bay, 4 miles south of largest river entering the bay, northeastern Bathurst Island, District of Franklin.

?*Hymenozonotriletes denticulatus* Naumova
Hypotype 15443
McGregor, D.C., 1967, Review Palaeobotany and Palynology, I, pl. 1, fig. 25.
Griper Bay Formation, Upper Devonian, northern Helena Island, 2.8 miles from north coast, in valley of stream entering the sea at Noel Point, District of Franklin.

Hymenozonotriletes deliquescens Naum.
Hypotype 29411
McGregor, D.C. and Uyeno, T.T., 1972, Geol. Surv. Can., Paper 71-13, pl. 4, fig. 3.
Griper Bay Formation, Upper Devonian, Beverley Inlet, southeastern Melville Island, District of Franklin.

Hymenozonotriletes deliquescens Naumova
Hypotype 73299
Richardson, J.B. and McGregor, D.C., 1986, Geol. Surv. Can., Bull. 364, pl. 17, fig. 6.
Fram Formation, Upper Devonian, east side of inner part of southeast arm of Bird Fiord, southwestern Ellesmere Island, District of Franklin.

Hymenozonotriletes denticulatus Naum.
Hypotype 29420
McGregor, D.C. and Uyeno, T.T., 1972, Geol. Surv. Can., Paper 71-13, pl. 4, fig. 12.
Griper Bay Formation, Upper Devonian, Beverley Inlet, southeastern Melville Island, District of Franklin.

Hymenozonotriletes denticulatus Naumova 1953
Hypotype 49850
McGregor, D.C., 1979, Palynology, vol. 3, pl. 4, fig. 15.
Griper Bay Formation, Upper Devonian, 2881-2900 feet, Uminmak H-07 well, Banks Island, District of Franklin.
=*"(?)Hymenozonotriletes denticulatus Naumova"*, Richardson, J.B. and McGregor, D.C., 1986, Geol. Surv. Can., Bull. 364, pl. 19, figs. 2, 3.

?*Hymenozonotriletes elegans* (Waltz) Naumova, 1953, sensu Kedo 1957
Hypotype 86114
McGregor, D.C. and McCutcheon, S.R., 1988, Can. J. Earth Sci., vol. 25, no. 9, p. 1353, pl. 2, fig. 20.
Carrow Formation, Piskahegan Group, Late Devonian, depth 72.85 m, drill hole M-6, lat. 45°34′43″N, long. 66°35′27″W, southwestern New Brunswick.

Hymenozonotriletes explanatus (Luber) Kedo
Hypotype 73313
Richardson, J.B. and McGregor, D.C., 1986, Geol. Surv. Can., Bull. 364, pl. 21, fig. 6.

Devonian, west side of St. George's Bay, north of McIsaac Point, lat. 45°44′24.4″N, long. 61°53′28.46″W, Antigonish County, Nova Scotia.

Hymenozonotriletes famenensis Kedo
Hypotypes 25443, 25454
McGregor, D.C., 1970, Colloque Stratig. Carbonifère, Congr. Colloques Univ. Liège, vol. 55, pl. 21, figs. 6, 16.
Bedford Formation, Upper Devonian, Courtright borehole, Courtright village, Lambton County, Ontario.
=*Rugospora flexuosa*, McGregor, D.C., 1979, Palynology, vol. 3, pl. 4, fig. 24 (25454).

?*Hymenozonotriletes famenensis* Kedo
Hypotypes 25463, 25480
McGregor, D.C., 1970, Colloque Stratig. Carbonifère, Congr. Colloques Univ. Liège, vol. 55, pl. 22, fig. 5; pl. 23, fig. 4.
Upper Devonian, Bedford Formation, Courtright borehole, Courtright village, Lambton County, Ontario, and Horton Group, west shore George Bay at mouth of McInnis Brook near Antigonish, Nova Scotia.

Hymenozonotriletes lepidophytus Kedo, 1957
Hypotypes 15478-15480
McGregor, D.C. and Owens, B., 1966, Geol. Surv. Can., Paper 66-30, pl. 29, figs. 6-8.
Kettle Point Formation, Upper Devonian, depth 295-299 feet, Courtright borehole, Courtright village, Lambton County, Ontario.

Hymenozonotriletes lepidophytus Kedo
Hypotypes 15486-15488
Owens, B. and Streel, M., 1967, Review Palaeobotany and Palynology, I, pl. 1, figs. D, F. G.
Upper Devonian, Port Lambton Formation, Courtright borehole, Courtright village, Lambton County, Ontario; Imperial Formation, Richardson Mountains region, Northwest Territories.

Hymenozonotriletes lepidophytus Kedo
Hypotype 15069 [25459]
McGregor, D.C., 1970, Geol. Surv. Can., Econom. Geol. Rept. No. 1, 5th Edit., pl. 31, fig. 6.
McGregor, D.C., 1970, Colloque Stratig. Carbonifère, Congr. Colloques Univ. Liège, vol. 55, pl. 22, fig. 2 (25459).
Bedford Formation, Upper Devonian, Courtright borehole, Courtright village, Lambton County, Ontario.
=*Retispora lepidophyta*, McGregor, D.C., 1979, Palynology, vol. 3, pl. 4, fig. 23 (25459).

Hymenozonotriletes lepidophytus Kedo
Hypotypes 25438, 25439, 25451, 25452, 25458, 25460, 25461, 25468-25470, 25477, 25486, 25487, 25493-25495
McGregor, D.C., 1970, Colloque Stratig. Carbonifère, Congr. Colloques Univ. Liège, vol. 55, pl. 21, figs. 1, 2, 14, 17; pl. 22, figs. 1, 4, 8, 11-13; pl. 23, figs. 1, 10, 11, 17, 18, 22.
Upper Devonian, Bedford Formation, Courtright borehole, Courtright village, Lambton County, and Sunbury Formation, Gibraltar 77 borehole, 6 miles south of Courthright village (25468-25470), Ontario; Horton Group, west shore George Bay at mouth of McInnis Brook near Antigonish, Nova Scotia (25477); Imperial Formation, lat. 66°27.5′N, long. 135°22′W, Yukon (25486, 25487, 25493-25495).

Hymenozonotriletes longus Arkh.
Hypotype 29383
McGregor, D.C. and Uyeno, T.T., 1972, Geol. Surv. Can., Paper 71-13, pl. 1, fig. 15.
Bird Fiord Formation, Middle Devonian, Young Inlet, northeastern Bathurst Island, District of Franklin.

?*Hymenozonotriletes longus* Arkhangelskaya, 1963
Hypotype 15277
McGregor, D.C. and Owens, B., 1966, Geol. Surv. Can., Paper 66-30, pl. 7, fig. 4.
Battery Point Formation, Lower Devonian, south side Gaspé Bay, one mile southeast of Douglastown, Gaspé, Québec.

Hymenozonotriletes sp. cf. *H. melanidus* Naumova *in* Kedo, 1955
Fig. spec. 15298
McGregor, D.C. and Owens, B., 1966, Geol. Surv. Can., Paper 66-30, pl. 10, fig. 6.
Weatherall Formation, Lower Devonian, south limb of Robertson Point anticline, 8 miles northeast of Beverley Inlet, Melville Island, District of Franklin.

Hymenozonotriletes parvimammatus (Naumova) Kedo
Hypotype 13628
Kerr, J.W., McGregor, D.C. and McLaren, D.J., 1965, Bull. Can. Petrol. Geol., vol. 13, no. 3, p. 425, pl. 4, fig. 20.
Griper Bay Formation, Upper Devonian, 3/4 mile from shore in stream valley entering Reindeer Bay, 4 miles south of largest river entering the bay, northeastern Bathurst Island, District of Franklin.

Hymenozonotriletes parvimammatus (Naumova) Kedo, 1957
Hypotypes 15439, 15440
McGregor, D.C. and Owens, B., 1966, Geol. Surv. Can., Paper 66-30, pl. 27, figs. 13, 14.
Griper Bay Formation, Upper Devonian, stream valley west of Sargeant Point, 2 miles from shore, northeastern Bathurst Island, District of Franklin.

?*Hymenozonotriletes polymorphus* Naumova of Ozolin'a, 1960
Hypotype 13627
Kerr, J.W., McGregor, D.C. and McLaren, D.J., 1965, Bull. Can. Petrol. Geol., vol. 13, no. 3, p. 428, pl. 4, fig. 19.
Griper Bay Formation, Upper Devonian, northern Helena Island, 2.8 miles from north coast, in valley of stream entering the sea at Noel Point, District of Franklin.

(?)*Hymenozonotriletes proteus* Naumova
Hypotype 16593
McGregor, D.C., 1967, Review Palaeobotany and Palynology, I, p. 180, pl. 1, fig. 16.
Hecla Bay Formation, Melville Island Group, Middle or Upper Devonian, Helena Island, District of Franklin.

Hymenozonotriletes pusillites Kedo, 1957
Hypotypes 15476, 15475
McGregor, D.C. and Owens, B., 1966, Geol. Surv. Can., Paper 66-30, pl. 29, figs. 1-4.
Kettle Point Formation, Upper Devonian, depth 295-299 feet, Courtright borehole, Courtright village, Lambton County, Ontario.

Hymenozonotriletes pusillites Kedo
Hypotypes 25440, 25456, 25465, 25471, 25478, 25479
McGregor, D.C., 1970, Colloque Stratig. Carbonifère, Congr. Colloques Univ. Liège, vol. 55, pl. 21, figs. 3, 19; pl. 22, figs. 7, 14; pl. 23, figs.2, 3.
Upper Devonian, Bedford Formation, Courtright borehole, Courtright village, Lambton County, and Sunbury Formation, Gibraltar 77 borehole, 6 miles south of Courtright village (25471), Ontario; Horton Group, west shore George Bay at mouth of McInnis Brook near Antiguonish, Nova Scotia (25478, 25479).
=*Vallatisporites pusillites*, McGregor, D.C., 1979, Palynology, vol. 3, pl. 4, fig. 22 (25465).

Hymenozonotriletes pusillites Kedo (?*Vallatisporites spleadens* Staplin and Jansonius)
Hypotype 25441
McGregor, D.C., 1970, Colloque Stratig. Carbonifère, Congr. Colloques Univ. Liège, vol. 55, pl. 21, fig. 4.
Bedford Formation, Upper Devonian, Courtright borehole, Courtright village, Lambton County, Ontario.

cf. *Hymenozonotriletes pusillites* Kedo
Hypotype 15481
McGregor, D.C. and Owens, B., 1966, Geol. Surv. Can., Paper 66-30, pl. 29, fig. 9.
Kettle Point Formation, Upper Devonian, depth 295-299 feet, Courtright borehole, Courtright village, Lambton County, Ontario.

Hymenozonotriletes sp. cf. *H. rigidispinus* Naumova, 1953
Hypotype 15310
McGregor, D.C. and Owens, B., 1966, Geol. Surv. Can., Paper 66-30, pl. 14, figs. 5, 6.
Weatherall Formation, Lower Devonian, south limb of Robertson Point anticline, 8 miles northeast of Beverley Inlet, Melville Island, District of Franklin.

?*Hymenozonotriletes rugosiusculus* Jushko
Hypotype 25481
McGregor, D.C., 1970, Colloque Stratig. Carbonifère, Congr. Colloques Univ. Liège, vol. 55, pl. 23, fig. 5.
Horton Group, Upper Devonian, west shore George Bay at mouth of McInnis Brook near Antigonish, Nova Scotia.

?*Hymenozonotriletes semilucensis* (Naumova) Kedo, 1957
Hypotype 15442
McGregor, D.C. and Owens, B., 1966, Geol. Surv. Can., Paper 66-30, pl. 27, figs. 17, 18.
Griper Bay Formation, Upper Devonian, stream valley west of Sargeant Point, 2 miles from shore, northeastern Bathurst Island, District of Franklin.

Hymenozonotriletes sp. cf. *H. tichomirovii* Naumova var. *devius* Chibrikova, 1959
Hypotype 15304
McGregor, D.C. and Owens, B., 1966, Geol. Surv. Can., Paper 66-30, pl. 12, figs. 4, 5.
Weatherall Formation, Lower Devonian, south limb of Robertson Point anticline, 8 miles northeast of Beverley Inlet, Melville Island, District of Franklin.

?*Hymenozonotriletes varius* Naumova, 1953, sensu Kedo, 1957
Hypotype 86108
McGregor, D.C. and McCutcheon, S.R., 1988, Can. J. Earth Sci., vol. 25, no. 9, p. 1353, pl. 2, fig. 12.
Carrow Formation, Piskahegan Group, Late Devonian, depth 73.5 m, drill hole M-6, lat. 45°34′43″N, long. 66°35′27″W, southwestern New Brunswick.

?*Hymenozonotriletes varius* var. *minor* Naumova, 1953
Hypotype 15472
McGregor, D.C. and Owens, B., 1966, Geol. Surv. Can., Paper 66-30, pl. 28, fig. 24.
Kettle Point Formation, Upper Devonian, depth 295-299 feet, Courtright borehole, Courtright village, Lambton County, Ontario.

Hymenozonotrilletes verrucosus Menéndez and Pöthe de Baldis, 1971
Hypotype 67187
McGregor, D.C., 1984, Academia Nacional de Ciencias, Misc. No. 69, p. 31, pl. 3, fig. 6.
Icla Formation, Lower Devonian, Laurani section, Chuquisaca Dept., northwest of Tarabuco, Cordillera Oriental, Bolivia.

Hymenozonotriletes versabilis Kedo
Hypotype 13602
Kerr, J.W., McGregor, D.C. and McLaren, D.J., 1965, Bull. Can. Petrol. Geol., vol. 13, no. 3, p. 425, pl. 3, fig. 20.
Griper Bay Formation, Upper Devonian, stream valley west of Sargeant Point, 2 miles from shore, northeastern Bathurst Island, District of Franklin.

Hymenozonotriletes sp.
Fig. specs. 15294, 15300, 15303, 15309
McGregor, D.C. and Owens, B., 1966, Geol. Surv. Can., Paper 66-30, pl. 10, fig. 1; pl. 11, fig. 2; pl. 12, figs. 2, 3; pl. 14, figs.3, 4.
Weatherall Formation, Middle Devonian, south limb of Robertson Point anticline, 8 miles northeast of Beverley Inlet , Melville Island, District of Franklin.

Hystricosporites bullosus Chi and Hills
Holotype 43696; paratypes 43697, 43698
Chi, B.I. and Hills, L.V., 1977, Bull. Can. Petrol. Geol., vol. 24, no. 4, p. 717, pl. 5, figs. 5-9.
Beverley Inlet Formation, Upper Devonian, Cape Terrace, west side Melville Island, District of Franklin.

Hystricosporites costatus Virgran 1964
Hypotypes 43690, 43691
Chi, B.I. and Hills, L.V., 1977, Bull. Can. Petrol. Geol., vol. 24, no. 4, p. 712, pl. 4, figs. 6, 7.
Beverley Inlet Formation, Upper Devonian, Cape Terrace, west side Melville Island, District of Franklin.

Hystricosporites costatus Vigran
Hypotype 66337
McGregor, D.C. and Camfield, M., 1982, Geol. Surv. Can., Bull. 348, p. 50, pl. 13, fig. 12.
Hecla Bay Formation, Middle Devonian, stream section 16.2 km east of east arm of Weatherall Bay, northeastern Melville Island, District of Franklin.

Hystricosporites delectabilis McGregor 1966
Hypotype 43689
Chi, B.I. and Hills, L.V., 1977, Bull. Can. Petrol. Geol., vol. 24, no. 4, p. 711, pl. 4, fig. 5.
Imperial Formation, Upper Devonian, Imperial River 35 miles west of Norman Wells, lat. 65°07′N, long. 127°51′W, District of Mackenzie.

Hystricosporites elongatus Chi and Hills
Holotype 43699; paratypes 43700, 43701
Chi, B.I. and Hills, L.V., 1977, Bull. Can. Petrol. Geol., vol. 24, no. 4, p. 720, pl. 5, figs. 10-12.
Weatherall Formation, Middle Devonian, Cape Terrace and South McCormick Inlet (43701), Melville Island, District of Franklin.

Hystricosporites expandus Chi and Hills
Holotype 43702; paratypes 43703, 43704
Chi, B.I. and Hills, L.V., 1977, Bull. Can. Petrol. Geol., vol. 24, no. 4, p. 721, pl. 6, fig. 1-5.
Beverley Inlet Formation, Upper Devonian, Cape Terrace, west side Melville Island, District of Franklin.

Hystricosporites furcatus Owens 1971
Hypotypes 43692, 43693
Chi, B.I. and Hills, L.V., 1977, Bull. Can. Petrol. Geol., vol. 24, no. 4, p. 714, pl. 4, figs. 8, 9.
Imperial Formation, Upper Devonian, Imperial River 35 miles west of Norman Wells, lat. 65°07′N, long. 127°51′W, District of Mackenzie.

Hystricosporites grandis Owens 1971
Hypotype 43694
Chi, B.I. and Hills, L.V., 1977, Bull. Can. Petrol. Geol., vol. 24, no. 4, p. 715, pl. 5, figs. 1, 2.
Beverley Inlet Formation, Upper Devonian, Cape Terrace, west side Melville Island, District of Franklin.

Hystricosporites gravis Owens 1971
Hypotype 43695
Chi, B.I. and Hills, L.V., 1977, Bull. Can. Petrol. Geol., vol. 24, no. 4, p. 716, pl. 5, figs. 3, 4.
Beverley Inlet Formation, Upper Devonian, Cape Terrace, west side Melville Island, District of Franklin.

Hystricosporites gravis Owens
Hypotypes 66340, 66341
McGregor, D.C. and Camfield, M., 1982, Geol. Surv. Can., Bull. 348, p. 50, pl. 13, figs. 15, 16.
Weatherall Formation, Middle Devonian, stream section 16.2 km east of east arm of Weatherall Bay, northeastern Melville Island, District of Franklin.

Hystricosporites multifurcatus (Winslow) Mortimer and Chaloner
Hypotype 73305
Richardson, J.B. and McGregor, D.C., 1986, Geol. Surv. Can., Bull. 364, pl. 18, fig. 10.
McAras Brook (?) Formation, Upper Devonian, Cape George Brook, Nova Scotia.

Hystricosporites multifurcatus (Windsor) Mortimer and Chaloner, 1967, sensu Allen and Robsen 1981
Hypotypes 86089, 86090
McGregor, D.C. and McCutcheon, S.R., 1988, Can. J. Earth Sci., vol. 25, no. 9, p. 1353, pl. 1, figs. 16-18.
Carrow Formation, Piskahegan Group, Late Devonian, depth 72.85 m, drill hole M-6, lat. 45°34′43″N, long. 66°35′27″W, southwestern New Brunswick.

Hystricosporites porcatus (Winslow) Allen, 1965
Hypotype 15396
McGregor, D.C. and Owens, B., 1966, Geol. Surv. Can., Paper 66-30, pl. 25, figs. 3, 4.
Escarpment Member, Hay River Formation, Upper Devonian, left bank Hay River at base of Louise Falls, Northwest Territories.

Hystricosporites ?richardsonii De Jersey
Fig. specs. 31979, 31980
McGregor, D.C., 1973, Palaeontographica, Abt. B, vol. 142, no. 1-3, p. 34, pl. 4. fig. 3, 4.
Battery Point Formation, Lower Devonian, 1300 yards south of second stream south of Bois Brulé Brook, south shore Gaspé Bay, Québec.

Hystricosporites setigerus? (Naumova ex Kedo) Arkhangelskaya
Hypotypes 66338, 66339
McGregor, D.C. and Camfield, M., 1982, Geol. Surv. Can., Bull. 348, p. 51, pl. 13, figs. 13, 14.
Hecla Bay and Weatherall formations, Middle Devonian, stream section 16.2 km east of east arm of Weatherall Bay, northeastern Melville Island, District of Franklin.

Hystricosporites spiralis Chi and Hills
Holotype 43705; paratypes 43706-43708
Chi, B.I. and Hills, L.V., 1977, Bull. Can. Petrol. Geol., vol. 24, no. 4, p. 722, pl. 6, figs. 6-9.
Weatherall Formation, Middle Devonian, South McCormick Inlet, north-central Melville Island, District of Franklin.

Hystricosporites validus Chi and Hills
Holotype 43709; paratypes 43710, 43711
Chi, B.I. and Hills, L.V., 1977, Bull. Can. Petrol. Geol., vol. 24, no. 4, p. 723, pl. 6, figs. 10-12.
Weatherall Formation, Middle Devonian, west side Mould Bay, Prince Patrick Island, District of Franklin.

Hystricosporites sp.
Fig. spec. 13608
Kerr, J.W., McGregor, D.C. and McLaren, D.J., 1965, Bull. Can. Petrol. Geol., vol. 13, no. 3, p. 425, pl. 3, fig. 30.
Griper Bay Formation, Upper Devonian, 3/4 mile from shore in stream entering Reindeer Bay, 4 miles south of largest river entering the bay, northeastern Bathurst Island, District of Franklin.

Hystricosporites sp.
Fig. specs. 15292, 15293, 15336-15339, 15395, 15483
McGregor, D.C. and Owens, B., 1966, Geol. Surv. Can., Paper 66-30, pl. 9, figs. 7, 8; pl. 21, figs. 1-6; pl. 25, figs. 1, 2; pl. 29, fig. 12.
Weatherall Formation, Middle Devonian (15292, 15293), and Griper Bay Formation, Upper Devonian, south limb of Robertson Point anticline, 8 miles northeast of Beverley Inlet, Melville Island, District of Franklin; Escarpment Member, Hay River Formation, Upper Devonian, left bank Hay River at base of Louise Falls, Northwest Territories (15395); Kettle Point Formation, Upper Devonian, depth 295-299 feet, Courtright borehole, Courtright village, Lambton County, Ontario (15483).

Hystricosporites sp.
Fig. spec. 29376
McGregor, D.C. and Uyeno, T.T., 1972, Geol. Surv. Can., Paper 71-13, pl. 1, fig. 8.
Eids Formation, Middle Devonian, Young Inlet, northeastern Bathurst Island, District of Franklin.

Hystricosporites sp.
Fig. specs. 86091, 86094
McGregor, D.C. and McCutcheon, S.R., 1988, Can. J. Earth Sci., vol. 25, no. 9, p. 1353, pl. 1, figs. 19, 23.
Carrow Formation, Piskahegan Group, Late Devonian, depth 72.85 m, drill hole M-6, lat. 45°34′43″N, long. 66°35′27″W, southwestern New Brunswick.

Hystricosporites sp. A of Owens
Fig. spec. 66342
McGregor, D.C. and Camfield, M., 1982, Geol. Surv. Can., Bull. 348, p. 51, pl. 14, fig. 1.
Hecla Bay Formation, Middle Devonian, stream section 16.2 km east of east arm of Weatherall Bay, northeastern Melville Island, District of Franklin.

Hystricosporites n.sp. 1, 2, 3
Fig. specs. 29399, 29394, 29410
McGregor, D.C. and Uyeno, T.T., 1972, Geol. Surv. Can., Paper 71-13, pl. 2, figs. 12, 7; pl. 4, fig. 2.
Weatherall and Griper Bay formations, Middle and Upper Devonian, Beverley Inlet, southeastern Melville Island, District of Franklin.

cf. *Iberoespora cantabrica* Cramer and Diez, 1975
Hypotype 67784
McGregor, D.C., 1984, Academia Nacional de Ciencias, Misc. No. 69, pl. 3, fig. 17.
Santa Rosa Formation, Lower Devonian, Laurani section, Chuquisaca Dept., northwest of Tarabuco, Cordillera Oriental, Bolivia.

cf. *Iberoespora guzmanii* Cramer and Diez, 1975
Hypotype 67782
McGregor, D.C., 1984, Academia Nacional de Ciencias, Misc. No. 69, pl. 3, fig. 15.
Santa Rosa Formation, Lower Devonian, Laurani section, Chuquisaca Dept., northwest of Tarabuco, Cordillera Oriental, Bolivia.

Klukisporites sp.
Fig. spec. 15360
McGregor, D.C. and Owens, B., 1966, Geol. Surv. Can., Paper 66-30, pl. 22, figs. 24, 25.
Escarpment Member, Hay River Formation, left bank Hay River at base of Louise Falls, Northwest Territories.

Knoxisporites dedaleus (Naumova) Streel
Hypotype 73306
Richardson, J.B. and McGregor, D.C., 1986, Geol. Surv. Can., Bull. 364, pl. 19, fig. 1.
Imperial Formation, Upper Devonian, about 48 km southwest of Fort McPherson, Vittrekwa River, lat. 67°7′N, long. 135°29′W, Yukon.

Knoxisporites dedaleus? (Naumova) Streel, 1977
Hypotype 86085
McGregor, D.C. and McCutcheon, S.R., 1988, Can. J. Earth Sci., vol. 25, no. 9, p. 1353, pl. 1, fig. 11.
Carrow Formation, Piskahegan Group, Late Devonian, depth 72.85 m, drill hole M-6, lat. 45°34′43″N, long. 66°35′27″W, southwestern New Brunswick.

Kraeuselisporites acerosus (Arkhangelskaya) n.comb.
Hypotypes 66347-66349
McGregor, D.C. and Camfield, M., 1982, Geol. Surv. Can., Bull. 348, p. 52, pl. 14, figs. 6, 7, 9, 10.
Richardson, J.B. and McGregor, D.C., 1986, Geol. Surv. Can., Bull. 364, pl. 11, figs. 3, 4 (66349).
Weatherall (66347) and Hecla Bay formations, Middle Devonian, stream section 16.2 km east of east arm of Weatherall Bay, northeastern Melville Island, District of Franklin.

Kraeuselisporites gaspesiensis McGregor
Holotype 15175; hypotype 32019
McGregor, D.C., 1973, Palaeontographica, Abt. B, vol. 142, no. 1-3, p. 55, pl. 7, figs. 11, 12.
Battery Point Formation, Lower Devonian, 480 yards northeast of first small brook south of Rivière de l'Anse-à-Brillant, and 550 yards north of Bois Brulé Brook, south shore Gaspé Bay, Québec.

Kraeuselisporites gaspiensis? McGregor
Hypotypes 61691, 61742
McGregor, D.C., 1979, Geol. Surv. Can., Paper 79-1B, pl. 22.2, fig. 45; pl. 22.3, fig. 97.
Lower Devonian, Daleje Shale, roadside outcrop near Karlštejn, 1 km north of Budnany in the Hluboké Valley; Trebotov Limestone, northwest side of railway, north side of Prokopské Valley at Hluboepy near Prague, Czechoslovakia.

Kraeuselisporites ollii? (Chibrikova) n.comb.
Hypotypes 66343-66346
McGregor, D.C. and Camfield, M., 1982, Geol. Surv. Can., Bull. 348, p. 52, pl. 14, figs. 2-5, 8.
Hecla Bay and Weatherall (66345) formations, Middle Devonian, stream section 16.2 km east of east arm of Weatherall Bay, northeastern Melville Island, District of Franklin.

Kraeuselisporites rugosus (Owens) n.comb
Hypotypes 66350, 66351
McGregor, D.C. and Camfield, M., 1982, Geol. Surv. Can., Bull. 348, p. 53, pl. 14, figs. 11, 12.
Hecla Bay Formation, Middle Devonian, stream section 16.2 km east of east arm of Weatherall Bay, northeastern Melville Island, District of Franklin.

Kraeuselisporites spinutissimus? (Naumova ex Kedo) n.comb.
Hypotype 66352
McGregor, D.C. and Camfield, M., 1982, Geol. Surv. Can., Bull. 348, p. 53, pl. 15, figs. 1, 2.
Cape De Bray Formation, Middle Devonian, stream section 16.2 km east of east arm of Weatherall Bay, northeastern Melville Island, District of Franklin.

?*Kraeuselisporites* sp.
Fig. spec. 15192
McGregor, D.C. and Owens, B., 1966, Geol. Surv. Can., Paper 66-30, pl. 4, fig. 13.
Battery Point Formation, Lower Devonian, north side Gaspé Bay, shore cliffs between Gaspé Peninsula and Laurencelle Road, eastern Gaspé Peninsula, Québec.

?*Kraeuselisporites* sp.
Fig. spec. 61674
McGregor, D.C., 1979, Geol. Surv. Can., Paper 79-1B, pl. 22.1, fig. 26.
Zlíchov Limestone, Lower Devonian, near Pekárek Mill west of village of Slopysky, Czechoslovakia.

Kraeuselisporites sp.
Fig. spec. 61724
McGregor, D.C., 1979, Geol. Surv. Can., Paper 79-1B, pl. 22.3, fig. 79.
Daleje Shale, Lower Devonian, upper part of southeast slope of Císařská rokle, southeast of Srbsko, Czechoslovakia.

Kraeuselisporites sp.
Fig. spec. 67788
McGregor, D.C., 1984, Academia Nacional de Ciencias, Misc. No. 69, pl. 4, fig. 4.
Huamampampa Formation, Lower Devonian, Laurani section, Chuquisaca Dept., northwest of Tarabuco, Cordillera Oriental, Bolivia.

Laevigatosporites sp.
Fig. specs. 25444, 25475
McGregor, D.C., 1970, Colloque Stratig. Carbonifère, Congr. Colloques Univ. Liège, vol. 55, pl. 21, fig. 7; pl. 22, fig. 18.
Bedford and Sunbury formations, Upper Devonian, Courtright borehole, Courtright village, Lambton County, and Gibraltar 77 borehole, 6 miles south of Courtright village, Ontario.

Laevigatosporites sp.
Fig. specs. 66353, 66354
McGregor, D.C. and Camfield, M., 1982, Geol. Surv. Can., Bull. 348, p. 53, pl. 15, figs. 3, 4.
Weatherall Formation, Middle Devonian, stream section 16.2 km east of east arm of Weatherall Bay, northeastern Melville Island, District of Franklin.

Lagenicula bulbosus Chi and Hills
Holotype 43726; paratypes 43727, 43728
Chi, B.I. and Hills, L.V., 1977, Bull. Can. Petrol. Geol., vol. 24, no. 4, p. 733, pl. 8, figs. 9-12.
Imperial Formation, Upper Devonian, Imperial River 35 miles west of Norman Wells, lat. 65°07′N, long. 127°51′W, District of Mackenzie.

Lagenicula constrictus Chi and Hills
Holotype 43729; paratypes 43730, 43731
Chi, B.I. and Hills, L.V., 1977, Bull. Can. Petrol. Geol., vol. 24, no. 4, p. 735, pl. 8, figs. 13-16; pl. 9, figs. 1, 2.
Griper Bay Subgroup, Upper Devonian, 0.5 miles west of Cape Vesey Hamilton, northern Banks Island, District of Franklin.

Lagenicula devonica var. *devonica* Chaloner 1959
Hypotypes 43720, 43721
Chi, B.I. and Hills, L.V., 1977, Bull. Can. Petrol. Geol., vol. 24, no. 4, p. 730, pl. 8, figs. 1-4.
Imperial Formation, Upper Devonian, Imperial River 35 miles west of Norman Wells, lat. 65°07′N, long. 127°51′W, District of Mackenzie.

Lagenicula devonica var. *reticulatus* Chi and Hills
Holotype 43722; paratypes 43723-43725
Chi, B.I. and Hills, L.V., 1977, Bull. Can. Petrol. Geol., vol. 24, no. 4, p. 732, pl. 8, figs. 5-8.
Weatherall Formation, Middle Devonian, 0.5 mile northwest from Mould Bay Weather Station, east side Mould Bay, Prince Patrick Island, District of Franklin.

Lagenicula sp. A, Streel in Becker et al. 1974
Fig. spec. 86123
McGregor, D.C. and McCutcheon, S.R., 1988, Can. J. Earth Sci., vol. 25, no. 9, p. 1353, pl. 3, fig. 12.
Carrow Formation, Piskahegan Group, Late Devonian, depth 72.85 m, drill hole M-6, lat. 45°34′43″N, long. 66°35′27″W, southwestern New Brunswick.

large bifurcate spine
Fig. spec. 25476
McGregor, D.C., 1970, Colloque Stratig. Carbonifère, Congr. Colloques Univ. Liège, vol. 55, pl. 22, fig. 19.
Sunbury Formation, Upper Devonian, Gibraltar 77 borehole, 6 miles south of Courtright village, Lambton County, Ontario.

Large spores, cf. *Calamospora* sp.
Fig. spec. 31943
McGregor, D.C., 1973, Palaeontographica, Abt. B, vol. 142, no. 1-3, p. 15, pl. 1, fig. 4.
Early Devonian, south side of Restigouche River near Dalhousie, New Brunswick.

Leiotriletes pagius Allen of Schultz, 1968
Hypotype 32811
McGregor, D.C., 1974, Can. J. Earth Sci., vol. 11, no. 1, pl. 1, fig. 29.
Vendom Fiord Formation, Early Devonian, lat. 79°28′40″N, long. 80°02′W, 12 miles south of head of Cañon Fiord, central Ellesmere Island, District of Franklin.

Leiotriletes priddyi (Berry) Potonié and Kremp, 1955
Hypotype 13206
McGregor, D.C., 1964, Geol. Surv. Can., Bull. 109, p. 5, pl. 2, fig. 16.
Ghost River Formation (sensu stricto), Middle or Upper Devonian, Phantom Crag, 4 miles northeast of eastern end of Lake Minnewanka, Alberta.

Lophotriletes devonicus (Naumova ex Chibrikova) n.comb.
Hypotypes 66355-66357
McGregor, D.C. and Camfield, M., 1982, Geol. Surv. Can., Bull. 348, p. 54, pl. 15, figs. 5-11.
Richardson, J.B. and McGregor, D.C., 1986, Geol. Surv. Can., Bull. 364, pl. 13, fig. 3 (66356).
Hecla Bay (66355) and Weatherall formations, Middle Devonian, stream section 16.2 km east of east arm of Weatherall Bay, northeastern Melville Island, District of Franklin.

Lophotriletes devonicus (Naumova ex Chibrikova) McGregor and Camfield, 1982
Hypotype 67188
McGregor, D.C., 1984, Academia Nacional de Ciencias, Misc. No. 69, p. 32, pl. 3, fig. 8.
Huamampampa Formation, Lower Devonian, Laurani section, Chuquisaca Dept., northwest of Tarabuco, Cordillera Oriental, Bolivia.

?*Lophotriletes evlanensis* Naumova, 1953
Hypotype 15457
McGregor, D.C. and Owens, B., 1966, Geol. Surv. Can., Paper 66-30, pl. 28, fig. 8.
Kettle Point Formation, Upper Devonian, depth 295-299 feet, Courtright borehole, Courtright village, Lambton County, Ontario.

Lophotriletes portentosus? Arkhangelskaya 1972
Hypotype 96778
Arkhangelskaya, A.D., McGregor, D.C. and Richardson, J.B., 1990, Geol. Surv. Can., Paper 90-1D, pl. 1, fig. h.
Strathcona Fiord Formation, Middle Devonian, east of Bird Fiord, Ellesmere Island, District of Franklin.

Lophotriletes salebrosus var. *famenensis* Naumova, 1953
Hypotype 15465
McGregor, D.C. and Owens, B., 1966, Geol. Surv. Can., Paper 66-30, pl. 28, fig. 16.
Kettle Point Formation, Upper Devonian, depth 295-299 feet, Courtright borehole, Courtright village, Lambton County, Ontario.

Lophotriletes sp.
Fig. spec. 29397
McGregor, D.C. and Uyeno, T.T., 1972, Geol. Surv. Can., Paper 71-13, pl. 2, figs. 10.
Weatherall Formation, Middle Devonian, Beverley Inlet, southeastern Melville Island, District of Franklin.

?*Lophozonotriletes crassus* Naumova
Hypotype 13595
Kerr, J.W., McGregor, D.C. and McLaren, D.J., 1965, Bull. Can. Petrol. Geol., vol. 13, no. 3, p. 425, pl. 4, fig. 11.
Griper Bay Formation, Upper Devonian, stream valley west of Sargeant Point, 2 miles from shore, northeastern Bathurst Island, District of Franklin.

Lophozonotriletes cristifer (Luber) Kedo, 1957
Hypotypes 15434, 15435, 15468
McGregor, D.C. and Owens, B., 1966, Geol. Surv. Can., Paper 66-30, pl. 27, figs. 6-8; pl. 28, fig. 20.
McGregor, D.C., 1967, Review Palaeobotany and Palynology, I, pl. 1, fig. 27 (15434).
Upper Devonian, Griper Bay Formation, stream valley west of Sargeant Point, 2 miles from shore, northeastern Bathurst Island, District of Franklin; Kettle Point Formation, depth 295-299 feet, Courtright borehole, Courtright village, Lambton County, Ontario (15468).

Lophozonotriletes cristifer (Luber) Kedo
Hypotypes 25489, 25497
McGregor, D.C., 1970, Colloque Stratig. Carbonifère, Congr. Colloques Univ. Liège, vol. 55, pl. 23, figs. 13, 20.
Imperial Formation, Upper Devonian, lat. 66°27.5′N, long. 135°22′W, Yukon.

Lophozonotriletes cristifer (Luber) Kedo 1957
Hypotype 49853
McGregor, D.C., 1979, Palynology, vol. 3, pl. 4, fig. 19.
Imperial Formation, Upper Devonian, along Vittrekwa River approximately 30 miles southwest of Fort McPherson, lat. 67°07′N, long. 135°29′W, Yukon.
=*Cyrtospora cristifera*, Richardson, J.B. and McGregor, D.C., 1986, Geol. Surv. Can., Bull. 364, pl. 18, fig. 7.

Lophozonotriletes cristifer (Luber) Kedo (=*Torispora*? *tiara* Staplin?)
Hypotype 13625
Kerr, J.W., McGregor, D.C. and McLaren, D.J., 1965, Bull. Can. Petrol. Geol., vol. 13, no. 3, p. 425, pl. 4, fig. 14.
Griper Bay Formation, Upper Devonian, 3/4 mile from shore in stream valley entering Reindeer Bay, 4 miles south of largest river entering the bay, northeastern Bathurst Island, District of Franklin.

Lophozonotriletes curvatus Naumova, 1953
Hypotypes 15419, 15420
McGregor, D.C. and Owens, B., 1966, Geol. Surv. Can., Paper 66-30, pl. 26, figs. 22, 27, 28.
Griper Bay Formation, Upper Devonian, stream valley west of Sargeant Point, 2 miles from shore, northeastern Bathurst Island, District of Franklin.

Lophozonotriletes dentatus Hughes and Playford
Hypotype 13974
Kerr, J.W., McGregor, D.C. and McLaren, D.J., 1965, Bull. Can. Petrol. Geol., vol. 13, no. 3, p. 425, pl. 3, fig. 2.
Griper Bay Formation, Upper Devonian, stream valley west of Sargeant Point, 2 miles from shore, northeastern Bathurst Island, District of Franklin.
=?*Lophozonotriletes dentatus*, McGregor, D.C. and Owens, B., 1966, Geol. Surv. Can., Paper 66-30, pl. 26, fig. 24.

Lophozonotriletes dentatus Hughes and Playford, 1961
Hypotype 15423
McGregor, D.C. and Owens, B., 1966, Geol. Surv. Can., Paper 66-10, pl. 26, fig. 35.
Griper Bay Formation, Upper Devonian, stream valley west of Sargeant Point, 2 miles from shore, northeastern Bathurst Island, District of Franklin.

?*Lophozonotriletes grandis* var. *minor* Naumova, 1953
Hypotype 15430
McGregor, D.C. and Owens, B., 1966, Geol. Surv. Can., Paper 66-30, pl. 27, fig. 2.
Griper Bay Formation, Upper Devonian, stream valley west of Sargeant Point, 2 miles from shore, northeastern Bathurst Island, District of Franklin.

Lophozonotriletes lebedianensis Naumova, 1953
Hypotype 15422
McGregor, D.C. and Owens, B., 1966, Geol. Surv. Can., Paper 66-30, pl. 26, figs. 32, 33.
McGregor, D.C., 1967, Review Palaeobotany and Palynology, I, p. 181, pl. 1, fig. 24.
Richardson, J.B. and McGregor, D.C., 1986, Geol. Surv. Can., Bull. 364, pl. 19, fig. 6.
Griper Bay Formation, Upper Devonian, stream valley west of Sargeant Point, 2 miles from shore, northeastern Bathurst Island, District of Franklin.

?*Lophozonotriletes lebedianensis* Naumova, 1953
Hypotype 15462
McGregor, D.C. and Owens, B., 1966, Geol. Surv. Can., Paper 66-30, pl. 28, fig. 13.
Kettle Point Formation, Upper Devonian, depth 295-299 feet, Courtright borehole, Courtright village, Lambton County, Ontario.

Lophozonotriletes macrogrumosus Kedo, 1957
Hypotype 15371
McGregor, D.C. and Owens, B., 1966, Geol. Surv. Can., Paper 66-30, pl. 23, fig. 10.
Escarpment Member, Hay River Formation, Upper Devonian, left bank Hay River at base of Louise Falls, Northwest Territories.

Lophozonotriletes malevkensis Naumova
Hypotype 25490
McGregor, D.C., 1970, Colloque Stratig. Carbonifère, Congr. Colloques Univ. Liège, vol. 55, pl. 23, fig. 14.
Imperial Formation, Upper Devonian, lat. 66°27.5′N, long. 135°22′W, Yukon.

Lophozonotriletes malevkensis Kedo 1963
Hypotype 49854
McGregor, D.C., 1979, Palynology, vol. 3, pl. 4, fig. 21.
Imperial Formation, Upper Devonian, Caribou River, Yukon.
=*Tumulispora malevkensis*, Richardson, J.B. and McGregor, D.C., 1986, Geol. Surv. Can., Bull. 364, pl. 21, fig. 4.

Lophozonotriletes media Taugourdeau-Lantz
Hypotype 73302
Richardson, J.B. and McGregor, D.C., 1986, Geol. Surv. Can., Bull. 364, pl. 17, fig. 10.
Weatherall Formation, Upper Devonian, southeastern Melville Island, District of Franklin.

Lophozonotriletes? *poecilomorphus* Richardson and Ioannides, 1973
Hypotype 67171
McGregor, D.C., 1984, Academia Nacional de Ciencias, Misc. No. 69, p. 32, pl. 1, fig. 3.
Santa Rosa Formation, Lower Devonian, Laurani section, Chuquisaca Dept., northwest of Tarabuco, Cordillera Oriental, Bolivia.

Lophozonotriletes rarituberculatus (Luber) Kedo
Hypotypes 25488, 25498
McGregor, D.C., 1970, Colloque Stratig. Carbonifère, Congr. Colloques Univ. Liège, vol. 55, pl. 23, figs. 12, 21.
McGregor, D.C., 1979, Palynology, vol. 3, pl. 4, fig. 20 (25498).
Imperial Formation, Upper Devonian, lat. 66°27.5′N, long. 135°22′W, Yukon.

?*Lophozonotriletes rarituberculatus* Naumova, 1953 *non* (Luber) Kedo, 1957
Hypotype 15425
McGregor, D.C. and Owens, B., 1966, Geol. Surv. Can., Paper 66-30, pl. 26, fig. 38.
Griper Bay Formation, Upper Devonian, stream valley west of Sargeant Point, 2 miles from shore, northeastern Bathurst Island, District of Franklin.

cf. *Lophozonotriletes rarituberculatus* (Luber) Kedo
Hypotype 13973
Kerr, J.W., McGregor, D.C. and McLaren, D.J., 1965, Bull. Can. Petrol. Geol., vol. 13, no. 3, p. 425, pl. 3, fig. 1.
McGregor, D.C. and Owens, B., 1966, Geol. Surv. Can., Paper 66-30, pl. 26, fig. 31.
Griper Bay Formation, Upper Devonian, stream valley west of Sargeant Point, 2 miles from shore, northeastern Bathurst Island, District of Franklin.

Lophozonotriletes tylophorus Naumova
Hypotype 13604
Kerr, J.W., McGregor, D.C. and McLaren, D.J., 1965, Bull. Can. Petrol. Geol., vol. 13, no. 3, p. 425, pl. 3, fig. 22.
Griper Bay Formation, Upper Devonian, near mouth of an easterly flowing stream, 4 miles south of head of Reindeer Bay, northeastern Bathurst Island, District of Franklin.

Lophozonotriletes sp.
Fig. specs. 15370, 15373, 15421, 13588, 15463
McGregor, D.C. and Owens, B., 1966, Geol. Surv. Can., Paper 66-30, pl. 23, figs. 9, 12; pl. 26, figs. 29, 30, 34; pl. 28, fig. 14..
Upper Devonian, Escarpment Member, Hay River Formation, left bank Hay River at base of Louise Falls, Northwest Territories; Griper Bay Formation, stream valley west of Sargeant Point, 2 miles from shore, northeastern Bathurst Island, District of Franklin (15421, 13588); and Kettle Point Formation, depth 295-299 feet, Courtright borehole, Courtright village, Lambton County, Ontario (15463).

Lycopodiacidites ogygius McGregor
Holotype 31982
McGregor, D.C., 1973, Palaeontographica, Abt. B, vol. 142, no. 1-3, p. 35, pl. 4, figs. 7, 8.
Battery Point Formation, Lower Devonian, 400 yards east of Rivière de l'Anse-à-Brillant, south shore Gaspé Bay, Québec.

Lycospora magnifica McGregor
=*Cymbosporites magnificus*, Richardson, J.B. and McGregor, D.C., 1986, Geol. Surv. Can., Bull. 364, pl. 14, fig. 11 (holotype 13036).

cf. *Lycospora magnifica* McGregor, 1960
Hypotype 15316
McGregor, D.C. and Owens, B., 1966, Geol. Surv. Can., Paper 66-30, pl. 15, figs. 8, 9.
Griper Bay Formation, Upper Devonian, south limb of Robertson Point anticline, 8 miles northeast of Beverley Inlet, Melville Island, District of Franklin.

Minutely sculptured spore with wall 4.5 thick
Fig. spec. 32827
McGregor, D.C., 1974, Can. J. Earth Sci., vol. 11, no. 1, pl. 2, fig. 52.
Vendon Fiord Formation, Early Devonian, lat. 79°28′40″N, long. 80°02′W, 12 miles south of head of Cañon Fiord, central Ellesmere Island, District of Franklin.

New genus n. sp.
Fig. spec. 29379
McGregor, D.C. and Uyeno, T.T., 1972, Geol. Surv. Can., Paper 71-13, pl. 1, fig. 11.
Weatherall Formation, Middle Devonian, Beverley Inlet, southeastern Melville Island, District of Franklin.

new species
Fig. spec. 27066
McGregor, D.C., Sanford, B.V. and Norris, A.W, 1970, Proc. Geol. Assoc. Can., vol. 22, pl. 1, fig. 7.
Kenogami River Formation, Lower Devonian, 1072 feet depth, ODM Jaab Lake No. 1 well, lat. 51°11.9′N, long. 82°57′W, Moose River Basin, Ontario.

Nikitinsporites canadensis Chaloner 1959
Hypotypes 43732, 43733
Chi, B.I. and Hills, L.V., 1977, Bull. Can. Petrol. Geol., vol. 24, no. 4, p. 737, pl. 9, figs. 3, 4.
Weatherall Formation, Middle Devonian, 0.5 mile northwest of Mould Bay Weather Station, east side Mould Bay, Prince Patrick Island, District of Farnklin.

Nikitinsporites canadensis Chaloner
Hypotype 73301
Richardson, J.B. and McGregor, D.C., 1986, Geol. Surv. Can., Bull. 364, pl. 17, fig. 8.
Beverley Inlet Formation, Upper Devonian, southeastern Melville Island, District of Franklin.

Nikitinsporites radiatus Chi and Hills
Holotype 43735; paratypes 43734, 43736-43738
Chi, B.I. and Hills, L.V., 1977, Bull. Can. Petrol. Geol., vol. 24, no. 4, p. 739, pl. 9, figs. 5-8; pl. 10, figs. 1-3.
Beverley Inlet Formation, Upper Devonian, Cape Terrace, west side Melville Island, District of Franklin.

Nikitinsporites simplex Chi and Hills
Holotype 43739; paratypes 43740, 43741
Chi, B.I. and Hills, L.V., 1977, Bull. Can. Petrol. Geol., vol. 24, no. 4, p. 741, pl. 10, figs. 4-9.
Beverley Inlet Formation, Upper Devonian, Cape Terrace, west side Melville Island, District of Franklin.

Nikitinsporites verrucatus Chi and Hills
Holotype 43742; paratypes 43743, 43744
Chi, B.I. and Hills, L.V., 1977, Bull. Can. Petrol. Geol., vol. 24, no. 4, p. 742, pl. 11, figs. 1-6.
Beverley Inlet Formation, Upper Devonian, Cape Terrace, west side Melville Island, District of Franklin.

Obligate tetrad
Fig. spec. 29369
McGregor, D.C. and Uyeno, T.T., 1972, Geol. Surv. Can., Paper 71-13, pl. 1, fig. 1.
Bathurst Island Formation, Lower Devonian, Young Inlet, northeastern Bathurst Island, District of Franklin.

Ocksisporites connatispinosus Hills, Chi and Sweet
Holotype 34067; paratypes 34126-34128, 34130; hypotypes 34125, 34129
Hills, L.V., Chi., B.I. and Sweet, A.R., 1975, Rev. Palaeobot. Palynol., vol. 19, p. 107, pl. II, figs. A-L.
Chi, B.I. and Hills, L.V., 1977, Bull. Can. Petrol. Geol., vol. 24, no. 4, p. 746, pl. 12, fig. 4 (34067).
Griper Bay Formation, Upper Devonian, 0.5 mile northwest from Mould Bay Weather Station, east side of Mould Bay, Prince Patrick Island, and east arm Weatherall Bay, lat. 76°14′N, long. 119°30′W, Melville Island (34130), District of Franklin.

Ocksisporites maclarenii Chaloner
Topotype 34069; hypotypes 34121-34124
Hills, L.V., Chi., B.I. and Sweet, A.R., 1975, Rev. Palaeobot. Palynol., vol. 19, p. 104, pl. I, figs. A-J.
Chi, B.I. and Hills, L.V., 1977, Bull. Can. Petrol. Geol., vol. 24, no. 4, p. 745, pl. 12, figs. 1-3 (34121).
Upper Devonian, Okse Bay Formation, Okse Bay, Ellesmere Island, District of Franklin; Imperial Formation, Imperial River 35 miles west of Norman Wells, lat. 65°07′N, long. 127°51′W, District of Mackenzie (34121); Griper Bay Formation, Mould Bay Weather Station (34122, 34123) and west side of Mould Bay (34124), Prince Patrick Island, District of Franklin.

Ocksisporites rugulatus Hills, Chi and Sweet
Holotype 34068; paratypes 34131, 34133, 34135; hypotypes 34132, 34134, 34136
Hills, L.V., Chi., B.I. and Sweet, A.R., 1975, Rev. Palaeobot. Palynol., vol. 19, p. 110, pl. III, figs. A-L.
Chi, B.I. and Hills, L.V., 1977, Bull. Can. Petrol. Geol., vol. 24, no. 4, p. 746, pl. 12, figs. 5-7 (34068).
Griper Bay Formation, Upper Devonian, east arm Weatherall Bay, lat. 75°43′N, long. 106°50′W, Melville Island, District of Franklin.

Ocksisporites rugulatus Hills, Chi and Sweet 1975
Hypotype 43747
Chi, B.I. and Hills, L.V., 1977, Bull. Can. Petrol. Geol., vol. 24, no. 4, p. 746, pl. 12, figs. 8-11.
Beverley Inlet Formation, Upper Devonian, east arm Weatherall Bay, lat. 75°43′N, long. 106°50′W, Melville Island, District of Franklin.

Periplecotriletes tortus Egorova
Hypotype 73290
Richardson, J.B. and McGregor, D.C., 1986, Geol. Surv. Can., Bull. 364, pl. 11, figs. 5, 6.
Strathcona Fiord Formation, Middle Devonian, east of Bird Fiord, southwestern Ellesmere Island, District of Franklin.

Periplecotriletes tortus Egorova 1974
Hypotype 96779
Arkhangelskaya, A.D., McGregor, D.C. and Richardson, J.B., 1990, Geol. Surv. Can., Paper 90-1D, pl. 1, fig. i.
Strathcona Fiord Formation, Middle Devonian east of Bird Fiord, Ellesmere Island, District of Franklin.

Perotrilites bifurcatus Richardson
Hypotypes 66360, 66361
McGregor, D.C. and Camfield, M., 1982, Geol. Surv. Can., Bull. 348, p.55, pl. 15, figs. 13-15.
Weatherall Formation, Middle Devonian, stream section 16.2 km east of east arm of Weatherall Bay, northeastern Melville Island, District of Franklin.

cf. *Perotrilites bifurcatus* Richardson
Hypotype 16589
McGregor, D.C., 1967, Review Palaeobotany and Palynology, I, p. 180, pl. 1, figs. 5, 6.
Bird Fiord Formation, Middle Devonian, Helena Island, District of Franklin.
=*Perotrilites bifurcatus*, McGregor, D.C., 1979, Palynology, vol. 3, pl. 2, fig. 8.

Perotrilites conatus Richardson
Hypotype 16590
McGregor, D.C., 1967, Review Palaeobotany and Palynology, I, p. 180, pl. 1, figs. 11, 12.
Bird Fiord Formation, Middle Devonian, Helena Island, District of Franklin.

Perotrilites conatus Richardson 1965
Hypotype 49841
McGregor, D.C., 1979, Palynology, vol. 3, pl. 3, fig. 10.
Weatherall Formation, Middle Devonian, creek section 5.5 to 10 miles north of head of Beverley Inlet, Melville Island, District of Franklin.

?*Perotrilites ergatus* Allen
Hypotype 29378
McGregor, D.C. and Uyeno, T.T., 1972, Geol. Surv. Can., Paper 71-13, pl. 1, fig. 10.
Stuart Bay Formation, Lower Devonian, Young Inlet, northeastern Bathurst Island, District of Franklin.

Perotrilites ergatus? Allen
Hypotypes 66358, 66359
McGregor, D.C. and Camfield, M., 1982, Geol. Surv. Can., Bull. 348, p.55, pl. 15, figs. 12, 16, 17.
Weatherall Formation, Middle Devonian, stream section 16.2 km east of east arm of Weatherall Bay, northeastern Melville Island, District of Franklin.

Perotrilites heclaensis McGregor and Camfield
Holotype 66362; paratypes 66363, 66364
McGregor, D.C. and Camfield, M., 1982, Geol. Surv. Can., Bull. 348, p. 55, pl. 15, figs. 18-21.
Richardson, J.B. and McGregor, D.C., 1986, Geol. Surv. Can., Bull 364, pl. 15, fig. 3 (66362).
Hecla Bay Formation, Middle Devonian, stream section 16.2 km east of east arm of Weatherall Bay, northeastern Malville Island, District of Franklin.

Perotrilites perinatus Hughes and Playford 1961
Hypotype 49855
McGregor, D.C., 1979, Palynology, vol. 3, pl. 4, fig. 25.
Imperial Formation, Upper Devonian, along Vitterkwa River, approximately 30 miles southwest of Fort McPhereson, lat. 67°07′N, long. 135°29′W, Yukon.

Perotrilites selectus (Arkhangelskaya)
Hypotype 41732
McGregor, D.C. and Camfield, M., 1976, Geol. Surv. Can., Bull. 263, p. 55, pl. 8, figs. 7, 8.
Williams Island Formation, Middle Devonian, 147-149 feet depth, ODM Jaab Lake no. 1 well, lat. 51°11′54″N, long. 82°56′W, Ontario.

Perotrilites selectus (Arkhangelskaya) McGregor and Camfield
Hypotypes 66365, 66366

McGregor, D.C. and Camfield, M., 1982, Geol. Surv. Can., Bull. 348, p. 56, pl. 15, figs. 22, 23.
Hecla Bay and Weatherall formations, Middle Devonian, 16.2 km east of east arm of Weatherall Bay, northeastern Melville Island, District of Franklin.

Perotrilites sp.
Fig. specs. 13601, 13615
Kerr, J.W., McGregor, D.C. and McLaren, D.J., 1965, Bull. Can. Petrol. Geol., vol. 13, no. 3, p. 425, pl. 3, fig. 19; pl. 4, fig. 7.
Griper Bay Formation, Upper Devonian, stream valley west of Sargeant Point, 2 miles from shore, and 3/4 mile from shore in stream valley entering Reindeer Bay, 4 miles south of largest river entering the bay, northeastern Bathurst Island, District of Franklin.

?*Perotrilites* sp.
Fig. specs. 15259, 15291, 15473
McGregor, D.C. and Owens, B., 1966, Geol. Surv. Can., Paper 66-30, pl. 5, fig. 13; pl. 9, fig. 6; pl. 28, fig. 5.
Battery Point Formation, Lower Devonian, south side Gaspé Bay, shore section from Bois Brulé Brook to Pointe Jaune, eastern Gaspé Peninsula, Québec; Weatherall Formation, Middle Devonian, south limb of Robertson Point Anticline, 8 miles northeast of Beverley Inlet, Melville Island, District of Franklin; Kettle Point Formation, Upper Devonian, depth 295-299 feet, Courtright borehole, Courtright village, Lambton County, Ontario.

Perotrilites sp.
Fig. specs. 15397, 15432
McGregor, D.C. and Owens, B., 1966, Geol. Surv. Can., Paper 66-30, pl. 25, fig. 5; pl. 22, fig. 4.
Upper Devonian, Escarpment Member, Hay River Formation, left bank Hay River at base of Louise Falls, Northwest Territories; Griper Bay Formation, stream valley west of Sargeant Point, 2 miles from shore, northeastern Bathurst Island, District of Franklin.

?*Perotrilites* sp.
Fig. spec. 86092
McGregor, D.C. and McCutcheon, S.R., 1988, Can. J. Earth Sci., vol. 25, no. 9, p. 1313, pl. 1, figs. 20, 21.
Carrow Formation, Piskahegan Group, Late Devonian, depth 72.85 m, drill hole M-6, lat. 45°34′43″N, long. 66°35′27″W, souuthwestern New Brunswick.

Phyllothecotriletes rotundus Streel, 1964
Hypotype 15251
McGregor, D.C. and Owens, B., 1966, Geol. Surv. Can., Paper 66-30, pl. 5, fig. 4.
Battery Point Formation, Lower Devonian, south side Gaspé Bay, shore section from Bois Brulé Brook to Pointe Jaune, eastern Gaspé Peninsula, Québec.

Phyllothecotriletes sp.
Fig. specs. 15344, 15349
McGregor, D.C. and Owens, B., 1966, Geol. Surv. Can., Paper 66-30, pl. 22, figs. 5, 10.
Escarpment Member, Hay River Formation, Upper Devonian, left bank Hay River at base of Louise Falls, Northwest Territories.

Pilasporites plurigenus Balme and Hennelly, 1956
Hypotype 13205
McGregor, D.C., 1964, Geol. Surv. Can., Bull. 109, p. 3, pl. 2, fig. 15.
Ghost River Formation (sensu stricto), Middle or Upper Devonian, Phantom Crag, 4 miles northeast of eastern end of Lake Minnewanka, Alberta.

Planisporites sp.
Fig. spec. 13210
McGregor, D.C., 1964, Geol. Surv. Can., Bull. 109, p. 15, pl. 2, fig. 19.
Ghost River Formation (sensu stricto), Middle or Upper Devonian, Phantom Crag, 4 miles northeast of eastern end of Lake Minnewanka, Alberta.

Planisporites sp.
Fig. specs. 15116, 15158
McGregor, D.C. and Owens, B., 1966, Geol. Surv. Can., Paper 66-30, pl. 1, fig. 16; pl. 2, fig. 27.
Shiphead Member, Grand Grève and Battery Point formations, Lower Devonian, Cape Gaspé, shore cliff below lighthouse, and south side Gaspé Bay, shore section from Anse-à-Brillant to Bois Brulé Brook, eastern Gaspé Peninsula, Québec.

Poorly preserved zonata-camerate spore
Fig. spec. 32836
McGregor, D.C., 1974, Can. J. Earth Sci., vol. 11, no. 1, pl. 2, fig. 61.
Vendom Fiord Formation, Early Devonian, lat. 79°28′40″N, long. 80°02′W, 12 miles south of head of Cânon Fiord, central Ellesmere Island, District of Franklin.

Procoronospora sp.
Fig. spec. 27083
McGregor, D.C., Sanford, B.V. and Norris, A.W., 1970, Proc. Geol. Assoc. Can., vol. 22, pl. 2, figs. 1, 2.
Stooping River Formation, Lower Devonian, 725 feet depth, ODM Jaab Lake No. 1 well, lat. 51°11.9′N, long. 82°57′W, Moose River Basin.
=*Camarozonotriletes sextantii*, Richardson, J.B. and McGregor, D.C., 1986, Geol. Surv. Can., Bull. 364, pl. 6, figs. 11, 12.

Procoronaspora sp.
Fig. spec. 29374
McGregor, D.C. and Uyeno, T.T., 1972, Geol. Surv. Can., Paper 71-13, pl. 1, fig. 6.
Eids Formation, Middle Devonian, Young Inlet, northeastern Bathurst Island, District of Franklin.

cf. *Procoronaspora* sp.
Fig. spec. 15161
McGregor, D.C. and Owens, B., 1966, Geol. Surv. Can., Paper 66-30, pl. 2, fig. 30.

Battery Point Formation, Lower Devonian, south side Gaspé Bay, shore section from Anse-à-Brillant to Bois Brulé Brook, easthern Gaspé Peninsula, Québec.

Punctatisporites inspissatus (Owens) n. comb.
Hypotypes 66367, 66368
McGregor, D.C. and Camfield, M., 1982, Geol. Surv. Can., Bull. 348, p. 56, pl. 16, figs. 1, 2.
Hecla Bay and Weatherall formations, Middle Devonian, 16.2 km east of east arm of Weatherall Bay, northeastern Melville Island, District of Franklin.

Punctatisporites? dilutus Hoffmeister, 1959
Hypotype 67189
McGregor, D.C., 1984, Academia Nacional de Ciensias, Misc. No. 69, p. 33, pl. 1, fig. 14.
Santa Rosa Formation, Lower Devonian, Laurani section, Chuquisaca Dept., northwest of Tarabuco, Cordillera Oriental, Bolivia.

?*Punctatisporites parvivermiculatus* Playford, 1962
Hypotype 15354
McGregor, D.C. and Owens, B., 1966, Geol. Surv. Can., Paper 66-30, pl. 22, fig. 16.
Escarpment Member, Hay River Formation, Upper Devonian, left bank Hay River at base of Louise Falls, Northwest Territories.

Punctatisporites sp. cf. *P. solidus* Hacquebard
Fig. spec. 66369
McGregor, D.C. and Camfield, M., 1982, Geol. Surv. Can., Bull. 348, p. 57, pl. 16, fig. 3.
Weatherall Formation, Middle Devonian, stream section 16.2 km east of east arm of Weatherall Bay, northeastern Melville Island, District of Franklin.

Punctatisporites sp.
Fig. specs. 15103, 15198, 15340, 15341, 15347
McGregor, D.C. and Owens, B., 1966, Geol. Surv. Can., Paper 66-30, pl. 1, fig. 1; pl. 22, figs. 1, 2, 7.
Forillon Member, Cape Bon Ami and Battery Point formations, north side Forillon Peninsula, Cape Rosier Cove, and south side Gaspé Bay, shore section from Bois Brulé Brook to Pointe Jaune, eastern Gaspé Peninsula, Québec; Escarpment Member, Hay River Formation, left bank Hay River at base of Louise Falls, northwest Territories.

Punctatisporites sp.
Fig. spec. 32817
McGregor, D.C., 1974, Can. J. Earth Sci., vol. 11, no. 1, pl. 1, fig. 36.
Vendom Fiord Formation, Early Devonian, lat. 79°28′40″N, long. 80°02′W, 12 miles south of head of Cañon Fiord, central Ellesmere Island, District of Franklin.

?*Punctatisporites* sp.
Fig. specs. 15115, 15456
McGregor, D.C. and Owens, B., 1966, Geol. Surv. Can., Paper 66-30, pl. 1, fig. 15; pl. 27, fig. 7.
Shiphead Member, Grand Grève Formation, Lower Devonian, Cape Gaspé, shore cliff below lighthouse, eastern Gaspé Peninsula, Québec; Kettle Point Formation, Upper Devonian, depth 295-299 feet, Courtright borehole, Courtright village, Lambton County, Ontario.

Punctatisporites sp. A, B
Fig. specs. 31946, 31947
McGregor, D.C., 1973, Palaeontographica, Ant. B, vol. 142, no.1-3, p. 17, pl. 1, figs. 10, 11.
Battery Point Formation, Middle and Lower Devonian, 100 yards south of most prominent point between Cap Rouge and Pointe Jaune, and 400 yards southeast of Tar Point dyke, south shore Gaspé Bay, Québec.

Punctatisporites spp. 1, 2
Fig. specs. 67190, 67774
McGregor, D.C., 1984, Academia Nacional de Ciencas, Misc. No. 69, p. 33, pl. 2, figs. 12, 13.
Santa Rosa and Huamampampa formations, Lower and Middle Devonian, Laurani section, Chuquisaca Dept., northwest of Tarabuco, Cordillera Oriental, Bolivia.

Pustulatisporites gibberosus (Hacquebard) Playford
Hypotype 13606
Kerr, J.W., McGregor, D.C. and McLaren, D.J., 1965, Bull. Can. Petrol. Geol., vol. 13, no. 3, p. 425, pl. 3, fig. 24.
Griper Bay Formation, Upper Devonian, 3/4 mile from shore in stream entering Reindeer Bay, 4 miles south of largest river entering the bay, northeastern Bathurst Island, District of Franklin.

Pustulatisporites sp.
Fig. specs. 15170, 15408, 15409, 15413
McGregor, D.C. and Owens, B., 1966, Geol. Surv. Can., Paper 66-30, pl. 3, fig. 5; pl. 26, figs. 10, 11, 16.
Battery Point Formation, Lower Devonian, south side Gaspé Bay, shore section from Anse-à-Brillant to Bois Brulé Brook, eastern Gaspé Peninsula, Québec (15170); Griper Bay Formation, Upper Devonian, stream valley west of Sargeant Point, 2 miles from shore, northeastern Bathurst Island, District of Franklin.

Pustulatisporites sp.
Fig. spec. 31981
McGregor, D.C., 1973, Palaeontographica, Abt. B, vol. 142, no. 1-3, p. 33, pl. 4, figs. 5, 6, 12.
Battery Point Formation, Middle Devonian, 100 yards south of most prominent point between Cap Rouge and Pointe Jaune, south shore Gaspé Bay, Québec.

Raistrickia aratra Allen
Hypotype 16594
McGregor, D.C., 1967, Review Palaeobotany and Palynology, I, p. 180, pl. 1, fig. 21.
Hecla Bay Formation, Middle or Upper Devonian, Helena Island, District of Franklin.

Raistrickia aratra Allen
Hypotype 41771
McGregor, D.C. and Camfield, M., 1976, Geol. Surv. Can., Bull. 263, p. 25, pl. 6, fig. 6.
Williams Island Formation, Middle Devonian, 159.5-159.5 feet depth, ODM Jaab Lake no. 1 well, lat. 51°11′54″N, long. 82°56′W, Ontario.

Raistrickia corynoges Sullivan
Hypotype 73315
Richardson, J.B. and McGregor, D.C., 1986, Geol. Surv. Can., Bull. 364, pl. 21, fig. 9.
Harrylock Formation, Upper Devonian, near Shoal Rock, Hook Head, southern Ireland.

Raistrickia variabilis Dolby and Neves
Hypotype 73312
Richardson, J.B. and McGregor, D.C., 1986, Geol. Surv. Can., Bull. 364, pl. 21, fig. 1.
Harrylock Formation, Upper Devonian, Hook Head, southern Ireland.

Raistrickia variabilis Dolby and Neves, 1970
Hypotype 86002
McGregor, D.C. and McCuthcheon, S.R., 1988, Can. J. Earth Sci., vol. 25, no. 9, p. 1353, pl. 1, figs. 7, 8.
Carrow Formation, Piskahegan Group, Late Devonian, depth 72.85 m, drill hole M-6, lat. 45°34′47″N, long. 66°35′27″W, southwestern New Brunswick.

Raistrickia sp.
Fig. spec. 31983
McGregor, D.C., 1973, Palaeontographica, Abt.B, vol. 142, no. 1-3, p. 35, pl. 4, figs. 9, 10.
Battery Point Formation, Lower Devonian, 550 yards north of Bois Brulé Brook, south shore Gaspé Bay, Québec.

?*Retialetes* sp.
Fig. spec. 15104
McGregor, D.C. and Owens, B., 1966, Geol. Surv. Can., Paper 66-30, pl. 1, fig. 2.
Forillon Member, Cape Bon Ami Formation, Lower Devonian, north side Forillon Peninsula, Cape Rosier Cave, eastern Gaspé Peninsula, Québec.

cf. *Reticulatisporites emsiensis* Allen, 1965
Hypotype 15067
McGregor, D.C. and Owens, B., 1966, Geol. Surv. Can., Paper 66-30, pl. 4, fig. 3.
Battery Point Formation, Lower Devonian, south side Gaspé Bay, shore section from Anse-à-Brillant to Bois Brulé Brook, eastern Gaspé Peninsula, Québec.

Reticulatisporites? *fimbriatus* Winslow
Hypotype 25446
McGregor, D.C., 1970, Colloque Stratig. Carbonifère, Congr, Colloques Univ. Liège, vol. 55, pl. 21, fig. 9.
Bedford Formation, Upper Devonian, Courtright borehole, Courtright village, Lambton County, Ontario.

?*Reticulatisporites fimbriatus* var. *spathulatus* Winslow, 1962
Hypotype 15474
McGregor, D.C. and Owens, B., 1966, Geol. Surv. Can., Paper 66-30, pl. 28, figs. 26, 27.
Kettle Point Formation, Upper Devonian, depth 295-299 feet, Courtright borehole, Courtright village, Lambton County, Ontario.

Reticulatisporites? *fimbriatus* var. *spathulatus* Winslow
Hypotype 25473
McGregor, D.C., 1970, Colloque Stratig. Carbonifère, Congr, Colloques Univ. Liège, vol. 55, pl. 22, fig. 16.
Sunbury Formation, Upper Devonian, Gibraltar 77 borehole, 6 miles south of Courtright village, Lambton County, Ontario.

Reticulatisporites sp. cf. *Dictyotriletes minor* Naumova *in* Allen, 1965
Fig. spec. 15180
McGregor, D.C. and Owens, B., 1966, Geol. Surv. Can., Paper 66-30, pl. 3, fig. 18.
Battery Point Formation, Lower Devonian, south side Gaspé Bay, shore section from Anse-à-Brillant to Bois Brulé Brook, eastern Gaspé Peninsula, Québec.

Reticulatisporites vulgaris (Arkhangelskaya) Arkhangelskaya 1985
Hypotype 96784
Arkhangelskaya, A.D., McGregor, D.C. and Richardson, J.B., 1990, Geol. Surv. Can., Paper 90-1D, pl. 1, fig. n.
Strathcona Fiord Formation, Middle Devonian, east of Bird Fiord, Ellesmere Island, District of Franklin.

Reticulatisporites sp.
Fig. spec. 13596
Kerr, J.W., McGregor, D.C. and McLaren, D.J., 1965, Bull. Can. Petrol. Geol., vol. 13, no. 3, p. 425, pl. 3, fig. 14.
Griper Bay Formation, Upper Devonian, stream valley west of Sargeant Point, 2 miles from shore, northeastern Bathurst Island, District of Franklin.

Reticulatisporites sp., *sensu* Potonié and Kremp, 1955
Fig. spec. 15188
McGregor, D.C. and Owens, B., 1966, Geol. Surv. Can., Paper 66-30, pl. 4, fig. 7.
Battery Point Formation, Lower Devonian, south side Gaspé Bay, shore section from Anse-à-Brillant to Bois Brulé Brook, eastern Gaspé Peninsula, Québec.

Reticulatisporites sp.
Fig. spec. 15131
McGregor, D.C. and Owens, B., 1966, Geol. Surv. Can., Paper 66-30, pl. 1, fig. 32.
Shiphead Member, Grand Grève Formation, Lower Devonian, Cape Gaspé, shore cliff below lighthouse, eastern Gaspé Peninsula, Québec.

Retispora archaelepidophyta (Kedo) n. comb.
Hypotypes 66370-66372
McGregor, D.C. and Camfield, M., 1982, Geol. Surv. Can., Bull. 348, p. 57, pl. 16, figs. 4-7, 10.
Richardson, J.B. and McGregor, D.C., 1986, Geol. Surv. Can., Bull. 364, pl. 14, fig. 4 (66372).
Hecla Bay and Weatherall (66371) formations, Middle Devonian, stream section 16.2 km east of east arm of Weatherall Bay, northeastern Melville Island, District of Franklin.

Retispora cassicula (Higgs) Higgs and Russell, 1981
Hypotype 86109
McGregor, D.C. and McCutcheon, S.R., 1988, Can. J. Earth Sci., vol. 25, no. 9, p. 1353, pl. 2, figs. 13, 14.
Carrow Formation, Piskahegan Group, Late Devonian, depth 72.85 m, drill hole M-6, lat. 45°34′43″N, long. 66°35′27″W, southwestern New Brunswick.

Retispora lepidophyta (Kedo) Playford
Hypotype 73310
Richardson, J.B. and McGregor, D.C., 1986, Geol. Surv. Can., Bull. 364, pl. 20, fig. 4.
Harrylock Formation, Upper Devonian, Hook Head, southern Ireland.

Retispora lepidophyta? (Kedo) Playford, 1976
Hypotype 86110
McGregor, D.C. and McCutcheon, S.R., 1988, Can. J. Earth Sci., vol. 25, no. 9, p. 1353, pl. 2, figs. 15, 16.
Carrow Formation, Piskahegan Group, Late Devonian, depth 72.85 m, drill hole M-6, lat. 45°34′43″N, long. 66°35′27″W, southwestern New Brunswick.

Retispora sp.
Fig. spec. 66373
McGregor, D.C. and Camfield, M., 1982, Geol. Surv. Can., Bull. 348, p. 57, pl. 16, figs. 8, 9.
Hecla Bay Formation, Middle Devonian, stream section 16.2 km east of east arm of Weatherall Bay, northeastern Melville Island, District of Franklin.

?*Retispora* sp.
Fig. spec. 86122
McGregor, D.C. and McCutcheon, S.R., 1988, Can. J. Earth Sci., vol. 25, no. 9, p. 1353, pl. 3, figs. 8, 9.
Carrow Formation, Piskahegan Group, Late Devonian, depth 72.85 m, drill hole M-6, lat. 45°34′43″N, long. 66°35′27″W, southwestern New Brunswick.

Retusotriletes actinomorphus Chibrikova, 1962
Hypotypes 15194
McGregor, D.C. and Owens, B., 1966, Geol. Surv. Can., Paper 66-30, pl. 4, fig. 15.
Battery Point Formation, Lower Devonian, south side Gaspé Bay, shore section from Anse-à-Brillant to Bois Brulé Brook, eastern Gaspé Peninsula, Québec.

Retusotriletes actinomorphus Chibrikova
Hypotypes 31948-31950
McGregor, D.C., 1973, Palaeontographica, Abt.B, vol. 142, no. 1-3, p. 18, pl. 1, figs. 12-14.
Battery Point Formation, Lower Devonian, 500 and 400 yards east of Rivière de l'Anse-à-Brillant, south shore Gaspé Bay, Québec.

?*Retusotriletes actinomorphus* Chibrikova
Hypotype 32812
McGregor, D.C., 1974, Can. J. Earth Sci., vol. 11, no. 1, pl. 1, fig. 31.
Vendom Fiord Formation, Early Devonian, lat. 79°28′40″N, long. 80°02′W, 12 miles south of head of Cañon Fiord, central Ellesmere Island, District of Franklin.

Retusotriletes sp. cf. *R. actinomorphus* Chibrikova
Fig. spec. 61650
McGregor, D.C., 1979, Geol. Surv. Can., Paper 79-1B, pl. 22, fig. 2.
Dvorce-Prokop Limestone, Lower Devonian, road-cut on west side of Vitava River below Barrandov, southern Prague, north of old quarry "Ukapliky", Czechoslovakia.

Retusotriletes ?*actinomorphus* Chibrikova 1962
Hypotypes 48176-48181
McGregor, D.C., 1977, Palaeontographica, Abt.B, vol. 163, p. 132, pl. 2, figs. 7-12.
York River Formation, Lower Devonian, north of Anse-à-Brillant, south shore Gaspé Bay, Québec.

Retusotriletes cf. *avonensis* Playford, of Kemp, 1972
Hypotype 67191
McGregor, D.C., 1984, Academia Nacional de Ciensias, Misc. No. 69, p. 34, pl. 1, fig. 16.
Santa Rosa Formation, Lower Devonian, Laurani section, Chuquisaca Dept., northwest of Tarabuco, Cordillera Oriental, Bolivia.

Retusotriletes biarealis McGregor
Holotype 13192; paratype 13193
McGregor, D.C., 1964, Geol. Surv. Can., Bull. 109, p. 7, pl. 1, figs. 13-15.
Ghost River Formation, Middle or Upper Devonian, Phantom Crag, 4 miles northeast of eastern end of Lake Minnewanka, Alberta.

?*Retusotriletes biornatus* Schultz
Hypotype 61703
McGregor, D.C., 1979, Geol. Surv. Can., Paper 79-1B, pl. 22.2, fig. 57.
Daleje Shale, Lower Devonian, upper part of southeast slope of Císařská rokle, southeast of Srbsko, Czechoslovakia.

Retusotriletes cf. *R. concinnus* Kedo of Lanninger, in Kemp, 1972
Hypotype 67192
McGregor, D.C., 1984, Academia Nacional de Ciensias, Misc. No. 69, p. 34, pl. 2, fig. 11.
Icla Formation, Middle Devonian, Laurani section, Chuquisaca Dept., northwest of Tarabuco, Cordillera Oriental, Bolivia.

Retusotriletes crassus Clayton in Clayton et al. 1980
Hypotype 86077
McGregor, D.C. and McCutcheon, S.R., 1988, Can. J. Earth Sci., vol. 25, no. 9, p. 1353, pl. 1, fig. 1.
Carrow Formation, Piskahegan Group, Late Devonian, depth 72.85 m, drill hole M-6, lat. 45°34′43″N, long. 66°35′27″W, southwestern New Brunswick.

Retusotriletes distinctus Richardson 1965
Hypotype 30697
Owens, B. and Richardson, J.B., 1972, 7th Internat. Congr. Carboniferous Strat. Geol. Compte Rendu, vol. 1, pl. 1, fig. 13.
Griper Bay Formation, Upper Devonian, Melville Island, District of Franklin.

Retusotriletes distinctus Richardson
Hypotype 32804
McGregor, D.C., 1974, Can. J. Earth Sci., vol. 11, no. 1, pl. 1, fig. 23.
Vendom Fiord Formation, Early Devonian, lat. 79°28′N, long. 80°02′W, 12 miles south of head of Cañon Fiord, central Ellesmere Island, District of Franklin.

Retusotriletes ?distinctus Richardson
Hypotype 41751
McGregor, D.C. and Camfield, M., 1976, Geol. Surv. Can., Bull. 263, p. 25, pl. 4, fig. 15.
Stooping River Formation, Lower Devonian, 815-817 feet depth, ODM Jaab Lake no. 1 well, lat. 51°11′54N, long. 82°56′W, Ontario.

Retusotriletes distinctus Richards 1965
Hypotype 43659
Chi, B.I. and Hills, L.V., 1977, Bull. Can. Petrol. Geol., vol. 24, no. 4, p. 692, pl. 1, fig. 2.
Weatherall Formation, Middle Devonian, Cape Terrace, west side Melville Island, District of Franklin.

Retusotriletes cf. *R. distinctus* Richardson, 1965
Hypotype 15311
McGregor, D.C. and Owens, B., 1966, Geol. Surv. Can., Paper 66-30, pl. 15, figs. 1, 2.
Griper Bay Formation, Upper Devonian, south limb of Robertson Point anticline, 8 miles northeast of Beverley Inlet, Melville Island, District of Franklin.

Retusotriletes cf. *distinctus* Richardson 1965
Hypotype 49820
McGregor, D.C., 1979, Palynology, vol. 3, pl. 1, fig. 17.
Williams Island Formation, Middle Devonian, southeast side of island in Nine Mile Rapids, Abitibi River, Ontario.

Retusotriletes dittonensis Richardson and Lister
Hypotype 27060
McGregor, D.C., Sanford, B.V. and Norris, A.W., 1970, Proc. Geol. Asoc. Can., vol. 22, pl. 1, fig. 1.
Kenogami River Formation, Lower Devonian, 1153 feet depth, ODM Jaab Lake No. 1, lat. 51°11.9′N, long. 82°57′W, Moose River Basin, Ontario.

Retusotriletes dubiosus McGregor
Hypotype 31951
McGregor, D.C., 1973, Palaeontographica, Abt.B, vol. 142, no. 1-3, p. 21, pl. 2, fig. 1.
Battery Point Formation, Lower Devonian, 1300 yards south of second stream south of Bois Brulé Brook, south shore Gaspé Bay, Québec.

Retusotriletes dubiosus McGregor
Hypotype 41735
McGregor, D.C. and Camfield, M., 1976, Geol. Surv. Can., Bull. 263, p. 25, pl. 3, fig. 10.
Stooping River Formation, Lower Devonian, 617-619 feet depth, ODM Jaab Lake No. 1 well, lat. 51°11′54″N, long. 82°56′W, Ontario.

Retusotriletes dubiosus McGregor
Hypotype 66374
McGregor, D.C. and Camfield, M., 1982, Geol. Surv. Can., Bull. 348, p. 57, pl. 16, fig. 11.
Hecla Bay Formation, Middle Devonian, stream section 16.2 km east of east arm of Weatherall Bay, northeastern Melville Island, District of Franklin.

Retusotriletes cf. *R. dubius* (Eisenack) Richardson
Hypotype 41740
McGregor, D.C. and Camfield, M., 1976, Geol. Surv. Can., Bull. 263, p. 25, pl. 3, fig. 17.
Stooping River Formation, Lower Devonian, 909-911 feet depth, ODM Jaab Lake No. 1 well, lat. 51°11′54″N, long. 82°56′W, Ontario.

Retusotriletes greggsii McGregor
Holotype 13184; paratypes 13185-13191
McGregor, D.C., 1964, Geol. Surv. Can., Bull. 109, p. 8, pl. 1, figs. 1-12.
Ghost River Formation, Middle or Upper Devonian, Phantom Crag, 4 miles northeast of eastern end of Lake Minnewanka, Alberta.
=*Aneurospora greggsii*, Richardson, J.B. and McGregor, D.C., 1986, Geol. Surv. Can., Bull. 364, pl. 15, fig. 10 (13184).

cf. *Retusotriletes greggsii* McGregor
Hypotype 15377
McGregor, D.C. and Owens, B., 1966, Geol. Surv. Can., Paper 66-30, pl. 23, fig. 16.
Escarpment Member, Hay River Formation, Upper Devonian, left bank Hay River at base of Louise Falls, Northwest Territories.

Retusotriletes incohatus Sullivan
Hypotypes 25464, 25484
McGregor, D.C., 1970, Colloque Stratig. Carbonifère, Congr. Colloques Univ. Liège, vol. 55, pl. 22, fig. 6; pl. 23, fig. 8.
Upper Devonian, Bedford Formation, Courtright borehole, Courtright village, Lambton County, Ontario, and Horton Group, west shore George Bay at mouth of McInnis Brook near Antigonish, Nova Scotia.

Retusotriletes incomptus McGregor
Holotype 31956
McGregor, D.C., 1973, Palaeontographica, Abt.B, vol. 142, no. 1-3, p. 22, pl. 2, fig. 6.
Battery Point Formation, Middle Devonian, 350 yards southeast of small brook south of Cap Rouge, 25 feet beyond second small point southeast of brook, south shore Gaspé Bay, Québec.

Retusotriletes maculatus McGregor and Campfield
Holotype 41706
McGregor, D.C. and Camfield, M., 1976, Geol. Surv. Can., Bull. 263, p. 26, pl. 1, fig. 6.
Stooping River Formation, Lower Devonian, 960-962 feet depth, ODM Jaab Lake No. 1 well, lat. 51°11′54″N, long. 82°56′W, Ontario.

Retusotriletes maculatus McGregor and Campfield, 1976
Hypotype 67196
McGregor, D.C., 1984, Academia Nacional de Ciencias, Misc. No. 69, p. 34, pl. 2, fig. 8.
Santa Rosa Formation, Lower Devonian, Laurani section, Chuquisaca Dept., northwest of Tarabuco, Cordillera Oriental, Bolivia.

Retusotriletes cf. *R. obliteratus* Chibrikova, 1962
Hypotype 15314
McGregor, D.C. and Owens, B., 1966, Geol. Surv. Can., Paper 66-30, pl. 15, fig. 6.
Griper Bay Formation, Upper Devonian, south limb of Robertson Point anticline, 8 miles northeast of Beverley Inlet, Melville Island, District of Franklin.

Retusotriletes ocellatus McGregor
Holotype 31955
McGregor, D.C., 1973, Palaeontographica, Abt.B, vol. 142, no. 1-3, p. 22, pl. 2, fig. 5.
York River Formation, Lower Devonian, 400 yards southeast of Tar Point dyke, south shore Gaspé Bay, Québec.

Retusotriletes ocellatus McGregor
Hypotype 61697
McGregor, D.C., 1979, Geol. Surv. Can., Paper 79-1B, pl. 22.2, fig. 51.
Daleje Shale, Lower Devonian, upper part southeast slope of Císařská rokle, southeast of Srbsko, Czechoslovakia.

?*Retusotriletes punctatus* Chibrikova
Hypotype 25466
McGregor, D.C., 1970, Colloque Stratig. Carbonifère, Congr. Colloques Univ. Liège, vol. 55, pl. 22, fig. 9.
Bedford Formation, Upper Devonian, Courtright borehole, Courtright village, Lambton County, Ontario.

Retusotriletes pychovii Naumova
Hypotype 31953
McGregor, D.C., 1973, Palaeontographica, Abt.B, vol. 142, no. 1-3, p. 19, pl. 2, fig. 3.
York River Formation, Lower Devonian, 150 yards southeast of Tar Point dyke, south shore Gaspé Bay, Québec.

Retusotriletes pychovii Naumova
Hypotype 32805 [not 32905]
McGregor, D.C., 1974, Can. J. Earth Sci., vol. 11, no. 1, pl. 1, fig. 24.
Vendom Fiord Formation, Early Devonian, lat. 79°28′40″N, long. 80°02′W, 12 miles south of head of Cañon Fiord, central Ellesmere Island, District of Franklin.

Retusotriletes pychovii? Naumova
Hypotypes 61661, 61682
McGregor, D.C., 1979, Geol. Surv. Can, Paper 79-1B, pl. 22.1, figs. 13, 34.
Lower Devonian, Zlíchov Limestone, near Pekárek mill west of village of Solopysky; Daleje Shale, roadside outcrop near Karlštejn, 1 km north of Budnany in the Hluboké Valley, Czechoslovakia.

Retusotriletes pychovii Naumova
Hypotype 66375
McGregor, D.C. and Camfield, M., 1982, Geol. Surv. Can., Bull. 348, p. 58, pl. 16, fig. 12.
Weatherall Formation, Middle Devonian, stream section 16.2 km east of east arm of Weatherall Bay, northeastern Melville Island, District of Franklin.

Retusotriletes cf. *R. psychovii* Naumova, 1953
Hypotype 13594
Kerr, J.W., McGregor, D.C. and McLaren, D.J., 1965, Bull. Can. Petrol. Geol., vol. 13, no. 3, p. 425, pl. 3, fig. 7.
McGregor, D.C. and Owens, B., 1966, Geol. Surv. Can., Paper 66-30, pl. 26, fig. 2.
Griper Bay Formation, Upper Devonian, stream valley west of Sargeant Point, 2 miles from shore, northeastern Bathurst Island, District of Franklin.

Retusotriletes cf. *R. pychovii* Naumova, 1953
Hypotype 15450
McGregor, D.C. and Owens, B., 1966, Geol. Surv. Can., Paper 66-30, pl. 28, fig. 1.
Kettle Point Formation, Upper Devonian, depth 295-299 feet, Courtright borehole, Courtright village, Lambton County, Ontario.

Retusotriletes cf. *R. pychovii* Naumova
Hypotype 31954
McGregor, D.C., 1973, Palaeontographica, Abt.B, vol. 142, no. 1-3, p. 20, pl. 2, fig. 4.
York River Formation, Lower Devonian, 500 yards north of southern extremity of cliff north of Rivière de l'Anse-à-Brillant, south shore Gaspé Bay, Québec.

Retusotriletes rotundus (Streel) Streel
Hypotype 15251
McGregor, D.C., 1973, Palaeontographica, Abt.B, vol. 142, no. 1-3, p. 20, pl. 2, fig. 9.
Malbaie Formation, Middle Devonian, 65 yards southeast of conglomerate of Pointe Jaune, south shore Gaspé Bay, Québec.

Retusotriletes rotundus (Streel) Streel
Hypotype 32806
McGregor, D.C., 1974, Can. J. Earth Sci., vol. 11, no. 1, pl. 1, fig. 25.
Vendom Fiord Formation, Early Devonian, lat. 79°28'40"N, long. 80°02'W, 12 miles south of head of Cañon Fiord, central Ellesmere Island, District of Franklin.

Retusotriletes rotundus (Streel) Streel
Hypotype 41762
McGregor, D.C. and Camfield, M., 1976, Geol. Surv. Can., Bull. 263, p. 26, pl. 5, fig. 9.
Williams Island Formation, Middle Devonian, 169.5-171.5 feet depth, ODM Jaab Lake No. 1 well, lat. 51°11'54"N, long. 82°56'W, Ontario.

Retusotrilites rotundus Streel
Hypotypes 61677, 61729
McGregor, D.C., 1979, Geol. Surv. Can., Paper 79-1B, pl. 22.1, fig. 29; pl. 22.3, fig. 84.
Lower Devonian, Daleje Shale, roadside outcrop near Kalštejn, 1 km north of Budnany in the Hluboké Valley; Trebotov Limestone, northwest side of railway, north side of Prokopské Valley at Hluboepy near Prague, Czechoslovakia.

Retusotriletes rotundus (Streel) Streel
Hypotype 66376
McGregor, D.C. and Camfield, M., 1982, Geol. Surv. Can., Bull. 348, p. 58, pl. 16, fig. 13.
Weatherall Formation, Middle Devonian, stream section 16.2 km east of east arm of Weatherall Bay, northeastern Melville Island, District of Franklin.

Retusotriletes rugulatus Riegel 1973
Hypotype 49827
McGregor, D.C., 1979, Palynology, vol. 3, pl. 2, fig. 3.
Hecla Bay Formation, Middle Devonian, creek flowing south, 10 miles east of Weatherall Bay, Melville Island, District of Franklin.

Retusotriletes rugulatus Riegel
Hypotypes 66377-66381
McGregor, D.C. and Camfield, M., 1982, Geol. Surv. Can., Bull. 348, p. 58, pl. 17, figs. 1-5.
Richardson, J.B. and McGregor, D.C., 1986, Geol. Surv. Can., Bull. 364, pl. 12, fig. 4 (66379).
Weatherall (66377, 66380) and Hecla Bay formations, Middle Devonian, stream section 16.2 km east of east arm of Weatherall Bay, northeastern Melville Island, District of Franklin.

Retusotriletes rugulatus Riegel 1973
Hypotype 96780
Arkhangelskaya, A.D., McGregor, D.C. and Richardson, J.B., 1990, Geol. Surv. Can., Paper 90-1D, pl. 1, fig. j.
Strathcona Fiord Formation, Middle Devonian, east of Bird Fiord, Ellesmere Island, District of Franklin.

Retusotriletes semizonalis McGregor
Holotype 13197; paratypes 13194-13196, 13198, 13199
McGregor, D.C., 1964, Geol. Surv. Can., Bull. 109, p. 10, pl. 2, figs. 1-5, 7, 8.
Ghost River Formation, Middle or Upper Devonian, Phantom Crag, 4 miles northeast of eastern end of Lake Minnewanka, Alberta.

Retusotriletes simplex Naumova, 1953
Hypotypes 15139, 15140
McGregor, D.C. and Owens, B., 1966, Geol. Surv. Can., Paper 66-30, pl. 2, figs. 4, 5.
Battery Point Formation, Lower Devonian, south side Gaspé Bay, shore section from Anse-à-Brillant to Bois Brulé Brook, eastern Gaspé Peninsula, Québec.

Retusotriletes simplex Naumova
Hypotype 31952
McGregor, D.C., 1973, Palaeontographica, Abt.B, vol. 142, no. 1-3, p. 19, pl. 2, fig. 2.
Battery Point Formation, Middle Devonian, 100 yards south of most prominent point between Cap Rouge and Pointe Jaune, south shore Gaspé Bay, Québec.

Retusotriletes simplex Naumova
Hypotype 32800
McGregor, D.C., 1974, Can. J. Earth Sci., vol. 11, no. 1, pl. 1, fig. 19.
Vendom Fiord Formation, Early Devonian, lat. 79°28'40"N, long. 80°02'W, 12 miles south of head of Cañon Fiord, central Ellesmere Island, District of Franklin.

cf. *Retusotriletes subgibberosus* Naumova
Hypotypes 32833-32835
McGregor, D.C., 1974, Can. J. Earth Sci., vol. 11, no. 1, pl. 2, figs. 58-60, 62, 63.
Vendom Fiord Formation, Early Devonian, lat. 79°28'40"N, long. 80°02'W, 12 miles south of head of Cañon Fiord, central Ellesmere Island, District of Franklin.

Retusotriletes cf. *R. subgibberosus* Naumova
Hypotype 13610
Kerr, J.W., McGregor, D.C. and McLaren, D.J., 1965, Bull. Can. Petrol. Geol., vol. 13, no. 3, p. 425, pl. 4, fig. 2.
Griper Bay Formation, Upper Devonian, 3/4 mile from shore in stream entering Reindeer Bay, 4 miles south of largest river entering the bay, northeastern Bathurst Island, District of Franklin.

Retusotriletes ?translaticus Chibrikova
Hypotypes 32802, 32803
McGregor, D.C., 1974, Can. J. Earth Sci., vol. 11, no. 1, pl. 1, figs. 21, 22.
Vendom Fiord Formation, Early Devonian, lat. 79°28'40"N, long. 80°02'W, 12 miles south of head of Cañon Fiord, central Ellesmere Island, District of Franklin.

Retusotriletes cf. *triangulatus* (Streel) Streel of Richardson and Lister, 1969
Hypotype 32791

McGregor, D.C., 1974, Can. J. Earth Sci., vol. 11, no. 1, pl. 1, fig. 6.
Unnamed formation, Early Devonian, lat. 79°28′40″N, long. 80°02′W, 12 miles south of head of Cañon Fiord, central Ellesmere Island, District of Franklin.

Retusotriletes warringtonii Richardson and Lister, 1969
Hypotype 67194
McGregor, D.C., 1984, Academia Nacional de Ciencias, Misc. No. 69, p. 35, pl. 35, pl. 1, fig. 28.
Santa Rosa Formation, Lower Devonian, Laureni section, Chuquisaca Dept., northwest of Tarabuco, Cordillera Oriental, Bolivia.

Retusotriletes sp.
Fig. spec. 13437
McGregor, D.C., 1963, Bull. Can. Petrol. Geol., vol. 11, no. 3, fig. 12.
Ghost River Formation (sensu stricto), Middle or Upper Devonian, Phantom Crag, 4 miles northeast of eastern end of Lake Minnewanka, Alberta.

Retusotriletes sp.
Fig. spec. 13200
McGregor, D.C., 1964, Geol. Surv. Can., Bull. 109, p. 11, pl. 2, fig. 6.
Ghost River Formation (sensu stricto), Middle or Upper Devonian, Phantom Crag, 4 miles northeast of eastern end of Lake Minnewanka, Alberta.

Retusotriletes sp.
Fig. specs. 15106, 15141, 15148, 15064, 15166, 15199, 15287-15289, 15401, 15452
McGregor, D.C. and Owens, B., 1966, Geol. Surv. Can., Paper 66-30, pl. 1, fig. 4; pl. 2, figs. 6, 10, 11, 16; pl. 3, fig. 1; pl. 5, fig. 2; pl. 9, figs. 2-4; pl. 26, fig. 1; pl. 28, fig. 3.
Lower Devonian, Forillon Member, Cape Bon Ami (15106), Battery Point and Malbaie (15199) formations, north side Forillon Peninsula, Cape Rosier Cove, south side Gaspé Bay, shore section from Anse-à-Brillant to Bois Brulé Brook and shore section from Bois Brulé to Pointe Jaune (15199), eastern Gaspé Peninsula, Québec; Weatherall Formation, Middle Devonian, south limb of Robertson Point anticline, 8 miles northeast of Beverley Inlet, Melville Island, District of Franklin (15287-15289); Griper Bay Formation, Upper Devonian, stream valley west of Sargeant Point, 2 miles from shore, northeastern Bathurst Island, District of Franklin (15401); Kettle Point Formation, Upper Devonian, depth 295-299 feet, Courtright borehole, Courtright village, Lambton County, Ontario (15452).

?*Retusotriletes* sp.
Fig. specs. 15451, 15454
McGregor, D.C. and Owens, B., 1966, Geol. Surv. Can., Paper 66-30, pl. 28, figs. 2, 5.
Kettle Point Formation, Upper Devonian, depth 295-299 feet, Courtright borehole, Courtright village, Lambton County, Ontario.

Retusotriletes sp.
Fig. specs. 25445, 25467
McGregor, D.C., 1970, Colloque Stratig. Carbonifère, Congr. Colloques Univ. Liège, vol. 55, pl. 21, fig. 8; pl. 22, fig. 10.
Bedford Formation, Upper Devonian, Courtright borehole, Courtright village, Lambton County, Ontario.

Retusotriletes sp.
Fig. spec. 30698
Owens, B. and Richardson, J.B., 1972, 7th Internat. Congr. Carboniferous Strat. Geol. Compte Rendu, vol. 1, pl. 2, fig. 1.
Griper Bay Formation, Upper Devonian, Melville Island, District of Franklin.

Retusotriletes sp.
Fig. spec. 15148
McGregor, D.C., 1973, Palaeontographica, Abt.B, vol. 142, no. 1-3, p. 22, pl. 2, fig. 7.
Battery Point Formation, Lower Devonian, at first point east of first small brook south of Rivière de l'Anse-à-Brillant, south shore Gaspé Bay, Québec.

Retusotriletes sp.
Fig. specs. 61649, 61660, 61662, 61680, 61681, 61698
McGregor, D.C., 1979, Geol. Surv. Can., Paper 79-1B, pl. 22.1, figs. 1, 12, 14, 32, 33; pl. 22.2, fig. 52.
Lower Devonian, Dvorce-Prokop Limestone, road-cut on west side of Vitava River below Barrandov, southern Prague, north of old quarry ″U Kapliky; Zlíchov Limestone, near Pekárek Mill west of village of Solopysky (61660, 61662); Daleje Shale, roadside outcrop near Karlštejn, 1 km north of Budnany in the Hluboké Valley (61680, 61681), and upper part of southeast slope of Císařská rokle, southeast of Srbsko (61698), Czechoslovakia.

Retusotriletes spp.
Fig. specs. 32792, 32807-32809
McGregor, D.C., 1974, Can. J. Earth Sci., vol. 11, no. 1, pl. 1, figs. 7, 26, 27, 30.
Unnamed and Vendom Fiord formations, Early Devonian, lat. 79°28′40″N, long. 80°02′W, 12 miles south of head of Cañon Fiord, central Ellesmere Island, District of Franklin.

Retusotriletes? sp. A Richardson and Lister 1969
Fig. spec. 41707
McGregor, D.C. and Camfield, M., 1976, Geol. Surv. Can., Bull. 263, p. 26, pl. 1, fig. 7.
Kenogami River Formation, Lower Devonian, 1153-1155 feet depth, ODM Jaab Lake No. 1 well, lat. 51°11′54″N, long. 82°56′W, Ontario.

Retusotriletes spp. 2, 3
Fig. specs. 67195, 67197
McGregor, D.C., 1984, Academia Nacional de Ciencias, Musc. No. 69, p. 36, pl. 1, fig. 15; pl. 2, fig. 18.
Santa Rosa end Icla formations, Lower and Middle Devonian, Laurani section, Chuquisaca Dept., northwest of Tarabuco, Cordillera Oriental, Bolivia.

Rhabdosporites langi Richardson, 1960
Hypotypes 15295, 15297
McGregor, D.C. and Owens, B., 1966, Geol. Surv. Can., Paper 66-30, pl. 10, figs. 2, 5.
Weatherall Formation, Middle Devonian, south limb of Robertson Point anticline, 8 miles northeast of Beverley Inlet, Melville Island, District of Franklin.

Rhabdosporites langi (Eisenack) Richardson
Hypotype 15276
McGregor, D.C. and Owens, B., 1966, Geol. Surv. Can., Paper 66-30, pl. 7, figs. 2, 3.
Malbaie Formation, Lower Devonian, south side Gaspé Bay, shore section between Pointe Jaune and hale Head, eastern Gaspé Peninsula, Québec.

Rhabdosporites langi (Eisenack) Richardson
Hypotype 27091
McGregor, D.C., Sanford, B.V. and Norris, A.W., 1970, Proc. Geol. Assoc. Can., vol. 22, pl. 2, fig. 12.
Williams Island Formation, Middle Devonian, 157 feet depth, ODM Jaab Lake No. 1 well, lat. 51°11.9′N, long. 82°57′W, Moose River Basin, Ontario.

Rhabdosporites langii (Eisenack) Richardson
Hypotypes 32038, 32039
McGregor, D.C., 1973, Palaeontographica, Abt.B, vol. 142, no. 1-3, p. 63, pl. 9, figs. 4, 9.
Malbaie Formation, Lower Devonian, 530 yards southeast of first stream south of Pointe Jaune; Battery Point Formation, Middle Devonian, 350 yards southeast of small brook south of Cap Rouge, 25 feet beyond second small point southeast of brook, south shore Gaspé Bay, Québec.

Rhadbdosporites langii (Eisenack) Richardson
Hypotype 41781
McGregor, D.C. and Camfield, M., 1976, Geol. Surv. Can., Bull. 263, p. 27, pl. 7, fig. 7.
Williams Island Formation, Middle Devonian, 157.5-159.5 feet depth, ODM Jaab Lake No. 1 well, lat. 51°11′54″N, long. 82°56′W, Ontario.

Rhabdosporites langii (Eisenack) Richardson 1960
Hypotype 49831
McGregor, D.C., 1979, Palynology, vol. 3, pl. 2, fig. 9.
Hecla Bay Formation, Upper Devonian, creek section 5.5 to 10 miles north of head of Beverley Inlet, Melville Island, District of Franklin.

Rhabdosporites langii (Eisenack) Richardson
Hypotypes 66382-66386
McGregor, D.C. and Camfield, M., 1982, Geol. Surv. Can., Bull. 348, p. 59, pl. 17, figs. 6-11, 13.
Weatherall and Hecla Bay (66383, 66386) formations, Middle Devonian, stream section 16.2 km east of east arm of Weatherall Bay, northeastern Melville Island, District of Franklin.

Rhabdosporites mirus Arkhangelskaya 1985
Hypotype 96783
Arkhangelskaya, A.D., McGregor, D.C. and Richardson, J.B., 1990, Geol. Surv. Can., Paper 90-1D, pl. 1, fig. m.
Strathcona Fiord Formation, Middle Devonian, east of Bird Fiord, Ellesmere Island, District of Franklin.

Rhabdosporites parvulus Richardson
Hypotype 41763
McGregor, D.C. and Camfield, M., 1976, Geol. Surv. Can., Bull. 263, p. 27, pl. 5, fig. 8.
Stooping River Formation, Lower Devonian, 589-591 feet depth, ODM Jaab Lake No. 1 well, lat. 51°11′54″N, long. 82°56′W, Ontario.

Rhabdosporites sp. cf. *R. parvulus* Richardson
Fig. spec. 66387
McGregor, D.C. and Camfield, M., 1982, Geol. Surv. Can., Bull. 348, p. 60, pl. 17, fig. 12.
Weatherall Formation, Middle Devonian, stream section 16.2 km east of east arm of Weatherall Bay, northeastern Melville Island, District of Franklin.

?*Rhabdosporites scamnus* Allen
Hypotype 29386
McGregor, D.C. and Uyeno, T.T., 1972, Geol. Surv. Can., Paper 71-13, pl. 1, fig. 18.
Weatherall Formation, Middle Devonian, Beverley Inlet, southeastern Melville Island, District of Franklin.

Rhabdosporites n. sp.
Fig. spec. 29404
McGregor, D.C. and Uyeno, T.T., 1972, Geol. Surv. Can., Paper 71-13, pl. 3, fig. 5.
Hecla Bay Formation, Middle Devonian, Beverley Inlet, southeastern Melville Island, District of Franklin.

Rhabdosporites sp.
Fig. spec. 61732
McGregor, D.C., 1979, Geol. Surv. Can., Paper 79-1B, pl. 22.3, fig. 87.
Trebotov Limestone, Lower Devonian, northwest side of railway, north side of Prokopské Valley at Hluboepy near Prague, Czechoslovakia.

Rhabdosporites sp.
Fig. specs. 66388, 66389
McGregor, D.C. and Camfield, M., 1982, Geol. Surv. Can., Bull. 348, p. 60, pl. 18, figs. 1-3.
Weatherall Formation, Middle Devonian, stream section 16.2 km of east of east arm of Weatherall Bay, northeastern Melville Island, District of Franklin.

?*Rhabdosporites* sp.
Fig. spec. 61744
McGregor, D.C., 1979, Geol. Surv. Can., Paper 79-1B, pl. 22.3, fig. 99.
Trebotov Limestone, Lower Devonian, northwest side of railway, north side of Prokopské Valley at Hluboepy near Prague, Czechoslovakia.

Rubinella sp.
Fig. spec. 15123
McGregor, D.C. and Owens, B., 1966, Geol. Surv. Can., Paper 66-30, pl. 1, fig. 23.
Shiphead Member, Grande Grève Formation, Lower Devonian, Cape Gaspé, shore cliff below lighthouse, eastern Gaspé Peninsula, Québec.

Rugospora flexuosa (Jushko) Streel
Hypotype 73308
Richardson, J.B. and McGregor, D.C., 1986, Geol. Surv. Can., Bull 364, pl. 19, fig. 7.
McAras Brook(?) Formation, Upper Devonian, Cape George Brook, Nova Scotia.

Rugulatisporites sp.
Fig. spec. 32009
McGregor, D.C., 1973, Palaeontographica, Abt.B, vol. 142, no. 1-3, p. 49, pl. 6, figs. 15, 16.
York River Formation, Lower Devonian, 400 yards southeast of Tar Point dyke, south shore Gaspé Bay, Québec.

Samarisporites concinnus Owens 1971
Hypotype 49829
McGregor, D.C., 1979, Palynology, vol. 3, pl. 2, fig. 6.
Weatherall Formation, Middle Devonian, creek section 5.5 to 10 miles north of head of Beverley Inlet, Melville Island, District of Franklin.

Samarisporites hesperus Allen, 1965
Hypotype 15273
McGregor, D.C. and Owens, B., 1966, Geol. Surv. Can., Paper 66-30, pl. 6, fig. 9.
Malbaie Formation, Lower Devonian, south side Gaspé Bay, shore section between Pointe Jaune and Whale Head, eastern Gaspé Peninsula, Québec.

Samarisporites inaequus (McGregor) Owens
Hypotype 30695
Owens, B. and Richardson, J.B., 1972, 7th Internat. Congr. Carboniferous Strat. Geol. Compte Rendu, vol. 1, pl. 1, fig. 11.
Griper Bay Formation, Upper Devonian?, Melville Island, District of Franklin.

Samarisporites ?megaformis Richardson
Hypotype 27094
McGregor, D.C., Sanford, B.V. and Norris, A.W., 1970, Proc. Geol. Assoc. Can., vol. 22, pl. 2, fig. 15.
Williams Island Formation, Middle Devonian, 157 feet depth, ODM Jaab Lake No. 1 well, lat. 51°11.9′N, long. 82°57′W, Moose River Basin, Ontario.

Samarisporite cf. S. *megaformis* Richardson, 1965
Fig. spec. 15279
McGregor, D.C. and Owens, B., 1966, Geol. Surv. Can., Paper 66-30, pl. 7, fig. 6.
Battery Point Formation, Lower Devonian, south side Gaspé Bay, shore section from Bois Brulé Brook to Pointe Jaune, Gaspé Peninsula, Québec.

Samarisporites triangulatus Allen
Hypotype 49842
McGregor, D.C., 1979, Palynology, vol. 3, pl. 3, fig. 11.
Devonian, depth 9250-9280 feet, Amoco Imp. Skelly Phalarope P-62 well, lat. 45°11.25′N, long. 51°24′14.40″W, Grand Banks.
=*Cristatisporites triangulatus*, Richardson, J.B. and McGregor, D.C., 1986, Geol. Surv. Can., Bull. 364, pl. 15, fig. 9.

Samarisporites sp.
Fig. spec. 15196
McGregor, D.C. and Owens, B., 1966, Geol. Surv. Can., Paper 66-30, pl. 4, fig. 18.
Battery Point Formation, Lower Devonian, south side Gaspé Bay, shore section from Anse-à-Brillant to Bois Brulé Brook, eastern Gaspé Peninsula, Québec.

Samarisporites sp.
Fig. spec. 29835
McGregor, D.C. and Uyeno, T.T., 1972, Geol. Surv. Can., Paper 71-13, pl. 1, fig. 17.
Bird Fiord Formation, Middle Devonian, Young Inlet, northeastern Bathurst Island, District of Franklin.

Samarisporites n. sp.
Fig. spec. 29390
McGregor, D.C. and Uyeno, T.T., 1972, Geol. Surv. Can., Paper 71-13, pl. 2, fig. 3.
Weatherall Formation, Middle Devonian, Beverley Inlet, southeastern Melville Island, District of Franklin.

Scabratisporites minor (Naumova) McGregor
Hypotype 31984
McGregor, D.C., 1973, Palaeontographica, Abt.B, vol. 142, no. 1-3, p. 36, pl. 4, figs. 11, 18.
Battery Point Formation, Lower Devonian, 420 yards north of Bois Brulé Brook, south shore Gaspé Bay, Québec.

Scabratisporites minor (Naumova) McGregor, 1973
Hypotype 67198
McGregor, D.C., 1984, Academia Nacional de Ciencias, Misc. No. 69, p. 36, pl. 1, fig. 27.
Santa Rosa Formation, Lower Devonian, Laurani section, Chuquisaca Dept., northwest of Tarabuco, Cordillera Oriental, Bolivia.

?*Schopfites minor* Schultz, 1968
Hypotype 67199
McGregor, D.C., 1984, Academia Nacional de Ciencias, Misc. No. 69, p. 37, pl. 2, fig. 6.
Santa Rosa Formation, Lower Devonian, Laurani section, Chuquisaca Dept., northwest of Tarabuco, Cordillera Oriental, Bolivia.

Sinuosisporis sinuosus (V.Umnova) Arkhangelskaya 1985
Hypotype 96782
Arkhangelskaya, A.D., McGregor, D.C. and Richardson, J.B., 1990, Geol. Surv. Can., Paper 90-1D, pl. 1, fig. l.
Strathcona Fiord Formation, Middle Devonian, east of Bird Fiord, Ellesmere Island, District of Franklin.

cf. *Sinuosisporis* sp.
Fig. spec. 86088
McGregor, D.C. and McCutcheon, S.R., 1988, Can. J. Earth Sci., vol. 25, no. 9, p. 1353, pl. 1, fig. 15.
Carrow Formation, Piskahegan Group, Late Devonian, depth 72.85 m, drill hole M-6, lat. 45°34′43″N, long. 66°35′27″W, southwestern New Brunswick.

Small spore, cf. *Calamospora pannucea* Richardson
Fig. spec. 31944
McGregor, D.C., 1973, Palaeontographica, Abt.B, vol. 142, no. 1-3, p. 15, pl. 1, fig. 5.
Early Devonian, south side of Restigouche River, Dalhousie, New Brunswick.

Spelaeotriletes resolutus Higgs, 1975
Hypotype 86116
McGregor, D.C. and McCutcheon, S.R., 1988, Can. J. Earth Sci., vol. 25, no. 9, p. 1353, pl. 2, figs. 22, 23.
Carrow Formation, Piskahegan Group, Late Devonian, depth 72.85 m, drill hole M-6, lat. 45°34′43″N, long. 66°35′27″W, southwestern New Brunswick.

?*Spelaeotriletes resolutus* Higgs, 1975
Hypotype 86086
McGregor, D.C. and McCutcheon, S.R., 1988, Can. J. Earth Sci., vol. 25, no. 9, p. 1353, pl. 1, figs. 12, 13.
Carrow Formation, Piskahegan Group, Late Devonian, depth 72.85 m, drill hole M-6, lat. 45°34′43″N, long. 66°35′27″W, southwestern New Brunswick.

cf. *Spelaeotriletes resolutus* Higgs, 1975
Hypotype 86120
McGregor, D.C. and McCutcheon, S.R., 1988, Can. J. Earth Sci., vol. 25, no. 9, p. 1353, pl. 3, fig. 6.
Carrow Formation, Piskahegan Group, Late Devonian, depth 72.85 m, drill hole M-6, lat. 45°34′43″N, long. 66°35′27″W, southwestern New Brunswick.

?*Spelaeotriletes* sp.
Fig. spec. 29395
McGregor, D.C. and Uyeno, T.T., 1972, Geol. Surv. Can., Paper 71-13, pl. 2, fig. 8.
Weatherall Formation, Middle Devonian, Beverley Inlet, southeastern Melville Island, District of Franklin.

spines of *Hystricosporites* sp. and ?*Discrospora multifurcata* Winslow
Fig. specs. 25482, 25483
McGregor, D.C., 1970, Colloque Stratig. Carbonifère, Congr. Colloques Univ. Liège, vol. 55, pl. 23, figs. 6, 7.
Horton Group, Upper Devonian, west shore George Bay at mouth of McInnis Brook near Antigonish, Nova Scotia.

Spinozonotriletes naumovii (Kedo) Richardson
Hypotype 27092
McGregor, D.C., Sanford, B.V. and Norris, A.W., 1970, Proc. Geol. Assoc. Can., vol 22, pl. 2, fig. 13.
Williams Island Formation, Middle Devonian, 157 feet depth, ODM Jaab Lake No. 1 well, lat. 51°11.9′N, long. 82°57′W, Moose River Basin, Ontario.

?*Spinozonotriletes* cf. *naumovii* (Kedo) Richardson, 1965
Fig. spec. 15278
McGregor, D.C. and Owens, B., 1966, Geol. Surv. Can., Paper 66-30, pl. 7, fig. 5.
Malbaie Formation, Lower Devonian, south side Gaspé Bay shore section between Pointe Jaune and Whale Head, eastern Gaspé Peninsula, Québec.

Spinozonotriletes uncatus Hacquebard
Hypotypes 25462, 25492
McGregor, D.C., 1970, Colloque Stratig. Carbonifère, Congr. Colloques Univ. Liège, vol. 55, pl. 22, fig. 3; pl. 23, fig. 16.
Upper Devonian, Bedford Formation, Courtright borehole, Courtright village, Lambton County, Ontario, and Imperial Formation, lat. 66°27.5′N, long. 135°22′W, Yukon.

Spinozonotriletes sp.
Fig. spec. 15330
McGregor, D.C. and Owens, B., 1966, Geol. Surv. Can., Paper 66-30, pl. 18, fig. 8.
Griper Bay Formation, Upper Devonian, south limb of Robertson Point anticline, 8 miles northeast of Beverley Inlet, Melville Island, District of Franklin.

?*Spinozonotriletes* n. sp.
Hypotype 29387
McGregor, D.C. and Uyeno, T.T., 1972, Geol. Surv. Can., Paper 71-13, pl. 1, figs. 19, 20.
Weatherall Formation, Middle Devonian, Beverley Inlet, southeastern Melville Island, District of Franklin.

Spinozonotriletes sp. A
Fig. spec. 29388
McGregor, D.C. and Uyeno, T.T., 1972, Geol. Surv. Can., Paper 71-13, pl. 2, fig. 1.
Bird Fiord Formation, Middle Devonian, Young Inlet, northeastern Bathurst Island, District of Franklin.

Spore from sporangium of *Sawdonia ornata* (Dawson) Hueber
Fig. spec. 31966
McGregor, D.C., 1973, Palaeontographica, Abt. B, vol. 142, no. 1-3, p. 27, pl. 2, fig. 20.
Sextant Formation, Lower Devonian, Abitibi River below Sextant Rapids, northern Ontario.

Spores from sporangium similar to *Dawsonites arcuatus* Halle
Fig.specs. 31963, 31964
McGregor, D.C., 1973, Palaeontographica, Abt.B, vol. 142, no. 1-4, p. 26, pl. 2, figs. 17, 18.
Sextant Formation, Lower Devonian, Abitibi River, northern Ontario.

Stenozonotriletes cf. *S. clarus* Ishchenko
Hypotype 13593
Kerr, J.W., McGregor, D.C. and McLaren, D.J., 1965, Bull. Can. Petrol. Geol., vol. 13, no. 3, p. 425, pl. 3, fig. 6.
Griper Bay Formation, Upper Devonian, stream valley west of Sargeant Point, 2 miles from shore, northeastern Bathurst Island, District of Franklin.

?*Stenozonotriletes definitus* Naumova
Hypotype 13605
Kerr, J.W., McGregor, D.C. and McLaren, D.J., 1965, Bull. Can. Petrol. Geol., vol. 13, no. 3, p. 425, pl. 3, fig. 23.
Griper Bay Formation, Upper Devonian, 3/4 mile from shore in stream valley entering Reindeer Bay, 4 miles south of largest river entering the bay, northeastern Bathurst Island, District of Franklin.

?*Stenozonotriletes furtivus* Allen
Hypotypes 41747, 41748
McGregor, D.C. and Camfield, M., 1976, Geol. Surv. Can., Bull. 263, p. 27, pl. 4, figs. 7, 8, 11, 12.
Stooping River Formation, Lower Devonian, 896-898 and 617-619 feet depths, ODM Jaab Lake No. 1 well, lat. 51°11′54″N, long. 82°56′W, Ontario.

cf. *Stenozotriletes furtivus* Allen 1965
Hypotype 48182
McGregor, D.C., 1977, Palaeontographica, Abt.B, vol. 163, p. 133, pl. 2, figs. 13-15.
York River Formation, Lower Devonian, north of Anse-à-Brillant, south shore Gaspé Bay, Québec.

cf. *Stenozotriletes insessus* Allen, 1965
Hypotype 15119
McGregor, D.C. and Owens, B., 1966, Geol. Surv. Can., Paper 66-30, pl. 1, fig. 19.
Shiphead Member, Grand Grève Formation, Lower Devonian, Cape Gaspé, shore cliff below lighthouse, eastern Gaspé Peninsula, Québec.

?*Stenozonotriletes irregularis* Schultz
Hypotype 29370
McGregor, D.C. and Uyeno, T.T., 1972, Geol. Surv. Can., Paper 71-13, pl. 1, fig. 2.
Bathurst Island Formation, Lower Devonian, Young Inlet, northeastern Bathurst Island, District of Franklin.

?*Stenozonotriletes irregularis* Schultz
Hypotype 32789
McGregor, D.C., 1974, Can. J. Earth Sci., vol. 11, no. 1, pl. 1, fig. 4.
Unnamed formation, Early Devonian, lat. 79°28′40″N, long. 80°02′W, 12 miles south of head of Cañon Fiord, central Ellesmere Island, District of Franklin.

Stenozonotriletes irregularis Schultz 1968
Hypotype 49815
McGregor, D.C., 1979, Palynology, vol. 3, pl. 1, fig. 5.
?Peel Sound Formation, Lower Devonian, east side of Smith Bay, Prince of Whales Island, District of Franklin.

?*Stenozonotriletes irregularis* Schultz, 1968
Hypotype 67200
McGregor, D.C., 1984, Academia Nacional de Ciencias, Misc. No. 69, p. 37, pl. 1, fig. 26.
Santa Rosa Formation, Lower Devonian, Laurani section, Chuquisaca Dept., northwest of Tarabuco, Cordillera Oriental, Bolivia.

Stenozonotriletes laevigatus Naumova, 1953
Hypotype 15345
McGregor, D.C. and Owens, B., 1966, Geol. Surv. Can., Paper 66-30, pl. 22, figs. 6, 9.
Escarpment Member, Hay River Formation, Upper Devonian, left bank Hay River at base of Louise Falls, Northwest Territories.

Stenozonotriletes minus McGregor and Campfield
Holotype 41730
McGregor, D.C. and Camfield, M., 1976, Geol. Surv. Can., Bull. 263, p. 27, pl. 3, figs. 3, 4.
Stooping River Formation, Lower Devonian, 909-911 feet depth, ODM Jaab Lake No. 1 well, lat. 51°11′54″N, long. 82°56′W, Ontario.

Stenozonotriletes simplex Naumova, 1953
Hypotypes 15427, 15428, 15458
McGregor, D.C. and Owens, B., 1966, Geol. Surv. Can., Paper 66-30, pl. 26, figs. 41, 42; pl. 28, fig. 9.
Griper Bay Formation, Upper Devonian, stream valley west of Sargeant Point, 2 miles from shore, northeastern Bathurst Island, District of Franklin; Kettle Point Formation, Upper Devonian, depth 295-299 feet, Courtright borehole, Courtright village, Lambton County, Ontario (15458).

Stenozonotriletes tersus (Waltz) Naumova *in* Kedo, 1957
Hypotype 15372
McGregor, D.C. and Owens, B., 1966, Geol. Surv. Can., Paper 66-30, pl. 23, fig. 11.
Escarpment Member, Hay River Formation, left bank Hay River at base of Louise Falls, Northwest Territories.

Stenozonotriletes sp.
Fig. spec. 15117
McGregor, D.C. and Owens, B., 1966, Geol. Surv. Can., Paper 66-30, pl. 1, fig. 17.
Shiphead Member, Grand Grève Formation, Lower Devonian, Cape Gaspé, shore cliff below lighthouse, eastern Gaspé, Peninsula, Québec.

?*Stenozonotriletes* sp.
Fig. specs. 15109, 15147, 15348, 15406, 15407, 15412
McGregor, D.C. and Owens, B., 1966, Geol. Surv. Can., Paper 66-30, pl. 1, fig. 7; pl. 2, fig. 9; pl. 22, fig. 8; pl. 26, figs. 8, 9, 15.
Lower Devonian, Forillon Member, Cape Bon Ami and Battery Point formations, north side Forillon Peninsula, Cape Rosier Cove, and north side Gaspé Bay, shore cliffs between Gaspé Peninsula and Laurencelle Road, eastern Gaspé Peninsula, Québec; Upper Devonian, Escarpment Member, Hay River Formation, left bank Hay River at base of Louise Falls, Northwest Territories (15348), and Griper Bay Formation, stream valley west of Sargeant Point, two miles from shore, northeastern Bathurst Island, District of Franklin (15406,15407).

?*Stenozonotriletes* sp.
Fig. spec. 27070
McGregor, D.C., Sanford, B.V. and Norris, A.W., 1970, Proc. Geol. Assoc. Can., vol. 22, pl. 1, fig. 16.

Stooping River Formation, Lower Devonian, 1027 feet depth, ODM Jaab Lake No. 1 well, lat. 51°11.9′N, long. 82°57′W, Moose River, Ontario.

cf. *Streelispora granulata* Richardson and Lister, 1969
Hypotype 67778
McGregor, D.C., 1984, Academia Nacional de Ciencias, Misc. No. 69, pl. 3, fig. 11.
Santa Rosa Formation, Lower Devonian, Laurani section, Chuquisaca Dept., northwest of Tarabuco, Cordillera Oriental, Bolivia.

Streelispora (?=Aneurospora) sp.
Fig. spec. 61671
McGregor, D.C., 1979, Geol. Surv. Can., Paper 79-1B, pl. 22, fig. 22.
Zlíchov Limestone, Lower Devonian, near Pekárek Mill west of village of Slopysky, Czechoslovakia.

Synorisporites libycus Richardson and Ioannides, 1973
Hypotype 67204
McGregor, D.C., 1984, Academia Nacional de Ciencias, Misc. No. 69, p. 38, pl. 1, fig. 2.
Santa Rosa Formation, Lower Devonian, Laurani section, Chuquisaca Dept., northwest of Tarabuco, Cordillera Oriental, Bolivia.

Synorisporites papillensis McGregor
Holotype 32013; hypotype 15163
McGregor, D.C., 1973, Palaeontographica, Abt. B, vol. 142, no. 1-3, p. 51, pl. 6, figs. 26-29.
Battery Point Formation, Lower Devonian, 500 and 400 yards east of Rivière de l'Anse-à-Brillant, north shore Gaspé Bay, Québec.

Synorisporites papillensis? McGregor, 1973
Hypotype 67201
McGregor, D.C., 1984, Academia Nacional de Ciencias, Misc. No. 69, p. 38, pl. 1, fig. 12.
Santa Rosa Formation, Lower Devonian, Laurani section, Chuquisaca Dept., northwest of Tarabuco, Cordillera Oriental, Bolivia.

Synorisporites tripapillatus Richardson and Lister
Hypotype 27063
McGregor, D.C., Sanford, B.V. and Norris, A.W., 1970, Proc. Geol. Assoc. Can., vol. 22, pl. 1, fig. 4.
Kenogami River Formation, Lower Devonian, ODM Jaab Lake No. 1 well, lat. 51°11′54″N, long. 82°56′W, Ontario.
=*Synorisporites ?papillensis*, McGregor, D.C. and Campfield, M., 1976, Geol. Surv. Can., Bull. 263, p. 28, pl. 1, figs. 24, 27.

Synorisporites verrucatus Richardson and Lister
Hypotype 32014
McGregor, D.C., 1973, Palaeontographica, Abt.A, vol. 142, no. 1-3, p. 51, pl. 6, figs. 30, 31.
Battery Point Formation, Lower Devonian, 1050 yards northwest of first small brook south of Rivière de l'Anse-à-Brillant, south shore Gaspé Bay, Québec.

Synorisporites verrucatus Richardson and Lister, 1969
Hypotype 67203
McGregor, D.C., 1984, Academia Nacional de Ciencias, Misc. No. 69, p. 36, pl. 1, fig. 10.
Santa Rosa Formation, Lower Devonian, Laurani section, Chuquisaca Dept., northwest of Tarabuco, Cordillera Oriental, Bolivia.

Synorisporites sp.
Fig. spec. 61658
McGregor, D.C., 1979, Geol. Surv. Can., Paper 79-1B, pl. 22.1, fig. 10.
Dvorce-Prokop Limestone, Lower Devonian, road-cut on west side of Vitava River below Barrandov, southern Prague, north of old quarry "U Kapliky", Czechoslovakia.

Tetrad of thin-walled spores
Fig. spec. 32837
McGregor, D.C., 1974, Can. J. Earth Sci., vol. 11, no. 1, pl. 2, fig. 64.
Vendom Fiord Formation, Early Devonian, lat. 79°28′40″N, long. 80°02′W, 12 miles south of head of Cañon Fiord, central Ellesmere Island, District of Franklin.

Tetraletes variabilis Cramer
Hypotypes 67772, 67773
McGregor, D.C., 1984, Academia Nacional de Ciencias, Misc. No. 69, pl. 2, figs. 9, 10.
Santa Rosa Formation, Lower Devonian, Laurani section, Chuquisaca Dept., northwest of Tarabuco, Cordillera Oriental, Bolivia.

Tholisporites chulus (Cramer) var. *chulus* (Richardson and Lister)
Hypotypes 32020, 32021
McGregor, D.C., 1973, Palaeontographica, Abt.B, vol. 142, no. 1-3, p. 56, pl. 7, figs. 13-15.
Battery Point Formation, Lower Devonian, 500 yards east of small stream 3/4 mile north of Bois Brulé Brook, and at tip of first point east of Rivière de l'Anse-à-Brillant, south shore Gaspé Bay, Québec.

Tholisporites chulus (Cramer) McGregor var. *chulus* (Richardson and Lister)
Hypotype 32832
McGregor, D.C., 1974, Can. J. Earth Sci., vol. 11, no. 1, pl. 2, fig. 57.
Vendom Fiord Formation, Early Devonian, lat. 79°28′40″N, long. 80°02′W, 12 miles south of head of Cañon Fiord, central Ellesmere Island, District of Franklin.

Tholisporites chulus (Cramer) McGregor var. *chulus* (Richardson and Lister)
Hypotype 41715
McGregor, D.C. and Camfield, M., 1976, Geol. Surv. Can., Bull. 263, p. 28, pl. 1, figs. 19, 20.
Stooping River Formation, Lower Devonian, 909-911 feet depth, ODM Jaab Lake No. 1 well, lat. 51°11′54″N, long. 82°56′W, Ontario.

Tholisporites chulus (Cramer) McGregor, 1973, var. *chulus*
Hypotype 67206
McGregor, D.C., 1984, Academia Nacional de Ciencias, Misc. No. 69, p. 39, pl. 2, fig. 14.
Santa Rosa Formation, Lower Devonian, Laurani section, Chuquisaca Dept., northwest of Tarabuco, Cordillera Oriental, Bolivia.

Tholisporites chulus (Cramer) var. *magnus* McGregor
Holotype 32023; hypotype 32022
McGregor, D.C., 1973, Palaeontographica, Abt.B, vol. 142, no. 1-3, p. 57, pl. 7, figs. 16-18.
McGregor, D.C., 1979, Palynology, vol. 3, pl. 1, fig. 2 (32022).
Battery Point Formation, Lower Devonian, 1050 yards northwest of first small brook south of, and at tip of first point east of Rivière de l'Anse-à-Brillant, south shore Gaspé Bay, Québec.

Tholisporites chulus (Cramer) McGregor var. *magnus*
Hypotype 61734
McGregor, D.C., 1979, Geol. Surv. Can., Paper 79-1B, pl. 22.3, fig. 89.
Trebotov Limestone, Lower Devonian, northwest side of railway, north side of Prokopské Valley at Hluboepy near Prague, Czechoslovakia.

Tholisporites chulus (Cramer) var. *nanus* (Richardson and Lister)
Hypotype 41705
McGregor, D.C. and Camfield, M., 1976, Geol. Surv. Can., Bull. 263, p. 28, pl. 1, figs. 4, 5.
Kenogami River Formation, Lower Devonian, 1153-1155 feet depth, ODM Jaab Lake No. 1 well, lat. 51°11'54"N, long. 82°56'W, Ontario.

Tholisporites chulus (Cramer) McGregor var. *nanus*
Hypotypes 61699, 61700
McGregor, D.C., 1979, Geol. Surv. Can., Paper 79-1B, pl. 22.2, figs. 53, 54.
Daleje Shale, Lower Devonian, upper part of southeast slope of Císařská rokle, southeast of Srbsko, Czechoslovakia.

Tholisporites chulus (Cramer) McGregor, 1973, var. nanus
Hypotype 67205
McGregor, D.C., 1984, Academia Nacional de Ciencias, Misc. No. 69, p. 40, pl. 2, fig. 7.
Santa Rosa Formation, Lower Devonian, Laurani section, Chuquisaca Dept., northwest of Tarabuco, Cordillera Oriental, Bolivia.

Tholisporites densus McGregor
Hypotype 16597
McGregor, D.C., 1967, Review Palaeobotany and Palynology, I, pl. 1, fig. 17.
Griper Bay Formation, Upper Devonian, northern Helena Island, 2.8 miles from north coast in valley of stream entering the sea at Noel Point, District of Franklin.

Tholisporites densus McGregor 1960
Hypotype 49849
McGregor, D.C., 1979, Palynology, vol. 3, pl. 4, fig. 14.
Griper Bay Formation, Upper Devonian, creek section 5.5 to 10 miles north of head of Beverley Inlet, Melville Island, District of Franklin.

Tholisporites sp.
Fig. spec. 13614
Kerr, J.W., McGregor, D.C. and McLaren, D.J., 1965, Bull. Can. Petrol. Geol., vol. 13, no. 3, p. 428, pl. 4, fig. 6.
Griper Bay Formation, Upper Devonian, northern Helena Island, 2.8 miles from north coast in valley of stream entering the sea at Noel Point, District of Franklin.

Tholisporites sp.
Fig. specs. 32026, 32027
McGregor, D.C., 1973, Palaeontographica, Abt.B, vol. 142, no. 1-3, p. 57, pl. 7, figs. 22-25.
York River Formation, Lower Devonian, 300 yards northwest of coal exposed 500 yards north of southern extremity of cliff north of Rivière de l'Anse-à-Brillant, south shore Gaspé Bay, Québec.

?*Tholisporites* sp.
Fig. spec. 61713
McGregor, D.C., 1979, Geol. Surv. Can., Paper 79-1B, pl. 22.2, fig. 68.
Daleje Shale, Lower Devonian, upper part of southeast slope of Císařská rokle, southeast of Srbsko, Czechoslovakia.

Triangulatisporites rootsii var. *irregularis* Chi and Hills
Holotype 43647; paratypes 43648-43650
Chi, B.I. and Hills, L.V., 1976, Can. J. Earth Sci., vol. 13, no. 6, p. 853, figs. 6A-L.
Upper Devonian, Griper Bay Formation, west side Mould Bay, Prince Patrick Island (43647), and Okse Bay Formation, Okse Bay, Ellesmere Island (43648), District of Franklin; Imperial Formation, Imperial River 35 miles west of Norman Wells, lat. 65°1'N, long. 127°51'W, District of Mackenzie.

Triangulatisporites rootsii var. *reticulatus* Chi and Hills
Holotype 43651; paratypes 43784, 43785
Chi, B.I. and Hills, L.V., 1976, Can. J. Earth Sci., vol. 13, no. 6, p. 854, figs. 8A-C.
Weatherall Formation, Middle Devonian, Cape Terrace, Melville Island, District of Franklin.

Triangulatisporites rootsii var. *rootsii* Chaloner
Topotype 43643; hypotypes 43644-43646
Chi, B.I. and Hills, L.V., 1976, Can. J. Earth Sci., vol. 13, no. 6, p. 852, figs. 4A-L.
Okse Bay and Imperial formations, Upper Devonian, east arm Okse Bay, Ellesmere Island, District of Franklin (43643) and Imperial River 35 miles west of Norman Wells, lat. 67°07'N, long. 127°51'W, District of Mackenzie.

Triangulatisporites rootsii var. *rugulatus* Chi and Hills
Holotype 43652; paratypes 43653, 43654
Chi, B.I. and Hills, L.V., 1976, Can. J. Earth Sci., vol. 13, no. 6, p. 854, figs. 8D-L.
Weatherall Formation, Middle Devonian, Cape Terrace, Melville Island, District of Franklin.

Triangulatisporites rootsii var.A.
Hypotypes 43655-43657
Chi, B.I. and Hills, L.V., 1976, Can. J. Earth Sci., vol. 13, no. 6, p. 856, figs. 11A-F.
Imperial Formation, Upper Devonian, Imperial River 35 miles west of Norman Wells, lat. 67°07′N, long. 127°51′W, District of Mackenzie.

Trichodosporites delicatus Chi and Hills
Holotype 43712; paratypes 43713-43715
Chi, B.I. and Hills, L.V., 1977, Bull. Can. Petrol. Geol., vol. 24, no. 4, p. 725, pl. 7, figs. 1-8.
Weatherall Formation, Middle Devonian, Cape Terrace, west side Melville Island, District of Franklin.

Trichodosporites minor Chi and Hills
Holotype 43716; paratype 43717
Chi, B.I. and Hills, L.V., 1977, Bull. Can. Petrol. Geol., vol. 24, no. 4, p. 727, pl. 7, figs. 9-12.
Wetherall Formation, Middle Devonian, Weatherall Bay and Robertson Point Anticline north of Beverley Inlet, Melville Island, District of Franklin.

Trichosporites divisus Chi and Hills
Holotype 43718; paratype 43719
Chi, B.I. and Hills, L.V., 1977, Bull. Can. Petrol. Geol., vol. 24, no. 4, p. 729, pl. 7, figs. 13-16.
Weatherall Formation, Middle Devonian, Cape Terrace, west side Melville Island, District of Franklin.

cf. *Trileites langi* Richardson, 1965
Hypotype 15164
McGregor, D.C. and Owens, B., 1966, Geol. Surv. Can., Paper 66-30, pl. 2, fig. 33.
Battery Point Formation, Lower Devonian, south side Gaspé Bay, shore section from Anse-à-Brillant to Bois Brulé Brook,eastern Gaspé Peninsula, Québec.

Trileites langii Richardson 1965
Hypotype 43660
Chi, B.I. and Hills, L.V., 1977, Bull. Can. Petrol. Geol., vol. 24, no. 4, p. 693, pl. 1, fig. 3.
Imperial Formation, Upper Devonian, lat. 65°07′N, long. 127°51′W, Imperial River, District of Mackenzie.

?Tripapillate spore
Fig. spec. 86080
McGregor, D.C. and McCutcheon, S.R., 1988, Can. J. Earth Sci., vol. 25, no. 9, p. 1353, pl. 1, fig. 4.
Carrow Formation, Piskahegan Group, Late Devonian, depth 73.5 m, drill hole M-6, lat. 45°34′47″N, long. 66°35′27″W, southwestern New Brunswick.

Unidentified
Fig. specs. 13583-13592, 13612
Kerr, J.W., McGregor, D.C. and McLaren, D.J., 1965, Bull. Can. Petrol. Geol., vol. 13, no. 1, pl. 3, figs. 4, 5, 8-10, 12, 13, 25, 26, 28, 29; pl. 4, fig. 4.
Griper Bay Formation, Upper Devonian, stream valley west of Sargeant Point, 2 miles from shore, and stream valley entering Reindeer Bay 3/4 mile from shore, 4 miles south of largest river entering the bay (13589, 13590-13592, 13612), northeastern Bathurst Island, District of Franklin.

Unidentified
Fig. specs. 15121, 15129, 15135, 15066, 15162, 15163, 15167, 15172, 15193, 15197, 15263, 15268, 15271, 15274
McGregor, D.C. and Owens, B., 1966, Geol. Surv. Can., Paper 66-30, pl. 1, figs. 21, 30, 36; pl. 2, figs. 26, 31, 32; pl. 3, figs. 2, 7; pl. 4, figs. 14, 19-21; pl. 5, figs. 3, 47, 10.
Lower Devonian, Shiphead Member, Grand Grève Formation, Cape Gaspé, shore cliff below lighthouse; south side Gaspé Bay, shore cliff 1/2 mile southeast of Tar Point (15066), shore section from Anse-à-Brillant to Bois Brulé Brook (15162, 15163, 15167, 15172, 15197), and shore section from Bois Brulé Point to Pointe Jaune (15263, 15268, 15271, 15274); north side Gaspé Bay, shore cliffs between Gaspé Peninsula and Laurencelle Road (15193), eastern Gaspé Peninsula, Québec.

Unidentified
Fig. specs. 15320, 15387, 15411, 15414, 15417, 15418, 15426, 15436, 15437, 15443-15446
McGregor, D.C. and Owens, B., 1966, Geol. Surv. Can., Paper 66-30, pl. 17, fig. 1; pl. 26, figs. 6, 14, 17, 21, 23, 39, 40; pl. 27, figs. 9, 10, 19-23.
Griper Bay Formation, Upper Devonian, south limb of Robertson Point anticline, 8 miles northeast of Beverley Inlet, Melville Island (15320), and stream valley west of Sargeant Point, 2 miles from shore, northeastern Bathurst Island, District of Franklin.

unidentified
Fig. specs. 15362-15364, 15365-15367, 15384, 15387, 15393, 15389, 15392
McGregor, D.C. and Owens, B., 1966, Geol. Surv. Can., Paper 66-30, pl. 22, figs. 27-32; pl. 23, figs. 1-5, 22, 23; pl. 24, figs. 2, 3, 5, 6, 10, 11.
Escarpment Member, Hay River Formation, Upper Devonian, left bank Hay River at base of Louise Falls, Northwest Territories.
=*Geminospora membrana*, McGregor, D.C., 1979, Palynology, vol. 3, pl. 4, fig. 2 (15387).

unidentified
Fig. specs. 15466, 15469, 15470, 15477, 15482, 15486, 15484
McGregor, D.C. and Owens, B., 1966, Geol. Surv. Can., Paper 66-30, pl. 28, figs. 17, 18, 21, 22; pl. 29, figs. 5, 10, 11, 13.
Kettle Point Formation, Upper Devonian, depth 295-299 feet, Courtright borehole, Courtright village, Lambton County, Ontario.

unidentified
Fig. spec. 27077
McGregor, D.C., Sanford, B.V. and Norris, A.W., 1970, Proc. Geol. Assoc. Can., vol. 22, pl. 1, figs. 21, 22.
Stooping River Formation, Lower Devonian, 909 feet depth, ODM Jaab Lake No. 1 well, lat. 51°11.9′N, long. 82°57′W, Moose River basin, Ontario.

unidentified
Fig. specs. 25447-25449, 25455, 25472, 25485, 25499
McGregor, D.C., 1970, Colloque Stratig. Carbonifère, Congr. Colloques Univ. Liège, vol. 55, pl. 21, figs. 10-12, 18; pl. 22, fig. 15; pl. 23, figs. 9, 23.
Upper Devonian, Bedford Formation, Courtright borehole, Courtright village, Lambton County, and Sunbury Formation, Gibraltar 77 borehole, 6 miles south of Courtright village (25472), Ontario; Horton Group, west shore George Bay at mouth of McInnis Brook near Antigonish, Nova Scotia (25485); Imperial Formation, lat. 66°27.5′N, long. 135°22′W, Yukon (25499).

Unidentified
Fig. spec. 86125
McGregor, D.C. and McCutcheon, S.R., 1988, Can. J. Earth Sci., vol. 25, no. 9, pl. 3, figs. 10, 11.
Carrow Formation, Piskahegan Group, Late Devonian, depth 72.85 m, drill hole M-6, lat. 45°34′43″N, long. 66°35′27″W, southwestern New Brunswick.

Unidentified spore
Fig. spec. 32801
McGregor, D.C., 1974, Can. J. Earth Sci., vol. 11, no. 1, pl. 1, fig. 20.
Vendom Fiord Formation, Early Devonian, lat. 79°28′40″N, long. 80°02′W, 12 miles south of head of Cañon Fiord, central Ellesmere Island, District of Franklin.

Vallatisporites pusillites (Kedo) Dolby and Neves, 1970 sensu lato
Hypotype 86121
McGregor, D.C. and McCutcheon, S.R., 1988, Can. J. Earth Sci., vol. 25, no. 9, p. 1353, pl. 3, fig. 7.
Carrow Formation, Piskahegan Group, Late Devonian, depth 72.85 m, drill hole M-6, lat. 45°34′43″N, long. 66°35′27″W, southwestern New Brunswick.

Verruciretusispora dubia (Eisenack) Richardson and Rasul
Hypotypes 66390-66393
McGregor, D.C. and Camfield, M., 1982, Geol. Surv. Can., Bull. 348, p. 60, pl. 18, figs. 4-7.
Weatherall and Hecla Bay (66392, 66293) formations, Middle Devonian, stream section 16.2 km east of east arm of Weatherall Bay, northeastern Melville Island, District of Franklin.

?*Verruciretussispora dubia* (Eisenack) Richardson and Rasul
Hypotype 61696
McGregor, D.C., 1979, Geol. Surv. Can., Paper 79-1B, pl. 22.2, fig. 50.
Daleje Shale, Lower Devonian, roadside outcrop near Karlštejn, 1 km north of Budnany in the Hluboké Valley, Czechoslovakia.

Verruciretusispora magnifica (McGregor) Owens
Hypotype 30691
Owens, B. and Richardson, J.B., 1972, 7th Internat. Congr. Carboniferous Strat. Geol. Compte Rendu, vol. 1, pl. 1, fig. 5.
Griper Bay Formation, Upper Devonian?, Melville Island, District of Franklin.

Verruciretusispora magnifica (McGregor) Owens 1971
Hypotype 49844
McGregor, D.C., 1979, Palynology, vol. 3, pl. 4, fig. 4.
Hecla Bay Formation, Upper Devonian, creek section 5.5 to 10 miles north of head of Beverley Inlet, Melville Island, District of Franklin.

Verruciretusispora multituberculata (Lanninger) McGregor
Hypotypes 31985, 31986
McGregor, D.C., 1973, Palaeontographica, Abt.B, vol. 142, no. 1-3, p. 36, pl. 4, figs. 13, 14.
Battery Point Formation, Middel Devonian, 550 yards southeast of small brook south of Cap Rouge, on point, south shore Gaspé Bay, Québec.

Verruciretusispora multituberculata (Lanninger) McGregor 1973
Hypotype 49825
McGregor, D.C., 1979, Palynology, vol. 3, pl. 1, fig. 25.
Weatherall Formation, Middle Devonian, creek section 5.5 to 10 miles north of head of Beverley Inlet, Melville Island, District of Franklin.

Verruciretusispora primus Chi and Hills
Holotype 43664; paratypes 43665, 43666
Chi, B.I. and Hills, L.V., 1977, Bull. Can. Petrol. Geol., vol. 24, no. 4, p. 697, pl. 1, figs. 7-11.
Hecla Bay Formation, Middle Devonian, Cape Terrace, west side Melville Island, District of Franklin.

Verruciretusispora sp.
Fig. specs. 61704, 61705
McGregor, D.C., 1979, Geol. Surv. Can., Paper 79-1B, pl. 22.2, figs. 58, 59.
Daleje Shale, Lower Devonian, upper part of southeast slope of Císařská rokle, southeast of Srbsko, Czechoslovakia.

Verrucisporites ellesmerensis var. *connatus* Chi and Hills
Holotype 43674; paratypes 43675-43677
Chi, B.I. and Hills, L.V., 1977, Bull. Can. Petrol. Geol., vol. 24, no. 4, p. 703, pl. 2, figs. 11-16.
Beverley Inlet Formation, Upper Devonian, Melville Island, Cape Terrace and Weatherall Bay, District of Franklin.

Verrucisporites ellesmerensis var. *ellesmerensis* (Chaloner)
Hypotype 43673
Chi, B.I. and Hills, L.V., 1977, Bull. Can. Petrol. Geol., vol. 24, no. 4, p. 703, pl. 2, figs. 9, 10.
Beverley Inlet Formation, Upper Devonian, Cape Terrace, west side Melville Island, District of Franklin.

Verrucisporites ellesmerensis var. *parvus* Chi and Hills
Holotype 43678; paratypes 43679-43681
Chi, B.I. and Hills, L.V., 1977, Bull. Can. Petrol. Geol., vol. 24, no. 4, p. 704, pl. 3, figs. 1-6.
Weatherall and Hecla Bay formations, Middle Devonian, Cape Terrace, west side Melville Island, District of Franklin.

Verrucisporites medius var. *magnus* Chi and Hills
Holotype 43671; paratype 43672
Chi, B.I. and Hills, L.V., 1977, Bull. Can. Petrol. Geol., vol. 24, no. 4, p. 700, pl. 2, figs. 5-8.
Beverley Inlet Formation, Upper Devonian, Cape Terrace, west side Melville Island, District of Franklin.

Verrucisporites medius var. *medius* Chi and Hills
Holotype 43670; paratypes 43671, 43672
Chi, B.I. and Hills, L.V., 1977, Bull. Can. Petrol. Geol., vol. 24, no. 4, p. 699, pl. 2, figs. 1-4.
Beverley Inlet Formation, Upper Devonian, Cape Terrace, west side Melville Island, District of Franklin.

Verrucisporites medius var. *minutus* Chi and Hills
Holotype 43667; paaratypes 43668, 43669
Chi, B.I. and Hills, L.V., 1977, Bull. Can. Petrol. Geol., vol. 24, no. 4, p. 699, pl. 1, figs. 12-17.
Beverley Inlet Formation, Upper Devonian, Cape Terrace, west side Melville Island, District of Franklin.

Verrucisporites submamillarius (McGregor)
Hypotype 43682
Chi, B.I. and Hills, L.V., 1977, Bull. Can. Petrol. Geol., vol. 24, no. 4, p. 705, pl. 3, figs. 7-8.
Beverley Inlet Formation, Upper Devonian, Cape Terrace, west side Melville Island, District of Franklin.

Verrucisporites sp.
Fig. spec. 86096
McGregor, D.C. and McCutcheon, S.R., 1988, Can. J. Earth Sci., vol. 25, no. 9, p. 1353, pl. 1, fig. 25.
Carrow Formation, Piskahegan Group, Late Devonian, depth 72.85 m, drill hole M-6, lat. 45°34′43″N, long. 66°35′27″W, southwestern New Brunswick.

Verrucosisporites bulliferus Richardson and McGregor
Hypotype 73297
Richardson. J.B. and McGregor, D.C.. 1986, Geol. Surv. Can., Bull. 364, pl. 17, fig. 2.
Long Rapids Formation, Upper Devonian, Abitibi River, Ontario.

Verrucosisporites cf. *confertus* Owens
Hypotype 30692
Owens, B. and Richardson, J.B., 1972, 7th Internat. Congr. Carboniferous Strat. Geol. Compte Rendu, vol. 1, pl. 1, fig. 7.
Griper Bay Formation, Upper Devonian, Prince Patrick Island, District of Franklin.

Verrucosisporites congestus Playford, 1964
Hypotype 15464
McGregor, D.C. and Owens, B., 1966, Geol. Surv. Can., Paper 66-30, pl. 28, fig. 15.
Kettle Point Formation, Upper Devonian, depth 295-299 feet, Courtright borehole, Courtright village, Lambton County, Ontario.

Verrucosisporites devonicus McGregor
Holotype 31988; hypotype 31987
McGregor, D.C., 1973, Palaeontographica, Abt.B, vol. 142, no. 1-3, p. 38, pl. 4, figs. 15, 21, 22.
York River and Battery Point formations, Lower Devonian, 400 yards southeast of Tar Point dyke, and 1/2 mile south of Seal Cove River, south shore Gaspé Bay, Québec.

Verrucosisporites nitidus (Naumova) Playford
Hypotype 73314
Richardson, J.B. and McGregor, D.C., 1986, Geol. Surv. Can., Bull. 364, pl. 21, fig. 7.
Fairfield Formation(?), Upper Devonian, depth 1750.5 m, Amax Ellendale well, northernmost tip of Western Australia.

Verrucosisporites ?polygonalis Lanninger
Hypotypes 15152, 31989
McGregor, D.C., 1973, Palaeontographica, Abt.B, vol. 142, no. 1-3, p. 37, pl. 4, figs. 16, 17, 19, 25, 26.
Battery Point and York River formations, Lower Devonian, 380 yards northwest of first small brook south of Rivière de l'Anse-à-Brillant, and 400 yards southeast of Tar Point dyke, south shore Gaspé Bay, Québec.

Verrucosisporites polygonalis Lanninger
Hypotype 41744
McGregor, D.C. and Camfield, M., 1976, Geol. Surv. Can., Bull. 263, p. 28, pl. 3, figs. 22-24.
Stooping River Formation, Lower Devonian, 960-962 feet depth, ODM Jaab Lake No. 1 well, lat. 51°11′54″N, long. 82°56′W, Ontario.

Verrucosisporites premnus Richardson
Hypotypes 66394, 66395
McGregor, D.C. and Camfield, M., 1976, Geol. Surv. Can., Bull. 348, p. 61, pl. 18, figs. 8, 9.
Richardson, J.B. and McGregor, D.C., 1986, Geol. Surv. Can., Bull. 364, pl. 14, figs. 5, 6 (66394).
Hecla Bay Formation, Middle Devonian, stream section 16.2 km east of east arm of Weatherall Bay, northeastern Melville Island, District of Franklin.

Verrucosisporites scurrus (Naumova) n.comb.
Hypotypes 66396-66400
McGregor, D.C. and Camfield, M., 1982, Geol. Surv. Can., Bull. 348, p. 61, pl. 18, figs. 10-17.
Richardson, J.B. and McGregor, D.C., 1986, Geol. Surv. Can., Bull. 364, pl. 14, fig. 2 (66396).
Weatherall Formation, Middle Devonian, stream section 16.2 km east of east arm of Weatherall Bay, northeastern Melville Island, District of Franklin.

Verrucosisporites scurrus (Naumova) n. comb., corroded, cf. *Geminospora verrucosa*
Hypotype 66405
McGregor, D.C. and Camfield, M., 1982, Geol. Surv. Can., Bull. 348, p. 61, pl. 18, fig. 22.

Weatherall Formation, Middle Devonian, stream section 16.2 km east of east arm of Weatherall Bay, northeastern Melville Island, District of Franklin.

Verrucosisporites tumulentis Clayton and Graham
Hypotypes 66401-66404
McGregor, D.C. and Camfield, M., 1982, Geol. Surv. Can., Bull. 348, p. 62, pl. 18, figs. 18-21.
Weatherall Formation, Middle Devonian, stream section 16.2 km east of east arm of Weatherall Bay, northeastern Melville Island, District of Franklin.

Verrucosisporites sp.
Fig. specs. 15124, 15133, 15145, 15152, 15156, 15165, 15255, 15264, 15267, 15415, 15459
McGregor, D.C. and Owens, B., 1966, Geol. Surv. Can., Paper 66-30, pl. 1, figs. 24, 25, 34; pl. 2, figs. 20, 22, 25, 34; pl. 5, figs. 8, 18; pl. 6, fig. 2; pl. 26, fig. 18; pl. 28, fig. 10.
Lower Devonian, Shiphead Member, Grand Grève and Battery Point formations, Cape Gaspé, shore cliff below lighthouse (15124, 15133), south side Gaspé Bay, shore section from Anse-à-Brillant to Bois Brulé Brook, and shore section from Bois Brulé Brook to Pointe Jaune (15255, 15264, 15267), eastern Gaspé Peninsula, Québec; Upper Devonian, Griper Bay Formation, stream valley west of Sargeant Point, 2 miles from shore, northeastern Bathurst Island, District of Franklin (15415), and Kettle Point Formation, depth 295-299 feet, Courtright borehole, Courtright village, Lambton County, Ontario (15459).

cf. *Verrucosisporites* sp. *in* Balme and Hassell, 1962
Fig. spec. 15369
McGregor, D.C. and Owens, B., 1966, Geol. Surv. Can., Paper 66-30, pl. 23, fig. 8.
Escarpment Member, Hay River Formation, left bank Hay River at base of Louise Falls, Northwest Territories.

Verrucosisporites sp.
Fig. spec. 61709
McGregor, D.C., 1979, Geol. Surv. Can., Paper 79-1B, pl. 22.2, fig. 64.
Daleje Shale, Lower Devonian, upper part of southeast slope of Císařská rokle, southeast of Srbsko, Czechoslovakia.

Verrucosisporites n. sp.
Fig. spec. 29419
McGregor, D.C. and Uyeno, T.T., 1972, Geol. Surv. Can., Paper 71-13, pl. 4, fig. 11.
Griper Bay Formation, Upper Devonian, Beverley Inlet, southeastern Melville Island, District of Franklin.

?*Verrucosisporites* sp.
Fig. spec. 15368
McGregor, D.C. and Owens, B., 1966, Geol. Surv. Can., Paper 66-30, pl. 23, figs. 6, 7.
Escarpment Member, Hay River Formation, left bank Hay River at base of Louise Falls, Northwest Territories.

?*Verrucosisporites*
Fig. spec. 32797
McGregor, D.C., 1974, Can. J. Earth Sci., vol. 11, no. 1, pl. 1, fig. 13.
Unnamed formation, Early Devonian, lat. 79°28′40″N, long. 80°02′W, 12 miles south of head of Cañon Fiord, central Ellesmere Island, District of Franklin.

Carboniferous-Permian

Acanthotriletes castanea Butterworth and Williams, 1958
Hypotype 15855
Barss, M.S., 1967, Geol. Surv. Can., Paper 67-11, pl. 7, fig. 21.
Mattson Formation, Mississippian, south face Twisted Mountain, about 15 miles northwest of South Nahanni, lat. 61°13′N. long. 123°37′W, Northwest Territories.

Acanthotriletes hacquebardii Playford
Holotype 13105; paratypes 13104, 13106, 13107
Playford, G., 1964, Geol. Surv. Can., Bull. 107, p. 20, pl. 4, figs. 1-4.
Barss, M.S., 1967, Geol. Surv. Can., Paper 67-11, pl. 2, fig. 5 (13105).
Cheverie Formation, Horton Group, Mississippian, west shore of Avon River, Nova Scotia.

Acanthotriletes hacquebardii Playford, 1964
Hypotype 73735
Utting, J., 1987, Geol. Surv. Can., Bull. 374, pl. 5, fig. 11.
Windsor Group, Lower Carboniferous, near Baddeck Bridge, lat. 46°9′N, long. 61°11′W, Nova Scotia.

Acanthotriletes sp. cf. *A. mirus* Ischchenko, 1956
Fig. spec. 92841
Utting, J., Jachowicz, M. and Jachowicz, A., 1989, Geol. Surv. Can., Bull. 396, pl. 6.1, fig. 2.
Emma Fiord Formation, Early Carboniferous, Lyall River area, Grinnell Peninsula, Devon Island, District of Franklin.

Acanthotriletes multisetus (Luber) Potonié and Kremp, 1955
Hypotype 15851
Barss, M.S., 1967, Geol. Surv. Can., Paper 67-11, pl. 7, fig. 17.
Mattson Formation, Mississippian, south face Twisted Mountain, about 15 miles northwest of South Nahanni, lat. 61°13′N. long. 123°37′W, Northwest Territories.

Acanthotriletes socraticus Neves and Ioannides, 1974
Hypotype 92842
Utting, J., Jachowicz, M. and Jachowicz, A., 1989, Geol. Surv. Can., Bull. 396, pl. 6.1, fig. 3.
Emma Fiord Formation, Early Carboniferous, Lyall River area, Grinnell Peninsula, Devon Island, District of Franklin.

Acanthotriletes cf. *A. tereteangulatus* Balme and Hennelly, 1956
Hypotype 16628
Barss, M.S., 1967, Geol. Surv. Can., Paper 67-11, pl. 38, fig. 26.
Sabine Bay Formation, Permian, vicinity of Tingmisut Lake, east side of Sabine Peninsula, Melville Island, District of Franklin.

Acanthotriletes sp.
Fig. specs. 15852, 15954
Barss, M.S., 1967, Geol. Surv. Can., Paper 67-11, pl. 7, fig. 18; pl. 12, fig. 7.
Mattson Formation, Mississippian, south face Twisted Mountain, about 15 miles northwest of South Nahanni, lat. 61°13′N, long. 123°37′W, Northwest Territories; Canso Group, Upper Carboniferous, Pomquet River south of Route 4, Antigonish County, Nova Scotia.

Acanthotriletes sp.
Fig. spec. 92840
Utting, J., Jachowicz, M. and Jachowicz, A., 1989, Geol. Surv. Can., Bull. 396, pl. 6.1, fig. 1.
Emma Fiord Formation, Early Carboniferous, Lyall River area, Grinnell Peninsula, Devon Island, District of Franklin.

Aculeispores variabilis Jansonius, 1962
Hypotype 16630
Barss, M.S., 1967, Geol. Surv. Can., Paper 67-11, pl. 38, fig. 28.
Sabine Bay Formation, Permian, vicinity of Tingmisut Lake, east side of Sabine Peninsula, Melville Island, District of Franklin.

Ahrensisporites cf. *A. beeleyensis* Neves, 1961
Hypotype 16126
Barss, M.S., 1967, Geol. Surv. Can., Paper 67-11, pl. 18, fig. 28.
Cumberland Group, Pennsylvanian, Jim Haley's mine on Chignecto coal seam, 1.1 miles west of Maccan, Cumberland County, Nova Scotia.

Ahrenisporites sp.
Fig. spec. 16064
Barss, M.S., 1967, Geol. Surv. Can., Paper 67-11, pl. 16, fig. 10.
Port Hood Formation, Riversdale Group, Pennsylvanian, 200 feet from mouth of adit of Harbourview coal mine, Port Hood, Inverness County, Nova Scotia.

Alatisporites cf. *A. pustulatus* Ibrahim, 1933
Hypotype 16260
Barss, M.S., 1967, Geol. Surv. Can., Paper 67-11, pl. 24, fig. 8.
Skinner Brook Member, Stellarton Group, Pennsylvanian, northwest of Westville, 2,000 feet east of Skinner Brook, Pictou coalfield, Pictou County, Nova Scotia.

Alatisporites tessellatus Staplin, 1960
Hypotype 15934
Barss, M.S., 1967, Geol. Surv. Can., Paper 67-11, pl. 11, fig. 8.
Mattson Formation, Mississippian, south face Twisted Mountain, about 15 miles northwest of South Nahanni, lat. 61°13′N. long. 123°37′W, Northwest Territories.

Alatisporites trialatus Kosanke, 1950
Hypotypes 16088, 16324
Barss, M.S., 1967, Geol. Surv. Can., Paper 67-11, pl. 17, fig. 1; pl. 27, fig 2.
Pennsylvanian, Port Hood Formation, Riversdale Group, 200 feet from mouth of adit of Harbourview coal mine, Port Hood, Inverness County; Mabou Mines Formation, eight-foot coal seam in drift from cliff section to old workings, south side of syncline, Mabou mines, Inverness County, Nova Scotia.

Alatisporites trialatus Kosanke, 1950
Hypotypes 16402, 16403
Barss, M.S., 1967, Geol. Surv. Can., Paper 67-11, pl. 29, figs. 23,24.
Morien Group, Pennsylvanian, Harbour coal seam, Dominion No. 12 colliery at face of No. 21 west level, Sydney coalfield, Cape Breton County, Nova Scotia.

Anapiculatisporites ampullaceus (Hacquebard) Playford
Hypotypes 13097, 13098
Playford, G., 1964, Surv. Can., Bull. 107, p. 16, pl. 3, figs. 16, 17.
Barss, M.S., 1967, Geol. Surv. Can., Paper 67-11, pl. 1, fig. 16 (13097).
Horton Group, Mississippian, Mill Brook, one mile north of East Gore, Nova Scotia.

Anapiculatisporites fundiensis Utting
Holotype 73736; hypotypes 73737, 73738
Utting, J., 1987, Geol. Surv. Can., Bull. 374, p. 26, pl. 1, figs. 12-14.
Windsor Group, Lower Carboniferous, north shore of Port Hood Island, lat. 46°1′N, long. 61°26′W, Cape Breton Island, Nova Scotia.

Anapiculatisporites hystricosus Playford
Holotype 13096; paratypes 13094, 13095
Playford, G., 1964, Geol. Surv. Can., Bull. 107, p. 16, pl. 3, figs. 13-15.
Barss, M.S., 1967, Geol. Surv. Can., Paper 67-11, pl. 1, fig. 20 (13094).
Cheverie Formation, Horton Group, Mississippian, west shore of Avon River, Nova Scotia.

Anapiculatisporites hystricosus Playford, 1964
Hypotype 89902
Utting, J., Keppie, J.D. and Giles, P.S., 1989, Geol. Surv. Can., Bull. 396, pl. 5.1, fig. 2.
Horton Bluff Formation, Horton Group, Lower Carboniferous, Hurd Creek, Avon River, Windsor Basin, Nova Scotia.

Anapiculatisporites minor Butterworth and Williams, 1958
Hypotype 15858
Barss, M.S., 1967, Geol. Surv. Can., Paper 67-11, pl. 7, fig. 24.
Mattson Formation, Mississippian, south face Twisted Mountain, about 15 miles northwest of South Nahanni, lat. 61°13′N. long. 123°37′W, Northwest Territories.

Anapiculatisporites spinosus (Kosanke) Potonié and Kremp, 1955
Hypotype 16207
Barss, M.S., 1967, Geol. Surv. Can., Paper 67-11, pl. 22, fig. 9.
Morien Group, Pennsylvanian, McAulay Cove coal seam, from outcrop on shore 1.75 miles south of Homeville, Sydney coalfield, Cape Breton County, Nova Scotia.

Anapiculatisporites spinosus (Kosanke) Potonié and Kremp, 1955
Hypotype 16281
Barss, M.S., 1967, Geol. Surv. Can., Paper 67-11, pl. 25, fig. 15.
Mabou Mines Formation, Pennsylvanian, coal seam immediately below fault north of Coal Mine Point, Mabou mines, Inverness County, Nova Scotia.

Anapiculatisporites spinosus (Kosanke) Potonié and Kremp, 1955
Hypotype 16357
Barss, M.S., 1967, Geol. Surv. Can., Paper 67-11, pl. 28, fig. 10.
Morien Group, Pennsylvanian, Harbour coal seam, Dominion No. 12 colliery at face of No. 21 west level, Sydney coal field, Cape Breton County, Nova Scotia.

Anapiculatisporites tersus Playford
Holotypes 13091; paratypes 13092, 13093
Playford, G., 1964, Surv. Can., Bull. 107, p. 17, pl. 3, figs. 10-12.
Barss, M.S., 1967, Geol. Surv. Can., Paper 67-11, pl. 1, fig. 12 (13091).
Horton Group, Mississippian, Mill Brook, one mile north of East Gore, Nova Scotia.

?*Anapiculatisporites* sp.
Fig. spec. 16119
Barss, M.S., 1967, Geol. Surv. Can., Paper 67-11, pl. 18, fig. 21.
Cumberland Group, Pennsylvanian, Pearly McFadden's coal mine on Chignecto coal seam, 0.3 miles east of Chignecto, Cumberland County, Nova Scotia.

Anaplanisporites sp.
Fig. spec. 62497
Utting, J., 1980, Can. J. Earth Sci., vol. 17, no. 8, pl. 2, fig. 3
Upper Windsor Group, Mississippian, United States Borax and Chemical Corporation borehole SB-1 near Stewiacke, Nova Scotia.

Anaplanisporites sp.
Fig. spec. 16624
Barss, M.S., 1967, Geol. Surv. Can., Paper 67-11, pl. 38, fig. 22.
Sabine Bay Formation, Permian, vicinity of Tingmisut Lake, east side of Sabine Peninsula, Melville Island, District of Franklin.

cf. *Anguisporites intonsus* Wilson, 1962
Hypotype 16616
Barss, M.S., 1967, Geol. Surv. Can., Paper 67-11, pl. 38, fig. 14.
Sabine Bay Formation, Permian, vicinity of Tingmisut Lake, east side of Sabine Peninsula, Melville Island, District of Franklin.

Anulatisporites labiatus Hughes and Playford, 1961
Hypotype 15903
Barss, M.S., 1967, Geol. Surv. Can., Paper 67-11, pl. 9, fig. 23.
Mattson Formation, Mississippian, south face Twisted Mountain, about 15 miles northwest of South Nahanni, lat. 61°13′N. long. 123°37′W, Northwest Territories.

Anulatisporites lycosporoides Butterworth and Williams, 1958
Hypotype 15997
Barss, M.S., 1967, Geol. Surv. Can., Paper 67-11, pl. 13, fig. 20.
Canso Group, Upper Carboniferous, Pomquet River south of Route 4, Antigonish County, Nova Scotia.

Anulatisporites sp.
Fig. spec. 16180
Barss, M.S., 1967, Geol. Surv. Can., Paper 67-11, pl. 20, fig. 20.
Cumberland Group, Pennsylvanian, Pearly McFadden's mine on Chignecto coal seam, 0.3 miles east of Chignecto, Cumberland County, Nova Scotia.

Apiculatisporis abditus? (Loose) Potonié and Kremp, 1955
Hypotype 16155
Barss, M.S., 1967, Geol. Surv. Can., Paper 67-11, pl. 19, fig. 23.
Cumberland Group, Pennsylvanian, bore-hole No. 1 east, in Roger's field near outcrop on No. 6 seam at 40 foot depth, Springhill, Cumberland County, Nova Scotia.

Apiculatisporis aculeatus Ibrahim, 1933
Hypotype 15960
Barss, M.S., 1967, Geol. Surv. Can., Paper 67-11, pl. 12, fig. 23.
Canso Group, Upper Carboniferous, Pomquet River south of Route 4, Antigonish County, Nova Scotia.

Apiculatisporis aculeatus (Ibrahim) Potonié and Kremp, 1955
Hypotype 16135
Barss, M.S., 1967, Geol. Surv. Can., Paper 67-11, pl. 19, fig. 3.
Cumberland Group, Pennsylvanian, Pearly McFadden's mine on Chignecto coal seam, 0.3 miles east of Chignecto, Cumberland County, Nova Scotia.

Apiculatisporis baccatus (Hoffmeister, Staplin and Malloy) Butterworth and Williams, 1958
Hypotype 15958
Barss, M.S., 1967, Geol. Surv. Can., Paper 67-11, pl. 12, fig. 21.
Canso Group, Upper Carboniferous, Pomquet River south of Route 4, Antigonish County, Nova Scotia.

Apiculatisporis baccatus var. *raistrickii* (Dybková and Jachowicz) Staplin, 1960
Hypotype 15959
Barss, M.S., 1967, Geol. Surv. Can., Paper 67-11, pl. 12, fig. 22.
Canso Group, Upper Carboniferous, Pomquet River south of Route 4, Antigonish County, Nova Scotia.

Apiculatisporis baccatus Staplin, 1960
Hypotypes 92843, 92844
Utting, J., Jachowicz, M. and Jachowicz, A., 1989, Geol. Surv. Can., Bull. 396, pl. 6.1, figs. 4, 5.
Emma Fiord Formation, Early Carboniferous, Lyall River area, Grinnell Peninsula, Devon Island, District of Franklin.

Apiculatisporis decorus Singh, 1964
Hypotype 16626
Barss, M.S., 1967, Geol. Surv. Can., Paper 67-11, pl. 38, fig. 24.
Sabine Bay Formation, Permian, vicinity of Tingmisut Lake, east side of Sabine Peninsula, Melville Island, District of Franklin.

Apiculatisporis grumosus (Ibrahim) Potonié and Kremp, 1955
Hypotype 16154
Barss, M.S., 1967, Geol. Surv. Can., Paper 67-11, pl. 19, fig. 22.
Cumberland Group, Pennsylvanian, Morrison's pit on Chignecto coal seam, 0.4 miles east of Chignecto, Cumberland County, Nova Scotia.

Apiculatisporis grumosus? (Ibrahim) Potonié and Kremp, 1955
Hypotypes 16297, 16298
Barss, M.S., 1967, Geol. Surv. Can., Paper 67-11, pl. 25, figs. 31, 32.
Pennsylvanian, Mabou Mines Formation, Angus MacPhee's pit on north side of syncline in cliff section, 112 feet above beach, and coal seam 95 feet above beach, cliff section northeast side of syncline, Coal Mine Point, Mabou mines, Inverness County, Nova Scotia.

Apiculatisporis imbricatus (Kosanke) Potonié and Kremp, 1955
Hypotype 16466
Barss, M.S., 1967, Geol. Surv. Can., Paper 67-11, pl. 32, fig. 15.
Pictou Group, Pennsylvanian, Shediac water well approximately 1,400 feet east of shore of harbour and 1,800 feet south of highway at Shediac, New Brunswick.

Apiculatisporis cf. *A. irregularis* (Kosanke) Potonié and Kremp, 1955
Hypotype 16465
Barss, M.S., 1967, Geol. Surv. Can., Paper 67-11, pl. 32, fig. 14.
Pictou Group, Pennsylvanian, Shediac water well approximately 1,400 feet east of shore of harbour and 1,800 feet south of highway at Shediac, New Brunswick.

Apiculatisporis irregularis (Kosanke) Potonié and Kremp, 1955
Hypotype 16241
Barss, M.S., 1967, Geol. Surv. Can., Paper 67-11, pl. 23, fig. 6.
Westville Member, Stellarton Group, Pennsylvanian, main coal seam in No. 1 mine, No. 2 north bord on No. 1 Balance, No. 7 north level 150 feet north of slope, Pictou coalfield, Pictou County, Nova Scotia.

Apiculatisporis latigranifer (Loose) Potonié and Kremp, 1956
Hypotype 16379
Barss, M.S., 1967, Geol. Surv. Can., Paper 67-11, pl. 28, fig. 33.
Pictou Group, Pennsylvanian, stripping operation one mile south of Beersville, on east side of Coal Branch River, New Brunswick.

Apiculatisporis cf. *A. priscus* (Kosanke) Potonié and Kremp, 1955
Hypotype 16370
Barss, M.S., 1967, Geol. Surv. Can., Paper 67-11, pl. 28, fig. 24.
Morien Group, Pennsylvanian, Harbour coal seam, Dominion No. 12 colliery at face of No. 21 west level, Sydney coalfield, Cape Breton County, Nova Scotia.

Apiculatisporis setulosus (Kosanke) Potonié and Kremp, 1955
Hypotype 16158
Barss, M.S., 1967, Geol. Surv. Can., Paper 67-11, pl. 19, fig. 26.
Cumberland Group, Pennsylvanian, No. 7 coal seam from 3,200 foot mine bord, 150 feet east of back auxiliary slope in No. 4 mine, Springhill, Cumberland County, Nova Scotia.

Apiculatisporis spinosaetosus (Loose) Potonié and Kremp, 1955
Hypotype 16058
Barss, M.S., 1967, Geol. Surv. Can., Paper 67-11, pl. 16, fig. 4.
Port Hood Formation, Riversdale Group, Pennsylvanian, 200 feet from mouth of adit of Harbourview coal mine, Port Hood, Inverness County, Nova Scotia.

Apiculatisporis cf. *A. spinosaetosus* (Loose) Potonié and Kremp, 1955
Hypotypes 16057, 16156
Barss, M.S., 1967, Geol. Surv. Can., Paper 67-11, pl. 15, fig. 31; pl. 19, fig. 24.
Pennsylvanian, Port Hood Formation, Riversdale Group, 200 feet from mouth of adit of Harbourview coal mine, Port Hood, Inverness County, and Cumberland Group, Morrison's pit on Chignecto seam, 0.4 miles east of Chignecto, Cumberland County, Nova Scotia.

Apiculatisporis cf. *A. variocorneus* Sullivan, 1964
Hypotype 16061
Barss, M.S., 1967, Geol. Surv. Can., Paper 67-11, pl. 16, fig. 7.

Port Hood Formation, Riversdale Group, Pennsylvanian, 200 feet from mouth of adit of Harbourview coal mine, Port Hood, Inverness County, Nova Scotia.

Apiculatisporis cf. *A. verrucifer* (Kosanke) Potonié and Kremp, 1955
Hypotype 16459
Barss, M.S., 1967, Geol. Surv. Can., Paper 67-11, pl. 32, fig. 8.
Pictou Group, Pennsylvanian, Shediac water well approximately 1,400 feet east of shore of harbour and 1,800 feet south of highway at Shediac, New Brunswick.

Apiculatisporis sp.
Fig. specs. 15961, 16287, 16307
Barss, M.S., 1967, Geol. Surv. Can., Paper 67-11, pl. 12, fig. 24; pl. 25, fig. 21.
Canso Group, Upper Carboniferous, Pomquet River south of Route 4, Antigonish County; Pennsylvanian, Cumberland Group, Tingley pit in Roger's field near southern extension of McDougall Street, Springhill, Cumberland County; Mabou Mines Formation, coal seam immediately below fault north of Coal Mine Point, Mabou mines, Inverness County, Nova Scotia.

Apiculatisporis sp.
Fig. specs. 16372, 16374, 16375
Barss, M.S., 1967, Geol. Surv. Can., Paper 67-11, pl. 28, figs. 26,28,29.
Morien Group, Pennsylvanian, Phalen coal seam, Dominion No. 18 colliery at face of No. 6 west level, 3,400 feet from main deep (16372), and Harbour coal seam, Dominion No. 12 colliery at face of No. 21 west level, Sydney coalfield, Cape Breton County, Nova Scotia.

Apiculatisporis sp.
Fig. spec. 16623
Barss, M.S., 1967, Geol. Surv. Can., Paper 67-11, pl. 38, fig. 21.
Sabine Bay Formation, Permian, vicinity of Tingmisut Lake, east side of Sabine Peninsula, Melville Island, District of Franklin.

?*Apiculatisporis* sp.
Fig. specs. 16059, 16161
Barss, M.S., 1967, Geol. Surv. Can., Paper 67-11, pl. 16, fig. 5; pl. 20, fig. 1.
Pennsylvanian, Port Hood Formation, Riversdale Group, 200 feet from mouth of adit of Harbourview coal mine, Port Hood, Inverness County; Cumberland Group, No. 1 coal seam from 5,400 feet east level in no. 2 mine, Springhill, Cumberland County, Nova Scotia.

?*Apiculatisporis* sp.
Fig. spec. 16620
Barss, M.S., 1967, Geol. Surv. Can., Paper 67-11, pl. 38, fig. 18.
Sabine Bay Formation, Permian, vicinity of Tingmisut Lake, east side of Sabine Peninsula, Melville Island, District of Franklin.

Apiculatasporites compactus Playford
Holotype 13102; paratype 13103
Playford, G., 1964, Geol. Surv. Can., Bull. 107, p. 18, pl. 3, figs. 21, 22.
Barss, M.S., 1967, Geol. Surv. Can., Paper 67-11, pl. 2, fig. 2 (13102).
Cheverie Formation, Horton Group, Mississippian, west shore of Avon River, Nova Scotia.

Apiculatasporites spinulistratus (Loose) Potonié and Kremp, 1955
Hypotype 16044
Barss, M.S., 1967, Geol. Surv. Can., Paper 67-11, pl. 15, fig. 18.
Port Hood Formation, Riversdale Group, Pennsylvanian, No. 5 coal seam, 1,300 feet east of Evans' coal mine on No. 2 seam, near Neil MacLeod Brook, St. Rose, Inverness County, Nova Scotia.

Apiculatasporites sp.
Fig. specs. 15793, 16060
Barss, M.S., 1967, Geol. Surv. Can., Paper 67-11, pl. 5, fig. 17; pl. 16, fig. 6.
Windsor Group, Mississippian, Frenchvale Brook (Balls Creek), Northwest Arm of Sydney Harbour, and Port Hood Formation, Riversdale Group, Pennsylvanian, 200 feet from mouth of adit of Harbourview coal mine, Port Hood, Inverness County, Nova Scotia.

Apiculatasporites sp.
Fig. spec. 16277
Barss, M.S., 1967, Geol. Surv. Can., Paper 67-11, pl. 25, fig. 11.
Mabou Mines Formation, Pennsylvanian, Angus MacPhee's pit on north side of syncline in cliff section, 112 feet above beach, Mabou mines, Inverness County, Nova Scotia.

Armatisporites cf. *A. castaneaformis* Dybová and Jachowicz, 1957
Hypotype 16654
Barss, M.S., 1967, Geol. Surv. Can., Paper 67-11, pl. 16, fig. 3.
Port Hood Formation, Riversdale Group, Pennsylvanian, 200 feet from mouth of adit of Harbourview coal mine, Port Hood, Inverness County, Nova Scotia.

Armatisporites sp.
Fig. spec. 16048
Barss, M.S., 1967, Geol. Surv. Can., Paper 67-11, pl. 15, fig. 22.
Port Hood Formation, Riversdale Group, Pennsylvanian, 200 feet from mouth of adit of Harbourview coal mine, Port Hood, Inverness County, Nova Scotia.

?*Armatisporites* sp.
Fig. spec. 16254
Barss, M.S., 1967, Geol. Surv. Can., Paper 67-11, pl. 24, fig. 2.
Morien Group, Pennsylvanian, LeCras coal seam, from pit 200 feet southwest side of Highway 22, 1.1 miles from intersection of Old Mira Road and Highway 22, Sydney coalfield, Cape Breton County, Nova Scotia.

Auroraspora macra Sullivan
Hypotype 62496
Utting, J., 1980, Can. J. Earth Sci., vol. 17, no. 8, pl. 2, fig. 2.
Upper Windsor Group, Mississippian, United States Borax and Chemical Corporation borehole SB-1 near Stewiacke, Nova Scotia.

Auroraspora macra Sullivan 1968
Hypotype 73826
Utting, J., 1987, Geol. Surv. Can., Bull. 374, p. 37, pl. 5, fig. 12.
Windsor Group, Lower Carboniferous, Frenchvale Brook, Balls Creek, northwest arm of Sydney Harbor, lat. 46°10′N, long. 60°18′W, Cape Breton Island, Nova Scotia.

Auroraspora macra Sullivan, 1968
Hypotype 76351
Utting, J., 1987, Palynology, vol. 11, p. 79, pl. 1, fig. 1.
Hiram Brook Member, Albert Formation, Lower Carboniferous, tributary from west, Boyd Creek, east of Gautreau Village, New Brunswick.

Auroraspora macra Sullivan, 1968
Hypotype 89903
Utting, J., Keppie, J.D., and Giles, P.S., 1989, Geol. Surv. Can., Bull. 396, pl. 5.1, fig. 3.
Unnamed unit, Horton Group, Lower Carboniferous, Lakevale cliff north of syncline, St. Georges Bay, Antigonish Basin, Nova Scotia.

Auroraspora macra Sullivan, 1968
Hypotype 92845
Utting, J., Jachowicz, M. and Jachowicz, A., 1989, Geol. Surv. Can., Bull. 396, pl. 6.1, fig. 6.
Emma Fiord Formation, Early Carboniferous, Lyall River area, Grinnell Peninsula, Devon Island, District of Franklin.

Auroraspora solisorta Hoffmeister, Staplin and Malloy, 1955
Hypotype 92846
Utting, J., Jachowicz, M. and Jachowicz, A., 1989, Geol. Surv. Can., Bull. 396, pl. 6.1, fig. 7.
Emma Fiord Formation, Early Carboniferous, Lyall River area, Grinnell Peninsula, Devon Island, District of Franklin.

Auroraspora solisorta Hoffmeister, Staplin and Malloy, 1955
Hypotype 94484
Utting, J., 1991, Geol. Surv. Can., Bull. 412, pl. 4.1, fig. 27.
Hart River Formation, Lower Carboniferous, north bank Peel River, northern Yukon.

Auroraspora solisortus Hoffmeister, Staplin and Malloy, 1955
Hypotype 16001
Barss, M.S., 1967, Geol. Surv. Can., Paper 67-11, pl. 13, fig. 24.
Canso Group, Upper Carboniferous, Pomquet River south of Route 4, Antigonish County, Nova Scotia.

Auroraspora solisortus Hoffmeister, Staplin and Malloy, 1955
Hypotype 73827
Utting, J., 1987, Geol. Surv. Can., Bull. 374, pl. 5, fig. 13.
Windsor Group, Lower Carboniferous, north shore of Port Hood Island, lat. 46°1′N, long. 61°26′W, Cape Breton Island, Nova Scotia.

Aurospora sp.
Fig. spec. 16084
Barss, M.S., 1967, Geol. Surv. Can., Paper 67-11, pl. 16, fig. 29.
Port Hood Formation, Riversdale Group, Pennsylvanian, 200 feet from mouth of adit of Harbourview coal mine, Port Hood, Inverness County, Nova Scotia.

Biannulatisporites simplex Neville
Hypotype 62505
Utting, J., 1980, Can. J. Earth Sci., vol. 17, no. 8, pl. 2, fig. 11.
Upper Windsor Group, Mississippian, United States Borax and Chemical Corporation borehole SB-1 near Stewiacke, Nova Scotia.

Biannulatisporites simplex Neville, 1973
Hypotypes 73831, 73832
Utting, J., 1987, Geol. Surv. Can., Bull. 374, p. 40, pl. 5, figs. 19, 20.
Windsor Group, Lower Carboniferous, west side of headland, northeast of Knoydart Point, lat. 45°42′5″N, long. 62°13′6″W, Nova Scotia.

Botryococcus sp.
Fig. spec. 76391
Utting, J., 1987, Palynology, vol. 11, pl. 3, fig. 12.
Hiram Brook Member, Albert Formation, Lower Carboniferous, Boyd Creek, northeast of Gautreau Village, New Brunswick.

Cadiospora magna Kosanke, 1950
Hypotype 16313
Barss, M.S., 1967, Geol. Surv. Can., Paper 67-11, pl. 26, fig. 14.
Mabou Mines Formation, Pennsylvanian, borehole No. 5, depth 136 feet, 5,690 feet south of Stack Point, Mabou mines, Inverness County, Nova Scotia.

Calamospora breviradiata Kosanke, 1950
Hypotype 16028
Barss, M.S., 1967, Geol. Surv. Can., Paper 67-11, pl. 15, fig. 2.
Port Hood Formation, Riversdale Group, Pennsylvanian, No. 5 coal seam, 1,300 feet east of Evans' coal mine on No. 2 seam, near Neil MacLeod Brook, St. Rose, Inverness County, Nova Scotia.

Calamospora breviradiata Kosanke, 1950
Hypotype 16502
Barss, M.S., 1967, Geol. Surv. Can., Paper 67-11, pl. 34, fig. 1.
Pictou Group, Permian, depth 350 feet, Gallas Point borehole No. 2 on property of H. Tweedy, Gallas Point, lat. 46°7′42″N, long. 63°57′28″W, Prince Edward Island.

Calamospora breviradiata Kosanke, 1950
Hypotype 92847
Utting, J., Jachowicz, M. and Jachowicz, A., 1989, Geol. Surv. Can., Bull. 396, pl. 6.1, fig. 8.
Emma Fiord Formation, Early Carboniferous, Lyall River area, Grinnell Peninsula, Devon Island, District of Franklin.

Calamospora breviradiata Kosanke, 1950
Hypotype 94464
Utting, J., 1991, Geol. Surv. Can., Bull. 412, pl. 4.1, fig. 7.
Hart River Formation, Lower Carboniferous, north bank Peel River, northern Yukon.

Calamospora hartungiana Schopf, in Schopf, Wilson and Bentall, 1944
Hypotypes 16099, 16100
Barss, M.S., 1967, Geol. Surv. Can., Paper 67-11, pl. 8, figs. 1, 2.
Cumberland Group, Pennsylvanian, Pearly McFadden's mine on Chignecto coal seam, 0.3 miles east of Chignecto, Cumberland County, and Tingley pit in Roger's field near southern extension of McDougal Street, Springhill, Cumberland County, Nova Scotia.

Calamospora liquida Kosanke, 1950
Hypotype 16209
Barss, M.S., 1967, Geol. Surv. Can., Paper 67-11, pl. 22, fig. 11.
Morien Group, Pennsylvanian, Beaver mine on Tracey coal seam east of MacDonald Lake and south of Morrison road, approximately 5.0 miles from Highway 22, Sydney coalfield, Cape Breton County, Nova Scotia.

Calamospora liquida Kosanke, 1950
Hypotype 16454
Barss, M.S., 1967, Geol. Surv. Can., Paper 67-11, pl. 32, fig. 3.
Pictou Group, Pennsylvanian, Shediac water well approximately 1,400 feet east of shore of harbour and 1,800 feet south of highway at Shediac, New Brunswick.

Calamospora microrugosa (Ibrahim) Schopf, Wilson and Bentall, 1944
Hypotypes 15800, 16656, 16027, 16101, 16102
Barss, M.S., 1967, Geol. Surv. Can., Paper 67-11, pl. 5, fig. 14; pl. 7, fig. 16; pl. 15, fig. 1; pl. 18, figs. 3, 4.
Mississippian, Windsor Group, Frenchvale Brook (Balls Creek), Northwest Arm of Sydney Harbour, Nova Scotia and Mattson Formation, south face Twisted Mountain, about 15 miles northwest of South Nahanni, lat. 61°13′N, long. 123°37′W, Northwest Territories; Pennsylvanian, Port Hood Formation, Riversdale Group, 200 feet from mouth of adit of Harbourview coal mine, Port Hood, Inverness County, and Cumberland Group, Tingley pit in Roger's field near southern extension of McDougal Street, Springhill, Cumberland County (16101, 16102), Nova Scotia.

Calamospora microrugosa (Ibrahim) Schopf, Wilson and Bentall, 1944
Hypotype 92848
Utting, J., Jachowicz, M. and Jachowicz, A., 1989, Geol. Surv. Can., Bull. 396, pl. 6.1, fig. 9.
Emma Fiord Formation, Early Carboniferous, Lyall River area, Grinnell Peninsula, Devon Island, District of Franklin.

Calamospora microrugosa (Ibrahim) Schopf, Wilson and Bentall, 1944
Hypotype 94465
Utting, J., 1991, Geol. Surv. Can., Bull. 412, pl. 4.1, fig. 8.
Hart River Formation, Lower Carboniferous, north bank Peel River, northern Yukon.

Calamospora minuta Bhardwaj, 1957
Hypotype 16107
Barss, M.S., 1967, Geol. Surv. Can., Paper 67-11, pl. 18, fig. 9.
Cumberland Group, Pennsylvanian, No. 6 coal seam from 5,400 foot west level in No. 4 mine, 35 feet from rock tunnel slope, Springhill, Cumberland County, Nova Scotia.

Calamospora cf. *C. minuta* Bhardwaj, 1957
Fig. spec. 16030
Barss, M.S., 1967, Geol. Surv. Can., Paper 67-11, pl. 15, fig. 4.
Port Hood Formation, Riversdale Group, Pennsylvanian, No. 5 coal seam, 1,300 feet east of Evans' coal mine on No. 2 seam, near Neil MacLeod Brook, St. Rose, Inverness County, Nova Scotia.

Calamospora pallida (Loose) Schopf, Wilson and Bentall, 1944
Hypotypes 16201, 16269
Barss, M.S., 1967, Geol. Surv. Can., Paper 67-11, pl. 22, fig. 3; pl. 25, fig. 3.
Pennsylvanian, Morien Group, Beaver mine on Tracey coal seam east of MacDonald Lake and south of Morrison road, approximately 5.0 miles from Highway 22, Sydney coalfield, Cape Breton County; Mabou Mines Formation, seven-foot coal seam in drift from cliff shore to old workings, south side of syncline, Mabou mines, Inverness County, Nova Scotia.

Calamospora pallida (Loose) Schopf, Wilson and Bentall, 1944
Hypotype 16351
Barss, M.S., 1967, Geol. Surv. Can., Paper 67-11, pl. 28, fig. 5.
Morien Group, Pennsylvanian, Phalen coal seam, Dominion No. 18 colliery at face of No. 6 west level, 3,400 feet from main deep, Sydney coalfield, Cape Breton County, Nova Scotia.

Calamospora cf. *C. pallida* (Loose) Schopf, Wilson and Bentall, 1944
Hypotype 16103
Barss, M.S., 1967, Geol. Surv. Can., Paper 67-11, pl. 18, fig. 5.
Cumberland Group, Pennsylvanian, Morrison's pit on Chignecto coal seam, 0.4 miles east of Chignecto, Cumberland County, Nova Scotia.

Calamospora parva Guennel, 1958
Hypotype 16029
Barss, M.S., 1967, Geol. Surv. Can., Paper 67-11, pl. 15, fig. 3.
Port Hood Formation, Riversdale Group, Pennsylvanian, No. 5 coal seam, 1,300 feet east of Evans' coal mine on No. 2 seam, near Neil MacLeod Brook, St. Rose, Inverness County, Nova Scotia.

Calamospora parva Guennel, 1958
Hypotype 16268
Barss, M.S., 1967, Geol. Surv. Can., Paper 67-11, pl. 25, fig. 2.
Mabou Mines Formation, Pennsylvanian, Angus MacPhee's pit, north side syncline in cliff section, 112 feet above beach, Mabou mines, Inverness County, Nova Scotia.

Calamospora parva Guennel, 1958
Hypotype 73727
Utting, J., 1987, Geol. Surv. Can., Bull. 374, pl. 1, fig. 1.
Windsor Group, Lower Carboniferous, Stewiacke borehole SB-1, Shubenacadie Basin, lat. 45°7'N, long. 63°20'W, Hants County, Nova Scotia.

Calamospora perrugosa (Loose) Schopf, Wilson and Bentall, 1944
Hypotype 16270
Barss, M.S., 1967, Geol. Surv. Can., Paper 67-11, pl. 25, fig. 4.
Mabou Mines Formation, Pennsylvanian, first coal seam northeast side syncline, 84 feet below Stack sandstone of Coal Mine Point, Mabou mines, Inverness County, Nova Scotia.

Calamospora sarriana Bhardwaj, 1957
Hypotype 15946
Barss, M.S., 1967, Geol. Surv. Can., Paper 67-11, pl. 12, fig. 9.
Canso Group, Upper Carboniferous, Pomquet River south of Route 4, Antigonish County, Nova Scotia.

Calamospora saariana Bhardwaj, 1957
Hypotype 16350
Barss, M.S., 1967, Geol. Surv. Can., Paper 67-11, pl. 28, fig. 4.
Morien Group, Pennsylvanian, Phalen coal seam, Dominion No. 18 colliery at face of No. 6 west level, 3,400 feet from main deep, Sydney coalfield, Cape Breton County, Nova Scotia.

Calamospora sp.
Fig. spec. 16104
Barss, M.S., 1967, Geol. Surv. Can., Paper 67-11, pl. 18, fig. 6.
Cumberland Group, Pennsylvanian, test pit on Golden coal seam, 0.6 mile southeast from Millers Corner on Leamington road, 100 feet east of road, Springhill, Cumberland County, Nova Scotia.

cf. *Calamospora* Type A Hoffmeister, Staplin and Malloy, 1955
Fig. spec. 15795
Barss, M.S., 1967, Geol. Surv. Can., Paper 67-11, pl. 5, fig. 9.
Windsor Group, Mississippian, Frenchvale Brook (Balls Creek), Northwest Arm of Sydney Harbour, Nova Scotia.

?*Camptotriletes corrugatus* (Ibrahim) Potonié and Kremp, 1955
Hypotype 16460
Barss, M.S., 1967, Geol. Surv. Can., Paper 67-11, pl. 32, fig. 9.
Pictou Group, Pennsylvanian, Shediac water well approximately 1,400 feet east of shore of harbour and 1,800 feet south of highway at Shediac, New Brunswick.

Camptotriletes cristatus Sullivan and Marshall, 1966
Hypotype 15969
Barss, M.S., 1967, Geol. Surv. Can., Paper 67-11, pl. 12, fig. 32.
Canso Group, Upper Carboniferous, Pomquet River south of Route 4, Antigonish County, Nova Scotia.

Camptotriletes superbus Neves, 1961
Hypotype 16050
Barss, M.S., 1967, Geol. Surv. Can., Paper 67-11, pl. 15, fig. 24.
Port Hood Formation, Riversdale Group, Pennsylvanian, 200 feet from mouth of adit of Harbourview coal mine, Port Hood, Inverness County, Nova Scotia.

Camptotriletes verrucosus Butterworth and Williams, 1958
Hypotypes 13167, 13168
Playford, G., 1964, Geol. Surv. Can., Bull. 107, p. 31, pl. 9, figs. 6, 7.
Barss, M.S., 1967, Geol. Surv. Can., Paper 67-11, pl. 3, fig. 6 (13167).
Horton Group, Mississippian, Mill Brook, one mile north of East Gore, Nova Scotia.

Camptotriletes sp.
Fig. spec. 13165
Playford, G., 1964, Geol. Surv. Can., Bull. 107, p. 32, pl. 9, fig. 4.
Horton Group, Mississippian, Mill Brook, one mile north of East Gore, Nova Scotia.

Camptotriletes sp.
Fig. specs. 15874, 15955, 16047, 16652, 16653, 16121, 16149, 16289
Barss, M.S., 1967, Geol. Surv. Can., Paper 67-11, pl. 8, fig. 11; pl. 12, fig. 18; pl. 15, fig. 21; pl. 16, figs. 1, 2; pl 18, fig. 23; pl. 119, fig. 17; pl. 25, fig. 23.
Carboniferous, Mississippian, Mattson Formation, south face Twisted Mountain, about 15 miles northwest of South Nahanni, lat. 61°13'N, long. 123°37'W, Northwest Territories; Pennsylvanian, Canso Group, Pomquet River south of Route 4, Antigonish County, Port Hood Formation, Riversdale Group, 200 feet from mouth of adit of Harbourview mine, Port Hood, Inverness County (16047, 16652, 16653), and Cumberland Group, Fenwick mine (16121) and Pearly McFadden's mine (16149) on Chignecto coal seam, 2.1 and 0.3 miles east of Chignecto, Cumberland County; Morien Group, Le Cras coal seam,

pit 200 feet southwest side of Highway 22, 1.1 miles from intersection of Old Mira Road, Sydney coalfield, Cape Breton County (16289), Nova Scotia.

Camptozonotriletes velatus (Waltz) Playford, 1963
Hypotype 15925
Barss, M.S., 1967, Geol. Surv. Can., Paper 67-11, pl. 10, fig. 20.
Mattson Formation, Mississippian, south face Twisted Mountain, about 15 miles northwest of South Nahanni, lat. 61°13′N. long. 123°37′W, Northwest Territories.

Camptozontriletes velatus (Waltz) Playford, 1963
Hypotype 92849
Utting, J., Jachowicz, M. and Jachowicz, A., 1989, Geol. Surv. Can., Bull. 396, pl. 6.1, fig. 10.
Emma Fiord Formation, Early Carboniferous, Lyall River area, Grinnell Peninsula, Devon Island, District of Franklin.

Chaetosphaerites pollenisimilis (Horst) Butterworth and Williams, 1958
Hypotype 15836
Barss, M.S., 1967, Geol. Surv. Can., Paper 67-11, pl. 7, fig. 1.
Mattson Formation, Mississippian, south face Twisted Mountain, about 15 miles northwest of South Nahanni, lat. 61°13′N. long. 123°37′W, Northwest Territories.

Chaetosphaerites sp.
Fig. spec. 15837
Barss, M.S., 1967, Geol. Surv. Can., Paper 67-11, pl. 7, fig. 2.
Mattson Formation, Mississippian, south face Twisted Mountain, about 15 miles northwest of South Nahanni, lat. 61°13′N. long. 123°37′W, Northwest Territories.

Chomotriletes multivittatus Playford, 1978
Hypotype 73830
Utting, J., 1987, Geol. Surv. Can., Bull. 374, pl. 5, fig. 18.
Windsor Group, Lower Carboniferous, northeast of Knoydart Point, lat. 45°42′5″N, long. 62°13′6″W, Nova Scotia.

Chomotriletes multivittatus Playford, 1978
Hypotype 89904
Utting, J., Keppie, J.D. and Giles, P.S., 1989, Geol. Surv. Can., Bull. 396, pl. 5.1, fig. 4.
Unnamed unit, Horton Group, Lower Carboniferous, Lakevale cliff north of syncline, St. Georges Bay, Antigonish Basin, Nova Scotia.

Chomotriletes multivittatus Playford, 1978
Hypotype 92850
Utting, J., Jachowicz, M. and Jachowicz, A., 1989, Geol. Surv. Can., Bull. 396, pl. 6.1, fig. 11.
Emma Fiord Formation, Early Carboniferous, Lyall River area, Grinnell Peninsula, Devon Island, District of Franklin.

Cincturasporites altilis Hacquebard and Barss, 1957
Hypotype 15894
Barss, M.S., 1967, Geol. Surv. Can., Paper 67-11, pl. 9, fig. 14.
Mattson Formation, Mississippian, south face Twisted Mountain, about 15 miles northwest of South Nahanni, lat. 61°13′N. long. 123°37′W, Northwest Territories.

Cingulizonates bialatus (Waltz) Smith and Butterworth, 1967
Hypotypes 92851, 92936-92938
Utting, J., Jachowicz, M. and Jachowicz, A., 1989, Geol. Surv. Can., Bull. 396, pl. 6.1, fig. 13; pl. 6.6, figs. 2-4.
Emma Fiord Formation, Early Carboniferous, Lyall River area, Grinnell Peninsula, Devon Island, and Svartevaeg Cliffs, northern Axel Heiberg Island, District of Franklin.

Cingulizonates bialatus (Waltz) Smith and Butterworth, 1967
Hypotype 94494
Utting, J., 1991, Geol. Surv. Can., Bull. 412, pl. 4.2, fig. 10.
Hart River Formation, Lower Carboniferous, north bank Peel River, northern Yukon.

Cirratriradites annulatus Kosanke and Brokaw, in Kosanke, 1950
Hypotypes 16395, 16396
Barss, M.S., 1967, Geol. Surv. Can., Paper 67-11, pl. 29, fig. 16,17.
Morien Group, Pennsylvanian, Harbour coal seam, Dominion No. 12 colliery at face of No. 21 west level, Sydney coalfield, Cape Breton County, Nova Scotia.

Cirratriradites elegans (Waltz) Potonié and Kremp, 1956
Hypotype 15936
Barss, M.S., 1967, Geol. Surv. Can., Paper 67-11, pl. 11, fig. 10.
Mattson Formation, Mississippian, south face Twisted Mountain, about 15 miles northwest of South Nahanni, lat. 61°13′N. long. 123°37′W, Northwest Territories.

Cirrotriradites flabelliformis Wilson and Kosanko, 1950
Hypotype 16264
Barss, M.S., 1967, Geol. Surv. Can., Paper 67-11, pl. 24, fig. 12.
Morien Group, Pennsylvanian, Beaver mine on Tracey coal seam east of MacDonald Lake and south of Morrison road, approximately 5.0 miles from Highway 22, Sydney coalfield, Cape Breton County, Nova Scotia.

cf. *Cirratriradites granulatipunctatus* Hofffmeister, Staplin and Malloy, 1955
Hypotypes 15814, 16000
Barss, M.S., 1967, Geol. Surv. Can., Paper 67-11, pl. 6, fig. 7; pl. 13, fig. 23.
Windsor Group, Mississippian, Frenchvale Brook (Balls Creek), Northwest Arm of Sydney Harbour; Canso Group, Upper Carboniferous, Pomquet River south of Route 4, Nova Scotia.

Cirratriradites ornatus (Luber and Waltz) Hart, 1965
Hypotype 16641
Barss, M.S., 1967, Geol. Surv. Can., Paper 67-11, pl. 38, fig. 39.
Sabine Bay Formation, Permian, vicinity of Tingmisut Lake, east side of Sabine Peninsula, Melville Island, District of Franklin.

Cirratriradites saturni (Ibrahim) Schopf, Wilson and Bentall, 1944
Hypotypes 16317, 16318
Barss, M.S., 1967, Geol. Surv. Can., Paper 67-11, pl. 26, figs. 18, 19.
Mabou Mines Formation, Pennsylvanian, coal seam 95 feet above beach, cliff section northeast side of syncline, Coal Mine Point, Mabou mines, Inverness County, Nova Scotia.

Cirratriradites solaris Hacquebard and Barss, 1957
Hypotype 15937
Barss, M.S., 1967, Geol. Surv. Can., Paper 67-11, pl. 11, fig. 11.
Mattson Formation, Mississippian, south face Twisted Mountain, about 15 miles northwest of South Nahanni, lat. 61°13′N. long. 123°37′W, Northwest Territories.

Cirratriradites splendens Balme and Hennelly, 1956
Hypotype 16511
Barss, M.S., 1967, Geol. Surv. Can., Paper 67-11, pl. 34, fig. 10.
Pictou Group, Permian, depth 332 feet, Gallas Point borehole No. 2 on property of H. Tweedy, Gallas Point, lat. 46°7′N, long. 63°57′28″W, Prince Edward Island.

Colatisporites decorus (Bharadwaj and Venkatachala) Williams, 1973
Hypotype 83360
Utting, J., 1987, Geol. Surv. Can., Bull. 374, p. 37, pl. 5, fig. 14.
Windsor Group, Lower Carboniferous, north shore of Port Hood Island, lat. 46°1′N, long. 61°26′W, Cape Breton Island, Nova Scotia.

Colatisporites decorus (Bharadwaj and Venkatachala) Williams, 1978
Hypotypes 89905, 89906
Utting, J., Keppie, J.D. and Giles, P.S., 1989, Geol. Surv. Can., Bull. 396, pl. 5.1, figs. 5, 7.
Wilkie Brook Formation, Horton Group, Lower Carboniferous, 380 m north of mouth of Wilkie Brook, western shore of St. Georges Bay, Antigonish Basin, Nova Scotia.

Colatisporites decorus (Bharadwaj and Venkachatala) Williams, 1973
Hypotypes 92852-92854
Utting, J., Jachowicz, M. and Jachowicz, A., 1989, Geol. Surv. Can., Bull. 396, pl. 6.1, figs. 12, 15, 16.
Emma Fiord Formation, Early Carboniferous, Lyall River area, Grinnell Peninsula, Devon Island, District of Franklin.

Colatisporites decorus (Bharadwaj and Venkatachala) emend. Williams, 1973
Hypotypes 94480, 94503
Utting, J., 1991, Geol. Surv. Can., Bull. 412, pl. 4.1, fig. 23; pl. 4.2, fig. 19.
Hart River Formation, Lower Carboniferous, north bank Peel River, northern Yukon.

Converrucosisporites parvinodosus Playford
Holotype 13089; paratypes 13088, 13090
Playford, G., 1964, Geol. Surv. Can., Bull. 107, p. 15, pl. 3, figs. 7-9.
Horton Group, Mississippian, Mill Brook, one mile north of East Gore, Nova Scotia.

Converrucosisporites parvinodosus Playford 1964
Hypotype 76352
Utting, J., 1987, Palynology, vol. 11, p. 79, pl. 1, fig. 2.
Hiram Brook Member, Albert Formation, Lower Carboniferous, Boyd Creek, east Gautreau Village area, New Brunswick.

Converrucosisporites sulcatus (Wilson and Kosanke) Potonié and Kremp, 1955
Hypotype 16368
Barss, M.S., 1967, Geol. Surv. Can., Paper 67-11, pl. 28, fig. 22.
Morien Group, Pennsylvanian, Harbour coal seam, Dominion No. 12 colliery at face of No. 21 west level, Sydney coalfield, Cape Breton County, Nova Scotia.

Converrucosisporites sp.
Fig. spec. 16037
Barss, M.S., 1967, Geol. Surv. Can., Paper 67-11, pl. 16, fig. 11.
Port Hood Formation, Riversdale Group, Pennsylvanian, 200 feet from mouth of adit of Harbourview coal mine, Port Hood, Inverness County, Nova Scotia.

Converrucosisporites sp.
Fig. spec. 16369
Barss, M.S., 1967, Geol. Surv. Can., Paper 67-11, pl. 28, fig. 23.
Morien Group, Pennsylvanian, Phalen coal seam, Dominion No. 18 colliery at face of No. 6 west level, 3,400 feet from main deep, Sydney coalfield, Cape Breton County, Nova Scotia.

?*Converrucosisporites* sp.
Fig. spec. 16062
Barss, M.S., 1967, Geol. Surv. Can., Paper 67-11, pl. 16, fig. 8.
Port Hood Formation, Riversdale Group, Pennsylvannian, 200 feet from mouth of adit of Harbourview coal mine, Port Hood, Inverness County, Nova Scotia.

Convolutispora ampla Hoffmeister, Staplin and Malloy, 1955
Hypotype 15869
Barss, M.S., 1967, Geol. Surv. Can., Paper 67-11, pl. 8, fig. 6.

Mattson Formation, Mississippian, south face Twisted Mountain, about 15 miles northwest of South Nahanni, lat. 61°13′N. long. 123°37′W, Northwest Territories.

Convolutispora ampla? Hoffmeister, Staplin and Malloy, 1955
Hypotype 16140
Barss, M.S., 1967, Geol. Surv. Can., Paper 67-11, pl. 19, fig. 8.
Cumberland Group, Pennsylvanian, No. 1 coal seam, 5,400 feet east level in No. 2 mine, Springhill, Cumberland County, Nova Scotia.

Convolutispora cerebra Butterworth and Williams, 1958
Hypotype 15865
Barss, M.S., 1967, Geol. Surv. Can., Paper 67-11, pl. 8, fig. 2.
Mattson Formation, Mississippian, south face Twisted Mountain, about 15 miles northwest of South Nahanni, lat. 61°13′N. long. 123°37′W, Northwest Territories.

Convolutispora circumvallata Clayton 1970
Hypotype 76353
Utting, J., 1987, Palynology, vol. 11, p. 79, pl. 1, fig. 3.
Frederick Brook Member, Albert Formation, Lower Carboniferous, depth 132.21 m, Petro-Canada Beaudreau No. 1 well, Highway 925 south of Beaudreau Village, New Brunswick.

Convolutispora crassa Playford, 1962
Hypotype 15856
Barss, M.S., 1967, Geol. Surv. Can., Paper 67-11, pl. 7, fig. 22.
Mattson Formation, Mississippian, south face Twisted Mountain, about 15 miles northwest of South Nahanni, lat. 61°13′N. long. 123°37′W, Northwest Territories.

Convolutisporis cf. *finis* Love, 1960
Hypotypes 13144, 13145
Playford, G., 1964, Geol. Surv. Can., Bull. 107, p. 27, pl. 8, figs. 1, 2.
Barss, M.S., 1967, Geol. Surv. Can., Paper 67-11, pl. 3, fig. 3 (13145).
Cheverie Formation, Horton Group, Mississippian, west shore of Avon River, Nova Scotia.

Convolutispora sp. cf. *C. finis* Love, 1960
Fig. spec. 89908
Utting, J., Keppie, J.D. and Giles, P.S., 1989, Geol. Surv. Can., Bull. 396, pl. 5.1, fig. 8.
Horton Bluff Formation, Horton Group, Lower Carboniferous, depth 75 m, Saarberg Interplan Canada Limited SI/8 drill core, north of west branch of Avon River, lat. 44°57′10′N, long. 64°15′50″W, Windsor Basin, Nova Scotia.

Convolutisporis flexuosa forma *major* Hacquebard, 1957
Holotype 4254
Barss, M.S., 1967, Geol. Surv. Can., Paper 67-11, pl. 3, fig. 1.
Horton Group, Mississippian, Gore Court House Hill, near Kennetcook, Nova Scotia.

Convolutispora flexuosa forma *major* Hacquebard 1957
Hypotype 76354
Utting, J., 1987, Palynology, vol. 11, p. 79, pl. 1, fig. 4.
Hiram Brook Member, Albert Formation, Lower Carboniferous, tributary from west, Boyd Creek, east of Gautreau Village, New Brunswick.

Convolutispora flexuosa forma *major* Hacquebard 1957
Hypotype 89907
Utting, J., Keppie, J.D. and Giles, P.S., 1989, Geol. Surv. Can., Bull. 396, pl. 5.1, fig. 1.
Wilkie Brook Formation, Horton Group, Lower Carboniferous, 380 m north of mouth of Wilkie Brook, western shore of St. Georges Bay, Antigonish Basin, Nova Scotia.

Convolutispora florida Hoffmeister, Staplin and Malloy, 1955
Hypotypes 15979, 15982
Barss, M.S., 1967, Geol. Surv. Can., Paper 67-11, pl. 13, figs. 2, 5.
Canso Group, Upper Carboniferous, Pomquet River south of Route 4, Antigonish County, Nova Scotia.

Convolutispora florida Hoffmeister, Staplin and Malloy, 1955
Hypotype 92855
Utting, J., Jachowicz, M. and Jachowicz, A., 1989, Geol. Surv. Can., Bull. 396, pl. 6.1, fig. 17.
Emma Fiord Formation, Early Carboniferous, Lyall River area, Grinnell Peninsula, Devon Island, District of Franklin.

Convolutispora labyrinthea Utting
Holotype 73758; hypotypes 73759, 73760
Utting, J., 1987, Geol. Surv. Can., Bull. 374, p. 29, pl. 2, figs. 5-7.
Windsor Group, Lower Carboniferous, north shore of Port Hood Island, lat. 46°1′N, long. 61°26′W, Cape Breton Island, Nova Scotia.

Convolutispora mellita Hoffmeister, Staplin and Malloy, 1955
Hypotypes 15859, 16148
Barss, M.S., 1967, Geol. Surv. Can., Paper 67-11, pl. 7, fig. 25; pl. 19, fig. 16.
Mattson Formation, Mississippian, south face Twisted Mountain, about 15 miles northwest of South Nahanni, lat. 61°13′N. long. 123°37′W, Northwest Territories; Cumberland Group, Pennsylvanian, Fenwick mine on Chignecto coal seam, 2.1 miles east of Chignecto, Cumberland County, Nova Scotia.

Convolutispora mellita Hoffmeister, Staplin and Malloy, 1955
Hypotype 73761
Utting, J., 1987, Geol. Surv. Can., Bull. 374, pl. 2, fig. 8.
Windsor Group, Lower Carboniferous, near Baddeck Bridge, lat. 46°9′N, long. 61°11′W, Cape Breton Island, Nova Scotia.

Convolutispora cf. *C. mellita* Hoffmeister, Staplin and Malloy, 1955
Hypotype 15807
Barss, M.S., 1967, Geol. Surv. Can., Paper 67-11, pl. 5, fig. 21.
Windsor Group, Mississippian, Frenchvale Brook (Balls Creek), Northwest Arm of Sydney Harbour, Nova Scotia.

Convolutispora tessellata Hoffmeister, Staplin and Malloy, 1955
Hypotype 15864
Barss, M.S., 1967, Geol. Surv. Can., Paper 67-11, pl. 8, fig. 1.
Mattson Formation, Mississippian, south face Twisted Mountain, about 15 miles northwest of South Nahanni, lat. 61°13′N. long. 123°37′W, Northwest Territories.

Convolutispora tessellata Hoffmeister, Staplin and Malloy, 1955
Hypotype 73762
Utting, J., 1987, Geol. Surv. Can., Bull. 374, pl. 2, fig. 9.
Windsor Group, Lower Carboniferous, Baddeck Bridge, Cape Breton Island, Nova Scotia.

Convolutispora tessellata Hoffmeister, Staplin and Malloy, 1955
Hypotype 92857
Utting, J., Jachowicz, M. and Jachowicz, A., 1989, Geol. Surv. Can., Bull. 396, pl. 6.1, fig. 19.
Emma Fiord Formation, Early Carboniferous, Lyall River area, Grinnell Peninsula, Devon Island, District of Franklin.

Convolutispora tuberculata (Waltz) Hoffmeister, Staplin and Malloy, 1955
Hypotype 15871
Barss, M.S., 1967, Geol. Surv. Can., Paper 67-11, pl. 8, fig. 8.
Mattson Formation, Mississippian, south face Twisted Mountain, about 15 miles northwest of South Nahanni, lat. 61°13′N. long. 123°37′W, Northwest Territories.

Convolutispora tuberculata (Waltz) Hoffmeister, Staplin and Malloy, 1955
Hypotype 92858
Utting, J., Jachowicz, M. and Jachowicz, A., 1989, Geol. Surv. Can., Bull. 396, pl. 6.1, fig. 18.
Emma Fiord Formation, Early Carboniferous, Lyall River area, Grinnell Peninsula, Devon Island, District of Franklin.

Convolutispora usitata Playford, 1962
Hypotype 15862
Barss, M.S., 1967, Geol. Surv. Can., Paper 67-11, pl. 7, fig. 28.
Mattson Formation, Mississippian, south face Twisted Mountain, about 15 miles northwest of South Nahanni, lat. 61°13′N. long. 123°37′W, Northwest Territories.

Convolutispora usitata Playford, 1962
Hypotype 92859
Utting, J., Jachowicz, M. and Jachowicz, A., 1989, Geol. Surv. Can., Bull. 396, pl. 6.1, fig. 20.
Emma Fiord Formation, Early Carboniferous, Lyall River area, Grinnell Peninsula, Devon Island, District of Franklin.

Convolutispora cf. *C. varicosa* Butterworth and Williams, 1958
Hypotype 15977
Barss, M.S., 1967, Geol. Surv. Can., Paper 67-11, pl. 12, fig. 40.
Canso Group, Upper Carboniferous, Pomquet River south of Route 4, Antigonish County, Nova Scotia.

Convolutispora venusta Hoffmeister, Staplin and Malloy, 1955
Hypotype 15861
Barss, M.S., 1967, Geol. Surv. Can., Paper 67-11, pl. 7, fig. 27.
Mattson Formation, Mississippian, south face Twisted Mountain, about 15 miles northwest of South Nahanni, lat. 61°13′N. long. 123°37′W, Northwest Territories.

Convolutispora venusta Hoffmeister, Staplin and Malloy, 1955
Hypotype 92860
Utting, J., Jachowicz, M. and Jachowicz, A., 1989, Geol. Surv. Can., Bull. 396, pl. 6.1, fig. 21.
Emma Fiord Formation, Early Carboniferous, Lyall River area, Grinnell Peninsula, Devon Island, District of Franklin.

Convolutispora cf. *C. venusta* Hoffmeister, Staplin and Malloy, 1955
Hypotype 15871
Barss, M.S., 1967, Geol. Surv. Can., Paper 67-11, pl. 8, fig. 7.
Mattson Formation, Mississippian, south face Twisted Mountain, about 15 miles northwest of South Nahanni, lat. 61°13′N. long. 123°37′W, Northwest Territories.

Convolutispora vermiformis Hughes and Playford, 1961
Hypotypes 13142, 13143, 4256
Playford, G., 1964, Geol. Surv. Can., Bull. 107, p. 27, pl. 7, figs. 8, 9.
Barss, M.S., 1967, Geol. Surv. Can., Paper 67-11, pl. 3, fig. 9 (13143). Micropaleontology, 1957, vol. 3, no.4, pl. 2, fig. 10 of *Convolutispora flexuosa* forma *minor* (Hacquebard) Playford.
Cheverie Formation, Horton Group, Mississippian, west shore of Avon River, Nova Scotia.

Convolutispora vermiformis Hughes and Playford, 1961
Hypotype 73763
Utting, J., 1987, Geol. Surv. Can., Bull. 374, pl. 2, fig. 10.
Windsor Group, Lower Carboniferous, north shore of Port Hood Island, lat. 46°1′N, long. 61°26′W, Cape Breton Island, Nova Scotia.

Convolutispora vermiformis Hughes and Playford, 1961
Hypotype 76355
Utting, J., 1987, Palynology, vol. 11, p. 79, pl. 1, fig. 5.

Hiram Brook Member, Albert Formation, Lower Carboniferous, tributary from west, Boyd Creek, east of Gautreau Village, New Brunswick.

Convolutispora vermiformis Hughes and Playford, 1961
Hypotype 89909
Utting, J., Keppie, J.D. and Giles, P.S., 1989, Geol. Surv. Can., Bull. 396, pl. 5.1, fig. 6.
Horton Bluff Formation, Horton Group, Lower Carboniferous, depth 108 m, Saarberg Interplan Canada Limited SI/8 drill core, north of west branch of Avon River, lat. 44°57'10'N, long. 64°15'50"W, Windsor Basin, Nova Scotia.

Convolutispora vermiformis Hughes and Playford, 1961
Hypotype 92861
Utting, J., Jachowicz, M. and Jachowicz, A., 1989, Geol. Surv. Can., Bull. 396, pl. 6.1, fig. 22.
Emma Fiord Formation, Early Carboniferous, Lyall River area, Grinnell Peninsula, Devon Island, District of Franklin.

Convolutispora vermiformis Hughes and Playford, 1961
Hypotype 94468
Utting, J., 1991, Geol. Surv. Can., Bull. 412, pl. 4.1, fig. 11.
Hart River Formation, Lower Carboniferous, north bank Peel River, northern Yukon.

Convolutispora sp.
Fig. specs. 15817, 15866, 15868, 15972-15975, 15984-15988, 16046, 16051, 16138, 16146, 16147, 16151, 16157, 16233, 16235, 16288, 16290, 16291, 16364-16367, 16467
Bass, M.S., 1967, Geol. Surv. Can., Paper 67-11, pl. 6, fig. 10; pl. 8, figs. 3, 5; pl. 12, figs. 35-38; pl. 13, figs. 7-11; pl. 15, figs. 20, 25; pl. 19, figs, 6, 14, 15, 19, 25; pl. 22, figs. 35, 37; pl. 25, figs. 22, 24, 25; pl. 28, figs. 18-21; pl. 32, fig. 16.
Mississippian, Windsor Group, Frenchvale Brook (Balls Creek), Northwest Arm of Sydney Harbour, Nova Scotia (15817); Mattson Formation, south face Twisted Mountain, about 15 miles northwest of South Nahanni, lat. 61°13'N, long. 123°37'W, Northwest Territories (15866, 15868); Canso Group, Upper Carboniferous, Pomquet River south of Route 4, Antigonish County; Pennsylvanian, Port Hood Formation, Riversdale Group, 200 feet from mouth of adit of Harbourview coal mine, Port Hood, Inverness County (16046, 16051), Cumberland Group, No. 7 coal seam from 3,200 foot mine bord, 150 feet east of back axiliary slope in No. 4 mine (16138), Pearly McFadden's mine on Chignecto coal seam (16146), Morrison's Pit on Chignecto coal seam (16147), Fenwick mine on Chignecto coal seam (16157), 0.3, 0.4 and 2.1 miles east of Chignecto, and pit on Park Road on Joggins Fall coal seam, Joggins (16151), Cumberland County; Morien Group, McAuley coal seam, on shore 1.75 miles south of Homeville and first seam below Tracey seam, 1 mile west of Homeville, Sydney Coalfield, Cape Breton County (16233, 16235); Mabou Mines Formation, Morien Group, Angus MacPhee's pit, north side of syncline in cliff section 112 feet above beach (16288) and south side of syncline (16290), Mabou Mines, Inverness County; Morien Group, Le Cras coal seam, pit 200 feet southwest side of Highway 22, 1.6 miles from intersection of Old Mira Road, Sydney Coalfield, Cape Breton County (16291), Phalen coal seam in Dominion No. 18 colliery, at face of No. 6 west level, 3400 feet from main deep (16364, 16366, 16367) and Harbour coal seam in Dominion No. 12 colliery, at face of No. 21 west level (16365), Sydney Coalfield, Cape Breton County, Nova Scotia; Pictou Group, Shediac water well, approximately 1400 feet east of shore of harbour and 1800 feet south of highway at Shediac, New Brunswick (16467).

Convolutispora sp.
Fig. spec. 62493
Utting, J., 1980, Can. J. Earth Sci., vol. 17, no. 8, pl. 1, fig. 14.
Upper Windsor Group, Mississippian, United States Borax and Chemical Corporation borehole SB-1 near Stewiacke, Nova Scotia.

Convolutispora sp.
Fig. spec. 92856
Utting, J., Jachowicz, M. and Jachowicz, A., 1989, Geol. Surv. Can., Bull. 396, pl. 6.1, fig. 14.
Emma Fiord Formation, Early Carboniferous, Lyall River area, Grinnell Peninsula, Devon Island, District of Franklin.

Convolutispora sp.
Fig. spec. 94490
Utting, J., 1991, Geol. Surv. Can., Bull. 412, pl. 4.2, fig. 6.
Hart River Formation, Lower Carboniferous, north bank Peel River, northern Yukon.

Convolutispora cf. sp. 2 in Peppers, 1964
Fig. spec. 16463
Barss, M.S., 1967, Geol. Surv. Can., Paper 67-11, pl. 32, fig. 12.
Pictou Group, Pennsylvanian, Scoudouc River 1 1/8 miles upstream from highway running south from Dorchester Crossing, New Brunswick.

cf. *Corbulispora subalveolaris* (Luber) Sullivan, 1964
Hypotype 15879
Barss, M.S., 1967, Geol. Surv. Can., Paper 67-11, pl. 8, fig. 16.
Mattson Formation, Mississippian, south face Twisted Mountain, about 15 miles northwest of South Nahanni, lat. 61°13'N. long. 123°37'W, Northwest Territories.

Cordaitina cf. *C. crenulata* (Wilson) Hart, 1965
Hypotype 16558
Barss, M.S., 1967, Geol. Surv. Can., Paper 67-11, pl. 36, fig. 15.
Permian, Tatonduck River, west limb of syncline 3 miles east of Yukon-Alaska border, lat. 64°58'N, long. 140°54'W, Yukon.

Cordaitina cf. *C. triangularis* (Mehta) Hart, 1965
Hypotype 16517
Barss, M.S., 1967, Geol. Surv. Can., Paper 67-11, pl. 34, fig. 16.

Pictou Group, Permian, depth 332 feet, Gallas Point borehole No. 2 on property of H. Tweedy, Gallas Point, lat. 46°7′42″N, long. 63°57′28″W, Prince Edward Island.

Cordaitina sp.
Fig. spec. 16344
Barss, M.S., 1967, Geol. Surv. Can., Paper 67-11, pl. 27, fig. 22.
Mabou Mines Formation, Pennsylvanian, seven-foot coal seam in drift from cliff shore to old workings, south side of syncline, Mabou mines, Inverness County, Nova Scotia.

Cordaitina sp.
Fig. specs. 16514, 16516, 16519, 16524
Barss, M.S., 1967, Geol. Surv. Can., Paper 67-11, pl. 34, figs. 13, 15; pl. 35, figs. 1, 6.
Pictou Group, Permian, depths 332, 335 (16519) and 350 (16516) feet, Gallas Point borehole No. 2 on property of H. Tweedy, Gallas Point, lat. 46°7′42″N, long. 63°57′28″W, Prince Edward Island.

?*Cordaitina* sp.
Fig. spec. 16487
Barss, M.S., 1967, Geol. Surv. Can., Paper 67-11, pl. 33, fig. 2.
Pictou Group, Pennsylvanian, Shediac water well approximately 1,400 feet east of shore of harbour and 1,800 feet south of highway at Shediac, New Brunswick.

Cordylosporites papillatus (Naumova) Playford and Satterthwait, 1985
Hypotypes 92913, 92959
Utting, J., Jachowicz, M. and Jachowicz, A., 1989, Geol. Surv. Can., Bull. 396, pl. 6.4, fig. 20; pl. 6.6, fig. 28.
Emma Fiord Formation, Early Carboniferous, Lyall River area, Grinnell Peninsula, Devon Island, and Svartevaeg Cliffs, northern Axel Hieberg Island, District of Franklin.

Cordylosporites sp. cf. *C. papillatus* Playford and Satterthwait, 1985
Fig. spec. 94496
Utting, J., 1991, Geol. Surv. Can., Bull. 412, pl. 4.2, fig. 12.
Hart River Formation, Lower Carboniferous, north bank Peel River, northern Yukon.

Cordylosporites sp.
Fig. spec. 92878
Utting, J., Jachowicz, M. and Jachowicz, A., 1989, Geol. Surv. Can., Bull. 396, pl. 6.2, fig. 18.
Emma Fiord Formation, Early Carboniferous, Lyall River area, Grinnell Peninsula, Devon Island, District of Franklin.

Crassispora kosankei (Potonié and Kremp) Bhardwaj, 1957
Hypotype 16134
Barss, M.S., 1967, Geol. Surv. Can., Paper 67-11, pl. 19, fig. 2.
Cumberland Group, Pennsylvanian, Morrison's pit on Chignecto coal seam, 0.4 miles east of Chignecto, Cumberland County, Nova Scotia.

Crassispora kosankei (Potonié and Kremp) Bhardwaj, 1957
Hypotypes 16222, 16223
Barss, M.S., 1967, Geol. Surv. Can., Paper 67-11, pl. 22, figs. 24, 25.
Morien Group, Pennsylvanian, Beaver mine on Tracey coal seam east of MacDonald Lake and south of Morrison road, approximately 5.0 miles from Highway 22 and McAulay Cove coal seam, from outcrop on shore 1.75 miles south of Homeville, Sydney coalfield, Cape Breton County, Nova Scotia.

Crassispora kosankei (Potonié and Kremp) Bhardwaj, 1957
Hypotype 16363
Barss, M.S., 1967, Geol. Surv. Can., Paper 67-11, pl. 28, fig. 17.
Morien Group, Pennsylvanian, Phalen coal seam, Dominion No. 18 colliery at face of No. 6 west level, 3,400 feet from main deep, Sydney coalfield, Cape Breton County, Nova Scotia.

Crassispora sp. cf. *C. maculosa* Sullivan and Marshall, 1966
Fig. spec. 94476
Utting, J., 1991, Geol. Surv. Can., Bull. 412, pl. 4.1, fig. 19.
Hart River Formation, Lower Carboniferous, north bank Peel River, northern Yukon.

Crassispora cf. *C. ovalis* (Bhardwaj) Bhardwaj, 1957
Hypotype 16285
Barss, M.S., 1967, Geol. Surv. Can., Paper 67-11, pl. 25, fig. 19.
Mabou Mines Formation, Pennsylvanian, Angus MacPhee's pit, on north side of syncline in cliff section, 112 feet above beach, Mabou mines, Inverness County, Nova Scotia.

Crassispora plicata Peppers, 1964
Hypotype 16284
Barss, M.S., 1967, Geol. Surv. Can., Paper 67-11, pl. 25, fig. 18.
Mabou Mines Formation, Pennsylvanian, seven-foot coal seam in drift from cliff shore to old workings, south side of syncline, Mabou mines, Inverness County, Nova Scotia.

Crassispora trychera Neves and Ioannides
Hypotype 62492
Utting, J., 1980, Can. J. Earth Sci., vol. 17, no. 8, pl. 1, fig. 13.
Upper Windsor Group, Mississippian, United States Borax and Chemical Corporation borehole SB-1 near Stewiacke, Nova Scotia.

Crassispora trychera Neves and Ioannides, 1974
Hypotypes 73783, 73785, 73786
Utting, J., 1987, Geol. Surv. Can., Bull. 374, p. 31, pl. 3, figs. 6-8.
Windsor Group, Lower Carboniferous, north shore of Port Hood Island, lat. 46°1′N, long. 61°26′W, Cape Breton Island (73783), and northeast of Knoydart Point, east side of headland, lat. 45°42′5″N, long. 62°13′6″W, Nova Scotia.

Crassispora trychera Neves and Ioannides, 1974
Hypotypes 89910, 89911
Utting, J., Keppie, J.D. and Giles, P.S., 1989, Geol. Surv. Can., Bull. 396, pl. 5.1, figs. 9, 10.
Wilkie Brook Formation, Horton Group, Lower Carboniferous, 380 m north of mouth of Wilkie Brook, western shore of St. Georges Bay, Antigonish Basin, Nova Scotia.

Crassispora trychera Neves and Ioannides, 1974
Hypotype 94474
Utting, J., 1991, Geol. Surv. Can., Bull. 412, pl. 4.1, fig. 17.
Hart River Formation, Lower Carboniferous, north bank Peel River, northern Yukon.

Crassisporites sp.
Fig. specs. 16328, 16331
Barss, M.S., 1967, Geol. Surv. Can., Paper 67-11, pl. 27, figs. 6, 9.
Pennsylvanian, Mabou Mines Formation, coal seam 95 feet above beach, cliff section northeast side of syncline, Coal Mine Point, Mabou mines, Inverness County; Morien Group, LeCras coal seam, from pit 200 feet southwest side of Highway 22, 1.1 miles from intersection of Old Mira Road and Highway 22, Sydney coalfield, Cape Breton County, Nova Scotia.

Cribrosporites cribellatus Sullivan, 1964
Hypotypes 92862, 92935
Utting, J., Jachowicz, M. and Jachowicz, A., 1989, Geol. Surv. Can., Bull. 396, pl. 6.2, fig. 1; pl. 6.6, fig. 1.
Emma Fiord Formation, Early Carboniferous, Lyall River area, Grinnell Peninsula, Devon Island, and Svartevaeg Cliffs, northern Axel Hieberg Island, District of Franklin.

Cristatisporites aculeatus (Hacquebard) Potonié 1960
Hypotype 76356
Utting, J., 1987, Palynology, vol. 11, p. 79, pl. 1, fig. 6.
Frederick Brook Member, Albert Formation, Lower Carboniferous, depth 129.21 m, Petro-Canada Beaudreau No. 1 well, Highway 925 south of Beaudreau Village, New Brunswick.

Cristatisporites aculeatus (Hacquebard) Potonié 1960
Hypotype 89912
Utting, J., Keppie, J.D. and Giles, P.S., 1989, Geol. Surv. Can., Bull. 396, pl. 5.1, fig. 13.
Wilkie Brook Formation, Horton Group, Lower Carboniferous, 380 m north of mouth of Wilkie Brook, western shore of St. Georges Bay, Antigonish Basin, Nova Scotia.

Cristatisporites aculeatus (Hacquebard) Potonié, 1960
Hypotypes 13181-13183
Playford, G., 1964, Geol. Surv. Can., Bull. 107, p. 36, pl. 11, figs. 3-5.
Barss, M.S., 1967, Geol. Surv. Can., Paper 67-11, pl. 4, fig. 9 (13182).
Horton Group, Mississippian, 3 1/2 miles south of Kennetcook, Mill Brook, 1 mile north of East Gore, and Little Ninemile River, 2 miles southeast of East Gore, Nova Scotia.

Cristatisporites apisaculeus Utting
Holotype 92863; Hypotypes 92864-92868
Utting, J., Jachowicz, M. and Jachowicz, A., 1989, Geol. Surv. Can., Bull. 396, p. 156, pl. 6.2, figs. 2-8.
Emma Fiord Formation, Early Carboniferous, Lyall River area, Grinnell Peninsula, Devon Island, District of Franklin.

Cristatisporites echninatus Playford
Hypotype 13178
Playford, G., 1964, Geol. Surv. Can., Bull. 107, p. 36, pl. 10, fig. 10.
Barss, M.S., 1967, Geol. Surv. Can., Paper 67-11, pl. 4, fig. 6.
Horton Group, Mississippian, Mill Brook, one mile north of East Gore, Nova Scotia.

Cristatisporites submarginatus (Playford) comb. nov.
Hypotype 73787
Utting, J., 1987, Geol. Surv. Can., Bull. 374, p. 32, pl. 3, fig. 9.
Windsor Group, Lower Carboniferous, north shore of Port Hood Island, lat. 46°1′N, long. 61°26′W, Cape Breton Island, Nova Scotia.

Cristatisporites submarginatus (Playford) Utting, 1987
Hypotype 89913
Utting, J., Keppie, J.D. and Giles, P.S., 1989, Geol. Surv. Can., Bull. 396, pl. 5.1, fig. 11.
Horton Bluff Formation, Horton Group, Lower Carboniferous, west end of Horton Bluff, Avon River, Windsor Basin, Nova Scotia.

Cristatisporites submarginatus (Playford) Utting, 1987
Hypotype 94501
Utting, J., 1991, Geol. Surv. Can., Bull. 412, pl. 4.2, fig. 17.
Hart River Formation, Lower Carboniferous, north bank Peel River, northern Yukon.

Cycadopites caperatus (Luber and Waltz) Hart, 1965
Hypotype 16642
Barss, M.S., 1967, Geol. Surv. Can., Paper 67-11, pl. 38, fig. 40.
Sabine Bay Formation, Permian, vicinity of Tingmisut Lake, east side of Sabine Peninsula, Melville Island, District of Franklin.

Cycadopites cf. *C. conjunctur* (Andreyeva) Hart, 1965
Hypotype 16631
Barss, M.S., 1967, Geol. Surv. Can., Paper 67-11, pl. 38, fig. 29.
Sabine Bay Formation, Permian, vicinity of Tingmisut Lake, east side of Sabine Peninsula, Melville Island, District of Franklin.

Cycadopites cf. *C. cymbatus* (Balme and Hennelly) Potonié and Lele, 1961
Hypotype 16640

Barss, M.S., 1967, Geol. Surv. Can., Paper 67-11, pl. 38, fig. 38.
Sabine Bay Formation, Permian, vicinity of Tingmisut Lake, east side of Sabine Peninsula, Melville Island, District of Franklin.

Cycadopites glaber (Luber and Waltz) Hart, 1965
Hypotype 16654
Barss, M.S., 1967, Geol. Surv. Can., Paper 67-11, pl. 36, fig. 11.
Permian, Tatonduck River, west limb of syncline 3 miles east of Yukon-Alaska border, lat. 64°58′N, long. 140°54′W, Yukon.

Cycadopites vetus (Balme and Hennelly) Hart, 1965
Hypotype 16553
Barss, M.S., 1967, Geol. Surv. Can., Paper 67-11, pl. 36, fig. 10.
Permian, Tatonduck River, west limb of syncline 3 miles east of Yukon-Alaska border, lat. 64°58′N, long. 140°54′W, Yukon.

Cycadopites sp.
Fig. spec. 16555
Barss, M.S., 1967, Geol. Surv. Can., Paper 67-11, pl. 36, fig. 12.
Permian, Tatonduck River, west limb of syncline 3 miles east of Yukon-Alaska border, lat. 64°58′N, long. 140°54′W, Yukon.

Cycadopites sp.
Fig. specs. 16633, 16636
Barss, M.S., 1967, Geol. Surv. Can., Paper 67-11, pl. 38, figs. 31, 4.
Sabine Bay Formation, Permian, vicinity of Tingmisut Lake, east side of Sabine Peninsula, Melville Island, District of Franklin.

Cyclogranisporites aureus (Loose) Potonié and Kremp, 1955
Hypotypes 16039, 16042
Barss, M.S., 1967, Geol. Surv. Can., Paper 67-11, pl. 16, figs. 13, 16.
Port Hood Formation, Riversdale Group, Pennsylvanian, 200 feet from mouth of adit of Harbourview coal mine, Port Hood, Inverness County, Nova Scotia.

Cyclogranisporites aureus (Loose) Potonié and Kremp, 1955
Hypotypes 16128, 16129
Barss, M.S., 1967, Geol. Surv. Can., Paper 67-11, pl. 18, figs. 30, 31.
Cumberland Group, Pennsylvanian, Tingley pit in Roger's field near southern extension of McDougal Street, Springhill, and Fenwick mine on Chignecto coal seam, 2.1 miles east of Chignecto, Cumberland County, Nova Scotia.

Cyclogranisporites aureus (Loose) Potonié and Kremp, 1955
Hypotype 16279
Barss, M.S., 1967, Geol. Surv. Can., Paper 67-11, pl. 25, fig. 13.
Mabou Mines Formation, Pennsylvanian, seven-foot coal seam in drift from cliff shore to old workings, south side of syncline, Mabou mines, Inverness County, Nova Scotia.

Cyclogranisporites cf. *C. aureus* (Loose) Potonié and Kremp, 1955
Hypotypes 16130-16132
Barss, M.S., 1967, Geol. Surv. Can., Paper 67-11, pl. 18, figs. 32-34.
Cumberland Group, Pennsylvanian, bore-hole No. 1 east, in Roger's field near outcrop No. 6 coal seam, at depth 40 feet, Springhill, and Jim Haley's mine on Chignecto coal seam, 1.1 miles west of Maccan (16132), Cumberland County, Nova Scotia.

Cyclogranisporites commodus Playford
Holotype 13069; paratypes 13070, 13071
Playford, G., 1964, Geol. Surv. Can., Bull. 107, p. 12, pl. 2, figs. 3-5.
Barss, M.S., 1967, Geol. Surv. Can., Paper 67-11, pl. 1, fig. 11 (13069)
Horton Group, Mississippian, Mill Brook, one mile north of East Gore, Nova Scotia.

Cyclogranisporites commodus Playford, 1964
Hypotype 15797
Barss, M.S., 1967, Geol. Surv. Can., Paper 67-11, pl. 5, fig. 11.
Windsor Group, Mississippian, Frenchvale Brook (Balls Creek), Northwest Arm of Sydney Harbour, Nova Scotia.

Cyclogranisporites commodus Playford 1964
Hypotype 76357
Utting, J., 1987, Palynology, vol. 11, p. 79, pl. 1, fig. 7.
Hiram Brook Member, Albert Formation, Lower Carboniferous, Boyd Creek, notheast of Gautreau Village, New Brunswick.

Cyclogranisporites commodus Playford, 1964
Hypotype 89914
Utting, J., Keppie, J.D. and Giles, P.S., 1989, Geol. Surv. Can., Bull. 396, pl. 5.1, fig. 14.
Wilkie Brook Formation, Lower Carboniferous, 380 m north of mouth of Wilkie Brook, western shore of St. Georges Bay, Antigonish Basin, Nova Scotia.

Cyclogranisporites cf. *C. flexuosa* Playford, 1962
Hypotype 15860
Barss, M.S., 1967, Geol. Surv. Can., Paper 67-11, pl. 7, fig. 26.
Mattson Formation, Mississippian, south face Twisted Mountain, about 15 miles northwest of South Nahanni, lat. 61°13′N. long. 123°37′W, Northwest Territories.

Cyclogranisporites lasius (Waltz) Playford, 1962
Hypotype 92869
Utting, J., Jachowicz, M. and Jachowicz, A., 1989, Geol. Surv. Can., Bull. 396, pl. 6.2, fig. 9.
Emma Fiord Formation, Early Carboniferous, Lyall River area, Grinnell Peninsula, Devon Island, District of Franklin.

Cyclogranisporites cf. *C. micaceus* (Imgrund) Potonié and Kremp, in Bhardwaj, 1957
Hypotype 16361
Barss, M.S., 1967, Geol. Surv. Can., Paper 67-11, pl. 28, fig. 15.

Morien Group, Pennsylvanian, Phalen coal seam, Dominion No. 18 colliery at face of No. 6 west level, 3,400 feet from main deep, Sydney coalfield, Cape Breton County, Nova Scotia.

Cyclogranisporites microgranus Bhardwaj, 1957
Hypotype 16219
Barss, M.S., 1967, Geol. Surv. Can., Paper 67-11, pl. 22, fig. 21.
Morien Group, Pennsylvanian, Beaver mine on Tracey coal seam east of MacDonald Lake and south of Morrison road, approximately 5.0 miles from Highway 22, Sydney coalfield, Cape Breton County, Nova Scotia.

cf. *Cyclogranisporites microgranus* Bhardwaj, 1957
Hypotype 16458
Barss, M.S., 1967, Geol. Surv. Can., Paper 67-11, pl. 32, fig. 7.
Pictou Group, Pennsylvanian, Shediac water well approximately 1,400 feet east of shore of harbour and 1,800 feet south of highway at Shediac, New Brunswick.

Cyclogranisporites orbicularis (Kosanke) Potonié and Kremp, 1955
Hypotype 16040
Barss, M.S., 1967, Geol. Surv. Can., Paper 67-11, pl. 15, fig. 14.
Port Hood Formation, Riversdale Group, Pennsylvanian, 200 feet from mouth of adit of Harbourview coal mine, Port Hood, Inverness County, Nova Scotia.

Cyclogranisporites orbicularis (Kosanke) Potonié and Kremp, 1955
Hypotype 16123
Barss, M.S., 1967, Geol. Surv. Can., Paper 67-11, pl. 18, fig. 25.
Cumberland Group, Pennsylvanian, bore-hole No. 1 east, in Roger's field near outcrop No. 6 seam, at depth 40 feet, Springhill, Cumberland County, Nova Scotia.

Cyclogranisporites orbicularis (Kosanke) Potonié and Kremp, 1955
Hypotype 16358
Barss, M.S., 1967, Geol. Surv. Can., Paper 67-11, pl. 28, fig. 12.
Morien Group, Pennsylvanian, Harbour coal seam, Dominion No. 12 colliery at face of No. 21 west level, Sydney coalfield, Cape Breton County, Nova Scotia.

Cyclogranisporites palaeophytus Neves and Ioannides
Hypotype 62490
Utting, J., 1980, Can. J. Earth Sci., vol. 17, no. 8, pl. 1, fig. 11.
Lower Windsor Group, Mississippian, United States Borax and Chemical Corporation borehole SB-1 near Stewiacke, Nova Scotia.

Cyclogranisporites sp. cf. *C. palaeophytus* Neves and Ioannides, 1974
Fig. spec. 92870
Utting, J., Jachowicz, M. and Jachowicz, A., 1989, Geol. Surv. Can., Bull. 396, pl. 6.2, fig. 10.
Emma Fiord Formation, Early Carboniferous, Lyall River area, Grinnell Peninsula, Devon Island, District of Franklin.

Cyclogranisporites cf. *C. provectus* (Kosanke) Potonié and Kremp, 1955
Hypotype 16220
Barss, M.S., 1967, Geol. Surv. Can., Paper 67-11, pl. 22, fig. 22.
Morien Group, Pennsylvanian, Beaver mine on Tracey coal seam east of MacDonald Lake and south of Morrison road, approximately 5.0 miles from Highway 22, Sydney coalfield, Cape Breton County, Nova Scotia.

Cyclogranisporites pergranulus Alpern, 1959
Hypotype 16359
Barss, M.S., 1967, Geol. Surv. Can., Paper 67-11, pl. 28, fig. 13.
Morien Group, Pennsylvanian, Phalen coal seam, Dominion No. 18 colliery at face of No. 6 west level, 3,400 feet from main deep, Sydney coalfield, Cape Breton County, Nova Scotia.

Cyclogranisporites cf. *C. provectus* (Kosanke) Potonié and Kremp, 1955
Hypotype 16278
Barss, M.S., 1967, Geol. Surv. Can., Paper 67-11, pl. 25, fig. 12.
Mabou Mines Formation, Pennsylvanian, first coal seam northeast side syncline, 84 feet below Stack sandstone of Coal Mine Point, Mabou mines, Inverness County, Nova Scotia.

Cyclogranisporites cf. *C. provectus* (Kosanke) Potonié and Kremp, 1955
Hypotype 16362
Barss, M.S., 1967, Geol. Surv. Can., Paper 67-11, pl. 28, fig. 16.
Morien Group, Pennsylvanian, coal seam 93 feet above Harbour seam on shore northwest of New Waterford, Sydney coalfield, Cape Breton County, Nova Scotia.

Cyclogranisporites vagus (Kosanke) Potonié and Kremp, 1955
Hypotype 16508
Barss, M.S., 1967, Geol. Surv. Can., Paper 67-11, pl. 34, fig. 7.
Pictou Group, Permian, depth 350 feet, Gallas Point borehole No. 2 on property of H. Tweedy, Gallas Point, lat. 46°7′42″N, long. 63°57′28″W, Prince Edward Island.

Cyclogranisporites sp.
Fig. specs. 16041, 16122, 16133
Barss, M.S., 1967, Geol. Surv. Can., Paper 67-11, pl. 15, fig. 15; pl. 18, fig. 24; pl. 19, fig. 1.
Pennsylvanian, Port Hood Formation, Riversdale Group, 200 feet from mouth of adit of Harbourview coal mine, Port Hood, Inverness County; Cumberland Group, No. 1 east, in Roger's field near outcrop No. 6 coal seam, at depth 40 feet, and No. 7 coal seam from 3,200 feet mine bord, 150 feet east of back auxilliary slope in No. 4 mine, Springhill, Cumberland County, Nova Scotia.

?*Cyclogranisporites* sp.
Fig. spec. 75759
Utting, J., 1985, Bull. Can. Petrol. Geol., vol. 33, no. 3, pl. 1, fig. 1.
Otto Fiord Formation, Upper Carboniferous, 3.5 km northeast of van Hauen Pass, north shore Hare Fiord, Ellesmere Island District of Franklin.

Cymbosporites acutus (Kedo) Byvscheva, 1985
Hypotype 89915
Utting, J., Keppie, J.D. and Giles, P.S., 1989, Geol. Surv. Can., Bull. 396, pl. 5.1, fig. 19.
Unnamed formation, Horton Group, Lower Carboniferous, Lakevale cliff south of syncline, St. Georges Bay, Antigonish Basin, Nova Scotia.

Densosporites anulatus (Loose) Smith and Butterworth, 1967
Hypotype 92871
Utting, J., Jachowicz, M. and Jachowicz, A., 1989, Geol. Surv. Can., Bull. 396, pl. 6.2, fig. 11.
Emma Fiord Formation, Early Carboniferous, Lyall River area, Grinnell Peninsula, Devon Island, District of Franklin.

Densosporites anulatus (Loose) Smith and Butterworth, 1967
Hypotype 94470
Utting, J., 1991, Geol. Surv. Can., Bull. 412, pl. 4.1, fig. 13.
Hart River Formation, Lower Carboniferous, north bank Peel River, northern Yukon.

Densosporites bialatus (Waltz) Potonié and Kremp, 1956
Hypotype 15897
Barss, M.S., 1967, Geol. Surv. Can., Paper 67-11, pl. 9, fig. 17.
Mattson Formation, Mississippian, south face Twisted Mountain, about 15 miles northwest of South Nahanni, lat. 61°13'N. long. 123°37'W, Northwest Territories.

Densosporites columbaris Utting
Holotype 73788; hypotypes 73789-73791
Utting, J., 1987, Geol. Surv. Can., Bull. 374, p. 32, pl. 3, figs. 10-13.
Windsor Group, Lower Carboniferous, Frenchvale Brook, Balls Creek, northwest arm of Sydney Harbour; Kaiser Celestite Ltd. borehole K112, Lake Eon, Richmond County, lat. 45°46'19"N, long. 60°34'37"W (73789); northeast of Knoydart Point, east (73780) and west (73781) sides of headland, lat. 45°42'5"N, long. 62°13'6"W, Nova Scotia.

Densosporites dentatus (Waltz) Potonié and Kremp, 1956
Hypotype 92939
Utting, J., Jachowicz, M. and Jachowicz, A., 1989, Geol. Surv. Can., Bull. 396, pl. 6.6, fig. 5.
Emma Fiord Formation, Early Carboniferous, Svartevaeg Cliffs, northern Axel Hieberg Island, District of Franklin.

Densosporites duplicatus (Naumova) Potonié and Kremp, 1956
Hypotype 15901
Barss, M.S., 1967, Geol. Surv. Can., Paper 67-11, pl. 9, fig. 21.
Mattson Formation, Mississippian, south face Twisted Mountain, about 15 miles northwest of South Nahanni, lat. 61°13'N. long. 123°37'W, Northwest Territories.

Densosporites duplicatus (Naumova) Potonié and Kremp, 1956
Hypotype 94471
Utting, J., 1991, Geol. Surv. Can., Bull. 412, pl. 4.1, fig. 14.
Hart River Formation, Lower Carboniferous, north bank Peel River, northern Yukon.

Densosporites intermedius Butterworth and Williams, 1958
Hypotype 16074
Barss, M.S., 1967, Geol. Surv. Can., Paper 67-11, pl. 16, fig. 19.
Port Hood Formation, Riversdale Group, Pennsylvanian, 200 feet from mouth of adit of Harbourview coal mine, Port Hood, Inverness County, Nova Scotia.

Densosporites cf. *D. landesii* Staplin, 1960
Hypotypes 15905, 16557
Barss, M.S., 1967, Geol. Surv. Can., Paper 67-11, pl. 9, fig. 25; pl. 10, fig. 1.
Mattson Formation, Mississippian, south face Twisted Mountain, about 15 miles northwest of South Nahanni, lat. 61°13'N. long. 123°37'W, Northwest Territories.

Densosporites cf. *D. lobatus* Kosanke, 1950
Hypotype 16414
Barss, M.S., 1967, Geol. Surv. Can., Paper 67-11, pl. 30, fig. 11.
Morien Group, Pennsylvanian, Harbour coal seam, Dominion No. 12 colliery at face of No. 21 west level, Sydney coalfield, Cape Breton County, Nova Scotia.

Densosporites pseudoannulatus Butterworth and Williams, 1958
Hypotype 15893
Barss, M.S., 1967, Geol. Surv. Can., Paper 67-11, pl. 9, fig. 13.
Mattson Formation, Mississippian, south face Twisted Mountain, about 15 miles northwest of South Nahanni, lat. 61°13'N. long. 123°37'W, Northwest Territories.

Densosporites rarispinosus Playford, 1963
Hypotype 15912
Barss, M.S., 1967, Geol. Surv. Can., Paper 67-11, pl. 10, fig. 17.
Mattson Formation, Mississippian, south face Twisted Mountain, about 15 miles northwest of South Nahanni, lat. 61°13'N. long. 123°37'W, Northwest Territories.

Densosporites rarispinosus Playford, 1963
Hypotypes 92872, 92940
Utting, J., Jachowicz, M. and Jachowicz, A., 1989, Geol. Surv. Can., Bull. 396, pl. 6.2, fig. 12; pl. 6.6, fig. 6.
Emma Fiord Formation, Early Carboniferous, Lyall River area, Grinnell Peninsula, Devon Island, and Svartevaeg Cliffs, northern Axel Hieberg Island, District of Franklin.

Densosporites rarispinosus Playford, 1963
Hypotype 94497
Utting, J., 1991, Geol. Surv. Can., Bull. 412, pl. 4.2, fig. 13.
Hart River Formation, Lower Carboniferous, north bank Peel River, northern Yukon.

Densosporites cf. *D. spinifer* Hoffmeister, Staplin and Malloy, 1955
Hypotype 15902
Barss, M.S., 1967, Geol. Surv. Can., Paper 67-11, pl. 9, fig. 22.
Mattson Formation, Mississippian, south face Twisted Mountain, about 15 miles northwest of South Nahanni, lat. 61°13′N. long. 123°37′W, Northwest Territories.

Densosporites spinosus Dybová and Jachowicz, 1957
Hypotype 16075
Barss, M.S., 1967, Geol. Surv. Can., Paper 67-11, pl. 16, fig. 20.
Port Hood Formation, Riversdale Group, Pennsylvanian, 200 feet from mouth of adit of Harbourview coal mine, Port Hood, Inverness County, Nova Scotia.

Densosporites spitsbergensis Playford, 1963
Hypotype 15906
Barss, M.S., 1967, Geol. Surv. Can., Paper 67-11, pl. 9, fig. 26.
Mattson Formation, Mississippian, south face Twisted Mountain, about 15 miles northwest of South Nahanni, lat. 61°13′N. long. 123°37′W, Northwest Territories.

Densosporites spitsbergensis Playford, 1963
Hypotype 92942
Utting, J., Jachowicz, M. and Jachowicz, A., 1989, Geol. Surv. Can., Bull. 396, pl. 6.6, fig. 8.
Emma Fiord Formation, Early Carboniferous, Svartevaeg Cliffs, northern Axel Hieberg Island, District of Franklin.

Densosporites spitsbergensis Playford, 1963
Hypotype 94499
Utting, J., 1991, Geol. Surv. Can., Bull. 412, pl. 4.2, fig. 15.
Hart River Formation, Lower Carboniferous, north bank Peel River, northern Yukon.

Densosporites subcrenatus (Waltz) Potonié and Kremp, 1956
Hypotypes 92873, 92941
Utting, J., Jachowicz, M. and Jachowicz, A., 1989, Geol. Surv. Can., Bull. 396, pl. 6.2, fig. 13; pl. 6.6, fig. 7.
Emma Fiord Formation, Early Carboniferous, Lyall River area, Grinnell Peninsula, Devon Island, and Svartevaeg Cliffs, northern Axel Hieberg Island, District of Franklin.

Densosporites subcrenatus (Waltz) Potonié and Kremp, 1956
Hypotype 94489
Utting, J., 1991, Geol. Surv. Can., Bull. 412, pl. 4.2, fig. 5.
Hart River Formation, Lower Carboniferous, north bank Peel River, northern Yukon.

Densosporites subserratus Hacquebard and Barss, 1957
Hypotype 15914
Barss, M.S., 1967, Geol. Surv. Can., Paper 67-11, pl. 10, fig. 9.
Mattson Formation, Mississippian, south face Twisted Mountain, about 15 miles northwest of South Nahanni, lat. 61°13′N. long. 123°37′W, Northwest Territories.

Densosporites tripapillatus Staplin, 1960
Hypotype 15888
Barss, M.S., 1967, Geol. Surv. Can., Paper 67-11, pl. 9, fig. 8.
Mattson Formation, Mississippian, south face Twisted Mountain, about 15 miles northwest of South Nahanni, lat. 61°13′N. long. 123°37′W, Northwest Territories.

Densosporites variabilis (Waltz) Potonié and Kremp, 1956
Hypotype 92874
Utting, J., Jachowicz, M. and Jachowicz, A., 1989, Geol. Surv. Can., Bull. 396, pl. 6.2, fig. 14.
Emma Fiord Formation, Early Carboniferous, Lyall River area, Grinnell Peninsula, Devon Island, District of Franklin.

Densosporites variabilis (Waltz) Potonié and Kremp, 1956
Hypotype 94475
Utting, J., 1991, Geol. Surv. Can., Bull. 412, pl. 4.1, fig. 18.
Hart River Formation, Lower Carboniferous, north bank Peel River, northern Yukon.

Densosporites variomarginatus Playford, 1963
Hypotype 15799
Barss, M.S., 1967, Geol. Surv. Can., Paper 67-11, pl. 5, fig. 13.
Windsor Group, Mississippian, Frenchvale Brook (Balls Creek), Northwest Arm of Sydney Harbour, Nova Scotia.

Densosporites sp.
Fig. specs. 15900, 15907-15909, 15911
Barss, M.S., 1967, Geol. Surv. Can., Paper 67-11, pl. 9, fig. 20; pl. 10, figs. 2-4, 6.
Mattson Formation, Mississippian, south face Twisted Mountain, about 15 miles northwest of South Nahanni, lat. 61°13′N. long. 123°37′W, Northwest Territories.

Densosporites sp.
Fig. specs. 16411-16413
Barss, M.S., 1967, Geol. Surv. Can., Paper 67-11, pl. 30, figs. 8-10.
Morien Group, Pennsylvanian, Harbour coal seam, Dominion No. 12 colliery at face of No. 21 west level, Sydney coalfield, Cape Breton County, Nova Scotia.

Diatomozonotriletes hughesii Playford, 1963
Hypotype 15919
Barss, M.S., 1967, Geol. Surv. Can., Paper 67-11, pl. 10, fig. 14.
Mattson Formation, Mississippian, south face Twisted Mountain, about 15 miles northwest of South Nahanni, lat. 61°13′N. long. 123°37′W, Northwest Territories.

Diatomozonotriletes hughesii Playford, 1963
Hypotypes 92875, 92943
Utting, J., Jachowicz, M. and Jachowicz, A., 1989, Geol. Surv. Can., Bull. 396, pl. 6.2, fig. 15; pl 6.6, fig. 9.
Emma Fiord Formation, Early Carboniferous, Lyall River area, Grinnell Peninsula, Devon Island, and Kleybolte Peninsula, northwestern Ellesmere Island, District of Franklin.

Diatomozonotriletes rarus Playford, 1963
Hypotype 89916
Utting, J., Keppie, J.D. and Giles, P.S., 1989, Geol. Surv. Can., Bull. 396, pl. 5.1, fig. 12.
Horton Bluff Formation, Horton Group, Lower Carboniferous, west end of Horton Bluff, Avon River, Windsor Basin, Nova Scotia.

Diatomozonotriletes rarus Playford, 1963
Hypotypes 92876, 92944
Utting, J., Jachowicz, M. and Jachowicz, A., 1989, Geol. Surv. Can., Bull. 396, pl. 6.2, fig. 16; pl 6.6, fig. 10.
Emma Fiord Formation, Early Carboniferous, Lyall River area, Grinnell Peninsula, Devon Island, and Kleybolte Peninsula, northwestern Ellesmere Island, District of Franklin.

Diatomozonotriletes saetosus (Hacquebard and Barss) Hughes and Playford, 1961
Hypotype 15923
Barss, M.S., 1967, Geol. Surv. Can., Paper 67-11, pl. 10, fig. 18.
Mattson Formation, Mississippian, south face Twisted Mountain, about 15 miles northwest of South Nahanni, lat. 61°13′N. long. 123°37′W, Northwest Territories.

Diatomozonotriletes saetosus (Hacquebard and Barss) Hughes and Playford, 1961
Hypotypes 92877, 92945
Utting, J., Jachowicz, M. and Jachowicz, A., 1989, Geol. Surv. Can., Bull. 396, pl. 6.2, fig. 17; pl 6.6, fig. 11.
Emma Fiord Formation, Early Carboniferous, Lyall River area, Grinnell Peninsula, Devon Island, and Kleybolte Peninsula, northwestern Ellesmere Island, District of Franklin.

Dictyotriletes admirabilis Playford
Holotype 13148; paratypes 13149-13155
Playford, G., 1964, Geol. Surv. Can., Bull. 107, p. 29, pl. 8, figs. 5-8.
Barss, M.S., 1967, Geol. Surv. Can., Paper 67-11, pl. 3, fig. 2 (13154).
Horton Group, Mississippian, Mill Brook, one mile north of East Gore, Nova Scotia.

Dictyotriletes cf. *D. clatriformis* (Artuz) Sullivan, 1964
Hypotype 15875
Barss, M.S., 1967, Geol. Surv. Can., Paper 67-11, pl. 8, fig. 12.
Mattson Formation, Mississippian, south face Twisted Mountain, about 15 miles northwest of South Nahanni, lat. 61°13′N. long. 123°37′W, Northwest Territories.

Dictyotriletes cf. *D. falsus* Potonié and Kremp, 1955
Hypotype 15981
Barss, M.S., 1967, Geol. Surv. Can., Paper 67-11, pl. 13, fig. 4.
Canso Group, Upper Carboniferous, Pomquet River south of Route 4, Antigonish County, Nova Scotia.

Dictyotriletes cf. *D. falsus* Potonié and Kremp, 1955
Fig. spec. 16169
Barss, M.S., 1967, Geol. Surv. Can., Paper 67-11, pl. 20, fig. 9.
Cumberland Group, Pennsylvanian, Pearly McFadden's mine on Chignecto coal seam, 0.3 miles east of Chignecto, Cumberland County, Nova Scotia.

?*Dictyotriletes maculatus* (Ibrahim) Potonié and Kremp, 1955
Hypotype 16232
Barss, M.S., 1967, Geol. Surv. Can., Paper 67-11, pl. 22, fig. 34.
Morien Group, Pennsylvanian, McAulay Cove coal seam, from outcrop on shore 1.75 miles south of Homeville, Sydney coalfield, Cape Breton County, Nova Scotia.

Dictyotriletes odontolophos Utting
Holotype 73764; hypotypes 73765, 73766
Utting, J., 1987, Geol. Surv. Can., Bull. 374, p. 29, pl. 2, figs. 11-13.
Windsor Group, Lower Carboniferous, north shore of Port Hood Island, lat. 46°1′N, long. 61°26′W, and Cape Dauphin cliff, lat. 46°21′N, long. 60°25′W (73766), Cape Breton Island, Nova Scotia.

Dictyotriletes submarginatus Playford, 1964
Holotype 13156; paratypes 13157-13160
Playford, G., 1964, Geol. Surv. Can., Bull. 107, p. 29, pl. 8, figs. 9-13.
Barss, M.S., 1967, Geol. Surv. Can., Paper 67-11, pl. 3, fig. 7 (13156).
Cheverie Formation, Horton Group, Mississippian, west shore of Avon River, Nova Scotia.

Dictyotriletes submarginatus Playford 1964
Hypotype 76358
Utting, J., 1987, Palynology, vol. 11, p. 79, pl. 1, fig. 8.
Hiram Brook Member, Albert Formation, Lower Carboniferous, tributary from west, Boyd Creek, east of Gautreau Village, New Brunswick.

Dictyotriletes trivialis Naumova *in* Kedo, 1963
Hypotype 89917
Utting, J., Keppie, J.D. and Giles, P.S., 1989, Geol. Surv. Can., Bull. 396, pl. 5.1, fig. 20.
Unnamed formation, Horton Group, Lower Carboniferous, Lakevale cliff north of syncline, St. Georges Bay, Antigonish Basin, Nova Scotia.

Dictyotriletes sp.
Fig. specs. 15992, 15993, 15998
Barss, M.S., 1967, Geol. Surv. Can., Paper 67-11, pl. 13, figs. 15, 16, 21.
Canso Group, Upper Carboniferous, Pomquet River south of Route 4, Antigonish County, Nova Scotia.

Dictyotriletes sp.
Fig. spec. 16506
Barss, M.S., 1967, Geol. Surv. Can., Paper 67-11, pl. 34, fig. 5.
Pictou Group, Permian, depth 350 feet, Gallas Point borehole No. 2 on property of H. Tweedy, Gallas Point, lat. 46°7′42″N, long. 63°57′28″W, Prince Edward Island.

?*Dictyotriletes* sp.
Fig. spec. 16150
Barss, M.S., 1967, Geol. Surv. Can., Paper 67-11, pl. 19, fig. 18.
Cumberland Group, Pennsylvanian, Pearly McFadden's mine on Chignecto coal seam, 0.3 miles east of Chignecto, Cumberland County, Nova Scotia.

Dictyotriletes sp.
Fig. spec. 94467
Utting, J., 1991, Geol. Surv. Can., Bull. 412, pl. 4.1, fig. 10.
Hart River Formation, Lower Carboniferous, north bank Peel River, northern Yukon.

?*Direticuloidispora* sp.
Fig. spec. 16512
Barss, M.S., 1967, Geol. Surv. Can., Paper 67-11, pl. 34, fig. 11.
Pictou Group, Permian, depth 335 feet, Gallas Point borehole No. 2 on property of H. Tweedy, Gallas Point, lat. 46°7′42″N, long. 63°57′28″W, Prince Edward Island.

Discernisporites barssii Utting
Holotype 73814; hypotypes 73815, 73816
Utting, J., 1987, Geol. Surv. Can., Bull. 374, p. 35, pl. 4, figs. 17-19.
Windsor Group, Lower Carboniferous, north shore of Port Hood Island, lat. 46°1′N, long. 61°26′W, and Cape Dauphin cliff, lat. 46°21′N, long. 60°25′W (73815), Cape Breton Island, Nova Scotia.

cf. *Discernisporites concentricus* Neves, 1958
Hypotype 16004
Barss, M.S., 1967, Geol. Surv. Can., Paper 67-11, pl. 13, fig. 27.
Canso Group, Upper Carboniferous, Pomquet River south of Route 4, Antigonish County, Nova Scotia.

Discernisporites crenulatus (Playford) Clayton 1970
Hypotype 76359
Utting, J., 1987, Palynology, vol. 11, p. 79, pl. 1, fig. 9.
Hiram Brook Member, Albert Formation, Lower Carboniferous, tributary from west, Boyd Creek, east of Gautreau Village, New Brunswick.

Discernisporites diaphanes Utting
Holotype 73817, hypotypes 73818, 73819
Utting, J., 1987, Geol. Surv. Can., Bull. 374, p. 36, pl. 5, figs. 1-3.
Windsor Group, Lower Carboniferous, north shore of Port Hood Island, lat. 46°1′N, long. 61°26′W, and Stewiacke borehole SB-1, Shubenacadie Basin, lat. 45°7′N, long. 63°20′W (73819), Cape Breton Island, Nova Scotia.

Discernisporites macromanifestus (Hacquebard) comb. nov.
Hypotype 62480
Utting, J., 1987, Geol. Surv. Can., Bull. 374, p. 36, pl. 5, fig. 4.
Windsor Group, Lower Carboniferous, Stewiacke borehole SB-1, Shubenacadie Basin, lat. 45°7′N, long. 63°20′W, Cape Breton Island, Nova Scotia.

Discernisporites macromanifestus (Hacquebard) Utting, 1987
Hypotype 89918
Utting, J., Keppie, J.D. and Giles, P.S., 1989, Geol. Surv. Can., Bull. 396, pl. 5.1, fig. 18.
Horton Bluff Formation, Horton Group, Lower Carboniferous, depth 29.25 m, Saarberg Interplan Canada Limited SI/8 drill core, north of west branch of Avon River, lat. 44°57′10″N, long. 64°15′50″W, Windsor Basin, Nova Scotia.

Discernisporites micromanifestus (Hacquebard 1957)
Hypotype 62489
Utting, J., 1980, Can. J. Earth Sci., vol. 17, no. 8, pl. 1, fig. 10.
Upper Windsor Group, Mississippian, United States Borax and Chemical Corporation borehole SB-1 near Stewiacke, Nova Scotia.

Discernisporites micromanifestus (Hacquebard) Sabry and Neves, 1971
Hypotypes 62489, 73820
Utting, J., 1987, Geol. Surv. Can., Bull. 374, p. 36, pl. 5, figs. 5, 6.
Windsor Group, Lower Carboniferous, Stewiacke borehole SB-1, Shubenacadie Basin, lat. 45°7′N, long. 63°20′W, and north shore of Port Hood Island, lat. 46°1′N, long. 61°26′W, Cape Breton Island, Nova Scotia.

Discernisporites micromanifestus (Hacquebard) Sabry and Neves 1971
Hypotype 76360
Utting, J., 1987, Palynology, vol. 11, p. 79, pl. 1, fig. 10.
Hiram Brook Member, Albert Formation, Lower Carboniferous, tributary from west, Boyd Creek, east of Gautreau Village, New Brunswick.

Discernisporites micromanifestus (Hacquebard) Sabry and Neves, 1971
Hypotype 89919
Utting, J., Keppie, J.D. and Giles, P.S., 1989, Geol. Surv. Can., Bull. 396, pl. 5.1, fig. 15.

Wilkie Brook Formation, Horton Group, Lower Carboniferous, 380 m north of mouth of Wilkie Brook, western shore of St. Georges Bay, Antigonish Basin, Nova Scotia.

Discernisporites micromanifestus (Hacquebard) Sabry and Neves, 1971
Hypotype 92881
Utting, J., Jachowicz, M. and Jachowicz, A., 1989, Geol. Surv. Can., Bull. 396, pl. 6.3, fig. 1.
Emma Fiord Formation, Early Carboniferous, Lyall River area, Grinnell Peninsula, Devon Island, District of Franklin.

Discernisporites micromanifestus (Hacquebard) Sabry and Neves, 1971
Hypotype 94479
Utting, J., 1991, Geol. Surv. Can., Bull. 412, pl. 4.1, fig. 22.
Hart River Formation, Lower Carboniferous, north bank Peel River, northern Yukon.

Discernisporites sp.
Fig. spec. 16006
Barss, M.S., 1967, Geol. Surv. Can., Paper 67-11, pl. 13, fig. 29.
Canso Group, Upper Carboniferous, Pomquet River south of Route 4, Antigonish County, Nova Scotia.

Discernisporites sp.
Fig. spec. 92879
Utting, J., Jachowicz, M. and Jachowicz, A., 1989, Geol. Surv. Can., Bull. 396, pl. 6.2, fig. 19.
Emma Fiord Formation, Early Carboniferous, Lyall River area, Grinnell Peninsula, Devon Island, District of Franklin.

(?)*Divarisaccus* sp.
Fig. spec. 45738
Davidson, A. and McGregor, D.C., 1976, Nature, vol. 262, no. 5567, fig. 2h.
Gilo sandstone, Permian, stream gully approximately 1 km north of Gilo River, 34 km southeast of Agenga, Ilubabor Province, southwestern Ethiopia.

Emphanispsorites rotatus McGregor emend. McGregor 1973
Hypotypes 89920, 89921, 89923
Utting, J., Keppie, J.D. and Giles, P.S., 1989, Geol. Surv. Can., Bull. 396, p. 132, pl. 5.1, figs. 16, 17; pl. 5.2, fig.1.
Unnamed unit, Horton Group, Lower Carboniferous, Lakevale cliff north of syncline, St. Georges Bay, Antigonish Basin, Nova Scotia.

Emphanispsorites sp. cf. *E. rotatus* McGregor, 1961
Fig. spec. 89922
Utting, J., Keppie, J.D. and Giles, P.S., 1989, Geol. Surv. Can., Bull. 396, p. 132, pl. 5.1, fig. 21.
Unnamed unit, Horton Group, Lower Carboniferous, Lakevale cliff north of syncline, St. Georges Bay, Antigonish Basin, Nova Scotia.

Endosporites formosus Kosanke, 1950
Hypotype 16482
Barss, M.S., 1967, Geol. Surv. Can., Paper 67-11, pl. 32, fig. 31.
Pictou Group, Pennsylvanian, small northerly tributary of Scoudouc River, 1/2 mile west of old dam and 1 1/2 miles south of town of Shediac, New Brunswick.

Endosporites globiformis (Ibrahim) Schopf, Wilson and Bentall, 1944
Hypotype 16257
Barss, M.S., 1967, Geol. Surv. Can., Paper 67-11, pl. 24, fig. 5.
Morien Group, Pennsylvanian, McAulay Cove coal seam, from outcrop on shore 1.75 miles south of Homeville, Sydney coalfield, Cape Breton County, Nova Scotia.

Endosporites macromanifestus Hacquebard, 1957
Hypotype 13179
Playford, G., 1964, Geol. Surv. Can., Bull. 107, p. 37, pl. 11, fig. 1.
Barss, M.S., 1967, Geol. Surv. Can., Paper 67-11, pl. 4, fig. 8.
Horton Group, Mississippian, Mill Brook, 1 mile north of East Gore, Nova Scotia.

Endosporites macromanifestus Hacquebard
Hypotype 62480
Utting, J., 1980, Can. J. Earth Sci., vol. 17, no. 8, pl. 1, fig. 1.
Lower Windsor Group, Mississippian, United States Borax and Chemical Corporation borehole SB-1 near Stewiacke, Nova Scotia.

Endosporites micromanifestus Hacquebard, 1957
Hypotype 13180
Playford, G., 1964, Geol. Surv. Can., Bull. 107, p. 37, pl. 11, fig. 2.
Barss, M.S., 1967, Geol. Surv. Can., Paper 67-11, pl. 4, fig. 4.
Horton Group, Mississippian, Little Ninemile River, 2 miles southeast of East Gore, Nova Scotia.

Endosporites micromanifestus Hacquebard, 1957
Hypotype 15930
Barss, M.S., 1967, Geol. Surv. Can., Paper 67-11, pl. 11, fig. 4.
Mattson Formation, Mississippian, south face Twisted Mountain, about 15 miles northwest of South Nahanni, lat. 61°13′N. long. 123°37′W, Northwest Territories.

Endosporites cf. *E. micromanifestus* Hacquebard, 1957
Hypotype 16005
Barss, M.S., 1967, Geol. Surv. Can., Paper 67-11, pl. 13, fig. 28.
Canso Group, Upper Carboniferous, Pomquet River south of Route 4, Antigonish County, Nova Scotia.

Endosporites minutus Hoffmeister, Staplin and Malloy, 1955
Hypotype 15810
Barss, M.S., 1967, Geol. Surv. Can., Paper 67-11, pl. 6, fig. 3.
Windsor Group, Mississippian, Frenchvale Brook (Balls Creek), Northwest Arm of Sydney Harbour, Nova Scotia.

Endosporites minutus Hoffmeister, Staplin and Malloy, 1955
Hypotype 16002
Barss, M.S., 1967, Geol. Surv. Can., Paper 67-11, pl. 13, fig. 25.
Canso Group, Upper Carboniferous, Pomquet River south of Route 4, Antigonish County, Nova Scotia.

Endosporites ornatus Wilson and Coe, 1940
Hypotype 16196
Barss, M.S., 1967, Geol. Surv. Can., Paper 67-11, pl. 21, fig. 5.
Cumberland Group, Pennsylvanian, Fenwick mine on Chignecto coal seam, 2.1 miles east of Chignecto, Cumberland County, Nova Scotia.

Endosporites ornatus Wilson and Coe, 1940
Hypotype 16256
Barss, M.S., 1967, Geol. Surv. Can., Paper 67-11, pl. 24, fig. 4.
Morien Group, Pennsylvanian, McAulay Cove coal seam, from outcrop on shore 1.75 miles south of Homeville, Sydney coalfield, Cape Breton County, Nova Scotia.

Endosporites ornatus Wilson and Coe, 1940
Hypotype 16325
Barss, M.S., 1967, Geol. Surv. Can., Paper 67-11, pl. 27, fig. 3.
Morien Group, Pennsylvanian, LeCras coal seam, from pit 200 feet southwest side of Highway 22, 1.1 miles from intersection of Old Mira Road and Highway 22, Sydney coalfield, Cape Breton County, Nova Scotia.

Endosporites sp.
Fig. specs. 15811, 15812, 15819, 15820, 15935, 16079, 16082
Barss, M.S., 1967, Geol. Surv. Can., Paper 67-11, pl. 6, figs. 4, 5, 12, 13; pl. 11, fig. 9; pl. 16, figs. 24, 27.
Mississippian, Windsor Group, Frenchvale Brook (Balls Creek), Northwest Arm of Sydney Harbour, Nova Scotia; Mattson Formation, Mississippian, south face Twisted Mountain, about 15 miles northwest of South Nahanni, lat. 61°13′N. long. 123°37′W, Northwest Territories (15935); Port Hood Formation, Riversdale Group, Pennsylvanian, 200 feet from mouth of adit of Harbourview coal mine, Port Hood, Inverness County, Nova Scotia.

Endosporites sp.
Fig. spec. 16557
Barss, M.S., 1967, Geol. Surv. Can., Paper 67-11, pl. 36, fig. 14.
Permian, Tatonduck River, west limb of syncline 3 miles east of Yukon-Alaska border, lat. 64°58′N, long. 140°54′W, Yukon.

?*Enzonalasporites* sp.
Fig. spec. 16625
Barss, M.S., 1967, Geol. Surv. Can., Paper 67-11, pl. 38, fig. 23.
Sabine Bay Formation, Permian, vicinity of Tingmisut Lake, east side of Sabine Peninsula, Melville Island, District of Franklin.

Eupunctisporites sp.
Fig. spec. 16627
Barss, M.S., 1967, Geol. Surv. Can., Paper 67-11, pl. 38, fig. 25.
Sabine Bay Formation, Permian, vicinity of Tingmisut Lake, east side of Sabine Peninsula, Melville Island, District of Franklin.

Florinites antiquus Schopf, in Schopf, Wilson and Bentall, 1944
Hypotype 16261
Barss, M.S., 1967, Geol. Surv. Can., Paper 67-11, pl. 24, fig. 9.
Morien Group, Pennsylvanian, McAulay Cove coal seam, from outcrop on shore 1.75 miles south of Homeville, Sydney coalfield, Cape Breton County, Nova Scotia.

Florinites antiquus Schopf, in Schopf, Wilson and Bentall, 1944
Hypotype 16342
Barss, M.S., 1967, Geol. Surv. Can., Paper 67-11, pl. 27, fig. 20.
Mabur Mines Formation, Pennsylvanian, seven-foot coal seam in drift from cliff shore to old workings, south side of syncline, Mabou mines, Inverness County, Nova Scotia.

Florinites antiquus Schopf, in Schopf, Wilson and Bentall, 1944
Hypotypes 16647, 16449
Barss, M.S., 1967, Geol. Surv. Can., Paper 67-11, pl. 31, figs. 10, 12.
Morien Group, Pennsylvanian, Phalen coal seam, Dominion No. 18 colliery at face of No. 6 west level, 3,400 feet from main deep, Sydney coalfield, Cape Breton County, Nova Scotia.

Florinites circularis Bhardwaj, 1957
Hypotype 16262
Barss, M.S., 1967, Geol. Surv. Can., Paper 67-11, pl. 24, fig. 10.
Morien Group, Pennsylvanian, McAulay Cove coal seam, from outcrop on shore 1.75 miles south of Homeville, Sydney coalfield, Cape Breton County, Nova Scotia.

Florinites circularis Bhardwaj, 1957
Hypotype 16648
Barss, M.S., 1967, Geol. Surv. Can., Paper 67-11, pl. 31, fig. 11.
Morien Group, Pennsylvanian, Harbour coal seam, Dominion No. 12 colliery at face of No. 21 west level, Sydney coalfield, Cape Breton County, Nova Scotia.

Florinites eremus Balme and Hennelly, 1955
Hypotype 16346
Barss, M.S., 1967, Geol. Surv. Can., Paper 67-11, pl. 27, fig. 24.
Mabou Mines Formation, Pennsylvanian, borehole No. 5, depth 136 feet, 690 feet south of Stack Point, Mabou mines, Inverness County, Nova Scotia.

Florinites eremus Balme and Hennelly, 1955
Hypotype 16485
Barss, M.S., 1967, Geol. Surv. Can., Paper 67-11, pl. 32, fig. 34.
Pictou Group, Pennsylvanian, Shediac water well approximately 1,400 feet east of shore of harbour and 1,800 feet south of highway at Shediac, New Brunswick.

Florinites eremus Balme and Hennelly, 1955
Hypotypes 16513, 16567
Barss, M.S., 1967, Geol. Surv. Can., Paper 67-11, pl. 34, fig. 12; pl. 37, fig. 1.
Permian, Pictou Group, depth 33 feet, Gallas Point borehole No. 2 on property of H. Tweedy, Gallas Point, lat. 46°7′42″N, long. 63°57′28″W, Prince Edward Island; Tatonduck River, west limb of syncline 3 miles east of Yukon-Alaska border, lat. 64°58′N, long. 140°54′W, Yukon.

Florinites cf. *F. diversiformis* Kosanke, 1950
Hypotype 16094
Barss, M.S., 1967, Geol. Surv. Can., Paper 67-11, pl. 17, fig. 7.
Port Hood Formation, Riversdale Group, Pennsylvanian, No. 5 coal seam, 1,300 feet east of Evans′ coal mine on No. 2 seam, near Neil MacLeod Brook, St. Rose, Inverness County, Nova Scotia.

Florinites junior Potonié and Kremp, 1956
Hypotype 16194
Barss, M.S., 1967, Geol. Surv. Can., Paper 67-11, pl. 21, fig. 3.
Cumberland Group, Pennsylvanian, No. 1 coal seam from 5,400 feet east level in no. 2 mine, Springhill, Cumberland County, Nova Scotia.

Florinites millotti Butterworth and Williams, 1954
Hypotype 16263
Barss, M.S., 1967, Geol. Surv. Can., Paper 67-11, pl. 24, fig. 11.
Morien Group, Pennsylvanian, Beaver mine on Tracey coal seam east of MacDonald Lake and south of Morrison road, approximately 5.0 miles from Highway 22, Sydney coalfield, Cape Breton County, Nova Scotia.

Florinites cf. *F. millotti* Butterworth and Williams, 1954
Hypotype 16445
Barss, M.S., 1967, Geol. Surv. Can., Paper 67-11, pl. 31, fig. 8.
Morien Group, Pennsylvanian, Phalen coal seam, Dominion No. 18 colliery at face of No. 6 west level, 3,400 feet from main deep, Sydney coalfield, Cape Breton County, Nova Scotia.

Florinites pumicosus (Ibrahim) Schopf, Wilson and Bentall, 1944
Hypotypes 16093, 16190, 16259
Barss, M.S., 1967, Geol. Surv. Can., Paper 67-11, pl. 17, fig. 6; pl. 20, fig. 30; pl. 24, fig. 7.
Pennsylvanian, Port Hood Formation, Riversdale Group, 200 feet from mouth of adit of Harbourview coal mine, Port Hood, Inverness County; Cumberland Group, Morrison′s pit on Chignecto seam, 0.4 miles east of Chignecto, Cumberland County; and Morien Group, McAulay Cove coal seam, from outcrop on shore 1.75 miles south of Homeville, Sydney coalfield, Cape Breton County, Nova Scotia.

Florinites similis Kosanke, 1950
Hypotype 16265
Barss, M.S., 1967, Geol. Surv. Can., Paper 67-11, pl. 24, fig. 13.
Skinner Brook Member, Stellarton Group, Pennsylvanian, northwest of Westville, 2,000 feet east of Skinner Brook, Pictou coalfield, Pictou County, Nova Scotia.

Florinites similis Kosanke, 1950
Hypotype 16450
Barss, M.S., 1967, Geol. Surv. Can., Paper 67-11, pl. 31, fig. 13.
Morien Group, Pennsylvanian, Phalen coal seam, Dominion No. 18 colliery at face of No. 6 west level, 3,400 feet from main deep, Sydney coalfield, Cape Breton County, Nova Scotia.

Florinites visendus (Ibrahim) Schopf, Wilson and Bentall, 1944
Hypotypes 16095, 16193
Barss, M.S., 1967, Geol. Surv. Can., Paper 67-11, pl. 17, fig. 8; pl. 21, fig. 2.
Pennsylvanian, Port Hood Formation, Riversdale Group, 200 feet from mouth of adit of Harbourview coal mine, Port Hood, Inverness County, and Cumberland Group, No. 1 coal seam from 5,400 feet east level in no. 2 mine, Springhill, Cumberland County, Nova Scotia.

Florinites visendus (Ibrahim) Schopf, Wilson and Bentall, 1944
Hypotype 16486
Barss, M.S., 1967, Geol. Surv. Can., Paper 67-11, pl. 33, fig. 1.
Pictou Group, Pennsylvanian, Shediac water well approximately 1,400 feet east of shore of harbour and 1,800 feet south of highway at Shediac, New Brunswick.

Florinites sp.
Fig. specs. 16089, 16092, 16096, 16198
Barss, M.S., 1967, Geol. Surv. Can., Paper 67-11, pl. 17, figs. 2, 5, 9; pl. 21, fig. 7.
Pennsylvanian, Port Hood Formation, Riversdale Group, 200 feet from mouth of adit of Harbourview coal mine, Port Hood, Inverness County, and Cumberland Group, Morrison′s pit on Chignecto seam, 0.4 miles east of Chignecto, Cumberland County (16198), Nova Scotia.

Florinites sp.
Fig. spec. 16340
Barss, M.S., 1967, Geol. Surv. Can., Paper 67-11, pl. 27, fig. 18.
Mabou Mines Formation, Pennsylvanian, eight-foot coal seam in drift from cliff shore to old workings, south side of syncline, Mabou mines, Inverness County, Nova Scotia.

Foveosporites insculptus Playford, 1962
Hypotype 15877
Barss, M.S., 1967, Geol. Surv. Can., Paper 67-11, pl. 8, fig. 14.
Mattson Formation, Mississippian, south face Twisted Mountain, about 15 miles northwest of South Nahanni, lat. 61°13′N. long. 123°37′W, Northwest Territories.

Foveosporites insculptus Playford, 1963
Hypotype 92882
Utting, J., Jachowicz, M. and Jachowicz, A., 1989, Geol. Surv. Can., Bull. 396, pl. 6.3, fig. 2.
Emma Fiord Formation, Early Carboniferous, Lyall River area, Grinnell Peninsula, Devon Island, District of Franklin.

Foveosporites cf. *F. insculptus* Playford, 1962
Hypotype 15980
Barss, M.S., 1967, Geol. Surv. Can., Paper 67-11, pl. 13, fig. 3.
Canso Group, Upper Carboniferous, Pomquet River south of Route 4, Antigonish County, Nova Scotia.

Foveosporites sp.
Fig. spec. 15983
Barss, M.S., 1967, Geol. Surv. Can., Paper 67-11, pl. 13, fig. 6.
Canso Group, Upper Carboniferous, Pomquet River south of Route 4, Antigonish County, Nova Scotia.

Foveosporites sp.
Fig. spec. 92946
Utting, J., Jachowicz, M. and Jachowicz, A., 1989, Geol. Surv. Can., Bull. 396, pl. 6.6, fig. 12.
Emma Fiord Formation, Early Carboniferous, Svartevaeg Cliffs, northern Axel Heiberg Island, District of Franklin.

cf. *Gardenasporites leonardi* Klause, 1963
Hypotype 16574
Barss, M.S., 1967, Geol. Surv. Can., Paper 67-11, pl. 37, fig. 8.
Tatonduck River, Permian, west limb of syncline 3 miles east of Yukon-Alaska border, lat. 64°58′N, long. 140°54′W, Yukon.

Grandispora echinata Hacquebard, 1957
Hypotype 13112
Playford, G., Geol. Surv. Can., Bull. 107, p. 22, pl. 5, fig. 2.
Barss, M.S., 1967, Geol. Surv. Can., Paper 67-11, pl. 2, fig. 9.
Horton Group, Mississippian, Mill Brook, 1 mile north of East Gore, Nova Scotia.

Grandispora echinata Hacquebard 1957
Hypotypes 76361, 76362
Utting, J., 1987, Palynology, vol. 11, p. 79, pl. 1, figs. 11, 12.
Hiram Brook Member, Albert Formation, Lower Carboniferous, tributary from west, Boyd Creek, east and northeast of Gautreau Village, New Brunswick.

Grandispsora spinosa Hoffmeister, Staplin and Malloy
Hypotype 62506
Utting, J., 1980, Can. J. Earth Sci., vol. 17, no. 8, pl. 2, fig. 12.
Upper Windsor Group, Mississippian, United States Borax and Chemical Corporation borehole SB-1 near Stewiacke, Nova Scotia.

Grandispsora spinosa Hoffmeister, Staplin and Malloy, 1955
Hypotype 73794
Utting, J., 1987, Geol. Surv. Can., Bull. 374, pl. 3, fig. 16.
Windsor Group, Lower Carboniferous, depth 34.75 m, Borehole Noranda Exploration 4, Lake Enon, Cape Breton Island, Nova Scotia.

Grandispsora uncata (Hacquebard) Playford
Hypotype 62481
Utting, J., 1980, Can. J. Earth Sci., vol. 17, no. 8, pl. 1, fig. 2.
Lower Windsor Group, Mississippian, United States Borax and Chemical Corporation borehole SB-1 near Stewiacke, Nova Scotia.

Grandispsora uncata (Hacquebard) Playford, 1971
Hypotype 62481
Utting, J., 1987, Geol. Surv. Can., Bull. 374, pl. 3, fig. 17.
Windsor Group, Lower Carboniferous, Stewiacke borehole SB-1, Shubenacadie Basin, lat. 45°7′N, long. 63°20′W, Hants County, Nova Scotia.

Grandispsora uncata (Hacquebard) Gupta 1969
Hypotype 76363
Utting, J., 1987, Palynology, vol. 11, p. 79, pl. 1, fig. 13.
Hiram Brook Member, Albert Formation, Lower Carboniferous, tributary from west, Boyd Creek, east of Gautreau Village, New Brunswick.

Grandispsora uncata (Hacquebard) Gupta, 1969
Hypotype 89924
Utting, J., Keppie, J.D. and Giles, P.S., 1989, Geol. Surv. Can., Bull. 396, pl. 5.2, fig. 6.
Horton Bluff Formation, Horton Group, Lower Carboniferous, depth 37.75 m, Saarberg Interplan Canada Limited SI/8 drill core, north of west branch of Avon River, lat. 44°57′10″N, long. 64°15′50″W, Windsor Basin, Nova Scotia.

Grandispsora uncata (Hacquebard) Gupta, 1969
Hypotype 92883
Utting, J., Jachowicz, M. and Jachowicz, A., 1989, Geol. Surv. Can., Bull. 396, pl. 6.3, fig. 7.

Emma Fiord Formation, Early Carboniferous, Lyall River area, Grinnell Peninsula, Devon Island, District of Franklin.

Grandispsora sp.
Fig. specs. 15931, 16016
Barss, M.S., 1967, Geol. Surv. Can., Paper 67-11, pl. 11, fig. 5; pl. 14, fig. 8.
Mattson Formation, Mississippian, south face Twisted Mountain, about 15 miles northwest of South Nahanni, lat. 61°13′N, long. 123°37′W, Northwest Territories; Canso Group, Upper Carboniferous, Pomquet River south of Route 4, Antigonish County, Nova Scotia.

Grandispsora sp.
Fig. spec. 92880
Utting, J., Jachowicz, M. and Jachowicz, A., 1989, Geol. Surv. Can., Bull. 396, pl. 6.2, fig. 20.
Emma Fiord Formation, Early Carboniferous, Lyall River area, Grinnell Peninsula, Devon Island, District of Franklin.

Grandispsora sp.
Fig. spec. 94483
Utting, J., 1991, Geol. Surv. Can., Bull. 412, pl. 4.1, fig. 26.
Hart River Formation, Lower Carboniferous, north bank Peel River, northern Yukon.

?*Grandispsora* sp.
Fig. specs. 15823, 16015
Barss, M.S., 1967, Geol. Surv. Can., Paper 67-11, pl. 6, fig. 16; pl. 14, fig. 7.
Windsor Group, Mississippian, Frenchvale Brook (Balls Creek), Northwest Arm of Sydney Harbour; Canso Group, Upper Carboniferous, Pomquet River south of Route 4, Antigonish County, Nova Scotia.

Granulatispora parvus (Ibrahim) Potonié and Kremp, 1955
Hypotypes 16034, 16110
Barss, M.S., 1967, Geol. Surv. Can., Paper 67-11, pl. 15, fig. 8; pl. 18, fig. 12.
Pennsylvanian, Port Hood Formation, Riversdale Group, 200 feet from mouth of adit of Harbourview coal mine, Port Hood, Inverness County, and Cumberland Group, Morrison's pit on Chignecto seam, 0.4 miles east of Chignecto, Cumberland County, Nova Scotia.

Granulatisporites crenulatus Playford
Holotype 13076; paratypes 13074, 13075
Playford, G., 1964, Geol. Surv. Can., Bull. 107, p. 11, pl. 2, figs. 8-10.
Barss, M.S., 1967, Geol. Surv. Can., Paper 67-11, pl. 1, fig. 14 (13075).
Horton Group, Mississippian, southeast Mabou River, 3 3/4 miles south at Brook village, Nova Scotia.

Granulatisporites crenulatus Playford, 1964
Hypotype 89925
Utting, J., Keppie, J.D. and Giles, P.S., 1989, Geol. Surv. Can., Bull. 396, pl. 5.2, fig. 2.
Horton Bluff Formation, Horton Group, Lower Carboniferous, depth 100 m, Saarberg Interplan Canada Limited SI/8 drill core, north of west branch of Avon River, lat. 44°57′10″N, long. 64°15′50″W, Windsor Basin, Nova Scotia.

Granulatisporites granulatus Ibrahim, 1933
Hypotype 15951
Barss, M.S., 1967, Geol. Surv. Can., Paper 67-11, pl. 12, fig. 14.
Canso Group, Upper Carboniferous, Pomquet River south of Route 4, Antigonish County, Nova Scotia.

Granulatisporites granulatus Ibrahm, 1933
Hypotype 16203
Barss, M.S., 1967, Geol. Surv. Can., Paper 67-11, pl. 22, fig. 5.
Morien Group, Pennsylvanian, McAulay Cove coal seam, from outcrop on shore 1.75 miles south of Homeville, Sydney coalfield, Cape Breton County, Nova Scotia.

Granulatisporites cf. G. *granulatus* Ibrahim, 1933
Hypotype 16272
Barss, M.S., 1967, Geol. Surv. Can., Paper 67-11, pl. 25, fig. 6.
Mabou Mines Formation, Pennsylvanian, coal seam in cliff section on north side of syncline, 100 feet above beach, Mabou mines, Inverness County, Nova Scotia.

Granulatisporites granulatus Ibrahim, 1933
Hypotypes 92884, 92947
Utting, J., Jachowicz, M. and Jachowicz, A., 1989, Geol. Surv. Can., Bull. 396, pl. 6.3, fig. 3; pl. 6.6, fig. 13.
Emma Fiord Formation, Early Carboniferous, Lyall River area, Grinnell Peninsula, Devon Island, and Kleybolte Peninsula, northwestern Ellesmere Island, District of Franklin.

Granulatisporites inspissatus Playford
Holotype 13072; paratype 13073
Playford, G., 1964, Geol. Surv. Can., Bull. 107, p. 11, pl. 2, figs. 6, 7.
Barss, M.S., 1967, Geol. Surv. Can., Paper 67-11, pl. 1, fig. 13 (13073).
Horton Group, Mississippian, Mill Brook, one mile north of East Gore, Nova Scotia.

Granulatisporites microgranifer Ibrahim, 1933
Hypotype 16202
Barss, M.S., 1967, Geol. Surv. Can., Paper 67-11, pl. 22, fig. 4.
Morien Group, Pennsylvanian, LeCras coal seam, from pit 200 feet southwest side of Highway 22, 1.1 miles from intersection of Old Mira Road and Highway 22, Sydney coalfield, Cape Breton County, Nova Scotia.

Granulatisporites microgranifer (Ibrahim) Potonié and Kremp 1954
Hypotype 76364
Utting, J., 1987, Palynology, vol. 11, p. 79, pl. 1, fig. 14.

Frederick Brook Member, Albert Formation, Lower Carboniferous, depth 132.21 m, Petro-Canada Boudreau No. 1 well, Highway 925 south of Boudreau Village, New Brunswick.

Granulatisporites micrograníifer (Ibrahim) Potonié and Kremp, 1955
Hypotype 89926
Utting, J., Keppie, J.D. and Giles, P.S., 1989, Geol. Surv. Can., Bull. 396, pl. 5.2, fig. 3.
Rights River Formation, Horton Group, Lower Carboniferous, depth 73.15 m, Drillhole DDH2, east of Beaver Settlement lat. 45°48′12″N, long. 61°55′8″W, Antigonish Basin, Nova Scotia.

Granulatisporites minutus Potonié and Kremp, 1955
Hypotypes 16106, 16109
Barss, M.S., 1967, Geol. Surv. Can., Paper 67-11, pl. 18, figs. 8, 11.
Cumberland Group, Pennsylvanian, No. 7 coal seam from 3,200 foot mine bord, 150 feet east of back auxiliary slope in No. 4 mine, and No. 6 coal seam from 5,400 foot west level in No. 4 mine, 35 feet from rock tunnel slope, Springhill, Cumberland County, Nova Scotia.

Granulatisporites minutus Potonié and Kremp, 1955
Hypotype 16455
Barss, M.S., 1967, Geol. Surv. Can., Paper 67-11, pl. 32, fig. 4.
Pictou Group, Pennsylvanian, Scoudouc River 1 1/8 miles upstream from highway running south from Dorchester Crossing, New Brunswick.

Granulatisporites cf. *G. pallidus* (Kosanke) Hoffmeister, Staplin and Malloy, 1955
Hypotype 15839
Barss, M.S., 1967, Geol. Surv. Can., Paper 67-11, pl. 7, fig. 4.
Mattson Formation, Mississippian, south face Twisted Mountain, about 15 miles northwest of South Nahanni, lat. 61°13′N. long. 123°37′W, Northwest Territories.

Granulatisporites parvus (Ibrahim) Potonié and Kremp, 1955
Hypotype 16618
Barss, M.S., 1967, Geol. Surv. Can., Paper 67-11, pl. 38, fig. 16.
Sabine Bay Formation, Permian, vicinity of Tingmisut Lake, east side of Sabine Peninsula, Melville Island, District of Franklin.

Granulatisporites piroformis Loose, 1934
Hypotype 16204
Barss, M.S., 1967, Geol. Surv. Can., Paper 67-11, pl. 22, fig. 6.
Morien Group, Pennsylvanian, Beaver mine on Tracey coal seam east of MacDonald Lake and south of Morrison road, approximately 5.0 miles from Highway 22, Sydney coalfield, Cape Breton County, Nova Scotia.

Granulatisporites triconvexus Staplin, 1960
Hypotype 15843
Barss, M.S., 1967, Geol. Surv. Can., Paper 67-11, pl. 7, fig. 8.
Mattson Formation, Mississippian, south face Twisted Mountain, about 15 miles northwest of South Nahanni, lat. 61°13′N. long. 123°37′W, Northwest Territories.

Granulatisporites tuberculatus Hoffmeister, Staplin and Malloy, 1955
Hypotype 73733
Utting, J., 1987, Geol. Surv. Can., Bull. 374, pl. 1, fig. 8.
Windsor Group, Lower Carboniferous, northeast of Knoydart Point, lat. 45°42′5″N, long. 62°13′6″W, Nova Scotia.

Granulatisporites sp. cf. *G. tuberculatus* Hoffmeister, Staplin and Malloy, 1955
Fig. spec. 73844
Utting, J., 1987, Geol. Surv. Can., Bull. 374, p. 26, pl. 1, fig. 9.
Windsor Group, Lower Carboniferous, north shore of Port Hood Island, lat. 46°1′N, long. 61°26′W, Cape Breton Island, Nova Scotia.

Granulatisporites sp.
Fig. spec. 16038
Barss, M.S., 1967, Geol. Surv. Can., Paper 67-11, pl. 15, fig. 12.
Port Hood Formation, Riversdale Group, Pennsylvanian, 200 feet from mouth of adit of Harbourview coal mine, Port Hood, Inverness County, Nova Scotia.

Granulatisporites sp.
Fig. spec. 16111
Barss, M.S., 1967, Geol. Surv. Can., Paper 67-11, pl. 18, fig. 13.
Cumberland Group, Pennsylvanian, Fenwick mine on Chignecto coal seam, 2.1 miles east of Chignecto, Cumberland County, Nova Scotia.

Granulatisporites sp.
Fig. specs. 16503, 16617
Barss, M.S., 1967, Geol. Surv. Can., Paper 67-11, pl. 34, fig. 2; pl. 38, fig. 15.
Permian, Pictou Group, Gallas Point borehole No. 2 on property of H. Tweedy, Gallas Point, lat. 46°7′42″N, long. 63°57′28″W, Prince Edward Island; Sabine Bay Formation, vicinity of Tingmisut Lake, east side of Sabine Peninsula, Melville Island, District of Franklin.

Gravisporites sphaerus (Butterworth and Williams) Bhardwaj, 1954
Hypotype 16310
Barss, M.S., 1967, Geol. Surv. Can., Paper 67-11, pl. 26, fig. 11.
Mabou Mines Formation, Pennsylvanian, depth 590 feet in borehole No. 2, 350 feet south of Stack Point, Mabou mines, Inverness County, Nova Scotia.

Gravisporites sphaerus (Butterworth and Williams) Bhardwaj, 1954
Hypotype 16477
Barss, M.S., 1967, Geol. Surv. Can., Paper 67-11, pl. 32, fig. 26.

Pictou Group, Pennsylvanian, Scoudouc River 1 1/8 miles upstream from highway running south from Dorchester Crossing, New Brunswick.

Gulisporites torpidus Playford
Holotype 13063; paratype 13064
Playford, G., 1964, Geol. Surv. Can., Bull. 107, p. 8, pl. 1, figs. 13, 14.
Barss, M.S., 1967, Geol. Surv. Can., Paper 67-11, pl. 1, fig. 10 (13063).
Horton Group, Mississippian, Mill Brook, one mile north of East Gore, Nova Scotia.

Gulisporites torpidus Playford 1964
Hypotype 76366
Utting, J., 1987, Palynology, vol. 11, p. 79, pl. 2, fig. 1.
Dawson Settlement Member, Albert Formation, Lower Carboniferous, depth 126.20 m, Petro-Canada Dover No. 1 well, north of Gautreau Village, lat. 45°59′15″N, long. 65°39′W, New Brunswick.

?*Gulisporites* sp.
Fig. spec. 15967
Barss, M.S., 1967, Geol. Surv. Can., Paper 67-11, pl. 12, fig. 30.
Canso Group, Upper Carboniferous, Pomquet River south of Route 4, Antigonish County, Nova Scotia.

Guthorlisporites magnificus Bhardwaj, 1954
Hypotype 16343
Barss, M.S., 1967, Geol. Surv. Can., Paper 67-11, pl. 27, fig. 17.
Morien Group, Pennsylvanian, LeCras coal seam, from pit 200 feet southwest side of Highway 22, 1.1 miles from intersection of Old Mira Road and Highway 22, Sydney coalfield, Cape Breton County, Nova Scotia.

Guthorlisporites sp.
Fig. spec. 16258
Barss, M.S., 1967, Geol. Surv. Can., Paper 67-11, pl. 24, fig. 6.
Morien Group, Pennsylvanian, McAulay Cove coal seam, from outcrop on shore 1.75 miles south of Homeville, Sydney coalfield, Cape Breton County, Nova Scotia.

Hamiapollenites sp.
Fig. specs. 16600, 16601
Barss, M.S., 1967, Geol. Surv. Can., Paper 67-11, pl. 37, figs. 21-23.
Permian, Tatonduck River, west limb of syncline 3 miles east of Yukon-Alaska border, lat. 64°58′N, long. 140°54′W, Yukon.

?*Hymenospora* sp.
Fig. spec. 15818
Barss, M.S., 1967, Geol. Surv. Can., Paper 67-11, pl. 6, fig. 11.
Windsor Group, Mississippian, Frenchvale Brook (Balls Creek), Northwest Arm of Sydney Harbour, Nova Scotia.

Hymenozonotriletes explanatus (Luber) Kedo, 1963
Hypotypes 89927, 89928
Utting, J., Keppie, J.D. and Giles, P.S., 1989, Geol. Surv. Can., Bull. 396, pl. 5.2, figs. 4, 7.
Unnamed formation, Horton Group, Lower Carboniferous, Lakevale cliff south of syncline, St. Georges Bay, Antigonish Basin, Nova Scotia.

Ibrahimispores magnificus Neves
Hypotype 62502
Utting, J., 1980, Can. J. Earth Sci., vol. 17, no. 8, pl. 2, fig. 8.
Upper Windsor Group, Mississippian, United States Borax and Chemical Corporation borehole SB-1 near Stewiacke, Nova Scotia.

Ibrahimispores magnificus Neves, 1961
Hypotype 73740
Utting, J., 1987, Geol. Surv. Can., Bull. 374, p. 26, pl. 1, fig. 16.
Windsor Group, Lower Carboniferous, near Baddeck Bridge, lat. 46°9′N, long. 61°11′W, Nova Scotia.

Illinites unicus Kosanke, 1950
Hypotype 16489
Barss, M.S., 1967, Geol. Surv. Can., Paper 67-11, pl. 33, fig. 4.
Pictou Group, Pennsylvanian, Shediac water well approximately 1,400 feet east of shore of harbour and 1,800 feet south of highway at Shediac, New Brunswick.

Illinites cf. *I. unicus* Kosanke, 1950
Hypotype 16339
Barss, M.S., 1967, Geol. Surv. Can., Paper 67-11, pl. 27, fig. 17.
Mabou Mines Formation, Pennsylvanian, Angus MacPhee's pit, on north side of syncline in cliff section, 112 feet above beach, Mabou mines, Inverness County, Nova Scotia.

Klausipollenites schaubergeri (Potonié and Klaus)
Hypotype 45733
Davidson, A. and McGregor, D.C., 1976, Nature, vol. 262, no. 5567, fig. 2c.
Kari sandstone, Permian, banks Kari River approximately 14 km upstream from its junction to the north with Akobo River, Kefa Province, southwestern Ethiopia.

cf. *Knoxisporites carnosus* (Knox) Butterworth and Williams, 1958
Hypotype 15929
Barss, M.S., 1967, Geol. Surv. Can., Paper 67-11, pl. 9, fig. 3.
Mattson Formation, Mississippian, south face Twisted Mountain, about 15 miles northwest of South Nahanni, lat. 61°13′N. long. 123°37′W, Northwest Territories.

Knoxisporites hederatus (Ischenko) Playford, 1963
Hypotype 19916
Barss, M.S., 1967, Geol. Surv. Can., Paper 67-11, pl. 10, fig. 11.

Mattson Formation, Mississippian, south face Twisted Mountain, about 15 miles northwest of South Nahanni, lat. 61°13′N. long. 123°37′W, Northwest Territories.

Knoxisporites literatus (Waltz) Playford, 1963
Hypotype 15917
Barss, M.S., 1967, Geol. Surv. Can., Paper 67-11, pl. 10, fig. 12.
Mattson Formation, Mississippian, south face Twisted Mountain, about 15 miles northwest of South Nahanni, lat. 61°13′N. long. 123°37′W, Northwest Territories.

Knoxisporites literatus (Waltz) Playford, 1962
Hypotype 73776
Utting, J., 1987, Geol. Surv. Can., Bull. 374, pl. 2, fig. 23.
Windsor Group, Lower Carboniferous, north shore of Port Hood Island, lat. 46°1′N, long. 61°26′W, Cape Breton Island, Nova Scotia.

Knoxisporites literatus (Waltz) Playford 1962
Hypotype 76365
Utting, J., 1987, Palynology, vol. 11, p. 79, pl. 1, fig. 15.
Hiram Brook Member, Albert Formation, Lower Carboniferous, tributary from west, Boyd Creek, east of Gautreau Village, New Brunswick.

Knoxisporites literatus (Waltz) Playford, 1962
Hypotype 89929
Utting, J., Keppie, J.D. and Giles, P.S., 1989, Geol. Surv. Can., Bull. 396, pl. 5.2, fig. 8.
Unnamed unit, Horton Group, Lower Carboniferous, Lakevale cliff north of syncline, St. Georges Bay, Antigonish Basin, Nova Scotia.

Knoxisporites literatus (Waltz) Playford, 1963
Hypotype 92885
Utting, J., Jachowicz, M. and Jachowicz, A., 1989, Geol. Surv. Can., Bull. 396, pl. 6.3, fig. 4.
Emma Fiord Formation, Early Carboniferous, Lyall River area, Grinnell Peninsula, Devon Island, District of Franklin.

Knoxisporites margarethae Hughes and Playford, 1961
Hypotype 15913
Barss, M.S., 1967, Geol. Surv. Can., Paper 67-11, pl. 10, fig. 8.
Mattson Formation, Mississippian, south face Twisted Mountain, about 15 miles northwest of South Nahanni, lat. 61°13′N. long. 123°37′W, Northwest Territories.

Knoxisporites probolos Utting
Holotype 73777; hypotypes 73778, 73779
Utting, J., 1987, Geol. Surv. Can., Bull. 374, p. 30, pl. 2, figs. 24-27.
Windsor Group, Lower Carboniferous, north shore of Port Hood Island, lat. 46°1′N, long. 61°26′W, Cape Breton Island, Nova Scotia; Cap Adèle Member, Havre aux Maisons Formation, Carboniferous, Anse à la Cabane, east of Cap du Sud-ouest, Île du Havre Aubert, Îles-de-la-Madeleine, Québec (73779).

Knoxisporites rotatus Hoffmeister, Staplin and Malloy, 1955
Hypotype 92886
Utting, J., Jachowicz, M. and Jachowicz, A., 1989, Geol. Surv. Can., Bull. 396, pl. 6.3, fig. 9.
Emma Fiord Formation, Early Carboniferous, Lyall River area, Grinnell Peninsula, Devon Island, District of Franklin.

Knoxisporites seniradiatus Neves, 1961
Hypotype 16011
Barss, M.S., 1967, Geol. Surv. Can., Paper 67-11, pl. 14, fig. 3.
Canso Group, Upper Carboniferous, Pomquet River south of Route 4, Antigonish County, Nova Scotia.

Knoxisporites stephanephorus Love, 1960
Hypotypes 16010, 16013, 16173
Barss, M.S., 1967, Geol. Surv. Can., Paper 67-11, pl. 14, figs. 2, 5; pl. 20, fig. 13.
Canso Group, Upper Carboniferous, Pomquet River south of Route 4, Antigonish County; Cumberland Group, Pennsylvanian, Pearly McFadden's mine on Chignecto coal seam, 0.3 mile east of Chignecto, Cumberland County (16173), Nova Scotia.

Knoxisporites stephanephorus Love
Hypotype 62491
Utting, J., 1980, Can. J. Earth Sci., vol. 17, no. 8, pl. 1, fig. 12.
Upper Windsor Group, Mississippian, United States Borax and Chemical Corporation borehole SB-1 near Stewiacke, Nova Scotia.

Knoxisporites stephanephorus Love, 1960
Hypotype 73780
Utting, J., 1987, Geol. Surv. Can., Bull. 374, pl. 3, fig. 1.
Windsor/Canso Group boundary, Lower Carboniferous, approximately 0.8 km downstream from Baddeck Bridge, lat. 46°9′N, long. 61°11′W, Cape Breton Island, Nova Scotia.

Knoxisporites stephanephorus Love, 1960
Hypotypes 92887, 92952
Utting, J., Jachowicz, M. and Jachowicz, A., 1989, Geol. Surv. Can., Bull. 396, pl. 6.3, fig. 10; pl. 6.6, fig. 18.
Emma Fiord Formation, Early Carboniferous, Lyall River area, Grinnell Peninsula, Devon Island, and Svartevaeg Cliffs, northern Axel Hieberg Island, District of Franklin.

Knoxisporites cf. *K. stephanephorus* Love, 1960
Hypotype 16014
Barss, M.S., 1967, Geol. Surv. Can., Paper 67-11, pl. 14, fig. 6.
Canso Group, Upper Carboniferous, Pomquet River south of Route 4, Antigonish County, Nova Scotia.

Knoxisporites cf. *K. stephanephorus* Love, 1960
Hypotypes 16091, 16177
Barss, M.S., 1967, Geol. Surv. Can., Paper 67-11, pl. 17, fig. 4; pl. 20, fig. 17.

Pennsylvanian, Port Hood Formation, Riversdale Group, 200 feet from mouth of adit of Harbourview coal mine, Port Hood, Inverness County, and Cumberland Group, Pearly McFadden's mine on Chignecto coal seam, 0.3 miles east of Chignecto, Cumberland County, Nova Scotia.

Knoxisporites cf. *K. stephanephorus* Love, 1960
Hypotypes 16308, 16309, 16392, 16393
Barss, M.S., 1967, Geol. Surv. Can., Paper 67-11, pl. 26, figs. 9, 10; pl. 29, figs. 13, 14.
Pennsylvanian, Morien Group, LeCras coal seam, from pit 200 feet southwest side of Highway 22, 1.1 miles from intersection of Old Mira Road and Highway 22, Ormond coal seam, pit approximately 600 feet southwest side of Highway 22 and 600 feet east of old road to Middle Lane, and Phalen coal seam, Dominion No. 18 colliery at face of No. 6 west level, 3,400 feet from main deep (16392), Sydney coalfield, Cape Breton County, Nova Scotia; Picton Group, stripping operation one mile south of Beersville on east side of Coal Branch River, New Brunswick (16393).

Knoxisporites triradiatus Hoffmeister, Staplin and Malloy, 1955
Hypotype 16009
Barss, M.S., 1967, Geol. Surv. Can., Paper 67-11, pl. 14, fig. 1.
Canso Group, Upper Carboniferous, Pomquet River south of Route 4, Antigonish County, Nova Scotia.

Knoxisporites triradiatus Hoffmeister, Staplin and Malloy
Hypotype 62503
Utting, J., 1980, Can. J. Earth Sci., vol. 17, no. 8, pl. 2, fig. 9.
Upper Windsor Group, Mississippian, United States Borax and Chemical Corporation borehole SB-1 near Stewiacke, Nova Scotia.

Knoxisporites triradiatus Hoffmeister, Staplin and Malloy, 1955
Hypotype 73781
Utting, J., 1987, Geol. Surv. Can., Bull. 374, pl. 3, fig. 2.
Windsor Group, Lower Carboniferous, northeast of Knoydart Point, east side, lat. 45°42′5″N, long. 62°13′6″W, Nova Scotia.

Knoxisporites triradiatus Hoffmeister, Staplin and Malloy, 1955
Hypotypes 92888-92891
Utting, J., Jachowicz, M. and Jachowicz, A., 1989, Geol. Surv. Can., Bull. 396, pl. 6.3, figs. 11-14.
Emma Fiord Formation, Early Carboniferous, Lyall River area, Grinnell Peninsula, Devon Island, District of Franklin.

Knoxisporites cf. *K. triradiatus* Hoffmeister, Staplin and Malloy, 1955
Hypotype 16012
Barss, M.S., 1967, Geol. Surv. Can., Paper 67-11, pl. 14, fig. 4.
Canso Group, Upper Carboniferous, Pomquet River south of Route 4, Antigonish County, Nova Scotia.

Knoxisporites sp.
Fig. spec. 16017
Barss, M.S., 1967, Geol. Surv. Can., Paper 67-11, pl. 14, fig. 9.
Canso Group, Upper Carboniferous, Pomquet River south of Route 4, Antigonish County, Nova Scotia.

Knoxisporites sp.
Fig. specs. 16080, 16242
Barss, M.S., 1967, Geol. Surv. Can., Paper 67-11, pl. 16, fig. 25; pl. 23, fig. 7.
Pennsylvanian, Port Hood Formation, Riversdale Group, 200 feet from mouth of adit of Harbourview coal mine, Port Hood, Inverness County and Skinner Brook Member, Stellarton Group, northwest of Westville, 2,000 feet east of Skinner Brook, Pictou coalfield, Pictou County, Nova Scotia.

Knoxisporites sp. A
Fig. spec. 73845
Utting, J., 1987, Geol. Surv. Can., Bull. 374, p. 31, pl. 3, fig. 3.
Windsor Group, Lower Carboniferous, north shore of Port Hood Island, lat. 46°1′N, long. 61°26′W, Cape Breton Island, Nova Scotia.

Kraeuselisporites sp.
Fig. spec. 92892
Utting, J., Jachowicz, M. and Jachowicz, A., 1989, Geol. Surv. Can., Bull. 396, pl. 6.3, figs. 5, 6.
Emma Fiord Formation, Early Carboniferous, Lyall River area, Grinnell Peninsula, Devon Island, District of Franklin.

Labiadensites fimbriatus (Waltz) Hacquebard and Barss, 1957
Hypotype 15926
Barss, M.S., 1967, Geol. Surv. Can., Paper 67-11, pl. 10, fig. 21.
Mattson Formation, Mississippian, south face Twisted Mountain, about 15 miles northwest of South Nahanni, lat. 61°13′N. long. 123°37′W, Northwest Territories.

Labiadensites fimbriatus (Waltz) Hacquebard and Barss, 1957
Hypotypes 92893, 92894
Utting, J., Jachowicz, M. and Jachowicz, A., 1989, Geol. Surv. Can., Bull. 396, pl. 6.3, figs. 15, 16.
Emma Fiord Formation, Early Carboniferous, Lyall River area, Grinnell Peninsula, Devon Island, District of Franklin.

cf. *Laevigatosporites crassus* Peppers, 1964
Hypotype 16481
Barss, M.S., 1967, Geol. Surv. Can., Paper 67-11, pl. 32, fig. 30.
Pictou Group, Pennsylvanian, Shediac water well approximately 1,400 feet east of shore of harbour and 1,800 feet south of highway at Shediac, New Brunswick.

Laevigatosporites desmoinensis (Wilson and Coe) Schopf, Wilson and Bentall, 1944
Hypotypes 16083, 16185, 16218

Barss, M.S., 1967, Geol. Surv. Can., Paper 67-11, pl. 16, fig. 28; pl. 20, fig. 25; pl. 22, fig. 20.
Pennsylvanian, Port Hood Formation, Riversdale Group, No. 5 coal seam, 1,300 feet east of Evans' coal mine on No. 2 seam, near Neil MacLeod Brook, St. Rose, Inverness County; Cumberland Group, Tingley pit in Roger's field near southern extension of McDougal Street, Springhill, Cumberland County; and Morien Group, Beaver mine on Tracey coal seam east of MacDonald Lake and south of Morrison road, approximately 5 miles from Highway 22, Sydney coalfield, Cape Breton County, Nova Scotia.

Laevigatosporites desmoinensis (Wilson and Coe) Schopf, Wilson and Bentall, 1944
Hypotype 16474
Barss, M.S., 1967, Geol. Surv. Can., Paper 67-11, pl. 32, fig. 23.
Pictou Group, Pennsylvanian, Shediac water well approximately 1,400 feet east of shore of harbour and 1,800 feet south of highway at Shediac, New Brunswick.

Laevigatosporites medius Kosanke, 1950
Hypotypes 16086, 16186
Barss, M.S., 1967, Geol. Surv. Can., Paper 67-11, pl. 16, fig. 31; pl. 20, fig. 26.
Pennsylvanian, Port Hood Formation, Riversdale Group, 200 feet from mouth of adit of Harbourview coal mine, Port Hood, Inverness County, and Cumberland Group, Tingley pit in Roger's field near southern extension of McDougal Street, Springhill, Cumberland County, Nova Scotia.

Laevigatosporites medius Kosanke, 1950
Hypotype 16424
Barss, M.S., 1967, Geol. Surv. Can., Paper 67-11, pl. 30, fig. 21.
Morien Group, Pennsylvanian, Harbour coal seam, Dominion No. 12 colliery at face of No. 21 west level, Sydney coalfield, Cape Breton County, Nova Scotia.

Laevigatosporites minimus (Wilson and Coe) Schopf, Wilson and Bentall, 1944
Hypotype 16085
Barss, M.S., 1967, Geol. Surv. Can., Paper 67-11, pl. 16, fig. 30.
Port Hood Formation, Riversdale Group, Pennsylvanian, 200 feet from mouth of adit of Harbourview coal mine, Port Hood, Inverness County, Nova Scotia.

Laevigatosporites minimus (Wilson and Coe) Schopf, Wilson and Bentall, 1944
Hypotype 16470
Barss, M.S., 1967, Geol. Surv. Can., Paper 67-11, pl. 32, fig. 19.
Pictou Group, Pennsylvanian, Shediac water well approximately 1,400 feet east of shore of harbour and 1,800 feet south of highway at Shediac, New Brunswick.

Laevigatosporites minimus (Wilson and Coe) Schopf, Wilson and Bentall, 1944
Hypotype 16329
Barss, M.S., 1967, Geol. Surv. Can., Paper 67-11, pl. 27, fig. 7.
Mabou Mines Formation, Pennsylvanian, coal seam 95 feet above beach, cliff section northwest side of syncline, Coal Mine Point, Mabou mines, Inverness County, Nova Scotia.

Laevigatosporites perminutus Alpern, 1959
Hypotype 16326
Barss, M.S., 1967, Geol. Surv. Can., Paper 67-11, pl. 27, fig. 4.
Mabou Mines Formation, Pennsylvanian, Colonel Snow's adit at beach level, northeast side of syncline at Coal Mine Point, approximately 440 feet west of gypsum fault, Mabou mines, Inverness County, Nova Scotia.

Laevigatosporites vulgaris Ibrahim, 1933
Hypotype 16475
Barss, M.S., 1967, Geol. Surv. Can., Paper 67-11, pl. 32, fig. 24.
Pictou Group, Pennsylvanian, Shediac water well approximately 1,400 feet east of shore of harbour and 1,800 feet south of highway at Shediac, New Brunswick.

Laevigatosporites vulgaris Ibrahim, 1933
Hypotypes 16336, 16425
Barss, M.S., 1967, Geol. Surv. Can., Paper 67-11, pl. 27, fig. 14; pl. 30, fig 22.
Pennsylvanian, Mabou Mines Formation, seven-foot coal seam in drift from cliff shore to old workings, south side syncline, Mabou mines, Inverness County, and Morien Group, Harbour coal seam, Dominion No. 12 colliery at face of No. 21 west level, Sydney coalfield, Cape Breton County, Nova Scotia.

Laevigatosporites vulgaris Ibrahim, 1933
Hypotype 16621
Barss, M.S., 1967, Geol. Surv. Can., Paper 67-11, pl. 38, fig. 19.
Sabine Bay Formation, Permian, vicinity of Tingmisut Lake, east side of Sabine Peninsula, Melville Island, District of Franklin.

?*Laevigatosporites* sp.
Fig. spec. 16622
Barss, M.S., 1967, Geol. Surv. Can., Paper 67-11, pl. 38, fig. 20.
Sabine Bay Formation, Permian, vicinity of Tingmisut Lake, east side of Sabine Peninsula, Melville Island, District of Franklin.

Latosporites sp.
Fig. spec. 16560
Barss, M.S., 1967, Geol. Surv. Can., Paper 67-11, pl. 36, fig. 17.
Permian, Tatonduck River, west limb of syncline 3 miles east of Yukon-Alaska border, lat. 64°58′N, long. 140°54′W, Yukon.

Latosporites sp.
Fig. spec. 89930
Utting, J., Keppie, J.D. and Giles, P.S., 1989, Geol. Surv. Can., Bull. 396, pl. 5.2, fig. 9.
Unnamed unit, Horton Group, Lower Carboniferous, Lakevale cliff north of syncline, St. Georges Bay, Antigonish Basin, Nova Scotia.

Latosporites sp.
Fig. spec. 92895
Utting, J., Jachowicz, M. and Jachowicz, A., 1989, Geol. Surv. Can., Bull. 396, pl. 6.3, fig. 8.
Emma Fiord Formation, Early Carboniferous, Lyall River area, Grinnell Peninsula, Devon Island, District of Franklin.

Leiotriletes adnatoides Potonié and Kremp, 1955
Hypotype 16349
Barss, M.S., 1967, Geol. Surv. Can., Paper 67-11, pl. 28, fig. 3.
Morien Group, Pennsylvanian, Harbour coal seam, Dominion No. 12 colliery at face of No. 21 west level, Sydney coalfield, Cape Breton County, Nova Scotia.

Leiotrilites adnatus (Kosanke) Potonié and Kremp, 1955
Hypotype 16199
Barss, M.S., 1967, Geol. Surv. Can., Paper 67-11, pl. 22, fig. 1.
Morien Group, Pennsylvanian, Beaver mine on Tracey coal seam east of MacDonald Lake and south of Morrison road, approximately 5.0 miles from Highway 22, Sydney coalfield, Cape Breton County, Nova Scotia.

Leiotriletes adnatus (Kosanke) Potonié and Kremp, 1955
Hypotype 16267
Barss, M.S., 1967, Geol. Surv. Can., Paper 67-11, pl. 25, fig. 1.
Morien Group, Pennsylvanian, LeCras coal seam, from pit 200 feet southwest side of Highway 22, 1.1 miles from intersection of Old Mira Road and Highway 22, Sydney coalfield, Cape Breton County, Nova Scotia.

Leiotriletes adnatus (Kosanke) Potonié and Kremp, 1955
Hypotypes 16347, 16348
Barss, M.S., 1967, Geol. Surv. Can., Paper 67-11, pl. 28, figs. 1, 2.
Morien Group, Pennsylvanian, lower Jubilee coal seam, first seam below Phalen coal seam, east shore of Sydney Harbour, and Harbour coal seam, Dominion No. 12 colliery at face of No. 21 west level, Sydney coalfield, Cape Breton County, Nova Scotia.

Leiotriletes curiosus Playford, 1962
Hypotype 15840
Barss, M.S., 1967, Geol. Surv. Can., Paper 67-11, pl. 7, fig. 5.
Mattson Formation, Mississippian, south face Twisted Mountain, about 15 miles northwest of South Nahanni, lat. 61°13′N. long. 123°37′W, Northwest Territories.

Leiotriletes gulaferus Potonié and Kremp, 1955
Hypotype 16452
Barss, M.S., 1967, Geol. Surv. Can., Paper 67-11, pl. 32, fig. 1.
Pictou Group, Pennsylvanian, Shediac water well approximately 1,400 feet east of shore of harbour and 1,800 feet south of highway at Shediac, New Brunswick.

Leiotriletes inermis (Waltz) Ischenko, 1952
Hypotype 73728
Utting, J., 1987, Geol. Surv. Can., Bull. 374, pl. 1, fig. 2.
Windsor Group, Lower Carboniferous, north shore of Port Hood Island, lat. 46°1′N, long. 61°26′W, Cape Breton Island, Nova Scotia.

Leiotriletes inermis (Waltz) Ishchenko, 1952
Hypotype 92948
Utting, J., Jachowicz, M. and Jachowicz, A., 1989, Geol. Surv. Can., Bull. 396, pl. 6.6, fig. 14.
Emma Fiord Formation, Early Carboniferous, Svartevaeg Cliffs, northern Axel Heiberg Island, District of Franklin.

Leiotriletes inflatus (Schemel) Potonié and Kremp, 1955
Hypotypes 15787, 16033
Barss, M.S., 1967, Geol. Surv. Can., Paper 67-11, pl. 5, fig. 1; pl. 15, fig. 7.
Windsor Group, Mississippian, Frenchvale Brook (Balls Creek), Northwest Arm of Sydney Harbour, and Port Hood Formation, Riversdale Group, Pennsylvanian, 200 feet from mouth of adit of Harbourview coal mine, Port Hood, Inverness County, Nova Scotia.

Leiotriletes inflatus Schemel, 1950
Hypotype 15939
Barss, M.S., 1967, Geol. Surv. Can., Paper 67-11, pl. 12, fig. 2.
Canso Group, Upper Carboniferous, Pomquet River south of Route 4, Antigonish County, Nova Scotia.

Leiotriletes notatus Hacquebard 1957
Hypotype 76367
Utting, J., 1987, Palynology, vol. 11, p. 79, pl. 2, fig. 2.
Hiram Brook Member, Albert Formation, Lower Carboniferous, depth 132.21 m, Petro-Canada Boudreau No. 1 well, Highway 925 south of Boudreau Village, New Brunswick.

Leiotriletes ornatus Ischenko, 1956
Hypotype 15789
Barss, M.S., 1967, Geol. Surv. Can., Paper 67-11, pl. 5, fig. 3.
Windsor Group, Mississippian, Frenchvale Brook (Balls Creek), Northwest Arm of Sydney Harbour, Nova Scotia.

Leiotriletes ornatus Ischenko, 1956
Hypotype 73729
Utting, J., 1987, Geol. Surv. Can., Bull. 374, pl. 1, fig. 3.
Windsor Group, Lower Carboniferous, St. Ann's Bay, North Gut of St. Ann's, Victoria County, lat. 46°12′19″N, long. 60°37′44″W, Cape Breton Island, Nova Scotia.

Leiotriletes ornatus Ishchenko 1956
Hypotype 76368
Utting, J., 1987, Palynology, vol. 11, p. 79, pl. 2, fig. 3.
Hiram Brook Member, Albert Formation, Lower Carboniferous, Boyd Creek east of Gautreau Village, New Brunswick.

Leiotriletes ornatus Ischenko 1956
Hypotype 89931
Utting, J., Keppie, J.D. and Giles, P.S., 1989, Geol. Surv. Can., Bull. 396, pl. 5.2, fig. 5.
Wilkie Brook Formation, Horton Group, Lower Carboniferous, 380 m north of mouth of Wilkie Brook, western shore of St. Georges Bay, Antigonish Basin, Nova Scotia.

Leiotriletes ornatus Ishchenko, 1956
Hypotypes 92896, 92949
Utting, J., Jachowicz, M. and Jachowicz, A., 1989, Geol. Surv. Can., Bull. 396, pl. 6.4, fig. 1; pl. 6.6, fig. 15.
Emma Fiord Formation, Early Carboniferous, Lyall River area, Grinnell Peninsula, Devon Island, and Svartevaeg Cliffs, northern Axel Heiberg Island, District of Franklin.

Leiotriletes cf. *L. parvus* Guennel, 1958
Hypotype 16105
Barss, M.S., 1967, Geol. Surv. Can., Paper 67-11, pl. 18, fig. 7.
Cumberland Group, Pennsylvanian, Pearly McFadden's mine on Chignecto coal seam, 0.3 miles east of Chignecto, Cumberland County, Nova Scotia.

Leiotriletes sphaerotriangulus (Loose) Potonié and Kremp, 1955
Hypotype 16108
Barss, M.S., 1967, Geol. Surv. Can., Paper 67-11, pl. 18, fig. 10.
Cumberland Group, Pennsylvanian, Morrison's pit on Chignecto coal seam, 0.4 miles east of Chignecto, Cumberland County, Nova Scotia.

cf. *Leiotriletes subintortus* (Waltz) Ischenko 1952 var. *rotundus* Waltz, 1941
Hypotype 15838
Barss, M.S., 1967, Geol. Surv. Can., Paper 67-11, pl. 7, fig. 3.
Mattson Formation, Mississippian, south face Twisted Mountain, about 15 miles northwest of South Nahanni, lat. 61°13′N. long. 123°37′W, Northwest Territories.

Leiotriletes sp.
Fig. specs. 15938, 16032
Barss, M.S., 1967, Geol. Surv. Can., Paper 67-11, pl. 12, fig. 1; pl. 15, fig. 6.
Canso Group, Upper Carboniferous, Pomquet River south of Route 4, Antigonish County, and Port Hood Formation, Riversdale Group, Pennsylvanian, No. 5 coal seam, 1300 feet east of Evans' mine on No. 2 coal seam, near Neil MacLeod Brook, St. Rose, Inverness County, Nova Scotia.

Leiotriletes sp.
Fig. specs. 16504, 16556
Barss, M.S., 1967, Geol. Surv. Can., Paper 67-11, pl. 34, fig. 3; pl. 36, fig. 13.
Permian, Pictou Group, depth 350 feet, Gallas Point borehole No. 2 on property of H. Tweedy, Gallas Point, lat. 46°7′42″N, long. 63°57′28″W, Prince Edward Island; Tatonduck River, west limb of syncline 3 miles east of Yukon-Alaska border, lat. 64°58′N, long. 140°54′W, Yukon.

?*Leiotriletes* sp.
Fig. spec. 16505
Barss, M.S., 1967, Geol. Surv. Can., Paper 67-11, pl. 34, fig. 4.
Pictou Group, Permian, depth 350 feet, Gallas Point borehole No. 2 on property of H. Tweedy, Gallas Point, lat. 46°7′42″N, long. 63°57′28″W, Prince Edward Island.

Leiotriletes sp. A *in* Hacquebard, 1957
Fig. spec. 15940
Barss, M.S., 1967, Geol. Surv. Can., Paper 67-11, pl. 12, fig. 3.
Canso Group, Upper Carboniferous, Pomquet River south of Route 4, Antigonish County, Nova Scotia.

Leiotriletes sp. A Utting
Fig. spec. 89932
Utting, J., Keppie, J.D. and Giles, P.S., 1989, Geol. Surv. Can., Bull. 396, p. 132, pl. 5.2, fig. 10.
Rights River Formation, Horton Group, Lower Carboniferous, depth 73.15 m, Drillhole DDH2, east of Beaver Settlement, lat. 45°48′12″N, long. 61°55′8″W, Antigonish Basin, Nova Scotia.

Leiozonotriletes insignitus Hacquebard, 1957
Hypotype 89933
Utting, J., Keppie, J.D. and Giles, P.S., 1989, Geol. Surv. Can., Bull. 396, pl. 5.2, fig. 12.
Horton Bluff Formation, Horton Group, Lower Carboniferous, depth 70 m, Saarberg Interplan Canada Limited SI/8 drill core, north of west branch of Avon River, lat. 44°57′10″N, long. 64°15′50″W, Windsor Basin, Nova Scotia.

Limitisporites monstruosus (Luber and Waltz) Hart, 1965
Hypotypes 16568, 16570, 16576
Barss, M.S., 1967, Geol. Surv. Can., Paper 67-11, pl. 37, figs. 2, 4, 10.
Permian, Tatonduck River, west limb of syncline 3 miles east of Yukon-Alaska border, lat. 64°58′N, long. 140°54′W, Yukon.

Limitisporites cf. *L. monstruosus* (Luber and Waltz) Hart, 1965
Hypotype 16492
Barss, M.S., 1967, Geol. Surv. Can., Paper 67-11, pl. 33, fig. 7.
Pictou Group, Pennsylvanian, Shediac water well approximately 1,400 feet east of shore of harbour and 1,800 feet south of highway at Shediac, New Brunswick.

Limitisporites sp.
Fig. specs. 16490, 16494
Barss, M.S., 1967, Geol. Surv. Can., Paper 67-11, pl. 33, figs. 5, 9.

Pictou Group, Pennsylvanian, Shediac water well approximately 1,400 feet east of shore of harbour and 1,800 feet south of highway at Shediac, New Brunswick.

Limitisporites sp.
Fig. spec. 16538
Barss, M.S., 1967, Geol. Surv. Can., Paper 67-11, pl. 35, fig. 21.
Pictou Group, Permian, depth 332 feet, Gallas Point borehole No. 2 on property of H. Tweedy, Gallas Point, lat. 46°7′42″N, long. 63°57′28″W, Prince Edward Island.

?*Limitisporites* sp.
Fig. spec. 16528
Barss, M.S., 1967, Geol. Surv. Can., Paper 67-11, pl. 35, fig. 10.
Pictou Group, Permian, depth 332 feet, Gallas Point borehole No. 2 on property of H. Tweedy, Gallas Point, lat. 46°7′42″N, long. 63°57′28″W, Prince Edward Island.

Lophotriletes commissuralis (Kosanke) Potonié and Kremp, 1955
Hypotypes 16035, 16112, 16205
Barss, M.S., 1967, Geol. Surv. Can., Paper 67-11, pl. 15, fig. 9; pl. 18, fig. 14; pl. 22, fig. 7.
Pennsylvanian, Port Hood Formation, Riversdale Group, 200 feet from mouth of adit of Harbourview coal mine, Port Hood, Inverness County; Cumberland Group, No. 1 coal seam from 5,400 foot east level in No. 2 mine, Springhill, Cumberland County; and Morien Group, Beaver mine on Tracey coal seam, east of MacDonald Lake and south of Morrison road, approximately 5 miles from Highway 22, Sydney coalfield, Cape Breton County, Nova Scotia.

Lophotriletes commissuralis (Kosanke) Pontié and Kremp, 1955
Hypotype 16275
Barss, M.S., 1967, Geol. Surv. Can., Paper 67-11, pl. 25, fig. 9.
Mabou Mines Formation, Pennsylvanian, coal seam 95 feet above beach, cliff section northwest side of syncline, Coal Mine Point, Mabou mines, Inverness County, Nova Scotia.

Lophotriletes microsaetosus (Loose) Potonié and Kremp, 1955
Hypotypes 16036, 16115, 16206, 16208
Barss, M.S., 1967, Geol. Surv. Can., Paper 67-11, pl. 15, fig. 10; pl. 18, fig. 17; pl. 22, figs. 8, 10.
Pennsylvanian, Port Hood Formation, Riversdale Group, No. 5 coal seam, 1,300 foot east of Evans' coal mine on No. 2 seam, near Neil MacLeod Brook, St. Rose, Inverness County, and Cumberland Group, Tingley pit in Roger's field near southern extension of McDougal Street, Springhill, Cumberland County (16115); Morien Group, McAulay Cove coal seam, from outcrop on shore 1.75 miles south of Homeville (16206) and Beaver mine on Tracey coal seam east of MacDonald Lake and south of Morrison road, approximately 5 miles from Highway 22, Sydney coalfield, Cape Breton County, Nova Scotia.

Lophotriletes microsaetosus (Loose) Pontié and Kremp, 1955
Hypotypes 16273, 16274
Barss, M.S., 1967, Geol. Surv. Can., Paper 67-11, pl. 25, fig. 7,8.
Mabou Mines Formation, Pennsylvanian, Colonel Snow's adit at beach level, northeast side of syncline at Coal Mine Point, approximately 440 feet west of gypsum fault, and seven-foot coal seam in drift from cliff shore to old workings, south side of syncline, Mabou mines, Inverness County, Nova Scotia.

Lophotriletes microsaetosus (Loose) Potonié and Kremp, 1955
Hypotypes 16354, 16355
Barss, M.S., 1967, Geol. Surv. Can., Paper 67-11, pl. 28, figs. 8, 11.
Morien Group, Pennsylvanian, Phalen coal seam, Dominion No. 18 colliery at face of No. 6 west level, 3,400 feet from main deep, and Harbour coal seam, Dominion No. 12 colliery at face of No. 21 west level, Sydney coalfield, Cape Breton County, Nova Scotia.

Lophotriletes cf. *L. mosaicus* Potonié and Kremp, 1955
Hypotype 15953
Barss, M.S., 1967, Geol. Surv. Can., Paper 67-11, pl. 12, fig. 16.
Canso Group, Upper Carboniferous, Pomquet River south of Route 4, Antigonish County, Nova Scotia.

Lophotriletes cf. *L. pseudaculeatus* Potonié and Kremp, 1955
Hypotypes 16116, 16118
Barss, M.S., 1967, Geol. Surv. Can., Paper 67-11, pl. 18, figs. 18, 20.
Cumberland Group, Pennsylvanian, bore-hole No. 1 east, in Roger's field near outcrop on No. 6 seam, at depth 40 feet, Springhill, and Pearly McFadden's mine on Chignecto coal seam, 0.3 miles east of Chignecto, Cumberland County, Nova Scotia.

Lophotriletes cf. *L. pseudaculeatus* Potonié and Kremp, 1955
Hypotype 16619
Barss, M.S., 1967, Geol. Surv. Can., Paper 67-11, pl. 38, fig. 17.
Sabine Bay Formation, Permian, vicinity of Tingmisut Lake, east side of Sabine Peninsula, Melville Island, District of Franklin.

Lophotriletes sp.
Fig. specs. 16113, 16117, 16127
Barss, M.S., 1967, Geol. Surv. Can., Paper 67-11, pl. 18, figs. 15, 19, 29.
Cumberland Group, Pennsylvanian, Pearly McFadden's mine on Chignecto coal seam, 0.3 miles east of Chignecto, Joggins bore-hole No. 1, at 1641 foot depth, on Chrislie Brook, 2 miles upstream from confluence with Herbert River, and No. 6 coal seam, from 5,400 foot west level in No. 4 mine, 35 feet from rock tunnel slope, Springhill, Cumberland County, Nova Scotia.

Lophotriletes sp.
Fig. specs. 16456, 16457
Barss, M.S., 1967, Geol. Surv. Can., Paper 67-11, pl. 32, figs. 5, 6.

Pictou Group, Pennsylvanian, Shediac water well approximately 1,400 feet east of shore of harbour and 1,800 feet south of highway at Shediac, New Brunswick.

Lophotriletes sp.
Fig. spec. 16559
Barss, M.S., 1967, Geol. Surv. Can., Paper 67-11, pl. 36, fig. 16.
Permian, Tatonduck River, west limb of syncline 3 miles east of Yukon-Alaska border, lat. 64°58′N, long. 140°54′W, Yukon.

Lophozonotriletes appendices (Hacquebard and Barss) Playford, 1963
Hypotype 15927
Barss, M.S., 1967, Geol. Surv. Can., Paper 67-11, pl. 11, fig. 1.
Mattson Formation, Mississippian, south face Twisted Mountain, about 15 miles northwest of South Nahanni, lat. 61°13′N. long. 123°37′W, Northwest Territories.

Lophozonotriletes malevkensis Naumova *in* Kedo, 1964
Hypotype 89934
Utting, J., Keppie, J.D. and Giles, P.S., 1989, Geol. Surv. Can., Bull. 396, pl. 5.2, fig. 11.
Unnamed limestone, Horton Group, Lower Carboniferous, Lakevale cliff north of syncline, St. Georges Bay, Antigonish Basin, Nova Scotia.

Lophozonotriletes sp.
Fig. spec. 92897
Utting, J., Jachowicz, M. and Jachowicz, A., 1989, Geol. Surv. Can., Bull. 396, pl. 6.4, fig. 2.
Emma Fiord Formation, Early Carboniferous, Lyall River area, Grinnell Peninsula, Devon Island, District of Franklin.

?*Lophozonotriletes* sp.
Fig. spec. 15990
Barss, M.S., 1967, Geol. Surv. Can., Paper 67-11, pl. 13, fig. 13.
Canso Group, Upper Carboniferous, Pomquet River south of Route 4, Antigonish County, Nova Scotia.

Lueckisporites cf. *L. virkkiae* Potonié and Klaus, 1954
Fig. spec. 16569
Barss, M.S., 1967, Geol. Surv. Can., Paper 67-11, pl. 37, fig. 3.
Permian, Tatonduck River, west limb of syncline 3 miles east of Yukon-Alaska border, lat. 64°58′N, long. 140°54′W, Yukon.

Lycospora brevijuga Kosanke, 1950
Hypotype 16182
Barss, M.S., 1967, Geol. Surv. Can., Paper 67-11, pl. 20, fig. 22.
Cumberland Group, Pennsylvanian, Tingley pit in Roger's field near southern extension of McDougal Street, Springhill, Cumberland County, Nova Scotia.

Lycospora brevijuga Kosanke, 1950
Hypotype 16215
Barss, M.S., 1967, Geol. Surv. Can., Paper 67-11, pl. 22, fig. 17.
Morien Group, Pennsylvanian, Beaver mine on Tracey coal seam east of MacDonald Lake and south of Morrison road, approximately 5.0 miles from Highway 22, Sydney coalfield, Cape Breton County, Nova Scotia.

Lycospora brevijuga Kosanke, 1950
Hypotype 16315
Barss, M.S., 1967, Geol. Surv. Can., Paper 67-11, pl. 26, fig. 16.
Mabou Mines Formation, Pennsylvanian, coal seam 95 feet above beach, cliff section northeast side of syncline, Coal Mine Point, Mabou mines, Inverness County, Nova Scotia.

Lycospora granulata Kosanke, 1950
Hypotype 16213
Barss, M.S., 1967, Geol. Surv. Can., Paper 67-11, pl. 22, fig. 15.
Morien Group, Pennsylvanian, McAulay Cove coal seam, from outcrop on shore 1.75 miles south of Homeville, Sydney coalfield, Cape Breton County, Nova Scotia.

Lycospora granulata Kosanke, 1950
Hypotype 16406
Barss, M.S., 1967, Geol. Surv. Can., Paper 67-11, pl. 30, fig. 3.
Morien Group, Pennsylvanian, Harbour coal seam, Dominion No. 12 colliery at face of No. 21 west level, Sydney coalfield, Cape Breton County, Nova Scotia.

Lycospora cf. *L. lobulata* Staplin, 1960
Hypotype 15898
Barss, M.S., 1967, Geol. Surv. Can., Paper 67-11, pl. 9, fig. 18.
Mattson Formation, Mississippian, south face Twisted Mountain, about 15 miles northwest of South Nahanni, lat. 61°13′N. long. 123°37′W, Northwest Territories.

Lycospora microgranulata Bhardwaj, 1957
Hypotypes 16408, 16409
Barss, M.S., 1967, Geol. Surv. Can., Paper 67-11, pl. 30, figs. 5, 6.
Morien Group, Pennsylvanian, Phalen coal seam, Dominion No. 18 colliery at face of No. 6 west level, 3,400 feet from main deep, Sydney coalfield, Cape Breton County, Nova Scotia.

Lycospora microgranulata Bhardwaj, 1957
Hypotype 16468
Barss, M.S., 1967, Geol. Surv. Can., Paper 67-11, pl. 32, fig. 17.
Pictou Group, Pennsylvanian, Shediac water well approximately 1,400 feet east of shore of harbour and 1,800 feet south of highway at Shediac, New Brunswick.

Lycospora noctuina Butterworth and Williams, 1958
Hypotype 92950
Utting, J., Jachowicz, M. and Jachowicz, A., 1989, Geol. Surv. Can., Bull. 396, pl. 6.6, fig. 16.
Emma Fiord Formation, Early Carboniferous, Svartevaeg Cliffs, northern Axel Heiberg Island, District of Franklin.

Lycospora noctuina Butterworth and Williams var *noctuina*
Hypotype 62483
Utting, J., 1980, Can. J. Earth Sci., vol. 17, no. 8, pl. 1, fig. 4.
Upper Windsor Group, Mississippian, United States Borax and Chemical Corporation borehole SB-1 near Stewiacke, Nova Scotia.

Lycospora noctuina Butterworth and Williams, 1958, var *noctuina*
Hypotype 73792
Utting, J., 1987, Geol. Surv. Can., Bull. 374, pl. 3, fig. 14.
Windsor Group, Lower Carboniferous, north shore of Port Hood Island, lat. 46°1′N, long. 61°26′W, Cape Breton Island, Nova Scotia.

Lycospora parva Kosanke, 1950
Hypotype 16405
Barss, M.S., 1967, Geol. Surv. Can., Paper 67-11, pl. 30, fig. 2.
Morien Group, Pennsylvanian, Harbour coal seam, Dominion No. 12 colliery at face of No. 21 west level, Sydney coalfield, Cape Breton County, Nova Scotia.

Lycospora pellucida (Wicher) Schopf, Wilson and Bentall, 1944
Hypotype 16212
Barss, M.S., 1967, Geol. Surv. Can., Paper 67-11, pl. 22, fig. 14.
Morien Group, Pennsylvanian, McAulay Cove coal seam, from outcrop on shore 1.75 miles south of Homeville, Sydney coalfield, Cape Breton County, Nova Scotia.

Lycospora pressoides? (Potonié and Kremp) Bhardwaj, 1957
Hypotype 16410
Barss, M.S., 1967, Geol. Surv. Can., Paper 67-11, pl. 30, fig. 7.
Morien Group, Pennsylvanian, Harbour coal seam, Dominion No. 12 colliery at face of No. 21 west level, Sydney coalfield, Cape Breton County, Nova Scotia.

Lycospora pseudoannulata Kosanke, 1950
Hypotypes 16072, 16181, 16210
Barss, M.S., 1967, Geol. Surv. Can., Paper 67-11, pl. 16, fig. 17; pl. 20, fig. 21; pl. 22, fig. 12.
Pennsylvanian, Port Hood Formation, Riversdale Group, 200 feet from mouth of adit of Harbourview coal mine, Port Hood, Inverness County; Cumberland Group, Pearly McFadden's mine on Chignecto coal seam, 0.3 miles east of Chignecto, Cumberland County; Morien Group, Beaver mine on Tracey coal seam east of MacDonald Lake and south of Morrison road, approximately 5 miles from Highway 22, Sydney coalfield, Cape Breton County, Nova Scotia.

Lycospora pseudoannulata? Kosanke, 1950
Hypotype 16316
Barss, M.S., 1967, Geol. Surv. Can., Paper 67-11, pl. 26, fig. 17.
Mabou Mines Formation, Pennsylvanian, coal seam 95 feet above beach, cliff section northeast side of syncline, Coal Mine Point, Mabou mines, Inverness County, Nova Scotia.

Lycospora punctata Kosanke, 1950
Hypotype 16183
Barss, M.S., 1967, Geol. Surv. Can., Paper 67-11, pl. 20, fig. 23.
Cumberland Group, Pennsylvanian, Styles mine on Chignecto coal seam, 5.4 miles east of Chignecto near Styles Brook, Cumberland County, Nova Scotia.

Lycospora punctata Kosanke, 1950
Hypotype 16211
Barss, M.S., 1967, Geol. Surv. Can., Paper 67-11, pl. 22, fig. 13.
Morien Group, Pennsylvanian, McAulay Cove coal seam, from outcrop on shore 1.75 miles south of Homeville, Sydney coalfield, Cape Breton County, Nova Scotia.

Lycospora punctata Kosanke, 1950
Hypotypes 16314, 16407
Barss, M.S., 1967, Geol. Surv. Can., Paper 67-11, pl. 26, fig. 15; pl. 30, fig. 4.
Pennsylvanian, Mabou Mines Formation, from 590 foot depth in borehole No. 2, 350 feet south of Stack Point, Mabou mines, Inverness County, Nova Scotia, and Pictou Group, stripping operation one mile south of Beersville, on east side of Coal Branch River, New Brunswick.

Lycospora punctata Kosanke, 1950
Hypotype 16507
Barss, M.S., 1967, Geol. Surv. Can., Paper 67-11, pl. 34, fig. 6.
Pictou Group, Permian, depth 350 feet, Gallas Point borehole No. 2 on property of H. Tweedy, Gallas Point, lat. 46°7′42″N, long. 63°57′28″W, Prince Edward Island.

Lycospora pusilla (Ibrahim) Schopf, Wilson and Bentall, 1944
Hypotype 16214
Barss, M.S., 1967, Geol. Surv. Can., Paper 67-11, pl. 22, fig. 16.
Morien Group, Pennsylvanian, Beaver mine on Tracey coal seam east of MacDonald Lake and south of Morrison road, approximately 5.0 miles from Highway 22, Sydney coalfield, Cape Breton County, Nova Scotia.

Lycospora pusilla (Ibrahim) Schopf, Wilson and Bentall, 1944
Hypotypes 92898, 92951
Utting, J., Jachowicz, M. and Jachowicz, A., 1989, Geol. Surv. Can., Bull. 396, pl. 6.4, fig. 3; pl. 6.6, fig. 17.
Emma Fiord Formation, Early Carboniferous, Lyall River area, Grinnell Peninsula, Devon Island, and Svartevaeg Cliffs, northern Axel Heiberg Island, District of Franklin.

Lycospora pusilla (Ibrahim) emend. Somers, 1972
Hypotype 94993
Utting, J., 1991, Geol. Surv. Can., Bull. 412, pl. 4.2, fig. 9.
Hart River Formation, Lower Carboniferous, north bank Peel River, northern Yukon.

Lycospora pusilla (Ibrahim) Schopf, Wilson and Bentall "tendance B2", Somers, 1972
Hypotype 73793
Utting, J., 1987, Geol. Surv. Can., Bull. 374, pl. 3, fig. 15.
Windsor Group, Lower Carboniferous, small bay 100m southwest of Hopewell Cape, lat. 45°48′56″N, long. 64°34′40″W, New Brunswick.

Lycospora cf. *L. pusilla* (Ibrahim) Schopf, Wilson and Bentall, 1944
Hypotype 16469
Barss, M.S., 1967, Geol. Surv. Can., Paper 67-11, pl. 32, fig. 18.
Pictou Group, Pennsylvanian, Shediac water well approximately 1,400 feet east of shore of harbour and 1,800 feet south of highway at Shediac, New Brunswick.

Lycospora uber Hoffmeister, Staplin and Malloy, 1955
Hypotype 15899
Barss, M.S., 1967, Geol. Surv. Can., Paper 67-11, pl. 9, fig. 19.
Mattson Formation, Mississippian, south face Twisted Mountain, about 15 miles northwest of South Nahanni, lat. 61°13′N. long. 123°37′W, Northwest Territories.

Lycospora uber (Hoffmeister, Staplin and Malloy) Staplin, 1960
Hypotype 15991
Barss, M.S., 1967, Geol. Surv. Can., Paper 67-11, pl. 13, fig. 14.
Canso Group, Upper Carboniferous, Pomquet River south of Route 4, Antigonish County, Nova Scotia.

Lycospora sp.
Fig. specs. 15804, 16184
Barss, M.S., 1967, Geol. Surv. Can., Paper 67-11, pl. 5, fig. 18; pl. 20, fig. 24.
Windsor Group, Mississippian, Frenchvale Brook (Balls Creek), Northwest Arm of Sydney Harbour; Cumberland Group, Pennsylvanian, Pearly McFadden's mine on Chignecto coal seam, 0.3 mile east of Chignecto, Cumberland County, Nova Scotia.

?*Lycospora* sp.
Fig. specs. 15808, 15809
Barss, M.S., 1967, Geol. Surv. Can., Paper 67-11, pl. 6, figs. 1, 2.
Windsor Group, Mississippian, Frenchvale Brook (Balls Creek), Northwest Arm of Sydney Harbour, Nova Scotia.

(?)*Marsupipollenites* sp.
Fig. spec. 45737
Davidson, A. and McGregor, D.C., 1976, Nature, vol. 262, no. 5567, fig. 2g.
Gilo sandstone, Permian, stream gully approximately 1 km north of Gilo River, 34 km southeast of Agenga, Ilubabor Province, southwestern Ethiopia.

Microreticulatisporites concavus Butterworth and Williams, 1958
Hypotype 92899
Utting, J., Jachowicz, M. and Jachowicz, A., 1989, Geol. Surv. Can., Bull. 396, pl. 6.4, fig. 4.
Emma Fiord Formation, Early Carboniferous, Lyall River area, Grinnell Peninsula, Devon Island, District of Franklin.

cf. *Microreticulatisporites concavus* Butterworth and Williams, 1958
Hypotype 15853
Barss, M.S., 1967, Geol. Surv. Can., Paper 67-11, pl. 7, fig. 19.
Mattson Formation, Mississippian, south face Twisted Mountain, about 15 miles northwest of South Nahanni, lat. 61°13′N. long. 123°37′W, Northwest Territories.

Microreticulatisporites fistulosus (Ibrahim) Knox, 1950
Hypotype 16283
Barss, M.S., 1967, Geol. Surv. Can., Paper 67-11, pl. 25, fig. 17.
Morien Group, Pennsylvanian, Ormond Coal seam, pit approximately 600 feet southwest side of Highway 22 and 600 feet east of old road to Middle Lake, Sydney coalfield, Cape Breton County, Nova Scotia.

Microreticulatisporites hacquebardii Utting
Holotype 73767; hypotypes 73768, 73769
Utting, J., 1987, Geol. Surv. Can., Bull. 374, p. 29, pl. 2, figs. 14-16.
Windsor Group, Lower Carboniferous, north shore of Port Hood Island, lat. 46°1′N, long. 61°26′W, Cape Breton Island, Nova Scotia; Lower Codroy Group, Carboniferous, Heatherton, St. George's Bay, Newfoundland (73768).

Microreticulatisporites hortonensis Playford, 1964
Hypotype 89935
Utting, J., Keppie, J.D. and Giles, P.S., 1989, Geol. Surv. Can., Bull. 396, pl. 5.2, fig. 13.
Wilkie Brook Formation, Horton Group, Lower Carboniferous, 380 m north of mouth of Wilkie Brook, western shore of St. Georges Bay, Antigonish Basin, Nova Scotia.

Microreticulatisporites cf. *M. hortonensis* Playford, 1964
Fig. spec. 16045
Barss, M.S., 1967, Geol. Surv. Can., Paper 67-11, pl. 15, fig. 19.
Port Hood Formation, Riversdale Group, Pennsylvanian, 200 feet from mouth of adit of Harbourview coal mine, Port Hood, Inverness County, Nova Scotia.

Microreticulatisporites lunatus Knox, 1950
Hypotype 15976
Barss, M.S., 1967, Geol. Surv. Can., Paper 67-11, pl. 12, fig. 39.
Canso Group, Upper Carboniferous, Pomquet River south of Route 4, Antigonish County, Nova Scotia.

?*Microreticulatisporites microreticulatus* Knox, 1950
Hypotype 16136
Barss, M.S., 1967, Geol. Surv. Can., Paper 67-11, pl. 19, fig. 4.
Cumberland Group, Pennsylvanian, Pearly McFadden's mine on Chignecto coal seam, 0.3 miles east of Chignecto, Cumberland County, Nova Scotia.

Microreticulatisporites cf. *M. novicus* Bhardwaj, 1957
Hypotype 16282
Barss, M.S., 1967, Geol. Surv. Can., Paper 67-11, pl. 25, fig. 16.
Mabou Mines Formation, Pennsylvanian, seven-foot coal seam in drift from cliff shore to old workings, south side of syncline, Mabou mines, Inverness County, Nova Scotia.

Microreticulatisporites punctatus Knox, 1950
Hypotype 15854
Barss, M.S., 1967, Geol. Surv. Can., Paper 67-11, pl. 7, fig. 20.
Mattson Formation, Mississippian, south face Twisted Mountain, about 15 miles northwest of South Nahanni, lat. 61°13'N. long. 123°37'W, Northwest Territories.

Monilospora dignata Playford, 1963
Hypotype 94492
Utting, J., 1991, Geol. Surv. Can., Bull. 412, pl. 4.2, fig. 8.
Hart River Formation, Lower Carboniferous, north bank Peel River, northern Yukon.

Monilospora moniliformis Hacquebard and Barss, 1957
Hypotype 15924
Barss, M.S., 1967, Geol. Surv. Can., Paper 67-11, pl. 10, fig. 19.
Mattson Formation, Mississippian, south face Twisted Mountain, about 15 miles northwest of South Nahanni, lat. 61°13'N. long. 123°37'W, Northwest Territories.

Monosaccate spore
Fig. specs. 75761, 75762
Utting, J., 1985, Bull. Can. Petrol. Geol., vol. 33, no. 3, pl. 1, figs. 3, 4.
Otto Fiord Formation, Upper Carboniferous, 3.5 km northeast of van Hauen Pass, north shore Hare Fiord, Ellesmere Island, District of Franklin.

Monosaccate spore with ornament of grana
Fig. specs. 75760
Utting, J., 1985, Bull. Can. Petrol. Geol., vol. 33, no. 3, pl. 1, fig. 2.
Otto Fiord Formation, Upper Carboniferous, 3.5 km northeast of van Hauen Pass, north shore Hare Fiord, Ellesmere Island, District of Franklin.

Mooreisporites sp.
Fig. spec. 16255
Barss, M.S., 1967, Geol. Surv. Can., Paper 67-11, pl. 24, fig. 3.
Morien Group, Pennsylvanian, LeCras coal seam, from pit 200 feet southwest side of Highway 22, 1.1 miles from intersection of Old Mira Road and Highway 22, Sydney coalfield, Cape Breton County, Nova Scotia.

Murospora aurita (Waltz) Playford, 1962
Hypotype 15891
Barss, M.S., 1967, Geol. Surv. Can., Paper 67-11, pl. 9, fig. 11.
Mattson Formation, Mississippian, south face Twisted Mountain, about 15 miles northwest of South Nahanni, lat. 61°13'N. long. 123°37'W, Northwest Territories.

Murospora aurita (Waltz) Playford, 1962
Hypotypes 92900, 92901, 92954, 92955
Utting, J., Jachowicz, M. and Jachowicz, A., 1989, Geol. Surv. Can., Bull. 396, pl. 6.4, figs. 6, 7; pl. 6.6, figs. 22, 23.
Emma Fiord Formation, Early Carboniferous, Lyall River area, Grinnell Peninsula, Devon Island, and Svartevaeg Cliffs, northern Axel Heiberg Island (92954, 92955), District of Franklin.

Murospora aurita (Waltz) Playford, 1962
Hypotypes 94463, 94488, 94495
Utting, J., 1991, Geol. Surv. Can., Bull. 412, pl. 4.1, fig. 6; pl. 4.2, figs. 4, 11.
Hart River Formation, Lower Carboniferous, north bank Peel River, northern Yukon.

Murospora conduplicata (Andreyeva) Playford, 1962
Hypotype 15892
Barss, M.S., 1967, Geol. Surv. Can., Paper 67-11, pl. 9, fig. 12.
Mattson Formation, Mississippian, south face Twisted Mountain, about 15 miles northwest of South Nahanni, lat. 61°13'N. long. 123°37'W, Northwest Territories.

Murospora conduplicata (Andrejeva) Playford, 1962
Hypotype 92902
Utting, J., Jachowicz, M. and Jachowicz, A., 1989, Geol. Surv. Can., Bull. 396, pl. 6.4, fig. 8.
Emma Fiord Formation, Early Carboniferous, Lyall River area, Grinnell Peninsula, Devon Island, District of Franklin.

Murospora dupla (Ischenko) Playford, 1962
Hypotype 15896
Barss, M.S., 1967, Geol. Surv. Can., Paper 67-11, pl. 9, fig. 16.
Mattson Formation, Mississippian, south face Twisted Mountain, about 15 miles northwest of South Nahanni, lat. 61°13'N. long. 123°37'W, Northwest Territories.

Murospora friendii Playford, 1962
Hypotype 15895
Barss, M.S., 1967, Geol. Surv. Can., Paper 67-11, pl. 9, fig. 15.
Mattson Formation, Mississippian, south face Twisted Mountain, about 15 miles northwest of South Nahanni, lat. 61°13'N. long. 123°37'W, Northwest Territories.

Murospora intorta (Waltz) Playford, 1962
Hypotype 15890
Barss, M.S., 1967, Geol. Surv. Can., Paper 67-11, pl. 9, fig. 10.
Mattson Formation, Mississippian, south face Twisted Mountain, about 15 miles northwest of South Nahanni, lat. 61°13'N. long. 123°37'W, Northwest Territories.

Murospora kosankei Somers, 1952
Hypotypes 16305, 16306, 16391
Barss, M.S., 1967, Geol. Surv. Can., Paper 67-11, pl. 26, figs. 6,7; pl. 29, fig. 12.
Pennsylvanian, Mabou Mines Formation, coal seam immediately below fault north of Coal Mine Point, Mabou mines, Inverness County; Morien Group, Pennsylvanian, LeCras coal seam, from pit 200 feet southwest side of Highway 22, 1.1 miles from intersection of Old Mira Road and Highway 22, and lower Jubilee coal seam, first seam below Phalen seam, east side of Sydney Harbour, Sydney coalfield, Cape Breton County, Nova Scotia.

Murospora strigata (Waltz) Playford, 1962
Hypotype 94491
Utting, J., 1991, Geol. Surv. Can., Bull. 412, pl. 4.2, fig. 7.
Hart River Formation, Lower Carboniferous, north bank Peel River, northern Yukon.

Murospora sublobata (Waltz) Playford, 1962
Hypotypes 92903, 92956
Utting, J., Jachowicz, M. and Jachowicz, A., 1989, Geol. Surv. Can., Bull. 396, pl. 6.4, fig. 9; pl. 6.6, fig. 24.
Emma Fiord Formation, Early Carboniferous, Lyall River area, Grinnell Peninsula, Devon Island, and Svartevaeg Cliffs, northern Axel Heiberg Island, District of Franklin.

Murospora tripulvinata Staplin, 1960
Hypotypes 92904, 92905
Utting, J., Jachowicz, M. and Jachowicz, A., 1989, Geol. Surv. Can., Bull. 396, pl. 6.4, figs. 10, 11.
Emma Fiord Formation, Early Carboniferous, Lyall River area, Grinnell Peninsula, Devon Island, District of Franklin.

Murospora sp.
Fig. spec. 94478
Utting, J., 1991, Geol. Surv. Can., Bull. 412, pl. 4.1, fig. 21.
Hart River Formation, Lower Carboniferous, north bank Peel River, northern Yukon.

?*Murospora* sp.
Fig. specs. 16076, 16077
Barss, M.S., 1967, Geol. Surv. Can., Paper 67-11, pl. 16, figs. 21, 22.
Port Hood Formation, Riversdale Group, Pennsylvanian, No. 5 coal seam, 1,300 feet east of Evans' coal mine on No. 2 seam, near Neil MacLeod Brook, St. Rose, Inverness County, and 200 feet from mouth of adit of Harbourview coal mine, Port Hood, Inverness County, Nova Scotia.

?*Neoraistrickia inconstans* Neves, 1961
Hypotype 16114
Barss, M.S., 1967, Geol. Surv. Can., Paper 67-11, pl. 18, fig. 16.
Cumberland Group, Pennsylvanian, No. 6 coal seam from 5,400 foot west level in No. 4 mine, 35 feet from rock tunnel slope, Springhill, Cumberland County, Nova Scotia.

Neoraistrickia logani (Winslow) Coleman and Clayton, 1987
Hypotypes 89936, 89937
Utting, J., Keppie, J.D. and Giles, P.S., 1989, Geol. Surv. Can., Bull. 396, pl. 5.2, figs. 16-18.
Horton Bluff Formation, Horton Group, Lower Carboniferous, depth 51m, Saarberg Interplan Canada Limited SI/8 drill core, north of west branch of Avon River, lat. 44°57′10″N, long. 64°15′50″W, Windsor Basin, Nova Scotia.

Neoraistrickia versiformia Utting
Holotype 73741; hypotypes 73742, 73743
Utting, J., 1987, Geol. Surv. Can., Bull. 374, p. 27, pl. 1, figs. 17-19.
Lower Carboniferous, Windsor Group, north shore of Port Hood Island, lat. 46°1′N, long. 61°26′W, Cape Breton Island, Nova Scotia; Bassin aux Huîtres Member, Havre aux Maisons Formation, Ile Boudreau, Iles-de-la-Madeleine, Québec.

(?)*Nuskoisporites gondwanensis* (Balme and Hassell)
Hypotype 45735
Davidson, A. and McGregor, D.C., 1976, Nature, vol. 262, no. 5567, fig. 2e.
Gilo sandstone, Permian, stream gully approximately 1 km north of Gilo River, 34 km southeast of Agenga, Ilubabor Province, southwestern Ethiopia.

Nuskoisporites sp.
Fig. spec. 16518
Barss, M.S., 1967, Geol. Surv. Can., Paper 67-11, pl. 34, fig. 17.
Pictou Group, Permian, depth 335 feet, Gallas Point borehole No. 2 on property of H. Tweedy, Gallas Point, lat. 46°7′42″N, long. 63°57′28″W, Prince Edward Island.

Pakhapites cf. *P. pachyrrhachis* (Andreyeva) Hart, 1965
Hypotypes 16638, 16639
Barss, M.S., 1967, Geol. Surv. Can., Paper 67-11, pl. 38, figs. 36, 37.
Sabine Bay Formation, Permian, vicinity of Tingmisut Lake, east side of Sabine Peninsula, Melville Island, District of Franklin.

Pakhapites sp.
Fig. spec. 16632
Barss, M.S., 1967, Geol. Surv. Can., Paper 67-11, pl. 38, fig. 30.
Sabine Bay Formation, Permian, vicinity of Tingmisut Lake, east side of Sabine Peninsula, Melville Island, District of Franklin.

(?)*Pallidosporites magnus* Schaar
Hypotype 45736
Davidson, A. and McGregor, D.C., 1976, Nature, vol. 262, no. 5567, fig. 2f.
Gilo sandstone, Permian, stream gully approximately 1 km north of Gilo River, 34 km southeast of Agenga, Ilubabor Province, southwestern Ethiopia.

Parasaccites sp.
Fig. spec. 45731
Davidson, A. and McGregor, D.C., 1976, Nature, vol. 262, no. 5567, fig. 2a.
Kari sandstone, Permian, banks Kari River approximately 14 km upstream from its junction to the north with Akobo River, Kefa Province, southwestern Ethiopia.

Perotrilites perinatus Hughes and Playford, 1961
Hypotypes 15815, 15889, 15995
Barss, M.S., 1967, Geol. Surv. Can., Paper 67-11, pl. 6, fig. 8; pl. 9, fig. 9; pl. 13, fig. 18.
Mississippian, Windsor Group, Frenchvale Brook (Balls Creek), Northwest Arm of Sydney Harbour, Nova Scotia; Mattson Formation, south face Twisted Mountain, about 15 miles northwest of South Nahanni, lat. 61°13′N. long. 123°37′W, Northwest Territories; Canso Group, Upper Carboniferous, Pomquet River south of Route 4, Antigonish County, Nova Scotia.

Perotriletes perinatus Hughes and Playford, 1961
Hypotype 92906
Utting, J., Jachowicz, M. and Jachowicz, A., 1989, Geol. Surv. Can., Bull. 396, pl. 6.4, fig. 5.
Emma Fiord Formation, Early Carboniferous, Lyall River area, Grinnell Peninsula, Devon Island, District of Franklin.

Perotriletes tessellatus (Staplin) Neville, 1973
Hypotypes 92907, 92957
Utting, J., Jachowicz, M. and Jachowicz, A., 1989, Geol. Surv. Can., Bull. 396, pl. 6.4, fig. 13; pl. 6.6, fig. 26.
Emma Fiord Formation, Early Carboniferous, Lyall River area, Grinnell Peninsula, Devon Island, and Kleybolte Peninsula, northwestern Ellesmere Island, District of Franklin.

Perotriletes tessellatus (Staplin) Neville, 1973
Hypotype 94481
Utting, J., 1991, Geol. Surv. Can., Bull. 412, pl. 4.1, fig. 24.
Hart River Formation, Lower Carboniferous, north bank Peel River, northern Yukon.

Perotrilites sp.
Fig. spec. 15813
Barss, M.S., 1967, Geol. Surv. Can., Paper 67-11, pl. 6, fig. 6.
Windsor Group, Mississippian, Frenchvale Brook (Balls Creek), Northwest Arm of Sydney Harbour, Nova Scotia.

?*Perotrilites* sp.
Fig. spec. 15999
Barss, M.S., 1967, Geol. Surv. Can., Paper 67-11, pl. 13, fig. 22.
Canso Group, Upper Carboniferous, Pomquet River south of Route 4, Antigonish County, Nova Scotia.

Phyllothecotriletes sp.
Fig. specs. 15944, 15947
Barss, M.S., 1967, Geol. Surv. Can., Paper 67-11, pl. 12, figs. 7, 10.
Canso Group, Upper Carboniferous, Pomquet River south of Route 4, Antigonish County, Nova Scotia.

Piceapollenites sp.
Fig. specs. 16521, 16522
Barss, M.S., 1967, Geol. Surv. Can., Paper 67-11, pl. 35, figs. 3, 4.
Pictou Group, Permian, depth 332 feet, Gallas Point borehole No. 2 on property of H. Tweedy, Gallas Point, lat. 46°7′42″N, long. 63°57′28″W, Prince Edward Island.

Piceapollenites sp.
Fig. specs. 16578, 16645
Barss, M.S., 1967, Geol. Surv. Can., Paper 67-11, pl. 37, fig. 12; pl. 38, fig. 43.
Permian, Tatonduck River, west limb of syncline 3 miles east of Yukon-Alaska border, lat. 64°58′N, long. 140°54′W, Yukon; Sabine Bay Formation, vicinity of Tingmisut Lake, east side of Sabine Peninsula, Melville Island, District of Franklin.

?*Piceaepollenites* sp.
Fig. specs. 75774, 75775
Utting, J., 1985, Bull. Can. Petrol. Geol., vol. 33, no. 3, pl. 2, figs. 7, 8.
Otto Fiord Formation, Upper Carboniferous, 3.5 km northeast of van Hauen Pass, north shore Hare Fiord, Ellesmere Island, District of Franklin.

Pityosporites papilionis (Potonié and Klaus) Hart, 1965
Hypotype 16527
Barss, M.S., 1967, Geol. Surv. Can., Paper 67-11, pl. 35, fig. 9.
Pictou Group, Permian, depth 350 feet, Gallas Point borehole No. 2 on property of H. Tweedy, Gallas Point, lat. 46°7′42″N, long. 63°57′28″W, Prince Edward Island.

Pityosporites sp.
Fig. spec. 16529
Barss, M.S., 1967, Geol. Surv. Can., Paper 67-11, pl. 35, fig. 11.
Pictou Group, Permian, depth 350 feet, Gallas Point borehole No. 2 on property of H. Tweedy, Gallas Point, lat. 46°7′42″N, long. 63°57′28″W, Prince Edward Island.

?*Pityosporites* sp.
Fig. spec. 75772
Utting, J., 1985, Bull. Can. Petrol. Geol., vol. 33, no. 3, pl. 2, fig. 5.
Otto Fiord Formation, Upper Carboniferous, 3.5 km northeast of van Hauen Pass, north shore Hare Fiord, Ellesmere Island, District of Franklin.

Potoniespores delicatus Playford, 1963
Hypotype 15915
Barss, M.S., 1967, Geol. Surv. Can., Paper 67-11, pl. 10, fig. 10.
Mattson Formation, Mississippian, south face Twisted Mountain, about 15 miles northwest of South Nahanni, lat. 61°13′N. long. 123°37′W, Northwest Territories.

Potonieisporites elegans (Wilson and Kosanke) Wilson and Venkachatala, 1964
Hypotypes 94485-94487
Utting, J., 1991, Geol. Surv. Can., Bull. 412, pl. 4.2, figs. 1-3.
Hart River Formation, Lower Carboniferous, north bank Peel River, northern Yukon.

Potonieisporites cf. *P. novicus* Bhardwaj, 1954
Hypotype 16500
Barss, M.S., 1967, Geol. Surv. Can., Paper 67-11, pl. 33, fig. 15.
Pictou Group, Pennsylvanian, Shediac water well approximately 1,400 feet east of shore of harbour and 1,800 feet south of highway at Shediac, New Brunswick.

Potonieisporites cf. *P. novicus* Bhardwaj, 1954
Hypotypes 16509, 16520, 16565
Barss, M.S., 1967, Geol. Surv. Can., Paper 67-11, pl. 34, fig. 8; pl. 35, fig. 2; pl. 36, fig. 22.
Permian, Pictou Group, depth 332 feet, Gallas Point borehole No. 2 on property of H. Tweedy, Gallas Point, lat. 46°7′42″N, long. 63°57′28″W, Prince Edward Island; Tatonduck River, west limb of syncline 3 miles east of Yukon-Alaska border, lat. 64°58′N, long. 140°54′W, Yukon (16565).

Potonieisporites cf. *P. simplex* Wilson, 1962
Hypotypes 16510, 16564
Barss, M.S., 1967, Geol. Surv. Can., Paper 67-11, pl. 34, fig. 9; pl. 36, fig. 21.
Permian, Pictou Group, depth 350 feet, Gallas Point borehole No. 2 on property of H. Tweedy, Gallas Point, lat. 46°7′42″N, long. 63°57′28″W, Prince Edward Island; Tatonduck River, west limb of syncline 3 miles east of Yukon-Alaska border, lat. 64°58′N, long. 140°54′W, Yukon.

Potonieisporites sp.
Fig. spec. 16192
Barss, M.S., 1967, Geol. Surv. Can., Paper 67-11, pl. 21, fig. 1.
Cumberland Group, Pennsylvanian, No. 3 coal seam from steep tunnel off 5,400 foot level in No. 2, Springhill, Cumberland County, Nova Scotia.

Potonieisporites sp.
Fig. spec. 16496
Barss, M.S., 1967, Geol. Surv. Can., Paper 67-11, pl. 33, fig. 11.
Pictou Group, Pennsylvanian, Shediac water well approximately 1,400 feet east of shore of harbour and 1,800 feet south of highway at Shediac, New Brunswick.

Potonieisporites sp.
Fig. spec. 16566
Barss, M.S., 1967, Geol. Surv. Can., Paper 67-11, pl. 36, fig. 23.
Permian, Tatonduck River, west limb of syncline 3 miles east of Yukon-Alaska border, lat. 64°58′N, long. 140°54′W, Yukon.

?*Potonieisporites* sp.
Fig. spec. 16189
Barss, M.S., 1967, Geol. Surv. Can., Paper 67-11, pl. 20, fig. 29.
Cumberland Group, Pennsylvanian, Morrison's pit on Chignecto coal seam, 0.4 miles east of Chignecto, Cumberland County, Nova Scotia.

?*Potonieisporites*
Fig. specs. 74763, 74764
Utting, J., 1985, Bull. Can. Petrol. Geol., vol. 33, no. 3, pl. 1, figs. 5, 6.
Otto Fiord Formation, Upper Carboniferous, 3.5 km northeast of van Hauen Pass, north shore Hare Fiord, Ellesmere Island, District of Franklin.

Probable aberranat form of *Murospora aurita* (cf. Playford, 1962, pl. 87, fig. 9.
Fig. spec. 92953
Utting, J., Jachowicz, M. and Jachowicz, A., 1989, Geol. Surv. Can., Bull. 396, pl. 6.6, fig. 21.
Emma Fiord Formation, Early Carboniferous, Svartevaeg Cliffs, northern Axel Heiberg Island, District of Franklin.

Protohaploxypinus cf. *P. globus* (Hart) Hart, 1964
Hypotype 16499
Barss, M.S., 1967, Geol. Surv. Can., Paper 67-11, pl. 33, fig. 14.
Pictou Group, Pennsylvanian, Shediac water well approximately 1,400 feet east of shore of harbour and 1,800 feet south of highway at Shediac, New Brunswick.

Protohaploxypinus cf. *P. globus* (Hart) Hart, 1964
Hypotypes 16540, 16541, 16545, 16580, 16581
Barss, M.S., 1967, Geol. Surv. Can., Paper 67-11, pl. 35, figs. 23, 24; pl. 36, fig. 2; pl. 37, figs. 14, 15.
Permian, Pictou Group, depth 332 feet, Gallas Point borehole No. 2 on property of H. Tweedy, Gallas Point, lat. 46°7′42″N, long. 63°57′28″W, Prince Edward Island; Tatonduck River, west limb of syncline 3 miles east of Yukon-Alaska border, lat. 64°58′N, long. 140°54′W, Yukon (16580, 16581).

Protohaploxypinus cf. *P. jacobii* (Jansonius) Hart, 1964
Hypotype 16536
Barss, M.S., 1967, Geol. Surv. Can., Paper 67-11, pl. 35, fig. 19.
Pictou Group, Permian, depth 350 feet, Gallas Point borehole No. 2 on property of H. Tweedy, Gallas Point, lat. 46°7′42″N, long. 63°57′28″W, Prince Edward Island.

Protohaploxypinus sewardi (Virkki) Hart, 1964
Hypotype 16493
Barss, M.S., 1967, Geol. Surv. Can., Paper 67-11, pl. 33, fig. 8.
Pictou Group, Pennsylvanian, Shediac water well approximately 1,400 feet east of shore of harbour and 1,800 feet south of highway at Shediac, New Brunswick.

Protohaploxypinus sewardi (Virkki) Hart, 1964
Hypotype 16534
Barss, M.S., 1967, Geol. Surv. Can., Paper 67-11, pl. 35, fig. 17.

Pictou Group, Permian, depth 332 feet, Gallas Point borehole No. 2 on property of H. Tweedy, Gallas Point, lat. 46°7′42″N, long. 63°57′28″W, Prince Edward Island.

Protohaploxypinus sp.
Fig. spec. 16498
Barss, M.S., 1967, Geol. Surv. Can., Paper 67-11, pl. 33, fig. 13.
Pictou Group, Pennsylvanian, Shediac water well approximately 1,400 feet east of shore of harbour and 1,800 feet south of highway at Shediac, New Brunswick.

Protohaploxypinus sp.
Fig. specs. 16531, 16532, 16544, 16562, 16584
Barss, M.S., 1967, Geol. Surv. Can., Paper 67-11, pl. 35, figs. 14, 15; pl. 36, figs. 1, 19; pl. 37, fig. 18.
Permian, Pictou Group, depths 350 and 335 (16532) feet, Gallas Point borehole No. 2 on property of H. Tweedy, Gallas Point, lat. 46°7′42″N, long. 63°57′28″W, Prince Edward Island; Tatonduck River, west limb of syncline 3 miles east of Yukon-Alaska border, lat. 64°58′N, long. 140°54′W, Yukon (16562, 16584).

Protohaploxypinus cf. *P. samoilovichii* (Jansonius) Hart, 1964
Hypotype 16533
Barss, M.S., 1967, Geol. Surv. Can., Paper 67-11, pl. 35, fig. 16.
Pictou Group, Permian, depth 332 feet, Gallas Point borehole No. 2 on property of H. Tweedy, Gallas Point, lat. 46°7′42″N, long. 63°57′28″W, Prince Edward Island.

?*Protohaploxypinus* sp.
Fig. spec. 16572
Barss, M.S., 1967, Geol. Surv. Can., Paper 67-11, pl. 37, fig. 6.
Permian, Tatonduck River, west limb of syncline 3 miles east of Yukon-Alaska border, lat. 64°58′N, long. 140°54′W, Yukon.

Protohaploxypinus sp.
Fig. spec. 16648
Barss, M.S., 1967, Geol. Surv. Can., Paper 67-11, pl. 38, fig. 46.
Sabine Bay Formation, Permian, vicinity of Tingmisut Lake, east side of Sabine Peninsula, Melville Island, District of Franklin.

Protohaploxypinus sp.
Fig. spec. 45732
Davidson, A. and McGregor, D.C., 1976, Nature, vol. 262, no. 5567, fig. 2b.
Kari sandstone, Permian, banks Kari River approximately 14 km upstream from its junction to the north with Akobo River, Kefa Province, southwestern Ethiopia.

?*Protohaploxypinus* sp.
Fig. spec. 75765
Utting, J., 1985, Bull. Can. Petrol. Geol., vol. 33, no. 3, pl. 1, fig. 7.
Otto Fiord Formation, Upper Carboniferous, 3.5 km northeast of van Hauen Pass, north shore Hare Fiord, Ellesmere Island, District of Franklin.

?*Protohaploxypinus* sp.; taeniae visible
Fig. specs. 75768-75770
Utting, J., 1985, Bull. Can. Petrol. Geol., vol. 33, no. 3, pl. 1, figs. 1-3.
Otto Fiord Formation, Upper Carboniferous, 3.5 km northeast of van Hauen Pass, north shore Hare Fiord, Ellesmere Island, District of Franklin.

Punctatisporites admirabilis (Playford) Clayton 1970
Hypotype 76369
Utting, J., 1987, Palynology, vol. 11, p. 79, pl. 2, fig. 4.
Frederick Brook Member, Albert Formation, Lower Carboniferous, depth 132.21 m, Petro-Canada Boudreau No. 1 well, Highway 925 south of Boudreau Village, New Brunswick.

Punctatisporites aerarius Butterworth and Williams, 1958
Hypotype 15847
Barss, M.S., 1967, Geol. Surv. Can., Paper 67-11, pl. 7, fig. 12.
Mattson Formation, Mississippian, south face Twisted Mountain, about 15 miles northwest of South Nahanni, lat. 61°13′N. long. 123°37′W, Northwest Territories.

Punctatisporites aerarius Butterworth and Williams, 1958
Hypotype 15948
Barss, M.S., 1967, Geol. Surv. Can., Paper 67-11, pl. 12, fig. 11.
Canso Group, Upper Carboniferous, Pomquet River south of Route 4, Antigonish County, Nova Scotia.

Punctatisporites glaber (Naumova) Playford, 1962
Hypotypes 15788, 15842, 15945
Barss, M.S., 1967, Geol. Surv. Can., Paper 67-11, pl. 5, fig. 2; pl. 7, fig. 7; pl. 12, fig. 8.
Mississippian, Windsor Group, Frenchvale Brook (Balls Creek), Northwest Arm of Sydney Harbour, Nova Scotia; Mattson Formation, south face Twisted Mountain, about 15 miles northwest of South Nahanni, lat. 61°13′N. long. 123°37′W, Northwest Territories; Canso Group, Upper Carboniferous, Pomquet River south of Route 4, Antigonish County, Nova Scotia.

Punctatisporites glaber (Naumova) Playford, 1962
Hypotype 89938
Utting, J., Keppie, J.D. and Giles, P.S., 1989, Geol. Surv. Can., Bull. 396, pl. 5.2, fig. 14.
Horton Bluff Formation, Horton Group, Lower Carboniferous, depth 37.75 m, Saarberg Interplan Canada Limited SI/8 drill core, north of west branch of Avon River, lat. 44°57′10″N, long. 64°15′50″W, Windsor Basin, Nova Scotia.

Punctatisporites glaber (Naumova) Playford, 1962
Hypotype 94459
Utting, J., 1991, Geol. Surv. Can., Bull. 412, pl. 4.1, fig. 2.
Hart River Formation, Lower Carboniferous, north bank Peel River, northern Yukon.

Punctatisporites glaber (Naumova) Playford, 1962
Hypotype 92908
Utting, J., Jachowicz, M. and Jachowicz, A., 1989, Geol. Surv. Can., Bull. 396, pl. 6.4, fig. 14.
Emma Fiord Formation, Early Carboniferous, Lyall River area, Grinnell Peninsula, Devon Island, District of Franklin.

Punctatisporites irrasus Hacquebard, 1957
Hypotype 15792
Barss, M.S., 1967, Geol. Surv. Can., Paper 67-11, pl. 5, fig. 6.
Windsor Group, Mississippian, Frenchvale Brook (Balls Creek), Northwest Arm of Sydney Harbour, Nova Scotia.

Punctatisporites irrasus Hacquebard, 1957
Hypotype 73731
Utting, J., 1987, Geol. Surv. Can., Bull. 374, pl. 1, fig. 5.
Windsor Group, Lower Carboniferous, north shore of Port Hood Island, lat. 46°1'N, long. 61°26'W, Cape Breton Island, Nova Scotia.

Punctatisporites irrasus Hacquebard, 1957
Hypotype 76370
Utting, J., 1987, Palynology, vol. 11, p. 79, pl. 2, fig. 5.
Hiram Brook Member, Albert Formation, Lower Carboniferous, Boyd Creek, northeast of Gautreau Village, New Brunswick.

Punctatisporites minutus Ibrahim 1933
Hypotypes 16187, 16200
Barss, M.S., 1967, Geol. Surv. Can., Paper 67-11, pl. 20, fig. 27; pl. 22, fig. 2.
Pennsylvanian, Cumberland Group, Tingley pit in Roger's field near southern extension of McDougal Street, Springhill, Cumberland County; Morien Group, McAulay Cove coal seam, from outcrop on shore 1.75 miles south of Homeville, Cape Breton County, Nova Scotia.

Punctatisporites minutus Kosanke, 1950
Hypotype 73730
Utting, J., 1987, Geol. Surv. Can., Bull. 374, pl. 1, fig. 4.
Windsor Group, Lower Carboniferous, north shore of Port Hood Island, lat. 46°1'N, long. 61°26'W, Cape Breton Island, Nova Scotia.

Punctatisporites cf. *P. minutus* Ibrahim, 1933
Hypotype 16217
Barss, M.S., 1967, Geol. Surv. Can., Paper 67-11, pl. 22, fig. 19.
Morien Group, Pennsylvanian, Beaver mine on Tracey coal seam east of MacDonald Lake and south of Morrison road, approximately 5.0 miles from Highway 22, Sydney coalfield, Cape Breton County, Nova Scotia.

Punctatisporites nahannensis Hacquebard and Barss, 1957
Hypotype 15850
Barss, M.S., 1967, Geol. Surv. Can., Paper 67-11, pl. 7, fig. 15.
Mattson Formation, Mississippian, south face Twisted Mountain, about 15 miles northwest of South Nahanni, lat. 61°13'N. long. 123°37'W, Northwest Territories.

Punctatisporites obliquus Kosanke, 1950
Hypotype 16031
Barss, M.S., 1967, Geol. Surv. Can., Paper 67-11, pl. 15, fig. 5.
Port Hood Formation, Riversdale Group, Pennsylvanian, No. 5 coal seam, 1,300 feet east of Evans' coal mine on No. 2 seam, near Neil MacLeod Brook, St. Rose, Inverness County, Nova Scotia.

Punctatisporites obliquus Kosanke, 1950
Hypotype 16276
Barss, M.S., 1967, Geol. Surv. Can., Paper 67-11, pl. 25, fig. 10.
Mabou Mines Formation, Pennsylvanian, Angus MacPhee's pit, on north side of syncline in cliff section, 112 feet above beach, Mabou mines, Inverness County, Nova Scotia.

Punctatisporites obliquus Kosanke, 1950
Hypotype 16453
Barss, M.S., 1967, Geol. Surv. Can., Paper 67-11, pl. 32, fig. 2.
Pictou Group, Pennsylvanian, Scoudouc River 1 1/8 miles upstream from highway running south from Dorchester Crossing, New Brunswick.

Punctatisporites planus Hacquebard, 1957
Hypotypes 15942, 15943
Barss, M.S., 1967, Geol. Surv. Can., Paper 67-11, pl. 12, figs. 5, 6.
Canso Group, Upper Carboniferous, Pomquet River south of Route 4, Antigonish County, Nova Scotia.

Punctatisporites planus Hacquebard
Hypotype 62495
Utting, J., 1980, Can. J. Earth Sci., vol. 17, no. 8, pl. 2, fig. 1.
Lower Windsor Group, Mississippian, United States Borax and Chemical Corporation borehole SB-1 near Stewiacke, Nova Scotia.

Punctatisporites planus Hacquebard, 1957
Hypotype 55421
Utting, J., 1987, Geol. Surv. Can., Bull. 374, pl. 1, fig. 6.
Windsor Group, Lower Carboniferous, north shore of Port Hood Island, lat. 46°1'N, long. 61°26'W, Cape Breton Island, Nova Scotia.

Punctatisporites planus Hacquebard, 1957
Hypotype 76371
Utting, J., 1987, Palynology, vol. 11, p. 79, pl. 2, fig. 6.
Frederick Brook Member, Albert Formation, Lower Carboniferous, depth 164.60 m, Arco Dover No. 1A well, northwest of Gautreau Village, New Brunswick.

Punctatisporites platirugosus (Waltz) Sullivan, 1964
Hypotype 15941
Barss, M.S., 1967, Geol. Surv. Can., Paper 67-11, pl. 12, fig. 4.
Canso Group, Upper Carboniferous, Pomquet River south of Route 4, Antigonish County, Nova Scotia.

Punctatisporites cf. *P. platirugosus* (Waltz) Sullivan, 1964
Hypotype 15791
Barss, M.S., 1967, Geol. Surv. Can., Paper 67-11, pl. 5, fig. 5.
Windsor Group, Mississippian, Frenchvale Brook (Balls Creek), Northwest Arm of Sydney Harbour, Nova Scotia.

Punctatisporites punctatus Ibrahim, 1933
Hypotypes 15790, 16125
Barss, M.S., 1967, Geol. Surv. Can., Paper 67-11, pl. 5, fig. 4; pl. 18, fig. 27.
Windsor Group, Mississippian, Frenchvale Brook (Balls Creek), Northwest Arm of Sydney Harbour; Cumberland Group, Pennsylvanian, Pearly McFadden's mine on Chignecto coal seam, 0.3 miles east of Chignecto, Cumberland County, Nova Scotia.

Punctatisporites cf. *P. punctatus* Ibrahim, 1933
Hypotype 16352
Barss, M.S., 1967, Geol. Surv. Can., Paper 67-11, pl. 28, fig. 6.
Pictou Group, Pennsylvanian, stripping operation one mile south of Beersville, on east side of Coal Branch River, New Brunswick.

?Punctatisporites staplini Peppers, 1964
Hypotype 16286
Barss, M.S., 1967, Geol. Surv. Can., Paper 67-11, pl. 25, fig. 20.
Morien Group, Pennsylvanian, LeCras coal seam, from pit 200 feet southwest side of Highway 22, 1.1 miles from intersection of Old Mira Road and Highway 22, Sydney coalfield, Cape Breton County, Nova Scotia.

?*Punctatisporites verrucifer* Kosanke, 1950
Hypotype 16142
Barss, M.S., 1967, Geol. Surv. Can., Paper 67-11, pl. 19, fig. 10.
Cumberland Group, Pennsylvanian, Tingley pit in Roger's field near southern extension of McDougal Street, Springhill, Cumberland County, Nova Scotia.

Punctatisporites sp.
Fig. specs. 15796, 15846
Barss, M.S., 1967, Geol. Surv. Can., Paper 67-11, pl. 5, fig. 10; pl. 7, fig. 11.
Mississippian, Windsor Group, Frenchvale Brook (Balls Creek), Northwest Arm of Sydney Harbour, Nova Scotia; Mattson Formation, south face Twisted Mountain, about 15 miles northwest of South Nahanni, lat. 61°13′N. long. 123°37′W, Northwest Territories.

Punctatisporites sp.
Fig. spec. 15949
Barss, M.S., 1967, Geol. Surv. Can., Paper 67-11, pl. 12, fig. 12.
Canso Group, Upper Carboniferous, Pomquet River south of Route 4, Antigonish County, Nova Scotia.

Punctatisporites sp.
Fig. specs. 16271, 16356
Barss, M.S., 1967, Geol. Surv. Can., Paper 67-11, pl. 25, fig. 5; pl. 28, fig. 9.
Morien Group, Pennsylvanian, LeCras coal seam, from pit 200 feet southwest side of Highway 22, 1.1 miles from intersection of Old Mira Road and Highway 22, and Harbour coal seam, Dominion No. 12 colliery at face of No. 21 west level, Sydney coalfield, Cape Breton County, Nova Scotia.

Punctatosporites minutus Ibrahim 1933
Hypotypes 16067, 16216
Barss, M.S., 1967, Geol. Surv. Can., Paper 67-11, pl. 16, fig. 12 (pars); pl. 22, fig. 18.
Pennsylvanian, Port Hood Formation, Riversdale Group, 200 feet from mouth of adit of Harbourview coal mine, Port Hood, Inverness County; Morien Group, Beaver mine on Tracey coal seam east of MacDonald Lake and south of Morrison road, approximately 5 miles from Highway 22, Sydney coalfield, Cape Breton County, Nova Scotia.

Puctatosporites minutus Ibrahim, 1933
Hypotypes 16327, 16420
Barss, M.S., 1967, Geol. Surv. Can., Paper 67-11, pl. 27, fig. 5; pl. 30, fig. 17.
Pennsylvanian, Mabou Mines Formation, first coal seam northeast side of syncline, 84 feet below Stack sandstone at Coal Mine Point, Mabou mines, Inverness County, and Morien Group, Harbour coal seam, Dominion No. 12 colliery at face of No. 21 west level, Sydney coalfield, Cape Breton County Nova Scotia.

Punctatosporites minutus Ibrahim, 1933
Hypotype 16473
Barss, M.S., 1967, Geol. Surv. Can., Paper 67-11, pl. 32, fig. 22.
Pictou Group, Pennsylvanian, Shediac water well approximately 1,400 feet east of shore of harbour and 1,800 feet south of highway at Shediac, New Brunswick.

Punctatosporites cf. *P. minutus* Ibrahim, 1933
Hypotype 16422
Barss, M.S., 1967, Geol. Surv. Can., Paper 67-11, pl. 30, fig. 19.
Morien Group, Pennsylvanian, Harbour coal seam, Dominion No. 12 colliery at face of No. 21 west level, Sydney coalfield, Cape Breton County, Nova Scotia.

Punctatosporites sp.
Fig. spec. 16087
Barss, M.S., 1967, Geol. Surv. Can., Paper 67-11, pl. 16, fig. 32.
Port Hood Formation, Riversdale Group, Pennsylvanian, 200 feet from mouth of adit of Harbourview coal mine, Port Hood, Inverness County, Nova Scotia.

Punctatosporites sp.
Fig. spec. 16480
Barss, M.S., 1967, Geol. Surv. Can., Paper 67-11, pl. 32, fig. 29.
Pictou Group, Pennsylvanian, Shediac water well approximately 1,400 feet east of shore of harbour and 1,800 feet south of highway at Shediac, New Brunswick.

Pustulatisporites gibberosus (Hacquebard) Playford, 1964
Hypotype 89940
Utting, J., Keppie, J.D. and Giles, P.S., 1989, Geol. Surv. Can., Bull. 396, pl. 5.2, fig. 19.
Horton Bluff Formation, Horton Group, Lower Carboniferous, depth 29.25 m, Saarberg Interplan Canada Limited SI/8 drill core, north of west branch of Avon River, lat. 44°57′10″N, long. 64°15′50″W, Windsor Basin, Nova Scotia.

Pustulatisporites multicapitis Bertelsen
Hypotype 62485
Utting, J., 1980, Can. J. Earth Sci., vol. 17, no. 8, pl. 1, fig. 6.
Upper Windsor Group, Mississippian, United States Borax and Chemical Corporation borehole SB-1 near Stewiacke, Nova Scotia.

Pustulatisporites multicapitis Bertelsen, 1972
Hypotype 73739
Utting, J., 1987, Geol. Surv. Can., Bull. 374, pl. 1, fig. 15.
Windsor Group, Lower Carboniferous, north shore of Port Hood Island, lat. 46°1′N, long. 61°26′W, Cape Breton Island, Nova Scotia.

Pustulatisporites sp.
Fig. spec. 15957
Barss, M.S., 1967, Geol. Surv. Can., Paper 67-11, pl. 12, fig. 20.
Canso Group, Upper Carboniferous, Pomquet River south of Route 4, Antigonish County, Nova Scotia.

Pustulatisporites sp.
Fig. spec. 16043
Barss, M.S., 1967, Geol. Surv. Can., Paper 67-11, pl. 15, fig. 17.
Port Hood Formation, Riversdale Group, Pennsylvanian, 200 feet from mouth of adit of Harbourview coal mine, Port Hood, Inverness County, Nova Scotia.

Raistrickia aculeolata Wilson and Kosanke, 1944
Hypotype 16295
Barss, M.S., 1967, Geol. Surv. Can., Paper 67-11, pl. 25, fig. 29.
Mabou Mines Formation, Pennsylvanian, Angus MacPhee's pit, on north side of syncline in cliff section, 112 feet above beach, Mabou mines, Inverness County, Nova Scotia.

Raistrickia cf. *R. aculeata* Kosanke, 1950
Hypotypes 16371, 16373
Barss, M.S., 1967, Geol. Surv. Can., Paper 67-11, pl. 28, figs. 25, 27.
Morien Group, Pennsylvanian, Harbour coal seam, Dominion No. 12 colliery at face of No. 21 west level, Sydney coalfield, Cape Breton County, Nova Scotia.

Raistrickia baculosa Hacquebard, 1957
Hypotype 73747
Utting, J., 1987, Geol. Surv. Can., Bull. 374, p. 27, pl. 1, fig. 23.
Windsor Group, Lower Carboniferous, approximately 0.8 km downstream from Baddeck Bridge, lat. 46°9′N, long. 61°11′W, Nova Scotia.

Raistrickia baculosa Hacquebard, 1957
Hypotype 89941
Utting, J., Keppie, J.D. and Giles, P.S., 1989, Geol. Surv. Can., Bull. 396, pl. 5.2, fig. 20.
Horton Bluff Formation, Horton Group, Lower Carboniferous, depth 60.5 m, Saarberg Interplan Canada Limited SI/8 drill core, north of west branch of Avon River, lat. 44°57′10″N, long. 64°15′50″W, Windsor Basin, Nova Scotia.

Raistrickia clavata Hacquebard, 1957
Hypotype 62484
Utting, J., 1980, Can. J. Earth Sci., vol. 17, no. 8, pl. 1, fig. 5.
Lower Windsor Group, Mississippian, United States Borax and Chemical Corporation borehole SB-1 near Stewiacke, Nova Scotia.

Raistrickia clavata Hacquebard emend. Playford, 1964
Hypotype 73748
Utting, J., 1987, Geol. Surv. Can., Bull. 374, p. 27, pl. 1, fig. 24.
Windsor Group, Lower Carboniferous, north shore of Port Hood Island, lat. 46°1′N, long. 61°26′W, Cape Breton Island, Nova Scotia.

Raistrickia clavata Hacquebard emend. Playford 1964
Hypotype 76372
Utting, J., 1987, Palynology, vol. 11, p. 79, pl. 2, fig. 7.
Frederick Brook Member, Albert Formation, Lower Carboniferous, depth 107.08 m, Petro-Canada Dover No. 2 well, northeast of Gautreau Village, New Brunswick.

Raistrickia clavata Hacquebard, 1957
Hypotype 89942
Utting, J., Keppie, J.D. and Giles, P.S., 1989, Geol. Surv. Can., Bull. 396, pl. 5.2, fig. 15.
Horton Bluff Formation, Horton Group, Lower Carboniferous, depth 100 m, Saarberg Interplan Canada Limited SI/8 drill core, north of west branch of Avon River, lat. 44°57′10″N, long. 64°15′50″W, Windsor Basin, Nova Scotia.

Raistrickia corynoges Sullivan 1968
Hypotype 76373
Utting, J., 1987, Palynology, vol. 11, p. 79, pl. 2, fig. 8.

Frederick Brook Member, Albert Formation, Lower Carboniferous, depth 107.08 m, Petro-Canada Dover No. 2 well, northeast of Gautreau Village, New Brunswick.

Raistrickia corynoges Sullivan, 1968
Hypotype 89943
Utting, J., Keppie, J.D. and Giles, P.S., 1989, Geol. Surv. Can., Bull. 396, pl. 5.2, fig. 21.
Unnamed unit, Horton Group, Lower Carboniferous, Lakevale cliff north of syncline, St. Georges Bay, Antigonish Basin, Nova Scotia.

Raistrickia cf. *R. crinita* Kosanke, 1950
Hypotype 16299
Barss, M.S., 1967, Geol. Surv. Can., Paper 67-11, pl. 25, fig. 33.
Mabou Mines Formation, Pennsylvanian, first coal seam northeast side of syncline, 84 feet below Stack sandstone at Coal Mine Point, Mabou mines, Inverness County, Nova Scotia.

Raistrickia crocea Kosanke, 1950
Hypotypes 16163, 16166
Barss, M.S., 1967, Geol. Surv. Can., Paper 67-11, pl. 20, figs. 3, 6.
Cumberland Group, Pennsylvanian, No. 1 coal seam from 5,400 feet east level in No. 2 mine, Springhill, Cumberland County, Nova Scotia.

Raistrickia crocea Kosanke, 1950
Hypotype 16237
Barss, M.S., 1967, Geol. Surv. Can., Paper 67-11, pl. 23, fig. 2.
Morien Group, Pennsylvanian, LeCras coal seam, from pit 200 feet southwest side of Highway 22, 1.1 miles from intersection of Old Mira Road and Highway 22, Sydney coalfield, Cape Breton County, Nova Scotia.

Raistrickia crocea Kosanke, 1950
Hypotype 16378
Barss, M.S., 1967, Geol. Surv. Can., Paper 67-11, pl. 28, fig. 32.
Morien Group, Pennsylvanian, Phalen coal seam, Dominion No. 18 colliery at face of No. 6 west level, 3,400 feet from main deep, Sydney coalfield, Cape Breton County, Nova Scotia.

Raistrickia cf. *R. crocea* Kosanke, 1950
Hypotype 16472
Barss, M.S., 1967, Geol. Surv. Can., Paper 67-11, pl. 32, fig. 21.
Pictou Group, Pennsylvanian, Shediac water well approximately 1,400 feet east of shore of harbour and 1,800 feet south of highway at Shediac, New Brunswick.

Raistrickia golatensis Staplin, 1960
Hypotype 92909
Utting, J., Jachowicz, M. and Jachowicz, A., 1989, Geol. Surv. Can., Bull. 396, pl. 6.4, fig. 19.
Emma Fiord Formation, Early Carboniferous, Lyall River area, Grinnell Peninsula, Devon Island, District of Franklin.

Raistrickia irregularis Kosanke, 1950
Hypotype 16167
Barss, M.S., 1967, Geol. Surv. Can., Paper 67-11, pl. 20, fig. 7.
Cumberland Group, Pennsylvanian, bore-hole No. 1 east, in Roger's field near outcrop No. 6 seam at 40 foot depth, Springhill, Cumberland County, Nova Scotia.

Raistrickia magdalena Utting
Holotype 73744; hypotypes 73745, 73746
Utting, J., 1987, Geol. Surv. Can., Bull. 374, p. 28, pl. 1, figs. 20-22.
Windsor Group, Lower Carboniferous, north shore of Port Hood Island, lat. 46°1′N, long. 61°26′W (73744), and Cape Dauphin, lat. 46°21′N, long. 60°25′W, Cape Breton Island, Nova Scotia.

Raistrickia nigra Love, 1960
Hypotype 73749
Utting, J., 1987, Geol. Surv. Can., Bull. 374, pl. 1, fig. 25.
Windsor Group, Lower Carboniferous, Cape Dauphin, lat. 46°21′N, long. 60°25′W, Cape Breton Island, Nova Scotia.

Raistrickia nigra Love, 1960
Hypotype 94466
Utting, J., 1991, Geol. Surv. Can., Bull. 412, pl. 4.1, fig. 9.
Hart River Formation, Lower Carboniferous, north bank Peel River, northern Yukon.

Raistrickia ponderosa Playford, 1964
Hypotype 76374
Utting, J., 1987, Palynology, vol. 11, p. 79, pl. 2, fig. 9.
Frederick Brook Member, Albert Formation, Lower Carboniferous, depth 164.60 m, Arco Dover No. 1A well, northwest of Gautreau Village, New Brunswick.

Raistrickia ponderosa Playford, 1964
Hypotype 89947
Utting, J., Keppie, J.D. and Giles, P.S., 1989, Geol. Surv. Can., Bull. 396, pl. 5.2, fig. 25.
Horton Bluff Formation, Horton Group, Lower Carboniferous, depth 108 m, Saarberg Interplan Canada Limited SI/8 drill core, north of west branch of Avon River, lat. 44°57′10″N, long. 64°15′50″W, Windsor Basin, Nova Scotia.

Raistrickia sp. cf. *R. ponderosa* Playford, 1964
Fig. spec. 92911
Utting, J., Jachowicz, M. and Jachowicz, A., 1989, Geol. Surv. Can., Bull. 396, pl. 6.4, fig. 18.
Emma Fiord Formation, Early Carboniferous, Lyall River area, Grinnell Peninsula, Devon Island, District of Franklin.

Raistrickia protensa Kosanke, 1950
Hypotype 16236
Barss, M.S., 1967, Geol. Surv. Can., Paper 67-11, pl. 23, fig. 1.

Morien Group, Pennsylvanian, LeCras coal seam, from pit 200 feet southwest side of Highway 22, 1.1 miles from intersection of Old Mira Road and Highway 22, Sydney coalfield, Cape Breton County, Nova Scotia.

?*Raistrickia rubida* Kosanke, 1950
Hypotypes 16292, 16296
Barss, M.S., 1967, Geol. Surv. Can., Paper 67-11, pl. 25, fig. 26,30.
Mabou Mines Formation, Pennsylvanian, Angus MacPhee's pit, on north side of syncline in cliff section, 112 feet above beach, and coal seam 95 feet above beach, cliff section northeast side of syncline, Coal Mine Point, Mabou mines, Inverness County, Nova Scotia.

Raistrickia saetosa (Loose) Schopf, Wilson and Bentall, 1944
Hypotype 16054
Barss, M.S., 1967, Geol. Surv. Can., Paper 67-11, pl. 15, fig. 28.
Port Hood Formation, Riversdale Group, Pennsylvanian, 200 feet from mouth of adit of Harbourview coal mine, Port Hood, Inverness County, Nova Scotia.

Raistrickia saetosa (Loose) Potonié and Kremp, 1955
Hypotypes 16162, 16165
Barss, M.S., 1967, Geol. Surv. Can., Paper 67-11, pl. 20, figs. 2, 5.
Cumberland Group, Pennsylvanian, Pearly McFadden's mine on Chignecto coal seam, 0.3 mile east of Chignecto, and Fenwick mine on Chignecto coal seam, 2.1 miles east of Chignecto, Cumberland County, Nova Scotia.

Raistrickia saetosa (Loose) Schopf, Wilson and Bentall, 1944
Hypotype 16294
Barss, M.S., 1967, Geol. Surv. Can., Paper 67-11, pl. 25, fig. 28.
Mabou Mines Formation, Pennsylvanian, Angus MacPhee's pit, on north side of syncline in cliff section, 112 feet above beach, Mabou mines, Inverness County, Nova Scotia.

Raistrickia saetosa (Loose) Schopf, Wilson and Bentall, 1944
Hypotype 16471
Barss, M.S., 1967, Geol. Surv. Can., Paper 67-11, pl. 32, fig. 20.
Pictou Group, Pennsylvanian, Shediac water well approximately 1,400 feet east of shore of harbour and 1,800 feet south of highway at Shediac, New Brunswick.

Raistrickia saetosa? (Loose) Potonié and Kremp, 1955
Hypotype 16160
Barss, M.S., 1967, Geol. Surv. Can., Paper 67-11, pl. 19, fig. 28.
Cumberland Group, Pennsylvanian, Roger's field near outcrop of No. 6 coal seam at depth 40 feet, Springhill, Cumberland County, Nova Scotia.

Raistrickia cf. *R. superba* (Ibrahim) Schopf, Wilson and Bentall, 1944
Hypotype 15964
Barss, M.S., 1967, Geol. Surv. Can., Paper 67-11, pl. 12, fig. 27.
Canso Group, Upper Carboniferous, Pomquet River south of Route 4, Antigonish County, Nova Scotia.

Raistrickia cf. *R. superba* (Ibrahim) Schopf, Wilson and Bentall, 1944
Hypotype 16164
Barss, M.S., 1967, Geol. Surv. Can., Paper 67-11, pl. 20, fig. 4.
Cumberland Group, Pennsylvanian, Pearly McFadden's mine on Chignecto coal seam, 0.3 mile east of Chignecto, Cumberland County, Nova Scotia.

Raistrickia sp.
Fig. specs. 15904, 15962, 15963, 15965, 15966, 16052, 16168, 16240, 16293, 16376, 16377, 16629
Barss, M.S., 1967, Geol. Surv, Can., Paper 67-11, pl. 9, fig. 24; pl. 12, figs. 25, 26, 28, 29; pl. 15, fig. 26; pl. 20, fig. 8; pl. 23, fig. 5; pl. 25, fig. 27; pl. 28, figs. 30, 31; pl. 38, fig. 27.
Mattson Formation, Mississippian, south face Twisted Mountain, about 15 miles northwest of South Nahanni, lat. 61°13′N, long. 123°37′W, Northwest Territories (15904); Canso Group, Upper Carboniferous, Pomquet River south of Route 4, Antigonish County, and Pennsylvanian, Port Hood Formation, Riversdale Group, 200 feet from mouth of adit of Harbourview coal mine, Port Hood, Inverness County (16052); Cumberland Group, Pearly McFadden's mine on Chignecto coal seam, 0.3 mile east of Chignecto, Cumberland County (16168); Morien Group, Tracey coal seam, Beaver mine, east of MacDonald Lake and south of Morrison Raod, approximately 5 miles from Highway 22, Sydney Coalfield, Cape Breton County (16240); Mabou Mines Formation, Angus MacPhee's pit on north side of syncline in cliff section, 112 feet above beach, Mabou mines, Inverness County (16293); Morien Group, Harbour coal seam in Dominion No. 12 colliery, at face of No. 21 west level, Sydney coalfield, Cape Breton County (16376, 16377), Nova Scotia; Sabine Bay Formation, Permian, Tingmisut Lake, east side Sabine Peninsula, Melville Island, District of Franklin (16629).

?*Raistrickia* sp.
Fig. spec. 15978
Barss, M.S., 1967, Geol. Surv. Can., Paper 67-11, pl. 13, fig. 1.
Canso Group, Upper Carboniferous, Pomquet River south of Route 4, Antigonish County, Nova Scotia.

?*Raistrickia* sp.
Fig. spec. 16238
Barss, M.S., 1967, Geol. Surv. Can., Paper 67-11, pl. 23, fig. 3.
Skinner Brook Member, Stellarton Group, Pennsylvanian, northwest of Westville, 2,000 feet east of Skinner Brook, Pictou coalfield, Pictou County, Nova Scotia.

Reinschospora speciosa (Loose) Schopf, Wilson and Bentall, 1944
Hypotype 16400
Barss, M.S., 1967, Geol. Surv. Can., Paper 67-11, pl. 29, fig. 21.
Morien Group, Pennsylvanian, Harbour coal seam, Dominion No. 12 colliery at face of No. 21 west level, Sydney coalfield, Cape Breton County, Nova Scotia.

Reinschospora triangularis Kosanke, 1950
Hypotype 16399
Barss, M.S., 1967, Geol. Surv. Can., Paper 67-11, pl. 29, fig. 20.
Morien Group, Pennsylvanian, Harbour coal seam, Dominion No. 12 colliery at face of No. 21 west level, Sydney coalfield, Cape Breton County, Nova Scotia.

Remnant of trilete(?) spore
Fig. specs. 76389, 76390
Utting, J., 1987, Palynology, vol. 11, pl. 3, figs. 10, 11.
Lower Carboniferous, Dawson Settlement Member, Albert Formation, depth 263.6 m, Petro-Canada Dover No. 2 well, and Frederick Brook Member, Albert Formation, depth 319 m, Petro-Canada Dover No. 2 well, northeast of Gautreau Village, New Brunswick.

Remysporites albertensis Staplin, 1960
Hypotype 15933
Barss, M.S., 1967, Geol. Surv. Can., Paper 67-11, pl. 11, fig. 7.
Mattson Formation, Mississippian, south face Twisted Mountain, about 15 miles northwest of South Nahanni, lat. 61°13′N. long. 123°37′W, Northwest Territories.

Reticulatisporites cancellatus (Waltz) Playford, 1962
Hypotypes 15873, 15876
Barss, M.S., 1967, Geol. Surv. Can., Paper 67-11, pl. 8, figs. 10, 13.
Mattson Formation, Mississippian, south face Twisted Mountain, about 15 miles northwest of South Nahanni, lat. 61°13′N. long. 123°37′W, Northwest Territories.

Reticulatisporites cancellatus (Waltz) Playford, 1962
Hypotypes 92912, 92958
Utting, J., Jachowicz, M. and Jachowicz, A., 1989, Geol. Surv. Can., Bull. 396, pl. 6.4, fig. 16; pl. 6.6, fig. 27.
Emma Fiord Formation, Early Carboniferous, Lyall River area, Grinnell Peninsula, Devon Island, and Svartevaeg Cliffs, northern Axel Heiberg Island, District of Franklin.

Reticulatisporites cancellatus (Waltz) Playford, 1962
Hypotype 94502
Utting, J., 1991, Geol. Surv. Can., Bull. 412, pl. 4.2, fig. 18.
Hart River Formation, Lower Carboniferous, north bank Peel River, northern Yukon.

Reticulatisporites decoratus Hoffmeister, Staplin and Malloy, 1955
Hypotype 15882
Barss, M.S., 1967, Geol. Surv. Can., Paper 67-11, pl. 9, fig. 2.
Mattson Formation, Mississippian, south face Twisted Mountain, about 15 miles northwest of South Nahanni, lat. 61°13′N. long. 123°37′W, Northwest Territories.

Reticulatisporites lacunosus Kosanke, 1950
Hypotype 16247
Barss, M.S., 1967, Geol. Surv. Can., Paper 67-11, pl. 23, fig. 13.
Morien Group, Pennsylvanian, LeCras coal seam, from pit 200 feet southwest side of Highway 22, 1.1 miles from intersection of Old Mira Road and Highway 22, Sydney coalfield, Cape Breton County, Nova Scotia.

Reticulatisporites cf. *R. lacunosus* Kosanke, 1950
Hypotype 16404
Barss, M.S., 1967, Geol. Surv. Can., Paper 67-11, pl. 30, fig. 1.
Morien Group, Pennsylvanian, Point Aconi coal seam, on shore of Point Aconi, Sydney coalfield, Cape Breton County, Nova Scotia.

Reticulatisporites mediareticulatus (Ibrahim) Potonié and Kremp, 1955
Hypotype 16655
Barss, M.S., 1967, Geol. Surv. Can., Paper 67-11, pl. 23, fig. 12.
Skinner Brook Member, Stellarton Group, Pennsylvanian, northwest of Westville, 2,000 feet east of Skinner Brook, Pictou coalfield, Pictou County, Nova Scotia.

Reticulatisporites muricatus Kosanke, 1950
Hypotype 16244
Barss, M.S., 1967, Geol. Surv. Can., Paper 67-11, pl. 23, fig. 9.
Westville Member, Stellarton Group, Pennsylvanian, No. 1 mine, No. 2 north bord on No. 1 Balance, No. 7 north level, 150 feet north of slope, Pictou coalfield, Pictou County, Nova Scotia.

Reticulatisporites peltatus Playford, 1962
Hypotype 15880
Barss, M.S., 1967, Geol. Surv. Can., Paper 67-11, pl. 8, fig. 17.
Mattson Formation, Mississippian, south face Twisted Mountain, about 15 miles northwest of South Nahanni, lat. 61°13′N. long. 123°37′W, Northwest Territories.

Reticulatisporites cf. *R. peltatus* Playford, 1962
Hypotype 15881
Barss, M.S., 1967, Geol. Surv. Can., Paper 67-11, pl. 9, fig. 11.
Mattson Formation, Mississippian, south face Twisted Mountain, about 15 miles northwest of South Nahanni, lat. 61°13′N. long. 123°37′W, Northwest Territories.

Reticulatisporites polygonalis (Ibrahim) Loose, 1934
Hypotypes 16078, 16081, 16245, 16246
Barss, M.S., 1967, Geol. Surv. Can., Paper 67-11, pl. 16, figs. 23, 26; pl. 23, figs. 10, 11.
Pennsylvanian, Port Hood Formation, Riversdale Group, 200 feet from mouth of adit of Harbourview coal mine, Port Hood, Inverness County, and Skinner Brook Member, Stellarton Group, northwest of Westville, 2,200 and 2,000 feet east of Skinner Brook, Pictou coalfield, Pictou County, Nova Scotia.

Reticulatisporites polygonalis (Ibrahim) Loose, 1934
Hypotypes 16311, 16312
Barss, M.S., 1967, Geol. Surv. Can., Paper 67-11, pl. 26, figs. 12,13.

Mabou Mines Formation, Pennsylvanian, coal seam 95 feet above beach, cliff section northeast side of syncline, Coal Mine Point, and coal seam immediately below fault north of Coal Mine Point, Mabou mines, Inverness County, Nova Scotia.

Reticulatisporites cf. *R. polygonalis* (Ibrahim) Loose, 1934
Hypotype 15989
Barss, M.S., 1967, Geol. Surv. Can., Paper 67-11, pl. 13, fig. 12.
Canso Group, Upper Carboniferous, Pomquet River south of Route 4, Antigonish County, Nova Scotia.

Reticulatisporites rudis Staplin, 1960
Hypotype 92914
Utting, J., Jachowicz, M. and Jachowicz, A., 1989, Geol. Surv. Can., Bull. 396, pl. 6.4, fig. 17.
Emma Fiord Formation, Early Carboniferous, Lyall River area, Grinnell Peninsula, Devon Island, District of Franklin.

Reticulatisporites sp.
Fig. spec. 15872
Barss, M.S., 1967, Geol. Surv. Can., Paper 67-11, pl. 8, fig. 9.
Mattson Formation, Mississippian, south face Twisted Mountain, about 15 miles northwest of South Nahanni, lat. 61°13′N. long. 123°37′W, Northwest Territories.

Reticulatisporites sp.
Fig. specs. 16178, 16243
Barss, M.S., 1967, Geol. Surv. Can., Paper 67-11, pl. 20, fig. 18; pl. 23, fig. 8.
Pennsylvanian, Cumberland Group, Fenwick mine on Chignecto coal seam, 2.1 miles east of Chignecto, Cumberland County; Morien Group, McAulay Cove coal seam, from outcrop on shore 1.75 miles south of Homeville, Sydney coalfield, Cape Breton County, Nova Scotia.

?*Reticulatisporites* sp.
Fig. spec. 15878
Barss, M.S., 1967, Geol. Surv. Can., Paper 67-11, pl. 8, fig. 15.
Mattson Formation, Mississippian, south face Twisted Mountain, about 15 miles northwest of South Nahanni, lat. 61°13′N. long. 123°37′W, Northwest Territories.

Retusotriletes avonensis Playford, 1964
Hypotype 89944
Utting, J., Keppie, J.D. and Giles, P.S., 1989, Geol. Surv. Can., Bull. 396, pl. 5.2, fig. 22.
Cheverie Formation, Horton Group, Lower Carboniferous, west shore of Avon River, Windsor Basin, Nova Scotia.

Retusotriletes crassus Clayton, *in* Clayton et al., 1980
Hypotype 89945
Utting, J., Keppie, J.D. and Giles, P.S., 1989, Geol. Surv. Can., Bull. 396, pl. 5.2, fig. 23.
Unnamed formation, Horton Group, Lower Carboniferous, Lakevale cliff north of syncline, St. Georges Bay, Antigonish Basin, Nova Scotia.

Retusotriletes incohatus Sullivan, 1964
Hypotypes 15798, 15950
Barss, M.S., 1967, Geol. Surv. Can., Paper 67-11, pl. 5, fig. 12; pl. 12, fig. 13.
Windsor Group, Mississippian, Frenchvale Brook (Balls Creek), Northwest Arm of Sydney Harbour, and Canso Group, Upper Carboniferous, Pomquet River south of Route 4, Antigonish County, Nova Scotia.

Retusotriletes incohatus Sullivan
Hypotype 62486
Utting, J., 1980, Can. J. Earth Sci., vol. 17, no. 8, pl. 1, fig. 7.
Lower Windsor Group, Mississippian, United States Borax and Chemical Corporation borehole SB-1 near Stewiacke, Nova Scotia.

Retusotriletes incohatus Sullivan, 1964
Hypotype 73732
Utting, J., 1987, Geol. Surv. Can., Bull. 374, pl. 1, fig. 7.
Windsor Group, Lower Carboniferous, St. Ann's Bay, North Gut of St. Ann's, Victoria County, lat. 46°12′19″N, long. 60°37′44″W, Cape Breton Island, Nova Scotia.

Retusotriletes incohatus Sullivan 1964
Hypotype 76375
Utting, J., 1987, Palynology, vol. 11, p. 79, pl. 2, fig. 10.
Hiram Brook Member, Albert Formation, Lower Carboniferous, tributary from west, Boyd Creek, east of Gautreau Village, New Brunswick.

Retusotriletes incohatus Sullivan, 1964
Hypotype 89946
Utting, J., Keppie, J.D. and Giles, P.S., 1989, Geol. Surv. Can., Bull. 396, pl. 5.2, fig. 24.
Horton Bluff Formation, Horton Group, Lower Carboniferous, depth 37.75 m, Saarberg Interplan Canada Limited SI/8 drill core, north of west branch of Avon River, lat. 44°57′10″N, long. 64°15′50″W, Windsor Basin, Nova Scotia.

Retusotriletes incohatus Sullivan, 1964
Hypotype 94461
Utting, J., 1991, Geol. Surv. Can., Bull. 412, pl. 4.1, fig. 4.
Hart River Formation, Lower Carboniferous, north bank Peel River, northern Yukon.

Retusotriletes triangulatus (Streel) Streel
Hypotype 89948
Utting, J., Keppie, J.D. and Giles, P.S., 1989, Geol. Surv. Can., Bull. 396, pl. 5.3, fig. 2.
Rights River Formation, Horton Group, Lower Carboniferous, depth 73.15m, Drillhole DDH2, east of Beaver Settlement, lat. 45°48′12″N, long. 61°55′8″W, Antigonish Basin, Nova Scotia.

Retusotriletes sp.
Fig. spec. 15794
Barss, M.S., 1967, Geol. Surv. Can., Paper 67-11, pl. 5, fig. 8.
Windsor Group, Mississippian, Frenchvale Brook (Balls Creek), Northwest Arm of Sydney Harbour, Nova Scotia.

?*Retusotriletes* sp.
Fig. spec. 15803
Barss, M.S., 1967, Geol. Surv. Can., Paper 67-11, pl. 5, fig. 17.
Windsor Group, Mississippian, Frenchvale Brook (Balls Creek), Northwest Arm of Sydney Harbour, Nova Scotia.

Retusotriletes sp. A of Higgs, 1975
Fig. spec. 76376
Utting, J., 1987, Palynology, vol. 11, p. 79, pl. 2, fig. 11.
Hiram Brook Member, Albert Formation, Lower Carboniferous, tributary from west, Boyd Creek, east of Gautreau Village, New Brunswick.

Rotaspora fracta (Schemel) Smith and Butterworth, 1967
Hypotypes 92915, 92960
Utting, J., Jachowicz, M. and Jachowicz, A., 1989, Geol. Surv. Can., Bull. 396, pl. 6.4, fig. 15; pl 6.6, fig. 19.
Emma Fiord Formation, Early Carboniferous, Lyall River area, Grinnell Peninsula, Devon Island, and Kleybolte Penninsula, northwestern Ellesmere Island, District of Franklin.

Rotaspora fracta (Schemel) Smith and Butterworth, 1967
Hypotype 94458
Utting, J., 1991, Geol. Surv. Can., Bull. 412, pl. 4.1, fig. 1.
Hart River Formation, Lower Carboniferous, north bank Peel River, northern Yukon.

Rotaspora cf. *R. fracta* Schemel, 1950
Hypotype 15857
Barss, M.S., 1967, Geol. Surv. Can., Paper 67-11, pl. 7, fig. 23.
Mattson Formation, Mississippian, south face Twisted Mountain, about 15 miles northwest of South Nahanni, lat. 61°13′N. long. 123°37′W, Northwest Territories.

Rugospora corporata var. *verrucosa* Neville
Hypotype 62499
Utting, J., 1980, Can. J. Earth Sci., vol. 17, no. 8, pl. 2, fig. 5.
Upper Windsor Group, Mississippian, United States Borax and Chemical Corporation borehole SB-1 near Stewiacke, Nova Scotia.

Rugospora corporata Neves and Owens, var. *verrucosa* Neville, 1968
Hypotype 73795
Utting, J., 1987, Geol. Surv. Can., Bull. 374, pl. 3, fig. 18.
Windsor Group, Lower Carboniferous, north shore of Port Hood Island, lat. 46°1′N, long. 61°26′W, Cape Breton Island, Nova Scotia.

Rugospora corporata var. *verrucosa* Neville, 1968
Hypotype 92916
Utting, J., Jachowicz, M. and Jachowicz, A., 1989, Geol. Surv. Can., Bull. 396, pl. 6.5, fig. 1.
Emma Fiord Formation, Early Carboniferous, Lyall River area, Grinnell Peninsula, Devon Island, District of Franklin.

Rugospora minuta Neves and Ioannides
Hypotype 62494
Utting, J., 1980, Can. J. Earth Sci., vol. 17, no. 8, pl. 1, fig. 15.
Lower Windsor Group, Mississippian, United States Borax and Chemical Corporation borehole SB-1 near Stewiacke, Nova Scotia.

Rugospora minuta Neves and Ioannides, 1974
Hypotypes 73796, 73797
Utting, J., 1987, Geol. Surv. Can., Bull. 374, pl. 3, figs. 19, 20.
Windsor Group, Lower Carboniferous, east and west sides of headland, Knoydart Point, lat. 45°42′5″N, long. 62°13′6″W, Nova Scotia.

Rugospora minuta Neves and Ioannides, 1974
Hypotype 89949
Utting, J., Keppie, J.D. and Giles, P.S., 1989, Geol. Surv. Can., Bull. 396, pl. 5.3, fig. 3.
Wilkie Brook Formation, Horton Group, Lower Carboniferous, 380 m north of mouth of Wilkie Brook, western shore of St. Georges Bay, Antigonish Basin, Nova Scotia.

Rugospora minuta Neves and Ioannides, 1974
Hypotype 94473
Utting, J., 1991, Geol. Surv. Can., Bull. 412, pl. 4.1, fig. 16.
Hart River Formation, Lower Carboniferous, north bank Peel River, northern Yukon.

Rugospora polyptycha Neves and Ioannides, 1974
Hypotype 73798
Utting, J., 1987, Geol. Surv. Can., Bull. 374, pl. 3, fig. 21.
Windsor Group, Lower Carboniferous, north shore of Port Hood Island, lat. 46°1′N, long. 61°26′W, Cape Breton Island, Nova Scotia.

Rugospora polyptycha Neves and Ioannides, 1974
Hypotype 89950
Utting, J., Keppie, J.D. and Giles, P.S., 1989, Geol. Surv. Can., Bull. 396, pl. 5.3, fig. 5.
Wilkie Brook Formation, Horton Group, Lower Carboniferous, 380 m north of mouth of Wilkie Brook, western shore of St. Georges Bay, Antigonish Basin, Nova Scotia.

Rugospora polyptycha Neves and Ioannides, 1974
Hypotype 92917
Utting, J., Jachowicz, M. and Jachowicz, A., 1989, Geol. Surv. Can., Bull. 396, pl. 6.5, fig. 4.
Emma Fiord Formation, Early Carboniferous, Lyall River area, Grinnell Peninsula, Devon Island, District of Franklin.

Rugospora polyptycha Neves and Ioannides, 1974
Hypotype 94500
Utting, J., 1991, Geol. Surv. Can., Bull. 412, pl. 4.2, fig. 16.
Hart River Formation, Lower Carboniferous, north bank Peel River, northern Yukon.

Rugospora sp.
Fig. specs. 15821, 15822, 15824, 15825
Barss, M.S., 1967, Geol. Surv. Can., Paper 67-11, pl. 6, figs. 14, 15, 17, 18.
Windsor Group, Mississippian, Frenchvale Brook (Balls Creek), Northwest Arm of Sydney Harbour, Nova Scotia.

Rugospora sp.
Fig. spec. 94498
Utting, J., 1991, Geol. Surv. Can., Bull. 412, pl. 4.2, fig. 14.
Hart River Formation, Lower Carboniferous, north bank Peel River, northern Yukon.

Savitrisporites asperatus Sullivan, 1964
Hypotype 16003
Barss, M.S., 1967, Geol. Surv. Can., Paper 67-11, pl. 13, fig. 26.
Canso Group, Upper Carboniferous, Pomquet River south of Route 4, Antigonish County, Nova Scotia.

Savitrisporites asperatus Sullivan, 1964
Hypotypes 16171, 16172
Barss, M.S., 1967, Geol. Surv. Can., Paper 67-11, pl. 20, figs. 11, 12.
Cumberland Group, Pennsylvanian, Pearly McFadden's mine and Fenwick mine on Chignecto coal seam, 0.3 and 2.1 miles east of Chignecto, Cumberland County, Nova Scotia.

Savitrisporites nux (Butterworth and Williams) Sullivan, 1964
Hypotypes 16068, 16069, 16175, 16176
Barss, M.S., 1967, Geol. Surv. Can., Paper 67-11, pl. 16, figs. 13, 14; pl. 20, figs. 15, 16.
Pennsylvanian, Port Hood Formation, Riversdale Group, No. 5 coal seam, 1,300 feet east of Evans' coal mine on No. 2 seam, near Neil MacLeod Brook, St. Rose, and 200 feet from mouth of adit of Harbourview coal mine, Port Hood, Inverness County; Cumberland Group, bore-hole No. 1 east, in Roger's field near outcrop on No. 6 seam at 40 foot depth, Springhill and Jim Haley's mine on Chignecto coal seam, 1.1 miles west of Maccan, Cumberland County, Nova Scotia.

Savitrisporites nux (Butterworth and Williams) Sulivan, 1964
Hypotype 16226
Barss, M.S., 1967, Geol. Surv. Can., Paper 67-11, pl. 22, fig. 28.
Morien Group, Pennsylvanian, LeCras coal seam, from pit 200 feet southwest side of Highway 22, 1.1 miles from intersection of Old Mira Road and Highway 22, Sydney coalfield, Cape Breton County, Nova Scotia.

Savitrisporites sp.
Fig. spec. 16174
Barss, M.S., 1967, Geol. Surv. Can., Paper 67-11, pl. 20, fig. 14.
Cumberland Group, Pennsylvanian, Pearly McFadden's mine on Chignecto coal seam, 0.3 miles east of Chignecto, Cumberland County, Nova Scotia.

Savitrisporites sp.
Fig. specs. 16390, 16394
Barss, M.S., 1967, Geol. Surv. Can., Paper 67-11, pl. 29, figs. 11, 15.
Pennsylvanian, Pictou Group, stripping operation one mile south of Beersville, on east side of Coal Branch River, New Brunswick; Morien Group, Point Aconi coal seam, on shore at Point Aconi, Sydney coalfield, Cape Breton County, Nova Scotia.

Schopfipollenites acadiensis Utting
Holotype 73821; hypotypes 73822-73825
Utting, J., 1987, Geol. Surv. Can., Bull. 374, p. 38, pl. 5, figs. 7-11.
Windsor Group, Lower Carboniferous, north shore of Port Hood Island, lat. 46°1′N, long. 61°26′W, Cape Breton Island, and west side of headland, northeast of Knoydart Point, lat. 45°42′5″N, long. 62°13′6″W (73825), Nova Scotia.

Schopfipollenites cf. *S. ellipsoides* (Ibrahim) Potonié and Kremp
Hypotypes 62500, 62501
Utting, J., 1980, Can. J. Earth Sci., vol. 17, no. 8, pl. 2, figs. 6, 7.
Upper Windsor Group, Mississippian, United States Borax and Chemical Corporation borehole SB-1 near Stewiacke, Nova Scotia.

Schopfipollenites ellipsoides var. *corporeus* Neves, 1961
Hypotype 16097
Barss, M.S., 1967, Geol. Surv. Can., Paper 67-11, pl. 17, fig. 10.
Port Hood Formation, Riversdale Group, Pennsylvanian, 200 feet from mouth of adit of Harbourview coal mine, Port Hood, Inverness County, Nova Scotia.

Schopfipollenites sp.
Fig. specs. 16021, 16023-16025
Barss, M.S., 1967, Geol. Surv. Can., Paper 67-11, pl. 14, figs. 13, 15-17.
Canso Group, Upper Carboniferous, Pomquet River south of Route 4, Antigonish County, Nova Scotia.

Schopfipollenites sp.
Fig. specs. 16197, 16266
Barss, M.S., 1967, Geol. Surv. Can., Paper 67-11, pl. 21, fig. 6; pl. 24, fig. 14.
Pennsylvanian, Cumberland Group, Pearly McFadden's mine on Chignecto coal seam, 0.3 miles east of Chignecto, Cumberland County; McAulay Cove coal seam, from outcrop on shore 1.75 miles south of Homeville, Sydney coalfield, Cape Breton County, Nova Scotia.

Schopfipollenites sp.
Fig. specs. 16345, 16451
Barss, M.S., 1967, Geol. Surv. Can., Paper 67-11, pl. 27, fig. 23; pl. 31, fig. 14.
Pennsylvanian, Mabou Mines Formation, seven-foot coal seam in drift from cliff shore to old workings, south side of syncline, Mabou mines, Inverness County; Morien Group, coal seam 12 feet above Point Aconi seam at Point Aconi, Sydney coalfield, Cape Breton County, Nova Scotia.

Schopfipollenites sp.
Fig. spec. 16501
Barss, M.S., 1967, Geol. Surv. Can., Paper 67-11, pl. 33, fig. 16.
Pictou Group, Pennsylvanian, Shediac water well approximately 1,400 feet east of shore of harbour and 1,800 feet south of highway at Shediac, New Brunswick.

?*Schopfipollenites* sp.
Fig. spec. 16098
Barss, M.S., 1967, Geol. Surv. Can., Paper 67-11, pl. 17, fig. 11.
Port Hood Formation, Riversdale Group, Pennsylvanian, 200 feet from mouth of adit of Harbourview coal mine, Port Hood, Inverness County, Nova Scotia.

Schopfites augustus Playford, 1964
Hypotype 89951
Utting, J., Keppie, J.D. and Giles, P.S., 1989, Geol. Surv. Can., Bull. 396, pl. 5.3, fig. 6.
Horton Bluff Formation, Horton Group, Lower Carboniferous, depth 29.25 m, Saarberg Interplan Canada Limited SI/8 drill core, north of west branch of Avon River, lat. 44°57′10″N, long. 64°15′50″W, Windsor Basin, Nova Scotia.

Schopfites sp. cf. *S. augustus* Playford 1964
Fig. spec. 73677
Utting, J., 1987, Palynology, vol. 11, p. 79, pl. 2, fig. 12.
Hiram Brook Member, Albert Formation, Lower Carboniferous, Boyd Creek, east of Gautreau Village, New Brunswick.

Schopfites claviger Sullivan
Hypotype 62482
Utting, J., 1980, Can. J. Earth Sci., vol. 17, no. 8, pl. 1, fig. 3.
Lower Windsor Group, Mississippian, United States Borax and Chemical Corporation borehole SB-1 near Stewiacke, Nova Scotia.

Schopfites claviger Sullivan, 1968
Hypotype 73757
Utting, J., 1987, Geol. Surv. Can., Bull. 374, pl. 2, fig. 4.
Windsor Group, Lower Carboniferous, northeast of Knoydart Point, lat. 45°42′5″N, long. 62°13′6″W, Antigonish County, Nova Scotia.

Schopfites claviger Sullivan, 1968
Hypotype 89952
Utting, J., Keppie, J.D. and Giles, P.S., 1989, Geol. Surv. Can., Bull. 396, pl. 5.3, fig. 4.
Wilkie Brook Formation, Horton Group, Lower Carboniferous, 380 m north of mouth of Wilkie Brook, western shore of St. Georges Bay, Antigonish Basin, Nova Scotia.

Schopfites colchesterensis Kosanke, 1950
Hypotype 16401
Barss, M.S., 1967, Geol. Surv. Can., Paper 67-11, pl. 29, fig. 22.
Morien Group, Pennsylvanian, Harbour coal seam, Dominion No. 12 colliery at face of No. 21 west level, Sydney coalfield, Cape Breton County, Nova Scotia.

Schopfites delicatus Higgs, 1975
Hypotype 89953
Utting, J., Keppie, J.D. and Giles, P.S., 1989, Geol. Surv. Can., Bull. 396, pl. 5.3, fig. 7.
Wilkie Brook Formation, Horton Group, Lower Carboniferous, 380 m north of mouth of Wilkie Brook, western shore of St. Georges Bay, Antigonish Basin, Nova Scotia.

Schopfites cf. *S. dimorphus* Kosanke, 1950
Hypotype 16397
Barss, M.S., 1967, Geol. Surv. Can., Paper 67-11, pl. 29, fig. 18.
Morien Group, Pennsylvanian, Harbour coal seam, Dominion No. 12 colliery at face of No. 21 west level, Sydney coalfield, Cape Breton County, Nova Scotia.

Schulzospora bilunata Utting
Holotype 62504; hypotypes 73828, 73829
Utting, J., 1987, Geol. Surv. Can., Bull. 374, p. 37, pl. 5, figs. 15-17.
Windsor Group, Lower Carboniferous, Stewiacke borehole SB-1, Shubenacadie Basin, lat. 45°7′N, long. 63°20′W, Cape Breton Island, Nova Scotia.

Schulzospora cf. *S. elongata* Hoffmeister, Staplin and Malloy
Fig. spec. 62504
Utting, J., 1980, Can. J. Earth Sci., vol. 17, no. 8, pl. 2, fig. 10.
Upper Windsor Group, Mississippian, United States Borax and Chemical Corporation borehole SB-1 near Stewiacke, Nova Scotia.

Schulzospora rara Kosanke, 1950
Hypotypes 16090, 16191
Barss, M.S., 1967, Geol. Surv. Can., Paper 67-11, pl. 17, fig. 3; pl. 20, fig. 31.
Pennsylvanian, Port Hood Formation, Riversdale Group, 200 feet from mouth of adit of Harbourview coal mine, Port Hood, Inverness County; Cumberland Group, Styles mine on Chignecto coal seam, 5.4 miles east of Chignecto near Styles Brook, Cumberland County, Nova Scotia.

Scutulum trisupplementum Utting
Holotype 73772; hypotypes 73773-73775
Utting, J., 1987, Geol. Surv. Can., Bull. 374, p. 30, pl. 2, figs. 19-22.

Windsor Group, Lower Carboniferous, Cape Dauphin Cliff, lat. 46°21'N, long. 60°25'W, and north shore of Port Hood Island, lat. 46°1'N, long. 61°26'W (73775) Cape Breton Island, Nova Scotia.

Secarisporites lobatus Neves, 1967
Hypotype 73770
Utting, J., 1987, Geol. Surv. Can., Bull. 374, pl. 2, fig. 17.
Windsor Group, Lower Carboniferous, Stewiacke borehole SB-1, Shubenacadie Basin, lat. 47°7'N, long. 63°20'W, Hants County, Nova Scotia.

cf. *Secarisporites lobatus* Neves, 1961
Hypotypes 16141, 16145
Barss, M.S., 1967, Geol. Surv. Can., Paper 67-11, pl. 19, figs. 9, 13.
Cumberland Group, Pennsylvanian, No. 1 coal seam, from 5,400 foot east level in No. 2 mine, Springhill, Cumberland County, Nova Scotia.

Secarisporites? *remotus* Neves
Hypotype 62498
Utting, J., 1980, Can. J. Earth Sci., vol. 17, no. 8, pl. 2, fig. 4.
Upper Windsor Group, Mississippian, United States Borax and Chemical Corporation borehole SB-1 near Stewiacke, Nova Scotia.

Secarisporites remotus Neves, 1961
Hypotype 73771
Utting, J., 1987, Geol. Surv. Can., Bull. 374, pl. 2, fig. 18.
Windsor Group, Lower Carboniferous, north shore of Port Hood Island, lat. 46°1'N, long. 61°26'W, Cape Breton Island, Nova Scotia.

Secarisporites remotus Neves, 1961
Hypotype 92918
Utting, J., Jachowicz, M. and Jachowicz, A., 1989, Geol. Surv. Can., Bull. 396, pl. 6.5, fig. 2.
Emma Fiord Formation, Early Carboniferous, Lyall River area, Grinnell Peninsula, Devon Island, District of Franklin.

cf. *Secarisporites remotus* Neves, 1961
Hypotype 16137
Barss, M.S., 1967, Geol. Surv. Can., Paper 67-11, pl. 19, fig. 5.
Cumberland Group, Pennsylvanian, No. 1 coal seam, from 5,400 foot east level in No. 2 mine, Springhill, Cumberland County, Nova Scotia.

?*Secarisporites* sp.
Fig. specs. 15826-15835
Barss, M.S., 1967, Geol. Surv. Can., Paper 67-11, pl. 6, figs. 19-28.
Windsor Group, Mississippian, Frenchvale Brook (Balls Creek), Northwest Arm of Sydney Harbour, Nova Scotia.

?*Secarisporites* sp.
Fig. spec. 15996
Barss, M.S., 1967, Geol. Surv. Can., Paper 67-11, pl. 13, fig. 19.
Canso Group, Upper Carboniferous, Pomquet River south of Route 4, Antigonish County, Nova Scotia.

?*Secarisporites* sp.
Fig. spec. 16139
Barss, M.S., 1967, Geol. Surv. Can., Paper 67-11, pl. 19, fig. 7.
Cumberland Group, Pennsylvanian, Pearly McFadden's mine on Chignecto coal seam, 0.3 mile east of Chignecto, Cumberland County, Nova Scotia.

Speciososporites laevigatus Alpern, 1959
Hypotype 16334
Barss, M.S., 1967, Geol. Surv. Can., Paper 67-11, pl. 27, fig. 12.
Mabou Mines Formation, Pennsylvanian, Angus MacPhee's pit, on north side of syncline in cliff section, 112 feet above beach, Mabou mines, Inverness County, Nova Scotia.

Speciososporites minutus Alpern, 1959
Hypotypes 16332, 16421
Barss, M.S., 1967, Geol. Surv. Can., Paper 67-11, pl. 27 fig. 10; pl. 30, fig. 18.
Pennsylvanian, Mabou Mines Formation, eight-foot coal seam in drift from cliff to old workings, south side of syncline, Mabou mines, Inverness County, and Morien Group, Harbour coal seam, Dominion No. 12 colliery at face of No. 21 west level, Sydney coalfield, Cape Breton County, Nova Scotia.

Speciososporites triletoides Alpern, 1959
Hypotype 16423
Barss, M.S., 1967, Geol. Surv. Can., Paper 67-11, pl. 30, fig. 20.
Inverness Formation, Pennsylvanian, thirty-four inch coal seam at Inverness near mine entrance, Inverness coalfield, Inverness County, Nova Scotia.

Spelaeotriletes arenaceus Neves and Owens, 1966
Hypotypes 92919, 92920, 92961
Utting, J., Jachowicz, M. and Jachowicz, A., 1989, Geol. Surv. Can., Bull. 396, pl. 6.5, figs. 5, 6; pl. 6.6, fig. 29.
Emma Fiord Formation, Early Carboniferous, Lyall River area, Grinnell Peninsula, Devon Island, and Svartevaeg Cliffs, northern Axel Heiberg Island (92961), District of Franklin.

Spelaeotriletes balteatus (Playford) Higgs, 1975
Hypotype 92921
Utting, J., Jachowicz, M. and Jachowicz, A., 1989, Geol. Surv. Can., Bull. 396, pl. 6.5, fig. 3.
Emma Fiord Formation, Early Carboniferous, Lyall River area, Grinnell Peninsula, Devon Island, District of Franklin.

Spelaeotriletes balteatus Playford, 1963
Hypotype 94472
Utting, J., 1991, Geol. Surv. Can., Bull. 412, pl. 4.1, fig. 15.
Hart River Formation, Lower Carboniferous, north bank Peel River, northern Yukon.

Spelaeotriletes cabottii Utting
Holotype 89954; hypotypes 89955-89960
Utting, J., Keppie, J.D. and Giles, P.S., 1989, Geol. Surv. Can., Bull. 396, p. 133, pl. 5.3, figs. 8-13, 15.
Unnamed unit, Horton Group, Lower Carboniferous, Lakevale cliff north of syncline, St. Georges Bay, lat. 45°47′55″N, long. 61°54′45″W, Antigonish Basin, Nova Scotia.

Spelaeotriletes echinatus Hacquebard comb. nov. and emend.
Holotype 73807; hypotypes 73808, 73809
Utting, J., 1987, Geol. Surv. Can., Bull. 374, p. 33, pl. 4, figs. 9-12.
Windsor Group, Lower Carboniferous, north shore of Port Hood Island, lat. 46°1′N, long. 61°26′W, Cape Breton Island, Nova Scotia.

Spelaeotriletes echinatus (Hacquebard) Utting, 1987
Hypotypes 89961, 89962
Utting, J., Keppie, J.D. and Giles, P.S., 1989, Geol. Surv. Can., Bull. 396, pl. 5.3, figs. 14, 16.
Wilkie Brook Formation, Horton Group, Lower Carboniferous, 380 m north of mouth of Wilkie Brook, western shore of St. Georges Bay, Antigonish Basin, Nova Scotia.

Spelaeotriletes pretiosus (Playford) Neves and Belt 1970
Hypotypes 76378, 76379
Utting, J., 1987, Palynology, vol. 11, p. 80, pl. 2, figs. 13, 14.
Hiram Brook Member, Albert Formation, Lower Carboniferous, Boyd Creek, east and northeast of Gautreau Village, New Brunswick.

Spelaeotriletes pretiosus (Playford) Neves and Belt emend. var. *bellii* Utting
Holotype 73800; hypotypes 73801, 73802
Utting, J., 1987, Geol. Surv. Can., Bull. 374, p. 34, pl. 4, figs. 2-4.
Windsor Group, Lower Carboniferous, Stewiacke borehole SB-1, Shubenacadie Basin, lat. 47°7′N, long. 63°20′W, stream southwest of North Gut of St′ Ann′s harbour, lat. 46°12′19″N, long. 60°37′44″W, and north shore of Port Hood Island, lat. 46°1′N, long. 61°26′W, Cape Breton Island, Nova Scotia.

Spelaeotriletes pretiosus (Playford) Neves and Belt var. *bellii* Utting, 1987
Hypotype 89967
Utting, J., Keppie, J.D. and Giles, P.S., 1989, Geol. Surv. Can., Bull. 396, pl. 5.4, fig. 3.
Chevrie Formation, Horton Group, Lower Carboniferous, west shore of Avon River, Windsor Basin, Nova Scotia.

Spelaeotriletes pretiosus (Playford) Neves and Belt emend. var. *pretiosus*
Hypotype 73799
Utting, J., 1987, Geol. Surv. Can., Bull. 374, p. 34, pl. 4, fig. 1.
Cheverie Formation, Horton Group, Lower Carboniferous, east shore of Avon River, Cheverie, lat. 45°9′N, long. 64°11′30″W, Nova Scotia.

Spelaeotriletes pretiosus (Playford) Neves and Belt var. *pretiosus*
Hypotypes 89963-89966
Utting, J., Keppie, J.D. and Giles, P.S., 1989, Geol. Surv. Can., Bull. 396, pl. 5.3, figs. 17, 18; pl. 5.4, figs. 1, 2.
Chevrie Formation, Horton Group, Lower Carboniferous, west shore of Avon River, Windsor Basin, Nova Scotia.

Spelaeotriletes pretiosus (Playford) Neves and Belt emend. var. *windsorensis* Utting
Holotype 73803; hypotypes 73802-73806
Utting, J., 1987, Geol. Surv. Can., Bull. 374, p. 34, pl. 4, figs. 5-8.
Windsor Group, Lower Carboniferous, north shore of Port Hood Island, lat. 46°1′N, long. 61°26′W, Cape Breton Island, and east and west of headland, Knoydart Point, lat. 45°42′5″N, long. 62°13′6″W, Nova Scotia.

Spelaeotriletes tuberosus Utting
Holotype 73810; hypotypes 73811-73813
Utting, J., 1987, Geol. Surv. Can., Bull. 374, p. 35, pl. 4, figs. 13-16.
Windsor Group, Lower Carboniferous, north shore of Port Hood Island, lat. 46°1′N, long. 61°26′W, Cape Breton Island, and Knoydart Point, lat. 45°42′5″N, long. 62°13′6″W, (73812, 73813), Nova Scotia.

Spelaeotriletes tuberosus Utting, 1987
Hypotype 92922
Utting, J., Jachowicz, M. and Jachowicz, A., 1989, Geol. Surv. Can., Bull. 396, pl. 6.5, fig. 7.
Emma Fiord Formation, Early Carboniferous, Lyall River area, Grinnell Peninsula, Devon Island, District of Franklin.

Spelaeotriletes tuberosus Utting, 1987
Hypotype 94477
Utting, J., 1991, Geol. Surv. Can., Bull. 412, pl. 4.1, fig. 20.
Hart River Formation, Lower Carboniferous, north bank Peel River, northern Yukon.

Spinozonotriletes balteatus Playford, 1963
Hypotype 15932
Barss, M.S., 1967, Geol. Surv. Can., Paper 67-11, pl. 11, fig. 6.
Mattson Formation, Mississippian, south face Twisted Mountain, about 15 miles northwest of South Nahanni, lat. 61°13′N. long. 123°37′W, Northwest Territories.

Spinozonotriletes uncatus Hacquebard, 1957
Hypotype 16026
Barss, M.S., 1967, Geol. Surv. Can., Paper 67-11, pl. 14, fig. 18.
Canso Group, Upper Carboniferous, Pomquet River south of Route 4, Antigonish County, Nova Scotia.

Stenozonotriletes clarus Ischenko, 1958
Hypotype 15885
Barss, M.S., 1967, Geol. Surv. Can., Paper 67-11, pl. 9, fig. 5.
Mattson Formation, Mississippian, south face Twisted Mountain, about 15 miles northwest of South Nahanni, lat. 61°13′N. long. 123°37′W, Northwest Territories.

Stenozonotriletes clarus Ishchenko, 1958
Hypotype 92923
Utting, J., Jachowicz, M. and Jachowicz, A., 1989, Geol. Surv. Can., Bull. 396, pl. 6.5, fig. 8.
Emma Fiord Formation, Early Carboniferous, Lyall River area, Grinnell Peninsula, Devon Island, District of Franklin.

Stenozonotriletes facilis Ischenko, 1956
Hypotype 15801
Barss, M.S., 1967, Geol. Surv. Can., Paper 67-11, pl. 5, fig. 15.
Windsor Group, Mississippian, Frenchvale Brook (Balls Creek), Northwest Arm of Sydney Harbour, Nova Scotia.

Stenozonotriletes facilis Ischenko, 1956
Hypotype 16007
Barss, M.S., 1967, Geol. Surv. Can., Paper 67-11, pl. 13, fig. 30.
Canso Group, Upper Carboniferous, Pomquet River south of Route 4, Antigonish County, Nova Scotia.

Stenozonotriletes cf. *S. perforatus* Playford, 1962
Hypotype 15886
Barss, M.S., 1967, Geol. Surv. Can., Paper 67-11, pl. 9, fig. 6.
Mattson Formation, Mississippian, south face Twisted Mountain, about 15 miles northwest of South Nahanni, lat. 61°13′N. long. 123°37′W, Northwest Territories.

Stenozonotriletes sp.
Fig. spec. 15802
Barss, M.S., 1967, Geol. Surv. Can., Paper 67-11, pl. 5, fig. 16.
Windsor Group, Mississippian, Frenchvale Brook (Balls Creek), Northwest Arm of Sydney Harbour, Nova Scotia.

Striatoabietites sp.
Fig. specs. 16535, 16537, 16539, 16542, 16543
Barss, M.S., 1967, Geol. Surv. Can., Paper 67-11, pl. 35, figs. 18, 20, 22, 25, 26.
Pictou Group, Permian, depths 332 (16535, 16543) and 335 feet, Gallas Point borehole No. 2 on property of H. Tweedy, Gallas Point, lat. 46°7′42″N, long. 63°57′28″W, Prince Edward Island.

Striatoabietites sp.
Fig. specs. 16571, 16583
Barss, M.S., 1967, Geol. Surv. Can., Paper 67-11, pl. 37, figs. 5, 17.
Permian, Tatonduck River, west limb of syncline 3 miles east of Yukon-Alaska border, lat. 64°58′N, long. 140°54′W, Yukon.

?*Striatoabietites* sp.
Fig. spec. 16526
Barss, M.S., 1967, Geol. Surv. Can., Paper 67-11, pl. 35, figs. 8.
Pictou Group, Permian, depth 332, Gallas Point borehole No. 2 on property of H. Tweedy, Gallas Point, lat. 46°7′42″N, long. 63°57′28″W, Prince Edward Island.

?*Striatoabieites* sp.
Fig. specs. 75771, 75773
Utting, J., 1985, Bull. Can. Petrol. Geol., vol. 33, no. 3, pl. 2, figs. 4, 6.
Otto Fiord Formation, Upper Carboniferous, 3.5 km northeast of van Hauen Pass, north shore Hare Fiord, Ellesmere Island, District of Franklin.

?*Striatoabieites* sp.; taenia visible
Fig. spec. 75776
Utting, J., 1985, Bull. Can. Petrol. Geol., vol. 33, no. 3, pl. 2, fig. 9.
Otto Fiord Formation, Upper Carboniferous, 3.5 km northeast of van Hauen Pass, north shore Hare Fiord, Ellesmere Island, District of Franklin.

Striomonosaccites sp.
Fig. spec. 16488
Barss, M.S., 1967, Geol. Surv. Can., Paper 67-11, pl. 33, fig. 3.
Pictou Group, Pennsylvanian, small northerly tributary of Scoudouc River, 1/2 mile west of old dam and 1 1/2 miles south of town of Shediac, New Brunswick.

Striomonosaccites sp.
Fig. spec. 16577
Barss, M.S., 1967, Geol. Surv. Can., Paper 67-11, pl. 37, fig. 11.
Permian, Tatonduck River, west limb of syncline 3 miles east of Yukon-Alaska border, lat. 64°58′N, long. 140°54′W, Yukon.

Striatopodocarpites cf. *S. cancellatus* (Balme and Hennelly) Hart, 1963
Hypotype 16586
Barss, M.S., 1967, Geol. Surv. Can., Paper 67-11, pl. 37, fig. 20.
Permian, Tatonduck River, west limb of syncline 3 miles east of Yukon-Alaska border, lat. 64°58′N, long. 140°54′W, Yukon.

Striatopodocarpites cf. *S. divaricata* (Wilson) Hart, 1964
Hypotype 16582
Barss, M.S., 1967, Geol. Surv. Can., Paper 67-11, pl. 37, fig. 16.
Permian, Tatonduck River, west limb of syncline 3 miles east of Yukon-Alaska border, lat. 64°58′N, long. 140°54′W, Yukon.

Striatopodocarpites sp.
Fig. spec. 16546
Barss, M.S., 1967, Geol. Surv. Can., Paper 67-11, pl. 36, fig. 3.
Pictou Group, Permian, depth 350 feet, Gallas Point borehole No. 2 on property of H. Tweedy, Gallas Point, lat. 46°7′42″N, long. 63°57′28″W, Prince Edward Island.

Striatopodocarpites sp.
Fig. specs. 16573, 16575
Barss, M.S., 1967, Geol. Surv. Can., Paper 67-11, pl. 37, figs. 7, 9.
Permian, Tatonduck River, west limb of syncline 3 miles east of Yukon-Alaska border, lat. 64°58′N, long. 140°54′W, Yukon.

Taeniaesporites sp.
Fig. spec. 16497
Barss, M.S., 1967, Geol. Surv. Can., Paper 67-11, pl. 33, fig. 12.
Pictou Group, Pennsylvanian, Shediac water well approximately 1,400 feet east of shore of harbour and 1,800 feet south of highway at Shediac, New Brunswick.

Tendosporites suborenatus (Waltz) Hacquebard and Barss, 1957
Hypotype 15918
Barss, M.S., 1967, Geol. Surv. Can., Paper 67-11, pl. 10, fig. 13.
Mattson Formation, Mississippian, south face Twisted Mountain, about 15 miles northwest of South Nahanni, lat. 61°13′N. long. 123°37′W, Northwest Territories.

Tetraporina horologia (Staplin) Playford, 1963
Hypotype 15928
Barss, M.S., 1967, Geol. Surv. Can., Paper 67-11, pl. 11, fig. 2.
Mattson Formation, Mississippian, south face Twisted Mountain, about 15 miles northwest of South Nahanni, lat. 61°13′N. long. 123°37′W, Northwest Territories.

cf. *Tholisporites foveolatus* Hughes and Playford, 1961
Hypotype 15910
Barss, M.S., 1967, Geol. Surv. Can., Paper 67-11, pl. 10, fig. 5.
Mattson Formation, Mississippian, south face Twisted Mountain, about 15 miles northwest of South Nahanni, lat. 61°13′N. long. 123°37′W, Northwest Territories.

Thymospora obscura (Kosanke) Wilson and Venkatchala, 1963
Hypotypes 16427, 16428
Barss, M.S., 1967, Geol. Surv. Can., Paper 67-11, pl. 30, figs. 24, 25.
Morien Group, Pennsylvanian, Harbour coal seam, Dominion No. 12 colliery at face of No. 21 west level, and Phalen coal seam, Dominion No. 18 colliery at face of No. 6 west level, Sydney coalfield, Cape Breton County, Nova Scotia.

Thymospora perverrucosa (Alpern) Wilson and Venkatchala, 1963
Hypotypes 16430, 16431
Barss, M.S., 1967, Geol. Surv. Can., Paper 67-11, pl. 30, figs. 27, 28.
Pennsylvanian, Pictou Group, stripping operation of Wasson Company, one mile north of Coal Branch Station, New Brunswick; Morien Group, Harbour coal seam, Dominion No. 12 colliery at face of No. 21 west level, Sydney coalfield, Cape Breton County, Nova Scotia.

Thymospora perverrucosa (Alpern) Wilson and Venkatachala, 1963
Hypotype 16476
Barss, M.S., 1967, Geol. Surv. Can., Paper 67-11, pl. 32, fig. 25.
Pictou Group, Pennsylvanian, Shediac water well approximately 1,400 feet east of shore of harbour and 1,800 feet south of highway at Shediac, New Brunswick.

Thymospora cf. *T. thiessenii* (Kosanke) Wilson and Venkatachala, 1963
Hypotype 16432
Barss, M.S., 1967, Geol. Surv. Can., Paper 67-11, pl. 30, fig. 29.
Morien Group, Pennsylvanian, Harbour coal seam, Dominion No. 12 colliery at face of No. 21 west level, Sydney coalfield, Cape Breton County, Nova Scotia.

Thymospora verrucosa (Alpern) Wilson and Venkatachala, 1963
Hypotype 16429
Barss, M.S., 1967, Geol. Surv. Can., Paper 67-11, pl. 30, fig. 26.
Morien Group, Pennsylvanian, Harbour coal seam, Dominion No. 12 colliery at face of No. 21 west level, Sydney coalfield, Cape Breton County, Nova Scotia.

Torispora cf. *T. granulatus* Alpern, 1959
Hypotype 16434
Barss, M.S., 1967, Geol. Surv. Can., Paper 67-11, pl. 30, fig. 31.
Morien Group, Pennsylvanian, Harbour coal seam, Dominion No. 12 colliery at face of No. 21 west level, Sydney coalfield, Cape Breton County, Nova Scotia.

Torispora laevigata Bhardwaj, 1957
Hypotypes 16333, 16433
Barss, M.S., 1967, Geol. Surv. Can., Paper 67-11, pl. 27, fig. 11; pl. 30, fig. 30.
Pennsylvanian, Mabou Mines Formation, Colonel Snow's adit at beach level, northeast side of syncline at Coal Mine Point, approximately 440 feet west of gypsum fault, Mabou mines, Inverness County, and Morien Group, Harbour coal seam, Dominion No. 12 colliery at face of No. 21 west level, Sydney coalfield, Cape Breton County, Nova Scotia.

Torispora securis (Balme) Alpern, Doubinger and Horst, 1965
Hypotypes 16335, 16658, 16435
Barss, M.S., 1967, Geol. Surv. Can., Paper 67-11, pl. 27, fig. 13; pl. 30, fig. 32.
Morien Group, Pennsylvanian, LeCras coal seam, from pit 200 feet southwest side of Highway 22, 1.1 miles from intersection of Old Mira Road and Highway 22, and

Harbour coal seam, Dominion No. 12 colliery at face of No. 21 west level (16435), Sydney coalfield, Cape Breton County, Nova Scotia.

Torispora verrucosus Alpern, 1959
Hypotypes 16436, 16437
Barss, M.S., 1967, Geol. Surv. Can., Paper 67-11, pl. 30, figs. 33, 34.
Morien Group, Pennsylvanian, Harbour coal seam, Dominion No. 12 colliery at face of No. 21 west level, Sydney coalfield, Cape Breton County, Nova Scotia.

Tricidarisporites arcuatus Neville, 1973
Hypotype 73734
Utting, J., 1987, Geol. Surv. Can., Bull. 374, p. 34, pl. 1, fig. 10.
Windsor Group, Lower Carboniferous, Stewiacke borehole SB-1, Shubenacadie Basin, lat. 47°7′N, long. 63°20′W, Hants County, Nova Scotia.

Tricidarisporites arcuatus Neville, 1973
Hypotypes 92924, 92925
Utting, J., Jachowicz, M. and Jachowicz, A., 1989, Geol. Surv. Can., Bull. 396, pl. 6.5, figs. 13, 14.
Emma Fiord Formation, Early Carboniferous, Lyall River area, Grinnell Peninsula, Devon Island, District of Franklin.

Tricidarisporites magnificus Neville, 1973
Hypotype 92962
Utting, J., Jachowicz, M. and Jachowicz, A., 1989, Geol. Surv. Can., Bull. 396, pl. 6.6, fig. 25.
Emma Fiord Formation, Early Carboniferous, Svartevaeg Cliffs, northern Axel Heiberg Island, District of Franklin.

Tripartites incisotrilobus (Naumova) Potonié and Kremp, 1956
Hypotype 15884
Barss, M.S., 1967, Geol. Surv. Can., Paper 67-11, pl. 9, fig. 4.
Mattson Formation, Mississippian, south face Twisted Mountain, about 15 miles northwest of South Nahanni, lat. 61°13′N. long. 123°37′W, Northwest Territories.

Tripartites incisotrilobus (Naumova) Potonié and Kremp, 1956
Hypotype 92927
Utting, J., Jachowicz, M. and Jachowicz, A., 1989, Geol. Surv. Can., Bull. 396, pl. 6.5, fig. 10.
Emma Fiord Formation, Early Carboniferous, Lyall River area, Grinnell Peninsula, Devon Island, District of Franklin.

Tripartites incisotrilobus (Naumova) Karczewska and Turnau, 1974
Hypotype 94469
Utting, J., 1991, Geol. Surv. Can., Bull. 412, pl. 4.1, fig. 12.
Hart River Formation, Lower Carboniferous, north bank Peel River, northern Yukon.

Tripartites pressuens (Ishchenko) Agrali and Kongali, 1969
Hypotype 92926
Utting, J., Jachowicz, M. and Jachowicz, A., 1989, Geol. Surv. Can., Bull. 396, pl. 6.5, fig. 9.
Emma Fiord Formation, Early Carboniferous, Lyall River area, Grinnell Peninsula, Devon Island, District of Franklin.

Triquitrites additus Wilson and Hoffmeister, 1956
Hypotypes 16301, 16388
Barss, M.S., 1967, Geol. Surv. Can., Paper 67-11, pl. 26, fig. 2; pl. 29, fig. 9.
Pennsylvanian, Mabou Mines Formation, first coal seam northeast side of syncline, 84 feet below Stack sandstone at Coal Mine Point, Mabou mines, Inverness County, and Morien Group, Phalen coal seam, Dominion No. 18 colliery at face of No. 6 west level, 3,400 feet from main deep, Sydney coalfield, Cape Breton County, Nova Scotia.

Triquitrites arculatus Wilson and Coe, 1940
Hypotypes 16063, 16065
Barss, M.S., 1967, Geol. Surv. Can., Paper 67-11, pl. 16, figs. 9, 11.
Port Hood Formation, Riversdale Group, Pennsylvanian, 200 feet from mouth of adit of Harbourview coal mine, Port Hood, Inverness County, Nova Scotia.

Triquitrites batillatus Hughes and Playford, 1961
Hypotype 15887
Barss, M.S., 1967, Geol. Surv. Can., Paper 67-11, pl. 9, fig. 7.
Mattson Formation, Mississippian, south face Twisted Mountain, about 15 miles northwest of South Nahanni, lat. 61°13′N. long. 123°37′W, Northwest Territories.

Triquitrites bransonii Wilson and Hoffmeister, 1956
Hypotypes 16066, 16224
Barss, M.S., 1967, Geol. Surv. Can., Paper 67-11, pl. 16, fig. 12; pl. 22, fig. 26.
Pennsylvanian, Port Hood Formation, Riversdale Group, 200 feet from mouth of adit of Harbourview coal mine, Port Hood, Inverness County, and Morien Group, first coal seam below Tracey seam one mile west of Homeville, approximately 1,300 feet south of Broughton-Homeville road, Sydney coalfield, Cape Breton County, Nova Scotia.

Triquitrites bransonii Wilson and Hoffmeister, 1956
Hypotypes 16303, 16385, 16389
Barss, M.S., 1967, Geol. Surv. Can., Paper 67-11, pl. 26, fig. 4; pl. 29, figs. 6,10.
Pennsylvanian, Mabou Mines Formation, first coal seam northeast side of syncline, 84 feet below Stack sandstone at Coal Mine Point, Mabou mines, Inverness County, and Morien Group, Phalen coal seam, Dominion No. 18 colliery at face of No. 6 west level, 3,400 feet from main deep, Sydney coalfield, Cape Breton County, Nova Scotia.

Triquitrites bransonii Wilson and Hoffmeister, 1956
Hypotypes 16461, 16462
Barss, M.S., 1967, Geol. Surv. Can., Paper 67-11, pl. 32, figs. 10, 11.

Pictou Group, Pennsylvanian, Scoudouc River 1 1/8 miles upstream from highway running south from Dorchester Crossing, and Shediac water well approximately 1,400 feet east of shore of harbour and 1,800 feet south of highway at Shediac, New Brunswick.

Triquitrites cf. *T. crassus* Kosanke, 1950
Hypotypes 16304, 16387
Barss, M.S., 1967, Geol. Surv. Can., Paper 67-11, pl. 26, fig. 5; pl. 29, fig. 8.
Pennsylvanian, Mabou Mines Formation, coal seam immediately below fault north of Coal Mine Point, Mabou mines; Inverness Formation, thirty-four inch coal seam at Inverness near mine entrance, Inverness coalfield, Inverness County, Nova Scotia.

Triquitrites cf. *T. crassus* Kosanke, 1950
Hypotype 16464
Barss, M.S., 1967, Geol. Surv. Can., Paper 67-11, pl. 32, fig. 13.
Pictou Group, Pennsylvanian, Scoudouc River 1 1/8 miles upstream from highway running south from Dorchester Crossing, New Brunswick.

Triquitrites dividuus Wilson and Hoffmeister, 1956
Hypotype 16229
Barss, M.S., 1967, Geol. Surv. Can., Paper 67-11, pl. 22, fig. 31.
Morien Group, Pennsylvanian, Beaver mine on Tracey coal seam east of MacDonald Lake and south of Morrison road, approximately 5.0 miles from Highway 22, Sydney coalfield, Cape Breton County, Nova Scotia.

Triquitrites cf. *T. novicus* Bhardwaj, 1957
Hypotype 16381
Barss, M.S., 1967, Geol. Surv. Can., Paper 67-11, pl. 29, fig. 2.
Morien Group, Pennsylvanian, Phalen coal seam, Dominion No. 18 colliery at face of No. 6 west level, 3,400 feet from main deep, Sydney coalfield, Cape Breton County, Nova Scotia.

Triquitrites pulvinatus Kosanke, 1950
Hypotypes 16302, 16380, 16386
Barss, M.S., 1967, Geol. Surv. Can., Paper 67-11, pl. 26, fig. 3; pl. 29, figs. 1,7.
Pennsylvanian, Mabou Mines Formation, slope of fifteen-foot coal seam, south side of syncline, 150 feet from entrance of mine, Mabou mines, Inverness County (16302); Morien Group, Phalen coal seam, Dominion No. 18 colliery at face of No. 6 west level, 3,400 feet from main deep, and coal seam 93 feet above Harbour seam on shore northwest of New Waterford, Sydney coalfield, Cape Breton County, Nova Scotia.

Triquitrites sculptilis? Balme, 1952
Hypotypes 16225, 16228, 16230
Barss, M.S., 1967, Geol. Surv. Can., Paper 67-11, pl. 22, figs. 27, 30, 32.
Morien Group, Pennsylvanian, Beaver mine on Tracey coal seam east of MacDonald Lake and south of Morrison road, approximately 5.0 miles from Highway 22, Sydney coalfield, Cape Breton County, Nova Scotia.

Triquitrites cf. *T. spinosus* Kosanke, 1943
Hypotype 16383
Barss, M.S., 1967, Geol. Surv. Can., Paper 67-11, pl. 29, fig. 4.
Morien Group, Pennsylvanian, Phalen coal seam, Dominion No. 18 colliery at face of No. 6 west level, 3,400 feet from main deep, Sydney coalfield, Cape Breton County, Nova Scotia.

Triquitrites sp.
Fig. specs. 16382, 16384
Barss, M.S., 1967, Geol. Surv. Can., Paper 67-11, pl. 29, figs. 3, 5.
Morien Group, Pennsylvanian, Phalen coal seam, Dominion No. 18 colliery at face of No. 6 west level, 3,400 feet from main deep, and Harbour coal seam, Dominion No. 12 colliery at face of No. 21 west level, Sydney coalfield, Cape Breton County, Nova Scotia.

?*Triquitrites* sp.
Fig. spec. 15883
Barss, M.S., 1967, Geol. Surv. Can., Paper 67-11, pl. 9, fig. 3.
Mattson Formation, Mississippian, south face Twisted Mountain, about 15 miles northwest of South Nahanni, lat. 61°13′N. long. 123°37′W, Northwest Territories.

Tumoripollenites sp.
Fig. spec. 16585
Barss, M.S., 1967, Geol. Surv. Can., Paper 67-11, pl. 37, fig. 19.
Permian, Tatonduck River, west limb of syncline 3 miles east of Yukon-Alaska border, lat. 64°58′N, long. 140°54′W, Yukon.

Umbonatisporites abstrusus (Playford) Clayton 1970
Hypotype 76380
Utting, J., 1987, Palynology, vol. 11, p. 80, pl. 3, fig. 1.
Hiram Brook Member, Albert Formation, Lower Carboniferous, Boyd Creek, northeast of Gautreau Village, New Brunswick.

Umbonatisporites abstrusus (Playford) Clayton 1970
Hypotype 89968
Utting, J., Keppie, J.D. and Giles, P.S., 1989, Geol. Surv. Can., Bull. 396, pl. 5.4, fig. 4.
Horton Bluff Formation, Horton Group, Lower Carboniferous, depth 51 m, Saarberg Interplan Canada Limited SI/8 drill core, north of west branch of Avon River, lat. 44°57′10″N, long. 64°15′50″W, Windsor Basin, Nova Scotia.

Umbonatisporites distinctus Clayton 1971
Hypotypes 76381, 76382
Utting, J., 1987, Palynology, vol. 11, p. 80, pl. 3, figs. 2, 3.
Hiram Brook Member, Albert Formation, Lower Carboniferous, Boyd Creek, northeast of Gautreau Village, New Brunswick.

Umbonatisporites distinctus Clayton, 1971
Hypotypes 89969, 89970
Utting, J., Keppie, J.D. and Giles, P.S., 1989, Geol. Surv. Can., Bull. 396, pl. 5.4, figs. 5, 6.
Lower Carboniferous, Horton Bluff Formation, Horton Group, depth 75 m, Saarberg Interplan Canada Limited SI/8 drill core, north of west branch of Avon River, lat. 44°57′10″N, long. 64°15′50″W, Windsor Basin, and unnamed unit, Lakevale cliff north of syncline, St. Georges Bay, Antigonish Basin, Nova Scotia.

Umbonatisporites sp. cf. *U. distinctus* Clayton 1971
Fig. spec. 92910
Utting, J., Jachowicz, M. and Jachowicz, A., 1989, Geol. Surv. Can., Bull. 396, pl. 6.4, fig. 12.
Emma Fiord Formation, Early Carboniferous, Lyall River area, Grinnell Peninsula, Devon Island, District of Franklin.

Unidentified
Fig. specs. 15816, 15971, 16008, 16018-16020, 16022, 16073, 16170, 16195, 16239, 16338, 16515, 16530, 16635, 16646, 16651
Barss, M.S., 1967, Geol. Surv, Can., Paper 67-11, pl. 6, fig. 9; pl. 12, fig. 34; pl. 13, fig. 31; pl. 14, figs. 10-12, 14; pl. 16, fig. 18; pl. 20, fig. 10; pl. 21, fig. 4; pl. 23, fig. 4; pl. 27, fig. 16; pl. 34, fig. 14; pl. 35, figs. 12, 13; pl. 38, figs. 33, 44, 49.
Windsor Group, Mississippian, Frenchvale Brook (Balls Creek), Northwest Arm of Sydney Harbour (15816); Canso Group, Upper Carboniferous, Pomquet River south of Route 4, Antigonish County; Pennsylvanian, Port Hood Formation, Riversdale Group, No. 5 coal seam, 1,300 feet east of Evans′ mine on No. 2 coal seam, near Neil MacLeod Brook, St. Rose, Inverness County (16073), Cumberland Group, Tingley pit in Roger′s field near southern extension of McDougal Street, Springfield (16170) and Morrison′s pit on Chignecto coal seam, 0.4 miles east of Chignecto (16195), Cumberland County; Morien Group, McAulay Cove coal seam from outcrop on shore, 1.75 miles south of Homeville, Sydney coalfield, Cape Breton County (16239) and Angus MacPhee′s pit in cliff section 112 feet above beach, Mabou mines, Inverness County (16338), Nova Scotia; Permian, Pictou Group, depth 335 and 332 feet, Gallas Point borehole No. 2, lat. 46°7′42″N, long. 63°57′28″W, Prince Edward Island (16515, 16530); Sabine Bay Formation, Tingmisut Lake, east side of Sabine Peninsula, Melville Island, District of Franklin (16635, 16646, 16651).

Unidentified
Fig. specs. 16338, 16398
Barss, M.S., 1967, Geol. Surv. Can., Paper 67-11, pl. 27, fig. 16; pl. 29, fig. 19.
Pennsylvanian, Mabou Mines Formation, Angus MacPhee′s pit, on north side of syncline in cliff section, 112 feet above beach, Mabou mines, Inverness County; Morien Group, coal seam 93 feet above Harbour seam on shore northwest of New Waterford, Sydney coalfield, Cape Breton County, Nova Scotia.

Unidentified
Fig. specs. 16515, 16530, 16635, 16646, 16651
Barss, M.S., 1967, Geol. Surv. Can., Paper 67-11, pl. 34, fig. 14; pl. 35, figs. 12, 13; pl. 38, figs. 33, 44, 49.
Pictou Group, Permian, depths 335 and 332 (16530) feet, Gallas Point borehole No. 2 on property of H. Tweedy, Gallas Point, lat. 46°7′42″N, long. 63°57′28″W, Prince Edward Island; Sabine Bay Formation, vicinity of Tingmisut Lake, east side of Sabine Peninsula, Melville Island, District of Franklin.

Vallatisporites ciliaris (Luber) Sullivan
Hypotype 62488
Utting, J., 1980, Can. J. Earth Sci., vol. 17, no. 8, pl. 1, fig. 9.
Lower Windsor Group, Mississippian, United States Borax and Chemical Corporation borehole SB-1 near Stewiacke, Nova Scotia.

Vallatisporites ciliaris (Luber) Sullivan, 1964
Hypotype 62488
Utting, J., 1987, Geol. Surv. Can., Bull. 374, pl. 3, fig. 4.
Windsor Group, Lower Carboniferous, Stewiacke borehole SB-1, Shubenacadie Basin, lat. 45°7′N, long. 63°20′W, Hants County, Nova Scotia.

Vallatisporites ciliaris (Luber) Sullivan, 1964
Hypotype 89971
Utting, J., Keppie, J.D. and Giles, P.S., 1989, Geol. Surv. Can., Bull. 396, pl. 5.4, fig. 12.
Wilkie Brook Formation, Horton Group, Lower Carboniferous, 380 m north of mouth of Wilkie Brook, western shore of St. Georges Bay, Antigonish Basin, Nova Scotia.

Vallatisporites galearis Sullivan 1964
Hypotype 76383
Utting, J., 1987, Palynology, vol. 11, p. 80, pl. 3, fig. 4.
Hiram Brook Member, Albert Formation, Lower Carboniferous, Boyd Creek, northeast of Gautreau Village, New Brunswick.

Vallatisporites vallatus Hacquebard 1957
Hypotype 76384
Utting, J., 1987, Palynology, vol. 11, p. 80, pl. 3, fig. 5.
Hiram Brook Member, Albert Formation, Lower Carboniferous, Boyd Creek, east of Gautreau Village, New Brunswick.

Vallatisporites vallatus Hacquebard, 1957
Hypotypes 89972, 89973
Utting, J., Keppie, J.D. and Giles, P.S., 1989, Geol. Surv. Can., Bull. 396, pl. 5.4, figs. 7, 8.
Lower Carboniferous, Horton Bluff Formation, Horton Group, depth 29.25 m, Saarberg Interplan Canada Limited SI/8 drill core, north of west branch of Avon River, lat. 44°57′10″N, long. 64°15′50″W, Windsor Basin, and Wilkie Brook Formation, Horton Group, 380 m north of mouth of Wilkie Brook, western shore of St. Georges Bay, Antigonish Basin, Nova Scotia.

Vallatisporites verrucosus Hacquebard, 1957
Hypotype 73782
Utting, J., 1987, Geol. Surv. Can., Bull. 374, pl. 3, fig. 5.
Windsor Group, Lower Carboniferous, Stewiacke borehole SB-1, Shubenacadie Basin, lat. 45°7′N, long. 63°20′W, Hants County, Nova Scotia.

Vallatisporites verrucosus Hacquebard, 1957
Hypotype 76385
Utting, J., 1987, Palynology, vol. 11, p. 80, pl. 3, fig. 6.
Hiram Brook Member, Albert Formation, Lower Carboniferous, Boyd Creek, northeast of Gautreau Village, New Brunswick.

Vallatisporites verrucosus Hacquebard 1957
Hypotype 89974
Utting, J., Keppie, J.D. and Giles, P.S., 1989, Geol. Surv. Can., Bull. 396, pl. 5.4, fig. 13.
Rights River Formation, Horton Group, Lower Carboniferous, depth 73.15 m, Beaver Settlement DDH2, lat. 45°48′12″N, long. 61°55′8″W, Antigonish Basin, Nova Scotia.

Vallatisporites sp.
Fig. specs. 92928, 92929
Utting, J., Jachowicz, M. and Jachowicz, A., 1989, Geol. Surv. Can., Bull. 396, pl. 6.5, figs. 11, 12.
Emma Fiord Formation, Early Carboniferous, Lyall River area, Grinnell Peninsula, Devon Island, District of Franklin.

Verrucosisporites congestus Playford 1964
Hypotype 76386
Utting, J., 1987, Palynology, vol. 11, p. 80, pl. 3, fig. 7.
Dawson Settlement Member, Albert Formation, Lower Carboniferous, depth 166.8 m, Petro-Canada Dover No. 1 well, north of Gautreau Village, lat. 45°59′15″N, long. 65°39′W, New Brunswick.

Verrucosisporites congestus Playford, 1964
Hypotype 89975
Utting, J., Keppie, J.D. and Giles, P.S., 1989, Geol. Surv. Can., Bull. 396, pl. 5.4, fig. 9.
Horton Bluff Formation, Horton Group, Lower Carboniferous, depth 84 m, Saarberg Interplan Canada Limited SI/8 drill core, north of west branch of Avon River, lat. 44°57′10″N, long. 64°15′50″W, Windsor Basin, Nova Scotia.

Verrucosisporites donarii Potonié and Kremp, 1955
Hypotype 16360
Barss, M.S., 1967, Geol. Surv. Can., Paper 67-11, pl. 28, fig. 14.
Morien Group, Pennsylvanian, Harbour coal seam, Dominion No. 12 colliery at face of No. 21 west level, Sydney coalfield, Cape Breton County, Nova Scotia.

Verrucosisporites cf. *V. donarii* Potonié and Kremp, 1955
Hypotype 15956
Barss, M.S., 1967, Geol. Surv. Can., Paper 67-11, pl. 12, fig. 19.
Canso Group, Upper Carboniferous, Pomquet River south of Route 4, Antigonish County, Nova Scotia.

Verrucosisporites sp. cf. *V. eximius* Playford, 1962
Fig. spec. 92930
Utting, J., Jachowicz, M. and Jachowicz, A., 1989, Geol. Surv. Can., Bull. 396, pl. 6.5, fig. 19.
Emma Fiord Formation, Early Carboniferous, Lyall River area, Grinnell Peninsula, Devon Island, District of Franklin.

Verrucosisporites cf. *V. microtuberosus* (Loose) Smith and Butterworth, in Smith, 1966
Hypotype 16143
Barss, M.S., 1967, Geol. Surv. Can., Paper 67-11, pl. 19, fig. 11.
Cumberland Group, Pennsylvanian, Morrison's pit on Chignecto coal seam, 0.4 mile east of Chignecto, Cumberland County, Nova Scotia.

Verrucosisporites cf. *V. microtuberosus* (Loose) Smith and Butterworth, in Smith, 1966
Hypotype 16221
Barss, M.S., 1967, Geol. Surv. Can., Paper 67-11, pl. 22, fig. 23.
Morien Group, Pennsylvanian, McAulay Cove coal seam, from outcrop on shore 1.75 miles south of Homeville, Sydney coalfield, Cape Breton County, Nova Scotia.

Verrucosisporites cf. *V. microtuberosus* (Loose) Smith and Butterworth, in Smith, 1966
Hypotype 16280
Barss, M.S., 1967, Geol. Surv. Can., Paper 67-11, pl. 25, fig. 14.
Morien Group, Pennsylvanian, LeCras coal seam, from pit 200 feet southwest side of Highway 22, 1.1 miles from intersection of Old Mira Road and Highway 22, Sydney coalfield, Cape Breton County, Nova Scotia.

Verrucosisporites morulatus (Knox) Smith and Butterworth, 1967
Hypotype 73754
Utting, J., 1987, Geol. Surv. Can., Bull. 374, pl. 2, fig 1.
Windsor Group, Lower Carboniferous, north shore of Port Hood Island, lat. 46°1′N, long. 61°26′W, Cape Breton Island, Nova Scotia.

Verrucosisporites nitidus Playford
Hypotype 62487
Utting, J., 1980, Can. J. Earth Sci., vol. 17, no. 8, pl. 1, fig. 8.
Lower Windsor Group, Mississippian, United States Borax and Chemical Corporation borehole SB-1 near Stewiacke, Nova Scotia.

Verrucosisporites nitidus (Naumova) Playford, 1964
Hypotype 73755
Utting, J., 1987, Geol. Surv. Can., Bull. 374, pl. 2, fig. 2.
Windsor Group, Lower Carboniferous, north shore of Port Hood Island, lat. 46°1′N, long. 61°26′W, Cape Breton Island, Nova Scotia.

Verrucosisporites nitidus Playford, 1964
Hypotype 76387
Utting, J., 1987, Palynology, vol. 11, p. 80, pl. 3, fig. 8.
Hiram Brook Member, Albert Formation, Lower Carboniferous, Boyd Creek, northeast of Gautreau Village, New Brunswick.

Verrucosisporites nitidus Playford, 1964
Hypotypes 89976, 89977
Utting, J., Keppie, J.D. and Giles, P.S., 1989, Geol. Surv. Can., Bull. 396, pl. 5.4, figs. 10, 14.
Unnamed unit, Horton Group, Lower Carboniferous, Lakevale cliff north of syncline, St. Georges Bay, lat. 45°47′55″N, long. 61°54′45″W, Antigonish Basin, Nova Scotia.

Verrucosisporites papulosus Hacquebard, 1957
Hypotypes 73750-73753
Utting, J., 1987, Geol. Surv. Can., Bull. 374, p. 28, pl. 1, figs. 26-29.
Windsor Group, Lower Carboniferous, north shore of Port Hood Island, lat. 46°1′N, long. 61°26′W, Cape Breton Island, and northeast of Knoydart Point, lat. 45°42′5″N, long. 62°13′6″W (73753), Nova Scotia.

Verrucosisporites papulosus Hacquebard, 1957
Hypotype 76388
Utting, J., 1987, Palynology, vol. 11, p. 80, pl. 3, fig. 9.
Hiram Brook Member, Albert Formation, Lower Carboniferous, Boyd Creek, northeast of Gautreau Village, New Brunswick.

Verrucosisporites papulosus Hacquebard, 1957
Hypotype 89978
Utting, J., Keppie, J.D. and Giles, P.S., 1989, Geol. Surv. Can., Bull. 396, pl. 5.4, fig. 11.
Unnamed unit, Horton Group, Lower Carboniferous, Lakevale cliff north of syncline, St. Georges Bay, Antigonish Basin, Nova Scotia.

Verrucosisporites papulosus Hacquebard, 1957
Hypotype 92931
Utting, J., Jachowicz, M. and Jachowicz, A., 1989, Geol. Surv. Can., Bull. 396, pl. 6.5, fig. 15.
Emma Fiord Formation, Early Carboniferous, Lyall River area, Grinnell Peninsula, Devon Island, District of Franklin.

Verrucosisporites cf. *V. perverrucosus* (Loose) Potonié and Kremp, 1955
Hypotype 16159
Barss, M.S., 1967, Geol. Surv. Can., Paper 67-11, pl. 19, fig. 27.
Cumberland Group, Pennsylvanian, Pearly McFadden′s mine on Chignecto coal seam, 0.3 mile east of Chignecto, Cumberland County, Nova Scotia.

Verrucosisporites cf. *V. sifati* (Ibrahim) Smith and Butterworth, in Smith, 1966
Hypotype 16300
Barss, M.S., 1967, Geol. Surv. Can., Paper 67-11, pl. 26, fig. 1.
Morien Group, Pennsylvanian, LeCras coal seam, from pit 200 feet southwest side of Highway 22, 1.1 miles from intersection of Old Mira Road and Highway 22, Sydney coalfield, Cape Breton County, Nova Scotia.

Verrucosisporites verrucosus (Ibrahim) Ibrahim, 1933
Hypotype 73756
Utting, J., 1987, Geol. Surv. Can., Bull. 374, pl. 2, fig. 3.
Windsor Group, Lower Carboniferous, north shore of Port Hood Island, lat. 46°1′N, long. 61°26′W, Cape Breton Island, Nova Scotia.

Verrucosisporites sp.
Fig. specs. 15806, 15863, 15867, 16049, 16053, 16055
Barss, M.S., 1967, Geol. Surv. Can., Paper 67-11, pl. 5, fig. 20; pl. 7, fig. 29; pl. 8, fig. 4; pl. 15, figs. 23, 27, 29, 30.
Mississippian, Windsor Group, Frenchvale Brook (Balls Creek), Northwest Arm of Sydney Harbour, Nova Scotia (15806); Mattson Formation, south face Twisted Mountain, about 15 miles northwest of South Nahanni, lat. 61°13′N. long. 123°37′W, Northwest Territories (15863, 15867); Port Hood Formation, Riversdale Group, Pennsylvanian, 200 feet from mouth of adit of Harbourview coal mine, Port Hood, Inverness County, Nova Scotia.

Verrucosisporites sp.
Fig. specs. 16152, 16153
Barss, M.S., 1967, Geol. Surv. Can., Paper 67-11, pl. 19, figs. 20, 21.
Cumberland Group, Pennsylvanian, bore-hole no. 1 east, in Roger′s field near outcrop of No. 6 coal seam, at depth 40 feet, Springhill, and Morrison′s pit on Chignecto coal seam, 0.4 miles esat of Chignecto, Cumberland County, Nova Scotia.

Verrucosisporites sp.
Fig. specs. 16231, 16234
Barss, M.S., 1967, Geol. Surv. Can., Paper 67-11, pl. 22, figs. 33, 36.
Pennsylvanian, Westville Member, Stellarton Group, No. 1 mine, No. 2 north bord on No. 1 Balance, No. 7 north level, 150 feet north of slope, Pictou coalfield, Pictou County, and Morien Group, McAulay Cove coal seam, from outcrop on shore 1.75 miles south of Homeville, Sydney coalfield, Cape Breton County, Nova Scotia.

Verrucosisporites sp.
Fig. spec. 94462
Utting, J., 1991, Geol. Surv. Can., Bull. 412, pl. 4.1, fig. 5.
Hart River Formation, Lower Carboniferous, north bank Peel River, northern Yukon.

?*Verrucosisporites* sp.
Fig. spec. 16144
Barss, M.S., 1967, Geol. Surv. Can., Paper 67-11, pl. 19, fig. 12.
Cumberland Group, Pennsylvanian, No. 7 coal seam, from 3,200 foot mine bord 150 feet east of back auxiliary slope in No. 4 mine, Springhill, Cumberland County, Nova Scotia.

?*Verrucososporites facierugosus* (Loose) Butterworth and Williams, 1954
Hypotype 15994
Barss, M.S., 1967, Geol. Surv. Can., Paper 67-11, pl. 13, fig. 17.
Genso Group, Upper Carboniferous, Pomquet River south of Route 4, Antigonish County, Nova Scotia.

Vesicaspora cf. *V. ovata* (Balme and Hennelly) Hart, 1960
Hypotype 16434
Barss, M.S., 1967, Geol. Surv. Can., Paper 67-11, pl. 32, fig. 33.
Pictou Group, Pennsylvanian, Shediac water well approximately 1,400 feet east of shore of harbour and 1,800 feet south of highway at Shediac, New Brunswick.

Vesicaspora sp.
Fig. specs. 16442, 16443
Barss, M.S., 1967, Geol. Surv. Can., Paper 67-11, pl. 31, figs. 5, 6.
Pennsylvanian, Pictou Group, stripping operation one mile south of Beersville, on east side of Coal Branch River, New Brunswick; Phalen coal seam, Dominion No. 18 colliery at face of No. 6 west level, 3,400 feet from main deep, Sydney coalfield, Cape Breton County, Nova Scotia.

Vesicaspora sp.
Fig. specs. 16491, 16495
Barss, M.S., 1967, Geol. Surv. Can., Paper 67-11, pl. 33, figs. 6, 10.
Pictou Group, Pennsylvanian, Scoudouc River 1 1/8 miles upstream from highway running south from Dorchester Crossing, New Brunswick.

Vesicaspora sp.
Fig. spec. 16525
Barss, M.S., 1967, Geol. Surv. Can., Paper 67-11, pl. 35, fig. 7.
Pictou Group, Permian, depth 332 feet, Gallas Point borehole No. 2 on property of H. Tweedy, Gallas Point, lat. 46°7′42″N, long. 63°57′28″W, Prince Edward Island.

?*Vesicaspora* sp.
Fig. spec. 16561
Barss, M.S., 1967, Geol. Surv. Can., Paper 67-11, pl. 36, fig. 18.
Permian, Tatonduck River, west limb of syncline 3 miles east of Yukon-Alaska border, lat. 64°58′N, long. 140°54′W, Yukon.

Vestigisporites balmei? Hart, 1960
Hypotype 16649
Barss, M.S., 1967, Geol. Surv. Can., Paper 67-11, pl. 38, fig. 47.
Sabine Bay Formation, Permian, vicinity of Tingmisut Lake, east side of Sabine Peninsula, Melville Island, District of Franklin.

Vestigisporites cf. *V. methoris* Hart, 1960
Hypotype 16563
Barss, M.S., 1967, Geol. Surv. Can., Paper 67-11, pl. 36, fig. 20.
Permian, Tatonduck River, west limb of syncline 3 miles east of Yukon-Alaska border, lat. 64°58′N, long. 140°54′W, Yukon.

Vestigisporites sp.
Fig. specs. 16341, 16441, 16444
Barss, M.S., 1967, Geol. Surv. Can., Paper 67-11, pl. 27, fig. 19; pl. 31, figs. 4,7.
Pennsylvanian, Mabou Mines Formation, coal seam immediately beneath fault north of Coal Mine Point, Mabou mines, Inverness County; Morien Group, Phalen coal seam, Dominion No. 18 colliery at face of No. 6 west level, 3,400 feet from main deep, Sydney coalfield, Cape Breton County, Nova Scotia.

?*Vestigisporites* sp.
Fig. spec. 16337
Barss, M.S., 1967, Geol. Surv. Can., Paper 67-11, pl. 27, fig. 15.
Mabou Mines Formation, Pennsylvanian, Colonel Snow's adit at beach level, northeast side of syncline at Coal Mine Point, approximately 440 feet west of gypsum fault, Mabou mines, Inverness County, Nova Scotia.

Vestigisporites sp.
Fig. spec. 16523
Barss, M.S., 1967, Geol. Surv. Can., Paper 67-11, pl. 35, fig. 5.
Pictou Group, Permian, depth 335 feet, Gallas Point borehole No. 2 on property of H. Tweedy, Gallas Point, lat. 46°7′42″N, long. 63°57′28″W, Prince Edward Island.

cf. *Vestigisporites* sp.
Fig. spec. 16634
Barss, M.S., 1967, Geol. Surv. Can., Paper 67-11, pl. 38, fig. 32.
Sabine Bay Formation, Permian, vicinity of Tingmisut Lake, east side of Sabine Peninsula, Melville Island, District of Franklin.

Vestispora costata (Balme) Bhardwaj, 1957
Hypotype 15073
Barss, M.S., 1967, Geol. Surv. Can., Paper 67-11, pl. 23, fig. 18.
Pictou Group, Pennsylvanian, Moncton Hump Yard, Canadian National Railways borehole, depth 986 feet, lat. 46°04′18″N, long. 64°52′32″W, Moncton, New Brunswick.

Vestispora fenestrata (Kosanke and Brokaw) Wilson and Venkatachala, 1963
Hypotype 16250
Barss, M.S., 1967, Geol. Surv. Can., Paper 67-11, pl. 23, fig. 16.
Morien Group, Pennsylvanian, Beaver mine on Tracey coal seam east of MacDonald Lake and south of Morrison road, approximately 5.0 miles from Highway 22, Sydney coalfield, Cape Breton County, Nova Scotia.

Vestispora fenestrata (Kosanke and Brokaw) Wilson and Venkatachala, 1963
Hypotypes 16320, 16417, 16418
Barss, M.S., 1967, Geol. Surv. Can., Paper 67-11, pl. 26, fig. 21; pl. 30, figs. 14,15.
Morien Group, Pennsylvanian, LeCras coal seam, from pit 200 feet southwest side of Highway 22, 1.1 miles from intersection of Old Mira Road and Highway 22, Phalen coal seam, Dominion No. 18 colliery at face of No. 6 west level, 3,400 feet from main deep and Harbour coal seam, Dominion No. 12 colliery at face of No. 21 west level, Sydney coalfield, Cape Breton County, Nova Scotia.

Vestispora fenestrata (Kosanke and Brokaw) Wilson and Venkatachala, 1963
Hypotype 16479
Barss, M.S., 1967, Geol. Surv. Can., Paper 67-11, pl. 32, fig. 28.
Pictou Group, Pennsylvanian, Scoudouc River 1 1/8 miles upstream from highway running south from Dorchester Crossing, New Brunswick.

Vestispora irregularis (Kosanke) Wilson and Venkatachala, 1963
Hypotype 16323
Barss, M.S., 1967, Geol. Surv. Can., Paper 67-11, pl. 27, fig. 1.
Mabou Mines Formation, Pennsylvanian, seven-foot coal seam in drift from cliff shore to old workings, south side of syncline, Mabou mines, Inverness County, Nova Scotia.

Vestispora laevigata Wilson and Venkatachala, 1963
Hypotype 16248
Barss, M.S., 1967, Geol. Surv. Can., Paper 67-11, pl. 23, fig. 14.
Morien Group, Pennsylvanian, McAulay Cove coal seam, from outcrop on shore 1.75 miles south of Homeville, Sydney coalfield, Cape Breton County, Nova Scotia.

Vestispora laevigata Wilson and Venkatachala, 1963
Hypotypes 16319, 16415
Barss, M.S., 1967, Geol. Surv. Can., Paper 67-11, pl. 26, fig. 20; pl. 30, fig. 12.
Morien Group, Pennsylvanian, LeCras coal seam, from pit 200 feet southwest side of Highway 22, 1.1 miles from intersection of Old Mira Road and Highway 22, and Point Aconi coal seam on shore at Point Aconi, Sydney coalfield, Cape Breton County, Nova Scotia.

Vestispora laevigata Wilson and Venkatachala, 1963
Hypotype 16478
Barss, M.S., 1967, Geol. Surv. Can., Paper 67-11, pl. 32, fig. 27.
Pictou Group, Pennsylvanian, small northerly tributary of Scoudouc River, 1/2 mile west of old dam and 1 1/2 miles south of town of Shediac, New Brunswick.

Vestispora magna (Butterworth and Williams) Wilson and Venkatachala, 1963
Hypotype 16249
Barss, M.S., 1967, Geol. Surv. Can., Paper 67-11, pl. 23, fig. 5.
Pictou Group, Pennsylvanian, Moncton Hump Yard, Canadian National Railways borehole, depth 986 feet, lat. 46°04′18″N, long. 64°52′32″W, Moncton, New Brunswick.

Vestispora profunda Wilson and Hoffmeister, 1956
Hypotype 16253
Barss, M.S., 1967, Geol. Surv. Can., Paper 67-11, pl. 24, fig. 1.
Morien Group, Pennsylvanian, McAulay Cove coal seam, from outcrop on shore 1.75 miles south of Homeville, Sydney coalfield, Cape Breton County, Nova Scotia.

Vestispora profunda Wilson and Hoffmeister, 1956
Hypotypes 16321, 16322, 16416
Barss, M.S., 1967, Geol. Surv. Can., Paper 67-11, pl. 26, figs. 22, 23; pl. 30, fig. 13.
Morien Group, Pennsylvanian, LeCras coal seam, from pit 200 feet southwest side of Highway 22, 1.1 miles from intersection of Old Mira Road and Highway 22, Ormond coal seam, pit approximately 600 feet southwest side of Highway 22 and 600 feet east of old road to Middle Lake, and Phalen coal seam, Dominion No. 18 colliery at face of No. 6 west level, 3,400 feet from main deep, Sydney coalfield, Cape Breton County, Nova Scotia.

Vestispora sp.
Fig. spec. 16419
Barss, M.S., 1967, Geol. Surv. Can., Paper 67-11, pl. 30, fig. 16.
Morien Group, Pennsylvanian, Harbour coal seam, Dominion No. 12 colliery at face of No. 21 west level, Sydney coalfield, Cape Breton County, Nova Scotia.

Vitreisporites pallidus (Reissinger) Nilsson, 1958
Hypotype 16643
Barss, M.S., 1967, Geol. Surv. Can., Paper 67-11, pl. 38, fig. 41.
Sabine Bay Formation, Permian, vicinity of Tingmisut Lake, east side of Sabine Peninsula, Melville Island, District of Franklin.

Vitreisporites signatus Leschik, 1955
Hypotype 16644
Barss, M.S., 1967, Geol. Surv. Can., Paper 67-11, pl. 38, fig. 42.
Sabine Bay Formation, Permian, vicinity of Tingmisut Lake, east side of Sabine Peninsula, Melville Island, District of Franklin.

?*Vitreisporites* sp.
Fig. spec. 16483
Barss, M.S., 1967, Geol. Surv. Can., Paper 67-11, pl. 32, fig. 32.
Pictou Group, Pennsylvanian, Shediac water well approximately 1,400 feet east of shore of harbour and 1,800 feet south of highway at Shediac, New Brunswick.

Vittatina costabilis Wilson, 1966
Hypotypes 16549, 16550
Barss, M.S., 1967, Geol. Surv. Can., Paper 67-11, pl. 36, figs. 6, 7.
Pictou Group, Permian, depths 335 and 332 feet, Gallas Point borehole No. 2 on property of H. Tweedy, Gallas Point, lat. 46°7′42″N, long. 63°57′28″W, Prince Edward Island.

Vittatina costabilis? Wilson, 1962
Hypotype 16609
Barss, M.S., 1967, Geol. Surv. Can., Paper 67-11, pl. 38, fig. 7.
Permian, Tatonduck River, west limb of syncline 3 miles east of Yukon-Alaska border, lat. 64°58′N, long. 140°54′W, Yukon.

Vittatina minima Jansonius, 1962
Hypotype 16637
Barss, M.S., 1967, Geol. Surv. Can., Paper 67-11, pl. 38, fig. 35.
Sabine Bay Formation, Permian, vicinity of Tingmisut Lake, east side of Sabine Peninsula, Melville Island, District of Franklin.

Vittatina saccifer Jansonius, 1962
Hypotype 16650
Barss, M.S., 1967, Geol. Surv. Can., Paper 67-11, pl. 38, fig. 48.
Sabine Bay Formation, Permian, vicinity of Tingmisut Lake, east side of Sabine Peninsula, Melville Island, District of Franklin.

Vittatina cf. *V. saccifer* Jansonius, 1962
Hypotype 16607
Barss, M.S., 1967, Geol. Surv. Can., Paper 67-11, pl. 38, fig. 5.
Permian, Tatonduck River, west limb of syncline 3 miles east of Yukon-Alaska border, lat. 64°58′N, long. 140°54′W, Yukon.

Vittatina simplex Jansonius, 1962
Hypotype 16613
Barss, M.S., 1967, Geol. Surv. Can., Paper 67-11, pl. 38, fig. 11.
Permian, Tatonduck River, west limb of syncline 3 miles east of Yukon-Alaska border, lat. 64°58′N, long. 140°54′W, Yukon.

Vittatina subsaccata Samoilovich, 1953
Hypotypes 16579, 16603
Barss, M.S., 1967, Geol. Surv. Can., Paper 67-11, pl. 37, fig. 13; pl. 38, fig. 1.
Permian, Tatonduck River, west limb of syncline 3 miles east of Yukon-Alaska border, lat. 64°58′N, long. 140°54′W, Yukon.

Vittatina vittafer (Luber) Samoilovich, 1953
Hypotypes 16604, 16611, 16614
Barss, M.S., 1967, Geol. Surv. Can., Paper 67-11, pl. 38, figs. 2, 9, 12.
Permian, Tatonduck River, west limb of syncline 3 miles east of Yukon-Alaska border, lat. 64°58′N, long. 140°54′W, Yukon.

Vittatina vittifer (Luber) forma *cinctutus* Samoilovich, 1953
Hypotype 16615
Barss, M.S., 1967, Geol. Surv. Can., Paper 67-11, pl. 38, fig. 13.
Permian, Tatonduck River, west limb of syncline 3 miles east of Yukon-Alaska border, lat. 64°58′N, long. 140°54′W, Yukon.

cf. *Vittatina wodehousei* (Jansonius) Hart, 1964
Hypotype 16602
Barss, M.S., 1967, Geol. Surv. Can., Paper 67-11, pl. 37, fig. 24.
Permian, Tatonduck River, west limb of syncline 3 miles east of Yukon-Alaska border, lat. 64°58′N, long. 140°54′W, Yukon.

Vittatina sp.
Fig. specs. 16547, 16548, 16551
Barss, M.S., 1967, Geol. Surv. Can., Paper 67-11, pl. 36, figs. 4, 5, 8.
Pictou Group, Permian, depths 335 (16547) and 332 feet, Gallas Point borehole No. 2 on property of H. Tweedy, Gallas Point, lat. 46°7′42″N, long. 63°57′28″W, Prince Edward Island.

Vittatina sp.
Fig. specs. 16605, 16608, 16610, 16612
Barss, M.S., 1967, Geol. Surv. Can., Paper 67-11, pl. 38, figs. 3, 6, 8, 10.
Permian, Tatonduck River, west limb of syncline 3 miles east of Yukon-Alaska border, lat. 64°58′N, long. 140°54′W, Yukon.

Vittatina sp.
Fig. spec. 45734
Davidson, A. and McGregor, D.C., 1976, Nature, vol. 262, no. 5567, fig. 2d.
Kari sandstone, Permian, banks Kari River approximately 14 km upstream from its junction to the north with Akobo River, Kefa Province, southwestern Ethiopia.

cf. *Vittatina* sp. *L.* in Jansonius, 1962
Fig. spec. 16606
Barss, M.S., 1967, Geol. Surv. Can., Paper 67-11, pl. 38, fig. 4. Permian, Tatonduck River, west limb of syncline 3 miles east of Yukon-Alaska border, lat. 64°58′N, long. 140°54′W, Yukon.

Waltzispora albertensis Staplin, 1960
Hypotype 15848
Barss, M.S., 1967, Geol. Surv. Can., Paper 67-11, pl. 7, fig. 13.

Mattson Formation, Mississippian, south face Twisted Mountain, about 15 miles northwest of South Nahanni, lat. 61°13′N. long. 123°37′W, Northwest Territories.

Waltzispora albertensis Staplin, 1960
Hypotype 94460
Utting, J., 1991, Geol. Surv. Can., Bull. 412, pl. 4.1, fig. 3.
Hart River Formation, Lower Carboniferous, north bank Peel River, northern Yukon.

Waltzispora lobophora (Waltz) Staplin, 1960
Hypotype 15849
Barss, M.S., 1967, Geol. Surv. Can., Paper 67-11, pl. 7, fig. 14.
Mattson Formation, Mississippian, south face Twisted Mountain, about 15 miles northwest of South Nahanni, lat. 61°13′N. long. 123°37′W, Northwest Territories.

Waltzispora planiangulata Sullivan, 1964
Hypotypes 92932, 92963
Utting, J., Jachowicz, M. and Jachowicz, A., 1989, Geol. Surv. Can., Bull. 396, pl. 6.5, fig. 16; pl. 6.6, fig. 20.
Emma Fiord Formation, Early Carboniferous, Lyall River area, Grinnell Peninsula, Devon Island, and Svartevaeg Cliffs, northern Axel Heiberg Island, District of Franklin.

Waltzispora sagittata Playford, 1962
Hypotype 92933
Utting, J., Jachowicz, M. and Jachowicz, A., 1989, Geol. Surv. Can., Bull. 396, pl. 6.5, fig. 17.
Emma Fiord Formation, Early Carboniferous, Lyall River area, Grinnell Peninsula, Devon Island, District of Franklin.

Waltzispora sp.
Fig. spec. 92934
Utting, J., Jachowicz, M. and Jachowicz, A., 1989, Geol. Surv. Can., Bull. 396, pl. 6.5, fig. 18.
Emma Fiord Formation, Early Carboniferous, Lyall River area, Grinnell Peninsula, Devon Island, District of Franklin.

?*Waltzispora* sp.
Fig. spec. 15844
Barss, M.S., 1967, Geol. Surv. Can., Paper 67-11, pl. 7, fig. 9.
Mattson Formation, Mississippian, south face Twisted Mountain, about 15 miles northwest of South Nahanni, lat. 61°13′N. long. 123°37′W, Northwest Territories.

Wilsonites vesicata Kosanke, 1950
Hypotype 16330
Barss, M.S., 1967, Geol. Surv. Can., Paper 67-11, pl. 27, fig. 8.
Mabou Mines Formation, Pennsylvanian, seven-foot coal seam in drift from cliff shore to old workings, south side of syncline, Mabou mines, Inverness County, Nova Scotia.

Wilsonites sp.
Fig. spec. 16440
Barss, M.S., 1967, Geol. Surv. Can., Paper 67-11, pl. 31, fig. 3.
Morien Group, Pennsylvanian, Phalen coal seam, Dominion No. 18 colliery at face of No. 6 west level, 3,400 feet from main deep, Sydney coalfield, Cape Breton County, Nova Scotia.

Triassic

?*Aculeispores variabilis* Jansonius, 1962
Hypotype 13648
McGregor, D.C., 1965, Geol. Surv. Can., Paper 64-55, pl. 1, fig. 26.
Bjorne Formation, Lower Triassic, 7 miles northeast of head of Tanquary Fiord, northern Ellesmere Island, District of Franklin.

Alisporites grandis (Cookson) Dettmann, 1963
Hypotype 13731
McGregor, D.C., 1965, Geol. Surv. Can., Paper 64-55, pl. 3, fig. 30.
Heiberg Formation, Upper Triassic, east side of Black Top Ridge near Eureka, western Ellesmere Island, District of Franklin.

Alisporites plicatus Jizba, 1962
Hypotype 13728
McGregor, D.C., 1965, Geol. Surv. Can., Paper 64-55, pl. 3, fig. 26.
Heiberg Formation, Upper Triassic, east side of Black Top Ridge near Eureka, western Ellesmere Island, District of Franklin.

cf. *Alisporites plicatus* Jizba, 1962
Hypotype 13725
McGregor, D.C., 1965, Geol. Surv. Can., Paper 64-55, pl. 3, fig. 21.
Heiberg Formation, Upper Triassic, east side of Black Top Ridge near Eureka, western Ellesmere Island, District of Franklin.

Alisporites sp.
Fig. spec. 13726
McGregor, D.C., 1965, Geol. Surv. Can., Paper 64-55, pl. 3, fig. 22.
Heiberg Formation, Upper Triassic, east side of Black Top Ridge near Eureka, western Ellesmere Island, District of Franklin.

Apiculatasporites sp.
Fig. spec. 13636
McGregor, D.C., 1965, Geol. Surv. Can., Paper 64-55, pl. 1, fig. 8.
Bjorne Formation, Lower Triassic, 7 miles northeast of head of Tanquary Fiord, northern Ellesmere Island, District of Franklin.

Baculatisporites camaumensis (Cookson) Potonié, 1956
Hypotypes 13677, 13678
McGregor, D.C., 1965, Geol. Surv. Can., Paper 64-55, pl. 2, figs. 13, 25, 17.
Heiberg Formation, Upper Triassic, east side of Black Top Ridge near Eureka, western Ellesmere Island, District of Franklin.

Calamospora diversiformis Balme and Hennelly, 1956
Hypotype 13666
McGregor, D.C., 1965, Geol. Surv. Can., Paper 64-55, pl. 2, fig. 1.
Heiberg Formation, Upper Triassic, east side of Black Top Ridge near Eureka, western Ellesmere Island, District of Franklin.

Camarozonosporites rudis (Leschick) Klaus, 1960
Hypotype 13682
McGregor, D.C., 1965, Geol. Surv. Can., Paper 64-55, pl. 2, figs. 24, 28.
Heiberg Formation, Upper Triassic, east side of Black Top Ridge near Eureka, western Ellesmere Island, District of Franklin.

Concavisporites infirmus Balme, 1957
Hypotype 13683
McGregor, D.C., 1965, Geol. Surv. Can., Paper 64-55, pl. 2, fig. 26.
Heiberg Formation, Upper Triassic, east side of Black Top Ridge near Eureka, western Ellesmere Island, District of Franklin.

Converrucosisporites sp.
Fig. spec. 13687
McGregor, D.C., 1965, Geol. Surv. Can., Paper 64-55, pl. 2, fig. 33.
Heiberg Formation, Upper Triassic, east side of Black Top Ridge near Eureka, western Ellesmere Island, District of Franklin.

?*Cordaitina* sp.
Fig. spec. 13732
McGregor, D.C., 1965, Geol. Surv. Can., Paper 64-55, pl. 3, fig. 32.
Heiberg Formation, Upper Triassic, east side of Black Top Ridge near Eureka, western Ellesmere Island, District of Franklin.

Cornutisporites seebergensis Schulz, 1962
Hypotype 13698
McGregor, D.C., 1965, Geol. Surv. Can., Paper 64-55, pl. 2, figs. 52, 53.
Heiberg Formation, Upper Triassic, east side of Black Top Ridge near Eureka, western Ellesmere Island, District of Franklin.

Cyathidites australis Couper, 1953
Hypotype 13671
McGregor, D.C., 1965, Geol. Surv. Can., Paper 64-55, pl. 2, fig. 6.
Heiberg Formation, Upper Triassic, east side of Black Top Ridge near Eureka, western Ellesmere Island, District of Franklin.

Cyathidites concavus (Bolkhovitina) Dettmann, 1963
Hypotype 13673
McGregor, D.C., 1965, Geol. Surv. Can., Paper 64-55, pl. 2, fig. 8.
Heiberg Formation, Upper Triassic, east side of Black Top Ridge near Eureka, western Ellesmere Island, District of Franklin.

Cyathidites sabuli Reinhardt, 1962
Hypotype 13672
McGregor, D.C., 1965, Geol. Surv. Can., Paper 64-55, pl. 2, fig. 7.
Heiberg Formation, Upper Triassic, east side of Black Top Ridge near Eureka, western Ellesmere Island, District of Franklin.

Cycadopites fragilis Singh, 1964
Hypotype 13711
McGregor, D.C., 1965, Geol. Surv. Can., Paper 64-55, pl. 3, fig. 4.
Heiberg Formation, Upper Triassic, east side of Black Top Ridge near Eureka, western Ellesmere Island, District of Franklin.

Cycadopites sp. (?*Bennettites medius* Bolkhovitina, 1956)
Fig. spec. 13176
McGregor, D.C., 1965, Geol. Surv. Can., Paper 64-55, pl. 3, figs. 9, 10.
Heiberg Formation, Upper Triassic, east side of Black Top Ridge near Eureka, western Ellesmere Island, District of Franklin.

cf. *Deltoidisporites neddeni* var. *torosus* (Pflug) Danzé-Cosin and Laveine, 1963
Hypotype 13680
McGregor, D.C., 1965, Geol. Surv. Can., Paper 64-55, pl. 2, figs. 19, 20.
Heiberg Formation, Upper Triassic, east side of Black Top Ridge near Eureka, western Ellesmere Island, District of Franklin.

Eucommiidites troedssonii Erdtman, 1948
Hypotype 13715
McGregor, D.C., 1965, Geol. Surv. Can., Paper 64-55, pl. 3, fig. 8.
Heiberg Formation, Upper Triassic, east side of Black Top Ridge near Eureka, western Ellesmere Island, District of Franklin.

?*Fimbriaesporties fimbriatus* Singh, 1964
Hypotype 13655
McGregor, D.C., 1965, Geol. Surv. Can., Paper 64-55, pl. 1, fig. 36.
Bjorne Formation, Lower Triassic, 7 miles northeast of head of Tanquary Fiord, northern Ellesmere Island, District of Franklin.

Foveosporites sc. *F. canilis* Balme, 1957
Hypotype 13684
McGregor, D.C., 1965, Geol. Surv. Can., Paper 64-55, pl. 2, fig. 27.
Heiberg Formation, Upper Triassic, east side of Black Top Ridge near Eureka, western Ellesmere Island, District of Franklin.

Ginkgocycadophycus nitidus (Balme) de Jersey, 1962 (?=*Cycadopites* fragilis Singh, 1964)
Hypotype 13640
McGregor, D.C., 1965, Geol. Surv. Can., Paper 64-55, pl. 1, fig. 18.
Bjorne Formation, Lower Triassic, 7 miles northeast of head of Tanquary Fiord, northern Ellesmere Island, District of Franklin.

Ginkgocycadophytus sp.
Fig. specs. 13641, 13642
McGregor, D.C., 1965, Geol. Surv. Can., Paper 64-55, pl. 1, figs. 19, 20.
Bjorne Formation, Lower Triassic, 7 miles northeast of head of Tanquary Fiord, northern Ellesmere Island, District of Franklin.

Gnetaceaepollenites sp.
Fig. specs. 13644, 13645
McGregor, D.C., 1965, Geol. Surv. Can., Paper 64-55, pl. 1, figs. 22, 25.
Bjorne Formation, Lower Triassic, 7 miles northeast of head of Tanquary Fiord, northern Ellesmere Island, District of Franklin.

?*Leschikisporis* sp.
Fig. spec. 13670
McGregor, D.C., 1965, Geol. Surv. Can., Paper 64-55, pl. 2, fig. 5.
Heiberg Formation, Upper Triassic, east side of Black Top Ridge near Eureka, western Ellesmere Island, District of Franklin.

Limbosporites lundbladii Nilsson, 1958
Hypotype 13689
McGregor, D.C., 1965, Geol. Surv. Can., Paper 64-55, pl. 2, figs. 41, 49.
Heiberg Formation, Upper Triassic, east side of Black Top Ridge near Eureka, western Ellesmere Island, District of Franklin.

cf. *Limitisporites delasaucei* (Potonié and Klaus) Schaarschmidt, 1963
Hypotype 13727
McGregor, D.C., 1965, Geol. Surv. Can., Paper 64-55, pl. 3, fig. 25.
Heiberg Formation, Upper Triassic, east side of Black Top Ridge near Eureka, western Ellesmere Island, District of Franklin.

Lophotriletes sp.
Fig. specs. 13634, 13635
McGregor, D.C., 1965, Geol. Surv. Can., Paper 64-55, pl. 1, figs. 5, 9.
Bjorne Formation, Lower Triassic, 7 miles northeast of head of Tanquary Fiord, northern Ellesmere Island, District of Franklin.

Lueckisporites sp.
Fig. spec. 13654
McGregor, D.C., 1965, Geol. Surv. Can., Paper 64-55, pl. 1, fig. 33.
Bjorne Formation, Lower Triassic, 7 miles northeast of head of Tanquary Fiord, northern Ellesmere Island, District of Franklin.

Lycopodiacidites kuepperi Klaus, 1960
Hypotype 13686
McGregor, D.C., 1965, Geol. Surv. Can., Paper 64-55, pl. 2, fig. 32.
Heiberg Formation, Upper Triassic, east side of Black Top Ridge near Eureka, western Ellesmere Island, District of Franklin.

Lycopodiumsporites cf. *L. austroclavatidites* (Cookson) Potonié, 1956
Hypotype 13688
McGregor, D.C., 1965, Geol. Surv. Can., Paper 64-55, pl. 2, figs. 38-40.
Heiberg Formation, Upper Triassic, east side of Black Top Ridge near Eureka, western Ellesmere Island, District of Franklin.

Lycopodiumsporites sp.
Fig. spec. 13694
McGregor, D.C., 1965, Geol. Surv. Can., Paper 64-55, pl. 2, fig. 45.
Heiberg Formation, Upper Triassic, east side of Black Top Ridge near Eureka, western Ellesmere Island, District of Franklin.

Marattisporites scabratus Couper in Nilsson, 1958
Hypotype 13708
McGregor, D.C., 1965, Geol. Surv. Can., Paper 64-55, pl. 3, fig. 1.
Heiberg Formation, Upper Triassic, east side of Black Top Ridge near Eureka, western Ellesmere Island, District of Franklin.

cf. *Microfoveolatispora directa* (Balme and Hennelly) Bharadwaj, 1960
Hypotype 13669
McGregor, D.C., 1965, Geol. Surv. Can., Paper 64-55, pl. 2, fig. 4.
Heiberg Formation, Upper Triassic, east side of Black Top Ridge near Eureka, western Ellesmere Island, District of Franklin.

cf. *Monocolpopollenites* sp. forme *minor* Lantz, 1958
Hypotype 13709
McGregor, D.C., 1965, Geol. Surv. Can., Paper 64-55, pl. 3, fig. 2.
Heiberg Formation, Upper Triassic, east side of Black Top Ridge near Eureka, western Ellesmere Island, District of Franklin.

Monosulcites cf. *M. minimus* Cookson in Jansonius, 1962
Hypotype 13643
McGregor, D.C., 1965, Geol. Surv. Can., Paper 64-55, pl. 1, fig. 21.
Bjorne Formation, Lower Triassic, 7 miles northeast of head of Tanquary Fiord, northern Ellesmere Island, District of Franklin.

cf. *Monosulcites subgranulosus* Couper, 1958
Hypotype 13714
McGregor, D.C., 1965, Geol. Surv. Can., Paper 64-55, pl. 3, fig. 7.
Heiberg Formation, Upper Triassic, east side of Black Top Ridge near Eureka, western Ellesmere Island, District of Franklin.

Monosulcites sp.
Fig. spec. 13639
McGregor, D.C., 1965, Geol. Surv. Can., Paper 64-55, pl. 1, fig. 17.
Bjorne Formation, Lower Triassic, 7 miles northeast of head of Tanquary Fiord, northern Ellesmere Island, District of Franklin.

Monosulcites sp.
Fig. spec. 13710
McGregor, D.C., 1965, Geol. Surv. Can., Paper 64-55, pl. 3, fig. 3.
Heiberg Formation, Upper Triassic, east side of Black Top Ridge near Eureka, western Ellesmere Island, District of Franklin.

Monosulcites sp. K in Jansonius, 1962
Fig. spec. 13647
McGregor, D.C., 1965, Geol. Surv. Can., Paper 64-55, pl. 1, fig. 24.
Bjorne Formation, Lower Triassic, 7 miles northeast of head of Tanquary Fiord, northern Ellesmere Island, District of Franklin.

cf. spores of *Naiadita*, in Dettmann, 1963
Fig. spec. 13638
McGregor, D.C., 1965, Geol. Surv. Can., Paper 64-55, pl. 1, fig. 15.
Bjorne Formation, Lower Triassic, 7 miles northeast of head of Tanquary Fiord, northern Ellesmere Island, District of Franklin.

Osmundacidites wellmanii Couper, 1953
Hypotype 13676
McGregor, D.C., 1965, Geol. Surv. Can., Paper 64-55, pl. 2, fig. 12.
Heiberg Formation, Upper Triassic, east side of Black Top Ridge near Eureka, western Ellesmere Island, District of Franklin.

cf. *Phyllothecites microdiscus* Kara-Murza, 1958
Hypotype 13679
McGregor, D.C., 1965, Geol. Surv. Can., Paper 64-55, pl. 2, fig. 14.
Heiberg Formation, Upper Triassic, east side of Black Top Ridge near Eureka, western Ellesmere Island, District of Franklin.

cf. *Phyllothecites microdiscus* f. *limbata* Kara-Murza, 1958
Hypotype 13692
McGregor, D.C., 1965, Geol. Surv. Can., Paper 64-55, pl. 2, fig. 43.
Heiberg Formation, Upper Triassic, east side of Black Top Ridge near Eureka, western Ellesmere Island, District of Franklin.

Pinuspollenites labdacus var. *arcuatus* Danzé-Corsin and Laveine, 1963
Hypotypes 13723, 13724
McGregor, D.C., 1965, Geol. Surv. Can., Paper 64-55, pl. 3, figs. 18-20.
Heiberg Formation, Upper Triassic, east side of Black Top Ridge near Eureka, western Ellesmere Island, District of Franklin.

cf. *Pityosporites zapfei* Potonié and Klaus (kleine Variante) in Klaus, 1955
Hypotype 13720
McGregor, D.C., 1965, Geol. Surv. Can., Paper 64-55, pl. 3, fig. 14.
Heiberg Formation, Upper Triassic, east side of Black Top Ridge near Eureka, western Ellesmere Island, District of Franklin.

Planisporites sp.
Fig. specs. 13630-13632
McGregor, D.C., 1965, Geol. Surv. Can., Paper 64-55, pl. 1, figs. 1-3.
Bjorne Formation, Lower Triassic, 7 miles northeast of head of Tanquary Fiord, northern Ellesmere Island, District of Franklin.

cf. *Pollenites apertus* Rogalska, 1954
Hypotype 13693
McGregor, D.C., 1965, Geol. Surv. Can., Paper 64-55, pl. 2, fig. 44.
Heiberg Formation, Upper Triassic, east side of Black Top Ridge near Eureka, western Ellesmere Island, District of Franklin.

Rogalskaisporites multicicatricosus Danzé-Corsin and Leveine, 1963
Hypotype 13681
McGregor, D.C., 1965, Geol. Surv. Can., Paper 64-55, pl. 2, fig. 23.
Heiberg Formation, Upper Triassic, east side of Black Top Ridge near Eureka, western Ellesmere Island, District of Franklin.

Rotaspora sp.
Fig. spec. 13696
McGregor, D.C., 1965, Geol. Surv. Can., Paper 64-55, pl. 2, fig. 48.
Heiberg Formation, Upper Triassic, east side of Black Top Ridge near Eureka, western Ellesmere Island, District of Franklin.

Rouseisporites sp.
Fig. spec. 13695
McGregor, D.C., 1965, Geol. Surv. Can., Paper 64-55, pl. 2, fig. 47.
Heiberg Formation, Upper Triassic, east side of Black Top Ridge near Eureka, western Ellesmere Island, District of Franklin.

Semiretisporis sp.
Fig. spec. 13697
McGregor, D.C., 1965, Geol. Surv. Can., Paper 64-55, pl. 2, fig. 50.

Heiberg Formation, Upper Triassic, east side of Black Top Ridge near Eureka, western Ellesmere Island, District of Franklin.

Sporites adriennis f. mesozoicus Thiergart in Rogalska, 1956 (cf. *Deltoidisporites hallei* var. *auriculus* Danzé-Crosin and Laveine, 1963)
Hypotypes 13674, 13675
McGregor, D.C., 1965, Geol. Surv. Can., Paper 64-55, pl. 2, figs. 9, 10.
Heiberg Formation, Upper Triassic, east side of Black Top Ridge near Eureka, western Ellesmere Island, District of Franklin.

Stereisporites sp.
Fig. spec. 13668
McGregor, D.C., 1965, Geol. Surv. Can., Paper 64-55, pl. 2, fig. 3.
Heiberg Formation, Upper Triassic, east side of Black Top Ridge near Eureka, western Ellesmere Island, District of Franklin.

Striatites richteri (Klaus) Potonié, 1958
Hypotypes 13652, 13653
McGregor, D.C., 1965, Geol. Surv. Can., Paper 64-55, pl. 1, figs. 30, 31.
Bjorne Formation, Lower Triassic, 7 miles northeast of head of Tanquary Fiord, northern Ellesmere Island, District of Franklin.

Striatites sp.
Fig. specs. 13649-13651
McGregor, D.C., 1965, Geol. Surv. Can., Paper 64-55, pl. 1, figs. 29, 32, 35.
Bjorne Formation, Lower Triassic, 7 miles northeast of head of Tanquary Fiord, northern Ellesmere Island, District of Franklin.

Taeniaesporites cf. *T. kraeuseli* Leschick, 1955
Hypotype 13730
McGregor, D.C., 1965, Geol. Surv. Can., Paper 64-55, pl. 3, figs. 28, 29.
Heiberg Formation, Upper Triassic, east side of Black Top Ridge near Eureka, western Ellesmere Island, District of Franklin.

Taeniaesporites ?noviaulensis Leschik, 1955
Hypotype 13729
McGregor, D.C., 1965, Geol. Surv. Can., Paper 64-55, pl. 3, figs. 27, 31.
Heiberg Formation, Upper Triassic, east side of Black Top Ridge near Eureka, western Ellesmere Island, District of Franklin.

Taeniaesporites sp.
Fig. spec. 13733
McGregor, D.C., 1965, Geol. Surv. Can., Paper 64-55, pl. 3, figs. 33, 34.
Heiberg Formation, Upper Triassic, east side of Black Top Ridge near Eureka, western Ellesmere Island, District of Franklin.

Taeniaesporites sp. U in Jansonius, 1962
Fig. spec. 13646
McGregor, D.C., 1965, Geol. Surv. Can., Paper 64-55, pl. 1, fig. 23.
Bjorne Formation, Lower Triassic, 7 miles northeast of head of Tanquary Fiord, northern Ellesmere Island, District of Franklin.

Triancoraesporites communis Schulz, 1962
Hypotypes 13690, 13691
McGregor, D.C., 1965, Geol. Surv. Can., Paper 64-55, pl. 2, figs. 42, 46.
Heiberg Formation, Upper Triassic, east side of Black Top Ridge near Eureka, western Ellesmere Island, District of Franklin.

cf. *Verrucosisporites trisecatus* Balme and Hennelly, 1956
Hypotype 13637
McGregor, D.C., 1965, Geol. Surv. Can., Paper 64-55, pl. 1, fig. 12.
Bjorne Formation, Lower Triassic, 7 miles northeast of head of Tanquary Fiord, northern Ellesmere Island, District of Franklin.

Verrucosisporites sp.
Fig. spec. 13633
McGregor, D.C., 1965, Geol. Surv. Can., Paper 64-55, pl. 1, fig. 4.
Bjorne Formation, Lower Triassic, 7 miles northeast of head of Tanquary Fiord, northern Ellesmere Island, District of Franklin.

Vitreisporites pallidus (Reissinger) Nilsson, 1958
Hypotypes 13718, 13719
McGregor, D.C., 1965, Geol. Surv. Can., Paper 64-55, pl. 3, figs. 13, 15.
Heiberg Formation, Upper Triassic, east side of Black Top Ridge near Eureka, western Ellesmere Island, District of Franklin.

Vitreisporites pallidus (Reissinger) Nilsson, 1958
Hypotype 13721
McGregor, D.C., 1965, Geol. Surv. Can., Paper 64-55, pl. 3, fig. 16.
Heiberg Formation, Upper Triassic, east side of Black Top Ridge near Eureka, western Ellesmere Island, District of Franklin.

cf. *Zebrasporites interscriptus* (Thiergart) Klaus, 1960
Hypotype 13685
McGregor, D.C., 1965, Geol. Surv. Can., Paper 64-55, pl. 2, figs. 30, 31.
Heiberg Formation, Upper Triassic, east side of Black Top Ridge near Eureka, western Ellesmere Island, District of Franklin.

Unidentified
Fig. specs. 13656-13665
McGregor, D.C., 1965, Geol. Surv. Can., Paper 64-55, pl. 1, figs. 6, 7, 10, 11, 13, 14, 16, 27, 28, 34.
Bjorne Formation, Lower Triassic, 7 miles northeast of head of Tanquary Fiord, northern Ellesmere Island, District of Franklin.

Unidentified
Fig. specs. 13699-13707, 13734-13736
McGregor, D.C., 1965, Geol. Surv. Can., Paper 64-55, pl. 2, figs. 11, 15, 16, 21, 22, 29, 34-37, 51, 54, 55; pl. 3, figs. 12, 23, 24.
Heiberg Formation, Upper Triassic, east side of Black Top Ridge near Eureka, western Ellesmere Island, District of Franklin.

Jurassic

?*Abietineaepollenites microalatus* Potonié in Couper, 1958
Hypotype 13780
McGregor, D.C., 1965, Geol. Surv. Can., Paper 64-55, pl. 4, figs. 37, 43.
Borden Island Formation, Lower Jurassic, west side of Black Top Ridge near Eureka, western Ellesmere Island, District of Franklin.

Abietineaepollenites minimus Couper, 1958
Hypotypes 13782, 13783
McGregor, D.C., 1965, Geol. Surv. Can., Paper 64-55, pl. 4, figs. 40, 42.
Borden Island Formation, Lower Jurassic, west side of Black Top Ridge near Eureka, western Ellesmere Island, District of Franklin.

Alisporites bisaccus Rouse, 1959
Hypotype 13785
McGregor, D.C., 1965, Geol. Surv. Can., Paper 64-55, pl. 4, fig. 44.
Borden Island Formation, Lower Jurassic, west side of Black Top Ridge near Eureka, western Ellesmere Island, District of Franklin.

Alisporites rotundus Rouse
Hypotype 13557
Pocock, S.A.J., 1964, Bull. Can. Petrol. Geol., vol. 12, Special Issue, pl., fig. 17.
Mutz Member, Kootenay Formation, Upper Jurassic, Grassy Mountain, sec. 25, tp. 8, rge. 4, W.5th mer., lat. 49°41′N, long. 114°25′W, southwestern Alberta.

Apiculatisporis sp.
Fig. spec. 13553
Pocock, S.A.J., 1964, Bull. Can. Petrol. Geol., vol. 12, Special Issue, pl., fig. 3.
Mutz Member, Kootenay Formation, Upper Jurassic, Grassy Mountain, sec. 25, tp. 8, rge. 4, W.5th mer., lat. 49°41′N, long. 114°25′W, southwestern Alberta.

Apiculatisporis sp.
Fig. spec. 13756
McGregor, D.C., 1965, Geol. Surv. Can., Paper 64-55, pl. 4, fig. 5.
Borden Island Formation, Lower Jurassic, west side of Black Top Ridge near Eureka, western Ellesmere Island, District of Franklin.

Araucariacites punctatus (Nilsson) Cornet and Traverse
Hypotype 75434
Davies, E.H., 1985, Palynology, vol. 9, pl. 1, fig. 10.
Late Pliensbachian, Liassic, Peniche Cliffs, Lusitanian Basin, west-central Portugal.

Auriculozonospora cf. *A. reticulata* Singh, Srivastava and Roy 1964
Hypotypes 48906-48908
Sweet, A.R., 1979, Am. Assoc. Strat. Palynol., Contrib. Ser. No. 5B, pl. 2, figs. 2-4
Kootenay Formation, Upper Jurassic-Lower Cretaceous, lat. 49°21′59″N, long. 114°11′05″W, Natal Ridge, Crowsnest area, British Columbia.

Baculatisporites comaumensis (Cookson) Potonié, 1956
Hypotype 13755
McGregor, D.C., 1965, Geol. Surv. Can., Paper 64-55, pl. 4, fig. 4.
Borden Island Formation, Lower Jurassic, west side of Black Top Ridge near Eureka, western Ellesmere Island, District of Franklin.

Baculatisporites comaumensis (Cookson) Potonié, 1956
Hypotype 13804
McGregor, D.C., 1965, Geol. Surv. Can., Paper 64-55, pl. 5, fig. 21.
Awingak Formation, Upper Jurassic, Black Top Ridge near Eureka, western Ellesmere Island, District of Franklin.

Baculatisporites sp.
Fig. specs. 13753, 13754
McGregor, D.C., 1965, Geol. Surv. Can., Paper 64-55, pl. 4, figs. 3, 10-12.
Borden Island Formation, Lower Jurassic, west side of Black Top Ridge near Eureka, western Ellesmere Island, District of Franklin.

Bennettiteaepollenites lucifer Thierg.
Hypotype 13576
Pocock, S.A.J., 1964, Bull. Can. Petrol. Geol., vol. 12, Special Issue, pl., fig. 25.
Kootenay Formation, Upper Jurassic, depth 7,689 feet below K.B., Royalite DeWinton No. 2 borehole, l.s.d. 13, sec. 24, tp. 21, rge. 1, W.5th mer., southwestern Alberta.

Brachysaccus microsaccus (Couper) Madler
Hypotype 75444
Davies, E.H., 1985, Palynology, vol. 9, pl. 1, fig. 20.
Early Pliensbachian, Liassic, Zambujal Section, south of Condeixa, Lusitanian Basin, west-central Portugal.

cf. *Bracteolina rotunda* Malyavkina, 1953
Hypotype 13759
McGregor, D.C., 1965, Geol. Surv. Can., Paper 64-55, pl. 4, figs. 8, 9.
Borden Island Formation, Lower Jurassic, west side of Black Top Ridge near Eureka, western Ellesmere Island, District of Franklin.

Callialasporites dampieri (Balme) Sukh Dev
Hypotype 13567
Pocock, S.A.J., 1964, Bull. Can. Petrol. Geol., vol. 12, Special Issue, pl., fig. 13.
Kootenay Formation, Upper Jurassic, depth 7,689 feet below K.B., Royalite DeWinton No. 2 borehole, l.s.d. 13, sec. 24, tp. 21, rge. 1, W.5th mer., southwestern Alberta.

Callialasporites segmentatus (Balme) Dev
Hypotype 75426
Davies, E.H., 1985, Palynology, vol. 9, pl. 1, fig. 2.
Aalenian, Liassic, Zambujal Section, south of Condeixa, Lusitanian Basin, west-central Portugal.

Callialasporites trilobatus (Balme) Sukh Dev
Hypotype 13571
Pocock, S.A.J., 1964, Bull. Can. Petrol. Geol., vol. 12, Special Issue, pl., fig. 19.
Kootenay Formation, Upper Jurassic, depth 7,689 feet below K.B., Royalite DeWinton No. 2 borehole, l.s.d. 13, sec. 24, tp. 21, rge. 1, W.5th mer., southwestern Alberta.

Callialasporites trilobatus Balme
Hypotype 76605
Davies, E.H., Akande, S.O. and Zentilli, M., 1984, Geol. Surv. Can., Paper 84-1A, pl. 48.1, fig. 16.
Late Jurassic-Early Cretaceous, Sample DH2-372, Gays River Mine, lat. 45°02′N, long. 63°22′W, Nova Scotia.

Callialasporites trilobatus (Balme) Dev
Hypotype 75425
Davies, E.H., 1985, Palynology, vol. 9, pl. 1, fig. 1.
Toarcian, Liassic, Brehna Road Section, Lusitanian Basin, west-central Portugal.

Callialasporites turbatus (Balme) Schulz
Hypotype 75429
Davies, E.H., 1985, Palynology, vol. 9, pl. 1, fig. 5.
Aalenian, Liassic, Brehna Road Section, Lusitanian Basin, west-central Portugal.

Cerebropollenites mesozoicus (Couper)
Hypotype 41909
Williams, G.L., 1975, Geol. Surv. Can., Paper 74-30, vol. 2, pl. 8, fig. 14.
Middle Jurassic, Shell Oneida 0-25 well, Scotian Shelf.

Cerebropollenites mesozoicus (Couper) Nillson
Fig. spec. 58928
Poulton, T.P., Leskiw, K. and Audretsch, A., 1982, Geol. Surv. Can., Bull. 325, pl. 6, fig. 22.
Almstrom Creek Formation, Lower Jurassic, Little Fish Creek northward from junction with Almstrom Creek, approximately lat. 68°25′40″N, long. 136°10′30″W, District of Mackenzie.

Chasmatosporites canadensis Pocock
Hypotype 75431
Davies, E.H., 1985, Palynology, vol. 9, pl. 1, fig. 7.
Early Pliensbachian, Liassic, Peniche Cliffs, Lusitanian Basin, west-central Portugal.

Chasmatosporites major Nilsson
Hypotype 75432
Davies, E.H., 1985, Palynology, vol. 9, pl. 1, fig. 8.
Early Pliensbachian, Liassic, Peniche Cliffs, Lusitanian Basin, west-central Portugal.

cf. *Cheiropleuria congregata* Bolkhovitina, 1956
Hypotype 13762
McGregor, D.C., 1965, Geol. Surv. Can., Paper 64-55, pl. 4, fig. 16.
Borden Island Formation, Lower Jurassic, west side of Black Top Ridge near Eureka, western Ellesmere Island, District of Franklin.

cf. *Cheiropleura congregata* Bolkhovitinia, 1956
Hypotype 13794
McGregor, D.C., 1965, Geol. Surv. Can., Paper 64-55, pl. 5, fig. 9.
Awingak Formation, Upper Jurassic, Black Top Ridge near Eureka, western Ellesmere Island, District of Franklin.

Chomotriletes sp.
Fig. spec. 13738
McGregor, D.C., 1965, Geol. Surv. Can., Paper 64-55, pl. 3, fig. 36.
Borden Island Formation, Lower Jurassic, west side of Black Top Ridge near Eureka, western Ellesmere Island, District of Franklin.

Cicatricosisporites abacus Burger 1966, emend. nov.
Hypotypes 72046-72052, 72058-72061
Fensome, R.A., 1987, Palaeontographica Canadiana No. 4, p. 14, pl. 1, figs. 5-7, 9-11, 13-16; pl. 2, figs. 2-5, 16, 19.
Jurassic-Cretaceous, Husky Formation, north bank of first creek south of Mt. Gifford, Martin Creek, east bank, lat. 68°12′N, long. 135°37′30″W (72050, 72052); McGuire Formation, Grizzly Gorge, south bank, lat. 68°15′49″N, long. 135°40′59″W (72048, 72049, 72061), District of Mackenzie.

Classopollis classoides Pflug
Hypotypes 13578, 13579
Pocock, S.A.J., 1964, Bull. Can. Petrol. Geol., vol. 12, Special Issue, pl., figs. 27, 28.
Kootenay Formation, Upper Jurassic, depth 7,689 feet below K.B., Royalite DeWinton No. 2 borehole, l.s.d. 13, sec. 24, tp. 21, rge. 1, W.5th mer., southwestern Alberta.

Classopollis classoides Pflug emend. Pocock and Jansonius, 1961
Hypotypes 13769, 13770
McGregor, D.C., 1965, Geol. Surv. Can., Paper 64-55, pl. 4, figs. 27, 28.
Borden Island Formation, Lower Jurassic, west side of Black Top Ridge near Eureka, western Ellesmere Island, District of Franklin.

Classopollis classoides Pflug
Hypotype 41903
Williams, G.L., 1975, Geol. Surv. Can., Paper 74-30, vol. 2, pl. 8, fig. 8.
Callovian, Jurassic, Amoco IOE Eider M-75 well, Grand Banks.

Classopollis classoides Pflug
Fig. specs. 58926, 58931
Poulton, T.P., Leskiw, K. and Audretsch, A., 1982, Geol. Surv. Can., Bull. 325, pl. 6, figs. 20, 25.
Almstrom Creek Formation, Lower Jurassic, Little Fish Creek northward from junction with Almstrom Creek, approximately lat. 68°25′40″N, long. 136°10′30″W, District of Mackenzie.

Concavissimisporites crassatus (Delcourt and Spumont 1955) Delcourt et al. 1963
Hypotypes 72085-72087
Fensome, R.A., 1987, Palaeontographica Canadiana No. 4, p. 21, pl. 4, figs. 1, 6, 7.
Husky Formation, Jurassic, Martin Creek, east bank, lat. 68°12′N, long. 135°37′20′W, and east bank of north-flowing creek 0.7 km south of Martin Creek, lat. 68°10′54″N, long. 135°37′50″W; McGuire Formation, Cretaceous, Grizzly Gorge, south bank, lat. 68°15′N, long. 135°40′59″W, District of Mackenzie.

Concavissimisporites exquisitus (Singh, C. 1971) comb. nov.
Hypotype 72092
Fensome, R.A., 1987, Palaeontographica Canadiana No. 4, p. 21, pl. 4, fig. 11.
Husky Formation, Jurassic, Martin Creek, east bank, lat. 68°12′N, long. 135°37′20′W, District of Mackenzie.

Concavissimisporites ferniensis (Pocock 1970) comb. nov.
Hypotype 72093
Fensome, R.A., 1987, Palaeontographica Canadiana No. 4, p. 22, pl. 4, fig. 12.
Husky Formation, Jurassic, Martin Creek, east bank, one mile southwest of small lake and about 0.3 miles up a tributary, lat. 68°11′20″N, long. 135°38′W, District of Mackenzie.

Concavissimisporites granulosa (Tralau 1968) comb. nov.
Hypotype 72080
Fensome, R.A., 1987, Palaeontographica Canadiana No. 4, p. 22, pl. 3, fig. 13.
Husky Formation, Jurassic, Martin Creek, east bank, lat. 68°12′N, long. 135°37′20′W, District of Mackenzie.

Concavissimisporites irroratus Paden Phillips and Felix, 1971 emend. nov.
Hypotypes 72081, 72082
Fensome, R.A., 1987, Palaeontographica Canadiana No. 4, p. 22, pl. 3, figs. 15, 16.
Husky Formation, Jurassic, east bank of north-flowing creek 0.7 km south of Martin Creek, lat. 68°10′54″N, long. 135°37′50″W, and Martin Creek, east bank, lat. 68°12′N, long. 135°37′20″W, District of Mackenzie.

Concavissimisporites montuosus (Döring 1964) comb. nov.
Hypotypes 72088, 72095, 72096
Fensome, R.A., 1987, Palaeontographica Canadiana No. 4, p. 23, pl. 4, fig. 2; pl. 5, figs. 1, 2.
Husky Formation, Jurassic, east bank of north-flowing creek 0.7 km south of Martin Creek, lat. 68°10′54″N, long. 135°37′50″W, and Martin Creek, east bank, lat. 68°12′N, long. 135°17′20″W (72096), District of Mackenzie.

Concavissimisporites robustus Dorhöfer 1977
Hypotype 72089
Fensome, R.A., 1987, Palaeontographica Canadiana No. 4, p. 23, pl. 4, fig. 3.
Husky Formation, Jurassic, east bank of north-flowing creek 0.7 km south of Martin Creek, lat. 68°10′54″N, long. 135°37′50″W, District of Mackenzie.

Concavissimisporites subgranulosus (Couper) Pocock
Hypotype 75451
Davies, E.H., 1985, Palynology, vol. 9, pl. 2, fig. 3.
Late Toarcian, Liassic, Zambujal Section, south of Condeixa, Lusitanian Basin, west-central Portugal.

Concavissimisporites subgranulosus (Couper 1961) Pocock 1970
Hypotype 72083
Fensome, R.A., 1987, Palaeontographica Canadiana No. 4, p. 24, pl. 3, fig. 17.
Husky Formation, Jurassic, east bank, lat. 68°12′N, long. 135°17′20″W, District of Mackenzie.

cf. *Coniopteris tajmyrensis* Kara-Murza in Samoilovich et al., 1961
Hypotype 13747
McGregor, D.C., 1965, Geol. Surv. Can., Paper 64-55, pl. 3, fig. 45.
Borden Island Formation, Lower Jurassic, west side of Black Top Ridge near Eureka, western Ellesmere Island, District of Franklin.

Contignisporites cooksonii (Balme) Dettmann
Hypotype 56779
Williams, G.L. and Bujak, J.P., 1980, Initial Repts. Deep Sea Drilling Project, vol. 50, pl. 10, fig. 10.
Late Jurassic, DSDP Site 416, lat. 32°50.18′N, long. 10°48.06′W, northwestern Moroccan continental Margin.

Contignisporites sp. cf. *C. glebulentus* Dettmann 1963
Fig. spec. 72053
Fensome, R.A., 1987, Palaeontographica Canadiana No. 4, p. 11, pl. 1, fig. 12.
Husky Formation, Jurassic, east bank of north-flowing creek 0.7 km south of Martin Creek, lat. 68°10′54″N, long. 135°37′50″W, District of Mackenzie.

Contignisporites sp.
Fig. spec. 47934
Bujak, J.P. and Williams, G.L., 1977, Strat. Micropal. Atlantic Basin and Borderlands, pl. 2, fig. 2.
Jurassic, 11,070-11,100 feet, Mobil-Tetco Cohasset D-42 well, Scotian Shelf, Atlantic Ocean.

Convolutispora sp.
Fig. spec. 75454
Davies, E.H., 1985, Palynology, vol. 9, pl. 2, fig. 6.
Early Toarcian, Liassic, Zambujal Section, south of Condeixa, Lusitanian Basin, west-central Portugal.

?*Coptospora* sp.
Fig. spec. 13765
McGregor, D.C., 1965, Geol. Surv. Can., Paper 64-55, pl. 4, figs. 21, 22.
Borden Island Formation, Lower Jurassic, west side of Black Top Ridge near Eureka, western Ellesmere Island, District of Franklin.

Corollina meyeriana (Klaus) Venkatachala and Góczán
Hypotype 479299
Bujak, J.P. and Williams, G.L., 1977, Strat. Micropal. Atlantic Basin and Borderlands, pl. 1, fig. 1.
Jurassic, 2840-2880 feet, Shell Eurydice P-36 well, Scotian Shelf, Atlantic Ocean.

Corollina meyeriana (Klaus) Venkatachala
Hypotype 75438
Davies, E.H., 1985, Palynology, vol. 9, pl. 1, fig. 13.
Early Toarcian, Liassic, Zambujal Section, south of Condeixa, Lusitanian Basin, west-central Portugal.

Corollina murphyi Cornet and Traverse
Hypotype 75437
Davies, E.H., 1985, Palynology, vol. 9, pl. 1, fig. 13.
Early Toarcian, Liassic, Zambujal Section, south of Condeixa, Lusitanian Basin, west-central Portugal.

Crassitudisporites problematicus (Couper 1958) Hiltmann 1967 emend. nov.
Hypotypes 72043-72045
Fensome, R.A., 1987, Palaeontographica Canadiana No. 4, p. 12, pl. 1, figs. 3, 4, 8.
Husky Formation, Jurassic, east bank, lat. 68°12′N, long. 135°17′20″W (72043), and one mile southwest of small lake and up a tributary about 0.3 miles, lat. 68°11′20″N, long. 135°38′W, District of Mackenzie.

Cyathidites austrialis Couper, 1953
Hypotypes 13788, 13789
McGregor, D.C., 1965, Geol. Surv. Can., Paper 64-55, pl. 5, figs. 1, 2.
Awingak Formation, Upper Jurassic, Black Top Ridge near Eureka, western Ellesmere Island, District of Franklin.

?*Cyathidites australis* Couper, 1953
Hypotype 13751
McGregor, D.C., 1965, Geol. Surv. Can., Paper 64-55, pl. 4, fig. 1.
Borden Island Formation, Lower Jurassic, west side of Black Top Ridge near Eureka, western Ellesmere Island, District of Franklin.

Cyathidites australis rimalis Balme, 1957
Hypotype 13744
McGregor, D.C., 1965, Geol. Surv. Can., Paper 64-55, pl. 3, fig. 42.
Borden Island Formation, Lower Jurassic, west side of Black Top Ridge near Eureka, western Ellesmere Island, District of Franklin.

Cyathidites dewintonensis Pocock
Hypotype 13561
Pocock, S.A.J., 1964, Bull. Can. Petrol. Geol., vol. 12, Special Issue, pl., fig. 2.
Kootenay Formation, Upper Jurassic, depth 7,689 feet below K.B., Royalite DeWinton No. 2 borehole, l.s.d. 13, sec. 24, tp. 21, rge. 1, W.5th mer., southwestern Alberta.

Cyathidites lineatus (Bolkh.) Pocock
Hypotype 13560
Pocock, S.A.J., 1964, Bull. Can. Petrol. Geol., vol. 12, Special Issue, pl., fig. 1.
Kootenay Formation, Upper Jurassic, depth 7,689 feet below K.B., Royalite DeWinton No. 2 borehole, l.s.d. 13, sec. 24, tp. 21, rge. 1, W.5th mer., southwestern Alberta.

Cycadopites fragilis Singh, 1964
Hypotype 13773
McGregor, D.C., 1965, Geol. Surv. Can., Paper 64-55, pl. 4, fig. 31.
Borden Island Formation, Lower Jurassic, west side of Black Top Ridge near Eureka, western Ellesmere Island, District of Franklin.

Cycadopites minor (Cookson) Pocock
Hypotypes 13574, 13575
Pocock, S.A.J., 1964, Bull. Can. Petrol. Geol., vol. 12, Special Issue, pl., figs. 23, 24.
Kootenay Formation, Upper Jurassic, depth 7,689 feet below K.B., Royalite DeWinton No. 2 borehole, l.s.d. 13, sec. 24, tp. 21, rge. 1, W.5th mer., southwestern Alberta.

Cycadopites sp.
Fig. spec. 13805
McGregor, D.C., 1965, Geol. Surv. Can., Paper 64-55, pl. 5, fig. 22.
Awingak Formation, Upper Jurassic, Black Top Ridge near Eureka, western Ellesmere Island, District of Franklin.

?*Cycadopites* spp. A, B
Fig. specs. 47936, 47938, 47939
Bujak, J.P. and Williams, G.L., 1977, Strat. Micropal. Atlantic Basin and Borderlands, pl. 2, figs. 4, 7, 8.
Jurassic, 5048, 2991 and 3229 feet, Shell Eurydice P-36 well, Scotian Shelf, Atlantic Ocean.

Deltoidisporites hallei var. *auriculus* Danzé-corsin and Laveine, 1963
Hypotype 13758
McGregor, D.C., 1965, Geol. Surv. Can., Paper 64-55, pl. 4, fig. 7.
Borden Island Formation, Lower Jurassic, west side of Black Top Ridge near Eureka, western Ellesmere Island, District of Franklin.

Deltoidospora sp.
Fig. spec. 13745
McGregor, D.C., 1965, Geol. Surv. Can., Paper 64-55, pl. 3, fig. 43.
Borden Island Formation, Lower Jurassic, west side of Black Top Ridge near Eureka, western Ellesmere Island, District of Franklin.

Dictyotriletes sp. 1 of Cornet and Traverse (1975)
Fig. spec. 75456
Davies, E.H., 1985, Palynology, vol. 9, pl. 2, fig. 8.
Early Toarcian, Liassic, Peniche Cliffs, Lusitanian Basin, west-central Portugal.

cf. *Disaccites (Phyllocladidites) rüei* Cookson, 1947
Hypotype 13814
McGregor, D.C., 1965, Geol. Surv. Can., Paper 64-55, pl. 5, fig. 34.
Awingak Formation, Upper Jurassic, Black Top Ridge near Eureka, western Ellesmere Island, District of Franklin.

Duplexisporites anagrammensis (Kara-Muza) Pocock
Hypotype 75459
Davies, E.H., 1985, Palynology, vol. 9, pl. 2, fig. 11.
Aalenian, Liassic, Buhna Road Section, Lusitanian Basin, west-central Portugal.

Echinitosporites cf. *iliacoides* Schulz and Krutzsch
Hypotypes 47931, 47932, 47940, 47941
Bujak, J.P. and Williams, G.L., 1977, Strat. Micropal. Atlantic Basin and Borderlands, pl. 1, figs. 3, 4; pl. 2, figs. 9, 12.
Jurassic, 2840-2880 feet and 2575 feet, Shell Eurydice P-36 well, Scotian Shelf, Atlantic Ocean.

Echinitosporites cf. *E. iliacoides* Schulz and Krutzsch sensu Bujak and Williams (1977)
Hypotype 75443
Davies, E.H., 1985, Palynology, vol. 9, pl. 1, fig. 19.
Aalenian, Liassic, Zambujal Section, south of Condeixa, Lusitanian Basin, west-central Portugal.

Echinitosporites sp. A
Fig. spec. 47942
Bujak, J.P. and Williams, G.L., 1977, Strat. Micropal. Atlantic Basin and Borderlands, pl. 2, figs. 10, 11.
Jurassic, 2840-2880 feet, Shell Eurydice P-36 well, Scotian Shelf, Atlantic Ocean.

Echinitosporites sp. A of Bujak and Williams (1977)
Fig. specs. 75441, 75442
Davies, E.H., 1985, Palynology, vol. 9, p. 128, pl. 1, figs. 17, 18.
Liassic, Early Toarcion, Peniche Cliffs, and Aalenian, Zambujal Section, south of Condeixa, Lusitanian Basin, west-central Portugal.

Echitriletes hispidus Marcinkiewicz 1960
Hypotype 48889
Sweet, A.R., 1979, Am. Assoc. Strat. Palynol., Contrib. Ser. No. 5B, pl. 1, figs. 1, 2.
Jaeger Formation, Lower-Middle Jurassic, lat. 77°41′50″N, long. 94°14′25″W, Cornwall Island, District of Franklin.

cf.*Ephedracites pseudoretectiniformis* Malyavkina, 1958
Hypotype 13772
McGregor, D.C., 1965, Geol. Surv. Can., Paper 64-55, pl. 4, fig. 30.
Borden Island Formation, Lower Jurassic, west side of Black Top Ridge near Eureka, western Ellesmere Island, District of Franklin.

Erlansonisporites sparassis (Murray) Potonié 1956
Hypotype 48896
Sweet, A.R., 1979, Am. Assoc. Strat. Palynol., Contrib. Ser. No. 5B, pl. 1, fig. 9.
Jaeger Formation, Lower-Middle Jurassic, lat. 77°41′50″N, long. 94°14′25″W, Cornwall Island, District of Franklin.

Eucommiidites troedssonii Erdtman, 1948
Hypotype 13771
McGregor, D.C., 1965, Geol. Surv. Can., Paper 64-55, pl. 4, fig. 29.
Borden Island Formation, Lower Jurassic, west side of Black Top Ridge near Eureka, western Ellesmere Island, District of Franklin.

Exesipollenites montanensis (Miner) Pocock
Hypotype 13577
Pocock, S.A.J., 1964, Bull. Can. Petrol. Geol., vol. 12, Special Issue, pl., fig. 26.
Kootenay Formation, Upper Jurassic, depth 7,689 feet below K.B., Royalite DeWinton No. 2 borehole, l.s.d. 13, sec. 24, tp. 21, rge. 1, W.5th mer., southwestern Alberta.

Exesipollenites tumulus Balme
Hypotype 13559
Pocock, S.A.J., 1964, Bull. Can. Petrol. Geol., vol. 12, Special Issue, pl., fig. 29.
Mutz Member, Kootenay Formation, Upper Jurassic, Grassy Mountain, sec. 25, tp. 8, rge. 4, W.5th mer., lat. 49°41′N, long. 114°25′W, southwestern Alberta.

Exesipollenites tumulus Balme
Hypotype 75436
Davies, E.H., 1985, Palynology, vol. 9, pl. 1, fig. 12.
Early Pliensbachian, Liassic, Zambujal Section, south of Condeixa, Lusitanian Basin, west-central Portugal.

Foveosporites vischeri van Erve
Hypotype 75460
Davies, E.H., 1985, Palynology, vol. 9, pl. 2, fig. 12.
Early Pliensbachian, Liassic, Zambujal Section, south of Condeixa, Lusitanian Basin, west-central Portugal.

cf. *Ginkocycadophytus caperatus* (Luber) Samoilovich, 1953
Hypotype 13808
McGregor, D.C., 1965, Geol. Surv. Can., Paper 64-55, pl. 5, fig. 25.
Awingak Formation, Upper Jurassic, Black Top Ridge near Eureka, western Ellesmere Island, District of Franklin.

Gonyaulacysta sp. cf. *G. cladophora* (Deflandre) Dodekova
Fig. specs. 58914, 58915
Poulton, T.P., Leskiw, K. and Audretsch, A., 1982, Geol. Surv. Can., Bull. 325, pl. 6, figs. 8, 9.
Richardson Mountains Formation, Middle Jurassic, Little Fish Creek northward from junction with Almstrom Creek, approximately lat. 68°25′10″N, long. 136°10′30″W, and end of flat-topped spur projecting east-southeastward, approximately lat. 68°01′30″N, long. 135°25′30″W, 1.5 km. from Jurassic Butte, District of Mackenzie.

Gonyaulax jurassica Defl.
Hypotype 13580
Pocock, S.A.J., 1964, Bull. Can. Petrol. Geol., vol. 12, Special Issue, pl., fig. 30.
Kootenay Formation, Upper Jurassic, depth 7,689 feet below K.B., Royalite DeWinton No. 2 borehole, l.s.d. 13, sec. 24, tp. 21, rge. 1, W.5th mer., southwestern Alberta.

Granulatisporites sp.
Fig. spec. 13740
McGregor, D.C., 1965, Geol. Surv. Can., Paper 64-55, pl. 3, fig. 38.
Borden Island Formation, Lower Jurassic, west side of Black Top Ridge near Eureka, western Ellesmere Island, District of Franklin.

Hologinella spinosigibberosa Brideaux and Fisher
Holotype 34157; paratypes 42376-42379
Brideaux, W.W. and Fisher, M.J., 1976, Geol. Surv. Can., Bull. 259, p. 22, pl. 4, fig. 9; pl. 5, figs. 1-16.
Upper Jurassic, 1710 (34157) and 1765 (42378, 42379) feet, Banff-Aquitaine-Arco Rat Pass K-35 well, 6 miles southeast of summit of Mount Goodenough, just south of Goodenough Creek, District of Mackenzie, and Mould Bay Formation, Upper Jurassic, 300-310 and 410-420 feet, Elf Jameson Bay C-31 well near Jameson Bay, lat. 76°40′12″N, long. 116°43′45″W, Prince Patrick Island, District of Franklin.

Horologinella sp.
Fig. spec. 58920
Poulton, T.P., Leskiw, K. and Audretsch, A., 1982, Geol. Surv. Can., Bull. 325, pl. 6, fig. 14.
Almstrom Creek Formation, Lower Jurassic, Little Fish Creek northward from junction with Almstrom Creek, approximately lat. 68°25′10″N, long. 136°10′30″W, District of Mackenzie.

Hughesisporites pustulatus Marcinkiewicz 1960
Hypotypes 48894, 48897
Sweet, A.R., 1979, Am. Assoc. Strat. Palynol., Contrib. Ser. No. 5B, pl. 1, figs. 7, 10.
Jaeger Formation, Lower-Middle Jurassic, lat. 77°41′50″N, long. 94°14′25″W, Cornwall Island, District of Franklin.

Ichyosporites sp. cf. *I. disjunctus* Singh, C. 1971
Fig. specs. 72072-72075
Fensome, R.A., 1987, Palaeontographica Canadiana No. 4, p. 18, pl. 3, figs. 7, 9, 11, 14.
Husky Formation, Jurassic, and Cretaceous (72073, 72075), Fault Creek Formation, Cretaceous, (72074), east bank of north-flowing creek, 0.7 km, south of Martin Creek, lat. 68°10′54″N, long. 135°37′50″W, north bank of first creek south of Mt. Gifford, Martin Creek, north bank, lat. 68°12′17″N, long. 135°34′38″W and south side, 0.8 miles north of small lake, lat. 68°12′20″N, long. 135°36′W, District of Mackenzie.

Ichyosporites pseudoreticulatus (Couper 1958) Döring 1965
Hypotypes 72077, 72078
Fensome, R.A., 1987, Palaeontographica Canadiana No. 4, p. 18, pl. 3, figs. 10, 12.
Husky Formation, Jurassic, east bank of north-flowing creek 0.7 km south of Martin Creek, lat. 68°10′54″N, long. 135°37′50″W, District of Mackenzie.

Inaperturopollenites cf. *I. australis* (Cookson) Pocock
Hypotype 13566
Pocock, S.A.J., 1964, Bull. Can. Petrol. Geol., vol. 12, Special Issue, pl., fig. 11.
Kootenay Formation, Upper Jurassic, depth 7,689 feet below K.B., Royalite DeWinton No. 2 borehole, l.s.d. 13, sec. 24, tp. 21, rge. 1, W.5th mer., southwestern Alberta.

Inaperturopollenites dubius (Potonié and Venitz) Thomson and Pflug
Hypotypes 75428, 75430
Davies, E.H., 1985, Palynology, vol. 9, pl. 1, figs. 4, 6.
Liassic, Early Toarcion, Peniche Cliffs, and Aalenian, Brehna Road Section, Lusitanian Basin, west-central Portugal.

Inaperturopollenites giganteus Coczan
Hypotype 75433
Davies, E.H., 1985, Palynology, vol. 9, pl. 1, fig. 9.
Late Pliensbachian, Liassic, Brenha Road Section, Lusitanian Basin, west-central Portugal.

Ishyosporites punctatus Cookson and Dettmann
Hypotype 75455
Davies, E.H., 1985, Palynology, vol. 9, pl. 2, fig. 7.
Early Pliensbachian, Liassic, Zambujal Section, south of Condeixa, Lusitanian Basin, west-central Portugal.

Ischyosporites sp. A
Fig. spec. 72076
Fensome, R.A., 1987, Palaeontographica Canadiana No. 4, p. 18, pl. 3, fig. 8.
Husky Formation, Jurassic, east bank of north-flowing creek 0.7 km south of Martin Creek, lat. 68°10′54″N, long. 135°37′50″W, District of Mackenzie.

Kraeuselisporites reissingeri (Harris) Morbey
Hypotypes 47930, 47937
Bujak, J.P. and Williams, G.L., 1977, Strat. Micropal. Atlantic
Basin and Borderlands, pl. 1, figs 2; pl. 2, figs. 5, 6.
Jurassic, 2840-2880 feet and 5048 feet, Shell Eurydice P-36 well, Scotian Shelf, Atlantic Ocean.

Kraeuselisporites reissingeri (Harris) Morbey
Hypotype 75458
Davies, E.H., 1985, Palynology, vol. 9, pl. 2, fig. 10.
Late Pliensbachian, Liassic, Brehna Road Section, Lusitanian Basin, west-central Portugal.

cf. *Leiotriletes lineatus* Bolkhovitina, 1956
Hypotypes 13790, 13791
McGregor, D.C., 1965, Geol. Surv. Can., Paper 64-55, pl. 5, figs. 3, 4; pl. 6, fig. 8.
Awingak Formation, Upper Jurassic, Black Top Ridge near Eureka, western Ellesmere Island, District of Franklin.

Leptolepidites bossus (Couper) Schulz
Hypotype 75453
Davies, E.H., 1985, Palynology, vol. 9, pl. 2, fig. 5.
Toarcian, Liassic, Brehna Road Section, Lusitanian Basin, west-central Portugal.

Lycopodiacidisporites cerniidites (Ross) Danzé-Corsin and Laveine, 1963
Hypotype 13761
McGregor, D.C., 1965, Geol. Surv. Can., Paper 64-55, pl. 4, fig. 15.
Borden Island Formation, Lower Jurassic, west side of Black Top Ridge near Eureka, western Ellesmere Island, District of Franklin.

Lycopodiacidites sp.
Fig. spec. 41910
Williams, G.L., 1975, Geol. Surv. Can., Paper 74-30, vol. 2, pl. 8, fig. 15.
Kimmeridgian-Tithonian, Upper Jurassic, Mobil Dauntless D-35 well, Scotian Shelf.

Lycopodiacidites sp.
Fig. spec. 75450
Davies, E.H., 1985, Palynology, vol. 9, p. 128, pl. 2, fig. 2.
Early Toarcian, Liassic, Zambujal Section, south of Condeixa, Lusitanian Basin, west-central Portugal.

Lycopodiumsporites austroclavatidites (Cookson) Potonié, 1956
Hypotype 13750
McGregor, D.C., 1965, Geol. Surv. Can., Paper 64-55, pl. 3, figs. 47, 48.
Borden Island Formation, Lower Jurassic, west side of Black Top Ridge near Eureka, western Ellesmere Island, District of Franklin.

Lycopodiumsporites austroclavatidites (Cookson) Potonié, 1956
Hypotype 13800
McGregor, D.C., 1965, Geol. Surv. Can., Paper 64-55, pl. 5, figs. 16, 17.
Awingak Formation, Upper Jurassic, Black Top Ridge near Eureka, western Ellesmere Island, District of Franklin.

Lycopodiumsporites ?austroclavatidites (Cookson) Potonié, 1956
Hypotype 13760
McGregor, D.C., 1965, Geol. Surv. Can., Paper 64-55, pl. 4, figs. 13, 14.
Borden Island Formation, Lower Jurassic, west side of Black Top Ridge near Eureka, western Ellesmere Island, District of Franklin.

Lycopodiumsporites marginatus Singh
Fig. spec. 58930
Poulton, T.P., Leskiw, K. and Audretsch, A., 1982, Geol. Surv. Can., Bull. 325, pl. 6, fig. 24.
Almstrom Creek Formation, Lower Jurassic, Little Fish Creek northward from junction with Almstrom Creek, approximately lat. 68°25′40″N, long. 136°10′30″W, District of Mackenzie.

Lycopodiumsporites pseudoannotinus Nilsson
Hypotypes 13562, 13565
Pocock, S.A.J., 1964, Bull. Can. Petrol. Geol., vol. 12, Special Issue, pl., figs. 6, 10.
Kootenay Formation, Upper Jurassic, depth 7,689 feet below K.B., Royalite DeWinton No. 2 borehole, l.s.d. 13, sec. 24, tp. 21, rge. 1, W.5th mer., southwestern Alberta.

Lycopodiumsporites semimurus Danzé-Corsin and Laveine, 1963
Hypotypes 13748, 13749
McGregor, D.C., 1965, Geol. Surv. Can., Paper 64-55, pl. 3, figs. 46, 49, 50.
Borden Island Formation, Lower Jurassic, west side of Black Top Ridge near Eureka, western Ellesmere Island, District of Franklin.

Lycopodiumsporites sp.
Fig. specs. 13801, 13802
McGregor, D.C., 1965, Geol. Surv. Can., Paper 64-55, pl. 5, figs. 18, 26.
Awingak Formation, Upper Jurassic, Black Top Ridge near Eureka, western Ellesmere Island, District of Franklin.

?*Lycopodiumsporites* sp.
Fig. spec. 13797
McGregor, D.C., 1965, Geol. Surv. Can., Paper 64-55, pl. 5, fig. 12.
Awingak Formation, Upper Jurassic, Black Top Ridge near Eureka, western Ellesmere Island, District of Franklin.

Mancodinium semitabulatum Morgenroth
Fig. specs. 58918, 58919
Poulton, T.P., Leskiw, K. and Audretsch, A., 1982, Geol. Surv. Can., Bull. 325, pl. 6, figs. 12, 13.
Almstrom Creek Formation, Lower Jurassic, Little Fish Creek northward from junction with Almstrom Creek, approximately lat. 68°25′10″N, long. 136°10′30″W, District of Mackenzie.

Membranipollenites chilcotinensis Pocock
Hypotype 13573
Pocock, S.A.J., 1964, Bull. Can. Petrol. Geol., vol. 12, Special Issue, pl., fig. 22.
Kootenay Formation, Upper Jurassic, depth 7,689 feet below K.B., Royalite DeWinton No. 2 borehole, l.s.d. 13, sec. 24, tp. 21, rge. 1, W.5th mer., southwestern Alberta.

Michrystridium sp.
Fig. spec. 58921
Poulton, T.P., Leskiw, K. and Audretsch, A., 1982, Geol. Surv. Can., Bull. 325, pl. 6, fig. 15.
Almstrom Creek Formation, Lower Jurassic, Little Fish Creek northward from junction with Almstrom Creek, approximately lat. 68°25′40″N, long. 136°10′30″W, District of Mackenzie.

Microplankton indet. sp. 1, 2
Fig. specs. 58908, 58909, 58912, 58913
Poulton, T.P., Leskiw, K. and Audretsch, A., 1982, Geol. Surv. Can., Bull. 325, pl. 6, figs. 1, 2, 5-7.
Richardson Mountains Formation, Middle Jurassic, Little Fish Creek northward from junction with Almstrom Creek, approximately lat. 68°25′10″N, long. 136°10′30″W, District of Mackenzie.

Monosulcites carpentieri Delcourt and Sprumont, 1975
Hypotypes 13806, 13807
McGregor, D.C., 1965, Geol. Surv. Can., Paper 64-55, pl. 5, figs. 23, 24.
Awingak Formation, Upper Jurassic, Black Top Ridge near Eureka, western Ellesmere Island, District of Franklin.

cf. *Monosulcites subgranulosus* Couper, 1958
Hypotype 13774
McGregor, D.C., 1965, Geol. Surv. Can., Paper 64-55, pl. 4, fig. 32.
Borden Island Formation, Lower Jurassic, west side of Black Top Ridge near Eureka, western Ellesmere Island, District of Franklin.

Nannoceratopsis sp.
Fig. specs. 58910, 58911
Poulton, T.P., Leskiw, K. and Audretsch, A., 1982, Geol. Surv. Can., Bull. 325, pl. 6, figs. 3, 4.
Almstrom Creek Formation, Lower Jurassic, Little Fish Creek northward from junction with Almstrom Creek, approximately lat. 68°25′10″N, long. 136°10′30″W, District of Mackenzie.

cf. *Orbicularia aliferiniformis* Donsk. in Malyavkina, 1956
Hypotype 13815
McGregor, D.C., 1965, Geol. Surv. Can., Paper 64-55, pl. 5, fig. 35.
Awingak Formation, Upper Jurassic, Black Top Ridge near Eureka, western Ellesmere Island, District of Franklin.

Osmundacidites wellmanii Couper
Hypotype 13554
Pocock, S.A.J., 1964, Bull. Can. Petrol. Geol., vol. 12, Special Issue, pl., fig. 4.
Mutz Member, Kootenay Formation, Upper Jurassic, Grassy Mountain, sec. 25, tp. 8, rge. 4, W.5th mer., lat. 49°41′N, long. 114°25′W, southwestern Alberta.

Osmundacidites sp.
Fig. spec. 58925
Poulton, T.P., Leskiw, K. and Audretsch, A., 1982, Geol. Surv. Can., Bull. 325, pl. 6, fig. 19.
Almstrom Creek Formation, Lower Jurassic, Little Fish Creek northward from junction with Almstrom Creek, approximately lat. 68°25′10″N, long. 136°10′30″W, District of Mackenzie.

Osmundacidites "designate T8" Rouse
Fig. spec. 13555
Pocock, S.A.J., 1964, Bull. Can. Petrol. Geol., vol. 12, Special Issue, pl., fig. 12.
Mutz Member, Kootenay Formation, Upper Jurassic, Grassy Mountain, sec. 25, tp. 8, rge. 4, W.5th mer., lat. 49°41′N, long. 114°25′W, southwestern Alberta.

Palaeoperidinium sp.
Fig. spec. 13581
Pocock, S.A.J., 1964, Bull. Can. Petrol. Geol., vol. 12, Special Issue, pl., fig. 31.
Kootenay Formation, Upper Jurassic, depth 7,689 feet below K.B., Royalite DeWinton No. 2 borehole, l.s.d. 13, sec. 24, tp. 21, rge. 1, W.5th mer., southwestern Alberta.

Pareodinia borealis Brideaux and Fisher
Holotype 34156; paratypes 42371-42375, 42388
Brideaux, W.W. and Fisher, M.J., 1976, Geol. Surv. Can., Bull. 259, p. 21, pl. 3, figs. 3, 4, 6-9; pl. 4, figs. 1-8.
Upper Jurassic, 1710 and 1765 (42371) feet, Banff-Aquitaine-Arco Rat Pass K-35 well, 6 miles southeast of summit of Mount Goodenough, just south of Goodenough Creek; Gilmore Lake Member, Langton Bay Formation, Lower Creteceous, 1530-1540 (42374) and 1760-1770 (42388) feet, Elf Horton River G-02 well, lat. 69°51′23″N, long. 127°15′57″W, southwest of mouth of Horton River; Husky Formation (42373), Upper Jurassic, lat. 68°12′N, long. 135°37′W, Martin Creek, District of Mackenzie; and Mould Bay Formation (42372), Upper Jurassic, 360 feet, Elf Jameson Bay C-31 well near Jameson Bay, lat. 76°40′12″N, long. 116°43′45″W, Prince Patrick Island, District of Franklin.

Pareodinia ceratiphora Delflandre
Fig. specs. 58917, 58922
Poulton, T.P., Leskiw, K. and Audretsch, A., 1982, Geol. Surv. Can., Bull. 325, pl. 6, figs. 11, 16.
Richardson Mountains Formation, Middle Jurassic, Little Fish Creek northward from junction with Almstrom Creek, approximately lat. 68°25′10″N, long. 136°10′30″W, District of Mackenzie.

Pareodinia sp.
Fig. spec. 58916
Poulton, T.P., Leskiw, K. and Audretsch, A., 1982, Geol. Surv. Can., Bull. 325, pl. 6, fig. 10.

Richardson Mountains Formation, Middle Jurassic, Little Fish Creek northward from junction with Almstrom Creek, approximately lat. 68°25′10″N, long. 136°10′30″W, District of Mackenzie.

Paxillitriletes cf. *P. fairlightensis* (Batten) Hall and Nicholson 1973
Hypotype 48905
Sweet, A.R., 1979, Am. Assoc. Strat. Palynol., Contrib. Ser. No. 5B, pl. 2, fig. 1.
Kootenay Formation, Upper Jurassic-Lower Cretaceous, lat. 49°21′59″N, long. 114°11′05″W, Natal Ridge, Crowsnest area, British Columbia.

Paxillitriletes granulatus (Mädler) Hall and Nicholson 1973
Hypotype 48902
Sweet, A.R., 1979, Am. Assoc. Strat. Palynol., Contrib. Ser. No. 5B, pl. 2, fig. 15.
Kootenay Formation, Upper Jurassic-Lower Cretaceous, lat. 49°21′59″N, long. 114°11′05″W, Natal Ridge, Crowsnest area, British Columbia.

Paxillitriletes reticulatus (Mädler) Hall and Nicholson 1973
Hypotypes 48903, 48904
Sweet, A.R., 1979, Am. Assoc. Strat. Palynol., Contrib. Ser. No. 5B, pl. 1, figs. 16, 17.
Kootenay Formation, Upper Jurassic-Lower Cretaceous, lat. 49°21′59″N, long. 114°11′05″W, Natal Ridge, Crowsnest area, British Columbia.

Paxillitriletes sp. A
Fig. specs. 48899, 48900, 48926, 48927
Sweet, A.R., 1979, Am. Assoc. Strat. Palynol., Contrib. Ser. No. 5B, pl. 1, figs. 12, 13; pl. 3, figs. 1, 2.
Kootenay Formation, Upper Jurassic-Lower Cretaceous, lat. 49°26′59″N, long. 114°11′05″W, Natal Ridge, Crowsnest area, British Columbia; Hassel Formation, Lower Cretaceous, lat. 79°47′24″N, long. 85°22′W, near Slidre Fiord, Ellesmere Island, Arctic.

Perotriletes sp.
Fig. spec. 13766
McGregor, D.C., 1965, Geol. Surv. Can., Paper 64-55, pl. 4, fig. 23.
Borden Island Formation, Lower Jurassic, west side of Black Top Ridge near Eureka, western Ellesmere Island, District of Franklin.

Perotrilites sp.
Fig. spec. 13803
McGregor, D.C., 1965, Geol. Surv. Can., Paper 64-55, pl. 5, figs. 19, 20.
Awingak Formation, Upper Jurassic, Black Top Ridge near Eureka, western Ellesmere Island, District of Franklin.

Pilasporites plurigenus Balme and Hennelly, 1956
Hypotype 13737
McGregor, D.C., 1965, Geol. Surv. Can., Paper 64-55, pl. 3, fig. 35.
Borden Island Formation, Lower Jurassic, west side of Black Top Ridge near Eureka, western Ellesmere Island, District of Franklin.

Pilosisporites sp. A
Fig. spec. 47933
Bujak, J.P. and Williams, G.L., 1977, Strat. Micropal. Atlantic Basin and Borderlands, pl. 2, fig, 1.
Jurassic, 10,570-10,600 feet, Mobil-Tetco Cohasset D-42 well, Scotian Shelf, Atlantic Ocean.

Pinuspollenites vancampoi Danzé-Corsin and Laveine, 1963
Hypotypes 13778, 13779
McGregor, D.C., 1965, Geol. Surv. Can., Paper 64-55, pl. 4, figs. 36, 39.
Borden Island Formation, Lower Jurassic, west side of Black Top Ridge near Eureka, western Ellesmere Island, District of Franklin.

?*Pinuspollenites vancampoi* Danzé-Corsin and Laveine, 1963
Hypotype 13781
McGregor, D.C., 1965, Geol. Surv. Can., Paper 64-55, pl. 4, fig. 38.
Borden Island Formation, Lower Jurassic, west side of Black Top Ridge near Eureka, western Ellesmere Island, District of Franklin.

?*Pinuspollenites*
Fig. spec. 13558
Pocock, S.A.J., 1964, Bull. Can. Petrol. Geol., vol. 12, Special Issue, pl., fig. 21.
Mutz Member, Kootenay Formation, Upper Jurassic, Grassy Mountain, sec. 25, tp. 8, rge. 4, W.5th mer., lat. 49°41′N, long. 114°25′W, southwestern Alberta.

Platysaccus lopsiensis (Mal.) Pocock
Hypotype 13570
Pocock, S.A.J., 1964, Bull. Can. Petrol. Geol., vol. 12, Special Issue, pl., fig. 18.
Kootenay Formation, Upper Jurassic, depth 7,689 feet below K.B., Royalite DeWinton No. 2 borehole, l.s.d. 13, sec. 24, tp. 21, rge. 1, W.5th mer., southwestern Alberta.

Platysaccus papilionis Pontonié and Klaus, 1954
Hypotype 13786
McGregor, D.C., 1965, Geol. Surv. Can., Paper 64-55, pl. 4, fig. 45.
Borden Island Formation, Lower Jurassic, west side of Black Top Ridge near Eureka, western Ellesmere Island, District of Franklin.

Podocarpidites canadensis Pocock
Hypotype 75440
Davies, E.H., 1985, Palynology, vol. 9, pl. 1, fig. 16.
Late Pliensbachian, Liassic, Zambujal Section, south of Condeixa, Lusitanian Basin, west-central Portugal.

Podocarpidites? canadensis Pocock, 1962
Hypotype 13811, 13812
McGregor, D.C., 1965, Geol. Surv. Can., Paper 64-55, pl. 5, figs. 30, 31.
Awingak Formation, Upper Jurassic, Black Top Ridge near Eureka, western Ellesmere Island, District of Franklin.

Podocarpidites cf. *P. ellipticus* Cookson, 1947
Hypotype 13776
McGregor, D.C., 1965, Geol. Surv. Can., Paper 64-55, pl. 4, fig. 34.
Borden Island Formation, Lower Jurassic, west side of Black Top Ridge near Eureka, western Ellesmere Island, District of Franklin.

?*Podocarpidites* cf. *P. ellipticus* Cookson in Dettmann, 1963
Hypotypes 13776, 13777
McGregor, D.C., 1965, Geol. Surv. Can., Paper 64-55, pl. 4, figs. 35, 41.
Borden Island Formation, Lower Jurassic, west side of Black Top Ridge near Eureka, western Ellesmere Island, District of Franklin.

Podocarpidites radiatus Brenner
Hypotype 76609
Davies, E.H., Akande, S.O. and Zentilli, M., 1984, Geol. Surv. Can., Paper 84-1A, pl. 48.1, fig. 20.
Late Jurassic-Early Cretaceous, sample DH2-372, Gays River Mine, lat. 45°02′N, long. 63°22′W, Nova Scotia.

Podocarpidites sp.
Fig. spec. 13809
McGregor, D.C., 1965, Geol. Surv. Can., Paper 64-55, pl. 5, fig. 28.
Awingak Formation, Upper Jurassic, Black Top Ridge near Eureka, western Ellesmere Island, District of Franklin.

cf. *Podocarpus elliptica* (Naumova) Kara-Murza in Pervuninskaya, 1961
Hypotype 13816
McGregor, D.C., 1965, Geol. Surv. Can., Paper 64-55, pl. 5, fig. 36.
Awingak Formation, Upper Jurassic, Black Top Ridge near Eureka, western Ellesmere Island, District of Franklin.

?*Podocarpus* cf. *P. ellipticus* Cookson in Dettmann, 1963
Hypotype 13789
McGregor, D.C., 1965, Geol. Surv. Can., Paper 64-55, pl. 4, fig. 41.
Borden Island Formation, Lower Jurassic, west side of Black Top Ridge near Eureka, western Ellesmere Island, District of Franklin.

Porcellispora longdonensis Morbey
Hypotype 75447
Davies, E.H., 1985, Palynology, vol. 9, pl. 1, fig. 23.
Early Pliensbachian, Liassic, Zambujal Section, south of Condeixa, Lusitanian Basin, west-central Portugal.

Porcellispora sp. A, B
Fig. specs. 75446, 75545
Davies, E.H., 1985, Palynology, vol. 9, p. 128, pl. 1, figs. 22, 21.
Liassic, Early Toarcian, Zambujal Section, south of Condeixa, and Early Pliensbachien, Brehna Road Section, Lusitanian Basin, west-central Portugal.

Protoconiferus microsaccus (Couper) Pocock
Hypotype 13556
Pocock, S.A.J., 1964, Bull. Can. Petrol. Geol., vol. 12, Special Issue, pl., fig. 15
Mutz Member, Kootenay Formation, Upper Jurassic, Grassy Mountain, sec. 25, tp. 8, rge. 4, W.5th mer., lat. 49°41′N, long. 114°25′W, southwestern Alberta.

Protopicea exilioides (Bolkh.) Pocock
Hypotype 13569
Pocock, S.A.J., 1964, Bull. Can. Petrol. Geol., vol. 12, Special Issue, pl., fig. 16.
Kootenay Formation, Upper Jurassic, depth 7,689 feet below K.B., Royalite DeWinton No. 2 borehole, l.s.d. 13, sec. 24, tp. 21, rge. 1, W.5th mer., southwestern Alberta.

cf. *Pteris multiformis* Sakhanova in Kurnosova, 1960
Hypotype 13746
McGregor, D.C., 1965, Geol. Surv. Can., Paper 64-55, pl. 3, fig. 44.
Borden Island Formation, Lower Jurassic, west side of Black Top Ridge near Eureka, western Ellesmere Island, District of Franklin.

Quadraeculina anellaeformis Maljavikina
Hypotypes 75435,75439
Davies, E.H., 1985, Palynology, vol. 9, pl. 1, figs. 11, 15.
Early Pliensbachian, Liassic, Brehna Road Section, and Zambujal Section, south of Condeixa, Lusitanian Basin, west-central Portugal.

Rogalskaisporites cicatricosus (Rogalska) Danzé-Corsin and Laveine, 1963 (=*Leiotriletes incertus* Bolkhovitina, 1956)
Hypotypes 13741, 13742
McGregor, D.C., 1965, Geol. Surv. Can., Paper 64-55, pl. 3, figs. 39, 41.
Borden Island Formation, Lower Jurassic, west side of Black Top Ridge near Eureka, western Ellesmere Island, District of Franklin.

Rogalskaisporites cicatricosus var. *rotundus* Danzé-Corsin and Laveine, 1963
Hypotype 13796
McGregor, D.C., 1965, Geol. Surv. Can., Paper 64-55, pl. 5, fig. 11.
Awingak Formation, Upper Jurassic, Black Top Ridge near Eureka, western Ellesmere Island, District of Franklin.

?*Rubinella media* var. *parvituberculata* Malyavkina, 1953
Hypotype 13813
McGregor, D.C., 1965, Geol. Surv. Can., Paper 64-55, pl. 5, fig. 33.
Awingak Formation, Upper Jurassic, Black Top Ridge near Eureka, western Ellesmere Island, District of Franklin.

Rugubivesiculites sp. A
Fig. spec. 75448
Davies, E.H., 1985, Palynology, vol. 9, p. 128, pl. 1, fig. 24.
Early Pliensbachian, Liassic, Zambujal Section, south of Condeixa, Lusitanian Basin, west-central Portugal.

Schizosporis cf. *S. cooksoni* Pocock, 1962
Hypotype 13768
McGregor, D.C., 1965, Geol. Surv. Can., Paper 64-55, pl. 4, fig. 26.
Borden Island Formation, Lower Jurassic, west side of Black Top Ridge near Eureka, western Ellesmere Island, District of Franklin.

?*Sciadopityspollenites* sp.
Fig. spec. 13767
McGregor, D.C., 1965, Geol. Surv. Can., Paper 64-55, pl. 4, figs. 24, 25.
Borden Island Formation, Lower Jurassic, west side of Black Top Ridge near Eureka, western Ellesmere Island, District of Franklin.

Selaginella obscura Bolkhovitina in Verbitskaya, 1962
Hypotype 13795
McGregor, D.C., 1965, Geol. Surv. Can., Paper 64-55, pl. 5, fig. 10.
Awingak Formation, Upper Jurassic, Black Top Ridge near Eureka, western Ellesmere Island, District of Franklin.

Stereisporites antiquasporites (Wilson and Webster) Dettmann, 1963
Hypotype 13743
McGregor, D.C., 1965, Geol. Surv. Can., Paper 64-55, pl. 3, fig. 40.
Borden Island Formation, Lower Jurassic, west side of Black Top Ridge near Eureka, western Ellesmere Island, District of Franklin.

Striatriletes excavatus (Marcinkiewicz) Sweet 1979
Hypotypes 48892, 48893
Sweet, A.R., 1979, Am. Assoc. Strat. Palynol., Contrib. Ser. No. 5B, p. 14, pl. 1, figs. 5, 6.
Jaeger Formation, Lower-Middle Jurassic, lat. 77°41′50″N, long. 94°14′25″W, Cornwall Island, District of Franklin.

Striatriletes reticulatus (Zerndt) Sweet 1979
Hypotypes 48890, 48891
Sweet, A.R., 1979, Am. Assoc. Strat. Palynol., Contrib. Ser. No. 5B, p. 15, pl. 1, figs. 3, 4.
Jaeger Formation, Lower-Middle Jurassic, lat. 77°41′50″N, long. 94°14′25″W, Cornwall Island, District of Franklin.

Styxisporites reissingeri Danzé-Corsin and Laveine, 1963
Hypotype 13764
McGregor, D.C., 1965, Geol. Surv. Can., Paper 64-55, pl. 4, figs. 19, 20.
Borden Island Formation, Lower Jurassic, west side of Black Top Ridge near Eureka, western Ellesmere Island, District of Franklin.

Taurocusporites reduncus (Bolkhovitine) Stover
Fig. spec. 58929
Poulton, T.P., Leskiw, K. and Audretsch, A., 1982, Geol. Surv. Can., Bull. 325, pl. 6, fig. 23.
Richardson Mountains Formation, Middle Jurassic, Little Fish Creek northward from junction with Almstrom Creek, approximately lat. 68°25′40″N, long. 136°10′30″W, District of Mackenzie.

Taurocusporites sp.
Fig. spec. 13763
McGregor, D.C., 1965, Geol. Surv. Can., Paper 64-55, pl. 4, figs. 17, 18.
Borden Island Formation, Lower Jurassic, west side of Black Top Ridge near Eureka, western Ellesmere Island, District of Franklin.

Taxodiacea indet.
Fig. specs. 58923, 58924
Poulton, T.P., Leskiw, K. and Audretsch, A., 1982, Geol. Surv. Can., Bull. 325, pl. 6, figs. 17, 18.
Murray Ridge Formation, Lower Jurassic, east side of hill overlooking the headwaters of next creek east of Bear Creek that flows south into Rat River, lat. 67°54′20″N, long. 135°55′W, and end of flat-topped spur projecting east-southeastward, approximately lat. 68°01′30″N, long. 135°25′30″W, District of Mackenzie.

Tetralete form
Fig. spec. 48898
Sweet, A.R., 1979, Am. Assoc. Strat. Palynol., Contrib. Ser. No. 5B, pl. 1, fig. 11.
Jaeger Formation, Lower-Middle Jurassic, lat. 77°41′50″N, long. 94°14′25″W, Cornwall Island, District of Franklin.

Tigrisporites jonkeri van Erve
Hypotype 75457
Davies, E.H., 1985, Palynology, vol. 9, pl. 2, fig. 9.
Early Toarcian, Liassic, Zambujal Section, south of Condeixa, Lusitanian Basin, west-central Portugal.

Todisporites major Couper
Hypotype 75449
Davies, E.H., 1985, Palynology, vol. 9, pl. 2, fig. 1.
Late Pliensbachian, Liassic, Zambujal Section, south of Condeixa, Lusitanian Basin, west-central Portugal.

Triangulopsis discoidalis Doring
Hypotype 13568
Pocock, S.A.J., 1964, Bull. Can. Petrol. Geol., vol. 12, Special Issue, pl., fig. 14.
Kootenay Formation, Upper Jurassic, depth 7,689 feet below K.B., Royalite DeWinton No. 2 borehole, l.s.d. 13, sec. 24, tp. 21, rge. 1, W.5th mer., southwestern Alberta.

Triangulopsis discoidalis Döring
Hypotype 75427
Davies, E.H., 1985, Palynology, vol. 9, pl. 1, fig. 3.
Aalenian, Liassic, Brehna Road Section, Lusitanian Basin, west-central Portugal.

Triletes turbanaeformis Harris 1961
Hypotype 48901
Sweet, A.R., 1979, Am. Assoc. Strat. Palynol., Contrib. Ser. No. 5B, pl. 1, fig. 14.
Kootenay Formation, Upper Jurassic-Lower Cretaceous, lat. 49°21′59″N, long. 114°11′05″W, Natal Ridge, Crowsnest area, British Columbia.

Tsugaepollenites mesozoicus Couper
Hypotype 13572
Pocock, S.A.J., 1964, Bull. Can. Petrol. Geol., vol. 12, Special Issue, pl., fig. 20.
Kootenay Formation, Upper Jurassic, depth 7,689 feet below K.B., Royalite DeWinton No. 2 borehole, l.s.d. 13, sec. 24, tp. 21, rge. 1, W.5th mer., southwestern Alberta.

cf. *Tsugaepollenites mesozoicus* Couper in Danzé-Corsin and Laveine, 1963
Hypotype 13757
McGregor, D.C., 1965, Geol. Surv. Can., Paper 64-55, pl. 4, fig. 6.
Borden Island Formation, Lower Jurassic, west side of Black Top Ridge near Eureka, western Ellesmere Island, District of Franklin.

Unidentified
Fig. specs. 13739, 13787
McGregor, D.C., 1965, Geol. Surv. Can., Paper 64-55, pl. 3, fig. 37; pl. 4, fig. 46.
Borden Island Formation, Lower Jurassic, west side of Black Top Ridge near Eureka, western Ellesmere Island, District of Franklin.

Uvaesporites argenteaeformis (Bolchovitina) Schulz
Hypotype 75452
Davies, E.H., 1985, Palynology, vol. 9, pl. 2, fig. 4.
Late Pliensbachian, Liassic, Zambujal Section, south of Condeixa, Lusitanian Basin, west-central Portugal.

Uvaesporites puzzlei Guy
Fig. spec. 58927
Poulton, T.P., Leskiw, K. and Audretsch, A., 1982, Geol. Surv. Can., Bull. 325, pl. 6, fig. 21.
Almstrom Creek Formation, Lower Jurassic, Little Fish Creek northward from junction with Almstrom Creek, approximately lat. 68°25′40″N, long. 136°10′30″W, District of Mackenzie.

Verrucosisporites manumii Pocock
Hypotype 13563
Pocock, S.A.J., 1964, Bull. Can. Petrol. Geol., vol. 12, Special Issue, pl., fig. 8.
Kootenay Formation, Upper Jurassic, depth 7,689 feet below K.B., Royalite DeWinton No. 2 borehole, l.s.d. 13, sec. 24, tp. 21, rge. 1, W.5th mer., southwestern Alberta.

Verrucosisporites sp.
Fig. spec. 13564
Pocock, S.A.J., 1964, Bull. Can. Petrol. Geol., vol. 12, Special Issue, pl., fig. 9.
Kootenay Formation, Upper Jurassic, depth 7,689 feet below K.B., Royalite DeWinton No. 2 borehole, l.s.d. 13, sec. 24, tp. 21, rge. 1, W.5th mer., southwestern Alberta.

Verrucosisporites sp. A
Fig. spec. 47935
Bujak, J.P. and Williams, G.L., 1977, Strat. Micropal. Atlantic Basin and Borderlands, pl. 2, fig 3.
Jurassic, 6600-6640 feet, Shell Eurydice P-36 well, Scotian Shelf, Atlantic Ocean.

Vitreisporites pallidus (Reissinger) Nilsson, 1958
Hypotype 13775
McGregor, D.C., 1965, Geol. Surv. Can., Paper 64-55, pl. 4, fig. 33.
Borden Island Formation, Lower Jurassic, west side of Black Top Ridge near Eureka, western Ellesmere Island, District of Franklin.

Cretaceous

Abietineaepollenites microreticulatus Brenner
Hypotype 76608
Davis, E.H., Akande, S.O. and Zentilli, M., 1984, Geol. Surv. Can., Paper 84-1A, pl. 48.1, fig. 19.
Early Cretaceous, sample DH2-409, Gays River mine, lat. 45°02′N, long. 63°22′W, Nova Scotia.

?*Abietineaepollenites minimus* Couper, 1958
Hypotypes 13961, 13962
McGregor, D.C., 1965, Geol. Surv. Can., Paper 64-55, pl. 10, figs. 33, 34, 36.
Lower Cretaceous, Peel Plateau, Stony Creek about 18.5 miles upstream from mouth, District of Mackenzie.

Abietineaepollenites sp.
Fig. specs. 38202, 38203
McIntyre, D.J., 1974, Geol. Surv. Can., Paper 74-14, pl. 14, figs. 33, 34.
Upper Cretaceous, lat. 69°27′30″N, long. 126°58′W, and lat. 69°28′N, long. 126°58′W, Horton River, District of Mackenzie.

Acanthotriletes varispinosus Pocock
Hypotype 23792
Hopkins, Jr., W.S., 1971, Geol. Surv. Can., Bull. 197, p. 114, pl. 20, fig. 6.
Isachsen Formation, Lower Cretaceous, lat. 76°20′N, long. 115°34′W, northwest Melville Island, District of Franklin.

Acanthotriletes varispinosus Pocock
Hypotype 42400
Brideaux, W.W. and McIntyre, D.J., 1975, Geol. Surv. Can., Bull. 252, pl. 1, fig. 13.

Crossley Lakes Member, Langton Bay Formation, Lower Cretaceous, lat. 69°04′N, long. 125°51′W, Horton River, District of Mackenzie.

Acanthotriletes cf. *A. varispinosus* Pocock
Hypotype 33948
Hopkins, Jr., W.S., 1974, Geol. Surv. Can., Paper 73-12, p. 9, pl. 1, fig. 9.
Christopher Formation, Lower Cretaceous, lat. 76°20′N, long. 115°34′W, northwest Melville Island, District of Franklin.

Acanthotriletes sp.
Fig. spec. 13918
McGregor, D.C., 1965, Geol. Surv. Can., Paper 64-55, pl. 8, fig. 15.
Isachsen Formation, Lower Cretaceous, south of Buchanan Lake anticline, eastern Axel Heiberg Island, District of Franklin.

Aequitriradites ornatus Uphaw, 1963
Hypotype 91334
Sweet, A.R. and McIntyre, D.J., 1988, Can. Soc. Petrol. Geol., Mem. 15, fig. 5.1.
Moosehound Member, Cardium Formation, Upper Cretaceous, south side of McLeod River, l.s.d. 16, sec. 16, tp. 47, rge. 23, W.5th mer., Alberta.

Aequitriradites spinulosus (Cookson and Dettmann)
Hypotype 25070
Hopkins, Jr., W.S., 1971, Geol. Surv. Can., Bull. 197, p. 117, pl. 20, fig. 25.
Isachsen Formation, Lower Cretaceous, lat. 76°20′N, long. 115°34′W, northwest Melville Island, District of Franklin.

Aequitriradites spinulosus (Cookson and Dettmann)
Hypotype 41904
Williams, G.L., 1975, Geol. Surv. Can., Paper 74-30, vol. 2, pl. 8, fig. 9.
Berriasian-Valanginian, Lower Cretaceous, Amoco IEO Puffiin B-90 well, Grand Banks.

Aequitriradites spinulosus (Cookson and Dettmann) Cookson and Dettmann
Hypotype 42485
Brideaux, W.W. and McIntyre, D.J., 1975, Geol. Surv. Can., Bull. 252, pl. 3, fig. 19.
Horton River, Lower Cretaceous, lat. 69°19′N, long. 126°50′W, Horton River, District of Mackenzie.

Aequitriradites verrucosus (Cookson and Dettmann) Cookson and Dettmann
Hypotypes 56771, 56772
Williams, G.L. and Bujak, J.P., 1980, Initial Repts. Deep Sea Drilling Project, vol. 50, pl. 10, figs. 1-3.
Early Cretaceous, DSDP site 416, lat. 32°50.18′N, long. 10°48.06′W, northwestern Moroccan continental margin.

Aequitriradites verrucosus (Cookson and Dettmann) Dettmann
Hypotype 76602
Davis, E.H., Akande, S.O. and Zentilli, M., 1984, Geol. Surv. Can., Paper 84-1A, pl. 48.1, fig. 13.
Early Cretaceous, sample AGR 285-230, Gays River mine, lat. 45°02′N, long. 63°22′W, Nova Scotia.

Alisporites bilateralis Rouse
Hypotype 38207
McIntyre, D.J., 1974, Geol. Surv. Can., Paper 74-14, pl. 14, fig. 38.
Upper Cretaceous, lat. 69°27′30″N, long. 126°58′W, Horton River, District of Mackenzie.

Alisporites bilateralis Rouse
Hypotype 42498
Brideaux, W.W. and McIntyre, D.J., 1975, Geol. Surv. Can., Bull. 252, pl. 3, fig. 32.
Horton River, Lower Cretaceous, lat. 69°19′N, long. 126°50′W, Horton River, District of Mackenzie.

Alisporites cf. *A. bilateralis* Rouse
Hypotype 41623
Hopkins, Jr., W.S. and Sweet, A.R., 1976, Geol. Surv. Can., Bull. 256, pl. 10, fig. 10.
Mattagami Formation, Lower Cretaceous, borehole 68-42, lat. 50°35′36″N, long. 82°30′12″W, Ontario.

Alisporites bisaccus Rouse, 1959
Hypotype 13963
McGregor, D.C., 1965, Geol. Surv. Can., Paper 64-55, pl. 10, fig. 35.
Lower Cretaceous, Peel Plateau, Stony Creek about 18.5 miles upstream from mouth, District of Mackenzie.

?*Alisporites grandis* (Cookson) Dettmann, 1963
Hypotype 13851
McGregor, D.C., 1965, Geol. Surv. Can., Paper 64-55, pl. 6, fig. 36.
Lower Cretaceous, Fault Creek Canyon, Aklavik Range, Richardson Mountains, District of Mackenzie.

Alisporites grandis (Cookson) Dettmann
Hypotype 42495
Brideaux, W.W. and McIntyre, D.J., 1975, Geol. Surv. Can., Bull. 252, pl. 3, fig. 29.
Crossley Lakes Member, Langton Bay Formation, Lower Cretaceous, lat. 69°04′N, long. 125°51′W, Horton River, District of Mackenzie.

Alisporites cf. *A. grandis* (Cookson) Dettmann, 1963
Hypotypes 13892, 13893
McGregor, D.C., 1965, Geol. Surv. Can., Paper 64-55, pl. 7, figs. 42, 43.
Lower Cretaceous, Donna River Canyon, Aklavik Range, Richardson Mountains, District of Mackenzie.

Alisporites minutus Rouse, 1959
Hypotype 13887
McGregor, D.C., 1965, Geol. Surv. Can., Paper 64-55, pl. 7, fig. 37.
Lower Cretaceous, Donna River Canyon, Aklavik Range, Richardson Mountains, District of Mackenzie.

Alisporites minutus Rouse
Hypotype 42497
Brideaux, W.W. and McIntyre, D.J., 1975, Geol. Surv. Can., Bull. 252, pl. 3, fig. 31.
Crossley Lakes Member, Langton Bay Formation, Lower Cretaceous, lat. 69°12′30″N, long. 126°17′W, Horton River, District of Mackenzie.

Alisporites thomasii (Couper) Nilsson, 1958
Hypotype 13889
McGregor, D.C., 1965, Geol. Surv. Can., Paper 64-55, pl. 7, fig. 39.
Lower Cretaceous, Donna River Canyon, Aklavik Range, Richardson Mountains, District of Mackenzie.

Alisporites sp.
Fig. specs. 13884, 13891
McGregor, D.C., 1965, Geol. Surv. Can., Paper 64-55, pl. 7, figs. 33, 41.
Lower Cretaceous, Donna River Canyon, Aklavik Range, Richardson Mountains, District of Mackenzie.

Alisporites sp.
Fig. spec. 41621
Hopkins, Jr., W.S. and Sweet, A.R., 1976, Geol. Surv. Can., Bull. 256, pl. 10, fig. 8.
Mattagami Formation, Lower Cretaceous, borehole 68-42, lat. 50°35′36″N, long. 82°30′12″W, Ontario.

?*Alisporites* sp.
Fig. spec. 13852
McGregor, D.C., 1965, Geol. Surv. Can., Paper 64-55, pl. 6, fig. 38.
Lower Cretaceous, Fault Creek Canyon, Aklavik Range, Richardson Mountains, District of Mackenzie.

Alnipollenites trina (Stanley) Norton, 1969
Hypotype 71393
Wilson, M.A., 1978, Palaeontographica, Abt. B, vol. 166, p. 145, pl. 10, fig. 15.
Cretaceous, Police Island, north bank Mackenzie River, 12 miles southeast of Fort Norman, District of Mackenzie.

Alnipollenites verus Potonié, 1931
Hypotypes 71410, 71411
Wilson, M.A, 1988, Palaeontographica, Abt. B, vol. 166, p. 145, pl.11, figs. 9, 10.
Cretaceous, IEO Atkinson H-25 well, north shore Tuktoyaktuk Peninsula, lat. 69°44′20″N, long. 131°50′06″W, District of Mackenzie.

Alnipollenites sp.
Fig. spec. 38293
McIntyre, D.J., 1974, Geol. Surv. Can., Paper 74-14, pl.22, fig.15.
Upper Cretaceous, lat. 69°58′30″N, long. 127°04′W, Horton River, District of Mackenzie.

Alterbidinium sp.
Fig. spec. 89728
Jerzykiewicz, T. and Sweet, A.R., 1988, Sedimentary Geol., vol. 59, no. 1/2, p. 58, pl. III, fig. 1.
St. Mary River Formation, Upper Cretaceous, Crowsnest River, lat. 49°30′N, long. 114°3′W, Alberta.

Anacolosidites sp.
Fig. specs. 38232-38234, 38237
McIntyre, D.J., 1974, Geol. Surv. Can., Paper 74-14, pl. 16, figs. 5-7; pl. 17, fig. 3.
Upper Cretaeous, lat. 69°28′N, long. 126°58′W, Horton River, District of Mackenzie.

Apiculatisporis sp.
Fig. spec. 31747
Hopkins, Jr., W.S. and Balkwill, H.R., 1973, Geol. Surv. Can., Paper 72-37, p. 11, pl. 1, fig. 8.
Hassel Formation, Cretaceous, Hoodoo Dome, lat. 78°11′N, long. 99°40′W, southern Ellef Ringnes Island, District of Franklin.

Appendicisporites sp. cf. *Anemia trichacantha* (Maljavkina) Markova
Fig. spec. 42446
Brideaux, W.W. and McIntyre, D.J., 1975, Geol. Surv. Can., Bull. 252, pl. 2, fig. 18.
Crossley Lakes Member, Langton Bay Formation, Lower Cretaceous, lat. 69°04′N, long. 125°51′W, Horton River, District of Mackenzie.

Appendicisporites auritus Agasie *in* Singh, 1983
Hypotype 91335
Sweet, A.R. and McIntyre, D.J., 1988, Can. Soc. Petrol. Geol., Mem. 15, fig. 5.2.
Sturrock Member, Cardium Formation, Upper Cretaceous, south side of McLeod River, l.s.d. 16, sec. 16, tp. 47, rge. 23, W.5th mer., Alberta.

Appendicisporites baconicus (Deak) Srivastava, 1975
Hypotype 91336
Sweet, A.R. and McIntyre, D.J., 1988, Can. Soc. Petrol. Geol., Mem. 15, figs. 5.3, 5.4.
Moosehound Member, Cardium Formation, Upper Cretaceous, south side of McLeod River, l.s.d. 16, sec. 16, tp. 47, rge. 23, W.5th mer., Alberta.

Appendicisporites bifurcatus Singh
Hypotype 42438
Brideaux, W.W. and McIntyre, D.J., 1975, Geol. Surv. Can., Bull. 252, pl. 2, fig. 10.
Crossley Lakes Member, Langton Bay Formation, Lower Cretaceous, lat. 69°04′N, long. 125°51′W, Horton River, District of Mackenzie.

Appendicisporites bilateralis Singh
Hypotype 42439
Brideaux, W.W. and McIntyre, D.J., 1975, Geol. Surv. Can., Bull. 252, pl. 2, fig. 11.
Horton River Formation, Lower Cretaceous, lat. 69°19′N, long. 126°50′W, Horton River, District of Mackenzie.

Appendicisporites concentricus Kemp
Hypotype 56780
Williams, G.L. and Bujak, J.P., 1980, Initial Repts. Deep Sea Drilling Project, vol. 50, pl. 10, fig. 11.
Early Cretaceous, DSDP Site 416, lat. 32°50.18′N, long. 10°48.06′W, northwestern Morrocan continental margin.

Appendicisporites cristatus (Markova) Pocock
Hypotypes 42442, 42443
Brideaux, W.W. and McIntyre, D.J., 1975, Geol. Surv. Can., Bull. 252, pl. 2, figs. 14, 15.
Horton River Formation, Lower Cretaceous, lat. 69°19′N, long. 126°50′W, Horton River, District of Mackenzie.

Appendicisporites erdtmanii Pocock
Hypotypes 42444, 42445
Brideaux, W.W. and McIntyre, D.J., 1975, Geol. Surv. Can., Bull. 252, pl. 2, figs. 16, 17.
Horton River Formation, Lower Cretaceous, lat. 69°19′N, long. 126°50′W, Horton River, District of Mackenzie.

Appendicisporites erdtmanii Pocock, 1964
Hypotype 91337
Sweet, A.R. and McIntyre, D.J., 1988, Can. Soc. Petrol. Geol., Mem. 15, figs. 5.5, 5.8.
Moosehound Member, Cardium Formation, Upper Cretaceous, south side of McLeod River, l.s.d. 16, sec. 16, tp. 47, rge. 23, W.5th mer., Alberta.

Appendicisporites insignis (Markova) Chlonova *in* Singh, 1983
Hypotype 91338
Sweet, A.R. and McIntyre, D.J., 1988, Can. Soc. Petrol. Geol., Mem. 15, figs. 5.6, 5.7.
Ram Member, Cardium Formation, Upper Cretaceous, north side of Highway 47 north of Cadomin, l.s.d. 14, sec. 16, tp. 47, rge. 23, W.5th mer., Alberta.

Appendicisporites irregularis Pocock, 1964
Hypotype 91345
Sweet, A.R. and McIntyre, D.J., 1988, Can. Soc. Petrol. Geol., Mem. 15, fig. 5.17.
Ram Member, Cardium Formation, Upper Cretaceous, north side of Highway 47 north of Cadomin, l.s.d. 14, sec. 16, tp. 47, rge. 23, W.5th mer., Alberta.

Appendicisporites jansonii Pocock
Hypotype 41896
Williams, G.L., 1975, Geol. Surv. Can., Paper 74-30, vol. 2, pl. 8, fig. 1.
Albian, Lower Cretaceous, Mobil Sable Island C-67, Scotian Shelf.

Appendicisporites ?jansonii Pocock, 1962
Hypotype 13921
McGregor, D.C., 1965, Geol. Surv. Can., Paper 64-55, pl. 8, fig. 21.
Isachsen Formation, Lower Cretaceous, south of Buchanan Lake anticline, eastern Axel Heiberg Island, District of Franklin.

Appendicisporites potomacensis Brenner
Hypotypes 42440, 42441
Brideaux, W.W. and McIntyre, D.J., 1975, Geol. Surv. Can., Bull. 252, pl. 2, figs. 12, 13.
Crossley Lakes Member, Langton Bay Formation, Lower Cretaceous, lat. 69°04′N, long. 125°51′W, Horton River, District of Mackenzie.

Appendicisporites potomacensis Brenner, 1963
Hypotype 91340
Sweet, A.R. and McIntyre, D.J., 1988, Can. Soc. Petrol. Geol., Mem. 15, fig. 5.10.
Moosehound Member, Cardium Formation, Upper Cretaceous, south side of McLeod River, l.s.d. 16, sec. 16, tp. 47, rge. 23, W.5th mer., Alberta.

Appendicisporites problematicus (Burger)
Hypotype 41897
Williams, G.L., 1975, Geol. Surv. Can., Paper 74-30, vol. 2, pl. 8, fig. 2.
Berriasian - Velanginian, Lower Cretaceous, Amoco IEO Puffin B-90 well, Grand Banks.

Appendicisporites tricornatus Weyland and Greifeld
Hypotype 33967
Hopkins, Jr., W.S., 1974, Geol. Surv. Can., Paper 73-12, p. 14, pl. 3, fig. 29.
Christopher Formation, Lower Cretaceous, lat. 76°20′N, long. 115°34′W, northwest Melville Island, District of Franklin.

Appendicisporites undosus Hedlund, 1966
Hypotype 91339
Sweet, A.R. and McIntyre, D.J., 1988, Can. Soc. Petrol. Geol., Mem. 15, figs. 5.9, 5.13.
Moosehound Member, Cardium Formation, Upper Cretaceous, south side of Grande Cache Highway 40, l.s.d. 15, sec. 13, tp. 57, rge. 7, W.5th mer., Alberta.

Appendicisporites sp.
Fig. spec. 31759
Hokins, Jr., W.S. and Balkwill, H.R., 1973, Geol. Surv. Can., Paper 72-37, p.13, pl. 1, fig. 21.
Hassel Formation, Cretaceous, Hoodoo Dome, lat. 78°11′N, long. 99°40′W, southern Ellef Ringnes Island, District of Franklin.

Appendicisporites sp.
Fig. spec. 41597
Hopkins, Jr., W.S. and Sweet, A.R., 1976, Geol. Surv. Can., Bull. 256, pl. 9, fig. 13.
Mattagami Formation, Lower Cretaceous, borehole 68-38, lat. 50°36′N, long. 82°31′12″W, Ontario.

?*Appendicisporites* sp.
Fig. spec. 13951
McGregor, D.C., 1965, Geol. Surv. Can., Paper 64-55, pl. 10, fig. 16.
Lower Cretaceous, Peel Plateau, Stony Creek about 18.5 miles upstream from mouth, District of Mackenzie.

Aquilapollenites amicus Srivastava, 1968
Hypotype 76017
Jerzykiewicz, T. and Sweet, A.R., 1986, Can. J. Earth Sci., vol.23, no. 9, pl. 2, fig. 1.
Late Cretaceous, right bank of Embarras River towards the south end of Robb, l.s.d. 14, sec. 11, tp. 49, rge. 21, W.5th mer., Alberta.

Aquilapollenites cf. *A. amicus* Srivastava, 1968
Hypotype 48964
Sweet, A.R., 1978, Geol. Surv. Can., Paper 78-1B, p. 32, pl. 6.1, fig. 4.
Cuesta Creek Member, Tent Island Formation, Upper Cretaceous, northwestern bank Big Fish River near its junction with Boundary Creek, lat. 68°30′30″N, long. 136°23′50″W, Yukon.

Aquilapollenites asper Mchedlishvili
Hypotypes 38252, 38253, 38260
McIntyre, D.J., 1974, Geol. Surv. Can., Paper 74-14, p. 19, figs. 1-3; pl. 20, fig. 1.
Upper Cretaceous, lat. 69°28′N, long. 126°58′W, Horton River, District of Mackenzie.

Aquilapollenites attenuatus Funkhouser, 1961
Hypotype 71346
Wilson, M.A., 1978, Palaeontographica, Abt. B, vol. 166, p. 130, pl. 7, figs. 14, 15.
Cretaceous, Police Island, north bank Mackenzie River, 12 miles southeast of Fort Norman, District of Mackenzie.

Aquilapollenites attenuatus Funkhouser, 1961
Hypotype 76016
Jerzykiewicz, T. and Sweet, A.R., 1986, Can. J. Earth Sci., vol. 23, no. 9, pl. 1, fig. 20.
Late Cretaceous, N.W. sec. 11, tp. 49, rge. 21, W.5th mer., west-central Alberta Foothills.

Aquilapollenites cf. *A. attenuatus* Funkhouser, 1961
Hypotype 48966
Sweet, A.R., 1978, Geol. Surv. Can., Paper 78-1B, p.32, pl. 6.1, fig. 6.
Cuesta Creek Member, Tent Island Formation, Upper Cretaceous, northwestern bank Big Fish River near its junction with Boundary Creek, lat. 68°30′30″N, long. 136°23′50″W, Yukon.

Aquilapollenites aucellatus Srivastava
Hypotypes 38251, 38256
McIntyre, D.J., 1974, Geol. Surv. Can., Paper 74-14, pl. 18, fig. 19; pl. 19, fig. 4.
Upper Cretaceous, lat. 69°28″N, long. 126°58′W, Horton River, District of Mackenzie.

Aquilapollenites augustus Srivastava, 1969
Hypotype 76018
Jerzykiewicz, T. and Sweet, A.R., 1986, Can. J. Earth Sci., vol. 23, no. 9, pl. 2, fig. 2.
Cretaceous, depth 29.35-29.62 m , GSC core hole C4-42-2-2, Coal Valley strip mine, l.s.d. 1, sec. 10, tp. 47, rge. 19, W.5th mer., Alberta.

Aquilapollenites augustus Srivastava, 1969
Hypotype 89707
Jerzykiewicz, T. and Sweet, A.R., 1988, Sedimentary Geol., vol. 59, no. 1/2, p. 58, pl. I, fig. 11.
St. Mary River Formation, Upper Cretaceous, Crowsnest River, lat. 49°30′N, long. 114°3′W, Alberta.

Aquilapollenites bertillonites Funkhouser, 1961
Hypotypes 48962, 48963
Sweet, A.R., 1978, Geol. Surv. Can., Paper 78-1B, p. 32, pl. 6.1, figs. 2, 3.
Cuesta Creek Member, Tent Island Formation, Upper Cretaceous, northwestern bank Big Fish River near its junction with Boundary Creek, lat. 68°30′30″N, long. 136°23′50″W, Yukon.

Aquilapollenites calvus Tschudy and Leopold, 1971
Hypotype 71350
Wilson, M.A., 1978, Palaeontographica, Abt. B, vol. 166, p. 130, pl. 8, fig. 1.
Cretaceous, Police Island, north bank Mackenzie River, 12 miles southeast of Fort Norman, District of Mackenzie.

Aquilapollenites clairireticulatus Samoilovitch 1965
Hypotype 89708
Jerzykiewicz, T. and Sweet, A.R., 1988, Sedimentary Geol., vol. 58, no. 1/2, pl. I, fig. 12.
Brazeau Formation, Upper Cretaceous, Blackstone River, lat. 52°43′20″N, long. 116°16′30″W, Alberta.

Aquilapollenites sp. cf. *A. clarireticulatus* (Samoilovich) B.D. Tschudy
Fig. spec. 27819
Brideaux, W.W., 1971, Geol. Surv. Can., Paper 71-1, pt. B, p. 89, pl. 1, fig. 4.
Basal bed, East Fork Formation, Late Cretaceous, Little Bear River, 24 miles southwest of Fort Norman, District of Mavkenzie.

Aquilapollenites conatus Norton 1965
Hypotype 48961
Sweet, A.R., 1978, Geol. Surv. Can., Paper 78-1B, p. 32, pl. 6.1, fig. 1.
Cuesta Creek Member, Tent Island Formation, Upper Cretaceous, northwestern bank Big Fish River near its junction with Boundary Creek, lat. 68°30′30″N, long. 136°23′50″W, Yukon.

Aquilapollenites conatus Norton, 1965
Hypotypes 71345, 71349
Wilson, M.A., 1978, Palaeontographica, Abt. B, vol. 166, p. 131, pl. 7, figs. 13, 19.
Cretaceous, Police Island, north bank Mackenzie River, 12 miles southeast of Fort Norman, District of Mackenzie.

Aquilapollenites conatus Norton, 1965
Hypotypes 76020, 76019
Jerzykiewicz, T. and Sweet, A.R., 1986, Can. J. Earth Sci., vol. 23, no. 9, pl. 2, figs. 3, 4.

Late Cretaceous, right bank of Embarras River towards the south end of Robb, l.s.d. 14, sec. 11, tp. 49, rge. 21, W.5th mer. and Coal Valley mine point, Alberta.

Aquilapollenites conatus Norton 1965
Hypotype 76142
Jerzykiewicz, T. and Sweet, A.R., 1986, Geol. Surv. Can., Paper 86-1B, Fig. 67.2, fig. 12.
William Creek Formation, Late Cretaceous, Castle River, l.s.d. 4-5, sec. 12, tp. 7, rge. 1, W.5th mer., Alberta.

Aquilapollenites conatus Norton, 1965
Hypotypes 93649, 89857
Sweet, A.R., Ricketts, B.D., Cameron, A.R. and Norris, D.K., 1989, Geol. Surv. Can., Paper 89-1G, pl. 1, figs. 19, 28.
Summit Creek Formation, Upper Cretaceous, north bank Mackenzie River immediately downstream from Police Island, lat. 64°52′42″N, long. 125°14′44″W, District of Mackenzie.

Aquilapollenites contiguus Tschudy, 1969
Hypotype 71343
Wilson, M.A., 1978, Palaeontographica, Abt. B, vol. 166, p. 131, pl. 7, figs. 9, 10.
Cretaceous, IEO Atkinson H-25 well, north shore Tuktoyaktuk Peninsula, lat. 69°44′20″N, long. 131°50′06″W, District of Mackenzie.

Aquilapollenites delicatus var. *collaris* Tschudy and Leopold, 1971
Hypotype 76021
Jerzykiewicz, T. and Sweet, A.R., 1986, Can. J. Earth Sci., vol. 23, no. 9, pl. 2, fig. 5.
Late Cretaceous, right bank of Embarras River towards the southend of Robb, l.s.d. 14, sec. 11, tp. 49, rge. 21, W.5th mer., Alberta.

Aquilapollenites delicatus var. *collaris* Tschudy and Leopold, 1971
Hypotype 93657
Sweet, A.R., Ricketts, B.D., Cameron, A.R. and Norris, D.K., 1989, Geol. Surv. Can., p. 98, pl. 2, fig. 8.
Summit Creek Formation, Upper Cretaceous, north bank Mackenzie River immediately downstream from Police Island, lat. 64°52′42″N, long. 125°14′44″W, District of Mackenzie.

Aquilapollenites delicatus var. *delicatus*
Hypotypes 76023, 76026
Jerzykiewicz, T. and Sweet, A.R., 1986, Can. J. Earth Sci., vol. 23, no. 9, pl. 2, figs. 7, 10.
Late Cretaceous, right bank of Embarras River towards the south end of Robb, l.s.d. 14, sec. 11, tp. 49, rge.21, W.5th mer., Alberta.

Aquilapollenites drumhellerensis Srivastava 1969
Hypotype 89709
Jerzykiewicz, T. and Sweet, A.R., 1988, Sedimentary Geol., vol. 59, no. 1/2, pl. I, fig. 13.
Brazeau Formation, Upper Cretaceous, Blackstone River, lat. 52°43′20″N, long. 116°30′W, Alberta.

Aquilapollenites formosus Srivastava and Rouse
Hypotypes 38259, 38261
McIntyre, D.J., 1974, Geol. Surv. Can., Paper 74-14, pl. 19, figs 7; pl. 20, fig. 2.
Upper Cretaceous, lat. 69°28′N, long. 126°58′W, Horton River, District of Mackenzie.

Aquilapollenites funkhouseri Srivastava 1968
Hypotype 89710
Jerzykiewicz, T. and Sweet, A.R., 1988, Sedimentary Geol., vol. 59, no. 1/2, p. 58, pl. I, fig. 14.
St. Mary River Formation, Upper Cretaceous, Crowsnest River, lat. 49°30′N, long. 114°3′W, Alberta.

Aquilapollenites insignis Mchedlishvili, 1961
Hypotype 93647
Sweet, A.R., Ricketts, B.D., Cameron, A.R. and Norris, D.K., 1989, Geol. Surv. Can., Paper 89-1G, p. 93, pl. 1, fig. 17.
Little Bear Formation, Upper Cretaceous, east bank Little Bear River, lat. 64°38′24″N, long. 126°19′W, District of Mackenzie.

Aquilapollenites leucocephalus Srivastava 1968
Hypotype 89730
Jerzykiewicz, T. and Sweet, A.R., 1988, Sedimentary Geol., vol. 59, no. 1/2, p. 58, pl. III, fig. 3.
St. Mary Formation, Upper Cretaceous, Crowsnest River, lat. 49°30′N, long. 114°3′W, Alberta.

Aquilapollenites sp. cf. A. oblatus Srivastava
Fig. spec. 27816
Brideaux, W.W., 1971, Geol. Surv. Can., Paper 71-1, pt. B, p.89, pl. 1, fig. 1.
Basal bed, East Fork Formation, Late Cretaceous, Little Bear River, 24 miles southwest of Fort Norman, District of Mackenzie.

Aquilapollenites psilatus Srivastava, 1966
Hypotype 71347
Wilson, M.A., 1978, Palaeontographica, Abt. B, vol. 166, p. 132, pl. 7, figs. 16, 17.
Cretaceous, Police Island, north bank Mackenzie River, 12 miles southeast of Fort Norman, District of Mackenzie.

Aquilapollenites quadicretaeus Chlonova, 1961
Hypotype 93657
Sweet, A.R., Ricketts, B.D., Cameron, A.R. and Norris, D.K., 1989, Geol. Surv. Can., Paper 89-1G, pl. 2, fig. 5.
Summit Creek Formation, Upper Cretaceous, north bank Mackenzie River immediately downstream from Police Island, lat. 64°52′42″N, long. 125°14′44″W, District of Mackenzie.

Aquilapollenites quadrilobus Rouse
Hypotypes 38257, 38258
McIntyre, D.J., 1974, Geol. Surv. Can., Paper 74-14, pl. 19, figs. 5, 6.
Upper Cretaceous, lat. 69°28′N, long. 126°58′W, Horton River, District of Mackenzie.

Aquilapollenites quadrilobus Rouse, 1957
Hypotype 76025
Jerzykiewicz, T. and Sweet, A.R., 1986, Can. J. Earth Sci., vol. 23, no. 9, pl. 2, fig. 9.
Late Cretaceous, Embarras River, Alberta.

Aquilapollenites quadrilobus Rouse 1957
Hypotype 89714
Jerzykiewicz, T. and Sweet, A.R., 1988, Sedimentary Geol., vol. 59, no. 1/2, p. 58, pl. I, fig. 18.
Brazeau Formation, Upper Cretaceous, Blackstone River, lat. 52°43′20″N, long. 116°16′30″W, Alberta.

Aquilapollenites rectus Tschudy 1969
Hypotype 89715
Jerzykiewicz, T. and Sweet, A.R., 1988, Sedimentary Geol., vol. 59, no. 1/2, p. 58, pl. I, fig. 19.
St. Mary River Formation, Upper Cretaceous, Crowsnest River, lat. 49°30′N, long. 114°3′W, Alberta.

Aquilapollenites reductus Norton, 1965
Hypotype 76029
Jerzykiewicz, T. and Sweet, A.R., 1986, Can. J. Earth Sci., vol. 23, no. 9, pl. 2, fig. 13.
Late Cretaceous, depth 29.35-29.62 m, GSC core hole CV-42-2-2, Coal Valley strip mine, l.s.d. 1, sec. 10, tp. 47, rge. 19, W.5th mer., Alberta.

Aquilapollenites reticulatus (Mchedlishvili) Tschudy and Leopold 1970
Hypotype 48965
Sweet, A.R., 1978, Geol. Surv. Can., Paper 78-1B, p. 34, pl. 6.1, fig. 5.
Tent Island Formation, Upper Cretaceous, northwestern bank Big Fish River near its junction with Boundary Creek, lat. 68°30′30″N, long. 136°23′50″W, Yukon.

Aquilapollenites rigidus Tschudy and Leopold, 1971
Hypotype 93634
Sweet, A.R., Ricketts, B.D., Cameron, A.R. and Norris, D.K., 1989, Geol. Surv. Can., Paper 89-1G, p. 93, pl. 1, fig. 4.
Little Bear Formation, Upper Cretaceous, east of McKay Range, East Little Bear River area, lat. 64°36′24″N, long. 125°35′W, District of Mackenzie.

Aquilapollenites rombicus Samoilovich, 1965
Hypotype 93633
Sweet, A.R., Ricketts, B.D., Cameron, A.R. and Norris, D.K., 1989, Geol. Surv. Can., Paper 89-1G, p. 93, pl. 1, fig. 3.
Little Bear Formation, Upper Cretaceous, east of McKay Range, East Little Bear River area, lat. 64°36′24″N, long. 125°35′W, District of Mackenzie.

Aquilapollenites senonicus (Mtchedlishvili) Tschudy and Leopold, 1971
Hypotype 71357
Wilson, M.A., 1978, Palaeontographica, Abt. B, vol. 166, p. 132, pl. 8, figs. 9, 10.
Cretaceous, Police Island, north bank Mackenzie River, 12 miles southeast of Fort Norman, District of Mackenzie.

Aquilapollenites senonicus Mchedlishvili *in* Samoilovitch and Mchedlishvili 1961
Hypotype 89713
Jerzykiewicz, T. and Sweet, A.R., 1988, Sedimentary Geol., vol. 59, no. 1/2, p. 59, pl. I, fig. 17.
Brazeau Formation, Upper Cretaceous, Blackstone River, lat. 52°43′20″N, long. 116°16′30″W, Alberta.

Aquilapollenites sentus Srivastava, 1969
Hypotype 76022
Jerzykiewicz, T. and Sweet, A.R., 1986, Can. J. Earth Sci., vol. 23, no. 9, pl. 2, fig. 6.
Late Cretaceous, Embarras River, Alberta.

Aquilapollenites sentus Srivastava, 1969
Hypotype 89722
Jerzykiewicz, T. and Sweet, A.R., 1988, Sedimentary Geol., vol. 59, no. 1/2, p. 58, pl. II, fig. 7.
St. Mary River Formation, Upper Cretaceous, Crowsnest River, lat. 49°30′N, long. 114°3′W, Alberta.

Aquilapollenites cf. *A. sentus* Srivastava 1969
Hypotypes 48970, 48971
Sweet, A.R., 1978, Geol. Surv. Can., Paper 78-1B, p. 34, pl. 6.1, figs. 10, 11.
Cuesta Creek Member, Tent Island Formation, Upper Cretaceous, northwestern bank Big Fish River near its junction with Boundary Creek, lat. 68°30′30″N, long. 136°23′50″W, Yukon.

Aquilapollenites spinulosus Funkhouser, 1961
Hypotype 71344
Wilson, M.A., 1978, Palaeontographica, Abt. B, vol. 166, p. 133, pl. 7, figs. 11, 12.
Cretaceous, IEO Atkinson H-25 well, north shore Tuktoyaktuk Peninsula, lat. 67°44′20″N, long. 131°50′06″W, District of Mackenzie.

Aquilapollenites trialatus Rouse
Hypotypes 27822-27824
Brideaux, W.W., 1971, Geol. Surv. Can., Paper 71-1, pt. B, p. 89, pl. 1, figs. 7, 9.
Basal bed, East Fork Formation, Late Cretaceous, Little Bear River, 24 miles southeast of Fort Norman, District of Mackenzie.

Aquilapollenites trialatus Rouse 1957
Hypotype 89712
Jerzykiewicz, T. and Sweet, A.R., 1988, Sedimentary Geol., vol. 59, no. 1/2, p. 58, pl. I, fig. 16.
St. Mary Formation, Upper Cretaceous, lat. 49°30′N, long. 114°3′W, Alberta.

Aquilapollenites trialatus Rouse, 1957
Hypotype 93640
Sweet, A.R., Ricketts, B.D., Cameron, A.R. and Norris, D.K., 1989, Geol. Surv. Can., Paper 89-1G, p. 93, pl. 1, fig. 10.
Little Bear Formation, Upper Cretaceous, east bank Little Bear River, lat. 64°38′24″N, long. 126°19′W, District of Mackenzie.

Aquilapollenites cf. *A. trialatus* Rousse, 1957
Hypotype 71362
Wilson, M.A., 1978, Palaeontographica, Abt. B, vol. 166, p. 133, pl. 8, fig. 15.
Tent Island Formation, Cretaceous, Little Fish (Cache) Creek, Big Fish River area, District of Mackenzie.

Aquilapollenites trialatus Rouse var. *trialatus* Tschudy and Leopold, 1971
Hypotype 71353
Wilson, M.A., 1978, Palaeontographica, Abt. B, vol. 166, p. 133, pl. 8, fig. 5.
Cretaceous, IEO Atkinson H-25 well, north shore Tuktoyaktuk Peninsula, lat. 69°44′20″N, long. 131°50′06″W, District of Mackenzie.

Aquilapollenites trialatus Rouse var. *uniformis* Tschudy and Leopold, 1971
Hypotype 71352
Wilson, M.A., 1978, Palaeontographica, Abt. B, vol. 166, p. 134, pl. 8, fig. 4.
Cretaceous, IEO Atkinson H-25 well, north shore Tuktoyaktuk Peninsula, lat. 69°44′20″N, long. 131°50′06″W, District of Mackenzie.

Aquilapollenites trialatus var. *uniformis* Tschudy and Leopold, 1971
Hypotype 93639
Sweet, A.R., Ricketts, B.D., Cameron, A.R. and Norris, D.K., 1989, Geol. Surv. Can., Paper 89-1G, p. 93, pl. 1, fig. 9.
Little Bear Formation, Upper Cretaceous, east of McKay Range, East Little Bear River area, lat. 64°36′24″N, long. 125°35′W, District of Mackenzie.

Aquilapollenites trialatus Rouse var. *variabilis* Tschudy and Leopold, 1971
Hypotype 71356
Wilson, M.A., 1978, Palaeontographica, Abt. B, vol. 166, p. 134, pl. 8, fig. 8.
Cretaceous, IEO Atkinson H-25 well, north shore Tuktoyaktuk Peninsula, lat. 69°44′20″N, long. 131°50′06″W, District of Mackenzie.

Aquilapollenites trialatus var. *variabilis* Tschudy and Leopold, 1971
Hypotype 93638
Sweet, A.R., Ricketts, B.D., Cameron, A.R. and Norris, D.K., 1989, Geol. Surv. Can., Paper 89-1G, p. 93, pl. 1, fig. 8.
Little Bear Formation, Upper Cretaceous, east bank Little Bear River, lat. 64°38′24″N, long. 126°19′W, District of Mackenzie.

Aquilapollenites turbidus Tschudy, 1973
Hypotype 93641
Sweet, A.R., Ricketts, B.D., Cameron, A.R. and Norris, D.K., 1989, Geol. Surv. Can., Paper 89-1G, p. 93, pl. 1, fig. 11.
Little Bear Formation, Upper Cretaceous, east Little Bear River lat. 64°38′24″N, long. 126°19′W, District of Mackenzie.

Aquilapollenites vulsus Sweet
Holotype 76093; paratypes 76094, 76095
Sweet, A.R., 1986, Can. J. Earth Sci., vol. 23, no. 9, p. 1378, pl. 1, figs. 3-5.
Late Cretaceous, depth 29.35-29.62 m, GSC core hole CV-42-2-2, Coal Valley strip mine, l.s.d. 1, sec. 10, tp. 47, rge. 19, W.5th mer., Alberta.

Aquilapollenites vulsus Sweet, 1986
Hypotype 76031
Jerzykiewicz, T. and Sweet, A.R., 1986, Can. J. Earth Sci., vol. 23, no. 9, pl. 2, fig. 15.
Late Cretaceous, depth 29.35-29.62 m, GSC core hole CV-42-2-2, Coal Valley strip mine, l.s.d. 1, sec. 10, tp. 47, rge. 19, W.5th mer., Alberta.

Aquilapollenites sp.
Fig. spec. 71360
Wilson, M.A., 1978, Palaeontographica, Abt. B, vol. 166, p. 134, pl. 8, fig. 13.
Cretaceous, IEO Atkinson H-25 well, north shore Tuktoyaktuk Peninsula, lat. 69°44′20″N, long. 131°50′06″W, District of Mackenzie.

Aquilapollenites sp.
Fig. specs. 93682, 93683
Sweet, A.R. and Braman, D.R., 1989, Geol. Surv. Can., Paper 89-8, p. 35, pl.1, figs. 17, 18.
Chungo Member, Wapiabi Formation, Upper Cretaceous, Trapp Creek near junction with Highwood River, sec. 36, tp. 17, rge. 4, W.5th mer., Alberta.

Aquilapollenites sp. A
Fig. specs. 48972, 48973
Sweet, A.R., 1978, Geol. Surv. Can., Paper 78-1B, p. 34, pl. 6.1, figs. 12, 13.
Cuesta Creek Member, Tent Island Formation, Upper Cretaceous, northwestern bank Big Fish River near its junction with Boundary Creek, lat. 68°30′30″N, long. 136°23′50″W, Yukon.

Araucariacites australis Cookson
Hypotype 25096
Hopkins, Jr., W.S., 1971, Geol. Surv. Can., Bull. 191, p. 121, pl. 22, fig. 5.
Isachsen Formation, Lower Cretaceous, lat. 76°20′N, long. 115°34′W, northwest Melville Island, District of Franklin.

Araucariacites australis Cookson
Hypotype 33999
Hopkins, Jr., W.S., 1974, Geol. Surv. Can., Paper 73-12, p. 22, pl. 5, fig. 66.

Christopher Formation, Lower Cretaceous, lat. 76°20′N, long. 115°34′W, northwest Melville Island, District of Franklin.

Araucariacites australis Cookson
Hypotype 42517
Brideaux, W.W. and McIntyre, D.J., 1975, Geol. Surv. Can., Bull. 252, pl. 4, fig. 15.
Horton River Formation, Lower Cretaceous, lat. 69°19′N, long. 126°50′W, Horton River, District of Mackenzie.

Araucariacites limbatus (Balme) Habib, 1969
Hypotype 71342
Wilson, M.A., 1978, Palaeontographica, Abt. B, vol. 166, p. 127, pl. 7, fig. 8.
Cretaceous, Police Island, north bank Mackenzie River, 12 miles southeast of Fort Norman, District of Mackenzie.

Arcellites disciformis Miner
Hypotypes 41644, 41645
Hopkins, Jr., W.S. and Sweet, A.R., 1976, Geol. Surv. Can., Bull. 256, pl. 12, figs. 1, 4.
Mattagami Formation, Lower Cretaceous, Onakawana lignite pit stock pile, lat. 50°36′N, long. 81°23′W, Ontario.

Arcellites disciformis Miner emend. Ellis and Tschudy 1964
Hypotype 48937
Sweet, A.R., 1979, Am. Assoc. Strat. Palynol., Contrib. Ser. No. 5B, pl. 3, fig. 12.
Hassel Formation, Lower Cretaceous, lat. 79°47′24″N, long. 85°22′W, near Slidre Fiord, Ellesmere Island, District of Franklin.

Arcellites incipens Singh
Hypotypes 41650, 41651
Hopkins, Jr., W.S. and Sweet, A.R., 1976, Geol. Surv. Can., Bull. 256, pl. 12, figs. 7, 8.
Mattagami Formation, Lower Cretaceous, Onakawana lignite pit stock pile, lat. 50°36′N, long. 81°23′W and borehole 68-39, lat. 50°35′48″N, long. 82°30′42″W, Ontario.

Arcellites incipiens Singh 1964
Hypotype 48913
Sweet, A.R., 1979, Am. Assoc. Strat. Palynol., Contrib. Ser. No. 5B, pl. 2, fig. 9.
Mattagami Formation, Early Cretaceous, lat. 50°36′N, long. 81°23′W, Hudson Bay region, northern Ontario.

Arcellites reticulatus (Cookson and Dettmann) Potter 1963
Hypotype 48938
Sweet, A.R., 1979, Am. Assoc. Strat. Palynol., Contrib. Ser. No. 5B, pl. 3, fig. 13.
Hassel Formation, Lower Cretaceous, lat. 79°47′24″N, long. 85°22′W, near Slidre Fiord, Ellesmere Island, District of Franklin.

Arecipites columellus Leffingwell, 1971
Hypotype 76034
Jerzykiewicz, T. and Sweet, A.R., 1986, Can. J. Earth Sci., vol. 23, no. 9, pl. 3, fig. 1.
Late Cretaceous, Coal Valley mine pit, Alberta.

Ariadnaesporites cristatus Tschudy 1966
Hypotype 48956
Sweet, A.R., 1979, Am. Assoc. Strat. Palynol., Contrib. Ser. No. 5B, pl. 4, fig. 14.
Eureka Sound Formation, Upper Cretaceous, lat. 78°17′45″N, long. 100°45′W, Ellef Ringnes Island, District of Franklin.

Aridnaesporites cf. *A. fluviatilis* Hall 1967
Hypotype 48948
Sweet, A.R., 1979, Am. Assoc. Strat. Palynol., Contrib. Ser. No. 5B, pl. 4, fig. 6.
Eureka Sound Formation, Upper Cretaceous, Meteorologist Peninsula, lat. 78°10′N, long. 99°30′W, Ellef Ringnes Island, District of Franklin.

Ariadnaesporites intermedius Hall 1967
Hypotype 48939
Sweet, A.R., 1979, Am. Assoc. Strat. Palynol., Contrib. Ser. No. 5B, pl. 3, fig. 14.
Hassel Formation, Lower Cretaceous, Investigator Point, lat. 74°10′N, long. 119°9′W, Banks Island, District of Franklin.

Arecipites columellus Leffingwell
Hypotypes 38346, 38347
McIntyre, D.J., 1974, Geol. Surv. Can., Paper 74-14, pl. 24, figs. 22, 23.
Upper Cretaceous, lat. 69°28′N, long. 126°58′W, Horton River, District of Mackenzie.

Artipollis praecox Ward, 1986
Hypotype 91383
Sweet, A.R. and McIntyre, D.J., 1988, Can. Soc. Petrol. Geol., Mem. 15, fig. 7.2.
Moosehound Member, Cardium Formation, Upper Cretaceous, south side of McLeod River, l.s.d. 16, sec. 16, tp. 47, rge. 23, W.5th mer., Alberta.

Asteropollis vulgaris (Groot and Singh) Singh, 1983
Hypotype 91384
Sweet, A.R. and McIntyre, D.J., 1988, Can. Soc. Petrol. Geol., Mem. 15, fig. 7.3.
Moosehound Member, Cardium Formation, Upper Cretaceous, south side of Grande Cache Highway 40, l.s.d. 15, sec. 13, tp. 57, rge. 7, W.5th mer., Alberta.

Atopospora cancellata (Burbridge and Felix) Jansonius and Hills, 1978
Hypotype 93642
Sweet, A.R., Ricketts, B.D., Cameron, A.R. and Norris, D.K., 1989, Geol. Surv. Can., Paper 89-1G, p. 93, pl. 1, fig. 12.
Little Bear Formation, Upper Cretaceous, east of McKay Range, East Little Bear River area, lat. 64°36′24″N, long. 125°35′W, District of Mackenzie.

Azolla barbata Snead 1969
Hypotype 76143
Jerzykiewicz, T. and Sweet, A.R., 1986, Geol. Surv. Can., Paper 86-1B, Fig. 67.2, fig. 15.
William Creek Formation, Late Cretaceous, Castle River, l.s.d. 4-5, sec. 12, tp. 7, rge. 1, W.5th mer., Alberta.

Azolla cretacea Stanley, 1965
Hypotype 71278
Wilson, M.A., 1978, Palaeontographica, Abt. B, vol. 166, p. 149, pl. 2, figs. 2, 5.
Cretaceous, Police Island, north bank Mackenzie River, 12 miles southeast of Fort Norman, District of Mackenzie.

Azolla cretacea Stanley, 1965
Hypotype 75999
Jerzykiewicz, T. and Sweet, A.R., 1986, Can. J. Earth Sci., vol. 23, no. 9, pl. 1, fig. 2.
Late Cretaceous, depth 29.35-29.62 m, GSC core hole C4-42-2-2, Coal Valley strip mine, l.s.d. 1, sec. 10, tp. 47, rge. 19, W.5th mer., Alberta.

Azollopsis (*Azollopsis*) *coccoides* Hall, 1968 Sweet and Hills
Hypotypes 30832, 30834-30836, 30838-30840
Sweet, A.R. and Hills, L.V., 1974, Can. J. Botany, vol. 52, no. 7, p. 1629, pl. 1, figs. 13-23.
Oldman Formation, Belly River Group, Upper Cretaceous, depths 1843-1849, 1852-1855 (30834-30836), 1878-1881.5 (30838), 1962-1966.5 (30839a-d, 30840) feet, Canadian Pacific Oil and Gas EV Strathmore well, lot. 7, sec. 12, tp. 25, rge. 25, W.4th mer., Alberta.

Azollopsis (Azollopsis) tomentosa Hall, 1968
Hypotypes 30842, 30843, 30845-30852
Sweet, A.R. and Hills, L.V., 1974, Can. J. Botany, vol. 52, no. 7, p. 1630, pl. 2, figs. 26-30; pl. 3, figs. 31-38; pl.4, fig. 39.
Upper Cretaceous, Horseshoe Canyom Formation, sec. 6, tp. 30, rge. 20, W.4th mer., Horsethief Canyon, NW. 1/4, sec. 26, tp. 28, rge. 21, W.4th mer., Horseshoe Canyon (30846a, b), and depth 757-759 feet, Canadian Pacific Oil and Gas EV Strathmore well, lot 7, sec. 12, tp. 25, rge. 25, W.4th mer. (30850a, b); Whitemud Formation, near Huxley along west side of Red Deer River, NE. 1/4, sec. 24, tp. 34, rge. 22, W.4th mer. (30852), Alberta.

Azollopsis massula
Fig. spec. 49004
Sweet, A.R., 1978, Geol. Surv. Can., Paper 78-1B, p. 36, pl. 6.2, fig. 18.
Tent Island Formation, Upper Cretaceous, northwestern bank Big Fish River near its junction with Boundary Creek, lat. 68°30′30″N, long. 136°23′50″W, Yukon.

Azollopsis (*Spiralopsis*) *intermedia* Sweet and Hills
Holotype 30854; paratypes 30855-30867
Sweet, A.R. and Hills, L.V., 1974, Can. J. Botany, vol. 52, no. 7, p. 1634, pl. 4, figs. 42-50; pl. 5, figs. 51-57.
Foremost Formation, Belly River Group, Upper Cretaceous, depth 2466-2470, 2491-2493.5 (30855), 2371-2378 (30857), 2270-2275 (30859, 30862, 30865, 30866), 2435.5-2439.5 (30861), 2255.5-2260 (30863) feet, Canadian Pacific Oil and Gas EV Strathmore well, lot 7, sec. 12, tp. 25, rge. 25, W.4th mer., Alberta.

Azollopsis (Spiralopsis) *pilata* (Snead, 1969)
Topotypes 30868-30874; hypotype 30875
Sweet, A.R. and Hills, L.V., 1974, Can. J. Botany, vol. 52, no. 7, p. 1635, pl. 5, figs. 58-61, pl. 6, figs. 62-67.
Scollard Member, Paskapoo Formation, Upper Cretaceous, 286-291 and 300-305 (30875) feet, Research Council of Alberta Wizard Lake corehole 65-1, l.s.d. 4, sec. 8, tp. 48, rge. 27, W.4th mer., Alberta.

Azollopsis (*Spiralopsis*) *pusilla* Sweet and Hills
Holotype 30876; paratypes 30877-30883, 30885-30888
Sweet, A.R. and Hills, L.V., 1974, Can. J. Botany, vol. 52, no. 7, p. 1637, pl. 6, figs. 68, 69; pl. 7, figs. 70-79.
Upper Cretaceous, Foremost Formation, Bateman Ridge in foothills west of Calgary, SW. 1/4, sec. 2, tp. 25, rge. 5, W.5th mer. and Canadian Pacific Oil and Gas EV Strathmore well, lot 7, sec. 12, tp. 15, rge. 25, W.4th mer. (30881); Oldman Formation, 2074.5-2083 feet, Canadian Pacific Oil and Gas EV Strathmore well, lot 7, sec. 12. tp. 15, rge. 25, W.4th mer. (30886b-30888), Alberta.

Azonia fabacea Samoilovitch
Hypotype 38222
McIntyre, D.J., 1974, Geol. Surv. Can., Paper 74-14, pl. 15, fig. 14.
Upper Cretaceous, lat. 69°28′N, long. 126°58′W, Horton River, District of Mackenzie.

Azonia sp. cf. *A. parva* Wiggins, 1976
Fig. spec. 93635
Sweet, A.R., Ricketts, B.D., Cameron, A.R. and Norris, D.K., 1989, Geol. Surv. Can., Paper 89-1G, pl. 1, fig. 5.
Little Bear Formation, Upper Cretaceous, east of McKay Range, East Little Bear River area, lat. 64°36′24″N, long. 125°35′W, District of Mackenzie.

Azonia sp. cf. *A. parva* Wiggins, 1976
Fig. spec. 93684
Sweet, A.R. and Braman, D.R., 1989, Geol. Surv. Can., Paper 89-8, p. 35, pl. 1, fig. 19.
Chungo Member, Wapiabi Formation, Upper Cretaceous, Highwood River No. 1 section, S.W. 1/4, sec. 7, tp. 18, rge. 3, W.5th mer., Alberta.

Azonia recta (Bolkhovitina) Samoilovich, 1961
Hypotype 93644
Sweet, A.R., Ricketts, B.D., Cameron, A.R. and Norris, D.K., 1989, Geol. Surv. Can., Paper 89-1G, p. 93, pl. 1, fig. 14.
Little Bear Formation, Upper Cretaceous, east bank Little Bear River, lat. 64°38′24″N, long. 126°19′W, District of Mackenzie.

Azonia sufflata Wiggins, 1976
Hypotype 93645
Sweet, A.R., Ricketts, B.D., Cameron, A.R. and Norris, D.K., 1989, Geol. Surv. Can., Paper 89-1G, p. 93, pl. 1, fig. 15.
Little Bear Formation, Upper Cretaceous, east bank Little Bear River, lat. 64°38′24″N, long. 126°19′W, District of Mackenzie.

Azonia sp. cf. *sufflata* Wiggins, 1976
Fig. spec. 93631
Sweet, A.R., Ricketts, B.D., Cameron, A.R. and Norris, D.K., 1989, Geol. Surv. Can., Paper 89-1G, p. 92, pl. 1, fig. 1.
Little Bear Formation, Upper Cretaceous, east bank Little Bear River, lat. 64°37′24″N, long. 126°18′W, District of Mackenzie.

Azonia sp.
Fig. spec. 91382
Sweet, A.R. and McIntyre, D.J., 1988, Can. Soc. Petrol. Geol., Mem. 15, fig. 7.1.
Moosehound Member, Cardium Formation, Upper Cretaceous, south side of Little Berland River, l.s.d. 1, sec. 6, tp. 54, rge. 2, W.5th mer., Alberta.

Baculatisporites comaumensis (Cookson) Potonié, 1956
Hypotype 13869
McGregor, D.C., 1965, Geol. Surv. Can., Paper 64-55, pl. 7, fig. 10.
Lower Cretaceous, Donna River Canyon, Aklavik Range, Richardson Mountains, District of Mackenzie.

Baculatisporites comaumensis (Cookson) Potonié
Hypotype 23796
Hopkins, Jr., W.S., 1971, Geol. Surv. Can., Bull. 197, p. 114, pl. 20, figs. 10, 11.
Isachsen Formation, Lower Cretaceous, lat. 76°20′N, long. 115°34′W, northwest Melville Island, District of Franklin.

Baculatisporites comaumensis (Cookson) Potonié
Hypotype 31755
Hopkins, Jr., W.S. and Balkwill, H.R., 1973, Geol. Surv. Can., Paper 72-37, p. 12, pl. 1, fig. 17.
Hassel Formation, Cretaceous, Hoodoo Doome, lat. 78°11′N, long. 99°40′W, southern Ellef Ringnes Island, District of Franklin.

Baculatisporites comaumensis (Cookson) Potonié
Hypotype 33963
Hopkins, Jr., W.S., 1974, Geol. Surv. Can., Paper 73-12, p. 13, pl. 2, fig. 25.
Christopher Formation, Lower Cretaceous, lat. 76°20′N, long. 115°34′W, northwest Melville Island, District of Franklin.

Baculatisporites comaumensis (Cookson)
Hypotype 38178
McIntyre, D.J., 1974, Geol. Surv. Can., Paper 74-14, pl. 14, fig. 9.
Upper Cretaceous, lat. 69°28′N, long. 126°58′W, Horton River, District of Mackenzie.

Baculatisporites comaumensis (Cookson) Potonié
Hypotype 42399
Brideaux, W.W. and McIntyre, D.J., 1975, Geol. Surv. Can., Bull. 252, pl. 1, fig. 12.
Horton River Formation, Lower Cretaceous, lat. 69°27′30″N, long. 126°54′W, Horton River, District of Mackenzie.

Baculatisporites comaumensis (Cookson)
Hypotype 41591
Hopkins, Jr., W.S. and Sweet, A.R., 1976, Geol. Surv. Can., Bull. 256, pl. 9, fig. 7.
Mattagami Formation, Lower Cretaceous, Onakawana lignite pit, lat. 50°36′N, long. 82°23′W, Ontario.

Baculatisporites comaumensis (Cookson) Potonié, 1956
Hypotype 71287
Wilson, M.A., 1978, Palaeontographica, Abt. B, vol. 166, p. 110, pl. 2, fig. 12.
Cretaceous, Police Island, north bank Mackenzie River, 12 miles southeast of Fort Norman, District of Mackenzie.

Baculatisporites sp.
Fig. spec. 13834
McGregor, D.C., 1965, Geol. Surv. Can., Paper 64-55, pl. 6, fig. 14.
Lower Cretaceous, Fault Creek Canyon, Aklavik Range, Richardson Mountains, District of Mackenzie.

Bacutriletes cf. *B. greenlandicus* (Miner) Potonié 1956
Hypotype 48931
Sweet, A.R., 1979, Am. Assoc. Strat. Palynol., Contrib. Ser. No. 5B, pl. 3, fig. 6.
Hassel Formation, Lower Cretaceous, lat. 77°9′N, long. 104°50′W, Lougheed Island, District of Franklin.

Bacutriletes cf. *B. onodios* (Harris)
Hypotypes 41625-41628
Hopkins, Jr., W.S. and Sweet, A.R., 1976, Geol. Surv. Can., Bull. 256, pl. 11, figs. 1-4.
Mattagami Formation, Lower Cretaceous, Onakawana lignite pit stock pile, lat. 50°36′N, long. 81°23′W and borehole 68-43, lat. 50°35′26″N, long. 82°28′56″W (41628), Ontario.

Bacutriletes sp. A
Fig. spec. 48959
Sweet, A.R., 1979, Am. Assoc. Strat. Palynol., Contrib. Ser. No. 5B, pl. 4, fig. 17.
Eureka Sound Formation, Upper Cretaceous, east of Hoodoo Dome, lat. 78°11′N, long. 99°43′W, Ellef Ringnes Island, District of Franklin.

Balmeiopsis sp.
Fig. spec. 91381
Sweet, A.R. and McIntyre, D.J., 1988, Can. Soc. Petrol. Geol., Mem. 15, fig. 6.30.
Moosehound Member, Cardium Formation, Upper Cretaceous, south side of Grande Cache Highway 40, l.s.d. 15, sec. 13, tp. 57, rge. 7, W.5th mer., Alberta.

cf. *Bennettites dilucides* Bolkhovitina, 1956
Hypotype 13926
McGregor, D.C., 1965, Geol. Surv. Can., Paper 64-55, pl. 8, fig. 26.
Isachsen Formation, Lower Cretaceous, south of Buchanan Lake anticline, eastern Axel Heiberg Island, District of Franklin.

Betula claripites Wodehouse, 1933
Hypotype 71392
Wilson, M.A., 1978, Palaeontographica, Abt. B, vol. 166, p. 142, pl. 10, fig. 14.
Cretaceous, Police Island, north bank Mackenzie River, 12 miles southeast of Fort Norman, District of Mackenzie.

Betulaceae - Myricacea pollen
Fig. spec. 76036
Jerzykiewicz, T. and Sweet, A.R., 1986, Can. J. Earth Sci., vol. 23, no. 9, pl. 3, fig. 3.
Early Paleocene, depth 25.15-25.25 m, GSC core hole CV-42-2-2-1, Coal Vaalley strip mine, l.s.d. 1, sec. 6, tp. 47, rge. 19, W.5th mer., Alberta.

Biretisporites potoniaei Delcourt and Sprumont
Hypotype 42389
Brideaux, W.W. and McIntyre, D.J., 1975, Geol. Surv. Can., Bull. 252, pl. 1, fig. 1.
Horton River Formation, Lower Cretaceous, lat. 69°19′N, long. 126°50′W, Horton River, District of Mackenzie.

Biretisporites potoniaei Delcourt and Sprumont, 1955
Hypotype 68192
Ricketts, B.D. Sweet, A.R., 1986, Geol. Surv. Can., Paper 84-15, p. 18, pl. 1, fig. 1.
Elk Formation, Cretaceous, Coal Creek, lat. 49°30′N, long. 114°59′W, British Columbia.

Biretisporites potoniaei Delcourt and Sprumont, 1955
Hypotype 91341
Sweet, A.R. and McIntyre, D.J., 1988, Can. Soc. Petrol. Geol., Mem. 15, fig. 5.11.
Leyland Member, Cardium Formation, Upper Cretaceous, south side of McLeod River, l.s.d. 16, sec. 16, tp. 47, rge. 23, W.5th mer., Alberta.

Bombacacidites nacimientoensis (Anderson) Elsik, 1968
Hypotypes 71354, 71358, 71361
Wilson, M.A., 1978, Palaeontographica, Abt. B, vol. 166, p. 135, pl. 8, figs. 6, 11, 14.
Cretaceous and Paleocene (71358), IEO Atkinson H-25 well, north shore Tuktoyaktuk Peninsula, lat. 69°44′20″N, long. 131°50′06″W, District of Mackenzie.

Calamospora sp. A.
Fig. spec. 76600
Davies, E.H., Arkande, S.O. and Zentilli, M., 1984, Geol. Surv. Can., Paper 84-1A, pl. 48.1, fig. 11.
Early Cretaceous, sample DH2-324, Gays River mine, lat. 45°02′N, long. 63°22′W, Nova Scotia.

Callialasporites dampieri (Balme)
Hypotype 41907
Williams, G.L., 1975, Geol. Surv. Can., Paper 74-30, vol. 2, pl. 8, fig. 12.
Hauterivian, Lower Cretaceous, Shell Mohican I-100 well, Scotian Shelf.

Callialasporites dampieri (Balme) Dev
Hypotype 42494
Brideaux, W.W. and McIntyre, D.J., 1975, Geol. Surv. Can., Bull. 252, pl. 3, fig. 28.
Gilmore Lake Member, Langton Bay Formation, Lower Cretacous, lat. 68°50′N, long. 125°24′W, Horton River, District of Mackenzie.

Callialasporites dampieri Purce, 1961
Hypotype 91373
Sweet, A.R. and McIntyre, D.J., 1988, Can. Soc. Petrol. Geol., Mem. 15, fig. 6.22.
Moosehound Member, Cardium Formation, Upper Cretaceous, west bank of Little Berland River, l.s.d. 1, sec. 6, tp. 54, rge. 2, W.5th mer., Alberta.

Callialasporites obrutus Norris, 1969
Hypotype 68823
Ricketts, B.D. and Sweet, A.R., 1986, Geol. Surv. Can., Paper 84-15, p.22, pl.4, fig. 6.
Elk Formation, Cretaceous, Coal Creek, lat. 49°30′N, long. 114°59′W, British Columbia.

Callialasporites segmentatus
Hypotype 91979
Banerjee, I. and Davies, E.H., 1988, Can. Soc. Petrol. Geol., Mem. 15, fig. 6.7.
Mannville Group, Lower Cretaceous, depth 5287 feet, well 11-08-41-24W4, Alberta.

Callialasporites trilobatus (Balme)
Hypotype 56787
Williams, G.L. and Bujak, J.P., 1980, Initial Repts. Deep Sea Drilling Project, vol. 50, pl. 11, fig. 6.
Early Cretaceous, DSDP site 416, lat. 32°50.18′N, long. 10°48.06′W, northwestern Moroccan Continental margin.

Callistopollenites radiostriatus (Mchedlishvili)
Hypotype 38274
McIntyre, D.J., 1974, Geol. Surv. Can., Paper 74-14, pl. 21, fig. 6.
Upper Cretaceous, lat. 69°28′N, long. 126°58′W, Horton River, District of Mackenzie.

Callistopollenites tumidoporous Srivastava, 1969
Hypotypes 71375, 71378
Wilson, M.A., 1978, Palaeontographica, Abt. B, vol. 166, p. 140, pl. 9, figs. 15, 19.
Cretaceous, Police Island, north bank Mackenzie River, 12 miles southeast of Fort Norman, District of Mackenzie.

Callistopollenites tumidoporus Srivastava, 1969
Hypotype 93651
Sweet, A.R., Ricketts, B.D., Cameron, A.R. and Norris, D.K., 1989, Geol. Surv. Can., Paper 89-1G, pl. 1, fig. 22.
Summit Creek Formation, Upper Cretaceous, north bank Mackenzie River immediately downstream from Police Island, lat. 64°52′42″N, long. 125°14′44″W, District of Mackenzie.

Callistosporites graniferus Singh 1971
Hypotype 48919
Sweet, A.R., 1979, Am. Assoc. Strat. Palynol., Contrib. Ser. No. 5B, pl. 2, fig. 15.
Christopher Formation, Lower Cretaceous, lat. 74°31′N, long. 121°10′W, Cape McClure, Banks Island, District of Franklin.

Callistosporites sp. A.
Fig. spec. 48918
Sweet, A.R., 1979, Am. Assoc. Strat. Palynol., Contrib. Ser. No. 5B, pl. 2, fig. 14.
Christopher Formation, Lower Cretaceous, lat. 74°31′N, long. 121°10′W, Cape McClure, Banks Island, District of Franklin.

Camarozonosporites ambigens (Fradkina) Playford
Hypotype 42453
Brideaux, W.W. and McIntyre, D.J., 1975, Geol. Surv. Can., Bull. 252, pl. 2, fig. 25.
Crossley Lakes Member, Langton Bay Formation, Lower Cretaceous, lat. 69°01′30″N, long. 126°02′W, Horton River, District of Mackenzie.

Camarozonosporites insignis Norris
Hypotype 31750
Hopkins, Jr., W.S. and Balkiwill, H.R., 1973, Geol. Surv. Can., Paper 72-37, p. 11, pl. 1, fig. 11.
Hassel Formation, Lower Cretaceous, Hoodoo Dome, lat. 78°11′N, long. 99°40′W, southern Ellef Ringnes Island, District of Franklin.

Camarozonosporites insignis Norris
Hypotypes 38190-38192
McIntyre, D.J., 1974, Geol. Surv. Can., Paper 74-14, pl. 14, figs. 21-23.
Upper Cretaceous, lat. 69°28′N, long. 126°58′W, Horton River, District of Mackenzie.

Camrozonosporites insignis Norris, 1967
Hypotype 91370
Sweet, A.R. and McIntyre, D.J., 1988, Can. Soc. Petrol. Geol., Mem. 15, fig. 6.18.
Moosehound Member, Cardium Formation, Upper Cretaceous, south side of Grande Cache Highway 40, l.s.d. 15, sec. 13, tp. 57, rge. 7, W.5th mer., Alberta.

Canningia? sp. 1
Fig. specs. 62638, 62639
Ioannides, N.S. and McIntyre, D.J., 1980, Geol. Surv. Can., Paper 80-1A, pl. 31.2, figs. 1, 3.
Upper Cretaceous, Caribou Hills, Mackenzie River, District of Mackenzie.

Canningia sp. CE
Fig. spec. 58599
McIntyre, D.J. and Brideaux, W.W., 1980, Geol. Surv. Can., Bull. 320, p.17, pl. 5, figs. 2, 3.
Lower Cretaceous, Martin Creek, lat. 68°12′10″N, long. 135°34′50″W, Richardson Mountains, District of Mackenzie.

Carpinites ancipites (Wodehouse) Srivastava, 1966
Hypotype 71380
Wilson, M.A., 1978, Palaeontographica, Abt. B, vol. 166, p. 142, pl. 10, fig. 2.
Cretaceous, Police Island, north bank Mackenzie River, 12 miles southeast of Fort Norman, District of Mackenzie.

Caryapollenites scabratus Groot and Groot, 1962
Hypotype 71389
Wilson, M.A., 1978, Palaeontographica, Abt. B, vol. 166, p. 143, pl. 10, fig. 11.
Cretaceous, IEO Atkinson H-25 well, north shore Tuktoyaktuk Peninsula, lat. 69°44′20″N, long. 131°50′06″W, District of Mackenzie.

Cedripites canadensis Pocock
Hypotype 42503
Brideaux, W.W. and McIntyre, D.J., 1975, Geol. Surv. Can., Bull. 252, pl. 3, fig. 37.
Crossley Lakes Member, Langton Bay Formation, Lower Cretaceous, lat. 69°04′N, long. 125°51′W, Horton River, District of Mackenzie.

Cedripites canadensis Pocock, 1962
Hypotype 91380
Sweet, A.R. and McIntyre, D.J., 1988, Can. Soc. Petrol. Geol., Mem. 15, figs. 6.29.
Leyland Member, Cardium Formation, Upper Cretaceous, south side of McLeod River, l.s.d. 16, sec. 16, tp. 47, rge. 23, W.5th mer., Alberta.

Cedripites cretaceus Pocock
Hypotype 38206
McIntyre, D.J., 1974, Geol. Surv. Can., Paper 74-14, pl. 14, fig. 37.
Upper Cretaceous, lat. 69°27′30″N, long. 126°58′W, Horton River, District of Mackenzie.

Cedripites cretaceus Pocock
Hypotypes 42499, 42500
Brideaux, .W.W. and McIntyre, D.J., 1975, Geol. Surv. Can., Bull. 252, pl. 3, figs. 33, 34.
Horton River Formation, Lower Cretaceous, lat. 69°27′30″N, long. 126°54′W, Horton River, District of Mackenzie.

cf. *Cedrus* sp.
Fig. spec. 31777
Hopkins, Jr., and Balkill, H.R., 1973, Geol. Surv. Can., Paper 72-37, p. 18, pl. 2, fig. 39.
Hassel Formation, Cretaceous, Hoodoo Dome, lat. 78°11′N, long. 99°40′W, southern Ellef Ringnes Island, District of Franklin.

Ceratosporites sp. cf. *C. couliensis* Srivastava
Hypotypes 38186, 38187
McIntyre, D.J., 1974, Geol. Surv. Can., Paper 74-14, pl. 14, figs. 17, 18.
Upper Cretaceous, lat. 69°28′N, long. 126°58′W, Horton River, District of Mackenzie.

Ceratosporites sp.
Fig. specs. 58557, 58558
McIntyre, D.J. and Brideaux, W.W., 1980, Geol. Surv. Can., Bull. 320, p. 9, pl. 1, figs. 1, 2.
Lower Cretaceous, 'Grizzly Gorge', lat. 68°15′49″N, long. 135°40′59″W, Richardson Mountains, District of Mackenzie.

Ceratosporites sp.
Fig. spec. 68198
Ricketts, B.D. and Sweet, A.R., 1986, Geol. Surv. Can., Paper 84-15, p. 18, pl. 1, fig. 9.
Elk Formation, Cretaceous, Coal Creek, lat. 49°30′N, long. 114°59′W, British Columbia.

Ceratosporites sp.
Fig. spec. 91353
Sweet, A.R. and McIntyre, D.J., 1988, Can. Soc. Petrol. Geol., Mem. 15, fig. 6.1.
Ram Member, Cardium Formation, Upper Cretaceous, north side of Highway 47 north of Cadomin, l.s.d. 14, sec. 16, tp. 47, rge. 23, W.5th mer., Alberta.

Cerebropollenites mesozoicus (Couper)
Hypotype 38219
McIntyre, D.J., 1974, Geol. Surv. Can., Paper 74-14, pl. 15, fig. 12.
Upper Cretaceous, lat. 69°27′30″N, long. 126°58′W, Horton River, District of Mackenzie.

Cerebropollenites mesozoicus (Couper) Nilsson
Hypotypes 42501, 42502
Brideaux, W.W. and McIntyre, D.J., 1975, Geol. Surv. Can., Bull 252, pl. 3, figs. 35, 36.
Gilmore Lake and Crossley Lakes members, Langton Bay Formation, Lower Cretaceous, lat. 68°50′N, long. 125°24′W and lat. 69°04′N, long. 125°51′W, Horton River, District of Mackenzie.

Chatangiella? biapertura (McIntyre) Lentin and Williams
Hypotype 62645
Ioannides, N.S. and McIntyre, D.J., 1980, Geol. Surv. Can., Paper 80-1A, pl. 31.2, fig. 8.
Upper Cretaceous, Caribou Hills, Mackenzie River, District of Mackenzie.

Chatangiella coronata (McIntyre) Lentin and Williams
Hypotypes 62630-62632
Ioannides, N.S. and McIntyre, D.J., 1980, Geol. Surv. Can., Paper 80-1A, pl. 31.3, figs. 1-3.
Upper Cretaceous, Caribou Hills, Mackenzie River, District of Mackenzie.

Chatangiella decorosa (McIntyre) Lentin and Williams
Hypotype 62635
Ioannides, N.S. and McIntyre, D.J., 1980, Geol. Surv. Can., Paper 80-1A, pl. 31.3, fig. 6.
Upper Cretaceous, Cariboo Hills, Mackenzie River, District of Mackenzie.

Chlamydophorella nyei Cookson and Eisenack
Hypotype 41558
Brideaux, W.W. and McIntyre, D.J., 1975, Geol. Surv. Can., Bull. 252, p. 38, pl. 14, fig. 9.
Horton River Formation, Lower Cretaceous, lat. 69°19′N, long. 126°50′W, Horton River, District of Mackenzie.

Chomotriletes fragilis Pocock
Hypotype 41601
Hopkins, Jr., W.S., 1976, Geol. Surv. Can., Bull. 256, pl. 9, fig. 17.
Mattagami Formation, Lower Cretaceous, borehole 68-42, lat. 50°35′36″N, long. 82°30′12″W, Ontario.

Cicatricosisporites sp. cf. *Anemia exiliodes* (Maljavkina) Bolkhovitina
Fig. specs. 42425, 42426
Brideaux, W.W. and McIntyre, D.J., 1975, Geol. Surv. Can., Bull. 252, pl. 1, figs. 39-41.
Horton River Formation, Lower Cretaceous, lat. 69°27′30″N, long. 126°54′W, and lat. 69°19′N, long. 126°50′W, Horton River, District of Mackenzie.

Cicatricosisporites angicanalis Döring, 1965
Hypotype 58567
McIntyre, D.J. and Brideaux, W.W., 1980, Geol. Surv. Can., Bull. 320, p. 9, pl. 1, fig. 12.
Lower Cretaceous, 'Grizzly Gorge', lat. 68°15′49″N, long. 135°40′59″W, Richardson Mountains, District of Mackenzie.

Cicatricosisporites annulatus Archangelsky and Gamerro
Hypotype 42420
Brideaux, W.W. and McIntyre, D.J., 1975, Geol. Surv. Can., Bull. 252, pl. 1, figs. 33, 34.
Crossley Lakes Member, Langton Bay Formation, Lower Cretaceous, lat. 69°04′N, long. 125°51′W, Horton River, District of Mackenzie.

Cicatricosisporites annulatus Archangelsky and Gamerro, 1966 *in* Singh, 1971
Hypotype 93668
Sweet, A.R. and Braman, D.R., 1989, Geol. Surv. Can., Paper 89-8, pl. 1, fig. 3.
Chungo Member, Wapiabi Formation, Upper Cretaceous, Trap Creek near junction with Highwood River, sec. 36, tp. 17, rge. 4, W.5th mer., Alberta.

Cicatricosisporites sp. cf. *C. aralica* (Bolkhovitina) Brenner, 1963
Fig. specs. 68193, 68194
Ricketts, B.D. and Sweet, A.R., 1986, Geol. Surv. Can., Paper 84-15, p. 18, pl. 1, figs. 2-4.

Cretaceous, Elk Formation, Coal Creek, lat. 49°30′N, long. 114°59′W, British Columbia, and Pocaterra Member Cadomin Formation, Highwood Pass, lat. 50°35′N, long. 115°1′W, Alberta.

Cicatricosisporites aralicus (Bolkhovitina 1961) Brenner 1963
Hypotypes 72062, 72063
Fensome, R.A., 1987, Palaeontographica Canadiana No. 4, p. 15, pl. 2, figs. 6, 7, 10, 11.
Husky Formation, Cretaceous, Martin Creek, south side, 0.8 miles north of small lake, lat. 68°12′20″N, long. 135°36′W, District of Mackenzie.

Cicatricosisporites aralicus (Bolkhovitina) Brenner
Hypotype 93669
Sweet, A.R. and Braman, D.R., 1989, Geol. Surv. Can., Paper 89-8, pl. 1, fig. 4.
Chungo Member, Wapiabi Formation, Upper Cretaceous, Highwood River No. 2 section, NE. 1/4, sec. 1, tp. 18, rge. 4, W.5th mer., Alberta.

Cicatricosisporites augustus Singh
Hypotypes 42421, 42422
Brideaux, W.W. and McIntyre, D.J., 1975, Geol. Surv. Can., Bull. 252, pl. 1, figs. 35, 36.
Crossley Lakes Member, Langton Bay Formation, Lower Cretaceous, lat. 69°04′N, long. 125°51′W, Horton River, District of Mackenzie.

Cicatricosisporites australiensis (Cookson)
Hypotypes 23797, 25519
Hopkins, Jr., W.S., 1971, Geol. Surv. Can., Bull. 197, p. 115, pl. 20, figs. 12, 13.
Isachsen Formation, Lower Cretaceous, lat. 76°20′N, long. 115°34′W, northwest Melville Island, District of Franklin.

Cicatricosisporites australiensis (Cookson) Potonié
Hypotype 33969
Hopkins, Jr., W.S., 1974, Geol. Surv. Can., Paper 73-12, p. 15, pl. 3, fig. 32.
Christopher Formation, Lower Cretaceous, lat. 76°20′N, long. 115°34′W, northwest Melville Island, District of Franklin.

Cicatricosisporites australiensis (Cookson) Potonié
Hypotype 42423
Brideaux, W.W. and McIntyre, D.J., 1975, Geol. Surv. Can., Bull. 252, pl. 1, fig. 37.
Gilmore Lake Member, Langton Bay Formation, Lower Cretaceous, lat. 68°50′N, long. 125°24′W, Horton River, District of Mackenzie.

Cicatricosisporites australiensis (Cookson) Potonié, 1956
Hypotype 68195, 68196
Ricketts, B.D. and Sweet, A.R., 1986, Geol. Surv. Can., Paper 84-15, p.18, pl. 1, figs. 5, 6.
Cretaceous, Pocaterra Member, Cadomin Formation, Highwood Pass, lat. 50°35′N, long. 115°1′W, Alberta and Elk Formation, Coal Creek, lat. 49°30′N, long. 114°59′W, British Columbia.

Cicatricosisporites australiensis (Cookson 1953) Potonié 1956
Hypotypes 72064, 72065
Fensome, R.A., 1987, Palaeontographica Canadiana No. 4, p. 16, pl. 2, figs. 8, 13.
Fault Creek and ?Lower Canyon formations, Cretaceous, Martin Creek, north bank, lat. 68°12′17″N, long. 135°34′38″W, District of Mackenzie.

Cicatricosisporites australiensis (Cookson) Potonié, 1956
Hypotype 91346
Sweet, A.R. and McIntyre, D.J., 1988, Can. Soc. Petrol. Geol., Mem. 15, fig. 5.18.
Moosehound Member, Cardium Formation, Upper Cretaceous, south side of McLeod River, l.s.d. 16, sec. 16, tp. 47, rge. 23, W.5th mer., Alberta.

Cicatricosisporites cf. *C. australiensis* (Cookson) Potonié
Hypotype 31757
Hopkins, W.S. and Balkwill, H.R., 1973, Geol. Surv. Can., Paper 72-37, p. 13, pl. 1, fig. 19.
Hassel Formation, Cretaceous, Hoodoo Dome, lat. 78°11′N, long. 99°40′W, southern Ellef Ringnes Island, District of Franklin.

Cicatrisosiporites crassistriatus Burger 1966
Hypotypes 72054-72057
Fensome, R.A., 1987, Palaeontographica Canadiana No. 4, p. 16, pl. 2, figs. 1, 9, 15, 17, 18.
Husky Formation, Cretaceous, Martin Creek, south side, 0.8 miles north of small lake, lat. 68°12′20″N, long. 135°36′W, and north bank of first creek south of Mt. Gifford (72055, 72056), District of Mackenzie.

Cicatricosisporites crassiterminatus Hedlund, 1966
Hypotype 91343
Sweet, A.R. and McIntyre, D.J., 1988, Can. Soc. Petrol. Geol., Mem. 15, figs. 5.14, 5.15.
Leyland Member, Cardium Formation, Upper Cretaceous, south side of McLeod River, l.s.d. 16, sec. 16, tp. 47, rge. 23, W.5th mer., Alberta.

Cicatricosisporites dorogensis Potonié and Gelletich, 1933
Hypotype 13953
McGregor, D.C., 1965, Geol. Surv. Can., Paper 64-55, pl. 10, fig. 20.
Lower Cretaceous, Peel Plateau, Stony Creek about 18.5 miles upstream from mouth, District of Mackenzie.

Cicatricosisporitees dorogensis Potonię and Gelletich
Hypotype 33970
Hopkins, Jr., W.S., 1974, Geol. Surv. Can., Paper 73-12, p. 15, pl. 3, fig. 33
Christopher Formation, Lower Cretaceous, lat. 76°20′N, long. 115°34′W, northwest Melville Island, District of Franklin.

Cicatricosisporites dorogensis Potonié and Gelletich, 1933
Hypotypes 71288, 71290
Wilson, M.A., 1978, Palaeontographica, Abt. B, vol. 166, p. 113, pl. 3, figs. 1, 2, 5.
Cretaceous, Police Island, north bank Mackenzie River, 12 miles southeast of Fort Norman, District of Mackenzie.

Cicatricosisporites exilioides (Maljavkina)
Hypotype 58570
McIntyre, D.J. and Brideaux, W.W., 1980, Geol. Surv. Can., Bull. 320, p.10, pl. 1, fig. 15.
Lower Cretaceous, 'Grizzly Gorge', lat. 68°15′49″N, long. 135°40′59″W, Richardson Mountains, District of Mackenzie.

Cicatricosisporites exilioides (Maljavkina) Bolchovitina
Hypotype 76594
Davies, E.H., Akande, S.O. and Zentilli, M., 1984, Geol. Surv. Can., Paper 84-1A, pl. 48.1, fig. 5.
Early Cretaceous, sample AGR 285-230, Gays River mine, lat. 45°02′N, long. 63°22′W, Nova Scotia.

Cicatricosisporites sp. cf. *grabowensis* Döring, 1965
Fig. spec. 68197
Ricketts, B.D. and Sweet, A.R., 1986, Geol. Surv. Can., Paper 84-15, p. 18, pl. 1, figs. 7, 8.
Elk Formation, Cretaceous, Coal Creek, lat. 49°30′N, long. 114°59′W, British Columbia.

Cicatricosisporites hallei Delcourt and Sprumont
Hypotype 23798
Hopkins, Jr., W.S., 1971, Geol. Surv. Can., Bull. 197, p. 115, pl. 20, fig. 14.
Isachsen Formation, Lower Cretaceous, lat. 76°20′N, long. 115°34′W, northwest Melville Island, District of Franklin.

Cicatricosisporites hallei Delcourt and Sprumont
Hypotype 33971
Hopkins, Jr., W.S., 1974, Geol. Surv. Can., Paper 73-12, p. 15, pl. 3, fig. 34.
Christopher Formation, Lower Cretaceous, lat. 76°20′N, long. 115°34′W, northwest Melville Island, District of Franklin.

Cicatricosisporites hallei Delcourt and Sprumont
Hypotype 42424
Brideaux, W.W. and McIntyre, D.J., 1975, Geol. Surv. Can., Bull. 252, pl. 1, fig. 38.
Crossley Lakes Member, Langton Bay Formation, Lower Cretaceous, lat. 69°04′N, long. 125°51′W, Horton River, District of Mackenzie.

Cicatricosisporites cf. *C. hallei* Del. and Sprum.
Hypotype 31756
Hopkins, Jr., W.S. and Balkwill, H.R., 1973, Geol. Surv. Can., Paper 72-37, p. 13, pl. 1, fig. 18.
Hassel Formation, Cretaceous, Hoodoo Dome, lat. 78°11′N, long. 99°40′W, southern Ellef Ringnes Island, District of Franklin.

Cicatricosisporites hughesii Dettmann
Hypotype 42427
Brideaux, W.W. and McIntyre, D.J., 1975, Geol. Surv. Can., Bull. 252, pl. 1, fig. 42.
Horton River Formation, Lower Cretaceous, lat. 69°19′N, long. 126°50′W, Horton River, District of Mackenzie.

Cicatricosisporites hughesi Dettmann
Hypotypes 56773, 56774
Williams, G.L. and Bujak, J.P., 1980, Initial Repts, Deep Sea Drilling Project, vol. 50, pl. 10, figs. 4, 5.
Early Cretaceous, DSDP site 416, lat. 32°50.18′N, long. 10°48.06′W, northwestern Moroccan continental margin.

Cicatricosisporites hughesii Dettmann, 1963
Hypotype 58569
McIntyre, D.J. and Brideaux, W.W., 1980, Geol. Surv. Can., Bull. 320, p.10, pl. 1, fig. 14.
Lower Cretaceous, 'Grizzly Gorge', lat. 68°15′49″N, long. 135°40′59″W, Richardson Mountains, District of Mackenzie.

Cicatricosisporites hughesi Dettmann, 1963
Hypotype 91350
Sweet, A.R. and McIntyre, D.J., 1988, Can. Soc. Petrol. Geol., Mem. 15, figs. 5.22, 5.23.
Moosehound Member, Cardium Formation, Upper Cretaceous, south side of McLeod River, l.s.d. 16, sec. 16, tp. 47, rge. 23, W.5th mer., Alberta.

Cicatricosisporites ?hughesi Dettmann, 1963
Hypotype 13917
McGregor, D.C., 1965, Geol. Surv. Can., Paper 64-55, pl. 8, figs. 13, 14.
Isachsen Formation, Lower Cretaceous, south of Buchanan Lake anticline, eastern Axel Heiberg Island, District of Franklin.

Cicatricosisporites ?ludbrooki Dettmann, 1963
Hypotype 13920
McGregor, D.C., 1965, Geol. Surv. Can., Paper 64-55, pl. 8, fig. 19.
Isachsen Formation, Lower Cretaceous, south of Buchanan Lake anticline, eastern Axel Heiberg Island, District of Franklin.

Cicatricosisporites ludbrooki Dettmann
Hypotypes 23799, 23800
Hopkins, Jr., W.S., 1971, Geol. Surv. Can., Bull. 197, p. 115, pl. 20, figs. 15-17.
Isachsen Formation, Lower Cretaceous, lat. 76°20′N, long. 115°34′W, northwest Melville Island, District of Franklin.

Cicatricosisporites ludbrooki Dettmann
Hypotype 33968
Hopkins, Jr., W.S., 1974, Geol. Surv. Can., Paper 73-12, p. 14, pl. 3, figs. 30, 31.
Christopher Formation, Lower Cretaceous, lat. 76°20′N, long. 115°34′W, northwest Melville Island, District of Franklin.

Cicatricosisporites ludbrooki Dettmann
Hypotype 42428
Brideaux, W.W. and McIntyre, D.J., 1975, Geol. Surv. Can., Bull. 252, pl. 1, fig. 43.
Crossley Lakes Member, Langton Bay Formation, Lower Cretaceous, lat. 69°04′N, long. 125°51′W, Horton River, District of Mackenzie.

Cicatricosisporites myrtellii Burger, 1966
Hypotype 58566
McIntyre, D.J. and Brideaux, W.W., 1980, Geol. Surv. Can., Bull. 320, p.10, pl. 1, figs. 10, 11.
Lower Cretaceous, 'Grizzly Gorge', lat. 68°15'49"N, long. 135°40'59"W, Richardson Mountains, District of Mackenzie.

Cicatricosisporites ornatus Srivastava 1972
Hypotype 89731
Jerzykiewicz, T. and Sweet, A.R., 1988, Sedimentary Geol., vol. 59, no. 1/2, p. 58, pl. III, fig. 4.
St. Mary River Formation, Upper Cretaceous, Crowsnest River, lat. 49°30'N, long. 114°3'W, Alberta.

Cicatricosisporites ornatus Srivastava, 1972
Hypotype 93670
Sweet, A.R. and Braman, D.R., 1989, Geol. Surv. Can., Paper 89-8, pl. 1, fig. 5.
Chungo Member, Wapiabi Formation, Upper Cretaceous, Highwood River No. 2 section, N.E. 1/4, sec. 1, tp. 18, rge. 4, W.5th mer., Alberta.

Cicatricosisporites pacificus (Bolchovitina) Chlonova, 1976
Hypotype 91347
Sweet, A.R. and McIntyre, D.J., 1988, Can. Soc. Petrol. Geol., Mem. 15, fig. 5.19.
Moosehound Member, Cardium Formation, Upper Cretaceous, south side of McLeod River, l.s.d. 16, sec. 16, tp. 47, rge. 23, W.5th mer., Alberta.

Cicatricosisporites potomacensis Brenner
Hypotype 42429
Brideaux, W.W. and McIntyre, D.J., 1975, Geol. Surv. Can., Bull. 252, pl. 1, fig. 44.
Crossley Lakes Member, Langton Bay Formation, Lower Cretaceous, lat. 69°04'N, long. 125°51'W, Horton River, District of Mackenzie.

Cicatricosisporites potomacensis Brenner, 1963
Hypotype 58568
McIntyre, D.J. and Brideaux, W.W., 1980, Geol. Surv. Can., Bull. 320, p.10, pl. 1, fig. 13.
Lower Cretaceous, Martin Creek, lat. 68°12'10"N, long. 135°34'50"W, Richardson Mountains, District of Mackenzie.

Cicatricosisporites potamacensis (Brenner)
Hypotype 76593
Davies, E.H., Akande, S.O. and Zentilli, M., 1984, Geol. Surv. Can., Paper 84-1A, pl. 48.1, fig. 4.
Early Cretaceous, sample AGR 285-230, Gays River mine, lat. 45°02'N, long. 63°22'W, Nova Scotia.

Cicatricosisporites pseudotripartitus (Bolkhovitina) Dettmann
Hypotype 23801
Hopkins, Jr., W.S., 1971, Geol. Surv. Can., Bull. 197, p. 115, pl. 20, fig. 18.
Isachsen Formation, Lower Cretaceous, lat. 76°20'N, long. 115°34'W, northwest Melville Island, District of Franklin.

Cicatricosisporites pseudotripartitus (Bolkhovitina) Dettmann
Hypotypes 42430, 42431
Brideaux, W.W. and McIntyre, D.J., 1975, Geol. Surv. Can., Bull. 252, pl. 2, figs. 1, 2.
Crossley Lakes Member, Langton Bay Formation, Lower Cretaceous, lat. 64°01'30"N, long. 126°02'W, and lat. 69°04'N, long. 125°51'W, Horton River, District of Mackenzie.

Cicatricosisporites cf. *C. pseudotripartitus* (Bolkhovitina)
Hypotype 33972
Hopkins, Jr., W.S., 1974, Geol. Surv. Can., Paper 73-12, p. 15, pl. 3, fig. 35, 36.
Christopher Formation, Lower Cretaceous, lat. 76°20'N, long. 115°34'W, northwest Melville Island, District of Franklin.

Cicatricosisporites purbekensis Norris, 1969
Hypotypes 72066, 72070, 72071
Fensome, R.A., 1987, Palaeontographica Canadiana No. 4, p. 16, pl. 2, fig. 12; pl. 3, figs. 4-6.
Lower Canyon, Husky and McGuire formations, Cretaceous, Martin Creek, south bank, lat. 68°12'N, long. 135°34'15"W, north bank of first creek south of Mt. Gifford, and Martin Creek, north bank, lat. 68°12'10"N, long. 135°34'50"W, District of Mackenzie.

Cicatricosisporites spiralis Singh, C. 1971
Hypotype 72067
Fensome, R.A., 1987, Palaeontographica Canadiana No. 4, p. 16, pl. 2, fig. 14.
Lower Canyon Formation, Cretaceous, Martin Creek, south bank, lat. 68°12'N, long. 135°34'15"W, District of Mackenzie.

Cicatricosisporites sp. cf. *C. subrotundus* Brenner
Fig. spec. 42432
Brideaux, W.W. and McIntyre, D.J., 1975, Geol. Surv. Can., Bull. 252, pl. 2, fig. 3.
Horton River Formation, Lower Cretaceous, lat. 69°19'N, long. 126°50'W, Horton River, District of Mackenzie.

Cicatricosisporites venustus Deák
Hypotypes 42433, 42434
Brideaux, W.W. and McIntyre, D.J., 1975, Geol. Surv. Can., Bull. 252, pl. 2, figs. 4, 5.
Crossley Lakes Member, Langton Bay Formation, Lower Cretaceous, lat. 69°04'N, long. 125°51'W, Horton River, District of Mackenzie.

Cicatricosisporites venustus Deák, 1963
Hypotype 91348
Sweet, A.R. and McIntyre, D.J., 1988, Can. Soc. Petrol. Geol., Mem. 15, fig. 5.20.
Moosehound Member, Cardium Formation, Upper Cretaceous, south side of McLeod River, l.s.d. 16, sec. 16, tp. 47, rge. 23, W.5th mer., Alberta.

Cicatricosisporites sp.
Fig. specs. 13836, 13837
McGregor, D.C., 1965, Geol. Surv. Can., Paper 64-55, pl. 6, figs. 17-19.
Lower Cretaceous, Fault Creek Canyon, Aklavik Range, Richardson Mountains, District of Mackenzie.

Cicatricosisporites sp.
Fig. spec. 13919
McGregor, D.C., 1965, Geol. Surv. Can., Paper 64-55, pl. 8, fig. 16.
Isachsen Formation, Lower Cretaceous, south of Buchanan Lake anticline, eastern Axel Heiberg Island, District of Franklin.

Cicatricosisporites sp.
Fig. spec. 13957
McGregor, D.C., 1965, Geol. Surv. Can., Paper 64-55, pl. 10, fig. 25.
Lower Cretaceous, Peel Plateau, Stony Creek about 18.5 miles upstream from mouth, District of Mackenzie.

Cicatricosisporites sp.
Fig. spec. 31758
Hopkins, W.S. and Balkwill, H.R., 1973, Geol. Surv. Can., Paper 72-37, p. 13, pl. 1, fig. 20.
Hassel Formation, Cretaceous, Hoodoo Dome, lat. 78°11′N, long. 99°40′W, southern Ellef Ringnes Island, District of Franklin.

Cicatricosisporites sp.
Fig. specs. 41595, 41596
Hopkins, Jr., W.S. and Sweet, A.R., 1976, Geol. Surv. Can., Bull. 256, pl. 9, figs. 11, 12.
Mattagami Formation, Lower Cretaceous, Onakawana lignite pit, lat. 50°36′N, long. 81°23′12″W and borehole 68-39, lat. 50°35′48″N, long. 82°30′42″W, Ontario.

Cicatricosisporites sp.
Fig. spec. 56775
Williams, G.L. and Bujak, J.P., 1980, Initial Repts. Deep Sea Drilling Project, vol. 50, pl. 10, fig. 6.
Early Cretaceous, DSDP Site 416, lat. 32°50.18′N, long. 10°48.06′W, northwestern Moroccan continental margin.

Cicatricosporites sp.
Fig. spec. 91344
Sweet, A.R. and McIntyre, D.J., 1988, Can. Soc. Petrol. Geol., Mem. 15, fig. 5.16.
Moosehound Member, Cardium Formation, Upper Cretaceous, south side of McLeod River, l.s.d. 16, sec. 16, tp. 47, rge. 23, W.5th mer., Alberta.

Cicatricosisporites sp. A
Fig. spec. 33973
Hopkins, Jr., W.S., 1974, Geol. Surv. Can., Paper 73-12, p. 16, pl. 3, fig. 37.
Christopher Formation, Lower Cretaceous, lat. 76°20′N, long. 115°34′W, northwest Melville Island, District of Franklin.

Cicatricosisporites sp. A
Fig. specs. 76591, 76592
Davies, E.H., Akande, S.O. and Zentilli, M., 1984, Geol. Surv. Can., Paper 84-1A, pl. 48.1, figs. 2, 3.
Early Cretaceous, sample P19715, Gays River mine, lat. 45°02′N, long. 63°22′W, Nova Scotia.

Cicatricosisporites sp. A
Fig. specs 72068, 72069
Fensome, R.A., 1987, Palaeontographica Canadiana No. 4, p. 17, pl. 3, figs. 1-3.
Husky Formation, Cretaceous, north bank of first creek south of Mt. Gifford, District of Mackenzie.

Cicatricososporites auritus Singh
Hypotypes 56777, 56778
Williams, G.L. and Bujak, J.P., 1980, Initial Repts. Deep Sea Drilling Project, vol. 50, pl. 10, figs. 8, 9.
Early Cretaceous, DSDP Site 416, lat. 32°50.18′N, long. 10°48.06′W, northwestern Moroccan Continental Margin.

Cingulatisporites distaverrucosus Brenner
Hypotype 33946
Hopkins, Jr., W.S., 1974, Geol. Surv. Can., Paper 73-12, p. 9, pl. 1, fig. 6.
Christopher Formation, Lower Cretaceous, lat. 76°20′N, long. 115°34′W, northwest Melville Island, District of Franklin.

Cingulatisporites cf. *C. radiatus* Stanley
Hypotype 31745
Hopkins, Jr., W.S. and Balkwill, H.R., 1973, Geol. Surv. Can., Paper 72-37, p. 10, pl. 1, fig. 6.
Hassel Formation, Cretaceous, near Cape Cairo, lat. 78°43′N, long. 100°06′W, northeastern Ellef Ringnes Island, District of Franklin.

?*Cingulatisporites rigidus* Couper, 1958
Hypotype 13911
McGregor, D.C., 1965, Geol. Surv. Can., Paper 64-55, pl. 8, fig. 7.
Isachsen Formation, Lower Cretaceous, south of Buchanan Lake anticline, eastern Axel Heiberg Island, District of Franklin.

Cingulatisporites scabratus Couper, 1958
Hypotypes 71283, 71284
Wilson, M.A., 1978, Palaeontographica, Abt. B, vol. 166, p. 116, pl. 2, figs. 8, 9.
Cretaceous, Police Island, north bank Mackenzie River, 12 miles southeast of Fort Norman, District of Mackenzie.

Cingulatisporites sp.
Fig. spec. 31746
Hopkins, Jr., W.S. and Balkwill, H.R., 1973, Geol. Surv. Can., Paper 72-37, p. 10, pl. 1, fig. 7.
Hassel Formation, Cretaceous, near Cape Cairo, lat. 78°43′N, long. 100°06′W, northeastern Ellef Ringnes Island, District of Franklin.

Cingutriletes clavus (Balme) Dettmann
Hypotype 33947
Hopkins, Jr., W.S., 1974, Geol. Surv. Can., Paper 73-12, p. 9, pl. 1, figs. 7, 8.
Christopher Formation, Lower Cretaceous, lat. 76°20′N, long. 115°34′W, northwest Melville Island, District of Franklin.

Cingutriletes clavus Dettmann
Hypotype 42462
Brideaux, W.W. and McIntyre, D.J., 1975, Geol. Surv. Can., Bull. 252, pl. 2, fig. 34.
Horton River Formation, Lower Cretaceous, lat. 69°19′N, long. 126°50′W, Horton River, District of Mackenzie.

Cingutriletes clavus (Balme)
Hypotype 41586
Hopkins, Jr., W.S. and Sweet, A.R., 1976, Geol. Surv. Can., Bull. 256, pl. 9, fig. 2.
Mattagami Formation, Lower Cretaceous, Onakawana lignite pit, lat. 50°36′N, long. 81°23′W, Ontario.

Circulina parva Brenner
Hypotypes 42528, 42529
Brideaux, W.W. and McIntyre, D.J., 1975, Geol. Surv. Can., Bull. 252, pl. 4, figs. 27, 28.
Crossley Lakes Member, Langton Bay Formation, Lower Cretaceous, lat. 69°04′N, long. 125°51′W, Horton River, District of Mackenzie.

Cirratriradites tetu Norris
Hypotype 42464
Brideaux, W.W. and McIntyre, D.J., 1975, Geol. Surv. Can., Bull. 252, pl. 2, fig. 36.
Crossley Lakes Member, Langton Bay Formation, Lower Cretaceous, lat. 69°12′30″N, long. 126°17′W, Horton River, District of Mackenzie.

Classopollis classoides (Pflug) Pocock and Jansonius
Hypotype 31784
Hopkins, Jr., W.S. and Balkwill, H.R., 1973, Geol. Surv. Can., Paper 72-37, p. 21, pl. 2, fig. 46.
Hassel Formation, Cretaceous, Hoodoo Dome, lat. 78°11′N, long. 99°40′W, southern Ellef Ringnes Island, District of Franklin.

Classopollis classoides Pflug emend. Pocock and Jansonius
Hypotypes 42526, 42527
Brideaux, W.W. and McIntyre, D.J., 1975, Geol. Surv. Can., Bull. 252, pl. 4, figs. 25, 26.
Crossley Lakes Member, Langton Bay Formation, Lower Cretaceous, lat. 69°04′N, long. 125°51′W and lat. 69°12′30″N, long. 126°17′W Horton River, District of Mackenzie.

Classopollis classoides (Pflug)
Hypotype 41610
Hopkins, Jr., W.S. and Sweet, A.R., 1976, Geol. Surv. Can., Bull. 256, pl. 9, fig. 26.
Mattagami Formation, Lower Cretaceous, borehole 68-42, lat. 50°35′36″N, long. 82°30′12″W, Ontario.

Classopollis torosus (Reissinger) Couper
Hypotypes 25105-25107
Hopkins, Jr., W.S., 1971, Geol. Surv. Can., Bull. 197, p. 123, pl. 22, figs. 144-16.
Isachsen Formation, Lower Cretaceous, lat. 76°20′N, long. 115°34′W, northwest Melville Island, District of Franklin.

Classopollis torosus (Reissinger) Couper
Hypotype 34005
Hopkins, Jr., W.S., 1974, Geol. Surv. Can., Paper 73-12, p. 24, pl. 6, figs. 72, 73.
Christopher Formation, Lower Cretaceous, lat. 76°20′N, long. 115°34′W, northwest Melville Island, District of Franklin.

Classopollis torosus (Reissinger) Couper, 1958
Hypotype 68826
Ricketts, B.D. and Sweet, A.R., 1986, Geol. Surv. Can., Paper 84-15, p. 22, pl. 4, fig. 9.
Elk Formation, Cretaceous, Coal Creek, lat. 49°30′N, long. 114°59′W, British Columbia.

Classopollis sp.
Fig. spec. 76145
Jerzykiewicz, T. and Sweet, A.R., 1986, Geol. Surv. Can., Paper 86-1B, Fig. 67.2, fig. 14.
Willow Creek Formation, Late Cretaceous, Oldman River bridge section, l.s.d. 3,4, sec. 35, tp. 6, rge. 1, W.5th mer., Alberta.

Clavabivesiculites pannosus Pierce, 1961
Hypotype 91372
Sweet, A.R. and McIntyre, D.J., 1988, Can. Soc. Petrol. Geol., Mem. 15, fig. 6.21.
Moosehound Member, Cardium Formation, Upper Cretaceous, south side of Grande Cache Highway 40, l.s.d. 15, sec. 13, tp. 57, rge. 7, W.5th mer., Alberta.

Clavatipollenites evittii Chumura, 1973
Hypotype 91385
Sweet, A.R. and McIntyre, D.J., 1988, Can. Soc. Petrol. Geol., Mem. 15, fig. 7.4.
Moosehound Member, Cardium Formation, Upper Cretaceous, south side of Grande Cache Highway 40, l.s.d. 15, sec. 13, tp. 57, rge. 7, W.5th mer., Alberta.

Clavatipollenites hughesii Couper emend. Kemp
Hypotype 42531
Brideaux, W.W. and McIntyre, D.J., 1975, Geol. Surv. Can., Bull. 252, pl. 4, fig. 30.
Crossley Lakes Member, Langton Bay Formation, Lower Cretaceous, lat. 69°04′N, long. 125°51′W, Horton River, District of Mackenzie.

Clavatipollenites cf. *C. hughesii* Couper
Hypotype 41619
Hopkins, Jr., W.S. and Sweet, A.R., 1976, Geol. Surv. Can., Bull. 256, pl. 10, fig. 6.
Mattagami Formation, Lower Cretaceous, Onakawana Lignite pit stock pile, lat. 50°36′N, long. 81°23′W, Ontario.

Clavatipollenites minutus Brenner, 1963
Hypotype 91419
Sweet, A.R. and McIntyre, D.J., 1988, Can. Soc. Petrol. Geol., Mem. 15, fig. 7.38.
Leyland Member, Cardium Formation, Upper Cretaceous, south side of McLeod River, l.s.d. 16, sec. 16, tp. 47, rge. 23, W.5th mer., Alberta.

Clavifera triplex (Bolkhovitina) Bolkhovitina
Hypotype 33975
Hopkins, Jr., W.S., 1974, Geol. Surv. Can., Paper 73-12, p. 16, pl. 4, fig. 39.
Christopher Formation, Lower Cretaceous, lat. 76°20′N, long. 115°34′W, northwest Melville Island, District of Franklin.

Complexipollis sp.
Fig. spec. 91388
Sweet, A.R. and McIntyre, D.J., 1988, Can. Soc. Petrol. Geol., Mem. 15, fig. 7.7.
Moosehound Member, Cardium Formation, Upper Cretaceous, south side of Grande Cache Highway 40, l.s.d. 15, sec. 13, tp. 57, rge. 7, W.5th mer., Alberta.

?*Concavisporites punctatus* Delcourt and Sprumont, 1955
Hypotype 13838
McGregor, D.C., 1965, Geol. Surv. Can., Paper 64-55, pl. 6, fig. 20.
Lower Cretaceous, Fault Creek Canyon, Aklavik Range, Richardson Mountains, District of Mackenzie.

Concavissimisporites apiverrucatus (Couper 1958) Döring 1965
Hypotype 72091
Fensome, R.A., 1987, Palaeontographica Canadiana No. 4, p. 20, pl. 4, fig. 5.
Martin Creek Formation, Cretaceous, Martin Creek, north bank, lat. 68°12′10″N, long. 135°34′50″W, District of Mackenzie.

Concavissimisporites parkinii (Pocock) Singh
Hypotype 25071
Hopkins, Jr., W.S., 1971, Geol. Surv. Can., Bull. 197, p. 117, pl. 20, fig. 26.
Isachsen Formation, Lower Cretaceous, lat. 76°20′N, long. 115°34′W, northwest Melville Island, District of Franklin.

Concavissimisporites parkini (Pocock) Singh
Hypotype 33976
Hopkins, Jr., W.S., 1974, Geol. Surv. Can., Paper 73-12, p. 16, pl. 4, fig. 40.
Christopher Formation, Lower Cretaceous, lat. 76°20′N, long. 115°34′W, northwest Melville Island, District of Franklin.

Concavissimisporites punctatus (Delcourt and Sprumont) Singh
Hypotype 25072
Hopkins, Jr., W.S., 1971, Geol. Surv. Can., Bull. 197, p. 117, pl. 20, fig. 27.
Isachsen Formation, Lower Cretaceous, lat. 76°20′N, long. 115°34′W, northwest Melville Island, District of Franklin.

Concavissimisporites punctatus (Delcourt and Sprumont) Brenner
Hypotype 42395
Brideaux, W.W. and McIntyre, D.J., 1975, Geol. Surv. Can., Bull. 252, pl. 1, fig. 7.
Horton River Formation, Lower Cretaceous, lat. 69°19′N, long. 126°50′W, Horton River, District of Mackenzie.

Concavissimisporites punctatus Delcourt and Sprumont
Hypotype 76601
Davies, E.H., Akande, S.O. and Zentilli, M., 1984, Geol. Surv. Can., Paper 84-1A, pl. 48.1, fig. 12.
Early Cretaceous, sample AGR 285-230, Gays River mine, lat. 45°02′N, long. 63°22′W, Nova Scotia.

Concavissimisporites punctatus (Delcourt and Sprumont) Brenner, 1963
Hypotypes 68200, 68201
Ricketts, B.D. and Sweet, A.R., 1986, Geol. Surv. Can., Paper 84-15, p. 19, pl. 1, figs. 12, 13.
Elk Formation, Cretaceous, Coal Creek, lat. 49°30′N, long. 114°59′W, British Columbia.

Concavissimisporites sp. cf. *C. uralensis* (Bolkhovitina 1961) comb. nov.
Fig. spec. 72084
Fensome, R.A., 1987, Palaeontographica Canadiana No. 4, p. 24, pl. 3, fig. 18.
Husky Formation, Cretaceous, Martin Creek, south side, 0.8 miles north of small lake, lat. 68°12′20″N, long. 135°36′W, District of Mackenzie.

Concavissimisporites variverrucatus (Couper) Singh
Hypotype 25073
Hopkins, Jr., W.S., 1971, Geol. Surv. Can., Bull. 197, p. 117, pl. 21, fig. 1.
Isachsen Formation, Lower Cretaceous, lat. 76°20′N, long. 115°34′W, northwest Melville Island, District of Franklin.

Concavissimisporites variverrucatus (Couper) Singh
Hypotype 33977
Hopkins, Jr., W.S., 1974, Geol. Surv. Can., Paper 73-12, p. 17, pl. 3, fig. 41.
Christopher Formation, Lower Cretaceous, lat. 76°20′N, long. 115°34′W, northwest Melville Island, District of Franklin.

Concavissimisporites variverrucatus (Couper) Brenner
Hypotype 42397
Brideaux, W.W. and McIntyre, D.J., 1975, Geol. Surv. Can., Bull. 252, pl. 1, fig. 10.
Crossley Lakes Member, Langton Bay Formation, Lower Cretaceous, lat. 69°04′N, long. 125°51′W, Horton River, District of Mackenzie.

Concavissimisporites variverrucatus (Couper) Brenner
Hypotype 76603
Davies, E.H., Akande, S.O. and Zentilli, M., 1984, Geol. Surv. Can., Paper 84-1A, pl. 48.1, fig. 14.
Early Cretaceous, sample AGR 285-230, Gays River mine, lat. 45°02′N, long. 63°22′W, Nova Scotia.

Concavissimisporites verrucosus Delcourt and Sprumont
Hypotype 33978
Hopkins, Jr., W.S., 1974, Geol. Surv. Can., Paper 73-12, p. 17, pl. 4, fig. 42.
Christopher Formation, Lower Cretaceous, lat. 76°20′N, long. 115°34′W, northwest Melville Island, District of Franklin.

Concavissimisporites cf. *C. verrucosus* Del. and Sprum.
Hypotype 25074
Hopkins, Jr., W.S., 1971, Geol. Surv. Can., Bull. 197, p. 117, pl. 21, fig. 1.
Isachsen Formation, Lower Cretaceous, lat. 76°20′N, long. 115°34′W, northwest Melville Island, District of Franklin.

Concentricystes sp.
Fig. spec. 76144
Jerzykiewicz, T. and Sweet, A.R., 1986, Geol. Surv. Can., Paper 86-1B, Fig. 67.2, fig. 13.
Willow Creek Formation, Late Cretaceous, Crowsnest River, l.s.d. 13, sec. 28, tp. 1, rge. 1, W.5th mer., Alberta.

Contignisporites cooksonii (Balme) Dettmann
Hypotype 33979
Hopkins, Jr., W.S., 1974, Geol. Surv. Can., Paper 73-12, p. 17, pl. 4, figs. 43, 44.
Christopher Formation, Lower Cretaceous, lat. 76°20′N, long. 115°34′W, northwest Melville Island, District of Franklin.

Contignisporites cooksonii (Balme)
Hypotype 41899
Williams, G.L., 1975, Geol. Surv. Can., Paper 74-30, vol. 2, pl. 8, fig. 4.
Hauterivian, Lower Cretaceous, Mobil Sable Island C-67 well, Scotian Shelf.

Contignisporites cooksoni (Balme) Dettmann
Hypotype 42468
Brideaux, W.W. and McIntyre, D.J., 1975, Geol. Surv. Can., Bull. 252, pl. 2, figs. 41, 42.
Gilmore Lake Member, Langton Bay Formation, Lower Cretaceous, lat. 68°50′N, long. 125°24′W, Horton River, District of Mackenzie.

Contignisporites cf. *C. cooksonii* (Balme) Dettmann, 1963
Hypotype 13835
McGregor, D.C., 1965, Geol. Surv. Can., Paper 64-55, pl. 6, figs. 15, 16.
Lower Cretaceous, Fault Creek Canyon, Aklavik Range, Richardson Mountains, District of Mackenzie.

Contignisporites fornicatus Dettmann
Hypotype 33980
Hopkins, Jr., W.S., 1974, Geol. Surv. Can., Paper 73-12, p. 17, pl. 4, figs. 45, 46.
Christopher Formation, Lower Cretaceous, lat. 76°20′N, long. 115°34′W, northwest Melville Island, District of Franklin.

Contignisporites glebulentus Dettmann
Hypotypes 42465, 42466
Brideaux, W.W. and McIntyre, D.J., 1975, Geol. Surv. Can., Bull. 252, pl. 2, figs. 37-39.
Crossley Lakes Member, Langton Bay Formation, Lower Cretaceous, lat. 64°01′30″N, long. 126°02′W and lat. 69°04′N, long. 125°51′W, Horton River, District of Mackenzie.

Contignisporites glebulentus
Hypotype 91978
Banerjee, I. and Davies, E.H., 1988, Can. Soc. Petrol. Geol., Mem. 15, fig. 6.6.
Mannville Group, Lower Cretaceous, depth 5295 feet, well 11-08-41-24W4, Alberta.

Converrucosisporites saskatchewanensis Pocock
Hypotypes 25075, 25076
Hopkins, Jr., W.S., 1971, Geol. Surv. Can., Bull. 197, p. 118, pl. 21, figs. 3, 4.
Isachsen Formation, Lower Cretaceous, lat. 76°20′N, long. 115°34′W, northwest Melville Island, District of Franklin.

Converrucosisporites cf. C. saskatchewanensis Pocock
Hypotype 33981
Hopkins, Jr., W.S., 1974, Geol. Surv. Can., Paper 73-12, p. 18, pl. 4, fig. 47.
Christopher Formation, Lower Cretaceous, lat. 76°20′N, long. 115°34′W, northwest Melville Island, District of Franklin.

Converrucosisporites sp.
Fig. spec. 41602
Hopkins, Jr., W.S. and Sweet, A.R., 1976, Geol. Surv. Can., Bull. 256, pl. 9, fig. 18.
Mattagami Formation, Lower Cretaceous, borehole 68-39, lat. 50°35′48″N, long. 82°30′42″W, Ontario.

Convolutriletes sp. cf. *C. rousei* Pocock, 1970
Fig. spec. 68205
Ricketts, B.D. and Sweet, A.R., 1986, Geol. Surv. Can., Paper 84-15, p. 19, pl. 1, fig. 18.
Elk Formation, Cretaceous, Coal Creek, lat. 49°30′N, long. 114°59′W, British Columbia.

Convolutispora sp.
Fig. spec. 13915
McGregor, D.C., 1965, Geol. Surv. Can., Paper 64-55, pl. 8, fig. 10.
Isachsen Formation, Lower Cretaceous, south of Buchanan Lake anticline, eastern Axel Heiberg Island, District of Franklin.

Cooksonites variabilis Pocock
Hypotypes 42486, 42487
Brideaux, W.W. and McIntyre, D.J., 1975, Geol. Surv. Can., Bull. 252, pl. 3, figs. 20, 21.
Crossley Lakes Member, Langton Bay Formation, Lower Cretaceous, lat. 69°04′N, long. 125°51′W, Horton River, District of Mackenzie.

Coptospora sp. cf. *C. williamsii* Playford
Fig. specs. 42488, 42489
Brideaux, W.W. and McIntyre, D.J., 1975, Geol. Surv. Can., Bull. 252, p. 17, pl. 3, figs. 22, 23.
Horton River Formation and Gilmore Lake Member, Langton Bay Formation, Lower Cretaceous, lat. 69°19′N, long. 126°50′W and lat. 68°50′N, long. 125°24′W Horton River, District of Mackenzie.

Coptospora sp.
Fig. specs. 42490, 42491
Brideaux, W.W. and McIntyre, D.J., 1975, Geol. Surv. Can., Bull. 252, p. 17, pl. 3, figs. 24, 25.
Gilmore Lake Member, Langton Bay Formation, Lower Cretaceous, lat. 68°50′N, long. 125°24′W, Horton River, District of Mackenzie.

Coronatispora valdensis (Couper) Dettmann
Hypotypes 42454, 42455
Brideaux, W.W. and McIntyre, D.J., 1975, Geol. Surv. Can., Bull. 252, pl. 2, figs. 26, 27.
Crossley Lakes Member, Langton Bay Formation and Horton River Formation, Lower Cretaceous, lat. 69°04′N, long. 125°51′W and lat. 69°19′N, long. 126°50′W, Horton River, District of Mackenzie.

Costatheca diskoensis (Miner) Hall 1967
Hypotype 48944
Sweet, A.R., 1979, Am. Assoc. Strat. Palynol., Contrib. Ser. No. 5B, pl. 4, fig. 2.
Eureka Sound Formation, Upper Cretaceous, Meteorologist Peninsula, lat. 78°10′N, long. 99°30′W, Ellef Ringnes Island, District of Franklin.

Costatheca levis (Vangerow) Hall 1967
Hypotype 48943
Sweet, A.R., 1979, Am. Assoc. Strat. Palynol., Contrib. Ser. No. 5B, pl. 4, fig. 1.
Eureka Sound Formation, Upper Cretaceous, east of Hoodoo Dome, lat. 78°11′N, long. 99°34′W, Ellef Ringnes Island, District of Franklin.

Costatoperforosporites sp. cf. *Anemia perforata* Baranov, Nemkova and Kondratiev
Fig. spec. 42436
Brideaux, W.W. and McIntyre, D.J., 1975, Geol. Surv. Can., Bull. 252, pl. 2, fig. 8.
Crossley Lakes Member, Langton Bay Formation, Lower Cretaceous, lat. 69°04′N, long. 125°51′W, Horton River, District of Mackenzie.

Costatoperforosporites fistulosus Deák
Hypotype 42435
Brideaux, W.W. and McIntyre, D.J., 1975, Geol. Surv. Can., Bull. 252, pl. 2, figs. 6, 7.
Crossley Lakes Member, Langton Bay Formation, Lower Cretaceous, lat. 69°04′N, long. 125°51′W, Horton River, District of Mackenzie.

Costatoperforosporites foveolatus Deák
Hypotype 41900
Williams, G.L., 1975, Geol. Surv. Can., Paper 74-30, vol. 2, pl. 8, fig. 5.
Barremian, Lower Cretaceous, Shell Naskapi N-30 well, Scotian Shelf.

Costatoperforosporites foveolatus Deák
Hypotype 42437
Brideaux, W.W. and McIntyre, D.J., 1975, Geol. Surv. Can., Bull. 252, pl. 2, fig. 9.
Gilmore Lake Member, Langton Bay Formation, Lower Cretaceous, lat. 68°50′N, long. 125°24′W, Horton River, District of Mackenzie.

Costatoperforosporites foveolatus Deák, 1962
Hypotype 91342
Sweet, A.R. and McIntyre, D.J., 1988, Can. Soc. Petrol. Geol., Mem. 15, fig. 5.12.
Sturrock Member, Cardium Formation, Upper Cretaceous, south side of McLeod River, l.s.d. 16, sec. 16, tp. 47, rge. 23, W.5th mer., Alberta.

Cranwellia rumseyensis Srivastava, 1966
Hypotypes 71355, 71359
Wilson, M.A., 1978, Palaeontographica, Abt. B, vol. 166, p. 135, pl. 8, figs. 7, 12.
Cretaceous, Police Island, north bank Mackenzie River, 12 miles southeast of Fort Norman, and Tent Island Formation, Little Fish (Cache) Creek, Big Fish River area, District of Mackenzie.

Cranwellia rumseyensis Srivastava, 1966
Hypotype 76039
Jerzykiewicz, T. and Sweet, A.R., 1986, Can. J. Earth Sci., vol. 23, no. 9, pl. 3, fig. 6.
Late Cretaceous, depth 29.35-29.62 m, GSC core hole CV-42-2-2, Coal Valley strip mine, l.s.d. 1, sec. 10, tp. 47, rge. 19, W.5th mer., Alberta.

Cranwellia rumseyensis Srivastava, 1966
Hypotype 76100
Sweet, A.R., 1986, Can. J. Earth. Sci., vol. 23, no. 9, pl. 1, fig. 10.
Late Cretaceous, left bank Embarras River near Coalspur, l.s.d. 10, sec. 33, tp. 48, rge. 21, W.5th mer., Alberta.

Cranwellia rumseyensis Srivastava 1966
Hypotype 76141
Jerzykiewicz, T. and Sweet, A.R., 1986, Geol. Surv. Can., Paper 86-1B, Fig. 67.2, fig. 11.
Willow Creek Formation, Late Cretaceous, Castle River, l.s.d. 4-5, sec. 12, tp. 7, rge. 1, W.5th mer., Alberta.

Cranwellia rumseyensis Srivastava 1966
Hypotype 89718
Jerzykiewicz, T. and Sweet, A.R., 1988, Sedimentary Geol., vol. 59, no. 1/2, p. 58, pl. II, fig. 3.
Brazeau Formation, Upper Cretaceous, Blackstone River, lat. 52°43′20″N, long. 116°16′30″W, Alberta.

Cranwellia sp.
Fig. specs. 38277-38279
McIntyre, D.J., 1974, Geol. Surv. Can., Paper 74-14, pl. 21, figs. 9, 10; pl. 22, fig. 1.
Upper Cretaceous, lat. 69°28′N, long. 126°58′W, Horton River, District of Mackenzie.

Cranwellia spp.
Fig. specs. 48974-48979
Sweet, A.R., 1978, Geol. Surv. Can., Paper 78-1B, p. 34, pl. 6.1, figs. 15-19.
Cuesta Creek Member, Tent Island Formation, Upper Cretaceous, northwestern bank Big Fish River near its junction with Boundary Creek, lat. 68°30′30″N, long. 136°23′50″W, Yukon.

cf. *Cranwellia* sp.
Fig. specs. 89698, 89699
Jerzykiewicz, T. and Sweet, A.R., 1988, Sedimentary Geol., vol. 59, no. 1/2, p. 48, pl. I, figs. 2, 3.
Belly River Formation, Upper Cretaceous, Crowsnest River valley west of Lundbreck, lat. 49°35′N, long. 114°11′W, Alberta.

cf. *Cranwellia* sp.
Fig. spec. 93685
Sweet, A.R. and Braman, D.R., 1989, Geol. Surv. Can., Paper 89-8, p. 35, pl. 1, fig. 20.
Chungo Member, Wapiabi Formation, Upper Cretaceous, Trap Creek near junction with Highwood River, sec. 36, tp. 17, rge. 4, W.5th mer., Alberta.

Crybelosporites vectensis Kremp
Hypotypes 42471-42473
Brideaux, W.W. and McIntyre, D.J., 1975, Geol. Surv. Can., Bull. 252, pl. 3, figs. 5-7.
Horton River Formation (42471) and Crossley Lakes Member, Langton Bay Formation, Lower Cretaceous, lat. 69°19′N, long. 126°50′W and lat. 69°04′N, long. 125°51′W, Horton River, District of Mackenzie.

Crybelosporites vectensis Kremp, 1970
Hypotypes 58562, 58563
McIntyre, D.J. and Brideaux, W.W., 1980, Geol. Surv. Can., Bull. 320, p.10, pl. 1, figs. 6, 7.
Lower Cretaceous, 'Grizzly Gorge', lat. 68°15′49″N, long. 135°40′59″W, Richardson Mountains, District of Mackenzie.

Cupuliferoidaepollenites levitas (Tschudy) Jameossansaie, 1987
Hypotypes 91386, 91387
Sweet, A.R. and McIntyre, D.J., 1988, Can. Soc. Petrol. Geol., Mem. 15, figs. 7.5, 7.6.
Moosehound Member, Cardium Formation, Upper Cretaceous, south side of Grande Cache Higway 40, l.s.d. 15, sec. 13, tp. 57, rge. 7, W.5th mer., Alberta.

Cupuliferoidaepollenites parvulus (Groot and Penny) Dettmann, 1973
Hypotypes 91389, 91390
Sweet, A.R. and McIntyre, D.J., 1988, Can. Soc. Petrol. Geol., Mem. 15, figs. 7.8, 7.9.
Moosehound Member, Cardium Formation, Upper Cretaceous, south side of Grande Cache Highway 40, l.s.d. 15, sec. 13, tp. 57, rge. 7, W.5th mer., Alberta.

Cyathidites australis Couper, 1953
Hypotypes 13907, 13908
McGregor, D.C., 1965, Geol. Surv. Can., Paper 64-55, pl. 8, figs. 3, 4.
Isachsen Formation, Lower Cretaceous, south of Buchanan Lake anticline, eastern Axel Heiberg Island, District of Franklin.

Cyathidites australis Couper
Hypotype 25068
Hopkins, Jr., W.S., 1971, Geol. Surv. Can., Bull. 197, p. 116, pl. 20, fig. 23.
Isachsen Formation, Lower Cretaceous, lat. 76°20′N, long. 115°34′W, northwest Melville Island, District of Franklin.

Cyathidites australis Couper
Hypotype 31764
Hopkins, Jr., W.S. and Balkwill, H.R., 1973, Geol. Surv. Can., Paper 72-37, p. 15, pl. 2, fig. 26.
Hassel Formation, Cretaceous, Hoodoo Dome, lat. 78°11′N, long. 99°40′W, southern Ellef Ringnes Island, District of Franklin.

Cyathidites australis Couper
Hypotype 33957
Hopkins, Jr., W.S., 1974, Geol. Surv. Can., Paper 73-12, p. 12, pl. 2, fig. 19.
Christopher Formation, Lower Cretaceous, lat. 76°20′N, long. 115°34′W, northwest Melville Island, District of Franklin.

Cyathidites australis Couper
Hypotype 42390
Brideaux, W.W. and McIntyre, D.J., 1975, Geol. Surv. Can., Bull. 252, pl. 1, fig. 2.
Crossley Lakes Member, Langton Bay Formation, Lower Cretaceous, lat. 69°04′N, long. 125°51′W, Horton River, District of Mackenzie.

Cyathidites cf. *C. australis* Couper, 1953
Hypotype 13944
McGregor, D.C., 1965, Geol. Surv. Can., Paper 64-55, pl. 10, fig. 7.
Lower Cretaceous, Peel Plateau, Stony Creek about 18.5 miles upstream from mouth, District of Mackenzie.

Cyathidites minor Couper, 1953
Hypotypes 13861, 13862
McGregor, D.C., 1965, Geol. Surv. Can., Paper 64-55, pl. 7, figs. 3, 4.
Lower Cretaceous, Donna River Canyon, Aklavik Range, Richardson Mountains, District of Mackenzie.

Cyathidites minor Couper
Hypotype 25069
Hopkins, Jr., W.S., 1971, Geol. Surv. Can., Bull. 197, p. 116, pl. 20, fig. 24.

Isachsen Formation, Lower Cretaceous, lat. 76°20′N, long. 115°34′W, northwest Melville Island, District of Franklin.

Cyathidites minor Couper
Hypotype 31765
Hopkins, Jr., W.S. and Balkwill, H.R., 1973, Geol. Surv. Can., Paper 72-37, p. 15, pl. 2, fig. 27.
Hassel Formation, Cretaceous, Hoodoo Dome, lat. 78°11′N, long. 99°40′W, southern Ellef Ringnes Island, District of Franklin.

Cyathidites minor Couper
Hypotype 33958
Hopkins, Jr., W.S., 1974, Geol. Surv. Can., Paper 73-12, p. 12, pl. 2, fig. 20.
Christopher Formation, Lower Cretaceous, lat. 76°20′N, long. 115°34′W, northwest Melville Island, District of Franklin.

Cyathidites minor Couper
Hypotype 38177
McIntyre, D.J., 1974, Geol. Surv. Can., Paper 74-14, pl. 14, fig. 8.
Upper Cretaceous, lat. 69°28′N, long. 126°58′W, Horton River, District of Mackenzie.

Cyathidites minor Couper
Hypotype 42391
Brideaux, W.W. and McIntyre, D.J., 1975, Geol. Surv. Can., Bull. 252, pl. 1, fig. 3.
Horton River Formation, Lower Cretaceous, lat. 69°27′30″N, long. 126°54′W, Horton River, District of Mackenzie.

Cyathidites minor Couper
Hypotype 41594
Hopkins, Jr., W.S. and Sweet, A.R., 1976, Geol. Surv. Can., Bull. 256, pl. 9, fig. 10.
Mattagami Formation, Lower Cretaceous, borehole 68-42, lat. 50°35′36″N, long. 82°30′12″W, Ontario.

Cyathidites sp.
Fig. spec. 13909
McGregor, D.C., 1965, Geol. Surv. Can., Paper 64-55, pl. 8, fig. 5.
Isachsen Formation, Lower Cretaceous, south of Buchanan Lake anticline, eastern Axel Heiberg Island, District of Franklin.

Cycadopites follicularis Wilson and Webster, 1946
Hypotype 13924
McGregor, D.C., 1965, Geol. Surv. Can., Paper 64-55, pl. 8, fig. 24.
Isachsen Formation, Lower Cretaceous, south of Buchanan Lake anticline, eastern Axel Heiberg Island, District of Franklin.

Cycadopites follicularis Wilson and Webster, 1946
Hypotype 71338
Wilson, M.A., 1978, Palaeontographica, Abt. B, vol. 166, p. 129, pl. 7, fig. 4.
Cretaceous, Police Island, north bank Mackenzie River, 12 miles southeast of Fort Norman, District of Mackenzie.

?*Cycadopites formosus* Singh, 1964
Hypotype 13880
McGregor, D.C., 1965, Geol. Surv. Can., Paper 64-55, pl. 7, fig. 28.
Lower Cretaceous, Donna River Canyon, Aklavik Range, Richardson Mountains, District of Mackenzie.

cf. *Cycadopites formosus* Singh, 1964
Hypotype 13925
McGregor, D.C., 1965, Geol. Surv. Can., Paper 64-55, pl. 8, fig. 25.
Isachsen Formation, Lower Cretaceous, south of Buchanan Lake anticline, eastern Axel Heiberg Island, District of Franklin.

Cycadopites fragilis Singh, 1964
Hypotype 13878
McGregor, D.C., 1965, Geol. Surv. Can., Paper 64-55, pl. 7, fig. 25.
Lower Cretaceous, Donna River Canyon, Aklavik Range, Richardson Mountains, District of Mackenzie.

cf. *Cycadopites fragilis* Singh, 1964
Hypotype 13875
McGregor, D.C., 1965, Geol. Surv. Can., Paper 64-55, pl. 7, fig. 21.
Lower Cretaceous, Donna River Canyon, Aklavik Range, Richardson Mountains, District of Mackenzie.

Cycadopites nitidus (Balme)
Hypotype 38216
McIntyre, D.J., 1974, Geol. Surv. Can., Paper 74-14, pl. 15, fig. 9.
Upper Cretaceous, lat. 69°27′30″N, long. 126°58′W, Horton River, District of Mackenzie.

Cycadopites nitidus (Balme) Norris
Hypotype 42525
Brideaux, W.W. and McIntyre, D.J., 1975, Geol. Surv. Can., Bull. 252, pl. 4, fig. 24.
Crossley Lakes Member, Langton Bay Formation, Lower Cretaceous, lat. 69°01′30″N, long. 126°02′W, Horton River, District of Mackenzie.

Cycadopites scabratus Stanley, 1965
Hypotype 71339
Wilson, M.A., 1978, Palaeontographica, Abt. B, vol. 166, p. 129, pl. 7, fig. 5.
Cretaceous, Police Island, north bank Mackenzie River, 12 miles southeast of Fort Norman, District of Mackenzie.

Cycadopites cf. *C. scabratus* Stanley
Hypotype 31773
Hopkins, Jr., W.S. and Balkwill, H.R., 1973, Geol. Surv. Can., Paper 72-37, p. 17, pl. 2, fig. 35.
Hassel Formation, Cretaceous, Hoodoo Dome, lat. 78°11′N, long. 99°40′W, southern Ellef Ringnes Island, District of Franklin.

Cycadopites cf. *C. scabratus* Stanley
Hypotype 41604
Hopkins, Jr., W.S. and Sweet, A.R., 1976, Geol. Surv. Can., Bull. 256, pl. 9, fig. 20.
Mattagami Formation, Lower Cretaceous, Onakawana Lignite pit, lat. 50°36′N, long. 81°23′W, Ontario.

Cycadopites sp.
Fig. spec. 31774
Hopkins, Jr., W.S. and Balkwill, H.R., 1973, Geol. Surv. Can., Paper 72-37, p. 17, pl. 2, fig. 36.
Hassel Formation, Cretaceous, Hoodoo Dome, lat. 78°11′N, long. 99°40′W, southern Ellef Ringnes Island, District of Franklin.

Cycadopites sp.
Fig. spec. 71337
Wilson, M.A., 1978, Palaeontographica, Abt. B, vol. 166, p. 129, pl. 7, fig. 3.
Cretaceous, IOE Atkinson H-25 well, north shore Tuktoyaktuk Peninsula, lat. 69°44′20″N, long. 131°50′06″W, District of Mackenzie.

Cycadopitys sp. A, B
Fig. specs. 33996, 33997
Hopkins, Jr., W.S., 1974, Geol. Surv. Can., Paper 73-12, p. 21, pl. 5, figs. 63, 64.
Christopher Formation, Lower Cretaceous, lat. 76°20′N, long. 115°34′W, northwest Melville Island, District of Franklin.

Deltoidospora cf. *D. hallei* Miner, 1935
Hypotype 13940
McGregor, D.C., 1965, Geol. Surv. Can., Paper 64-55, pl. 10, fig. 2.
Lower Cretaceous, Peel Plateau, Stony Creek about 18.5 miles upstream from mouth, District of Mackenzie.

Deltoidospora hallii Miner
Hypotype 25085
Hopkins, Jr., W.S., 1971, Geol. Surv. Can., Bull. 197, p. 119, pl. 21, fig. 13.
Isachsen Formation, Lower Cretaceous, lat. 76°20′N, long. 115°34′W, northwest Melville Island, District of Franklin.

Deltoidospora juncta (Kara-Murza) Singh, 1964
Hypotype 13826
McGregor, D.C., 1965, Geol. Surv. Can., Paper 64-55, pl. 6, fig. 4.
Lower Cretaceous, Fault Creek Canyon, Aklavik Range, Richardson Mountains, District of Mackenzie.

Deltoidospora juncta (Kara-Murza) Singh
Hypotype 33983
Hopkins, Jr., W.S., 1974, Geol. Surv. Can., Paper 73-12, p. 18, pl. 4, fig. 49.
Christopher Formation, Lower Cretaceous, lat. 76°20′N, long. 115°34′W, northwest Melville Island, District of Franklin.

Deltoidospora junctum (Kara-Murza) Singh
Hypotype 25084
Hopkins, Jr., W.S., 1971, Geol. Surv. Can., Bull. 197, p. 119, pl. 21, fig. 12.
Isachsen Formation, Lower Cretaceous, lat. 76°20′N, long. 115°34′W, northwest Melville Island, District of Franklin.

Deltoidospora junctum (Kara-Murza) Singh
Hypotype 31766
Hopkins, Jr., W.S. and Balkwill, H.R., 1973, Geol. Surv. Can., Paper 72-37, p. 15, pl. 2, fig. 28.
Hassel Formation, Cretaceous, Hoodoo Dome, lat. 78°11′N, long. 99°40′W, southern Ellef Ringnes Island, District of Franklin.

Deltoidospora juncta (Kara-Murza) Singh
Hypotype 42392
Brideaux, W.W. and McIntyre, D.J., 1975, Geol. Surv. Can., Bull. 252, pl. 1, fig. 4.
Horton River Formation, Lower Cretaceous, lat. 69°27′30″N, long. 126°54′W, Horton River, District of Mackenzie.

Deltoidospora psilostoma Rouse
Hypotype 33982
Hopkins, Jr., W.S., 1974, Geol. Surv. Can., Paper 73-12, p. 18, pl. 4, fig. 48.
Christopher Formation, Lower Cretaceous, lat. 76°20′N, long. 115°34′W, northwest Melville Island, District of Franklin.

Deltoidospora sp.
Fig. spec. 13824
McGregor, D.C., 1965, Geol. Surv. Can., Paper 64-55, pl. 6, fig. 2.
Lower Cretaceous, Fault Creek Canyon, Aklavik Range, Richardson Mountains, District of Mackenzie.

Deltoidospora sp.
Fig. spec. 13905
McGregor, D.C., 1965, Geol. Surv. Can., Paper 64-55, pl. 8, fig. 1.
Isachsen Formation, Lower Cretaceous, south of Buchanan Lake anticline, eastern Axel Heiberg Island, District of Franklin.

Deltoidospora sp.
Fig. spec. 31767
Hopkins, Jr., W.S. and Balkwill, H.R., 1973, Geol. Surv. Can., Paper 72-37, p. 16, pl. 2, fig. 29.
Hassel Formation, Cretaceous, Hoodoo Dome, lat. 78°11′N, long. 99°40′W, southern Ellef Ringnes Island, District of Franklin.

Densoisporites cf. *D. crassus* Tralau
Hypotypes 25088, 25089
Hopkins, Jr., W.S., 1971, Geol. Surv. Can., Bull. 197, p. 120, pl. 21, figs. 16, 17.
Isachsen Formation, Lower Cretaceous, lat. 76°20′N, long. 115°34′W, northwest Melville Island, District of Franklin.

Densoisporites microrugulatus Brenner
Hypotype 42470
Brideaux, W.W. and McIntyre, D.J., 1975, Geol. Surv. Can., Bull. 252, pl. 3, fig. 4.
Crossley Lakes Member, Langton Bay Formation, Lower Cretaceous, lat. 69°04′N, long. 125°51′W, Horton River, District of Mackenzie.

Densoisporites velatus Weyland and Krieger, 1953
Hypotype 68202
Ricketts, B.D. and Sweet, A.R., 1986, Geol. Surv. Can., Paper 84-15, p. 19, pl. 1, fig. 14.
Elk Formation, Cretaceous, Coal Creek, lat. 49°30′N, long. 114°59′W, British Columbia.

Densoisporites sp.
Fig. spec. 41603
Hopkins, Jr., W.S. and Sweet, A.R., 1976, Geol. Surv. Can., Bull. 256, pl. 9, fig. 19.
Mattagami Formation, Lower Cretaceous, borehole 68-39, lat. 50°35′48″N, long. 82°30′42″W, Ontario.

Densosporites annulatus (Loose) Schopf, Wilson and Bentall, 1944
Hypotypes 71297, 71298
Wilson, M.A., 1978, Palaeontographica, Abt. B, vol. 166, p. 158, pl. 4, figs. 1, 2.
Cretaceous, Police Island, north bank Mackenzie River, 12 miles southeast of Fort Norman, District of Mackenzie.

Densosporites cf. *D. irregularis* Hacquebard and Barss, 1957
Hypotype 71299
Wilson, M.A., 1978, Palaeontographica, Abt. B, vol. 166, p. 158, pl. 4, fig. 3.
Cretaceous, Police Island, north bank Mackenzie River, 12 miles southeast of Fort Norman, District of Mackenzie.

Densosporites cf. *D. spitsbergensis* Playford, 1963
Hypotype 71296
Wilson, M.A., 1978, Palaeontographica, Abt. B, vol. 166, p. 159, pl. 3, figs. 12, 13.
Cretaceous, Police Island, north bank Mackenzie River, 12 miles southeast of Fort Norman, District of Mackenzie.

Diatomozonotriletes saetosus (Hacquebard and Barss) Hughes and Playford, 1961
Hypotype 71270
Wilson, M.A., 1978, Palaeontographica, Abt. B, vol. 166, p. 156, pl. 1, fig. 8.
Cretaceous, Police Island, north bank Mackenzie River, 12 miles southeast of Fort Norman, District of Mackenzie.

Diconodinium articum Manum and Cookson
Hypotypes 38075, 38076
McIntyre, D.J., 1974, Geol. Surv. Can., Paper 74-14, pl. 6, figs. 12, 13.
Upper Cretaceous, lat. 69°27′30″N, long. 126°58′W, Horton River, District of Mackenzie.

Diconodinium firmum Harland
Hypotypes 38072, 38073
McIntyre, D.J., 1974, Geol. Surv. Can., Paper 74-14, pl. 6, figs. 9, 10.
Upper Cretaceous, lat. 69°28′N, long. 126°58′W, Horton River, District of Mackenzie.

Diconodinium sp. 2
Fig. specs. 38061-38063
McIntyre, D.J., 1974, Geol. Surv. Can., Paper 74-14, pl. 5, figs. 12-14.
Upper Cretaceous, lat. 69°28′N, long. 126°58′W, Horton River, District of Mackenzie.
=*Laciniadinium biconiculum*, McIntyre, D.J., 1975, Geoscience and Man, vol. 11, p.71, pl. 4, figs. 5, 6 (holotype 38063), 8 (paratype 38062).

Dictyophyllidites sp.
Fig. spec. 31762
Hopkins, Jr., W.S. and Balkwill, H.R., 1973, Geol. Surv. Can., Paper 72-37, p. 14, pl. 1, fig. 24.
Hassel Formation, Cretaceous, Hoodoo Dome, lat. 78°11′N, long. 99°40′W, southern Ellef Ringnes Island, District of Franklin.

Dictyophyllidites sp.
Fig. spec. 33956
Hopkins, Jr., W.S., 1974, Geol. Surv. Can., Paper 73-12, p. 11, pl. 2, fig. 18.
Christopher Formation, Lower Cretaceous, lat. 76°20′N, long. 115°34′W, northwest Melville Island, District of Franklin.

Dictyophyllidites sp.
Fig. spec. 91355
Sweet, A.R. and McIntyre, D.J., 1988, Can. Soc. Petrol. Geol., Mem. 15, figs. 6.3.
Ram Member, Cardium Formation, Upper Cretaceous, north side of Highway 47 north of Cadomin, l.s.d. 14, sec. 16, tp. 47, rge. 23, W.5th mer., Alberta.

?*Dictyophyllidites* sp.
Fig. spec. 13945
McGregor, D.C., 1965, Geol. Surv. Can., Paper 64-55, pl. 10, fig. 8.
Lower Cretaceous, Peel Plateau, Stony Creek about 18.5 miles upstream from mouth, District of Mackenzie.

Dictyotriletes granulatus Pocock
Hypotypes 42404, 42405
Brideaux, W.W. and McIntyre, D.J., 1975, Geol. Surv. Can., Bull. 252, pl. 1, figs. 17, 18.
Horton River Formation, Lower Cretaceous, lat. 69°19′N, long. 126°50′W, Horton River, District of Mackenzie.

Dijkstraisporites sp. A
Fig. spec. 48940
Sweet, A.R., 1979, Am. Assoc. Strat. Palynol., Contrib. Ser. No. 5B, pl. 3, fig. 15.
Hassel Formation, Lower Cretaceous, lat. 78°11′N, long. 99°43′W, Ellef Ringnes Island, District of Franklin.

Diptycha sp.
Fig. spec. 93636
Sweet, A.R., Ricketts, B.D., Cameron, A.R. and Norris, D.K., 1989, Geol. Surv. Can., Paper 89-1G, p. 93, pl. 1, fig. 6.
Little Bear Formation, Upper Cretaceous, east of McKay Range, East Little Bear River area, lat. 64°36′24″N, long. 125°35′W, District of Mackenzie.

cf. *Disaccites grandis* Cookson, 1953
Hypotype 13966
McGregor, D.C., 1965, Geol. Surv. Can., Paper 64-55, pl. 10, fig. 40.
Lower Cretaceous, Peel Plateau, Stony Creek about 18.5 miles upstream from mouth, District of Mackenzie.

cf. *Disaccites* (*Phyllocladidites*) Rüei Cookson, 1947
Hypotypes 13876, 13877
McGregor, D.C., 1965, Geol. Surv. Can., Paper 64-55, pl. 7, figs. 22, 27.
Lower Cretaceous, Donna River Canyon, Aklavik Range, Richardson Mountains, District of Mackenzie.

Distaltriangulisporites perplexus (Singh) Singh
Hypotype 42467
Brideaux, W.W. and McIntyre, D.J., 1975, Geol. Surv. Can., Bull. 252, pl. 2, fig. 40.
Horton River Formation, Lower Cretaceous, lat. 69°19′N, long. 126°50′W, Horton River, District of Mackenzie.

Distaltriangulisporites perplexus (Singh) Singh, 1971
Hypotype 91351
Sweet, A.R. and McIntyre, D.J., 1988, Can. Soc. Petrol. Geol., Mem. 15, figs. 5.24, 5.25.
Kaskapau Formation, Upper Cretaceous, south side of Grande Cache Highway 40, l.s.d. 15, sec. 13, tp. 57, rge. 7, W.5th mer., Alberta.

Distaltriangilisporites sp.
Fig. spec. 93671
Sweet, A.R. and Braman, D.R., 1989, Geol. Surv. Can., Paper 89-8, pl. 1, fig. 6.
Chungo Member, Wapibiabi Formation, Upper Cretaceous, Trap Creek near junction with Highwood River, sec. 36, tp. 17, rge. 4, W.5th mer., Alberta.

Dorocysta litotes Davey
Hypotypes 38107-38109
McIntyre, D.J., 1974, Geol. Surv. Can., Paper 74-14, pl. 8, figs. 13-15.
Upper Cretaceous, lat. 69°27′30″N, long. 126°58′W, Horton River, District of Mackenzie.

Echinatisporis varispinosus (Pocock) Srivastava, 1975
Hypotype 68206
Ricketts, B.D. and Sweet, A.R., 1986, Geol. Surv. Can., Paper 84-15, p. 19, pl. 1, figs. 19, 20.
Pocaterra Member, Cadomin Formation, Highwood Pass, lat. 50°35′N, long. 115°1′W, Alberta.

Echinatisporis varispinosus (Pocock) Srivastava, 1975
Hypotype 91354
Sweet, A.R. and McIntyre, D.J., 1988, Can. Soc. Petrol. Geol., Mem. 15, fig. 6.2.
Moosehound Member, Cardium Formation, Upper Cretaceous, south side of McLeod River, l.s.d. 16, sec. 16, tp. 47, rge. 23, W.5th mer., Alberta.

Echinatisporis sp.
Fig. spec. 71267
Wilson, M.A., 1978, Palaeontographica, Abt. B, vol. 166, p. 111, pl. 1, figs. 5, 14.
Cretaceous, Police Island, north bank Mackenzie River, 12 miles southeast of Fort Norman, District of Mackenzie.

Echiperiporites sp.
Fig. spec. 76041
Jerzykiewicz, T. and Sweet, A.R., 1986, Can. J. Earth Sci., vol. 23, no. 9, pl. 3, fig. 8.
Late Cretaceous, depth 27.41-27.43, GSC core hole CV-42-2-2, Coal Valley strip mine, l.s.d. 1, sec. 10, tp. 47, rge. 19, W.5th mer., Alberta.

Echitriletes lanatus (Diykstra) Potonié 1956
Hypotype 48941
Sweet, A.R., 1979, Am. Assoc. Strat. Palynol., Contrib. Ser. No. 5B, pl. 3, figs. 16, 17.
Hassel Formation, Upper Cretaceous, lat. 79°47′24″N, long. 85°22′W, near Slidre Fiord, Ellesmere Island, District of Franklin.

Echitriletes mineri Sweet
Holotype 48932
Sweet, A.R., 1979, Am. Assoc. Strat. Palynol., Contrib. Ser. No. 5B, pl. 3, p. 15, pl. 3, fig. 7.
Hassel Formation, Lower Cretaceous, lat. 77°9′N, long. 104°50′W, Lougheed Island, District of Franklin.

Echitriletes cf. *E. mineri* Sweet 1979
Hypotype 48955
Sweet, A.R., 1979, Am. Assoc. Strat. Palynol., Contrib. Ser. No. 5B, pl. 4, fig. 13.
Eureka Sound Formation, Upper Cretaceous, Meteorologist Peninsula, lat. 78°10′N, long. 99°30′W, Ellef Ringnes Island, District of Franklin.

Echitriletes sp. A
Fig. spec. 48960
Sweet, A.R., 1979, Am. Assoc. Strat. Palynol., Contrib. Ser. No. 5B, pl. 4, fig. 18.
Eureka Sound Formation, Upper Cretaceous, east of Hoodoo Dome, lat. 78°11′N, long. 99°43′W, Ellef Ringnes Island, District of Franklin.

Ellipsodinum rugulosum Clarke and Verdier
Hypotype 38074
McIntyre, D.J., 1974, Geol. Surv. Can., Paper 74-14, pl. 6, fig. 11.
Upper Cretaceous, lat. 69°27′30″N, long. 126°58′W, Horton River, District of Mackenzie.

Ephedra sp.
Fig. spec. 31775
Hopkins, Jr., W.S. and Balkwill, H.R., 1973, Geol. Surv. Can., Paper 72-37, p. 18, pl. 2, fig. 37.
Hassel Formation, Cretaceous, Hoodoo Dome, lat. 78°11′N, long. 99°40′W, southern Ellef Ringnes Island, District of Franklin.

Ephedra sp.
Fig. spec. 71335
Wilson, M.A., 1978, Palaeontographica, Abt. B, vol. 166, p. 125, pl. 7, fig. 1.
Cretaceous, IEO Atkinson H-25 well, north shore Tuktoyaktuk Peninsula, lat. 69°44′20″N, long. 131°50′06″W, District of Mackenzie.

Ephedripites ambiguus Hedlund, 1966
Hypotype 91376
Sweet, A.R. and McIntyre, D.J., 1988, Can. Soc. Petrol. Geol., Mem. 15, fig. 6.25.
Moosehound Member, Cardium Formation, Upper Cretaceous, south side of McLeod River, l.s.d. 16, sec. 16, tp. 47, rge. 23, W.5th mer., Alberta.

Ephedripites multicostatus Brenner, 1963
Hypotype 68232
Ricketts, B.D. and Sweet, A.R., 1986, Geol. Surv. Can., Paper 84-15, p. 22, pl. 4, fig. 3.
Elk Formation, Cretaceous, Coal Creek, lat. 49°30′N, long. 114°59′W, British Columbia.

Ephedripites multicostatus Brenner, 1963
Hypotype 91375
Sweet, A.R. and McIntyre, D.J., 1988, Can. Soc. Petrol. Geol., Mem. 15, fig. 6.24.
Kaskapau Formation, Upper Cretaceous, south side of Grande Cache Highway 40, l.s.d. 15, sec. 13, tp. 57, rge. 7, W.5th mer., Alberta.

Ephedripites multipartitus (Chlonova) Gao and Zhav, 1976
Hypotype 76011
Jerzykiewicz, T. and Sweet, A.R., 1986, Can. J. Earth Sci., vol. 23, no. 9, pl. 1, fig. 14.
Late Cretaceous, right bank of Embarras River towards the south end of Robb, l.s.d. 14, sec. 11, tp. 49, rge. 21, W.5th mer., Alberta.

Ephedripites sp. cf. *E. patapscoensis* Brenner, 1963
Fig. spec. 68233
Ricketts, B.D. and Sweet, A.R., 1986, Geol. Surv. Can., Paper 84-15, p. 22, pl. 4, fig. 4.
Elk Formation, Cretaceous, Coal Creek, lat. 49°30′N, long. 114°59′W, British Columbia.

Ephedrpites sp.
Fig. spec. 91377
Sweet, A.R. and McIntyre, D.J., 1988, Can. Soc. Petrol. Geol., Mem. 15, fig. 6.26.
Ram Member, Cardium Formation, Upper Cretaceous, north side of Highway 47 north of Cadomin, l.s.d. 14, sec. 16, tp. 47, rge. 23, W.5th mer., Alberta.

Ephedripites sp. A
Fig. spec. 56776
Williams, G.L. and Bujak, J.P., 1980, Initial Repts. Deep Sea Drilling Project, vol. 50, pl. 10, fig. 7.
Early Cretaceous, DSDP Site 416, lat. 32°50.18′N, long. 10°48.06′W, northwestern Moroccan continental margin.

Equisetosporites amabilis Srivastava 1968
Hypotype 89697
Jerzykiewicz, T. and Sweet, A.R., 1988, Sedimentary Geol., vol. 59, no. 1/2, p. 48, pl. I, fig. 1.
Belly River Formation, Upper Cretaceous, Crowsnest River valley west of Lundbreck, lat. 49°35′N, long. 114°11′W, Alberta.

Equisetosporites lajwantis Srivastava, 1968
Hypotype 76012
Jerzykiewicz, T. and Sweet, A.R., 1986, Can. J. Earth Sci., vol. 23, no. 9, pl. 1, fig. 15.
Late Cretaceous, right bank of Embarras River towards the south end of Robb, l.s.d. 14, sec. 11, tp. 49, rge. 21, W.5th mer., Alberta.

Equisetosporites multicostatus (Brenner)
Hypotype 42518
Brideaux, W.W. and McIntyre, D.J., 1975, Geol. Surv. Can., Bull. 252, pl. 4, fig. 16.
Crossley Lakes Member, Langton Bay Formation, Lower Cretaceous, lat. 69°04′N, long. 125°51′W, Horton River, District of Mackenzie.

Equisetosporites spp.
Fig. specs. 93677, 93678
Sweet, A.R. and Braman, D.R., 1989, Geol. Surv. Can., Paper 89-8, p. 35, pl. 1, figs. 12, 13.
Chungo Member, Wapiabi Formation, Upper Cretaceous, Trap Creek near junction with Highwood River, sec. 36, tp. 17, rge. 4, W.5th mer., Alberta.

Erdtmanipollis pachysandroides Krutzsch, 1962
Hypotype 71400
Wilson, M.A., 1978, Palaeontographica, Abt. B, vol. 166, p. 147, pl. 10, fig. 23.
Cretaceous, Police Island, north bank Mackenzie River, 12 miles southeast of Fort Norman, District of Mackenzie.

Erdtmanipollis procumbentiformis (Samoilovitch)
Hypotype 38275
McIntyre, D.J., 1974, Geol. Surv. Can., Paper 74-14, pl. 21, fig. 7.
Upper Cretaceous, lat. 69°28′N, long. 126°58′W, Horton River, District of Mackenzie.

Erdtmanipollis procumbentiformis (Samoilovich) Krutzsch, 1966
Hypotype 76043
Jerzykiewicz, T. and Sweet, A.R., 1986, Can. J. Earth Sci., vol. 23, no. 9, pl. 3, fig. 10.
Late Cretaceous, depth 31.1-31.32, GSC core hole CV-42-2-2, Coal Valley strip mine, l.s.d. 1, sec. 10, tp. 47, rge. 19, W.5th mer., Alberta.

Erdtmanipollis procumbentiformis (Samoilovitch) Krutzsch, 1966
Hypotype 89717
Jerzykiewicz, T. and Sweet, A.R., 1988, Sedimentary Geol., vol. 59, no. 1/2, pl. II, fig. 2.
Brazeau Formation, Upper Cretaceous, Blackstone River, lat. 52°43′20′N, long. 116°16′30″W, Alberta.

Ericaceoipollenites rallus Stanley, 1965
Hypotype 76042
Jerzykiewicz, T. and Sweet, A.R., 1986, Can. J. Earth Sci., vol. 23, no. 9, pl. 3, fig. 9.
Late Cretaceous, depth 27.41-27.43 m, GSC core hole CV-42-2-2, Coal Valley strip mine, l.s.d. 1, sec. 10, tp. 47, rge. 19, W.5th mer., Alberta.

Erlansonisporites sparassis (Murray)
Hypotypes 41629-41631
Hopkins, Jr., W.S. and Sweet, A.R., 1976, Geol. Surv. Can., Bull. 256, pl. 11, figs. 5-7.
Mattagami Formation, Lower Cretaceous, borehole 68-38, lat. 50°36′N, long. 82°31′12″W (41629), and borehole 68-39, lat. 50°35′48″N, long. 82°30′42″W, Ontario.

Eucommiidites minor Groot and Penny
Hypotype 34006
Hopkins, Jr., W.S., 1974, Geol. Surv. Can., Paper 73-12, p. 24, pl. 6, fig. 74.
Christopher Formation, Lower Cretaceous, lat. 76°20′N, long. 115°34′W, northwest Melville Island, District of Franklin.

Eucommiidites minor Groot and Penny
Hypotype 41911
Williams, G.L, 1975, Geol. Surv. Can., Paper 74-30, vol. 2, pl. 8, fig. 16.
Aptian, Lower Cretaceous, Mobil Dauntless D-35 well, Scotian Shelf.

Eucommiidites minor Groot and Penny
Hypotypes 42519, 42520
Brideaux, W.W. and McIntyre, D.J., 1975, Geol. Surv. Can., Bull. 252, pl. 4, figs. 17, 18.
Horton River Formation, Lower Cretaceous, lat. 69°27′30″N, long. 126°54′W and lat. 69°19′N, long. 126°50′W, Horton River, District of Mackenzie.

Eucommiidites minor Groot and Penny, 1960
Hypotype 91378
Sweet, A.R. and McIntyre, D.J., 1988, Can. Soc. Petrol. Geol., Mem. 15, fig. 6.27.
Ram Member, Cardium Formation, Upper Cretaceous, north side of Highway 47 north of Cadomin, l.s.d. 14, sec. 16, tp. 47, rge. 23, W.5th mer., Alberta.

Eucommiidites minor Groot and Penny, 1960
Hypotype 93679
Sweet, A.R. and Braman, D.R., 1989, Geol. Surv. Can., Paper 89-8, p. 35, pl. 1, fig. 14.
Chungo Member, Wapiabi Formation, Upper Cretaceous, Highwood River No. 2 section, N.E. 1/4, sec. 1, tp. 18, rge. 4, W.5th mer., Alberta.

Eucommiidites troedssonii Erdtman
Hypotypes 25500-25502
Hopkins, Jr., W.S., 1971, Geol. Surv. Can., Bull. 197, p. 123, pl. 22, figs. 17-19.
Isachsen Formation, Lower Cretaceous, lat. 76°20′N, long. 115°34′W, northwest Melville Island, District of Franklin.

Eucommiidites troedssonii Erdtman
Hypotype 31785
Hopkins, Jr., W.S. and Balkwill, H.R., 1973, Geol. Surv. Can., Paper 72-37, p. 21, pl. 2, fig. 47.
Hassel Formation, Cretaceous, Hoodoo Dome, lat. 78°11′N, long. 99°40′W, southern Ellef Ringnes Island, District of Franklin.

Eucommiidites troedssonii Erdtman
Hypotypes 41611, 41612
Hopkins, Jr., W.S. and Sweet, A.R., 1976, Geol. Surv. Can., Bull. 256, pl. 9, figs. 27, 28.
Mattagami Formation, Lower Cretaceous, borehole 68-39, lat. 50°35′48″N, long. 82°30′42″W, Ontario.

Eucommiidites troedssonii Erdtman, 1948
Hypotype 68231
Ricketts, B.D. and Sweet, A.R., 1986, Geol. Surv. Can., Paper 84-15, p. 23, pl. 4, fig. 2.
Elk Formation, Cretaceous, Coal Creek, lat. 49°30′N, long. 114°59′W, British Columbia.

?*Eucommiidites* sp.
Fig. spec. 13874
McGregor, D.C., 1965, Geol. Surv. Can., Paper 64-55, pl. 7, fig. 20.
Lower Cretaceous, Donna River Canyon, Aklavik Range, Richardson Mountains, District of Mackenzie.

Exesipollenites tumulus Balme
Hypotype 42530
Brideaux, W.W. and McIntyre, D.J., 1975, Geol. Surv. Can., Bull. 252, pl. 4, fig. 29.
Horton River Formation, Lower Cretaceous, lat. 69°27′30″N, long. 126°54′W, Horton River, District of Mackenzie.

Exochosphaeridium striolatum (Deflandre) Davey
Hypotype 62648
Ioannides, N.S. and McIntyre, D.J., 1980, Geol. Surv. Can., Paper 80-1A, pl. 31.2, fig. 11.
Upper Cretaceous, Caribou Hills, Mackenzie River, District of Mackenzie.

Exochosphaeridium sp. cf. *E. striolatum* (Deflandre)
Hypotypes 38160, 38161
McIntyre, D.J., 1974, Geol. Surv. Can., Paper 74-14, pl. 13, figs. 5, 6.
Upper Cretaceous, lat. 69°27′30″N, long. 126°58′W, Horton River, District of Mackenzie.

Expressipollis sp. cf. *E. barbatus* Chlonova 1961
Fig. spec. 89723
Jerzykiewicz, T. and Sweet, A.R., 1988, Sedimentary Geol., vol. 59, no. 1/2, p. 58, pl. II, fig. 8.
Brazeau Formation, Upper Cretaceous, Blackstone River, lat. 52°43′20″N, long. 116°16′30″W, Alberta.

Expressipollis spp.
Fig. specs. 48983-48986
Sweet, A.R., 1978, Geol. Surv. Can., Paper 78-1B, p. 34, pl. 6.1, figs. 23-26.
Cuesta Creek Member, Tent Island Formation, Upper Cretaceous, northwestern bank Big Fish River near its junction with Boundary Creek, lat. 68°30′30″N, long. 136°23′50″W, Yukon.

Expressipollis sp. 1, 2
Fig. specs. 38235, 38236, 38242
McIntyre, D.J., 1974, Geol. Surv. Can., Paper 74-14, pl. 17, figs. 1, 2, 8.
Upper Cretaceous, lat. 69°28′N, long. 126°58′W, Horton River, District of Mackenzie.

Extratriporopollenites sp. 1
Fig. specs. 38282, 38284
McIntyre, D.J., 1974, Geol. Surv. Can., Paper 74-14, pl. 22, figs. 5, 6.
Upper Cretaceous, lat. 69°28′N, long. 126°58′W, Horton River, District of Mackenzie.

Extratriporopollenites sp. 2
Fig. specs. 38280-38282
McIntyre, D.J., 1974, Geol. Surv. Can., Paper 74-14, pl. 22, figs. 2-4.
Upper Cretaceous, lat. 69°28′N, long. 126°58′W, Horton River, District of Mackenzie.

Fibulapollis sp.
Fig. specs. 91392, 91393
Sweet, A.R. and McIntyre, D.J., 1988, Can. Soc. Petrol. Geol., Mem. 15, figs. 7.11, 7.12.
Moosehound Member, Cardium Formation, Upper Cretaceous, south side of Grande Cache Highway 40, l.s.d. 15, sec. 13, tp. 57, rge. 7, W.5th mer., Alberta.

cf. *Fibulapollis* sp.
Fig. spec. 38318
McIntyre, D.J., 1974, Geol. Surv. Can., Paper 74-14, pl. 23, figs. 14.
Upper Cretaceous, lat. 69°28′N, long. 126°58′W, Horton River, District of Mackenzie.

Foraminisporis asymmetrica
Hypotype 91984
Banerjee, I. and Davies, E.H., 1988, Can. Soc. Petrol. Geol., Mem. 15, fig. 6.12.
Mannville Group, Lower Cretaceous, depth 4335 feet, well 15-21-47-23W4, Alberta.

Foraminisporis asymmetricus (Cookson and Dettmann) Dettmann
Hypotypes 42458, 42459
Brideaux, W.W. and McIntyre, D.J., 1975, Geol. Surv. Can., Bull. 252, pl. 2, figs. 30, 31.
Horton River Formation and Crossley Lakes Member, Langton Bay Formation, Lower Cretaceous, lat. 69°19′N, long. 126°50′W and lat. 69°04′N, long. 125°51′W, Horton River, District of Mackenzie.

Foraminisporis dailyi (Cookson and Dettmann) Dettmann, 1963
Hypotype 68204
Ricketts, B.D. and Sweet, A.R., 1986, Geol. Surv. Can., Paper 84-15, p. 19, pl. 1, fig. 17.
Elk Formation, Cretaceous, Coal Creek, lat. 49°30′N, long. 114°59′W, British Columbia.

Foraminisporis dailyi (Cookson and Dettmann) Dettmann, 1963
Hypotype 913557
Sweet, A.R. and McIntyre, D.J., 1988, Can. Soc. Petrol. Geol., Mem. 15, fig. 6.5.
Moosehound Member, Cardium Formation, Upper Cretaceous, south side of McLeod River, l.s.d. 16, sec. 16, tp. 47, rge. 23, W.5th mer., Alberta.

Foraminisporis undulatus Leffingwell, 1971
Hypotype 91371
Sweet, A.R. and McIntyre, D.J., 1988, Can. Soc. Petrol. Geol., Mem. 15, figs. 6.19, 6.20.
Moosehound Member, Cardium Formation, Upper Cretaceous, south side of McLeod River, l.s.d. 16, sec. 16, tp. 47, rge. 23, W.5th mer., Alberta.

Foraminisporis undulosa Leffingwell, 1971
Hypotype 76000
Jerzykiewicz, T. and Sweet, A.R., 1986, Can. J. Earth Sci., vol. 23, no. 9, pl. 1, fig. 3.
Late Cretaceous, depth 29.35-29.62 m, GSC core hole C4-42-2-2, Coal Valley strip mine, l.s.d. 1, sec. 10, tp. 47, rge. 19, W.5th mer., Alberta.

Foraminisporis wonthaggiensis (Cookson and Dettmann) Dettmann
Hypotypes 42460, 42461
Brideaux, W.W. and McIntyre, D.J., 1975, Geol. Surv. Can., Bull. 252, pl. 2, figs. 32, 33.
Crossley Lakes Member, Langton Bay Formation, Lower Cretaceous, lat. 69°12′30″N, long. 126°17′W and lat. 69°04′N, long. 125°51′W, Horton River, District of Mackenzie.

Foraminisporis wonthaggiensis (Cookson and Dettmann) Dettmann, 1963
Hypotype 68203
Ricketts, B.D. and Sweet, A.R., 1986, Geol. Surv. Can., Paper 84-15, p. 20, pl. 1, figs. 15, 16.
Elk Formation, Cretaceous, Coal Creek, lat. 49°30′N, long. 114°59′W, British Columbia.

Foraminisporis sp.
Fig. spec. 91356
Sweet, A.R. and McIntyre, D.J., 1988, Can. Soc. Petrol. Geol., Mem. 15, fig. 6.4.

Moosehound Member, Cardium Formation, Upper Cretaceous, south side of McLeod River, l.s.d. 16, sec. 16, tp. 47, rge. 23, W.5th mer., Alberta.

Foraminisporis sp.
Fig. spec. 93673
Sweet, A.R. and Braman, D.R., 1989, Geol. Surv. Can., Paper 89-8, p. 35, pl. 1, fig. 8.
Chungo Member, Wapiabi Formation, Upper Cretaceous, Trap Creek near junction with Highwood River, sec. 36, tp. 17, rge. 4, W.5th mer., Alberta.

Foraminisporis #EA
Fig. spec. 91980
Banerjee, I. and Davies, E.H., 1988, Can. Soc. Petrol. Geol., Mem. 15, fig. 6.8.
Mannville Group, Lower Cretaceous, depth 4414 feet, well 3-22-46-23W4, Alberta.

Foveosporites canalis Balme
Hypotype 25090
Hopkins, Jr., W.S., 1971, Geol. Surv. Can., Bull. 197, p. 120, pl. 21, fig. 18.
Isachsen Formation, Lower Cretaceous, lat. 76°20′N, long. 115°34′W, northwest Melville Island, District of Franklin.

Foveosporites cenomanicus (Chlonova) Schvetzova, 1960
Hypotype 91352
Sweet, A.R. and McIntyre, D.J., 1988, Can. Soc. Petrol. Geol., Mem. 15, fig. 5.26.
Leyland Member, Cardium Formation, Upper Cretaceous, south side of McLeod River, l.s.d. 16, sec. 16, tp. 47, rge. 23, W.5th mer., Alberta.

Foveosporites pantostikos Phillips and Felix, 1971
Hypotype 91358
Sweet, A.R. and McIntyre, D.J., 1988, Can. Soc. Petrol. Geol., Mem. 15, fig. 6.6.
Moosehound Member, Cardium Formation, Upper Cretaceous, south side of McLeod River, l.s.d. 16, sec. 16, tp. 47, rge. 23, W.5th mer., Alberta.

Foveosporites pantostiktos Phillips and Felix, 1971
Hypotype 89732
Jerzykiewicz, T. and Sweet, A.R., 1986, Sedimentary Geol., vol. 59, no. 1/2, p. 58, pl. III, fig. 5.
St. Mary River Formation, Upper Cretaceous, Crownest River, lat. 49°30′N, long. 114°3′W, Alberta.

Foveotetradites dignatus Singh, 1983
Hypotype 91401
Sweet, A.R. and McIntyre, D.J., 1988, Can. Soc. Petrol. Geol., Mem. 15, fig. 7.20.
Moosehound Member, Cardium Formation, Upper Cretaceous, north side of Highway 47 north of Cadomin, l.s.d. 14, sec. 16, tp. 47, rge. 23, W.5th mer., Alberta.

Foveotetradites sp.
Fig. spec. 93689
Sweet, A.R. and Braman, D.R., 1989, Geol. Surv. Can., Paper 89-8, p. 35, pl. 1, fig. 24.
Chungo Member, Wapiabi Formation, Upper Cretaceous, Trap Creek near junction with Highwood River, sec. 36, tp. 17, rge. 4, W.5th mer., Alberta.

Foveotriletes subtriangularis Brenner
Hypotype 42416
Brideaux, W.W. and McIntyre, D.J., 1975, Geol. Surv. Can., Bull. 252, pl. 1, fig. 29.
Horton River Formation, Lower Cretaceous, lat. 69°19′N, long. 126°50′W, Horton River, District of Mackenzie.

Fraxinoipollenites variabilis Stanley, 1965
Hypotypes 91398, 91399
Sweet, A.R. and McIntyre, D.J., 1988, Can. Soc. Petrol. Geol., Mem. 15, figs. 7.17, 7.18.
Moosehound Member, Cardium Formation, Upper Cretaceous, north side of Highway 47 north of Cadomin, l.s.d. 14, sec. 16, tp. 47, rge. 23, W.5th mer., and south side of Grande Cache Highway 40, l.s.d. 15, sec. 13, tp. 57, rge. 7, .5th mer., Alberta.

Fraxinoipollenites variabilis Stanley, 1965
Hypotype 93692
Sweet, A.R. and Braman, D.R., 1989, Geol. Surv. Can., Paper 89-8, pl. 1, fig. 27.
Chungo Member, Wapiabi Formation, Upper Cretaceous, Trap Creek near junction with Highwood River, sec. 36, tp. 17, rge. 4, W.5th mer., Alberta.

Fraxinoipollenites venustus Singh
Hypotype 42534, 42535
Brideaux, W.W. and McIntyre, D.J., 1975, Geol. Surv. Can., Bull. 252, pl. 4, figs. 35-37.
Horton River Formation, Lower Cretaceous, lat. 69°19′N, long. 126°50′W, Horton River, District of Mackenzie.

Fromea amphora Cookson and Eisenack
Hypotype 38112
McIntyre, D.J., 1974, Geol. Surv. Can., Paper 74-14, pl. 8, fig. 18.
Upper Cretaceous, lat. 69°27′30″N, long. 126°58′W, Horton River, District of Mackenzie.

Fromea amphora Cookson and Eisenack, 1958
Hypotype 96495
Leckie, D.A. and Singh, C., 1991, J. Sed. Petrology, vol. 61, no. 5, fig. 32.
Paddy Member, Peace River Formation, Lower Cretaceous, near town of Peace River, Alberta.

Fungal spore
Fig. spec. 47894
Williams, G.L., 1978, Initial Repts. Deep Sea Drilling Project, vol. 41, pl. 6, fig. 3.
Early Cretaceous, DSDP Site 370, lat. 32°50.2′N, long. 10°46.6′W, northwestern Moroccan continental margin.

Fungal spore
Fig. spec. 49002
Sweet, A.R., 1978, Geol. Surv. Can., Paper 78-1B, pl. 6.2, fig. 16.

Cuesta Creek Member, Tent Island Formation, Upper Cretaceous, northwestern bank Big Fish River near its junction with Boundary Creek, lat. 68°30′30″N, long. 136°23′50″W, Yukon.

Fungal spore sp.
Fig. spec. 71408
Wilson, M.A., 1978, Palaeontographica, Abt. B, vol. 166, p. 150, pl. 11, fig. 7.
Moose Channel Formation, Cretaceous, Aklak Creek, Big Fish River area, District of Mackenzie.

Fusiformisporites cf. *F. pseudocrabbii* Elsik, 1968
Hypotype 71397
Wilson, M.A., 1978, Palaeontographica, Abt. B, vol. 166, p. 149, pl. 10, figs. 19, 20.
Cretaceous, IEO Atkinson H-25 well, north shore Tuktoyaktuk Peninsula, lat. 69°44′20″N, long. 131°50′06″W, District of Mackenzie.

Fusiformisporites sp.
Fig. spec. 49000
Sweet, A.R., 1978, Geol. Surv. Can. Paper 78-1B, p. 36, pl. 6.2, fig. 14.
Cuesta Creek Member, Tent Island Formation, Upper Cretaceous, northwestern bank Big Fish River near its junction with Boundary Creek, lat. 68°30′30″N, long. 136°23′50″W, Yukon.

Gabonisporis cristata (Stanley) comb. nov.
Hypotype 76092
Sweet, A.R., 1986, Can. J. Earth Sci., vol. 23, no. 9, p.1375, pl. 1, figs. 1, 2.
Late Cretaceous, depth 29.35-29.62 m, GSC core hole CV-42-2-2, Coal Valley strip mine, l.s.d. 1, sec. 10, tp. 47, rge. 19, W.5th mer., Alberta.

Gabonisporis cristata (Stanley) Sweet, 1986
Hypotype 76002
Jerzykiewicz, T. and Sweet, A.R., 1986, Can. J. Earth Sci., vol. 23, no. 9, pl. 1, fig. 5.
Late Cretaceous, right bank of Embarras River towards the south end of Robb, l.s.d. 14, sec. 11, tp. 49, rge. 21, W.5th mer., Alberta.

Gemmatriletes clavatus Brenner, 1968 *in* Singh, 1983
Hypotype 91363
Sweet, A.R. and McIntyre, D.J., 1988, Can. Soc. Petrol. Geol., Mem. 15, fig. 6.11.
Moosehound Member, Cardium Formation, Upper Cretaceous, south side of Grande Cache Highway 40, l.s.d. 15, sec. 13, tp. 59, rge. 7, W.5th mer., Alberta.

Ghoshispora sp.
Fig. spec. 76003
Jerzykiewicz, T. and Sweet, A.R., 1986, Can. J. Earth Sci., vol. 23, no. 9, pl. 1, fig. 6.
Late Cretaceous, depth 29.35-29.62 m, GSC core CV-42-2-2, Coal Valley strip mine, l.s.d. 1, sec. 10, tp. 47, rge. 19, W.5th mer., Alberta.

Gillinia hymenophora Cookson and Eisenack
Hypotypes 38086, 38087
McIntyre, D.J., 1974, Geol. Surv. Can., Paper 74-14, pl. 7, figs, 8, 9.
Upper Cretaceous, lat. 69°28′N, long. 126°58′W, Horton River, District of Mackenzie.

?*Ginkgocycadophytus nitidus* (Balme) de Jersey in Dettmann, 1963
Hypotype 13881
McGregor, D.C., 1965, Geol. Surv. Can., Paper 64-55, pl. 7, fig. 29.
Lower Cretaceous, Donna River Canyon, Aklavik Range, Richardson Mountains, District of Mackenzie.

?*Ginkgocycadophytus nitidus* (Balme) de Jersey, 1962
Hypotype 13956
McGregor, D.C., 1965, Geol. Surv. Can., Paper 64-55, pl. 10, fig. 24.
Lower Cretaceous, Peel Plateau, Stony Creek about 18.5 miles upstream from mouth, District of Mackenzie.

Ginkgocycadophytus sp.
Fig. spec. 13844
McGregor, D.C., 1965, Geol. Surv. Can., Paper 64-55, pl. 6, fig. 27.
Lower Cretaceous, Fault Creek Canyon, Aklavik Range, Richardson Mountains, District of Mackenzie.

Ginkgocycadophytus sp.
Fig. spec. 93681
Sweet, A.R. and Braman, D.R., 1989, Geol. Surv. Can., Paper 89-8, pl. 1, fig. 16.
Chungo Member, Wapiabi Formation, Upper Cretaceous, Trap Creek near junction with Highwood River, sec. 36, tp. 17, rge. 4, W.5th mer., Alberta.

?*Ginkgocycadophytus* sp.
Fig. spec. 13883
McGregor, D.C., 1965, Geol. Surv. Can., Paper 64-55, pl. 7, fig. 32.
Lower Cretaceous, Donna River Canyon, Aklavik Range, Richardson Mountains, District of Mackenzie.

Gleicheniidites cf. *G. cercinidites* (Cookson) Dettmann, 1963
Hypotype 13906
McGregor, D.C., 1965, Geol. Surv. Can., Paper 64-55, pl. 8, fig. 2.
Isachsen Formation, Lower Cretaceous, south of Buchanan Lake anticline, eastern Axel Heiberg Island, District of Franklin.

Gleicheniidites cf. *G. cercinidites* (Cookson) Dettmann, 1963
Hypotype 13942
McGregor, D.C., 1965, Geol. Surv. Can., Paper 64-55, pl. 10, fig. 5.
Lower Cretaceous, Peel Plateau, Stony Creek about 18.5 miles upstream from mouth, District of Mackenzie.

Gleicheniidites senonicus Ross, 1949
Hypotype 13943
McGregor, D.C., 1965, Geol. Surv. Can., Paper 64-55, pl. 10, fig. 6.
Lower Cretaceous, Peel Plateau, Stony Creek about 18.5 miles upstream from mouth, District of Mackenzie.

Gleicheniidites senonicus Ross
Hypotypes 25065-25067
Hopkins, Jr., W.S., 1971, Geol. Surv. Can., Bull. 197, p. 116, pl. 20, figs. 20-22.
Isachsen Formation, Lower Cretaceous, lat. 76°20′N, long. 115°34′W, northwest Melville Island, District of Franklin.

Gleicheniidites senonicus Ross
Hypotype 31761
Hopkins, Jr., W.S. and Balkwill, H.R., 1973, Geol. Surv. Can., Paper 72-37, p. 14, pl. 1, fig. 23.
Hassel Formation, Cretaceous, Hoodoo Dome, lat. 78°11′N, long. 99°40′W, southern Ellef Ringnes Island, District of Franklin.

Gleicheniidites senonicus Ross
Hypotype 33960
Hopkins, Jr., W.S., 1974, Geol. Surv. Can., Paper 73-12, p. 12, pl. 2, fig. 22.
Christopher Formation, Lower Cretaceous, lat. 76°20′N, long. 115°34′W, northwest Melville Island, District of Franklin.

Gleicheniidites senonicus Ross
Hypotypes 38173, 38174
McIntyre, D.J., 1974, Geol. Surv. Can., Paper 74-14, pl. 14, figs, 4, 5.
Upper Cretaceous, lat. 69°58′30″N, long. 127°04′W, and lat. 69°28′N, long. 126°58′W, Horton River, District of Mackenzie.

Gleicheniidites senonicus Ross
Hypotype 41592
Hopkins, Jr., W.S. and Sweet, A.R., 1976, Geol. Surv. Can., Bull. 256, pl. 9, fig. 8.
Mattagami Formation, Lower Cretaceous, borehole Onakawana lignite pit, lat. 50°36′N, long. 81°23′W, Ontario.

Gleicheniidites senonicus Ross
Hypotype 42456
Brideaux, W.W. and McIntyre, D.J., 1976, Geol. Surv. Can., Bull. 252, pl. 2, fig. 28.
Horton River Formation, Lower Cretaceous, lat. 69°27′30″N, long. 126°54′W, Horton River, District of Mackenzie.

Gleicheniidites senonicus Ross, 1949
Hypotype 71293
Wilson, M.A., 1978, Palaeontographica, Abt. B, vol. 166, p. 115, pl. 3, fig. 8.
Cretaceous, Police Island, north bank Mackenzie River, 12 miles southeast of Fort Norman, District of Mackenzie.

Gleicheniidites senonicus Ross, 1949
Hypotype 68207
Ricketts, B.D. and Sweet, A.R., 1986, Geol. Surv. Can., Paper 84-15, p. 20, pl. 1, fig. 21.
Elk Formation, Cretaceous, Coal Creek, lat. 49°30′N, long. 114°59′W, British Columbia.

Gleicheniidites senonicus Ross, 1969
Hypotype 91369
Sweet, A.R. and McIntyre, D.J., 1988, Can. Soc. Petrol. Geol., Mem. 15, fig. 6.17.
Ram Member, Cardium Formation, Upper Cretaceous, south side of Grande Cache Highway 40, l.s.d. 15, sec. 13, tp. 57, rge. 7, W.5th mer., Alberta.

?*Gleicheniidites triplex* (Bolkhovitina) Grigoryeva in Samoilovich et al., 1961
Hypotype 13941
McGregor, D.C., 1965, Geol. Surv. Can., Paper 64-55, pl. 10, fig. 3.
Lower Cretaceous, Peel Plateau, Stony Creek about 18.5 miles upstream from mouth, District of Mackenzie.

Gleicheniidites sp.
Fig. spec. 13828
McGregor, D.C., 1965, Geol. Surv. Can., Paper 64-55, pl. 6, fig. 6.
Lower Cretaceous, Fault Creek Canyon, Aklavik Range, Richardson Mountains, District of Mackenzie.

?*Gleicheniidites* sp.
Fig. specs. 13827, 13863
McGregor, D.C., 1965, Geol. Surv. Can., Paper 64-55, pl. 6, fig. 5; pl. 7, fig. 5.
Lower Cretaceous, Fault Creek Canyon, Aklavik Range, Richardson Mountains, District of Mackenzie.

Glyptostrobus cf. *G. vacuipites* Wodehouse
Hypotype 31779
Hopkins, Jr., W.S. and Balkwill, H.R., 1973, Geol. Surv. Can., Paper 72-37, p. 19, pl. 2, fig. 41.
Hassel Formation, Cretaceous, Hoodoo Dome, lat. 78°11′N, long. 99°40′W, southern Ellef Ringnes Island, District of Franklin.

Glyptostrobus sp.
Fig. spec. 41608
Hopkins, Jr., W.S. and Sweet, A.R., 1976, Geol. Surv. Can., Bull. 256, pl. 9, fig. 24.
Mattagami Formation, Lower Cretaceous, borehole 68-39, lat. 50°35′48″N, long. 82°30′42″W, Ontario.

Gnetaceaepollenites ventrosus Srivastava, 1968
Hypotype 76013
Jerzykiewicz, T. and Sweet, A.R., 1986, Can. J. Earth Sci., vol. 23, no. 9, pl. 1, fig. 16.
Late Cretaceous, Embarras River, Alberta.

Granabivesiculites inchoatus Pierce, 1961
Hypotype 91379
Sweet, A.R. and McIntyre, D.J., 1988, Can. Soc. Petrol. Geol., Mem. 15, fig. 6.28.
Leyland Member, Cardium Formation, Upper Cretaceous, south side of McLeod River, l.s.d. 16, sec. 16, tp. 47, rge. 23, W.5th mer., Alberta.

Granulatisporites cf. *G.dailyi* Cookson and Dettmann
Hypotype 33984
Hopkins, Jr., W.S., 1974, Geol. Surv. Can., Paper 73-12, p. 18, pl. 4, fig. 50.

Christopher Formation, Lower Cretaceous, lat. 76°20′N, long. 115°34′W, northwest Melville Island, District of Franklin.

Gunnaripollis superba Srivastava, 1969
Hypotype 76044
Jerzykiewicz, T. and Sweet, A.R., 1986, Can. J. Earth Sci., vol. 23, no. 9, pl. 3, fig. 11.
Late Cretaceous, right bank of Embarras River towards the south end of Robb, l.s.d. 14, sec. 11, tp. 49, rge. 21, W.5th mer., Alberta.

Gunnera microreticulata (Belsky, Boltenhagen and Potonié) Leffingwell, 1971
Hypotype 76045
Jerzykiewicz, T. and Sweet, A.R., 1986, Can. J. Earth Sci., vol. 23, no. 9, pl. 3, fig. 12.
Late Cretaceous, Embarras River, Alberta.

Hamulatisporis amplus Stanley
Hypotypes 38193, 38194
McIntyre, D.J., 1974, Geol. Surv. Can., Paper 74-14, pl. 14, figs, 24, 25.
Upper Cretaceous, lat. 69°28′N, long. 126°58′W, Horton River, District of Mackenzie.

Hamulatisporis hamulatis Krutzsch, 1959
Hypotype 71289
Wilson, M.A., 1978, Palaeontographica, Abt. B, vol. 166, p. 114, pl. 3, figs. 3, 4.
Cretaceous, Police Island, north bank Mackenzie River, 12 miles southeast of Fort Norman, District of Mackenzie.

Hazaria sheoparii Srivastava
Hypotypes 38195, 38196
McIntyre, D.J., 1974, Geol. Surv. Can., Paper 74-14, pl. 14, figs, 26, 27.
Upper Cretaceous, lat. 69°28′N, long. 126°58′W, Horton River, District of Mackenzie.

Heliosporites altmarkensis Schulz, 1962
Hypotype 76007
Jerzykiewicz, T. and Sweet, A.R., 1986, Can. J. Earth Sci., vol. 23, no. 9, pl. 1, fig. 10.
Late Cretaceous, depth 29.35-29.62 m, GSC core hole C4-42-2-2, Coal Valley strip mine, l.s.d. 11, sec. 10, tp. 47, rge. 19, W.5th mer., Alberta.

Heliosporites cf. *H. altmarkensis* Schulz, 1962
Hypotypes 71275, 71280
Wilson, M.A., 1978, Palaeontographica, Abt. B, vol. 166, p. 111, pl. 1, fig. 13; pl. 2, fig. 4.
Cretaceous, Police Island, north bank Mackenzie River, 12 miles southeast of Fort Norman, District of Mackenzie.

Heliosporites kemensis (Chlonova) Srivastava, 1972
Hypotype 93674
Sweet, A.R. and Braman, D.R., 1989, Geol. Surv. Can., Paper 89-8, pl. 1, fig. 9.
Chungo Member, Wapiabi Formation, Upper Cretaceous, Trap Creek near junction with Highwood River, sec. 36, tp. 17, rge. 4, W.5th mer., Alberta.

Henrisporites cf. *H. affinis* (Dijkstra) Potonié 1956
Hypotypes 48924, 48925
Sweet, A.R., 1979, Am. Assoc. Strat. Palynol., Contrib. Ser. No. 5B, pl. 2, figs. 20, 21.
Christopher(?) Formation, Lower Cretaceous, lat. 72°22′N, long. 118°54′W, northeastern Banks Island, District of Franklin.

Henrisporites sp. A
Fig. specs. 48921, 48922
Sweet, A.R., 1979, Am. Assoc. Strat. Palynol., Contrib. Ser. No. 5B, pl. 2, figs. 17, 18.
Christopher Formation, Lower Cretaceous, lat. 74°31′N, long. 121°10′W, Cape McClure, Banks Island, District of Franklin.

Henrisporites sp. B
Fig. specs. 48945, 48946
Sweet, A.R., 1979, Am. Assoc. Strat. Palynol., Contrib. Ser. No. 5B, pl. 4, figs. 3, 5.
Eureka Sound Formation, Upper Cretaceous, lat. 77°20′45″N, long. 104°57′W, Lougheed Island, and Meteorologist Peninsula, lat. 78°10′N, long. 99°30′W, Ellef Ringnes Island, District of Franklin.

Hepiticae spores
Fig. specs. 76005, 76006
Jerzykiewicz, T. and Sweet, A.R., 1986, Can. J. Earth Sci., vol. 23, no. 9, pl. 1, figs. 8, 9.
Late Cretaceous, depth 29.35-29.62 m, GSC core hore C4-42-2-2, Coal Valley strip mine, l.s.d. 1, sec. 10, tp. 47, rge. 19, W.5th mer., and Paskapoo Formation, Early Paleocene, roadcut along Highway 17, Coalspur, Alberta.

Heteropolar, spinose forms of *Aquilapollenites*
Fig. specs. 48967-48969
Sweet, A.R., 1978, Geol. Surv. Can., Paper 78-1B, p. 32, pl. 6.1, figs. 7, 9.
Cuesta Creek Member, Tent Island Formation, Upper Cretaceous, northwestern bank Big Fish River near its junction with Boundary Creek, lat. 68°30′30″N, long. 136°23′50″W, Yukon.

Hexagonifera chlamydata Cookson and Eisenack
Hypotypes 38141-38143
McIntyre, D.J., 1974, Geol. Surv. Can., Paper 74-14, pl. 11, figs, 7, 9.
Upper Cretaceous, lat. 69°27′30″N, long. 126°58′W, and lat. 69°28′N, long. 126°58′W, Horton River, District of Mackenzie.

Horologinella apiculata Coolson and Eisenack
Hypotype 38106
McIntyre, D.J., 1974, Geol. Surv. Can., Paper 74-14, pl. 8, fig. 12.
Upper Cretaceous, lat. 69°28′N, long. 126°58′W, Horton River, District of Mackenzie.

Horologinella sp.
Fig. spec. 49007
Sweet, A.R., 1978, Geol. Surv. Can., Paper 78-1B, pl. 6.2, fig. 21.

Cuesta Creek Member, Tent Island Formation, Upper Cretaceous, northwestern bank Big Fish River near its junction with Boundary Creek, lat. 68°30′30″N, long. 136°23′50″W, Yukon.

Hymenozonotriletes pseudoalveolatus (Couper) Singh, 1964
Hypotype 13847
McGregor, D.C., 1965, Geol. Surv. Can., Paper 64-55, pl. 16, fig. 30.
Lower Cretaceous, Fault Creek Canyon, Aklavik Range, Richardson Mountains, District of Mackenzie.

Hymenozonotriletes cf.*H. pseudoalveolatus* (Couper) Singh
Fig. spec. 25091
Hopkins, Jr., W.S., 1971, Geol. Surv. Can., Bull. 197, p. 120, pl. 21, fig. 19.
Isachsen Formation, Lower Cretaceous, lat. 76°20′N, long. 115°34′W, northwest Melville Island, District of Franklin.

Hystrichosphaeridium sp. 2, *in* McIntyre 1974
Fig. spec. 62636
Ioannides, N.S. and McIntyre, D.J., 1980, Geol. Surv. Can., Paper 80-1A, pl. 31.3, fig. 7.
Upper Cretaceous, Caribou Hills, Mackenzie River, District of Mackenzie.

Hystrichosphaeropsis sp.
Fig. specs. 38150-38152
McIntyre, D.J., 1974, Geol. Surv. Can., Paper 74-14, pl. 12, figs, 7-9.
Upper Cretaceous, lat. 69°28′N, long. 126°58′W, Horton River, District of Mackenzie.

Impardecispora apiverrucata (Couper) Venkatachela et al.
Hypotype 76599
Davies, E.H., Akande, S.O. and Zentilli, M., 1984, Geol. Surv. Can., Paper 84-1A, pl. 48.1, fig. 10.
Early Cretaceous, sample AGR 285-230, Gays River mine, lat. 45°02′N, long. 63°22′W, Nova Scotia.

Impardecispora minor (Pocock) Venkatchala, Kar and Raza, 1968
Hypotype 68213
Ricketts, B.D. and Sweet, A.R., 1986, Geol. Surv. Can., Paper 84-15, p. 20, pl. 2, fig. 7.
Elk Formation, Cretaceous, Coal Creek, lat. 49°30′N, long. 114°59′W, British Columbia.

Impardecispora uralensis (Bolkhovitina) Venkatachala, Kar and Raza, 1968
Hypotype 68219
Ricketts, B.D. and Sweet, A.R., 1986, Geol. Surv. Can., Paper 84-15, p. 20, pl. 3, fig. 1.
Elk Formation, Cretaceous, Coal Creek, lat. 49°30′N, long. 114°59′W, British Columbia.

Imperdecispora sp. A
Fig. specs. 68212, 68215
Ricketts, B.D. and Sweet, A.R., 1986, Geol. Surv. Can., Paper 84-15, p. 20, pl. 2, figs. 5, 6, 9, 10.
Elk Formation, Cretaceous, Coal Creek, lat. 49°30′N, long. 114°59′W, British Columbia, and Pocaterra Creek Member, Cadomin Formation, Highwood Pass, lat. 50°35′N, long. 115°1′W, Alberta.

Impletosphaeridium sp. cf. *I. whitei* (Deflandre and Courtville)
Fig. spec. 38155
McIntyre, D.J., 1974, Geol. Surv. Can., Paper 74-14, pl. 12, fig. 12.
Upper Cretaceous, lat. 69°27′30″N, long. 126°58′W, Horton River, District of Mackenzie.

Inaperturopollenites atlanticus Groot, Penny and Groot, 1961
Hypotype 71341
Wilson, M.A., 1978, Palaeontographica, Abt. B, vol. 166, p. 128, pl. 7, fig. 7.
Cretaceous, Police Island, north bank Mackenzie River, 12 miles southeast of Fort Norman, District of Mackenzie.

Inaperturopollenites dubius (Potonié and Venitz) Thomson and Pflug, 1953
Hypotype 71329
Wilson, M.A., 1978, Palaeontographica, Abt. B, vol. 166, p. 127, pl. 6, fig. 8.
Cretaceous, Police Island, north bank Mackenzie River, 12 miles southeast of Fort Norman, District of Mackenzie.

Inaperturopollenites limbatus Balme
Hypotype 42511
Brideaux, W.W. and McIntyre, D.J., 1975, Geol. Surv. Can., Bull. 252, pl. 4, fig. 8.
Crossley Lakes Member, Langton Bay Formation, Lower Cretaceous, lat. 69°04′N, long. 125°51′W, Horton River, District of Mackenzie.

Inaperturopollenites sp.
Fig. specs. 25503-25505
Hopkins, Jr., W.S., 1971, Geol. Surv. Can., Bull. 197, p. 124, pl. 22, figs. 20-22.
Isachsen Formation, Lower Cretaceous, lat. 76°20′N, long. 115°34′W, northwest Melville Island, District of Franklin.

Inaperturopollenites sp.
Fig. spec. 31783
Hopkins, Jr., W.S. and Balkwill, H.R., 1973, Geol. Surv. Can., Paper 72-37, p. 20, pl. 2, fig. 45.
Hassel Formation, Upper Cretaceous, lat. 78°11′N, long. 99°40′W, southern Ellef Ringnes Island, District of Franklin.

Inaperturopollenites sp.
Fig. spec. 34003
Hopkins, Jr., W.S., 1974, Geol. Surv. Can., Paper 73-12, p. 23, pl. 6, fig. 70.
Christopher Formation, Lower Cretaceous, lat. 76°20′N, long. 115°34′W, northwest Melville Island, District of Franklin.

Inaperturopollenites sp.
Fig. spec. 41615
Hopkins, Jr., W.S. and Sweet, A.R., 1976, Geol. Surv. Can., Bull. 256, pl. 10, fig. 2.
Mattagami Formation, Lower Cretaceous, borehole Okanawana lignite pit, lat. 50°36′N, long. 81°23′W, Ontario.

Inaperturotetrdites scabratus Tschudy 1973
Hypotype 89716
Jerzykiewicz, T. and Sweet, A.R., 1988, Sedimentary Geol., vol. 59, no. 1/2, p. 58, pl. II, fig. 1.
St. Mary River Formation, Upper Cretaceous, Crownest River, lat. 49°30′N, long. 114°3′W, Alberta.

Integricorpus clarireticulatus Samoilovitch
Hypotypes 38268-38269
McIntyre, D.J., 1974, Geol. Surv. Can., Paper 74-14, pl. 20, figs, 9, 10.
Upper Cretaceous, 69°28′N, long. 126°58′W, Horton River, District of Mackenzie.

Integricorpus venustus (Srivastava)
Hypotypes 38266, 38267
McIntyre, D.J., 1974, Geol. Surv. Can., Paper 74-14, pl. 20, figs, 7, 8.
Upper Cretaceous, lat. 69°28′N, long. 126°58′W, Horton River, District of Mackenzie.

Integricorpus (*Intergricorpus* sp. 1 of Samoilovitch, 1967)
Fig. specs. 38270, 38271
McIntyre, D.J., 1974, Geol. Surv. Can., Paper 74-14, pl. 21, figs, 1-3.
Upper Cretaceous, lat. 69°28′N, long. 126°58′W, Horton River, District of Mackenzie.

Interulobites sp.
Fig. spec. 93672
Sweet, A.R. and Braman, D.R., 1989, Geol. Surv. Can., Paper 89-8, p. 35, pl. 1, fig. 7.
Chungo Member, Wapiabi Formation, Upper Cretaceous, Trap Creek near junction with Highwood River, sec. 36, tp. 17, rge. 4, W.5th mer., Alberta.

Isabelidinium microarmum (McIntyre) Lentin and Williams
Hypotypes 62640, 62646
Ioannides, N.S. and McIntyre, D.J., 1980, Geol. Surv. Can., Paper 80-1A, pl. 31.2, figs. 2, 9.
Upper Cretaceous, Caribou Hills, Mackenzie River, District of Mackenzie.

Ischyosporites estherae Deak
Hypotype 76595
Davies, E.H., Akande, S.O. and Zentilli, M., 1984, Geol. Surv. Can., Paper 84-1A, pl. 48.1, fig. 6.
Early Cretaceous, sample AGR 285-230, Gays River mine, lat. 45°02′N, long. 63°22′W, Nova Scotia.

Ischysporites punctatus Cookson and Dettmann, 1958
Hypotype 68199
Ricketts, B.D. and Sweet, A.R., 1986, Geol. Surv. Can., Paper 84-15, p. 20, pl. 1, figs. 10, 11.
Elk Formation, Cretaceous, Coal Creek, lat. 49°30′N, long. 114°59′W, British Columbia.

Januasporites spiniferus Singh
Hypotypes 42492, 42493
Brideaux, W.W. and McIntyre, D.J., 1975, Geol. Surv. Can., Bull. 252, pl. 3, figs. 26, 27.
Horton River Formation, Lower Cretaceous, lat. 69°19′N, long. 126°50′W, Horton River, District of Mackenzie.

Jiangupollis major Song *in* Song Zhi-Chen, Zhen Ya-hui et al., 1980
Hypotype 89849
Sweet, A.R., Ricketts, B.D., Cameron, A.R. and Norris, D.K., 1989, Geol. Surv. Can., Paper 89-1G, p. 98, pl. 2, fig. 2.
Summit Creek Formation, Upper Cretaceous, north bank Mackenzie River immediately downstream from Police Island, lat. 64°52′42″N, long. 125°14′44″W, District of Mackenzie.

cf. Juglandeceae
Fig. spec. 38300
McIntyre, D.J., 1974, Geol. Surv. Can., Paper 74-14, pl. 22, fig. 22.
Upper Cretaceous, lat. 69°28′N, long. 126°58′W, Horton River, District of Mackenzie.

Klukisporites areolatus Singh
Hypotype 42414
Brideaux, W.W. and McIntyre, D.J., 1975, Geol. Surv. Can., Bull. 252, pl. 1, fig. 27.
Gilmore Lake Member, Langton Bay Formation, Lower Cretaceous, lat. 68°50′N, long. 125°24′W, Horton River, District of Mackenzie.

Klukisporites areolaus (Singh)
Hypotype 76590
Davies, E.H., Akande, S.O. and Zentilli, M., 1984, Geol. Surv. Can., Paper 84-1A, pl. 48.1, fig. 1.
Early Cretaceous, sample DH2-409, Gays River mine, lat. 45°02′N, long. 63°22′W, Nova Scotia.

Klukisporites foveolatus Pocock, 1964
Hypotype 93666
Sweet, A.R. and Braman, D.R., 1989, Geol. Surv. Can., Paper 89-8, p. 35, pl. 1, fig. 1.
Chungo Member, Wapiabi Formation, Upper Cretaceous, Trap Creek near junction with Highwood River, sec. 36, tp. 17, rge. 4, W.5th mer., Alberta.

Klukisporites pseudoreticulatus Couper
Hypotype 42415
Brideaux, W.W. and McIntyre, D.J., 1975, Geol. Surv. Can., Bull. 252, pl. 1, fig. 28.
Crossley Lakes Member, Langton Bay Formation, Lower Cretaceous, lat. 69°12′30″N, long. 126°17′W, Horton River, District of Mackenzie.

Klukisporites cf. *K. pseudoreticulatus* Couper
Hypotype 23802
Hopkins, Jr., W.S., 1971, Geol. Surv. Can., Bull. 197, p. 116, pl. 20, fig. 19.
Isachsen Formation, Lower Cretaceous, lat. 76°20′N, long. 115°34′W, northwest Melville Island, District of Franklin.

Klukisporites sp. A
Fig. specs. 56792, 56793
Williams, G.L. and Bujak, J.P., 1980, Initial Repts. Deep Sea Drilling Project, vol. 50, pl. 11, figs. 11, 12.
Early Cretaceous and Late Jurassic, DSDP Site 416, lat. 32°50.18′N, long. 10°48.06′W, northwestern Moroccan continental margin.

Knoxisporites cf. *K. literatus* (Waltz) Playford, 1963
Hypotype 71305
Wilson, M.A., 1978, Palaeontographica, Abt. B, vol. 166, p. 159, pl. 4, figs. 9, 10.
Cretaceous, Police Island, north bank Mackenzie River, 12 miles southeast of Fort Norman, District of Mackenzie.

Krauselisporites hastilobatus Playford
Hypotype 42469
Brideaux, W.W. and McIntyre, D.J., 1975, Geol. Surv. Can., Bull. 252, pl. 3, figs. 1-3.
Crossley Lakes Member, Langton Bay Formation, Lower Cretaceous, lat. 69°04′N, long. 125°51′W, Horton River, District of Mackenzie.

Kraeuselisporites hastilobatus Playford, 1971
Hypotype 58559
McIntyre, D.J. and Brideaux, W.W., 1980, Geol. Surv. Can., Bull. 320, p.10, pl. 1, fig. 3.
Lower Cretaceous, 'Grizzly Gorge', lat. 68°15′49″N, long. 135°40′59″W, Richardson Mountains, District of Mackenzie.

Kurtzipites andersonii Srivastava 1981
Hypotype 89735
Jerzykiewicz, T. and Sweet, A.R., 1988, Sedimentary Geol., vol. 59, no. 1/2, pl. III, fig. 8.
Brazeau Formation, Upper Cretaceous, Blackstone River, lat. 52°43′20″N, long. 116°16′30″W, Alberta.

Kurtzipites trispissatus Anderson, 1960
Hypotype 71396
Wilson, M.A., 1978, Palaeontographica, Abt. B, vol. 166, p. 141, pl. 10, fig. 18.
Cretaceous, Police Island, north bank Mackenzie River, 12 miles southeast of Fort Norman, District of Mackenzie.

Kurtzipites trispissatus Anderson, 1960
Hypotype 76047
Jerzykiewicz, T. and Sweet, A.R., 1986, Can. J. Earth Sci., vol. 23, no. 9, pl. 3, fig. 14.
Late Cretaceous, depth 27.41-27.43 m, GSC core hole CV-42-2-2, Coal Valley strip mine, l.s.d. 1, sec. 10, tp. 47, rge. 19, W.5th mer., Alberta.

Kurtzipites sp.
Fig. specs. 38296, 38297
McIntyre, D.J., 1974, Geol. Surv. Can., Paper 74-14, pl. 22, figs, 18, 19.
Upper Cretaceous, lat. 69°28′N, long. 126°58′W, Horton River, District of Mackenzie.

Kuylisporites lunaris Cookson and Dettmann
Hypotype 33959
Hopkins, Jr., W.S., 1974, Geol. Surv. Can., Paper 73-12, p. 12, pl. 2, fig. 21.
Christopher Formation, Lower Cretaceous, lat. 76°20′N, long. 115°34′W, northwest Melville Island, District of Franklin.

Kuylisporites lunaris Cookson and Dettmann
Hypotype 42403
Brideaux, W.W. and McIntyre, D.J., 1975, Geol. Surv. Can., Bull. 252, pl. 1, fig. 16.
Crossley Lakes Member, Langton Bay Formation, Lower Cretaceous, lat. 69°04′N, long. 125°51′W, Horton River, District of Mackenzie.

Kuylisporites lunaris Cookson and Dettmann
Hypotype 41593
Hopkins, Jr., W.S., 1971, Geol. Surv. Can., Bull. 256, pl. 9, fig. 9.
Mattagami Formation, Lower Cretaceous, borehole 68-38, lat. 50°36′N, long. 82°31′12″W, Ontario.

Kuylisporites scutatus Newman, 1965
Hypotype 91367
Sweet, A.R. and McIntyre, D.J., 1988, Can. Soc. Petrol. Geol., Mem. 15, figs. 6.15.
Moosehound Member, Cardium Formation, Upper Cretaceous, south side of Grande Cache Highway 40, l.s.d. 15, sec. 13, tp. 57, rge. 7, W.5th mer., Alberta.

Laciniadinium biconiculum McIntyre
Paratypes 34036, 34037
McIntyre, D.J., 1975, Geoscience and Man, vol. 11, p. 71, pl. 4, figs. 7, 9.
Upper Cretaceous, lat. 69°28′N, long. 126°58′W, Horton River, District of Mackenzie.

Laciniadinium orbiculatum McIntyre
Paratype 34038
McIntyre, D.J., 1975, Geoscience and Man, vol. 11, p. 70, pl. 4, fig. 11.
Upper Cretaceous, lat. 69°27′30″N, long. 126°58′W, Horton River, District of Mackenzie.

Laevigatosporites haardti (Potonié and Venitz) Thomson and Pflug, 1953
Hypotypes 71303, 71304
Wilson, M.A., 1978, Palaeontographica, Abt. B, vol. 166, p. 117, pl. 4, figs. 7, 8.
Cretaceous, Police Island, north bank Mackenzie River, 12 miles southeast of Fort Norman, District of Mackenzie.

Laevigatosporites ovatus Wilson and Webster
Hypotype 38200
McIntyre, D.J., 1974, Geol. Surv. Can., Paper 74-14, pl. 14, fig. 31.
Upper Cretaceous, lat. 69°28′N, long. 126°58′W, Horton River, District of Mackenzie.

Laevigatosporites ovatus Wilson and Webster
Hypotype 31763
Hopkins, Jr., W.S. and Balkwill, H.R., 1973, Geol. Surv. Can., Paper 72-37, p. 15, pl. 1, fig. 25.
Hassel Formation, Cretaceous, Hoodoo Dome, lat. 78°11′N, long. 99°40′W, southern Ellef Ringnes Island, District of Franklin.

Laevigatosporites ovatus Wilson and Webster
Hypotype 33966
Hopkins, Jr., W.S., 1974, Geol. Surv. Can., Paper 73-12, p. 14, pl. 3, fig. 28.
Christopher Formation, Lower Cretaceous, lat. 76°20′N, long. 115°34′W, northwest Melville Island, District of Franklin.

Laevigatosporites ovatus Wilson and Webster
Hypotype 42474
Brideaux, W.W. and McIntyre, D.J., 1975, Geol. Surv. Can., Bull. 252, pl. 3, fig. 8.
Horton River Formation, Lower Cretaceous, lat. 69°27′30″N, long. 126°54′W, Horton River, District of Mackenzie.

Laevigatosporites sp.
Fig. spec. 41589
Hopkins, Jr., W.S. and Sweet, A.R., 1976, Geol. Surv. Can., Bull. 256, pl. 9, fig. 5.
Mattagami Formation, Lower Cretaceous, borehole 68-39, lat. 50°35′48″N, long. 82°30′42″W, Ontario.

Laevigatosporites sp. *in* Orlansky, 1971
Fig. spec. 91360
Sweet, A.R. and McIntyre, D.J., 1988, Can. Soc. Petrol. Geol., Mem. 15, fig. 6.8.
Moosehound Member, Cardium Formation, Upper Cretaceous, south side of McLeod River, l.s.d. 16, sec. 16, tp. 47, rge. 23, W.5th mer., Alberta.

Laricoidites gigantus Brenner
Hypotype 25097
Hopkins, Jr., W.S., 1971, Geol. Surv. Can., Bull. 197, p. 122, pl. 22, fig. 6.
Isachsen Formation, Lower Cretaceous, lat. 76°20′N, long. 115°34′W, northwest Melville Island, District of Franklin.

Laricoidites gigantus Brenner, 1963
Hypotype 71315
Wilson, M.A., 1978, Palaeontographica, Abt. B, vol. 166, p. 126, pl. 5, fig. 7.
Cretaceous, IEO Atkinson H-25 well, north shore Tuktoyaktuk Peninsula, lat. 69°44′20″N, long. 131°50′06″W, District of Mackenzie.

Laricoidites magnus (Potonié) Potonié, Thomson and Thiergart
Hypotype 25098
Hopkins, Jr., W.S., 1971, Geol. Surv. Can., Bull. 197, p. 122, pl. 22, fig. 7.
Isachsen Formation, Lower Cretaceous, lat. 76°20′N, long. 115°34′W, northwest Melville Island, District of Franklin.

Laricoidites magnus (Potonié) Potonié, Thomson and Thiergart
Hypotype 34000
Hopkins, Jr., W.S., 1974, Geol. Surv. Can., Paper 73-12, p. 22, pl. 6, fig. 67.
Christopher Formation, Lower Cretaceous, lat. 76°20′N, long. 115°34′W, northwest Melville Island, District of Franklin.

Laricoidites magnus (Potonié) Potonié, Thomson and Thiergart
Hypotype 42510
Brideaux, W.W. and McIntyre, D.J., 1975, Geol. Surv. Can., Bull. 252, pl. 4, fig. 7.
Horton River Formation, Lower Cretaceous, lat. 69°27′30″N, long. 126°54′W, Horton River, District of Mackenzie.

Laricoidites magnus (Potonié) Potonié, Thomson and Thiergart, 1950
Hypotype 71317
Wilson, M.A., 1978, Palaeontographica, Abt. B, vol. 166, p. 125, pl. 5, fig. 9.
Cretaceous, Police Island, north bank Mackenzie River, 12 miles southeast of Fort Norman, District of Mackenzie.

Laricoidites cf. *L. magnus* (Potonié)
Hypotype 41620
Hopkins, Jr., W.S. and Sweet, A.R., 1976, Geol. Surv. Can., Bull. 256, pl. 10, fig. 7.
Mattagami Formation, Lower Cretaceous, borehole 68-42, lat. 50°35′36″N, long. 82°30′12″W, Ontario.

cf. *Leiotriletes lineatus* Bolkhovitina, 1956
Hypotype 13830
McGregor, D.C., 1965, Geol. Surv. Can., Paper 64-55, pl. 6, fig. 8.
Lower Cretaceous, Fault Creek Canyon, Aklavik Range, Richardson Mountains, District of Mackenzie.

Leptolepidites major Couper
Hypotype 33985
Hopkins, Jr., W.S., 1974, Geol. Surv. Can., Paper 73-12, p. 19, pl. 4, fig. 51.
Christopher Formation, Lower Cretaceous, lat. 76°20′N, long. 115°34′W, northwest Melville Island, District of Franklin.

Leptolepidites psarorus Norris
Hypotype 56790
Williams, G.L. and Bujak, J.P., 1980, Initial Repts. Deep Sea Drilling Project, vol. 50, pl. 11, fig. 9.

Early Cretaceous, DSDP Site 416, lat. 32°50.18'N, long. 10°48.06'W, northwestern Moroccan continental margin.

Leptolepidites verrucatus Couper
Hypotype 42396
Brideaux, W.W. and McIntyre, D.J., 1975, Geol. Surv. Can., Bull. 252, pl. 1, figs. 8, 9.
Crossley Lakes Member, Langton Bay Formation, Lower Cretaceous, lat. 69°04'N, long. 125°51'W, Horton River, District of Mackenzie.

Leptolepidites verrucatus Couper
Hypotype 56788
Williams, G.L. and Bujak, J.P., 1980, Initial Repts. Deep Sea Drilling Project, vol. 50, pl. 11, fig. 7.
Early Cretaceous, DSDP Site 416, lat. 32°50.18'N, long. 10°48.06'W, northwestern Moroccan continental margin.

Leptolepidites cf. *L. verrucatus* Couper
Hypotype 25077
Hopkins, Jr., W.S., 1971, Geol. Surv. Can., Bull. 197, p. 118, pl. 21, fig. 5.
Isachsen Formation, Lower Cretaceous, lat. 76°20'N, long. 115°34'W, northwest Melville Island, District of Franklin.

Leptopecopites pocockii (Srivastava) Srivastava, 1978
Hypotype 76048
Jerzykiewicz, T. and Sweet, A.R., 1986, Can. J. Earth Sci., vol. 23, no. 9, pl. 3, fig. 15.
Late Cretaceous, depth 29.35-29.62 m, GSC core hole CV-42-2-2, Coal Valley mine strip, l.s.d. 1, sec. 10, tp. 47, rge. 19, W.5th mer., Alberta.

Liburnisporis adnaca Srivastava, 1972
Hypotype 76009
Jerzykiewicz, T. and Sweet, A.R., 1986, Can. J. Earth Sci., vol. 23, no. 9, pl. 1, fig. 12.
Late Cretaceous, depth 80.9 m, Manalta Coal Limited core hole ME-111-81CD, l.s.d. 3, sec. 17, tp. 48, rge. 21, W.5th mer., Alberta.

Liliacidites altimurus Leffingwell, 1971
Hypotype 76049
Jerzykiewicz, T. and Sweet, A.R., 1986, Can. J. Earth Sci., vol. 23, no. 9, pl. 3, figs. 16, 17.
Late Cretaceous, depth 25.82-25.95 m, GSC core hole CV-42-2-1, l.s.d. 1, sec. 6, tp. 47, rge. 19, W.5th mer., Alberta.

Liliacidites complexus (Stanley)
Hypotypes 38348, 38349
McIntyre, D.J., 1974, Geol. Surv. Can., Paper 74-14, pl. 24, figs, 24, 25.
Upper Cretaceous, lat. 69°28'N, long. 126°58'W, Horton River, District of Mackenzie.

Liliacidites complexus (Stanley) Leffinwfell, 1971
Hypotype 76053
Jerzykiewicz, T. and Sweet, A.R., 1986, Can. J. Earth Sci., vol. 23, no. 9, pl. 3, fig. 21.
Late Cretaceous, right bank Embarras River towards south end of Robb, l.s.d. 14, sec. 11, tp. 49, rge. 21, W.5th mer., Alberta.

Liliacidites crassatus Singh, 1971
Hypotype 91407
Sweet, A.R. and McIntyre, D.J., 1988, Can. Soc. Petrol. Geol., Mem. 15, fig. 7.26.
Ram Member, Cardium Formation, Upper Cretaceous, north side of Highway 47 north of Cadomin, l.s.d. 14, sec. 16, tp. 47, rge. 23, W.5th mer., Alberta.

Liliacidites dividus Singh, 1971
Hypotype 91406
Sweet, A.R. and McIntyre, D.J., 1988, Can. Soc. Petrol. Geol., Mem. 15, fig. 7.25.
Moosehound Member, Cardium Formation, Upper Cretaceous, south side of McLeod River, l.s.d. 16, sec. 16, tp. 47, rge. 23, W.5th mer., Alberta.

Liliacidites doylei Ward 1986
Hypotype 89706
Jerzykiewicz, T. and Sweet, A.R., 1988, Sedimentary Geol., vol. 59, no. 1/2, p. 46, pl. I, fig. 10.
Belly River Formation, Upper Cretaceous, Crownest River valley west of Lundbreck, lat. 49°35'N, long. 114°11'W, Alberta.

Liliacidites dolyei Ward, 1986
Hypotype 91403
Sweet, A.R. and McIntyre, D.J., 1988, Can. Soc. Petrol. Geol., Mem. 15, fig. 7.22.
Ram Member, Cardium Formation, Upper Cretaceous, north side of Highway 47 north of Cadomin, l.s.d. 14, sec. 16, tp. 47, rge. 23, W.5th mer., Alberta.

Liliacidites mirus Srivastava, 1969
Hypotype 89720
Jerzykiewicz, T. and Sweet, A.R., 1988, Sedimentary Geol., vol. 59, no. 1/2, p. 59, pl. II, fig. 5.
St. Mary Formation, Upper Cretaceous, Crownest River, lat. 49°30'N, long. 114°3'W, Alberta.

Liliacidites sp.
Fig. specs. 38255, 38302
McIntyre, D.J., 1974, Geol. Surv. Can., Paper 74-14, pl. 24, figs, 26, 27.
Upper Cretaceous, lat. 69°28'N, long. 126°58'W, Horton River, District of Mackenzie.

Liliacidites sp.
Fig. spec. 48990
Sweet, A.R., 1978, Geol. Surv. Can., Paper 78-1B, pl. 6.2, fig. 4.
Cuesta Creek Member, Tent Island Formation, Upper Cretaceous, northwestern bank Big Fish River near its junction with Boundary Creek, lat. 68°30'30"N, long. 136°23'50"W, Yukon.

Liliacidites sp. (spinose)
Fig. spec. 76050
Jerzykiewicz, T. and Sweet, A.R., 1986, Can. J. Earth Sci., vol. 23, no. 9, pl. 3, fig. 18.
Late Cretaceous, depth 29.35-29.62 m, GSC core hole CV-42-2-2, Coal Valley strip mine, l.s.d. 1, sec. 10, tp. 47, rge. 19, W.5th mer., Alberta.

Liliacidites sp.
Fig. spec. 93686
Sweet, A.R. and Braman, D.R., 1989, Geol. Surv. Can., Paper 89-8, p. 35, pl. 1, fig. 21.
Chungo Member, Wapiabi Formation, Upper Cretaceous, Trap Creek near junction with Highwood River, sec. 36, tp. 17, rge. 4, W.5th mer., Alberta.

Liriodendrow cf. *L. psilopites* Wodehouse, 1933
Hypotypes 71332, 71333
Wilson, M.A., 1978, Palaeontographica, Abt. B, vol. 166, p. 130, pl. 6, figs. 13, 14.
Cretaceous, IEO Atkinson H-25 well, north shore Tuktoyaktuk Peninsula, lat. 69°44′20″N, long. 131°50′06″W, and Police Island, north bank Mackenzie River, 12 miles southeast of Fort Norman, District of Mackenzie.

Lophozonotriletes cf. *L. rarituberculatus* (Luber) Kedo, 1957
Hypotype 71300
Wilson, M.A., 1978, Palaeontographica, Abt. B, vol. 166, p. 160, pl. 4, fig. 4.
Cretaceous, Police Island, north bank Mackenzie River, 12 miles southeast of Fort Norman, District of Mackenzie.

Loranthacites pilatus Mchedlishvili
Hypotypes 38240, 38241
McIntyre, D.J., 1974, Geol. Surv. Can., Paper 74-14, pl. 17, figs, 6, 7.
Upper Cretaceous, lat. 69°28′N, long. 126°58′W, Horton River, District of Mackenzie.

Loranthacites sp.
Fig. specs. 93690, 93691
Sweet, A.R. and Braman, D.R., 1989, Geol. Surv. Can., Paper 89-8, p. 35, pl. 1, figs. 25, 26.
Chungo Member, Wapiabi Formation, Upper Cretaceous, Highwood River No. 1 section, SW. 1/4, sec. 7, tp. 18, rge. 3, and Trap Creek near junction with Highwood River, sec. 36, tp. 17, rge. 4, W.5th mer., Alberta.

Loranchacites sp. 1
Fig. specs. 38238, 38239
McIntyre, D.J., 1974, Geol. Surv. Can., Paper 74-14, pl. 17, figs, 4, 5.
Upper Cretaceous, lat. 69°28′N, long. 126°58′W, Horton River, District of Mackenzie.

Loranthacites sp. 2
Fig. specs. 38228, 38229
McIntyre, D.J., 1974, Geol. Surv. Can., Paper 74-14, pl. 16, figs, 1, 2.
Upper Cretaceous, lat. 69°28′N, long. 126°58′W, Horton River, District of Mackenzie.

Loranthacites sp. 3
Fig. specs. 38230, 38231
McIntyre, D.J., 1974, Geol. Surv. Can., Paper 74-14, pl. 16, figs, 3, 4.
Upper Cretaceous, lat. 69°28′N, long. 126°58′W, Horton River, District of Mackenzie.

? *Lugodium multituberculatum* Bolkhovitina, 1961
Hypotype 13846
McGregor, D.C., 1965, Geol. Surv. Can., Paper 64-55, pl. 6, fig. 29.
Lower Cretaceous, Fault Creek Canyon, Aklavik Range, Richardson Mountains, District of Mackenzie.

Lunatadinium dissolutum Brideaux and McIntyre
Holotype 33931; hypotypes 33929, 33930, 33932-33940
Brideaux, W.W. and McIntyre, D.J., 1973, Bull. Can. Petrol. Geol., vol. 21, no. 3, p. 396, pl. 1, figs. 1-13.
Lower Cretaceous, Western Minerals Standard Oil of British Columbia Shaeffer Creek 0-22 well, 1948 feet, lat. 66°41′54″N, long. 137°19′40″W, Eagle Plain, and lat. 67°23′N, long. 137°47′W, Porcupine River, Yukon; lat. 68°12′N, long. 135°25′-40′W, Martin Creek, Richardson Mountains, Banff Aquitaine Arco Rat Pass K-35 well, 70-1930 feet, lat. 67°54′43″N, long. 135°21′57″W, and lat. 68°17′N, long. 135°46′W, Canyon Creek, Richrdson Mountains, District of Mackenzie.

Lycopidites sp.
Fig. spec. 31752
Hopkins, Jr., W.S. and Balkwill, H.R., 1973, Geol. Surv. Can., Paper 72-37, p. 12, pl. 1, fig. 13.
Hassel Formation, Cretaceous, Hoodoo Dome, lat. 78°11′N, long. 99°40′W, southern Ellef Ringnes Island, District of Franklin.

Lycopodiacidites caniculatus Singh
Hypotype 33955
Hopkins, Jr., W.S., 1974, Geol. Surv. Can., Paper 73-12, p. 11, pl. 2, fig. 17.
Christopher Formation, Lower Cretaceous, lat. 76°20′N, long. 115°34′W, northwest Melville Island, District of Franklin.

Lycopodiacidites caperatus Singh
Hypotype 33954
Hopkins, Jr., W.S., 1974, Geol. Surv. Can., Paper 73-12, p. 11, pl. 2, fig. 16.
Christopher Formation, Lower Cretaceous, lat. 76°20′N, long. 115°34′W, northwest Melville Island, District of Franklin.

Lycopodiacidites cf. *L. caperatus* Singh
Hypotype 31751
Hopkins, Jr., W.S. and Balkwill, H.R., 1973, Geol. Surv. Can., Paper 72-37, p. 11, pl. 1, fig. 12.
Hassel Formation, Cretaceous, Hoodoo Dome, lat. 78°11′N, long. 99°40′W, southern Ellef Ringnes Island, District of Franklin.

Lycopodiacidites intraverrucatus Brenner
Hypotypes 42406, 42407
Brideaux, W.W. and McIntyre, D.J., 1975, Geol. Surv. Can., Bull. 252, pl. 1, figs. 19, 20.
Crossley Lakes Member, Langton Bay Formation, Lower Cretaceous, lat. 69°04′N, long. 125°51′W, and lat. 64°01′30″N, long. 126°02′W, Horton River, District of Mackenzie.

Lycopodiumsporites austroclavatidites (Cookson) Pontonié, 1956
Hypotype 13870
McGregor, D.C., 1965, Geol. Surv. Can., Paper 64-55, pl. 7, fig. 11.
Lower Cretaceous, Donna River Canyon, Aklavik Range, Richardson Mountains, District of Mackenzie.

Lycopodiumsporites austroclavatidites (Cookson) Potonié. 1956
Hypotypes 13949, 13950
McGregor, D.C., 1965, Geol. Surv. Can., Paper 64-55, pl. 10, figs. 13-15.
Lower Cretaceous, Peel Plateau, Stony Creek about 18.5 miles upstream from mouth, District of Mackenzie.

Lycopodiumsporites austroclavatidites (Cookson) Potonié
Hypotype 23790
Hopkins, Jr., W.S., 1971, Geol. Surv. Can., Bull. 197, p. 113, pl. 20, fig. 4.
Isachsen Formation, Lower Cretaceous, lat. 76°20′N, long. 115°34′W, northwest Melville Island, District of Franklin.

Lycopodiumsporites autroclavatidites (Cookson) Potonié
Hypotype 31748
Hopkins, Jr., W.S. and Balkwill, H.R., 1973, Geol. Surv. Can., Paper 72-37, p. 11, pl. 1, fig. 9.
Hassel Formation, Cretaceous, Hoodoo Dome, lat. 78°11′N, long. 99°40′W, southern Ellef Ringnes Island, District of Franklin.

Lycopodiumsporites austroclavatidites (Cookson) Potonié
Hypotype 33952
Hopkins, Jr., W.S., 1974, Geol. Surv. Can., Paper 73-12, p. 10, pl. 2, fig. 14.
Christopher Formation, Lower Cretaceous, lat. 76°20′N, long. 115°34′W, northwest Melville Island, District of Franklin.

Lycopodiumsporites austroclavatidites (Cookson)
Hypotype 38181
McIntyre, D.J., 1974, Geol. Surv. Can., Paper 74-14, pl. 14, fig. 12.
Upper Cretaceous, lat. 69°28′N, long. 126°58′W, Horton River, District of Mackenzie.

Lycopodiumsporites austroclavatidites (Cookson) Potonié
Hypotype 42408
Brideaux, W.W. and McIntyre, D.J., 1975, Geol. Surv. Can., Bull. 252, pl. 1, fig. 21.
Horton River Formation, Lower Cretaceous, lat. 69°19′N, long. 126°50′W, Horton River, District of Mackenzie.

Lycopodiumsporites autroclavatidites (Cookson) Potonié, 1956
Hypotype 71285
Wilson, M.A., 1978, Palaeontographica, Abt. B, vol. 166, p. 114, pl. 2, figs. 10, 13.
Cretaceous, Police Island, north bank Mackenzie River, 12 miles southeast of Fort Norman, District of Mackenzie.

Lycopodiumsporites cf. *L. cerniidites* (Ross) Del. and Sprum.
Hypotype 23791
Hopkins, Jr., W.S., 1971, Geol. Surv. Can., Bull. 197, p. 114, pl. 20, fig. 5.
Isachsen Formation, Lower Cretaceous, lat. 76°20′N, long. 115°34′W, northwest Melville Island, District of Franklin.

Lycopodiumsporites facetus Dettmann
Hypotype 33953
Hopkins, Jr., W.S., 1974, Geol. Surv. Can., Paper 73-12, p. 10, pl. 2, fig. 15.
Christopher Formation, Lower Cretaceous, lat. 76°20′N, long. 115°34′W, northwest Melville Island, District of Franklin.

Lycopodiumsporites marginatus Singh
Hypotype 42409
Brideaux, W.W. and McIntyre, D.J., 1975, Geol. Surv. Can., Bull. 252, pl. 1, fig. 22.
Crossley Lakes Member, Langton Bay Formation, Lower Cretaceous, lat. 69°04′N, long. 125°51′W, Horton River, District of Mackenzie.

Lycopodiumsporites marginatus Singh, 1964
Hypotypes 58564, 58565
McIntyre, D.J. and Brideaux, W.W., 1980, Geol. Surv. Can., Bull. 320, p.10, pl. 1, figs. 8, 9.
Lower Cretaceous, Martin Creek, lat. 68°12′10″N, long. 135°34′50″W, Richardson Mountains, District of Mackenzie.

Lycopodiumsporites sp.
Fig. spec. 13912
McGregor, D.C., 1965, Geol. Surv. Can., Paper 64-55, pl. 8, fig. 8.
Isachsen Formation, Lower Cretaceous, south of Buchanan Lake anticline, eastern Axel Heiberg Island, District of Franklin.

Lycopodiumsporites sp.
Fig. spec. 31749
Hopkins, Jr., W.S. and Balkwill, H.R., 1973, Geol. Surv. Can., Paper 72-37, p. 11, pl. 1, fig. 10.
Hassel Formation, Cretaceous, Hoodoo Dome, lat. 78°11′N, long. 99°40′W, southern Ellef Ringnes Island, District of Franklin.

Lycopodiumsporites sp.
Fig. spec. 41587
Hopkins, Jr., W.S. and Sweet, A.R., 1976, Geol. Surv. Can., Bull. 256, pl. 9, fig. 3.
Mattagami Formation, Lower Cretaceous, borehole Onakawana lignite pit, lat. 50°36′N, long. 81°23′W, Ontario.

Lycospora sp.
Fig. spec. 33951
Hopkins, Jr., W.S., 1974, Geol. Surv. Can., Paper 73-12, p. 10, pl. 1, fig. 13.
Christopher Formation, Lower Cretaceous, lat. 76°20′N, long. 115°34′W, northwest Melville Island, District of Franklin.

Lygodioisporites sp. (sensu Singh, 1964)
Fig. spec. 13866
McGregor, D.C., 1965, Geol. Surv. Can., Paper 64-55, pl. 7, fig. 8.
Lower Cretaceous, Donna River Canyon, Aklavik Range, Richardson Mountains, District of Mackenzie.

cf. *Lygodium crispaeformis* (Bolkhovitina) Bolkhovitina, 1961
Hypotype 13832
McGregor, D.C., 1965, Geol. Surv. Can., Paper 64-55, pl. 6, fig. 11.
Lower Cretaceous, Fault Creek Canyon, Aklavik Range, Richardson Mountains, District of Mackenzie.

Lygodium cf. *L. reticulosporites* Rouse
Hypotype 41600
Hopkins, Jr., W.S. and Sweet, A.R., 1976, Geol. Surv. Can., Bull. 256, pl. 9, fig. 16.
Mattagami Formation, Lower Cretaceous, borehole 68-39, lat. 50°35′48″N, long. 82°30′42″W, Ontario.

Lygodiumsporites sp.
Fig. spec. 33974
Hopkins, Jr., W.S., 1974, Geol. Surv. Can., Paper 73-12, p. 16, pl. 3, fig. 38.
Christopher Formation, Lower Cretaceous, lat. 76°20′N, long. 115°34′W, northwest Melville Island, District of Franklin.

Maculatisporites granulatus (Ivanova) Döring, 1964
Hypotypes 68208, 68209
Ricketts, B.D. and Sweet, A.R., 1986, Geol. Surv. Can., Paper 84-15, p. 20, pl. 2, figs. 1, 2.
Elk Formation, Cretaceous, Coal Creek, lat. 49°30′N, long. 114°59′W, British Columbia.

Maculatisporites microverrucatus Döring, 1964
Hypotype 68210
Ricketts, B.D. and Sweet, A.R., 1986, Geol. Surv. Can., Paper 84-15, p. 20, pl. 2, fig. 3.
Elk Formation, Cretaceous, Coal Creek, lat. 49°30′N, long. 114°59′W, British Columbia.

Maculatisporites parkinii (Pocock) Dörhöfer, 1979
Hypotype 68211
Ricketts, B.D. and Sweet, A.R., 1986, Geol. Surv. Can., Paper 84-15, p. 21, pl. 2, fig. 4.
Elk Formation, Cretaceous, Coal Creek, lat. 49°30′N, long. 114°59′W, British Columbia.

Maexisporites sp. A
Fig. spec. 48934
Sweet, A.R., 1979, Am. Assoc. Strat. Palynol., Contrib. Ser. No. 5B, pl. 3, fig. 9.
Hassel Formation, Lower Cretaceous, lat. 77°9′N, long. 104°50′W, Lougheed Island, District of Franklin.

Mancicorpus albertensis Srivastava, 1968
Hypotype 89719 (missing)
Jerzykiewicz, T. and Sweet, A.R., 1988, Sedimentary Geol., vol. 59, no. 1/2, pl. II, fig. 4.
Brazeau Formation, Upper Cretaceous, Blackstone River, lat. 52°43′20″N, long. 116°16′30″W, Alberta.

Mancicorpus calvus (Tschudy and Leopold) Tschudy, 1973
Hypotype 93643
Sweet, A.R., Ricketts, B.D., Cameron, A.R. and Norris, D.K., 1989, Geol. Surv. Can., Paper 89-1G, pl. 1, fig. 13.
Little Bear Formation, Upper Cretaceous, east Little Bear River, lat. 64°38′24″N, long. 126°19′W, District of Mackenzie.

Mancicorpus gibbus Srivastava 1968
Hypotype 89737
Jerzykiewicz, T. and Sweet, A.R., 1988, Sedimentary Geol., vol. 59, no. 1/2, pl. III, fig. 10.
Brazeau Formation, Upper Cretaceous, Blackstone River, lat. 52°42′55″N, long. 116°18′30″W, Alberta.

Mancicorpus gibbus Srivastava, 1968
Hypotype 93652
Sweet, A.R., Ricketts, B.D., Cameron, A.R. and Norris, D.K., 1989, Geol. Surv. Can., Paper 89-1G, p. 97, pl. 1, fig. 23.
Summit Creek Formation, Upper Cretaceous, north bank Mackenzie River immediately downstream from Police Island, lat. 64°52′42″N, long. 125°14′44″W, District of Mackenzie.

Mancicorpus cf. *M. gibbus* Srivastava 1968
Hypotype 48974
Sweet, A.R., 1978, Geol. Surv. Can., Paper 78-1B, p. 32, pl. 6.1, fig. 14.
Cuesta Creek Member, Tent Island Formation, Upper Cretaceous, northwestern bank Big Fish River near its junction with Boundary Creek, lat. 68°30′30″N, long. 136°23′50″W, Yukon.

Mancicorpus sp. cf. *M. gibbus* Srivastava, 1968
Fig. spec. 89865
Sweet, A.R., Ricketts, B.D., Cameron, A.R. and Norris, D.K., 1989, Geol. Surv. Can., Paper 89-1G, pl. 2, fig. 1.
Summit Creek Formation, Upper Cretaceous, north bank Mackenzie River immediately downstream from Police Island, lat. 64°52′42″N, long. 125°14′44″W, District of Mackenzie.

Mancicorpus glaber (Chlonova) Krutzsch 1969
Hypotype 89721
Jerzykiewicz, T. and Sweet, A.R., 1988, Sedimentary Geol., vol. 59, no. 1/2, p. 58, pl. II, fig. 6.
Brazeau Formation, Upper Cretaceous, Blackstone River, lat. 52°43′20″N, long. 116°16′30″W, Alberta.

Mancicorpus rostratus Srivastava, 1968
Hypotype 93754
Sweet, A.R., Ricketts, B.D., Cameron, A.R. and Norris, D.K., 1989, Geol. Surv. Can., Paper 89-1G, p. 97, pl. 1, fig. 26.
Summit Creek Formation, Upper Cretaceous, tributary of East Little Bear River, lat. 64°33′20″N, long. 125°49′45″W, District of Mackenzie.

Mancicorpus senonicum Mchedlishvili
Hypotypes 38262, 38263
McIntyre, D.J., 1974, Geol. Surv. Can., Paper 74-14, pl. 20, figs. 3, 4.
Upper Cretaceous, lat. 69°28′N, long. 126°58′W, Horton River, District of Mackenzie.

Mancicorpus trapeziforme Mchedlishvili
Hypotypes 38264, 38265
McIntyre, D.J., 1974, Geol. Surv. Can., Paper 74-14, pl. 20, figs, 5, 6.
Upper Cretaceous, lat. 69°28′N, long. 126°58′W, Horton River, District of Mackenzie.

Mancicorpus tripodiformis (Tschudy and Leopold) Tschudy 1973
Hypotype 89725
Jerzykiewicz, T. and Sweet, A.R., 1988, Sedimentary Geol., vol. 59, no. 1/2, p. 58, pl. II, fig. 10.
Brazeau Formation, Upper Cretaceous, Blackstone River, lat. 52°43′20″N, long. 116°16′30″W, Alberta.

Mancicorpus sp.
Fig. spec. 38276
McIntyre, D.J., 1974, Geol. Surv. Can., Paper 74-14, pl. 21, fig. 8.
Upper Cretaceous, lat. 69°58′30″N, long. 127°04′W, Horton River, District of Mackenzie.

Margocolporites sp.
Fig. specs. 91416, 91417
Sweet, A.R. and McIntyre, D.J., 1988, Can. Soc. Petrol. Geol., Mem. 15, figs. 7.35, 7.36.
Ram Member, Cardium Formation, Upper Cretaceous, north side of Highway 47 north of Cadomin, l.s.d. 14, sec. 16, tp. 47, rge. 23, W.5th mer., Alberta.

Marsypiletis cretaceus Robertson emend. Robertson and Elsik, 1978
Hypotype 76054
Jerzykiewicz, T. and Sweet, A.R., 1986, Can. J. Earth Sci., vol. 23, no. 9, pl. 3, fig. 22.
Late Cretaceous, depth 27.41-27.43 cm, GSC core hole, CV-42-2-2, Coal Valley strip mine, l.s.d. 1, sec. 10, tp. 47, rge. 19, W.5th mer., Alberta.

Matonisporites phlebopteroides Couper
Hypotype 33965
Hopkins, Jr., W.S., 1974, Geol. Surv. Can., Paper 73-12, p. 14, pl. 3, fig. 27.
Christopher Formation, Lower Cretaceous, lat. 76°20′N, long. 115°34′W, northwest Melville Island, District of Franklin.

cf. *Matonisporites* sp.
Fig. specs. 13859, 13860
McGregor, D.C., 1965, Geol. Surv. Can., Paper 64-55, pl. 7, figs. 1, 2.
Lower Cretaceous, Donna River Canyon, Aklavik Range, Richardson Mountains, District of Mackenzie.

Membranosphaera maastrichtia Samoilovitch ex Norris and Sarjeant, 1965
Hypotypes 38070, 38071
McIntyre, D.J., 1974, Geol. Surv. Can., Paper 74-14, pl. 6, figs, 7, 8.
Upper Cretaceous, lat. 69°27′30″N, long. 126°58′W, and lat. 69°28′N, long. 126°58′W, Horton River, District of Mackenzie.

cf. *Metasequoia* sp.
Fig. spec. 31780
Hopkins, Jr., W.S. and Balkwill, H.R., 1973, Geol. Surv. Can., Paper 72-37, p. 19, pl. 2, fig. 42.
Hassel Formation, Cretaceous, Hoodoo Dome, lat. 78°11′N, long. 99°40′W, southern Ellef Ringnes Island, District of Franklin.

Microfoveolatosporis cf. *M. fromensis* (Cookson) Harris, 1965
Hypotype 71306
Wilson, M.A., 1978, Palaeontographica, Abt. B, vol. 166, p. 119, pl. 4, fig. 11.
Cretaceous, Police Island, north bank Mackenzie River, 12 miles southeast of Fort Norman, District of Mackenzie.

Microfoveolatosporis pseudoreticulatus (Hedlund) Singh, 1983
Hypotypes 91361, 91362
Sweet, A.R. and McIntyre, D.J., 1988, Can. Soc. Petrol. Geol., Mem. 15, figs. 6.9, 6.10.
Moosehound Member, Cardium Formation, Upper Cretaceous, south side of McLeod River, l.s.d. 16, sec. 16, tp. 47, rge. 23, W.5th mer., Alberta.

Microreticulatisporites lunatus Knox, 1950
Hypotypes 71291, 71292
Wilson, M.A., 1978, Palaeontographica, Abt. B, vol. 166, p. 157, pl. 3, figs. 6, 7.
Cretaceous, Police Island, north bank Mackenzie River, 12 miles southeast of Fort Norman, District of Mackenzie.

Microreticulatisporites uniformis Singh
Hypotypes 42410, 42411
Brideaux, W.W. and McIntyre, D.J., 1975, Geol. Surv. Can., Bull. 252, pl. 1, figs. 23, 24.

Crossley Lakes Member, Langton Bay Formation, Lower Cretaceous, lat. 69°04′N, long. 125°51′W, Horton River, District of Mackenzie.

Microreticulatisporites uniformis Singh, 1964
Hypotype 91349
Sweet, A.R. and McIntyre, D.J., 1988, Can. Soc. Petrol. Geol., Mem. 15, fig. 5.21.
Moosehound Member, Cardium Formation, Upper Cretaceous, west bank Little Berland River, l.s.d. 1, sec. 6, tp. 54, rge. 2, W.5th mer., Alberta.

Minerisporites cf. *M. alius* Batten
Hypotypes 41632-41635
Hopkins, Jr., W.S. and Sweet, A.R., 1976, Geol. Surv. Can., Bull. 256, pl. 11, figs. 8-11.
Mattagami Formation, Lower Cretaceous, borehole 68-42, lat. 50°35′36″N, long. 82°30′12″W, Ontario.

Minerisporites cf. *M. alius* Batten 1969
Hypotype 48916
Sweet, A.R., 1979, Am. Assoc. Strat. Palynol., Contrib. Ser. No. 5B, pl. 2, fig. 12.
Christopher Formation, Lower Cretaceous, lat. 74°31′N, long. 121°10′W, Cape McClure, Banks Island, District of Franklin.

Minerisporites borealis (Miner) Potonié 1956
Hypotype 48935
Sweet, A.R., 1979, Am. Assoc. Strat. Palynol., Contrib. Ser. No. 5B, pl. 3, fig. 10.
Christopher Formation, Lower Cretaceous, lat. 79°47′24″N, long. 85°22′W, near Slidre Fiord, Ellesmere Island, District of Franklin.

Minerisporites irregularis Gunther and Hills 1973
Hypotype 48953
Sweet, A.R., 1979, Am. Assoc. Strat. Palynol., Contrib. Ser. No. 5B, pl. 4, fig. 11.
Eureka Sound Formation, Upper Cretaceous, east of Hoodoo Dome, lat. 78°1′N, long. 99°43′W, Ellef Ringnes Island, District of Franklin.

Minerisporites cf. *M. macroreticulatus* Singh 1964
Hypotypes 48909, 48910
Sweet, A.R., 1979, Am. Assoc. Strat. Palynol., Contrib. Ser. No. 5B, pl. 2, figs. 5, 6.
Lower Cretaceous, Martin Creek, lat. 68°12′20″N, long. 135°35′W, District of Mackenzie.

Minerisporites marginatus (Dijkstra) Potonié 1956
Hypotypes 48914, 48917
Sweet, A.R., 1979, Am. Assoc. Strat. Palynol., Contrib. Ser. No. 5B, pl. 2, figs. 10, 13.
Christopher(?) Formation, Lower Cretaceous, lat. 73°22′N, long. 118°54′W, and Cape McClure, lat. 74°31′N, long. 121°10′W, Banks Island, District of Franklin.

Minerisporites cf. *M. marginatus* (Dijkstra) Potonié 1956
Hypotype 48923
Sweet, A.R., 1979, Am. Assoc. Strat. Palynol., Contrib. Ser. No. 5B, pl. 2, fig. 19.
Christopher(?) Formation, Lower Cretaceous, lat. 73°28′N, long. 118°08′W, northeastern Banks Island, District of Franklin.

Minerisporites richardsonii (Murray) Pontonié emend. Harris 1961
Hypotype 48895
Sweet, A.R., 1979, Am. Assoc. Strat. Palynol., Contrib. Ser. No. 5B, pl. 1, fig. 8.
Jaeger Formation, Lower-Middle Jurassic, lat. 77°41′50″N, long. 94°14′25″W, Cornwall Island, District of Franklin.

Minerisporites venustus Singh
Hypotype 48928
Sweet, A.R., 1979, Am. Assoc. Strat. Palynol., Contrib. Ser. No. 5B, pl. 3, fig. 3.
Hassel Formation, Lower Cretaceous, lat. 77°9′N, long. 104°50′W, Lougheed Island, District of Franklin.

Minerisporites sp. A
Fig. spec. 48920
Sweet, A.R., 1979, Am. Assoc. Strat. Palynol., Contrib. Ser. No. 5B, pl. 2, fig. 16.
Christopher Formation, Lower Cretaceous, lat. 74°31′N, long. 121°10′W, Cape McClure, Banks Island, District of Franklin.

Minerisporites sp. B
Fig. spec. 48954
Sweet, A.R., 1979, Am. Assoc. Strat. Palynol., Contrib. Ser. No. 5B, pl. 4, fig. 12.
Eureka Sound Formation, Upper Cretaceous, east of Hoodoo Dome, lat. 78°11′N, long. 99°43′W, Ellef Ringnes Island, District of Franklin.

Molaspora lobata (Dijkstra) Hall 1963
Hypotype 48950
Sweet, A.R., 1979, Am. Assoc. Strat. Palynol., Contrib. Ser. No. 5B, pl. 4, fig. 8.
Eureka Sound Formation, Upper Cretaceous, Meteorologist Peninsula, lat. 78°10′N, long. 99°30′W, Ellef Ringnes Island, District of Franklin.

Momipites cf. *M. inaequalis* Anderson, 1960
Hypotype 71390
Wilson, M.A., 1978, Palaeontographica, Abt. B, vol. 166, p. 144, pl. 10, fig. 12.
Cretaceous, IEO Atkinson H-25 well, north shore Tuktoyaktuk Peninsula, lat. 69°44′20″N, long. 131°50′06″W, District of Mackenzie.

Momipites sp.
Fig. specs. 38298, 38299
McIntyre, D.J., 1974, Geol. Surv. Can., Paper 74-14, pl. 22, figs, 20, 21.
Upper Cretaceous, lat. 69°58′30″N, long. 127°04′W, and lat. 69°28′N, long. 126°58′W, Horton River, District of Mackenzie.

?*Monosulcites couperi* Sukh Dev, 1961
Hypotype 13879
McGregor, D.C., 1965, Geol. Surv. Can., Paper 64-55, pl. 7, fig. 26.
Lower Cretaceous, Donna River Canyon, Aklavik Range, Richardson Mountains, District of Mackenzie.

Monosulcites sp.
Fig. spec. 38335
McIntyre, D.J., 1974, Geol. Surv. Can., Paper 74-14, pl. 24, fig. 28.
Upper Cretaceous, lat. 69°28′N, long. 126°58′W, Horton River, District of Mackenzie.

Monosulcites spp.
Fig. specs. 25094, 25095, 25520
Hopkins, Jr., W.S., 1971, Geol. Surv. Can., Bull. 197, p. 121, pl. 22, figs. 2-4.
Isachsen Formation, Lower Cretaceous, lat. 76°20′N, long. 115°34′W, northwest Melville Island, District of Franklin.

Montanapollis endannulatus Tschudy 1971
Hypotype 89727
Jerzykiewicz, T. and Sweet, A.R., 1988, Sedimentary Geol., vol. 59, no. 1/2, p. 58, pl. II, fig. 12.
St. Mary River Formation, Upper Cretaceous, Crowsnest River, lat. 49°30′N, long. 114°3′W, Alberta.

Montanapollis sp. (disporate)
Fig. spec. 76059
Jerzykiewicz, T. and Sweet, A.R., 1986, Can. J. Earth Sci., vol. 23, no. 9, pl. 3, fig. 27.
Late Cretaceous, right bank of Embarras River towards south end of Robb, l.s.d. 14, sec. 11, tp. 49, rge. 21, W.5th mer., Alberta.

Multinodisporites sp.
Fig. spec. 76014
Jerzykiewicz, T. and Sweet, A.R., 1986, Can. J. Earth Sci., vol. 23, no. 9, pl. 1, figs. 17, 18.
Late Cretaceous, Embarras River, Alberta.

Murospora mesozoica Pocock, 1961
Hypotype 13946
McGregor, D.C., 1965, Geol. Surv. Can., Paper 64-55, pl. 10, fig. 9.
Lower Cretaceous, Peel Plateau, Stony Creek about 18.5 miles upstream from mouth, District of Mackenzie.

Myricipites sp.
Fig. specs. 38294, 38295
McIntyre, D.J., 1974, Geol. Surv. Can., Paper 74-14, pl. 22, figs, 16, 17.
Upper Cretaceous, lat. 69°28′N, long. 126°58′W, and lat. 69°58′30″N, long. 126°58′W, Horton River, District of Mackenzie.

Myrtaceidites solidus Sweet, 1986
Hypotype 76055
Jerzykiewicz, T. and Sweet, A.R., 1986, Can. J. Earth Sci., vol. 23, no. 9, pl. 3, fig. 23.
Late Cretaceous, depth 29.35-29.62 m, GSC core hole CV-42-2-2, oal Valley strip mine, l.s.d. 1, sec. 10, rge. 19, W.5th mer, Alberta.

Myrtaceidites sp.
Fig. spec. 71372
Wilson, M.A., 1978, Palaeontographica, Abt. B, vol. 166, p. 137, pl. 9, fig. 11.
Cretaceous, Police Island, north bank Mackenzie River, 12 miles southeast of Fort Norman, District of Mackenzie.

Myrtipites scabratus Norton *in* Norton and Hall, 1969
Hypotypes 76104, 76105
Sweet, A.R., 1986, Can.J. Earth Sci., vol. 23, no. 9, pl. 1, figs. 14, 15.
Late Cretaceous, depth 29.35-29.62 m, GSC core hole CV-42-2-2, Coal Valley strip mine, l.s.d. 1, sec. 10, tp. 47, rge. 19, W.5th mer., Alberta.

Myrtipites scabratus Norton *in* Norton and Hall, 1969
Hypotype 76056
Jerzykiewicz, T. and Sweet, A.R., 1986, Can. J. Earth Sci., vol. 23, no. 9, pl. 3, fig. 24.
Late Cretaceous, depth 25.82-25.95 m, GSC core hole CV-42-2-1, Coal Valley strip mine, l.s.d. 1, sec. 6, tp. 47, rge. 19, W.5th mer., Alberta.

Myrtipites scabratus Norton *in* Norton and Hall, 1969
Hypotype 89856
Sweet, A.R., Ricketts, B.D., Cameron, A.R. and Norris, D.K., 1989, Geol. Surv. Can., Paper 89-1G, p. 98, pl. 1, fig. 30.
Summit Creek Formation, Upper Cretaceous, north bank Mackenzie River immediately downstream from Police Island, lat. 64°52′42″N, long. 125°14′44″W, District of Mackenzie.

Nelsoniella aceras Cookson and Eisenack, 1960
Hypotype 38052
McIntyre, D.J., 1974, Geol. Surv. Can., Paper 74-14, pl. 5, fig. 3.
Upper Cretaceous, lat. 69°28′N, long. 126°58′W, Horton River, District of Franklin.

Nelsoniella sp. AE
Fig. spec. 45803
Brideaux, W.W., 1976, Geol. Surv. Can., Paper 76-1B, p. 251, pl. 44.1, fig. 7.
Boundary Creek Formation, Upper Cretaceous, 6630-6660 feet, Gulf Mobil Parsons N-10 well, ca. 1.5 miles east-northeast of Parsons Lake, lat. 69°59′48.52″N, long. 133°31′50.33″W, District of Mackenzie.

Neoraistrickia sp. cf. *N. truncata* (Cookson) Potonié, 1956
Hypotypes 38182, 38183
McIntyre, D.J., 1974, Geol. Surv. Can., Paper 74-14, pl. 14, figs. 13, 14.
Upper Cretaceous, lat. 69°28′N, long. 126°58′W, Horton River, District of Franklin.

Neoraistrickia cf. *N. robusta* Brenner
Hypotype 33949
Hopkins, Jr., W.S., 1974, Geol. Surv. Can., Paper 73-12, p. 10, pl. 1, fig. 10.
Christoper Formation, Lower Cretaceous, lat. 76°20′N, long. 115°34′W, northwestern Melville Island, District of Franklin.

Neoraistrickia sp.
Fig. spec. 13948
McGregor, D.C., 1965, Geol. Surv. Can., Paper 64-55, pl. 10, figs. 11, 12.
Lower Cretaceous, Peel Plateau, Stony Creek about 18.5 miles upstream from mouth, District of Mackenzie.

Neoraistrickia sp.
Fig. specs. 71279, 71281
Wilson, M.A., 1978, Palaeontographica, Abt. B, vol. 166, p. 112, pl. 2, figs. 3, 6.
Cretaceous, Police Island, north bank Mackenzie River, 12 miles southeast of Fort Norman, District of Mackenzie.

cf. *Neoraistrickia* sp.
Fig. spec. 33950
Hopkins, Jr., W.S., 1974, Geol. Surv. Can., Paper 73-12, p. 10, pl. 1, figs. 11, 12.
Christoper Formation, Lower Cretaceous, lat. 76°20′N, long. 115°34′W, northwestern Melville Island, District of Franklin.

Nyssapollenites albertensis Singh, 1971
Hypotype 91414
Sweet, A.R. and McIntyre, D.J., 1988, Can. Soc. Petrol. Geol., Mem. 15, fig. 7.33.
Ram Member, Cardium Formation, Upper Cretaceous, north side of Highway 47 north of Cadomin, l.s.d. 14, sec. 16, tp. 47, rge. 23, W.5th mer., Alberta.

Nyssapollenites sp.
Fig. spec. 91391
Sweet, A.R. and McIntyre, D.J., 1988, Can. Soc. Petrol. Geol., Mem. 15, fig. 7.10.
Moosehound Member, Cardium Formation, Upper Cretaceous, south side of Grande Cache Highway 40, l.s.d. 15, sec. 13, tp. 57, rge. 7, W.5th mer., Alberta.

Nyssapollenites sp.
Fig. spec. 93695
Sweet, A.R. and Braman, D.R., 1989, Geol. Surv. Can., Paper 89-8, pl. 1, fig. 30.
Chungo Member, Wapiabi Formation, Upper Cretaceous, Trap Creek near junction with Highwood River, sec. 36, tp. 17, rge. 4, W.5th mer., Alberta.

Nyssoidites anulatus (Chlonova) Sweet, 1986
Hypotype 76057
Jerzykiewicz, T. and Sweet, A.R., 1986, Can. J. Earth Sci., vol. 23, no. 9, pl. 3, fig. 25.
Late Cretaceous, Embarras River, Alberta.

Oblate?, tricolpate angiosperm pollen
Fig. specs. 93698-93700
Sweet, A.R. and Braman, D.R., 1989, Geol. Surv. Can., Paper 89-8, pl. 1, figs. 33-35.
Chungo Member, Wapiabi Formation, Upper Cretaceous, Trap Creek near junction with Highwood River, sec. 36, tp. 17, rge. 4, W.5th mer., Alberta.

cf. *Oblatinella rotundiformis* Malyavkina, 1949
Hypotype 13936
McGregor, D.C., 1965, Geol. Surv. Can., Paper 64-55, pl. 9, fig. 8.
Isachsen Formation, Lower Cretaceous, south of Buchanan Lake anticline, eastern Axel Heiberg Island, District of Franklin.

Onychiites incertus Kara-Murza, 1960
Hypotype 13833
McGregor, D.C., 1965, Geol. Surv. Can., Paper 64-55, pl. 6, fig. 13.
Lower Cretaceous, Fault Creek Canyon, Aklavik Range, Richardson Mountains, District of Mackenzie.

cf. *Onychium amplectiformis* Kara-Murza(?) in Bolkhovitina, 1956
Hypotype 13954
McGregor, D.C., 1965, Geol. Surv. Can., Paper 64-55, pl. 10, fig. 21.
Lower Cretaceous, Peel Plateau, Stony Creek about 18.5 miles upstream from mouth, District of Mackenzie.

Operculites sp.
Fig. spec. 49005
Sweet, A.R. 1978, Geol. Surv. Can., Paper 78-1B, pl. 6.2, fig. 19.
Cuesta Creek Member, Trent Island Formation, Upper Cretaceous, northwestern bank Big Fish River near its junction with Boundary Creek, lat. 68°30′30″N, long. 136°23′50″W, Yukon.

Ophiobolus lapidaris O. Wetzel, 1933
Hypotype 38162
McIntyre, D.J., 1974, Geol. Surv. Can., Paper 74-14, pl. 13, fig. 7.
Upper Cretaceous, lat. 69°27′30″N, long. 126°58′W, Horton River, District of Franklin.

Orbiculapollis globosus (Chlonova) Chlonova, 1961
Hypotypes 38243, 38244, 38254
McIntyre, D.J., 1974, Geol. Surv. Can., Paper 74-14, pl. 17, figs. 9-11.
Upper Cretaceous, lat. 69°28′N, long. 126°58′W, Horton River, District of Franklin.

Orbiculapollis globosus Chlonova 1961
Hypotype 89736
Jerzykiewicz, T. and Sweet, A.R., 1988, Sedimentary Geol., vol. 59, no. 1/2, pl. III, fig. 9.
Brazeau Formation, Upper Cretaceous, Blackstone River, lat. 52°43′20″N, long. 116°16′30″W, Alberta.

Orbiculapollis lucida Chlonova, 1961
Hypotype 76052
Jerzykiewicz, T. and Sweet, A.R., 1986, Can. J. Earth Sci., vol. 23, no. 9, pl. 3, fig. 20.
Late Cretaceous, depth 27.41-27.43 m, GSC core hole CV-42-2-2, Coal Valley strip mine, l.s.d. 1, sec. 10, tp. 47, rge. 19, W.5th mer., Alberta.

Orbiculapollis sp.
Fig. spec. 48982
Sweet, A.R., 1978, Geol. Surv. Can., Paper 78-1B, p. 34, pl. 6.1, fig. 22.
Cuesta Creek Member, Trent Island Formation, Upper Cretaceous, northwestern bank Big Fish River near its junction with Boundary Creek, lat. 68°30′30″N, long. 136°23′50″W, Yukon.

Orbiculapollis sp. 2
Fig. specs. 38245, 38246
McIntyre, D.J., 1974, Geol. Surv. Can., Paper 74-14, pl. 18, figs. 1, 2.
Upper Cretaceous, lat. 69°28′N, long. 126°58′W, Horton River, District of Franklin.

Ornamentifera baculata Singh
Hypotype 33961
Hopkins, Jr., W.S., 1974, Geol. Surv. Can., Paper 73-12, p. 13, pl. 2, fig. 23.
Christoper Formation, Lower Cretaceous, lat. 76°20′N, long. 115°34′W, northwestern Melville Island, District of Franklin.

Ornamentifera baculata Ross
Hypotype 42457
Brideaux, W.W. and McIntyre, D.J., 1975, Geol. Surv. Can., Bull. 252, pl. 2, fig. 29.
Crossley Lakes Member, Langton Bay Formation, Lower Cretaceous, lat. 69°07′30″N, long. 126°20′W, west bank Horton River opposite mouth of Coal Creek, District of Mackenzie.

Ornamentifera tuberculata (Grigorjeva) Bolkhovitina, 1968
Hypotypes 38175, 38176
McIntyre, D.J., 1974, Geol. Surv. Can., Paper 74-14, pl. 14, figs. 6, 7.
Upper Cretaceous, lat. 69°27′30″N, long. 126°58′W, and lat. 69°28′N, long. 126°58′W, Horton River, District of Franklin.

Osmundacidites alpinus Klaus, 1960
Hypotype 13829
McGregor, D.C., 1965, Geol. Surv. Can., Paper 64-55, pl. 6, fig. 7.
Lower Cretaceous, Fault Creek Canyon, Aklavik Range, Richardson Mountains, District of Mackenzie.

Osmundacidites major Döring, 1965
Hypotype 68216
Ricketts, B.D. and Sweet, A.R., 1986, Geol. Surv. Can., Paper 84-15, p. 21, pl. 2, fig. 11.
Elk Formation, Cretaceous, Coal Creek, lat. 49°30′N, long. 114°59′W, British Columbia.

Osmundacidites wellmanii Couper
Hypotypes 23793, 23794
Hopkins, Jr., W.S., 1971, Geol. Surv. Can., Bull. 197, p. 114, pl. 20, figs. 7, 8.
Isachsen Formation, Lower Cretaceous, lat. 76°20′N, long. 115°34′W, northwest Melville Island, District of Franklin.

Osmundacidites wellmanii Couper
Hypotype 31754
Hopkins, Jr., W.S. and Balkwill, H.R., 1973, Geol. Surv. Can., Paper 72-37, p. 12, pl. 1, figs. 15, 16.
Hassel Formation, Cretaceous, Hoodoo Dome, lat. 78°11′N, long. 99°40′W, northern Ellef Ringnes Island, District of Franklin.

Osmundacidites wellmanii Couper
Hypotype 33962
Hopkins, Jr., W.S., 1974, Geol. Surv. Can., Paper 73-12, p. 13, pl. 2, fig. 24.
Christoper Formation, Lower Cretaceous, lat. 76°20′N, long. 115°34′W, northwestern Melville Island, District of Franklin.

Osmundacidites wellmanii Couper, 1953
Hypotypes 38179, 38180
McIntyre, D.J., 1974, Geol. Surv. Can., Paper 74-14, pl. 14, figs. 10, 11.
Upper Cretaceous, lat. 69°28′N, long. 126°58′W, Horton River, District of Franklin.

Osmundacidites wellmanii Couper
Hypotype 42398
Brideaux, W.W. and McIntyre, D.J., 1975, Geol. Surv. Can., Bull. 252, pl. 1, fig. 11.
Crossley Lakes Member, Langton Bay Formation, Lower Cretaceous, lat. 64°01′30″N, long. 126°02′W, Horton River, District of Mackenzie.

Osmundacidites wellmanii Couper
Hypotype 41590
Hopkins, Jr., W.S. and Sweet, A.R., 1976, Geol. Surv. Can., Bull. 256, pl. 9, fig. 6.
Mattagami Formation, Lower Cretaceous, Onakawana lignite pit, lat. 50°36′N, long. 81°23′W, Ontario.

Osmunda-sporites elongatus Rouse
Hypotype 23795
Hopkins, Jr., W.S., 1971, Geol. Surv. Can., Bull. 197, p. 114, pl. 20, fig. 9.
Isachsen Formation, Lower Cretaceous, lat. 76°20′N, long. 115°34′W, northwest Melville Island, District of Franklin.

Osmunda-sporites sp. in Rouse, 1959
Fig. spec. 13865
McGregor, D.C., 1965, Geol. Surv. Can., Paper 64-55, pl. 7, fig. 7.
Lower Cretaceous, Donna River Canyon, Aklavik Range, Richardson Mountains, District of Mackenzie.

Ovoid cysts
Fig. specs. 62642, 62643
Ioannides, N.S. and McIntyre, D.J., 1980, Geol. Surv. Can., Paper 80-1A, pl. 31.2, figs. 5 6.
Smoking Hills Formation, Upper Cretaceous (reworked), Douglas Creek, lat. 68°33′N, long. 133°50′W, Caribou Hills, District of Mackenzie.

Ovoidites ligneolus (Potonié) Thomson and Pflug 1953
Hypotype 48994
Sweet, A.R. 1978, Geol. Surv. Can., Paper 78-1B, pl. 6.2, fig. 8.
Cuesta Creek Member, Trent Island Formation, Upper Cretaceous, northwestern bank Big Fish River near its junction with Boundary Creek, lat. 68°30′30″N, long. 136°23′50″W, Yukon.

Palaeostomocystis chytra Drugg, 1967
Hypotypes 38098-38100
McIntyre, D.J., 1974, Geol. Surv. Can., Paper 74-14, pl. 8, figs. 4-6.
Upper Cretaceous, lat. 69°28′N, long. 126°58′W, Horton River, District of Mackenzie.

Palaeostomocystis fragilis Cookson and Eisenack, 1962
Hypotype 38114
McIntyre, D.J., 1974, Geol. Surv. Can., Paper 74-14, pl. 9, fig. 1.
Upper Cretaceous, lat. 69°58′30″N, long. 127°04′W, Horton River, District of Mackenzie.

Palaeostomocystis laevigata Drugg, 1967
Hypotypes 38093, 38094
McIntyre, D.J., 1974, Geol. Surv. Can., Paper 74-14, pl. 7, figs. 15, 16.
Upper Cretaceous, lat. 69°27′30″N, long. 126°58′W, and lat. 69°28′N, long. 126°58′W, Horton River, District of Mackenzie.

Palaeostomocystis sp. 2
Fig. spec. 38105
McIntyre, D.J., 1974, Geol. Surv. Can., Paper 74-14, pl. 8, fig. 11.
Upper Cretaceous, lat. 69°27′30″N, long. 126°58′W, Horton River, District of Mackenzie.

Palambages Form A Manum and Cookson, 1964
Fig. spec. 38167
McIntyre, D.J., 1974, Geol. Surv. Can., Paper 74-14, pl. 13, fig. 12.
Upper Cretaceous, lat. 69°27′30″N, long. 126°58′W, Horton River, District of Mackenzie.

Palambages Form A Manum and Cookson, 1964
Fig. spec. 38166
McIntyre, D.J., 1974, Geol. Surv. Can., Paper 74-14, pl. 13, figs. 11.
Upper Cretaceous, lat. 69°28′N, long. 126°58′W, Horton River, District of Mackenzie.

Paliurus triplicatus Anderson, 1960 *(Scabratriporites)*
Hypotypes 93711, 93712
Sweet, A.R. and Braman, D.R., 1989, Geol. Surv. Can., Paper 89-8, p. 35, pl. 1, figs. 46, 47.
Chungo Member, Wapiabi Formation, Upper Cretaceous, Trap Creek near junction with Highwood River, sec. 36, tp. 17, rge. 4, and Highwood River No. 2 section, N.E. 1/4 sec. 1, tp. 18, rge. 4, W.5th mer., Alberta.

Pandaniidites typicus (Norton)
Hypotype 76120
Sweet, A.R., 1986, Can. J. Earth Sci., vol. 23, no. 9, p. 1382, pl. 2, fig. 12.
Late Cretaceous, depth 90-100 feet, Cypress 25R test hole, l.s.d. 13, sec. 36, tp. 6, rge. 20, W.3rd mer., Saskatchewan.

Pandaniidites typicus (Norton) Sweet, 1986
Hypotype 76062
Jerzykiewicz, T. and Sweet, A.R., 1986, Can. J. Earth Sci., vol. 23, no. 9, pl. 4, fig. 2.
Late Cretaceous, depth 29.35-29.62 m, GSC core hole CV-42-2-2, Coal Valley strip mine, l.s.d. 1, sec. 10, tp. 47, rge. 19, W.5th mer., Alberta.

Paraalnipollenites confusus (Zaklinskaia) Hills and Wallace, 1969
Hypotype 38292
McIntyre, D.J., 1974, Geol. Surv. Can., Paper 74-14, pl. 22, fig. 14.
Upper Cretaceous, lat. 69°58′30″N, long. 127°04′W, Horton River, District of Mackenzie.

Paraalnipollenites confusus (Zaklinskaia) Hills and Wallace, 1969
Hypotype 71402
Wilson, M.A., 1978, Palaeontographica, Abt. B, vol. 166, p. 144, pl. 10, figs. 25.
Cretaceous, Police Island, north bank Mackenzie River, 12 miles southeast of Fort Norman, District of Mackenzie.

Pareodinia capillosa Brideaux and Fisher
Holotype 34155; paratypes 42365-42370
Brideaux, W.W. and Fisher, M.J., 1976, Geol. Surv. Can., Bull. 259, p. 20, pl. 2, figs. 3-10; pl. 3, figs. 1, 2, 5.
Gilmore Lake Member, Langton Bay Formation, Lower Cretaceous, 1530-1540 (34155, 42365), 1650-1670 (42369), 1680-1700 (42367) and 1760-1770 (42366) feet, Elf Horton River G-02 well, lat. 69°51′23″N, long. 127°15′57″W, southwest of mouth of Horton River, and Upper Jurassic, 1710 feet, Banff-Aquitaine-Arco Rat Pass K-35 well, 6 miles southeast of summit of Mount Goodenough, just south of Goodenough Creek, District of Mackenzie.

Parvisaccites amplus Brenner
Hypotype 76606
Davies, E.H., Akande, S.O. and Zentilli, M., 1984, Geol. Surv. Can., Paper 84-1A, pl. 48.1, fig. 17.
Early Cretaceous, Gays River mine, lat. 45°02′N, long. 63°22′W, Nova Scotia.

Parvisaccites cf. *P. enigmaticus* Couper, 1958
Hypotype 13958
McGregor, D.C., 1965, Geol. Surv. Can., Paper 64-55, pl. 10, figs. 28, 29.
Lower Cretaceous, Peel Plateau, Stony Creek about 18.5 miles upstream from mouth, District of Mackenzie.

Parvisaccites hortonensis Brideaux and McIntyre
Holotype 34159; paratypes 42514-42516
Brideaux, W.W. and McIntyre, D.J., 1975, Geol. Surv. Can., Bull. 252, p. 17, pl. 4, figs. 11-14.
Gilmore Lake Member, Langton Bay Formation, Lower Cretaceous, lat. 68°50′N, long. 125°24′W, Horton River, District of Mackenzie.

Parvisaccites radiatus Couper
Hypotype 42512
Brideaux, W.W. and McIntyre, D.J., 1975, Geol. Surv. Can., Bull. 252, pl. 4, fig. 9.
Crossley Lakes Member, Langton Bay Formation, Lower Cretaceous, lat. 69°04′N, long. 125°51′W, Horton River, District of Mackenzie.

Parvisaccites cf. *P. radiatus* Couper, 1958
Hypotype 13960
McGregor, D.C., 1965, Geol. Surv. Can., Paper 64-55, pl. 10, figs. 31, 37.
Lower Cretaceous, Peel Plateau, Stony Creek about 18.5 miles upstream from mouth, District of Mackenzie.

Parvisaccites rugulatus Brenner
Hypotype 42513
Brideaux, W.W. and McIntyre, D.J., 1975, Geol. Surv. Can., Bull. 252, pl. 4, fig. 10.
Crossley Lakes Member, Langton Bay Formation, Lower Cretaceous, lat. 69°04′N, long. 125°51′W, Horton River, District of Mackenzie.

Parvisaccites rugulatus Brenner
Hypotype 76607
Davies, E.H., Akande, S.O. and Zentilli, M., 1984, Geol. Surv. Can., Paper 84-1A, pl. 48.1, fig. 18.
Early Cretaceous, sample AGR 285-230, Gays River mine, lat. 45°02′N, long. 63°22′W, Nova Scotia.

cf. *Parvisaccites* sp. in Delcourt and Sprumont, 1959
Fig. spec. 13964
McGregor, D.C., 1965, Geol. Surv. Can., Paper 64-55, pl. 10, fig. 38.
Lower Cretaceous, Peel Plateau, Stony Creek about 18.5 miles upstream from mouth, District of Mackenzie.

Paxillitriletes fairlightensis (Batten) Hall and Nicholson 1973
Hypotypes 48911, 48912
Sweet, A.R., 1979, Am. Assoc. Strat. Palynol., Contrib. Ser. No. 5B, pl. 2, figs. 7, 8.
Lower Cretaceous, Martin Creek, lat. 68°12′20″N, long. 135°35′W, District of Mackenzie.

Paxillitriletes cf. *P. fairlightensis* (Batten)
Hypotypes 41640-41643
Hopkins, Jr., W.S. and Sweet, A.R., 1976, Geol. Surv. Can., Bull. 256, pl. 11, fig. 16-19.
Mattagami Formation, Lower Cretaceous, borehole 68-39, lat. 50°35′48″N, long. 82°30′42″W (41640, 41643) and borehole 68-42, lat. 50°35′36″N, long. 82°30′12″W, Ontario.

Pediastrum sp.
Fig. spec. 49006
Sweet, A.R. 1978, Geol. Surv. Can., Paper 78-1B, pl. 6.2, fig. 20.
Cuesta Creek Member, Trent Island Formation, Upper Cretaceous, northwestern bank Big Fish River near its junction with Boundary Creek, lat. 68°30′30″N, long. 136°23′50″W, Yukon.

Penetetrapites inconspicuus Sweet
Holotype 76106; paratypes 76107, 76108
Sweet, A.R., 1986, Can. J. Earth Sci., vol. 23, no. 9, p. 1382, pl. 1, figs. 16-19.
Late Cretaceous, right bank of Embarras River towards south end of Robb, l.s.d. 14, sec. 11, tp. 49, rge. 21, W.5th mer., Alberta and depth 90-100 feet, Cypress 25R coal test hole, l.s.d. 13, sec. 36, tp. 6, rge. 20, W.3rd mer., Saskatchewan (76107, 76108).

Penetetrapites inconspicuus Sweet, 1986
Hypotype 76060
Jerzykiewicz, T. and Sweet, A.R., 1986, Can. J. Earth Sci., vol. 23, no. 9, pl. 3, fig. 28.
Late Cretaceous, right bank of Embarras River towards the end of Robb, l.s.d. 14, sec. 11, tp. 49, rge. 21, W.5th mer., Alberta.

Pentetrapites mollis Hedlund and Norris, 1968
Hypotype 91409
Sweet, A.R. and McIntyre, D.J., 1988, Can. Soc. Petrol. Geol., Mem. 15, fig. 7.28.
Ram Member, Cardium Formation, Upper Cretaceous, north side of Highway 47 north of Cadomin, l.s.d. 14, sec. 16, tp. 47, rge. 23, W.5th mer., Alberta.

Perinopollenites elatoides Couper
Hypotypes 42523, 42524
Brideaux, W.W. and McIntyre, D.J., 1975, Geol. Surv. Can., Bull. 252, pl. 4, figs. 22, 23.
Horton River Formation and Crossley Lakes Member, Langton Bay Formation, Lower Cretaceous, lat. 69°27′30″N, long. 126°54′W and lat. 69°04′N, long. 125°51′W, Horton River, District of Mackenzie.

Perinopollenites cf. *P. elatoides* Couper
Hypotype 41614
Hopkins, Jr., W.S. and Sweet, A.R., 1976, Geol. Surv. Can., Bull. 256, pl. 10, fig. 1.
Mattagami Formation, Lower Cretaceous, borehole 68-39, lat. 50°35′48″N, long. 82°30′42″W, Ontario.

Peromonolites subengelmanii (Elsik) Jameossanie, 1987
Hypotype 91368
Sweet, A.R. and McIntyre, D.J., 1988, Can. Soc. Petrol. Geol., Mem. 15, fig. 6.16.
Moosehound Member, Cardium Formation, Upper Cretaceous, south side of McLeod River, l.s.d. 16, sec. 16, tp. 47, rge. 23, W.5th mer., Alberta.

Peromonolites sp.
Fig. spec. 93676
Sweet, A.R. and Braman, D.R., 1989, Geol. Surv. Can., Paper 89-8, pl. 1, fig. 11.
Chungo Member, Wapiabi Formation, Upper Cretaceous, Highwood River No. 2 section, N.E. 1/4, sec. 1, tp. 18, rge. 4, W.5th mer., Alberta.

Pesavis sp.
Fig. spec. 49001
Sweet, A.R., 1978, Geol. Surv. Can., Paper 78-1B, p. 36, pl. 6.2, fig. 15.
Cuesta Creek Member, Tent Island Formation, Upper Cretaceous, northwestern bank Big Fish River near its junction with Boundary Creek, lat. 68°30′30″N, long. 136°23′50″W, Yukon.

Pesavis sp.
Fig. spec. 76004
Jerzykiewicz, T. and Sweet, A.R., 1986, Can. J. Earth Sci., vol. 23, no. 9, pl. 1, fig. 7.
Late Cretaceous, depth 80.9 m, Manalta Coal Limited, core hole ME-111-81CD, l.s.d. 3, sec. 17, tp. 48, rge. 21, W.5th mer., Alberta.

Phyllocladidites inchoatus (Pierce) Norris, 1967
Hypotypes 38214, 38213
McIntyre, D.J., 1974, Geol. Surv. Can., Paper 74-14, pl. 15, figs. 7, 8.
Upper Cretaceous, lat. 69°28′N, long. 126°58′W, and lat. 69°27′30″N, long. 126°58′W, Horton River, District of Mackenzie.

Phyllocladidites microreticulatus Brenner, 1963
Hypotypes 38212, 38213
McIntyre, D.J., 1974, Geol. Surv. Can., Paper 74-14, pl. 15, figs. 5, 6.
Upper Cretaceous, lat. 69°27′30″N, long. 126°58′W, Horton River, District of Mackenzie.

Phyllocladidites microreticulatus Brenner
Hypotype 76604
Davies, E.H., Akande, S.O. and Zentilli, M., 1984, Geol. Surv. Can., Paper 84-1A, pl. 48.1, fig. 15.
Early Cretaceous, sample AGR 285-230, Gays River Mine, lat. 45°02′N, long. 63°22′W, Nova Scotia.

cf. *Phyllocladidites* sp.
Fig. spec. 41622
Hopkins, Jr., W.S. and Sweet, A.R., 1976, Geol. Surv. Can., Bull. 256, pl. 10, fig. 9.
Mattagami Formation, Lower Cretaceous, borehole 68-42, lat. 50°35′36″N, long. 82°30′12″W, Ontario.

Piceapollenites sp. 1
Fig. spec. 38210
McIntyre, D.J., 1974, Geol. Surv. Can., Paper 74-14, pl. 15, fig. 3.
Upper Cretaceous, lat. 69°58′30″N, long. 127°04′W, Horton River, District of Mackenzie.

Piceapollenites sp. 2
Fig. spec. 38211
McIntyre, D.J., 1974, Geol. Surv. Can., Paper 74-14, pl. 15, fig. 4.
Upper Cretaceous, lat. 69°28′N, long. 126°58′W, Horton River, District of Mackenzie.

Pilosisporites brecibaculatus Döring, 1965
Hypotype 68214
Ricketts, B.D. and Sweet, A.R., 1986, Geol. Surv. Can., Paper 84-15, p. 21, pl. 2, fig. 8.
Elk Formation, Cretaceous, Coal Creek, lat. 49°30′N, long. 114°59′W, British Columbia.

Pilosisporites delicatulus Norris 1969
Hypotype 72090
Fensome, R.A., Palaeontographica Canadian No. 4, p. 27, pl. 4, figs. 4, 8.
McGuire Formation, Cretaceous, Grizzly Gorge, south bank, lat. 68°15′49″N, long. 135°40′59″W, District of Mackenzie.

Pilosisporites trichopapillosus (Thiergart) Del. and Sprum.
Hypotype 25078
Hopkins, Jr., W.S., 1971, Geol. Surv. Can., Bull. 197, p. 118, pl. 21, fig. 6.
Isachsen Formation, Lower Cretaceous, lat. 76°20′N, long. 115°34′W, northwest Melville Island, District of Franklin.

Pilosisporites trichopapillosus (Thiergart) Delcourt and Sprumont
Hypotype 33987
Hopkins, Jr., W.S., 1974, Geol. Surv. Can., Paper 73-12, p. 19, pl. 5, fig. 53.
Christopher Formation, Lower Cretaceous, lat. 76°20′N, long. 115°34′W, northwest Melville Island, District of Franklin.

Pilosisporites trichopapillosus (Thiergart) Delcourt and Sprumont
Hypotype 42401
Brideaux, W.W. and McIntyre, D.J., 1975, Geol. Surv. Can., Bull. 252, pl. 1, fig. 14.
Horton River Formation, Lower Cretaceous, lat. 69°19′N, long. 126°50′W, Horton River, District of Mackenzie.

Pilosisporites trichopapillosus (Thiergart)
Hypotype 41901
Williams, G.L., 1975, Geol. Surv. Can., Paper 74-30, vol. 2, pl. 8, fig. 6.
Aptian, Lower Cretaceous, Mobil Dauntless D-35 well, Scotian Shelf.

Pilosisporites trichopapillosus (Thiergart) Delcourt and Sprumont, 1955
Hypotypes 68217, 68218
Ricketts, B.D. and Sweet, A.R., 1986, Geol. Surv. Can., Paper 84-15, p. 21, pl. 2, figs. 12, 13.

Cretaceous, Elk Formation, Coal Creek, lat. 40°30′N, long. 114°59′W, British Columbia, and Pocaterra Creek Member, Cadomin Formation, Highwood Pass, lat. 50°35′N, long. 115°1′W, Alberta

Pilosisporites trichopapillosus (Thiergart 1949) Delcourt and Sprumont, 1955, emend. Döring, 1965
Hypotype 72094
Fensome, R.A., 1987, Palaeontographica Canadiana No. 4, p. 28, pl. 4, fig. 13.
Martin Creek Formation, Cretaceous, Martin Creek, north bank, lat. 68°12′10″N, long. 135°34′50″W, District of Mackenzie.

Pilosisporites verus Delcourt and Sprumont
Hypotype 33986
Hopkins, Jr., W.S., 1974, Geol. Surv. Can., Paper 73-12, p. 19, pl. 4, fig. 52.
Christopher Formation, Lower Cretaceous, lat. 76°20′N, long. 115°34′W, northwest Melville Island, District of Franklin.

Pilosisporites verus Delcourt and Sprumont
Hypotype 42402
Brideaux, W.W. and McIntyre, D.J., 1975, Geol. Surv. Can., Bull. 252, pl. 1, fig. 15.
Horton River Formation, Lower Cretaceous, lat. 69°19′N, long. 126°50′W, Horton River, District of Mackenzie.

Pilosisporites verus
Hypotype 91983
Banerjee, I. and Davies, E.H., 1988, Can. Soc. Petrol. Geol., Mem. 15, fig. 6.11.
Mannville Group, Lower Cretaceous, depth 4441 feet, well 3-22-46-23W4, Alberta.

Pilosisporites sp. A
Fig. specs. 56782-56784
Williams, G.L. and Bujak, J.P., 1980, Initial Repts. Deep Sea Drilling Project, vol. 50, pl. 11, figs. 1-3.
Early Cretaceous, DSDP Site 416, lat. 32°50.18′N, long. 10°48.06′W, northwestern Moroccan continental margin.

Pinuspollenites cf. *P. semicircularis* (Stanley)
Hypotype 71328
Wilson, M.A., 1978, Palaeontographica, Abt. B, vol. 166, p. 122, pl. 6, fig. 7.
Cretaceous, Police Island, north bank Mackenzie River, 12 miles southeast of Fort Norman, District of Mackenzie.

Pistillipollenites macgregorii Rouse, 1962
Hypotypes 71384, 71385
Wilson, M.A., 1978, Palaeontographica, Abt. B, vol. 166, p. 145, pl. 10, figs. 6, 7.
Cretaceous, IEO Atkinson H-35 well, north shore Tuktoyaktuk Peninsula, lat. 69°44′20″N, long. 131°50′06″W, District of Mackenzie.

Pistillipollenites mcgregori Rouse
Hypotypes 62613-62617
Ioannides, N.S. and McIntyre, D.J., 1980, Geol. Surv. Can., Paper 80-1A, pl. 31.5, figs. 1-5.
Upper Cretaceous (62615, 62617), Upper Paleocene-(?) Lower Eocene (62613, 62616), Eocene, Caribou Hills, Mackenzie River, District of Mackenzie.

?*Pityosporites constrictus* Singh, 1964
Hypotype 13965
McGregor, D.C., 1965, Geol. Surv. Can., Paper 64-55, pl. 10, fig. 39.
Lower Cretaceous, Peel Plateau, Stony Creek about 18.5 miles upstream from mouth, District of Mackenzie.

cf. *Pityosporites* sp.
Fig. spec. 13886
McGregor, D.C., 1965, Geol. Surv. Can., Paper 64-55, pl. 7, fig. 36.
Lower Cretaceous, Donna River Canyon, Aklavik Range, Richardson Mountains, District of Mackenzie.

?*Platysaccus papilionis* Potonié and Klaus, 1954
Hypotype 13853
McGregor, D.C., 1965, Geol. Surv. Can., Paper 64-55, pl. 6, fig. 39.
Lower Cretaceous, Fault Creek Canyon, Aklavik Range, Richardson Mountains, District of Mackenzie.

Pleuricellaesporites sp.
Fig. spec. 31740
Hopkins, Jr., W.S. and Balkwill, H.R., 1973, Geol. Surv. Can., Paper 72-37, p. 9, pl. 1, fig. 1.
Hassel Formation, Cretaceous, Hoodoo Dome, lat. 78°11′N, long. 99°40′W, southern Ellef Ringnes Island, District of Franklin.

Pleurospermaepollenites sp.
Fig. spec. 93717
Sweet, A.R. and Braman, D.R., 1989, Geol. Surv. Can., Paper 89-8, pl. 1, fig. 52.
Chungo Member, Wapiabi Formation, Upper Cretaceous, Trap Creek near junction with Highwood River sec. 36, tp. 17, rge. 4, W.5th mer., Alberta.

Pleurospermaepollenites sp. (*Tricolporopollenites* sp. 1 *in* Norris et al., 1975)
Fig. specs. 93715, 93716
Sweet, A.R. and Braman, D.R., 1989, Geol. Surv. Can., Paper 89-8, p. 35, pl. 1, figs. 50, 51.
Chungo Member, Wapiabi Formation, Upper Cretaceous, Trap Creek near junction with Highwood River, sec. 36, tp. 17, rge. 4, W.5th mer., Alberta.

Plicatella pseudomacrorhyza
Hypotypes 91976, 91977
Banerjee, I. and Davies, E.H., 1988, Can. Soc. Petrol. Geol., Mem. 15, figs. 6.4, 6.5.
Mannville Group, Lower Cretaceous, depths 5276 and 4414 feet, well 11-28-40-24W4 and 3-22-46-23W4, Alberta.

Plicatella undosa (Hedlund) Davies, 1985
Hypotype 93667
Sweet, A.R. and Braman, D.R., 1989, Geol. Surv. Can., Paper 89-8, p. 35, pl. 1, fig. 2.
Chungo Member, Wapiabi Formation, Upper Cretaceous, Highwood River No. 1, section, S.W. 1/4 sec. 7, tp. 18, rge. 3, W.5th mer., Alberta.

Plicatella #ES
Fig. specs. 91974, 91975
Banerjee, I. and Davies, E.H., 1988, Can. Soc. Petrol. Geol., Mem. 15, figs. 6.1-6.3.
Mannville Group, Lower Cretaceous, depths 5081 and 5287 feet, well 11-08-41-24W4, Alberta.

Pluricellaesporites psilatus Clarke, 1965
Hypotypes 71381, 71398, 71399
Wilson, M.A., 1978, Palaeontographica, Abt. B, vol. 166, p. 149, pl. 10, figs. 3, 21, 22.
Cretaceous, IOE Atkinson H-25 well, north shore Tuktoyaktuk Peninsula, lat. 69°44′20″N, long. 131°50′06″W (71381); Aklak Member, Reindeer Formation, Paleocene, Aklak Cree, Big Fish River area, District of Mackenzie.

Pluricellaesporites sp.
Fig. spec. 49003
Sweet, A.R., 1978, Geol. Surv. Can., Paper 78-1B, pl. 6.2, fig. 17.
Tent Island Formation, Upper Cretaceous, northwestern bank Big Fish River near its junction with Boundary Creek, lat. 68°30′30″N, long. 136°23′50″W, Yukon.

Podacarpidites biformis Rouse, 1957
Hypotype 13849
McGregor, D.C., 1965, Geol. Surv. Can., Paper 64-55, pl. 6, fig. 32.
Lower Cretaceous, Fault Creek Canyon, Aklavik Range, Richardson Mountains, District of Mackenzie.

Podocarpidites biformis Rouse, 1957
Hypotypes 38208, 38209
McIntyre, D.J., 1974, Geol. Surv. Can., Paper 74-14, pl. 15, figs. 1, 2.
Upper Cretaceous, lat. 69°28′N, long. 126°58′W, Horton River, District of Mackenzie.

Podocarpidites biformis Rouse
Hypotypes 42508, 42509
Brideaux, W.W. and McIntyre, D.J., 1975, Geol. Surv. Can., Bull. 252, pl. 4, figs. 5, 6.
Horton River Formation, Lower Cretaceous, lat. 69°19′N, long. 126°50′W Horton River, District of Mackenzie.

Podocarpidites canadensis Pocock, 1962
Hypotype 13890
McGregor, D.C., 1965, Geol. Surv. Can., Paper 64-55, pl. 7, fig. 40.
Lower Cretaceous, Donna River Canyon, Aklavik Range, Richardson Mountains, District of Mackenzie.

Podocarpidites canadensis Pocock, 1962
Hypotype 13934
McGregor, D.C., 1965, Geol. Surv. Can., Paper 64-55, pl. 9, fig. 6.
Isachsen Formation, Lower Cretaceous, south of Buchanan Lake anticline, eastern Axel Heiberg Island, District of Franklin.

Podocarpidites canadensis Pocock
Hypotype 42505
Brideaux, W.W. and McIntyre, D.J., 1975, Geol. Surv. Can., Bull. 252, pl. 4, fig. 2.
Gilmore Lake Member, Langton Bay Formation, Lower Cretaceous, lat. 69°50′N, long. 125°24′W, Horton River, District of Mackenzie.

Podacarpidites cf. *P. ellipticus* Cookson, 1947
Hypotype 13850
McGregor, D.C., 1965, Geol. Surv. Can., Paper 64-55, pl. 6, fig. 35.
Lower Cretaceous, Fault Creek Canyon, Aklavik Range, Richardson Mountains, District of Mackenzie.

Podocarpidites maximus (Stanley) Norton, 1969
Hypotype 71334
Wilson, M.A., 1978, Palaeontographica, Abt. B, vol. 166, p. 123, pl. 6, figs. 15, 16.
Cretaceous, Police Island, north bank Mackenzie River, 12 miles southeast of Fort Norman, District of Mackenzie.

Podocarpidites multesimus (Bolkhovitina) Pocock, 1962
Hypotypes 38204, 38205
McIntyre, D.J., 1974, Geol. Surv. Can., Paper 74-14, pl. 14, figs. 35, 36.
Upper Cretaceous, lat. 69°27′30″N, long. 126°58′W and lat. 69°28′N, long. 126°58′W, Horton River, District of Mackenzie.

Podocarpidites multesimus (Bolkhovitina) Pocock
Hypotypes 42506, 42507
Brideaux, W.W. and McIntyre, D.J., 1975, Geol. Surv. Can., Bull. 252, pl. 4, figs. 3, 4.
Horton River Formation and Crossley Lakes Member, Langton Bay Formation, Lower Cretaceous, lat. 69°27′30″N, long. 126°54′W and lat. 69°07′30″N, long. 125°20′W, Horton River, District of Mackenzie.

Podocarpidites otagoensis Couper, 1953
Hypotype 71336
Wilson, M.A., 1978, Palaeontographica, Abt. B, vol. 166, p. 124, pl. 7, fig. 2.
Cretaceous, Police Island, north bank Mackenzie River, 12 miles southeast of Fort Norman, District of Mackenzie.

Podocarpidites sp.
Fig. specs. 41616, 41617
Hopkins, Jr., W.S. and Sweet, A.R., 1976, Geol. Surv. Can., Bull. 256, pl. 10, figs. 3, 4.
Mattagami Formation, Lower Cretaceous, Onakawana lignite pit stock pile, lat. 50°36′N, long. 81°23′W, Ontario.

?*Podocarpidites* sp.
Fig. spec. 13935
McGregor, D.C., 1965, Geol. Surv. Can., Paper 64-55, pl. 9, fig. 7.
Isachsen Formation, Lower Cretaceous, south of Buchanan Lake anticline, eastern Axel Heiberg Island, District of Franklin.

cf. *Podocarpidites* sp.
Fig. spec. 31782
Hopkins, Jr., W.S. and Balkwill, H.R., 1973, Geol. Surv. Can., Paper 72-37, p. 20, pl. 2, fig. 44.
Hassel Formation, Cretaceous, Hoodoo Dome, lat. 78°11′N, long. 99°40′W, southern Ellef Ringnes Island, District of Franklin.

cf. *Podocarpidites* sp.
Fig. spec. 34004
Hopkins, Jr., W.S., 1974, Geol. Surv. Can., Paper 73-12, p. 23, pl. 6, fig. 71.
Christopher Formation, Lower Cretaceous, lat. 76°20′N, long. 115°34′W, northwest Melville Island, District of Franklin.

Podocarpidites sp. A, B
Fig. specs. 25103, 25104
Hopkins, Jr., W.S., 1971, Geol. Surv. Can., Bull. 197, p. 123, pl. 22, figs. 12, 13.
Isachsen Formation, Lower Cretaceous, lat. 76°20′N, long. 115°34′W, northwest Melville Island, District of Franklin.

Podocarpites sp.
Fig. spec. 91374
Sweet, A.R. and McIntyre, D.J., 1988, Can. Soc. Petrol. Geol., Mem. 15, fig. 6.23.
Leyland Member, Cardium Formation, Upper Cretaceous, south side of McLeod River, l.s.d. 16, sec. 16, tp. 47, rge. 23, W.5th mer. Alberta.

cf. *Podocarpus tricocca* (Malyavkina) Bolkhovitina, 1953
Hypotype 13932
McGregor, D.C., 1965, Geol. Surv. Can., Paper 64-55, pl. 9, fig. 3.
Isachsen Formation, Lower Cretaceous, south of Buchanan Lake anticline, eastern Axel Heiberg Island, District of Franklin.

cf. *Podocarpus tricocca* (Maljavkina) Bolkhovitina
Hypotype 42504
Brideaux, W.W. and McIntyre, D.J., 1975, Geol. Surv. Can., Bull. 252, pl. 4, fig. 1.
Crossley Lakes Member, Langton Bay Formation, Lower Cretaceous, lat. 69°04′N, long. 125°51′W, Horton River, District of Mackenzie.

Podocarpus cf. *P. wallichianus* Presl.
Hypotype 27838
Hopkins, Jr., W.S., Ruther, N.W. and Rouse, G.E, 1972, Can, J. Earth Sci, vol. 9, No. 4, pl. 1, fig. 7.
Oligocene, east side Parsnip River 2 1/2 miles north of John Hart Highway bridge, lat. 55°09′N, long. 122°59′W, British Columbia.

Pollen sp. A
Fig. spec. 71406
Wilson, M.A., 1978, Palaeontographica, Abt. B, vol. 166, p. 140, pl. 11, fig. 5.
Cretaceous, IOE Atkinson H-25 well, north shore Tuktoyaktuk Peninsula, lat. 69°44′20″N, long. 131°50′06″W, District of Mackenzie.

cf. *Pollenites verrucosus* Rogalska, 1954
Hypotype 13952
McGregor, D.C., 1965, Geol. Surv. Can., Paper 64-55, pl. 10, figs. 17, 18.
Lower Cretaceous, Peel Plateau, Stony Creek about 18.5 miles upstream from mouth, District of Mackenzie.

Polotricolporites rotundus var. *rotundus* Sweet
Holotype 76115; paratypes 76109-76112, 76114
Sweet, A.R., 1986, Can. J. Earth Sci., vol. 23, p. 1383, pl. 1, figs. 20, 21; pl. 2, figs. 1-3, 5, 6.
Late Cretaceous, Highway 37 north side of Frenchman River Valley, l.s.d. 14, sec. 31, tp. 4, rge. 18, W.3rd mer., and depth 495 feet, Quantock test hole, l.s.d. 13, sec. 35, tp. 2, rge. 1, W.3rd mer. (76110), Saskatchewan.

Polycingulatisporites reduncus (Bolkhovitina) Playford and Dettmann, 1965
Hypotypes 38188, 38189
McIntyre, D.J., 1974, Geol. Surv. Can., Paper 74-14, pl. 14, figs. 19, 20.
Upper Cretaceous, lat. 69°28′N, long. 126°58′W, Horton River, District of Mackenzie.

Polycingulatisporites reduncus (Bolkhovitina) Playford and Dettmann, 1965
Hypotypes 71301, 71302
Wilson, M.A., 1978, Palaeontographica, Abt. B, vol. 166, p. 117, pl. 4, figs. 5, 6.
Cretaceous, Police Island, north bank Mackenzie River, 12 miles southeast of Fort Norman, and Moose Channel Formation, Aklak Creek, Big Fish River area, District of Mackenzie.

Polypodiidites sp.
Fig. specs. 38197, 38198
McIntyre, D.J., 1974, Geol. Surv. Can., Paper 74-14, pl. 14, figs. 28, 29.
Upper Cretaceous, lat. 69°28′N, long. 126°58′W, Horton River, District of Mackenzie.

Porosipollis porosus (Mchedlishvili) Krutzsch, 1969
Hypotype 89842
Sweet, A.R., Ricketts, B.D., Cameron, A.R. and Norris, D.K., 1989, Geol. Surv. Can., Paper 89-1G, p. 97, pl. 1, fig. 29.
Summit Creek Formation, Upper Cretaceous, north bank Mackenzie River immediately dowstream from Police Island, lat. 64°52′42″N, long. 125°14′44″W, District of Mackenzie.

Proteacidites auratus Srivastava 1969
Hypotype 48981
Sweet, A.R., 1978, Geol. Surv. Can., Paper 78-1B, p. 34, pl. 6.1, fig. 21.

Cuesta Creek Member, Tent Island Formation, Upper Cretaceous, northwestern bank Big Fish River near its junction with Boundary Creek, lat. 68°30′30″N, long. 136°23′50″W, Yukon.

Proteacidites sp. cf. *P. globosiporus* Samoilovich, 1961
Fig. spec. 93655
Sweet, A.R., Ricketts, B.D., Cameron, A.R. and Norris, D.K., 1989, Geol. Surv. Can., Paper 89-1G, p. 98, pl. 2, fig. 3.
Summit Creek Formation, Upper Cretaceous, tributary of East Little Bear River, lat. 64°33′20″N, long. 125°49′45″W, District of Mackenzie.

Proteacidites thalmanni Anderson
Hypotype 38303
McIntyre, D.J., 1974, Geol. Surv. Can., Paper 74-14, pl. 22, fig. 24.
Upper Cretaceous, 69°58′30″N, long. 127°04′W, Horton River, District of Mackenzie.

Proteacidites thalmanni Anderson, 1960
Hypotypes 93706-93710
Sweet, A.R. and Braman, D.R., 1989, Geol. Surv. Can., Paper 89-8, p. 35, pl. 1, figs. 41-45.
Chungo Member, Wapiabi Formation, Upper Cretaceous, Trap Creek near junction with Highwood River, sec. 36, tp. 17, rge. 4, W.5th mer., Alberta.

Proteacidites sp.
Fig. specs. 93704, 93705
Sweet, A.R. and Braman, D.R., 1989, Geol. Surv. Can., Paper 89-8, p. 35, pl. 1, figs. 39, 40.
Chungo Member, Wapiabi Formation, Upper Cretaceous, Trap Creek near junction with Highwood River, sec. 36, tp. 17, rge. 4, W.5th mer., Alberta.

Proteacidites sp. 1
Fig. specs. 38301, 38305
McIntyre, D.J., 1974, Geol. Surv. Can., Paper 74-14, pl. 22, fig. 23; pl. 23, fig. 2.
Upper Cretaceous, lat. 69°28′N, long. 126°58′W, Horton River, District of Mackenzie.

cf. *Proteacidites* sp.
Fig. spec. 38304
McIntyre, D.J., 1974, Geol. Surv. Can., Paper 74-14, pl. 23, fig. 1.
Upper Cretaceous, lat. 69°28′N, long. 126°58′W, Horton River, District of Mackenzie.

cf. *Proteacidites* sp.
Fig. spec. 91413
Sweet, A.R. and McIntyre, D.J., 1988, Can. Soc. Petrol. Geol., Mem. 15, fig. 7.32.
Leyland Member, Cardium Formation, Upper Cretaceous, south side of McLeod River, l.s.d. 16, sec. 16, tp. 47, rge. 23m, W.5th mer., Alberta.

cf. *Protoconiferus funarius* (Naumova?) Bolkhovitina, 1956
Hypotype 13923
McGregor, D.C., 1965, Geol. Surv. Can., Paper 64-55, pl. 8, fig. 23.
Isachsen Formation, Lower Cretaceous, south of Buchanan Lake anticline, eastern Axel Heiberg Island, District of Franklin.

?*Protopicea* sp.
Fig. spec. 41624
Hopkins, Jr., W.S. and Sweet, A.R., 1976, Geol. Surv. Can., Bull. 256, pl. 10, fig. 11.
Mattagami Formation, Lower Cretaceous, borehole 68-38, lat. 50°36′N, long. 82°13′12″W, Ontario.

?*Protopodocarpus monstrificaliformis* Bolkhovitina in Verbitskaya, 1963
Hypotype 13931
McGregor, D.C., 1965, Geol. Surv. Can., Paper 64-55, pl. 9, fig. 1.
Isachsen Formation, Lower Cretaceous, south of Buchanan Lake anticline, eastern Axel Heiberg Island, District of Franklin.

"Psaligonyaulax apatela" auct. non (Cookson and Eisenack) Sarjeant, 1966
Hypotype 45815
Brideaux, W.W., 1976, Geol. Surv. Can., Paper 76-1B, p. 256, pl. 44.2, fig. 6.
Lower Cretaceous, 8400-8420 feet, Gulf Mobil Parsons N-10 well, ca. 1.5 miles east-northeast of Parsons Lake, lat. 69°59′48.52″N, long. 133°31′50.33″W, District of Mackenzie.

Psaligonyaulax dualis Brideaux and Fisher
Holotype 34154; paratypes 42359-42364
Brideaux, W.W. and Fisher, M.J., 1976, Geol. Surv. Can., Bull. 259, p. 18, pl. 1, figs. 4-6, 8-12; pl. 2, figs. 1, 2.
Gilmore Lake Member, Langton Bay Formation, Lower Cretaceous, 1800-1810 and 1760-1770 (42359) feet, Elf Horton River G-02 well, lat. 69°51′23″N, long. 127°15′57″W, southwest of mouth of Horton River; Husky Formation (42362), Upper Jurassic, lat. 68°12′N, long. 135°37′W, Martin Creek; and Upper Jurassic, 1900-1910 (42363) and 1765 (42364) feet, Banff-Aquitaine-Arco Rat Pass K-35 well, 6 miles southeast of summit of Mount Goodernough, just north of Goodenough Creek, District of Mackenzie.

Pseudointegricorpus sp.
Fig. specs. 93713, 93714
Sweet, A.R. and Braman, D.R., 1989, Geol. Surv. Can., Paper 89-8, p. 35. pl. 1, figs. 48, 49.
Chungo Member, Wapiabi Formation, Upper Cretaceous, Trap Creek near junction with Highwood River, sec. 36, tp. 17, rge. 4, W.5th mer., Alberta.

?*Pseudopicea monstruosa* Bolkhovitina, 1956
Hypotype 13933
McGregor, D.C., 1965, Geol. Surv. Can., Paper 64-55, pl. 9, fig. 5.
Isachsen Formation, Lower Cretaceous, south of Buchanan Lake anticline, eastern Axel Heiberg Island, District of Franklin.

cf. *Pseudopinus prolongata* Kara-Murza, 1960
Hypotype 13894
McGregor, D.C., 1965, Geol. Surv. Can., Paper 64-55, pl. 7, fig. 44.
Lower Cretaceous, Donna River Canyon, Aklavik Range, Richardson Mountains, District of Mackenzie.

Pseudoplicapollis newmannii Nichols and Jacobson, 1982
Hypotypes 93702, 93703
Sweet, A.R. and Braman, D.R., 1989, Geol. Surv. Can., Paper 89-8, p. 35, pl. 1, figs. 37, 38.
Chungo Member, Wapiabi Formation, Upper Cretaceous, Trap Creek near junction with Highwood River, sec. 36, tp. 17, rge. 4, W.5th mer., Alberta.

Pseudotricolpites reticulatus Stanley, 1965
Hypotype 71340
Wilson, M.A., 1978, Palaeontographica, Abt. B, vol. 166, p. 128, pl. 7, fig. 6.
Cretaceous, Police Island, north bank Mackenzie River, 12 miles southeast of Fort Norman, District of Mackenzie.

Psilatricolporites subtilis (Groot and Penny) Singh, 1963
Hypotype 91415
Sweet, A.R. and McIntyre, D.J., 1988, Can. Soc. Petrol. Geol., Mem. 15, fig. 7.34.
Moosehound Member, Cardium Formation, Upper Cretaceous, south side of Grande Cache Highway 40, l.s.d. 15, sec. 13, tp. 57, rge. 7, W.5th mer., Alberta.

Pterodinium sp. 2, *in* McIntyre 1974
Fig. spec. 62637
Ioannides, N.S. and McIntyre, D.J., 1980, Geol. Surv. Can., Paper 80-1A, pl. 31.3, fig. 8.
Upper Cretaceous, Caribou Hills, Mackenzie River, District of Mackenzie.

?*Pteruchipollenites microsaccus* Couper, 1958
Hypotype 13882
McGregor, D.C., 1965, Geol. Surv. Can., Paper 64-55, pl. 7, fig. 31.
Lower Cretaceous, Donna River Canyon, Aklavik Range, Richardson Mountains, District of Mackenzie.

cf. *Pteruchipollenites microsaccus* Couper, 1958
Hypotype 13895
McGregor, D.C., 1965, Geol. Surv. Can., Paper 64-55, pl. 7, fig. 45.
Lower Cretaceous, Donna River Canyon, Aklavik Range, Richardson Mountains, District of Mackenzie.

Pulcheripollenites inrasus Srivastava, 1969
Hypotype 93653
Sweet, A.R., Ricketts, B.D., Cameron, A.R. and Norris, D.K., 1989, Geol. Surv. Can., Paper 89-1G, p. 97, pl. 1, fig. 25.
Summit Creek Formation, Upper Cretaceous, north bank Mackenzie River immediately dowstream from Police Island, lat. 64°52′42″N, long. 125°14′44″W, District of Mackenzie.

Pulcheripollenites krempii Srivastava 1969
Hypotype 89724
Jerzykiewicz, T. and Sweet, A.R., 1988, Sedimentary Geol., vol. 59, no. 1/2, p. 58, pl. II, fig. 9.
Brazeau Formation, Upper Cretaceous, Blackstone River, lat. 52°43′20″N, long. 116°16′30″W, Alberta.

Pulcheripollenites spp.
Fig. specs. 48980, 48991
Sweet, A.R., 1978, Geol. Surv. Can., Paper 78-1B, pl. 6.1, fig. 20; pl. 6.2, fig. 5.
Cuesta Creek Member, Tent Island Formation, Upper Cretaceous, northwestern bank Big Fish River near its junction with Boundary Creek, lat. 68°30′30″N, long. 136°23′50″W, Yukon.

Punctodiporites sp.
Fig. spec. 62629
Ioannides, N.S. and McIntyre, D.J., 1980, Geol. Surv. Can., Paper 80-1A, pl. 31.5, fig. 18.
Upper Cretaceous, Caribou Hills, Mackenzie River, District of Mackenzie.

Quadripollis krempii Drugg, 1967
Hypotype 93675
Sweet, A.R. and Braman, D.R., 1989, Geol. Surv. Can., Paper 89-8, p. 35, pl. 1, fig. 10.
Chungo Member, Wapiabi Formation, Upper Cretaceous, Trap Creek near junction with Highwood River, sec. 36, tp. 17, rge. 4, W.5th mer., Alberta.

Quercoidites genustriatus Stanley, 1965
Hypotype 76068
Jerzykiewicz, T. and Sweet, A.R., 1986, Can. J. Earth Sci., vol. 23, no. 9, pl. 4, fig. 8.
Late Cretaceous, depth 78.1 m, Manalta Coal Limited, core hole ME-111-81CD, l.s.d. 3, sec. 17, tp. 48, rge. 21, W.5th mer., Alberta.

Racemonocolpites formosus Sweet
Holotype 76118; paratype 76119
Sweet, A.R., 1978, Can. J. Earth Sci., vol. 23, no. 9, p. 1384, pl. 2, figs. 9-11.
Late Cretaceous, depth 29.35-29.62 m, GSC core hole CV-42-2-2, Coal Valley strip mine, l.s.d. 1, sec. 10, tp. 47, rge. 19, W.5th mer., Alberta.

Racemonocolpites formosus Sweet, 1986
Hypotype 76067
Jerzykiewicz, T. and Sweet, A.R., 1986, Can. J. Earth Sci., vol. 23, no. 9, pl. 4, fig. 7.
Late Cretaceous, depth 29.35-29.62 m, GSC core hole CV-42-2-2, Coal Valley strip mine, l.s.d. 1, sec. 10, tp. 47, rge. 19, W.5th mer., Alberta.

Radialisporis radiatus (Krutzsch) Krutzsch, 1967
Hypotypes 38184, 38185
McIntyre, D.J., 1974, Geol. Surv. Can., Paper 74-14, pl. 14, figs. 15, 16.
Upper Cretaceous, lat. 69°28′N, long. 126°58′W, Horton River, District of Mackenzie.

Raistrickia sp.
Fig. spec. 13872
McGregor, D.C., 1965, Geol. Surv. Can., Paper 64-55, pl. 7, fig. 17.
Lower Cretaceous, Donna River Canyon, Aklavik Range, Richardson Mountains, District of Mackenzie.

Retibrevitrocolporites beccus Sweet
Holotype 76121
Sweet, A.R., 1986, Can. J. Earth Sci., vol. 23, no. 9, p. 1385, pl. 2, fig. 13.
Late Cretaceous, abandoned mine working, Coalspur, l.s.d. 10, sec. 33, tp. 48, rge. 21, W.5th mer., Alberta.

Retibrevitricolporites beccus Sweet, 1986
Hypotype 76063
Jerzykiewicz, T. and Sweet, A.R., 1986, Can. J. Earth Sci., vol. 23, no. 9, pl. 4, fig. 3.
Late Cretaceous, left bank of Embarras River, l.s.d. 10, sec. 33, tp. 48, rge. 21, W.5th mer., Alberta.

Reticulatasporites dupliexinous Brenner
Hypotype 42522
Brideaux, W.W. and McIntyre, D.J., 1975, Geol. Surv. Can., Bull. 252, pl. 4, figs. 20, 21.
Crossley Lakes Member, Langton Bay Formation, Lower Cretaceous, lat. 69°04′N, long. 125°51′W, Horton River, District of Mackenzie.

Reticulatisporites sp.
Fig. spec. 31768
Hopkins, Jr., W.S. and Balkwill, H.R., 1973, Geol. Surv. Can., Paper 72-37, p. 16, pl. 2, fig. 30.
Hassel Formation, Cretaceous, Hoodoo Dome, lat. 78°11′N, long. 99°40′W, southern Ellef Ringnes Island, District of Franklin.

Reticulisporites vermiformis Kremp
Hypotypes 42412, 42413
Brideaux, W.W. and McIntyre, D.J., 1975, Geol. Surv. Can., Bull. 252, pl. 1, figs. 25, 26.
Crossley Lakes Member, Langton Bay Formation and Horton River Formation, Lower Cretaceous, lat. 69°12′30″N, long. 126°17′W and lat. 69°19′N, long. 126°50′W, Horton River, District of Mackenzie.

Reticuloidosporites dentatus Pflug, 1953
Hypotypes 71307, 71308
Wilson, M.A., 1978, Palaeontographica, Abt. B, vol. 166, p. 119, pl. 4, figs. 12, 13.
Cretaceous, Police Island, north bank Mackenzie River, 12 miles southeast of Fort Norman, District of Mackenzie.

Reticuloidosporites sp.
Fig. spec. 38201
McIntyre, D.J., 1974, Geol. Surv. Can., Paper 74-14, pl. 14, fig. 32.
Upper Cretaceous, lat. 69°28′N, long. 126°58′W, Horton River, District of Mackenzie.

Retimonocolpites excelsus Ward, 1986
Hypotype 91405
Sweet, A.R. and McIntyre, D.J., 1988, Can. Soc. Petrol. Geol., Mem. 15, fig. 7.24.
Ram Member, Cardium Formation, Upper Cretaceous, north side of Highway 47 north of Cadmin, l.s.d. 14, sec. 16, tp. 47, rge. 23, W.5th mer., Alberta.

Retitricolpites crassus Samoilovich, 1965
Hypotype 76064
Jerzykiewicz, T. and Sweet, A.R., 1986, Can. J. Earth Sci., vol. 23, no. 9, pl. 4, fig. 4.
Early Paleocene, right bank of Embarras River towards south end of Robb, l.s.d. 14, sec. 11, tp. 49, rge. 21, .5th mer., Alberta.

Retitricolpites georgensis Brenner
Hypotype 42532
Brideaux, W.W. and McIntyre, D.J., 1975, Geol. Surv. Can., Bull. 252, pl. 4, figs. 31, 32.
Horton River Formation, Lower Cretaceous, lat. 69°19′N, long. 126°50′W, Horton River, District of Mackenzie.

Retitricolpites prosimilis Norris
Hypotype 42533
Brideaux, W.W. and McIntyre, D.J., 1975, Geol. Surv. Can., Bull. 252, pl. 4, figs. 33, 34.
Crossley Lakes Member, Langton Bay Formation, Lower Cretaceous, lat. 69°12′30″N, long. 126°17′W, Horton River, District of Mackenzie.

Retitricolpites sp.
Fig. specs. 31789, 31790
Hopkins, Jr., W.S. and Balkwill, H.R., 1973, Geol. Surv. Can., Paper 72-37, p. 22, pl. 2, figs. 51, 52.
Hassel Formation, Cretaceous, Hoodoo Dome, lat. 78°11′N, long. 99°40′W, southern Ellef Ringnes Island, District of Franklin.

Retitricolporites pristinus Singh, 1983
Hypotypes 91394, 91395
Sweet, A.R. and McIntyre, D.J., 1988, Can. Soc. Petrol. Geol., Mem. 15, figs. 7.13, 7.14.
Moosehound Member, Cardium Formation, Upper Cretaceous, north side of Highway 47 north of Cadomin, l.s.d. 14, sec. 16, tp. 47, rge. 23, W.5th mer., and south side of McLeod River, l.s.d. 16, sec. 16, tp. 47, rge. 23, W.5th mer., Alberta.

Retitriletes cf. *R. saxatilis* Srivastava, 1972
Hypotypes 71282, 71286
Wilson, M.A., 1978, Palaeontographica, Abt. B, vol. 166, p. 115, pl. 2, figs. 7, 11.
Cretaceous, Police Island, north bank Mackenzie River, 12 miles southeast of Fort Norman, District of Mackenzie.

Rogalskaisporites cicatricosus (Rogalska) Danzé-Corsin and Laveine
Hypotype 42463
Brideaux, W.W. and McIntyre, D.J., 1975, Geol. Surv. Can., Bull. 252, pl. 2, fig. 35.

Crossley Lakes Member, Langton Bay Formation, Lower Cretaceous, lat. 64°01′30″N, long. 126°02′W, Horton River, District of Mackenzie.

Rousea brenneri Singh, 1983
Hypotype 91404
Sweet, A.R. and McIntyre, D.J., 1988, Can. Soc. Petrol. Geol., Mem. 15, fig. 7.23.
Moosehound Member, Cardium Formation, Upper Cretaceous, south side of Grande Cache Highway 40, l.s.d. 15, sec. 13, tp. 57, rge. 7, W.5th mer., Alberta.

Rousea georgensis (Brenner) Dettmann, 1975
Hypotypes 91396, 91397
Sweet, A.R. and McIntyre, D.J., 1988, Can. Soc. Petrol. Geol., Mem. 15, figs. 7.15, 7.16.
Moosehound Member, Cardium Formation, Upper Cretaceous, south side of Grande Cache Highway 40, l.s.d. 15, sec. 13, tp. 57, rge. 7, W.5th mer., Alberta.

Rousea sp.
Fig. spec. 91400
Sweet, A.R. and McIntyre, D.J., 1988, Can. Soc. Petrol. Geol., Mem. 15, fig. 7.19.
Moosehound Member, Cardium Formation, Upper Cretaceous, south side of Grande Cache Highway 40, l.s.d. 15, sec. 13, tp. 57, rge. 7, W.5th mer., Alberta.

Rouseisporites reticulatus Pocock
Hypotype 56791
Williams, G.L. and Bujak, J.P., 1980, Initial Repts. Deep Sea Drilling Project, vol. 50, pl. 11, fig. 10.
Early Cretaceous, DSDP Site 416, lat. 32°50.18′N, long. 10°48.06′W, northwestern Moroccan continental margin.

Rubinella major (Couper) Norris, 1969
Hypotype 58560
McIntyre, D.J., and Brideaux, W.W. 1980, Geol. Surv. Can., Bull. 320, p. 10, pl. , fig. 4.
Lower Cretaceous, Martin Creek, lat. 68°12′10″N, long. 135°34′50″W, Richardson Mountains, District of Mackenzie.

Rugubivesiculites reductus Pierce, 1961
Hypotype 96497
Leckie, D.A. and Singh, C., 1991, J. Sed. Petrology, vol. 61, no. 5, fig. 3N.
Paddy Member, Peace River Formation, Lower Cretaceous, near town of Peace River, Alberta.

Rugubivesiculites rugosus Pierce, 1961
Hypotype 84751
Leckie, D.A. et al., 1990, J. Sed. Petrology, vol. 60, no. 1, p. 101, fig. 6N.
Shaftesbury Formation, Lower Cretaceous, west bank Peace River near town of Peace River, Alberta.

Rugulatisporites sp.
Fig. spec. 25087
Hopkins, Jr., W.S., 1971, Geol. Surv. Can., Bull. 197, p. 120, pl. 21, fig. 15.
Isachsen Formation, Lower Cretaceous, lat. 76°20′N, long. 115°34′W, northwest Melville Island, District of Franklin.

Rutihesperipites trochuensis (Srivastava) Srivastava, 1977
Hypotypes 76069, 76070
Jerzykiewicz, T. and Sweet, A.R., 1986, Can. J. Earth Sci., vol. 23, no. 9, pl. 4, figs. 9, 10.
Late Cretaceous, depth 29.35-29.62 m, GSC core hole CV-42-2-2, Coal Valley strip mine, l.s.d. 1, sec. 10, tp. 47, rge. 19, W.5th mer. and depth 25.6-25.75 m, GSC core hole CV-42-2-1, l.s.d. 1, sec. 6, tp. 47, rge. 19, W.5th mer., Alberta.

Rutihesperipites sp.
Fig. spec. 76123
Sweet, A.R., 1986, Can. J. Earth Sci., vol. 23, no. 9, p. 1385, pl. 2, fig. 16.
Late Cretaceous, depth 29.35-29.62 m, GSC core hole CV-42-2-2, Coal Valley strip mine, l.s.d. 1, sec. 10, tp. 47, rge. 19, W.5th mer., Alberta.

Rutihesperipites sp.
Fig. spec. 76072
Jerzykiewicz, T. and Sweet, A.R., 1986, Can. J. Earth Sci., vol. 23, no. 9, pl. 4, fig. 12.
Late Cretaceous, depth 29.35-29.62 m, GSC core hole CV-42-2-2, Coal Valley strip mine, l.s.d. 1, sec. 10, tp. 47, rge. 19, W.5th mer., Alberta.

Sabalpollenites dividuus Kimyai, 1966
Hypotype 91411
Sweet, A.R. and McIntyre, D.J., 1988, Can. Soc. Petrol. Geol., Mem. 15, fig. 7.30.
Moosehound Member, Cardium Formation, Upper Cretaceous, south side of Grande Cache Highway 40, l.s.d. 15, sec. 13, tp. 57, rge. 7, W.5th mer., Alberta.

Sabalpollenites sp.
Fig. spec. 91412
Sweet, A.R. and McIntyre, D.J., 1988, Can. Soc. Petrol. Geol., Mem. 15, fig. 7.31.
Ram Member, Cardium Formation, Upper Cretaceous, north side of Highway 47 north of Cadomin, l.s.d. 14, sec. 16, tp. 47, rge. 23, W.5th mer., Alberta.

Scabrastephanocolpites albertensis Srivastava, 1969
Hypotype 76074
Jerzykiewicz, T. and Sweet, A.R., 1986, Can. J. Earth Sci., vol. 23, no. 9, pl. 4, fig. 14.
Late Cretaceous, Embarras River, Alberta.

?*Sciadopityspollenites serratus* (Potonié and Venitz) Raatz in Manum, 1962
Hypotype 13955
McGregor, D.C., 1965, Geol. Surv. Can., Paper 64-55, pl. 10, fig. 23.
Lower Cretaceous, Peel Plateau, Stony Creek about 18.5 miles upstream from mouth, District of Mackenzie.

Schizaea sp. A
Fig. spec. 71309
Wilson, M.A., 1978, Palaeontographica, Abt. B, vol. 166, p. 119, pl. 5, fig. 1.

Cretaceous, Police Island, north bank Mackenzie River, 12 miles southeast of Fort Norman, District of Mackenzie.

Schizaea spp. C, B
Fig. specs. 71310, 71318
Wilson, M.A., 1978, Palaeontographica, Abt. B, vol. 166, p. 120, pl. 5, figs. 2, 10.
Cretaceous and Paleocene, IOE Atkinson H-25 well, north shore Tuktoyaktuk Peninsula, lat. 69°44′20″N, long. 131°50′06″W, District of Mackenzie.

Schizocystia laevigate Cookson and Eisenack
Hypotype 38165
McIntyre, D.J., 1974, Geol. Surv. Can., Paper 74-14, pl. 13, fig. 10.
Upper Cretaceous, lat. 69°28′N, long. 126°58′W, Horton River, District of Mackenzie.

Scizocystia laevigata Cookson and Eisenack, 1962
Hypotype 96496
Leckie, D.A. and Singh, C., 1991, J. Sed. Petrology, vol. 61, no. 5, fig. 3M.
Paddy Member, Peace River Formation, Lower Cretaceous, near town of Peace River, Alberta.

Schizophacus parvus (Cookson and Dettmann) Pierce 1976
Hypotype 48995
Sweet, A.R., 1978, Geol. Surv. Can., Paper 78-1B, pl. 6.2, fig. 9.
Cuesta Creek Member, Tent Island Formation, Upper Cretaceous, northwestern bank Big Fish River near its junction with Boundary Creek, lat. 68°30′30″N, long. 136°23′50″W, Yukon.

Schizosporis microreticulatus Brenner, 1963
Hypotype 91410
Sweet, A.R. and McIntyre, D.J., 1988, Can. Soc. Petrol. Geol., Mem. 15, fig. 7.29.
Moosehound Member, Cardium Formation, Upper Cretaceous, west bank of Little Berland River, l.s.d. 1, sec. 6, tp. 54, rge. 2, W.5th mer., Alberta.

Schizosporis parvus Cookson and Dettmann
Hypotype 33994
Hopkins, Jr., W.S., 1974, Geol. Surv. Can., Paper 73-12, p. 21, pl. 5, fig. 61.
Christopher Formation, Lower Cretaceous, lat. 76°20′N, long. 115°34′W, northwestern Melville Island, District of Franklin.

Scizosporis reticulatus Cookson and Dettmann
Hypotype 25506
Hopkins, Jr., W.S., 1971, Geol. Surv. Can., Bull. 197, p. 124, pl. 22, fig. 23.
Isachsen Formation, Lower Cretaceous, lat. 76°20′N, long. 115°34′W, northwest Melville Island, District of Franklin.

Schizosporis reticulatus Cookson and Dettmann
Hypotype 76610
Davies, E.H., Akande, S.O. and Zentilli, M., 1984, Geol. Surv. Can., Paper 84-1A, pl. 48.1, fig. 21.
Early Cretaceous, Gays River Mine, lat. 45°02′N, long. 63°22′W, Nova Scotia.

Schizosporis sp.
Fig. spec. 41618
Hopkins, Jr., W.S. and Sweet, A.R., 1976, Geol. Surv. Can., Bull. 256, pl. 10, fig. 5.
Mattagami Formation, Lower Cretaceous, Onakawana lignite pit, lat. 50°36′N, long. 81°23′W, Ontario.

Schizosporis sp.
Fig. spec. 93680
Sweet, A.R. and Braman, D.R., 1989, Geol. Surv. Can., Paper 89-8, pl. 1, fig. 15.
Chungo Member, Wapiabi Formation, Upper Cretaceous, Trap Creek near junction with Highwood River, sec. 36, tp. 17, rge. 4, W.5th mer., Alberta.

Schizosporis sp. 1 (cf. *Disulcites magnus* Srivastava, 1969)
Fig. spec. 38168
McIntyre, D.J., 1974, Geol. Surv. Can., Paper 74-14, pl. 13, fig. 13.
Upper Cretaceous, lat. 69°28′N, long. 126°58′W, Horton River, District of Mackenzie.

Schizosporis sp. 2
Fig. spec. 38169
McIntyre, D.J., 1974, Geol. Surv. Can., Paper 74-14, pl. 13, fig. 14.
Upper Cretaceous, lat. 69°28′N, long. 126°58′W, Horton River, District of Mackenzie.

Sciadopitys sp.
Fig. spec. 31781
Hopkins, Jr., W.S. and Balkwill, H.R., 1973, Geol. Surv. Can., Paper 72-37, p. 19, pl. 2, fig. 43.
Hassel Formation, Cretaceous, Hoodoo Dome, lat. 78°11′N, long. 99°40′W, southern Ellef Ringnes Island, District of Franklin.

Sciadopitys sp.
Fig. spec. 41609
Hopkins, Jr., W.S. and Sweet, A.R., 1976, Geol. Surv. Can., Bull. 256, pl. 9, fig. 25.
Mattagami Formation, Lower Cretaceous, Onakawana lignite pit, lat. 50°36′N, long. 81°23′W, Ontario.

Scollardia trapaformis Srivastava, 1966
Hypotypes 89738, 89739
Jerzykiewicz, T. and Sweet, A.R., 1988, Sedimentary Geol., vol. 59, no. 1/2, pl. III, figs. 11, 12.
Brazeau Formation, Upper Cretaceous, Blackstone River, lat. 52°42′55″N, long. 116°18′30″W, Alberta.

Scollardia trapaformis Srivastava, 1966
Hypotype 89851
Sweet, A.R., Ricketts, B.D., Cameron, A.R. and Norris, D.K., 1989, Geol. Surv. Can., Paper 89-1G, p. 97, pl. 1, fig. 24.

Summit Creek Formation, Upper Cretaceous, north bank Mackenzie River immediately dowstream from Police Island, lat. 64°52′42″N, long. 125°14′44″W, District of Mackenzie.

Scortea tecta
Hypotype 91982
Banerjee, I. and Davies, E.H., 1988, Can. Soc. Petrol. Geol., Mem. 15, fig. 6.10.
Mannville Group, Lower Cretaceous, depth 5276 feet, well 11-28-40-24W4, Alberta.

Senipites drumhellerensis Srivastava, 1969
Hypotypes 76065, 76066
Jerzykiewicz, T. and Sweet, A.R., 1986, Can. J. Earth Sci., vol. 23, no. 9, pl. 4, figs. 5, 6.
Late Cretaceous, depth 80.9 m, Manalta Coal Limited core hole ME-111-81CD, l.s.d. 3, sec. 17, tp. 48, rge. 21, W.5th mer. and Embarras River, Alberta.

Sequoiapollenites papillapollenites (Rouse) Srivastava, 1966
Hypotypes 71325, 71326
Wilson, M.A., 1978, Palaeontographica, Abt. B, vol. 166, p. 127, pl. 6, figs. 3, 4.
Cretaceous, Police Island, north bank Mackenzie River, 12 miles southeast of Fort Norman, and IOE Atkinson H-25 well, north shore Tuktoyaktuk Peninsula, lat. 69°44′20″N, long. 131°50′06″W, District of Mackenzie.

Sestrosporites pseudoalveolatus (Couper) Dettmann
Hypotype 31753
Hopkins, Jr., W.S. and Balkwill, H.R., 1973, Geol. Surv. Can., Paper 72-37, p. 12, pl. 1, fig. 14.
Hassel Formation, Cretaceous, Hoodoo Dome, lat. 78°11′N, long. 99°40′W, southern Ellef Ringnes Island, District of Franklin.

Sestrosporites pseudoalveolatus (Couper)
Hypotype 41588
Hopkins, Jr., W.S. and Sweet, A.R., 1976, Geol. Surv. Can., Bull. 256, pl. 9, fig. 4.
Mattagami Formation, Lower Cretaceous, Onakawa lignite pit, lat. 50°36′N, long. 81°23′W, Ontario.

Sestrosporites pseudoalveolatus (Couper)
Hypotype 42452
Brideaux, W.W. and McIntyre, D.J., 1975, Geol. Surv. Can., Bull. 252, pl. 2, fig. 24.
Crossley Lakes Member, Langton Bay Formation, Lower Cretaceous, lat. 69°07′30″N, long. 126°20′W, west bank Horton River opposite mouth of Coal Creek, District of Mackenzie.

Sestrosporites sp.
Fig. spec. 13916
McGregor, D.C., 1965, Geol. Surv. Can., Paper 64-55, pl. 8, fig. 11.
Isachsen Formation, Lower Cretaceous, south of Buchanan Lake anticline, eastern Axel Heiberg Island, District of Franklin.

Siberiapollis montanensis Tschudy 1971
Hypotype 89726
Jerzykiewicz, T. and Sweet, A.R., 1988, Sedimentary Geol., vol. 59, no. 1/2, pl. II, fig. 11.
Brazeau Formation, Upper Cretaceous, Blackstone River, lat. 52°43′20″N, long. 116°16′30″W, Alberta.

Siberiapollis sp. cf. *S. occulata* (Samoilovich) Tschudy, 1971
Fig. spec. 76078
Jerzykiewicz, T. and Sweet, A.R., 1986, Can. J. Earth Sci., vol. 23, no. 9, pl. 4, fig. 18.
Late Cretaceous, depth 80.9 m, Manalta Coal Limited core hole ME-111-81CD, l.s.d. 3, sec. 17, tp. 48, rge. 21, W.5th mer., Alberta.

Siberiapollis sp.
Fig. spec. 89700
Jerzykiewicz, T. and Sweet, A.R., 1988, Sedimentary Geol., vol. 59, no. 1/2, p. 48, pl. I, fig. 4.
Belly River Formation, Upper Cretaceous, Crownest River Valley west of Lundbreck, lat. 49°35′N, long. 114°11′W, Alberta.

Sigmopollis sp.
Fig. spec. 76073
Jerzykiewicz, T. and Sweet, A.R., 1986, Can. J. Earth Sci., vol. 23, no. 9, pl. 4, fig. 13.
Late Cretaceous, Coal Valley strip mine pit, Alberta.

Singhia urwashii Srivastava, 1968
Hypotype 76008
Jerzykiewicz, T. and Sweet, A.R., 1986, Can. J. Earth Sci., vol. 23, no. 9, pl. 1, fig. 11.
Late Cretaceous, right bank of Embarras River towards south end of Robb, l.s.d. 14, sec. 11, tp. 49, rge. 21, W.5th mer., Alberta.

Singularia aculeata Samoilovitch, 1961
Hypotype 38225
McIntyre, D.J., 1974, Geol. Surv. Can., Paper 74-14, pl. 15, fig. 17.
Upper Cretaceous, lat. 69°28′N, long. 126°58′W, Horton River, District of Mackenzie.

Sparganium sp.
Fig. spec. 76075
Jerzykiewicz, T. and Sweet, A.R., 1986, Can. J. Earth Sci., vol. 23, no. 9, pl. 4, fig. 15.
Late Cretaceous, left bank Embarras River, l.s.d. 10, sec. 33, tp. 48, rge. 21, W.5th mer., Alberta.

Spermatites ellipticus Miner 1935
Hypotype 48936
Sweet, A.R., 1979, Am. Assoc. Strat. Palynol, Contrib. Ser. No. 5B, pl. 3, fig. 11.
Hassel Formation, Lower Cretaceous, lat. 77°9′N, long. 104°50′W, Lougheed Island, District of Franklin.

Spermatites cf. *S. ellipticus* Miner
Hypotype 41649
Hopkins, Jr., W.S. and Sweet, A.R., 1976, Geol. Surv. Can., Bull. 256, pl. 12, fig. 6.
Mattagami Formation, Lower Cretaceous, borehole 68-42, lat. 50°35′36″N, long. 82°30′12″W, Ontario.

Sphagnum antiquasporites Wilson and Webster
Hypotype 23787
Hopkins, Jr., W.S., 1971, Geol. Surv. Can., Bull. 197, p. 113, pl. 20, fig. 1.
Isachsen Formation, Lower Cretaceous, lat. 76°20′N, long. 115°34′W, northwest Melville Island, District of Franklin.

Sphagnum antiquasporites Wilson and Webster
Hypotype 31741
Hopkins, Jr., W.S. and Balkwill, H.R., 1973, Geol. Surv. Can., Paper 72-37, p. 10, pl. 1, fig. 2.
Hassel Formation, Cretaceous, Hoodoo Dome, lat. 78°11′N, long. 99°40′W, southern Ellef Ringnes Island, District of Franklin.

Sphagnum antiquasporites Wilson and Webster
Hypotype 33943
Hopkins, Jr., W.S., 1974, Geol. Surv. Can., Paper 73-12, p. 8, pl. 1, fig. 3.
Christopher Formation, Lower Cretaceous, lat. 76°20′N, long. 115°34′W, northwestern Melville Island, District of Franklin.

Sphagnum cf. *S. australe* (Cookson) Drozh.
Hypotype 31742
Hopkins, Jr., W.S. and Balkwill, H.R., 1973, Geol. Surv. Can., Paper 72-37, p. 10, pl. 1, fig. 3.
Hassel Formation, Cretaceous, Hoodoo Dome, lat. 78°11′N, long. 99°40′W, southern Ellef Ringnes Island, District of Franklin.

Sphagnum cf. *S. australe parvum* (Cookson) Elsik
Hypotype 31744
Hopkins, Jr., W.S. and Balkwill, H.R., 1973, Geol. Surv. Can., Paper 72-37, p. 10, pl. 1, fig. 5.
Hassel Formation, Cretaceous, Hoodoo Dome, lat. 78°11′N, long. 99°40′W, southern Ellef Ringnes Island, District of Franklin.

Sphagnum bimammatus (Naumova ex Bolkhovitina) Elsik
Hypotypes 23788, 23789
Hopkins, Jr., W.S., 1971, Geol. Surv. Can., Bull. 197, p. 113, pl. 20, figs. 2, 3.
Isachsen Formation, Lower Cretaceous, lat. 76°20′N, long. 115°34′W, northwest Melville Island, District of Franklin.

Sphagnum buchenauensis (Krutzsch) Elsik
Hypotype 33944
Hopkins, Jr., W.S., 1974, Geol. Surv. Can., Paper 73-12, p. 8, pl. 1, fig. 4.
Christopher Formation, Lower Cretaceous, lat. 76°20′N, long. 115°34′W, northwest Melville Island, District of Franklin.

Sphagnum regium Drozhastchich
Hypotype 33945
Hopkins, Jr., W.S., 1974, Geol. Surv. Can., Paper 73-12, p. 9, pl. 1, fig. 5.
Christopher Formation, Lower Cretaceous, lat. 76°20′N, long. 115°34′W, northwestern Melville Island, District of Franklin.

Sphagnum cf. *S. regium* Drozh.
Hypotype 31743
Hopkins, Jr., W.S. and Balkwill, H.R., 1973, Geol. Surv. Can., Paper 72-37, p. 10, pl. 1, fig. 4.
Hassel Formation, Cretaceous, Hoodoo Dome, lat. 78°11′N, long. 99°40′W, southern Ellef Ringnes Island, District of Franklin.

Sphagnumsporites psilatus (Ross) Couper, 1958
Hypotype 13823
McGregor, D.C., 1965, Geol. Surv. Can., Paper 64-55, pl. 6, fig. 1.
Lower Cretaceous, Fault Creek Canyon, Aklavik Range, Richardson Mountains, District of Mackenzie.

Sphagnumsporites psilatus (Ross) Couper, 1958
Hypotype 13864
McGregor, D.C., 1965, Geol. Surv. Can., Paper 64-55, pl. 7, fig. 6.
Lower Cretaceous, Donna River Canyon, Aklavik Range, Richardson Mountains, District of Mackenzie.

Sphagnumsporites psilatus (Ross) Couper, 1958
Hypotype 13939
McGregor, D.C., 1965, Geol. Surv. Can., Paper 64-55, pl. 10, fig. 1.
Lower Cretaceous, Peel Plateau, Stony Creek about 18.5 miles upstream from mouth, District of Mackenzie.

Spheripollenites psilatus Couper, 1958
Hypotype 13959
McGregor, D.C., 1965, Geol. Surv. Can., Paper 64-55, pl. 10, fig. 30.
Lower Cretaceous, Peel Plateau, Stony Creek about 18.5 miles upstream from mouth, District of Mackenzie.

?*Spheripollenites* sp.
Fig. spec. 41613
Hopkins, Jr., W.S. and Sweet, A.R., 1976, Geol. Surv. Can., Bull. 256, pl. 9, fig. 29.
Mattagami Formation, Lower Cretaceous, Onakawana lignite pit stock pile, lat. 50°36′N, long. 81°23′W, Ontario.

Spongodinium delitiense (Ehrenberg) Deflandre, 1936
Hypotypes 38144, 38145
McIntyre, D.J., 1974, Geol. Surv. Can., Paper 74-14, pl. 12, figs. 1, 2.
Upper Cretaceous, lat. 69°27′30″N, long. 126°58′W, Horton River, District of Mackenzie.

Sporites adriennis forma *mesozoicus* Thiergart in Rogalska, 1956
Hypotype 13910
McGregor, D.C., 1965, Geol. Surv. Can., Paper 64-55, pl. 8, fig. 6.
Isachsen Formation, Lower Cretaceous, south of Buchanan Lake anticline, eastern Axel Heiberg Island, District of Franklin.

Staplinisporites caminus (Balme) Pocock
Hypotype 25086
Hopkins, Jr., W.S., 1971, Geol. Surv. Can., Bull. 197, p. 119, pl. 21, fig. 14.

Isachsen Formation, Lower Cretaceous, lat. 76°20′N, long. 115°34′W, northwest Melville Island, District of Franklin.

Stereisporites antiquasporites (Wilson and Webster) Dettmann, 1963
Hypotype 38170
McIntyre, D.J., 1974, Geol. Surv. Can., Paper 74-14, pl. 14, fig. 1.
Upper Cretaceous, lat. 69°28′N, long. 126°58′W, Horton River, District of Mackenzie.

Stereisporites antiquasporites (Wilson and Webster) Dettmann
Hypotype 42394
Brideaux, W.W. and McIntyre, D.J., 1975, Geol. Surv. Can., Bull. 252, pl. 1, fig. 6.
Horton River Formation, Lower Cretaceous, lat. 69°27′30″N, long. 126°54′W, Horton River, District of Mackenzie.

Stereisporites antiquasporites (Wilson and Webster)
Hypotype 41585
Hopkins, Jr., W.S. and Sweet, A.R., 1976, Geol. Surv. Can., Bull. 256, pl. 9, fig. 1.
Mattagami Formation, Lower Cretaceous, Onakawana lignite pit, lat. 50°36′N, long. 81°23′W, Ontario.

Stereisporites antiquasporites (Wilson and Webster) Dettmann, 1963
Hypotypes 71269, 71273
Wilson, M.A., 1978, Palaeontographica, Abt. B, vol. 166, p. 108, pl. 1, figs. 7, 11.
Cretaceous, Police Island, north bank Mackenzie River, 12 miles southeast of Fort Norman, District of Mackenzie.

Stereisporites regium (Drozhastchich) Drugg, 1967
Hypotypes 38171, 38172
McIntyre, D.J., 1974, Geol. Surv. Can., Paper 74-14, pl. 14, figs. 2, 3.
Upper Cretaceous, lat. 69°28′N, long. 126°58′W, Horton River, District of Mackenzie.

Stereisporites regium (Drozhastchich) Drugg, 1967
Hypotype 71266
Wilson, M.A., 1978, Palaeontographica, Abt. B, vol. 166, p. 109, pl. 1, fig. 4.
Cretaceous, IOE Atkinson H-25 well, north shore Tuktoyaktuk Peninsula, lat. 69°44′20″N, long. 131°50′06″W, District of Mackenzie.

Striadipores sp.
Fig. spec. 62628
Ioannides, N.S. and McIntyre, D.J., 1980, Geol. Surv. Can., Paper 80-1A, pl. 31.5, fig. 17.
Upper Cretaceous, Caribou Hills, Mackenzie River, District of Mackenzie.

Striatellipollis radiata Sweet
Holotype 76127; paratypes 76124-76126
Sweet, A.R., 1986, Can. J. Earth Sci., vol. 23, p. 1385, pl. 2, figs. 17-20.
Late Cretaceous, depth 78.1 m, Manalta Coal Limited core hole ME-111-81CD, l.s.d. 3, sec. 17, tp. 48, rge. 21, W.5th mer. and depth 29.35-29.62 m, GSC core hole CV-42-2-2, Coal Valley strip mine, l.s.d. 1, sec. 10, tp. 47, rge. 19, W.5th mer. (76124), Alberta.

Striatellipollis radiata Sweet 1986
Hypotype 76084
Jerzykiewicz, T. and Sweet, A.R., 1986, Can. J. Earth Sci., vol. 23, no. 9, pl. 4, fig. 24.
Late Cretaceous, depth 78.1 m, Manalta Coal Limited core hole ME-111-81CD, l.s.d. 3, sec. 17, tp. 48, rge. 21, W.5th mer. Alberta.

Striatellipollis radiata Sweet, 1986
Hypotype 89855
Sweet, A.R., Ricketts, B.D., Cameron, A.R. and Norris, D.K., 1989, Geol. Surv. Can., Paper 89-1G, p. 98, pl. 1, fig. 31.
Summit Creek Formation, Upper Cretaceous, north bank Mackenzie River immediately dowstream from Police Island, lat. 65°52′42″N, long. 125°14′44″W, District of Mackenzie.

Striatellipollis striatella (Mchedlishvili) Krutzsch, 1969
Hypotypes 76128-76130
Sweet, A.R., 1986, Can. J. Earth Sci., vol. 23, no. 9, p. 1386, pl. 2, figs. 21-23.
Late Cretaceous, depth 90-100 feet, Cypress 25R test hole, l.s.d. 13, sec. 36, tp. 6, rge. 20, W.3rd mer., Saskatchewan.

Striatellipollis striatella (Mchedlishvili) Krutzsch 1969
Hypotype 76087
Jerzykiewicz, T. and Sweet, A.R., 1986, Can. J. Earth Sci., vol. 23, no. 9, pl. 4, fig. 27.
Late Cretaceous, right bank of Embarras River towards the south end of Robb, l.s.d. 14, sec. 11, tp. 49, rge. 21, W.5th mer. Alberta.

Striatopollis sp.
Fig. spec. 89704
Jerzykiewicz, T. and Sweet, A.R., 1988, Sedimentary Geol., vol. 59, no. 1/2, p. 46, pl. I, fig. 8.
Belly River Formation, Upper Cretaceous, Crownest River valley west of Lundbreck, lat. 49°35′N, long. 114°11′W, Alberta.

Striatriletes radialis (Gunther and Hills) Sweet 1979
Hypotype 48951
Sweet, A.R., 1979, Am. Assoc. Strat. Palynol., Contrib. Ser. No. 5B, p. 15, pl. 4, fig. 9.
Eureka Sound Formation, Upper Cretacous, Meteorologist Peninsula, lat. 78°10′N, long. 99°30′W, Ellef Ringnes Island, Arctic.

Striatriletes sulcatus (Dijkstra) Potonié 1956
Hypotype 48915
Sweet, A.R., 1979, Am. Assoc. Strat. Palynol., Contrib. Ser. No. 5B, pl. 2, fig. 11.
Christopher Formation, Lower Cretaceous, lat. 74°31′N, long. 121°10′W, Cape McClure, Banks Island, District of Franklin.

Styx canadensis (Srivastava and Binda) Sweet 1979
Hypotype 48949
Sweet, A.R., 1979, Am. Assoc. Strat. Palynol., Contrib. Ser. No. 5B, p. 15, pl. 4, fig. 7.
Eureka Sound Formation, Upper Cretacous, Meteorologist Peninsula, lat. 78°10′N, long. 99°30′W, Ellef Ringnes Island, District of Franklin.

Styx of *S. kondinskayae* (Srivastava and Binda) Sweet 1979
Hypotype 48952
Sweet, A.R., 1979, Am. Assoc. Strat. Palynol., Contrib. Ser. No. 5B, p. 15, pl. 4, fig. 10.
Eureka Sound Formation, Upper Cretacous, Meteorologist Peninsula, lat. 78°10′N, long. 99°30′W, Ellef Ringnes Island, District of Franklin.

Styx sp.
Fig. specs. 48997-48999
Sweet, A.R., 1978, Geol. Surv. Can., Paper 78-1B, p. 36, pl. 6.2, figs. 11-13.
Cuesta Creek and mudstone (48998) members, Tent Island Formation, Upper Cretaceous, northwestern bank Big Fish River near its junction with Boundary Creek, lat. 68°30′30″N, long. 136°23′50″W, Yukon.

Subtriporopollenites alpinus (Wolf) Tschudy 1973
Hypotype 89701
Jerzykiewicz, T. and Sweet, A.R., 1988, Sedimentary Geol., vol. 59, no. 1/2, p. 46, pl. I, fig. 5.
Belly River Formation, Upper Cretaceous, Crowsnest River valley west of Lundbreck, lat. 49°35′N, long. 114°11′W, Alberta.

Symplocoipollenites cf. *S. vestibulum* (Potonié) Potonié, 1951
Hypotype 71373
Wilson, M.A., 1978, Palaeontographica, Abt. B, vol. 166, p. 139, pl. 9, figs. 12, 13.
Cretaceous, Police Island, north bank Mackenzie River, 12 miles southeast of Fort Norman, District of Mackenzie.

Symplocoipollenites sp.
Fig. spec. 71379
Wilson, M.A., 1978, Palaeontographica, Abt. B, vol. 166, p. 139, pl. 10, fig. 1.
Cretaceous, IOE Atkinson H-25 well, north shore Tuktoyaktuk Peninsula, lat. 69°44′20″N, long. 131°50′06″W, District of Mackenzie.

Syncolpites porosus Mchedlishvili, 1961
Hypotypes 38272, 38273
McIntyre, D.J., 1974, Geol. Surv. Can., Paper 74-14, pl. 21, figs. 4, 5.
Upper Cretaceous, lat. 69°28′N, long. 126°58′W, Horton River, District of Mackenzie.

Syncolporate sp.
Fig. spec. 71407
Wilson, M.A., 1978, Palaeontographica, Abt. B, vol. 166, p. 141, pl. 11, fig. 6.
Cretaceous, Police Island, north bank Mackenzie River, 12 miles southeast of Fort Norman, District of Mackenzie.

Syncolporites minimus Leffingwell, 1971
Hypotype 71394
Wilson, M.A., 1978, Palaeontographica, Abt. B, vol. 166, p. 141, pl. 10, fig. 16.
Cretaceous, IOE Atkinson H-25 well, north shore Tuktoyaktuk Peninsula, lat. 69°44′20″N, long. 131°50′06″W, District of Mackenzie.

Syncolporites minimus Leffingwell 1971
Hypotypes 76076, 76077
Jerzykiewicz, T. and Sweet, A.R., 1986, Can. J. Earth Sci., vol. 23, no. 9, pl. 4, figs. 16, 17.
Late Cretaceous, depth 78.1 m, Manalta Coal Limited core hole ME-111-81CD, l.s.d. 3, sec. 17, tp. 48, rge. 21, W.5th mer. and depth 25.75-25.755 m, GSC core hole CV-42-2-1, Coal Valley strip mine, l.s.d. 1, sec. 6, tp. 47, rge. 19, W.5th mer., Alberta.

Taurocusporites segmentatus Stover
Hypotype 41906
Williams, G.L., Geol. Surv. Can., Paper 74-30, vol. 2, pl. 8, fig. 11.
Barremian, Lower Cretaceous, Mobil Sable Island C-67 well, Scotian Shelf.

Taxodiaceaepollenites hiatus (Potonié) Kremp, 1949
Hypotypes 38217, 38218
McIntyre, D.J., 1974, Geol. Surv. Can., Paper 74-14, pl. 15, figs. 10, 11.
Upper Cretaceous, lat. 69°27′30″N, long. 126°58′W, and lat. 69°28′N, long. 126°58′W, Horton River, District of Mackenzie.

Taxodiaceaepollenites hiatus (Potonié) Kremp
Hypotype 42521
Brideaux, W.W. and McIntyre, D.J., 1975, Geol. Surv. Can., Bull. 252, pl. 4, fig. 19.
Crossley Lakes Member, Langton Bay Formation, Lower Cretaceous, lat. 69°04′N, long. 125°51′W, Horton River, District of Mackenzie.

Taxodiaceaepollenites hiatus (Potonié) Kremp, 1949
Hypotype 71320
Wilson, M.A., 1978, Palaeontographica, Abt. B, vol. 166, p. 126, pl. 5, fig. 12.
Cretaceous, IOE Atkinson H-25 well, north shore Tuktoyaktuk Peninsula, lat. 69°44′20″N, long. 131°50′06″W, District of Mackenzie.

Taxodium cf. *T. hiatipites* Wodehouse
Hypotype 31778
Hopkins, Jr., W.S. and Balkwill, H.R., 1973, Geol. Surv. Can., Paper 72-37, p. 19, pl. 2, fig. 40.
Hassel Formation, Cretaceous, Hoodoo Dome, lat. 78°11′N, long. 99°40′W, southern Ellef Ringnes Island, District of Franklin.

Taxodium cf. *T. hiatipites* Wodehouse
Hypotype 34002
Hopkins, Jr., W.S., 1974, Geol. Surv. Can., Paper 73-12, p. 23, pl. 6, fig. 69.

Christopher Formation, Lower Cretaceous, lat. 76°20′N, long. 115°34′W, northwest Melville Island, District of Franklin.

Tenellisporites spinatus Peake
Hypotypes 41636-41639
Hopkins, Jr., W.S. and Sweet, A.R., 1976, Geol. Surv. Can., Bull. 256, pl. 11, figs. 12-15.
Mattagami Formation, Lower Cretaceous, borehole 68-42, lat. 50°35′36″N, long. 82°30′12″W, Ontario.

Tenellisporites sp. B
Fig. spec. 48930
Sweet, A.R., 1979, Am. Assoc. Strat. Palynol., Contrib. Ser. No. 5B, pl. 3, fig. 5.
Christopher Formation, Lower Cretaceous, lat. 79°47′24″N, long. 85°22′W, near Slidre Fiord, Ellesmere Island, District of Franklin.

Tenellisporites sp. C
Fig. spec. 48942
Sweet, A.R., 1979, Am. Assoc. Strat. Palynol., Contrib. Ser. No. 5B, pl. 3, figs. 18, 19.
Hassel Formation, Upper Cretaceous, lat. 79°47′24″N, long. 85°22′W, near Slidre Fiord, Ellesmere, District of Franklin.

Tenellisporites sp. D
Fig. spec. 48947
Sweet, A.R., 1979, Am. Assoc. Strat. Palynol., Contrib. Ser. No. 5B, pl. 4, fig. 4.
Eureka Sound Formation, Upper Cretacous, lat. 77°20′45″N, long. 104°57′W, Lougheed Island, District of Franklin.

Tetracolpites sp.
Fig. spec. 27817
Brideaux, W.W., 1976, Geol. Surv. Can., Paper 71-1, pt.B, p. 89, pl. 1, fig. 2.
Basal bed, East Fork Formation, Late Cretaceous, Little Bear River, 24 miles southwest of Fort Norman, District of Mackenzie.

Tetraporate
Fig. spec. 93701
Sweet, A.R. and Braman, D.R., 1989, Geol. Surv. Can., Paper 89-8, pl. 1, fig. 36.
Chungo Member, Wapiabi Formation, Upper Cretaceous, Trap Creek near junction with Highwood River, sec. 36, tp. 17, rge. 4, W.5th mer., Alberta.

Tetraporites sp. 1 (cf. *Extratriporollenites* sp.)
Fig. specs. 38287, 38288
McIntyre, D.J., 1974, Geol. Surv. Can., Paper 74-14, pl. 22, figs. 9, 10.
Upper Cretaceous, lat. 69°28′N, long. 126°58′W, Horton River, District of Mackenzie.

Tetraporites sp. 2 (cf. Haloragaceae)
Fig. specs. 38289, 38290
McIntyre, D.J., 1974, Geol. Surv. Can., Paper 74-14, pl. 22, figs. 11, 12.
Upper Cretaceous, lat. 69°28′N, long. 126°58′W, Horton River, District of Mackenzie.

?*Thomsonipollis* sp.
Fig. spec. 76071
Jerzykiewicz, T. and Sweet, A.R., 1986, Can. J. Earth Sci., vol. 23, no. 9, pl. 4, fig. 11.
Late Cretaceous, depth 29.35-29.62 m, GSC core hole CV-42-2-2, Coal Valley strip mine, l.s.d. 1, sec. 10, rge. 19, W.5th, Alberta.

Tigrisporites scurrandus Norris
Hypotype 33988
Hopkins, Jr., W.S., 1974, Geol. Surv. Can., Paper 73-12, p. 19, pl. 5, fig. 54.
Christopher Formation, Lower Cretaceous, lat. 76°20′N, long. 115°34′W, northwest Melville Island, District of Franklin.

Todisporites major Couper
Hypotype 33964
Hopkins, Jr., W.S., 1974, Geol. Surv. Can., Paper 73-12, p. 13, pl. 2, fig. 26.
Christopher Formation, Lower Cretaceous, lat. 76°20′N, long. 115°34′W, northwestern Melville Island, District of Franklin.

Todisporites major Couper, 1958
Hypotypes 71264, 71265
Wilson, M.A., 1978, Palaeontographica, Abt. B, vol. 166, p. 110, pl. 1, figs. 2, 3.
Cretaceous, Police Island, north bank Mackenzie River, 12 miles southeast of Fort Norman, District of Mackenzie.

Todisporites minor Couper, 1958
Hypotype 71274
Wilson, M.A., 1978, Palaeontographica, Abt. B, vol. 166, p. 110, pl. 1, fig. 12.
Cretaceous, Police Island, north bank Mackenzie River, 12 miles southeast of Fort Norman, District of Mackenzie.

Todisporites minor Couper
Hypotype 42393
Brideaux, W.W. and McIntyre, D.J., 1975, Geol. Surv. Can., Bull. 252, pl. 1, fig. 5.
Crossley Lakes Member, Langton Bay Formation, Lower Cretaceous, lat. 69°04′N, long. 125°51′W, Horton River, District of Mackenzie.

Trichicola scollardiana Srivastava, 1972
Hypotype 91365
Sweet, A.R. and McIntyre, D.J., 1988, Can. Soc. Petrol. Geol., Mem. 15, fig. 6.13.
Moosehound Member, Cardium Formation, Upper Cretaceous, south side of Grande Cache Highway 40, l.s.d. 15, sec. 13, tp. 57, rge. 7, W.5th mer., Alberta.

Tricoloporoidites subtilis Pacltová, 1971
Hypotype 84752
Leckie, D.A. et al., 1990, J. Sed. Petrology, vol. 60, no. 1, p. 101, fig. 60.
Shaftesbury Formation, Lower Cretaceous, west bank Peace River near town of Peace River, Alberta.

Tricolpate pollen
Fig. specs. 48992, 48993
Sweet, A.R., 1978, Geol. Surv. Can., Paper 78-1B, pl. 6.2, figs. 6, 7.
Cuesta Creek Member, Tent Island Formation, Upper Cretaceous, northwestern bank Big Fish River near its junction with Boundary Creek, lat. 68°30′30″N, long. 136°23′50″W, Yukon.

Tricolpites bathyreticulatus Stanley, 1965
Hypotypes 71376, 71377
Wilson, M.A., 1978, Palaeontographica, Abt. B, vol. 166, p. 137, pl. 9, figs. 16-18.
Cretaceous, Police Island, north bank Mackenzie River, 12 miles southeast of Fort Norman, District of Mackenzie.

Tricolpites cellatus Norton *in* Norton and Hall 1967
Hypotype 76085
Jerzykiewicz, T. and Sweet, A.R., 1986, Can. J. Earth Sci., vol. 23, no. 9, pl. 4, fig. 25.
Late Cretaceous, depth 25.82-25.95 m, GSC core hole CV-42-2-1, Coal Valley strip mine, l.s.d. 1, sec. 6, tp. 47, rge. 19, W.5th mer., Alberta.

Tricolpites hians Stanley, 1965
Hypotype 71367
Wilson, M.A., 1978, Palaeontographica, Abt. B, vol. 166, p. 138, pl. 9, figs. 5, 6.
Cretaceous, Police Island, north bank Mackenzie River, 12 miles southeast of Fort Norman, District of Mackenzie.

Tricolpites interangulus Newman, 1965
Hypotype 89847
Sweet, A.R., Ricketts, B.D., Cameron, A.R. and Norris, D.K., 1989, Geol. Surv. Can., Paper 89-1G, p. 97, pl. 1, fig. 27.
Summit Creek Formation, Upper Cretaceous, north bank Mackenzie River immediately dowstream from Police Island, lat. 64°52′42″N, long. 125°14′44″W, District of Mackenzie.

Tricolpites micromunus (Groot and Penny) Singh, 1971
Hypotype 91418
Sweet, A.R. and McIntyre, D.J., 1988, Can. Soc. Petrol. Geol., Mem. 15, fig. 7.37.
Moosehound Member, Cardium Formation, Upper Cretaceous, south side of Grande Cache Highway 40, l.s.d. 15, sec. 13, tp. 57, rge. 7, W.5th mer., Alberta.

Tricolpites parvistriatus Norton 1967
Hypotype 76079
Jerzykiewicz, T. and Sweet, A.R., 1986, Can. J. Earth Sci., vol. 23, no. 9, pl. 4, fig. 19.
Late Cretaceous, depth 27.41-27.43 m, GSC core hole CV-42-2-2, Coal Valley strip mine, l.s.d. 1, sec. 10, tp. 47, rge. 19, W.5th mer., Alberta.

Tricolpites parvistriatus Norton *in* Norton and Hall, 1967
Hypotype 89862
Sweet, A.R., Ricketts, B.D., Cameron, A.R. and Norris, D.K., 1989, Geol. Surv. Can., Paper 89-1G, pl. 2, fig. 4.
Summit Creek Formation, Upper Cretaceous, north bank Mackenzie River immediately dowstream from Police Island, lat. 64°52′42″N, long. 125°14′44″W, District of Mackenzie.

Tricolpites sp.
Fig. spec. 31791
Hopkins, Jr., W.S. and Balkwill, H.R., 1973, Geol. Surv. Can., Paper 72-37, p. 22, pl. 2, fig. 53.
Hassel Formation, Cretaceous, near Cape Cairo, lat. 78°43′N, long. 100°06′W, northeastern Ellef Ringnes Island, District of Franklin.

Tricolpites sp.
Fig. spec. 93694
Sweet, A.R. and Braman, D.R., 1989, Geol. Surv. Can., Paper 89-8, pl. 1, fig. 29.
Chungo Member, Wapiabi Formation, Upper Cretaceous, Trap Creek near junction with Highwood River, sec. 36, tp. 17, rge. 4, W.5th mer., Alberta.

Tricolpites sp. 1
Fig. specs. 38344, 38345
McIntyre, D.J., 1974, Geol. Surv. Can., Paper 74-14, pl. 24, figs. 20, 21.
Upper Cretaceous, lat. 69°27′30″N, long. 126°58′W, Horton River, District of Mackenzie.

Tricolpites sp. 2
Fig. specs. 38317, 38329
McIntyre, D.J., 1974, Geol. Surv. Can., Paper 74-14, pl. 13, fig. 13; pl. 24, fig. 5.
Upper Cretaceous, lat. 69°28′N, long. 126°58′W, Horton River, District of Mackenzie.

Tricolpites sp. 3
Fig. specs. 38309, 38327
McIntyre, D.J., 1974, Geol. Surv. Can., Paper 74-14, pl. 23, fig. 6; pl. 24, fig. 3.
Upper Cretaceous, lat. 69°27′30″N, long. 126°58′W, and lat. 69°28′N, long. 126°58′W, Horton River, District of Mackenzie.

Tricolpites sp. 4
Fig. spec. 38315
McIntyre, D.J., 1974, Geol. Surv. Can., Paper 74-14, pl. 23, fig. 11.
Upper Cretaceous, lat. 69°27′30″N, long. 126°58′W, Horton River, District of Mackenzie.

Tricolpites sp. 5
Fig. specs. 38311-38314
McIntyre, D.J., 1974, Geol. Surv. Can., Paper 74-14, pl. 23, figs. 8-10, 17.
Upper Cretaceous, lat. 69°27′30″N, long. 126°58′W, and lat. 69°28′N, long. 126°54′W(38312, 38314), Horton River, District of Mackenzie.

Tricolpites sp. 6
Fig. specs. 38325, 38326
McIntyre, D.J., 1974, Geol. Surv. Can., Paper 74-14, pl. 24, figs. 2, 8.
Upper Cretaceous, lat. 69°27′30″N, long. 126°58′W, Horton River, District of Mackenzie.

Tricolpites sp. 7, 10
Fig. specs. 38328, 38337
McIntyre, D.J., 1974, Geol. Surv. Can., Paper 74-14, pl. 24, figs. 4, 12.
Upper Cretaceous, lat. 69°28′N, long. 126°58′W, Horton River, District of Mackenzie.

Tricolpites sp. 8
Fig. spec. 38324
McIntyre, D.J., 1974, Geol. Surv. Can., Paper 74-14, pl. 24, fig. 1.
Upper Cretaceous, lat. 69°28′N, long. 126°58′W, Horton River, District of Mackenzie.

Tricolpites sp. 9
Fig. specs. 38330-38333
McIntyre, D.J., 1974, Geol. Surv. Can., Paper 74-14, pl. 24, figs. 6, 9, 10, 13.
Upper Cretaceous, lat. 69°28′N, long. 126°58′W, Horton River, District of Mackenzie.

Tricolpites sp. 11
Fig. specs. 38322, 38323
McIntyre, D.J., 1974, Geol. Surv. Can., Paper 74-14, pl. 23, figs. 19, 20.
Upper Cretaceous, lat. 69°58′30″N, long. 127°04′W, Horton River, District of Mackenzie.

Tricolpites sp. 12
Fig. spec. 38316
McIntyre, D.J., 1974, Geol. Surv. Can., Paper 74-14, pl. 23, fig. 12.
Upper Cretaceous, lat. 69°58′30″N, long. 127°04′W, Horton River, District of Mackenzie.

Tricolpites sp. 13
Fig. specs. 38319, 38320, 38336
McIntyre, D.J., 1974, Geol. Surv. Can., Paper 74-14, pl. 23, figs. 15, 16; pl. 24, fig. 7.
Upper Cretaceous, lat. 69°28′N, long. 126°58′W, Horton River, District of Mackenzie.

Tricolpites sp. "A" Oltz 1969
Fig. spec. 76086
Jerzykiewicz, T. and Sweet, A.R., 1986, Can. J. Earth Sci., vol. 23, no. 9, pl. 4, fig. 26.
Late Cretaceous, right bank of Embarras River towards the south end of Robb, l.s.d. 14, sec. 11, tp. 49, rge. 21, W.5th mer., Alberta.

Tricolpites sp. A *in* Nichols and Jacobson, 1982
Fig. spec. 93693
Sweet, A.R. and Braman, D.R., 1989, Geol. Surv. Can., Paper 89-8, pl. 1, fig. 28.
Chungo Member, Wapiabi Formation, Upper Cretaceous, Trap Creek near junction with Highwood River, sec. 36, tp. 17, rge. 4, W.5th mer., Alberta.

Tricolpites sp. A, B
Fig. specs. 34007-34009
Hopkins, Jr., W.S., 1974, Geol. Surv. Can., Paper 73-12, p. 24, pl. 6, figs. 75-77.
Christopher Formation, Lower Cretaceous, lat. 76°20′N, long. 115°34′W, northwestern Melville Island, District of Franklin.

Tricolpopollenites cf. *T. crassimurus* Groot and Penny
Hypotype 31786
Hopkins, Jr., W.S. and Balkwill, H.R., 1973, Geol. Surv. Can., Paper 72-37, p. 21, pl. 2, fig. 48.
Hassel Formation, Cretaceous, Hoodoo Dome, lat. 78°11′N, long. 99°40′W, southern Ellef Ringnes Island, District of Franklin.

Tricolpopollenites cf. *T. elongatus* Groot and Groot
Hypotype 31787
Hopkins, Jr., W.S. and Balkwill, H.R., 1973, Geol. Surv. Can., Paper 72-37, p. 22, pl. 2, fig. 49.
Hassel Formation, Cretaceous, Hoodoo Dome, lat. 78°11′N, long. 99°40′W, southern Ellef Ringnes Island, District of Franklin.

Tricolpopollenites cf. *T. parvulus* Groot and Penny
Hypotype 31788
Hopkins, Jr., W.S. and Balkwill, H.R., 1973, Geol. Surv. Can., Paper 72-37, p. 22, pl. 2, fig. 50.
Hassel Formation, Cretaceous, Hoodoo Dome, lat. 78°11′N, long. 99°40′W, southern Ellef Ringnes Island, District of Franklin.

Tricolpopollenites sp.
Fig. spec. 89705
Jerzykiewicz, T. and Sweet, A.R., 1988, Sedimentary Geol., vol. 59, no. 1/2, p. 46, pl. I, fig. 9.
Belly River Formation, Upper Cretaceous, Crownest River valley west of Lundbreck, lat. 49°35′N, long. 114°11′W, Alberta.

Tricolporate angiosperm pollen
Fig. specs. 93696, 93697
Sweet, A.R. and Braman, D.R., 1989, Geol. Surv. Can., Paper 89-8, p. 35, pl. 1, figs. 31, 32.
Chungo Member, Wapiabi Formation, Upper Cretaceous, Trap Creek near junction with Highwood River, sec. 36, tp. 17, rge. 4, W.5th mer., Alberta.

Tricolporites spp.
Fig. specs. 89702, 89703
Jerzykiewicz, T. and Sweet, A.R., 1988, Sedimentary Geol., vol. 59, no. 1/2, p. 46, pl. I, figs. 6, 7.
Belly River Formation, Upper Cretaceous, Crownest River valley west of Lundbreck, lat. 49°35′N, long. 114°11′W, Alberta.

Tricolporopollenites sp. 1, 3, 4, 6
Fig. specs. 38334, 38338, 38341, 38343
McIntyre, D.J., 1974, Geol. Surv. Can., Paper 74-14, pl. 24, figs. 11, 14, 17, 19.
Upper Cretaceous, lat. 69°28′N, long. 126°58′W, Horton River, District of Mackenzie.

Tricolporopollenites sp. 2 *in* Norris et al. (1975)
Fig. spec. 89711
Jerzykiewicz, T. and Sweet, A.R., 1988, Sedimentary Geol., vol. 59, no. 1/2, p.58, pl. I, fig. 15.
St. Mary River Formation, Upper Cretaceous, lat. 49°30′N, long. 114°3′W, Alberta.

Tricolporopollenites sp. 5
Fig. specs. 38339, 38340
McIntyre, D.J., 1974, Geol. Surv. Can., Paper 74-14, pl. 24, figs. 15, 16.
Upper Cretaceous, lat. 69°28′N, long. 126°58′W, and lat. 69°58′30″N, long. 127°04′W, Horton River, District of Mackenzie.

Tricolporopollenites sp. 7
Fig. spec. 38342
McIntyre, D.J., 1974, Geol. Surv. Can., Paper 74-14, pl. 24, fig. 18.
Upper Cretaceous, lat. 69°58′30″N, long. 127°04′W, Horton River, District of Mackenzie.

Trifossapollenites ellipticus Rouse, 1957
Hypotype 38310
McIntyre, D.J., 1974, Geol. Surv. Can., Paper 74-14, pl. 23, fig. 7.
Upper Cretaceous, lat. 69°27′30″N, long. 126°58′W, Horton River, District of Mackenzie.

Trigonopyxidia ginella (Cookson and Eisenack) Cookson and Eisenack, 1961
Hypotype 38113
McIntyre, D.J., 1974, Geol. Surv. Can., Paper 74-14, pl. 8, fig. 19.
Upper Cretaceous, lat. 69°27′30″N, long. 126°58′W, Horton River, District of Mackenzie.

Trigrisporites scurrandus Norris
Hypotypes 42417-42419
Brideaux, W.W. and McIntyre, D.J., 1975, Geol. Surv. Can., Bull. 252, pl. 1, figs. 30-32.
Horton River Formation, Lower Cretaceous, lat. 69°19′N, long. 126°50′W, Horton River, District of Mackenzie.

Trilete A
Fig. spec. 25092
Hopkins, Jr., W.S., 1971, Geol. Surv. Can., Bull 197, p. 120, pl. 21, fig. 21.
Isachsen Formation, Lower Cretaceous, lat. 76°20′N, long. 115°34′W, northwest Melville Island, District of Franklin.

Triletes equatibossus Couper, 1958
Hypotype 13947
McGregor, D.C., 1965, Geol. Surv. Can., Paper 64-55, pl. 10, fig. 10.
Lower Cretaceous, Peel Plateau, Stony Creek about 18.5 miles upstream from mouth, District of Mackenzie.

Triletes lanaris Vangerow
Hypotype 41648
Hopkins, Jr., W.S. and Sweet, A.R., 1976, Geol. Surv. Can., Bull. 256, pl. 12, fig. 5.
Mattagami Formation, Lower Cretaceous, borehole 68-42, lat. 50°35′36″N, long. 82°30′12″W, Ontario.

Triletes pileatus Dijkstra 1959
Hypotype 48929
Sweet, A.R., 1979, Am. Assoc. Strat. Palynol., Contrib. Ser. No. 5B, pl. 3, fig. 4.
Christopher Formation, Lower Cretaceous, lat. 79°47′24″N, long. 85°22′W, near Slidre Fiord, Ellesmere Island, District of Franklin.

?*Triletes verrucatus* Couper, 1963
Hypotype 13873
McGregor, D.C., 1965, Geol. Surv. Can., Paper 64-55, pl. 7, fig. 18.
Lower Cretaceous, Donna River Canyon, Aklavik Range, Richardson Mountains, District of Mackenzie.

cf. *Triletes verrucatus* Couper in Reyre, 1964
Hypotypes 13867, 13868
McGregor, D.C., 1965, Geol. Surv. Can., Paper 64-55, pl. 7, figs. 9, 15.
Lower Cretaceous, Donna River Canyon, Aklavik Range, Richardson Mountains, District of Mackenzie.

Triletes sp. (sensu Dettmann, 1963)
Fig. spec. 13871
McGregor, D.C., 1965, Geol. Surv. Can., Paper 64-55, pl. 7, fig. 16.
Lower Cretaceous, Donna River Canyon, Aklavik Range, Richardson Mountains, District of Mackenzie.

Triletes sp.
Fig. spec. 31771
Hopkins, Jr., W.S. and Balkwill, H.R., 1973, Geol. Surv. Can., Paper 72-37, p. 16, pl. 2, fig. 33.
Hassel Formation, Cretaceous, Hoodoo Dome, lat. 78°11′N, long. 99°44′W, southern Ellef Ringnes Island, District of Franklin.

Triletes sp.
Fig. specs. 41646, 41647
Hopkins, Jr., W.S. and Sweet, A.R., 1976, Geol. Surv. Can., Bull. 256, pl. 11, figs. 2, 3.
Mattagami Formation, Lower Cretaceous, borehole 68-43, lat. 50°35′26″N, long. 82°28′56″W and borehole 68-38, lat. 50°36′N, long. 82°31′12″W, , Ontario.

Triletes sp. A
Fig. specs. 48957, 48958
Sweet, A.R., 1979, Am. Assoc. Strat. Palynol., Contrib. Ser. No. 5B, pl. 4, figs. 15, 16.
Eureka Sound Formation, Upper Cretacous, Meteorologist Peninsula, lat. 78°10′N, long. 99°30′W, Ellef Ringnes Island, Arctic.

Triletes sp. A
Fig. spec. 56785
Williams, G.L. and Bujak, J.P., 1980, Initial Repts. Deep Sea Drilling Project, vol. 50, pl. 11, fig. 4.

Early Cretaceous, DSDP Site 416, lat. 32°50.18′N, long. 10°48.06′W, northwestern Moroccan continental margin.

cf. *Trilites sinuatus* Couper, 1953
Hypotype 13825
McGregor, D.C., 1965, Geol. Surv. Can., Paper 64-55, pl. 6, fig. 3.
Lower Cretaceous, Fault Creek Canyon, Aklavik Range, Richardson Mountains, District of Mackenzie.

Trilites verrucatus Couper, 1953
Hypotype 13843
McGregor, D.C., 1965, Geol. Surv. Can., Paper 64-55, pl. 6, fig. 25.
Lower Cretaceous, Fault Creek Canyon, Aklavik Range, Richardson Mountains, District of Mackenzie.

Trilobosporites apiverrucatus Couper, 1958
Hypotypes 13841, 13842
McGregor, D.C., 1965, Geol. Surv. Can., Paper 64-55, pl. 6, figs. 23, 26.
Lower Cretaceous, Fault Creek Canyon, Aklavik Range, Richardson Mountains, District of Mackenzie.

Trilobosporites bernissartensis (Delcourt and Sprumont) Potonié, 1956
Hypotypes 13839, 13840
McGregor, D.C., 1965, Geol. Surv. Can., Paper 64-55, pl. 6, figs. 21, 22, 24.
Lower Cretaceous, Fault Creek Canyon, Aklavik Range, Richardson Mountains, District of Mackenzie.

Trilobosporites sp. cf. *T. aequiverrucosus* Dörhöfer 1977
Fig. spec. 72106
Fensome, R.A., 1987, Palaeontographica Canadiana No. 4, p. 25, pl. 5, fig. 12.
Husky Formation, Cretaceous, north bank of first creek south of Mt. Gifford, District of Mackenzie.

Trilobosporites aornatus Döring, 1965
Hypotype 58561
McIntyre, D.J. and Brideaux, W.W., 1980, Geol. Surv. Can., Bull. 320, p. 11, pl. 1, fig. 5.
Lower Cretaceous, 'Grizzly Gorge', lat. 68°15′49″N, long. 135°40′59″W, Richardson Mountains, District of Mackenzie.

Trilobosporites aornatus Döring, 1965
Hypotype 68226
Ricketts, B.D. and Sweet, A.R., 1986, Geol. Surv. Can., Paper 84-15, p. 21, pl. 3, fig. 9.
Elk Formation, Cretaceous, Coal Creek, lat. 40°30′N, long. 114°59′W, British Columbia.

Trilobosporites aornatus Döring, 1965
Hypotypes 72099, 72100
Fensome, R.A., 1987, Palaeontographica Canadiana No. 4, p. 25, pl. 5, figs. 4, 7.
Husky Formation, Cretaceous, north bank of first creek south of Mt. Gifford, District of Mackenzie.

Trilobosporites apibaculatus Brenner
Hypotypes 42447
Brideaux, W.W. and McIntyre, D.J., 1975, Geol. Surv. Can., Bull. 252, pl. 2, fig. 19.
Crossley Lakes Member, Langton Bay Formation, Lower Cretaceous, lat. 69°04′N, long. 125°51′W, Horton River, District of Mackenzie.

Trilobosporites apiverrucatus Couper
Hypotypes 25079, 25080
Hopkins, Jr., W.S., 1971, Geol. Surv. Can., Bull. 197, p. 118, pl. 21, figs. 7, 8.
Isachsen Formation, Lower Cretaceous, lat. 76°20′N, long. 115°34′W, northwest Melville Island, District of Franklin.

Trilobosporites apiverrucatus Couper
Hypotype 31760
Hopkins, Jr., W.S. and Balkwill, H.R., 1973, Geol. Surv. Can., Paper 72-37, p. 13, pl. 1, fig. 22.
Hassel Formation, Cretaceous, near Cape Cairo, lat. 78°43′N, long. 100°06′W, northeastern Ellef Ringnes Island, District of Franklin.

Trilobosporites apiverrucatus Couper
Hypotype 33989
Hopkins, Jr., W.S., 1974, Geol. Surv. Can., Paper 73-12, p. 20, pl. 5, fig. 55.
Christopher Formation, Lower Cretaceous, lat. 76°20′N, long. 115°34′W, northwestern Melville Island, District of Franklin.

Trilobosporites apiverrucatus Couper
Hypotype 41902
Williams, G.L., 1975, Geol. Surv. Can., Paper 74-30, vol. 2, pl. 8, fig. 7.
Aptian, Lower Cretaceous, Shell Oneida O-25 well, Scotian Shelf.

Trilobosporites apiverrucatus Couper
Hypotypes 41598, 41599
Hopkins, Jr., W.S. and Sweet, A.R., 1976, Geol. Surv. Can., Bull. 256, pl. 9, figs. 14, 15.
Mattagami Formation, Lower Cretaceous, Onakawana lignite pit, lat. 50°36′N, long. 81°23′W, and borehole 68-39, lat. 50°35′48″N, long. 82°30′42″W, Ontario.

Trilobosporites bernissartensis (Del. and Sprum.) Potonié
Hypotype 25081
Hopkins, Jr., W.S., 1971, Geol. Surv. Can., Bull. 197, p. 118, pl. 21, fig. 9.
Isachsen Formation, Lower Cretaceous, lat. 76°20′N, long. 115°34′W, northwest Melville Island, District of Franklin.

cf. *Trilobosporites bernissartensis* (Delcourt and Sprumont) Potonié in Delcourt et al., 1963
Hypotype 13922
McGregor, D.C., 1965, Geol. Surv. Can., Paper 64-55, pl. 8, fig. 22.
Isachsen Formation, Lower Cretaceous, south of Buchanan Lake anticline, eastern Axel Heiberg Island, District of Franklin.

Trilobosporites sp. cf. *T. bernissartensis* (Delcourt and Sprumont) Potonié, 1956
Fig. spec. 68222
Ricketts, B.D. and Sweet, A.R., 1986, Geol. Surv. Can., Paper 84-15, p. 21, pl. 3, fig. 5.
Elk Formation, Cretaceous, Coal Creek, lat. 40°30′N, long. 114°59′W, British Columbia.

Trilobosporites canadensis Pocock
Hypotype 25082
Hopkins, Jr., W.S., 1971, Geol. Surv. Can., Bull. 197, p. 119, pl. 21, fig. 10.
Isachsen Formation, Lower Cretaceous, lat. 76°20′N, long. 115°34′W, northwest Melville Island, District of Franklin.

Trilobosporites canadensis Pocock
Hypotype 33990
Hopkins, Jr., W.S., 1974, Geol. Surv. Can., Paper 73-12, p. 20, pl. 5, fig. 56.
Christopher Formation, Lower Cretaceous, lat. 76°20′N, long. 115°34′W, northwest Melville Island, District of Franklin.

Trilobosporites? canadensis Pocock 1962
Hypotype 72101
Fensome, R.A., 1988, Palaeontographica Canadiana No. 4, p. 25, pl. 5, fig. 5.
Husky Formation, Cretaceous, north bank of first creek south Mt. Gifford, District of Mackenzie.

Trilosboporites canadensis Pocock, 1962
Hypotypes 68223-68225
Ricketts, B.D. and Sweet, A.R., 1986, Geol. Surv. Can., Paper 84-15, p. 21, pl. 3, figs. 6-8.
Elk Formation, Cretaceous, Coal Creek, lat. 40°30′N, long. 114°59′W, British Columbia.

Trilobosporites crassus Brenner
Hypotype 76596
Davies, E.H., Akande, S.O. and Zentilli, M., 1984, Geol. Surv. Can., Paper 84-1A, pl. 48.1, fig. 7.
Early Cretaceous, sample AGR 285-230, Gays River Mine, lat. 45°02′N, long. 63°22′W, Nova Scotia.

Trilobosporites granulatus Döring 1965
Hypotype 72104
Fensome, R.A., 1988, Palaeontographica Canadiana No. 4, p. 26, pl. 5, fig. 10.
Lower Canyon Formation, Cretaceous, Martin Creek, south bank, lat. 68°12′N, long. 135°34′W, District of Mackenzie.

Trilobosporites sp. cf. *T. granulatus* Döring, 1965
Fig. spec. 68229
Ricketts, B.D. and Sweet, A.R., 1986, Geol. Surv. Can., Paper 84-15, p. 22, pl. 3, fig. 12.
Elk Formation, Cretaceous, Coal Creek, lat. 40°30′N, long. 114°59′W, British Columbia.

Trilobosporites hannonicus (Delcourt and Sprumont) Potonié, 1956
Hypotypes 68227, 68228
Ricketts, B.D. and Sweet, A.R., 1986, Geol. Surv. Can., Paper 84-15, p. 22, pl. 3, figs. 10, 11.
Elk Formation, Cretaceous, Coal Creek, lat. 40°30′N, long. 114°59′W, British Columbia.

Trilobosporites sp. cf. *T. humilis*
Fig. spec. 91981
Banerjee, I. and Davies, E.H., 1988, Can. Soc. Petrol. Geol., Mem. 15, fig. 6.9.
Mannville Group, Lower Cretaceous, depth 4356 feet, well 3-22-46-23W4, Alberta.

Trilobosporites jurassicus Pocock
Hypotype 56786
Williams, G.L. and Bujak, J.P., 1980, Initial Repts. Deep Sea Drilling Project, vol. 50, pl. 11, fig. 5.
Early Cretaceous, DSDP Site 416, lat. 32°50.18′N, long. 10°48.06′W, northwestern Moroccan continental margin.

Trilobosporites marylandensis Brenner
Hypotype 42448
Brideaux, W.W. and McIntyre, D.J., 1975, Geol. Surv. Can., Bull. 252, pl. 2, fig. 20.
Crossley Lakes Member, Langton Bay Formation, Lower Cretaceous, lat. 69°04′N, long. 125°51′W, Horton River, District of Mackenzie.

Trilobosporites obsitus Norris, 1969
Hypotypes 68220, 68221
Ricketts, B.D. and Sweet, A.R., 1986, Geol. Surv. Can., Paper 84-15, p. 22, pl. 3, figs. 2-4.
Elk Formation, Cretaceous, Coal Creek, lat. 40°30′N, long. 114°59′W, British Columbia.

Trilobosporites perverulentus (Verbitskaya) Dettmann
Hypotypes 42449, 42450
Brideaux, W.W. and McIntyre, D.J., 1975, Geol. Surv. Can., Bull. 252, pl. 2, figs. 21, 22.
Horton River Formation and Crossley Lakes Member, Langton Bay Formation, Lower Cretaceous, lat. 69°19′N, long. 126°50′W and lat. 69°04′N, long. 125°51′W, Horton River, District of Mackenzie.

Trilobosporites purverulentus (Verbitskaya)
Hypotype 41905
Williams, G.L., 1975, Geol. Surv. Can., Paper 74-30, vol. 2, pl. 8, fig. 10.
Barremian, Lower Cretaceous, Shell Naskapi N-30 well, Scotian Shelf.

Trilobosporites purverulentus (Verbitskaja) Dettmann
Hypotype 76598
Davies, E.H., Akande, S.O. and Zentilli, M., 1984, Geol. Surv. Can., Paper 84-1A, pl. 48.1, fig. 9.
Early Cretaceous, sample DH2-409, Gays River Mine, lat. 45°02′N, long. 63°22′W, Nova Scotia.

Trilobosporites cf. *T. purverulentus* (Verbitskaya) Dettmann
Hypotype 33991
Hopkins, Jr., W.S., 1974, Geol. Surv. Can., Paper 73-12, p. 20, pl. 5, fig. 57.

Christopher Formation, Lower Cretaceous, lat. 76°20′N, long. 115°34′W, northwestern Melville Island, District of Franklin.

Trilobosporites trioreticulatus Cookson and Dettmann
Hypotype 76597
Davies, E.H., Akande, S.O. and Zentilli, M., 1984, Geol. Surv. Can., Paper 84-1A, pl. 48.1, fig. 8.
Early Cretaceous, Gays River mine, lat. 45°02′N, long. 63°22′W, Nova Scotia.

Trilobosporites trioreticulosus Cookson and Dettmann
Hypotype 42451
Brideaux, W.W. and McIntyre, D.J., 1975, Geol. Surv. Can., Bull. 252, pl. 2, figs. 23.
Crossley Lakes Member, Langton Bay Formation, Lower Cretaceous, lat. 69°04′N, long. 125°51′W, Horton River, District of Mackenzie.

Trilobosporites sp. cf. *T. verrucatus* (Lantz 1985) comb. nov.
Fig. spec. 72103
Fensome, R.A., 1988, Palaeontographica Canadiana No. 4, p. 26, pl. 5, fig. 9.
Husky Formation, Cretaceous, Martin Creek, south side, 0.8 miles north of small lake, lat. 68°12′20″N, long. 135°36′W, District of Mackenzie.

Trilobosporites sp.
Fig. spec. 25083
Hopkins, Jr., W.S., 1974, Geol. Surv. Can., Bull. 197, p. 119, pl. 21, fig. 11.
Isachsen Formation, Lower Cretaceous, lat. 76°20′N, long. 115°34′W, northwest Melville Island, District of Franklin.

Trilobosporites sp.
Fig. spec. 33992
Hopkins, Jr., W.S., 1974, Geol. Surv. Can., Paper 73-12, p. 20, pl. 5, figs. 58, 59.
Christopher Formation, Lower Cretaceous, lat. 76°20′N, long. 115°34′W, northwest Melville Island, District of Franklin.

Trilobosporites? sp. indet.
Fig. specs. 72097, 72098
Fensome, R.A., 1988, Palaeontographica Canadiana No. 4, pl. 5, figs. 3, 6.
Fossil Creek Formation, Cretaceous, Martin Creek, north bank, lat. 68°12′17″N, long. 135°34′38″W; Husky Formation, north bank of first creek south of Mt. Gifford, District of Mackenzie.

Trilobosporites sp. A
Fig. spec. 72105
Fensome, R.A., 1988, Palaeontographica Canadiana No. 4, p. 27, pl. 5, fig. 11.
Husky Formation, Cretaceous, north bank of first south of Mt. Gifford, District of Mackenzie.

Trilobosporites sp. B
Fig. spec. 72102
Fensome, R.A., 1988, Palaeontographica Canadiana No. 4, p. 27, pl. 5, fig. 8.
Martin Creek Formation, Cretaceous, Martin Creek, north bank, lat. 68°12′17″N, long. 135°34′38″W, District of Mackenzie.

Tripartites incisotrilobus (Naumova) Potonié and Kremp, 1956
Hypotype 71294
Wilson, M.A., 1978, Palaeontographica, Abt. B, vol. 166, p. 157, pl. 3, fig. 9.
Cretaceous, Police Island, north bank Mackenzie River, 12 miles southeast of Fort Norman, District of Mackenzie.

triporate(?)
Fig. spec. 91408
Sweet, A.R. and McIntyre, D.J., 1988, Can. Soc. Petrol. Geol., Mem. 15, fig. 7.27.
Moosehound Member, Cardium Formation, Upper Cretaceous, north side of Highway 47 north Cadomin, l.s.d. 14, sec. 16, tp. 47, rge. 23, W.5th mer., Alberta.

Triporoletes cenomanianus (Agasie) Srivastava, 1975
Hypotype 91359
Sweet, A.R. and McIntyre, D.J., 1988, Can. Soc. Petrol. Geol., Mem. 15, fig. 6.7.
Moosehound Member, Cardium Formation, Upper Cretaceous, north side of Highway 47, l.s.d. 14, sec. 16, tp. 47, rge. 23, W.5th mer., Alberta.

Triporoletes incertus McIntyre and Brideaux
Holotype 58571; paratypes 58572-58576
McIntyre, D.J. and Brideaux, W.W., 1980, Geol. Surv. Can., Bull. 320, p. 11, pl. 1, figs. 16-26.
Lower Cretaceous, 'Grizzly Gorge', lat. 68°15′49″N, long. 135°40′59″W and Martin Creek, lat. 68°12′10″N, long. 135°34′50″W (58574, 58575), Richardson Mountains, District of Mackenzie.

Triporoletes radiatus (Dettmann) Playford
Hypotype 33941
Hopkins, Jr., W.S., 1974, Geol. Surv. Can., Paper 73-12, p. 8, pl. 1, fig. 1.
Christopher Formation, Lower Cretaceous, lat. 76°20′N, long. 115°34′W, northwestern Melville Island, District of Franklin.

Triporoletes radiatus (Dettmann) Playford
Hypotypes 42476-42478
Brideaux, W.W. and McIntyre, D.J., 1975, Geol. Surv. Can., Bull. 252, pl. 3, figs. 10-12.
Horton River Formation, Lower Cretaceous, lat. 69°19′N, long. 126°50′W, Horton River, District of Mackenzie.

Triporoletes reticulatus (Pocock) Playford
Hypotype 33942
Hopkins, Jr., W.S., 1974, Geol. Surv. Can., Paper 73-12, p. 8, pl. 1, fig. 2.
Christopher Formation, Lower Cretaceous, lat. 76°20′N, long. 115°34′W, northwest Melville Island, District of Franklin.

Triporoletes reticulatus (Pocock) Playford
Hypotypes 42479, 42480
Brideaux, W.W. and McIntyre, D.J., 1975, Geol. Surv. Can., Bull. 252, pl. 3, figs. 13, 14.
Gilmore Lake Member, Langton Bay Formation and Horton River Formation, Lower Cretaceous, lat. 68°50′N, long. 125°24′W and lat. 69°19′N, long. 126°50′W, Horton River, District of Mackenzie.

Triporoletes reticulatus (Pocock) Playford, 1971
Hypotype 91364
Sweet, A.R. and McIntyre, D.J., 1988, Can. Soc. Petrol. Geol., Mem. 15, fig. 6.12.
Moosehound Member, Cardium Formation, Upper Cretaceous, west bank of Little Berland River, l.s.d. 1, sec. 6, tp. 54, rge. 2, W.5th mer., Alberta.

Triporoletes simplex (Cookson and Dettmann) PLayford
Hypotype 42481
Brideaux, W.W. and McIntyre, D.J., 1975, Geol. Surv. Can., Bull. 252, pl. 3, fig. 15.
Horton River Formation, Lower Cretaceous, lat. 69°19′N, long. 126°50′W, Horton River, District of Mackenzie.

Triporoletes singularis Mtchedlishvili
Hypotypes 42482-42484
Brideaux, W.W. and McIntyre, D.J., 1975, Geol. Surv. Can., Bull. 252, pl. 3, figs. 16-18.
Horton River Formation, Lower Cretaceous, lat. 69°19′N, long. 126°50′W, Horton River, District of Mackenzie.

Triporopollenites sp.
Fig. spec. 27818
Brideaux, W.W., 1971, Geol. Surv. Can., Paper 71-1, pt.B, p. 89, pl. 1, fig. 3.
Basal bed, East Fork Formation, Late Cretaceous, Little Bear River, 24 miles southwest of Fort Norman, District of Mackenzie.

Triporopollenites sp. 1 (cf. *Betula)*
Fig. spec. 38291
McIntyre, D.J., 1974, Geol. Surv. Can., Paper 74-14, pl. 22, fig. 13.
Upper Cretaceous, lat. 69°58′30″N, long. 127°04′W, Horton River, District of Mackenzie.

Triporopollenites sp. 2
Fig. spec. 38308
McIntyre, D.J., 1974, Geol. Surv. Can., Paper 74-14, pl. 23, fig. 5.
Upper Cretaceous, lat. 69°27′30″N, long. 126°58′W, Horton River, District of Mackenzie.

Triporopollenites sp. 3
Fig. specs. 38285, 38286
McIntyre, D.J., 1974, Geol. Surv. Can., Paper 74-14, pl. 22, figs. 7, 8.
Upper Cretaceous, lat. 69°28′N, long. 126°58′W, Horton River, District of Mackenzie.

Triporopollenites sp. 4, 5
Fig. specs. 38306, 38321
McIntyre, D.J., 1974, Geol. Surv. Can., Paper 74-14, pl. 23, figs. 3, 18.
Upper Cretaceous, lat. 69°28′N, long. 126°58′W, Horton River, District of Mackenzie.

Triprojectus magnus (Mchedlishvili) Stanley, 1970
Hypotypes 38247-38249
McIntyre, D.J., 1974, Geol. Surv. Can., Paper 74-14, pl. 18, figs. 3-5.
Upper Cretaceous, lat. 69°28′N, long. 126°58′W, Horton River, District of Mackenzie.

Triprojectus sp. 2
Fig. spec. 38250
McIntyre, D.J., 1974, Geol. Surv. Can., Paper 74-14, pl. 18, fig. 6.
Upper Cretaceous, lat. 69°28′N, long. 126°58′W, Horton River, District of Mackenzie.

Triprojectus unicus (Chlonova) Mchedlishvili, 1961
Hypotype 93648
Sweet, A.R., Ricketts, B.D., Cameron, A.R. and Norris, D.K., 1989, Geol. Surv. Can., Paper 89-1G, p. 93, pl. 1, fig. 18.
East Fork Formation, Upper Cretaceous, near junction of East Little Bear and Little Bear rivers, lat. 64°46′55″N, long. 126°2′17″W, District of Mackenzie.

Triquitrites batillatus Hughes and Playford, 1961
Hypotype 71295
Wilson, M.A., 1978, Palaeontographica, Abt. B, vol. 166, p. 158, pl. 3, figs. 10, 11.
Cretaceous, Police Island, north bank Mackenzie River, 12 miles southeast of Fort Norman, District of Mackenzie.

Trithyrodinium ornatum Felix and Burbridge, 1973
Hypotypes 38053-38056
McIntyre, D.J., 1974, Geol. Surv. Can., Paper 74-14, pl. 5, figs. 4-7.
McIntyre, D.J., 1975, Geoscience and Man, vol. 11, p. 69, pl. 1, figs. 9(38054), 10(38053).
Upper Cretaceous, lat. 69°27′30″N, long. 126°58′W, Horton River, District of Mackenzie.

Trithyrodinium suspectum (Manum and Cookson) Davey, 1969
Hypotypes 38042-38044
McIntyre, D.J., 1974, Geol. Surv. Can., Paper 74-14, pl. 3, figs. 8-10.
Upper Cretaceous, lat. 69°27′30″N, long. 126°58′W, Horton River, District of Mackenzie.

Trithyrodinium suspectum (Manum and Cookson)
Hypotype 34016, 34017
McIntyre, D.J., 1975, Geoscience and Man, vol. 11, p. 70, pl. 1, figs. 11, 12.
Upper Cretaceous, lat. 69°27′30″N, long. 126°58′W, Horton River, District of Mackenzie.

Trudopollis sp.
Fig. spec. 93646
Sweet, A.R., Ricketts, B.D., Cameron, A.R. and Norris, D.K., 1989, Geol. Surv. Can., Paper 89-1G, p. 93, pl. 1, fig. 16.
Little Bear Formation, Upper Cretaceous, east of McKay Range, East Little Bear River area, lat. 64°36′24″N, long. 125°35′W, District of Mackenzie.

Tsugaepollenites cf. *T. dampieri* (Balme) Dettmann
Hypotype 25102
Hopkins, Jr., W.S., 1971, Geol. Surv. Can., Bull. 197, p. 122, pl. 22, fig. 11.
Isachsen Formation, Lower Cretaceous, lat. 76°20′N, long. 115°34′W, northwest Melville Island, District of Franklin.

Tsugaepollenites mesozoicus Couper, 1958
Hypotypes 13848, 13885
McGregor, D.C., 1965, Geol. Surv. Can., Paper 64-55, pl. 6, fig. 31; pl. 7, fig. 34.
Lower Cretaceous, Fault Creek Canyon and Donna River Canyon, Aklavik Range, Richardson Mountains, District of Mackenzie.

Tsugaepollenites mesozoicus Couper, 1958
Hypotypes 13913, 13914
McGregor, D.C., 1965, Geol. Surv. Can., Paper 64-55, pl. 8, figs. 9, 20.
Isachsen Formation, Lower Cretaceous, south of Buchanan Lake anticline, eastern Axel Heiberg Island, District of Franklin.

Tsugaepollenites mesozoicus Couper
Hypotypes 25099-25101
Hopkins, Jr., W.S., 1971, Geol. Surv. Can., Bull. 197, p. 122, pl. 22, figs. 8-10.
Isachsen Formation, Lower Cretaceous, lat. 76°20′N, long. 115°34′W, northwest Melville Island, District of Franklin.

Tsugaepollenites mesozoicus Couper
Hypotype 34001
Hopkins, Jr., W.S., 1974, Geol. Surv. Can., Paper 73-12, p. 22, pl. 6, fig. 68.
Christopher Formation, Lower Cretaceous, lat. 76°20′N, long. 115°34′W, northwest Melville Island, District of Franklin.

Tsugaepollenites mesozoicus Couper
Hypotype 41607
Hopkins, Jr., W.S. and Sweet, A.R., 1976, Geol. Surv. Can., Bull. 256, pl. 9, fig. 23.
Mattagami Formation, Lower Cretaceous, borehole 68-42, lat. 50°35′36″N, long. 82°30′12″W, Ontario.

?*Tsugaepollenites* sp.
Fig. spec. 31776
Hopkins, Jr., W.S. and Balkwill, H.R., 1973, Geol. Surv. Can., Paper 72-37, p. 18, pl. 2, fig. 38.
Hassel Formation, Cretaceous, Hoodoo Dome, lat. 78°11′N, long. 99°40′W, southern Ellef Ringnes Island, District of Franklin.

Ulmoideipites krempi Anderson emend. Elsik, 1968
Hypotype 71409
Wilson, M.A., 1978, Palaeontographica, Abt. B, vol. 166, p. 147, pl. 11, fig. 8.
Cretaceous, IOE Atkinson H-25 well, north shore Tuktoyaktuk Peninsula, lat. 69°44′20″N, long. 131°50′06″W, District of Mackenzie.

Ulmoideipites tricostatus Anderson 1960
Hypotype 76082
Jerzykiewicz, T. and Sweet, A.R., 1986, Can. J. Earth Sci., vol. 23, no. 9, pl. 4, fig. 22.
Late Cretaceous, Coal Valley strip pit, Alberta.

Umbosporites callosus Newman, 1965
Hypotype 38199
McIntyre, D.J., 1974, Geol. Surv. Can., Paper 74-14, pl. 14, fig. 30.
Upper Cretaceous, lat. 69°27′30″N, long. 126°58′W, Horton River, District of Mackenzie.

Umbosporites callosus Newman, 1965
Hypotype 93632
Sweet, A.R., Ricketts, B.D., Cameron, A.R. and Norris, D.K., 1989, Geol. Surv. Can., Paper 89-1G, p. 92, pl. 1, fig. 2.
Little Bear Formation, Upper Cretaceous, east bank of Little Bear River, lat. 64°37′48″N, long. 126°18′W, District of Mackenzie.

Undulatisporites undulapolus Brenner
Hypotype 33995
Hopkins, Jr., W.S., 1974, Geol. Surv. Can., Paper 73-12, p. 21, pl. 5, fig. 62.
Christopher Formation, Lower Cretaceous, lat. 76°20′N, long. 115°34′W, northwest Melville Island, District of Franklin.

Undulatisporites sp.
Fig. spec. 31769
Hopkins, Jr., W.S. and Balkwill, H.R., 1973, Geol. Surv. Can., Paper 72-37, p. 16, pl. 2, fig. 31.
Hassel Formation, Cretaceous, Hoodoo Dome, lat. 78°11′N, long. 99°40′W, southern Ellef Ringnes Island, District of Franklin.

Undulatisporites sp.
Fig. spec. 71268
Wilson, M.A., 1978, Palaeontographica, Abt. B, vol. 166, p. 113, pl. 1, fig. 6.
Cretaceous, Police Island, north bank Mackenzie River, 12 miles southeast of Fort Norman, District of Mackenzie.

Unidentified
Fig. specs. 13854-13858, 13896-13904
McGregor, D.C., 1965, Geol. Surv. Can., Paper 64-55, pl. 6, figs. 10, 12, 33, 34, 37; pl. 7, figs. 12-14, 19, 23, 24, 30, 35, 46.
Lower Cretaceous, Fault Creek Canyon and Donna River Canyon (13896-13904), Aklavik Range, Richardson Mountains, District of Mackenzie.

Unidentified
Fig. specs. 13927-13930, 13937, 13938
McGregor, D.C., 1965, Geol. Surv. Can., Paper 64-55, pl. 8, figs. 12, 17, 18, 27; pl. 9, figs, 2, 4.
Isachsen Formation, Lower Cretaceous, south of Buchanan Lake anticline, eastern Axel Heiberg Island, District of Franklin.

Unidentified
Fig. specs. 13967-13972
McGregor, D.C., 1965, Geol. Surv. Can., Paper 64-55, pl. 10, figs. 4, 19, 22, 26, 27, 32.
Lower Cretaceous, Peel Plateau, Stony Creek about 18.5 miles upstream from mouth, District of Mackenzie.

Verrucatosporites choloneri (Brenner) Kremp
Hypotype 42475
Brideaux, W.W. and McIntyre, D.J., 1975, Geol. Surv. Can., Bull. 252, pl. 3, fig. 9.
Horton River Formation, Lower Cretaceous, lat. 69°27′30″N, long. 126°54′W, Horton River, District of Mackenzie.

Verrucosisporites dentonianus Wingate, 1980
Hypotype 91366
Sweet, A.R. and McIntyre, D.J., 1988, Can. Soc. Petrol. Geol., Mem. 15, fig. 7.14.
Moosehound Member, Cardium Formation, Upper Cretaceous, south side of Grande Cache Highway 40, l.s.d. 15, sec. 13, tp. 57, rge. 7, W.5th mer., Alberta.

Verrucosisporites rotundus Singh
Hypotype 33993
Hopkins, Jr., W.S., 1974, Geol. Surv. Can., Paper 73-12, p. 20, pl. 5, fig. 60.
Christopher Formation, Lower Cretaceous, lat. 76°20′N, long. 115°34′W, northwest Melville Island, District of Franklin.

Verrucosisporites sp.
Fig. spec. 31770
Hopkins, Jr., W.S. and Balkwill, H.R., 1973, Geol. Surv. Can., Paper 72-37, p. 16, pl. 2, fig. 32.
Hassel Formation, Cretaceous, near Cape Cairo, lat. 78°43′N, long. 100°06′W, northeastern Ellef Ringnes Island, District of Franklin.

Verrucosisporites sp.
Fig. specs. 56781, 56789
Williams, G.L. and Bujak, J.P., 1980, Initial Repts. Deep Sea Drilling Project, vol. 50, pl. 10, fig. 12; pl. 11, fig. 8.
Early Cretaceous, DSDP Site 416, lat. 32°50.18′N, long. 10°48.06′W, northwestern Moroccan continental margin.

Verrutriletes papillosus (Dijkstra) Potonié 1956
Hypotype 48933
Sweet, A.R., 1979, Am. Assoc. Strat. Palynol., Contrib. Ser. No. 5B, pl. 3, fig. 8.
Christopher Formation, Lower Cretaceous, lat. 79°47′24″N, long. 85°22′W, near Slidre Fiord, Ellesmere Island, District of Franklin.

Virgo amiantopollis (Srivastava) Ward, 1986
Hypotype 93688
Sweet, A.R. and Braman, D.R., 1989, Geol. Surv. Can., Paper 89-8, pl. 1, fig. 23.
Chungo Member, Wapiabi Formation, Upper Cretaceous, Highwood River No. 2, section, N.E. 1/4 sec. 1, tp. 18, rge. 4, W.5th mer., Alberta.

Vitreisporites pallidus (Reissinger) Nilsson, 1958
Hypotype 13888
McGregor, D.C., 1965, Geol. Surv. Can., Paper 64-55, pl. 7, fig. 38.
Lower Cretaceous, Donna River Canyon, Aklavik Range, Richardson Mountains, District of Mackenzie.

Vitreisporites pallidus (Reissinger) Nilsson
Hypotype 25093
Hopkins, Jr., W.S., 1971, Geol. Surv. Can., Bull. 197, p. 121, pl. 21, fig. 20; pl. 22, fig. 1.
Isachsen Formation, Lower Cretaceous, lat. 76°20′N, long. 115°34′W, northwest Melville Island, District of Franklin.

Vitreisporites pallidus (Reissinger) Nilsson
Hypotype 31772
Hopkins, Jr., W.S. and Balkwill, H.R., 1973, Geol. Surv. Can., Paper 72-37, p. 16, pl. 2, fig. 34.
Hassel Formation, Cretaceous, Hoodoo Dome, lat. 78°11′N, long. 99°40′W, southern Ellef Ringnes Island, District of Franklin.

Vitreisporites pallidus (Reissinger) Nilsson
Hypotype 33998
Hopkins, Jr., W.S., 1974, Geol. Surv. Can., Paper 73-12, p. 21, pl. 5, fig. 65.
Christopher Formation, Lower Cretaceous, lat. 76°20′N, long. 115°34′W, northwest Melville Island, District of Franklin.

Vitreisporites pallidus (Reissinger) Nilsson
Hypotype 42496
Brideaux, W.W. and McIntyre, D.J., 1975, Geol. Surv. Can., Bull. 252, pl. 3, fig. 30.
Horton River Formation, Lower Cretaceous, lat. 69°27′30″N, long. 126°54′W, Horton River, District of Mackenzie.

Vitreisporites pallidus (Reissinger)
Hypotypes 41605, 41606
Hopkins, Jr., W.S. and Sweet, A.R., 1976, Geol. Surv. Can., Bull. 256, pl. 9, figs. 21, 22.
Mattagami Formation, Lower Cretaceous, Onakawana lignite pit, lat. 50°36′N, long. 81°23′W and borehole 68-38, lat. 50°36′N, long. 82°31′12″W, Ontario.

Vitreisporites pallidus (Reissinger) Nilsson, 1958
Hypotypes 71321, 71322
Wilson, M.A., 1978, Palaeontographica, Abt. B, vol. 166, p. 124, pl. 5, figs. 13, 14.
Cretaceous, Police Island, north bank Mackenzie River, 12 miles southeast of Fort Norman and IOE Atkinson H-25 well, north shore Tuktoyaktuk Peninsula, lat. 69°44′20″N, long. 131°50′06″W, District of Mackenzie.

Wallodinium anglicum (Cookson and Hughes) Lentin and Williams, 1973
Hypotypes 38115, 38116
McIntyre, D.J., 1974, Geol. Surv. Can., Paper 74-14, pl. 9, figs. 2, 3.
Upper Cretaceous, lat. 69°27′30″N, long. 126°58′W, Horton River, District of Mackenzie.

Weylandipollis retiformis Takahashi, 1964
Hypotype 93637
Sweet, A.R., Ricketts, B.D., Cameron, A.R. and Norris, D.K., 1989, Geol. Surv. Can., Paper 89-1G, pl. 1, fig. 7.
Little Bear Formation, Upper Cretaceous, east of McKay Range, East Little Bear River area, lat. 64°36′24″N, long. 125°35′W, District of Mackenzie.

Weylandipollis sp.
Fig. spec. 93687
Sweet, A.R. and Braman, D.R., 1989, Geol. Surv. Can., Paper 89-8, p. 35, pl. 1, fig. 22.
Chungo Member, Wapiabi Formation, Upper Cretaceous, Trap Creek near junction with Highwood River, sec. 36, tp. 17, rge. 4, W.5th mer., Alberta.

Williamsidinium banksianum Lentin
Holotype 56597; paratypes 56598-56601, 73725, 73726
Lentin, J.K., 1983, Palynology, vol. 7, p. 150, pl. 1, figs. 1-7.
? Eureka Sound Formation, Upper Cretaceous, lat. 73°26′55″N, long. 123°56′45″W, Banks Island, District of Franklin.

Wodehouseia edmontonicola Wiggins, 1976
Hypotypes 89733, 89734
Jerzykiewicz, T. and Sweet, A.R., 1988, Sedimentary Geol., vol. 59, no. 1/2, pl. III, figs. 6, 7.
Brazeau Formation, Upper Cretaceous, Blackstone River, lat. 52°43′20″N, long. 116°16′30″W, Alberta.

Wodehouseia fimbriata Stanley, 1961
Hypotype 38227
McIntyre, D.J., 1974, Geol. Surv. Can., Paper 74-14, pl. 15, fig. 20.
Upper Cretaceous, lat. 69°28′N, long. 126°58′W, Horton River, District of Mackenzie.

Wodehouseia fimbriata var. *constricta* Wiggins, 1976
Hypotype 89866
Sweet, A.R., Ricketts, B.D., Cameron, A.R. and Norris, D.K., 1989, Geol. Surv. Can., Paper 89-1G, p. 98, pl. 2, fig. 7.
Summit Creek Formation, Upper Cretaceous, tributary of East Little Bear River, lat. 64°33′20″N, long. 125°49′45″W, District of Mackenzie.

Wodehouseia gracile (Samoilovitch) Pokrovskaya, 1966
Hypotypes 38223, 38224
McIntyre, D.J., 1974, Geol. Surv. Can., Paper 74-14, pl. 15, figs. 15, 16.
Upper Cretaceous, lat. 69°28′N, long. 126°58′W, Horton River, District of Mackenzie.

Wodehouseia gracile (Samoilovich) Pokrovskaya, 1966
Hypotype 93650
Sweet, A.R., Ricketts, B.D., Cameron, A.R. and Norris, D.K., 1989, Geol. Surv. Can., Paper 89-1G, p. 97, pl. 1, fig. 20.
Summit Creek Formation, Upper Cretaceous, north bank Mackenzie River immediately dowstream from Police Island, lat. 64°52′42″N, long. 125°14′44″W, District of Mackenzie.

Wodehouseia jactuense (Samoilovitch) Samoilovitch, 1967
Hypotypes 38220, 38221
McIntyre, D.J., 1974, Geol. Surv. Can., Paper 74-14, pl. 15, figs. 13, 19.
Upper Cretaceous, lat. 69°28′N, long. 126°58′W, Horton River, District of Mackenzie.

Wodehouseia octospina Wiggins, 1976
Hypotype 76090
Jerzykiewicz, T. and Sweet, A.R., 1986, Can. J. Earth Sci., vol. 23, no. 9, pl. 4, fig. 30.
Late Cretaceous, depth 26.15-26.25 m, GSC core hole CV-42-2-1, Coal Valley strip mine, l.s.d. 1, sec. 6, tp. 47, rge. 19, W.5th mer., Alberta.

Wodehouseia quadrispina Wiggins, 1976
Hypotype 89863
Sweet, A.R., Ricketts, B.D., Cameron, A.R. and Norris, D.K., 1989, Geol. Surv. Can., Paper 89-1G, p. 98, pl. 2, fig. 6.
Summit Creek Formation, Upper Cretaceous, north bank Mackenzie River immediately dowstream from Police Island, lat. 64°52′42″N, long. 125°14′44″W, District of Mackenzie.

Wodehouseia spinata Stanley, 1961
Hypotype 38226
McIntyre, D.J., 1974, Geol. Surv. Can., Paper 74-14, pl. 15, fig. 18.
Upper Cretaceous, lat. 69°28′N, long. 126°58′W, Horton River, District of Mackenzie.

Wodehouseia spinata Stanley 1961
Hypotype 48987
Sweet, A.R., 1978, Geol. Surv. Can., Paper 78-1B, p. 34, pl. 6.2, fig. 1.
Cuesta Creek Member, Tent Island Formation, Upper Cretaceous, northwestern bank Big Fish River near its junction with Boundary Creek, lat. 68°30′30″N, long. 136°23′50″W, Yukon.

Wodehouseia spinata Stanley, 1961
Hypotype 71417
Wilson, M.A., 1978, Palaeontographica, Abt. B, vol. 166, p. 148, pl. 11, fig. 16.
Cretaceous, Police Island, north bank Mackenzie River, 12 miles southeast of Fort Norman, District of Mackenzie.

Wodehouseia spinata Stanley 1961
Hypotype 76091
Jerzykiewicz, T. and Sweet, A.R., 1986, Can. J. Earth Sci., vol. 23, no. 9, pl. 4, fig. 31.
Late Cretaceous, Coal Valley strip mine pit, Alberta.

Wodehouseia spinata Stanley, 1961
Hypotype 89854
Sweet, A.R., Ricketts, B.D., Cameron, A.R. and Norris, D.K., 1989, Geol. Surv. Can., Paper 89-1G, pl. 1, fig. 21.
Summit Creek Formation, Upper Cretaceous, north bank Mackenzie River immediately dowstream from Police Island, lat. 64°52′42″N, long. 125°14′44″W, District of Mackenzie.

Wodehouseia sp.
Fig. spec. 71413
Wilson, M.A., 1978, Palaeontographica, Abt. B, vol. 166, p. 148, pl. 11, fig. 12.
Cretaceous, Police Island, north bank Mackenzie River, 12 miles southeast of Fort Norman, District of Mackenzie.

Wodehouseia sp. A
Fig. specs. 48988, 48989
Sweet, A.R., 1978, Geol. Surv. Can., Paper 78-1B, p. 36, pl. 6.2, figs. 2, 3.
Cuesta Creek Member, Tent Island Formation, Upper Cretaceous, northwestern bank Big Fish River near its junction with Boundary Creek, lat. 68°30′30″N, long. 136°23′50″W, Yukon.

Xenikoon sp.
Fig. spec. 62644
Ioannides, N.S. and McIntyre, D.J., 1980, Geol. Surv. Can., Paper 80-1A, pl. 31.2, fig. 7.
Upper Cretaceous, Caribou Hills, Mackenzie River, District of Mackenzie.

Zlivisporis sp.
Fig. spec. 48996
Sweet, A.R., 1978, Geol. Surv. Can., Paper 78-1B, p. 36, pl. 6.2, fig. 10.
Cuesta Creek Member, Tent Island Formation, Upper Cretaceous, northwestern bank Big Fish River near its junction with Boundary Creek, lat. 68°30′30″N, long. 136°23′50″W, Yukon.

Zonosulcites scollardensis Srivastava 1969
Hypotype 76083
Jerzykiewicz, T. and Sweet, A.R., 1986, Can. J. Earth Sci., vol. 23, no. 9, pl. 4, fig. 23.
Late Cretaceous, Coal Valley strip mine pit, Alberta.

Zonosulcites sp.
Fig. spec. 91402
Sweet, A.R. and McIntyre, D.J., 1988, Can. Soc. Petrol. Geol., Mem. 15, fig. 7.21.
Kaskapau Formation, Upper Cretaceous, south side of Grande Cache Highway 40, l.s.d. 15, sec. 13, tp. 57, rge. 7, W.5th mer., Alberta.

Tertiary-Holocene

?*Abies* sp.
Fig. spec. 96310
McIntyre, D.J., 1991, Geol. Surv. Can., Bull. 403, pl. 2, fig. 2.
Buchanan Lake Formation, Eureka Sound Group, Eocene, 25 km east of Geodetic Hills, lat. 79°55′N, long. 88°58′W, northeastern Axel Heiberg Island, District of Franklin.

Abietineaepollenites latisulcatus Norton, 1969
Hypotypes 71312, 71313
Wilson, M.A., 1978, Palaeontographica, Abt. B., vol. 166, p. 120, pl. 5, figs. 4, 5.
Paleocene, IOE Atkinson H-25 well, north shore Tuktoyaktuk Peninsula, lat. 69°44′20″N, long. 131°50′06″W, District of Mackenzie.

Abietineaepollenites cf. *A. microalatus microalatus* Potonié, 1958
Hypotype 71311
Wilson, M.A., 1978, Palaeontographica, Abt. B., vol. 166, p. 120, pl. 5, fig. 3.
Paleocene, IOE Atkinson H-25 well, north shore Tuktoyaktuk Peninsula, lat. 69°44′20″N, long. 131°50′06″W, District of Mackenzie.

Acer sp.
Fig. specs. 96364-96367
McIntyre, D.J., 1991, Geol. Surv. Can., Bull. 403, pl. 4, figs. 12-15.
Buchanan Lake Formation, Eureka Sound Group, Eocene, 25 km east of Geodetic Hills, lat. 79°55′N, long. 88°58′W, northeastern Axel Heiberg Island, District of Franklin.

cf. *Acer* sp.
Fig. spec. 93600
McIntyre, D.J., 1989, Geol. Surv. Can., Paper 89-1G, pl. 1, fig. 9.
Eureka Sound Group, Paleocene, valley southeast of Cunningham Inlet between lat. 73°54′N, long. 92°35′W and lat. 74°2′N, long. 93°33′W, north end of Somerset Island, District of Franklin.

Albies pollen
Fig. spec. 68781
Dixon, J. et al., 1984, Geol. Surv. Can., Paper 82-13, pl. 5, fig. 10.
Pliocene to Holocene, 488-515 m interval, Dome Gulf et al. Hunt Kopanoar M-13 well, lat. 70°22′55.1″N, long. 135°05′33.9″W, Beaufort Sea.

Alisporites bilateralis Rouse
Hypotype 80484
Burden, E. and Holloway, D., 1985, Can. J. Earth Sci., vol. 22, no. 10, pl. 1, fig. 3.
Quaternary, Scott Inlet area, Baffin Island, District of Franklin.

Alisporites grandis (Cookson)
Hypotype 38723
Haskell, T.R. and Wilson, G.T., 1975, Initial Repts. Deep Sea Drilling Project, vol. 29, pl. 4, fig. 16.
Middle-Upper Eocene, DSDP Site 280, lat. 48°57.44′S, long. 147°14.08′E, South Tasman Rise, south of Tasmania.

Alisporites grandis (Cookson) Dettmann, 1963
Hypotypes 71316, 71319
Wilson, M.A., 1978, Palaeontographica, Abt. B., vol. 166, p. 121, pl. 5, figs. 8, 11.
Paleocene and Cretaceous, IOE Atkinson H-25 well, north shore Tuktoyaktuk Peninsula, lat. 69°44′20″N, long. 131°50′06″W, District of Mackenzie.

Alnipollenites verus Potonié, 1931
Hypotype 93665
Sweet, A.R., Ricketts, B.D., Cameron, A.R. and Norris, D.K., 1989, Geol. Surv. Can., Paper 89-1G, pl. 2, fig. 16.
Summit Creek Formation, mid-Paleocene, west bank Mackenzie River opposite north end of Seagull Island, lat. 64°45′N, long. 125°7′20″W, District of Mackenzie.

Alnipollenites sp.
Fig. spec. 62608
Ioannides, N.S. and McIntyre, D.J., 1980, Geol. Surv. Can., Paper 80-1A, pl. 31.4, fig. 16.
Eocene, Caribou Hills, Mackenzie River, District of Mackenzie.

Alnus trina Stanley, 1965
Hypotype 76015
Jerzykiewicz, T. and Sweet, A.R., 1986, Can. J. Earth Sci., vol. 23, no. 9, pl. 1, fig. 19.
Early Paleocene, abandoned open pit on hill about 800 m southeast of Highway 47, southeast of Coalspur, Alberta.

Alnus sp.
Fig. spec. 27850
Hopkins, Jr., W.S., Rutter, N.W. and Rouse, G.E., 1972, Can. J. Earth Sci., vol. 9, no. 4, pl. 2, fig. 22.
Oligocene, east side Parsnip River 2 1/2 miles north of John Hart Highway bridge, lat. 55°09′N, long. 122°59′W, British Columbia.

Alnus sp.
Fig. spec. 92034
Dietrich, J.R. et al., 1989, Geol. Surv. Can., Paper 89-1G, pl. 3, fig. 4.
Paleocene, depth 8200-8290 feet, Dome Pacific et al. PEX Natsek E-56 well, lat. 69°45′21.46″N, long. 139°44′34.58″W, western Beafort Sea.

Alnus sp.
Fig. spec. 93613
McIntyre, D.J., 1989, Geol. Surv. Can., Paper 89-1G, p. 195, pl. 2, fig. 4.
Eureka Sound Group, Paleocene, valley southeast of Cunningham Inlet between lat. 73°54′N, long. 92°35′W and lat. 74°2′N, long. 93°33′W, north end of Somerset Island, District of Franklin.

Alnus sp.
Fig. spec. 94358
White, J.M., 1990, Can. J. Earth Sci., vol. 27, no. 4, p. 535, pl. 1, fig. 3.
Early Eocene to Early Oligocene, float in Mud Bay Creek, lat. 51°12′42″N, long. 132°15′23″W, north shore Skidegate Inlet, Queen Charlotte Islands, British Columbia.

Alnus sp.
Fig. specs. 96316, 96320-96322
McIntyre, D.J., 1991, Geol. Surv. Can., Bull. 403, pl. 2, figs. 8, 12-14.
Buchanan Lake Formation, Eureka Sound Group, Eocene, 25 km east of Geodetic Hills, lat. 79°55′N, long. 88°58′W, northeastern Axel Heiberg Island, District of Franklin.

Alnus sp.
Fig. spec. 97325
Kalgutkar, R.M and McIntyre, D.J., 1991, Can. J. Earth Sci., vol. 28, no. 3, p. 370, pl. 2, fig. 12.
Iceberg Bay Formation, Eureka Sound Group, Early Eocene, north shore Strand Fiord, 26 km from head, lat. 79°14′N, long. 91°20′W, Axel Heiberg Island, District of Franklin.

Alterbia obscura (Drugg) Lentin and Williams, 1976
Hypotype 45802
Brideaux, W.W., 1976, Geol. Surv. Can., Paper 76-1B, pl. 44.1, fig. 6.
Reindeer Formation, Paleocene, 1400-1430 feet, Gulf Mobil Parsons N-10 well, ca. 1.5 miles east-northeast of Parsons Lake, lat. 69°59′48.52″N, long. 133°31′50.33″W, District of Mackenzie.

Alterbia sp.
Fig. spec. 62652
Ioannides, N.S. and McIntyre, D.J., 1980, Geol. Surv. Can., Paper 80-1A, pl. 31.2, fig. 15.
Eocene, Caribou Hills, Mackenzie River, District of Mackenzie.

Ambrosia sp.
Fig. spec. 41912
Williams, G.L., 1975, Geol. Surv. Can., Paper 74-30, vol. 2, pl. 8, fig. 17.
Pleistocene, Shell Mohican I-100 well, Scotian Shelf.

Anacolosidites sp. *A. reklawensis* Elsik
Fig. specs. 96347, 96348
McIntyre, D.J., 1991, Geol. Surv. Can., Bull. 403, pl. 3, figs. 26, 27.
Buchanan Lake Formation, Eureka Sound Group, Eocene, 25 km east of Geodetic Hills, lat. 79°55′N, long. 88°58′W, northeastern Axel Heiberg Island, District of Franklin.

Annutriporites tripollenites (Rouse)
Hypotypes 68095-68098
Norris, G., 1986, Geol. Surv. Can., Bull. 340, p. 39, pl. 10, figs. 20-23.

Richards Formation, Eocene to Early Oligocene and Arnak Member, Kugmallit Formation, Late Oligocene (68096), depths 11,300-11,400, 3300-3400, 9200-9300 and 7800-8000 feet, Imperial Nuktak C-22 well, lat. 69°41′07″N, long. 134°51′30″W, District of Mackenzie.

Annutriporites sp. A
Fig. specs. 68099-68102
Norris, G., 1986, Geol. Surv. Can., Bull. 340, p. 39, pl. 10, figs. 24-27
Richards Formation, Middle and Late Eocene, depths 10,000-10,100, 9600-9700, 9300-9400 and 10,300-10,400 feet, Imperial Nuktak C-22 well, lat. 69°41′07″N, long. 134°51′30″W, District of Mackenzie.

Apectodinium parvum (Alberti) Lentin and Williams
Hypotypes 62663-62666
Ioannides, N.S. and McIntyre, D.J., 1980, Geol. Surv. Can., Paper 80-1A, pl. 31.1, figs. 10-13.
Eocene, Caribou Hills, Mackenzie River, District of Mackenzie.

Aquilapollenites immiser Sweet
Holotype 76096; paratypes 76097-76099
Sweet, A.R., 1986, Can. J. Earth Sci., vol. 23, no. 9, p. 1378, pl. 1, figs. 6-9.
Early Paleocene and Late Cretaceous, depth 27.38-27.41 m (76096) and 29.35-29.62 m, GSC core hole CV-42-2-2, Coal Valley strip mine, l.s.d. 1, sec. 10, tp. 47, rge. 19, W.5th mer., Alberta.

Aquilapollenites immiser Sweet, 1986
Hypotype 76024
Jerzykiewicz, T. and Sweet, A.R., 1986, Can. J. Earth Sci., vol. 23, no. 9, pl. 2, fig. 8.
Early Paleocene, depth 79.5 m, Manalta Coal Limited core hole ME-111-81CD, l.s.d. 3, sec. 17, tp. 48, rge. 21, W.5th mer., Alberta.

Aquilapollenites polaris Funkhouser, 1961
Hypotype 71348
Wilson, M.A., 1978, Palaeontographica, Abt. B., vol. 166, p. 131, pl. 7, fig. 18.
Paleocene, Police Island, north bank Mackenzie River, 12 miles southeast of Fort Norman, District of Mackenzie.

Aquilapollenites sp. cf. *A. quadrilobus* Rouse, 1957
Hypotype 76030
Jerzykiewicz, T. and Sweet, A.R., 1986, Can. J. Earth Sci., vol. 23, no. 9, pl. 2, fig. 14.
Early Paleocene, depth 26.62-26.71 m, GSC core hole CV-42-2-2, Coal Valley strip mine, l.s.d. 1, sec. 10, tp. 47, rge. 19, W.5th mer., Alberta.

Aquilapollenites reticulatus Stanley, 1961
Hypotype 71351
Wilson, M.A., 1978, Palaeontographica, Abt. B., vol. 166, p. 132, pl. 8, figs. 2, 3.
Paleocene, IOE Atkinson H-25 well, north shore Tuktoyaktuk Peninsula, lat. 69°44′20″N, long. 131°50′06″W, District of Mackenzie.

Aquilapollenites reticulatus (Mchedlishvili) Tschudy and Leopold, 1971
Hypotype 76027
Jerzykiewicz, T. and Sweet, A.R., 1986, Can. J. Earth Sci., vol. 23, no. 9, pl. 2, fig. 11.
Early Paleocene, Coal Valley mine pit, Alberta.

Aquilapollenites spinulosus Funkhouser, 1961
Hypotype 76028
Jerzykiewicz, T. and Sweet, A.R., 1986, Can. J. Earth Sci., vol. 23, no. 9, pl. 2, fig. 12.
Paskapoo Formation, Early Paleocene, road cut along Highway 47 near Coalspur, Alberta.

Aquilapollenites spinulosus Funkhouser 1961
Hypotype 76131
Jerzykiewicz, T. and Sweet, A.R., 1986, Geol. Surv. Can., Paper 86-1B, Fig. 67.2, fig. 1.
Porcupine Hills Formation, Middle Paleocene, sec. 9, tp. 15, rge. 1, W.5th mer., Alberta.

Aquilapollenites spinulosus Funkhouser, 1961
Hypotype 93660
Sweet, A.R., Ricketts, B.D., Cameron, A.R. and Norris, D.K., 1989, Geol. Surv. Can., Paper 89-1G, p. 98, pl. 2, fig. 11.
Summit Creek Formation, Mid-Paleocene, south bank Mackenzie River opposite the downstream end of Seagull Island, lat. 64°46′N, long. 125°7′30″W, District of Mackenzie.

Arecipites tenuiexinous Leffingwell, 1971
Hypotype 76035
Jerzykiewicz, T. and Sweet, A.R., 1986, Can. J. Earth Sci., vol. 23, no. 9, pl. 3, fig. 2.
Early Paleocene, abandoned open pit on hill about 800 m southeast of Highway 47, southeast of Coalspur, Alberta.

Arecipites tenuiexinous Leffingwell
Hypotype 93598
McIntyre, D.J., 1989, Geol. Surv. Can., Paper 89-1G, p. 195, pl. 1, fig. 7.
Eureka Sound Group, Paleocene, valley southeast of Cunningham Inlet between lat. 73°54′N, long. 92°35′W and lat. 74°2′N, long. 93°33′W, north end of Somerset Island, District of Franklin.

Artemisia sp.
Fig. spec. 41913
Williams, G.L., 1975, Geol. Surv. Can., Paper 74-30, vol. 2, pl. 8, fig. 18.
Pleistocene, Shell Mohican I-100 well, Scotian Shelf.

Azolla areolata Sweet and Hills
Holotype 31078; paratypes 31079-31086
Sweet. A.R. and Hills, L.V, 1976, Can. J. Botany, vol. 54, nos. 3 and 4, p. 345, figs. 48-74.
Eureka Sound Formation, Paleocene-Eocene, 25 miles south of Cape Wrottesley, lat. 74°24′N, long. 120°47′W, north-central Banks Island, District of Franklin.

Azolla bulbosa Snead
Topotypes 31027-31031, 31033, 31034, 31072
Sweet. A.R. and Hills, L.V, 1976, Can. J. Botany, vol. 54, nos. 3 and 4, p. 338, figs. 3-24.
Paskapoo Formation, Paleocene, depth 80-87 feet, Research Council of Alberta corehole 65-1, Wizard Lake, l.s.d. 4, sec. 8, tp. 48, rge. 27, W.4th mer., Alberta.

Azolla schopfii Dijkstra 1961
Hypotype 76140
Jerzykiewicz, T. and Sweet, A.R., 1986, Geol. Surv. Can., Paper 86-1B, Fig. 67.2, fig. 10.
Willow Creek Formation, Early Paleocene, Castle River, l.s.d. 4-5, sec. 12, tp. 7, rge. 1, W.5th mer., Alberta.

Azolla stanleyi Jain and Hall
Hypotypes 31036-31038, 31040, 31041c, 31073-31077
Sweet. A.R. and Hills, L.V, 1976, Can. J. Botany, vol. 54, nos. 3 and 4, p. 344, figs. 25-39, 42, 43, 46, 47.
Paskapoo and Ravenscrag formations, Paleocene, depth 45-50, 13-20 (31040), and 32-44 (31041c) feet, Research Council of Alberta corehole 65-1, Wizard Lake, l.s.d. 4, sec. 8, tp. 48, rge. 27, W.4th mer., Alberta; depths 63-64.5 (31073, 31074), 35-40 (31075a, n, 31077), and 80-90 (31076) feet, South Saskatachewan Federal-Provincial testholes Estevan 14, N.E.14, sec. 3, tp. 2, rge. 6, W.2nd mer, Estevan 115R, N.W.13, sec. 23, tp. 2, rge. 7, W.2nd mer, and Estevan 76, N.W. l.s.d. 5, sec. 32, tp. 1, rge. 8, W.2nd mer., Saskatchewan.

Azolla sp. A
Fig. spec. 68040
Norris, G., 1986, Geol. Surv. Can., Bull. 340, p. 33, pl. 8, fig. 7.
Richards Formation, Late Eocene, depth 9500-9600 feet, Imperial Nuktak C-22 well, lat. 69°41′07″N, long. 134°51′30″W, District of Mackenzie.

Azolla sp.
Fig. spec. 92053
Dietrich, J.R. et al., 1989, Geol. Surv. Can., Paper 89-1G, p. 139, pl. 3, fig. 23.
Early to Middle Eocene, depth 1200-1290 feet, Dome Pacific et al. PEX Natsek E-56 well, lat. 69°45′21.46″N, long. 139°44′34.58″W, western Beafort Sea.

Baculatisporites comaumensis (Cookson) Potonié
Hypotype 68021
Norris, G., 1986, Geol. Surv. Can., Bull. 340, p. 32, pl. 7, fig. 15.
Arnak Member, Kugmallit Formation, Late Oligocene, depth 3300-3400 feet, Imperial Nuktak C-22 well, lat. 69°41′07″N, long. 134°51′30″W, District of Mackenzie.

Baculatisporites crassiprimarius (Krutzsch)
Hypotypes 68022-68025
Norris, G., 1986, Geol. Surv. Can., Bull. 340, p. 32, pl. 7, figs. 16, 19-21.
Richards Formation, Early Oligocene to late Eocene, Arnak Member, Late Oligocene (68023) and Mackenzie Bay Formation, Miocene (68025), depths 8200-8300, 4100-4200, 9400-9500 and 2400-2500 feet, Imperial Nuktak C-22 well, lat. 69°41′07″N, long. 134°51′30″W, District of Mackenzie.

Baculatisporites quintus (Thomson and Pflug) Krutzsch
Hypotypes 68027, 68028
Norris, G., 1986, Geol. Surv. Can., Bull. 340, p. 32, pl. 7, figs. 23, 24.
Arnak Member, Kugmallit Formation, Late Oligocene, depths 4000-4100 and 3300-3400 feet, Imperial Nuktak C-22 well, lat. 69°41′07″N, long. 134°51′30″W, District of Mackenzie.

Beaupreaeidites eleganisformis Cookson
Hypotype 38708
Haskell, T.R. and Wilson, G.T., 1975, Initial Repts. Deep Sea Drilling Project, vol. 29, pl. 4, fig. 1.
Eocene, DSDP Site 282, lat. 42°14.76′S, long. 143°29.18′E, west of Tasmania.

Betula sp.
Fig. spec. 92033
Dietrich, J.R. et al., 1989, Geol. Surv. Can., Paper 89-1G, pl. 3, fig. 3.
Early to Middle Eocene, depth 600-690 feet, Dome Pacific et al. PEX Natsek E-56 well, lat. 69°45′21.46″N, long. 139°44′34.58″W, western Beafort Sea.

Betula sp.
Fig. spec. 96318
McIntyre, D.J., 1991, Geol. Surv. Can., Bull. 403, pl. 2, fig. 10.
Buchanan Lake Formation, Eureka Sound Group, Eocene, 25 km east of Geodetic Hills, lat. 79°55′N, long. 88°58′W, northeastern Axel Heiberg Island, District of Franklin.

Betula sp.
Fig. spec. 97326
Kalgutkar, R.M. and McIntyre, D.J., 1991, Can. J. Earth Sci., vol. 28, no. 3, p. 370, pl. 2, fig. 13.
Iceberg Bay Formation, Eureka Sound Group, Early Eocene, north shore Strand Fiord, 26 km from head, lat. 79°14′N, long. 91°20′W, Axel Heiberg Island, District of Franklin.

Birestisporites potoniaei Delcourt and Sprumont
Hypotype 68015
Norris, G., 1986, Geol. Surv. Can., Bull. 340, p. 31, pl. 7, fig. 10.
Richards Formation, Middle Eocene, depth 11,000-11,100 feet, Imperial Nuktak C-22 well, lat. 69°41′07″N, long. 134°51′30″W, District of Mackenzie.

Bohemiasporis sp.
Fig. spec. 75998
Jerzykiewicz, T. and Sweet, A.R., 1986, Can. J. Earth Sci., vol. 23, no. 9, pl. 1, fig. 1.
Early Paleocene, depth 26.71-26.81 m, GSC core hole CV-42-2-2, Coal Valley strip mine, l.s.d. 1, sec. 10, tp. 47, rge. 19, W.5th mer., Alberta.

Boisduvalia clavatites Piel
Hypotype 27859
Hopkins, Jr., W.S., Rutter, N.W. and Rouse, G.E., 1972, Can. J. Earth Sci., vol. 9, no. 4, pl. 2, figs. 31, 32.
Oligocene, east side Parsnip River 2 1/2 miles north of John Hart Highway bridge, lat. 55°09′N, long. 122°59′W, British Columbia.

Boisduvalia clavatites Piel
Hypotype 62619
Ioannides, N.S. and McIntyre, D.J., 1980, Geol. Surv. Can., Paper 80-1A, pl. 31.5, fig. 7.
Oligocene(?), Caribou Hills, Mackenzie River, District of Mackenzie.

Bombacacidites sp. A
Fig. spec. 41908
Williams, G.L., 1975, Geol. Surv. Can., Paper 74-30, vol. 2, pl. 8, fig. 13.
Oligocene, Mobil Sable Island C-67 well, Scotian Shelf.

Bombacacidites sp. A Williams and Brideaux 1975
Fig. spec. 49353
Williams, G.L. and Bujak, J. P., 1977, Am. Assoc. Strat. Palynol., Contr. Strat. Palynol., vol. 1, Contr. Ser. 5A, pl. 4, fig. 1.
Miocene, depth 2610-2640 feet, Tenneco et al. Leif E-38 well, lat. 54°17′29.87″N, long. 55°5′52.17″W, Labrador Shelf.

Brachysporisporites cotalis (Elsik and Jansonius)
Hypotypes 67933-67935
Norris, G., 1986, Geol. Surv. Can., Bull. 340, p. 24, pl. 2, figs. 45-47.
Richards Formation, Eocene, depths 11,800-11,900, 9000-9100 and 12,000-12,100 feet, Imperial Nuktak C-22 well, lat. 69°41′07″N, long. 134°51′30″W, District of Mackenzie.

Brachysporisporites sp. cf. *B. cotalis* (Elsik and Jansonius)
Fig. specs. 67936, 67937
Norris, G., 1986, Geol. Surv. Can., Bull. 340, p. 24, pl. 2, figs. 48, 49.
Richards Formation, Late Eocene, depth 9200-9300 feet, Imperial Nuktak C-22 well, lat. 69°41′07″N, long. 134°51′30″W, District of Mackenzie.

Brachysporisporites opimus (Elsik and Jansonius)
Hypotypes 67938, 67939
Norris, G., 1986, Geol. Surv. Can., Bull. 340, p. 24, pl. 2, figs. 50, 51.
Richards Formation, Middle Eocene, depths 12,500-12,600 and 12,200-12,300 feet, Imperial Nuktak C-22 well, lat. 69°41′07″N, long. 134°51′30″W, District of Mackenzie.

Brachysporisporites pyriformis Lange and Smith 1971
Hypotype 93710
Kalgutkar, R.M and McIntyre, D.J., 1991, Can. J. Earth Sci., vol. 28, no. 3, p. 368, pl. 1, fig. 20.
Iceberg Bay Formation, Eureka Sound Group, Early Eocene, north shore Strand Fiord, 26 km from head, lat. 79°14′N, long. 91°20′W, Axel Heiberg Island, District of Franklin.

Brachysporisporites sp.
Fig. specs. 92049, 92051
Dietrich, J.R. et al., 1989, Geol. Surv. Can., Paper 89-1G, pl. 3, figs. 19, 21.
Early to Middle Eocene, depths 3400-3490 and 6500-6590 feet, Dome Pacific et al. PEX Natsek E-56 well, lat. 69°45′21.46″N, long. 139°44′34.58″W, western Beafort Sea.

Brevicolporites colpella Anderson, 1960
Hypotypes 76037, 76038
Jerzykiewicz, T. and Sweet, A.R., 1986, Can. J. Earth Sci., vol. 23, no. 9, pl. 3, figs. 4, 5.
Early Paleocene, depth 27.34-27.38 m, GSC core hole CV-42-2-2, Coal Valley strip mine, l.s.d. 1, sec. 10, tp. 47, rge. 19, W.5th mer., Alberta.

Brevicolporites colpella Anderson, 1960
Hypotype 76138
Jerzykiewicz, T. and Sweet, A.R., 1986, Geol. Surv. Can., Paper 86-1B, Fig. 67.2, fig. 8.
Porcupine Hills Formation, Early Paleocene, Oldman River, sec. 17, tp. 7, rge. 30, W.4th mer., Alberta.

Callimothallus pertusus Dilcher
Hypotypes 67998-68002
Norris, G., 1986, Geol. Surv. Can., Bull. 340, p. 30, pl. 6, figs. 4, 6-9.
Richards Formation, Middle Eocene, depths 12,200-12,300, 12,000-12,100, 12,300-12,400, 11,700-11,800 and 12,600-12,650 feet, Imperial Nuktak C-22 well, lat. 69°41′07″N, long. 134°51′30″W, District of Mackenzie.

Carpinipites sp. cf. *Carpinipites spackmaniana* (Traverse) Zhou
Fig. specs. 68106-68108
Norris, G., 1986, Geol. Surv. Can., Bull. 340, p. 39, pl. 10, figs. 31-33.
Arnak Member, Kugmallit Formation, Late Oligocene and Richards Formation, Middle Eocene (68107), depths 4500-4600, 12,500-12,600, and 3600-3700 feet, Imperial Nuktak C-22 well, lat. 69°41′07″N, long. 134°51′30″W, District of Mackenzie.

Carpinus sp.
Fig. spec. 27849
Hopkins, Jr., W.S., Rutter, N.W. and Rouse, G.E., 1972, Can. J. Earth Sci., vol. 9, no. 4, pl. 2, fig. 21.
Oligocene, east side Parsnip River 2 1/2 miles north of John Hart Highway bridge, lat. 55°09′N, long. 122°59′W, British Columbia.

Carya juxtaporipites (Wodehouse) Rouse, 1962
Hypotype 71386
Wilson, M.A., 1978, Palaeontographica, Abt. B., vol. 166, p. 142, pl. 10, fig. 8.

Paleocene, IOE Atkinson H-25 well, north shore Tuktoyaktuk Peninsula, lat. 69°44′20″N, long. 131°50′06″W, District of Mackenzie.

Carya veripites Wilson and Webster, 1946
Hypotype 71387
Wilson, M.A., 1978, Palaeontographica, Abt. B., vol. 166, p. 143, pl. 10, fig. 9.
Paleocene, IOE Atkinson H-25 well, north shore Tuktoyaktuk Peninsula, lat. 69°44′20″N, long. 131°50′06″W, District of Mackenzie.

Carya veripites Wilson and Webster
Hypotype 96327
McIntyre, D.J., 1991, Geol. Surv. Can., Bull. 403, pl. 3, fig. 5.
Buchanan Lake Formation, Eureka Sound Group, Eocene, 25 km east of Geodetic Hills, lat. 79°55′N, long. 88°58′W, northeastern Axel Heiberg Island, District of Franklin.

Carya sp. cf. *C. viridifluminipites* Wodehouse
Fig. spec. 96326
McIntyre, D.J., 1991, Geol. Surv. Can., Bull. 403, pl. 3, fig. 4.
Buchanan Lake Formation, Eureka Sound Group, Eocene, 25 km east of Geodetic Hills, lat. 79°55′N, long. 88°58′W, northeastern Axel Heiberg Island, District of Franklin.

Carya sp.
Fig. spec. 27851
Hopkins, Jr., W.S., Rutter, N.W. and Rouse, G.E., 1972, Can. J. Earth Sci., vol. 9, no. 4, pl. 2, fig. 23.
Oligocene, east side Parsnip River 2 1/2 miles north of John Hart Highway bridge, lat. 55°09′N, long. 122°59′W, British Columbia.

Carya sp.
Fig. specs. 71382, 71383
Wilson, M.A., 1978, Palaeontographica, Abt. B., vol. 166, p. 143, pl. 10, figs. 4, 5.
Paleocene, IOE Atkinson H-25 well, north shore Tuktoyaktuk Peninsula, lat. 69°44′20″N, long. 131°50′06″W, District of Mackenzie.

Carya sp.
Fig. specs. 92022, 92030
Dietrich, J.R. et al., 1989, Geol. Surv. Can., Paper 89-1G, p. 139, pl. 2, figs. 16, 22.
Early to Middle Eocene, depths 6500-6590 and 750-790 feet, Dome Pacific et al. PEX Natsek E-56 well, lat. 69°45′21.46″N, long. 139°44′34.58″W, western Beafort Sea.

Carya sp.
Fig. spec. 94367
White, J.M., 1990, Can. J. Earth Sci., vol. 27, no. 4, p. 535, pl. 1, fig. 12.
Early Eocene to Early Oligocene, depth 1452.4-1458.5 m, Union Port Louis well, west coast of Graham Island, lat. 53°41′18″N, long. 132°58′23″W, Queen Charlotte Islands, British Columbia.

Carya sp.
Fig. spec. 96328
McIntyre, D.J., 1991, Geol. Surv. Can., Bull. 403, pl. 3, fig. 6.
Buchanan Lake Formation, Eureka Sound Group, Eocene, 25 km east of Geodetic Hills, lat. 79°55′N, long. 88°58′W, northeastern Axel Heiberg Island, District of Franklin.

Caryapollenites imparalis Nichols and Ott
Hypotype 93616
McIntyre, D.J., 1989, Geol. Surv. Can., Paper 89-1G, p. 195, pl. 2, fig. 7.
Eureka Sound Group, Paleocene, valley southeast of Cunningham Inlet between lat. 73°54′N, long. 92°35′W and lat. 74°2′N, long. 93°33′W, north end of Somerset Island, District of Franklin.

Caryapollenites imparalis Nichols and Ott 1978
Hypotype 97333
Kalgutkar, R.M and McIntyre, D.J., 1991, Can. J. Earth Sci., vol. 28, no. 3, p. 370, pl. 2, fig. 20.
Iceberg Bay Formation, Eureka Sound Group, Early Eocene, north shore Strand Fiord, 26 km from head, lat. 79°14′N, long. 91°20′W, Axel Heiberg Island, District of Franklin.

Caryapollenites veripites (Wislon and Webster) Nichols and Ott
Hypotype 68130
Norris, G., 1986, Geol. Surv. Can., Bull. 340, p. 41, pl. 10, fig. 58.
Richards Formation, Middle Eocene, depth 10,890 feet, Imperial Nuktak C-22 well, lat. 69°41′07″N, long. 134°51′30″W, District of Mackenzie.

Caryapollenites veripites (Wislon and Webster) Nichols and Ott
Hypotype 93617
McIntyre, D.J., 1989, Geol. Surv. Can., Paper 89-1G, p. 195, pl. 2, fig. 8.
Eureka Sound Group, Paleocene, valley southeast of Cunningham Inlet between lat. 73°54′N, long. 92°35′W and lat. 74°2′N, long. 93°33′W, north end of Somerset Island, District of Franklin.

Caryapollenites wodehousei Nichols and Ott
Hypotype 63164
McNeil, D.H, Ioannides, N.S. and Dixon, J., 1982, Geol. Surv. Can., Paper 80-32, pl. 4, fig. 2.
Oligocene, 2225 m below Kelly Bushing, Dome Gulf et al. Ukalerk C-50 well, lat. 70°8′5.598″N, long. 132°44′8.492″W, Beaufort Sea.

Caryapollenites wodehousei Nichols and Ott
Hypotype 93615
McIntyre, D.J., 1989, Geol. Surv. Can., Paper 89-1G, p. 195, pl. 2, fig. 6.
Eureka Sound Group, Paleocene, valley southeast of Cunningham Inlet between lat. 73°54′N, long. 92°35′W and lat. 74°2′N, long. 93°33′W, north end of Somerset Island, District of Franklin.

Caryapollenites wodehousei Nichols and Ott, 1978
Hypotypes 97331, 97332
Kalgutkar, R.M and McIntyre, D.J., 1991, Can. J. Earth Sci., vol. 28, no. 3, p. 370, pl. 2, figs. 18, 19.
Iceberg Bay Formation, Eureka Sound Group, Early Eocene, north shore Strand Fiord, 26 km from head, lat. 79°14′N, long. 91°20′W, Axel Heiberg Island, District of Franklin.

Caryapollenites sp. A
Fig. specs. 62593, 62594
Ioannides, N.S. and McIntyre, D.J., 1980, Geol. Surv. Can., Paper 80-1A, pl. 31.4, figs. 1, 2.
Eocene and Upper Paleocene-(?)Lower Eocene, Caribou Hills, Mackenzie River, District of Mackenzie.

Caryapollenites sp. A of Ioannides and McIntyre (1980)
Fig. spec. 63169
McNeil, D.H, Ioannides, N.S. and Dixon, J., 1982, Geol. Surv. Can., Paper 80-32, pl. 4, fig. 7.
Oligocene, 2042 m below Kelly Bushing, Dome Gulf et al. Ukalerk C-50 well, lat. 70°8′5.598″N, long. 132°44′8.492″W, Beaufort Sea.

Caryapollenites sp. B
Fig. spec. 62595
Ioannides, N.S. and McIntyre, D.J., 1980, Geol. Surv. Can., Paper 80-1A, pl. 31.4, fig. 3.
Eocene, Caribou Hills, Mackenzie River, District of Mackenzie.

Caryapollenites sp. B of Ioannides and McIntyre (1980)
Fig. spec. 63163
McNeil, D.H, Ioannides, N.S. and Dixon, J., 1982, Geol. Surv. Can., Paper 80-32, pl. 4, fig. 1.
Oligocene, 2012 m below Kelly Bushing, Dome Gulf et al. Ukalerk C-50 well, lat. 70°8′5.598″N, long. 132°44′8.492″W, Beaufort Sea.

Castanea sp.
Fig. spec. 27846
Hopkins, Jr., W.S., Rutter, N.W. and Rouse, G.E., 1972, Can. J. Earth Sci., vol. 9, no. 4, pl. 2, fig. 17.
Oligocene, east side Parsnip River 2 1/2 miles north of John Hart Highway bridge, lat. 55°09′N, long. 122°59′W, British Columbia.

Castanea sp.
Fig. spec. 96378
McIntyre, D.J., 1991, Geol. Surv. Can., Bull. 403, pl. 4, fig. 26.
Buchanan Lake Formation, Eureka Sound Group, Eocene, 25 km east of Geodetic Hills, lat. 79°55′N, long. 88°58′W, northeastern Axel Heiberg Island, District of Franklin.

Cedripites parvus Norton, 1969
Hypotype 71314
Wilson, M.A., 1978, Palaeontographica, Abt. B., vol. 166, p. 121, pl. 5, fig. 6.
Paleocene, Police Island, north bank of Mackenzie River, 12 miles southeast of Fort Norman, District of Mackenzie.

Cedrus perialata Martin and Rouse
Hypotype 27904
Hopkins, Jr., W.S., Rutter, N.W. and Rouse, G.E., 1972, Can. J. Earth Sci., vol. 9, no. 4, pl. 1, fig. 11.
Oligocene, east side Parsnip River 2 1/2 miles north of John Hart Highway bridge, lat. 55°09′N, long. 122°59′W, British Columbia.

Centonites sp. A
Fig. spec. 67967
Norris, G., 1986, Geol. Surv. Can., Bull. 340, p. 27, pl. 3, fig. 26.
Arnak Member, Kugmallit Formation, Late Oligocene, depth 3600-3700 feet, Imperial Nuktak C-22 well, lat. 69°41′07″N, long. 134°51′30″W, District of Mackenzie.

Cercidiphyllum sp.
Fig. specs. 96345, 96346
McIntyre, D.J., 1991, Geol. Surv. Can., Bull. 403, pl. 3, figs. 24, 25.
Buchanan Lake Formation, Eureka Sound Group, Eocene, 25 km east of Geodetic Hills, lat. 79°55′N, long. 88°58′W, northeastern Axel Heiberg Island, District of Franklin.

cf. *Cercidiphyllum* sp.
Fig. spec. 93623
McIntyre, D.J., 1989, Geol. Surv. Can., Paper 89-1G, p. 195, pl. 2, fig. 14.
Eureka Sound Group, Paleocene, valley southeast of Cunningham Inlet between lat. 73°54′N, long. 92°35′W and lat. 74°2′N, long. 93°33′W, north end of Somerset Island, District of Franklin.

Chenopodiaceae
Fig. spec. 92044
Dietrich, J.R. et al., 1989, Geol. Surv. Can., Paper 89-1G, p. 135, pl. 3, fig. 14.
Pliocene, depth 500-590 feet, Dome Pacific et al. PEX Natsek E-56 well, lat. 69°45′21.46″N, long. 139°44′34.58″W, western Beafort Sea.

Chenopodipollis nuktakensis Norris
Holotype 68140; hypotype 68139
Norris, G., 1986, Geol. Surv. Can., Bull. 340, p. 42, pl. 11, figs. 10-12.
Arnak Member, Kugmallit Formation, Late Oligocene and Nuktak Formation, Pliocene or Late Miocene, depths 4100-4200 and 1400-1500 feet, Imperial Nuktak C-22 well, lat. 69°41′07″N, long. 134°51′30″W, District of Mackenzie.

Chenopodipollis sp. A
Fig. specs. 68141-68143
Norris, G., 1986, Geol. Surv. Can., Bull. 340, p. 42, pl. 11, figs. 13-16.
Mackenzie Bay Formation, Miocene, Ivik Member, Kugmallit Formation, Middle to Early Oligocene and Nuktak Formation, Pliocene or Late Miocene, depths 2300-2400 and 5100-5200 and 1200-1300 feet, Imperial Nuktak C-22 well, lat. 69°41′07″N, long. 134°51′30″W, District of Mackenzie.

Chenopodipollis sp. B
Fig. specs. 68144, 68145
Norris, G., 1986, Geol. Surv. Can., Bull. 340, p. 42, pl. 11, figs. 17-19.
Nuktak Formation, Poliocene or Late Miocene, depth 1400-1500 feet, Imperial Nuktak C-22 well, lat. 69°41′07″N, long. 134°51′30″W, District of Mackenzie.

Chiropteridium dispersum Gocht
Hypotype 41810
Williams, G.L., 1975, Geol. Surv. Can., Paper 74-30, vol. 2, pl. 3, fig. 1.
Oligocene, core hole 2, Grand Banks.

Cicatricosisporites australiensis (Cookson)
Hypotype 38703
Haskell, T.R. and Wilson, G.T., 1975, Initial Repts. Deep Sea Drilling Project, vol. 29, pl. 3, fig. 11.
Middle Eocene, DSDP Site 283, lat. 43°54.6′S, long. 154°16.96′E, east of Tasmania.

Cicatricosisporites australiensis (Cookson)
Hypotype 41898
Williams, G.L., 1975, Geol. Surv. Can., Paper 74-30, vol. 2, pl. 8, fig. 3.
Late Miocene (reworked), Amoco IOE Puffin B-90 well, Grand Banks.

Cicatricosisporites cicatricosoidess Krutzsch
Hypotype 93593
McIntyre, D.J., 1989, Geol. Surv. Can., Paper 89-1G, p. 195, pl. 1, fig. 2.
Eureka Sound Group, Paleocene, valley southeast of Cunningham Inlet between lat. 73°54′N, long. 92°35′W and lat. 74°2′N, long. 93°33′W, north end of Somerset Island, District of Franklin.

Cicatricosisporites sp.
Fig. spec. 80482
Burden, E. and Holloway, D., 1985, Can. J. Earth Sci., vol. 22, no. 10, pl. 1, fig. 1.
Quaternary, Scott Inlet area, Baffin Island, District of Franklin.

Coronatispora perforata Dettman
Hypotype 38728
Haskell, T.R. and Wilson, G.T., 1975, Initial Repts. Deep Sea Drilling Project, vol. 29, pl. 4, fig. 21.
Lower Oligocene, DSDP Site 282, lat. 42°14.76′S, long. 143°29.18′E, west of Tasmania.

Corsinipollenites triangulatus (Zaklinskaya) Ke et Shi ex Sung et al.
Hypotypes 68127, 68128
Norris, G., 1986, Geol. Surv. Can., Bull. 340, p. 40, pl. 10, figs. 54-56.
Arnak and Ivik members, Kugmallit Formation, Oligocene, depths 3450 and 6490 feet, Imperial Nuktak C-22 well, lat. 69°41′07″N, long. 134°51′30″W, District of Mackenzie.

Corylus sp.
Fig. spec. 93619
McIntyre, D.J., 1989, Geol. Surv. Can., Paper 89-1G, p. 195, pl. 2, fig. 10.
Eureka Sound Group, Paleocene, valley southeast of Cunningham Inlet between lat. 73°54′N, long. 92°35′W and lat. 74°2′N, long. 93°33′W, north end of Somerset Island, District of Franklin.

Corylus sp.
Fig. spec. 96317
McIntyre, D.J., 1991, Geol. Surv. Can., Bull. 403, pl. 2, fig. 9.
Buchanan Lake Formation, Eureka Sound Group, Eocene, 25 km east of Geodetic Hills, lat. 79°55′N, long. 88°58′W, northeastern Axel Heiberg Island, District of Franklin.

Cranwellia subtilis Sweet
Holotype 76103; paratypes 76101, 76102
Sweet, A.R., 1986, Can. J. Earth Sci., vol. 23, no. 9, p. 1379, pl. 1, figs. 11-13.
Early Paleocene, depth 26.62-26.71 m, GSC core hole CV-42-2-2, Coal Valley strip mine, l.s.d. 1, sec. 10, tp. 47, rge. 19 and depth 24.93-24.95 m (76102), GSC core hole CV-42-2-2, Coal Valley strip mine, l.s.d. 1, sec. 6, tp. 47, rge. 19, W.5th mer., Alberta.

Cranwellia subtilis Sweet, 1986
Hypotype 76040
Jerzykiewicz, T. and Sweet, A.R., 1986, Can. J. Earth Sci., vol. 23, no. 9, pl. 3, fig. 7.
Early Paleocene, depth 26.62-26.71 m, GSC core hole CV-42-2-2, Coal Valley strip mine, l.s.d. 1, sec. 10, tp. 47, rge. 19, W.5th mer., Alberta.

Cranwellia sp.
Fig. spec. 71363
Wilson, M.A., 1978, Palaeontographica, Abt. B., vol. 166, p. 136, pl. 9, fig. 1.
Paleocene, Police Island, north bank Mackenzie River, 12 miles southeast of Fort Norman, District of Mackenzie.

?*Cryptogramma* sp.
Fig. spec. 96295
McIntyre, D.J., 1991, Geol. Surv. Can., Bull. 403, pl. 1, fig. 8.
Buchanan Lake Formation, Eureka Sound Group, Eocene, 25 km east of Geodetic Hills, lat. 79°55′N, long. 88°58′W, northeastern Axel Heiberg Island, District of Franklin.

Ctenosporites eskerensis Elsik and Jansonius emend. Smith 1978
Hypotype 94361
White, J.M., 1990, Can. J. Earth Sci., vol. 27, no. 4, p. 535, pl. 1, fig. 6.
Early Eocene to Early Oligocene, depth 1452.4-1458.5 m, Union Port Louis well, west coast of Graham Island, lat. 53°41′18″N, long. 132°58′23″W, Queen Charlotte Islands, British Columbia.

Ctenosporites eskerensis Elsik and Jansonius 1974 emend. Smith 1978
Hypotype 97300
Kalgutkar, R.M and McIntyre, D.J., 1991, Can. J. Earth Sci., vol. 28, no. 3, p. 368, pl. 1, fig. 10.
Iceberg Bay Formation, Eureka Sound Group, Early Eocene, north shore Strand Fiord, 26 km from head, lat. 79°14′N, long. 91°20′W, Axel Heiberg Island, District of Franklin.

Ctenosporites wolfei Elsik and Jansonius
Hypotype 67962
Norris, G., 1986, Geol. Surv. Can., Bull. 340, p. 27, pl. 3, fig. 22.
Richards Formation, Late Eocene, depth 10,300-10,400 feet, Imperial Nuktak C-22 well, lat. 69°41′07″N, long. 134°51′30″W, District of Mackenzie.

Ctenosporites wolfei Elsik and Jansonius
Hypotype 92055
Dietrich, J.R. et al., 1989, Geol. Surv. Can., Paper 89-1G, pl. 3, fig. 25.
Early to Middle Eocene, depth 4500-4590 feet, Dome Pacific et al. PEX Natsek E-56 well, lat. 69°45′21.46″N, long. 139°44′34.58″W, western Beafort Sea.

Cupanieidites sp. A
Fig. spec. 49354
Williams, G.L. and Bujak, J. P., 1977, Am. Assoc. Strat. Palynol., Contr. Strat. Palynol., vol. 1, Contr. Ser. 5A, pl. 4, fig. 2.
Late Eocene, depth 2460-2490 feet, Eastcan et al. Freydis B-87 well, lat. 53°56′13.39″N, long. 54°42′39.75″W, Labrador Shelf.

Cupuliferoipollemites oviformis (Potonié) Potonié
Hypotypes 68077-68079
Norris, G., 1986, Geol. Surv. Can., Bull. 340, p. 37, pl. 9, figs. 17-19.
Richards Formation, Middle Eocene and Mackenzie Bay Formation, Micoene, depths 11,700-11,800, 3000-3100 and 11,000-11,100 feet, Imperial Nuktak C-22 well, lat. 69°41′07″N, long. 134°51′30″W, District of Mackenzie.

Cupuliferoipollemites pusillus (Potonié) Potonié
Hypotypes 68075, 68076
Norris, G., 1986, Geol. Surv. Can., Bull. 340, p. 36, pl. 9, figs. 15, 16.
Richards Formation, Middle Eocene, depth 10,600-10,700 and 12,500-12,600 feet, Imperial Nuktak C-22 well, lat. 69°41′07″N, long. 134°51′30″W, District of Mackenzie.

Cyathidites minor Couper
Hypotype 38704
Haskell, T.R. and Wilson, G.T., 1975, Initial Repts. Deep Sea Drilling Project, vol. 29, pl. 3, fig. 12.
Eocene, DSDP Site 282, lat. 42°14.76′S, long. 143°29.18′E, west of Tasmania.

Cyathidites minor Couper, 1953
Hypotypes 71263, 71272, 71277
Wilson, M.A., 1978, Palaeontographica, Abt. B., vol. 166, p. 107, pl. 1, figs. 1, 10; pl. 2, fig. 1.
Paleocene and Cretaceous (71277), Police Island, north bank Mackenzie River, 12 miles southeast of Fort Korman,, District of Mackenzie.

Cyathidites minor Couper
Hypotype 68013
Norris, G., 1986, Geol. Surv. Can., Bull. 340, p. 30, pl. 7, fig. 8.
Mackenzie Bay Formation, Late Oligocene, depth 3000-3100 feet, Imperial Nuktak C-22 well, lat. 69°41′07″N, long. 134°51′30″W, District of Mackenzie.

Dacrydiumites
Fig. spec. 38705
Haskell, T.R. and Wilson, G.T., 1975, Initial Repts. Deep Sea Drilling Project, vol. 29, pl. 3, fig. 13.
Eocene, DSDP Site 282, lat. 42°14.76′S, long. 143°29.18′E, west of Tasmania.

Deltoidospora hallii Miner
Hypotype 68014
Norris, G., 1986, Geol. Surv. Can., Bull. 340, p. 31, pl. 7, fig. 9.
Richards Formation, Middle Eocene, depth 11,000-11,100 feet, Imperial Nuktak C-22 well, lat. 69°41′07″N, long. 134°51′30″W, District of Mackenzie.

Deltoidospora sp.
Fig. spec. 97317
Kalgutkar, R.M and McIntyre, D.J., 1991, Can. J. Earth Sci., vol. 28, no. 3, p. 370, pl. 2, fig. 4.
Iceberg Bay Formation, Eureka Sound Group, Early Eocene, north shore Strand Fiord, 26 km from head, lat. 79°14′N, long. 91°20′W, Axel Heiberg Island, District of Franklin.

Deltoidospora sp.
Fig. spec. 96297
McIntyre, D.J., 1991, Geol. Surv. Can., Bull. 403, pl. 1, fig. 10.
Buchanan Lake Formation, Eureka Sound Group, Eocene, 25 km east of Geodetic Hills, lat. 79°55′N, long. 88°58′W, northeastern Axel Heiberg Island, District of Franklin.

Densoisporites velatus Weyland and Krieger
Hypotype 38727
Haskell, T.R. and Wilson, G.T., 1975, Initial Repts. Deep Sea Drilling Project, vol. 29, pl. 4, fig. 20.
Middle Eocene, DSDP Site 282, lat. 43°54.6′S, long. 154°16.96′E, west of Tasmania.

Dicellaesporites aculeolatus Sheffy and Dilcher
Hypotype 67897
Norris, G., 1986, Geol. Surv. Can., Bull. 340, p. 21, pl. 2, fig. 8.
Richards Formation, Middle Eocene, depth 11,400-11,500 feet, Imperial Nuktak C-22 well, lat. 69°41′07″N, long. 134°51′30″W, District of Mackenzie.

Dicellaesporites obnixus Norris
Holotype 67894; hypotypes 67892, 67893, 67895, 67896
Norris, G., 1986, Geol. Surv. Can., Bull. 340, p. 21, pl. 2, figs. 3-7.
Richards Formation, Middle Eocene, depths 12,000-12,100, 12,400-12,500 and 11,800-11,900 feet, Imperial Nuktak C-22 well, lat. 69°41′07″N, long. 134°51′30″W, District of Mackenzie.

Dicellaesporites popovii Elsik
Hypotypes 67890, 67891
Norris, G., 1986, Geol. Surv. Can., Bull. 340, p. 21, pl. 2, figs. 1, 2.
Richards Formation, Middle Eocene, depths 11,700-11,800 and 12,400-12,500 feet, Imperial Nuktak C-22 well, lat. 69°41′07″N, long. 134°51′30″W, District of Mackenzie.

Dicellaesporites sp.
Fig. specs. 97301, 97302
Kalgutkar, R.M and McIntyre, D.J., 1991, Can. J. Earth Sci., vol. 28, no. 3, p. 368, pl. 1, figs. 11, 12.
Iceberg Bay Formation, Eureka Sound Group, Early Eocene, north shore Strand Fiord, 26 km from head, lat. 79°14′N, long. 91°20′W, Axel Heiberg Island, District of Franklin.

Didymosporisporonites ovatus Ke et Shi ex Sung et al.
Hypotypes 67913-67916
Norris, G., 1986, Geol. Surv. Can., Bull. 340, p. 23, pl. 2, figs. 25-28.
Richards Formation, Late Eocene, depths 9100-9200, 10,600-10,700 (67914) and 10,300-10,400 (67915) feet, Imperial Nuktak C-22 well, lat. 69°41′07″N, long. 134°51′30″W, District of Mackenzie.

Diervilla echinata Piel
Hypotype 62624
Ioannides, N.S. and McIntyre, D.J., 1980, Geol. Surv. Can., Paper 80-1A, pl. 31.5, fig. 12.
Oligocene(?), Caribou Hills, Mackenzie River, District of Mackenzie.

Diervilla sp.
Fig. specs. 96353, 96354
McIntyre, D.J., 1991, Geol. Surv. Can., Bull. 403, pl. 4, figs. 1, 2.
Buchanan Lake Formation, Eureka Sound Group, Eocene, 25 km east of Geodetic Hills, lat. 79°55′N, long. 88°58′W, northeastern Axel Heiberg Island, District of Franklin.

Diporicellaesporites bellulus Ke et Shi ex Sung et al.
Hypotypes 67920-67922
Norris, G., 1986, Geol. Surv. Can., Bull. 340, p. 23, pl. 2, figs. 32-34.
Arnak Member, Kugmallit Formation, Late Oligocene (67920) and Richards Formation, Eocene, depths 4000-4100, 12,600-12,650 and 9200-9300 feet, Imperial Nuktak C-22 well, lat. 69°41′07″N, long. 134°51′30″W, District of Mackenzie.

Diporicellaesporites sp. cf. *D. bellulus* Ke et Shi ex Sung et al.
Fig. specs. 67917-67919
Norris, G., 1986, Geol. Surv. Can., Bull. 340, p. 23, pl. 2, figs. 29-31.
Richards Formation, Late Eocene, depths 9300-9400, 11,800-11,900 and 9200-9300 feet, Imperial Nuktak C-22 well, lat. 69°41′07″N, long. 134°51′30″W, District of Mackenzie.

Diporicellaesporites laevigataeformis Ke et Shi ex Sung et al.
Hypotypes 67923, 67924
Norris, G., 1986, Geol. Surv. Can., Bull. 340, p. 23, pl. 2, figs. 35, 42.
Richards Formation, Middle Eocene, depths 12,300-12,400 and 11,700-11,800 feet, Imperial Nuktak C-22 well, lat. 69°41′07″N, long. 134°51′30″W, District of Mackenzie.

Diporicellaesporites laevigataeformis Ke et Shi ex Sung et al.
Hypotype 92054
Dietrich, J.R. et al., 1989, Geol. Surv. Can., Paper 89-1G, pl. 3, fig. 24.
Early to Middle Eocene, depth 3400-3490 feet, Dome Pacific et al. PEX Natsek E-56 well, lat. 69°45′21.46″N, long. 139°44′34.58″W, western Beafort Sea.

Diporicellaesporites sp. B, A
Fig. specs. 97307, 97308
Kalgutkar, R.M and McIntyre, D.J., 1991, Can. J. Earth Sci., vol. 28, no. 3, p. 368, pl. 1, figs. 17, 18.
Iceberg Bay Formation, Eureka Sound Group, Early Eocene, north shore Strand Fiord, 26 km from head, lat. 79°14′N, long. 91°20′W, Axel Heiberg Island, District of Franklin.

Diporisporites communis Ke et Shi ex Sung et al.
Hypotypes 67870, 67871
Norris, G., 1986, Geol. Surv. Can., Bull. 340, p. 20, pl. 1, figs. 18, 19.
Richards Formation, Early Oligocene and Late Eocene, depths 7900-8000, and 10,600-10,700 feet, Imperial Nuktak C-22 well, lat. 69°41′07″N, long. 134°51′30″W, District of Mackenzie.

Diporisporites sp.
Fig. spec. 67872
Norris, G., 1986, Geol. Surv. Can., Bull. 340, p. 20, pl. 1, fig. 20.
Richards Formation, Early Oligocene, depth 8300-8400 feet, Imperial Nuktak C-22 well, lat. 69°41′07″N, long. 134°51′30″W, District of Mackenzie.

Dyadosporites oblongatus (Ke et Shi ex Sung et al.) comb. nov.
Hypotypes 67901-67905
Norris, G., 1986, Geol. Surv. Can., Bull. 340, p. 22, pl. 2, figs. 12-16.
Richards Formation, Middle and Late Eocene, depths 11,300-11,400, 9200-9300, 10,600-10,700 and 12,600-12,650 feet, Imperial Nuktak C-22 well, lat. 69°41′07″N, long. 134°51′30″W, District of Mackenzie.

Dyadosporites sp. cf. *D. schwabii* (Elsik)
Fig. spec. 67898
Norris, G., 1986, Geol. Surv. Can., Bull. 340, p. 22, pl. 2, fig. 9.
Arnak Member, Kugmallit Formation, Late Oligocene, depth 4100-4200 feet, Imperial Nuktak C-22 well, lat. 69°41′07″N, long. 134°51′30″W, District of Mackenzie.

Dyadosporites sp. A
Fig. specs. 67906-67908
Norris, G., 1986, Geol. Surv. Can., Bull. 340, p. 22, pl. 2, figs. 17-19.
Richards Formation, Late Eocene, depths 9400-9500 and 9300-9400 (67908) feet, Imperial Nuktak C-22 well, lat. 69°41′07″N, long. 134°51′30″W, District of Mackenzie.

Echinatisporis sp. A
Fig. specs. 68031, 68032
Norris, G., 1986, Geol. Surv. Can., Bull. 340, p. 32, pl. 7, figs. 27-30.
Nuktak Formation, Pliocene or Late Miocene and Mackenzie Bay Formation, Miocene, depths 1400-1500 and 2800-2900 feet, Imperial Nuktak C-22 well, lat. 69°41′07″N, long. 134°51′30″W, District of Mackenzie.

Eleocharis cf. *E. pauciflora* (Lightf.) Link
Hypotype 46939
Prest, V.J., Terasmae, J., Matthews, Jr., J.V., and Lichti-Federovich, S., 1976, Maritime Sediments, vol. 12, no. 2, pl. 1, fig. 1b.
Holocene, Portage-du-Cap, Amherst Island, Magdalen Islands, Québec.

Engelhardtia sp.
Fig. spec. 92025
Dietrich, J.R. et al., 1989, Geol. Surv. Can., Paper 89-1G, pl. 2, fig. 13.
Early to Middle Eocene, depth 1300-1390 feet, Dome Pacific et al. PEX Natsek E-56 well, lat. 69°45′21.46″N, long. 139°44′34.58″W, western Beafort Sea.

Engelhardtia sp.
Fig. specs. 96329, 96330
McIntyre, D.J., 1991, Geol. Surv. Can., Bull. 403, pl. 3, figs. 7, 8.
Buchanan Lake Formation, Eureka Sound Group, Eocene, 25 km east of Geodetic Hills, lat. 79°55′N, long. 88°58′W, northeastern Axel Heiberg Island, District of Franklin.

Ericaceae
Fig. spec. 93599
McIntyre, D.J., 1989, Geol. Surv. Can., Paper 89-1G, p. 195, pl. 1, fig. 8.
Eureka Sound Group, Paleocene, valley southeast of Cunningham Inlet between lat. 73°54′N, long. 92°35′W and lat. 74°2′N, long. 93°33′W, north end of Somerset Island, District of Franklin.

Ericaceae
Fig. spec. 92027
Dietrich, J.R. et al., 1989, Geol. Surv. Can., Paper 89-1G, pl. 2, fig. 19.
Paleocene, depth 7200-7290 feet, Dome Pacific et al. PEX Natsek E-56 well, lat. 69°45′21.46″N, long. 139°44′34.58″W, western Beafort Sea.

Eriraceae
Fig. specs. 96314, 96315
McIntyre, D.J., 1991, Geol. Surv. Can., Bull. 403, pl. 2, figs. 6, 7.
Buchanan Lake Formation, Eureka Sound Group, Eocene, 25 km east of Geodetic Hills, lat. 79°55′N, long. 88°58′W, northeastern Axel Heiberg Island, District of Franklin.

Ericaceae
Fig. spec. 97330
Kalgutkar, R.M and McIntyre, D.J., 1991, Can. J. Earth Sci., vol. 28, no. 3, p. 370, pl. 2, fig. 17.
Iceberg Bay Formation, Eureka Sound Group, Early Eocene, north shore Strand Fiord, 26 km from head, lat. 79°14′N, long. 91°20′W, Axel Heiberg Island, District of Franklin.

Ericaceoipollenites rallus Stanley, 1965
Hypotypes 71364, 71365
Wilson, M.A., 1978, Palaeontographica, Abt. B., vol. 166, p. 136, pl. 9, figs. 2, 3.
Paleocene, IOE Atkinson H-25 well, north shore Tuktoyaktuk Peninsula, lat. 69°44′20″N, long. 131°50′06″W, District of Mackenzie.

Ericipites antecursoroides Norris
Holotype 68087; hypotypes 68085, 68086, 68088, 68089
Norris, G., 1986, Geol. Surv. Can., Bull. 340, p. 38, pl. 10, figs. 8-13.
Richards Formation, Late and Middle Eocene, depths 9000-9100, 10,900-11,000 (68086), 9300-9400 (68088) and 10,000-10,100 (68089) feet, Imperial Nuktak C-22 well, lat. 69°41′07″N, long. 134°51′30″W, District of Mackenzie.

Ericipites compactipolliniatus (Traverse)
Hypotypes 68083, 68084
Norris, G., 1986, Geol. Surv. Can., Bull. 340, p. 37, pl. 10, figs. 6, 7.
Arnak Member, Kugmallit Formation, Late Oligocene and Mackenzie Bay Formation, Late Oligocene, depths 3500-3600 and 3100-3200 feet, Imperial Nuktak C-22 well, lat. 69°41′07″N, long. 134°51′30″W, District of Mackenzie.

Extratriporopollenites sp.
Fig. specs. 41894, 41895
Williams, G.L., 1975, Geol. Surv. Can., Paper 74-30, vol. 2, pl. 7, figs. 15, 16.
Late Paleocene, Mobil Dauntless D-35 well, Scotian Shelf.

Extratriporopollenites sp. A
Fig. spec. 49355
Williams, G.L. and Bujak, J. P., 1977, Am. Assoc. Strat. Palynol., Contrib. Strat. Palynol., vol. 1, Contr. Ser. 5A, pl. 4, fig. 3.

Late Eocene, depth 5480-5510 feet, Eastcan et al. Gudrid H-55 well, lat. 55°54′20.06″N, long. 55°52′31.04″W, Labrador Shelf.

Fagus sp.
Fig. specs. 96373, 96374
McIntyre, D.J., 1991, Geol. Surv. Can., Bull. 403, pl. 4, figs. 21, 22.
Buchanan Lake Formation, Eureka Sound Group, Eocene, 25 km east of Geodetic Hills, lat. 79°55′N, long. 88°58′W, northeastern Axel Heiberg Island, District of Franklin.

cf. *Fagus* sp.
Fig. spec. 27847
Hopkins, Jr., W.S., Rutter, N.W. and Rouse, G.E., 1972, Can. J. Earth Sci., vol. 9, no. 4, pl. 2, figs. 18, 19.
Oligocene, east side Parsnip River 2 1/2 miles north of John Hart Highway bridge, lat. 55°09′N, long. 122°59′W, British Columbia.

Fisheripollis undulatus Krutzsch
Hypotype 62625
Ioannides, N.S. and McIntyre, D.J., 1980, Geol. Surv. Can., Paper 80-1A, pl. 31.5, fig. 13.
Oligocene(?), Caribou Hills, Mackenzie River, District of Mackenzie.

Fisheripollis undulatus Krutzsch
Hypotype 63174
McNeil, D.H, Ioannides, N.S. and Dixon, J., 1982, Geol. Surv. Can., Paper 80-32, pl. 4, fig. 12.
Miocene, 1707 m below Kelly Bushing, Dome Gulf et al. Ukalerk C-50 well, lat. 70°8′5.598″N, long. 132°44′8.492″W, Beaufort Sea.

Fractisporonites sp. cf. *F. canalis* Clarke
Hypotype 67948
Norris, G., 1986, Geol. Surv. Can., Bull. 340, p. 25, pl. 3, fig. 9.
Richards Formation, Late Eocene, depth 8800-8900 feet, Imperial Nuktak C-22 well, lat. 69°41′07″N, long. 134°51′30″W, District of Mackenzie.

Fractisporonites spp. B, A
Fig. specs. 67953-67956
Norris, G., 1986, Geol. Surv. Can., Bull. 340, p. 26, pl. 3, figs. 14-16, 28.
Richards Formation, Middle Eocene, depths 11,700-11,800 (67953), 12,500-12,600 and 12,600-12,650 (67955) feet, Imperial Nuktak C-22 well, lat. 69°41′07″N, long. 134°51′30″W, District of Mackenzie.

Fraxinus sp.
Fig. specs. 96358, 96359
McIntyre, D.J., 1991, Geol. Surv. Can., Bull. 403, pl. 4, figs. 6, 7.
Buchanan Lake Formation, Eureka Sound Group, Eocene, 25 km east of Geodetic Hills, lat. 79°55′N, long. 88°58′W, northeastern Axel Heiberg Island, District of Franklin.

?*Fraxinus* sp.
Fig. specs. 97339, 97340
Kalgutkar, R.M and McIntyre, D.J., 1991, Can. J. Earth Sci., vol. 28, no. 3, p. 370, pl. 2, figs. 26, 27.
Iceberg Bay Formation, Eureka Sound Group, Early Eocene, north shore Strand Fiord, 26 km from head, lat. 79°14′N, long. 91°20′W, Axel Heiberg Island, District of Franklin.

Fungal hyphea type A
Fig. specs. 67968-67972
Norris, G., 1986, Geol. Surv. Can., Bull. 340, p. 27, pl. 4, figs. 1-5.
Nuktak Formation, Pliocene-Late Miocene, Mackenzie Bay Formation (67971) and Arnak Member, Kugmallit Formation (67972), Late Oligocene, depths 1700-1800, 600-700, 900-1000 , 3000-3100 and 3600-3700 feet, Imperial Nuktak C-22 well, lat. 69°41′07″N, long. 134°51′30″W, District of Mackenzie.

Fungal hyphea type B
Fig. specs. 67973-67975
Norris, G., 1986, Geol. Surv. Can., Bull. 340, p. 27, pl. 4, figs. 6-8.
Mackenzie Bay Formation, Miocene and Arnak Member, Kugmallit Formation, Late Oligocene (67975), depths 2800-2900 and 4500-4600 feet, Imperial Nuktak C-22 well, lat. 69°41′07″N, long. 134°51′30″W, District of Mackenzie.

Fungal hyphea type C
Fig. specs. 67976, 67977
Norris, G., 1986, Geol. Surv. Can., Bull. 340, p. 27, pl. 4, figs. 9, 10.
Mackenzie Bay Formation, and Arnak Member, Kugmallit Formation, Late Oligocene, depths 3000-3100 and 3600-3700 feet, Imperial Nuktak C-22 well, lat. 69°41′07″N, long. 134°51′30″W, District of Mackenzie.

Fungal hyphea type D
Fig. specs. 67978-67981
Norris, G., 1986, Geol. Surv. Can., Bull. 340, p. 28, pl. 4, figs. 11-14.
Arnak and Ivik (67981) members, Kugmallit Formation, Oligocene, depths 3600-3700, 4200-4300 and 7200-7300 feet, Imperial Nuktak C-22 well, lat. 69°41′07″N, long. 134°51′30″W, District of Mackenzie.

Fungal hyphea type E
Fig. specs. 67982, 67983
Norris, G., 1986, Geol. Surv. Can., Bull. 340, p. 28, pl. 4, figs. 15, 16.
Arnak Member, Kugmallit Formation, Late Oligocene, depth 4200-4300 feet, Imperial Nuktak C-22 well, lat. 69°41′07″N, long. 134°51′30″W, District of Mackenzie.

Fungal hyphea type F
Fig. spec. 67984
Norris, G., 1986, Geol. Surv. Can., Bull. 340, p. 28, pl. 4, fig. 17.
Arnak Member, Kugmallit Formation, Late Oligocene, depth 4200-4300 feet, Imperial Nuktak C-22 well, lat. 69°41′07″N, long. 134°51′30″W, District of Mackenzie.

Fungal hyphea type G
Fig. specs. 67985-67998
Norris, G., 1986, Geol. Surv. Can., Bull. 340, p. 28, pl. 4, figs. 18-22.
Arnak Member, Kugmallit Formation, Late Oligocene, depths 4200-4300 (67985), 4900-5000 (67986) and 4700-4800 feet, Imperial Nuktak C-22 well, lat. 69°41′07″N, long. 134°51′30″W, District of Mackenzie.

Fusiformisporites microstriatus Hopkins
Hypotype 67899
Norris, G., 1986, Geol. Surv. Can., Bull. 340, p. 22, pl. 2, fig. 10.
Richards Formation, Middle Eocene, depth 12,000-12,100 feet, Imperial Nuktak C-22 well, lat. 69°41′07″N, long. 134°51′30″W, District of Mackenzie.

Fusiformisporites sp. A
Fig. spec. 67900
Norris, G., 1986, Geol. Surv. Can., Bull. 340, p. 21, pl. 2, fig. 11.
Richards Formation, Middle Eocene, depth 11,800-11-900 feet, Imperial Nuktak C-22 well, lat. 69°41′07″N, long. 134°51′30″W, District of Mackenzie.

Glaphyrocysta sp. cf. *G. ordinata* (Williams and Downie) Stover and Evitt
Fig. spec. 62634
Ioannides, N.S. and McIntyre, D.J., 1980, Geol. Surv. Can., Paper 80-1A, pl. 31.3, fig. 5.
Upper Paleocene-(?)Lower Eocene, Caribou Hills, Mackenzie River, District of Mackenzie.

Glyptostrobus sp.
Fig. spec. 27844
Hopkins, Jr., W.S., Rutter, N.W. and Rouse, G.E., 1972, Can. J. Earth Sci., vol. 9, no. 4, pl. 1, fig. 15.
Oligocene, east side Parsnip River 2 1/2 miles north of John Hart Highway bridge, lat. 55°09′N, long. 122°59′W, British Columbia.

Glyptostrobus sp.
Fig. specs. 96301, 96304
McIntyre, D.J., 1991, Geol. Surv. Can., Bull. 403, pl. 1, figs. 14, 17.
Buchanan Lake Formation, Eureka Sound Group, Eocene, 25 km east of Geodetic Hills, lat. 79°55′N, long. 88°58′W, northeastern Axel Heiberg Island, District of Franklin.

Gothanipollis sp. cf.*G. cockfieldensis* Engelhardt 1964
Fig. spec. 94365
White, J.M., 1990, Can. J. Earth Sci., vol. 27, no. 4, p. 535, pl. 1, fig. 10.
Early Eocene to Early Oligocene, depth 1392.9-1396 m, Union Port Louis well, west crest of Graham Island, lat. 53°41′18″N, long. 132°58′23″W, Queen Charlotte Islands, British Columbia.

Gothanipollis sp.
Fig. specs. 27857, 27858
Hopkins, Jr., W.S., Rutter, N.W. and Rouse, G.E., 1972, Can. J. Earth Sci., vol. 9, no. 4, pl. 2, figs. 29, 30.
Oligocene, east side Parsnip River 2 1/2 miles north of John Hart Highway bridge, lat. 55°09′N, long. 122°59′W, British Columbia.

Gothanipollis sp.
Fig. spec. 96343
McIntyre, D.J., 1991, Geol. Surv. Can., Bull. 403, pl. 3, fig. 22.
Buchanan Lake Formation, Eureka Sound Group, Eocene, 25 km east of Geodetic Hills, lat. 79°55′N, long. 88°58′W, northeastern Axel Heiberg Island, District of Franklin.

Gramineae
Fig. spec. 92045
Dietrich, J.R. et al., 1989, Geol. Surv. Can., Paper 89-1G, p. 135, pl. 3, fig. 15.
Pliocene, depth 500-590 feet, Dome Pacific et al. PEX Natsek E-56 well, lat. 69°45′21.46″N, long. 139°44′34.58″W, western Beafort Sea.

Graminidites sp. A
Fig. specs. 68093, 68094
Norris, G., 1986, Geol. Surv. Can., Bull. 340, p. 38, pl. 10, figs. 18, 19.
Arnak Member, Kugmallit Formation, Late Oligocene and Nuktak Formation, Pliocene or Late Miocene, depths 3300-3400 and 1400-1500 feet, Imperial Nuktak C-22 well, lat. 69°41′07″N, long. 134°51′30″W, District of Mackenzie.

Hazaria sheopiarii Srivastava
Hypotype 76001
Jerzykiewicz, T. and Sweet, A.R., 1986, Can. J. Earth Sci., vol. 23, no. 9, pl. 1, fig. 4.
Early Paleocene, Coal Valley strip mine, Alberta.

Helicoön - Helicodendron type
Fig. specs. 97291-97283
Kalgutkar, R.M and McIntyre, D.J., 1991, Can. J. Earth Sci., vol. 28, no. 3, p. 365, pl. 1, figs. 1-3.
Iceberg Bay Formation, Eureka Sound Group, Early Eocene, north shore Strand Fiord, 26 km from head, lat. 79°14′N, long. 91°20′W, Axel Heiberg Island, District of Franklin.

Helicosporium sp.
Fig. specs. 97294-97296
Kalgutkar, R.M and McIntyre, D.J., 1991, Can. J. Earth Sci., vol. 28, no. 3, p. 366, pl. 1, figs. 4-6.
Iceberg Bay Formation, Eureka Sound Group, Early Eocene, north shore Strand Fiord, 26 km from head, lat. 79°14′N, long. 91°20′W, Axel Heiberg Island, District of Franklin.

Hypoxylonites sp.
Fig. spec. 97305
Kalgutkar, R.M and McIntyre, D.J., 1991, Can. J. Earth Sci., vol. 28, no. 3, p. 368, pl. 1, fig. 15.
Iceberg Bay Formation, Eureka Sound Group, Early Eocene, north shore Strand Fiord, 26 km from head, lat. 79°14′N, long. 91°20′W, Axel Heiberg Island, District of Franklin.

Hystrichokolpoma sp.
Fig. specs 62650, 62651
Ioannides, N.S. and McIntyre, D.J., 1980, Geol. Surv. Can., Paper 80-1A, pl. 31.2, figs. 13, 14.
Upper Paleocene-(?)Lower Eocene and Lower and Middle Eocene, Caribou Hills, Mackenzie River, District of Mackenzie.

Ilex sp.
Fig. specs. 96337, 96338
McIntyre, D.J., 1991, Geol. Surv. Can., Bull. 403, pl. 3, figs. 16, 17.
Buchanan Lake Formation, Eureka Sound Group, Eocene, 25 km east of Geodetic Hills, lat. 79°55′N, long. 88°58′W, northeastern Axel Heiberg Island, District of Franklin.

Ilexpollenites sp.
Fig. specs. 62611, 62612
Ioannides, N.S. and McIntyre, D.J., 1980, Geol. Surv. Can., Paper 80-1A, pl. 31.4, figs. 19, 20.
Oligocene(?), Caribou Hills, Mackenzie River, District of Mackenzie.

Ilexpollenites sp.
Fig. spec. 63179
McNeil, D.H, Ioannides, N.S. and Dixon, J., 1982, Geol. Surv. Can., Paper 80-32, pl. 4, fig. 17.
Miocene, 1615 m below Kelly Bushing, Dome Gulf et al. Ukalerk C-50 well, lat. 70°8′5.598″N, long. 132°44′8.492″W, Beaufort Sea.

Ilexpollenites sp. A of Williams and Brideaux, 1975
Fig. spec. 68770
Dixon, J. et al., 1984, Geol. Surv. Can., Paper 82-13, pl. 4, fig. 11.
Late Middle to Late Miocene, 2637 m interval, Dome Gulf et al. Hunt Kopanoar M-13 well, lat. 70°22′55.1″N, long. 135°05′33.9″W, Beaufort Sea.

Imprimospora tankensis Norris
Holotype 67909; hypotypes 67910-67912
Norris, G., 1986, Geol. Surv. Can., Bull. 340, p. 22, pl. 2, figs. 20-24.
Richards Formation, Late Eocene, depths 9000-9100 (67909) and 9200-9300 feet, Imperial Nuktak C-22 well, lat. 69°41′07″N, long. 134°51′30″W, District of Mackenzie.

Inapertisporites circularis Sheffy and Dilcher
Hypotypes 67856, 67857
Norris, G., 1986, Geol. Surv. Can., Bull. 340, p. 18, pl. 1, figs. 3, 4.
Richards Formation, Middle Eocene, depth 12,400-12,500 feet, Imperial Nuktak C-22 well, lat. 69°41′07″N, long. 134°51′30″W, District of Mackenzie.

Inapertisporites sp. cf. *I. subovoideus* Sheffy and Dilcher
Fig. spec. 67858
Norris, G., 1986, Geol. Surv. Can., Bull. 340, p. 18, pl. 1, fig. 5.
Richards Formation, Middle Eocene, depth 11,900-12,000 feet, Imperial Nuktak C-22 well, lat. 69°41′07″N, long. 134°51′30″W, District of Mackenzie.

"*Inapertisporites*"sp. cf. I. vittatus Sheffy and Dilcher
Fig. specs. 67854, 67855
Norris, G., 1986, Geol. Surv. Can., Bull. 340, p. 18, pl. 1, figs. 1, 2.
Richards Formation, Middle Eocene, depths 12,300-12,400 and 11,700-11,800 feet, Imperial Nuktak C-22 well, lat. 69°41′07″N, long. 134°51′30″W, District of Mackenzie.

"Inapertisporites" sp. A
Fig. spec. 67862
Norris, G., 1986, Geol. Surv. Can., Bull. 340, p. 18, pl. 1, figs. 9, 10.
Richards Formation, Middle Eocene, depth 10,600-10,700 feet, Imperial Nuktak C-22 well, lat. 69°41′07″N, long. 134°51′30″W, District of Mackenzie.

Indeterminate peltate fruiting bodies
Fig. specs. 68004, 68005
Norris, G., 1986, Geol. Surv. Can., Bull. 340, p. 30, pl. 6, figs. 10, 11.
Ivik Member, Kugmallit Formation, Middle to Early Oligocene and Richards Formation, Late Eocene, depths 6100-6200 and 9200-9300 feet, Imperial Nuktak C-22 well, lat. 69°41′07″N, long. 134°51′30″W, District of Mackenzie.

Insulapollenites cf. I. regulatus Leffingwell, 1971
Hypotypes 71370, 71371
Wilson, M.A., 1978, Palaeontographica, Abt. B., vol. 166, p. 136, pl. 9, figs. 9, 10.
Paleocene, IOE Atkinson H-25 well, north shore Tuktoyaktuk Peninsula, lat. 69°44′20″N, long. 131°50′06″W, District of Mackenzie.

Insulapollenites rugulatus Leffingwell, 1971
Hypotype 93664
Sweet, A.R., Ricketts, B.D., Cameron, A.R. and Norris, D.K., 1989, Geol. Surv. Can., Paper 89-1G, p. 99, pl. 2, fig. 15.
Summit Creek Formation, mid-Paleocene, west bank Mackenzie River opposite north end of Seagull Island, lat. 64°45′N, long. 125°7′20″W, District of Mackenzie.

Insulapollenites rugulatus Leffingwell, 1971
Hypotypes 93605, 93606
McIntyre, D.J., 1989, Geol. Surv. Can., Paper 89-1G, p. 195, pl. 1, figs. 14, 15.
Eureka Sound Group, Paleocene, valley southeast of Cunningham Inlet between lat. 73°54′N, long. 92°35′W and lat. 74°2′N, long. 93°33′W, north end of Somerset Island, District of Franklin.

Integricorpus sp. A
Fig. specs. 68072-68074
Norris, G., 1986, Geol. Surv. Can., Bull. 340, p. 36, pl. 9, figs. 12-14.
Richards Formation, Middle to Late Eocene, depths 9500-9600, 9300-9400 and 8800-8900 feet, Imperial Nuktak C-22 well, lat. 69°41′07″N, long. 134°51′30″W, District of Mackenzie.

Intratriporopollenites crassipites (Wodehouse)
Hypotypes 68117-68119
Norris, G., 1986, Geol. Surv. Can., Bull. 340, p. 40, pl. 10, figs. 43-45, 50.
Mackenzie Bay Formation, Late Oligocene, Richards Formation, Middle Eocene and Ivik Member, Kugmallit Formation, Middle to Early Oligocene, depths 3000-3100, 11,700-11,800 and 6100-6200 feet, Imperial Nuktak C-22 well, lat. 69°41′07″N, long. 134°51′30″W, District of Mackenzie.

Intratriporopollenites minimus Mai
Hypotypes 68125, 68126
Norris, G., 1986, Geol. Surv. Can., Bull. 340, p. 41, pl. 10, figs. 51, 52.
Richards Formation, Middle Eocene, depths 12,400-12,500 and 12,500-12,600 feet, Imperial Nuktak C-22 well, lat. 69°41′07″N, long. 134°51′30″W, District of Mackenzie.

Intratriporopollenites sp.
Fig. specs. 96335, 96336
McIntyre, D.J., 1991, Geol. Surv. .Can., Bull. 403, pl. 3, figs. 14, 15.
Buchanan Lake Formation, Eureka Sound Group, Eocene, 25 km east of Geodetic Hills, lat. 79°55′N, long. 88°58′W, northeastern Axel Heiberg Island, District of Franklin.

Involutisporites foraminus Clarke 1965
Hypotypes 97299, 97306
Kalgutkar, R.M and McIntyre, D.J., 1991, Can. J. Earth Sci., vol. 28, no. 3, p. 368, pl. 1, figs. 9, 16.
Iceberg Bay Formation, Eureka Sound Group, Early Eocene, north shore Strand Fiord, 26 km from head, lat. 79°14′N, long. 91°20′W, Axel Heiberg Island, District of Franklin.

Involutisporites sp. A
Fig. spec. 97304
Kalgutkar, R.M and McIntyre, D.J., 1991, Can. J. Earth Sci., vol. 28, no. 3, p. 368, pl. 1, fig. 14.
Iceberg Bay Formation, Eureka Sound Group, Early Eocene, north shore Strand Fiord, 26 km from head, lat. 79°14′N, long. 91°20′W, Axel Heiberg Island, District of Franklin.

Juglans sp.
Fig. spec. 27852
Hopkins, Jr., W.S., Rutter, N.W. and Rouse, G.E., 1972, Can. J. Earth Sci., vol. 9, no. 4, pl. 2, fig. 24.
Oligocene, east side Parsnip River 2 1/2 miles north of John Hart Highway bridge, lat. 55°09′N, long. 122°59′W, British Columbia.

Juglans sp.
Fig. spec. 92028
Dietrich, J.R. et al., 1989, Geol. Surv. Can., Paper 89-1G, p. 139, pl. 2, fig. 20.
Early to Middle Eocene, depth 700-790 feet, Dome Pacific et al. PEX Natsek E-56 well, lat. 69°45′21.46″N, long. 139°44′34.58″W, western Beaufort Sea.

Juglans sp.
Fig. spec. 96323
McIntyre, D.J., 1991, Geol. Surv. Can., Bull. 403, pl. 3, fig. 1.
Buchanan Lake Formation, Eureka Sound Group, Eocene, 25 km east of Geodetic Hills, lat. 79°55′N, long. 88°58′W, northeastern Axel Heiberg Island, District of Franklin.

Juglanspollenites sp.
Fig. specs. 62596, 62597
Ioannides, N.S. and McIntyre, D.J., 1980, Geol. Surv. Can., Paper 80-1A, pl. 31.4, figs. 4, 5.
Eocene, Caribou Hills, Mackenzie River, District of Mackenzie.

Juglanspollenites sp.
Fig. spec. 63168
McNeil, D.H, Ioannides, N.S. and Dixon, J., 1982, Geol. Surv. Can., Paper 80-32, pl. 4, fig. 6.
Miocene, 1707 m below Kelly Bushing, Dome Gulf et al. Ukalerk C-50 well, lat. 70°8′5.598″N, long. 132°44′8.492″W, Beaufort Sea.

Jussiaea sp.
Fig. specs. 27856, 27860
Hopkins, Jr., W.S., Rutter, N.W. and Rouse, G.E., 1972, Can. J. Earth Sci., vol. 9, no. 4, pl. 2, figs. 28, 33.
Oligocene, east side Parsnip River 2 1/2 miles north of John Hart Highway bridge, lat. 55°09′N, long. 122°59′W, British Columbia.

Kurtzipites circularis (Norton) Srivastava, 1981
Hypotype 76046
Jerzykiewicz, T. and Sweet, A.R., 1986, Can. J. Earth Sci., vol. 23, no. 9, pl. 3, fig. 13.
Early Paleocene, Coal Valley strip mine, Alberta.

Kurtzipites trispissatus Anderson 1960
Hypotype 76137
Jerzykiewicz, T. and Sweet, A.R., 1986, Geol. Surv. Can., Paper 86-1B, Fig. 67.2, fig. 7.
Willow Creek Formation, Early Paleocene, Castle River, l.s.d. 4-5, sec. 12, tp. 7, rge. 1, W.5th mer., Alberta.

Lacrimasporionites sp. A
Fig. spec. 67866
Norris, G., 1986, Geol. Surv. Can., Bull. 340, p. 19, pl. 1, fig. 14.
Richards Formation, Late Eocene, depth 9000-9100 feet, Imperial Nuktak C-22 well, lat. 69°41′07″N, long. 134°51′30″W, District of Mackenzie.

Laevigatosporites novus Norris
Holotype 68042; hypotypes 68041, 68043
Norris, G., 1986, Geol. Surv. Can., Bull. 340, p. 33, pl. 8, figs. 8-10.
Nuktak Formation, Pliocene and Arnak Member, Kugmallit Formation, Late Oligocene, depths 1100-1200, 600-700 and 4620-4700 feet, Imperial Nuktak C-22 well, lat. 69°41′07″N, long. 134°51′30″W, District of Mackenzie.

Laevigatosporites ovatus Wilson and Webster
Hypotype 68046
Norris, G., 1986, Geol. Surv. Can., Bull. 340, p. 33, pl. 8, fig. 13.
Arnak Member, Kugmallit Formation, Late Oligocene, depth 3500-3600 feet, Imperial Nuktak C-22 well, lat. 69°41′07″N, long. 134°51′30″W, District of Mackenzie.

Laevigatosporites sp.
Fig. spec. 96300
McIntyre, D.J., 1991, Geol. Surv. Can., Bull. 403, pl. 1, fig. 13.
Buchanan Lake Formation, Eureka Sound Group, Eocene, 25 km east of Geodetic Hills, lat. 79°55′N, long. 88°58′W, northeastern Axel Heiberg Island, District of Franklin.

Laevigatosporites sp.
Fig. spec. 97318
Kalgutkar, R.M and McIntyre, D.J., 1991, Can. J. Earth Sci., vol. 28, no. 3, p. 370, pl. 2, fig. 5.
Iceberg Bay Formation, Eureka Sound Group, Early Eocene, north shore Strand Fiord, 26 km from head, lat. 79°14′N, long. 91°20′W, Axel Heiberg Island, District of Franklin.

?*Larix* sp.
Fig. spec. 96305
McIntyre, D.J., 1991, Geol. Surv. Can., Bull. 403, pl. 1, fig. 18.
Buchanan Lake Formation, Eureka Sound Group, Eocene, 25 km east of Geodetic Hills, lat. 79°55′N, long. 88°58′W, northeastern Axel Heiberg Island, District of Franklin.

Leptolepidites sp. A
Fig. spec. 68026
Norris, G., 1986, Geol. Surv. Can., Bull. 340, p. 31, pl. 7, fig. 22.
Richards Formation, Late Eocene, depth 9500-9600 feet, Imperial Nuktak C-22 well, lat. 69°41′07″N, long. 134°51′30″W, District of Mackenzie.

Liliacidites sp.
Fig. spec. 93595
McIntyre, D.J., 1989, Geol. Surv. Can., Paper 89-1G, p. 195, pl. 1, fig. 4.
Eureka Sound Group, Paleocene, valley southeast of Cunningham Inlet between lat. 73°54′N, long. 92°35′W and lat. 74°2′N, long. 93°33′W, north end of Somerset Island, District of Franklin.

Liliacidites sp.
Fig. spec. 92048
Dietrich, J.R. et al., 1989, Geol. Surv. Can., Paper 89-1G, pl. 3, fig. 18.
Paleocene, depth 7100-7190 feet, Dome Pacific et al. PEX Natsek E-56 well, lat. 69°45′21.46″N, long. 139°44′34.58″W, western Beaufort Sea.

Liliacidites sp.
Fig. specs. 96384-96386
McIntyre, D.J., 1991, Geol. Surv. Can., Bull. 403, pl. 4, figs. 31-34.
Buchanan Lake Formation, Eureka Sound Group, Eocene, 25 km east of Geodetic Hills, lat. 79°55′N, long. 88°58′W, northeastern Axel Heiberg Island, District of Franklin.

Liliacidites sp.
Fig. spec. 97345
Kalgutkar, R.M and McIntyre, D.J., 1991, Can. J. Earth Sci., vol. 28, no. 3, p. 370, pl. 2, fig. 32.
Iceberg Bay Formation, Eureka Sound Group, Early Eocene, north shore Strand Fiord, 26 km from head, lat. 79°14′N, long. 91°20′W, Axel Heiberg Island, District of Franklin.

Lingulodinium machaerophorum (Deflandre and Cookson) Wall
Hypotype 63180
McNeil, D.H, Ioannides, N.S. and Dixon, J., 1982, Geol. Surv. Can., Paper 80-32, pl. 4, fig. 18.
Miocene, 1707 m below Kelly Bushing, Dome Gulf et al. Ukalerk C-50 well, lat. 70°8′5.598″N, long. 132°44′8.492″W, Beaufort Sea.

Liquidambar sp.
Fig. spec. 92041
Dietrich, J.R., et al., 1989, Geol. Surv. Can., Paper 89-1G, p. 139, pl. 3, fig. 11.
Early to Middle Eocene, depth 1400-1490 feet, Dome Pacific et al. PEX Natsek E-56 well, lat. 69°45′21.46″N, long. 139°44′34.58″W, western Beafort Sea.

Liquidambar sp.
Fig. spec. 93604
McIntyre, D.J., 1989, Geol. Surv. Can., Paper 89-1G, p. 195, pl. 1, fig. 13.
Eureka Sound Group, Paleocene, valley southeast of Cunningham Inlet between lat. 73°54′N, long. 92°35′W and lat. 74°2′N, long. 93°33′W, north end of Somerset Island, District of Franklin.

Liquidambar sp.
Fig. spec. 96339
McIntyre, D.J., 1991, Geol. Surv. Can., Bull. 403, pl. 3, fig. 18.
Buchanan Lake Formation, Eureka Sound Group, Eocene, 25 km east of Geodetic Hills, lat. 79°55′N, long. 88°58′W, northeastern Axel Heiberg Island, District of Franklin.

Liquidambarpollenites sp.
Fig. spec. 62610
Ioannides, N.S. and McIntyre, D.J., 1980, Geol. Surv. Can., Paper 80-1A, pl. 31.4, fig. 18.
Upper Paleocene-(?)Lower Eocene, Caribou Hills, Mackenzie River, District of Mackenzie.

Lonicera sp.
Fig. spec. 96355
McIntyre, D.J., 1991, Geol. Surv. Can., Bull. 403, pl. 4, fig. 3.
Buchanan Lake Formation, Eureka Sound Group, Eocene, 25 km east of Geodetic Hills, lat. 79°55′N, long. 88°58′W, northeastern Axel Heiberg Island, District of Franklin.

Lonicerapollis spiniformis (Ke et Shi ex Sung et al.) comb. nov.
Hypotype 68080
Norris, G., 1986, Geol. Surv. Can., Bull. 340, p. 37, pl. 10, figs. 1, 2.
Arnak Member, Kugmallit Formation, Late Oligocene, depth 3450 feet, Imperial Nuktak C-22 well, lat. 69°41′07″N, long. 134°51′30″W, District of Mackenzie.

Lycopodium cf. *L. annotinioides* (Krutzsch) Martin and Rouse
Hypotypes 27834, 27835
Hopkins, Jr., W.S., Rutter, N.W. and Rouse, G.E., 1972, Can. J. Earth Sci., vol. 9, no. 4, pl. 1, figs. 1, 2.
Oligocene, east side Parsnip River 2 1/2 miles north of John Hart Highway bridge, lat. 55°09′N, long. 122°59′W, British Columbia.

Lycopodium annotinum
Hypotypes 96288, 96289
McIntyre, D.J., 1991, Geol. Surv. Can., Bull. 403, pl. 1, figs. 1, 2.
Buchanan Lake Formation, Eureka Sound Group, Eocene, 25 km east of Geodetic Hills, lat. 79°55′N, long. 88°58′W, northeastern Axel Heiberg Island, District of Franklin.

Lycopodium sp. cf. *L. annotinum* Linnaeus 1753
Fig. spec. 97315
Kalgutkar, R.M and McIntyre, D.J., 1991, Can. J. Earth Sci., vol. 28, no. 3, p. 370, pl. 2, fig. 2.
Iceberg Bay Formation, Eureka Sound Group, Early Eocene, north shore Strand Fiord, 26 km from head, lat. 79°14′N, long. 91°20′W, Axel Heiberg Island, District of Franklin.

Lycopodiumsporites austroclavatidites (Cookson)
Hypotype 38729
Haskell, T.R. and Wilson, G.J., 1975, Initial Repts. Deep Sea Drilling Project, vol. 29, pl. 4, fig. 22.
Lower Oligocene, DSDP Site 282, lat. 42°14.76′S, long. 143°29.18′E, west of Tasmania.

Lycopodium inundatum
Hypotype 96290
McIntyre, D.J., 1991, Geol. Surv. Can., Bull. 403, pl. 1, fig. 3.
Buchanan Lake Formation, Eureka Sound Group, Eocene, 25 km east of Geodetic Hills, lat. 79°55′N, long. 88°58′W, northeastern Axel Heiberg Island, District of Franklin.

Lycopodium selago
Hypotype 96291
McIntyre, D.J., 1991, Geol. Surv. Can., Bull. 403, pl. 1, fig. 4.
Buchanan Lake Formation, Eureka Sound Group, Eocene, 25 km east of Geodetic Hills, lat. 79°55′N, long. 88°58′W, northeastern Axel Heiberg Island, District of Franklin.

Manicorpus sp. cf. *M. tripodiformis* (Tschudy and Leopold) Tschudy, 1973
Fig. specs. 76032, 76033
Jerzykiewicz, T. and Sweet, A.R., 1986, Can. J. Earth Sci., vol. 23, no. 9, pl. 2, figs. 16, 17.
Early Paleocene, depth 24.2-24.43 m., GSC core CV-42-2-2, Coal Valley strip mine, l.s.d. 1, sec. 10, tp. 47, rge. 19, W.5th mer. and Late Cretaceous, Embarras River, Alberta.

Margocolporites stenosus Ke et Shi ex Sung et al.
Hypotype 68081
Norris, G., 1986, Geol. Surv. Can., Bull. 340, p. 37, pl. 10, figs. 3, 4.
Arnak Member, Kugmallit Formation, Late Oligocene, depth 4700-4800 feet, Imperial Nuktak C-22 well, lat. 69°41′07″N, long. 134°51′30″W, District of Mackenzie.

Matonisporites sp.
Fig. spec. 76010
Jerzykiewicz, T. and Sweet, A.R., 1986, Can. J. Earth Sci., vol. 23, no. 9, pl. 1, fig. 13.
Early Paleocene, depth 27.2-27.34 m, GSC core hole C 4-42-2-2, Coal Valley strip mine, l.s.d. 1, sec. 10, tp. 47, rge. 19, W.5th mer., Alberta.

Megaspore sp. 1
Fig. spec. 64263
Gradstein, F.M. and Agterberg, F.P., 1982, Quantitative Stratigraphic Correlation, pl. 7, fig. 20.
Early Eocene, depth 6260 feet, Eastcan et al Bjarni H-81, Labrador Shelf, East Coast Canada.

Metasequoia papillapollenites Rouse
Hypotype 27843
Hopkins, W.S., Jr., Rutter, N.W. and Rouse, G.E., 1972, Can. J. Earth Sci., vol. 9, no. 4, pl. 1, fig. 14.
Oligocene, east side Parsnip River 2 1/2 miles north of John Hart Highway bridge, lat. 55°09′N, long. 122°59′W, British Columbia.

Metasequoia sp.
Fig. specs. 96302, 96303, 96306, 96307
McIntyre, D.J., 1991, Geol. Surv. Can., Bull. 403, pl. 1, figs. 15, 16, 19, 20.
Buchanan Lake Formation, Eureka Sound Group, Eocene, 25 km east of Geodetic Hills, lat. 79°55′N, long. 88°58′W, northeastern Axel Heiberg Island, District of Franklin.

Metasequoia sp.
Fig. specs. 97321, 97322
Kalgutkar, R.M. and McIntyre, D.J., 1991, Can. J. Earth Sci., vol. 28, no. 3, p. 370, pl. 2, figs. 8, 9.

Iceberg Bay Formation, Eureka Sound Group, Early Eocene, north shore Strand Fiord, 26 km from head, lat. 79°14′N, long. 91°20′W, Axel Heiberg Island, District of Franklin.

Michrystridium? frasera Piel
Hypotype 62622
Ioannides, N.S. and McIntyre, D.J., 1980, Geol. Surv. Can., Paper 80-1A, pl. 31.5, fig. 10.
Eocene, Caribou Hills, Mackenzie River, District of Mackenzie.

Microthallites sp. cf. *M. lutosus* Dilcher
Fig. spec. 68003
Norris, G., 1986, Geol. Surv. Can., Bull. 340, p. 29, pl. 6, fig. 5.
Richards Formation, Late Eocene, depth 9100-9200 feet, Imperial Nuktak C-22 well, lat. 69°41′07″N, long. 134°51′30″W, District of Mackenzie.

Microthallites sp.
Fig. spec. 94357
White, J.M., 1990, Can. J. Earth Sci., vol. 27, no. 4, p. 535, pl. 1, fig. 2.
Early Eocene to Early Oligocene, float in Mud Bay Creek, lat. 51°12′42″N, long. 132°15′23″W, north shore Skidegate Inlet, Queen Charlotte Islands, British Columbia.

Microthyriacites sp. A
Fig. specs. 67994
Norris, G., 1986, Geol. Surv. Can., Bull. 340, p. 29, pl. 5, figs. 6, 7.
Ivik Member, Kugmallit Formation, Middle to Early Oligocene, depth 7200-7300 feet, Imperial Nuktak C-22 well, lat. 69°41′07″N, long. 134°51′30″W, District of Mackenzie.

Microthyrites sp. A
Fig. spec. 67997
Norris, G., 1986, Geol. Surv. Can., Bull. 340, p. 29, pl. 6, fig. 3.
Richards Formation, Late Eocene, depth 10,300-10,400 feet, Imperial Nuktak C-22 well, lat. 69°41′07″N, long. 134°51′30″W, District of Mackenzie.

Milfordia sp.
Fig. spec. 93630
McIntyre, D.J., 1989, Geol. Surv. Can., Paper 89-1G, p. 195, pl. 2, fig. 21.
Eureka Sound Group, Paleocene, valley southeast of Cunningham Inlet between lat. 73°54′N, long. 92°35′W and lat. 74°2′N, long. 93°33′W, north end of Somerset Island, District of Franklin.

Momipites actinus Nichols and Ott, 1978
Hypotypes 93662, 93663
Sweet, A.R., Ricketts, B.D., Cameron, A.R. and Norris, D.K., 1989, Geol. Surv. Can., Paper 89-1G, p. 99, pl. 2, figs. 13, 14.
Summit Creek Formation, mid-Paleocene, west bank Mackenzie River opposite north end of Seagull Island, lat. 64°45′N, long. 125°7′20″W, District of Mackenzie.

Momipites amplus (Leffingwell) Nichols
Hypotype 63165
McNeil, D.H., Ioannides, N.S. and Dixon, J., 1982, Geol. Surv. Can., Paper 80-32, pl. 4, fig. 3.
Oligocene, 2012 m below Kelly Bushing, Dome Gulf et al. Ukalerk C-50 well, lat. 70°8′5.598″N, long. 132°44′8.492″W, Beaufort Sea.

Momipites anellus Nichols and Ott 1978
Hypotype 76134
Jerzykiewicz, T. and Sweet, A.R., 1986, Geol. Surv. Can., Paper 86-1B, Fig. 67.2, fig. 4.
Porcupine Hills Formation, Middle Paleocene, sec. 9, tp. 15, rge. 1, W.5th mer., Alberta.

Momipites anellus Nichols and Ott
Hypotype 93610
McIntyre, D.J., 1989, Geol. Surv. Can., Paper 89-1G, p. 195, pl. 2, fig. 1.
Eureka Sound Group, Paleocene, valley southeast of Cunningham Inlet between lat. 73°54′N, long. 92°35′W and lat. 74°2′N, long. 93°33′W, north end of Somerset Island, District of Franklin.

Momipites leboensis (Leffingwell) Nichols
Hypotype 63167
McNeil, D.H., Ioannides, N.S. and Dixon, J., 1982, Geol. Surv. Can., Paper 80-32, pl. 4, fig. 5.
Oligocene, 2195 m below Kelly Bushing, Dome Gulf et al. Ukalerk C-50 well, lat. 70°8′5.598″N, long. 132°44′8.49″W, Beaufort Sea.

Momipites microfoveolatus (Stanley) Nichols, 1973
Hypotype 76051
Jerzykiewicz, T. and Sweet, A.R., 1986, Can. J. Earth Sci., vol. 23, no. 9, pl. 3, fig. 19.
Early Paleocene, abandoned open pit on hill about 800 m southeast of Highway 47, southeast of Coalspur, Alberta.

Momipites tenuipolus Anderson
Hypotype 68103
Norris, G., 1986, Geol. Surv. Can., Bull. 340, p. 39, pl. 10, figs. 28, 29.
Arnak Member, Kugmallit Formation, Late Oligocene, depth 4602 feet, Imperial Nuktak C-22 well, lat. 69°41′07″N, long. 134°51′30″W, District of Mackenzie.

Momipites ventifluminis Nichols and Ott 1978
Hypotype 76132
Jerzykiewicz, T. and Sweet, A.R., 1986, Geol. Surv. Can., Paper 86-1B, Fig. 67.2, fig. 2.
Porcupine Hills Formation, Middle Paleocene, sec. 9, tp. 15, rge. 1, W.5th mer., Alberta.

Momipites ventifluminis Nichols and Ott
Hypotype 93611
McIntyre, D.J., 1989, Geol. Surv. Can., Paper 89-1G, p. 195, pl. 2, fig. 2.
Eureka Sound Group, Paleocene, valley southeast of Cunningham Inlet between lat. 73°54′N, long. 92°35′W and lat. 74°2′N, long. 93°33′W, north end of Somerset Island, District of Franklin.

Momipites waltmanensis Nichols and Ott 1978
Hypotype 76136
Jerzykiewicz, T. and Sweet, A.R., 1986, Geol. Surv. Can., Paper 86-1B, Fig. 67.2, fig. 6.
Porcupine Hills Formation, Early Paleocene, Oldman River, sec. 17, tp. 7, rge. 30, W.4th mer., Alberta.

Momipites wyomingensis Nichols and Ott 1978
Hypotype 76135
Jerzykiewicz, T. and Sweet, A.R., 1986, Geol. Surv. Can., Paper 86-1B, Fig. 67.2, fig. 5.
Porcupine Hills Formation, Middle Paleocene, sec. 9, tp. 15, rge. 1, W.5th mer., Alberta.

Momipites wyomingensis Nichols and Ott
Hypotype 93612
McIntyre, D.J., 1989, Geol. Surv. Can., Paper 89-1G, p. 195, pl. 2, fig. 3.
Eureka Sound Group, Paleocene, valley southeast of Cunningham Inlet between lat. 73°54′N, long. 92°35′W and lat. 74°2′N, long. 93°33′W, north end of Somerset Island, District of Franklin.

Momipites sp.
Fig. spec. 63166
McNeil, D.H., Ioannides, N.S. and Dixon, J., 1982, Geol. Surv. Can., Paper 80-32, pl. 4, fig. 4.
Oligocene, 2225 m below Kelly Bushing, Dome Gulf et al. Ukalerk C-50 well, lat. 70°8′5.598″N, long. 132°44′8.492″W, Beaufort Sea.

Momipites sp.
Fig. spec. 92026
Dietrich, J.R. et al., 1989, Geol. Surv. Can., Paper 89-1G, pl. 2, fig. 18.
Early Eocene, depth 6500-6590 feet, Dome Pacific et al. PEX Natsek E-56 well, lat. 69°45′21.46″N, long. 139°44′34.58″W, western Beaufort Sea.

cf. *Momipites* sp.
Fig. spec. 94366
White, J.M., 1990, Can. J. Earth Sci., vol. 27, no. 4, p. 535, pl. 1, fig. 11.
Early Eocene to Early Oligocene, depth 1452.4-1458.5 m, Union Port Louis well, west coast of Graham Island, lat. 53°41′18″N, long. 132°58′23″W, Queen Charlotte Islands, British Columbia.

Monocolpopollenites sp. ?*Magnolia*
Fig. spec. 96383
McIntyre, D.J., 1991, Geol. Surv. Can., Bull. 403, pl. 1, fig. 31.
Buchanan Lake Formation, Eureka Sound Group, Eocene, 25 km east of Geodetic Hills, lat. 79°55′N, long. 88°58′W, northeastern Axel Heiberg Island, District of Franklin.

Monoporisporites abruptus Sheffy and Dilcher
Hypotypes 67863-67865
Norris, G., 1986, Geol. Surv. Can., Bull. 340, p. 19, pl. 1, figs. 11-13.
Richards Formation, Late to Middle Eocene, depths 9400-9500, 11,900-12,000 and 11,800-11,900 feet, Imperial Nuktak C-22 well, lat. 69°41′07″N, long. 134°51′30″W, District of Mackenzie.

Monoporisporites sp. cf. *M. culpuliformis* Sheffy and Dilcher
Fig. specs. 67868, 67869
Norris, G., 1986, Geol. Surv. Can., Bull. 340, p. 19, pl. 1, figs. 16, 17.
Arnak and Ivik members, Kugmallit Formation, Oligocene, depths 4820-4900 and 7600-7700 feet, Imperial Nuktak C-22 well, lat. 69°41′07″N, long. 134°51′30″W, District of Mackenzie.

Monoporisporites singularis Sheffy and Dilcher
Hypotypes 67859-67861
Norris, G., 1986, Geol. Surv. Can., Bull. 340, p. 19, pl. 1, figs. 6-8.
Mackenzie Bay Formation and Arnak Member, Kugmallit (67861) Formation, Late Oligocene and Miocene, depths 3000-3100, 2800-2900 and 3600-3700 feet, Imperial Nuktak C-22 well, lat. 69°41′07″N, long. 134°51′30″W, District of Mackenzie.

Monoporisporites sp. A
Fig. specs. 67873, 67874
Norris, G., 1986, Geol. Surv. Can., Bull. 340, p. 19, pl. 1, figs. 21, 22.
Richards Formation, Late Eocene, depths 10,100-10,200 and 9500-9600 feet, Imperial Nuktak C-22 well, lat. 69°41′07″N, long. 134°51′30″W, District of Mackenzie.

Multicellaesporites compactilis Ke et Shi ex Sung et al.
Hypotypes 67943-67945
Norris, G., 1986, Geol. Surv. Can., Bull. 340, p. 25, pl. 3, figs. 4-6.
Richards Formation, Eocene, depths 11,700-11,800, 10,600-10,700 and 9200-9300 feet, Imperial Nuktak C-22 well, lat. 69°41′07″N, long. 134°51′30″W, District of Mackenzie.

Multicellaesporites sp. cf. *M. conicus* Ke et Shi ex Sung et al.
Fig. spec. 67950
Norris, G., 1986, Geol. Surv. Can., Bull. 340, p. 25, pl. 3, fig. 11.
Richards Formation, Middle Eocene, depth 12,400-12,500 feet, Imperial Nuktak C-22 well, lat. 69°41′07″N, long. 134°51′30″W, District of Mackenzie.

Multicellaesporites sp. cf. *M. conspicuus* Ke et Shi ex Sung et al.
Fig. spec. 67951
Norris, G., 1986, Geol. Surv. Can., Bull. 340, p. 25, pl. 3, fig. 12.
Richards Formation, Middle Eocene, depth 12,500-12,600 feet, Imperial Nuktak C-22 well, lat. 69°41′07″N, long. 134°51′30″W, District of Mackenzie.

Multicellaesporites lanceolatus Ke et Shi ex Sung et al.
Hypotype 67949
Norris, G., 1986, Geol. Surv. Can., Bull. 340, p. 25, pl. 3, fig. 10.

Richards Formation, Middle Eocene, depth 11,500-11,600 feet, Imperial Nuktak C-22 well, lat. 69°41′07″N, long. 134°51′30″W, District of Mackenzie.

Multicellaesporites leptaleus Ke et Shi ex Sung et al.
Hypotypes 67940-67942
Norris, G., 1986, Geol. Surv. Can., Bull. 340, p. 24, pl. 3, figs. 1-3.
Richards Formation, Eocene, depths 12,000-12,100, 9200-9300 and 11,900-12,000 feet, Imperial Nuktak C-22 well, lat. 69°41′07″N, long. 134°51′30″W, District of Mackenzie.

Multicellaesporites margaritus Ke et Shi ex Sung et al.
Hypotypes 67946, 67947
Norris, G., 1986, Geol. Surv. Can., Bull. 340, p. 25, pl. 3, figs. 7, 8.
Richards Formation, Early Oligocene, depths 8300-8400 and 9100-9200 feet, Imperial Nuktak C-22 well, lat. 69°41′07″N, long. 134°51′30″W, District of Mackenzie.

"Multicellaesporites" sp. A
Fig. spec. 67952
Norris, G., 1986, Geol. Surv. Can., Bull. 340, p. 25, pl. 3, fig. 13.
Arnak Member, Kugmallit Formation, Late Oligocene, depth 4500-4600 feet, Imperial Nuktak C-22 well, lat. 69°41′07″N, long. 134°51′30″W, District of Mackenzie.

Myrica annulites Martin and Rouse, 1966
Hypotype 71401
Wilson, M.A., 1978, Palaeontographica, Abt. B, vol. 166, p. 144, pl. 10, fig. 24.
Paleocene, Police Island, north bank Mackenzie River, 12 miles southeast of Fort Norman, District of Mackenzie.

Myrica sp.
Fig. spec. 96319
McIntyre, D.J., 1991, Geol. Surv. Can., Bull. 403, pl. 2, fig. 11.
Buchanan Lake Formation, Eureka Sound Group, Eocene, 25 km east of Geodetic Hills, lat. 79°55′N, long. 88°58′W, northeastern Axel Heiberg Island, District of Franklin.

Myricipites annulites (Martin and Rouse)
Hypotypes 68104, 68105
Norris, G., 1986, Geol. Surv. Can., Bull. 340, p. 39, pl. 10, figs. 30, 34.
Arnak Member, Kugmallit Formation, Late Oligocene, depths 4300-4400 and 4200-4300 feet, Imperial Nuktak C-22 well, lat. 69°41′07″N, long. 134°51′30″W, District of Mackenzie.

Myrtaceidites parvus Cookson and Pike
Hypotype 38715
Haskell, T.P. and Wilson, G.J., 1975, Initial Repts. Deep Sea Drilling Project, vol. 29, pl. 4, fig. 8.
Oligocene, DSDP Site 282, lat. 42°14.76′S, long. 143°29.18′E, west of Tasmania.

Nothofagidites flemingii Couper
Hypotype 38719
Haskell, T.P. and Wilson, G.J., 1975, Initial Repts. Deep Sea Drilling Project, vol. 29, pl. 4, fig. 12.
Eocene, DSDP Site 282, lat. 42°14.76′S, long. 143°29.18′E, west of Tasmania.

Nothofagidites sp. (*fusca* group)
Fig. spec. 38714
Haskell, T.P. and Wilson, G.J., 1975, Initial Repts. Deep Sea Drilling Project, vol. 29, pl. 4, fig. 7.
Eocene, DSDP Site 282, lat. 42°14.76′S, long. 143°29.18′E, west of Tasmania.

Novemprojectus traversii Choi
Holotype 73416
Choi, D.K., 1984, Review of Palaeobotany and Palynology, vol. 43, no. 4, p. 338, pl. 1, fig. 1.
Eureka Sound Group, Middle Eocene, Kanguk Formation, Axel Heiberg Island, Canadian Arctic Archipelago.

Nudopollis sp.
Fig. spec. 93625
McIntyre, D.J., 1989, Geol. Surv. Can., Paper 89-1G, p. 175, pl. 2, fig. 16.
Eureka Sound Group, Paleocene, valley southeast of Cunningham Inlet between lat. 73°54′N, long. 92°35′W and lat. 74°2′N, long. 93°33′W, north end of Somerset Island, District of Franklin.

Nyssa sp.
Fig. spec. 92021
Dietrich, J.R. et al., 1989, Geol. Surv. Can., Paper 89-1G, pl. 2, fig. 15.
Early to Middle Eocene, depth 1300-1390 feet, Dome Pacific et al. PEX Natsek E-56 well, lat. 69°45′21.46″N, long. 139°44′34.58″W, western Beaufort Sea.

Nyssa sp.
Fig. specs. 96370, 96371
McIntyre, D.J., 1991, Geol. Surv. Can., Bull. 403, pl. 4, figs. 18, 19.
Buchanan Lake Formation, Eureka Sound Group, Eocene, 25 km east of Geodetic Hills, lat. 79°55′N, long. 88°58′W, northeastern Axel Heiberg Island, District of Franklin.

?*Nyssa* spp.
Fig. specs. 97338, 97342, 97344
Kalgutkar, R.M. and McIntyre, D.J., 1991, Can. J. Earth Sci., vol. 28, no. 3, p. 370, pl. 2, figs. 25, 29, 31.
Iceberg Bay Formation, Eureka Sound Group, Early Eocene, north shore Strand Fiord, 26 km from head, lat. 79°14′N, long. 91°20′W, Axel Heiberg Island, District of Franklin.

Nyssapollenites sp. A
Fig. specs. 49356, 49357
Williams, G.L. and Bujak, J.P., 1977, Am. Assoc. Strat. Palynol., Contr. Strat. Palynol., vol. 1, Contr. Ser. 5A, pl. 4, figs. 4, 5.

Late Eocene, depth 2960-2990 feet, Eastcan et al. Freydis B-87 well, lat. 53°56′13.39″N, long. 54°42′39.75″W, and Eastcan et al. Gudrid H-55 well, lat. 54°54′20.06″N, long. 55°52′31.04″W, Labrador Shelf.

Nyssoidites anulatus (Chlonova)
Hypotype 76122
Sweet, A.R., 1986, Can. J. Earth Sci., vol. 23, no. 9, p. 1380, pl. 2, figs. 14, 15.
Early Paleocene, depth 130-160 feet, Cypress 30R test hole, l.s.d. 4, sec. 24, tp. 9, rge. 18, W.3rd mer., Saskatchewan.

Onagracous pollen
Fig. spec. 62618
Ioannides, N.S. and McIntyre, D.J., 1980, Geol. Surv. Can., Paper 80-1A, pl. 31.5, fig. 6.
Eocene, Caribou Hills, Mackenzie River, District of Mackenzie.

Onagracous pollen
Fig. spec. 63172
McNeil, D.H., Ioannides, N.S. and Dixon, J., 1982, Geol. Surv. Can., Paper 80-32, pl. 4, fig. 10.
Oligocene, 1951 m below Kelly Bushing, Dome Gulf et al. Ukalerk C-50 well, lat. 70°8′5.598″N, long. 132°44′8.492″W, Beaufort Sea.

Operculodinium centrocarpum (Deflandre and Cookson) Wall
Hypotype 63182
McNeil, D.H., Ioannides, N.S. and Dixon, J., 1982, Geol. Surv. Can., Paper 80-32, pl. 4, fig. 20.
Miocene, 1768 m below Kelly Bushing, Dome Gulf et al. Ukalerk C-50 well, lat. 70°8′5.598″N, long. 132°44′8.492″W, Beaufort Sea.

Osmunda claytonites Graham
Hypotype 27836
Hopkins, Jr., W.S., Rutter, N.W. and Rouse, G.E., 1972, Can. J. Earth Sci., vol. 9, no. 4, pl. 1, figs. 3, 4.
Oligocene, east side Parsnip River 2 1/2 miles north of John Hart Highway bridge, lat. 55°09′N, long. 122°59′W, British Columbia.

Osmunda irregulites Martin and Rouse
Hypotype 27837
Hopkins, Jr., W.S., Rutter, N.W. and Rouse, G.E., 1972, Can. J. Earth Sci., vol. 9, no. 4, pl. 1, figs. 5, 6.
Oligocene, east side Parsnip River 2 1/2 miles north of John Hart Highway bridge, lat. 55°09′N, long. 122°59′W, British Columbia.

Osmunda sp.
Fig. specs. 96298, 96299
McIntyre, D.J., 1991, Geol. Surv. Can., Bull. 403, pl. 1, figs. 11, 12.
Buchanan Lake Formation, Eureka Sound Group, Eocene, 25 km east of Geodetic Hills, lat. 79°55′N, long. 88°58′W, northeastern Axel Heiberg Island, District of Franklin.

Osmunda sp.
Fig. spec. 97316
Kalgutkar, R.M. and McIntyre, D.J., 1991, Can. J. Earth Sci., vol. 28, no. 3, p. 370, pl. 2, fig. 3.
Iceberg Bay Formation, Eureka Sound Group, Early Eocene, north shore Strand Fiord, 26 km from head, lat. 79°14′N, long. 91°20′W, Axel Heiberg Island, District of Franklin.

Osmundacidites richardsii Norris
Holotype 68018; hypotypes 68017, 68019, 68020
Norris, G., 1986, Geol. Surv. Can., Bull. 340, p. 31, pl. 7, figs. 12-14, 17, 18.
Richards Formation, Late Eocene, depths 10,000-10,200, 8900-9000, 9300-9400 and 10,300-10,400 feet, Imperial Nuktak C-22 well, lat. 69°41′07″N, long. 134°51′30″W, District of Mackenzie.

Osmundacidites wellmanii Couper, 1953
Hypotypes 71271, 71276
Wilson, M.A., 1978, Palaeontographica, Abt. B, vol. 166, p. 112, pl. 1, figs. 9, 15.
Paleocene, IOE Atkinson H-25 well, north shore Tuktoyaktuk Peninsula, lat. 69°44′20″N, long. 131°50′06″W; Cretaceous, Police Island, north bank Mackenzie River, 12 miles southeast of Fort Norman, District of Mackenzie.

Osmundacidites wellmanii Couper
Hypotype 68016
Norris, G., 1986, Geol. Surv. Can., Bull. 340, p. 31, pl. 7, fig. 11.
Arnak Member, Kugmallit Formation, Late Oligocene, depth 3300-3400 feet, Imperial Nuktak C-22 well, lat. 69°41′07″N, long. 134°51′30″W, District of Mackenzie.

Osmundacidites sp. A
Fig. spec. 49358
Williams, G.L. and Bujak, J.P., 1977, Am. Assoc. Strat. Palynol., Contr. Strat. Palynol., vol. 1, Contr. Ser. 5A, pl. 4, fig. 6.
Miocene, Eastcan et al. Leif M-48 well, lat. 54°17′45.92″N, long. 55°7′20.17″W, Labrador Shelf.

Ostryoipollenites sp. cf. *O. rhenanus* (Thomson) Potonié ex Potonié
Fig. specs. 68109-68111
Norris, G., 1986, Geol. Surv. Can., Bull. 340, p. 40, pl. 10, figs. 35-37.
Mackenzie Bay Formation and Arnak Member, Kugmallit Formation, Late Oligocene, depths 3000-3100 and 3500-3600 (68110) feet, Imperial Nuktak C-22 well, lat. 69°41′07″N, long. 134°51′30″W, District of Mackenzie.

Pachysandra sp.
Fig. spec. 93602
McIntyre, D.J., 1989, Geol. Surv. Can., Paper 89-1G, p. 195, pl. 1, fig. 11.
Eureka Sound Group, Paleocene, valley southeast of Cunningham Inlet between lat. 73°54′N, long. 92°35′W and lat. 74°2′N, long. 93°33′W, north end of Somerset Island, District of Franklin.

Pachysandra sp.
Fig. spec. 96356
McIntyre, D.J., 1991, Geol. Surv. Can., Bull. 403, pl. 4, fig. 4.
Buchanan Lake Formation, Eureka Sound Group, Eocene, 25 km east of Geodetic Hills, lat. 79°55′N, long. 88°58′W, northeastern Axel Heiberg Island, District of Franklin.

Palaeoperidinium? sp.
Fig. spec. 62649
Ioannides, N.S. and McIntyre, D.J., 1980, Geol. Surv. Can., Paper 80-1A, pl. 31.2, fig. 12.
Eocene, Caribou Hills, Mackenzie River, District of Mackenzie.

Pandaniidites radicus Leffingwell
Hypotype 93629
McIntyre, D.J., 1989, Geol. Surv. Can., Paper 89-1G, p. 195, pl. 2, fig. 20.
Eureka Sound Group, Paleocene, valley southeast of Cunningham Inlet between lat. 73°54′N, long. 92°35′W and lat. 74°2′N, long. 93°33′W, north end of Somerset Island, District of Franklin.

Paraalnipollenites alterniporus (Simpson) Rouse and Srivastava
Hypotype 93614
McIntyre, D.J., 1989, Geol. Surv. Can., Paper 89-1G, p. 195, pl. 2, fig. 5.
Eureka Sound Group, Paleocene, valley southeast of Cunningham Inlet between lat. 73°54′N, long. 92°35′W and lat. 74°2′N, long. 93°33′W, north end of Somerset Island, District of Franklin.

Paraalnipollenites alterniporus (Simpson) Srivastava
Hypotype 62609
Ioannides, N.S. and McIntyre, D.J., 1980, Geol. Surv. Can., Paper 80-1A, pl. 31.4, fig. 17.
Eocene, Caribou Hills, Mackenzie River, District of Mackenzie.

Paraalnipollenites alterniporus (Simpson) Srivastava, 1975
Hypotype 76061
Jerzykiewicz, T. and Sweet, A.R., 1986, Can. J. Earth Sci., vol. 23, no. 9, pl. 4, fig. 1.
Val d'Or coal zone, Early Paleocene, roadcut along Highway 47, Coalspur, Alberta.

Paraalnipollenites alterniporus (Simpson) Srivastava, 1975
Hypotype 76139
Jerzykiewicz, T. and Sweet, A.R., 1986, Geol. Surv. Can., Paper 86-1B, Fig. 67.2, fig. 9.
Porcupine Hills Formation, Early Paleocene, Oldman River, sec. 17, tp. 7, rge. 30, W.4th mer., Alberta.

Paraalnipollenites alterniporus (Simpson) Srivastava, 1977
Hypotype 93659
Sweet, A.R., Ricketts, B.D., Cameron, A.R. and Norris, D.K., 1989, Geol. Surv. Can., Paper 89-1G, p. 98, pl. 2, fig. 10.
Summit Creek Formation, earliest Paleocene, north bank Mackenzie River immediately downstream from Police Island, lat. 64°52′42″N, long. 125°14′44″W, District of Mackenzie.

Paraalnipollenites alterniporus (Simpson) Srivastava
Hypotypes 92031, 92032
Dietrich, J.R. et al., 1989, Geol. Surv. Can., Paper 89-1G, p. 142, pl. 3, figs. 1, 2.
Paleocene to Early Eocene, depths 6700-6790 and 2750-2770 feet, Dome Pacific et al. PEX Natsek E-56 well, lat. 69°45′21.46″N, long. 139°44′34.58″W, western Beaufort Sea.

Paraalnipollenites alterniporus (Simpson 1936) Srivastava 1975
Hypotype 97324
Kalgutkar, R.M. and McIntyre, D.J., 1991, Can. J. Earth Sci., vol. 28, no. 3, p. 370, pl. 2, fig. 11.
Iceberg Bay Formation, Eureka Sound Group, Early Eocene, north shore Strand Fiord, 26 km from head, lat. 79°14′N, long. 91°20′W, Axel Heiberg Island, District of Franklin.

Pediastrum sp.
Fig. spec. 68775
Dixon, J. et al., 1984, Geol. Surv. Can., Paper 82-13, pl. 5, fig. 4.
Pliocene to Holocene, 488-515 m interval, Dome Gulf et al. Hunt Kopanoar M-13 well, lat. 70°22′55.1″N, long. 135°05′33.9″W, Beaufort Sea.

Peridiniacean cyst
Fig. specs. 68776, 68777, 68779, 68780
Dixon, J. et al., 1984, Geol. Surv. Can., Paper 82-13, pl. 5, figs. 5, 6, 8, 9.
Pliocene to Holocene, 610-637, 762-789 (68779), and 1524-1551 (68780) m interval, Dome Gulf et al. Hunt Kopanoar M-13 well, lat. 70°22′55.1″N, long. 135°05′33.9″W, Beaufort Sea.

Peridinium cyst type D of Norris and McAndrews 1970
Fig. spec. 68764
Dixon, J. et al., 1984, Geol. Surv. Can., Paper 82-13, pl. 4, fig. 1.
Pliocene to Holocene, 488-515 m interval, Dome Gulf et al. Hunt Kopanoar M-13 well, lat. 70°22′55.1″N, long. 135°05′33.9″W, Beaufort Sea.

Pesavis tagluensis Elsik and Jansonius
Hypotype 62627
Ioannides, N.S. and McIntyre, D.J., 1980, Geol. Surv. Can., Paper 80-1A, pl. 31.5, fig. 16.
Upper Paleocene-(?)Lower Eocene, Caribou Hills, Mackenzie River, District of Mackenzie.

Pesavis tagluensis Elsik and Jansonius
Hypotypes 67965, 67966
Norris, G., 1986, Geol. Surv. Can., Bull. 340, p. 27, pl. 3, figs. 24, 25.

Richards Formation, Late to Middle Eocene, depths 9200-9300 and 11,700-11,800 feet, Imperial Nuktak C-22 well, lat. 69°41′07″N, long. 134°51′30″W, District of Mackenzie.

Pesavis tagluensis Elsik and Jansonius
Hypotype 92052
Dietrich, J.R. et al., 1989, Geol. Surv. Can., Paper 89-1G, p. 141, pl. 3, fig. 22.
Early to Middle Eocene, depth 3500-3590 feet, Dome Pacific et al. PEX Natsek E-56 well, lat. 69°45′21.46″N, long. 139°44′34.58″W, western Beaufort Sea.

Pesavis tagluensis Elsik and Jansonius emend. Kalgutkar and Sweet 1988
Hypotypes 94356, 94360
White, J.M., 1990, Can. J. Earth Sci., vol. 27, no. 4, p. 535, pl. 1, figs. 1, 5.
Early Eocene to Early Oligocene, float in Mud Bay Creek, lat. 51°12′42″N, long. 132°15′23″W, north shore Skidegate Inlet, and depth 902.2-908.3 m, Union Port Louis well, west coast of Graham Island, lat. 53°41′18″N, long. 132°58′23″W, Queen Charlotte Islands, British Columbia.

Pesavis sp.
Fig. spec. 62626
Ioannides, N.S. and McIntyre, D.J., 1980, Geol. Surv. Can., Paper 80-1A, pl. 31.5, fig. 15.
Upper Paleocene-(?)Lower Eocene, Caribou Hills, Mackenzie River, District of Mackenzie.

Phragmothyrites eocaenica Edwards 1922 emend. Kar and Saxena 1976
Hypotypes 97297, 97298
Kalgutkar, R.M. and McIntyre, D.J., 1991, Can. J. Earth Sci., vol. 28, no. 3, p. 368, pl. 1, figs. 7, 8.
Iceberg Bay Formation, Eureka Sound Group, Early Eocene, north shore Strand Fiord, 26 km from head, lat. 79°14′N, long. 91°20′W, Axel Heiberg Island, District of Franklin.

Phragmothyrites sp. cf. *P. eocaenicus* Edwards
Fig. spec. 67992, 67993
Norris, G., 1986, Geol. Surv. Can., Bull. 340, p. 28, pl. 5, figs. 4, 5.
Richards Formation, Middle Eocene, depths 12,300-12,400 and 11,500-11,600 feet, Imperial Nuktak C-22 well, lat. 69°41′07″N, long. 134°51′30″W, District of Mackenzie.

Phyllocladidites mawsonii Cookson
Hypotype 38702
Haskell, T.P. and Wilson, G.J., 1975, Initial Repts. Deep Sea Drilling Project, vol. 29, pl. 3, fig. 10.
Eocene, DSDP Site 282, lat. 42°14.76′S, long. 143°29.18′E, west of Tasmania.

Picea sp.
Fig. spec. 27840
Hopkins, Jr., W.S., Rutter, N.W. and Rouse, G.E., 1972, Can. J. Earth Sci., vol. 9, no. 4, pl. 1, fig. 10.
Oligocene, east side Parsnip River 2 1/2 miles north of John Hart Highway bridge, lat. 55°09′N, long. 122°59′W, British Columbia.

Picea sp.
Fig. specs. 96311, 96313
McIntyre, D.J., 1991, Geol. Surv. Can., Bull. 403, pl. 2, figs. 3, 5.
Buchanan Lake Formation, Eureka Sound Group, Eocene, 25 km east of Geodetic Hills, lat. 79°55′N, long. 88°58′W, northeastern Axel Heiberg Island, District of Franklin.

Picea sp.
Fig. spec. 97320
Kalgutkar, R.M. and McIntyre, D.J., 1991, Can. J. Earth Sci., vol. 28, no. 3, p. 370, pl. 2, fig. 7.
Iceberg Bay Formation, Eureka Sound Group, Early Eocene, north shore Strand Fiord, 26 km from head, lat. 79°14′N, long. 91°20′W, Axel Heiberg Island, District of Franklin.

Piceaepollenites grandivescipites (Wodehouse)
Hypotypes 68057-68059, 68061
Norris, G., 1986, Geol. Surv. Can., Bull. 340, p. 35, pl. 8, figs. 24-26; pl. 9, fig. 1.
Richards Formation, Middle Eocene and Nuktak Formation,Pliocene (68058), depths 10,300-10,400, 600-700 (68058) and 12,600-12,650 feet, Imperial Nuktak C-22 well, lat. 69°41′07″N, long. 134°51′30″W, District of Mackenzie.

Piceaepollenites sp. A
Fig. specs. 68062-68064
Norris, G., 1986, Geol. Surv. Can., Bull. 340, p. 35, pl. 9, figs. 2, 4, 5.
Richards Formation, Late Eocene and Arnak Member, Kugmallit Formation, Late Oligocene, depths 9000-9100, 9100-9200 and 4745 feet, Imperial Nuktak C-22 well, lat. 69°41′07″N, long. 134°51′30″W, District of Mackenzie.

Pinus sp.
Fig. spec. 27839
Hopkins, Jr., W.S., Rutter, N.W. and Rouse, G.E., 1972, Can. J. Earth Sci., vol. 9, no. 4, pl. 1, figs. 8, 9.
Oligocene, east side Parsnip River 2 1/2 miles north of John Hart Highway bridge, lat. 55°09′N, long. 122°59′W, British Columbia.

Pinus sp.
Fig. spec. 97319
Kalgutkar, R.M. and McIntyre, D.J., 1991, Can. J. Earth Sci., vol. 28, no. 3, p. 370, pl. 2, fig. 6.
Iceberg Bay Formation, Eureka Sound Group, Early Eocene, north shore Strand Fiord, 26 km from head, lat. 79°14′N, long. 91°20′W, Axel Heiberg Island, District of Franklin.

Pinus sp.
Fig. spec. 96312
McIntyre, D.J., 1991, Geol. Surv. Can., Bull. 403, pl. 2, fig. 4.

Buchanan Lake Formation, Eureka Sound Group, Eocene, 25 km east of Geodetic Hills, lat. 79°55′N, long. 88°58′W, northeastern Axel Heiberg Island, District of Franklin.

Pinuspollenites elongatus Norton, 1969
Hypotype 71323
Wilson, M.A., 1978, Palaeontographica, Abt. B, vol. 166, p. 122, pl. 5, fig. 15.
Paleocene, IOE Atkinson H-25 well, north shore Tuktoyaktuk Peninsula, lat. 69°44′20″N, long. 131°50′06″W, District of Mackenzie.

Pinuspollenites labdacus (Potonié) Raatz ex Potonié, 1958
Hypotypes 68047, 68048
Norris, G., 1986, Geol. Surv. Can., Bull. 340, p. 34, pl. 8, figs. 14, 15.
Arnak Member, Kugmallit Formation, Late Oligocene, depths 3300-3400 and 4620-4700 feet, Imperial Nuktak C-22 well, lat. 69°41′07″N, long. 134°51′30″W, District of Mackenzie.

Pinuspollenites parvus Norton, 1969
Hypotype 71324
Wilson, M.A., 1978, Palaeontographica, Abt. B, vol. 166, p. 122, pl. 6, figs. 1, 2.
Paleocene, IOE Atkinson H-25 well, north shore Tuktoyaktuk Peninsula, lat. 69°44′20″N, long. 131°50′06″W, District of Mackenzie.

Pinuspollenites ruginosa (Stanley) Oltz, 1969
Hypotype 71331
Wilson, M.A., 1978, Palaeontographica, Abt. B, vol. 166, p. 122, pl. 6, figs. 11, 12.
Paleocene, IOE Atkinson H-25 well, north shore Tuktoyaktuk Peninsula, lat. 69°44′20″N, long. 131°50′06″W, District of Mackenzie.

Pinuspollenites sp. A
Fig. specs. 68049, 68050
Norris, G., 1986, Geol. Surv. Can., Bull. 340, p. 34, pl. 8, figs. 16, 17.
Arnak Member, Kugmallit Formation, Late Oligocene and Nuktak Formation, Pliocene, depths 3500-3600 and 600-700 feet, Imperial Nuktak C-22 well, lat. 69°41′07″N, long. 134°51′30″W, District of Mackenzie.

Pistillipollenites mcgregorii Rouse
Hypotype 49359
Williams, G.L. and Bujak, J.P., 1977, Am. Assoc. Strat. Palynol., Contr. Strat. Palynol., vol. 1, Contr. Ser. 5A, pl. 4, fig. 7.
Late Eocene, depth 2860-2890 feet, Eastcan et al. Freydis B-87 well, lat. 53°56′13.39″N, long. 54°42′39.75″W, Labrador Shelf.

Pistillipollenites mcgregorii Rouse
Hypotype 68129
Norris, G., 1986, Geol. Surv. Can., Bull. 340, p. 41, pl. 10, fig. 57.
Richards Formation, Middle Eocene, depth 11,900-12,000 feet, Imperial Nuktak C-22 well, lat. 69°41′07″N, long. 134°51′30″W, District of Mackenzie.

Pistillipollenites mcgregorii Rouse
Hypotypes 92039, 92040
Dietrich, J.R. et al., 1989, Geol. Surv. Can., Paper 89-1G, pl. 3, figs. 9, 10.
Early to Middle Eocene, depths 6500-6590 and 6700-6790 feet, Dome Pacific et al. PEX Natsek E-56 well, lat. 69°45′21.46″N, long. 139°44′34.58″W, western Beaufort Sea.

Pistillipollenites mcgregorii Rouse
Hypotypes 96349-96351
McIntyre, D.J., 1991, Geol. Surv. Can., Bull. 403, pl. 3, figs. 28-30.
Buchanan Lake Formation, Eureka Sound Group, Eocene, 25 km east of Geodetic Hills, lat. 79°55′N, long. 88°58′W, northeastern Axel Heiberg Island, District of Franklin.

?*Pistillipollenites mcgregorii* Rouse
Hypotype 96352
McIntyre, D.J., 1991, Geol. Surv. Can., Bull. 403, pl. 3, fig. 31.
Buchanan Lake Formation, Eureka Sound Group, Eocene, 25 km east of Geodetic Hills, lat. 79°55′N, long. 88°58′W, northeastern Axel Heiberg Island, District of Franklin.

Pistillipollenites? sp.
Fig. spec. 63178
McNeil, D.H., Ioannides, N.S. and Dixon, J., 1982, Geol. Surv. Can., Paper 80-32, pl. 4, fig. 16.
Oligocene, 2073 m below Kelly Bushing, Dome Gulf et al. Ukalerk C-50 well, lat. 70°8′5.598″N, long. 132°44′8.492″W, Beaufort Sea.

?*Planera* sp.
Fig. spec. 96332
McIntyre, D.J., 1991, Geol. Surv. Can., Bull. 403, pl. 3, fig. 10.
Buchanan Lake Formation, Eureka Sound Group, Eocene, 25 km east of Geodetic Hills, lat. 79°55′N, long. 88°58′W, northeastern Axel Heiberg Island, District of Franklin.

Platycarya sp.
Fig. specs. 92023, 92024
Dietrich, J.R. et al., 1989, Geol. Surv. Can., Paper 89-1G, p. 143, pl. 2, figs. 12, 17.
Early Eocene, depth 6500-6590 feet, Dome Pacific et al. PEX Natsek E-56 well, lat. 69°45′21.46″N, long. 139°44′34.58″W, western Beaufort Sea.

Platycaryapollenites sp. A
Fig. spec. 49360
Williams, G.L. and Bujak, J.P., 1977, Am. Assoc. Strat. Palynol., Contr. Strat. Palynol., vol. 1, Contr. Ser. 5A, pl. 4, fig. 8.
Middle-Late Eocene, depth 6080-6110 feet, Eastcan et al. Gudrid H-55 well, lat. 54°54′20.06″N, long. 55°52′31.04″W, Labrador Shelf.

Plochmopeltinites masonii Cookson 1947
Hypotype 97311
Kalgutkar, R.M. and McIntyre, D.J., 1991, Can. J. Earth Sci., vol. 28, no. 3, p. 368, pl. 1, fig. 21.
Iceberg Bay Formation, Eureka Sound Group, Early Eocene, north shore Strand Fiord, 26 km from head, lat. 79°14′N, long. 91°20′W, Axel Heiberg Island, District of Franklin.

Plochmopeltinites masonii Cookson
Hypotypes 67989-67991
Norris, G., 1986, Geol. Surv. Can., Bull. 340, p. 28, pl. 5, figs. 1-3.
Arnak Member, Kugmallit Formation, Late Oligocene and Richards Formation, Late and Middle Eocene, depths 4900-5000, 10,100-10,200 and 12,100-12,200 feet, Imperial Nuktak C-22 well, lat. 69°41′07″N, long. 134°51′30″W, District of Mackenzie.

Pluricellaesporites sp. B, A
Fig. specs. 97303, 97309
Kalgutkar, R.M. and McIntyre, D.J., 1991, Can. J. Earth Sci., vol. 28, no. 3, p. 368, pl. 1, figs. 13, 19.
Iceberg Bay Formation, Eureka Sound Group, Early Eocene, north shore Strand Fiord, 26 km from head, lat. 79°14′N, long. 91°20′W, Axel Heiberg Island, District of Franklin.

Podocarpidites ellipticus Cookson
Hypotype 38707
Haskell, T.P. and Wilson, G.J., 1975, Initial Repts. Deep Sea Drilling Project, vol. 29, pl. 3, fig. 15.
Cenozoic, DSDP Site 284, lat. 40°30.48′S, long. 167°40.81′E, Lord Howe Rise west of New Zealand.

Podocarpidites exiguus Harris
Hypotype 38706
Haskell, T.P. and Wilson, G.J., 1975, Initial Repts. Deep Sea Drilling Project, vol. 29, pl. 3, fig. 14.
Cenozoic, DSDP Site 284, lat. 40°30.48′S, long. 167°40.81′E, Lord Howe Rise west of New Zealand.

Podocarpidites biformis Rouse, 1957
Hypotype 71327
Wilson, M.A., 1978, Palaeontographica, Abt. B, vol. 166, p. 123, pl. 6, figs. 5, 6.
Paleocene, IOE Atkinson H-25 well, north shore Tuktoyaktuk Peninsula, lat. 69°44′20″N, long. 131°50′06″W, District of Mackenzie.

Podocarpidites cf. *P. major* Couper, 1953
Hypotype 71330
Wilson, M.A., 1978, Palaeontographica, Abt. B, vol. 166, p. 123, pl. 6, figs. 9, 10.
Paleocene, IOE Atkinson H-25 well, north shore Tuktoyaktuk Peninsula, lat. 69°44′20″N, long. 131°50′06″W, District of Mackenzie.

Podocarpidites sp.
Fig. spec. 49361
Williams, G.L. and Bujak, J.P., 1977, Am. Assoc. Strat. Palynol., Contr. Strat. Palynol., vol. 1, Contr. Ser. 5A, pl. 4, fig. 9.
Miocene (reworked?), depth 2610-2640 feet, Tennaco et al. Leif E-38 well, lat. 54°17′29.87″N, long. 55°5′52.17″W, Labrador Shelf.

cf. *Polemonium* sp.
Fig. spec. 63175
McNeil, D.H., Ioannides, N.S. and Dixon, J., 1982, Geol. Surv. Can., Paper 80-32, pl. 4, fig. 13.
Late Miocene-Pliocene, 1097 m below Kelly Bushing, Dome Gulf et al. Ukalerk C-50 well, lat. 70°8′5.598″N, long. 132°44′8.492″W, Beaufort Sea.

Pollen of Compositae
Fig. spec. 63176
McNeil, D.H., Ioannides, N.S. and Dixon, J., 1982, Geol. Surv. Can., Paper 80-32, pl. 4, fig. 14.
Late Miocene-Pliocene, 1280 m below Kelly Bushing, Dome Gulf et al. Ukalerk C-50 well, lat. 70°8′5.598″N, long. 132°44′8.492″W, Beaufort Sea.

Polotricolporites rotundus var. *pumilus* Sweet
Holotype 76113; paratypes 76116, 76117
Sweet, A.R., 1986, Can. J. Earth Sci., vol. 23, no. 9, p. 1384, pl. 2, figs. 4, 7, 8.
Early Paleocene, depth 534 feet, Scout Lake test hole, l.s.d. 2, sec. 15, tp. 5, rge. 1, W.3rd mer., Quontock test hole, l.s.d. 13, sec. 35, tp. 2, rge. 1, W.3rd mer., and Late Cretaceous, Highway 37 north side of Frenchman River valley, l.s.d. 14, sec. 31, tp. 4, rge. 18, W.3rd mer., Saskatchewan.

Polotricolporites rotundus var. *pumilus* Sweet, 1986
Hypotype 76058
Jerzykiewicz, T. and Sweet, A.R., 1986, Can. J. Earth Sci., vol. 23, no. 9, pl. 3, fig. 26.
Early Paleocene, depth 26.62-26.71 m, GSC core hole CV-42-2-2, Coal Valley strip mine, l.s.d. 1, sec. 10, tp. 47, rge. 19, W.5th mer., Alberta.

Polyadosporites suesae van der Hammen 1954
Hypotypes 97312, 97313
Kalgutkar, R.M. and McIntyre, D.J., 1991, Can. J. Earth Sci., vol. 28, no. 3, p. 368, pl. 1, figs. 22, 23.
Iceberg Bay Formation, Eureka Sound Group, Early Eocene, north shore Strand Fiord, 26 km from head, lat. 79°14′N, long. 91°20′W, Axel Heiberg Island, District of Franklin.

Polyatriopollenites stellatus (Potonié) Pflug
Hypotypes 68134-68136
Norris, G., 1986, Geol. Surv. Can., Bull. 340, p. 42, pl. 11, figs. 5-7.
Arnak Member, Kugmallit Formation, Late Oligocene and Richards Formation, Middle and Late Eocene, depths 4100-4200, 10,700-10,800 and 94000-9500 feet, Imperial Nuktak C-22 well, lat. 69°41′07″N, long. 134°51′30″W, District of Mackenzie.

Polybrevicolporites sp. A
Fig. spec. 68092
Norris, G., 1986, Geol. Surv. Can., Bull. 340, p. 38, pl. 10, figs. 15, 16.

Ivik Member, Kugmallit Formation, Middle to Early Oligocene, depth 7485 feet, Imperial Nuktak C-22 well, lat. 69°41′07″N, long. 134°51′30″W, District of Mackenzie.

Polypodiidites sp.
Fig. spec. 80483
Burden, E. and Holloway, D., 1985, Can. J. Earth Sci., vol. 22, no. 1, pl. 1, fig. 2.
Quaternary, Scott Inlet area, Baffin Island, District of Franklin.

?*Polypodium* sp.
Fig. specs. 96293, 96294
McIntyre, D.J., 1991, Geol. Surv. Can., Bull. 403, pl. 1, figs. 6, 7.
Buchanan Lake Formation, Eureka Sound Group, Eocene, 25 km east of Geodetic Hills, lat. 79°55′N, long. 88°58′W, northeastern Axel Heiberg Island, District of Franklin.

Polyvestibulopollenites trinus (Stanley)
Hypotypes 68120, 68121
Norris, G., 1986, Geol. Surv. Can., Bull. 340, p. 41, pl. 10, figs. 46, 53.
Mackenzie Bay Formation and Arnak Member, Kugmallit Formation, Late Oligocene, depths 3000-3100 and 3500-3600 feet, Imperial Nuktak C-22 well, lat. 69°41′07″N, long. 134°51′30″W, District of Mackenzie.

Polyvestibulopollenites verus (Potonié) Thomson and Pflug
Hypotypes 68132, 68133
Norris, G., 1986, Geol. Surv. Can., Bull. 340, p. 41, pl. 11, figs. 3, 4.
Arnak Member, Kugmallit Formation, Late Oligocene and Richards Formation, Middle Eocene, depths 4700-4800 and 10,600-10,700 feet, Imperial Nuktak C-22 well, lat. 69°41′07″N, long. 134°51′30″W, District of Mackenzie.

Potamogeton cf. *P. diversifolius* Raf.
Hypotype 46938
Prest, V.K., Terasmae, J., Matthews, Jr., J.V. and Lichti-Federovich, S., 1976, Maritime Sediments, vol. 12, no. 2, pl. 1, fig. 1a.
Holocene, Portage-du-Cap, Amherst Island, Magdalen Islands, Québec.

Proteacidites annularis Cookson
Hypotypes 38709, 38710
Haskell, T.P. and Wilson, G.J., 1975, Initial Repts. Deep Sea Drilling Project, vol. 29, pl. 4, figs. 2, 3.
Miocene and Eocene, DSDP Site 282, lat. 42°14.76′S, long. 143°29.18′E, west of Tasmania.

Proteacidites clintonensis Harris
Hypotype 38718
Haskell, T.P. and Wilson, G.J., 1975, Initial Repts. Deep Sea Drilling Project, vol. 29, pl. 4, fig. 11.
Lower Oligocene, DSDP Site 282, lat. 42°14.76′S, long. 143°29.18′E, west of Tasmania.

Proteacidites franktonensis Couper
Hypotype 38717
Haskell, T.P. and Wilson, G.J., 1975, Initial Repts. Deep Sea Drilling Project, vol. 29, pl. 4, fig. 10.
Lower Oligocene, DSDP Site 282, lat. 42°14.76′S, long. 143°29.18′E, west of Tasmania.

Proteacidites incurvatus Cookson
Hypotype 38713
Haskell, T.P. and Wilson, G.J., 1975, Initial Repts. Deep Sea Drilling Project, vol. 29, pl. 4, fig. 6.
Middle Eocene, DSDP Site 283, lat. 43°54.6′S, long. 154°16.96′E, east of Tasmania.

Proteacidites pachypolas Cookson and Pike
Hypotype 38721
Haskell, T.P. and Wilson, G.J., 1975, Initial Repts. Deep Sea Drilling Project, vol. 29, pl. 4, fig. 14.
Middle Eocene, DSDP Site 283, lat. 43°54.60′S, long. 154°16.96′E, east of Tasmania.

Proteacidites cf. *retiformis* Couper
Hypotype 38720
Haskell, T.P. and Wilson, G.J., 1975, Initial Repts. Deep Sea Drilling Project, vol. 29, pl. 4, fig. 13.
Eocene, DSDP Site 282, lat. 42°14.76′S, long. 143°29.18′E, west of Tasmania.

Proteacidites cf. *scaboratus* Couper
Hypotype 38716
Haskell, T.P. and Wilson, G.J., 1975, Initial Repts. Deep Sea Drilling Project, vol. 29, pl. 4, fig. 9.
Eocene, DSDP Site 282, lat. 42°14.76′S, long. 143°29.18′E, west of Tasmania.

Pterocarya sp.
Fig. spec. 27853
Hopkins, Jr., W.S., Rutter, N.W. and Rouse, G.E., 1972, Can. J. Earth Sci., vol. 9, no. 4, pl. 2, fig. 25.
Oligocene, east side Parsnip River 2 1/2 miles north of John Hart Highway bridge, lat. 55°09′N, long. 122°59′W, British Columbia.

Pterocarya sp.
Fig. spec. 92029
Dietrich, J.R. et al., 1989, Geol. Surv. Can., Paper 89-1G, p. 139, pl. 2, fig. 21.
Early to Middle Eocene, depth 1300-1390 feet, Dome Pacific et al. PEX Natsek E-56 well, lat. 69°45′21.46″N, long. 139°44′34.58″W, western Beaufort Sea.

Pterocarya sp.
Fig. spec. 93618
McIntyre, D.J., 1989, Geol. Surv. Can., Paper 89-1G, p. 195, pl. 2, fig. 9.
Eureka Sound Group, Paleocene, valley southeast of Cunningham Inlet between lat. 73°54′N, long. 92°35′W and lat. 74°2′N, long. 93°33′W, north end of Somerset Island, District of Franklin.

Pterocarya sp.
Fig. specs. 96324, 96325
McIntyre, D.J., 1991, Geol. Surv. Can., Bull. 403, pl. 3, figs. 2, 3.
Buchanan Lake Formation, Eureka Sound Group, Eocene, 25 km east of Geodetic Hills, lat. 79°55′N, long. 88°58′W, northeastern Axel Heiberg Island, District of Franklin.

Pterocarya sp.
Fig. spec. 97323
Kalgutkar, R.M. and McIntyre, D.J., 1991, Can. J. Earth Sci., vol. 28, no. 3, p. 370, pl. 2, fig. 10.
Iceberg Bay Formation, Eureka Sound Group, Early Eocene, north shore Strand Fiord, 26 km from head, lat. 79°14′N, long. 91°20′W, Axel Heiberg Island, District of Franklin.

Pterocaryapollenites stellatus (Potonié) Raatz, 1937
Hypotypes 71368, 71391, 71395
Wilson, M.A., 1978, Palaeontographica, Abt. B, vol. 166, p. 146, pl. 9, fig. 7; pl. 10, figs. 13, 17.
Paleocene and Cretaceous (71391), IOE Atkinson H-25 well, north shore Tuktoyaktuk Peninsula, lat. 69°44′20″N, long. 131°50′06″W, District of Mackenzie.

Pterocaryapollenites sp.
Fig. spec. 62598
Ioannides, N.S. and McIntyre, D.J., 1980, Geol. Surv. Can., Paper 80-1A, pl. 31.4, fig. 6.
Eocene, Caribou Hills, Mackenzie River, District of Mackenzie.

Pterocaryapollenites sp. A
Fig. spec. 49362
Williams, G.L. and Bujak, J.P., 1977, Am. Assoc. Strat. Palynol., Contr. Strat. Palynol., vol. 1, Contr. Ser. 5A, pl. 4, fig. 10.
Oligocene, depth 5010-5040 feet, BP Columbia Bonavista C-99 well, lat. 49°8′6.48″N, long. 51°14′27.85″W, northeast Newfoundland Basin.

?*Quadrapollenites* sp.
Fig. spec. 93597
McIntyre, D.J., 1989, Geol. Surv. Can., Paper 89-1G, p. 195, pl. 1, fig. 6.
Eureka Sound Group, Paleocene, valley southeast of Cunningham Inlet between lat. 73°54′N, long. 92°35′W and lat. 74°2′N, long. 93°33′W, north end of Somerset Island, District of Franklin.

Quercoidites microhenrica (Potonié) Potonié
Hypotype 68070
Norris, G., 1986, Geol. Surv. Can., Bull. 340, p. 36, pl. 9, fig. 10.
Arnak Member, Kugmallit Formation, Late Oligocene, depth 3300-3400 feet, Imperial Nuktak C-22 well, lat. 69°41′07″N, long. 134°51′30″W, District of Mackenzie.

Quercoidites sp. A
Fig. spec. 68071
Norris, G., 1986, Geol. Surv. Can., Bull. 340, p. 36, pl. 9, fig. 11.
Arnak Member, Kugmallit Formation, Late Oligocene, depth 4500-4600 feet, Imperial Nuktak C-22 well, lat. 69°41′07″N, long. 134°51′30″W, District of Mackenzie.

Quercus sp.
Fig. spec. 27848
Hopkins, Jr., W.S., Rutter, N.W. and Rouse, G.E., 1972, Can. J. Earth Sci., vol. 9, no. 4, pl. 2, fig. 20.
Oligocene, east side Parsnip River 2 1/2 miles north of John Hart Highway bridge, lat. 55°09′N, long. 122°59′W, British Columbia.

Quercus sp.
Fig. spec. 92047
Dietrich, J.R. et al., 1989, Geol. Surv. Can., Paper 89-1G, pl. 3, fig. 17.
Middle Eocene, depth 750-790 feet, Dome Pacific et al. PEX Natsek E-56 well, lat. 69°45′21.46″N, long. 139°44′34.58″W, western Beaufort Sea.

Quercus sp.
Fig. specs. 96360-96363, 96373
McIntyre, D.J., 1991, Geol. Surv. Can., Bull. 403, pl. 4, figs. 8-11, 20.
Buchanan Lake Formation, Eureka Sound Group, Eocene, 25 km east of Geodetic Hills, lat. 79°55′N, long. 88°58′W, northeastern Axel Heiberg Island, District of Franklin.

cf. *Quercus* sp.
Fig. spec. 93626
McIntyre, D.J., 1989, Geol. Surv. Can., Paper 89-1G, p. 175, pl. 2, fig. 17.
Eureka Sound Group, Paleocene, valley southeast of Cunningham Inlet between lat. 73°54′N, long. 92°35′W and lat. 74°2′N, long. 93°33′W, north end of Somerset Island, District of Franklin.

Radialisporis radiatus (Krutzsch) Krutzsch
Hypotype 96296
McIntyre, D.J., 1991, Geol. Surv. Can., Bull. 403, pl. 1, fig. 9.
Buchanan Lake Formation, Eureka Sound Group, Eocene, 25 km east of Geodetic Hills, lat. 79°55′N, long. 88°58′W, northeastern Axel Heiberg Island, District of Franklin.

Reduviasporonites anangus Norris
Holotype 67925; hypotypes 67926-67928
Norris, G., 1986, Geol. Surv. Can., Bull. 340, p. 23, pl. 2, figs. 36-39.
Arnak Member, Kugmallit Formation and Richards Formation, Oligocene, depths 3600-3700, 4700-4800 (67926) and 7800-7900 (67927) feet, Imperial Nuktak C-22 well, lat. 69°41′07″N, long. 134°51′30″W, District of Mackenzie.

Reduviasporonites sp. cf. *R. catenulatus* Wilson
Fig. specs. 67929-67931
Norris, G., 1986, Geol. Surv. Can., Bull. 340, p. 23, pl. 2, figs. 40, 41, 43.

Arnak Member, Kugmallit Formation, Late Oligocene, depths 3600-3700 and 4100-4200 (67931) feet, Imperial Nuktak C-22 well, lat. 69°41′07″N, long. 134°51′30″W, District of Mackenzie.

Reduviasporonites sp. A
Fig. spec. 67932
Norris, G., 1986, Geol. Surv. Can., Bull. 340, p. 24, pl. 2, fig. 44.
Arnak Member, Kugmallit Formation, Late Oligocene, depth 3300-3400 feet, Imperial Nuktak C-22 well, lat. 69°41′07″N, long. 134°51′30″W, District of Mackenzie.

Retitricolpites promiscuus Laing
Hypotype 80485
Burden, E. and Holloway, D., 1985, Can. J. Earth Sci., vol. 22, no. 10, pl. 1, fig. 4.
Quaternary, Scott Inlet area, Baffin Island, District of Franklin.

Retitriletes annotinioides Krutzsch
Hypotypes 68033, 68034
Norris, G., 1986, Geol. Surv. Can., Bull. 340, p. 33, pl. 7, figs. 31, 32.
Mackenzie Bay Formation, Miocene, depths 2400-2500 and 2800-2900 feet, Imperial Nuktak C-22 well, lat. 69°41′07″N, long. 134°51′30″W, District of Mackenzie.

Retitriletes sp. cf. *R. novomexicanus* (Anderson)
Fig. specs. 68035-68038
Norris, G., 1986, Geol. Surv. Can., Bull. 340, p. 33, pl. 8, figs. 1-5.
Richards Formation, Eocene to Early Oligocene (68036), depths 9500-9600, 7900-8000, 10,000-10,100 and 11,300-11,400 feet, Imperial Nuktak C-22 well, lat. 69°41′07″N, long. 134°51′30″W, District of Mackenzie.

Retitriletes sp. cf. *R. oligocenicus* Krutzsch
Fig. specs. 68029, 68030, 68039
Norris, G., 1986, Geol. Surv. Can., Bull. 340, p. 32, pl. 7, figs. 25, 26; pl. 8, fig. 6.
Mackenzie Bay Formation, Miocene and Richards Formation, Late Eocene (68030), depths 2400-2500, 10,000-10,100 and 2300-2400 feet, Imperial Nuktak C-22 well, lat. 69°41′07″N, long. 134°51′30″W, District of Mackenzie.

Rhizophagites cerasiformis Norris
Holotype 67877; hypotypes 67875, 67876, 67878, 67879
Norris, G., 1986, Geol. Surv. Can., Bull. 340, p. 20, pl. 1, figs. 23-27.
Nuktak Formation and Arnak Member, Kugmallit Formation (67876), Pliocene, Late Miocene and Late Oligocene, depths 600-700, 1400-1500, 4100-4200 and 900-1000 feet, Imperial Nuktak C-22 well, lat. 69°41′07″N, long. 134°51′30″W, District of Mackenzie.

Rhoipites sp.
Fig. spec. 71369
Wilson, M.A., 1978, Palaeontographica, Abt. B, vol. 166, p. 138, pl. 9, fig. 8.
Paleocene, IOE Atkinson H-25 well, north shore Tuktoyaktuk Peninsula, lat. 69°44′20″N, long. 131°50′06″W, District of Mackenzie.

?*Rhus* sp.
Fig. spec. 96379
McIntyre, D.J., 1991, Geol. Surv. Can., Bull. 403, pl. 4, fig. 27.
Buchanan Lake Formation, Eureka Sound Group, Eocene, 25 km east of Geodetic Hills, lat. 79°55′N, long. 88°58′W, northeastern Axel Heiberg Island, District of Franklin.

?Rosaceae
Fig. specs. 93607, 93608
McIntyre, D.J., 1989, Geol. Surv. Can., Paper 89-1G, p. 195, pl. 1, figs. 16, 17.
Eureka Sound Group, Paleocene, valley southeast of Cunningham Inlet between lat. 73°54′N, long. 92°35′W and lat. 74°2′N, long. 93°33′W, north end of Somerset Island, District of Franklin.

?Rosaceae
Fig. spec. 96377
McIntyre, D.J., 1991, Geol. Surv. Can., Bull. 403, pl. 4, fig. 25.
Buchanan Lake Formation, Eureka Sound Group, Eocene, 25 km east of Geodetic Hills, lat. 79°55′N, long. 88°58′W, northeastern Axel Heiberg Island, District of Franklin.

Salix sp.
Fig. spec. 96357
McIntyre, D.J., 1991, Geol. Surv. Can., Bull. 403, pl. 4, fig. 5.
Buchanan Lake Formation, Eureka Sound Group, Eocene, 25 km east of Geodetic Hills, lat. 79°55′N, long. 88°58′W, northeastern Axel Heiberg Island, District of Franklin.

Salixpollenites discoloripites (Wodehouse) Srivastava
Hypotype 68069
Norris, G., 1986, Geol. Surv. Can., Bull. 340, p. 36, pl. 9, fig. 9.
Arnak Member, Kugmallit Formation, Late Oligocene, depth 3300-3400 feet, Imperial Nuktak C-22 well, lat. 69°41′07″N, long. 134°51′30″W, District of Mackenzie.

Saxonipollis saxonicus Krutzsch
Hypotype 62620
Ioannides, N.S. and McIntyre, D.J., 1980, Geol. Surv. Can., Paper 80-1A, pl. 31.5, figs. 8, 14.
Oligocene(?), Caribou Hills, Mackenzie River, District of Mackenzie.

Saxonipollis saxonicus Krutzsch
Hypotype 63173
McNeil, D.H., Ioannides, N.S. and Dixon, J., 1982, Geol. Surv. Can., Paper 80-32, pl. 4, fig. 11.
Miocene, 1707 m below Kelly Bushing, Dome Gulf et al. Ukalerk C-50 well, lat. 70°8′5.598″N, long. 132°44′8.492″W, Beaufort Sea.

Saxonipollis sp.
Fig. spec. 62623
Ioannides, N.S. and McIntyre, D.J., 1980, Geol. Surv. Can., Paper 80-1A, pl. 31.5, fig. 11.
Upper Paleocene-(?)Lower Eocene, Caribou Hills, Mackenzie River, District of Mackenzie.

Saxonipollis sp. A of Ioannides and McIntyre
Fig. spec. 93609
McIntyre, D.J., 1989, Geol. Surv. Can., Paper 89-1G, p. 195, pl. 1, fig. 18.
Eureka Sound Group, Paleocene, valley southeast of Cunningham Inlet between lat. 73°54′N, long. 92°35′W and lat. 74°2′N, long. 93°33′W, north end of Somerset Island, District of Franklin.

Schizaea triangula Stanley
Hypotype 93592
McIntyre, D.J., 1989, Geol. Surv. Can., Paper 89-1G, p. 195, pl. 1, fig. 1.
Eureka Sound Group, Paleocene, valley southeast of Cunningham Inlet between lat. 73°54′N, long. 92°35′W and lat. 74°2′N, long. 93°33′W, north end of Somerset Island, District of Franklin.

Sequoiapollenites polyformosus Thiergart
Hypotypes 68051-68053
Norris, G., 1986, Geol. Surv. Can., Bull. 340, p. 35, pl. 8, figs. 18-20.
Richards Formation, Middle Eocene and Arnak Member, Kugmallit Formation, Late Oligocene, depths 11,700-11,800, 4620-4700 and 3300-3400 feet, Imperial Nuktak C-22 well, lat. 69°41′07″N, long. 134°51′30″W, District of Mackenzie.

Sequoiapollenites sp.
Fig. spec. 93594
McIntyre, D.J., 1989, Geol. Surv. Can., Paper 89-1G, p. 195, pl. 1, fig. 3.
Eureka Sound Group, Paleocene, valley southeast of Cunningham Inlet between lat. 73°54′N, long. 92°35′W and lat. 74°2′N, long. 93°33′W, north end of Somerset Island, District of Franklin.

Sigmopollis psilatus Piel
Hypotype 62621
Ioannides, N.S. and McIntyre, D.J., 1980, Geol. Surv. Can., Paper 80-1A, pl. 31.5, fig. 9.
Upper Paleocene-(?)Lower Eocene, Caribou Hills, Mackenzie River, District of Mackenzie.

Sigmopollis psilatus Piel
Hypotypes 68054, 68055
Norris, G., 1986, Geol. Surv. Can., Bull. 340, p. 35, pl. 8, figs. 21, 22.
Nuktak Formation, Pliocene or Late Miocene and Arnak Member, Kugmallit Formation, Late Oligocene, depths 1400-1500 and 3600-3700 feet, Imperial Nuktak C-22 well, lat. 69°41′07″N, long. 134°51′30″W, District of Mackenzie.

Sparganiaceaepollenites neogenicus Krutzsch
Hypotypes 68090, 68091
Norris, G., 1986, Geol. Surv. Can., Bull. 340, p. 38, pl. 10, figs. 14, 17.
Arnak Member, Kugmallit Formation, Late Oligocene, depths 3300-3400 and 3600-3700 feet, Imperial Nuktak C-22 well, lat. 69°41′07″N, long. 134°51′30″W, District of Mackenzie.

Sparganium sp.
Fig. spec. 93596
McIntyre, D.J., 1989, Geol. Surv. Can., Paper 89-1G, p. 195, pl. 1, fig. 5.
Eureka Sound Group, Paleocene, valley southeast of Cunningham Inlet between lat. 73°54′N, long. 92°35′W and lat. 74°2′N, long. 93°33′W, north end of Somerset Island, District of Franklin.

Sphagnum sp.
Fig. spec. 96292
McIntyre, D.J., 1991, Geol. Surv. Can., Bull. 403, pl. 1, fig. 5.
Buchanan Lake Formation, Eureka Sound Group, Eocene, 25 km east of Geodetic Hills, lat. 79°55′N, long. 88°58′W, northeastern Axel Heiberg Island, District of Franklin.

Sphagnum sp.
Fig. spec. 97314
Kalgutkar, R.M. and McIntyre, D.J., 1991, Can. J. Earth Sci., vol. 28, no. 3, p. 370, pl. 2, fig. 1.
Iceberg Bay Formation, Eureka Sound Group, Early Eocene, north shore Strand Fiord, 26 km from head, lat. 79°14′N, long. 91°20′W, Axel Heiberg Island, District of Franklin.

Spiniferites ramosus (Ehrenberg) Loeblich and Loeblich
Hypotype 63181
McNeil, D.H., Ioannides, N.S. and Dixon, J., 1982, Geol. Surv. Can., Paper 80-32, pl. 4, fig. 19.
Miocene, 1768 m below Kelly Bushing, Dome Gulf et al. Ukalerk C-50 well, lat. 70°8′5.598″N, long. 132°44′8.492″W, Beaufort Sea.

Spongodinium delitiense (Ehrenberg) Deflandre
Hypotype 62647
Ioannides, N.S. and McIntyre, D.J., 1980, Geol. Surv. Can., Paper 80-1A, pl. 31.2, fig. 10.
Eocene, Caribou Hills, Mackenzie River, District of Mackenzie.

Staphlosporonites sp. cf. *S. conoideus* Sheffy and Dilcher
Fig. specs. 67963, 67964
Norris, G., 1986, Geol. Surv. Can., Bull. 340, p. 26, pl. 3, figs. 23, 27.
Richards Formation, Late Eocene, depths 9000-9100 and 8900-9000 feet, Imperial Nuktak C-22 well, lat. 69°41′07″N, long. 134°51′30″W, District of Mackenzie.

Staphlosporonites delumbus Norris
Holotype 67958; hypotypes 67957, 67959, 67960
Norris, G., 1986, Geol. Surv. Can., Bull. 340, p. 26, pl. 3, figs. 17-20.

Arnak Member, Kugmallit Formation, Late Oligocene and Richards Formation, Late Eocene (67959), depths 3000-3600, 4400-4500, 9100-9200 and 4700-4800 feet, Imperial Nuktak C-22 well, lat. 69°41′07″N, long. 134°51′30″W, District of Mackenzie.

Staphlosporonites sp. A
Fig. spec. 67961
Norris, G., 1986, Geol. Surv. Can., Bull. 340, p. 26, pl. 3, fig. 21.
Arnak Member, Kugmallit Formation, Late Oligocene, depth 3500-3600 feet, Imperial Nuktak C-22 well, lat. 69°41′07″N, long. 134°51′30″W, District of Mackenzie.

Stereisporites microgranulus Krutzsch
Hypotypes 68010-68012
Norris, G., 1986, Geol. Surv. Can., Bull. 340, p. 30, pl. 7, figs. 5-7.
Arnak Member, Kugmallit Formation, Late Oligocene, depths 3300-3400, 4100-4200 and 4620-4700 feet, Imperial Nuktak C-22 well, lat. 69°41′07″N, long. 134°51′30″W, District of Mackenzie.

Stereisporites minor (Raatz) Krutzsch
Hypotypes 68008, 68009
Norris, G., 1986, Geol. Surv. Can., Bull. 340, p. 30, pl. 7, figs. 3, 4.
Nuktak Formation, Pliocene and Mackenzie Bay Formation, Miocene, depths 900-1000 and 2400-2500 feet, Imperial Nuktak C-22 well, lat. 69°41′07″N, long. 134°51′30″W, District of Mackenzie.

Stereisporites stereoides (Potonié and Venitz) Pflug *in* Thomson and Pflug
Hypotypes 68006, 68007
Norris, G., 1986, Geol. Surv. Can., Bull. 340, p. 30, pl. 7, figs. 1, 2.
Mackenzie Bay Formation, Miocene and Arnak Member, Kugmallit Formation, Late Oligocene, depths 2400-2500 and 3500-3600 feet, Imperial Nuktak C-22 well, lat. 69°41′07″N, long. 134°51′30″W, District of Mackenzie.

Striacolporites sp. A
Fig. spec. 68082
Norris, G., 1986, Geol. Surv. Can., Bull. 340, p. 37, pl. 10, fig. 5.
Arnak Member, Kugmallit Formation, Late Oligocene, depth 3300-3400 feet, Imperial Nuktak C-22 well, lat. 69°41′07″N, long. 134°51′30″W, District of Mackenzie.

Striadiporites bistriatus (Ke et Shi ex Sung et al.) comb. nov.
Hypotypes 67885-67887
Norris, G., 1986, Geol. Surv. Can., Bull. 340, p. 20, pl. 1, figs. 32-34.
Richards Formation, Late to Middle (67887) Eocene, depths 10,600-10,700, 9000-9100 and 11,500-11,600 feet, Imperial Nuktak C-22 well, lat. 69°41′07″N, long. 134°51′30″W, District of Mackenzie.

Striadiporites inflexus (Ke et Shi ex Sung et al.) comb. nov.
Hypotypes 67882-67884
Norris, G., 1986, Geol. Surv. Can., Bull. 340, p. 21, pl. 1, figs. 30, 31, 35.
Richards Formation, Middle and Late Eocene, depths 10,300-10,400, 9000-9100 and 9200-9300 feet, Imperial Nuktak C-22 well, lat. 69°41′07″N, long. 134°51′30″W, District of Mackenzie.

Striadiporites multistriatus (Ke et Shi ex Sung et al.) comb. nov.
Hypotypes 67888, 67889
Norris, G., 1986, Geol. Surv. Can., Bull. 340, p. 20, pl. 1, figs. 36, 37.
Richards Formation, Late Eocene, depths 9200-9300 and 8900-9000 feet, Imperial Nuktak C-22 well, lat. 69°41′07″N, long. 134°51′30″W, District of Mackenzie.

Striadiporites sanctaebarbarae Elsik and Jansonius
Hypotypes 67880, 67881
Norris, G., 1986, Geol. Surv. Can., Bull. 340, p. 20, pl. 1, figs. 28, 29.
Richards Formation, Middle and Late Eocene, depths 11,800-11,900 and 9000-9100 feet, Imperial Nuktak C-22 well, lat. 69°41′07″N, long. 134°51′30″W, District of Mackenzie.

Striadiporites sp.
Fig. spec. 92050
Dietrich, J.R. et al., 1989, Geol. Surv. Can., Paper 89-1G, p. 141, pl. 3, fig. 20.
Early to Middle Eocene, depth 1100-1190 feet, Dome Pacific et al. PEX Natsek E-56 well, lat. 69°45′21.46″N, long. 139°44′34.58″W, western Beaufort Sea.

Striadiporites sp.
Fig. spec. 94362
White, J.M., 1990, Can. J. Earth Sci., vol. 27, no. 4, p. 535, pl. 1, fig. 7.
Early Eocene to Early Oligocene, depth 1435.6-1440.2 m, Union Port Louis well, west coast of Graham Island, lat. 53°41′18″N, long. 132°58′23″W, Queen Charlotte Islands, British Columbia.

Striatites spp.
Fig. specs. 38724, 38725
Haskell, T.P. and Wilson, G.J., 1975, Initial Repts. Deep Sea Drilling Project, vol. 29, pl. 4, figs. 17, 18.
Miocene, DSDP Site 282, lat. 42°14.76′S, long. 143°29.18′E, west of Tasmania.

?Syncolpate sp. A
Fig. spec. 94364
White, J.M., 1990, Can. J. Earth Sci., vol. 27, no. 4, p. 535, pl. 1, fig. 9.
Early Eocene to Early Oligocene, depth 1452.4-1458.5 m, Union Port Louis well, west coast of Graham Island, lat. 53°41′18″N, long. 132°58′23″W, Queen Charlotte Islands, British Columbia.

Systematophora? sp.
Fig. spec. 63183
McNeil, D.H., Ioannides, N.S. and Dixon, J., 1982, Geol. Surv. Can., Paper 80-32, pl. 4, fig. 21.
Miocene, 1707 m below Kelly Bushing, Dome Gulf et al. Ukalerk C-50 well, lat. 70°8′5.598″N, long. 132°44′8.492″W, Beaufort Sea.

Taraxacum sp.
Fig. spec. 41914
Williams, G.L., 1975, Geol. Surv. Can., Paper 74-30, vol. 2, pl. 8, fig. 19.
Plio-Pleistocene, Amoco IOE Puffin B-90 well, Grand Banks.

Taxodiaceaepollenites hiatus Potonié ex Potonié
Hypotype 68056
Norris, G., 1986, Geol. Surv. Can., Bull. 340, p. 35, pl. 8, fig. 23.
Mackenzie Bay Formation, Miocene, depth 2800-2900 feet, Imperial Nuktak C-22 well, lat. 69°41′07″N, long. 134°51′30″W, District of Mackenzie.

Tetrapollis conspectus Manum
Hypotype 93627
McIntyre, D.J., 1989, Geol. Surv. Can., Paper 89-1G, p. 195, pl. 2, fig. 18.
Eureka Sound Group, Paleocene, valley southeast of Cunningham Inlet between lat. 73°54′N, long. 92°35′W and lat. 74°2′N, long. 93°33′W, north end of Somerset Island, District of Franklin.

Tilia danei Anderson 1960
Hypotype 76133
Jerzykiewicz, T. and Sweet, A.R., 1986, Geol. Surv. Can., Paper 86-1B, Fig. 67.2, fig. 3.
Porcupine Hills Formation, Middle Paleocene, sec. 9, tp. 15, rge. 1, W.5th mer., Alberta.

Tilia vescipites Wodehouse 1933
Hypotype 94359
White, J.M., 1990, Can. J. Earth Sci., vol. 27, no. 4, p. 535, pl. 1, fig. 4.
Early Eocene to Early Oligocene, float in Mud Bay Creek, lat. 51°12′42″N, long. 132°15′23″W, north shore Skidegate Inlet, Queen Charlotte Islands, British Columbia.

Tilia sp. *(T. vescipites* Wodehouse)
Fig. spec. 93603
McIntyre, D.J., 1989, Geol. Surv. Can., Paper 89-1G, p. 195, pl. 1, fig. 12.
Eureka Sound Group, Paleocene, valley southeast of Cunningham Inlet between lat. 73°54′N, long. 92°35′W and lat. 74°2′N, long. 93°33′W, north end of Somerset Island, District of Franklin.

Tilia sp. *(T. vescipites* Wodehouse 1933)
Fig. specs. 97328, 97329
Kalgutkar, R.M. and McIntyre, D.J., 1991, Can. J. Earth Sci., vol. 28, no. 3, p. 370, pl. 2, figs. 15, 16.
Iceberg Bay Formation, Eureka Sound Group, Early Eocene, north shore Strand Fiord, 26 km from head, lat. 79°14′N, long. 91°20′W, Axel Heiberg Island, District of Franklin.

Tilia sp.
Fig. spec. 27855
Hopkins, Jr., W.S., Rutter, N.W. and Rouse, G.E., 1972, Can. J. Earth Sci., vol. 9, no. 4, pl. 2, fig. 27.
Oligocene, east side Parsnip River 2 1/2 miles north of John Hart Highway bridge, lat. 55°09′N, long. 122°59′W, British Columbia.

Tilia sp.
Fig. specs. 92036-92038
Dietrich, J.R. et al., 1989, Geol. Surv. Can., Paper 89-1G, pl. 3, figs. 6-8.
Early to Middle Eocene and Paleocene (92038), depths 1300-1390 and 8400-8490 feet, Dome Pacific et al. PEX Natsek E-56 well, lat. 69°45′21.46″N, long. 139°44′34.58″W, western Beaufort Sea.

Tilia sp.
Fig. spec. 94363
White, J.M., 1990, Can. J. Earth Sci., vol. 27, no. 4, p. 535, pl. 1, fig. 8.
Early Eocene to Early Oligocene, depth 1452.4-1458.5 m, Union Port Louis well, west coast of Graham Island, lat. 53°41′18″N, long. 132°58′23″W, Queen Charlotte Islands, British Columbia.

Tilia sp.
Fig. specs. 96333, 96334
McIntyre, D.J., 1991, Geol. Surv. Can., Bull. 403, pl. 3, figs. 11-13.
Buchanan Lake Formation, Eureka Sound Group, Eocene, 25 km east of Geodetic Hills, lat. 79°55′N, long. 88°58′W, northeastern Axel Heiberg Island, District of Franklin.

Tiliapollenites - Bombacacidites complex
Fig. specs. 62599-62604
Ioannides, N.S. and McIntyre, D.J., 1980, Geol. Surv. Can., Paper 80-1A, pl. 31.4, figs. 7-12.
Eocene, Caribou Hills, Mackenzie River, District of Mackenzie.

Tiliapollenites - Bombacacidites complex
Fig. specs. 63170, 63171
McNeil, D.H., Ioannides, N.S. and Dixon, J., 1982, Geol. Surv. Can., Paper 80-32, pl. 4, figs. 8, 9.
Miocene, 1768 m below Kelly Bushing, and Oligocene, 2073 m below Kelly Bushing, Dome Gulf et al. Ukalerk C-50 well, lat. 70°8′5.598″N, long. 132°44′8.492″W, Beaufort Sea.

Tiliaepollenites sp.
Fig. spec. 49364
Williams, G.L. and Bujak, J.P., 1977, Am. Assoc. Strat. Palynol., Contr. Strat. Palynol., vol. 1, Contr. Ser. 5A, pl. 4, fig. 12.
Miocene, depth 1840-1870 feet, Eastcan et al. Freydis B-87 well, lat. 53°56′13.39″N, long. 54°42′39.75″W, Labrador Shelf.

Triatriopollenites sp. A
Fig. spec. 49363
Williams, G.L. and Bujak, J.P., 1977, Am. Assoc. Strat. Palynol., Contr. Strat. Palynol., vol. 1, Contr. Ser. 5A, pl. 4, fig. 11.
Pliocene-Pleistocene (reworked), Eastcan et al. Freydis B-87 well, lat. 53°56′13.39″N, long. 54°42′39.75″W, Labrador Shelf.

Trichothyrites sp. A
Fig. specs. 67995, 67996
Norris, G., 1986, Geol. Surv. Can., Bull. 340, p. 29, pl. 6, figs. 1, 2.
Ivik Member, Kugmallit Formation, Middle to Early Oligocene and Richards Formation, Late Eocene, depths 7600-7700 and 9450 feet, Imperial Nuktak C-22 well, lat. 69°41′07″N, long. 134°51′30″W, District of Mackenzie.

Tricolpites hians Stanley
Hypotypes 68067, 68068
Norris, G., 1986, Geol. Surv. Can., Bull. 340, p. 35, pl. 9, figs. 7, 8.
Richards Formation, Middle Eocene and Arnak Member, Kugmallit Formation, Late Oligocene, depths 9500-9600 and 3300-3400 feet, Imperial Nuktak C-22 well, lat. 69°41′07″N, long. 134°51′30″W, District of Mackenzie.

Tricolpites sabulosus Dettman and Playford
Hypotypes 38711, 38712
Haskell, T.P. and Wilson, G.J., 1975, Initial Repts. Deep Sea Drilling Project, vol. 29, pl. 4, figs. 4, 5.
Paleocene, DSDP Site 282, lat. 42°14.76′S, long. 143°29.18′E, west of Tasmania.

Tricolpites sp.
Fig. spec. 96382
McIntyre, D.J., 1991, Geol. Surv. Can., Bull. 403, pl. 4, fig. 30.
Buchanan Lake Formation, Eureka Sound Group, Eocene, 25 km east of Geodetic Hills, lat. 79°55′N, long. 88°58′W, northeastern Axel Heiberg Island, District of Franklin.

Tricolpites sp.
Fig. specs. 97335-97337
Kalgutkar, R.M. and McIntyre, D.J., 1991, Can. J. Earth Sci., vol. 28, no. 3, p. 370, pl. 2, figs. 22-24.
Iceberg Bay Formation, Eureka Sound Group, Early Eocene, north shore Strand Fiord, 26 km from head, lat. 79°14′N, long. 91°20′W, Axel Heiberg Island, District of Franklin.

Tricolporate pollen
Fig. spec. 63177
McNeil, D.H., Ioannides, N.S. and Dixon, J., 1982, Geol. Surv. Can., Paper 80-32, pl. 4, fig. 15.
Oligocene, 2133 m below Kelly Bushing, Dome Gulf et al. Ukalerk C-50 well, lat. 70°8′5.598″N, long. 132°44′8.492″W, Beaufort Sea.

Tricolporopollenites cingulum (Potonié) Thomson and Pflug, 1953
Hypotype 71374
Wilson, M.A., 1978, Palaeontographica, Abt. B, vol. 166, p. 139, pl. 9, fig. 14.
Paleocene, IOE Atkinson H-25 well, north shore Tuktoyaktuk Peninsula, lat. 69°44′20″N, long. 131°50′06″W, District of Mackenzie.

Tricolporopollenites kruschii (Potonié) Thomson and Pflug
Hypotype 92046
Dietrich, J.R. et al., 1989, Geol. Surv. Can., Paper 89-1G, p. 139, pl. 3, fig. 16.
Early to Middle Eocene, depth 2000-2090 feet, Dome Pacific et al. PEX Natsek E-56 well, lat. 69°45′21.46″N, long. 139°44′34.58″W, western Beaufort Sea.

Tricolporopollenites kruschii (Potonié) Thomson and Pflug
Fig. specs. 96368, 96369
McIntyre, D.J., 1991, Geol. Surv. Can., Bull. 403, pl. 4, figs. 16, 17.
Buchanan Lake Formation, Eureka Sound Group, Eocene, 25 km east of Geodetic Hills, lat. 79°55′N, long. 88°58′W, northeastern Axel Heiberg Island, District of Franklin.

Tricolporopollenites kruschii (Potonié 1934) Thomson and Pflug 1953
Hypotype 97341
Kalgutkar, R.M. and McIntyre, D.J., 1991, Can. J. Earth Sci., vol. 28, no. 3, p. 370, pl. 2, fig. 28.
Iceberg Bay Formation, Eureka Sound Group, Early Eocene, north shore Strand Fiord, 26 km from head, lat. 79°14′N, long. 91°20′W, Axel Heiberg Island, District of Franklin.

Tricolporopollenites sp.
Fig. specs. 96380, 96381
McIntyre, D.J., 1991, Geol. Surv. Can., Bull. 403, pl. 4, figs. 28, 29.
Buchanan Lake Formation, Eureka Sound Group, Eocene, 25 km east of Geodetic Hills, lat. 79°55′N, long. 88°58′W, northeastern Axel Heiberg Island, District of Franklin.

Tricolporopollenites sp.
Fig. specs. 97334, 97343
Kalgutkar, R.M. and McIntyre, D.J., 1991, Can. J. Earth Sci., vol. 28, no. 3, p. 370, pl. 2, figs. 21, 30.
Iceberg Bay Formation, Eureka Sound Group, Early Eocene, north shore Strand Fiord, 26 km from head, lat. 79°14′N, long. 91°20′W, Axel Heiberg Island, District of Franklin.

Tricolporopollenites sp. A
Fig. spec. 96344
McIntyre, D.J., 1991, Geol. Surv. Can., Bull. 403, pl. 3, fig. 23.
Buchanan Lake Formation, Eureka Sound Group, Eocene, 25 km east of Geodetic Hills, lat. 79°55′N, long. 88°58′W, northeastern Axel Heiberg Island, District of Franklin.

Tricolporopollenites sp. ?Acanthaceae
Fig. specs. 96375, 96376
McIntyre, D.J., 1991, Geol. Surv. Can., Bull. 403, pl. 4, figs. 23, 24.
Buchanan Lake Formation, Eureka Sound Group, Eocene, 25 km east of Geodetic Hills, lat. 79°55′N, long. 88°58′W, northeastern Axel Heiberg Island, District of Franklin.

Triporisporonites versus (Ke et Shi ex Sung et al.) comb. nov.
Hypotype 67867
Norris, G., 1986, Geol. Surv. Can., Bull. 340, p. 19, pl. 1, fig. 15.
Richards Formation, Middle Eocene, depth 12,500-12,600 feet, Imperial Nuktak C-22 well, lat. 69°41′07″N, long. 134°51′30″W, District of Mackenzie.

Triporopollenites mullensis (Simpson) Rouse and Srivastava
Hypotype 92035
Dietrich, J.R. et al., 1989, Geol. Surv. Can., Paper 89-1G, p. 144, pl. 3, fig. 5.
Paleocene, depth 7000-7090 feet, Dome Pacific et al. PEX Natsek E-56 well, lat. 69°45′21.46″N, long. 139°44′34.58″W, western Beaufort Sea.

Triporopollenites mullensis Simpson
Hypotype 93620
McIntyre, D.J., 1989, Geol. Surv. Can., Paper 89-1G, p. 195, pl. 2, fig. 11.
Eureka Sound Group, Paleocene, valley southeast of Cunningham Inlet between lat. 73°54′N, long. 92°35′W and lat. 74°2′N, long. 93°33′W, north end of Somerset Island, District of Franklin.

Triporopollenites plektosus Anderson 1960
Hypotypes 76080, 76081
Jerzykiewicz, T. and Sweet, A.R., 1986, Can. J. Earth Sci., vol. 23, no. 9, pl. 4, figs. 20, 21.
Early Paleocene, depths 27.34-27.38 m and 27.38-27.41 m, GSC core hole CV-42-2-2, Coal Valley strip mine, l.s.d. 1, sec. 10, tp. 47, rge. 19, W.5th mer., Alberta.

Triporopollenites sp.
Fig. specs. 93621, 93622
McIntyre, D.J., 1989, Geol. Surv. Can., Paper 89-1G, p. 195, pl. 2, figs. 12, 13.
Eureka Sound Group, Paleocene, valley southeast of Cunningham Inlet between lat. 73°54′N, long. 92°35′W and lat. 74°2′N, long. 93°33′W, north end of Somerset Island, District of Franklin.

Triprojectus magnus (Mtcheldlishvili) Stanley, 1970
Hypotype 71366
Wilson, M.A., 1978, Palaeontographica, Abt. B, vol. 166, p. 138, pl. 9, fig. 4.
Paleocene, IOE Atkinson H-25 well, north shore Tuktoyaktuk Peninsula, lat. 69°44′20″N, long. 131°50′06″W, District of Mackenzie.

Trivestibulopollenites betuloides Pflug, *in* Thomson and Pflug
Hypotypes 68112-68116
Norris, G., 1986, Geol. Surv. Can., Bull. 340, p. 40, pl. 10, figs. 38-42.
Nuktak Formation, Pliocene (68112) and Mackenzie Bay Formation, Miocene, depths 100-200, 2300-2400, 2800-2900 and 2100-2200 feet, Imperial Nuktak C-22 well, lat. 69°41′07″N, long. 134°51′30″W, District of Mackenzie.

Trivestibulopollenites claripites (Wodehouse)
Hypotypes 68122-68124
Norris, G., 1986, Geol. Surv. Can., Bull. 340, p. 40, pl. 10, figs. 47-49.
Nuktak Formation, Pliocene or Late Miocene, Mackenzie Bay Formation, Miocene and Arnak Member, Kugmallit Formation, Late Oligocene, depths 1200-1300, 2400-2500 and 3300-3400 feet, Imperial Nuktak C-22 well, lat. 69°41′07″N, long. 134°51′30″W, District of Mackenzie.

Trudopollis barentsii Manum
Hypotype 93628
McIntyre, D.J., 1989, Geol. Surv. Can., Paper 89-1G, p. 195, pl. 2, fig. 19.
Eureka Sound Group, Paleocene, valley southeast of Cunningham Inlet between lat. 73°54′N, long. 92°35′W and lat. 74°2′N, long. 93°33′W, north end of Somerset Island, District of Franklin.

Trudopollis sp., Type B of Manum
Fig. spec. 93624
McIntyre, D.J., 1989, Geol. Surv. Can., Paper 89-1G, p. 195, pl. 2, fig. 15.
Eureka Sound Group, Paleocene, valley southeast of Cunningham Inlet between lat. 73°54′N, long. 92°35′W and lat. 74°2′N, long. 93°33′W, north end of Somerset Island, District of Franklin.

Tsuga heterophyllites Martin and Rouse
Hypotype 27842
Hopkins, Jr., W.S., Rutter, N.W. and Rouse, G.E., 1972, Can. J. Earth Sci., vol. 9, no. 4, pl. 1, fig. 13.
Oligocene, east side Parsnip River 2 1/2 miles north of John Hart Highway bridge, lat. 55°09′N, long. 122°59′W, British Columbia.

Tsuga minisacca Martin and Rouse
Hypotype 27841
Hopkins, Jr., W.S., Rutter, N.W. and Rouse, G.E., 1972, Can. J. Earth Sci., vol. 9, no. 4, pl. 1, fig. 12.
Oligocene, east side Parsnip River 2 1/2 miles north of John Hart Highway bridge, lat. 55°09′N, long. 122°59′W, British Columbia.

Tsuga sp.
Fig. specs. 96308, 96309
McIntyre, D.J., 1991, Geol. Surv. Can., Bull. 403, pl. 1, fig. 21; pl. 2, fig. 1.
Buchanan Lake Formation, Eureka Sound Group, Eocene, 25 km east of Geodetic Hills, lat. 79°55′N, long. 88°58′W, northeastern Axel Heiberg Island, District of Franklin.

Tsugaepollenites igniculus (Potonié) Potonié and Venitz
Hypotype 68066
Norris, G., 1986, Geol. Surv. Can., Bull. 340, p. 34, pl. 9, fig. 6.
Mackenzie Bay Formation, Miocene, depth 2730 feet, Imperial Nuktak C-22 well, lat. 69°41′07″N, long. 134°51′30″W, District of Mackenzie.

Tsugaepollenites viridifluminipites (Wodehouse)
Hypotypes 68060, 68065
Norris, G., 1986, Geol. Surv. Can., Bull. 340, p. 34, pl. 8, fig. 27; pl. 9, fig. 3.
Richards Formation, Middle Eocene and Mackenzie Bay Formation, Miocene, depths 10,900-11,000 and 3000-3100 feet, Imperial Nuktak C-22 well, lat. 69°41′07″N, long. 134°51′30″W, District of Mackenzie.

Typha sp.
Fig. spec. 27845
Hopkins, Jr., W.S., Rutter, N.W. and Rouse, G.E., 1972, Can. J. Earth Sci., vol. 9, no. 4, pl. 1, fig. 16.
Oligocene, east side Parsnip River 2 1/2 miles north of John Hart Highway bridge, lat. 55°09′N, long. 122°59′W, British Columbia.

Ulmipollenites undulosus Wolfe
Hypotypes 62606, 62607
Ioannides, N.S. and McIntyre, D.J., 1980, Geol. Surv. Can., Paper 80-1A, pl. 31.4, figs. 14, 15.
Eocene, Caribou Hills, Mackenzie River, District of Mackenzie.

Ulmipollenites undulosus Wolff
Hypotypes 68137, 68138
Norris, G., 1986, Geol. Surv. Can., Bull. 340, p. 42, pl. 11, figs. 8, 9.
Arnak Member, Kugmallit Formation, Late Oligocene and Mackenzie Bay Formation, Miocene, depths 4000-4100 and 2800-2900 feet, Imperial Nuktak C-22 well, lat. 69°41′07″N, long. 134°51′30″W, District of Mackenzie.

Ulmipollenites sp.
Fig. spec. 62605
Ioannides, N.S. and McIntyre, D.J., 1980, Geol. Surv. Can., Paper 80-1A, pl. 31.4, fig. 13.
Eocene, Caribou Hills, Mackenzie River, District of Mackenzie.

Ulmoideipites herbridicus (Simpson) Sweet 1986
Hypotype 76088
Jerzykiewicz, T. and Sweet, A.R., 1986, Can. J. Earth Sci., vol. 23, no. 9, pl. 4, fig. 28.
Early Paleocene, Coal Valley strip mine, Alberta.

Ulmoideipites krempii Anderson 1960
Hypotype 76089
Jerzykiewicz, T. and Sweet, A.R., 1986, Can. J. Earth Sci., vol. 23, no. 9, pl. 4, fig. 29.
Early Paleocene, depth 26.62-26.7 m, GSC core hole CV-42-2-2, Coal Valley strip mine, l.s.d. 1, sec. 10, tp. 47, rge. 19, W.5th mer., Alberta.

Ulmoideipites tricostatus Anderson, 1960
Hypotype 68131
Norris, G., 1986, Geol. Surv. Can., Bull. 340, p. 42, pl. 11, figs. 1, 2.
Arnak Member, Kugmallit Formation, depth 3300-3400 feet, Imperial Nuktak C-22 well, lat. 69°41′07″N, long. 134°51′30″W, District of Mackenzie.

Ulmus sp. (*Ulmipollenites undulosus* Wolff 1934)
Fig. spec. 97327
Kalgutkar, R.M. and McIntyre, D.J., 1991, Can. J. Earth Sci., vol. 28, no. 3, p. 370, pl. 2, fig. 14.
Iceberg Bay Formation, Eureka Sound Group, Early Eocene, north shore Strand Fiord, 26 km from head, lat. 79°14′N, long. 91°20′W, Axel Heiberg Island, District of Franklin.

Ulmus sp.
Fig. spec. 27854
Hopkins, Jr., W.S., Rutter, N.W. and Rouse, G.E., 1972, Can. J. Earth Sci., vol. 9, no. 4, pl. 2, fig. 26.
Oligocene, east side Parsnip River 2 1/2 miles north of John Hart Highway bridge, lat. 55°09′N, long. 122°59′W, British Columbia.

Ulmus sp.
Fig. specs. 92042, 92043
Dietrich, J.R. et al., 1989, Geol. Surv. Can., Paper 89-1G, p. 139, pl. 3, figs. 12, 13.
Early to Middle Eocene, depth 1400-1490 and 6500-6590 feet, Dome Pacific et al. PEX Natsek E-56 well, lat. 69°45′21.46″N, long. 139°44′34.58″W, western Beaufort Sea.

Ulmus sp.
Fig. spec. 93601
McIntyre, D.J., 1989, Geol. Surv. Can., Paper 89-1G, p. 195, pl. 1, fig. 9.
Eureka Sound Group, Paleocene, valley southeast of Cunningham Inlet between lat. 73°54′N, long. 92°35′W and lat. 74°2′N, long. 93°33′W, north end of Somerset Island, District of Franklin.

Ulmus sp.
Fig. spec. 96331
McIntyre, D.J., 1991, Geol. Surv. Can., Bull. 403, pl. 3, fig. 9.
Buchanan Lake Formation, Eureka Sound Group, Eocene, 25 km east of Geodetic Hills, lat. 79°55′N, long. 88°58′W, northeastern Axel Heiberg Island, District of Franklin.

Verrucatosporites favus (Potonié) Thomson and Pflug
Hypotypes 68044, 68045
Norris, G., 1986, Geol. Surv. Can., Bull. 340, p. 34, pl. 8, figs. 11, 12.
Richards Formation, Late and Middle Eocene, depths 9200-9300 and 10,500-10,600 feet, Imperial Nuktak C-22 well, lat. 69°41′07″N, long. 134°51′30″W, District of Mackenzie.

?*Viburnum* sp. cf. *V. cassinoides* L.
Fig. specs. 96340, 96341
McIntyre, D.J., 1991, Geol. Surv. Can., Bull. 403, pl. 3, figs. 19, 20.
Buchanan Lake Formation, Eureka Sound Group, Eocene, 25 km east of Geodetic Hills, lat. 79°55′N, long. 88°58′W, northeastern Axel Heiberg Island, District of Franklin.

?*Viburnum* sp.
Fig. spec. 96342
McIntyre, D.J., 1991, Geol. Surv. Can., Bull. 403, pl. 3, fig. 21.
Buchanan Lake Formation, Eureka Sound Group, Eocene, 25 km east of Geodetic Hills, lat. 79°55′N, long. 88°58′W, northeastern Axel Heiberg Island, District of Franklin.

Vitreisporites pallidus (Reissinger)
Hypotype 38722
Haskell, T.P. and Wilson, G.J., 1975, Initial Repts. Deep Sea Drilling Project, vol. 29, pl. 4, fig. 15.
Eocene, DSDP Site 280, lat. 48°57.44′S, long. 147°14.08′E, South Tasman Rise, south of Tasmania.

Wetzeliella sp. aff. *W. articulata* Eisenack
Fig. spec. 62654
Ioannides, N.S. and McIntyre, D.J., 1980, Geol. Surv. Can., Paper 80-1A, pl. 31.1, fig. 1.
Eocene, Caribou Hills, Mackenzie River, District of Mackenzie.

Wetzeliella articulata articulata Eisenack, 1938
Hypotype 45797
Brideaux, W.W., 1976, Geol. Surv. Can., Paper 76-1B, pl. 44.1, fig. 1.
Reindeer Formation, Eocene, 900-1000 feet, Gulf Mobil Parsons N-10 well, ca. 1.5 miles east-northeast of Parsons Lake, lat. 69°59′48.52″N, long. 133°31′50.33″W, District of Mackenzie.

Wetzeliella articulata coniopa Williams and Downie, 1966
Hypotype 45809
Brideaux, W.W., 1976, Geol. Surv. Can., Paper 76-1B, pl. 44.1, fig. 13.
Reindeer Formation, Eocene, 1000-1030 feet, Gulf Mobil Parsons N-10 well, ca. 1.5 miles east-northeast of Parsons Lake, lat. 69°59′48.52″N, long. 133°31′50.33″W, District of Mackenzie.

Wetzeliella articulata subsp. *conopia* (Williams and Downie *in* Davey et al., 1966) Lentin and Williams
Hypotype 62655
Ioannides, N.S. and McIntyre, D.J., 1980, Geol. Surv. Can., Paper 80-1A, pl. 31.1, fig. 3.
Eocene, Caribou Hills, Mackenzie River, District of Mackenzie.

Wetzeliella homomorpha homomorpha Deflandre and Cookson, 1955
Hypotype 45798
Brideaux, W.W., 1976, Geol. Surv. Can., Paper 76-1B, pl. 44.1, fig. 2.
Reindeer Formation, Eocene, 900-1000 feet, Gulf Mobil Parsons N-10 well, ca. 1.5 miles east-northeast of Parsons Lake, lat. 69°59′48.52″N, long. 133°31′50.33″W, District of Mackenzie.

Wetzeliella parva Alberti, 1961
Hypotype 45799
Brideaux, W.W., 1976, Geol. Surv. Can., Paper 76-1B, pl. 44.1, fig. 3.
Reindeer Formation, Eocene, 900-1000 feet, Gulf Mobil Parsons N-10 well, ca. 1.5 miles east-northeast of Parsons Lake, lat. 69°59′48.52″N, long. 133°31′50.33″W, District of Mackenzie.

Wetzeliella sp.
Fig. specs. 62656-62662
Ioannides, N.S. and McIntyre, D.J., 1980, Geol. Surv. Can., Paper 80-1A, pl. 31.1, figs. 2, 4-9.
Eocene, Caribou Hills, Mackenzie River, District of Mackenzie.

Wetzeliella sp. AE
Fig. spec. 45804
Brideaux, W.W., 1976, Geol. Surv. Can., Paper 76-1B, p. 251, pl. 44.1, fig. 8.
Reindeer Formation, Eocene, 1000-1030 feet, Gulf Mobil Parsons N-10 well, ca. 1.5 miles east-northeast of Parsons Lake, lat. 69°59′48.52″N, long. 133°31′50.33″W, District of Mackenzie.

Wetzeliella sp. BE
Fig. spec. 45800
Brideaux, W.W., 1976, Geol. Surv. Can., Paper 76-1B, p. 251, pl. 44.1, fig. 4.
Reindeer Formation, Eocene, 1000-1030 feet, Gulf Mobil Parsons N-10 well, ca. 1.5 miles east-northeast of Parsons Lake, lat. 69°59′48.52″N, long. 133°31′50.33″W, District of Mackenzie.

Wodehouseia capillata Wiggins, 1976
Hypotype 93661
Sweet, A.R., Richetts, B.D., Cameron, A.R. and Norris, D.K., 1989, Geol. Surv. Can., Paper 89-1G, p. 98, pl. 2, fig. 12.
Summit Creek Formation, mid-Paleocene, west bank Mackenzie River between Birch Creek and Seagull Island, lat. 64°44′30″N, long. 125°7′30″W, District of Mackenzie.

Wodehouseia fimbriata Stanley, 1981
Hypotype 93658
Sweet, A.R., Richetts, B.D., Cameron, A.R. and Norris, D.K., 1989, Geol. Surv. Can., Paper 89-1G, p. 98, pl. 2, fig. 9.
Summit Creek Formation, earliest Paleocene, north bank Mackenzie River immediately downstream from Police Island, lat. 64°52′42″N, long. 125°14′44″W, District of Mackenzie.

Zonalapollenites trilobatus Balme
Hypotype 38726
Haskell, T.P. and Wilson, G.J., 1975, Initial Repts. Deep Sea Drilling Project, vol. 29, pl. 4, fig. 19.
Upper Eocene, DSDP Site 282, lat. 42°14.76′S, long. 143°29.18′E, west of Tasmania.

DINOFLAGELLATA

Acanthaulax aceras Gitmez and Sarjeant
Hypotype 76533
Davies, E.H. and Poulton, T.P., 1986, Geol. Surv. Can., Paper 86-1B, pl. 55.4, fig. 7.
Fernie Formation, Upper Jurassic, Christina Falls, lat. 56°40′N, long. 123°4′W, northeastern British Columbia.

Acanthaulax sp.
Fig. spec. 45958
Brideaux, W.W., 1977, Geol. Surv. Can., Bull. 281, p. 5, pl. 1, fig. 1.
Husky Formation, Jurassic, IOE Stoney Core Hole F-42, depth 1000-1010 feet, Stoney Creek about 18 miles west-southwest of Fort McPherson, lat. 67°21′23″N, long. 135°38′43″W, District of Mackenzie.

Achmosphaera ramulifera (Deflandre) Evitt, 1963
Hypotype 38127
McIntyre, D.J., 1974, Geol. Surv. Can., Paper 74-14, pl. 10, fig. 3.
Upper Cretaceous, lat. 69°27′30″N, long. 126°58′W, Horton River, District of Mackenzie.

Achomosphaera andalousiense Jan du Chêne, 1977
Hypotype 55307
Mudie, P.J, 1986, Initial Repts. Deep Sea Drilling Project, vol. XCIV, p. 801, pl. 2, fig. 6.
Upper Miocene, hole 611C, depth 445.2 m, lat. 52°50.15′N, long. 30°18.1′W, in the Gardar Drift on southwestern flank of Reykjanes Ridge, North Atlantic Ocean.

Achomosphaera neptuni (Eisenack)
Hypotype 41788
Williams, G.L., 1975, Geol. Surv. Can., Paper 74-30, vol. 2, pl. 1, fig. 8.
Berriasian-Valanginian, Lower Cretaceous, Amoco-IOE Puffin B-90 well, lat. 44°39′12.73″N, long. 53°42′28.35″W, Grand Banks.

Achomosphaera neptuni (Eisenack) Davey and Williams
Hypotypes 56744, 56745, 56767
Williams, G.L. and Bujack, J.P., 1980, Initial Repts. Deep Sea Drilling Project, vol. 50, pl. 5, figs. 7, 8; pl. 9, fig. 3.
Early Cretaceous, DSDP Site 416, lat. 32°50.18′N, long. 10°48.06′W, northwestern Morrocan continental margin.

Achomosphaera neptunii (Eisenack) Davey and Williams
Hypotype 46013
Brideaux, W.W., 1977, Geol. Surv. Can., Bull. 281, p. 21, pl. 8, fig. 7.
Cretaceous, Banff-Aquitaine-Arco Rat Pass K-35 well, depth 1025 feet, 6 miles southeast of Mount Goodenough, lat. 67°54′42.55″N, long. 135°21′56.83″W, District of Mackenzie.

Achomosphaera ramulifera (Deflandre) Evitt, 1963
Hypotype 55304
Mudie, P.J, 1986, Initial Repts. Deep Sea Drilling Project, vol. XCIV, p. 802, pl. 2, fig. 3.
Upper Miocene, hole 607, depth 246.5 m, lat. 41°0.07′N, long. 32°57.44′W, at base of upper western flank of the Mid-Atlantic Ridge, North Atlantic Ocean.

Achomosphaera ramulifera (Deflandre) Evitt, 1963
Hypotype 65106
Ioannides, N.S., 1986, Geol. Surv. Can., Bull. 371, p. 12, pl. 1, fig. 10.
Kanguk Formation, Upper Cretaceous, south coast Bylot Island, lat. 72°53′50″N, long. 78°16′W, District of Franklin.

Adnatosphaeridium caulleryi (Deflandre)
Hypotype 41787
Williams, G.L., 1975, Geol. Surv. Can., Paper 74-30, vol. 2, pl. 1, fig. 7.
Berriasian-Valanginian (reworked), Lower Cretaceous, Amoco IOE Puffin B-90 well, lat. 44°39′12.73″N, long. 53°42′28.35″W, Grand Banks.

?*Adnatosphaeridium patulum* Williams and Downie
Hypotype 49346
Williams, G.L. and Bujak, J.P., 1977, Am. Assoc. Strat. Palynol, Contr. Strat. Palynol, vol. 1, Contr. Ser. 5A, pl. 2, fig. 5.
Early to Middle Eocene, depth 1040-1070 feet, Amoco Imperial Skelly Pelican J-49 well, lat. 45°28′35.17″N, long. 52°36′41.71″W, Grand Banks.

?*Adnatosphaeridium patulum* Williams and Downie
Hypotype 47852
Williams, G.L., 1978, Initial Repts. Deep Sea Drilling Project, vol. 41, pl. 1, fig. 8.
Middle Eocene, DSDP site 370, lat. 32°50.2′N, long. 10°46.6′W, northwestern Moroccan continental margin.

?*Adnatosphaeridium* cf. *patulum* Williams and Downie
Hypotype 49345
Williams, G.L. and Bujak, J.R., 1977, Am. Assoc. Strat. Palynol, Contr. Strat. Palynol, vol. 1, Contr. Ser. 5A, pl. 2, fig. 4.
Early Oligocene, depth 1230-1260 feet, Mobil Tetco Esperanto K-78 well, lat. 44°47′31.26″N, long. 58°11′19.24″W, Scotian Shelf.

Adnatosphaeridium sp.
Fig. specs. 38134, 38135
McIntyre, D.J., 1974, Geol. Surv. Can., Paper 74-14, pl. 10, fig. 10; pl. 11, fig. 1.
Upper Cretaceous, lat. 69°28′N, long. 126°58′W, Horton River, District of Mackenzie.

Adnastosphaeridium sp.
Fig. spec. 91473
Sweet, A.R. and McIntyre, D.J., 1988, Can. Soc. Petrol. Geol., Mem. 15, fig. 9.18.
Sturrock Member, Cardium Formation, Upper Cretaceous, south side of McLeod River, l.s.d. 16, sec. 16, tp. 47, rge. 23, W.5th mer., Alberta.

Adnatosphaeridium sp.
Fig. specs. 71414, 71427
Wilson, M.A., 1978, Palaeontographica, Abt. B, vol. 166, p. 150, pl. 11, fig. 13; pl. 12, fig. 10.
Paleocene and Cretaceous, IOE Atkinson H-25 well, north shore Tuktoyaktuk Peninsula, lat. 69°44′20″N, long. 131°50′06″W,
District of Mackenzie.

Adnatosphaeridium? sp.
Fig. specs. 65114, 65115
Ioannides, N.S., 1986, Geol. Surv. Can., Bull. 371, p. 12, pl. 2, figs. 2, 3.
Kanguk Formation, Upper Cretaceous, south coast Bylot Island, lat. 72°53′30″N, long. 78°16′W, District of Franklin.

Adnatosphaeridium sp. A
Fig. spec. 49337
Williams, G.L. and Bujak, J.P., 1977, Am. Assoc. Strat. Palynol, Contr. Strat. Palynol, vol. 1, Contr. Ser. 5A, pl. 1, fig. 2.
Early-Middle Eocene, depth 1130-1160 feet, Amoco Imperial Skelly Pelican J-49 well, lat. 45°28′35.17″N, long. 52°36′41.71″W, Grand Banks.

Adnatosphaeridium sp. A
Fig. spec. 46016
Brideaux, W.W., 1977, Geol. Surv. Can., Bull. 281, p. 21, pl. 8, figs. 10, 11.
Cretaceous, Banff-Aquitaine-Arco Rat Pass K-35 well, depth 1025 feet, 6 miles southeast of Mount Goodenough, lat. 67°54′42.55″N, long. 135°21′56.83″W, District of Mackenzie.

Alterbia sp. cf. *A. foliacea* (Eisenack and Cookson) Lentin and Williams, 1976
Fig. spec. 65105
Ioannides, N.S., 1986, Geol. Surv. Can., Bull. 371, p. 12, pl. 1, fig. 9.
Kanguk Formation, Upper Cretaceous, south coast Bylot Island, lat. 72°53′50″N, long. 78°16′W, District of Franklin.

Alterbia minor (Alberti) Lentin and Williams, 1976
Hypotypes 65103, 65104
Ioannides, N.S., 1986, Geol. Surv. Can., Bull. 371, p. 12, pl. 1, figs. 7, 8.
Kanguk Formation, Upper Cretaceous, south coast Bylot Island, lat. 72°53′30″N, long. 78°16′W and "Twosnout Creek", lat. 73°14′30″N, long. 79°49′W, District of Franklin.

Alterbia sp.
Fig. spec. 71419
Wilson, M.A., 1978, Palaeontographica, Abt. B, vol. 166, p. 151, pl. 12, figs. 1, 3.
Cretaceous, IOE Atkinson H-25 well, north shore Tuktoyaktuk Peninsula, lat. 69°44′20″N, long. 131°50′06″W, District of Mackenzie.

Amiculosphaera umbracula Harland, 1979
Hypotypes 55303, 55366
Mudie, P.J, 1986, Initial Repts. Deep Sea Drilling Project, vol. XCIV, p. 802, pl. 2, figs. 5, 8.
Upper Miocene, hole 607, depth 246.5 m and 215.6 m, lat. 41°0.07′N, long. 32°57.44′W, at base upper western flank of the Mid-Atlantic Ridge, North Atlantic Ocean.

Amphorula metaelliptica Dodekova
Hypotype 56725
Williams, G.L. and Bujak, J.P, 1980, Initial Repts. Deep Sea Drilling Project, vol. 50, pl.3, fig. 10.
Late Jurassic, DSDP Site 416, lat. 32°50.18′N, long. 10°48.06′W, northwestern Moroccan continental margin.

Apectodinium homomorphum (Deflandre and Cookson) Lentin and Williams emend. Harland
Hypotype 92002
Dietrich, J.R. et al., 1989, Geol. Surv. Can., Paper 85-1G, pl. 1, fig. 6.
Early to Middle Eocene, depth 4900-4990 feet, Dome Pacific et al. PEX Natsek E-56 well, lat. 69°45′21.46″N, long. 139°44′34.58″W, western Beaufort Sea.

Aptea polymorpha Eisenack
Hypotypes 47848, 47849
Williams, G.L., 1978, Initial Repts. Deep Sea Drilling Project, vol. 41, pl. 1, figs. 4, 5.
Early Cretaceous, DSDP site 370, lat. 32°50.2′N, long. 10°46.6′W, northwestern Moroccan continental margin.

Apteodinium apiatum McIntyre and Brideaux
Holotype 58577; paratypes 58578-58580
McIntyre, D.J. and Brideaux, W.W., 1980, Geol. Surv. Can., Bull. 320, p. 11, pl. 2, figs. 1-7.
Lower Cretaceous, Martin Creek, lat. 68°12′10″N, long. 135°34′50″W, Richardson Mountains, District of Mackenzie.

Apteodinium australiense (Deflandre and Cookson)
Hypotype 47847
Williams, G.L., 1978, Initial Repts. Deep Sea Drilling Project, vol. 41, pl. 1, fig. 3.
Late Eocene, DSDP site 370, lat. 32°50.2′N, long. 10°46.6′W, northwestern Moroccan continental margin.

Apteodinium bucculiatum Davies
Holotype 69362
Davies, E.H., 1983, Geol. Surv. Can., Bull. 359, p. 20, pl. 4. figs. 13, 14.
Wilkie Point Formation, Upper Jurassic, depth 1840-1850, Panarctic Sandy Point L-46 well, northwestern Melville Island, District of Franklin.

Apteodinium bucculiatum
Hypotype 76294
Davies, E.H. and Poulton, T.P., 1986, Geol. Surv. Can., Paper 86-1B, pl. 55.5, fig. 5.
Fernie Formation, Upper Jurassic, Christina Falls, lat. 56°40′N, long. 123°4′W, northeastern British Columbia.

Apteodinium maculatum Eisenack and Cookson
Hypotype 41475
Brideaux, W.W. and McIntyre, D.J., 1975, Geol. Surv. Can., Bull. 252, p. 20, pl. 5, figs. 1, 2.
Horton River Formation, Lower Cretaceous, lat. 69°19′N, long. 126°50′W, Horton River, District of Mackenzie.

Apteodinium reticulatum Singh, 1971
Hypotype 84745
Leckie, D.A. et al., J. Sed. Petrology, vol. 60, no. 1, fig. 6H.
Shaftesbury Formation, Lower Cretaceous, west bank Peace River near town of Peace River, Alberta.

Apteodinium spongiosum McIntyre and Brideaux
Holotype 58584; paratypes 58581-58583
McIntyre, D.J. and Brideaux, W.W., 1980, Geol. Surv. Can., Bull. 320, p. 12, pl. 2, figs. 8-12.
Lower Cretaceous, Martin Creek, lat. 68°12′10″N, long. 135°34′50″W, Richardson Mountains, District of Mackenzie.

Apteodinium sp. B
Fig. spec. 41783
Williams, G.L., 1975, Geol. Surv. Can., Paper 74-30, vol. 2, pl. 1, figs. 2, 3.
Lower Miocene, core hole 2, lat. 45°8′N, long. 52°38′W, Grand Banks.

Areoligera medusettiformis (O. Wetzel) Lejeune-Carpentier
Hypotype 47850
Williams, G.L., 1978, Initial Repts. Deep Sea Drilling Project, vol. 41, pl. 1, fig. 6.
Early Eocene, DSDP site 370, lat. 32°50.2′N, long. 10°46.6′W, northwestern Moroccan continental margin.

Areoligera senonensis Lejeune-Carpentier sensu Gocht 1969
Hypotype 49336
Williams, G.L. and Bujak, J.R., 1977, Am. Assoc. Strat. Palynol, Contr. Strat. Palynol, vol. 1, Contr. Ser. 5A, pl. 1, fig. 1.
Early-Middle Eocene, depth 1040-1070 feet, Amoco Imperial Skelly Pelican J-49 well, lat. 48°28′35.17″N, long. 52°36′41.71″W, Grand Banks.

Areoligera senonensis Lejeune-Carpentier, 1938
Hypotypes 65107, 65108
Ioannides, N.S., 1986, Geol. Surv. Can., Bull. 371, p. 13, pl. 1, figs. 11, 12.
Eureka Sound Formation, Late Paleocene to (?)Early Eocene, Bylot Island south coast, lat. 72°52′N, long. 78°31′W and "Twosnout Creek", lat. 73°13′N, long. 79°53′W, District of Franklin.

Areoligera senonensis Lejeune-Carpentier, 1938
Hypotype 91470
Sweet, A.R. and McIntyre, D.J., 1988, Can. Soc. Petrol. Geol., Mem. 15, fig. 9.15.
Wapiabi Formation, Upper Cretaceous, south side of McLeod River, l.s.d. 16, sec. 16, tp. 47, rge. 23, W.5th mer., Alberta.

Areoligera sp.
Fig specs. 38095, 38096
McIntyre, D.J., 1974, Geol. Surv. Can., Paper 74-14, pl. 8, figs. 1, 2.
Upper Cretaceous, lat. 69°27′30″N, long. 126°58′W, Horton River, District of Mackenzie.

Areoligera sp.
Fig. spec. 41784
Williams, G.L., 1975, Geol. Surv. Can., Paper 74-30, vol. 2, pl. 1, fig. 4.
Turonian, Upper Cretaceous, Shell Naskapi N-30 well, lat. 43°29′46.29″N, long. 62°34′0.63″W, Scotian Shelf.
=*Areoligera* sp. A, Bujak, J.P. and Williams, G.L., 1978, Geol. Surv..Can., Bull. 297, pl.1, figs. 1, 2.

?*Areoligera* sp.
Fig. spec. 80488
Burden, E. and Holloway, D., 1985, Can. J. Earth Sci. vol. 22, no. 10, pl. 1, fig. 7.
Quaternary, Scott Inlet area, Baffin Island, District of Franklin.

Areoligera sp. a
Fig. spec. 53688
Bujak, J.P. and Williams, G.L., 1978, Geol. Surv. Can. Bull. 297, pl. 2, fig. 2.
Late Cretaceous, swc 4926 feet, Mobil Sable Island C-67 well, lat. 43°56′4.91″N, long. 59°55′1.38″W, Scotian Shelf.

Areosphaeridium arcuatum Eaton
Hypotype 41785
Williams, G.L., 1975, Geol. Surv. Can., Paper 79-30, vol. 2, pl. 1, fig. 5.
Middle Eocene, core hole 4, lat. 45°31′N, long. 52°22′W, Grand Banks.

Areosphaeridium sp. A
Fig. spec. 49339
Williams, G.L. and Bujak, J.P., 1977, Am. Assoc. Strat. Palynol, Contr. Strat. Palynol, vol. 1, Contr. Ser. 5A, pl. 1, fig. 4.
Tertiary, depth 3620-3650 feet, Elf et al. Emerillon C-56 well, lat. 45°15′4.79″N, long. 54°23′16.85″W, Grand Banks.

Astrocysta cretacea Pocock ex Davey
Hypotype 41552
Brideaux, W.W. and McIntyre, D.J., 1975, Geol. Surv. Can., Bull. 252, p. 36, pl. 14, fig. 3.
Horton River Formation, Lower Cretaceous, lat. 69°27′30″N, long. 126°54′W, Horton River, District of Mackenzie.

Astrocysta cretacea
Hypotype 91990
Banerjee, I. and Davies, E.H., 1988, Can. Soc. Petrol. Geol., Mem. 15, fig. 7.6.
Mannville Group, Lower Cretaceous, depth 5064 feet, well 11-08-41-24W4, Alberta.

Aterbidinium sp.
Fig. specs. 91432, 91433
Sweet, A.R. and McIntyre, D.J., 1988, Can. Soc. Petrol. Geol., Mem. 15, figs. 8.4, 8.5.
Leyland Member, Cardium Formation, Upper Cretaceous, south side of McLeod River, l.s.d. 16, sec. 16, tp. 47, rge. 23, W.5th mer., Alberta.

Atopodinium prostatum Drugg
Hypotypes 76285, 76530
Davies, E.H. and Poulton, T.P., 1986, Geol. Surv. Can., Paper 86-1B, pl. 55.3, figs. 5, 6.
Fernie Formation, Upper Jurassic, Christina Falls, lat. 56°40′N, long. 123°4′W, northeastern British Columbia.

Australiella victoriensis (Cookson and Manum)
Hypotype 41782
Williams, G.L., 1975, Geol. Surv. Can., Paper 74-30, vol. 2, pl. 1, fig. 1.
Santonian, Upper Cretaceous, core hole 5, lat. 45°48′N, long. 52°13′W, Grand Banks.

Balmula tripenta
Hypotype 91986
Banerjee, I. and Davies, E.H., 1988, Can. Soc. Petrol. Geol., Mem. 15, fig. 7.2.
Mannville Group, Lower Cretaceous, depth 5122 feet, well 3-8-40-23W4, Alberta.

Batioladinium jaegeri (Alberti) Brideaux
Hypotypes 41485, 41486
Brideaux, W.W. and McIntyre, D.J., 1975, Geol. Surv. Can., Bull. 252, p. 24, pl. 6, figs. 7, 10, 11.
Horton River Formation, Lower Cretaceous, lat. 69°19′N, long. 126°50′W, Horton River, District of Mackenzie.

Baltioladinium jaegeri (Alberti) Brideaux, 1975
Hypotype 58601
McIntyre, D.J. and Brideaux, W.W., 1980, Geol. Surv. Can., Bull. 320, p. 17, pl. 5, figs. 6, 8, 11.
Lower Cretaceous, Martin Creek, lat. 68°12′10″N, long. 135°34′50″W, Richardson Mountains, District of Mackenzie.

Batioladinium micropodum (Alberti) Brideaux
Hypotype 45974
Brideaux, W.W., 1977, Geol. Surv. Can., Bull. 281, p. 10, pl. 2, fig. 7.
Cretaceous, IOE Stoney Core Hole, depth 790-795 feet, Stoney Creek about 18 miles west-southwest of Fort McPherson, lat. 67°21′23″N, long. 135°38′43″W, District of Mackenzie.

Batioladinium sp. A
Fig. spec. 53671
Bujak, J.P. and Williams, G.L., 1978, Geol. Surv. Can., Bull. 297, pl. 1, fig. 3.
Early Cretaceous, 10460-10490 feet, Amoco-IOE Puffin B-90 well, lat. 44°39′12.73″N, long. 53°42′28.35″W, Grand Banks.

Biorbifera johnewingi Habib
Hypotypes 41796, 41797
Williams, G.L., 1975, Geol. Surv. Can., Paper 74-30, vol. 2, pl. 2, figs. 1, 2.
Berriasian-Valanginian reworked, Lower Cretaceous, Amoco-IOE Puffin B-90 well, lat. 44°39′12.73″N, long. 53°42′28.35″W, Grand Banks.

Biorbifera sp. A
Fig. specs. 53672, 53673, 53689
Bujak, J.P. and Williams, G.L., 1978, Geol. Surv. Can., Bull. 297, pl. 1, figs. 4, 5; pl. 2, fig. 3.
Early Cretaceous, swc 10,874 feet, Shell Cree E-35 well, lat. 43°44′20.71″N, long. 60°35′55.9″W, Scotian Shelf.

Bourkidinium sp. AE
Fig. spec. 58605
McIntyre, D.J. and Brideaux, W.W., 1980, Geol. Surv. Can., Bull. 320, p. 18, pl. 6, figs. 2, 3.
Lower Cretaceous, Martin Creek, lat. 68°12′10″N, long. 135°34′50″W, Richardson Mountains, District of Mackenzie.

Brigantedinium auranteum Reid, 1974
Hypotype 55350
Mudie, P.J, 1986, Initial Repts. Deep Sea Drilling Project, vol. XCIV, p. 802, pl. 5, fig. 16.
Pleistocene, hole 607, depth 0.5 m, lat. 41°0.07′N, long. 32°57.44′W, at the base of upper western flank of the Mid-Atlantic Ridge, North Atlantic Ocean.

Brigantedinium cariacoense Reid, 1974
Hypotype 55349
Mudie, P.J, 1986, Initial Repts. Deep Sea Drilling Project, vol. XCIV, p. 802, pl. 5, fig. 22.
Pleistocene, hole 607, depth 0.5 m, lat. 41°0.07′N, long. 32°57.44′W, at the base of upper western flank of the Mid-Atlantic Ridge, North Atlantic Ocean.

Brigantedinium simplex Reid, 1974
Hypotypes 55348, 55352
Mudie, P.J, 1986, Initial Repts. Deep Sea Drilling Project, vol. XCIV, p. 802, pl. 5, figs. 4, 9.
Pleistocene, hole 607, depth 0.5 m, lat. 41°0.07′N, long. 32°57.44′W, at the base of upper western flank of the Mid-Atlantic Ridge, North Atlantic Ocean.

Brigantedinium sp.
Fig. spec. 55359
Mudie, P.J, 1986, Initial Repts. Deep Sea Drilling Project, vol. XCIV, p. 802, pl. 5, fig. 20.
Pleistocene, hole 607, depth 0.5 m, lat. 41°0,07′N, long. 32°57.44′W, at the base of upper western flank of the Mid-Atlantic Ridge, North Atlantic Ocean.

Brigantodinium sp.
Fig. spec. 68765
Dixon, J. et al., 1984, Geol. Surv. Can., Paper 82-13, pl. 4, fig. 2.
Oligocene, 2749 m interval, Dome Gulf et al. Hunt Koponoar M-13 well, lat. 70°22′55.1″N, long. 135°05′33.9″W, Beaufort Sea.

Broomea sp.
Fig. spec. 41786
Williams, G.L., 1975, Geol. Surv. Can., Paper 74-30, vol. 2, pl. 1, fig. 6.
Barremian, Lower Cretaceous, Amoco-IOE Puffin B-90 well, lat. 44°39′12.73″N, long. 53°42′28.35″W, Grand Banks.
=*Batioladinium* sp. A, Bujak, J.P. and Williams, G.L., 1978, Geol. Surv. Can., Bull. 297, pl. 2, fig. 1.

"*Broomea*" sp. 1 (cf. *Canningia* sp. Manum and Cookson, 1964)
Fig. specs. 38103, 38104
McIntyre, D.J., 1974, Geol. Surv. Can., Paper 74-14, pl. 8, figs. 9, 10.
Upper Cretaceous, lat. 69°27′30″N, long. 126°58′W, Horton River, District of Mackenzie.

"*Broomea*" sp. 2
Fig. specs. 38110, 38111
McIntyre, D.J., 1974, Geol. Surv. Can., Paper 74-14, pl. 8, figs. 16, 17.
Upper Cretaceous, lat. 69°27′30″N, long. 126°58′W, Horton River, District of Mackenzie.

Caligodinium aceras (Manum and Cookson) Lantin and Williams
Hypotype 75474
Davies, E.H., 1985, Palynology, vol. 9, pl. 3, fig. 4.
Toarcian, Liassic, Brehna Road Section, Lusitanian Basin, west-central Portugal.

Caligodinium aceras (Manum and Cookson) Lentin and Williams, 1973
Hypotype 65116
Ioannides, N.S., 1986, Geol. Surv. Can., Bull. 371, p. 13, pl. 2, fig. 4.
Kanguk Formation, Upper Cretaceous, south coast Bylot Island, lat. 72°53′30″N, long. 78°16′W, District of Franklin.

Caligodinium? sp.
Fig. specs. 65117, 65118
Ioannides, N.S., 1986, Geol. Surv. Can., Bull. 371, p. 13, pl. 2, figs. 5, 6.
Eureka Sound Formation, Late Paleocene to (?)Early Eocene, south coast Bylot Island, lat. 72°52′N, long. 78°35′W, District of Franklin.

Callaiosphaeridium asymmetricum (Deflandre and Courteville) Davey and Williams
Hypotype 46014
Brideaux, W.W., 1977, Geol. Surv. Can., Bull. 281, p. 22, pl. 8, fig. 8.
Cretaceous, Banff-Aquitaine-Arco Rat Pass K-35 well, depth 1025 feet, 6 miles southeast of Mount Goodenough, lat. 67°54′42.55″N, long. 135°21′56.83″W, District of Mackenzie.

Callaiosphaeridium asymmetricum (Deflandre and Courteville)
Hypotype 53692
Bujak, J.P. and Williams, G.L., 1978, Geol. Surv. Can., Bull. 297, pl. 2, fig. 6.
Early Cretaceous, 10960-10990 feet, Amoco-IOE Puffin B-90 well, lat. 44°39′12.73″N, long. 53°42′28.35″W, Grand Banks.

Callaiosphaeridium asymmetricum (Deflandre and Courteville) subsp. A
Hypotypes 53690, 53691
Bujak, J.P. and Williams, G.L., 1978, Geol. Surv. Can., Bull. 297, pl. 2, figs. 4, 5.
Early Cretaceous, locality not available, and 11560-11590 feet, Shell Primrose 1(a)A-41 well, lat. 44°5.68′N, long. 54°6′18.26″W, Scotian Shelf.

Canningia apera Singh
Hypotypes 45971-45973
Brideaux, W.W., 1977, Geol. Surv. Can., Bull. 281, p. 11, pl. 2, figs. 6, 8, 9.
Cretaceous, depths 795-800, 145-150 and 310-320 feet, IOE Stoney Core Hole F-42, Stoney Creek about 18 miles west-southeast of Fort McPherson, lat. 67°21′23″N, long. 135°38′43″W, District of Mackenzie.

Canningia colliveri Cookson and Eisenack
Hypotype 41792
Williams, G.L., 1975, Geol. Surv. Can., Paper 74-30, vol. 2, pl. 1, fig. 12.
Aptian, Lower Cretaceous, Mobil Tetco Dauntless D-35 well, lat. 44°44′8.26″N, long. 57°20′46.62″W, Scotian Shelf.

Canningia sp. cf. *C. colliveri* Cookson and Eisenack
Fig. spec. 45975
Brideaux, W.W., 1977, Geol. Surv. Can., Bull. 281, p. 11, pl. 2, fig. 10.
Cretaceous, Banff-Aquitaine-Arco Rat Pass K-35 well, depth 945 feet, 6 miles southeast of Mount Goodenough, lat. 67°54′42.55″N, long. 135°21′56.83″W, District of Mackenzie.

Canningia? minor Cookson and Hughes, 1964
Hypotypes 65120, 65121
Ioannides, N.S., 1986, Geol. Surv. Can., Bull. 371, p. 13, pl. 2, figs. 10, 15.
Kanguk Formation, Upper Cretaceous, south coast Bylot Island, lat. 72°53′30″N, long. 78°16′W, District of Franklin.

Canningia minor Cookson and Hugues
Hypotype 41487
Brideaux, W.W. and McIntyre, D.J., 1975, Geol. Surv. Can., Bull. 252, p.24, pl. 6, fig. 8.
Horton River Formation, Lower Cretaceous, lat. 69°19′N, long. 126°50′W, Horton River, District of Mackenzie.

Canningia sp. cf. *C. minor* Cookson and Hugues
Fig. specs. 38084, 38085
McIntyre, D.J., 1974, Geol. Surv. Can., Paper 74-14, pl. 7, figs. 6, 7.
Upper Cretaceous, lat. 69°28′N, long. 126°58′W, Horton River, District of Mackenzie.

Canningia palliata Brideaux
Holotype 45976; paratypes 45977-45979
Brideaux, W.W., 1977, Geol. Surv. Can., Bull. 281, p. 11, pl. 3, figs. 1-8.
Cretaceous, IOE Stoney Core Hole F-42, depths 770-775, 795-800 and 610-620 feet, Stoney Creek about 18 miles west-southwest of Fort McPherson, lat. 67°21′23″N, long. 135°38′43″W, District of Mackenzie.

Canningia ringnesii Manum and Cookson
Hypotype 41488
Brideaux, W.W. and McIntyre, D.J., 1975, Geol. Surv. Can. Bull. 252, p. 24, pl. 6, fig. 9.
Horton River Formation, Lower Cretaceous, lat. 69°19′N, long. 126°50′W, Horton River, District of Mackenzie.

Canningia? ringnesiorum Manum and Cookson, 1964
Hypotype 65119
Ioannides, N.S., 1986, Geol. Surv. Can., Bull. 371, p. 13, pl. 2, fig. 7.
Kanguk Formation, Upper Cretaceous, south coast Bylot Island, lat. 72°53′30″N, long. 78°16′W, District of Franklin.

Canningia spumosa Brideaux
Holotype 45980; paratypes 45981, 45982
Brideaux, W.W., 1977, Geol. Surv. Can., Bull. 281, p. 12, pl. 3, figs. 9-14.
Cretaceous, IOE Stoney Core Hole F-42, depth 310-320 feet, Stoney Creek about 18 miles west-southwest of Fort McPherson, lat. 67°21′23″N, long. 135°38′43″W, District of Mackenzie.

"*Canningia*" turrita Brideaux
Holotype 45983; paratypes 45984-45986
Brideaux, W.W., 1977, Geol. Surv. Can., Bull. 281, p. 13, pl. 4, figs. 1-9.
Cretaceous, IOE Stoney Core Hole F-42, depths 780-785, 770-775, 740-745 and 845-850 feet, Stoney Creek about 18 miles west-southwest of Fort McPherson, lat. 67°21′23″N, long. 135°38′43″W, District of Mackenzie.

Canningia sp.
Fig. specs. 45929, 45930
Brideaux, W.W., 1976, Geol. Surv. Can., Paper 76-1C, pl. 23.1, figs. 8, 14.
Husky Formation, Lower Cretaceous, south bank Martin Creek, District of Mackenzie.

?*Canningia* sp.
Fig. spec. 45935
Brideaux, W.W., 1976, Geol. Surv. Can., Paper 76-1C, pl. 23.2, figs. 5, 6.
Husky Formation, Lower Cretaceous, south bank Martin Creek, District of Mackenzie.

Canningia sp. 1
Fig. specs. 38101, 38102
McIntyre, D.J., 1974, Geol. Surv. Can., Paper 74-14, pl. 8, figs. 7, 8.
Upper Cretaceous, lat. 69°28′N, long. 126°58′W, Horton River, District of Mackenzie.

"*Canningia*" sp. 1 of McIntyre
Fig. spec. 65445
Ioannides, N.S., 1986, Geol. Surv. Can., Bull. 371, pl. 25, fig. 5.
Kanguk Formation, Upper Cretaceous, south coast Bylot Island, lat. 72°53′50″N, long. 78°16′W, District of Franklin.

Canningia sp. A
Fig. specs. 56751, 56752
Williams, G.L. and Bujak, J.P., 1980, Initial Repts. Deep Sea Drilling Project, vol. 50, pl. 6, figs. 5, 6.
Early Cretaceous, DSDP Site 416, lat. 32°50.18′N, long. 10°48.06′W, northwestern Moroccan continental margin.

Canningia sp. AE
Fig. spec. 45817
Brideaux, W.W., 1976, Geol. Surv. Can., Paper 76-1B, p. 256, pl. 44.2, fig. 8.
Lower Cretaceous, 8270-8290 feet, Gulf Mobil Parson N-10 well, ca. 1.5 miles east-northeast of Parsons Lake, lat. 69°59′48.52″N, long. 133°31′50.33″W, District of Mackenzie.

Canningia sp. BE
Fig. spec. 45821
Brideaux, W.W., 1976, Geol. Surv. Can., Paper 76-1B, p. 256, pl. 44.2, fig. 12.
Lower Cretaceous, 8300-8320 feet, Gulf Mobil Parsons N-10 well, ca. 1.5 miles east-northeast of Parsons Lake, lat. 69°59′48.52″N, long. 133°31′50.33″W, District of Maackenzie.

Cannosphaeropsis utinensis Wetzel
Hypotype 41795
Williams, G.L., 1975, Geol. Surv. Can., Paper 74-30, vol. 2, pl. 1, fig. 15.
Maastrichtian, Upper Cretaceous, Mobil Sable Island C-67 well, lat. 43°56′4.91″N, long. 59°55′1.38″W, Scotian Shelf.

Cannosphaeropsis sp. A
Fig. spec. 41811
Williams, G.L., 1975, Geol. Surv. Can., Paper 74-30, vol. 2, pl. 3, fig. 2.
Late Miocene, Amoco IOE Kittiwake P-11 well, lat. 44°40′49.43″N, long. 53°31′45.65″W, Grand Banks.

Cannosphaeropsis sp. A Williams and Brideaux, 1975
Fig. spec. 47851
Williams, G.L., 1978, Initial Repts. Deep Sea Drilling Project, vol. 41, pl. 1, fig. 7.
Oligocene, DSDP Site 370, lat. 32°50.2′N, long. 10°46.6′W, northwestern Moroccan continental margin.

Cannosphaeropsis sp. A of Williams and Brideaux, 1975
Fig. specs. 68768, 68769, 68783
Dixon, J. et al., 1984, Geol. Surv. Can., Paper 82-13, pl. 4, figs. 8, 9; pl. 6, figs. 1, 2.
Late Middle to Late Miocene, 2621-2649, 2643 and 2613 m interval, Dome Gulf et al. Hunt Kopanoar M-13 well, lat. 70°22′55.1″N, long. 135°05′33.9″W, Beaufort Sea.

?*Cannosphaeropsis* sp. 1
Fig. specs. 55320, 55358
Mudie, P.J, 1986, Initial Repts. Deep Sea Drilling Project, vol. XCIV, p. 802, pl. 3, fig. 6; pl. 5, fig. 14.
Upper Miocene, hole 611C, depth 441.28 and 374.53 m, lat. 52°50.15′N, long. 30°18.1′W, in the Gardar Drift on southwestern flank of Reykjanes Ridge, North Atlantic Ocean.

Cantulodinium speciosum
Hypotypes 76273, 76274
Davies, E.H. and Poulton, T.P., 1986, Geol. Surv. Can., Paper 86-1B, pl. 55.2, figs. 7, 8.
Fernie Formation, Upper Jurassic, Christina Falls, lat. 56°40′N, long. 123°4′W, northeastern British Columbia.

Cassiculosphaeridia sp.
Fig. specs. 96493, 96494
Leckie, D.A. and Singh, C., 1991, J. Sed. Petrology, vol. 61, no. 5, figs. 3J, 3K.
Paddy Member, Peace River Formation, Lower Cretaceous, near town of Peace River, Alberta.

Cassidium fragilis (Harris)
Hypotype 38695
Haskell, T.R. and Wilson, G.J., 1975, Initial Repts. Deep Sea Drilling Project, vol. 29, pl. 3, fig. 3.
Eocene, DSDP Site 282, lat. 42°14.76′S, long. 143°29.18′E, west of Tasmania.

Catastomocystis spinosa Singh, 1983
Hypotype 91464
Sweet, A.R. and McIntyre, D.J., 1988, Can. Soc. Petrol. Geol., Mem. 15, fig. 9.9.
Leyland Member, Cardium Formation, Upper Cretaceous, south side of McLeod River, l.s.d. 16, sec. 16, tp. 47, rge. 23, W.5th mer., Alberta.

Catastomocystis spinosa Singh, 1983
Hypotype 84746
Leckie, D.A. et al., 1990, J. Sed. Petrology, vol. 60, no. 1, fig. 6I.
Shaftesbury Formation, Lower Cretaceous, west bank Peace River near town of Peace River, Alberta.

Ceratiopsis sp. J-43 of Staplin
Fig. specs. 92012, 92016
Dietrich, J.R. et al., 1989, Geol. Surv. Can., Paper 89-1G, p. 141, pl. 2, fugs. 4, 8.
Early to Middle Eocene, depths 4900-4990 and 3000-3090 feet, Dome Pacific et al. PEX Natsek E-56 well, lat. 69°45′21.46″N, long. 139°44′34.58″W, western Beaufort Sea.

Chálamydophorella sp.
Fig. spec. 47868
Williams, G.L., 1978, Initial Repts. Deep Sea Drilling Project, vol. 41, pl. 3, fig. 5.
Late Jurassic, DSDP Site 367, lat. 12°29.2′N, long. 20°2.8′W, Cape Verde Basin west of Senegal.

Chlamydophorella sp. A
Fig. spec. 75481
Davies, E.H., 1985, Palynology, vol. 9, p. 129, pl. 3, figs. 13, 14.
Early Toarecian, Liassic, Zambujal Section, south of Condeixa, Lusitanian Basin, west-central Portugal.

Chatangiella ditissima (McIntyre) Lestin and Williams, 1976
Hypotypes 65133-65137
Ioannides, N.S., 1986, Geol. Surv. Can., Bull. 371, p. 14, pl. 3, figs. 5-9.
Kanguk Formation, Upper Cretaceous, lat. 72°53′30″N, long. 78°16′W and Eureka Sound Formation, Late Paleocene to ?Early Eocene, lat. 72°52′N, long. 78°35′W (65136), south coast Bylot Island, District of Franklin.

Chatangiella ditissima (McIntyre) Lentin and Williams, 1976
Hypotype 91429
Sweet, A.R. and McIntyre, D.J., 1988, Can. Soc. Petrol. Geol., Mem. 15, fig. 8.1.
Leyland Member, Cardium Formation, Upper Cretaceous, south side of McLeod River, l.s.d. 16, sec. 16, tp. 47, rge. 23, W.5th mer., Alberta.

Chatangiella granulifera (Manum) Lentin and Williams, 1976
Hypotypes 65138-65143
Ioannides, N.S., 1986, Geol. Surv. Can., Bull. 371, p. 14, pl. 3, figs. 10, 11; pl. 4, figs. 1, 2, 4, 9.
Kanguk Formation, Upper Cretaceous, south coast Bylot Island, lat. 72°53′30″N, long. 78°16′W, District of Franklin.

Chantangiella granulifera (Manum) Lentin and Williams, 1976
Hypotype 91430
Sweet, A.R. and McIntyre, D.J., 1988, Can. Soc. Petrol. Geol., Mem. 15, fig. 8.2.
Leyland Member, Cardium Formation, Upper Cretaceous, south side of McLeod River, l.s.d. 16, sec. 16, tp. 47, rge. 23, W.5th mer., Alberta.

Chatangiella madura Lentin and Williams, 1976
Hypotypes 65149-65151
Ioannides, N.S., 1986, Geol. Surv. Can., Bull. 371, p. 14, pl. 4, fig. 10-12.
Kanguk Formation, Upper Cretaceous, south coast Bylot Island, lat. 72°53′30″N, long. 78°16′W, District of Franklin.

Chatangiella spectabilis (Alberti) Lentin and Williams, 1976
Hypotypes 65147, 65148
Ioannides, N.S., 1986, Geol. Surv. Can., Bull. 371, p. 15, pl. 4, figs. 7, 8.

Kanguk Formation, Upper Cretaceous, Bylot Island, south coast, lat. 72°53′30″N, long. 78°16′W, and ?"Twosnout Creek", lat. 73°14′30″N, long. 79°49′W, District of Franklin.

Chantangiella spectabilis (Alberti) Lentin and Williams, 1976
Hypotype 91435
Sweet, A.R. and McIntyre, D.J., 1988, Can. Soc. Petrol. Geol., Mem. 15, fig. 8.7.
Leyland Member, Cardium Formation, Upper Cretaceous, south side of McLeod River, l.s.d. 16, sec. 16, tp. 47, rge. 23, W.5th mer., Alberta.

Chatangiella verrucosa (Manum) Lentin and Williams, 1976
Hypotypes 65132, 65144-65146
Ioannides, N.S., 1986, Geol. Surv. Can., Bull. 371, p. 15, pl. 3, fig. 4; pl.4, figs. 3, 5, 6.
Kanguk Formation, Upper Cretaceous, south coast Bylot Island, lat. 72°53′30″N, long. 78°16′W, District of Franklin.

Chatangiella sp. cf. *C. victoriensis* (Cookson and Manum) Lentin and Williams, 1976
Fig. spec. 91437
Sweet, A.R. and McIntyre, D.J., 1988, Can. Soc. Petrol. Geol., Mem. 15, fig. 8.9.
Wapiabi Formation, Upper Cretaceous, south side of McLeod River, l.s.d. 16, sec. 16, tp. 47, rge. 23, W.5th mer., Alberta.

Chatangiella sp. cf. *C. vnigri* (Vozzhennikova) Lentin and Williams, 1976
Fig. spec. 91434
Sweet, A.R. and McIntyre, D.J., 1988, Can. Soc. Petrol. Geol., Mem. 15, fig. 8.6.
Leyland Member, Cardium Formation, Upper Cretaceous, south side of McLeod River, l.s.d. 16, sec. 16, tp. 47, rge. 23, W.5th mer., Alberta.

Chatangiella sp.
Fig. spec. 91431
Sweet, A.R. and McIntyre, D.J., 1988, Can. Soc. Petrol. Geol., Mem. 15, fig. 8.3.
Leyland Member, Cardium Formation, Upper Cretaceous, south side of McLeod River, l.s.d. 16, sec. 16, tp. 47, rge. 23, W.5th mer., Alberta.

Chenopodipollis multiplex Krutzsch
Hypotype 68771
Dixon, J. et al., 1884, Geol. Surv. Can., Paper 82-13, pl. 4, fig. 12.
Late Middle to Late Miocene, 2646 m interval, Dome Gulf et al. Hunt Kopanoar M-13 well, lat. 70°22′55.1″N, long. 135°05′33.9″W, Beaufort Sea.

Chichaouadinium vestitum
Hypotype 91994
Banerjee, I. and Davies, E.H., 1988, Can. Soc. Petrol. Geol., Mem. 15, fig. 7.10.
Mannville Group, Lower Cretaceous, depth 4395 feet, well 3-22-46-23W4, Alberta.

Chicaouadinium vestitum (Brideaux, 1971) Bujack and Davis, 1983
Hypotypes 96490-96492
Leckie, D.A. and Singh, C., 1991, J. Sed. Petrology, vol. 61, no. 5, figs. 3G-3I.
Paddy Member, Peace River Formation, Lower Cretaceous, near town of Peace River, Alberta.

Chiropteridium conispinum Williams
Holotype 47855; paratypes 47856, 47857
Williams, G.L., 1978, Initial Repts. Deep Sea Drilling Project, vol. 41, p. 794, pl. 2, figs. 1-5.
Oligocene, DSDP site 370, lat. 32°50.2′N, long. 10°46.6′W, northwestern Moroccan continental margin.

Chiropteridium dispersum Gocht
Hypotype 47858
Williams, G.L., 1978, Initial Repts. Deep Sea Drilling Project, vol. 41, pl. 2, fig. 7.
Oligocene, DSDP site 370, lat. 32°50.2′N, long. 10°46.6′W, northwestern Moroccan continental margin.

Chlamydophorella discreta Clarke and Verdier, 1967
Hypotype 65152
Ioannides, N.S., 1986, Geol. Surv. Can., Bull. 371, p. 16, pl. 5, fig. 1.
Kanguk Formation, Upper Cretaceous, south coast Bylot Island, lat. 72°53′30″N, long. 78°16′W, District of Franklin.

Chlamydophorella? grossa Manun and Cookson, 1964
Hypotypes 65158-65165, 65167
Ioannides, N.S., 1986, Geol. Surv. Can., Bull. 371, p. 16, pl. 5, figs. 6, 9-14, 16; pl. 6, fig. 1.
Kanguk Formation, Upper Cretaceous, south coast Bylot Island, lat. 72°53′30″N, long. 78°16′W, District of Franklin.

Chlamydophorella? grossa - Scriniodinium? obscurum Manum and Cookson, 1964
Hypotype 65168
Ioannides, N.S., 1986, Geol. Surv. Can., Bull. 371, p. 15, pl. 6, fig. 2.
Kanguk Formation, Upper Cretaceous, south coast Bylot Island, lat. 72°53′30″N, long. 78°16′W, District of Franklin.

Chlamydophorella? grossa - Scriniodinium? obscurum
Fig. specs. 65179, 65188
Ioannides, N.S., 1986, Geol. Surv. Can., Bull. 371, pl. 7, fig. 2; pl. 8, fig. 1.
Kanguk Formation, Upper Cretaceous, south coast Bylot Island, lat. 72°53′30″N, long. 78°16′W, and "Twosnout Creek", lat. 73°14′30″N, long. 79°49′W, District of Franklin.

Chlamydophorella? grossa - Scriniodinium? obscurum complex
Fig. specs. 65166, 65174-65176
Ioannides, N.S., 1986, Geol. Surv. Can., Bull. 371, p. 15, pl. 5, fig. 15; pl. 6, figs. 8-10.

Kanguk Formation, Upper Cretaceous, lat. 72°53′30″N, long. 78°16′W and Eureka Sound Formation, Late Paleocene to (?)Early Eocene, lat. 72°52′N, long. 78°31′W (65176), south coast Bylot Island; Upper Cretaceous, lat. 75°26′N, long. 91°35′W, Devon Island (65174, 65175), District of Franklin.

Chlamydophorella nyei Cookson and Eisenack
Hypotype 38133
McIntyre, D.J., 1974, Geol. Surv. Can., Paper 74-14, pl. 10, fig. 9.
Upper Cretaceous, lat. 69°27′30″N, long. 126°58′W, Horton River, District of Mackenzie.

Chlamydophorella nyei Cookson and Eisenack, 1958
Hypotypes 65153-65157
Ioannides, N.S., 1986, Geol. Surv. Can., Bull. 371, p. 17, pl. 5, figs. 2-5, 8.
Kanguk Formation, Upper Cretaceous, south coast Bylot Island, lat. 72°53′30″N, long. 78°16′W, District of Franklin.

Chlamydophorella nyei Cookson and Eisenack, 1958
Hypotype 91450
Sweet, A.R. and McIntyre, D.J., 1988, Can. Soc. Petrol. Geol., Mem. 15, fig. 8.22.
Kaskapau Formation, Upper Cretaceous, west bank of Little Berland River, l.s.d. 1, sec. 6, tp. 54, rge. 7, W.5th mer., Alberta.

Chlamydophorella sp.
Fig. spec. 53675
Bujak, J.P. and Williams, G.L., 1978, Geol. Surv. Can., Bull. 297, pl. 1, fig. 7.
Late Cretaceous, swc 4964 feet, Mobil-Tetco Dauntless D-35 well, lat. 44°44′8.26″N, long. 57°20′46.62″W, Scotian Shelf.

Chlamydophorella sp. A
Fig. spec. 53677
Bujak, J.P. and Williams, G.L., 1978, Geol. Surv. Can., Bull. 297, pl. 1, fig. 9; pl. 2, fig. 8.
Late Cretaceous, swc 4635 feet, Mobil Sable Island C-67 well, lat. 43°56′4.9″N, long. 59°55′1.38″W, Scotian Shelf.

Chlamydophorella sp. A
Fig. specs. 56723, 56724
Williams, G.L. and Bujak, J.P., Initial Repts. Deep Sea Drilling Project, vol. 50, pl. 3, figs. 8, 9.
Early Cretaceous, DSDP Site 416, lat. 32°50.18′N, long. 10°48.06′W, northwestern Moroccan continental margin.

Chlamydophorella sp. AE
Fig. spec. 46046
Brideaux, W.W., 1977, Geol. Surv. Can., Bull. 281, p. 33, pl. 14, fig. 8.
Cretaceous, Banff-Aquitaine-Arco Rat Pass K-35 well, depth 1025 feet, 6 miles southeast of Mount Goodenough, lat. 67°54′42.55″N, long. 135°21′56.83″W, District of Mackenzie.

Chytroeisphaerida sp.
Fig. specs. 58341-58343
Tan, J.T. and Hills, L.V., 1978, Geol. Surv. Can., Paper 78-1C, p.66, pl. 12.1, figs. 1-3.
Savik Formation, Upper Jurassic, lat. 77°48′N, long. 112°20′W, Mackenzie King Island, District of Franklin.

Chytroeisphaeridia minor (Cookson and Hughes) Morgan, 1980
Hypotype 91446
Sweet, A.R. and McIntyre, D.J., 1988, Can. Soc. Petrol. Geol., Mem. 15, fig. 8.18.
Leyland Member, Cardium Formation, Upper Cretaceous, south side of McLeod River, l.s.d. 16, sec. 16, tp. 47, rge. 23, W.5th mer., Alberta.

Chytroeisphaeridia ringnesiorum (Manum and Cookson) Morgan, 1980
Hypotype 91439
Sweet, A.R. and McIntyre, D.J., 1988, Can. Soc. Petrol. Geol., Mem. 15, fig. 8.11.
Leyland Member, Cardium Formation, Upper Cretaceous, south side of McLeod River, l.s.d. 16, sec. 16, tp. 47, rge. 23, W.5th mer., Alberta.

Chytroeisphaeridia sp.
Fig. spec. 75479
Davies, E.H., 1985, Palynology, vol. 9, p.129, pl. 3, fig. 11.
Early Toarcian, Liassic, Peniche Cliffs, Lusitanian Basin, west-central Portugal.

Cleistosphaeridium aciculare Davey, 1969
Hypotype 91456
Sweet, A.R. and McIntyre, D.J., 1988, Can. Soc. Petrol. Geol., Mem. 15, fig. 9.1.
Leyland Member, Cardium Formation, Upper Cretaceous, south side of McLeod River, l.s.d. 16, sec. 16, tp. 47, rge. 23, W.5th mer., Alberta.

Cleistosphaeridium? aciculare Davey
Hypotype 41496
Brideaux, W.W. and McIntyre, D.J., 1975, Geol. Surv. Can., Bull. 252, p. 26, pl. 7, fig. 7.
Horton River Formation, Lower Cretaceous, lat. 69°27′30″N, long. 126°54′W, Horton River, District of Mackenzie.

Cleistosphaeridium? aciculare Davey
Hypotype 46015
Brideaux, W.W., 1977, Geol. Surv. Can., Bull. 281, p. 22, pl. 8, fig. 9.
Cretaceous, IOE Stoney Core Hole F-42, depth 795-800 feet, Stoney Creek about 18 miles west-southwest of Fort McPherson, lat. 67°21′23″N, long. 135°38′43″W, District of Mackenzie.

Cleistosphaeridium? aciculare Davey, 1969
Hypotypes 65129-65131
Ioannides, N.S., 1986, Geol. Surv. Can., Bull. 371, p. 14, pl. 3, figs. 1-3.
Kanguk Formation, Upper Cretaceous, south coast Bylot Island, lat. 72°53′30″N, long. 78°16′W, District of Franklin.

Cleistosphaeridium sp. cf. *C. aciculare* Davey
Fig. specs. 38153, 38154
McIntyre, D.J., 1974, Geol. Surv. Can., Paper 74-14, pl. 12, figs. 10, 11.
Upper Cretaceous, lat. 69°28′N, long. 126°58′W, Horton River, District of Mackenzie.

Cleistosphaeridium ancoriferum (Cookson and Eisenack) Davey et al.
Hypotype 41499
Brideaux, W.W. and McIntyre, D.J., 1975, Geol. Surv. Can., Bull. 252, p. 27, pl. 7, figs. 10, 11.
Horton River Formation, Lower Cretaceous, lat. 69°19′N, long. 126°50′W, Horton River, District of Mackenzie.

Cleistosphaeridium araneosum Brideaux
Paratype 46017
Brideaux, W.W., 1977, Geol. Surv. Can., Bull. 281, p. 22, pl. 9, fig. 3.
Cretaceous, Banff-Aquitaine-Arco Rat Pass K-35 well, depth 1180 feet, 6 miles southeast of Mount Goodenough, lat. 67°54′42.55″N, long. 135°21′56.83″W, District of Mackenzie.

Cleistosphaeridium araneosum Brideaux, 1977
Hypotype 58608
Brideaux, W.W. and McIntyre, D.J., 1975, Geol. Surv. Can., Bull. 320, p. 19, pl. 6, fig. 6.
Lower Cretaceous; Martin Creek, lat. 68°12′10″N, long. 135°34′50″W, Richardson Mountains, District of Mackenzie.

Cleistosphaeridium huguonioti (Valensi)
Hypotype 41791
Williams, G.L., 1975, Geol. Surv. Can., Paper 74-30, vol. 2, pl. 1, fig. 11.
Turonian, Upper Cretaceous, Amoco IOE Kittiwake P-11 well, lat. 44°40′49.43″N, long. 53°31′45.65″W, Grand Banks.

Cleistosphaeridium huguonioti (Valensi) Davey
Hypotype 47860
Williams, G.L., 1978, Initial Repts. Deep Sea Drilling Project, vol. 41, pl. 2, fig. 9.
Late Eocene, DSDP site 370, lat. 32°50.2′N, long. 10°46.6′W, northwestern Moroccan continental margin.

Cleistosphaeridium multispinosum (Singh) Brideaux
Hypotype 41500
Brideaux, W.W. and McIntyre, D.J., 1975, Geol. Surv. Can., Bull. 252, p. 27, pl. 7, fig. 12.
Horton River Formation, Lower Cretaceous, lat. 69°19′N, long. 126°50′W, Horton River, District of Mackenzie.

Cleistosphaeridium multispinosum (Singh) Brideaux, 1971
Hypotype 71423
Wilson, M.A., 1978, Palaeontographica, Abt. B, vol. 166, p. 160, pl. 12, fig. 6.
Cretaceous, Police Island, north bank Mackenzie River, 12 miles southeast of Fort Norman, District of Mackenzie.

Cleistosphaeridium polypes (Cookson and Eisenack)
Hypotype 41790
Williams, G.L., 1975, Geol. Surv. Can., Paper 74-30, vol. 2, pl. 1, fig. 10.
Albian, Lower Cretaceous, Mobil Sable Island C-67 well, lat. 43°56′4.91″N, long. 59°55′1.38″W, Scotian Shelf.

Cleistospharidium sp. cf. *C. polypes* (Cookson and Eisenack) Davey, 1969
Fig. spec. 38157
McIntyre, D.J., 1974, Geol. Surv. Can., Paper 74-14, pl. 13, fig. 2.
Upper Cretaceous, lat. 69°27′30″N, long. 126°58′W, Horton River, District of Mackenzie.

Cleistosphaeridium polypes clavulum (Davey) Lentin and Williams
Hypotype 41497
Brideaux, W.W. and McIntyre, D.J., 1975, Geol. Surv. Can., Bull. 252, p. 27, pl. 7, fig. 8.
Horton River Formation, Lower Cretaceous, lat. 69°19′N, long. 126°50′W, Horton River, District of Mackenzie.

Cleistosphaeridium polypes subsp. *clavulum* Davey
Hypotype 45931
Brideaux, W.W., 1977, Geol. Surv. Can., Paper 76-1C, p. 118, pl. 23.1, fig. 9.
Husky Formation, Lower Cretaceous, south bank Martin Creek, District of Mackenzie.

Cleistosphaeridium polypes polypes (Cookson and Eisenack)
Hypotype 41498
Brideaux, W.W. and McIntyre, D.J., 1975, Geol. Surv. Can., Bull. 252, p. 27, pl. 7, fig. 9.
Horton River Formation, Lower Cretaceous, lat. 69°19′N, long. 126°50′W, Horton River, District of Mackenzie.

Cleistosphaeridium polypes subsp. A
Hypotype 41789
Williams, G.L., 1975, Geol. Surv. Can., Paper 74-30, vol. 2, pl. 1, fig. 9.
Bujak, J.P. and Williams, G.L., 1978, Geol. Surv. Can., Bull. 297, pl. 2, fig. 9.
Cenomanian, Upper Cretaceous, Shell Mohawk B-93 well, lat. 42°42′10.72″N, long. 64°43′53.04″W, Scotian Shelf.

Cleistosphaeridium polypes (Cookson and Eisenack) subsp. A sensu Williams, 1975
Hypotypes 53694, 53695
Bujak, J.P. and Williams, G.L., 1978, Geol. Surv. Can., Bull. 297, pl. 2, figs. 10, 11.
Late Cretaceous, swc 1950 feet, Shell Wyandot E-53 well, lat. 44°52′19.78″N, long. 59°23′55.7″W, and 3398-3430 feet, Shell Mic Mac J-77 well, lat. 44°36′42.8″N, long. 59°26′10.9″W, Scotian Shelf.

Cleistosphaeridium separatum McIntyre and Brideaux
Holotype 58606; paratype 58607
McIntyre, D.J. and Brideaux, W.W., 1980, Geol. Surv. Can., Bull. 320, p. 19, pl. 6, figs. 4, 5, 7, 8.

Lower Cretaceous, sidewall core 487.7 m, Banff-Aquitaine-Arco Rat Pass K-35 well, 6 miles southeast of Mount Goodenough, lat. 67°54′42.55″N, long. 135°21′56.83″W, and Martin Creek, lat. 68°12′10″N, long. 135°34′50″W, Richardson Mountains, District of Mackenzie.

Cleistosphaeridium spissum McIntyre and Brideaux
Holotype 58609; paratypes 58610, 58611
McIntyre, D.J. and Brideaux, W.W., 1980, Geol. Surv. Can., Bull. 320, p. 20, pl. 7, figs. 1-9.
Lower Cretaceous, 'Grizzly Gorge', lat. 68°15′49″N, long. 135°40′59″W, Richardson Mountains, District of Mackenzie.

Cleistosphaeridium sp.
Fig. specs. 47859, 47863, 47864, 47866
Williams, G.L., 1978, Initial Repts. Deep Sea Drilling Project, vol. 41, pl. 2, fig. 8; pl. 3, figs. 1, 2, 4.
Early Cretaceous, DSDP Site 370, lat. 32°50.2′N, long. 10°46.6′W, northwestern Moroccan continental margin.

?*Cleistosphaeridium* sp.
Fig. specs. 45936, 45939
Brideaux, W.W., 1976, Geol. Surv. Can., Paper 76-1C, pl. 23.2, figs. 8, 11.
Husky Formation, Lower Cretaceous, south bank Martin Creek, District of Mackenzie.

Cleitosphaeridium sp. A
Fig. specs. 69368, 69369
Davies, E.H., 1983, Geol. Surv. Can., Bull. 359, p. 26, pl. 8, figs. 15-18.
Middle-Upper Jurssic, depth 550-580 feet, Panartic Sandy Point L-46 well, northwestern Melville Island, District of Franklin.

Cleistosphaeridium sp. JE
Fig. spec. 45925
Brideaux, W.W., 1976, Geol. Surv. Can., Paper 76-1C, p. 118, pl. 23.1, figs. 5, 6.
Husky Formation, Lower Cretaceous, south bank Martin Creek, District of Mackenzie.

Cleistosphaeridium sp. KE
Fig. specs. 58612-58615
McIntyre, D.J. and Brideaux, W.W., 1980, Geol. Surv. Can., Bull. 320, p. 20, pl. 8, figs. 1-5.
Lower Cretaceous, lat. 68°12′10″N, long. 135°34′50″W, Richardson Mountains, District of Mackenzie.

Cleistosphaeridium? sp. AE
Fig. specs. 41501, 41502
Brideaux, W.W. and McIntyre, D.J., 1975, Geol. Surv. Can., Bull. 252, p. 27, pl. 7, figs. 13, 14.
Horton River Formation, Lower Cretaceous, lat. 69°19′N, long. 126°50′W, Horton River, District of Mackenzie.

Cleistosphaeridium? sp. BE
Fig. specs. 46019, 46020
Brideaux, W.W., 1977, Geol. Surv. Can., Bull. 281, p. 24, pl. 9, figs. 7, 8, 10, 11.
Cretaceous, Banff-Aquitaine-Arco Rat Pass K-35 well, depth 1600 feet, 6 miles southeast of Mount Goodenough, lat. 67°54′42.55″N, long. 135°21′56.83″W, District of Mackenzie.

Comasphaeridium cf. *comestes* (Valensi)
Hypotype 41805
Williams, G.L., 1975, Geol. Surv. Can., Paper 74-30, vol. 2, pl. 2, fig. 10.
Early Eocene, core hole 4, lat. 45°31′N, long. 52°22′W, Grand Banks.

Cometodinium whithei (Deflandre and Courteville) Stover and Evitt, 1978
Hypotype 91462
Sweet, A.R. and McIntyre, D.J., 1988, Can. Soc. Petrol. Geol., Mem. 15, fig. 9.7.
Leyland Member, Cardium Formation, Upper Cretaceous, south side of McLeod River, l.s.d. 16, sec. 16, tp. 47, rge. 23, W.5th mer., Alberta.

Cometodinium sp. A (Habib, 1972)
Fig. specs. 56726, 56727, 56768, 56770
Williams, G.L. and Bujak, J.P., 1980, Initial Repts. Deep Sea Drilling Project, vol. 50, pl. 3, figs. 11, 12; pl. 9, figs. 4, 6.
Early Cretaceous, DSDP Site 416, lat. 32°50.18′N, long. 10°48.06′W, northwestern Moroccan contimental margin.

Cometodinium sp. A of Habib (1972)
Fig. spec. 76535
Davies, E.H. and Poulton, T.P., 1986, Geol. Surv. Can., Paper 86-1B, pl. 55.5, fig. 9.
Fernie Formation, Upper Jurassic, Christina Falls, lat. 56°40′N, long. 123°4′W, northeastern British Columbia.

Comparodinium perpunctatum Wille and Gocht, 1979
Hypotype 69360
Davies, E.H., 1983, Geol. Surv. Can., Bull. 359, pl. 2, fig. 16.
Lower-Middle Jurassic, depth 2330-2360 feet, Panarctic Sandy Point L-46 well, northwestern Melville Island, District of Franklin.

Comparodium spiculatum Willie and Gocht
Hypotypes 75482, 75483
Davies, E.H., 1985, Palynology, vol. 9, pl. 3, figs. 15, 16.
Late Pliensbachian, Liassic, Zambujal Section, south of Condeixa, Lusitanian Basin, west-central Portugal.

Concentricystes rubinus Rossignol
Hypotype 47853
Williams, G.L., 1978, Initial Repts. Deep Sea Drilling Project, vol. 41, pl. 1, fig. 9.
Plio-Pleistocene, DSDP Site 367, lat. 12°29.2′N, long. 20°2.8′W, Cape Verde Basin west of Senegal.

Conosphaeridium sp. A
Fig. spec. 41504
Brideaux, W.W. and McIntyre, D.J., 1975, Geol. Surv. Can., Bull. 252, p. 28, pl. 7, figs. 17, 18.
Horton River Formation, Lower Cretaceous, lat. 69°19′N, long. 126°50′W, District of Mackenzie.

Conosphaeridium sp. B
Fig. spec. 46018
Brideaux, W.W., 1977, Geol. Surv. Can., Bull. 281, p. 24, pl. 9, figs. 4-6.
Cretaceous, Banff-Aquitaine-Arco Rat Pass K-35 well, depth 1445 feet, 6 miles southeast of Mount Goodenough, lat. 67°54′42.55″N, long. 135°21′56.83″W, District of Mackenzie.

Cordosphaeridium eoinodes (Eisenack)
Hypotype 41793
Williams, G.L., 1975, Geol, Surv. Can., Paper 74-30, vol. 2, pl. 1, fig. 13.
Hauterivian, Lower Cretaceous, Mobil Sable Island C-67 well, lat. 43°56′4.91″N, long. 59°55′1.38″W, Scotian Shelf.

Cordosphaeridium eoinodes (Eisenack) Eisenack
Hypotype 46021
Brideaux, W.W., 1977, Geol. Surv. Can., Bull. 281, p. 24, pl. 9, fig. 9.
Cretaceous, Banff-Aquitaine-Arco Rat Pass K-35 well, depth 1130 feet, 6 miles southeast of Mount Goodenough, lat. 67°54′42.55″N, long. 135°21′56.83″W, District of Mackenzie.

Cordosphaeridium exilimurum Davey and Williams *in* Davey et al. 1966
Hypotype 62641
Ioannides, N.S. and McIntyre, D.J., 1980, Geol. Surv. Can., Paper 80-1A, pl. 31.2, fig. 4.
Eocene, Caribou Hills, Mackenzie River, District of Mackenzie.

Cordosphaeridium exilimurum Davey and Williams, *in* Davey et al., 1966
Hypotype 65195
Ioannides, N.S., 1986, Geol. Surv. Can., Bull. 371, p. 17, pl. 8, fig. 8.
Eureka Sound Formation, Late Paleocene to ?Early Eocene, south coast Bylot Island, lat. 72°52′N, long. 78°31′W, District of Franklin.

Cordosphaeridium funiculatum Morgenroth
Hypotype 41794
Williams, G.L., 1975, Geol. Surv. Can., Paper 74-30, vol. 2, pl. 1, fig. 14.
Early Oligocene, Amoco IOE Puffin B-90 well, lat. 44°39′12.73″N, long. 53°42′28.35″W, Grand Banks.

Cordosphaeridium funiculatum Morgenroth
Hypotype 49347
Williams, G.L. and Bujak, J.P., 1977, Am. Assoc. Strat. Palynol, Contr. Strat. Palynol, vol. 1, Contr. Ser. 5A, pl. 2, fig. 6.
Early Oligocene, depth 1230-1260 feet, Mobil Tetco Esperanto K-78 well, lat. 44°47′31.26″N, long. 58°11′19.24″W, Scotian Shelf.

Cordosphaeridinium gracile (Eisenack) Davey and Williams
Hypotype 68147
Norris, G., 1986, Geol. Surv. Can., Bull. 340, p. 43, pl. 11, fig. 21.
Richards Formation, Middle Eocene, depth 11, 880 feet, Imperial Nuktak C-22 well, lat. 69°41′07″N, long. 134°51′30″W, District of Mackenzie.

Cordosphaeridium gracilis (Eisenack)
Hypotype 41804
Williams, G.L., 1975, Geol. Surv. Can., Paper 74-30, vol. 2, pl. 2, fig. 9.
Early Eocene, core hole 4, lat. 45°31′N, long. 52°22′W, Grand Banks.

Cordosphaeridium cf. *gracilis* (Eisenack)
Hypotype 41800
Williams, G.L., 1975, Geol. Surv. Can., Paper 74-30, vol. 2, pl. 2, fig. 5.
Maastrichtian, Upper Cretaceous, Shell Onondaga E-84 well, lat. 43°43′16.14″N, long. 60°14′17.18″W, Scotian Shelf.

Corrudinium harlandii Matsuoka, 1983
Hypotype 55323
Mudie, P.J, 1986, Initial Repts. Deep Sea Drilling Project, vol. XCIV, p. 803, pl. 3, figs. 9a, b.
Lower Pliocene, hole 611C, depth 196.69 m, lat. 52°50.15′N, long. 30°18.1′W, in the Gardar Drift on southwestern flank of Reykjanes Ridge, North Atlantic Ocean.

Cordosphaeridium inodes (Klumpp) Eisenack, 1963
Hypotypes 65193, 65194
Ioannides, N.S., 1986, Geol. Surv. Can., Bull. 371, p. 17, pl. 8, figs. 6, 7.
Eureka Sound Formation, Late Paleocene to ?Early Eocene, "Twosnout Creek" lat. 73°13′N, long. 79°53′W, Bylot Island, District of Franklin.

Cordosphaeridium inodes (Klumpp) Eisenack emend. Morgenroth
Hypotype 92004
Dietrich, J.R. et al., 1989, Geol. Surv. Can., Paper 89-1G, p. 141, pl. 1, fig. 8.
Early to Middle Eocene, depth 1200-1290 feet, Dome Pacific et al. PEX Natsek E-56 well, lat. 69°45′21.46″N, long. 139°44′34.58″W, western Beaufort Sea.

Cordosphaeridium sp.
Fig. spec. 71426
Wilson, M.A., 1978, Palaeontographica, Abt. B, vol. 166, p. 151, pl. 12, fig. 9.
Cretaceous, Police Island, north bank Mackenzie River, 12 miles southeast of Fort Norman, District of Mackenzie.

Coronifera oceanica Cookson and Eisenack
Hypotype 41814
Williams, G.L., 1975, Geol. Surv. Can., Paper 74-30, vol. 2, pl. 3, fig. 5.
Santonian, Upper Cretaceous, core hole 5, lat. 45°48′N, long. 52°13′W, Grand Banks.

Coronifera oceanica Cookson and Eisenack, 1958
Hypotype 65199
Ioannides, N.S., 1986, Geol. Surv. Can., Bull. 371, p. 18, pl. 8, fig. 10.
Kanguk Formation, Upper Cretaceous, south coast Bylot Island, lat. 72°53′30″N, long. 78°16′W, District of Franklin.

Cribroperidinium edwardsii (Cookson and Eisenack) Davey, 1969
Hypotypes 65196-65198
Ioannides, N.S., 1986, Geol. Surv. Can., Bull. 371, p. 18, pl. 8, figs. 9, 11, 12.
Kanguk Formation, Upper Cretaceous, Bylot Island, "Twosnout Creek", lat. 73°14′30″N, long. 79°49′W, and south coast, lat. 72°53′30″N, long. 78°16′W (65197, 65198), District of Franklin.

Cribroperidinium intricatum Davey
Hypotype 41809
Williams, G.L., 1975, Geol. Surv. Can., Paper 74-30, vol. 2, pl. 2, fig. 14.
Albian, Lower Cretaceous, Shell Naskapi N-30 well, lat. 43°29′46.29″N, long. 62°34′0.63″W, Scotian Shelf.

Cribroperidinium intricatum Davey, 1969
Hypotype 84742
Leckie, D.A. et al., 1990, J. Sed. Petrology, vol. 60, no. 1, fig. 6E.
Shaftesbury Formation, Lower Cretaceous, west bank Peace River near town of Peace River, Alberta.

Cribroperidinium jubaris (Davies) Lentin and Williams
Hypotypes 76289, 76290
Davies, E.H. and Poulton, T.P., 1986, Geol. Surv. Can., Paper 86-1B, pl. 55.4, figs. 1, 4.
Fernie Formation, Upper Jurassic, Christina Falls, lat. 56°40′N, long. 123°4′W, northeastern British Columbia.

Cribroperidinium muderongense (Cookson and Eisenack) Davey
Hypotypes 47861, 47862
Williams, G.L., 1978, Initial Repts. Deep Sea Drilling Project, vol. 41, pl. 2, figs. 10, 11.
Early Cretaceous, DSDP site 370, lat. 32°50.2′N, long. 10°46.6′W, northwestern Moroccan continental margin.

Cribroperidinium muderongensis (Cookson and Eisenack) Davey
Hypotype 45962
Brideaux, W.W., 1977, Geol. Surv. Can., Bull. 281, p. 5, pl. 1, fig. 5.
Cretaceous, Banff-Aquitaine-Arco Rat Pass K-35 well, depth 1025 feet, 6 miles southeast of Mount Goodenough, lat. 67°54′42.55″N, long. 135°21′56.83″W, District of Mackenzie.

Cribroperidinium orthoceras (Eisenack)
Hypotype 41807
Williams, G.L., 1975, Geol. Surv. Can., Paper 74-30, vol. 2, pl. 2, fig. 12.
Hauterivian, Lower Cretaceous, Mobil Sable Island C-67 well, lat. 43°56′4.91″N, long. 59°55′1.38″W, Scotian Shelf.

Cribroperidinium orthoceras (Eisenack) Davey
Hypotype 47865
Williams, G.L., 1978, Initial Repts. Deep Sea Drilling Project, vol. 41, pl. 3, fig. 3.
Early Cretaceous, DSDP site 370, lat. 32°50.2′N, long. 10°46.6′W, northwestern Moroccan continental margin.

Cribroperidinium perforans
Hypotypes 76300, 76301
Davies, E.H. and Poulton, T.P., 1986, Geol. Surv. Can., Paper 86-1B, pl. 55.3, figs. 2, 3.
Fernie Formation, Upper Jurassic, Christina Falls, lat. 56°40′N, long. 123°4′W, northeastern British Columbia.

Cribroperidinium sepimentum Neale and Sargeant
Hypotype 41808
Williams, G.L., 1975, Geol. Surv. Can., Paper 74-30, vol. 2, pl. 2, fig. 13.
Barremian, Lower Cretaceous, Amoco IOE Puffin B-90 well, lat. 44°39′12.73″N, long. 53°42′28.35″W, Grand Banks.

Cribroperidinium sp.
Fig. specs. 38139, 38140
McIntyre, D.J., 1974, Geol. Surv. Can., Paper 74-14, pl. 11, figs. 5, 6.
Upper Cretaceous, lat. 69°28′N, long. 126°58′W, Horton River, District of Mackenzie.

Cribroperidinium sp.
Fig. spec. 91458
Sweet, A.R. and McIntyre, D.J., 1988, Can. Soc. Petrol. Geol., Mem. 15, fig. 9.3.
Leyland Member, Cardium Formation, Upper Cretaceous, south side of McLeod River, l.s.d. 16, sec. 16, tp. 47, rge. 23, W.5th mer., Alberta.

Cribroperidinium sp. A
Fig. specs. 56715, 56716
Williams, G.L. and Bujak, J.P., 1980, Initial Repts. Deep Sea Drilling Project, vol. 50, pl. 2, figs. 8, 9.
Early Cretaceous and Late Jurassic, DDDP Site 416, lat. 32°50.18′N, long. 10°48.06′W, northwestern Moroccan continental margin.

Cribroperidinium sp. A
Fig. spec. 76302
Davies, E.H. and Poulton, T.P., 1986, Geol. Surv. Can., Paper 86-1B, p. 525, pl. 55.5, fig. 7.
Fernie Formation, Upper Jurassic, Christina Falls, lat. 56°40′N, long. 123°4′W, northeastern British Columbia.

Cribroperidinium sp. A
Fig. specs. 65202, 65203
Ioannides, N.S., 1986, Geol. Surv. Can., Bull. 371, p. 18, pl. 9, figs. 3, 4.

Kanguk Formation, Upper Cretaceous, south coast Bylot Island, lat. 72°53′30″N, long. 78°16′W, District of Franklin.

Cribroperidinium sp. B
Fig. specs. 65200, 65201
Ioannides, N.S., 1986, Geol. Surv. Can., Bull. 371, p. 18, pl. 9, figs. 1, 2.
Upper Cretaceous, Devon Island, lat. 75°26′N, long. 91°35′W and Kanguk Formation, south coast Bylot Island, lat. 72°53′30″N, long. 78°16′W, District of Franklin.

Cribroperidinium (alias *Gonyaulacysta*) sp. F of Gitnez and Sargeant (1972)
Fig. specs. 76299, 76534
Davies, E.H. and Poulton, T.P., 1986, Geol. Surv. Can., Paper 86-1B, pl. 55.5, figs. 1, 4.
Fernie Formation, Upper Jurassic, Christina Falls, lat. 56°40′N, long. 123°4′W, northeastern British Columbia.

Ctenidodinium culmulum (Norris)
Hypotype 41812
Williams, G.L., 1975, Geol. Surv. Can., Paper 74-30, vol. 2, pl. 3, fig. 3.
Tithonian, Upper Jurassic, Shell Naskapi N-30 well, lat. 43°29′46.29″N, long. 62°34′0.63″W, Scotian Shelf.

Ctenidodinium elegantulum Millioud
Hypotype 41813
Williams, G.L., 1975, Geol. Surv. Can., Paper 74-30, vol. 2, pl. 3, fig. 4.
Hauterivian, Lower Cretaceous, Shell Oneida 0-25 well, lat. 43°14′57.49″N, long. 61°33′36.38″W, Scotian Shelf.

Ctenidodinium elegantulum Millioud
Hypotype 56721
Williams, G.L. and Bujak, J.P., 1980, Initial Repts. Deep Sea Drilling Project, vol. 50, pl. 3, fig. 6.
Early Cretaceous, DSDP Site 416, lat. 32°50.18′N, long. 10°48.06′W, northwestern Moroccan continental margin.

Ctenidodinium ornatum (Eisenack)
Hypotype 41806
Williams, G.L., 1975, Geol. Surv. Can., Paper 74-30, vol. 2, pl. 2, fig. 11.
Callovian, Jurassic, Shell Mohican I-100 well, lat. 42°59′39.04″N, long. 62°28′51.32″W, Scotian Shelf.

Ctenidodinium ornatum (Eisenack) Deflandre
Hypotype 75503
Davies, E.H., 1985, Palynology, vol. 9, pl. 4, fig. 15.
Early Bathonian, Liassic, Brehna Road Section, Lusitanian Basin, west-central Portugal.

Ctenidodinium pachydermum Gocht
Hypotypes 75501, 75502
Davies, E.H., 1985, Palynology, vol. 9, pl. 4, figs. 13, 14.
Early Bathonian, Liassic, Brehna Road Section, Lusitanian Basin, west-central Portugal.

Ctenidodinium panneum (Norris)
Hypotype 41815
Williams, G.L., 1975, Geol. Surv. Can., Paper 74-30, vol. 2, pl. 3, fig. 6.
Tithonian, Upper Jurassic, Shell Naskapi N-30 well, lat. 43°29′46.29″N, long. 63°24′0.63″W, Scotian Shelf.

Ctenidodinium scissum McIntyre and Brideaux
Holotype 58616; paratypes 58617-58620
McIntyre, D.J. and Brideaux, W.W., 1980, Geol. Surv. Can., Bull. 320, p. 21, pl. 8, figs. 6-9; pl. 9, figs. 1-6.
Lower Cretaceous, Martin Creek, lat. 68°12′10″N, long. 135°34′50″W, Richardson Mountains, District of Mackenzie.

Ctenidodinium #ES
Fig. specs. 91995, 91996
Banerjee, I. and Davies, E.H., 1988, Can. Soc. Petrol. Geol., Mem. 15, figs. 7.11, 7.12.
Mannville Group, Lower Cretaceous, depths 5260 and 5270 feet, well 11-28-40-24W4, Alberta.

Cf. *Ctenidodinium* sp. AE
Fig. spec. 45820
Brideaux, W.W., 1976, Geol. Surv. Can., Paper 76-1B, p. 251, pl. 44.2, fig. 11.
Lower Cretaceous, 7500-7530 feet, Gulf Mobil Parsons N-10 well, ca. 1.5 miles east-northeast of Parsons Lake, lat. 69°59′48.52″N, long. 133°31′50.33″W, District of Mackenzie.

Cyclonephelium attadalicum Cookson and Eisenack
Hypotypes 41798, 41799
Williams, G.L., 1975, Geol. Surv. Can., Paper 74-30, vol. 2, pl. 2, figs. 3, 4.
Aptian, Lower Cretaceous, Shell Oneida 0-25 well, lat. 43°14′57.49″N, long. 61°33′36.38″W, Scotian Shelf.

Cyclonephelium attadalicum Cookson and Eisenack
Hypotype 47867
Williams, G.L., 1978, Initial Repts. Deep Sea Drilling Project, vol. 41, pl. 3, fig. 6.
Early Cretaceous, DSDP Site 370, lat. 32°50.2′N, long. 10°46.6′W, northwestern Moroccan continental margin.

Cyclonephelium attadalicum Cookson and Eisenack
Hypotypes 53679, 53693
Bujak, J.P. and Williams, G.L., 1978, Geol. Surv. Can., Bull. 297, pl. 1, figs. 11, 12; pl. 2, fig. 7.
Early Cretaceous, depth 10,960-10,990 feet, Shell Primrose 1(a) A-41 well, lat. 44°5.68′N, long. 59°6′18.62″W, and depth 8350-8380 feet, Shell Cree E-35 well, lat. 43°44′20.71″N, long. 60°35′35.9″W, Scotian Shelf.

Cyclonephelium cf. *attadalicum* Cookson and Eisenack
Hypotype 41802
Williams, G.L., 1975, Geol. Surv. Can., Paper 74-30, vol. 2, pl. 2, fig. 7.
Aptian, Lower Cretaceous, Shell Mohawk B-93 well, lat. 42°42′10.72″N, long. 64°43′53.04″W, Scotian Shelf.

Cyclonephelium brevispinum
Hypotype 91988
Banerjee, I. and Davies, E.H., 1988, Can. Soc. Petrol. Geol., Mem. 15, fig. 7.4.
Mannville Group, Lower Cretaceous, depth 5038 feet, well 11-08-41-24W4, Alberta.

Cyclonephelium compactum Deflandre and Cookson
Hypotype 41527
Brideaux, W.W. and McIntyre, D.J., 1975, Geol. Surv. Can., Bull. 252, p. 32, pl. 9, fig. 17.
Horton River Formation, Lower Cretaceous, lat. 69°19′N, long. 126°50′W, Horton River, District of Mackenzie.

Cyclonephelium distinctum Deflandre and Cookson
Hypotypes 38091, 38092, 38097
McIntyre, D.J., 1974, Geol. Surv. Can., Paper 74-14, pl. 7, figs. 13, 14; pl. 8, fig. 3.
Upper Cretaceous, lat. 69°28′N, long. 126°58′W, Horton River, District of Mackenzie.

Cyclonephelium distinctum Deflandre and Cookson
Hypotype 41528
Brideaux, W.W. and McIntyre, D.J., 1975, Geol. Surv. Can., Bull. 252, p. 32, pl. 9, fig. 18.
Horton River Formation, Lower Cretaceous, lat. 69°27′30″N, long. 126°54′W, Horton River, District of Mackenzie.

Cyclonephelium distinctum Deflandre and Cookson
Hypotype 53676
Bujak, J.P. and Williams, G.L., 1978, Geol. Surv. Can., Bull. 297, pl. 1, fig. 8.
Early Cretaceous, depth 11,350-11,380 feet, Shell Cree E-35 well, lat. 43°44′20.7″N, long. 60°35′55.9″W, Scotian Shelf.

Cyclonephelium distinctum Deflandre and Cookson
Hypotypes 56753, 56756
Williams, G.L. and Bujak, J.P., 1980, Initial Repts. Deep Sea Drilling Project, vol. 50, pl. 7, figs. 1, 4.
Early Cretaceous, DSDP Site 416, lat. 32°50.18′N, long. 10°48.06′W, northwestern Moroccan continental margin.

Cyclonephelium distinctum Deflandre and Cookson, 1955
Hypotypes 65204-65206
Ioannides, N.S., 1986, Geol. Surv. Can., Bull. 371, p. 19, pl. 9, figs. 5-7.
Kanguk Formation, Upper Cretaceous, south coast Bylot Island, lat. 72°53′30″N, long. 78°16′W, District of Franklin.

Cyclonephelium distinctum Deflandre and Cookson, 1955
Hypotype 91452
Sweet, A.R. and McIntyre, D.J., 1988, Can. Soc. Petrol. Geol., Mem. 15, fig. 8.24.
Leyland Member, Cardium Formation, Upper Cretaceous, south side of McLeod River, l.s.d. 16, sec. 16, tp. 47, rge. 23, W.5th mer., Alberta.

Cyclonephelium distinctum subsp. *brevispinatum* (Millioud) Lentin and Williams
Hypotype 56750
Williams, G.L. and Bujak, J.P., 1980, Initial Repts. Deep Sea Drilling Project, vol. 50, pl. 6, figs. 3, 4.
Early Cretaceous, DSDP Site 416, lat. 32°50.18′N, long. 10°48.06′W, northwestern Moroccan continental margin.

Cyclonephelium distinctum Deflandre and Cookson subsp. A
Hypotype 53698
Bujak, J.P. and Williams, G.L., 1978, Geol. Surv. Can., Bull. 297, pl. 2, fig. 14.
Early Cretaceous, 8500-8530 feet, Shell Oneida 0-25 well, lat. 43°14′57.49″N, long. 61°33′36.38″W, Scotian Shelf.

Cyclonephelium eisenacki Davey
Hypotype 47869
Williams, G.L., 1978, Initial Repts. Deep Sea Drilling Project, vol. 41, pl. 4, fig. 1.
Early Cretaceous, DSDP site 370, lat. 32°50.2′N, long. 10°46.6′W, northwestern Moroccan continental margin.

Cyclonephelium membraniforum Cookson and Eisenack, 1962
Hypotype 65207
Ioannides, N.S., 1986, Geol. Surv. Can., Bull. 371, p. 19, pl. 9, fig. 8.
Kanguk Formation, Upper Cretaceous, south coast Bylot Island, lat. 72°53′30″N, long. 78°16′W, District of Franklin.

Cyclonephelium ordinatum Williams and Downie
Hypotype 45808
Brideaux, W.W., 1976, Geol. Surv. Can., Paper 76-1B, pl. 44.1, fig. 9.
Reindeer Formation, Eocene, 1200-1230 feet, Gulf Mobil Parsons N-10 well, ca. 1.5 miles east-northeast of Parsons Lake, lat. 69°59′48.52″N, long. 133°31′50.33″W, District of Mackenzie.

Cyclonephelium paucispinum Davey
Hypotype 53713
Bujak, J.P. and Williams, G.L., 1978, Geol. Surv. Can., Bull. 297, pl. 3, fig. 16.
Early Cretaceous, 8550-8580 feet, Mobil Sable Island C-67 well, lat. 43°56′4.91″N, long. 59°55′1.38″W, Scotian Shelf.

Cyclonephelium retiintextum Cookson
Hypotype 47871
Williams, G.L., 1978, Initial Repts. Deep Sea Drilling Project, vol. 41, pl. 4, fig. 3.
Middle Eocene, DSDP site 370, lat. 32°50.2′N, long. 10°46.6′W, northwestern Moroccan continental margin.

Cyclonephelium vannophorum Davey
Hypotype 41801
Williams, G.L., 1975, Geol. Surv. Can., Paper 74-30, vol. 2, pl. 2, fig. 6.

Cenomanian, Upper Cretaceous, Mobil Tetco Dauntless D-35 well, lat. 44°44′8.26″N, long. 57°20′46.62″W, Scotian Shelf.

Cyclonephelium vannophorum Davey
Hypotype 47870
Williams, G.L., 1978, Initial Repts. Deep Sea Drilling Project, vol. 41, pl. 4, fig. 2.
Late Cretaceous, DSDP site 370, lat. 32°50.2′N, long. 10°46.6′W, northwestern Moroccan continental margin.

Cyclonephelium vannophorum Davey, 1969
Hypotype 91453
Sweet, A.R. and McIntyre, D.J., 1988, Can. Soc. Petrol. Geol., Mem. 15, fig. 8.25.
Leyland Member, Cardium Formation, Upper Cretaceous, south side of McLeod River, l.s.d. 16, sec. 16, tp. 47, rge. 23, W.5th mer., Alberta.

Cyclonephelium sp.
Fig. spec. 41803
Williams, G.L., 1975, Geol. Surv. Can., Paper 74-30, vol. 2, pl. 2, fig. 8.
Late Eocene, Mobil Sable Island C-67 well, lat. 43°56′4.91″N, long. 59°55′1.38″W, Scotian Shelf.

Cyclonephelium sp. D
Fig. spec. 49338
Williams, G.L. and Bujak, J.P., 1977, Am. Assoc. Strat. Palynol, Contr. Strat. Palynol, vol. 1, Contr. Ser. 5A, pl. 1, fig. 3.
Early Oligocene, depth 1230-1260 feet, Mobil Tetco Esperanto K-78 well, lat. 44°47′31.26″N, long. 58°11′19.24″W, Scotian Shelf.

Cyclonephelium sp. GE
Fig. spec. 45822
Brideaux, W.W., 1976, Geol. Surv. Can., Paper 76-1B, p. 251, pl. 44.2, fig. 13.
Lower Cretaceous, 7500-7530 feet, Gulf Mobil Parsons N-10 well, ca. 1.5 miles east-northeast of Parsons Lake, lat. 69°59′48.52″N, long. 133°31′50.33″W, District of Mackenzie.

Cyclopsiella cf. *elliptica* Drugg and Loeblich
Hypotype 47872
Williams, G.L., 1978, Initial Repts. Deep Sea Drilling Project, vol. 41, pl. 4, fig. 4.
Late Eocene, DSDP site 370, lat. 32°50.2′N, long. 10°46.6′W, northwestern Moroccan continental margin.

Cyclopsiella sp.
Fig. spec. 53697
Bujak, J.P. and Williams, G.L., 1978, Geol. Surv. Can., Bull. 297, pl. 2, fig. 13.
Early Cretaceous, 12050-12080 feet, Shell Cree E-35 well, lat. 43°44′20.71″N, long. 60°35′55.9″W, Scotaian Shelf.

Dapcodinium coalitum Davies
Hypotypes 75494, 75495
Davies, E.H, 1985, Palynology, vol. 9, pl. 4, figs. 5, 6.
Early and Late Pliensbachian, Liassic, Zambujal Section, south of Condeixa, Lusitanian Basin, west-central Portugal.

Dapcodinium holotabulatum Davies
Holotype 69359
Davies, E.H., 1983, Geol. Surv. Can., Bull. 359, p. 14, pl. 1, figs. 9-11; text-figs. 7A, B.
Savik Formation, Lower-Middle Jurassic, depth 1880 feet, Panarctic Sandy Point L-46 well, northwestern Melville Island, District of Franklin.

Dapcodinium holotabulatum Davies
Hypotype 75492
Davies, E.H., 1985, Palynology, vol. 9, pl. 4, fig. 3.
Early Pliensbachian, Liassic, Zambujal Section, south of Condeixa, Lusitanian Basin, west-central Portugal.

Decussosporites microreticulatus Brenner
Hypotype 41836
Williams, G.L., 1975, Geol. Surv. Can., Paper 74-30, vol. 2, pl. 4, fig. 10.
Aptian, Lower Cretaceous, Mobil Tetco Dauntless D-35 well, lat. 44°44′8.26″N, long. 57°20′46.62″W, Scotian Shelf.

Dapcodinium priscum Evitt
Hypotype 75491
Davies, E.H., 1985, Palynology, vol. 9, pl. 4, fig. 2.
Early Pliensbachian, Liassic, Zambujal Section, south of Condeixa, Lusitanian Basin, west-central Portugal.

Dapcodinium semitabulatum (Morgenroth) Dörhöfer and Davies
Hypotypes 75496, 75497
Davies, E.H., 1985, Palynology, vol. 9, pl. 4, figs. 7, 8.
Late and Early Pliensbachian, Liassic, Zambujal Section, south of Condeixa, Lusitanian Basin, west-central Portugal.

Dapcodinium sp. A
Fig. spec. 75490
Davies, E.H., 1985, Palynology, vol. 9, p. 129, pl. 4, fig. 1.
Early Pliensbachian, Liassic, Zambujal Section, south of Condeixa, Lusitanian Basin, west-central Portugal.

Deflandrea acuminata Cookson and Eisenack, 1958
Hypotype 38017
McIntyre, D.J., 1974, Geol. Surv. Can., Paper 74-14, pl. 1, fig. 15.
Upper Cretaceous, lat. 69°27′30″N, long. 126°58′W, Horton River, District of Mackenzie.

Deflandrea acuminata Cookson and Eisenack
Hypotypes 34026, 34027
McIntyre, D.J., 1975, Geoscience and Man, vol. 11, p. 66, pl. 2, figs. 11, 13.
Upper Cretaceous, lat. 69°27′30″N, long. 126°58′W, Horton River, District of Mackenzie.

Deflandrea acuminata Cookson and Eisenack
Hypotype 41828
Williams, G.L., 1975, Geol. Surv. Can., Paper 74-30, vol. 2, pl. 4, fig. 2.
Paleocene, Mobil Tetco Dauntless D-35 well, lat. 44°44′8.26″N, long. 57°20′46.62″W, Scotian Shelf.

Deflandrea amphiata McIntyre
Holotype 34022; paratype 34023
McIntyre, D.J., 1974, Geoscience and Man, vol. 11, p. 65, pl. 2, figs. 5, 6, 8.
Upper Cretaceous, lat. 69°28′N, long. 126°58′W, Horton River, District of Mackenzie.

Deflandrea sp. cf. *D. belfastensis* Cookson and Eisenack, 1961
Fig. specs. 38021, 38022
McIntyre, D.J., 1974, Geol. Surv. Can., Paper 74-14, pl. 2, figs. 1, 2.
Upper Cretaceous, lat. 69°27′30″N, long. 126°58′W, Horton River, District of Mackenzie.

Deflandrea biapertura McIntyre
Paratypes 34031, 34032
McIntyre, D.J., 1975, Geodcience and Man, vol. 11, p. 66, pl. 3, figs. 7, 8.
Upper Cretaceous, lat. 69°28′N, long. 126°58′W, Horton River, District of Mackenzie.

Deflandrea sp. cf. *D. cooksonii* Alberti, 1959
Fig. spec. 38018
McIntyre, D.J., 1974, Geol. Surv. Can., Paper 74-14, pl. 1, fig. 16.
Upper Cretaceous, lat. 69°28′N, long. 126°58′W, Horton River, District of Mackenzie.
=*Deflandrea acuminata*, McIntyre, D.J., 1975, Geoscience and Man, vol. 11, p. 66, pl. 2, fig. 12.

Deflandrea coronata McIntyre
Holotype 34028; paratypes 34029, 34030
McIntyre, D.J., 1975, Geoscience and Man, vol. 11, p. 64, pl. 1, figs. 1-4.
Upper Cretaceous, lat. 69°28′N, long. 126°58′W, Horton River, District of Mackenzie.

Deflandrea decorosa McIntyre
Holotype 34020; paratype 34021
McIntyre, D.J., 1975, Geoscience and Man, vol. 11, p. 63, pl. 2, figs. 1, 3.
Upper Cretaceous, lat. 69°28′N, long. 126°58′W, Horton River, District of Mackenzie.

Deflandrea delineata Cookson and Eisenack
Hypotype 41831
Williams, G.L., 1975, Geol. Surv. Can., Paper 74-30, vol. 2, pl. 4, fig. 5.
Danian, Mobil Sable Island C-67 well, lat. 43°56′4.91″N, long. 59°55′1.38″W, Scotian Shelf.

Deflandrea cf. *delineata* Cookson and Eisenack
Hypotype 41816
Williams, G.L., 1975, Geol. Surv. Can., Paper 74-30, vol. 2, pl. 3, fig. 7.
Late Paleocene-early Eocene, Shell Mohawk B-93 well, lat. 42°42′10.72″N, long. 64°43′53.04″W, Scotian Shelf.

Deflandrea diebeli Alberti, 1959
Hypotypes 38048, 38049
McIntyre, D.J., 1974, Geol. Surv. Can., Paper 74-14, pl. 4, figs. 4, 5.
Upper Cretaceous, lat. 69°28′N, long. 126°58′W, Horton River, District of Mackenzie.

Deflandrea diebili Alberti
Hypotypes 34034, 34035
McIntyre, D.J., 1975, Geoscience and Man, vol. 11, p. 67, pl. 4, figs. 1, 2.
Upper Cretaceous, lat. 69°58′30″N, long. 121°04′W, Horton River, District of Mackenzie.

Deflandrea diebili Alberti
Hypotype 41827
Williams, G.L., 1975, Geol. Surv. Can., Paper 74-30, vol. 2, pl. 4, fig. 1.
Danian, Eastcan et al. Bjarni H-81 well, lat. 55°30′29.35″N, long. 57°42′5.52″W, Labrador Shelf.

Deflandrea diebeli Alberti
Hypotype 49365
Williams, G.L. and Bujak, J.P., 1977, Am. Assoc. Strat. Palynol, Contr. Strat. Palynol, vol. 1, Contr. Ser. 5A, pl. 5, fig. 1.
Late Cretaceous, depth 8040-8070, Eastcan et al. Gudrid H-55 well, lat. 55°54′20.06″N, long. 53°52′31.04″W, Labrador Shelf.

Deflandrea diebeli Alberti, 1959
Hypotype 71416
Wilson, M.A., 1978, Palaeontographica, Abt. B, vol. 166, p. 151, pl. 111, fig. 15.
Tent Island Formation, Cretaceous, Little Fish (Cache) Creek, Big Fish River area, District of Mackenzie.

Deflandrea diebelli Alberti, 1959
Hypotypes 65249-65252
Ioannides, N.S., 1986, Geol. Surv. Can., Bull. 371, p. 19, pl. 11, figs. 6, 7, 10, 11.
Eureka Sound Formation, Late Paleocene to ?Early Eocene, south coast Bylot Island, lat. 72°53′30″N, long. 78°16′W, District of Franklin.

Deflandrea ditissima McIntyre
Paratypes 34011, 34012
McIntyre, D.J., 1975, Geoscience and Man, vol. 11, p. 62, pl. 1, figs. 3, 4.
Upper Cretaceous, lat. 69°27′30″N, long. 126°58′W, Horton River, District of Mackenzie.

Deflandrea granulifera Manum
Hypotype 38023
McIntyre, D.J., 1974, Geol. Surv. Can., Paper 74-14, pl. 2, fig. 3.

McIntyre, D.J., 1975, Geoscience and Man, vol. 11, p. 67, pl. 2, figs. 14.
Upper Cretaceous, lat. 69°27′30″N, long. 126°58′W, Horton River, District of Mackenzie.

Deflandrea heterophlycta Deflandre and Cookson
Hypotype 38700
Haskell, T.R. and Wilson, G.J., 1975, Initial Repts. Deep Sea Drilling Project, vol. 29, pl. 3, fig. 8.
Lower Oligocene, DSDP Site 282, lat. 42°14.76′S, long. 143°29.18′E, west of Tasmania.

Deflandrea heterophlycta Deflandre and Cookson
Hypotype 41833
Williams, G.L., 1975, Geol. Surv. Can., Paper 74-30, vol. 2, pl. 4, fig. 7.
Early Oligocene, Amoco IOE Puffin B-90 well, lat. 44°39′12.73″N, long. 53°42′28.35″W, Grand Banks.

Deflandrea microarma McIntyre
Holotype 34013; paratypes 34014, 34015
McIntyre, D.J., 1975, Geoscience and Man, vol. 11, p. 65, pl. 1, figs. 5-8.
Upper Cretaceous, lat. 69°27′30″N, long. 126°58′W, Horton River, District of Mackenzie.

Deflandrea magnifica Stanley
Hypotype 41832
Williams, G.L., 1975, Geol. Surv. Can., Paper 74-30, vol. 2, pl. 4, fig. 6.
Danian, Eastcan *et al.* Bjarni H-81 well, lat. 55°30′29.35″N, long. 57°42′5.52″W, Labrador Shelf.

Deflandrea minor Alberti, 1959
Hypotypes 38024, 38025
McIntyre, D.J., 1974, Geol. Surv. Can., Paper 74-14, pl. 2, figs. 4, 5.
Upper Cretaceous, lat. 69°27′30″N, long. 126°58′W, Horton River, District of Mackenzie.

Deflandrea minor Alberti
Hypotypes 34024, 34025
McIntyre, D.J., 1975, Geoscience and Man, vol. 11, p. 66, pl. 2, figs. 9, 10.
Upper Cretaceous, lat. 69°27′30″N, long. 126°58′W, Horton River, District of Mackenzie.

Deflandrea sp. cf. *D. obscura* Drugg, 1967
Fig. specs. 38032, 38033
McIntyre, D.J., 1974, Geol. Surv. Can., Paper 74-14, pl. 2, figs. 12, 13.
Upper Cretaceous, lat. 69°27′30″N, long. 126°58′W, Horton River, District of Mackenzie.
=*Deflandrea* sp. 1, McIntyre, D.J., 1975, Geoscience and Man, vol. 11, p. 67, pl. 1, fig. 13 (fig. spec. 38032).

Deflandrea oebisfeldensis Alberti
Hypotype 92009
Dietrich, J.R. et al., 1989, Geol. Surv. Can., Paper 89-1G, p. 141, pl. 2, fig. 1.
Early to Middle Eocene, depth 4100-4190 feet, Dome Pacific et al. PEX Natsek E-56 well, lat. 69°45′21.46″N, long. 139°44′34.58″W, western Beaufort Sea.

Deflandrea sp. cf. *D. pannucea* Stanley, 1965
Fig. spec. 38031
McIntyre, D.J., 1974, Geol. Surv. Can., Paper 74-14, pl. 2, fig. 11.
Upper Cretaceous, lat. 69°58′30″N, long. 127°04′W, Horton River, District of Mackenzie.

Deflandrea perlucida Alberti
Hypotypes 41817, 41829
Williams, G.L., 1975, Geol. Surv. Can., Paper 74-30, vol. 2, pl. 3, fig. 8; pl. 4, fig. 3.
Aptian, Lower Cretaceous, Mobil Sable Island C-67 well, lat. 43°56′4.91″N, long. 59°55′1.38″W, Scotian Shelf, and Barremian, Lower Cretaceous, Amoco IOE Puffin B-90 well, lat. 44°39′12.73′N, long. 53°42′28.35″W, Grand Banks.

Deflandrea phosphoritica Eisenack
Hypotype 41834
Williams, G.L., 1975, Geol. Surv. Can., Paper 74-30, vol. 2, pl. 4, fig. 8.
Oligocene, Mobil Sable Island C-67 well, lat. 43°56′4.9″N, long. 59°55′1.38″W, Scotian Shelf.

Deflandrea cf. *phosphoritica* Eisenack
Hypotype 49366
Williams, G.L. and Bujak, J.P., 1977, Am. Assoc. Strat. Palynol, Contr. Strat. Palynol, vol. 1, Contr. Ser. 5A, pl. 5, fig. 2.
Middle Eocene, swc. 6800 feet, Eastcan et al. Snorri J-90 well, lat. 57°19′44.52″N, long. 59°57′44.37″W, Labrador Shelf.

Deflandrea pirnaensis Alberti
Hypotypes 41820, 41821
Williams, G.L., 1975, Geol. Surv. Can., Paper 74-30, vol. 2, pl. 3, figs. 11, 12.
Aptian, Lower Cretaceous, Shell Fox I-22 well lat. 45°21′33.58″N, long. 59°33′16.03″W, and Mobil Tetco Dauntless D-35 well, lat. 44°44′8.26″N, long. 57°20′46.62″W, Scotian Shelf.

Deflandrea speciosa Alberti
Hypotype 41830
Williams, G.L., 1975, Geol. Surv. Can., Paper 74-30, vol. 2, pl. 4, fig. 4.
Paleocene, Mobil Tetco Dauntless D-35 well, lat. 44°44′8.26″N, long. 57°20′46.62″W, Scotian Shelf.

Deflandrea speciosa Alberti, 1959
Hypotype 65253
Ioannides, N.S., 1986, Geol. Surv. Can., Bull. 371, p. 19, pl. 11, fig. 9.
Eureka Sound Formation, Late Paleocene to ?Early Eocene, south coast Bylot Island, lat. 72°52′N, long. 78°35′W, District of Franklin.

Deflandrea spectabilis Alberti, 1959
Hypotypes 38013-38016
McIntyre, D.J., 1974, Geol. Surv. Can., Paper 74-14, pl. 1, figs. 8, 9, 13, 14.
Upper Cretaceous, lat. 69°27′30″N, long. 126°58′W, Horton River, District of Mackenzie.

Deflandrea spectabilis Alberti
Hypotype 34019
McIntyre, D.J., 1975, Geoscience and Man, vol. 11, p. 67, pl. 1, fig. 15.
Upper Cretaceous, lat. 69°27′30″N, long. 126°58′W, Horton River, District of Mackenzie.

Deflandrea sp. cf. *D. spectabilis* Alberti
Fig. spec. 27821
Brideaux, W.W., 1971, Geol. Surv. Can., Paper 71-1, pt.B, p. 89, pl. 1, fig. 6.
Little Bear Formation, Late Cretaceous, Little Bear River, 24 miles southwest of Fort Norman, District of Mackenzie.

Deflandrea sp. cf. *D. victoriensis* Cookson and Manum, 1964
Fig. specs. 38034, 38035
McIntyre, D.J., 1974, Geol. Surv. Can., Paper 74-14, pl. 2, fig. 14; pl. 3, fig. 1.
Upper Cretaceous, lat. 69°27′30″N, long. 126°58′W, Horton River, District of Mackenzie.
=*Deflandrea ditissima*, McIntyre, D.J., 1975, Geoscience and Man, vol. 11, p. 62, pl. 1, figs. 1, 2 (holotype 38035).

Deflandrea sp.
Fig. spec. 27820
Brideaux, W.W., 1971, Geol. Surv. Can., Paper 71-1, pt. B, p. 89, pl. 1, fig. 5.
Little Bear Formation, Late Cretaceous, Little Bear River, 24 miles southwest of Fort Norman, District of Mackenzie.

Deflandrea spp. B, C
Fig. specs. 49367, 49368
Williams, G.L. and Bujak, J.P., 1977, Am. Assoc. Strat. Palynol, Contr. Strat. Palynol, vol. 1, Contr. Ser. 5A, pl. 5, figs. 3, 4.
Early Eocene, depth 5190-5200 feet, Eastcan et al. Leif M-48 well, lat. 54°17′45.92″N, long. 55°7′20.17″W; Early Oligocene, depth 8100-8130 feet, BP Columbia Bonavista C-99 well, lat. 49°8′6.48″N, long. 51°14′27.85″W, northeast Newfoundland Basin.

Deflandrea sp. 1
Fig. spec. 34018
McIntyre, D.J., 1975, Geoscience and Man, vol. 11, p. 67, pl. 1, fig. 14.
Upper Cretaceous, lat. 69°27′30″N, long. 126°58′W, Horton River, District of Mackenzie.

Deflandrea sp. 2
Fig. specs. 38028-38030
McIntyre, D.J., 1974, Geol. Surv. Can., Paper 74-14, pl. 2, figs. 8-10.
McIntyre, D.J., 1975, Geoscience and Man, vol. 11, p. 68, pl. 3, figs. 10 (38029), 11 (38030).
Upper Cretaceous, lat. 69°28′N, long. 126°58′W, Horton River, District of Mackenzie.

Deflandrea sp. 2
Fig. spec. 34033
McIntyre, D.J., 1975, Geoscience and Man, vol. 11, p. 68, pl. 3, fig. 9.
Upper Cretaceous, lat. 69°28′N, long. 126°58′W, Horton River, District of Mackenzie.

Deflandrea sp. 3
Fig. specs. 38039, 38040
McIntyre, D.J., 1974, Geol. Surv. Can., Paper 74-14, pl. 3, figs. 5, 6.
Upper Cretaceous, lat. 69°28′N, long. 126°58′W, Horton River, District of Mackenzie.
=*Deflandrea decorosa*, McIntyre, D.J., 1975, Geoscience and Man, vol. 11, p. 63, pl. 2, figs. 2 (paratype 38040), 4 (paratype 38039).

Deflandrea sp. 4
Fig. specs. 38041, 38045
McIntyre, D.J., 1974, Geol. Surv. Can., Paper 74-14, pl. 3, fig. 7; pl. 4, fig. 1.
Upper Cretaceous, lat. 69°28′N, long. 126°58′W, Horton River, District of Mackenzie.

Deflandrea sp. 5
Fig. specs. 38019, 38020
McIntyre, D.J., 1974, Geol. Surv. Can., Paper 74-14, pl. 1, figs. 19, 20.
Upper Cretaceous, lat. 69°28′N, long. 126°58′W, Horton River, District of Mackenzie.
=*Deflandre biapertura*, McIntyre, D.J., 1975, Geoscience and Man, vol. 11, p. 66, pl. 3, figs. 5, 6 (holotype 38020).

Deflandrea sp. 6
Fig. specs. 38050, 38051
McIntyre, D.J., 1974, Geol. Surv. Can., Paper 74-14, pl. 5, figs. 1, 2.
Upper Cretaceous, lat. 69°58′30″N, long. 127°04′W, Horton River, District of Mackenzie.

Deflandrea sp. 8
Fig. specs. 38026, 38027
McIntyre, D.J., 1974, Geol. Surv. Can., Paper 74-14, pl. 2, figs. 6, 7.
Upper Cretaceous, lat. 69°28′N, long. 126°58′W, Horton River, District of Mackenzie.
=*Deflandrea amphiata*, McIntyre, D.J., 1975, Geoscience and Man, vol. 11, p. 65, pl. 2, fig. 7 (paratype 38027).

Deflandreoid Form 2
Fig. specs. 38057-38060
McIntyre, D.J., 1974, Geol. Surv. Can., Paper 74-14, pl. 5, figs. 8-11.
Upper Cretaceous, lat. 69°27′30″N, long. 126°58′W, Horton River, District of Mackenzie.
=*Laciniadinium orbiculatum*, McIntyre, D.J., 1975, Geoscience and Man, vol. 11, p. 70, pl. 4, figs. 10 (paratype 38058), 12, 13 (holotype 38060).

Diconodinium arcticum Manum and Cookson
Hypotype 62653
Ioannides, N.S. and McIntyre, D.J., 1980, Geol. Surv. Can., Paper 80-1A, pl. 31.2, fig. 16.
Upper Cretaceous, Caribou Hills, Mackenzie River, District of Mackenzie.

Diconodinium arcticum Manum and Cookson, 1964
Hypotype 71403
Wilson, M.A, 1978, Palaeontographica, Abt. B, vol. 166, p. 152, pl. 11, fig. 1.
Tent Island Formation, Cretaceous, Little Fish (Cache) Creek, Big Fish River area, District of Mackenzie.

Diconodinium arcticum Manum and Cookson, 1964
Hypotype 65221
Ioannides, N.S., 1986, Geol. Surv. Can, Bull. 371, p. 20, pl. 10, fig. 7.
Kangunk Formation, Upper Cretaceous, south coast Bylot Island, lat. 72°53′30″N, long. 78°16′W, District of Franklin.

Diconodinium sp. ? *D. arcticum* Manum and Cookson, 1964
Fig. specs. 65222-65230, 65243
Ioannides, N.S., 1986, Geol. Surv. Can., Bull. 371, p. 21, pl. 10, figs. 8-15, 25; pl. 11, fig. 1.
Kanguk Formation, Upper Cretaceous, Bylot Island, south coast, lat. 72°53′30″N, long. 78°16′W and "Twosnout Creek", lat. 73°14′30″N, long. 79°49′W (65229, 65243); Eureka Sound Formation, Late Paleocene to ?Early Eocene, Bylot Island, "Twosnout Creek", lat. 73°13′N, long. 79°53′W, District of Franklin.

Diconodinium multispinulum (Deflandre and Cookson) Eisenack and Cookson
Hypotype 68766
Dixon, J. et al., 1984, Geol. Surv. Can., Paper 82-13, pl. 4, fig. 3
Cretaceous (reworked), 2749 m interval, Dome Gulf et al. Hunt Kopanvar M-13 well, lat. 70°22′55.1″N, long. 135°05′33.9″W, Beaufort Sea.

Diconodinium - Spinidinium group
Fig. spec. 65212
Ioannides, N.S., 1986, Geol. Surv. Can., Bull. 371, p. 21, pl. 9, fig. 13.
Kanguk Formation, Upper Cretaceous, "Twosnout Creek", lat. 73°14′30″N, long. 79°49′W, Bylot Island, District of Franklin.

Diconodinium sp. A
Fig. specs. 65213, 65214
Ioannides, N.S., 1986, Geol. Surv. Can., Bull. 371, pl. 9, figs. 14, 15.
Eureka Sound Formation, Late Paleocene to ?Early Eocene, Bylot Island, "Twosnout Creek", lat. 73°9′30″N, long. 79°42′30″W, District of Franklin.

Dictyopxidia circulata Clarke and Verdier
Hypotype 38077
McIntyre, D.J., 1974, Geol. Surv. Can., Paper 74-14, pl. 6, fig. 14.
Upper Cretaceous, lat. 69°27′30″N, long. 126°58′W, Horton River, District of Mackenzie.

Dictyopyxidia imperfecta Brideaux and McIntyre
Holotype 34149; paratypes 41493, 41494
Brideaux, W.W. and McIntyre, D.J., 1975, Geol. Surv. Can., Bull. 252, p. 25, pl. 7, figs. 1-5.
Horton River Formation, Lower Cretaceous, lat. 69°19′N, long. 126°50′W, Horton River, District of Mackenzie.

?Dictyopyxidia sp.
Fig. spec. 53027
van Helden, B.G.T., 1977, Geol. Surv. Can., Paper 77-1B, p. 170, pl. 33.1, figs. 15, 16.
Wilkie Point Formation, Jurassic, east side Intrepid Inlet, Prince Patrick Island, District of Franklin.

Dimidiadinium uncinatum Brideaux
Holotype 46050; paratypes 46051, 46052
Brideaux, W.W., 1977, Geol. Surv. Can., Bull. 281, p. 38, pl. 15, figs. 2-5, 7, 8.
Cretaceous, Banff-Aqitaine-Arco-Rat Pass K-35 well, depths 1380 and 1180 feet, 6 miles southeast of Mount oodenough, lat. 67°54′42.55″N, long. 135°21′56.83″W, and IOE Stoney Core Hole F-42, depth 845-850 feet, Stoney Creek about 18 miles west-southwest of Fort McPherson, lat. 67°21′23″N, long. 135°38′43″W, District of Mackenzie.

Dingodinum cerviculum Cookson and Eisenack
Hypotype 38701
Haskell, T.R. and Wilson, G.J., 1975, Initial Repts. Deep Sea Drilling Project, vol. 29, pl. 3, fig. 9.
Lower Oligocene, DSDP Site 282, lat. 42°14.76′S, long. 143°29.18′E, west of Tasmania.

Dingodinium cerviculum Cookson and Eisenack
Hypotype 451554
Brideaux, W.W. and McIntyre, D.J., 1975, Geol. Surv. Can., Bull. 252, p. 38, pl. 14, fig. 5.
Horton River Formation, Lower Cretaceous, lat. 69°19′N, long. 126°50′W, Horton River, District of Mackenzie.

Dingodinium cerviculum Cookson and Eisenack
Hypotype 41837
Williams, G.L., 1975, Geol. Surv. Can., Paper 74-30, vol. 2, pl. 4, fig. 11.
Aptian, Lower Cretaceous, Mobil Sable Island C-67 well, lat. 43°56′4.91″N, long. 59°55′1.38″W, Scotian Shelf.

Dingodinium cerviculum Cookson and Eisenack
Hypotype 56743
Williams, G.L. and Bujak, J.P., 1980, Initial Repts. Deep Sea Drilling Project, vol. 50, pl. 5, fig. 6.
Early Cretaceous, DSDP Site 416, lat. 32°50.18′N, long. 10°48.06′W, northwestern Moroccan Continental margin.

Dingodinium cerviculum Cookson and Eisenack, 1958
Hypotype 58637
McIntyre, D.J. and Brideaux, W.W., 1980, Geol. Surv. Can., Bull. 320, p. 24, pl. 12, fig. 2.
Lower Cretaceous, Martin Creek, lat. 68°12′10″N, long. 135°34′50″W, Richardson Mountains, District of Mackenzie.

Dinoflagellate sp. S-1 of Staplin
Fig. spec. 92015
Dietrich, J.R. et al., 1989, Geol. Surv. Can., Paper 89-1G, p.141, pl. 2, fig. 7.
Early to Middle Eocene, depth 1100-1190 feet, Dome pacific et al. PEX Natsek E-56 well, lat. 69°45′21.46″N, long. 139°44′34.58″W, western Beaufort Sea.

Dinoflagellate Form 1
Fig. spec. 38078
McIntyre, D.J., 1974, Geol. Surv. Can., Paper 74-14, pl. 6, fig. 15.
Upper Cretaceous, lat. 69°28′N, long. 126°58′W, Horton River, District of Mackenzie.

Dinoflagellate type A
Fig. specs. 65423-65426
Ioannides, N.S., 1986, Geol. Surv. Can., Bull. 371, p. 41, pl. 23, fig. 17; pl. 24, figs. 1, 2, 6.
Kanguk Formation, Upper Cretaceous, south coast Bylot Island, lat. 72°53′30″N, long. 78°16′W (65423), District of Franklin, and IOE Martin House L50, lat. 66°49′42″N, long. 133°24′8″W, Northwest Territories.

Dinoflagellate type B
Fig. specs. 65409-65415
Ioannides, N.S., 1986, Geol. Surv. Can., Bull. 371, p. 41, pl. 23, figs. 2, 7-12.
Kanguk Formation, Upper Cretaceous, south coast Bylot Island, lat. 72°53′30″N, long. 78°16′W, District of Franklin.

Dinoflagellate type C
Fig. specs. 65432-65437
Ioannides, N.S., 1986, Geol. Surv. Can., Bull. 371, p. 41, pl. 24, figs. 5, 7-9, 11, 12.
Eureka Sound Formation, Late Paleocene to ?Early Eocene, Bylot Island, "Twosnout Creek", lat. 73°9′30″N, long. 79°42′30″W, District of Franklin.

Dinoflagellate? type D
Fig. specs. 65440-65444
Ioannides, N.S., 1986, Geol. Surv. Can., Bull. 371, p. 42, pl. 25, figs. 1-4, 6.
Kanguk Formation, Upper Cretaceous, south coast Bylot Island, lat. 72°53′30″N, long. 78°16′W, District of Franklin.

Dinoflagellate type E
Fig. specs. 65420-65422
Ioannides, N.S., 1986, Geol. Surv. Can., Bull. 371, p. 42, pl. 23, figs. 13-16.
Kanguk Formation, Upper Cretaceous, south coast Bylot Island, lat. 72°53′30″N, long. 78°16′W, District of Franklin.

Dinogymnium euclaense Cookson and Eisenack, 1970
Hypotypes 65097-65100
Ioannides, N.S., 1986, Geol. Surv. Can., Bull. 371, p. 11, pl. 1, figs. 1-4.
Kanguk Formation, Upper Cretaceous, south coast Bylot Island, lat. 72°53′30″N, long. 78°16′W, District of Franklin.

Dinogymnium eucalaensis Cookson and Eisenack
Hypotypes 38006, 38007
McIntyre, D.J., 1974, Geol. Surv. Can., Paper 74-14, pl. 1, figs. 6, 7.
Upper Cretaceous, lat. 69°27′30″N, long. 126°58′W, Horton River, District of Mackenzie.

Dinogymnium euclaensis Cookson and Eisenack
Hypotype 41823
Williams, G.L., 1975, Geol. Surv. Can., Paper 74-30, vol. 2, pl. 3, fig. 14.
Maastrichtian, Upper Cretaceous, core hole 23, lat. 45°38′N, long. 53°33′W, Grand Banks.

Dinogymnium sp. cf. *D. kasachstanicum* (Vozzhennikova) Lentin and Williams, 1973
Fig. spec. 65101
Ioannides, N.S., 1986, Geol. Surv. Can., Bull. 371, p. 11, pl. 1, fig. 5.
Kanguk Formation, Upper Cretaceous, south coast Bylot Island, lat. 72°53′30″N, long. 78°16′W, District of Franklin.

Dinogymnium sp. cf. *Gymnodinium kasachstanicum* Vozzhennikova
Fig. specs. 38011, 38012
McIntyre, D.J., 1974, Geol. Surv. Can., Paper 74-14, pl. 1, figs. 17, 18.
Upper Cretaceous, lat. 69°27′30″N, long. 126°58′W, Horton River, District of Mackenzie.

Dinogymnium sp. cf. *Gymnodinium sibiricum* Vozzhennikova
Hypotypes 38008-38010
McIntyre, D.J., 1974, Geol. Surv. Can., Paper 74-14, pl. 1, figs. 10-12.
Upper Cretaceous, lat. 69°27′30″N, long. 126°58′W, Horton River, District of Mackenzie.

Dinogymnium undulosum Cookson and Eisenack
Hypotype 41840
Williams, G.L., 1975, Geol. Surv. Can., Paper 74-30, vol.2, pl. 4, fig. 14.
Campanian, Upper Cretaceous, Shell Onondaga E-84 well, lat. 43°43′16.14″N, long. 60°14′17.18″W, Scotian Shelf.

Dinogymnium sp. cf. *D. westralium* (Cookson and Eisenack)
Fig. specs. 38004, 38005
McIntyre, D.J., 1974, Geol. Surv. Can., Paper 74-14, pl. 1, figs. 4, 5.
Upper Cretaceous, lat. 69°27′30″N, long. 126°58′W, Horton River, District of Mackenzie.

Dinogymnium sp. 1
Fig. specs. 38001-38003
McIntyre, D.J., 1974, Geol. Surv. Can., Paper 74-14, pl. 1, figs. 1-3.
Upper Cretaceous, lat. 69°27′30″N, long. 126°58′W, Horton River, District of Mackenzie.

Dinogymnium sp. 1 of McIntyre, 1974
Fig. spec. 65102
Ioannides, N.S., 1986, Geol. Surv. Can., Bull. 371, p. 11, pl. 1, fig. 6.
Kanguk Formation, Upper Cretaceous, south coast Bylot Island, lat. 72°53′30″N, long. 78°16′W, District of Franklin.

Dinogymnium sp. AE
Fig. spec. 45818
Brideaux, W.W., 1976, Geol. Surv. Can., Paper 76-1B, pl. 44.2, fig. 9.
Boundary Creek Formation, Upper Cretaceous, 6380-6400 feet, Gulf Mobil Parsons N-10 well, ca. 1.5 miles east-northeast of Parsons Lake, lat. 69°59′48.52″N, long. 133°31′50.33″W, District of Mackenzie.

Dinopterygium sp.
Fig.specs. 65208, 65209
Ioannides, N.S., 1986, Geol. Surv. Can., Bull. 371, p. 21, pl. 9, figs. 9, 10.
Kanguk Formation, Upper Cretaceous, south coast Bylot Island, lat. 72°53′30″N, long. 78°16′W, District of Franklin.

Dioxya(?) pignerata Norris
Holotype 68183; hypotypes 68179-68182, 68184, 68185
Norris, G., 1986, Geol. Surv. Can., Bull. 340, p. 46, pl. 14, figs. 8-14.
Richards Formation, Late Eocene, depths 10, 300-15, 400 and 10,600-10,700 (68181) feet, Imperial Nuktak C-22 well, lat. 69°41′07″N, long. 134°51′30″W, District of Mackenzie.

Dioxya(?) pignerata Norris
Hypotype 91999
Dietrich, J.R. et al, 1989, Geol. Surv. Can., Paper 89-1G, pl. 1, fig. 3.
Early to Middle Eocene, depth 1200-1290 feet, Dome Pacific et al. PEX Natsek E-56 well, lat. 69°45′21.46″N, long. 139°44′34.58″W, western Beaufort Sea.

Diphyes colligerum (Deflandre and Cookson)
Hypotype 41839
Williams, G.L., 1975, Geol. Surv. Can., Paper 74-30, vol. 2, pl. 4, fig. 13.
Late Eocene, core hole 2, lat. 45°08′N, long. 52°38′W, Grand Banks.

Diphyes colligerum (Deflandre and Cookson) Cookson
Hypotype 45801
Brideaux, W.W., 1976, Geol. Surv. Can., Paper 76-1B, pl. 44.1, fig. 5.
Reindeer Formation, Paleocene, 4260-4290 feet, Gulf Mobil Parsons N-10 well, ca. 1.5 miles east-northeast of Parsons Lake, lat. 69°59′48.52″N, long. 133°31′50.33″W, District of Mackenzie.

Diphyes sp.
Fig. spec. 91468
Sweet, A.R. and McIntyre, D.J., 1988, Can. Soc. Petrol. Geol., Mem. 15, fig. 9.13.
Kaskapau Formation, Upper Cretaceous, west bank of Little Berland River, l.s.d. 1, sec. 6, tp. 54, rge. 2, W.5th mer., Alberta.

Distatodinium sp. A
Fig. spec. 68146
Norris, G., 1986, Geol. Surv. Can., Bull. 340, p. 43, pl. 11, fig. 20.
Richards Formation, Middle Eocene, depth 11,700-11,800 feet, Imperial Nuktak C-22 well, lat. 69°41′07″N, long. 134°51′30″W, District of Mackenzie.

Doidyx anaphrisa Sarjeant, 1966
Hypotypes 71424, 71429
Wilson, M.A., 1978, Palaeontographica, Abt.B, vol. 166, p. 161, pl. 12, figs. 7, 12.
Tent Island Formation, Cretaceous, Little Fish (Cache) Creek, Big Fish aera, District of Mackenzie.

Dorocysta litotes Davey, 1970
Hypotypes 65210, 65211
Ioannides, N.S., 1986, Geol. Surv. Can., Bull. 371, p. 21, pl. 9, figs. 11, 12.
Kanguk Formation, Upper Cretaceous, south coast Bylot Island, lat. 72°53′30″N, long. 78°16′W, District of Franklin.

Dorocysta sp. A
Fig. specs. 53680, 53702
Bujak, J.P. and Williams, G.L., 1978, Geol. Surv. Can., Bull. 297, pl. 1, fig. 13; pl. 2, fig. 20.
Late Cretaceous, 5300-5330 feet, Amoco-Imp Gannet O-54 well, lat. 45°3′54.55″N, long. 52°38′9.72″W, Grand Banks.

Druggidium? sp. AE
Fig. spec. 58585
McIntyre, D.J. and Brideaux, W.W., 1980, Geol. Surv. Can., Bull. 320, p. 12, pl. 3, fig. 1.
ower Cretaceous, Martin Creek, lat. 68°12′10″N, long. 135°34′50″W, Richardson Mountains, District of Mackenzie.

Duosphaeridium nudum (Cookson)
Hypotype 41826
Williams, G.L., 1975, Geol. Surv. Can., Paper 74-30, vol. 2, pl. 3, fig. 17.
Early Eocene, Mobil Sable Island C-67 well, lat. 43°56′4.91″N, long. 59°55′1.38″W, Scotian Shelf.

Eisenackia circumtabulata Drugg
Hypotype 41835
Williams, G.L., 1975, Geol. Surv. Can., Paper 74-30, vol. 2, pl. 4, fig. 9.
Danian, Eastcan et al. Leif M-48 well, lat. 54°17′45.92″N, long. 55°7′20.17″W, Labrador Shelf.

Eisenackia circumtabulata Drugg
Hypotypes 49369, 49371
Williams, G.L. and Bujak, J.P., 1977, Am. Assoc. Strat. Palynol, Contr. Strat. Palynol, vol. 1, Contr. Ser. 5A, pl. 5, figs. 5, 7.
Middle Paleocene, Eastcan et al Leif M-48 well, lat. 54°17′45.92″N, long. 55°7′20.17″W; Early Paleocene, depth 6460-6490 feet, Eastcan et al., Bjarni H-81 well, lat. 55°30′29.35″N, long. 57°42′5.52″W, Labrador Shelf.

Eisenackia crassitabulata Deflandre and Cookson
Hypotype 38697
Haskell, T.R. and Wilson, G.J., 1975, Initial Repts. Deep Sea Drilling Project, vol. 29, pl. 3, fig. 5.
Paleocene, DSDP Site 283, lat. 43°54.60′S, long. 154°16.96′E, east of Tasmania.

Eisenackia sp.
Fig. spec. 47914
Williams, G.L., 1978, Initial Repts. Deep Sea Drilling Project, vol. 41, pl. 7, fig. 12.
Late Cretaceous, DSDP Site 370, lat. 32°50.2′N, long. 10°46.6′W, northwestern Moroccan Continental Margin.

Ellipsoidictyon sp. A
Fig. spec. 75504
Davies, E.H., 1985, Palynology, vol. 9, p. 129, pl. 4, fig. 16.
Early Bajocian, Liassic, Brehna Road Section, Lusitanian Basin, west-central Portugal.

Elystrocysta druggi Stover and Evitt, 1978
Hypotypes 65255-65259
Ioannides, N.S., 1986, Geol. Surv. Can., Bull. 371, p. 22, pl. 12, figs. 1-4, 8.
Eureka Sound Formation, Late Paleocene to ?Early Eocene, Bylot Island, lat. 72°53′30″N, long. 78°25′W, and "Twosnout Creek", lat. 73°14′30″N, long. 79°49′W (65259), District of Franklin.

Endoceratium sp.
Fig. spec. 65254
Ioannides, N.S., 1986, Geol. Surv. Can., Bull. 371, p. 22, pl. 11, fig. 12.
Kanguk Formation, Upper Cretaceous, south coast Bylot Island, lat. 72°53′30″N, long. 78°16′W, District of Franklin.

Endoscrinium campanula (Gocht) Vozzhennikova
Hypotype 45922
Brideaux, W.W., 1976, Geol. Surv. Can., Paper 76-1C, p. 118, pl. 23.1, fig. 2.
Lower Cretaceous, south bank Martin Creek, District of Mackenzie.

Endoscrinium campanula (Gocht) Vozzennikova, 1967
Hypotype 91459
Sweet, A.R. and McIntyre, D.J., 1988, Can. Soc. Petrol. Geol., Mem. 15, fig. 9.4.
Kaskapau Formation, Upper Cretaceous, west bank of Little Berland River, l.s.d. 1, sec. 6, tp. 54, rge. 2, W.5th mer., Alberta.

Endoscrinium campanulum (Gocht) Vozzhennikova
Hypotype 47919
Williams, G.L., 1978, Initial Repts. Deep Sea Drilling Project, vol. 41, pl. 8, fig. 5.
Early Cretaceous, DSDP Site 367, lat. 12°29.2′N, long. 20°2.8′W, Cape Verde Basin west of Senegal.

Endoscrinium luridum (Deflandre)
Hypotype 41819
Williams, G.L., 1975, Geol. Surv. Can., Paper 74-30, vol. 2, pl. 3, fig. 10.
Kimmeridgian, Upper Jurassic, Shell Mohican I-100 well, lat. 42°59′39.04″N, long. 62°28′51.32″W, Scotian Shelf.

Endoscrinium luridum (Deflandre) Gocht
Hypotype 45816
Brideaux, W.W., 1976, Geol. Surv. Can., Paper 76-1B, pl. 44.2, fig. 7.
Husky Formation, Upper Jurassic, Gulf Mobil Parsons N-10 well, ca. 1.5 miles east-northeast of Parsons Lake, lat. 69°54′48.52″N, long. 133°31′50.33″W, District of Mackenzie.

Endoscrinium luridum (Deflandre) Gocht
Hypotype 46044
Brideaux, W.W., 1977, Geol. Surv. Can., Bull 281, p. 34, pl. 14, fig. 6.
Husky Formation, Jurassic, IOE Stoney Core Hole F-42, depth 1000-1010 feet, Stoney Creek about 18 miles west-southwest of Fort McPherson, lat. 67°21′23″N, long. 135°38′43″W, District of Mackenzie.

Eopseudoceratium sp. cf. *E. gochtii* (Neale and Sarjeant) Lenti and Williams
Fig. spec. 45994
Brideaux, W.W., 1977, Geol. Surv. Can., Bull 281, p. 14, pl. 5, fig. 4.
Cretaceous, IOE Stoney Core Hole F-42, Stoney Creek about 18 miles west-southwest of Fort McPherson, lat. 67°21′23″N, long. 135°38′43″W, District of Mackenzie.

Epelidosphaeridia spinosa (Cookson and Hughes)
Hypotype 41838
Williams, G.L., 1975, Geol. Surv. Can., Paper 74-30, vol. 2, pl. 4, fig. 12.
Cenomanian, Upper Cretaceous, Shell Onondaga E-84 well, lat. 43°43′16.14″N, long. 60°14′17.18″W, Scotian Shelf.

Escharisphaerida rudis Davies
Hypotype 76281
Davies, E.H. and Poulton, T.P., 1986, Geol. Surv. Can., Paper 86-1B, pl. 55.2, fig. 5.
Fernie Formation, Upper Jurassic, Christina Falls, lat. 56°40′N, long. 123°4′W, northeastern British Columbia.

Escharisphaeridia sp.
Fig. specs. 65122, 65123, 65125
Ioannides, N.S., 1986, Geol. Surv. Can., Bull. 371, p. 22, pl. 2, figs. 9, 11, 14.
Kanguk Formation, Upper Cretaceous, south coast Bylot Island, lat. 72°53′30″N, long. 78°16′W, District of Franklin.

Escharisphaeridia? sp.
Fig. specs. 65124, 65126-65128
Ioannides, N.S., 1986, Geol. Surv. Can., Bull. 371, p. 22, pl. 2, figs. 8, 12, 13, 16.
Kanguk Formation, Upper Cretaceous, south coast Bylot Island, lat. 72°53′30″N, long. 78°16′W, District of Franklin.

Evittosphaerula paratabulata Munum
Hypotype 65089
Bujak, J.P. and Davies, E.H., 1982, Bull. Can. Petrol. Geol., vol. 29, no. 3, 1981, p. 422, pl.1, fig. 5.
Dixon, J. et al., 1984, Geol. Surv. Can., Paper 82-13, pl. 4, fig. 4.
Late Middle to Late Miocene, 2643 m interval, Dome Gulf et al. Hunt Kopanoar M-13 well, lat. 70°22′55.1″N, long. 135°05′33.9″W, Beaufort Sea.

Exochosphaeridium bifidum (Clarke and Verdier) Clarke et al., 1968
Hypotype 65267
Ioannides, N.S., 1986, Geol. Surv. Can., Bull. 371, p. 22, pl. 12, fig. 1.
Eureka Sound Formation, Late Paleocene to ?Early Eocene, south coast Bylot Island, lat. 72°52′N, long. 78°35′W, District of Franklin.

Exochosphaeridium bifidum (Clarke and Verdier) Clarke et al., 1968
Hypotype 91460
Sweet, A.R. and McIntyre, D.J., 1988, Can. Soc. Petrol. Geol., Mem. 15, fig. 9.5.
Sturrock Member, Cardium Formation, Upper Cretaceous, south side of McLeod River, l.s.d. 16, sec. 16, tp. 47, rge. 23, W.5th mer., Alberta.

Exochosphaeridium striolatum (Deflandre) Davey, 1969
Hypotypes 65264-65266
Ioannides, N.S., 1986, Geol. Surv. Can., Bull. 371, p. 22, pl. 12, figs. 6, 9, 10.
Kanguk Formation, Upper Cretaceous, south coast Bylot Island, lat. 79°53′30″N, long. 78°16′W, District of Franklin.

Exochosphaeridium sp.
Fig. spec. 91461
Sweet, A.R. and McIntyre, D.J., 1988, Can. Soc. Petrol. Geol., Mem. 15, fig. 9.6.
Leyland Member, Cardium Formation, Upper Cretaceous, south side of McLeod River, l.s.d. 16, sec. 16, tp. 47, rge. 23, W.5th mer., Alberta.

Eyachia prisca Gocht
Hypotype 75498
Davies, E.H., 1985, Palynology, vol. 9, pl. 4, figs. 9, 10.
Late Toarican, Liassic, west-central Portugal.

Fibrocysta? cf. *F. vectensis* (Eaton) Stover and Evitt, 1978
Fig. specs. 65260-65626
Ioannides, N.S., 1986, Geol. Surv. Can., Bull. 371, p. 23, pl. 12, figs. 5, 17, 12, 13.
Eureka Sound Formation, Late Paleocene to ?Early Eocene, Bylot Island, south coast, lat. 72°52′N, long. 78°35′W (65260), and "Twosnout Creek", lat. 73°13′N, long. 79°53′W, District of Franklin.

Filisphaera filifera Bujak, 1984
Hypotype 55334
Mudie, P.J., 1986, Initial Repts. Deep Sea Drilling Project, vol. XCIV, p. 803, pl. 4, fig. 6.
Upper Pliocene, hole 607, depth 105.5 m, lat. 41°0.07′N, long. 32°57.44′W, at base of upper western flank of the mid-Atlantic Ridge, North Atlantic Ocean.

Florentinia cooksoniae (Singh) Duxbury, 1980
Hypotype 91474
Sweet, A.R. and McIntyre, D.J., 1988, Can. Soc. Petrol. Geol., Mem. 15, fig. 9.19.
Leyland Member, Cardium Formation, Upper Cretaceous, south side of McLeod River, l.s.d. 16, sec. 16, tp. 47, rge. 23, W.5th mer., Alberta.

Florentinia? mantellii (Davey and Williams, *in* Davey et al.) Davey and Verdier, 1973
Hypotypes 65268-65270
Ioannides, N.S., 1986, Geol. Surv. Can., Bull. 371, p. 23, pl. 12, figs. 14-16.
Kanguk Formation, Upper Cretaceous, south coast Bylot Island, lat. 72°53′30″N, long. 78°16′W, District of Franklin; Upper Cretaceous, Dome Pacific Natsek E-56 well, lat. 69°45′21″N, long. 139°44′34.5″W, District of Mackenzie (65269).

Florentinia mantellii (Davey and Williams) Davey and Verdier, 1973
Hypotype 91475
Sweet, A.R. and McIntyre, D.J., 1988, Can. Soc. Petrol. Geol., Mem. 15, fig. 9.20.
Blackstone Formation, Upper Cretaceous, north side of Highway 47 north of Cadomin, l.s.d. 14, sec. 16, tp. 47, rge. 23, W.5th mer., Alberta.

Florentinia radiculata (Davey and Williams, 1966) Davey and Verdier, 1973
Hypotype 84744
Leckie, D.A. et al., 1990, J. Sed. Petrology, vol. 60, no. 1, fig. 6 G.
Shaftesbury Formation, Lower Cretaceous, west bank Peace River near town of Peace River, Alberta.

Florentinia? sp.
Fig. specs. 65271, 65277
Ioannides, N.S., 1986, Geol. Surv. Can., Bull. 371, pl. 12, fig. 17; pl. 13, fig. 6.

Kanguk Formation, Upper Cretaceous, "Twosnout Creek", Bylot Island, lat. 73°14′30″N, long. 79°49′W, District of Franklin.

Fromea amphora Cookson and Eisenack
Hypotype 41495
Brideaux, W.W. and McIntyre, D.J., 1975, Geol. Surv. Can., Bull. 252, p. 26, pl. 7, fig. 6.
Horton River Formation, Lower Cretaceous, lat. 69°19′N, long. 126°50′W, Horton River, District of Mackenzie.

Fromea amphora Cookson and Eisenack, 1958
Hypotype 58600
McIntyre, D.J. and Brideaux, W.W., 1980, Geol. Surv. Can., Bull. 320, p. 17, pl. 5, fig. 5.
Lower Cretaceous, Martin Creek, lat. 68°12′10″N, long. 135°34′50″W, Richardson Mountains, District of Mackenzie.

Fromea amphora Cookson and Eisenack, 1958
Hypotype 91448
Sweet, A.R. and McIntyre, D.J., 1988, Can. Soc. Petrol. Geol., Mem. 15, fig. 8.20.
Kaskapau Formation, Upper Cretaceous, west bank of Little Berland River, l.s.d. 1, sec. 6, tp. 54, rge. 2, W.5th mer., Alberta.

Fromea chytra (Drugg) Stover and Evitt, 1958
Hypotypes 65272-65275
Ioannides, N.S., 1986, Geol. Surv. Can., Bull. 371, p. 23, pl. 13, figs. 1, 2, 4, 5.
Kanguk Formation, Upper Cretaceous, Bylot Island, "Twosnout Creek", lat. 73°14′30″N, long. 79°49′W; Eureka Sound Formation, Late Paleocene to ?Early Eocene, south coast, lat. 72°53′30″N, long. 78°25′W (65275), District of Franklin.

Fromea complicata Brideaux
Holotype 4600; paratypes 46001, 46002
Brideaux, W.W., 1977, Geol. Surv. Can., Bull 281, p. 18, pl. 7, figs. 1-6.
Cretaceous, Horton River Formation, Horton River, lat. 69°27′30″N, long. 126°54′W, District of Mackenzie; Joli Fou Formation, depth 5447-5452 feet, Imperial Battle Lake No. 1 well, Alberta (missing); and Banff-Aquitaine-Arco Rat Pass K-35 well, depth 190-210 feet, 6 miles southeast of Mount Goodenough, lat. 67°54′42.55″N, long. 132°21′56.83″W, District of Mackenzie.

Fromea fragilis (Cookson and Eisenack) Stover and Evitt, 1978
Hypotypes 65279-65282
Ioannides, N.S., 1986, Geol. Surv. Can., Bull. 371, p. 24, pl. 13, figs. 8-10, 12.
Eureka Sound Formation, Late Paleocene to ?Early Eocene, "Twosnout Creek", Bylot Island, lat. 73°13′N, long. 79°53′W (65279); Kanguk Formation, Upper Cretaceous, south coast Bylot Island, lat. 72°53′30″N, long. 78°16′W; Upper Cretaceous, souhwestern Devon Island, lat. 75°26′N, long. 91°35′W (65282), District of Franklin.
=*Fromea nicosia*, Jansonius, J., 1989, Review Palaeobotany and Palynology, vol. 61, no. 1/2, p. 67, pl. 1, figs. 2(65279), 3(65280), 4(holotype 65281).

Fromea fragilis (Cookson and Eisenack) Stover and Evitt, 1978
Hypotype 91447
Sweet, A.R. and McIntyre, D.J., 1988, Can. Soc. Petrol. Geol., Mem. 15, fig. 8.19.
Wapiabi Formation, Upper Cretaceous, south side of McLeod River, l.s.d. 16, sec. 16, tp. 47, rge. 23, W.5th mer., Alberta.

Fromea sp. cf. *F. fragilis* (Cookson and Eisenack) Stover and Evitt, 1978
Fig. spec. 58603
McIntyre, D.J. and Brideaux, W.W., 1980, Geol. Surv. Can., Bull. 320, p. 17, pl. 5, fig. 9.
Lower Cretaceous, Martin Creek, lat. 68°12′10″N, long. 135°34′50″W, Richardson Mountains, District of Mackenzie.
=*Fromea grandis*, Jansonius, J., 1989, Review Palaeobotany and Palynology, vol. 61, no. 1/2, p. 65, pl. 1, fig. 1 (holotype).

Galeacornea sp.
Fig. spec. 47923
Williams, G.L., 1978, Initial Repts. Deep Sea Drilling Project, vol. 41, pl. 8, fig. 9.
Late Cretaceous, DSDP Site 370, lat. 32°50.2′N, long. 10°46.6′W, northwestern Moroccan continental margin.

Gardodinium deflandrei Clarke and Verdier
Hypotype 41818
Williams, G.L., 1975, Geol. Surv. Can., Paper 74-30, vol. 2, pl. 3, fig. 9.
Campanian, Upper Cretaceous, Shell Mohawk B-93 well, lat. 42°42′10.72″N, long. 64°43′53.04″W, Scotian Shelf.

Gardodinium eisenackii Alberti
Hypotypes 41529-41532
Brideaux, W.W. and McIntyre, D.J., 1975, Geol. Surv. Can., Bull. 252, p. 33, pl. 10, figs. 1-4.
Horton River Formation, Lower Cretaceous, lat. 69°19′N, long. 126°50′W, Horton River, District of Mackenzie.

Gardodinium sp. cf. *G. eisenackii* Alberti
Fig. spec. 45924
Brideaux, W.W., 1976, Geol. Surv. Can., Paper 76-1C, p. 118, pl. 23.1, fig. 4.
Lower Cretaceous, south bank Martin Creek, District of Mackenzie.

Gardodinium sp.
Fig.spec. 53674
Bujak, J.P and Williams, G.L., 1978, Geol. Surv. Can., Bull. 297, pl. 1, fig. 6
Late Cretaceous, swc 7200 feet, Amoco-IOE Heron H-37 well, lat. 44°2′26.65″N, long. 52°25′40.58″W, Grand Banks.

Gardoninium? sp.
Fig. spec. 45937
Brideaux, W.W., 1976, Geol. Surv. Can., Paper 76-1C, pl. 23.2, fig. 9.
Husky Formation, Lower Cretaceous, south bank Martin Creek, District of Mackenzie.

Gillinia hymenophora Cookson and Eisenack
Hypotype 53681
Bujak, J.P. and Williams, G.L., 1978, Geol. Surv. Can., Bull. 297, pl. 1, fig. 14.
Late Cretaceous, swc 4898 feet, Shell Primrose A-41 well, lat. 44°5.68"'N, long. 59°6′18.26″W, Scotian Shelf.

Gillinia hymenophora Cookson and Eisenack, 1960
Hypotype 65276
Ioannides, N.S., 1986, Geol. Surv. Can., Bull. 371, p. 24, pl. 13, fig. 3.
Kanguk Formation, Upper Cretaceous, "Twosnout Creek", Bylot Island, lat. 73°14′30″N, long. 79°49′W, District of Franklin.

Glaphyrocysta divaricata (Williams and Downie) Stover and Evitt
Hypotype 92005
Dietrich, J.R. et al., 1989, Geol. Surv. Can., Paper 89-1G, p. 141, pl. 1, fig. 9.
Early to Middle Eocene, depth 3900-3990 feet, Dome Pacific et al. PEX Natsek E-56 well, lat. 69°45′21.46″N, long. 139°44′34.58″W, western Beaufort Sea.

Glaphyrocysta ordinata Williams and Downie, *in* Davey et al.) Stover and Evitt, 1978
Hypotypes 65109-65113
Ioannides, N.S., 1986, Geol. Surv. Can., Bull. 371, p. 24, pl. 1, figs. 13-16; pl. 2, fig. 1.
Eureka Sound Formation, Late Paleocene to ?Early Eocene, Bylot Island, "Twosnout Creek", lat. 73°13′N, long. 79°53′W, District of Franklin.

Glaphyrocysta ordinata (Williams and Downie *in* Davey et al.) Stover and Evitt
Hypotypes 68148, 68151, 68152
Norris, G., 1986, Geol. Surv. Can., Bull 340, p. 43, pl. 11, fig. 22; pl. 12, figs. 1, 6.
Richards Formation, Middle Eocene, depths 11,800-11,900 and 11,700-11,800 (68151) feet, Imperial Nuktak C-22 well, lat. 69°41′07″N, long. 134°51′30″W, District of Mackenzie.

Glaphyrocysta ordinata (Williams and Downie) Stover and Evitt
Hypotype 92008
Dietrich, J.R. et al., 1989, Geol. Surv. Can., Paper 89-1G, p. 141, pl. 1, fig. 12.
Early to Middle Eocene, depth 2000-2090 feet, Dome Pacific et al. PEX Natsek E-56 well, lat. 69°45′21.46″N, long. 139°44′34.58″W, western Beaufort Sea.

Glenodinium sp. A
Fig. spec. 55354
Mudie, P.J., 1986, Initial Repts. Deep Sea Drilling Project, vol. XCIV, p. 803, pl. 5, figs. 10a, b.
Pleistocene, hole 607, depth 42.8 m, lat. 41°0.07′N, long. 32°57.44′W, at base of upper western flank of the Mid-Atlantic Ridge, North Atlantic Ocean.

Glomodinium evittii (Pocock) Davies
Hypotype 75473
Davies, E.H., 1985, Palynology, vol. 9, pl. 3, fig. 3.
Late Pliensbachian, Liassic, Zumbujal Section, south of Condeixa, Lusitanian Basin, west-central Portugal.

Gochteodinia judilentinae McIntyre and Brideaux
Holotype 58632; paratypes 58633, 58634
McIntyre, D.J. and Brideaux, W.W., 1980, Geol. Surv. Can., Bull. 320, p. 23, pl. 11, figs. 4-9.
Lower Cretaceous, Martin Creek, lat. 68°12′10″N, long. 135°34′50″W, Richardson Mountains, District of Mackenzie.

Gochteodinia villosa (Vozzhennikova) Norris, 1978
Hypotypes 58635, 58636
McIntyre, D.J. and Brideaux, W.W., 1980, Geol. Surv. Can., Bull. 320, p. 23, pl. 11, fig. 10; pl. 12, fig. 1.
Lower Cretaceous, Martin Creek, lat. 68°12′10″N, long. 135°34′50″W, and "Grizzly Gorge", lat. 68°15′49″N, long. 135°40′59″W, Richardson Mountains, District of Mackenzie.

Gonyaulacysta cladophora (Deflandre)
Hypotype 41842
Williams, G.L., 1975, Geol. Surv. Can., Paper 74-30, vol. 2, pl. 4, fig. 16.
Kimmeridgian, Upper Jurassic, Shell Mohican I-100 well, lat. 42°59′39.04″N, long. 62°28′51.32″W, Scotian Shelf.

Gonyaulacysta cladophora (Deflandre) Dodekova, 1976
Hypotype 58344
Tan, J.T. and Hills, L.V., 1978, Geol. Surv. Can., Paper 78-1C, p. 66, pl. 12.1, fig. 4.
Savik Formation, Upper Jurassic, lat. 77°48′N, long. 112°20′W, Mackenzie King Island, District of Franklin.

Gonyaulacysta sp. cf. *G. cladophora* (Deflandre) Dodekova
Fig. specs. 42356-42358
Brideaux, W.W. and Fisher, M.J., 1976, Geol. Surv. Can., Bull. 259, p. 18, pl. 1, figs. 1-3, 7.
Upper Jurassic, 1920 feet, Banff-Aquitaine-Arco Rat Pass K-35 well, 6 miles southeast of summit of Mount Goodenough, just south of Goodenough Creek, District of Mackenzie.

Gonyaulacysta cretacea (Neale and Sarjeant) Sarjeant
Hypotype 45960
Brideaux, W.W., 1977, Geol. Surv. Can., Bull 281, p. 6, pl. 1, fig. 3.
Cretaceous, IOE Stoney Core Hole F-42, depth 795-800 feet, Stoney Creek about 18 miles west-southwest of Fort McPherson, lat. 67°21′23″N, long. 135°38′43″W, District of Mackenzie.

Gonyaulacysta dualis (Brideaux and Fisher) Stover and Evitt
Hypotype 76304
Davies, E.H. and Poulton, T.P., 1986, Geol. Surv. Can., Paper 86-1B, pl. 55.3, fig. 1.
Fernie Formation, Upper Jurassic, Christina Falls, lat. 56°40′N, long. 123°4′W, northeastern British Columbia.

Gonyaulacysta episoma Sarjeant
Hypotype 45959
Brideaux, W.W., 1977, Geol. Surv. Can., Bull 281, p. 6, pl. 1, fig. 2.
Cretaceous, IOE Stoney Core Hole F-42, depth 1000-1010 feet, Stoney Creek about 18 miles west-southwest of Fort McPherson, lat. 67°21′23″N, long. 135°38′43″W, District of Mackenzie.

Gonyaulacysta fastigiata Duxbury
Hypotypes 56765, 56766
Williams, G.L. and Bujak, J.P., 1980, Initial Repts. Deep Sea Drilling Project, vol. 50, pl. 9, figs. 1, 2.
Early Cretaceous, DSDP Site 416, lat. 32°50.18‴N, long. 10°48.06′W, northwestern Moroccan continental margin.

Gonyaulacysta helicoidea (Eisenack and Cookson) Sarjeant
Hypotype 41476
Brideaux, W.W. and McIntyre, D.J., 1975, Geol. Surv. Can., Bull. 252, p. 20, pl. 5, fig. 3.
Horton River Formation, Lower Cretaceous, lat. 69°19′N, long. 126°50′W, Horton River, District of Mackenzie.

Gonyaulacysta helicoidea (Eisenack and Cookson) Sarjeant
Hypotype 45963
Brideaux, W.W., 1977, Geol. Surv. Can., Bull 281, p. 6, pl. 1, fig. 6.
Cretaceous, Banff-Aquitaine-Arco Rat Pass K-35 well, depth 1025 feet, 6 mile southeast of Mount Goodenough, lat. 67°54′42.55″N, long. 135°21′56.83″W, District of Mackenzie.

Gonyaulacysta hyalodermopsis (Cookson and Eisenack) Sarjeant
Hypotypes 41477, 41478
Brideaux, W.W. and McIntyre, D.J., 1975, Geol. Surv. Can., Bull. 252, p. 20, pl. 5, figs. 4, 5.
Horton River Formation, Lower Cretaceous, lat. 69°19′N, long. 126°50′W, Horton River, District of Mackenzie.

Gonyaulacysta hyalodermopsis (Cookson and Eisenack) Sarjeant
Hypotype 45961
Brideaux, W.W., 1977, Geol. Surv. Can., Bull 281, p. 6, pl. 1, fig. 4.
Cretaceous, Banff-Aquitaine-Arco Rat Pass K-35 well, depth 1025 feet, 6 miles southeast of Mount Goodernough, lat. 67°54′42.55″N, long. 135°21′56.83″W, District of Mackenzie.

Gonyaulacysta jurassica (Deflandre)
Hypotype 41844
Williams, G.L., 1975, Geol. Surv. Can., Paper 74-30, vol. 2, pl. 5, fig. 1.
Oxfordian, Upper Jurassic, Shell Naskapi N-30 well, lat. 43°29′46.29″N, long. 62°34′0.63″W, Scotian Shelf.

Gonyaulacysta jurassic (Deflandre, 1938) Norris and Sarjeant, 1967
Hypotype 58345
Tan, J.T. and Hills, L.V., 1978, Geol. Surv. Can., Paper 78-1C, p. 66, pl. 12.1, fig. 5.
Savik Formation, Upper Jurassic, lat. 77°48′N, long. 112°20′W, Mackenzie King Island, District of Franklin.

Gonyaulacysta jurassica subsp. longicornis (Deflandre, 1938) Lentin and Williams, 1973
Hypotype 58346
Tan, J.T. and Hills, L.V., 1978, Geol. Surv. Can., Paper 78-1C, p. 66, pl. 12.1, fig. 6.
Savik Formation, Upper Jurassic, lat. 77°48′N, long. 112°20′W, Mackenzie King Island, District of Franklin.

Gonyaulacysta sp. cf. *G. kostromiensis* (Vozzhennikova) Sarjeant, 1969
Fig. spec. 58593
McIntyre, D.J. and Brideaux, W.W., 1980, Geol. Surv. Can., Bull. 320, p. 13, pl. 4, figs. 2-5.
Lower Cretaceous, Martin Creek, lat. 68°12′10″N, long. 135°34′50″W, Richardson Mountains, District of Mackenzie.

Gonyaulacysta longicornis (Downie)
Hypotype 41845
Williams, G.L., 1975, Geol. Surv. Can., Paper 74-30, vol. 2, pl. 5, fig. 2.
Berriasian-Valanginian, Lower Cretaceous, Amoco IOE Puffin B-90 well, lat. 44°39′12.73″N, long. 53°42′28.35″W, Grand Banks.

Gonyaulacysta longicornis (Downie, 1957) Sarjeant, 1969
Hypotype 58347
Tan, J.T. and Hills, L.V., 1978, Geol. Surv. Can., Paper 78-1C, p. 66, pl. 12.1, fig. 7.
Deer Bay Formation, Upper Jurassic, lat. 77°48′N, long. 112°20′W, Mackenzie King Island, District of Franklin.

Gonyaulacysta spp. (*cretacea/helicoïdea* group)
Fig. spec. 58590
McIntyre, D.J. and Brideaux, W.W., 1980, Geol. Surv. Can., Bull. 320, p. 13, pl. 3, figs. 5, 6.
Lower Cretaceous, Martin Creek, lat. 68°12′10″N, long. 135°34′50″W, Richardson Mountains, District of Mackenzie.

Gonyaulacysta sp. A
Fig. spec. 45964
Brideaux, W.W., 1977, Geol. Surv. Can., Bull 281, p. 7, pl. 1, fig. 7.
Cretaceous, Banff-Aquitaine-Arco Rat Pass K-35 well, depth 1180 feet, 6 miles southeast of Mount Goodnough, lat. 67°54′42.55″N, long. 135°21′56.83″W, District of Mackenzie.

Hemicystodinium zoharyi (Rossignol)
Hypotype 41848
Williams, G.L., 1975, Geol. Surv. Can., Paper 74-30, vol. 2, pl. 5, fig. 5.
Late Eocene, core hole 4, lat. 45°31′N, long. 52°22′W, Grand Banks.

Hemicystodinium sp.
Fig. specs. 41822, 41847
Williams, G.L., 1975, Geol. Surv. Can., Paper 74-30, vol. 2, pl. 3, fig. 13; pl. 5, fig. 4.
Late and Middle Miocene, Amoco IOE Puffin B-90 well, lat. 44°39′12.73″N, long. 53°42′28.35″W, Grand Banks.

"*Hemicystodinium*" sp. CA
Fig spec. 45819
Brideaux, W.W., 1976, Geol. Surv. Can., Paper 76-1B, p. 251, pl. 44.2, fig. 10.
Lower Cretaceous, 6850-6880 feet, Gulf Mobil Parsons N-10 well, ca. 1.5 miles east-northeast of Parsons Lake, lat. 69°59′48.52″N, long. 133°31′50.33″W, District of Mackenzie.

Heslertonia heslertonensis (Neale and Sarjeant) Sarjeant
Hypotypes 46022, 46023
Brideaux, W.W., 1977, Geol. Surv. Can., Bull 281, p. 25, pl. 10, figs. 1-4.
Cretaceous, IOE Stoney Core Hole F-42, depths 780-785 and 805-810 feet, Stoney Creek about 18 miles west-southwest of Fort McPherson, lat. 67°21′23″N, long. 135°38′43″W, District of Mackenzie.

Heslertonia sp. BE
Fig. spec. 45940
Brideaux, W.W., 1976, Geol. Surv. Can., Paper 76-1C, p. 118, pl. 23.1, fig. 12.
Husky Formation, Lower Cretaceous, south bank Martin Creek, District of Mackenzie.

Heteraulacacysta leptalea Eaton
Hypotype 92003
Dietrich, J.R. et al., 1989, Geol. Surv. Can., Paper 89-1G, pl. 1, fig. 7.
Early to Middle Eocene, depth 5000-5090 feet, Dome Pacific et al. PEX Natsek E-56 well, lat. 69°45′21.46″N, long. 139°44′34.58″W, western Beaufort Sea.

Heterosphaeridium difficile (Manum and Cookson)
Hypotypes 65283-65288
Ioannides, N.S., 1986, Geol. Surv. Can., Bull. 371, p. 24, pl. 13, figs. 11, 13,-16; pl. 14, fig. 1.
Kanguk Formation, Upper Cretaceous, south coast Bylot Island, lat. 72°53′30″N, long. 78°16′W, District of Franklin.

Heterosphaeridium difficile (Manum and Cookson) Ioannides, 1980
Hypotypes 91466, 91471
Sweet, A.R. and McIntyre, D.J., 1988, Can. Soc. Petrol. Geol., Mem. 15, figs. 9.11, 9.16.
Upper Cretaceous, Wapiabi Formation, south side of McLeod River, l.s.d.16, sec. 16, tp. 47, rge. 23, W.5th mer., Moosehound Member, Cardium Formation, west bank of Little Berland River, l.s.d. 1, sec. 6, tp. 54, rge. 2, W.5th mer., Alberta.

Homotryblium aculeatum Williams
Holotype 47873; paratypes 47874-47876
Williams, G.L., 1978, Initial Repts. Deep Sea Drilling Project, vol. 41, p. 797, pl. 4, figs. 5, 6, 8, 9.
Middle and Late Eocene, DSDP Site 370, lat. 32°50.2′N, long. 10°46.6′W, northwestern Moroccan continental margin.

Homotryblium plectilum Drugg and Loeblich
Hypotype 41846
Williams, G.L., 1975, Geol. Surv. Can., Paper 74-30, vol. 2, pl. 5, fig. 3.
Oligocene, Mobil Sable Island C-67 well, lat. 43°56′4.91″N, long. 59°55′1.38″W, Scotian Shelf.

Homotryblium tenuispinosum Davey and Williams
Hypotype 41843
Williams, G.L., 1975, Geol. Surv. Can., Paper 74-30, vol. 2, pl. 4, fig. 17.
Early Eocene, core hole 4, lat. 45°31′N, long. 52°22′W, Grand Banks.

Hologinella spinosigibberosa Brideaux and Fisher, 1976
Hypotype 58348
Tan, J.T. and Hills, L.V., 1978, Geol. Surv. Can., Paper 78-1C, p. 66, pl. 12.1, fig. 8.
Ringnes Formation, Upper Jurassic, lat. 71°48′N, long. 112°20′W, Mackenzie King Island, District of Franklin.

Hystrichodinum pulchrum Deflandre
Hypotype 47877
Williams, G.L., 1978, Initial Repts. Deep Sea Drilling Project, vol. 41, pl. 4, fig. 7.
Late Cretaceous, DSDP Site 370, lat. 32°50.2′N, long. 10°46.6′W, northwestern Moroccan continental margin.

Hystrichodinium sp.
Fig. spec. 46024
Brideaux, W.W., 1977, Geol. Surv. Can., Bull 281, p. 25, pl. 10, fig. 5.
Cretaceous, IOE Stoney Core Hole F-42, depth 310-320 feet, Stoney Creek about 18 miles west-southwest of Fort McPherson, lat. 67°21′23″N, long. 135°38′43″W, District of Mackenzie.

Hystrichodinium sp. A (Ioannides et al., 1977)
Fig. specs. 56746, 56747
Williams, G.L. and Bujak, J.P., 1980, Initial Repts. Deep Sea Drilling Project, vol. 50, pl. 5, figs. 9, 10.
Early Cretaceous and Late Jurassic, DSDP Site 416, lat. 32°50.18′N, long. 10°48.06′W, northwestern Moroccan continental margin.

Hystrichokolpoma eisenacki Williams and Downie
Hypotype 49341
Williams, G.L. and Bujak, J.P., 1977, Am. Assoc. Strat. Palynol, Contr. Strat. Palynol, vol. 1, Contr. Ser. 5A, pl. 1, fig. 6.

Early-Middle Eocene, depth 1040-1070 feet, Amoco Imperial Skelly Pelican J-49 well, lat. 45°28′35.17″N, long. 52°36′41.71″W, Grand Banks.

Hystrichokolpoma eisenacki Williams and Downie
Hypotype 47878
Williams, G.L., 1978, Initial Repts. Deep Sea Drilling Project, vol. 41, pl. 4, fig. 10.
Oligocene, DSDP Site 370, lat. 32°50.2′N, long. 10°46.6′W, northwestern Moroccan continental margin.

Hystrichokolpoma ferox (Deflandre) Davey
Hypotype 41503
Brideaux, W.W. and McIntyre, D.J., 1975, Geol. Surv. Can., Bull. 252, p. 28, pl. 7, figs. 15, 16.
Horton River Formation, Lower Cretaceous, lat. 69°19′N, long. 126°50′W, Horton River, District of Mackenzie.

Hystrichokolpoma sp. 1
Fig. specs. 55338, 55339
Mudie, P.J, 1986, Initial Repts. Deep Sea Drilling Project, vol. XCIV, p. 803, pl. 4, figs. 10a, b, 11.
Lower Pliocene, hole 611C, depth 268.49 m, lat. 52°50.15′N, long. 38°18.1′W, in the Gardar Drift on southwestern flank of Reykjanes Ridge, North Atlantic Ocean.

Hystrichosphaeridium albertense Pocock
=*Oligosphaeridium albertense*, Brideaux, W.W., 1977, Geol. Surv. Can., Bull. 281, p. 27, 55, pl. 11, figs. 3, 4, 8-10 (holotype [missing] and paratype 14358).

Hystrichosphaeridium cooksonii Singh
Hypotype 46026
Brideaux, W.W., 1977, Geol. Surv. Can., Bull. 281, p. 25, pl. 10, fig. 7.
Cretaceous, Banff-Aquitainee-Arco Rat Pass K-35 well, depth 1080-1090 feet, 6 miles southeast of Mount Goodenough, lat. 67°54′42.55″N, long. 135°21′56.83″W, District of Mackenzie.

Hystrichosphaeridium costatum Davey and Williams
Hypotype 46025
Brideaux, W.W., 1977, Geol. Surv. Can., Bull 281, p. 26, pl. 10, fig. 6.
Cretaceous, Banff-Aquitaine-Arco Rat Pass K-35 well, depth 1600 feet, 6 miles southeast of Mount Goodenough, lat. 67°54′42.55″N, long. 135°21′56.83″W, District of Mackenzie.

Hystrichosphaeridium sp. cf. *H. difficile* Manum and Cookson
Fig. specs. 38128-38130
McIntyre, D.J., 1974, Geol. Surv. Can., Paper 74-14, pl. 10, figs. 4-6.
Upper Cretaceous, lat. 69°28′N, long. 126°58′W, and lat. 69°27′30″N, long. 126°58′W (38130), Horton River, District of Mackenzie.

Hystrichosphaeridium sp. *H.? palmatum* (White ex Brown) Downie and Sarjeant, 1965
Fig. spec. 65278
Ioannides, N.S., 1986, Geol. Surv. Can., Bull. 371, p. 25, pl. 13, fig. 7.
Kanguk Formation, Upper Cretaceous, south coast Bylot Island, lat. 72°53′30″N, long. 78°16′W, District of Franklin.

Hystrichosphaeridium paracostatum Cookson and Eisenack
Hypotypes 53678, 53696
Bujak, J.P. and Williams, G.L., 1978, Geol. Surv. Can., Bull. 297, pl. 1, fig. 10; pl. 2, fig. 12.
Late Cretaceous, swc 9380 feet, Amoco-IOE Heron H-73 well, lat. 44°2′26.65″N, long. 52°25′40.58″W, Grand Banks.

Hystrichosphaeridium pseudorecurvatum Morgenroth
Hypotypes 41824, 41825
Williams, G.L., 1975, Geol. Surv. Can., Paper 74-30, vol. 2, pl. 3, figs. 15, 16.
Late Miocene, Amoco IOE Puffin B-90 well, lat. 44°39′12.73″N, long. 53°42′28.35″W, Grand Banks.

Hystrichosphaeridium pseudorecurvatum Morgenroth
Hypotype 49340
Williams, G.L. and Bujak, J.P., 1977, Am. Assoc. Strat. Palynol, Contr. Strat. Palynol., vol. 1, Contr. Ser. 5A, pl. 1, fig. 5.
Early-Middle Eocene, depth 1130-1160 feet, Amoco Imperial Skelly Pelican J-49 well, lat. 48°28′35.17″N, long. 52°36′41.71″W, Grand Banks.

Hystrichosphaeridium recurvatum (White) Davey and Williams
Hypotype 56707
Williams, G.L. and Bujak, J.P., 1980, Initial Repts. Deep Sea Drilling Project, vol. 50, pl. 1, fig. 12.
Early Cretaceous, DSDP Site 416, lat. 32°50.18′N, long. 10°48.06′W, northwestern Moroccan continental margin.

Hystrichosphaeridium salpingophorum Deflandre emend. Davey and Williams, *in* Davey et al., 1966
Hypotypes 65291, 65292
Ioannides, N.S., 1986, Geol. Surv. Can., Bull. 371, p. 26, pl. 14, figs. 3, 9.
Eureka Sound Formation, Late Paleocene to ?Early Eocene, Bylot Island, south coast, lat. 72°53′30″N, long. 78°25′W, and "Twosnout Creek", lat. 73°13′N, long. 79°53′W, District of Franklin.

Hystrichosphaeridium stellatum Maier
Hypotype 38156
McIntyre, D.J., 1974, Geol. Surv. Can., Paper 74-14, pl. 13, fig. 1.
Upper Cretaceous, lat. 69°28′N, long. 126°58′W, Horton River, District of Mackenzie.

Hystrichosphaeridium truncigerum Deflandre
Hypotype 41841
Williams, G.L., 1975, Geol. Surv. Can., Paper 74-30, vol. 2, pl. 4, fig. 15.
Santonian, Upper Cretaceous, Core hole 5, lat. 45°48′N, long. 52°13′W, Grand Banks.

Hystrichosphaeridium tubiferum (Ehrenberg) Deflandre emend. Davey and Williams, *in* Davey et al., 1966
Hypotypes 65289, 65290
Ioannides, N.S., 1986, Geol. Surv. Can., Bull. 371, p. 25, pl. 14, figs. 2, 7.
Eureka Sound Formation, Late Paleocene to ?Early Eocene, south coast Bylot Island, lat. 72°52′N, long. 78°31′W, District of Franklin.

Hystrichosphaeridium tubiferum subsp. *brevispinum* Davey and Williams, *in* Davey et al., 1966
Hypotypes 65293-65295
Ioannides, N.S., 1986, Geol. Surv. Can., Bull. 371, p. 25, pl. 14, figs. 4-6.
Eureka Sound Formation, Late Paleocene to ?Early Eocene, Bylot Island, south coast, lat. 72°53′30″N, long. 78°25′W (65293), and "Twosnout Creek", lat. 73°13′N, long. 79°53′W, District of Franklin.

Hystrichosphaeridium sp. *?H. tubiferum* (Ehrenberg) Deflandre emend. Davey and Williams, *in* Davey et al., 1966
Hypotype 65296
Ioannides, N.S., 1986, Geol. Surv. Can., Bull. 371, p. 25, pl. 14, fig. 8.
Eureka Sound Formation, Late Paleocene to ?Early Eocene, "Twosnout Creek", Bylot Island, lat. 73°13′N, long. 79°53′W, District of Franklin.

Hystrichosphaeridium sp. 2
Fig. specs. 38136-38138
McIntyre, D.J., 1974, Geol. Surv. Can., Paper 74-14, pl. 11, figs. 2-4.
Upper Cretaceous, lat. 69°28′N, long. 126°58′W, Horton River, District of Mackenzie.

Hystrichosphaeridium sp. 3(cf. *H. dificile*)
Fig. specs. 38131, 38132
McIntyre, D.J., 1974, Geol. Surv. Can., Paper 74-14, pl. 10, figs. 7, 8.
Upper Cretaceous, lat. 69°28′N, long. 126°58′W, Horton River, District of Mackenzie.

Hystrichosphaeridium sp. A
Fig. spec. 49370
Williams, G.L. and Bujak, J.P., 1980, Am. Strat. Palynol., Contr. Strat. palynol, vol. 1, Contr. Ser. 5A, pl. 8, fig. 6.
Early Paleocene, depth 7860-7890 feet, Eastcan et al Gudrid H-55 well, lat. 55°54′20.06″N, long. 53°52′31.04″W, Labrador Shelf.

Hystrichosphaeridium sp. A
Fig. specs. 53682, 53700
Bujak, J.P. and Williams, G.L., 1978, Geol. Surv. Can., Bull. 297, pl. 1, fig. 15; pl. 2, figs. 16, 17.
Early and Late Cretaceous, 5880-5920 feet, Shell Mic Mac J-77 well, lat. 44°36′42.8″N, long. 59°26′10.9″W, and swc 3807, Shell Mohawk B-93 well, lat. 42°42′10.72″N, long. 64°43′53.04″W, Scotian Shelf.

Hystrichosphaeridium sp. A Habib, 1972
Fig. spec. 47881
Williams, G.L., 1978, Initial Repts. Deep Sea Drilling Project, vol. 41, pl. 5, fig. 1.
Late Jurassic, DSDP Site 367, lat. 12°29.2′N, long. 20°2.8′W, Cape Verde Basin west of Senegal.

Hystrichosphaeridium sp. AE
Fig. spec. 45812
Brideaux, W.W., 1976, Geol. Surv. Can., Paper 76-1B, p. 256, pl. 44.2, fig. 3.
Lower Cretaceous, depth 7500-7530 feet, Gulf Mobil Parsons N-10 well, ca. 1.5 miles east-northeast of Parsons Lake, lat. 69°59′48.52″N, long. 133°31′50.33″W, District of Mackenzie.

Hystrichosphaeridium sp. AE
Fig. specs. 46027, 46028
Brideaux, W.W., 1977, Geol. Surv. Can., Bull 281, p. 25, pl. 10, figs. 8-11.
Cretaceous, IOE Stoney Core Hole F-42, depth 825-830 feet, Stoney Creek about 18 miles west-southwest of Fort McPherson, lat. 67°21′23″N, long. 135°38′43″W and depth 1025 feet, Banff-Aquitaine-Arco Rat Pass K-35 well, 6 miles southeast of Mount Goodenough, lat. 67°54′42.55″N, long. 135°21′56.83″W, District of Mackenzie.

Hystrichosphaeridium sp.B
Fig. specs. 53683, 53684, 53701
Bujak, J.P. and Williams, G.L., 1978, Geol. Surv. Can., Bull. 297, pl. 1, figs. 16, 17; pl. 2, figs. 18, 19.
Late Cretaceous, depth 7860-7890 feet, Eastcan et al Gudrid H-55 well, lat. 54°54′20.06″N, long. 55°52′31.04″W (53683), and depths 5170-5200 and 5230-5260 feet, Eastcan et al. Freydis B-87 well, lat. 53°56′13.39″N, long. 54°42′39.75″W, Labrador Shelf.

Hystrichosphaeridium sp. BE
Fig. specs. 46029, 46030
Brideaux, W.W., 1977, Geol. Surv. Can., Bull 281, p. 26, pl. 11, figs. 1, 2.
Cretaceous, Banff-Aquitaine-Arco Rat Pass K-35 well, depths 1080-1090 and 1470-1480 feet, 6 miles southeast of Mount Goodenough, lat. 67°54′42.55″N, long. 135°21′56.83″W, District of Mackenzie.

Hystrichosphaeropsis sp.
Fig. specs. 65297, 65298, 65305
Ioannides, N.S., 1986, Geol. Surv. Can., Bull. 371, p. 26, pl. 14, fig. 10, 11; pl. 15, fig. 3.
Kanguk Formation, Upper Cretaceous, south coast Bylot Island, lat. 72°53′30″N, long. 78°16′W; Eureka Sound Formation, Late Paleocene to ?Early Eocene, "Twosnout Creek", Bylot Island, lat. 73°13′N, long. 79°53′W (65298), District of Franklin.

Imbatodinium kondratjevii Vozzhennikova
Hypotype 56711
Williams, G.L. and Bujak, J.P., 1980, Initial Repts. Deep Sea Drilling Project, vol. 50, pl. 2, fig. 4.

Late Jurassic, DSDP Site 416, lat. 32°50.18′N, long. 10°48.06′W, northwestern Moroccan continental margin.

Imbatodinium villosum Vozzhennikova
Hypotype 41849
Williams, G.L., 1975, Geol. Surv. Can., Paper 74-30, vol. 2, pl. 5, fig. 6.
Kimmeridgian, Upper Jurassic, Shell Oneida O-25 well, lat. 43°14′57.49″N, long. 61°33′36.38″W, Scotian Shelf.

Imbatodinium villosum Vozzhennikova
Hypotype 45813
Brideaux, W.W., 1976, Geol. Surv. Can., Paper 76-1B, pl. 44.2, fig. 4.
Husky Formation, Lower Cretaceous, depth 9480-9295 feet, Gulf Mobil Parsons N-10 well, ca. 1.5 miles east-northeast of Parsons Lake, lat. 69°54′48.52″N, long. 133°31′50.33″W, District of Mackenzie.

Imbatodinium sp. A
Fig. specs. 41525, 41526
Brideaux, W.W. and McIntyre, D.J., 1975, Geol. Surv. Can., Bull. 252, p. 32, pl. 9, figs. 14-16.
Horton River Formation, Lower Cretaceous, lat. 69°19′N, long. 126°50′W and lat. 69°27′30″N, long. 126°54′W, Horton River, District of Mackenzie.

Impagidinium aculeatum (Wall) Lentin and Williams, 1981
Hypotype 55332
Mudie, P.J., 1986, Initial Repts. Deep Sea Drilling Project, vol. XCIV, p. 803, pl. 4, fig. 4.
Pleistocene, hole 607, depth 16.1 m, lat. 41°0.07′N, long. 32°57.44′W, at base of upper western flank of the Mid-Atlantic Ridge, North Atlantic Ocean.

Impagidinium aliferum Mudie
Holotype 55329
Mudie, P.J., 1986, Initial Repts. Deep Sea Drilling Project, vol. XCIV, p. 803, pl. 4, figs. 1a, b.
Lower Pliocene, hole 607, depth 215.6 m, lat. 41°0.07′N, long. 32°57.44′W, at base of upper western flank of the Mid-Atlantic Ridge, North Atlantic Ocean.

Impagidinum sp. cf. *I. dispertitum* (Cookson and Eisenack) Stover and Evitt, 1978
Fig. specs. 65314, 65315
Ioannides, N.S., 1986, Geol. Surv. Can., Bull. 371, p. 26, pl. 15, figs. 11, 12.
Eureka Sound Formation, Late Paleocene to ?Early Eocene, south coast Bylot Island, lat. 72°53′30″N, long. 78°25′W, District of Franklin.

Impagidinium japonicum Matsuko, 1983
Hypotype 55330
Mudie, P.J., 1986, Initial Repts. Deep Sea Drilling Project, vol. XCIV, p. 803, pl. 4, figs. 2a, b.
Upper Pliocene, hole 607, depth 76.7 m, lat. 41°0.07′N, long. 32°57.44′W, at base of upper western flank of the Mid-Atlantic Ridge, North Atlantic Ocean.

Impagidinium multiplexum (Wall and Dale) Lentin and Williams, 1981
Hypotype 55364
Mudie, P.J., 1986, Initial Repts. Deep Sea Drilling Project, vol. XCIV, p. 803, pl. 5, fig. 15.
Pleistocene, hole 611, depth 49.49 m, lat. 52°50.47′N, long. 30°18.58′W, in the Gardar Drift on southwestern flank of Reykjanes Ridge, North Atlantic Ocean.

Impagidinium(?) pallidum Bujak 1984
Hypotype 55316
Mudie, P.J., 1986, Initial Repts. Deep Sea Drilling Project, vol. XCIV, p. 803, pl. 3, fig. 2.
Upper Miocene, hole 611C, depth 441.28 m, lat. 52°50.15′N, long. 30°18.1′W, in the Gardar Drift on southwestern flank of Reykjanes Ridge, North Atlantic Ocean.

Impagidinium patulum (Wall) Stover and Evitt, 1978
Hypotype 55331
Mudie, P.J., 1986, Initial Repts. Deep Sea Drilling Project, vol. XCIV, p. 803, pl. 4, fig. 3.
Lower Pliocene, hole 607, depth 215.6 m, lat. 41°0.07′N, long. 32°57.44′W, at base of upper western flank of the Mid-Atlantic Ridge, North Atlantic Ocean.

Impagidinium strialatum (Wall) Stover and Evitt, 1978
Hypotype 55333
Mudie, P.J., 1986, Initial Repts. Deep Sea Drilling Project, vol. XCIV, p. 804, pl. 4, fig. 5.
Pleistocene, hole 607, depth 5 m, lat. 41°0.07′N, long. 32°57.44′W, at base of upper western flank of the Mid-Atlantic Ridge, North Atlantic Ocean.

Impagidinium velorum Bujak, 1984
Hypotype 55319
Mudie, P.J., 1986, Initial Repts. Deep Sea Drilling Project, vol. XCIV, p. 804, pl. 3, fig. 5a, b.
Upper Pliocene, hole 607, depth 92.9 m, lat. 41°0.07′N, long. 32°57.44′W, at base of upper western flank of the Mid-Atlantic Ridge, North Atlantic Ocean.

Impagidinium sp.
Fig. specs. 58591, 58592
McIntyre, D.J. and Brideaux, W.W., 1980, Geol. Surv. Can., Bull. 320, p. 14, pl. 4, fig. 1, 6, 7.
Lower Cretaceous, Martin Creek, lat. 68°12′10″N, long. 135°34′50″W, Richardson Mountains, District of Mackenzie.

Impagidinium sp.
Fig. spec. 65090
Bujak, J.P. and Davies, E.H., 1982, Bull. Can. Petrol. Geol., vol. 49, no. 3, 1981, p. 422, pl. 1, fig. 6.
Late Middle to Late Miocene, DSDP Site 183, Aleutian Abyssal Plain, Pacific Ocean.
=*Impagidinium* sp. A, Dixon, J. et al., 1984, Geol. Surv. Can., Paper 82-13, pl. 4, fig. 5.

Impagidinium spp. A, B
Fig. specs. 55335, 55336
Mudie, P.J., 1986, Initial Repts. Deep Sea Drilling Project, vol. XCIV, p. 803, pl. 4, figs. 7a, b, 8a, b.

Lower Pliocene and Upper Miocene, hole 607, depths 200 m and 275. 34 m, lat. 41°0.07′N, long. 32°57.44′W, at base of upper western flank of the Mid-Atlantic Ridge, North Atlantic Ocean.

Impagidinium sp. C
Fig. spec. 55342
Mudie, P.J., 1986, Initial Repts. Deep Sea Drilling Project, vol. XCIV, p. 804, pl. 3, figs. 14a, b.
Upper Miocene, hole 611C, depth 313.49 m, lat. 52°50.15′N, long. 30°18.1′W, on the Gardar Drift on southwestern flank of Reykjanes Ridge, North Atlantic Ocean.

Impletosphaeridium sp.
Fig. specs. 65312, 65313
Ioannides, N.S., 1986, Geol. Surv. Can., Bull. 371, p. 26, pl. 15, figs. 10, 13.
Kanguk Formation, Upper Cretaceous, south coast Bylot Island, lat. 72°53′30″N, long. 78°16′W, District of Franklin.

Impletosphaeridium spp.
Fig. specs. 55356, 55357
Mudie, P.J., 1986, Initial Repts. Deep Sea Drilling Project, vol. XCIV, p. 804, pl. 5, figs. 12a, b, 13a, b.
Upper Pliocene, hole 611C, depth 166.82 m, lat. 52°50.15′N, long. 30°18.1′W, in the Gardar Drift on southwestern flank of Reykjanes Ridge, and hole 607, depth 105.5 m, lat. 41°0.07′N, long. 32°57.44′W, at base upper western flank of the Mid-Atlantic Ridge, North Atlantic Ocean.

Incertae sedis sp. I of Edwards (1984)
Fig. spec. 55344
Mudie, P.J., 1986, Initial Repts. Deep Sea Drilling Project, vol. XCIV, p. 806, pl. 4, fig. 16.
Lower Pliocene, hole 607, depth 177.2 m, lat. 41°0.07′N, long. 32°57.44′W, at base of upper western flank of the Mid-Atlantic Ridge, North Atlantic Ocean.

Invertocysta lactrmosa Edwards, 1984
Hypotype 55306
Mudie, P.J., 1986, Initial Repts. Deep Sea Drilling Project, vol. XCIV, p. 804, pl. 2, fig. 14.
Lower Pleiocene, hole 607, depth 215.6 m, lat. 41°0.07′N, long. 32°57.44′W, at base of upper western flank of the Mid-Atlantic Ridge, North Atlantic Ocean.

Isabelidinium acuminatum (Cookson and Eisenack) Stover and Evitt, 1978
Hypotypes 65299, 65300, 65303, 65304
Ioannides, N.S., 1986, Geol. Surv. Can., Bull. 371, p. 26, pl. 14, figs. 12, 13; pl. 15, figs. 1, 2.
Kanguk Formation, Upper Cretaceous, Bylot Island, south coast, lat. 72°53′30″N, long. 78°16′W, and "Twosnout Creek", lat. 73°14′30″N, long. 79°49′W (65303), District of Franklin.

Isabelidinium bakerii (Deflandre and Cookson) Lentin and Williams, 1977
Hypotypes 65301, 65302
Ioannides, N.S., 1986, Geol. Surv. Can., Bull. 371, p. 27, pl. 14, figs. 14, 15.
Eureka Sound Formation, Late Paleocene to ?Early Eocene, Bylot Island, "Twosnout Creek", lat. 73°9′30″N, long. 79°42′30″W, and south coast, lat. 72°52′N, long. 78°35′W, District of Franklin.

Isabelidinium cf. *belfastense* (Cookson and Eisenack)
Hypotype 53703
Bujak, J.P. and Williams, G.L., 1978, Geol. Surv. Can., Bull. 297, pl. 3, fig. 1.
Late Cretaceous, swc. 3400 feet, Shell Cree E-35 well, lat. 43°44′20.71″N, long. 60°35′55.9″W, Scotian Shelf.

Isabelidinum cooksoniae (Alberti) Lentin and Williams, 1977
Hypotype 65306
Ioannides, N.S., 1986, Geol. Surv. Can., Bull. 371, p. 27, pl. 15, fig. 4.
Eureka Sound Formation, Late Paleocene to ?Early Eocene, "Twosnout Creek", Bylot Island, lat. 73°13′N, long. 79°53′W, District of Franklin.

Isabelidinium cooksoniae (Alberti) Lentin and Williams, 1977
Hypotype 91436
Sweet, A.R. and McIntyre, D.J., 1988, Can. Soc. Petrol. Geol., Mem. 15, fig. 8.8.
Leyland Member, Cardium Formation, Upper Cretaceous, south side of McLeod River, l.s.d. 16, sec. 16, tp. 47, rge. 23, W.5th mer., Alberta.

Isabelidinium cretaceum (Cookson) Lentin and Williams, 1977
Hypotypes 65307-65309
Ioannides, N.S., 1986, Geol. Surv. Can., Bull. 371, p. 27, pl. 15, figs. 5-7.
Eureka Sound Formation, Late Paleocene to ?Early Eocene, south coast Bylot Island, lat. 72°53′30″N, long. 78°25′W, District of Franklin.

Isabelidinium microarmum (McIntyre) Lentin and Williams, 1977
Hypotype 65311
Ioannides, N.S., 1986, Geol. Surv. Can., Bull. 371, p. 28, pl. 15, fig. 9.
Kanguk Formation, Upper Cretaceous, south coast Bylot Island, lat. 72°53′30″N, long. 78°16′W, District of Franklin.

Kallosphaeridium ? agglutinatum McIntyre and Brideaux
Holotype 58642; paratypes 58640, 58641, 58643-58645
McIntyre, D.J. and Brideaux, W.W., 1980, Geol. Surv. Can., Bull. 320, p. 25, pl. 12, figs. 5-12.
Lower Cretaceous, Martin Creek, lat. 68°12′10″N, long. 135°34′50″W, and "Grizzly Gorge", lat. 68°15′49″N, long. 135°40′59″W (58640, 58643), Richardson Mountains, District of Mackenzie.

Kallosphaeridium sp. A
Fig. spec. 56718
Williams, G.L. and Bujak, J.P., 1980, Initial Repts. Deep Sea Drilling Project, vol. 50, p. 3, fig. 3.

Early Cretaceous, DSDP Site 416, lat. 32°50.18′N, long. 10°48.06′W, northwestern Moroccan continental margin.

Kallosphaeridium sp. AE
Fig. specs. 58597, 58598
McIntyre, D.J. and Brideaux, W.W., 1980, Geol. Surv. Can., Bull. 320, p. 25, pl. 5, figs. 1, 4.
Lower Cretaceous, Martin Creek, lat. 68°12′10″N, long. 135°34′50″W, Richardson Mountains, District of Mackenzie.

Kallosphaeridium sp. BE
Fig. spec. 58602
McIntyre, D.J. and Brideaux, W.W., 1980, Geol. Surv. Can., Bull. 320, p. 25, pl. 5, figs. 7, 10.
Lower Cretaceous, Martin Creek, lat. 68°12′10″N, long. 135°34′50″W, Richardson Mountains, District of Mackenzie.

Kalyptea aceras Manum and Cookson
Hypotypes 38117, 38118
McIntyre, D.J., 1974, Geol. Surv. Can., Paper 74-14, pl. 9, figs. 4, 5.
Upper Cretaceous, lat. 69°28′N, long. 126°58′W, Horton River, District of Mackenzie.

Kalyptea monoceras Cookson and Eisenack
Hypotype 38119
McIntyre, D.J., 1974, Geol. Surv. Can., Paper 74-14, pl. 9, fig. 6.
Upper Cretaceous, lat. 69°27′30″N, long. 126°58′W, Horton River, District of Mackenzie.

Kalyptea monoceras Cookson and Eisenack
Hypotype 41555
Brideaux, W.W. and McIntyre, D.J., 1975, Geol. Surv. Can., Bull. 252, p. 38, pl. 14, fig. 6.
Horton River Formation, Lower Cretaceous, lat. 69°19′N, long. 126°50′W, Horton River, District of Mackenzie.

Kalyptea monoceras Cookson and Eisenack
Hypotype 56748
Williams, G.L. and Bujak, J.P., 1980, Initial Repts. Deep Sea Drilling Project, vol. 50, pl. 5, fig. 11.
Early Cretaceous, DSDP Site 416, lat. 32°50.18′N, long. 10°48.06′W, northwestern Moroccan continental margin.

Kalyptea sp. A Brideaux, 1971
Fig. spec. 41557
Brideaux, W.W. and McIntyre, D.J., 1975, Geol. Surv. Can., Bull. 252, p. 38, pl. 14, fig. 8.
Horton River Formation, Lower Cretaceous, lat. 69°19′N, long. 126°50′W, Horton River, District of Mackenzie.

Kisselovia tenuivirgula (Williams and Downie) Lentin and Williams
Hypotype 92000
Dietrich, J.R. et al., 1989, Geol. Surv. Can., Paper 89-1G, p. 141, pl. 1, fig. 4.
Early to Middle Eocene, depth 5000-5090 feet, Dome Pacific et al. PEX Natsek E-56 well, lat. 69°45′21.46″N, long. 139°44′34.58″W, western Beaufort Sea.

Kleithriasphaeridium eoinodes (Eisenack) Davey
Hypotypes 56729, 56734
Williams, G.L. and Bujak, J.P., 1980, Initial Repts. Deep Sea Drilling Project, vol. 50, pl. 4, figs. 3, 9.
Early Cretaceous, DSDP Site 416, lat. 32°50.18′N, long. 10°48.06′W, northwestern Moroccan continental margin.

Labyrinthodinium truncatum Piasecki, 1980
Hypotypes 55326-55328
Mudie, P.J., 1986, Initial Repts. Deep Sea Drilling Project, vol. XCIV, p. 804, pl. 3, figs. 12a, b, 13; pl. 4, figs. 14a, b.
Upper Miocene and Lower Pliocene (55328), hole 611c, depths 510.09 m and 212.39m, lat. 52°50.15′N, long. 30°18.1′W, in the Gardar Drift on southwestern flank of Reykjanes Ridge, North Atlantic Ocean.

Laciniadinium arcticum (Manum and Cookson) Lentin and Williams, 1980
Hypotype 91451
Sweet, A.R. and McIntyre, D.J., 1988, Can. Soc. Petrol. Geol., Mem. 15, fig. 8.23.
Leyland Member, Cardium Formation, Upper Cretaceous, south side McLeod River, l.s.d.16, sec. 16, tp. 47, rge. 23, W.5th mer., Alberta.

Laciniadinium williamsi Ioannides
Holotype 65215; paratypes 62516-62520, 65248
Ioannides, N.S., 1986, Geol. Surv. Can., Bull. 371, p. 28, pl. 10, figs. 1-6; pl. 11, fig. 5.
Kanguk Formation, Upper Cretaceous, south coast Bylot Island, lat. 72°53′30″N, long. 78°16′W, District of Franklin.

Lanterna saturnalis Brideaux and Fisher
Holotype 34158; paratypes 42380-42387
Brideaux, W.W. and Fisher, M.J., 1976, Geol. Surv. Can., Bull. 259, p. 24, pl. 6, figs. 1-10; pl. 7, figs. 1-13.
Gilmore Lake Member, Langton Bay Formation, Lower Cretaceous, depths 1680-1700, 1760-1770 (42380, 42383, 42386), and 1800-1810 (42381, 42384) feet, Elf Horton River G-02 well, lat. 69°51′23″N, long. 127°15′57″W, southwest of mouth of Horton River, District of Mackenzie.

Lanterna saturnalis Brideaux and Fisher, 1976
Hypotypes 58349, 58352
Tan, J.T. and Hills, L.V, 1978, Geol. Surv. Can., Paper 78-1C, p. 66, pl. 12.1, figs. 9, 12.
Deer Bay Formation, Upper Jurassic, lat. 77°48′N, long. 112°20′W, Mackenzie King Island, District of Franklin.

Lanterna sportula Dodekova
Hypotype 41850
Williams, G.L., 1975, Geol. Surv. Can., Paper 74-30, vol. 2, pl. 5, fig. 7.

Kimmeridgian (caved), Upper Jurassic, Shelf Oneida O-25 well, lat. 43°14′57.49″N, long. 61°33′36.38″W, Scotian Shelf.

Lanternosphaeridium radiatum Morgenroth
Hypotype 47880
Williams, G.L., 1978, Initial Repts. Deep Sea Drilling Project, vol. 41, pl. 4, fig. 12.
Middle Eocene, DSDP Site 370, lat. 32°50.2′N, long. 10°46.6′W, northwestern Moroccan continental margin.

Lanthernosphaeridium sp. BE
Fig. specs. 45806, 45807
Brideaux, W.W., 1976, Geol. Surv. Can., Paper 76-1B, p. 251, pl. 44.1, figs. 10, 12.
Reindeer Formation, Eocene, depth 1200-1230 feet, Gulf Mobil Parsons N-10 well, ca. 1.5 miles east-northeast of Parsons Lake, lat. 69°59′48.52″N, long. 133°31′50.33″W, District of Mackenzie.

Leberidocysta chlamydata (Cookson and Eisenack) Stover and Evitt, 1978
Hypotypes 65321, 65322
Ioannides, N.S., 1986, Geol. Surv. Can., Bull. 371, p. 28, pl. 16, figs. 1, 2.
Eureka Sound Formation, Late Paleocene to ?Early Eocene, south coast Bylot Island, lat. 72°53′30″N, long. 78°25′W, District of Franklin.

Leberidocysta chlamydata (Cookson and Eisenack) Stover and Evitt, 1978
Hypotype 91442
Sweet, A.R. and McIntyre, D.J., 1988, Can. Soc. Petrol. Geol., Mem. 15, fig. 8.14.
Moosehound Member, Cardium Formation, Upper Cretaceous, west bank of Little Berland River, l.s.d. 1, sec. 6, tp. 54, rge. 2, W.5th mer., Alberta.

Lecaniella sp.
Fig. spec. 91463
Sweet, A.R. and McIntyre, D.J., 1988, Can. Soc. Petrol. Geol., Mem. 15, fig. 9.8.
Leyland Member, Cardium Formation, Upper Cretaceous, south side of McLeod River, l.s.d.16, sec. 16, tp. 47, rge. 23, W.5th mer., Alberta.

Lecaniella sp. AE
Fig. spec. 45932
Brideaux, W.W., 1976, Geol. Surv. Can., Paper 76-1C, p. 118, pl. 23.1, fig. 11.
Lower Cretaceous, south bank Martin Creek, District of Mackenzie.

Leiosphaera sp.
Fig. spec. 68782
Dixon, J. et al., 1984, Geol, Surv. Can., Paper 82-13, pl. 5, figs. 11, 12.
Pliocene to Holocene, 610-637 m interval, Dome Gulf et al. Hunt Kopanoar M-13 well, lat. 70°22′55.1″N, long. 135°05′33.9″W, Beaufort Sea.

Leiosphaera sp. A of Mudie and Short (1985)
Fig. spec. 55351
Mudie, P.J., 1986, Initial Repts. Deep Sea Drilling Project, vol. XCIV, p. 803, pl. 5, fig. 7.
Recent, Baffin Bay, District of Franklin.

Lejeuncysta sp.
Fig. spec. 65310
Ioannides, N.S., 1986, Geol. Surv. Can., Bull. 371, p. 28, pl. 15, fig. 8.
Kanguk Formation, Upper Cretaceous, south coast Bylot Island, lat. 72°53′30″N, long. 78°16′W, District of Franklin.

Lejeunia fallax Morgenroth
Hypotype 41851
Williams, G.L., 1975, Geol. Surv. Can., Paper 74-30, vol. 2, pl. 5, fig. 8.
Middle Miocene, Amoco IOE Puffin B-90 well, lat. 44°39′12.73″N, long. 53°42′28.35″W, Grand Banks.

Lejeunia hyalina Gerlach
Hypotype 38694
Haskell, T.R. and Wilson, G.J., 1975, Initial Repts. Deep Sea Drilling Project, vol. 29, pl. 3, fig. 2.
Upper Eocene, DSDP Site 282, lat. 42°14.76′S, long. 143°29.18′E, west of Tasmania.

Lejeunia hyalina Gerlach, 1961 emend. Kjellström, 1972
Hypotype 71425
Wilson, M.A., 1978, Palaeontographica, Abt.B, vol. 166, p. 152, pl. 12, fig. 8.
Tent Island Formation, Cretaceous, Little Fish (Cache) Creek, Big Fish River area, District of Mackenzie.

Lejeunia kozlowskii Gorka
Hypotypes 38046, 38047
McIntyre, D.J., 1974, Geol. Surv. Can., Paper 74-14, pl. 4, figs. 2, 3.
McIntyre, D.J., 1975, Geoscience and Man, vol. 11, p. 68, pl. 4, figs. 3 (38047), 4 (38046).
Upper Cretaceous, lat. 69°28′N, long. 126°58′W, Horton River, District of Mackenzie.

Lejeunia paratenella Benedek
Hypotype 41861
Williams, G.L., 1975, Geol. Surv. Can., Paper 74-30, vol. 2, pl. 5, fig. 18.
Middle Miocene, Amoco IOE Puffin B-90 well, lat. 44°39′12.73″N, long. 53°42′28.35″W, Grand Banks.

Lejeunia sp.
Fig. spec. 49372
Williams, G.L. and Bujak, J.P., 1977, Am. Assoc. Strat. Palynol, Contr. Strat. Palynol, vol. 1, Contr. Ser. 5A, pl. 5, fig 8.
Miocene, depth 1210-1290 feet, Tenneco et al. Leif E-38 well, lat. 54°17′29.87″N, long. 55°5′52.17″W, Labrador Shelf.

Leptodinium cancellatum Brideaux and McIntyre
Holotype 34146; paratypes 41480, 41481
Brideaux, W.W. and McIntyre, D.J., 1975, Geol. Surv. Can., Bull. 252, p. 21, pl. 5, figs. 7-9.

Horton River Formation, Lower Cretaceous, lat. 69°19′N, long. 126°50′W, Horton River, District of Mackenzie.

Leptodinium delicatum (Davey) Sarjeant
Hypotype 41479
Brideaux, W.W. and McIntyre, D.J., 1975, Geol. Surv. Can., Bull. 252, p. 22, pl. 5, fig. 6.
Horton River Formation, Lower Cretaceous, lat. 69°19′N, long. 126°50′W, Horton River, District of Mackenzie.

Leptodinium dispertitum Cookson and Eisenack
Hypotype 38698
Haskell, T.R. and Wilson, G.J., 1975, Initial Repts. Deep Sea Drilling Project, vol. 29, pl. 3, fig. 6.
Paleocene, DSDP Site 283, lat. 43°54.60′S, long. 154°16.96′E, east of Tasmania.

Leptodinium modicum Brideaux and McIntyre
Holotype 34147; paratype 41482
Brideaux, W.W. and McIntyre, D.J., 1975, Geol. Surv. Can., Bull. 252, p. 22, pl. 5, figs. 10-15.
Horton River Formation, Lower Cretaceous, lat. 69°19′N, long. 126°50′W, Horton River, District of Mackenzie.

Leptodinium modicum denticulatum Brideaux
Holotype 45967
Brideaux, W.W., 1977, Geol. Surv. Can., Bull 281, p. 8, pl. 1, figs. 10, 11.
Cretaceous, Banff-Aquitaine-Arco Rat Pass K-35 well, depth 1445 feet, 6 miles southeast of Mount Goodenough, lat. 67°54′42.55″W, long. 135°21′56.83″W, District of Mackenzie.

Leptodinium modicum modicum Brideaux and McIntyre
Hypotype 45965
Brideaux, W.W., 1977, Geol. Surv. Can., Bull 281, p. 7, pl. 1, fig. 8.
Cretaceous, Banff-Aquitaine-Arco Rat Pass K-35 well, depth 1380 feet, 6 miles southeast of Mount Goodenough, lat. 67°54′42.55″N, long. 135°21′56.83″W, District of Mackenzie.

Leptodinium vitorianum Cookson and Eisenack
Hypotype 38693
Haskell, T.R. and Wilson G.T., 1975, Initial Repts. Deep Sea Drilling Project, vol. 29, pl. 3, fig. 1.
Cenozoic, DSDP Site 284, lat. 40°30.48′S, long. 167°40.81′E, Lord Howe Rise west of New Zewland.

Leptodinium sp.
Fig. spec. 47884
Williams, G.L., 1978, Initial Repts. Deep Sea Drilling Project, vol. 41, pl. 5, fig. 3.
Early Cretaceous, DSDP Site 367, lat. 12°29.2′N, long. 20°2.8′W, Cape Verde Basin west of Senegal.

Leptodinium sp. A
Fig. spec. 56714
Williams, G.L. and Bujak, J.P., 1980, Initial Repts. Deep Sea Drilling Project, vol. 50, pl. 2, fig. 7.
Early Cretaceous, DSDP Site 416, lat. 32°50.18′N, long. 10°48.06′W, northwestern Moroccan continental margin.

Leptodinium sp. EE
Fig. spec. 45921
Brideaux, W.W., 1976, Geol. Surv. Can., Paper 76-1C, p. 118, pl. 23.1, fig. 1.
Lower Cretaceous, south bank Martin Creek, District of Mackenzie.

Lingulodinium machaerophorum (Deflandre and Cookson) Wall, 1967
Hypotype 55360
Mudie, P.J., 1986, Initial Repts. Deep Sea Drilling Project, vol. XCIV, p. 804, pl. 5, fig. 18.
Upper Miocene, hole 607, depth 249.5 m, lat. 41°0.07′N, long. 32°57.44′W, at base of upper western flank of the Mid-Atlantic Ridge, North Atlantic Ocean.

Litodinia serrulata Davies
Paratypes 69370, 69371
Davies, E.H., 1983, Geol. Surv. Can., Bull. 359, p. 26, pl. 9, figs. 1, 2, 6.
Lower-Middle Jurassic, depth 2330-2360 feet, Panarctic Sandy Point L-46 well, northwestern Melville Island, District of Franklin.

Lithodinia sp. A
Fig. spec. 75493
Davies, E.H., 1985, Palynology, vol. 9, p. 129, pl. 4, fig. 4.
Early Bathonian, Liassic, Brehna Road Section, Lusitanian Basin, west-central Portugal.

?*Lithodinia* sp. B
Fig. spec. 69372
Davies, E.H., 1983, Geol. Surv. Can., Bull. 359, p. 27, pl. 9, figs. 7, 8; text-figs. 23 A, B.
Lower Cretaceous, depth 430-460 feet, Panarctic Sandy Point L-46 well, northwestern Melville Island, District of Franklin.

Litosphaeridium siphoniphorum (Cookson and Eisenack)
Hypotype 41880
Williams, G.L., 1975, Geol. Surv. Can., Paper 74-30, vol. 2, pl. 7, fig. 1.
Turonian, Upper Cretaceous, Amoco IOE Kittiwake P-11 well, lat. 44°40′49.43″N, long. 53°31′45.65″W, Grand Banks.

Litosphaeridium siphoniphorum (Cookson and Eisenack) sensu Warren, 1967
Hypotype 53686
Bujak, J.P. and Williams, G.L., 1978, Geol. Surv. Can., Bull. 297, pl. 1, fig. 19.
Early Cretaceous. 12150-12180 feet, Shell Cree E-35 well, lat. 43°44′20.71″N, long. 60°35′55.9″W, Scotian Shelf.

Litosphaeridium siphoniphorum (Cookson and Eisenack) Davey and Williams
Hypotype 47879

Williams, G.L., 1978, Initial Repts. Deep Sea Drilling Project, vol. 41, pl. 4, fig. 11.
Late Cretaceous, DSDP Site 370, lat. 32°50.2′N, long. 10°46.6′W, northwestern Moroccan Continental Margin.

Luehndea sp. A
Fig. spec. 75475
Davies, E.H., 1985, Palynology, vol. 9, p. 129, pl. 3, figs. 5, 6.
Late Pliensbachian, Liassic, Brehna Road Section, Lusitanian Basin, west-central Portugal.

Luxadinium primulum Brideaux and McIntyre
Holotype 34153; paratypes 41548-41551
Brideaux, W.W. and McIntyre, D.J., 1975, Geol. Surv. Can., Bull. 252, p. 37, pl. 12, figs. 9-12; pl. 13, figs. 1-8.
Horton River Formation, Lower Cretaceous, lat. 69°19′N, long. 126°50′W and lat. 69°27′30″N, long. 126°54′W (41551), Horton River, District of Mackenzie.

Luxadinium sp. cf. *L. primulum* Brideaux and McIntyre, 1975
Fig. specs. 96484-96488
Leckie, D.A. and Singh, C., 1991, J. Sed. Petrology, vol. 61, no. 5, figs. 3A-3E.
Paddy Member, Peace River Formation, Lower Cretaceous, near town of Peace River, Alberta.

Luxadinium propatulum Brideaux and McIntyre, 1975
Hypotype 91438
Sweet, A.R. and McIntyre, D.J., 1988, Can. Soc. Petrol. Geol., Mem. 15, fig. 8.10.
Leyland Member, Cardium Formation, Upper Cretaceous, south side of McLeod River, l.s.d. 16, sec. 16, tp. 47, rge. 23, W.5th mer., Alberta.

Luxadinium propatulum Brideaux and McIntyre, 1975
Hypotypes 84749, 84750
Leckie, D.A. et al., 1990, J. Sed. Petrology, vol. 60, no. 1, p. 101, figs. 6L, M.
Shaftesbury Formation, Lower Cretaceous, west bank Peace River near town of Peace River, Alberta.

Maduradinium turpis Norris
Holotype 68167; hypotypes 68172-68175
Norris, G., 1986, Geol. Surv. Can., Bull. 340, p. 44, pl. 13, fig. 7; pl. 14, figs. 1-4.
Richards Formation, Middle Eocene, depths 12,600-12,650, 11,970, 12,300-12,400 and 12,600 (68174, 68175) feet, Imperial Nuktak C-22 well, lat. 69°41′07″N, long. 134°51′30″W, District of Mackenzie.

Meiourogonyaulax cf. *M. deflandrei* Sarjeant 1968
Hypotypes 53029, 53042
van Helden, B.G.T., 1977, Geol. Surv. Can., Paper 77-1B, p. 170, pl. 33.1, fig. 18; pl. 33.2, figs. 14, 15.
Wilkie Point Formation, Jurassic, east side Inteprid Inlet, Prince Patrick Island, District of Franklin.

Meiourogonyaulax stoveri Millioud
Hypotype 41854
Williams, G.L., 1975, Geol. Surv. Can., Paper 74-30, vol. 2, pl. 5, fig. 11.
Berriasian-Valanginian, Lower Cretaceous, Mobil Tetco Dauntless D-35 well, lat. 44°44′8.26″N, long. 57°20′46.62″W, Scotian Shelf.

Meiourogonyaulax stoveri Millioud
Hypotypes 47882, 47883
Williams, G.L., 1978, Initial Repts. Deep Sea Drilling Project, vol. 41, pl. 5, figs. 2, 5.
Early Cretaceous, DSDP Site 370, lat. 32°50.2′N, long. 10°46.6′W, northwestern Moroccan continental margin.

Meiourogonyaulax sp.
Fig. spec. 41881
Williams, G.L., 1975, Geol. Surv. Can., Paper 74-30, vol. 2, pl. 7, fig. 2.
Callovian, Jurassic, Shell Mohican I-100 well, lat. 42°59′39.04″N, long. 62°58′51.32″W, Scotian Shelf.

Meiourogonyaulax sp.
Fig. spec. 47886
Williams, G.L., 1978, Initial Repts. Deep Sea Drilling Project, vol. 41, pl. 5, fig. 6.
Early Cretaceous, DSDP Site 370, lat. 32°50.2′N, long. 10°46.6′W, northwestern Moroccan continental margin.

Meiourogonyaulax sp.
Fig. spec. 58351
Tan, J.T. and Hills. L.V., 1978, Geol. Surv. Can., Paper 78-1C, p. 66, pl. 12.1, fig. 11.
Savik Formation, Upper Jurassic, lat. 77°48′N, long. 112°20′W, Mackenzie King Island, District of Franklin.

Meiourogonyaulax sp. AE
Fig. spec. 58594
McIntyre, D.J. and Brideaux, W.W., 1980, Geol. Surv. Can., Bull. 320, p. 14, pl. 4, figs. 8, 9, 11.
Lower Cretaceous, Martin Creek, lat. 68°12′10″N, long. 135°34′50″W, Richardson Mountains, District of Mackenzie.

Mendicodinium caperatum Brideaux
Holotype 46003; paratypes 46004, 46005
Brideaux, W.W., 1977, Geol. Surv. Can., Bull. 281, p. 19, pl. 7, figs. 7-11.
Cretaceous, IOE Stoney Core Hole F-42, depth 310-320 feet, Stoney Creek about 18 miles west-southwest of Fort McPherson, lat. 67°21′23″N, long. 135°38′43″W, District of Mackenzie.

Mendicodinium reticulatum Morgenroth
Hypotype 75487
Davies, E.H., 1985, Palynology, vol. 9, pl. 3, fig. 21.
Early Pliensbachian, Liassic, Zambujal Section, south of Condeixa, Lusitanian Basin, west-central Portugal.

Mendicodinium woodhamense Drugg
Hypotype 75486
Davies, E.H., 1985, Palynology, vol. 9, pl. 3, fig. 20.
Early Pliensbachian, Liassic, Zambujal Section, south of Condeixa Lusitanian Basin, west-central Portugal.

Mendicodinium sp. C, A
Fig. specs. 75485, 75498
Davies, E.H., 1985, Palynology, vol. 9, p. 129, pl. 3, figs. 18, 19, 22, 23.
Liassic, Early Pliensbachian, Zambujal Section, south of Condeixa, and Early Toarcian, Peniche Cliffs, Lusitanian Basin, west-central Portugal.

Microdinium? crinitum Davey
Hypotype 41484
Brideaux, W.W. and McIntyre, D.J., 1975, Geol. Surv. Can., Bull. 252, p. 23, pl. 6, fig. 6.
Horton River Formation, Lower Cretaceous, lat. 69°19′N, long. 126°50′W, Horton River, District of Mackenzie.

Microdinium sp. cf. *M. irregulare* Clarke and Verdier
Fig. specs. 38079, 38080
McIntyre, D.J., 1974, Geol. Surv. Can., Paper 74-14, pl. 7, figs. 1, 2.
Upper Cretaceous, lat. 69°27′30″N, long. 126°58′W, Horton River, District of Mackenzie.

Microdinium ornatum Cookson and Eisenack
Hypotypes 38081-38083
McIntyre, D.J., 1974, Geol. Surv. Can., Paper 74-14, pl. 7, figs. 3-5.
Upper Cretaceous, lat. 69°27′30″N, long. 126°58′W, Horton River, District of Mackenzie.

Microdinium ornatum Cookson and Eisenack, 1960
Hypotype 65320
Ioannides, N.S., 1986, Geol. Surv. Can., Bull. 371, p. 29, pl. 15, fig. 18.
Eureka Sound Formation, Late Paleocene to ?Early Eocene, "Twosnout Creek", Bylot Island, lat. 73°9′30″N, long. 79°42′30″W, District of Franklin.

Microdinium spinosum Brideaux and McIntyre
Holotype 34148; paratype 41483
Brideaux, W.W. and McIntyre, D.J., 1975, Geol. Surv. Can., Bull. 252, p. 23, pl. 6, figs. 3-5.
Crossley Lakes Member, Langton Bay Formation, Lower Cretaceous, lat. 64°01′30″N, long. 126°02′W, Horton River, District of Mackenzie.

Microdinium sp. A
Fig. specs. 65316-65318
Ioannides, N.S., 1986, Geol. Surv. Can., Bull. 317, p. 29, pl. 15, figs. 14, 15, 17.
Kanguk Formation, Upper Cretaceous, Bylot Island, south coast, lat. 72°53′30″N, long. 78°16′W, and "Twosnout Creek", lat. 73°14′30″N, long. 79°49′W; Eureka Sound Formation, Late Paleocene to ?Early Eocene, south coast Bylot Island, lat. 72°53′30″N, long. 78°25′W, District of Franklin.

Microdinium sp. B
Fig. spec. 65319
Ioannides, N.S., 1986, Geol. Surv. Can., Bull. 371, p. 29, pl. 15, fig. 16.
Kanguk Formation, Upper Cretaceous, south coast Bylot Island, lat. 72°53′30″N, long. 78°16′W, District of Franklin.

Millioudinium saetigerum McIntyre and Brideaux
Holotype 58586
McIntyre, D.J. and Brideaux, W.W., 1980, Geol. Surv. Can., Bull. 320, p. 15, pl. 3, figs. 2, 3, 7.
Lower Cretaceous, Martin Creek, lat. 68°12′10″N, long. 135°34′50″W, Richardson Mountains, District of Mackenzie.

Millioudodinium spinoreticulatum McIntyre and Brideaux
Holotype 58587; paratypes 58588, 58589
McIntyre, D.J. and Brideaux, W.W., 1980, Geol. Surv. Can., Bull. 320, p. 15, pl. 3, figs. 4, 8-12.
Lower Cretaceous, Martin Creek, lat. 68°12′10″N, long. 135°34′50″W (58587) and "Grizzly Gorge", lat. 68°15′49″N, long. 135°40′59″W, Richardson Mountains, District of Mackenzie.

Moesiodinium raileanui Antonescu, 1974
Hypotype 69367
Davies, E.H., 1983, Geol. Surv. Can., Bull 359, pl. 6, fig. 23.
Lower Cretaceous, Panarctic Sandy Point L-46 well, northwestern Melville Island, District of Franklin.

Moesiodinium sp. A
Fig. specs. 75477, 75478
Davies, E.H., 1985, Palynology, vol. 9, p. 129, pl. 3, figs. 9, 10.
Early and Late Pliensbachian, Liassic, Zambujal Section, south of Condeixa, and Brehna Road Section, Lusitanian Basin, west-central Portugal.

Muderongia asymmetrica Brideaux
Holotype 46054; paratype 46055
Brideaux, W.W., 1977, Geol. Surv. Can., Bull 281, p. 40, pl. 15, figs. 9, 10; pl. 16, fig. 1.
Cretaceous, IOE Stoney Core Hole F-42, depths 390-400 and 510-520 feet, Stoney Creek about 18 miles west-southwest of Fort McPherson, lat. 67°21′23″N, long. 135°38′45″W, District of Mackenzie.

Muderongia asymmetrica Brideaux, 1977
Hypotypes 91443, 91449
Sweet, A.R. and McIntyre, D.J., 1988, Can. Soc. Petrol. Geol., Mem. 15, figs. 8.15, 8.21.
Upper Cretaceous, Kaskapau Formation, south side of Grande Cache Highway 40, l.s.d.15, sec. 13, tp. 57, rge. 7, W.5th mer.; Moosehound Member, Cardium Formation, west bank of Little Berland River, l.s.d. 1, sec. 6, tp. 54, rge. 7, W.5th mer., Alberta.

Muderongia perforata Alberti
Hypotype 41860
Williams, G.L., 1975, Geol. Surv. Can., Paper 74-30, vol. 2, pl. 5, fig. 17.
Barremian, Lower Cretaceous, Amoco-IOE Puffin B-90 well, lat. 44°39′12.73″N, long. 53°42′28.35″W, Grand Banks.

Muderongia simplex Alberti
Hypotype 41852
Williams, G.L., 1975, Geol. Surv. Can., Paper 74-30, vol. 2, pl. 5, fig. 9.
Barremian, Lower Cretaceous, Amoco-IOE Puffin B-90 well, lat. 44°39′12.73″N, long. 53°42′28.35″W, Grand Banks.

Muderongia simplex Alberti
Hypotype 45810
Brideaux, W.W., 1976, Geol. Surv. Can., Paper 76-1B, pl. 44.2, fig. 1.
Lower Cretaceous, depth 8100-8120 feet, Gulf Mobil Parsons N-10 well, ca. 1.5 miles east-northeast of Parsons Lake, lat. 69°59′48.52″N, long. 133°31′50.33″W, District of Mackenzie.

Muderongia simplex Alberti
Hypotypes 56739, 56741
Williams, G.L. and Bujak, J.P., 1980, Initial Repts. Deep Sea Drilling Project, vol. 50, pl. 5, figs. 2, 4.
Early Cretaceous, DSDP Site 416, lat. 32°50.18′N, long. 10°48.06′W, northwestern Moroccan continental margin.

Muderongia sp. cf. *M. simplex* Alberti
Fig. specs. 56740, 56742
Williams, G.L. and Bujak, J.P., 1980, Initial Repts. Deep Sea Drilling Project, vol. 50, pl. 5, figs. 3, 5.
Early Cretaceous, DSDP Site 416, lat. 32°50.18′N, long. 10°48.06′W, northwestern Moroccan continental margin.

Muderongia sp. cf. *M. simplex* Alberti, 1961
Fig. spec. 58639
McIntyre, D.J. and Brideaux, W.W., 1980, Geol. Surv. Can., Bull. 320, p. 25, pl. 12, fig. 4.
Lower Cretaceous, 'Grizzly Gorge', lat. 68°15′49″N, long. 135°40′59″W, Richardson Mountains, District of Mackenzie.

Muderongia tetracantha (Gocht) Alberti
Hypotype 41541
Brideaux, W.W. and McIntyre, D.J., 1975, Geol. Surv. Can., Bull. 252, p. 34, pl. 11, fig. 6.
Crossley Lakes Member, Langton Bay Formation, Lower Cretaceous, lat. 64°01′30″N, long. 126°02′W, Horton River, District of Mackenzie.

Muderongia tetracantha (Gocht) Alberti
Hypotype 46057
Brideaux, W.W., 1977, Geol. Surv. Can., Bull 281, p. 41, pl. 16, fig. 2.
Cretaceous, IOE Stoney Core Hole F-42, depth 875-880 feet, Stoney Creek about 18 miles west-southwest of Fort McPherson, lat. 67°21′23″N, long. 135°38′43″W, District of Mackenzie.

Muderongia sp. A
Fig. specs. 41539, 41540
Brideaux, W.W. and McIntyre, D.J., 1975, Geol. Surv. Can., Bull. 252, p. 34, pl. 11, figs. 4, 5.
Crossley Lakes Member, Langton Bay Formation, Lower Cretaceous, lat. 64°01′30″N, long. 126°02′W, Horton River, District of Mackenzie.

Muderongia sp. A Brideaux and McIntyre 1976
Fig. spec. 45814
Brideaux, W.W., 1976, Geol. Surv. Can., Paper 76-1B, pl. 44.2, fig. 5.
Lower Cretaceous, depth 7040-7070 feet, Gulf Mobil Parsons N-10 well, ca. 1.5 miles east-northeast of Parsons Lake, lat. 69°59′48.52″N, long. 133°31′50.33″W, District of Mackenzie.

Multispinula minuta Harland and Reid *in* Harland et al., 1980
Hypotype 55368
Mudie, P.J., 1986, Initial Repts. Deep Sea Drilling Project, vol. XCIV, p. 804, pl. 5, fig. 8.
Pleistocene, hole 607, depth 0.5 m, lat. 41°0.07′N, long. 32°57.44′W, at base of upper western flank of the Mid-Atlantic Ridge, North Atlantic Ocean.

Multispinula sp. A
Fig. spec. 55361
Mudie, P.J., 1986, Initial Repts. Deep Sea Drilling Project, vol. XCIV, p. 804, pl. 5, fig. 19.
Recent, grab sample from Lancaster Sound, Baffin Bay, District of Franklin.

Nannoceratopsis gracilis Alberti 1961 emend.
Hypotypes 53023-53026, 53030-53034
van Helden, B.G.T., 1977, Geol. Surv. Can., Paper 77-1B, p. 165, pl. 33.1, figs. 10-14; pl. 33.2, figs. 1-13.
Wilkie Point Formation, Jurassic, 1.6 to 8 km north and east of Mould Bay Weather Station (53023), east side Intrepid Inlet, Prince Patrick Island, District of Franklin.

Nannoceratopsis pellucida Deflandre, 1938, emend. Evitt, 1961
Hypotype 58350
Tan, J.T. and Hills, L.V., 1978, Geol. Surv. Can., Paper 78-1C, p. 66, pl. 12.1, fig. 10.
Savik Formation, Upper Jurassic, lat. 77°48′N, long. 112°20′W, Mackenzie King Island, Arctic.

Nannoceratopsis senex van Helden
Holotype 53017; paratypes 53018-53022
van Helden, B.G.T., 1977, Geol. Surv. Can., Paper 77-1B, p. 165, pl. 33.1, figs. 1-9.
Wilkie Point Formation, Jurassic, depth 4280-4300 feet, B.P. et al. Panarctic Emerald K-33 well, Emerald Island (53017) and 1.6 to 8 km north and east of Mould Bay Weather Station, Prince Patrick Island, District of Franklin.

Nannoceratopsis senex van Helden
Hypotypes 75505, 75506
Davies, E.H., 1985, Palynology, vol. 9, pl. 4, figs. 17, 18.
Early Pliensbachian and Aalenian, Liassic, Zambujal Section, south of Condeixa, Lusitanian Basin, west-central Portugal.

Nannoceratopsis senx van Helden, 1977 var. B
Hypotype 69366
Davies, E.H., 1983, Geol. Surv. Can., Bull. 359, p. 30, pl. 6, fig. 19.
Lower-Middle Jurassic, depth 1880 feet, Panarctic Sandy Point L-46 well, northwestern Melville Island, District of Franklin.

Nannoceratopsis? sp. A
Fig. specs. 76529, 76306
Davies, E.H. and Poulton, T.P., 1986, Geol. Surv. Can., Paper 86-1B, p. 525, pl. 55.3, figs. 4, 7.
Fernie Formation, Upper Jurassic, Christina Falls, lat. 56°40′N, long. 123°4′W, northeastern British Columbia.

Nematosphaeropsis aquaeducta Piasecki, 1980
Hypotypes 55321, 55325
Mudie, P.J., 1986, Initial Repts. Deep Sea Drilling Project, vol. XCIV, p. 804, pl. 3, figs. 7, 11a, b.
Lower Pliocene, hole 611C, depth 225.59 m, lat. 52°50.15′N, long. 30°18.1′W, in the Gardar Drift on southwestern flank of Reykjanes Ridge; Upper Miocene, hole 607A, depth 283 m, lat. 41°0.07′N, long. 32°57.44′W, at base of upper western flank of the Mid-Atlantic Ridge, North Atlantic Ocean.

Nematosphaeropsis labyrinthea (Ostenfeld) Reid
Hypotypes 65086, 65087
Bujak, J.P. and Davies, E.H., 1982, Bull. Can. Petrol. Geol., vol. 29, no. 3, 1981, p. 422, pl. 1, figs. 1, 2.
Dixon, J. et al., 1984, Geol. Surv. Can., Paper 82-13, pl. 4, fig. 10(65087).
Miocene, DSDP Site 406, core 1, sec. 1 (90-94 cm), Rockall Plateau, North Atlantic, and Gulf Dome et al. Hunt Kopanoar M-13 well, sidewall core 8603 feet, Beaufort Sea, Canada.

Nematosphaeropsis labyrinthea (Ostenfeld) Reid, 1974
Hypotype 55315
Mudie, P.J., 1986, Initial Repts. Deep Sea Drilling Project, vol. XCIV, p. 804, pl. 3, figs. 1a, b.
Upper Miocene, hole 611C, depth 313.49 m, lat. 52°50.15′N, long. 30°18.1′W, in the Gardar Drift on southwestern flank of Reykjanes Ridge, North Atlantic Ocean.

Nematosphaeropsis oblonga Mudie
Holotype 55318
Mudie, P.J., 1986, Initial Repts. Deep Sea Drilling Project, vol. XCIV, p. 804, pl. 3, figs. 4a, b.
Upper Miocene, hole 611C, depth 313.49 m, lat. 52°50.15′N, long. 30°18.1′W, in the Gardar Drift on southwestern flank of Reykjanes Ridge, North Atlantic Ocean.

Nematosphaeropsis sp.
Fig. spec. 91998
Dietrich, J.R. et al., 1989, Geol. Surv. Can., Paper 89-1G, pl. 1, fig. 2.
Early to Middle Eocene, depth 5100-5190 feet, Dome Pacific et al. PEX Natsek E056 well, lat. 69°45′21.46″N, long. 139°44′34.58″W, western Beaufort Sea.

Nematosphaeropsis sp. 1
Fig. spec. 55317
Mudie, P.J., 1986, Initial Repts. Deep Sea Drilling Project, vol. XCIV, p. 804, pl. 3, fig. 3.
Upper Miocene, hole 611C, depth 313.49 m, lat. 52°50.15′N, long. 30°18.1′W, in the Gardar Drift on southwestern flank of Reykjanes Ridge, North Atlantic Ocean.

Nematosphaeropsis sp. B of Williams and Brideaux, 1975
Fig. spec. 65088
Bujak, J.P. and Davies, E.H., 1982, Bull. Can. Petrol. Geol., vol. 29, no. 3, 1981, p. 422, pl. 1, fig. 3.
Dixon, J. et al, 1984, Geol. Surv. Can., Paper 82-13, pl. 4, fig. 7.
Miocene, Dome Gulf et al. Hunt Kopanoar M-13 well, sidewall core 8672 feet, Beaufort Sea, Canada.

Netrelytron sp. A
Fig. specs. 76275-76277, 76280, 76287
Davies, E.H. and Poulton, T.P., 1986, Geol. Surv. Can., Paper 86-1B, pl. 55.1, figs. 4-8.
Fernie and Monteith (76287) formations, Upper Jurassic, Christina Falls, lat. 56°40′N, long. 123°4′W, northeastern British Columbia.

Nyktericysta arachnion
Hypotypes 91987, 91989
Banerjee, I. and Davies, E.H., 1988, Can. J. Petrol. Geol., Mem. 15, figs. 7.3, 7.5.
Mannville Group, Lower Cretaceous, depth 4266 feet, well 15-21-47-23W4, Alberta.

Nyktericysta davissii Bint, 1986
Hypotype 84738
Leckie, D.A. et al., 1990, J. Sed. Petrology, vol. 60, no. 1, p. 105, fig. 6A.
Shaftesbury Formation, Lower Cretaceous, west bank Peace River near town of Peace River, Alberta.

Occisucysta monoheuriskos Gitmez
Hypotypes 76292, 76531, 76532
Davies, E.H. and Poulton, T.P., 1986, Geol. Surv. Can., Paper 86-1B, pl. 55.4, figs. 2, 5, 8.
Fernie Formation, Upper Jurassic, Christina Falls, lat. 56°40′N, long. 123°4′W, northeastern British Columbia.

Occisucysta sp. A
Fig. specs. 53685, 53704
Bujak, J.P. and Williams, G.L., 1978, Geol. Surv. Can., Bull. 297, pl. 1, fig. 18; pl. 3, figs. 2-4.
Early Cretaceous, depth 13460-13490 feet, Amoco-IOE Puffin B-90 well, lat. 44°39′12.73″N, long. 53°42′28.35″W, Grand Banks, and 9550-9580 feet, Shell Cree E-35 well, lat. 43°44′20.71″N, long. 60°35′55.9″W, Scotian Shelf.

Occisucysta sp. A
Fig. spec. 47885
Williams, G.L., 1978, Initial Repts. Deep Sea Drilling Project, vol. 41, pl. 5, fig. 4.
Early Cretaceous, DSDP Site 370, lat. 32°50.2′N, long. 10°46.6′W, northwestern Moroccan continental margin.

Occisucysta sp. A
Fig. specs. 76297, 76298
Davies, E.H. and Poulton, T.P., 1986, Geol. Surv. Can., Paper 86-1B, p. 525, pl. 55.5, figs. 2, 8.
Fernie Formation, Upper Jurassic, Christina Falls, lat. 56°40′N, long. 123°4′W, northeastern British Columbia.

Odontochitina costata (Alberti)
Hypotypes 38089, 38090
McIntyre, D.J., 1974, Geol. Surv. Can., Paper 74-14, pl. 7, figs. 11, 12.
Upper Cretaceous, lat. 69°27′30″N, long. 126°58′W, Horton River, District of Mackenzie.

Odontochitina costata Alberti
Hypotype 62633
Ioannides, N.S. and McIntyre, D.J., 1980, Geol. Surv. Can., Paper 80-1A, pl. 31.3, fig. 4.
Upper Cretaceous, Caribou Hills, Mackenzie River, District of Mackenzie.

Odontochitina costata Alberti, 1961
Hypotype 65328
Ioannides, N.S., 1986, Geol. Surv. Can., Bull. 371, p. 29, pl. 16, fig. 16.
Kanguk Formation, Upper Cretaceous, south coast Bylot Island, lat. 72°53′30″N, long. 78°16′W, District of Franklin.

Odontochitina operculata (Wetzel)
Hypotype 38088
McIntyre, D.J., 1974, Geol. Surv. Can., Paper 74-14, pl. 7, fig. 10.
Upper Cretaceous, lat. 69°27′30″N, long. 126°58′W, Horton River, District of Mackenzie.

Odontochitina operculata (Wetzel)
Hypotype 41857
Williams, G.L., 1975, Geol. Surv. Can., Paper 74-30, vol. 2, pl. 5, fig. 14.
Santonian, Upper Cretaceous, core hole 5, lat. 45°28′N, long. 50°5′W, Grand Banks.

Odontochitina operculata (Wetzel) Deflandre and Cookson, 1955
Hypotype 71415
Wilson, M.A., 1978, Palaeontographica, Abt.B, vol. 166, p. 153, pl. 11, fig. 14.
paleocene, IOE Atkinson H-25 well, north shore Tuktoyaktuk Peninsula, lat. 69°44′20″N, long. 131°50′06″W, District of Mackenzie.

Odontochitina operculata (O.Wetzel) Deflandre and Cookson, 1955
Hypotypes 65323-65327
Ioannides, N.S., 1986, Geol. Surv. Can., Bull. 371, p. 30, pl. 16, figs. 3-5, 7, 10.
Kanguk Formation, Upper Cretaceous, Bylot Island, south coast, lat. 72°53′30″N, long. 78°16′W, and "Twosnout Creek", lat. 73°14′30″N, District of Franklin.

Odontochitina operculata (O.Wetzel) Deflandre and Cookson, 1955
Hypotype 91455
Sweet, A.R. and McIntyre, D.J., 1988, Can. Soc. Petrol. Geol., Mem. 15, fig. 8.27.
Wapiabi Formation, Upper Cretaceous, south side of McLeod River, l.s.d.16, sec. 16, tp. 47, rge. 23, W.5th mer., Alberta.

Odontochitina singhii Morgan, 1980
Hypotype 84748
Leckie, D.A., 1990, J. Sed. Petrology, vol. 60, no. 1, fig. 6K.
Shaftesbury Formation, Lower Cretaceous, west bank Peace River near town of Peace River, Alberta.

Odontochitina sp.
Fig. spec. 46058
Brideaux, W.W., 1977, Geol. Surv. Can., Bull 281, p. 41, pl. 16, fig. 3.
Husky Formation, Jurassic, IOE Stoney Core Hole F-42, depth 1000-1010 feet, Stoney Creek about 18 miles west-southwest of Fort McPherson, lat. 67°21′23″N, long. 135°38′43″W, District of Mackenzie.

Odontochitina sp.
Fig. spec. 71388
Wilson, M.A., 1978, Palaeontographica, Abt.B, vol. 166, p. 153, pl. 10, fig. 10.
Cretaceous, Police Island, north bank Mackenzie River, 12 miles southeast of Fort Norman, District of Mackenzie.

Odontochitina sp.
Fig. spec. 65330
Ioannides, N.S., 1986, Geol. Surv. Can., Bull. 371, pl. 16, fig. 9.
Kanguk Formation, Upper Cretaceous, "Twosnout Creek", Bylot Island, lat. 73°14′30″N, long. 79°49′W, District of Franklin.

Oligosphaeridium albertense (Pocock) Davey and Williams
Hypotypes 46031, 46032
Brideaux, W.W., 1977, Geol. Surv. Can., Bull 281, p. 27, pl. 11, figs. 5-7; pl. 12, figs. 1-3.
Cretaceous, IOE Stoney Core Hole F-42, depth 760-765 feet, Stoney Creek about 18 miles west-southwest of Fort McPherson, lat. 67°21′23″N, long. 135°38′43″W, and Banff-Aquitaine-Arco-Rat Pass K-35 well, depth 1025 feet, 6 miles southeast of Mount Goodenough, lat. 67°54′42.55″N, long. 135°21′56.83″W, District of Mackenzie.

Oligosphaeridium anthophorum (Cookson and Eisenack)
Hypotype 38126
McIntyre, D.J., 1974, Geol. Surv. Can., Paper 74-14, pl. 10, fig. 2.
Upper Cretaceous, lat. 69°28′N, long. 126°58′W, Horton River, District of Mackenzie.

Oligosphaeridium anthophorum (Cookson and Eisenack) Davey
Hypotype 41515
Brideaux, W.W. and McIntyre, D.J., 1975, Geol. Surv. Can., Bull. 252, p. 28, pl. 8, figs. 12, 13.

Crossley Lakes Member, Langton Bay Formation, Lower Cretaceous, lat. 69°04′N, long. 125°51′W, Horton River, District of Mackenzie.

Oligosphaeridium anthophorum (Cookson and Eisenack) Davey, 1969
Hypotypes 65340, 65341
Ioannides, N.S., 1986, Geol. Surv. Can., Bull. 371, p. 30, pl. 18, figs. 4, 7.
Kanguk Formation, Upper Cretaceous, Bylot Island, south coast, lat. 72°53′30″N, long. 78°16′W, and "Twosnout Creek", lat. 73°14′30″N, long. 79°49′W, District of Franklin.

Oligosphaeridium asterigerum (Gocht) Davey and Williams, 1969
Hypotype 58626
McIntyre, D.J. and Brideaux, W.W., 1980, Geol. Surv. Can., Bull. 320, p. 21, pl. 10, fig. 5.
Lower Cretaceous, Martin Creek, lat. 68°12′10″N, long. 135°34′50″W, Richardson Mountains, District of Mackenzie.

Oligosphaeridium asterigium (Gocht) Davey and Williams
Hypotype 41505
Brideaux, W.W. and McIntyre, D.J., 1975, Geol. Surv. Can., Bull. 252, p. 28, pl. 8, fig. 1.
Horton River Formation, Lower Cretaceous, lat. 69°19′N, long. 126°50′W, Horton River, District of Mackenzie.

Oligosphaeridium complex (White)
Hypotype 38123
McIntyre, D.J., 1974, Geol. Surv. Can., Paper 74-14, pl. 9, fig. 10.
Upper Cretaceous, lat. 69°27′30″N, long. 126°58′W, Horton River, District of Mackenzie.

Oligosphaeridium complex (White) Davey and Williams
Hypotype 41506
Brideaux, W.W. and McIntyre, D.J., 1975, Geol. Surv. Can., Bull. 252, p. 28, pl. 8, fig. 2.
Horton River Formation, Lower Cretaceous, lat. 69°19′N, long. 126°50′W, Horton River, District of Mackenzie.

Oligosphaeridium complex (White)
Hypotype 41859
Williams, G.L., 1975, Geol. Surv. Can., Paper 74-30, vol. 2, pl. 5, fig. 16.
Aptian, Lower Cretaceous, Shell Mohican I-100 well, lat. 42°59′39.04″N, long. 62°28′51.32″W, Scotian Shelf.

Oligosphaeridium complex (White) Davey and Williams
Hypotype 47888
Williams, G.L., 1978, Initial Repts. Deep Sea Drilling Project, vol. 41, pl. 5, fig. 9.
Early Cretaceous, DSDP Site 370, lat. 32°50.2′N, long. 10°46.6′W, northwestern Moroccan continental margin.

Oligosphaeridium cf. *complex* (White) Davey and Williams
Hypotypes 47887, 47889, 47890
Williams, G.L., 1978, Initial Repts. Deep Sea Drilling Project, vol. 41, p. 797, pl. 5, figs. 7, 10, 11.
Early Cretaceous, DSDP Site 370, lat. 32°50.2′N, long. 10°46.6′W, northwestern Moroccan continental margin.

Oligosphaeridium sp. cf. *O. complex* (White) Davey and Williams
Fig. spec. 45823
Brideaux, W.W., 1976, Geol. Surv. Can., Paper 76-1B, p. 256, pl. 44.2, fig. 14.
Lower Cretaceous, depth 8400-8420 feet, Gulf Mobil Parsons N-10 well, ca. 1.5 miles east-northeast of Parsons Lake, lat. 69°59′48.52″N, long. 133°31′50.33″W, District of Mackenzie.

Oligosphaeridium sp. cf. *O. complex* (White) Davey and Williams
Fig. spec. 46033
Brideaux, W.W., 1977, Geol. Surv. Can., Bull 281, p. 29, pl. 12, figs. 4, 5.
Cretaceous, Banff-Aquitaine-Arco Rat Pass K-35 well, 6 miles southeast of Mount Goodenough, lat. 67°54′42.55″N, long. 135°21′56.83″W, District of Mackenzie.

Oligosphaeridium sp. cf. *O. complex (White) Davey and Williams, in* Davey et al., 1966
Hypotype 65339
Ioannides, N.S., 1986, Geol. Surv. Can., Bull. 371, p. 31, pl. 18, fig. 3.
Kanguk Formation, Upper Cretaceous, south coast Bylot Island, lat. 72°53′30″N, long. 78°16′W, District of Franklin.

Oligosphaeridium dividuum Williams
Holotype 47891
Williams, G.L., 1978, Initial Repts. Deep Sea Drilling Project, vol. 41, p. 797, pl. 5, fig. 8.
Late Cretaceous, DSDP Site 370, lat. 32°50.2′N, long. 10°46.6′W, northwestern Moroccan continental margin.

Oligosphaeridium irregulare (Pocock) Davey and Williams
Hypotypes 41507, 41508
Brideaux, W.W. and McIntyre, D.J., 1975, Geol. Surv. Can., Bull. 252, p. 28, pl. 8, figs. 3, 6.
Crossley Lakes Member, Langton Bay Formation, Lower Cretaceous, lat. 69°04′N, long. 125°51′W, Horton River, District of Mackenzie.

Oligosphaeridium nannum Davey
Hypotype 46034
Brideaux, W.W., 1977, Geol. Surv. Can., Bull 281, p. 29, pl. 12, figs. 6, 7.
Cretaceous, Banff-Aquitaine-Arco Rat Pass K-35 well, depth 1130 feet, 6 miles southeast of Mount Goodenough, lat. 67°54′42.55″N, long. 135°21′56.83″W, District of Mackenzie.

Oligosphaeridium pulcherrimum (Deflandre and Cookson)
Hypotypes 38124, 38125
McIntyre, D.J., 1974, Geol. Surv. Can., Paper 74-14, pl. 9, fig. 11; pl. 10, fig. 1.
Upper Cretaceous, lat. 69°27′30″N, long. 126°58′W, Horton River, District of Mackenzie.

Oligosphaeridium pulcherrimum (Deflandre and Cookson) Davey and Williams
Hypotype 41509
Brideaux, W.W. and McIntyre, D.J., 1975, Geol. Surv. Can., Bull. 252, p. 29, pl. 8, fig. 4.
Crossley Lakes Member, Langton Bay Formation, Lower Cretaceous, lat. 69°04′N, long. 125°51′W, Horton River, District of Mackenzie.

Oligosphaeridium pulcherrimum (Deflandre and Cookson)
Hypotype 41858
Williams, G.L., 1975, Geol. Surv. Can., Paper 74-30, vol. 2, pl. 5, fig. 15.
Hauterivian, Lower Cretaceous, Shell Mohican I-100 well, lat. 42°59′39.04″N, long. 62°28′51.32″W, Scotian Shelf.

Oligosphaeridium pulcherrimum (Deflandre and Cookson) Davey and Williams, *in* Davey et al., 1966
Hypotypes 65337, 65338
Ioannides, N.S., 1986, Geol. Surv. Can., Bull. 371, p. 31, pl. 18, figs. 1, 2.
Kanguk Formation, Upper Cretaceous, south coast Bylot Island, lat. 72°53′30″N, long. 78°16′W, District of Franklin.

Oligosphaeridium pulcherrimum (Deflandre and Cookson) Davey and Williams, 1966
Hypotype 91472
Sweet, A.R. and McIntyre, D.J., 1988, Can. Soc. Petrol. Geol., Mem. 15, fig. 9.17.
Leyland member, Cardium Formation, Upper Cretaceous, south side of McLeod River, l.s.d.16, sec. 16, tp. 47, rge. 23, W.5th mer., Alberta.

Oligosphaeridium totum totum Brideaux
Hypotype 41513
Brideaux, W.W. and McIntyre, D.J., 1975, Geol. Surv. Can., p. 29, pl. 8, fig. 10.
Crossley Lakes Member, Langton Bay Formation, Lower Cretaceous, lat. 64°01′30″N, long. 126°02′W, Horton River, District of Mackenzie.

Oligosphaeridium vasiformum (Neale and Sarjeant) Davey and Williams, 1966
Hypotypes 58621-58625
McIntyre, D.J. and Brideaux, W.W., 1980, Geol. Surv. Can., Bull. 320, p. 21, pl. 10, figs. 1-4, 7, 8, 12.
Lower Cretaceous, Martin Creek, lat. 68°12′10″N, long. 135°34′50″W, Richardson Mountains, District of Mackenzie.

Oligosphaeridium sp.
Fig. spec. 71405
Wilson, M.A., 1978, Palaeontographica, Abt.B, vol. 166, p. 154, pl. 11, fig. 4.
Cretaceous, Police Island, north bank Mackenzie River, 12 miles southeast of Fort Norman, District of Mackenzie.

Oligosphaeridium sp. A
Fig. spec. 41882
Williams, G.L., 1975, Geol. Surv. Can., Paper 74-30, vol. 2, pl. 7, fig. 3.
Santonian, Upper Cretaceous, Mobil Sable Island C-67 well, lat. 43°56′4.91″N, long. 59°55′1.38″W, Scotian Shelf.

Oligosphaeridium sp. A
Fig. spec. 56728
Williams, G.L. and Bujak, J.P., 1980, Initial Repts. Deep Sea Drilling Project, vol. 50, pl. 4, figs. 1, 2.
Early Cretaceous, DSDP Site 416, lat. 32°50.18′N, long. 10°48.06′W, northwestern Moroccan continental margin.

Oligosphaeridium? sp. A Brideaux, 1971
Fig. spec. 41511
Brideaux, W.W. and McIntyre, D.J., 1975, Geol. Surv. Can., Bull. 252, p. 29, pl. 8, fig. 7.
Gilmore Lake Member, Langton Bay Formation, Lower Cretaceous, lat. 68°50′N, long. 125°24′W, Horton River, District of Mackenzie.

Oligosphaeridium sp. A
Fig. specs. 76282, 76283
Davies, E.H. and Poulton, T.P., 1986, Geol. Surv. Can., Paper 86-1B, p. 525, pl. 55.2, figs. 1, 4.
Fernie Formation, Upper Jurassic, Christina Falls, lat. 56°40′N, long. 123°4′W, northeastern British Columbia.

Oligosphaeridium sp. AB
Fig. specs. 41510, 41512
Brideaux, W.W. and McIntyre, D.J., 1975, Geol. Surv. Can., Bull. 252, p. 29, pl. 8, figs. 5, 8, 9.
Horton River Formation, Lower Cretaceous, lat. 69°19′N, long. 126°50′W, Horton River, District of Mackenzie.

Oligosphaeridium? sp. B Brideaux, 1971
Fig. spec. 41514
Brideaux, W.W. and McIntyre, D.J., 1975, Geol. Surv. Can., Bull. 252, p. 29, pl. 8, fig. 11.
Horton River Formation, Lower Cretaceous, lat. 69°27′30″N, long. 126°54′W, Horton River, District of Mackenzie.

Oligosphaeridium sp. FE
Fig. spec. 45941
Brideaux, W.W., 1976, Geol. Surv. Can., Paper 76-1C, p. 119, pl. 23.2, fig. 7.
Lower Cretaceous, south bank Martin Creek, District of Mackenzie.

Oligosphaeridium sp. GE
Fig. specs. 58627-58630
McIntyre, D.J. and Brideaux, W.W., 1980, Geol. Surv. Can., Bull. 320, p. 22, pl. 10, figs. 6, 9-11.

Lower Cretaceous, Martin Creek, lat. 68°12′10″N, long. 135°34′50″W, Richardson Mountains, District of Mackenzie.

Oodnadattia tuberculata Eisenack and Cookson
Hypotype 47892
Williams, G.L., 1978, Initial Repts. Deep Sea Drilling Project, vol. 41, pl. 6, fig. 1.
Late Cretaceous, DSDP Site 370, lat. 32°50.2′N, long. 10°46.6′W, northwestern Moroccan continental margin.

Operculodinium centrocarpium (Deflandre and Cookson) Wall, 1967
Hypotype 55305
Mudie, P.J., 1986, Initial Repts. Deep Sea Drilling Project, vol. XCIV, p. 804, pl. 2, fig. 4.
Upper Pliocene, hole 611, depth 118.19 m, lat. 52°50.47′N, long. 30°18.58′W, in the Gardar Drift on southwestern flank of Reykjanes Ridge, North Atlantic Ocean.

Operculodinium crassum Harland, 1979
Hypotypes 55309, 55367
Mudie, P.J., 1986, Initial Repts. Deep Sea Drilling Project, vol. XCIV, p. 805, pl. 2, fig. 9; pl. 5, fig. 6.
Lower Pliocene, hole 611C, depths 209.39 m and 225.59 m, lat. 52°50.15′N, long. 30°18.1′W, in the Gardar Drift on southwestern flank of Reykjanes Ridge, North Atlantic Ocean.

?*Operculodinium echigoense* Matsuoka, 1983
Hypotype 55308
Mudie, P.J., 1986, Initial Repts. Deep Sea Drilling Project, vol. XCIV, p. 804, pl. 2, fig. 7.
Upper Miocene, hole 611C, depth 510.09 m, lat. 52°50.15′N, long. 30°18.1′W, in the Gardar Drift on southwestern flank of Reykjanes Ridge, North Atlantic Ocean.

Operculodinium israelianum (Rossignol)
Hypotype 38696
Haskell, T.R. and Wilson, G.J., 1975, Initial Repts. Deep Sea Drilling Project, vol. 29, pl. 3, fig. 4.
Cenozoic, DSDP Site 284, lat. 40°30.48′S, long. 167°40.81′E, Lord House Rise west of New Zealand.

Operculodinium longispinigerum Matsuoka, 1983
Hypotype 55314
Mudie, P.J., 1986, Initial Repts. Deep Sea Drilling Project, vol. XCIV, p. 805, pl. 2, fig. 13.
Upper Miocene, hole 607, depth 251 m, lat. 41°0.07′N, long. 32°57.44′W, at base of upper western flank of the Mid-Atlantic Ridge, North Atlantic Ocean.

Operculodinium? spinigerum Brideaux
Holotype 46035; paratypes 46036-46039
Brideaux, W.W., 1977, Geol. Surv. Can., Bull 281, p. 30, pl. 12, figs. 8, 9; pl. 13, figs. 1-11.
Cretaceous, Banff-Aquitaine-Arco Rat Pass K-35 well, depth 1130 feet, 6 miles southeast of Mount Goodenough, lat. 67°54′42.55″N, long. 135°21′56.83″W, District of Mackenzie.

Operculodinium wallii Matsuoka, 1983
Hypotype 55302
Mudie, P.J., 1986, Initial Repts. Deep Sea Drilling Project, vol. XCIV, p. 805, pl. 2, fig. 1.
Upper Miocene, hole 607, depth 251 m, lat. 41°0.07′N, long. 32°57.44′W, at base of upper western flank of the Mid-Atlantic Ridge, North Atlantic Ocean.

Operculodinium sp. of Jan du Chêne (1977)
Fig. spec. 55313
Mudie, P.J., 1986, Initial Repts. Deep Sea Drilling Project, vol. XCIV, p. 805, pl. 2, figs. 12a, b.
Lower Pliocene, hole 607, depth 212.60 m, lat. 41°0.07′N, long. 32°57.44′W, at base of upper western flank of the Mid-Atlantic Ridge, North Atlantic Ocean.

Operculodinium sp. of Piasecki (1980)
Fig. spec. 55312
Mudie, P.J., 1986, Initial Repts. Deep Sea Drilling Project, vol. XCIV, p. 805, pl. 2, figs. 11a, b.
Upper Miocene, hole 607, depth 275.34 m, lat. 41°0.07′N, long. 32°57.44′W, at base of upper western flank of the Mid-Atlantic Ridge, North Atlantic Ocean.

?*Operculodinium* sp. 1
Fig. spec. 55337
Mudie, P.J., 1986, Initial Repts. Deep Sea Drilling Project, vol. XCIV, p. 805, pl. 4, figs. 9a, b.
Lower Pliocene, hole 607, depth 215.6 m, lat. 41°0.07′N, long. 32°57.44′W, at base of upper western flank of the Mid-Atlantic Ridge, North Atlantic Ocean.

"Organism" sp AE
Fig. specs. 45926-45928
Brideaux, W.W., 1976, Geol. Surv. Can., Paper 76-1C, p. 119, pl. 23.1, figs. 7, 10, 13.
Husky Formation, Lower Cretaceous, south bank Martin Creek, District of Mackenzie.

Ovoidinium scabrosum (Cookson and Hughes)
Hypotypes 47854, 47893
Williams, G.L., 1978, Initial Repts. Deep Sea Drilling Project, vol. 41, pl. 1, fig. 10; pl. 6, fig. 2.
Late Cretaceous, DSDP Site 370, lat. 32°50.2′N, long. 10°46.6′W, northwestern Moroccan continental margin.

Palaeocystodinium benjaminii Drugg
Hypotype 41886
Williams, G.L., 1975, Geol. Surv. Can., Paper 74-30, vol. 2, pl. 7, fig. 7.
Danian, Mobil Tetco Dauntless D-35 well, lat. 44°44′8.26″N, long. 57°20′46.62″W, Scotian Shelf.

Palseocystodinium bulliformum Ioannides
Holotype 65334; paratypes 65332, 65333, 65335
Ioannides, N.S., 1986, Geol. Surv. Can., Bull. 371, p. 31, pl. 17, figs. 2-5.
Eureka Sound Formation, Late Paleocene to ?Early Eocene, Bylot Island, south coast, lat. 72°52′N, long. 78°35′W, and lat. 72°52′N, long. 78°31′W (65335); "Twosnout Creek", lat. 73°13′N, long. 79°53′W (65333), District of Franklin.

Palseocystodinium golzowense Alberti, 1961
Hypotype 65336
Ioannides, N.S., 1986, Geol. Surv. Can., Bull. 371, p. 31, pl. 17, fig. 6.
Eureka Sound Formation, Late Paleocene to ?Early Eocene, "Twosnout Creek", Bylot Island, lat. 73°9′30″N, long. 79°42′30″W, District of Franklin.

Palaeohystrichophora brevispinosa Pocock
= *Tenua brevispinosa*, Brideaux, W.W., 1977, Geol. Surv. Can., Bull. 281, p. 15, pl. 6, fig. 1 (holotype 14358).

Palaeohystrichophora infusorioides Deflandre
Hypotype 41862
Williams, G.L., 1975, Geol. Surv. Can., Paper 74-30, vol. 2, pl. 6, fig. 1.
Turonian, Upper Cretaceous, Amoco IOE Kittiwake P-11 well, lat. 44°40′49.43″N, long. 53°31′45.65″W, Grand Banks.

Palaeohystrichophora infusorioides Deflandre, 1935
Hypotypes 65352-65354
Ioannides, N.S., 1986, Geol. Surv. Can., Bull. 371, p. 32, pl. 19, figs. 1-3.
Kanguk Formation, Upper Cretaceous, south coast of Bylot Island, lat. 72°53′30″N, long. 78°16′W, District of Franklin.

Palaeohystrichophora infusorioides Deflandre, 1935
Hypotype 91445
Sweet, A.R. and McIntyre, D.J., 1988, Can. Soc. Petrol. Geol, Mem. 15, fig. 8.17.
Leyland Member, Cardium Formation, Upper Cretaceous, south side of McLeod River, l.s.d. 16, sec. 16, tp. 47, rge. 23, W. 5th mer., Alberta.

Palaeohystrichophora sp.
Fig spec. 47904
Williams, G.L., 1978, Initial Repts. Deep Sea Drilling Project, vol. 41, pl. 7, fig. 2.
Late Cretaceous, DSDP Site 370, lat. 32°50.2′N, long. 10°46.6′W, northwestern Moroccan continental margin.

Palaeoperidinium ariadnae Norris
Holotypes 68162; hypotypes 68158-68161, 68163, 68164
Norris, G., 1986, Geol. Surv. Can., Bull. 340, p. 45, pl. 12, figs. 10-12; pl. 13, figs. 1-4.
Richards Formation, Middle Eocene, depths 10,500-10,600, 10,700-10,800, 11,700-11,800, 11,100-11,200, 11,800-11,900, 12,300-12,400 and 12,200-12,300 feet, Imperial Nuktak C-22 well, lat. 69°41′07″N, long. 134°51′30″W, District of Mackenzie.

Palaeoperidinium basilium Drugg
Hypotypes 38067-38069
McIntyre, D.J., 1974, Geol. Surv. Can., Paper 74-14, pl. 6, figs. 4-6.
Upper Cretaceous, lat. 69°48′N, long. 126°58′W, Horton River, District of Mackenzie.

Palaeoperidinium cretaceum Pocock, 1962
Hypotype 96489
Leckie, D.A. and Singh, C., 1991, J. Sed. Petrology, vol. 61, no. 5, fig. 3F.
Paddy Member, Peace River Formation, Lower Cretaceous, near town of Peace River, Alberta.

Palaeoperidinium pyrophorum (Ehrenberg)
Hypotypes 38064-38066
McIntyre, D.J., 1974, Geol. Surv. Can., Paper 74-14, pl. 6, figs. 1-3.
Upper Cretaceous, lat. 69°28′N, long. 126°58′W, Horton River, District of Mackenzie.

Palaeoperidinium pyrophorum (Ehrenberg)
Hypotype 41883
Williams, G.L., 1975, Geol. Surv. Can., Paper 74-30, vol. 2, pl. 7, fig. 4.
Danian, Shell Oneida O-25 well, lat. 43°14′57.49″N, long. 61°35′36.38″W, Scotian Shelf.

Palaeoperidinium pyrophorum (Ehrenberg) Sarjeant, 1967
Hypotypes 65329, 65331, 65345-65347
Ioannides, N.S., 1986, Geol. Surv. Can., Bull. 371, p. 32, pl. 16, fig. 8; pl. 17, fig. 1; pl. 18, figs. 8-10.
Kanguk Formation, Upper Cretaceous, south coast of Bylot Island, lat. 72°53′30″N, long. 78°16′W and Eureka Sound Formation, Late Paleocene to ?Early Eocene, lat. 72°53′30″N, long. 78°25′W (65345-65347), District of Franklin.

Palaeoperidinium sp. A Brideaux, 1971
Fig. spec. 41553
Brideaux, W.W. and McIntyre, D.J., 1975, Geol. Surv. Can., Bull. 252, p. 38, pl. 14, fig. 4.
Horton River Formation, Lower Cretaceous, lat. 69°19′N, long. 126°50′W, Horton River, District of Mackenzie.

Palaeoperidinium sp. A
Fig. specs. 53705, 53706
Bujak, J.P. and Williams, G.L., 1978, Geol. Surv. Can., Bull. 297, pl. 3, figs. 5-7.
Late Cretaceous, depth 2240-2270 feet, Shell Wyandot E-53 well, lat. 44°52′19.78″N, long. 59°23′55.7″W, Scotian Shelf.

Palaeoperidinium sp. A
Fig. specs. 68165, 68166
Norris, G., 1986, Geol. Surv. Can., Bull. 340, p. 45, pl. 13, figs. 5,6.
Richards Formation, Middle Eocene, depths 12,200-12,300 and 11,700-11,800 feet, Imperial Nuktak C-22 well, lat. 69°41′07″N, long. 134°51′30″W, District of Mackenzie.

Palaeostomocystis expolita Brideaux
Holotype 46006; paratypes 46007, 46008
Brideaux, W.W., 1977, Geol. Surv. Can., Bull. 281, p. 20. pl. 7, figs. 12-14.
Cretaceous, IOE Stoney Core Hole F-42, Stoney Creek about 18 miles west-southwest of Fort McPherson, lat. 67°21′23″N, long. 135°38′43″W, District of Mackenzie.

Palaeostomocystis fragilis Cookson and Eisenack
=*Fromea nicosia*, Jansonius, J., 1989, Review Palaeobotany and Palynology, vol. 61, no. 1/2, p. 67, pl. 1, figs. 5-7 (hypotype 38114).

Palaeostomocystis senilis McIntyre and Brideaux
Holotype 58604
McIntyre, D.J. and Brideaux, W.W., 1980, Geol. Surv. Can., Bull. 320, p. 18. pl. 5, fig. 12; pl. 6, fig.1.
Lower Cretaceous, Martin Creek, lat. 68°12′10″N, long. 135°34′50″W, Richardson Mountains, District of Mackenzie.
= *Fromea senilis*, Jansonius, J., 1989, Review Palaeobotany and Palynology, vol. 61, no. 1/2, p. 67, pl. 1, fig. 22 (not GSC 50871).

Palaeostomocystis triquetra Brideaux
Holotype 46009; paratypes 46010-46012
Brideaux, W.W., 1977, Geol. Surv. Can., Bull. 281, p. 20. pl. 87, figs. 1-6.
Cretaceous, Banff-Aquitaine-Arco Rat Pass K-35 well, depths 1025, 1010-1020 and 990-1000 feet, 6 miles southeast of Mount Goodenough, lat. 67°54′42.55″N, long. 135°21′56.83″W, District of Mackenzie.

Palaeostomocystis sp. of Baltes (1971)
Fig. spec. 55363
Mudie, P.J., 1986, Initial Repts. Deep Sea Drilling Project, vol. XCIV, p. 805, pl. 5, fig. 21.
Upper Miocene, hole 611C, depth 445.2m, lat. 52°50.15′N, long. 30°18.1′W, in the Gardar Drift on southwestern flank of Reykjanes Ridge, North Atlantic Ocean.

Palaeotetradinium silicorum Deflandre, 1936
Hypotypes 65348, 65349
Ioannides, N.S., 1986, Geol. Surv. Can., Bull. 371, p. 32, pl. 18, figs. 11, 12.
Kanguk Formation, Upper Cretaceous, south coast of Bylot Island, lat. 72°53′30″N, long. 78°16′W, District of Franklin.

Palambages spp.
Fig. specs. 65427-65429, 65446, 65447
Ioannides, N.S., 1986, Geol. Surv. Can., Bull. 371, p. 42, pl. 24, figs. 3, 13, 16; pl. 25, figs. 7, 8.
Kanguk Formation, Upper Cretaceous, Bylot Island, south coast, lat. 72°53′30″N, long. 78°16′W, and "Twosnout Creek", lat. 73°14′30″N, long. 79°49′W (65546); Eureka Sound Formation, Late Paleocene to ?Early Eocene, "Twosnout Creek", Bylot Island, lat. 73°9′30″N, long. 79°42′30″W (65429), District of Franklin.

Paragonyaulacysta borealis (Brideaux and Fisher)
Hypotype 76271
Davies, E.H. and Poulton, T.P., 1986, Geol. Surv. Can., Paper 86-1B, pl. 55.1, fig. 1.
Fernie Formation, Upper Jurassic, Christina Falls, lat. 56°40′N, long. 123°4′W, northeastern British Columbia.

Paragonyaulacysta sp. cf. *P. borealis* (Brideaux and Fisher)
Fig. spec. 76272
Davies, E.H. and Poulton, T.P., 1986, Geol. Surv. Can., Paper 86-1B, pl. 55.1, figs. 2, 3.
Fernie Formation, Upper Jurassic, Christina Falls, lat. 56°40′N, long. 123°4′W, northeastern British Columbia.

Paragonyaulacysta capillosa (Brideaux and Fisher) Stover and Evitt
Hypotypes 76278, 76279
Davies, E.H. and Poulton, T.P., 1986, Geol. Surv. Can., Paper 86-1B, pl. 55.2, figs. 3, 6.
Fernie Formation, Upper Jurassic, Christina Falls, lat. 56°40′N, long. 123°4′W, northeastern British Columbia.

Paragonyaulacysta sp
Fig. spec. 58353
Tan, J.T. and Hills, L.V., 1978, Geol. Surv. Can., Paper 78-1C, p. 67, pl. 12.2, figs. 1, 2.
Ringnes Formation, Upper Jurassic, lat. 77°48′N, long. 112°20′W, Mackenzie King Island, District of Franklin.

Paralecaniella indentata (Deflandre and Cookson) Cookson and Eisenack emend. Elsik, 1975
Hypotypes 65430, 65431
Ioannides, N.S., 1986, Geol. Surv. Can., Bull. 371, p. 42, pl. 24, figs. 4, 10.
Kanguk Formation, Upper Cretaceous, south coast Bylot Island, lat. 72°53′30″N, long. 78°16′W, District of Franklin.

Pareodinia borealis Brideaux and Fisher, 1976
Hypotype 58356
Tan, J.T. and Hills, L.V., 1978, Geol. Surv. Can., Paper 78-1C, p. 67, pl. 12.2, fig. 5.
Ringnes Formation, Upper Jurassic, lat. 77°48′N, long. 112°20′W, Mackenzie King Island, District of Franklin.

Pareodinia capillosa Brideaux and Fisher, 1976
Hypotype 58357
Tan, J.T. and Hills, L.V., 1978, Geol. Surv. Can., Paper 78-1C, p. 67, pl. 12.2, fig. 6.
Ringnes Formation, Upper Jurassic, lat. 77°48′N, long. 112°20′W, Mackenzie King Island, District of Franklin.

Pareodinia ceratophora Deflandre, 1947, emend. Gocht, 1970
Hypotype 58354
Tan, J.T. and Hills, L.V., 1978, Geol. Surv. Can., Paper 78-1C, p. 67, pl. 12.2, fig. 3.
Savik Formation, Upper Jurassic, lat. 77°48′N, long. 112°20′W, Mackenzie King Island, District of Franklin.

Pareodinia ceratophora scopaeus (Sarjeant, 1972) Lentin and Williams, 1973
Hypotype 58355
Tan, J.T. and Hills, L.V., 1978, Geol. Surv. Can., Paper 78-1C, p. 67, pl. 12.2, fig. 4.
Savik Formation, Upper Jurassic, lat. 77°48′N, long. 112°20′W, Mackenzie King Island, District of Franklin.

Pareodinia? ceratophora subsp. *scopaea* (Sarjeant) Lentin and Williams, 1973
Hypotypes 65342-65344
Ioannides, N.S., 1986, Geol. Surv. Can., Bull. 371, p. 32, pl. 18, figs. 5, 6, 15.
Kanguk Formation, Upper Cretaceous, south coast Bylot Island, lat. 72°53′30″N, long. 78°16′W, District of Franklin.

Parvocavatus tuberosus Gitmez
Hypotype 41868
Williams, G.L., 1975, Geol. Surv. Can., Paper 74-30, vol. 2, pl. 6, fig. 7.
Kimmeridgian, Upper Jurassic, Shell Mohican I-100 well, lat. 42°59′39.04″N, long. 62°28′51.32″W, Scotian Shelf.

Pediastrum sp.
Fig. spec. 58646
McIntyre, D.J. and Brideaux, W.W., 1980, Geol. Surv. Can., Bull. 320, p. 26. pl. 12, fig. 13.
Lower Cretaceous, 'Grizzly George', lat. 68°15′49″N, long. 135°40′59″W, Richardson Mountains, District of Mackenzie.

Pentadinium laticinctum Gerlach
Hypotype 41856
Williams, G.L., 1975, Geol. Surv. Can., Paper 74-30, vol. 2, pl. 5, fig. 13.
Middle Miocene, core hole 2, lat. 45°08′N, long. 52°38′W, Grand Banks.

Pentadinium laticinctum Gerlach
Hypotype 47898
Williams, G.L., 1978, Initial Repts. Deep Sea Drilling Project, vol. 41, pl. 6, fig. 7.
Oligocene, DSDP Site 370, lat. 32°50.2′N, long. 10°46.6′W, northwestern Moroccan continental margin.

Pentadinium laticinctum granulatum Gocht
Hypotype 41853
Williams, G.L., 1975, Geol. Surv. Can., Paper 74-30, vol. 2, pl. 5, fig. 10.
Oligocene, core hole 2, lat. 45°08′N, long. 52°38′W, Grand Banks.

Peridinioid sp. A
Fig. specs. 92013, 92020
Dietrich, J.R. et al., 1989, Geol. Surv. Can., Paper 89-1G, p. 141, pl. 2, figs. 5, 14.
Early to Middle Eocene, depths 1100-1190 and 2500-2590 feet, Dome Pacific et al. PEX Natsek E-56 well, lat. 69°45′21.46″N, long. 139°44′34.58″W, western Beaufort Sea.

Peridinioid sp. A (cf. *Palaeoperidinium ariaduae* Norris)
Fig. spec. 92014
Dietrich, J.R. et al., 1989, Geol. Surv. Can., Paper 89-1G, pl. 2, fig. 6.
Early to Middle Eocene, depth 3200-3290 feet, Dome Pacific et al. PEX Natsek E-56 well, lat. 69°45′21.46″N, long. 139°44′34.58″W, western Beaufort Sea.

Peridinioid sp. B
Fig. spec. 92019
Dietrich, J.R. et al., 1989, Geol. Surv. Can., Paper 89-1G, p. 141, pl. 2, fig. 11.
Early to Middle Eocene, depths 2500-2590 feet, Dome Pacific et al. PEX Natsek E-56 well, lat. 69°45′21.46″N, long. 139°44′34.58″W, western Beaufort Sea.

Peridinioid sp. C (cf. *Maduradinium turpis* Norris)
Fig. specs. 92017, 92018
Dietrich, J.R. et al., 1989, Geol. Surv. Can., Paper 89-1G, pl. 2, figs. 9, 10.
Early to Middle Eocene, depths 2800-2890 and 2400-2490 feet, Dome Pacific et al. PEX Natsek E-56 well, lat. 69°45′21.46″N, long. 139°44′34.58″W, western Beaufort Sea.

Perisseiasphaeridium sp. A, B
Fig. specs. 56735, 56737
Williams, G.L. and Bujak, J.P., 1980, Initial Repts. Deep Sea Drilling Project, vol. 50, pl. 4, figs. 10, 12.
Early Cretaceous, DSDP Site 416, lat. 32°50.18′N, long. 10°48.06′W, northwestern Moroccan continental margin.

Phoberocysta neocomica (Gocht)
Hypotype 41863
Williams, G.L., 1975, Geol. Surv. Can., Paper 74-30, vol. 2, pl. 6, fig. 2.
Berriasian-Valanginian, Lower Cretaceous, Amoco IOE Puffin B-90 well, lat. 44°39′12.73″N, long. 53°42′28.35″W, Grand Banks.

Phoberocysta sp. cf. *P. neocomica* (Gocht) Millioud
Fig. spec. 56738
Williams, G.L. and Bujak, J.P., 1980, Initial Repts. Deep Sea Drilling Project, vol. 50, pl. 5, fig. 1.
Early Cretaceous, DSDP Site 416, lat. 32°50.18′N, long. 10°48.06′W, northwestern Moroccan continental margin.

Phoberocysta sp.
Fig. spec. 53707
Bujak, J.P. and Williams, G.L., 1978, Geol. Surv. Can., Bull. 297, pl. 3, fig. 8.
Early Cretaceous, depth 12,160-15,425 feet, Amoco IOE Puffin B-90 well?, Grand Banks.

Phoberocysta sp. A
Fig. spec. 53687
Bujak, J.P. and Williams, G.L., 1978, Geol. Surv. Can., Bull. 297, pl. 1, fig. 20.
Early Cretaceous, depth 12,450-12,480 feet, Shell Cree E-35 well, lat. 43°44′20.71″N, long. 60°35′55.9″W, Scotian Shelf.

Phthanoperidinium sp. A
Fig. spec. 49373
Williams, G.L. and Bujak, J.P., 1977, Am. Assoc. Strat. Palynol., Contr. Strat. Palynol., vol. 1, Contr. Ser. 5A, pl. 5, fig. 9.

Miocene, depth 5370-5400 feet, BP Columbia Bonavista C-99 well, lat. 49°8′6.45″N, long. 51°14′27.85″W, northeast Newfoundland Basin.

Phthanoperidinium sp.
Fig. specs. 92006, 92007
Dietrich, J.R. et al., 1989, Geol. Surv. Can., Paper 89-1G, p. 141, pl. 1, figs. 10, 11.
Early to Middle Eocene, depths 2300-2390 and 2100-2190 feet, Dome Pacific et al. PEX Natsek E-56 well, lat. 69°45′21.46″N, long. 139°44′34.58″W, western Beaufort Sea.

Piperodinium perplexum Mudie *in* Scott et al., 1984
Holotype 55369
Scott, D.B., Mudie, P.J., Vilks, G. and Younger, D.C., 1984, Marine Micropaleontol., vol. 9, p. 214, pl. 1, figs, 1-3.
Mudie, P.J., 1986, Initial Repts, Deep Sea Drilling Project, vol. XCIV, p. 805.
Recent, upper slope of the Nova Scotian Shelf, Atlantic Ocean.

Piperodinium perplexum Mudie *in* Scott et al., 1984
Hypotype 55355
Mudie, P.J., 1986, Initial Repts, Deep Sea Drilling Project, vol. XCIV, p. 805, pl. 5, fig. 11.
Pleistocene, hole 607, depth 42.8m, lat. 41°0.07′N, long. 32°57.44′W, at base of upper western flank of the Mid-Atlantic Ridge, North Atlantic Ocean.

Polysphaeridium laminaspinosum Davey and Williams
Hypotype 41516
Brideaux, W.W. and McIntyre, D.J., 1975, Geol. Surv. Can., Bull. 252, p. 30, pl. 8, fig. 14.
Horton River Formation, Lower Cretaceous, lat. 69°19′N, long. 126°50′W, Horton River, District of Mackenzie.

Polysphaeridium multispinosum Davey
Hypotypes 56699, 56700, 56759
Williams, G.L. and Bujak, J.P., 1980, Initial Repts. Deep Sea Drilling Project, vol. 50, pl. 1, figs. 3, 4; pl. 8, fig. 1.
Early Cretaceous, DSDP Site 416, lat. 32°50.18′N, long. 10°48.06′W, northwestern Moroccan continental margin.

Polysphaeridium sp. cf. *P. subtile* Davey and Williams
Fig. specs. 38158, 38159
McIntyre, D.J., 1974, Geol. Surv. Can., Paper 74-14, pl. 13, figs. 3, 4.
Upper Cretaceous, lat. 69°28′N, long. 126°58′W, Horton River, District of Mackenzie.

Polysphaeridium warrenii Habib
Hypotype 56697
Williams, G.L. and Bujak, J.P., 1980, Initial Repts. Deep Sea Drilling Project, vol. 50, pl. 1, fig. 1.
Early Cretaceous, DSDP Site 416, lat. 32°50.18′N, long. 10°48.06′W, northwestern Moroccan continental margin.

Polysphaeridium sp. A
Fig. specs. 56698, 56760, 56762, 56763
Williams, G.L. and Bujak, J.P., 1980, Initial Repts. Deep Sea Drilling Project, vol. 50, pl. 1, fig. 2; pl. 8, figs. 2, 4, 5.
Early Cretaceous, DSDP Site 416, lat. 32°50.18′N, long. 10°48.06′W, northwestern Moroccan continental margin.

Polysphaeridium sp. AE
Fig. spec. 58631
McIntyre, D.J. and Brideaux, W.W., 1980, Geol. Surv. Can., Bull. 320, p. 22. pl. 11, figs. 1-3.
Lower Cretaceous, Martin Creek, lat. 68°12′10″N, long. 135°34′50″W, Richardson Mountains, District of Mackenzie.

Polystephanephorus sarjeantii Gitmez
Hypotype 41864
Williams, G.L., 1975, Geol. Surv. Can., Paper 74-30, vol. 2, pl. 6, fig. 3.
?Tithonian, Upper Jurassic, Shell Naskapi N-30 well, lat. 43°29′46.29″N, long. 62°34′0.63″W, Scotian Shelf.

Prionodinium alaskense Leffingwell and Morgan
Hypotypes 46064, 45966, 45968, 45969
Brideaux, W.W., 1977, Geol. Surv. Can., Bull. 281, p. 9. pl. 1, figs. 9, 12; pl. 2, figs 1-3.
Cretaceous, Banff-Aquitaine-Arco Rat Pass K-35 well, depths 1380 and 1445 feet, 6 miles southeast of Mount Goodenough, lat. 67°54′42.55″N, long. 135°21′56.83″W, District of Mackenzie.

Prionodinium alveolatum Leffingwell and Morgan
Hypotype 45970
Brideaux, W.W., 1977, Geol. Surv. Can., Bull. 281, p. 9, pl. 2, figs. 4, 5.
Cretaceous, Banff-Aquitaine-Arco Rat Pass K-35 well, depth 1600 feet, 6 miles southeast of Mount Goodenough, lat. 67°54′42.55″N, long. 135°21′56.83″W, District of Mackenzie.

Prolixosphaeridium granulosum (Deflandre) Davey et al.
Hypotype 47896
Williams, G.L., 1978, Initial Repts. Deep Sea Drilling Project, vol. 41, pl. 6, fig. 5.
Late Cretaceous, DSDP Site 370, lat. 32°50.2′N, long. 10°46.6′W, northwestern Moroccan continental margin.

Prolixosphaeridium granulosum (Deflandre) Davey et al.
Hypotypes 56706, 56764
Williams, G.L. and Bujak, J.P., 1980, Initial Repts. Deep Sea Drilling Project, vol. 50, pl. 1, fig. 11; pl. 8, fig. 6.
Early Cretaceous, DSDP Site 416, lat. 32°50.18′N, long. 10°48.06′W, northwestern Moroccan continental margin.

Prolixosphaeridium granulosum (Deflandre) Davey, Downie, Sarjeant, and Williams, *in* Davey et al., 1966
Hypotype 65351
Ioannides, N.S., 1986, Geol. Surv. Can., Bull. 371, p. 33, pl. 18, fig. 4.

Kanguk Formation, Upper Cretaceous, "Twosnout Creek", Bylot Island, lat. 73°14′30″N, long. 79°49′W, District of Franklin.

Prolixosphaeridium parvispinum (Deflandre) Davey, Downie, Sarjeant and Williams
Hypotype 46040
Brideaux, W.W., 1977, Geol. Surv. Can., Bull. 281, p. 31, pl. 14, fig. 1.
Cretaceous, Banff-Aquitaine-Arco Rat Pass K-35 well, depth 1130 feet, 6 miles southeast of Mount Goodenough, lat. 67°54′42.55″N, long. 135°21′56.83″W, District of Mackenzie.

Prolixosphaeridium spissum (McIntyre and Brideaux) Lentin and Williams
Hypotype 76286
Davies, E.H. and Poulton, T.P., 1986, Geol. Surv. Can., Paper 86-1B, pl. 55.5, fig. 6.
Fernie Formation, Upper Jurassic, Christina Falls, lat. 56°40′N, long. 123°4′W, northeastern British Columbia.

?*Prolixosphaeridium torynum* (Cookson and Eisenack) Eisenack and Kjellström
Hypotype 56722
Williams, G.L. and Bujak, J.P., 1980, Initial Repts. Deep Sea Drilling Project, vol. 50, pl. 3, fig. 7.
Late Jurassic, DSDP Site 416, lat. 32°50.18′N, long. 10°48.06′W, northwestern Moroccan continental margin.

Prolixosphaeridium sp.
Fig. spec. 65355
Ioannides, N.S., 1986, Geol. Surv. Can., Bull. 371, p. 33, pl. 19, fig. 4.
Eureka Sound Formation, Late Paleocene to ?Early Eocene, south coast of Bylot Island, lat. 72°53′30″N, long. 78°25′W, District of Franklin.

Protoellipsodinium sp.
Fig. spec. 41893
Williams, G.L., 1975, Geol. Surv. Can., Paper 74-30, vol. 2, pl. 7, fig. 14.
Aptian, Lower Cretaceous, Shell Oneida O-25 well, lat. 43°14′57.49″N, long. 61°33′36.38″W, Scotian Shelf.

Psaligonyaulax duales Brideaux and Fisher, 1976
Hypotypes 58358, 58359
Tan, J.T. and Hills, L.V., 1978, Geol. Surv. Can., Paper 78-1C, p. 67, pl. 12.2, figs. 7, 8.
Ringnes Formation, Upper Jurassic, lat. 77°48′N, long. 112°20′W, Mackenzie King Island, District of Franklin.

Pseudoceratim pelliferum
Hypotype 91993
Banerjee, I. and Davies, E.H., 1988, Can. Soc. Petrol. Geol., Mem. 15, fig. 7.9.
Mannville Group, Lower Cretaceous, depth 4211 feet, well 15-14-48-23W4, Alberta.

Pseudoceratim regium
Hypotypes 91991, 91992
Banerjee, I. and Davies, E.H., 1988, Can. Soc. Petrol. Geol., Mem. 15, figs. 7.7, 7.8.
Mannville Group, Lower Cretaceous, depths 4176 and 5064 feet, wells 15-14-48-23W4 and 11-08-41-24W4, Alberta.

Pseudoceratium sp. cf. *P. gochtii* Neale and Sarjeant
Fig. spec. 45811
Brideaux, W.W., 1978, Geol. Surv. Can., Paper 76-1B, pl. 44.2, fig. 2.
Lower Cretaceous, depth 7300-7320 feet, Gulf Mobil Parsons N-10 well, ca. 1.5 miles east-northeast of Parsons Lake, lat. 69°59′48.52″N, long. 133°31′50.33″W, District of Mackenzie.

Pseudoceratium nudum Gocht
Hypotype 45993
Brideaux, W.W., 1977, Geol. Surv. Can., Bull. 281, p. 14. pl. 5, fig. 3.
Cretaceous, IOE Stoney Core Hole F-42, depth 795-800 feet, Stoney Creek about 18 miles west-southwest of Fort McPherson, lat. 67°21′23″N, long. 135°38′43″W, District of Mackenzie.

Pseudoceratium pelliferum Gocht
Hypotype 41855
Williams, G.L., 1975, Geol. Surv. Can., Paper 74-30, vol. 2, pl. 5, fig. 12.
Valanginian, Lower Cretaceous, Amoco IOE Puffin B-90, lat. 44°39′12.73″N, long. 53°42′28.35″W, Grand Banks.

Pseudoceratium pelliferum Gocht
Hypotypes 56708-56710, 56769
Williams, G.L. and Bujak, J.P., 1980, Initial Repts. Deep Sea Drilling Project, vol. 50, pl. 2, figs. 1-3; pl. 9, fig. 5.
Early Creatacenes, DSDP Site 416, lat. 32°50.18′N, long. 10°48.06′W, northwestern Moroccan continental margin.

Pseudoceratium polymorphum (Eisenack) Bint, 1966
Hypotype 91454
Sweet, A.R. and McIntyre, D.J., 1988, Can. Soc. Petrol. Geol, Mem. 15, fig. 8.26.
Leyland Member, Cardium Formation, Upper Cretaceous, south side of McLeod River, l.s.d. 16, sec. 16, tp. 47, rge. 23, W. 5th mer., Alberta.

Pseudoceratium retusum Brideaux
Holotype 45987; paratypes 45988-45992
Brideaux, W.W., 1977, Geol. Surv. Can., Bull. 281, p. 14, pl. 4, figs. 10-12; pl. 5, figs. 1, 2, 5-10.
Cretaceous, Banff-Aquitaine-Arco Rat Pass K-35 well, depths 1140-1100, 1010-1020 and 1025 feet, 6 miles southeast of Mount Goodenough, lat. 67°54′42.55″N, long. 135°21′56.83″W, and IOE Stoney Core Hole (45989, 45991), depths 795-800 and 825-830 feet, about 18 miles west-southwest of Fort McPherson, lat. 67°21′23″N, long. 135°38′43″W, District of Mackenzie.

Pseudoceratium sp.
Fig. spec. 47900
Williams, G.L., 1978, Initial Repts. Deep Sea Drilling Project, vol. 41, pl. 6, fig. 9.
Late Cretaceous, DSDP Site 370, lat. 32°50.2′N, long. 10°46.6′W, northwestern Moroccan continental margin.

Pterodinium sp. cf. *P. aliferum* Eisenack
Fig. spec. 41517
Brideaux, W.W. and McIntyre, D.J., 1975, Geol. Surv. Can., Bull. 252, p. 30, pl. 8, figs. 15, 16.
Horton River Formation, Lower Cretaceous, lat. 69°19′N, long. 126°50′W, Horton River, District of Mackenzie.

Pterodinium verrucosum Brideaux and McIntyre
Holotype 34150; paratypes 41518, 41519
Brideaux, W.W. and McIntyre, D.J., 1975, Geol. Surv. Can., Bull. 252, p. 30, pl. 9, figs. 1-8.
Horton River Formation, Lower Cretaceous, lat. 69°19′N, long. 126°50′W, Horton River, District of Mackenzie.

Pterodinium sp.
Fig. spec. 65350
Ioannides, N.S., 1986, Geol. Surv. Can., Bull. 371, p. 33, pl. 18, fig. 13.
Kanguk Formation, Upper Cretaceous, south coast Bylot Island, lat. 72°53′30″N, long. 78°16′W, District of Franklin.

Pterodinium sp. 1
Fig. specs. 38147, 38148
McIntyre, D.J., 1974, Geol. Surv. Can., Paper 74-14, pl. 12, figs. 4, 5.
Upper Cretaceous, lat. 69°27′30″N, long. 126°58′W, Horton River, District of Mackenzie.

Pterodinium sp. 2
Fig. spec. 38149
McIntyre, D.J., 1974, Geol. Surv. Can., Paper 74-14, pl. 12, fig. 6.
Upper Cretaceous, lat. 69°28′N, long. 126°58′W, Horton River, District of Mackenzie.

Pterodinium sp. A
Fig. spec. 41520
Brideaux, W.W. and McIntyre, D.J., 1975, Geol. Surv. Can., Bull. 252, p. 31, pl. 9, fig. 9.
Horton River Formation, Lower Cretaceous, lat. 69°19′N, long. 126°50′W, Horton River, District of Mackenzie.

Pterospermella sp.
Fig. spec. 68774
Dixon, J. et al., 1984, Geol. Surv. Can., Paper 82-13, pl. 5, fig. 3.
Pliocene to Holocene, 610-637m interval, Dome Gulf et al. Hunt Kopanoar M-13 well, lat. 70°22′55.1″N, long. 135°05′33.9″W, Beaufort Sea.

Pyxidiella sp.
Fig. spec. 41865
Williams, G.L., 1975, Geol. Surv. Can., Paper 74-30, vol. 2, pl. 6, fig. 4.
Kimmeridgian, Upper Jurassic, Shell Mohawk B-93 well, lat. 42°42′10.72″N, long. 64°43′53.04″W, Scotian Shelf.

Pyxidiella sp. 1
Fig. specs. 55340, 55365
Mudie, P.J., 1986, Initial Repts. Deep Sea Drilling Project, vol. XCIV, p. 805, pl. 4, figs. 12a, b.
Upper Miocene, hole 611C, depth 313.40m, lat. 52°50.15′N, long. 30°18.1′W, in the Gardar Drift on southwestern flank of Reykjanes Ridge, North Atlantic Ocean.

Pyxidiella sp. A Habib, 1972
Fig. specs. 47901, 47902
Williams, G.L., 1978, Initial Repts. Deep Sea Drilling Project, vol. 41, pl. 6, figs. 10, 11.
Late Jurassic, DSDP Site 367, lat. 12°29.2′N, long. 20°2.8′W, Cape Verde Basin west of Senegal.

Pyxidiella sp. A (Habib, 1972)
Fig. spec. 56713
Williams, G.L. and Bujak, J.P., 1980, Initial Repts. Deep Sea Drilling Project, vol. 50, pl. 2, fig. 6.
Early Createacenes, DSDP Site 416, lat. 32°50.18′N, long. 10°48.06′W, northwestern Moroccan continental margin.

Pyxidiella sp. A
Fig. specs. 68176-68178
Norris, G., 1986, Geol. Surv. Can., Bull. 340, p. 46, pl. 14, figs. 5-7.
Invik Member, Kugmallit Formation, Middle to Early Oligocene and Nuktak Formation, Pliocene or Late Miocene (68177), depths 7050 and 1650 feet, Imperial Nuktak C-22 well, lat. 69°41′07″N, long. 134°51′30″W, District of Mackenzie.

Rottnestia amphicavata Harland et al.
Hypotype 68778
Dixon, J. et al., 1984, Geol. Surv. Can., Paper 82-13, pl. 5, fig. 7.
Pliocene to Holocene, 610-637m interval, Dome Gulf et al. Hunt Kopanoar M-13 well, lat. 70°22′55.1″N, long. 135°05′33.9″W, Beaufort Sea.

Rottnestia cf. *borussica* (Eisenack)
Hypotype 49342
Williams, G.L. and Bujak, J.P., 1977, Am. Assoc. Strat. Palynol., Contr. Strat. Palynol., vol. 1, Contr. Ser. 5A, pl. 2, fig. 1.
Early-Middle Eocene, depth 1040-1070 feet, Amoco Imperial Skelly Pelican J-49 well, lat. 45°28′35.17″N, long. 52°36′41.71″W, Grand Banks.

Saeptodinium eurypylum (Manum and Cookson) Stover and Evitt, 1978
Hypotypes 65356, 65357

Ioannides, N.S., 1986, Geol. Surv. Can., Bull. 371, p. 33, pl. 19, figs. 5, 6.
Kanguk Formation, Upper Cretaceous, south coast Bylot Island, lat. 72°53′30″N, long. 78°16′W, District of Franklin.

Samlandia chlamydophora Eisenack
Hypotype 41887
Williams, G.L., 1975, Geol. Surv. Can., Paper 74-30, vol. 2, pl. 7, fig. 8.
Late Eocene, core hole 2, lat. 45°8′N, long. 52°38″W, Grand Banks.

Schematophora speciosa Deflandre and Cookson
Hypotype 47897
Williams, G.L., 1978, Initial Repts. Deep Sea Drilling Project, vol. 41, pl. 6, fig. 6.
Late Eocene, DSDP Site 370, lat. 32°50.2′N, long. 10°46.6′W, northwestern Moroccan continental margin.

Schizocystia laevigata Cookson and Eisenack, 1962
Hypotypes 65438, 65439
Ioannides, N.S., 1986, Geol. Surv. Can., Bull. 371, p. 43, pl. 24, figs. 14, 15.
Kanguk Formation, Upper Cretaceous, south coast Bylot Island, lat. 72°53′30″N, long. 78°16′W, District of Franklin.

Scriniocassis dictyotum (Cookson and Eisenack) Beju
Hypotypes 47895, 47899
Williams, G.L., 1978, Initial Repts. Deep Sea Drilling Project, vol. 41, pl. 6, figs. 4, 8.
Late Jurassic, DSDP Site 367, lat. 12°29.2′N, long. 20°2.8′W, Cape Verde Basin west of Senegal.

Scriniocassis dictyotus (Cookson and Eisenack) Beju (sensu Habib, 1972)
Hypotype 56717
Williams, G.L. and Bujak, J.P., 1980, Initial Repts. Deep Sea Drilling Project, vol. 50, pl. 3, figs. 1, 2.
Early Createacenes, DSDP Site 416, lat. 32°50.18′N, long. 10°48.06′W, northwestern Moroccan continental margin.

Scriniocassis weberi Gocht
Hypotype 75476
Davies, E.H., 1985, Palynology, vol. 9, pl. 3, figs. 7, 8.
Early Toarcian, Liassic, Peniche Cliffs, Lusitanian Basin, west-central Portugal.

Scriniocassis sp. A
Fig. spec. 75489
Davies, E.H., 1985, Palynology, vol. 9, p. 129, pl. 3, fig. 24.
Early Pliensbachian, Liassic, Zambujal Section, south of Condeixa, Lusitanian Basin, west-central Portugal.

Scriniodinium campanula Gocht
Hypotype 41533
Brideaux, W.W. and McIntyre, D.J., 1975, Geol. Surv. Can., Bull. 252, p. 33, pl. 10, fig. 5.
Horton River Formation, Lower Cretaceous, lat. 69°27′30″N, long. 126°54′W, Horton River, District of Mackenzie.

Scriniodinium? campanula Gocht, 1959
Hypotypes 68378, 65379
Ioannides, N.S., 1986, Geol. Surv. Can., Bull. 371, p. 34, pl. 20, figs. 2, 3.
Kanguk Formation, Upper Cretaceous, south coast Bylot Island, lat. 72°53′30″N, long. 78°16′W, District of Franklin.

Scriniodinium crystallinum (Deflandre) Klement
Hypotype 46053
Brideaux, W.W., 1977, Geol. Surv. Can., Bull. 281, p. 39, pl. 15, fig. 6.
Husky Formation, Jurassic, IOE Stoney Core Hole F-42, depth 1000-1010 feet, Stoney Creek about 18 miles west-southwest of Fort McPherson, lat. 67°21′23″N, long. 135°38′43″W, District of Mackenzie.

Scriniodinium crystallinum (Deflandre) Klement
Hypotypes 76295, 76296
Davies, E.H. and Poulton, T.P., 1986, Geol. Surv. Can., Paper 86-1B, pl. 55.4, figs. 3, 6.
Fernie Formation, Upper Jurassic, Christina Falls, lat. 56°40′N, long. 123°4′W, northeastern British Columbia.

Scriniodinium? obscurum Manum and Cookson, 1964
Hypotypes 65169-65173
Ioannides, N.S., 1986, Geol. Surv. Can., Bull. 371, p. 34, pl. 6, figs. 3-7.
Kanguk Formation, Upper Cretaceous, lat. 72°53′30″N, long. 78°16′W, and Eureka Sound Formation, Late Paleocene to ?Early Eocene, lat. 72°53′30″N, long. 78°25′W (65170, 65173), south coast of Bylot Island, District of Franklin.

Scriniodinium rostratum Brideaux and McIntyre
Holotype 34151; paratypes 41534-41538
Brideaux, W.W. and McIntyre, D.J., 1975, Geol. Surv. Can., Bull. 252, p. 33, pl. 10, figs. 6-14; pl. 11, figs. 1-3.
Horton River Formation, Lower Cretaceous, lat. 69°19′N, long. 126°50′W, Horton River, District of Mackenzie.

Scriniodinium? sp.
Fig. specs. 65177, 65178
Ioannides, N.S., 1986, Geol. Surv. Can., Bull. 371, pl. 7, figs. 1, 3.
Kanguk Formation, Upper Cretaceous, south coast Bylot Island, lat. 72°53′30″N, long. 78°16′W, District of Franklin.

Selenopemphix nephroides Benedek
Hypotype 47905
Williams, G.L., 1978, Initial Repts. Deep Sea Drilling Project, vol. 41, pl. 7, fig. 3.
Plio-Pleistocene, reworked, DSDP Site 370, lat. 32°50.2′N, long. 10°46.6′W, northwestern Moroccan continental margin.

Selenopemphix nephroides Benedek, 1972, emend. Bujak, 1980
Hypotype 55345
Mudie, P.J., 1986, Initial Repts. Deep Sea Drilling Project, vol. XCIV, p. 805, pl. 5, fig. 1.
Pleistocene, hole 607, depth 17.6m, lat. 41°0.07′N, long. 32°57.44′W, at base of upper western flank of the Mid-Atlantic Ridge, North Atlantic Ocean.

Senegalinium sp. cf. *S. microgranulatum* (Stanley) Stover and Evitt, 1978
Fig. specs. 65358, 65359
Ioannides, N.S., 1986, Geol. Surv. Can., Bull. 371, p. 35, pl. 19, figs. 7, 8.
Eureka Sound Formation, Late Paleocene to ?Early Eocene, Bylot Island, "Twosnout Creek", lat. 73°9′30″N, long. 79°42′30″W, and Kanguk Formation, Upper Cretaceous, south coast, lat. 72°53′30″N, long. 78°16′W, District of Franklin.

Senoniasphaera sp. cf. *S. jurassica* (Gitmez and Sarjeant) Lentin and Williams
Fig. spec. 56719
Williams, G.L. and Bujak, J.P., 1980, Initial Repts. Deep Sea Drilling Project, vol. 50, pl. 3, fig. 4.
Late Jurassic, DSDP Site 416, lat. 32°50.18′N, long. 10°48.06′W, northwestern Moroccan continental margin.

Senoniasphaera microreticulata Brideaux and McIntyre
Holotype 34152; paratypes 41542-41545, 41547
Brideaux, W.W. and McIntyre, D.J., 1975, Geol. Surv. Can., Bull. 252, p. 35, pl. 11, figs. 7-12; pl. 12, figs. 1-8.
Horton River Formation and Gilmore Lake Member, Langton Bay Formation (41543), Lower Cretaceous, lat. 69°19′N, long. 126°50′W and lat. 68°50′N, long. 125°24′W, Horton River, District of Mackenzie.

Senoniasphaera sp. A
Fig. specs. 68149, 68150, 68153, 68154
Norris, G., 1986, Geol. Surv. Can., Bull. 340, p. 44, pl. 11, figs. 23, 24; pl. 12, figs. 2-4.
Richards Formation, Middle Eocene, depths 12,300-12,400, 12,400-12,500 (68150) and 11,900-12,000 (68153) feet, Imperial Nuktak C-22 well, lat. 69°41′07″N, long. 134°51′30″W, District of Mackenzie.

Sentusidinium asymmetricum (Fenton, Neves and Piel) Lentin and Williams
Hypotypes 75499, 75500
Davies, E.H., 1985, Palynology, vol. 9, pl. 4, figs. 11, 12.
Liassic, Early Bajocian, Brehna Road Section, and Aalenian, Zambujal Section, south of Condeixa, Lusitanian Basin, west-central Portugal.

Sentusidinium baculatum (Dodekova) Stover and Evitt
Hypotype 76293
Davies, E.H. and Poulton, T.P., 1986, Geol. Surv. Can., Paper 86-1B, pl. 55.2, fig. 2.
Fernie Formation, Upper Jurassic, Christina Falls, lat. 56°40′N, long. 123°4′W, northeastern British Columbia.

Sentusidinium filiatum Davies
Hypotype 76528
Davies, E.H. and Poulton, T.P., 1986, Geol. Surv. Can., Paper 86-1B, pl. 55.2, fig. 9.
Fernie Formation, Upper Jurassic, Christina Falls, lat. 56°40′N, long. 123°4′W, northeastern British Columbia.

Sirmiodinium grossii Alberti emend. Warren
Hypotype 46056
Brideaux, W.W., 1977, Geol. Surv. Can., Bull. 281, p. 40, pl. 15, fig. 11.
Cretaceous, Banff-Aquitaine-Arco Rat Pass K-35 well, depth 1660 feet, 6 miles southeast of Mount Goodenough, lat. 67°54′42.55″N, long. 135°21′56.83″W, District of Mackenzie.

Sirmiodinium grossii Alberti sensu Gitmez and Sarjeant, 1972
Hypotypes 47907, 47908
Williams, G.L., 1978, Initial Repts. Deep Sea Drilling Project, vol. 41, pl. 7, figs. 5, 6.
Late Jurassic, DSDP Site 367, lat. 12°29.2′N, long. 20°2.8′W, Cape Verde Basin west of Senegal.

Sirmiodinium grossii Alberti, 1961, emend. Warren, 1973
Hypotype 58360
Tan, J.T. and Hills, L.V., 1978, Geol. Surv. Can., Paper 78-1C, p. 67, pl. 12.2, fig. 9.
Savik Formation, Upper Jurassic, lat. 77°48′N, long. 112°20′W, Mackenzie King Island, District of Franklin.

Specimens intermediate
Fig. specs. 56705, 56755
Williams, G.L. and Bujak, J.P., 1980, Initial Repts. Deep Sea Drilling Project, vol. 50, pl. 1, fig. 10; pl. 7, fig. 3.
Early Cretaceous, DSDP Site 416, lat. 32°50.18′N, long. 10°48.06′W, northwestern Moroccan continental margin.

Spinidinium balmei (Cookson and Eisenack)
Hypotypes 65361, 65362
Ioannides, N.S., 1986, Geol. Surv. Can., Bull. 371, p. 35, pl. 19, figs. 10, 11.
Kanguk Formation, Upper Cretaceous, south coast Bylot Island, lat. 72°53′30″N, long. 78°16′W, District of Franklin.

Spinidinium clavum Harland
Hypotypes 38036-38038
McIntyre, D.J., 1974, Geol. Surv. Can., Paper 74-14, pl. 3, figs. 2-4.
Upper Cretaceous, lat. 69°27′30″N, long. 126°58′W, Horton River, District of Mackenzie.

Spinidinium densispinatum Stanley sensu Benedek, 1927
Hypotypes 47916, 47921
Williams, G.L., 1978, Initial Repts. Deep Sea Drilling Project, vol. 41, pl. 8, figs. 2, 7.
Oligocene, DSDP Site 370, lat. 32°50.2′N, long. 10°46.6′W, northwestern Moroccan continental margin.

Spinidinium echinoideum (Cookson and Eisenack) Lentin and Williams, 1976
Hypotype 91444
Sweet, A.R. and McIntyre, D.J., 1988, Can. Soc. Petrol. Geol, Mem. 15, fig. 8.16.
Sturrock Member, Cardium Formation, Upper Cretaceous, south side of McLeod River, l.s.d. 16, sec. 16, tp. 47, rge. 23, W. 5th mer., Alberta.

Spinidinium sagittulum (Drugg) Lentin and Williams
Hypotype 92010
Dietrich, J.R. et al., 1989, Geol. Surv. Can., Paper 89-1G, p. 141, pl. 2, fig. 2.
Early to Middle Eocene, depth 1700-1790 feet, Dome Pacific et al. PEX Natsek E-56 well, lat. 69°45′21.46″N, long. 139°44′34.58″W, western Beaufort Sea.

Spinidinium sp. cf. *S. sagittulum* (Drugg) Lentin and Williams
Fig. specs. 68155-68157
Norris, G., 1986, Geol. Surv. Can., Bull. 340, p. 44, pl. 12, figs. 5, 7-9.
Richards Formation, Middle Eocene, depths 12,200-12,300 and 12,300-12,400(68156) feet, Imperial Nuktak C-22 well, lat. 69°41′07″N, long. 134°51′30″W, District of Mackenzie.

Spinidinium sp. cf. *S. sagittulum* (Drugg) Lentin and Williams
Fig. spec. 92011
Dietrich, J.R. et al., 1989, Geol. Surv. Can., Paper 89-1G, pl. 2, fig. 3.
Early to Middle Eocene, depth 1800-1890 feet, Dome Pacific et al. PEX Natsek E-56 well, lat. 69°45′21.46″N, long. 139°44′34.58″W, western Beaufort Sea.

Spinidinium uncinatum May, 1980
Hypotypes 65237-65242, 65244-65247
Ioannides, N.S., 1986, Geol. Surv. Can., Bull. 371, p. 35, pl. 10, figs. 24-24, 27, 28; pl. 11, figs. 2-4, 8.
Eureka Sound Formation, Late Paleocene to ?Early Eocene, south coast Bylot Island, lat. 72°53′30″N, long. 78°25′W, District of Franklin.

Spinidinium cf. *vestitum* Brideaux
Hypotypes 41866, 41870, 41871
Williams, G.L., 1975, Geol. Surv. Can., Paper 74-30, vol. 2, pl. 6, figs. 5, 9, 10.
Albian, Lower Cretaceous, Mobil Sable Island C-67 well, lat. 43°56′4.91″N, long. 59°55′1.38″W, Scotian Shelf.

Spinidinium cf. *vestitum* Brideaux sensu Williams, 1975
Hypotypes 53708-53710
Bujak, J.P. and Williams, G.L., 1978, Geol. Surv. Can., Bull. 297, pl. 3, figs. 9-12.
Early Cretaceous, depths 6150-6180 and 8050-8080 (53710) feet, Mobil Sable Island C-67 well, lat. 43°56′4.91″N, long. 59°55′1.38″W, Scotian Shelf.

Spinidinium sp.
Fig. specs. 65231-65236
Ioannides, N.S., 1986, Geol. Surv. Can., Bull. 371, p. 21, pl. 10, figs. 16-20, 26.
Kanguk Formation, Upper Cretaceous, Bylot Island, south coast, lat. 72°53′30″N, long. 78°16′W, and "Twosnout Creek", lat. 73°14′30″N, long. 79°49′W (65234), District of Franklin.

Spiniferites? *cingulatus* (O. Wetzel) Sarjeant, 1970
Hypotype 65364
Ioannides, N.S., 1986, Geol. Surv. Can., Bull. 371, p. 36, pl. 19, fig. 13.
Kanguk Formation, Upper Cretaceous, south coast Bylot Island, lat. 72°53′30″N, long. 78°16′W, District of Franklin.

Spiniferites elongatus Reed 1974
Hypotype 55343
Mudie, P.J., 1986, Initial Repts. Deep Sea Drilling Project, vol. XCIV, p. 805, pl. 4, fig. 15.
Pleistocene, hole 607, depth 3.5m, lat. 41°0.07′N, long. 32°57.44′W, at base of upper western flank of the Mid-Atlantic Ridge, North Atlantic Ocean.

Spiniferites mirabilis (Rossignol) Sarjeant, 1970
Hypotype 55311
Mudie, P.J., 1986, Initial Repts. Deep Sea Drilling Project, vol. XCIV, p. 806, pl. 2, fig. 10.
Upper Pliocene, hole 607, depth 140.9m, lat. 41°0.07′N, long. 32°57.44′W, at base of upper western flank of the Mid-Atlantic Ridge, North Atlantic Ocean.

Spiniferites nodosus (Wall)
Hypotype 41874
Williams, G.L., 1975, Geol. Surv. Can., Paper 74-30, vol. 2, pl. 6, fig. 13.
Late Miocene, Amoco IOE Puffin B-90, lat. 44°39′12.73″N, long. 53°42′28.35″W, Grand Banks.

Spiniferites porosus (Manum and Cookson) Harland, 1973
Hypotypes 65368, 65369
Ioannides, N.S., 1986, Geol. Surv. Can., Bull. 371, p. 36, pl. 19, figs. 16, 18.
Eureka Sound Formation, Late Paleocene to ?Early Eocene, Bylot Island, "Twosnout Creek", lat. 73°13′N, long. 79°53′W, and Kanguk Formation, Upper Cretaceous, lat. 72°53′30″N, long. 78°16′W, District of Franklin.

Spiniferites ramosus (Ehrenberg)
Hypotypes 38120, 38121
McIntyre, D.J., 1974, Geol. Surv. Can., Paper 74-14, pl. 9, figs. 7, 9.
Upper Cretaceous, lat. 69°27′30″N, long. 126°58′W, Horton River, District of Mackenzie.

Spiniferites ramosus (Ehrenberg) Mantell
Hypotype 41521
Brideaux, W.W. and McIntyre, D.J., 1975, Geol. Surv. Can., Bull. 252, p. 35, pl. 9, fig. 10.

Horton River Formation, Lower Cretaceous, lat. 69°19′N, long. 126°50′W, Horton River, District of Mackenzie.

Spiniferites ramosus (Ehrenberg) Loeblich and Loeblich, 1966
Hypotype 65363
Ioannides, N.S., 1986, Geol. Surv. Can., Bull. 371, p. 36, pl. 19, fig. 12.
Kanguk Formation, Upper Cretaceous, south coast Bylot Island, lat. 72°53′30″N, long. 78°16′W, District of Franklin.

Spiniferites ramosus (Ehrenberg) Loeblich and Loeblich, 1966
Hypotype 91467
Sweet, A.R. and McIntyre, D.J., 1988, Can. Soc. Petrol. Geol, Mem. 15, fig. 9.12.
Leyland Member, Cardium Formation, Upper Cretaceous, south side of McLeod River, l.s.d. 16, sec. 16, tp. 47, rge. 23, W. 5th mer., Alberta.

Spiniferites ramosus var. *granomembranaceus* (Davey and Williams) Lentin and Williams, 1973
Hypotype 71420
Wilson, M.A., 1978, Palaeontographica, Abt. B, vol. 166, p. 154, pl. 12, fig. 2.
Cretaceous, Police Island, north bank Mackenzie River, 12 miles southeast of Fort Norman, District of Mackenzie.

Spiniferites ramosus (Ehrenberg) subsp. *granosus* Davey and Williams
Hypotype 49344
Williams, G.L. and Bujak, J.P., 1977, Am. Assoc. Strat. Palynol., Contr. Strat. Palynol., vol. 1, Contr. Ser. 5A, pl. 2, fig. 3.
Early-Middle Eocene, depth 1220-1250 feet, Amoco Imperial Skelly Pelican J-49 well, lat. 45°28′35.17″N, long. 52°36′41.71″W, Grand Banks.

Spiniferites ramosus var. *multibrevis* (Davey and Williams) Davey and Verdier, 1971
Hypotypes 71418, 71428
Wilson, M.A., 1978, Palaeontographica, Abt. B, vol. 166, p. 154, pl. 11, fig. 17; pl. 12, fig. 11.
Cretaceous, Police Island, north bank Mackenzie River, 12 miles southeast of Fort Norman, District of Mackenzie.

Spiniferites sp. cf. *S. scabrosus* (Clarke and Verdier) Lentin and Williams, 1975
Fig. specs. 65365, 65366
Ioannides, N.S., 1986, Geol. Surv. Can., Bull. 371, p. 36, pl. 19, figs. 14, 17.
Kanguk Formation, Upper Cretaceous, Bylot Island, south coast, lat. 72°53′30″N, long. 78°16′W, and "Twosnout Creek", lat. 73°14′30″N, long. 79°49′W, District of Franklin.

?*Spiniferites splendidus* Harland 1979
Hypotype 55341
Mudie, P.J., 1986, Initial Repts. Deep Sea Drilling Project, vol. XCIV, p. 806, pl. 4, fig. 13.
Pleistocene, hole 607, depth 5m, lat. 41°0.07′N, long. 32°57.44′W, at base of upper western flank of the Mid-Atlantic Ridge, North Atlantic Ocean.

Spiniferites wetzelii (Deflandre) Sarjeant, 1970
Hypotype 65380
Ioannides, N.S., 1986, Geol. Surv. Can., Bull. 371, p. 36, pl. 20, fig. 4.
Kanguk Formation, Upper Cretaceous, south coast of Bylot Island, lat. 72°53′30″N, long. 78°16′W, District of Franklin.

Spiniferites sp.
Fig. spec. 41869
Williams, G.L., 1975, Geol. Surv. Can., Paper 74-30, vol. 2, pl. 6, fig. 8.
Late Miocene, Amoco IOE Puffin B-90 well, lat. 44°39′12.73″N, long. 53°42′28.35″W, Grand Banks.

Spiniferites spp.
Fig. specs. 55310, 55353
Mudie, P.J., 1986, Initial Repts. Deep Sea Drilling Project, vol. XCIV, p. 806, pl. 2, fig. 2; pl. 5, fig. 5.
Upper Miocene, hole 611C, depth 510.09m, lat. 52°50.15′N, long. 30°18.10′W, in the Gardar Drift on southwestern flank of Reykjanes Ridge, and hole 607A, depth 305.2m, lat. 41°0.07′N, long. 32°57.44′W, at base of upper western flank of the Mid-Atlantic Ridge, North Atlantic Ocean.

Spiniferites sp. 1
Fig. spec. 38122
McIntyre, D.J., 1974, Geol. Surv. Can., Paper 74-14, pl. 9, fig. 8.
Upper Cretaceous, lat. 69°27′30″N, long. 126°58′W, Horton River, District of Mackenzie.

Spiniferites sp. A
Fig. spec. 49343
Williams, G.L. and Bujak, J.P., 1977, Am. Assoc. Strat. Palynol., Contr. Strat. Palynol., vol. 1, Contr. Ser. 5A, pl. 2, fig. 2.
Early-Middle Eocene, depth 1220-1250 feet, Amoco Imperial Skelly Pelican J-49 well, lat. 45°28′35.17″N, long. 52°36′41.71″W, Grand Banks.

Spiniferites sp. A
Fig. spec. 68767
Dixon, J. et al., 1984, Geol. Surv. Can., Paper 82-13, pl. 4, fig. 6.
Late Middle to Late Miocene, 2646m interval, Dome Gulf et al. Hunt Kopanoar M-13 well, lat. 70°22′55.1″N, long. 135°05′33.9″W, Beaufort Sea.

Spiniferites sp. CE
Fig. spec. 45808
Brideaux, W.W., 1976, Geol. Surv. Can., Paper 76-1B, pl. 44.1, fig. 11.
Boundary Creek Formation, Upper Cretaceous, depth 6380-6400 feet, Gulf Mobile Parsons N-10 well, ca. 1.5 miles east-northeast of Parsons Lake, lat. 69°59′48.52″N, long. 133°31′50.33″W, District of Mackenzie.

Spongodinium? sp. (cf. *S. delitiense*)
Fig. specs. 65180-65187, 65189-65192
Ioannides, N.S., 1986, Geol. Surv. Can., Bull. 371, p. 37, pl. 7, figs. 4-11; pl. 8, figs. 2-5.
Eureka Sound Formation, Late Paleocene to ?Early Eocene, lat. 72°53′30″N, long. 78°25′W and lat. 72°52′N, long. 78°31′W (65189-65192); Kanguk Formation, Upper Cretaceous, lat. 72°53′30″N, long. 78°16′W (65184, 65186), south coast Bylot Island; Upper Cretaceous, lat. 75°17′N, long. 90°35′W, Devon Island (65187), District of Franklin.

Stephodinium coronatum Deflandre, 1936
Hypotype 65360
Ioannides, N.S., 1986, Geol. Surv. Can., Bull. 371, p. 37, pl. 19, fig. 9.
Kanguk Formation, Upper Cretaceous, south coast Bylot Island, lat. 72°53′30″N, long. 78°16′W, District of Franklin.

Subtilisphaera paeminosa (Drugg) Bujak and Davies
Hypotype 76291
Davies, E.H. and Poulton, T.P., 1986, Geol. Surv. Can., Paper 86-1B, pl. 55.1, fig. 9.
Fernie Formation, Upper Jurassic, Christina Falls, lat. 56°40′N, long. 123°4′W, northeastern British Columbia.

Subtilisphaera perlucida (Alberti) Jain and Millepied
Hypotypes 46059, 46060
Brideaux, W.W., 1977, Geol. Surv. Can., Bull. 281, p. 42, pl. 16, figs. 4, 5.
Cretaceous, Banff-Aquitaine-Arco Rat Pass K-35 well, depths 1350-1360 and 990-1000 feet, 6 miles southeast of Mount Goodenough, lat. 67°54′42.55″N, long. 135°21′56.83″W, District of Mackenzie.

Subtilisphaera cf. *perlucida* (Alberti) Jain and Millepied
Hypotype 47903
Williams, G.L., 1978, Initial Repts. Deep Sea Drilling Project, vol. 41, pl. 7, fig. 1.
Early Cretaceous, DSDP Site 370, lat. 32°50.2′N, long. 10°46.6′W, northwestern Moroccan continental margin.

Subtilisphaera cf. *perlucida* Alberti
Hypotype 53711
Bujak, J.P. and Williams, G.L., 1978, Geol. Surv. Can., Bull. 297, pl. 3, figs. 13, 14.
Early Cretaceous, depth 9460-9490 feet, Amoco IOE Puffin B-90 well, lat. 44°39′12.73″N, long. 53°42′28.35″W, Grand Banks.

Sumatradinium sp.
Fig. spec. 47913
Williams, G.L., 1978, Initial Repts. Deep Sea Drilling Project, vol. 41, pl. 7, fig. 11.
Plio-Pleistocene, DSDP Site 370, lat. 32°50.2′N, long. 10°46.6′W, northwestern Moroccan continental margin.

Surculosphaeridium longifurcatum (Firtion)
Hypotype 41885
Williams, G.L., 1975, Geol. Surv. Can., Paper 74-30, vol. 2, pl. 7, fig. 6.
Turonian, Upper Cretaceous, Shell Oneida O-25 well, lat. 43°14′57.49″N, long. 61°33′36.38″W, Scotian Shelf.

Surculosphaeridium longifurcatum (Firtion) Davey et al.
Hypotypes 80486, 80487
Burden, E. and Holloway, D., 1985, Can. J. Earth Sci., vol. 22, no. 10, pl. 1, figs. 5, 6.
Quaternary, Scott Inlet area, Baffin Island, District of Franklin.

Surculosphaeridium longifurcatum (Firtion) Davey et al., 1966
Hypotype 91465
Sweet, A.R. and McIntyre, D.J., 1988, Can. Soc. Petrol. Geol, Mem. 15, fig. 9.10.
Kaskapau Formation, Upper Cretaceous, south side of Grande Cache Highway 40, l.s.d. 15, sec. 13, tp. 57, rge. 7, W. 5th mer., Alberta.

Surculosphaeridium? *longifurcatum* (Firtion) Davey et al., *in* Davey et al., 1966
Hypotype 65367
Ioannides, N.S., 1986, Geol. Surv. Can., Bull. 371, p. 37, pl. 19, fig. 15.
Kanguk Formation, Upper Cretaceous, south coast Bylot Island, lat. 72°53′30″N, long. 78°16′W, District of Franklin.

Surculosphaeridium cf. *longifurcatum* (Firtion)
Hypotype 41889
Williams, G.L., 1975, Geol. Surv. Can., Paper 74-30, vol. 2, pl. 7, fig. 10.
Hauterivian, Lower Cretaceous, Amoco IOE Puffin B-90 well, lat. 44°39′12.73″N, long. 53°42′28.35″W, Grand Banks.

Surculosphaeridium cf. *longifurcatum* (Firtion) sensu Williams, 1975
Hypotypes 53699, 53712
Bujak, J.P. and Williams, G.L., 1978, Geol. Surv. Can., Bull. 297, pl. 2, fig. 15; pl. 3, fig. 15.
Early Cretaceous, swc 3807 feet, Shell Mohawk B-93 well, lat. 42°42′10.72″N, long. 64°43′53.04″W, and depth 13150-13180 feet, Mobil Sable Island C-67 well, lat. 43°56′4.91″N, long. 59°55′1.38″W, Scotian Shelf.

Susadinium scroforides Dörhöfer and Davies
Hypotype 75472
Davies, E.H., 1985, Palynology, vol. 9, pl. 3, figs. 1, 2.
Late Pliensbachian, Liassic, Zambujal Section, south of Condeixa, Lusitanian Basin, west-central Portugal.

Systematophora sp. cf. *S. areolata* Klement
Fig. specs. 56730, 56758
Williams, G.L. and Bujak, J.P., 1980, Initial Repts. Deep Sea Drilling Project, vol. 50, pl. 4, fig. 4; pl. 7, fig. 6.

Early Cretaceous, DSDP Site 416, lat. 32°50.18′N, long. 10°48.06′W, northwestern Moroccan continental margin.

Systematophora complicata Neale and Sarjeant
Hypotype 41872
Williams, G.L., 1975, Geol. Surv. Can., Paper 74-30, vol. 2, pl. 6, fig. 11.
Berriasian-Valanginian, Lower Cretaceous, Amoco IOE Puffin B-90 well, lat. 44°39′12.73″N, long. 53°42′28.35″W, Grand Banks.

Systematophora sp. cf. *S. complicata* Neale and Sarjeant
Fig. spec. 46042
Brideaux, W.W., 1977, Geol. Surv. Can., Bull. 281, p. 31, pl. 14, fig. 4.
Cretaceous, Banff-Aquitaine-Arco Rat Pass K-35 well, depth 1130 feet, 6 miles southeast of Mount Goodenough lat. 67°54′42.55″N, long. 135°21′56.83″W, District of Mackenzie.

Systematophora fasciculigera Klement
Hypotype 41890
Williams, G.L., 1975, Geol. Surv. Can., Paper 74-30, vol. 2, pl. 7, fig. 11.
Oxfordian, Upper Jurassic, Shell Mohican I-100 well, lat. 42°59′39.04″N, long. 62°28′51.32″W, Scotian Shelf.

Systematophora fasciculigera Klement
Hypotype 56732
Williams, G.L. and Bujak, J.P., 1980, Initial Repts. Deep Sea Drilling Project, vol. 50, pl. 4, fig. 7.
Late Jurassic, DSDP Site 416, lat. 32°50.18′N, long. 10°48.06′W, northwestern Moroccan continental margin.

Systematophora orbifera Klement
Hypotype 41884
Williams, G.L., 1975, Geol. Surv. Can., Paper 74-30, vol. 2, pl. 7, fig. 5.
Berriasian-Valanginian, Lower Cretaceous, Amoco IOE Puffin B-90 well, lat. 44°39′12.73″N, long. 53°42′28.35″W, Grand Banks.

Systematophora placacantha Deflandre and Cookson
Hypotype 38699
Haskell, T.R. and Wilson, G.J., 1975, Initial Repts. Deep Sea Drilling Project, vol. 29, pl. 3, fig. 7.
Cenozoic, DSDP Site 284, lat. 40°30.48′S, long. 167°40.81′E, Lord Howe Rise west of New Zealand.

Systematophora schindewolfi (Alberti)
Hypotype 41875
Williams, G.L., 1975, Geol. Surv. Can., Paper 74-30, vol. 2, pl. 6, fig. 14.
Hauterivian, Lower Cretaceous, Shell Mohican I-100 well, lat. 42°59′39.04″N, long. 62°28′51.32″W, Scotian Shelf.

Systematophora schindewolfi (Alberti) Downie and Sarjeant
Hypotype 46045
Brideaux, W.W., 1977, Geol. Surv. Can., Bull. 281, p. 31, pl. 14, figs. 7, 9.
Cretaceous, Banff-Aquitaine-Arco Rat Pass K-35 well, depth 1130 feet, 6 miles southeast of Mount Goodenough, lat. 67°54′42.55″N, long. 135°21′56.83″W, District of Mackenzie.

Systematophora schindewolfi (Alberti) Downie and Sarjeant
Hypotype 56731
Williams, G.L. and Bujak, J.P., 1980, Initial Repts. Deep Sea Drilling Project, vol. 50, pl. 4, figs. 5, 6.
Early Cretaceous, DSDP Site 416, lat. 32°50.18′N, long. 10°48.06′W, northwestern Moroccan continental margin.

Systematophora sp.
Fig. spec. 56736
Williams, G.L. and Bujak, J.P., 1980, Initial Repts. Deep Sea Drilling Project, vol. 50, pl. 4, fig. 11.
Early Cretaceous, DSDP Site 416, lat. 32°50.18′N, long. 10°48.06′W, northwestern Moroccan continental margin.

Systematophora sp. A (Habib, 1972)
Fig. spec. 56733
Williams, G.L. and Bujak, J.P., 1980, Initial Repts. Deep Sea Drilling Project, vol. 50, pl. 4, fig. 8.
Early Cretaceous, DSDP Site 416, lat. 32°50.18′N, long. 10°48.06′W, northwestern Moroccan continental margin.

Systematophora sp. AE
Fig. spec. 46041
Brideaux, W.W., 1977, Geol. Surv. Can., Bull. 281, p. 31, pl. 14, figs. 2, 3.
Cretaceous, Banff-Aquitaine-Arco Rat Pass K-35 well, depth 1010-1020 feet, 6 miles southeast of Mount Goodenough, lat. 67°54′42.55″N, long. 135°21′56.83″W, District of Mackenzie.

Tanyosphaeridium boletum Davey
Hypotypes 56701-56704, 56761
Williams, G.L. and Bujak, J.P., 1980, Initial Repts. Deep Sea Drilling Project, vol. 50, pl. 1, figs. 5-9; pl. 8, fig. 3.
Early Cretaceous, DSDP Site 416, lat. 32°50.18′N, long. 10°48.06′W, northwestern Moroccan continental margin.

Tanyosphaeridium variecalamum Davey and Williams, *in* Davey et al., 1966
Hypotype 65373
Ioannides, N.S., 1986, Geol. Surv. Can., Bull. 371, p. 37, pl. 19, fig. 22.
Eureka Sound Formation, Late Paleocene to ?Early Eocene, "Twosnout Creek", Bylot Island, lat. 73°13′N, long. 79°53′W, District of Franklin.

Tanyosphaeridium sp. A
Fig. spec. 41876
Williams, G.L., 1975, Geol. Surv. Can., Paper 74-30, vol. 2, pl. 6, fig. 15.
Oligocene, Amoco IOE Murre G-67 well, lat. 46°6′20.4″N, long. 49°9′38.2″W, Grand Banks.

Tanyosphaeridium spp. A, B Brideaux, 1971
Fig. specs. 41522, 41523
Brideaux, W.W. and McIntyre, D.J., 1975, Geol. Surv. Can., Bull. 252, p. 31, pl. 9, figs. 11, 12.
Horton River Formation, Lower Cretaceous, lat. 69°27′30″N, long. 126°54′W, Horton River, District of Mackenzie.

Tanyosphaeridium sp. C
Fig. spec. 41524
Brideaux, W.W. and McIntyre, D.J., 1975, Geol. Surv. Can., Bull. 252, p. 32, pl. 9, fig. 13.
Horton River Formation, Lower Cretaceous, lat. 69°19′N, long. 126°50′W, Horton River, District of Mackenzie.

Tanyosphaeridium sp. DE
Fig. spec. 46043
Brideaux, W.W., 1977, Geol. Surv. Can., Bull. 281, p. 32, pl. 14, fig. 5.
Cretaceous, Banff-Aquitaine-Arco Rat Pass K-35 well, depth 1130 feet, 6 miles southeast of Mount Goodenough, lat. 67°54′42.55″N, long. 135°21′56.83″W, District of Mackenzie.

Tectatodinium laminatum Davies
Holotype 69365; paratypes 69363, 69364
Davies, E.H., 1983, Geol. Surv. Can., Bull. 359, p. 21, pl. 6, figs. 1-3, 21; text-fig. 16.
Wilkie Point Formation, Upper Jurassic, depth 1840-1850 feet, Panarctic Sandy Point L-46 well, northwestern Melville Island, District of Franklin.

Tectatodinium laminatum Davies
Hypotype 76284
Davies, E.H. and Poulton, T.P., 1986, Geol. Surv. Can., Paper 86-1B, pl. 55.5, fig. 3.
Fernie Formation, Upper Jurassic, Christina Falls, lat. 56°40′N, long. 123°4′W, northeastern British Columbia.

Tectatodinium simplex (Harland) Edwards, 1984
Hypotypes 55322, 55324
Mudie, P.J., 1986, Initial Repts. Deep Sea Drilling Project, vol. XCIV, p. 806, pl. 3, figs. 8a, b, 10.
Lower Pliocene and Upper Miocene, hole 611C, depths 291.29m and 441.28m, lat. 52°50.15′N, long. 30°18.1′W, in the Gardar Drift on southwestern flank of Reykjanes Ridge, North Atlantic Ocean.

Tenua anaphrissa (Sarjeant)
Hypotype 41888
Williams, G.L., 1975, Geol. Surv. Can., Paper 74-30, vol. 2, pl. 7, fig. 9.
Barremian, Lower Cretaceous, Shell Naskapi N-30 well, lat. 43°29′46.29″N, long. 62°34′0.63″W, Scotian Shelf.

Tenua brevispinosa (Pocock)
Hypotypes 45995, 45996
Brideaux, W.W., 1977, Geol. Surv. Can., Bull. 281, p. 15, pl. 6, figs. 2-5.
Cretaceous, IOE Stoney Core Hole F-42, depth 550-560 feet, Stoney Creek about 18 miles west-southwest of Fort McPherson, lat. 67°21′23″N, long. 135°38′43″W, and Banff-Aquitaine-Arco Rat Pass K-35 well, depth 370-390 feet, 6 miles southeast of Mount Goodenough, lat. 67°54′42.55″N, long. 135°21′56.83″W, District of Mackenzie.

Tenua echinata Gitmez and Sarjeant
Hypotype 45997
Brideaux, W.W., 1977, Geol. Surv. Can., Bull. 281, p. 16, pl. 6, fig. 6.
Husky Formation, Jurassic, IOE Stoney Core Hole F-42, depth 1000-1010 feet, Stoney Creek about 18 miles west-southwest of Fort McPherson, lat. 67°21′23″N, long. 135°38′43″W, District of Mackenzie.

Tenua hystrix Eisenack
Hypotype 41489
Brideaux, W.W. and McIntyre, D.J., 1975, Geol. Surv. Can., Bull. 252, p. 24, pl. 6, figs. 12, 13.
Crossley Lakes Member, Langton Bay Formation, Lower Cretaceous, lat. 69°04′N, long. 125°51′W, Horton River, District of Mackenzie.

Tenua hystrix Eisenack
Hypotypes 56720, 56754, 56757
Williams, G.L. and Bujak, J.P., 1980, Initial Repts. Deep Sea Drilling Project, vol. 50, pl. 3, fig. 5; pl. 7, figs. 2, 5.
Early Cretaceous, DSDP Site 416, lat. 32°50.18′N, long. 10°48.06′W, northwestern Moroccan continental margin.

Tenua rioulti Sarjeant
Hypotype 41873
Williams, G.L., 1975, Geol. Surv. Can., Paper 74-30, vol. 2, pl. 6, fig. 12.
Berriasian-Valanginian (reworked), Lower Cretaceous, Amoco IOE Puffin B-90 well, lat. 44°39′12.73″N, long. 53°42′28.35″W, Grand Banks.

Tenua sp. cf. *T. rioultii* Sarjeant
Fig. spec. 45998
Brideaux, W.W., 1977, Geol. Surv. Can., Bull. 281, p. 16, pl. 6, fig. 7.
Husky Formation, Jurassic, IOE Stoney Core Hole F-42, depth 980-990 feet, Stoney Creek about 18 miles west-southwest of Fort McPherson, lat. 67°21′23″N, long. 135°38′43″W, District of Mackenzie.

Tenua verrucosa Sarjeant
Hypotype 47917
Williams, G.L., 1978, Initial Repts. Deep Sea Drilling Project, vol. 41, pl. 8, fig. 3.
Early Cretaceous, DSDP Site 367, lat. 12°29.2′N, long. 20°2.8′W, Cape Verde Basin west of Senegal.

Tenua sp.
Fig. spec. 45933
Brideaux, W.W., 1976, Geol. Surv. Can., Paper 76-1C, pl. 23.2, fig. 1.
Lower Cretaceous, south bank Martin Creek, District of Mackenzie.

Tenua sp.
Fig. spec. 47915
Williams, G.L., 1978, Initial Repts. Deep Sea Drilling Project, vol. 41, pl. 8, fig. 1.
Early Cretaceous, DSDP Site 367, lat. 12°29.2′N, long. 20°2.8′W, Cape Verde Basin west of Senegal.

Tenua sp.
Fig. spec. 71404
Wilson, M.A., 1978, Palaeontographica, Abt. B, vol. 166, p. 155, pl. 11, figs. 2, 3.
Cretaceous, Police Island, north bank Mackenzie River, 12 miles southeast of Fort Norman, District of Mackenzie.

?*Tenua* sp.
Fig. specs. 45934, 45938
Brideaux, W.W., 1976, Geol. Surv. Can., Paper 76-1C, pl. 23.2, figs. 2-4, 10.
Husky Formation, Lower Cretaceous, south bank Martin Creek, District of Mackenzie.

Tenua sp. A
Fig. specs. 41490-41492
Brideaux, W.W. and McIntyre, D.J., 1975, Geol. Surv. Can., Bull. 252, p. 25, pl. 6, figs. 14-16.
Horton River Formation (41490) and Crossley Lakes Member, Langton Bay Formation, Lower Cretaceous, lat. 69°19′N, long. 126°50′W, lat. 64°01′30″N, long. 126°02′W, and west bank opposite mouth of Coal Creek, lat. 69°07′30″N, long. 126°20′W, Horton River, District of Mackenzie.

Tenua sp. A
Fig. spec. 56749
Williams, G.L. and Bujak, J.P., 1980, Initial Repts. Deep Sea Drilling Project, vol. 50, pl. 6, figs. 1, 2.
Early Cretaceous, DSDP Site 416, lat. 32°50.18′N, long. 10°48.06′W, northwestern Moroccan continental margin.

Tenua? sp. B
Fig. spec. 45999
Brideaux, W.W., 1977, Geol. Surv. Can., Bull. 281, p. 17, pl. 6, figs. 8-10.
Cretaceous, Banff-Aquitaine-Arco Rat Pass K-35 well, depth 70-90 feet, 6 miles southeast of Mount Goodenough, lat. 67°54′42.55″N, long. 135°21′56.83″W, District of Mackenzie.

Tetranguladinium? sp. A
Fig. spec. 76303
Davies, E.H. and Poulton, T.P., 1986, Geol. Surv. Can., Paper 86-1B, p. 525, pl. 55.3, fig. 9.
Fernie Formation, Upper Jurassic, Christina Falls, lat. 56°40′N, long. 123°4′W, northeastern British Columbia.

Thalassiphora? sp. cf. *T. delicata* Williams and Downie, *in* Davey et al., 1966 emend. Eaton, 1976
Fig. specs. 65406, 65407
Ioannides, N.S., 1986, Geol. Surv. Can., Bull. 371, p. 38, pl. 22, figs. 6, 8.
Kanguk Formation, Upper Cretaceous, south shore Bylot Island, lat. 72°53′30″N, long. 78°16′W, District of Franklin.

Thalassiphora pelagica (Eisenack)
Hypotype 41877
Williams, G.L., 1975, Geol. Surv. Can., Paper 74-30, vol. 2, pl. 6, fig. 16.
Early Eocene, core hole 4, lat. 45°31′N, long. 52°22′W, Grand Banks.

Thalassiphora pelagica (Eisenack) Eisenack and Gocht, 1960
Hypotypes 65397-65405
Ioannides, N.S., 1986, Geol. Surv. Can., Bull. 371, p. 38, pl. 21, figs. 1-7; pl. 22, figs. 1-5, 7.
Eureka Sound Formation, Paleocene, south coast Bylot Island, lat. 72°52′N, long. 78°31′W, District of Franklin.

cf. *Thalassiphora* sp.
Fig. spec. 71421
Wilson, M.A., 1978, Palaeontographica, Abt. B, vol. 166, p. 155, pl. 12, fig. 4.
Tent Island Formation, Cretaceous, Boundary Creek, Big Fish River area, Yukon.

Trichodinium castaneum (Deflandre)
Hypotype 38146
McIntyre, D.J., 1974, Geol. Surv. Can., Paper 74-14, pl. 12, fig. 3.
Upper Cretaceous, lat. 69°27′30″N, long. 126°58′W, Horton River, District of Mackenzie.

Trichodinium castaneum (Deflandre)
Hypotype 41892
Williams, G.L., 1975, Geol. Surv. Can., Paper 74-30, vol. 2, pl. 7, fig. 13.
Cenomanian, Upper Cretaceous, Shell Oneida O-25 well, lat. 43°14′57.49″N, long. 61°33′36.38″W, Scotian Shelf.

Trichodinium castaneum (Deflandre) Clarke and Verdier, 1967
Hypotype 58595
McIntyre, D.J., and Brideaux, W.W., 1980, Geol. Surv. Can., Bull. 320, p. 16. pl. 4, figs. 10, 13.
Lower Cretaceous, Martin Creek, lat. 68°12′10″N, long. 135°34′50″W, Richardson Mountains, District of Mackenzie.

Trichodinium castaneum (Deflandre) Clarke and Verdier, 1967
Hypotypes 65374, 65381
Ioannides, N.S., 1986, Geol. Surv. Can., Bull. 371, p. 38, pl. 19, fig. 23; pl. 20, fig. 5.
Kanguk Formation, Upper Cretaceous, south coast Bylot Island, lat. 72°53′30″N, long. 78°16′W, District of Franklin.

Trichodinium spinosum Singh, 1971
Hypotype 84743
Leckie, D.A., 1990, J. Sed. Petrology, vol. 60, no. 1, fig. 6F.
Shaftesbury Formation, Lower Cretaceous, west bank Peace River near town of Peace River, Alberta.

Trichodinium sp. BE
Fig. spec. 58596
McIntyre, D.J., and Brideaux, W.W., 1980, Geol. Surv. Can., Bull. 320, p. 16. pl. 4, fig. 12.
Lower Cretaceous, Martin Creek, lat. 68°12′10″N, long. 135°34′50″W, Richardson Mountains, District of Mackenzie.

Trigonopyxidia ginella (Cookson and Eisenack) Downie and Sarjeant, 1965
Hypotypes 65370-65372
Ioannides, N.S., 1986, Geol. Surv. Can., Bull. 371, p. 38, pl. 19, figs. 19-21.
Kanguk Formation, Upper Cretaceous, south coast Bylot Island, lat. 72°53′30″N, long. 78°16′W, District of Franklin.

Trithyrodinium fragile Davey, 1969
Hypotypes 65392-65396
Ioannides, N.S., 1986, Geol. Surv. Can., Bull. 371, p. 39, pl. 20, figs. 10, 12, 17, 21, 22.
Kanguk Formation, Upper Cretaceous, south coast Bylot Island, lat. 72°53′30″N, long. 78°16′W, District of Franklin.

Trithyrodinium pentagonum May, 1980
Hypotypes 65375-65377
Ioannides, N.S., 1986, Geol. Surv. Can., Bull. 371, p. 39, pl. 20, figs. 1, 8, 16.
Kanguk Formation, Upper Cretaceous, Bylot Island, south coast, lat. 72°53′30″N, long. 78°16′W, District of Franklin.

Trithyrodinium suspectum (Manum and Cookson) Davey, 1969
Hypotypes 65382-65387
Ioannides, N.S., 1986, Geol. Surv. Can., Bull. 371, p. 39, pl. 20, figs. 6, 7, 11, 14, 15, 18.
Kanguk Formation, Upper Cretaceous, south coast Bylot Island, lat. 72°53′30″N, long. 78°16′W, and Eureka Sound Formation, Late Paleocene to ?Early Eocene, "Twosnout Creek", lat. 73°13′N, long. 79°53′W (65387), District of Franklin.

Trithyrodinium suspectum (Manum and Cookson) Davey, 1969
Hypotype 91440
Sweet, A.R. and McIntyre, D.J., 1988, Can. Soc. Petrol. Geol, Mem. 15, fig. 8.12.
Leyland Member, Cardium Formation, Upper Cretaceous, south side of McLeod River, l.s.d. 16, sec. 16, tp. 47, rge. 23, W. 5th mer., Alberta.

Trithyrodinium sp.
Fig. specs. 65388-65391 [65390 and 65391 same specimen]
Ioannides, N.S., 1986, Geol. Surv. Can., Bull. 371, p. 40, pl. 20, figs. 9, 13, 19, 20.
Kanguk Formation, Upper Cretaceous, south coast Bylot Island, lat. 72°53′30″N, long. 78°16′W, District of Franklin.

Trithyrodinium sp.
Fig. spec. 91441
Sweet, A.R. and McIntyre, D.J., 1988, Can. Soc. Petrol. Geol, Mem. 15, fig. 8.13.
Leyland Member, Cardium Formation, Upper Cretaceous, south side of McLeod River, l.s.d. 16, sec. 16, tp. 47, rge. 23, W. 5th mer., Alberta.

Tuberculodinium vancampoae (Rossignol)
Hypotype 41879
Williams, G.L., 1975, Geol. Surv. Can., Paper 74-30, vol. 2, pl. 6, fig. 18.
Late Eocene, core hole 2, lat. 45°8′N, long. 52°38″W, Grand Banks.

Tuberculodinium vancampoae (Rossignol) Wall, 1967
Hypotype 55362
Mudie, P.J., 1986, Initial Repts. Deep Sea Drilling Project, vol. XCIV, p. 806, pl. 5, fig. 17.
Upper Miocene, hole 611C, depth 508.69m, lat. 52°50.15′N, long. 30°18.1′W, in the Gardar Drift on southwestern flank of Reykjanes Ridge, North Atlantic Ocean.

Tubotuberella egemenii (Gitmez, 1970) Stover and Evitt, 1978
Hypotype 69361
Davies, E.H., 1983, Geol. Surv. Can., Bull. 359, pl. 4, fig. 5.
Upper Jurassic, depth 1220-1250 feet, Panarctic Sandy Point L-46 well, northwestern Melville Island, District of Franklin.

Tubotuberella rhombiformis Vozzhennikova emend.
Hypotypes 46047-46049
Brideaux, W.W., 1977, Geol. Surv. Can., Bull. 281, p. 36, pl. 14, figs. 10, 11; pl. 15, fig. 1.
Husky Formation, Jurassic, IOE Stoney Core Hole F-42, Stoney Creek about 18 miles west-southwest of Fort McPherson, lat. 67°21′23″N, long. 135°38′43″W, and Banff-Aquitaine-Arco Rat Pass K-35 well, depth 1710 feet, 6 miles southeast of Mount Goodenough, lat. 67°54′42.55″N, long. 135°21′56.83″W (46049), District of Mackenzie.

Tubotuberella rhombiformis Vozzhennikova emend. Brideaux, 1977
Hypotype 58638
McIntyre, D.J., and Brideaux, W.W., 1980, Geol. Surv. Can., Bull. 320, p. 24. pl. 12, fig. 3.
Lower Cretaceous, Martin Creek, lat. 68°12′10″N, long. 135°34′50″W, Richardson Mountains, District of Mackenzie.

Tubotuberella rhombiformis
Hypotype 76305
Davies, E.H. and Poulton, T.P., 1986, Geol. Surv. Can., Paper 86-1B, pl. 55.3, fig. 8.
Fernie Formation, Upper Jurassic, Christina Falls, lat. 56°40′N, long. 123°4′W, northeastern British Columbia.

Tubotuberella sp. cf. *T. rhombiformis* Vozzhennikova
Fig. spec. 45923
Brideaux, W.W., 1976, Geol. Surv. Can., Paper 76-1C, pl. 23.1, fig. 3.
Lower Cretaceous, south bank Martin Creek, District of Mackenzie.

Tubulospina oblongata Davey, 1970
Hypotype 84747
Leckie, D.A. et al., 1990, J. Sed. Petrology, vol. 60, no. 1, fig. 6J.
Shaftesbury Formation, Lower Cretaceous, west bank Peace River near town of Peace River, Alberta.

Turbiosphaera filosa (Wilson) Archangelsky, 1969
Hypotype 65408
Ioannides, N.S., 1986, Geol. Surv. Can., Bull. 371, p. 40, pl. 23, fig. 1.
Eureka Sound Formation, Late Paleocene to ?Early Eocene, "Twosnout Creek", Bylot Island, lat. 73°13′N, long. 79°53′W, District of Franklin.

Valensiella vermiculata Gocht
Hypotype 41891
Williams, G.L., 1975, Geol. Surv. Can., Paper 74-30, vol. 2, pl. 7, fig. 12.
Bathonian, Middle Jurassic, Amoco IOE Eider M-75 well, lat. 45°34′54.97″N, long. 51°56′41.52″W, Grand Banks.

Valvaedinium sp. A
Fig. spec. 75484
Davies, E.H., 1985, Palynology, vol. 9, p. 129, pl. 3, fig. 17.
Late Pliensbachian, Liassic, Zambujal Section, south of Condeixa, Lusitanian Basin, west-central Portugal.

Veryhachium sp.
Fig. spec. 91469
Sweet, A.R. and McIntyre, D.J., 1988, Can. Soc. Petrol. Geol, Mem. 15, fig. 9.14.
Sturrock Member, Cardium Formation, Upper Cretaceous, south side of McLeod River, l.s.d. 16, sec. 16, tp. 47, rge. 23, W. 5th mer., Alberta.

Veryhachium sp. A
Fig. spec. 76288
Davies, E.H. and Poulton, T.P., 1986, Geol. Surv. Can., Paper 86-1B, p. 525, pl. 55.4, fig. 9.
Fernie Formation, Upper Jurassic, Christina Falls, lat. 56°40′N, long. 123°4′W, northeastern British Columbia.

Vesperopsis mayi
Hypotype 91985
Banerjee, I. and Davies, E.H., 1988, Can. Svc. Petrol. Geol., Mem. 15, fig. 7.1.
Mannville Group, Lower Cretaceous, depth 5295 feet, well 11-28-40-24W4, Alberta.

Votadinium calvum Reid, 1977
Hypotype 55346
Mudie, P.J., 1986, Initial Repts. Deep Sea Drilling Project, vol. XCIV, p. 806, pl. 5, fig. 2.
Pleistocene, hole 607, depth 0.5m, lat. 41°0.07′N, long. 32°57.44′W, at base of upper western flank of the Mid-Atlantic Ridge, North Atlantic Ocean.

?*Vozzhennikovia elegantula* Williams
Holotype 47910; paratypes 47906, 47909, 47911
Williams, G.L., 1978, Initial Repts. Deep Sea Drilling Project, vol. 41, p. 797, pl. 7, figs. 4, 7-9.
Early Eocene, DSDP Site 370, lat. 32°50.2′N, long. 10°46.6′W, northwestern Moroccan continental margin.

Vozzhennikovia tenella (Morgenroth) Lentin and Williams sensu Morgenroth, 1966
Hypotypes 47912, 47920
Williams, G.L., 1978, Initial Repts. Deep Sea Drilling Project, vol. 41, pl. 7, fig. 10; pl. 8, fig. 6.
Oligocene, DSDP Site 370, lat. 32°50.2′N, long. 10°46.6′W, northwestern Moroccan continental margin.

Wallodinium krutzschi (Alberti) Habib
Hypotype 47918
Williams, G.L., 1978, Initial Repts. Deep Sea Drilling Project, vol. 41, pl. 8, fig. 4.
Early Cretaceous, DSDP Site 367, lat. 12°29.2′N, long. 20°2.8′W, Cape Verde Basin west of Senegal.

Wallodinium krutzschi (Alberti) Habib
Hypotype 56712
Williams, G.L. and Bujak, J.P., 1980, Initial Repts. Deep Sea Drilling Project, vol. 50, pl. 2, fig. 5.
Early Cretaceous, DSDP Site 416, lat. 32°50.18′N, long. 10°48.06′W, northwestern Moroccan continental margin.

Wallodinium luna (Cookson and Eisenack) Lentin and Williams
Hypotypes 46061, 46062
Brideaux, W.W., 1977, Geol. Surv. Can., Bull. 281, p. 43, pl. 16, figs. 6, 7.
Cretaceous, Banff-Aquitaine-Arco Rat Pass K-35 well, depth 1470-1480 feet, 6 miles southeast of Mount Goodenough, lat. 67°54′42.55″N, long. 135°21′56.83″W, and IOE Stoney Core Hole F-42, depth 780-785 feet, Stoney Creek about 18 miles west-southwest of Fort McPherson, lat. 67°21′23″N, long. 135°38′43″W, District of Mackenzie.

Wallodinium lunum (Cookson and Eisenack) Lentin and Williams, 1973
Hypotypes 65418, 65419
Ioannides, N.S., 1986, Geol. Surv. Can., Bull. 371, p. 40, pl. 23, figs. 5, 6.
Eureka Sound Formation, Late Paleocene to ?Early Eocene, "Twosnout Creek", Bylot Island, lat. 73°13′N, long. 79°53′W, and Kanguk Formation, Upper Cretaceous, south coast, lat. 72°53′30″N, long. 78°16′W, District of Franklin.

?*Wanaea* sp. A
Fig. spec. 75480
Davies, E.H., 1985, Palynology, vol. 9, p. 129, pl. 3, fig. 12.
Late Pliensbachian, Liassic, Brehna Road Section, Lusitanian Basin, west-central Portugal.

Wetzeliella articulata Eisenack, 1938
Hypotype 71422
Wilson, M.A., 1978, Palaeontographica, Abt. B, vol. 166, p. 155, pl. 12, fig. 5.
Cretaceous, IOE Atkinson H-25 well, north shore Tuktoyaktuk Peninsula, lat. 69°44′20″N, long. 131°50′06″W, District of Mackenzie.

Wetzeliella articulata Eisenack
Hypotype 91997
Dietrich, J.R. et al., 1989, Geol. Surv. Can., Paper 89-1G, p. 141, pl. 1, fig. 1.
Early to Middle Eocene, depths 3600-3690 feet, Dome Pacific et al. PEX Natsek E-56 well, lat. 69°45′21.46″N, long. 139°44′34.58″W, western Beaufort Sea.

Wetzeliella articulata (cf. *Wetzeliella hampdensis* Wilson)
Hypotype 92001
Dietrich, J.R. et al., 1989, Geol. Surv. Can., Paper 89-1G, pl. 1, fig. 5.
Early to Middle Eocene, depths 1200-1290 feet, Dome Pacific et al. PEX Natsek E-56 well, lat. 69°45′21.46″N, long. 139°44′34.58″W, western Beaufort Sea.

Wetzeliella condylos Williams and Downie
Hypotype 49375
Williams, G.L. and Bujak, J.P., 1977, Am. Assoc. Strat. Palynol., Contr. Strat. Palynol., vol. 1, Contr. Ser. 5A, pl. 5, fig. 11.
Early Eocene, depth 7100-7130 feet, Eastcan et al. Gudrid H-55 well, lat. 55°54′20.06″N, long. 53°52′31.04″W, Labrador Shelf.

Wetzeliella cf. *condylos* Williams and Downie
Hypotype 49374
Williams, G.L. and Bujak, J.P., 1977, Am. Assoc. Strat. Palynol., Contr. Strat. Palynol., vol. 1, Contr. Ser. 5A, pl. 5, fig. 10.
Early Eocene, depth 7100-7130 feet, Eastcan et al. Gudrid H-55 well, lat. 55°54′20.06″N, long. 53°52′31.04″W, Labrador Shelf.

Wetzeliella sp. cf. *W. hampdenensis* Wilson
Fig. specs. 68170, 68171
Norris, G., 1986, Geol. Surv. Can., Bull. 340, p. 45, pl. 13, figs. 10, 11.
Richards Formation, Middle Eocene, depths 12,600 and 12,400-12,500 feet, Imperial Nuktak C-22 well, lat. 69°41′07″N, long. 134°51′30″W, District of Mackenzie.

Wetzeliella homomorpha Deflandre and Cookson
Hypotype 49352
Williams, G.L. and Bujak, J.P., 1977, Am. Assoc. Strat. Palynol., Contr. Strat. Palynol., vol. 1, Contr. Ser. 5A, pl. 3, fig. 6.
Early-Middle Eocene, depth 1130-1160 feet, Amoco Imperial Skelly Pelican J-49 well, lat. 45°28′35.17″N, long. 52°36′41.71″W, Grand Banks.

Wetzeliella sp. A
Fig. spec. 49350
Williams, G.L. and Bujak, J.P., 1977, Am. Assoc. Strat. Palynol., Contr. Strat. Palynol., vol. 1, Contr. Ser. 5A, pl. 3, figs. 3, 4.
Early-Middle Eocene, depth 1230-1260 feet, Mobile Tetco Esperanto K-78 well, lat. 44°47′31.26″N, long. 58°11′19.24″W, Scotian Shelf.

Wetzeliella sp. B
Fig. spec. 49348
Williams, G.L. and Bujak, J.P., 1977, Am. Assoc. Strat. Palynol., Contr. Strat. Palynol., vol. 1, Contr. Ser. 5A, pl. 3, fig. 1.
Tertiary, depth 3620-3650 feet, Elf et al. Emerillon C-56 well, lat. 45°15′4.79″N, long. 54°23′16.85″W, Grand Banks.

Wetzeliella spp. D,C
Fig. specs. 49349, 49351
Williams, G.L. and Bujak, J.P., 1977, Am. Assoc. Strat. Palynol., Contr. Strat. Palynol., vol. 1, Contr. Ser. 5A, pl. 3, figs. 2, 5.
Early-Middle Eocene, depths 1040-1070 and 1130-1160 feet, Amoco Imperial Skelly Pelican J-49 well, lat. 45°28′35.17″N, long. 52°36′41.71″W, Grand Banks.

Wigginsiella grandstandica Lucus-Clark, 1987
Hypotypes 84739-87741
Leckie, D.A. et al., 1990, J. Sed. Petrology, vol. 60, no. 1, p. 101, figs. 6B-D.
Shaftesbury Formation, Lower Cretaceous, west bank Peace River near town of Peace River, Alberta.

Xenascus ceratioides (Deflandre)
Hypotype 41878
Williams, G.L., 1975, Geol. Surv. Can., Paper 74-30, vol. 2, pl. 6, fig. 17.
Santonian, Upper Cretaceous, core hole 5, lat. 45°48′N, long. 51°13′W, Grand Banks.

Xenascus ceratioides (Deflandre) Lentin and Williams, 1973
Hypotype 65416
Ioannides, N.S., 1986, Geol. Surv. Can., Bull. 371, p. 40, pl. 23, fig. 3.
Kanguk Formation, Upper Cretaceous, south coast Bylot Island, lat. 72°53′30″N, long. 78°16′W, District of Franklin.

Xenicodinium sp.
Fig. spec. 41867
Williams, G.L., 1975, Geol. Surv. Can., Paper 74-30, vol. 2, pl. 6, fig. 6.
Plio-Pleistocene, Amoco IOE Puffin B-90 well, lat. 44°39′12.73″N, long. 53°42′28.35″W, Grand Banks.

Xiphophoridium alatum (Cookson and Eisenack) Sarjeant
Hypotype 47922
Williams, G.L., 1978, Initial Repts. Deep Sea Drilling Project, vol. 41, pl. 8, fig. 8.
Late Cretaceous, DSDP Site 370, lat. 32°50.2′N, long. 10°46.6′W, northwestern Moroccan continental margin.

Xiphophoridium alatum (Cookson and Eisenack) Sarjeant, *in* Davey et al., 1966
Hypotype 65417
Ioannides, N.S., 1986, Geol. Surv. Can., Bull. 371, p. 41, pl. 23, fig. 4.
Kanguk Formation, Upper Cretaceous, ″Twosnout Creek″, Bylot Island, lat. 73°14′30″N, long. 79°49′W, District of Franklin.

Xiphophoridium alatum (Cookson and Eisenack) Sarjeant, 1966
Hypotype 91457
Sweet, A.R. and McIntyre, D.J., 1988, Can. Soc. Petrol. Geol, Mem. 15, fig. 9.2.
Leyland Member, Cardium Formation, Upper Cretaceous, south side of McLeod River, l.s.d. 16, sec. 16, tp. 47, rge. 23, W. 5th mer., Alberta.

Zyabikodinium lenticulatum Wall and Dale, 1968, emend. Bujak and Davies, 1983.
Hypotype 55347
Mudie, P.J., 1986, Initial Repts. Deep Sea Drilling Project, vol. XCIV, p. 806, pl. 5, fig. 3.
Pleistocene, hole 611, depth 7.49m, lat. 52°50.47′N, long. 30°18.58′W, in the Gardar Drift on southwestern flank of Reykjanes Ridge, North Atlantic Ocean.

ACRITARCHA

Precambrian

Acritarchs
Fig. specs. 42766
Hofmann, H.J., 1976, J. Pal., vol. 50, no. 6, p. 1072, pl. 3, figs. 8, 9.
Kasegalik Formation, Precambrian, unnamed island northern end of Churchill Sound, lat. 56°29.5′N, long. 79°30.2′W, Belcher Islands, Hudson Bay.

Archaeodiscina? sp.
Fig. spec. 76558
Hofmann, H.J., 1984, Geol. Surv. Can., Paper 84-1B, p. 287, pl. 32.1, fig. W.
Vampire Formation, Precambrian-Cambrian, lat. 64°38′30″N, long. 132°53′30″W, Wernecke Mountains, Yukon.

Archaeodiscina? sp. B
Fig. specs. 75707-75710
Baudet, D., Aitken, J.D. and Vauguestaine, M., 1989, Can. J. Earth Sci., vol. 26, no. 1, p. 137, pl. 1, figs. 7-10.
Sheepbed Formation, Late Proterozoic, near Palmer Lake, lat. 64°28′50″N, long. 129°23′30″W, and Blueflower Formation, Sekwi Brook, lat. 63°21′10″N, long. 128°24′52″W (75710), Northwest Territories.

Archaeorestis? sp.
Fig. spec. 76567
Hofmann, H.J., 1984, Geol. Surv. Can., Paper 84-1B, p. 293, pl. 32.2, fig. G.
Vampire Formation, Precambrian-Cambrian, lat. 64°33′-64°33′45″N, long. 132°59′-133°00′15″W, Wernecke Mountains, Yukon.

Archaeotrichion sp.
Fig. spec. 43586
Hofmann, H.J., 1976, J. Pal., vol. 50, no. 6, p. 1052, pl. 1, figs. 1, 2.
Kasegalik Formation, Precambrian, unnamed island northern end of Churchill Sound, lat. 56°29.5′N, long. 79°30.2′W, Belcher Islands, Hudson Bay.

Archaeotrichion sp.
Fig. specs. 76581, 76585
Hofmann, H.J., 1984, Geol. Surv. Can., Paper 84-1B, p. 290, pl. 32.3, figs. I, M.
Vampire Formation, Precambrian-Cambrian, lat. 64°33′-64°33′45″N, long. 132°59′-133°00′15″W, Wernecke Mountains, Yukon.

Bavlinella sp.
Fig. specs. 76536-76542
Hofmann, H.J., 1984, Geol. Surv. Can., Paper 84-1B, p. 289, pl. 32.1, figs. A-G.
Precambrian, lat. 64°39′-64°38′30″N, long. 132°53′30″W, lat. 64°33′-64°33′45″N, long. 132°59′-133°00′15″W (76541); Vampire Formation, Precambrian-Cambrian, lat. 64°25′30″-64°25′45″N, long. 132°21′45″-132°22′30″W (76539), lat. 64°39′15″-64°39′30″N, long. 132°19′45″-132°19′W (76540), lat. 64°33′-64°33′45″N, long. 132°59′-133°00′15″W (76542), Wernecke Mountains, Yukon.

Biocatenoides sphaerula Schopf
Hypotypes 42766
Hofmann, H.J., 1976, J. Pal., vol. 50, no. 6, p. 1053, pl. 1, figs. 8, 10.
Kasegalik Formation, Precambrian, unnamed island northern end of Churchill Sound, lat. 56°29.5′N, long. 79°30.2′W, Belcher Islands, Hudson Bay.

Carysphaeroides sp.
Fig. specs. 42773, 42775
Hofmann, H.J., 1976, J. Pal., vol. 50, no. 6, p. 1062, pl. 9, figs. 5, 7, 10.
McLeary Formation, Precambrian, Sanikiluag harbour, Flaherty Island, Belcher Islands, Hudson Bay.

Chuaria circularis Walcott
Hypotypes 42779a-j
Hofmann, H.J., 1977, Precambrian Res., vol. 4, p. 3, figs. 2a, j.
Red Pine Formation, Uninta Mountain Group, Precambrian, lat. 40°36′49″N, long. 110°22′04″W, Yellowstone Creek, Uninta Mountains, Utah, U.S.A.

Chuaria circularis Walcott
Hypotypes 57890, 57897, 57899
Hofmann, H.J. and Aitken, J.D, 1979, Can. J. Earth Sci., vol. 16, no. 1, p. 157, figs. 13A, I, K-M.
Little Dal Group, Precambrian, lat. 64°47′N, long. 129°40′W, and lat. 64°58′N, long. 130°36′N (57899), Mackenzie Mountains, District of Mackenzie.

Cymatiosphaera? sp.
Fig. specs. 98945-98948
Hofmann, H.J. and Jackson, G.D., 1991, J. Palentol., vol. 65, no. 3, p. 378, figs. 13.15-13.18.
Victor Bay Formation, Uluksan Group, Proterozoic, Eclipse Sound, lat. 72°33′N, long. 79°78′W, northwestern Baffin Island, District of Franklin.

Eoentophysalis belcherensis Hofmann
Holotype 42770; paratypes 42766, 43589-43591
Hofmann, H.J., 1976, J. Pal., vol. 50, no. 6, p. 1070, pl. 4, figs. 1-5.
Golubic, S. and Hofmann, H.J., 1976, ibid., p. 1079, pl. 1, figs. 7-10; pl. 2, figs. 4-6.
Kasegalik Formation, Precambrian, unnamed island northern end of Churchill Sound, lat. 56°29.5′N, long. 79°30.2′W, Belcher Islands, Hudson Bay.

Eoentophysalis belcherensis capsulata Hofmann
Holotype 42770; paratypes 42770, 42773
Hofmann, H.J., 1976, J. Pal., vol. 50, no. 6, p. 1070, pl. 5, fig. 3; pl. 6, figs. 2, 10, 11.
Kasegalik and McLeary formations, Precambrian, unnamed island northern end of Churchill Sound, lat. 56°29.5′N, long. 79°30.2′W, and Sanikiluaq Harbour, Flaherty Island (42773), Belcher Islands, Hudson Bay.

Eoentophysalis belcherensis capsulopunctata Hofmann
Holotype 42770; paratypes 42770, 42774, 43589, 43592
Hofmann, H.J., 1976, J. Pal., vol. 50, no. 6, p. 1070, pl. 5, figs. 4, 6; pl. 6, figs. 1, 3-6, 12-14.
Kasegalik and McLeary formations, Precambrian, unnamed island northern end of Churchill Sound, lat. 56°29.5′N, long. 79°30.2′W, and Sanikiluaq Harbour, Flahrty Island (42774, 43592) Belcher Islands, Hudson Bay.

Eoentophysalis belcherensis punctata Hofmann
Holotype 42770; paratypes 42770, 43590
Hofmann, H.J., 1976, J. Pal., vol. 50, no. 6, p. 1070, pl. 5, fig. 5; pl. 6, figs. 7-9.
Kasegalik Formation, Precambrian, unnamed island northern end of Churchill Sound, lat. 56°29.5′N, long. 79°30.2′W, Belcher Islands, Hudson Bay.

Eomicrhystridium? sp.
Fig. specs. 61593a-1, 2
Hofmann, H.J., Hill, J. and King, A.F., 1979, Geol. Surv. Can., Paper 79-1B, p. 92, pl. 12.3, figs. 16, 17.
Connecting Point Group, Precambrian, west side of road-cut Trans Canada Highway 0.9 km south of junction of Southern Harbour road, lat. 47°43′53″N, long. 53°56′03″W, southeastern Newfoundland.

Eomicrhystridium? sp.
Fig. spec. 76559
Hofmann, H.J., 1984, Geol. Surv. Can., Paper 84-1B, p. 289, pl. 32.1, fig. X.
Precambrian, lat. 64°39′15″-64°39′30″255DN, long. 132°19′45″-132°19′W, Wernecke Mountains, Yukon.

Eomycetopsis filiformis Schopf
Hypotypes 42766, 43590, 43592
Hofmann, H.J., 1976, J. Pal., vol. 50, no. 6, p. 1052, pl. 1, figs. 3-5.
Kasegalik and McLeary formations, Precambrian, unnamed island northern end of Churchill Sound, lat. 56°29.5′N, long. 79°30.2′W, and 5 km east-southeast of Sanikiluaq and 2.5 km west-southwest of Katuk Passage, Belcher Islands, Hudson Bay.

Eomycetopsis filiformis Schopf or *Biocatenoides sphaerula* Schopf
Hypotypes 42778, 42766
Hofmann, H.J., 1976, J. Pal., vol. 50, no. 6, p. 1052, pl. 1, figs. 6, 7.
McLeary and Kasegalik formations, Precambrian, Sanikiluaq Harbour and unnamed island northern end of Churchill Sound, lat. 56°29.5′N, long. 79°30.2′W, Belcher Islands, Hudson Bay.

Eomycetopsis spp.
Fig. specs. 76572-76574, 76587, 76588
Hofmann, H.J., 1984, Geol. Surv. Can., Paper 84-1B, p. 290, pl. 32.3, fig. L; pl. 32.3, figs. A, B, G, O, P.
Vampire Formation, Precambrian-Cambrian, lat. 64°33′-64°33′45″N, long. 132°59′-133°00′15; Precambrian, lat. 64°32′15″-64°32′30″2N, long. 132°57′45″-132°58′30″W (76573), and lat. 64°32′15″N, long. 132°56′45″W (76574), Wernecke Mountains, Yukon.

Eomycetopsis? sp
Fig. specs. 76577, 76578, 76583
Hofmann, H.J., 1984, Geol. Surv. Can., Paper 84-1B, p. 290, pl. 32.3, figs. E, F, X.
Precambrian, lat. 64°39′-64°38′30″N, long 132°55′-132°53′30″W; Precambrian and Vampire Formation, Precambrian-Cambrian, lat. 64°33′-64°33′45″N, long. 132°59′-133°00′15″W, Wernecke Mountains, Yukon.

Eosynechococcus grandis Hofmann
Holotype 42775; paratypes 42771, 42777
Hofmann, H.J., 1976, J. Pal., vol. 50, no. 6, p. 1058, pl. 2, figs. 11-14.
McLeary Formation, Precambrian, 0.5 km northwest of Sanikiluaq Harbour, Flaherty Island, Belcher Island, Hudson Bay.

Eosynechococcus medius Hofman
Holotype and paratype 43588
Hofmann, H.J., 1976, J. Pal., vol. 50, no. 6, p. 1058, pl. 2, fig. 9, 10.
Kasegalik Formation, Precambrian, unnamed island northern end of Churchill Sound, lat. 56°29.5′N, long. 79°30.2′W, Belcher Islands, Hudson Bay.

Eosynechococcus moorei Hofmann
Holotype 42770; paratypes 42766, 42770, 43589
Hofmann, H.J., 1976, J. Pal., vol. 50, no. 6, p. 1057, pl. 2, figs. 1-7.
Kasegalik Formation, Precambrian, unnamed island northern end of Churchill Sound, lat. 56°29.5′N, long. 79°30.2′W, Belcher Islands, Hudson Bay.

Eosynechococcus moorei Hofmann or *Sphaerophycus parvum* Schopf
Hypotype 43589
Hofmann, H.J., 1976, J. Pal., vol. 50, no. 6, p. 1057, pl. 2, fig. 8.
Kasegalik Formation, Precambrian, unnamed island northern end of Churchill Sound, lat. 56°29.5′N, long. 79°30.2′W, Belcher Islands, Hudson Bay.

Eozygion minutum Schopf
Hypotype 42775
Hofmann, H.J., 1976, J. Pal., vol. 50, no. 6, p. 1062, pl. 7, fig. 9.
McLeary Formation, Precambrian, Sanikiluaq Harbour, Flaherty Island, Belcher Island, Hudson Bay.

Eupoikilofusa? sp.
Fig. specs. 98877 (partim)-98879(partim)
Hofmann, H.J. and Jackson, G.D., 1991, J. Palentol., vol. 65, no. 3, p. 378, figs. 7.11-7.13.
Victor Bay Formation, Uluksan Group, Proterozoic, Eclipse Sound, lat. 72°33′N, long. 79°78′W, northwestern Baffin Island, District of Franklin.

Glenobotrydion majorinum Schopf and Blacic
Hypotype 42770
Hofmann, H.J., 1976, J. Pal., vol. 50, no. 6, p. 1065, pl. 8, fig. 12.
McLeary Formation, Precambrian, Sanikiluaq Harbour, Flaherty Island, Belcher Islands, Hudson Bay.

Globophycus sp.
Fig. specs. 42773, 42775, 43590
Hofmann, H.J., 1976, J. Pal., vol. 50, no. 6, p. 1062, pl. 9, figs. 2-4, 6, 9.
McLeary and Kasegalik (43590) formations, Precambrian, Sanikiluaq Harbour, Flaherty Island, and unnamed island northern end of Churchill Sound, lat. 56°29.5′N, long. 79°30.2′W, Belcher Islands, Hudson Bay.

Halythrix sp.
Fig. specs. 42769
Hofmann, H.J., 1976, J. Pal., vol. 50, no. 6, p. 1054, pl. 1, figs. 18, 19.
Kasegalik Formation, Precambrian, unnamed island northern end of Churchill Sound, lat. 56°29.5′N, long. 79°30.2′W, Belcher Islands, Hudson Bay.

Kakabekia? sp.
Fig. spec. 42774
Hofmann, H.J., 1976, J. Pal., vol. 50, no. 6, p. 1072, pl. 7, fig. 10.
McLeary Formation, Precambrian, Sanikiluaq Harbour, Flaherty Island, Belcher Islands, Hudson Bay.

Kildinella sp.
Fig. spec. 57896b
Hofmann, H.J. and Aitken, J.D., 1979, Can. J. Earth Sci., vol. 16, no. 1, p. 157, fig. 9K.
Little Dal Group, Precambrian, headwaters Gayna River, lat. 69°48′N, long. 129°47′W, Mackenzie Mountains, District of Mackenzie.

Leiosphaeridia sp.
Fig. specs. 76551, 76553-76556, 76560-76564
Hofmann, H.J., 1984, Geol. Surv. Can., Paper 84-1B, p. 287, pl. 32.1, figs. P, R-U, Y (?), Z; pl. 32.2, figs. A-D.
Vampire Formation, Precambrian-Cambrian, lat. 64°33′-64°33′45″N, long. 132°59′-133°00′15″W, lat. 64°26′N, long. 132°24′30″-132°23′45″N (76656), lat. 64°39′15″-64°39′30″N, long. 132°19′45″-13°19′W (76560); Precambrian, lat. 64°39′15″-64°39′30″N, long. 132°19′45″-132°19′W (76561, 76563, 76564), Wernecke Mountains, Yukon.

Melasmatosphaera magna Hofmann
Holotype 42775; paratype 43589
Hofmann, H.J., 1976, J. Pal., vol. 50, no. 6, p. 1066, pl. 8, figs. 1a, b, 2.
McLeary and Kasegalik formations, Precambrian, 0.5 km northwest of Sanikiluaq Harbour, Flaherty Island, and unnamed island northern end of Churchill Sound, lat. 56°29.5′N, long. 79°30.2′W, Belcher Islands, Hudson Bay.

Melasmatosphaera media Hofmann
Holotype 42776; paratypes 42773, 42775, 42776, 43589, 43590
Hofmann, H.J., 1976, J. Pal., vol. 50, no. 6, p. 1066, pl. 8, figs. 3-6, 9-11, 13; pl. 9, fig. 8.
McLeary and Kasegalik formations, Precambrian, 0.8 km north-northwest of Sanikiluaq Harbour, Flaherty Island, and unnamed island northern end of Churchill Sound, lat. 56°29.5′N, long. 79°30.2′W, Belcher Islands, Hudson Bay.

Melasmatosphaera media Hofmann or *Glenobotydion majorinum* Schopf and Blacic
Hypotype 42775

Hofmann, H.J., 1976, J. Pal., vol. 50, no. 6, p. 1066, pl. 8, fig. 14.
McLeary Formation, Precambrian, Sanikiluaq Harbour, Flaherty Island, Belcher Islands, Hudson Bay.

Melasmatosphaera parva Hofmann
Holotype 42774; paratype 42770
Hofmann, H.J., 1976, J. Pal., vol. 50, no. 6, p. 1066, pl. 8, figs. 7, 8.
McLeary and Kasegalik formations, Precambrian, 0.5 km northwest of Sanikiluaq Harbour, Flaherty Island, and unnamed island northern end of Churchill Sound, lat. 56°29.5′N, long. 79°30.2′W, Belcher Islands, Hudson Bay.

? *Micrhystridium tornatum* Volkova 1968
Hypotype 75746
Baudet, D., Aitken, J.D. and Vanguestaine, M., 1989, Can. J. Earth Sci., vol. 26, no. 1, p. 142, pl. 2, fig. 14.
Sheepbed Formation, Proterozoic, near Palmer Lake, lat. 64°28′50″N, long. 129°23′30″W, Northwest Territories.

Myxococcoides inornata Schopf
Hypotype 42772
Hofmann, H.J., 1976, J. Pal., vol. 50, no. 6, p. 1061, pl. 7, fig. 3.
McLeary Formation, Precambrian, Sanikiluaq Harbour, Flaherty Island, Belcher Islands, Hudson Bay.

Myxococcoides minor Schopf
Hypotypes 42773, 43589
Hofmann, H.J., 1976, J. Pal., vol. 50, no. 6, p. 1061, pl. 7, figs. 4-8.
McLeary and Kasegalik formations, Precambrian, Sanikiluaq Harbour, Flaherty Island, and unnamed island northern end of Churchill Sound, lat. 56°29.5′N, long. 79°30.2′W, Belcher Islands, Hudson Bay.

Myxococcoides sp.
Fig. specs. 42775
Hofmann, H.J., 1976, J. Pal., vol. 50, no. 6, p. 1061, pl. 7, figs. 1, 2.
McLeary Formation, Precambrian, Sanikiluaq Harbour, Flaherty Island, Belcher Islands, Hudson Bay.

Myxococcoides? sp.
Fig. specs. 42773
Hofmann, H.J., 1976, J. Pal., vol. 50, no. 6, p. 1061, pl. 9, fig. 11.
McLeary Formation, Precambrian, Sanikiluaq Harbour, Flaherty Island, Belcher Islands, Hudson Bay.

Nostocomorpha sp.
Fig. spec. 76586
Hofmann, H.J., 1984, Geol. Surv. Can., Paper 84-1B, p. 293, pl. 32.3, fig. N.
Vampire Formation, Precambrian-Cambrian, lat. 64°33′-64°33′45″N, long. 132°59′-133°00′15″W, Wernecke Mountains, Yukon.

Nucellosphaeridium spp.
Fig. spec. 57896C
Hofmann, H.J. and Aitken, J.D., 1979, Can. J. Earth Sci., vol. 16, no. 1, p. 156, figs. 12 A-C.
Little Dal Group, Precambrian, headwaters Gayna River, lat. 64°48′N, long. 129°47′W, Mackenzie Mounatins, District of Mackenzie.

Nucellosphaeridium sp.
Fig. specs. 76552, 76557
Hofmann, H.J., 1984, Geol. Surv. Can., Paper 84-1B, p. 287, pl. 32.1, figs. Q, V.
Precambrian, lat. 64°39′15″-64°39′30″N, long. 132°19′45″-132°19′W, Wernecke Mountains, Yukon.

Oscillatoriopsis psilata
Hypotype 76589
Hofmann, H.J., 1984, Geol. Surv. Can., Paper 84-1B, p. 291, pl. 32.3, figs. Q.
Vampire Formation, Precambrian-Cambrian, lat. 64°33′-64°33′45″N, long. 132°59′-133°00′15″W, Wernecke Mountains, Yukon.

Palaeoanacystis vulgaris Schopf
Hypotypes 43592, 42775
Hofmann, H.J., 1976, J. Pal., vol. 50, no. 6, p. 1061, pl. 5, figs. 1, 2.
McLeary Formation, Precambrian, 5 km east-southeast of Sanikiluaq and 2.5 km west-southwest of Katuk passage, and Sanikiluaq Harbour, Flaherty Island, Belcher Islands, Hudson Bay.

Pleurocapsa? sp.
Fig. spec. 42770
Hofmann, H.J., 1976, J. Pal., vol. 50, no. 6, p. 1072, pl. 3, fig. 7.
Kasegalik Formation, Precambrian, unnamed island northern end of Churchill Sound, lat. 56°29.5′N, long. 79°30.2′W, Belcher Islands, Hudson Bay.

Rhicnonema antiquum Hofmann or *Halythrix* sp.
Hypotype 42769
Hofmann, H.J., 1976, J. Pal., vol. 50, no. 6, p. 1053, pl. 1, fig. 15.
Kasegalik Formation, Precambrian, unnamed island northern end of Churchill Sound, lat. 56°29.5′N, long. 79°30.2′W, Belcher Islands, Hudson Bay.

Rhicnonema antiquum catenoides Hofmann
Holotype and paratype 42766
Hofmann, H.J., 1976, J. Pal., vol. 50, no. 6, p. 1053, pl. 1, figs. 9, 16.
Kasegalik Formation, Precambrian, unnamed island northern end of Churchill Sound, lat. 56°29.5′N, long. 79°30.2′W, Belcher Islands, Hudson Bay.

Rhicnonema antiquum spiraliforme Hofmann
Holotype 42766; paratypes 42766, 43587
Hofmann, H.J., 1976, J. Pal., vol. 50, no. 6, p. 1053, pl. 1, figs. 12-14.
Kasegalik Formation, Precambrian, unnamed island northern end of Churchill Sound, lat. 56°29.5′N, long. 79°30.2′W, Belcher Islands, Hudson Bay.

Rhicnonema antiquum velatum Hofmann
Holotype and paratype 42766
Hofmann, H.J., 1976, J. Pal., vol. 50, no. 6, p. 1053, pl. 1, figs. 11, 17.

Kasegalik Formation, Precambrian, unnamed island northern end of Churchill Sound, lat. 56°29.5′N, long. 79°30.2′W, Belcher Islands, Hudson Bay.

Siphonophycus spp.
Fig. specs. 76575, 76576, 76580, 76584
Hofmann, H.J., 1984, Geol. Surv. Can., Paper 84-1B, p. 291, pl. 32.3, figs. C, D, H, L.
Precambrian, lat. 64°39′-64°38′30″N, long. 132°55′-132°53′30″W, and lat. 64°39′15″-64°39′30″N, long. 132°19′45″-132°19′W; Vampire Formation, Precambrian-Cambrian, lat. 64°33′-64°33′45″N, long. 132°59′-133°00′15″W, and lat. 64°26′N, long. 132°24′30″-132°23′45″W, Wernecke Mountains, Yukon.

Siphonophycus? sp.
Fig. specs. 76582
Hofmann, H.J., 1984, Geol. Surv. Can., Paper 84-1B, p. 291, pl. 32.3, fig. J.
Precambrian, lat. 64°39′15″-64°39′30″N, long. 132°19′45″-132°19′W, Wernecke Mountains, Yukon.

Sphaeromorphe Evitt 1963
Fig. spec. 75749
Baudet, D., Aitken, J.D. and Vanguestaine, M., 1989, Can. J. Earth Sci., vol. 26, no. 1, p. 143, pl. 2, figs. 16.
Sheepbed Formation, Late Proterozoic, Palmer Lake, lat. 64°28′50″N, long. 129°23′30″W, Northwest Territories.

Sphaerophycus parvum Schopf
Hypotypes 42770, 42773, 42776, 43589, 43592
Hofmann, H.J., 1976, J. Pal., vol. 50, no. 6, p. 1058, pl. 3, figs. 1-6.
Kasegalik and McLeary formations, Precambrian, unnamed island northern end of Churchill Sound, lat. 56°29.5′N, long. 79°30.2′W; Sanikiluaq Harbour (42773, 42776); and 5 km east-southeast of Sanikiluaq Harbour and 2.5 km west-southwest of Katuk Passage (43592), Belcher Islands, Hudson Bay.

Spheroidal compressions
Fig. specs. 61588, 61590, 61597, 61605-61610
Hofmann, H.J., Hill, J. and King, A.F., 1979, Geol. Surv. Can., Paper 79-1B, p. 87, text-fig. 12.3.
Precambrian, Fermeuse Formation, road-cut at western outskirts of Port Kirwan, lat. 46°58′11″N, long. 52°55′W, east side of highway 6.5 km south of Cappahayden, lat. 46°50′N, long. 52°59′20″W (61590), and beach point southeast of causeway to Powles Peninsula, Mutton Bay, lat. 46°43′30″N, long. 53°22′W (61606); Connecting Point Group, northwest side of road 3 km southwest of Salvage outcrop, Wild Cove, lat. 48°40′48″N, long. 53°40′43″W (61597), road-cut south side of road, 7 km southwest of Salvage, Dark Cove, lat. 48°39′53″N, long. 53°42′36″W (61607), west side Hayward's Cove, Cottel Island, lat. 48°52′12″N, long. 53°39′W (61608), next to ferry wharf, Shalloway Cove, Cottel Island, lat. 48°50′24″N, long. 53°41′15″W (61609), and south side of road to Little Heart's Ease, Leonard's Cove, lat. 48°00′07″N, long. 53°51′W (61610); and Snows Pond Formation, west side of Green Harbour, lat. 47°39′N, long. 53°31′W (61605), southeastern Newfoundland.

"Spiromorphes"
Fig. specs. 75752-75757
Baudet, D., Aitken, J.D. and Vanguestaine, M., 1989, Can. J. Earth Sci., vol. 26, no. 1, p. 144, pl. 2, figs. 20-25.
Blueflower Formation, Late Proterozoic, Sekwi Brook, lat. 63°21′10″N, long. 128°24′52″W, Northwest Territories.

Trachysphaeridium spp.
Fig. spec. 57896b
Hofmann, H.J. and Aitken, J.D., 1979, Can. J., Earth Sci., vol. 16, no. 1, p. 156, figs. 9I, J.
Little Dal Group, Precambrian, headwaters Gayna River, lat. 64°48′N, long. 129°47′W, Mackenzie Mountains, District of Mackenzie.

Trachysphaeridium spp.
Fig. specs. 61573-1, 2, 61577-1 to 4, 61578, 61579a-2, 3, b, 61581-1, 2, 61582-1, 2, 61583-1, 61584, 61586, 61587-1 to 8, 61589, 61592, 61593b-1, 2, 61594-61596, 61598-61602
Hofmann, H.J., Hill, J. and King, A.F., 1979, Geol. Surv. Can., Paper 79-1B, p. 90, pl. 12.1, fig. 23; pl. 12.2, figs. 1, 2, 5-14; pl. 12.3, figs. 1, 2, 4-14, 18-28.
Precambrian, Musgravetown Group, 1.5 km east of Davis Cove, lat. 47°38′30″N; Snows Pond Formation, north side Placentia Highway, 8 km south of Trans Canada Highway junction, lat. 47°23′24″N, long. 53°36′14″W (61577-1 to 4), and Trans Canada Highway, 0.5 km west of junction to Placentia-Argentia, lat. 47°27′N, long. 53°34′24″W (61578); Halls Town Formation, 2 km southwest of Harricott, lat. 47°09′30″N, long. 53°33′24″W (61579a-2, 3, b, 61581-1, 2); Gibbett Hill and Cappahayden formations, Burnt Point, 1.2 km north of Cappahayden at end of road, lat. 46°51′45″N, long. 52°56′23″W (61582-1, 2, 61584); Cappahayden Formation, ca. 200 m west-northwest of lighthouse, Bear Cove point, lat. 46°56′27″N, long. 52°53′44″W (61583-1); Renews Head Formation, southside Southside Road opposite Gordon Forwarders and below overpass foundation, St. John's, lat. 47°33′12″N, long. 52°42′54″W (61586); Fermeuse Formation, road-cut opposite NLPG transformer station, Mobile, lat. 47°14′51″N, long. 52°50′41″W (61587-1 to 8), and coastal exposure opposite Crown Island, Ferryland, lat. 47°00′40″N, long. 52°53′35″W (61589); Connecting Point Group, west side of road-cut Trans Canada Highway 0.2 and 0.9 km south of junction of Southern Harbour road, lat. 47°44′09″N, long. 53°56′05″W (61592) and lat. 47°43′53″N, long. 53°56′03″W (61593b-1, 2), east side Trans Canada Highway, 1.5 km southwest of junction of Southern Harbour road, lat. 47°44′N, long. 54°55′24″W (61594), roads end southwest side Little Harbour east, lat. 47°38′30″N, long. 54°56′20″W, (61595), next to ferry wharf, Burnside, lat.

48°43′N, long. 53°48′12″W (61596), 1.5 km northwest of St. Brendan's bride northermost wharf, lat. 48°53′33″N, long. 53°39′55″W (61595), 0.5 km south of Cutwater Head, Canning's Cove, lat. 48°26′45″N, long. 53°50′40″W (61599), northwest side of road beside Long Pond, 3.5 km west of Hickman's harbour, Random Island, lat. 48°06′25″N, long. 53°46′30″W (61600), north side Cabot Highway 2 km west of Southern Bay Station, Lethbridge, lat. 48°22′N, long. 53°42′W (61601), and south side of road on east side of brook, Queen's Cove, lat. 48°00′40″N, long. 53°56′W (61602), southeastern Newfoundland.

Trachysphaeridium? sp. - *Eomichrystridium?* sp.
Fig. specs. 61579a-1, 61580, 61583-2
Hofmann, H.J., Hill, J. and King, A.E., 1979, Geol. Surv. Can., Paper 79-1B, pl. 12.3, figs. 3, 4; pl. 12.3, fig. 3.
Precambrian, Halls Town Formation, 2 km southwest of Harricott, lat. 47°09′30″N, long. 53°33′24″W; Cappahayden Formation, ca. 200 m west-northwest of lighthouse, Bear Cove point, lat. 46°56′27″N, long. 52°53′44″W (61583-2), southeastern Newfoundland.

Trematosphaeridium holtedahlii Timofeev
Hypotype 61592
Hofmann, H.J., Hill, J. and King, A.., 1979, Geol. Surv. Can., Paper 79-1B, p. 90, pl. 12.3, fig. 15.
Connecting Point Group, Precambrian, west side of road-cut on Trans Canada Highway 0.2 km south of junction of Southern Harbour road, lat. 47°44′09″N, long. 53°56′05″W, southeastern Newfoundland.

Unnamed form A, B
Fig. specs. 98902, 98930
Hofmann, H.J. and Jackson, G.D., 1991, J. Paleontol., vol. 65, no. 3, p. 378, 340, figs. 8.14, 12.10.
Society Cliffs Formation, Uluksan Group, Proterozoic, Eclipse Sound, lat. 72.25°N, long. 79°97′W, northwestern Baffin Island, and Navy Board Inlet, lat. 73°38′N, long. 80.39°W, Bylot Island, District of Franklin.

Veryhachium(?) sp.
Hofmann, H.J. and Schopf, J.W., 1983, Earth's Earliest Biosphere, Phot. 14-1T (fig. spec. 24380c).

Veryhachium? sp. and *Huroniospora macroreticulata* Barghoorn
Fig. spec. 24380c
Hofmann, H.J., 1971, J.Pal., vol. 45, No. 3, p. 522, pl. 67, figs. 1, 2, 4, 5; text-fig. 1.
Gunlint Formation, Precambrian, 6.4 km west of Schreiber, Ontario.

Paleozoic

Cambrian

Acanthodiacrodium ubui Martin, 1969
Hypotype 57831
Martin, F. and Dean, W.T., 1981, Geol. Surv. Can., Bull. 343, p. 13, pl. 14, figs. 2, 4.
Elliott Cove Formation, Upper Cambrian, south of Promontory VIII, south of Elliott's Cove, Random Island, eastern Newfoundland.

Acritarch gen. et sp. nov.
Fig. specs. 72935-72938
Martin, F. and Dean, W.T., 1984, Geol. Surv. Can., Paper 84-1A, p. 433, pl. 57.1, figs. 10, 17-19.
Chamberlains Brook Formation, Middle Cambrian, west coast Random Island north-northwest of Weybridge, Newfoundland.

Acritarch gen. et sp. nov. Martin *in* Martin and Dean, 1984
Fig. spec. 83123
Martin, F. and Dean, W.T., 1988, Geol. Surv. Can., Bull. 381, pl. 11, fig. 4.
Manuels River Formation, Middle Cambrian, west bank of Manuels River, eastern Newfoundland.

Actinotodissus achrasi (Martin, 1973)
Hypotypes 83124-83127
Martin, F. and Dean, W.T., 1988, Geol. Surv. Can., Bull. 381, p. 35, pl. 10, figs. 1-5.
Elliott Cove Formation, Upper Cambrian, northwest side of Promontory VII, coastal section south of Elliotts Cove, northwest Random Island, eastern Newfoundland.

Actinotodissus cf. *A. ubui* (Martin, 1969)
Hypotypes 83128, 83129
Martin, F. and Dean, W.T., 1988, Geol. Surv. Can., Bull. 381, p. 35, pl. 11, figs. 1, 2.
Elliott Cove Formation, Upper Cambrian, northwest side of Promontory VII, coastal section south of Elliotts Cove, northwest Random Island, eastern Newfoundland.

Adara alea Martin
Holotype 57782; paratypes 57781, 57783, 57835-57837
Martin, F. and Dean, W.T., 1981, Geol. Surv. Can., Bull. 343, p. 16, pl. 1, figs. 20-22, pl. 4, figs. 7, 9, 10.
Manuels River Formation, Middle Cambrian, south of Promontory I, south of Elliott's Cove, Random Island, eastern Newfoundland.

Adara alea Martin *in* Martin and Dean, 1981
Hypotypes 72933, 72934
Martin, F. and Dean, W.T., 1984, Geol. Surv. Can., Paper 84-1A, pl. 57.3, figs. 3, 11.
Manuels River Formation, Middle Cambrian, west coast of Random Island north-northwest of Weybridge, Newfoundland.

Adara alea Martin *in* Martin and Dean, 1981
Hypotypes 83130-83133
Martin, F. and Dean, W.T., 1988, Geol. Surv. Can., Bull. 381, pl. 11, figs. 3, 5, 6, 8.
Manuels River Formation, Middle Cambrian, west bank of Manuels River, eastern Newfoundland.

Adara sp. A
Fig. specs. 75729, 75730
Baudet, D., Aitken, J.D. and Vanguestaine, M., 1989, Can. J. Earth Sci., vol. 26, no. 1, p. 136, pl. 1, figs. 29, 30.
Vampire Formation, Early Cambrian, June Lake, lat. 63°29′30″N, long. 128°40′15′-41′30″W, Sekwi Mountain, Northwest Territories.

Annulum squamaceum (Volkova, 1968) comb. n.
Hypotypes 69618-69622
Martin, F. and Dean, W.T., 1983, Geol. Surv. Can., Paper 83-1B, p. 359, pl. 43.1, figs. 4, 7-12.
Chamberlains Brook Formation, Middle Cambrian, and Brigus Formation, Lower Cambrian (69619, 69620), west bank Manuels River, west-northwest of road bridge at Manuels, eastern Newfoundland.

?*Aranidium* sp.
Fig. spec. 75727
Baudet, D., Aitken, J.D. and Vanguestaine, M., 1989, Can. J. Earth Sci., vol. 26, no. 1, p. 136, pl. 1, fig. 27.
Vampire Formation, Early Cambrian, June Lake, lat. 63°29′30″N, long. 128°40′15″-41′30″W, Sekwi Mountain, Northwest Territories.

Arbusculidium rommelaerei Martin
Holotype 57818; paratypes 57817, 57819, 57820, 57839
Martin, F. and Dean, W.T., 1981, Geol. Surv. Can., Bull. 343, p. 16, pl. 3, figs. 6, 8, 21, 23; pl. 5, fig. 1.
Martin, F., 1982, Nat. Mus. Wales, Geol. Ser. 3, pl. 1, fig. 16 (57819).
Elliott Cove Formation, Upper Cambrian, lower part of cliff approximately 1800m southeast of southeast end of Bounds Mead, and south of Promontory VIII (57817), south of Elliott's Cove, Random Island, eastern Newfoundland.

Arbusculidium rommelaerei Martin *in* Martin and Dean, 1981
Hypotypes 83134-83137
Martin, F. and Dean, W.T., 1988, Geol. Surv. Can., Bull. 381, p. 36, pl. 11, figs. 7, 9-13.
Elliott Cove Formation, Upper Cambrian, cliff south of Promontory VIII, coastal section south of Elliotts Cove, northwest Random Island, eastern Newfoundland.

Archaeodiscina aff. *umbonulata* Volkova 1968
Hypotype 75703
Baudet, D., Aitken, J.D. and Vanguestaine, M., 1989, Can. J. Earth Sci., vol. 26, no. 1, p. 136, pl. 1, fig. 3.
Vampire Formation, Early Cambrian, June Lake, lat. 63°29′30″N, long. 128°40′15″-41′30″W, Sekwi Mountain, Northwest Territories.

Archaeodiscina cf. *umbonulata* Volkova 1968
Hypotypes 75701, 75702
Baudet, D., Aitken, J.D. and Vanguestaine, M., 1989, Can. J. Earth Sci., vol. 26, no. 1, p. 136, pl. 1, figs. 1, 2.
Vampire Formation, Early Cambrian, June Lake, lat. 63°29′30″N, long. 128°40′15″-41′30″W, Sekwi Mountain, Northwest Territories.

Archaeodiscina sp. A
Fig. specs. 75704-75706
Baudet, D., Aitken, J.D. and Vanguestaine, M., 1989, Can. J. Earth Sci., vol. 26, no. 1, p. 136, pl. 1, figs. 4-6.
Vampire Formation, Early Cambrian, June Lake, lat. 63°29′30″N, long. 128°40′15″-41′30″W, Sekwi Mountain; Sheepbed Formation, Proterozoic, near Palmer Lake, lat. 64°28é50″N, long. 129°23′30″W (75706), Northwest Territories.

cf. *Baltisphaeridium cerinum* Volkova 1968
Hypotype 75725
Baudet, D., Aitken, J.D. and Vanguestaine, M., 1989, Can. J. Earth Sci., vol. 26, no. 1, p. 138, pl. 1, fig. 25.
Vampire Formation, Early Cambrian, June Lake, lat. 63°29′30″N, long. 128°40′15″-41′30″W, Sekwi Mountain, Northwest Territories.

Comasphaeridium mackenziana Baudet, Aitken and Vanguestaine
Holotype 75715; paratypes 75714, 75716, 75717
Baudet, D., Aitken, J.D. and Vanguestaine, M., 1989, Can. J. Earth Sci., vol. 26, no. 1, p. 138, pl. 1, figs. 14-17.
Vampire and Sekwi (75717) formations, Early Cambrian, June Lake, lat. 63°29′30″N, long. 128°40′15″-41′30″W, Sekwi Mountain, Northwest Territories.

Comasphaeridium sp. A dans Downie 1982
Fig. specs. 75718, 75719
Baudet, D., Aitken, J.D. and Vanguestaine, M., 1989, Can. J. Earth Sci., vol. 26, no. 1, p. 140, pl. 1, figs. 18, 19.
Vampire Formation, Early Cambrian, June Lake, lat. 63°29′30″N, long. 128°40′15″-41′30″W, Sekwi Mountain, Northwest Territories.

Cristallinium cambriense (Slavíková) Vanguestaine, 1978
Hypotypes 57813-57816, 57842, 57843
Martin, F. and Dean, W.T., 1981, Geol. Surv. Can., Bull. 343, p. 17, pl. 3, figs. 4, 5, 9, 11; pl. 5, figs. 3, 5, 8, 11.
Manuels River Formation, Middle Cambrian, south of Promontory I, south of Elliott's Cove, Random Island, eastern Newfoundland.

Cristallinium cambriense (Slavíková) Vanguestaine, 1978
Hypotypes 72939-72944
Martin, F. and Dean, W.T., 1984, Geol. Surv. Can., Paper 84-1A, p. 433, pl. 57.1, figs. 1-7, 9, 13.
Manuels River Formation, Middle Cambrian, west coast Random Island north-northwest of Weybridge, Newfoundland.

Cristallinium cambriense (Slavíková) Vanguestaine, 1978
Hypotypes 83138, 83139
Martin, F. and Dean, W.T., 1988, Geol. Surv. Can., Bull. 381, p. 36, pl. 12, figs. 1, 2.
Manuels River Formation, Middle Cambrian, west bank of Manuels River, eastern Newfoundland.

Cristallinium ovillense (Cramer and Diez de Cramer)
Hypotype 57826
Martin, F. and Dean, W.T., 1981, Geol. Surv. Can., Bull. 343, p. 17, pl. 3, fig. 16.
Elliott Cove Formation, Upper Cambrian, cliff between Promontories IV and V, south of Elliott's Cove, Random Island, eastern Newfoundland.

Cristallinium randomense Martin
Holotype 57806; paratypes 57805, 57807-57811, 57851, 57852
Martin, F. and Dean, W.T., 1981, Geol. Surv. Can., Bull. 343, p. 18, pl. 3, figs. 2, 10, 12, 17, 20, 24, 26; pl. 6, figs, 4, 6.
Martin, F., 1982, Nat. Mus. Wales, Geol. Ser. 3, pl. 1, fig. 18 (57806).
Elliott Cove Formation, Upper Cambrian, south of Promontory VIII (57806, 57811) and Promontory V, south of Elliott's Cove, Random Island, eastern Newfoundland.

Cristallinium randomense Martin *in* Martin and Dean, 1981, emend.
Hypotypes 83140-83154
Martin, F. and Dean, W.T., 1988, Geol. Surv. Can., Bull. 381, p. 36, pl. 13, figs. 1-17.
Elliott Cove Formation, Upper Cambrian, at and just south of Promontory VII, coastal section south of Elliotts Cove, northwest Random Island, and low cliff in west bank of Manuels River (81341), eastern Newfoundland.

Cymatiogalea aspergillum Martin
Holotype 83157; paratypes 83155, 83156, 83158-83161
Martin, F. and Dean, W.T., 1988, Geol. Surv. Can., Bull. 381, p. 37, pl. 14, figs. 1-7, 9.
Elliott Cove Formation, Upper Cambrian, at and just south of, and northwest side (83159) of Promontory VII, coastal section south of Elliotts Cove, northwest Random Island, and low cliff in west bank of Manuels River (83156), eastern Newfoundland.

Cymatiogalea bouvardi Martin, 1973
Hypotypes 57824, 57834
Martin, F. and Dean, W.T., 1981, Geol. Surv. Can., Bull. 343, pl. 14, fig. 14; pl. 4, fig. 6.
Elliott Cove Formation, Upper Cambrian, south of Promontory VIII, south of Elliott's Cove, Random Island, eastern Newfoundland.

Cymatiogalea cf. *C. cylindrata* Rasul, 1974
Hypotypes 57821, 57822
Martin, F. and Dean, W.T., 1981, Geol. Surv. Can., Bull. 343, p. 18, pl. 3, figs. 7, 22.
Elliott Cove Formation, Upper Cambrian, south of Promontory VIII, south of Elliott's Cove, Random Island, eastern Newfoundland.

Cymatiogalea cf. *C. cylindrata* Rasul, 1974
Hypotype 83162
Martin, F. and Dean, W.T., 1988, Geol. Surv. Can., Bull. 381, pl. 14, fig. 8.
Elliott Cove Formation, Upper Cambrian, northwest side of Promontory VII, coastal section south of Elliotts Cove, northwest Random Island, eastern Newfoundland.

Cymatiogalae virgulta Martin
Holotype 83166; paratypes 83163-83165
Martin, F. and Dean, W.T., 1988, Geol. Surv. Can., Bull. 381, p. 38, pl. 14, figs. 10, 13, 14, 17.
Elliott Cove Formation, Upper Cambrian, at and just south of Promontory VII, coastal section south of Elliotts Cove, northwest Random Island, eastern Newfoundland.

Cymatiosphaera crameri Slavikova, 1968
Hypotype 69643
Martin, F. and Dean, W.T., 1983, Geol. Surv. Can., Paper 83-1B, pl. 43.2, fig. 1.
Chamberlains Brook Formation, Middle Cambrian, west bank Manuels River, west-northwest of road bridge at Manuels, eastern Newfoundland.

Cymatiosphaera crameri (Slavíková), 1968
Hypotype 72945
Martin, F. and Dean, W.T., 1984, Geol. Surv. Can., Paper 84-1A, pl. 57.1, fig. 8.
Chamberlains Brook Formation, Middle Cambrian, west coast Random Island north-northwest of Weybridge, Newfoundland.

Cymatiosphaera cf. *minuta* Jankauskas dans Volkova et al. 1979
Hypotype 75733
Baudet, D., Aitken, J.D. and Vanguestaine, M., 1989, Can. J. Earth Sci., vol. 26, no. 1, p. 140, pl. 2, fig. 1.
Vampire Formation, Early Cambrian, June Lake, lat. 63°29′30″N, long. 128°40′15″-41′30″W, Sekwi Mountain, Northwest Territories.

Cymatiosphaera spp.
Fig. specs. 75735-75737
Baudet, D., Aitken, J.D. and Vanguestaine, M., 1989, Can. J. Earth Sci., vol. 26, no. 1, p. 142, pl. 2, figs. 3-5.
Early Cambrian, Sekwi (75735) and Vampire formations, June Lake, lat. 63°29′30″N, long. 128°40′15″-41′30″W, Sekwi Mountain, Northwest Territories.

Cymatiosphaera sp. A dans Vauguestaine 1977
Fig. spec. 75734
Baudet, D., Aitken, J.D. and Vanguestaine, M., 1989, Can. J. Earth Sci., vol. 26, no. 1, p. 142, pl. 2, fig. 2.
Vampire Formation, Early Cambrian, June Lake, lat. 63°29′30″N, long. 128°40′15″-41′30″W, Sekwi Mountain, Northwest Territories.

Cymatiosphaeroides kullingii Knoll, 1984
Hypotype 81902
Allison, C.W. and Awrimak, S.M., 1989, Precambrian Res., vol. 43, no. 4, p. 287, figs. 11.1-11.3.

Tindir Group, earliest Cambrian or latest Precambrian, headwaters of Tindir Creek between, lat. 65°15′N and 65°17′N, long. 141°56′W and long. 141°57′W, Yukon.

Dasydiacrodium caudatum Vanguestaine, 1973
Hypotypes 83167, 83168
Martin, F. and Dean, W.T., 1988, Geol. Surv. Can., Bull. 381, pl. 16, figs. 5, 8.
Elliott Cove Formation, Upper Cambrian, at and just south of Promontory VII, coastal section south of Elliotts Cove, northwest Random Island, eastern Newfoundland.

Dasydiacrodium obsonum Martin
Holotype 83169; paratypes 83170-83175
Martin, F. and Dean, W.T., 1988, Geol. Surv. Can., Bull. 381, p. 38, pl. 10, figs. 6, 7, 10, 11, 13-15.
Elliott Cove Formation, Upper Cambrian, northwest side of Promontory VII, and at and just south of Promontory VII (83171, 83173), coastal section south of Elliotts Cove, northwest Random Island, eastern Newfoundland.

Eliasum cf. *E. asturicum* Fombella, 1977
Hypotype 57800
Martin, F. and Dean, W.T., 1981, Geol. Surv. Can., Bull. 343, p. 19, pl. 2, fig. 6.
Manuels River Formation, Middle Cambrian, south of Promontory I, south of Elliott's Cove, Random Island, eastern Newfoundland.

Eliasum cf. *E. asturicum* Fombella, 1977
Hypotype 83176
Martin, F. and Dean, W.T., 1988, Geol. Surv. Can., Bull. 381, pl. 15, fig. 3.
Manuels River Formation, Middle Cambrian, west bank of Manuels River, esatern Newfoundland.

Eliasum? *hutchinsonii* Martin
Holotype 72949; paratypes 72946-72948
Martin, F. and Dean, W.T., 1984, Geol. Surv. Can., Paper 84-1A, p. 435, pl. 57.2, figs. 1-5, 7, 9.
Manuels River Formation, Middle Cambrian, west coast of Random Island north-northwest of Weybridge, Newfoundland.

Eliasum jennessii Martin
Holotype 72953; paratypes 72950-72952, 97954-97957
Martin, F. and Dean, W.T., 1984, Geol. Surv. Can., Paper 84-1A, p. 435, pl. 57.3, figs. 1, 2, 4-7, 12-15.
Chamberlains Brook Formation, Middle Cambrian, west coast of Random Island north-northwest of Weybridge, Newfoundland.

Eliasum jennessii Martin *in* Martin and Dean, 1984
Hypotypes 83177, 83178
Martin, F. and Dean, W.T., 1988, Geol. Surv. Can., Bull. 381, pl. 15, figs. 1, 2.
Chamberlains Brook Formation, Middle Cambrian, Manuels River, eastern Newfoundland.

Eliasum llaniscum Fombella, 1977
Hypotype 57803
Martin, F. and Dean, W.T., 1981, Geol. Surv. Can., Bull. 343, p. 19, pl. 2, fig. 14.
Manuels River Formation, Middle Cambrian, south of Promontory I, south of Elliott's Cove, Random Island, eastern Newfoundland.

Eliasum llaniscum Fombella, 1977
Hypotype 69644
Martin, F. and Dean, W.T., 1983, Geol. Surv. Can., Paper 83-1B, pl. 43.2, fig. 8.
Chamberlains Brook Formation, Middle Cambrian, west bank Manuels River, west-northwest of road bridge at Manuels, eastern Newfoundland.

Eliasum llaniscum Fombella, 1978
Hypotypes 72958-72965
Martin, F. and Dean, W.T., 1984, Geol. Surv. Can., Paper 84-1A, pl. 57.2, figs. 6, 8, 10-15.
Chamberlains Brook and Manuels River (72959, 72965) formations, Middle Cambrian, west coast of Random Island north-northwest of Weybridge, Newfoundland.

Eliasum llaniscum Fombella, 1977
Hypotypes 83179, 83180
Martin, F. and Dean, W.T., 1988, Geol. Surv. Can., Bull. 381, pl. 15, figs. 8, 9.
Manuels River Formation, Middle Cambrian, west bank of Manuels River, eastern Newfoundland.

Goniosphaeridium aff. *G. akrochodermum* (Rasul, 1979)
Hypoypes 65667-65669
Dean, W.T. and Martin, F., 1982, Nat. Mus. Wales, Geol. Ser. 3, p. 138, pl.1, figs. 8, 12, 14.
Basal silty member, Survey Peak Formation, Upper Cambrian, southeastern end Wilcox Pass, Jasper Park, Alberta.

Hemisphaerium? sp.
Fig. spec. 75747
Baudet, D., Aitken, J.D. and Vanguestaine, M., 1989, Can. J. Earth Sci., vol. 26, no. 1, p. 142, pl. 2, fig. 15.
Vampire Formation, Early Cambrian, June Lake, lat. 63°29′30″N, long. 128°40′15″-41′30″W, Sekwi Mountain, Northwest Territories.

Impluviculus sp. A
Fig. specs. 83181-83185
Martin, F. and Dean, W.T., 1988, Geol. Surv. Can., Bull. 381, p. 39, pl. 15, figs. 4-6, 10, 11.
Elliott Cove Formation, Upper Cambrian, at and just south of Promontory VII, coastal section south of Elliotts Cove, northwest Random Island, eastern Newfoundland.

Indeterminé
Fig. specs. 75758
Baudet, D., Aitken, J.D. and Vanguestaine, M., 1989, Can. J. Earth Sci., vol. 26, no. 1, pl. 2, fig. 26.
Vampire Formation, Early Cambrian, June Lake, lat. 63°29′30″N, long. 128°40′15″-41′30″5DW, Sekwi Mountain, Northwest Territories.

Kildinella sp.
Fig. spec. 69645
Martin, F. and Dean, W.T., 1983, Geol. Surv. Can., Paper 83-1B, pl. 43.2, fig. 17.

Brigues Formation, Lower Cambrian, west bank Manuels River, west-northwest of road bridge at Manuels, eastern Newfoundland.

Kildinella sp.
Fig. spec. 72966
Martin, F. and Dean, W.T., 1984, Geol. Surv. Can., Paper 84-1A, pl. 57.3, fig. 9.
Manuels River Formation, Middle Cambrian, west coast of Random Island north-northwest of Weybridge, Newfoundland.

Leiofusa stoumonensis Vanguestaine, 1973
Hypotypes 57778, 57779
Martin, F. and Dean, W.T., 1981, Geol. Surv. Can., Bull. 343, p. 19, pl. 1, figs. 16, 17.
Elliott Cove Formation, Upper Cambrian, Promontory V, and immediately north of Promontory VI, south of Elliott's Cove, Random Island, eastern Newfoundland.

Leiofusa stoumonensis Vanguestaine, 1973
Hypotypes 83186-83188
Martin, F. and Dean, W.T., 1988, Geol. Surv. Can., Bull. 381, pl. 10, figs. 8, 9, 12.
Elliott Cove Formation, Upper Cambrian, at and just south of Promontory VII, coastal section south of Elliotts Cove, northwest Random Island, eastern Newfoundland.

Leiosphaeridia asperata (Naumova) Lindgren, 1982
Hypotype 81901
Allison, C.W. and Awrimak, S.M., 1989, Precambrian Res., vol. 43, no. 4, p. 287, fig. 11.4.
Tindir Group, earliest Cambrian or latest Precambrian, headwaters of Tindir Creek between lat. 65°15′N and 65°17′N, long. 141°56′W and long. 141°57′W, Yukon.

Leiosphaeridia fumiana Vanguestaine 1974
Hypotype 75728
Baudet, D., Aitken, J.D. and Vanguestaine, M., 1989, Can. J. Earth Sci., vol. 26, no. 1, p. 142, pl. 1, fig. 28.
Vampire Formation, Early Cambrian, June Lake, lat. 63°29′30″N, long. 128°40′15″-41′30″W, Sekwi Mountain, Northwest Territories.

Leiosphaeridia sp.
Fig. spec. 57812
Martin, F. and Dean, W.T., 1981, Geol. Surv. Can., Bull. 343, pl. 3, fig. 3.
Manuels River Formation, Middle Cambrian, south of Promontory I, south of Elliott's Cove, Random Island, eastern Newfoundland.

Lophosphaeridium tentativum Volkova 1968
Hypotype 75726
Baudet, D., Aitken, J.D. and Vanguestaine, M., 1989, Can. J. Earth Sci., vol. 26, no. 1, p. 142, pl. 1, fig. 26.
Vampire Formation, Early Cambrian, June Lake, lat. 63°29′30″N, long. 128°40′15″-41′30″W, Sekwi Mountain, Northwest Territories.

Micrhystridium lanatum Volkova 1969
Hypotype 75738
Baudet, D., Aitken, J.D. and Vanguestaine, M., 1989, Can. J. Earth Sci., vol. 26, no. 1, p. 142, pl. 2, fig. 6.
Vampire Formation, Early Cambrian, June Lake, lat. 63°29′30″N, long. 128°40′15″-41′30″W, Sekwi Mountain, Northwest Territories.

Micrhystridium pallidum Volkova 1968
Hypotypes 75739-75741
Baudet, D., Aitken, J.D. and Vanguestaine, M., 1989, Can. J. Earth Sci., vol. 26, no. 1, p. 142, pl. 2, figs.7-9.
Vampire Formation, Early Cambrian, June Lake, lat. 63°29′30″N, long. 128°40′15″-41′30″W, Sekwi Mountain, Northwest Territories.

Micrhystridium tornatum Volkova 1968
Hypotypes 75742-75744
Baudet, D., Aitken, J.D. and Vanguestaine, M., 1989, Can. J. Earth Sci., vol. 26, no. 1, p. 142, pl. 2, figs. 10-12.
Early Cambrian, Vampire and Sekwi (75744) formations, June Lake, lat. 63°29′30″N, long. 128°40′15″-41′30″W, Sekwi Mountain, Northwest Territories.

Micrhystridium cf. *tornatum* Volkova 1968
Hypotype 75745
Baudet, D., Aitken, J.D. and Vanguestaine, M., 1989, Can. J. Earth Sci., vol. 26, no. 1, p. 142, pl. 2, fig. 13.
Vampire Formation, Early Cambrian, June Lake, lat. 63°29′30″N, long. 128°40′15″-41′30″W, Sekwi Mountain, Northwest Territories.

Multiplicisphaeridium sp.
Fig. spec. 72967
Martin, F. and Dean, W.T., 1984, Geol. Surv. Can., Paper 84-1A, p. 437, pl. 57.3, fig. 8.
Manuels River Formation, Middle Cambrian, west coast of Random Island north-northwest of Weybridge, Newfoundland.

Nucellosphaeridium sp. A
Fig. spec. 69623
Martin, F. and Dean, W.T., 1983, Geol. Surv. Can., Paper 83-1B, p. 360, pl. 43.2, fig. 15.
Brigus Formation, Lower Cambrian, west bank Manuels River, west-northwest of road bridge at Manuels, eastern Newfoundland.

Ooidium? sp. A
Fig. spec. 57775
Martin, F. and Dean, W.T., 1981, Geol. Surv. Can., Bull. 343, p. 20, p. 20, pl. 1, fig. 13.
Elliott Cove Formation, Upper Cambrian, lower part of cliff approximately 1800m southeast of the south-eastern end of Bounds Mead, Random Island, eastern Newfoundland.

Peramorpha manuelsensis Martin
Holotype 69624; paratypes 69625-69628
Martin, F. and Dean, W.T., 1983, Geol. Surv. Can., Paper 83-1B, p. 360, pl. 43.1, figs. 1-3, 5, 6.
Brigus Formation, Lower Cambrian, west bank Manuels River west-northwest of road bridge at Manuels, eastern Newfoundland.

Pterospermopsimorpha sp.
Fig. spec. 57823
Martin, F. and Dean, W.T., 1981, Geol. Surv. Can., Bull. 343, p. 20, pl. 3, fig. 13.
Manuels River Formation, Middle Cambrian, south of Promontory I, south of Elliott's Cove, Random Island, eastern Newfoundland.

Pterospermopsimorpha sp.
Fig. spec. 69646
Martin, F. and Dean, W.T., 1983, Geol. Surv. Can., Paper 83-1B, pl. 43.2, fig. 10.
Brigus Formation, Lower Cambrian, west bank Manuels River, west-northwest of road bridge at Manuels, eastern Newfoundland.

Pterospermopsimorpha spp.
Fig. specs. 75711-75713
Baudet, D., Aitken, J.D. and Vanguestaine, M., 1989, Can. J. Earth Sci., vol. 26, no. 1, p. 142, pl. 1, figs. 11-13.
Vampire Formation, Early Cambrian, June Lake, lat. 63°29′30″N, long. 128°40′15″W, Sekwi Mountain; Late Proterozoic, Blueflower Formation, Sekwi Brook, lat. 63°21′10″N, long. 128°24′52″W, and Sheepbed Formation, near Palmer Lake, lat. 64°28′50″N, long. 129°23′30″W, Northwest Territories.

Retisphaeridium dichamerum Staplin, Jansonius and Pocock, 1965
Hypotypes 69629-69633
Martin, F. and Dean, W.T., 1983, Geol. Surv. Can., Paper 83-1B, p. 361, pl. 43.2, figs. 4, 9, 13, 14, 20.
Chamberlains Brook Formation, Middle Cambrian, west bank Manuels River, west-northwest of road bridge at Manuels, eastern Newfoundland.

Retisphaeridium dichamerum Staplin, Jansonius and Pocock, 1965
Hypotypes 72968-72970
Martin, F. and Dean, W.T., 1984, Geol. Surv. Can., Paper 84-1A, pl. 57.1, figs. 11, 12, 15.
Chamberlains Brook Formation, Middle Cambrian, west coast Random Island north-northwest of Weybridge, Newfoundland.

Retisphaeridium dichamerum Staplin, Jansonius and Pocock, 1965
Hypotypes 83189-83191
Martin, F. and Dean, W.T., 1988, Geol. Surv. Can., Bull. 381, pl. 12, figs. 7-9.
Manuels River Formation, Middle Cambrian, west bank of Manuels River, eastern Newfoundland.

Retisphaeridium dichamerum Staplin, Jansonius et Pocock 1965
Hypotype 75731
Baudet, D., Aitken, J.D. and Vanguestaine, M., 1989, Can. J. Earth Sci., vol. 26, no. 1, p. 143, pl. 1, fig. 31.
Vampire Formation, Early Cambrian, June Lake, lat. 63°29′30″N, long. 128°40′15″-41′30″W, Sekwi Mountain, Northwest Territories.

Retisphaeridium howelli Martin
Holotype 69634; paratypes 69635-69642
Martin, F. and Dean, W.T., 1983, Geol. Surv. Can., Paper 83-1B, p. 361, pl. 43.2, figs. 2, 3, 5, 6, 11, 12, 16, 18, 19.
Chamberlains Brook Formation, Middle Cambrian, west bank Manuels River, west-northwest of road bridge at Manuels, eastern Newfoundland.

Retisphaeridium howellii Martin *in* Martin and Dean, 1983
Hypotypes 72971, 72972
Martin, F. and Dean, W.T., 1984, Geol. Surv. Can., Paper 84-1A, pl. 57.1, figs. 14, 16.
Manuels River and Chamberlains Brook formations, Middle Cambrian, west coast Random Island north-northwest of Weybridge, Newfoundland.

Retisphaeridium howellii Martin *in* Martin and Dean, 1983
Hypotypes 83192, 83193
Martin, F. and Dean, W.T., 1988, Geol. Surv. Can., Bull. 381, pl. 12, figs. 3, 4.
Manuels River Formation, Middle Cambrian, west bank of Manuels River, eastern Newfoundland.

Retisphaeridium? sp.
Fig. spec. 75732
Baudet, D., Aitken, J.D. and Vanguestaine, M., 1989, Can. J. Earth Sci., vol. 26, no. 1, p. 143, pl. 1, fig. 32.
Vampire Formation, Early Cambrian, June Lake, lat. 63°29′30″N, long. 128°40′15″-41′30″W, Sekwi Mountain, Northwest Territories.

Rugasphaera terranovana Martin
Holotype 83198; paratypes 83194-83197, 83199-83201
Martin, F. and Dean, W.T., 1988, Geol. Surv. Can., Bull. 381, p. 39, pl. 17, figs. 1-11.
Manuels River Formation, Middle Cambrian, west bank of manuels River, eastern Newfoundland.

Skiagia compressa (Volkova 1968) Downie 1982
Hypotypes 75720, 75721
Baudet, D., Aitken, J.D. and Vanguestaine, M., 1989, Can. J. Earth Sci., vol. 26, no. 1, p. 143, pl. 1, figs. 20, 21.
Vampire Formation, Early Cambrian, June Lake, lat. 63°29′30″N, long. 128°40′15″-41′30″W, Sekwi Mountain, Northwest Territories.

Skiagia orbiculare (Volkova 1968) Downie 1982
Hypotypes 75722, 75723
Baudet, D., Aitken, J.D. and Vanguestaine, M., 1989, Can. J. Earth Sci., vol. 26, no. 1, p. 143, pl. 1, figs. 22, 23.
Vampire Formation, Early Cambrian, June Lake, lat. 63°29′30″N, long. 128°40′15″-41′30″W, Sekwi Mountain, Northwest Territories.

Skiagia scottica Downie 1982
Hypotype 75724
Baudet, D., Aitken, J.D. and Vanguestaine, M., 1989, Can. J. Earth Sci., vol. 26, no. 1, p. 143, pl. 1, fig. 24.
Vampire Formation, Early Cambrian, June Lake, lat. 63°29′30″N, long. 128°40′15″-41′30″W, Sekwi Mountain, Northwest Territories.

Stellichinatum uncinatum (Downie, 1958)
Hypotypes 83202-83204
Martin, F. and Dean, W.T., 1988, Geol. Surv. Can., Bull. 381, p. 40, pl. 17, figs. 12, 15.
Elliott Cove Formation, Upper Cambrian, at and just south of Promontory VII, coastal section south of Elliotts Cove, northwest Random Island, eastern Newfoundland.

Stelliferidium pingiculum Martin
Holotype 83205; paratypes 83206-83208
Martin, F. and Dean, W.T., 1988, Geol. Surv. Can., Bull. 381, p. 40, pl. 14, figs. 11, 12, 15, 16, 18, 19.
Elliott Cove Formation, Upper Cambrian, cliff in west bank of Manuels River, eastern Newfoundland.

Stictosphaeridium brayense Gardiner and Vauguestaine, 1971
Hypotype 69647
Martin, F. and Dean, W.T., 1983, Geol. Surv. Can., Paper 83-1B, pl. 43.2, fig. 7.
Chamberlains Brook Formation, Middle Cambrian, west bank Manuels River, west-northwest of road bridge at Manuels, eastern Newfoundland.

Timofeevia lancarae (Cramer and Diez de Cramer) Vanguestaine, 1978
Hypotypes 57784-57792, 57846, 57848-57850
Martin, F. and Dean, W.T., 1981, Geol. Surv. Can., Bull. 343, p. 20, pl. 2, figs. 1-3, 8, 9, 11, 12, 19, 20; pl. 6, figs. 1,3, 5, 7, 8.
Martin, F. 1982, Nat. Mus. Wales, Geol. Ser. 3, pl. 1, fig. 20 (57789).
Elliott Cove Formation, Upper Cambrian, Promontory III (57784, 57787-57789, 57791) and south side of Promontory I, south of Elliott's Cove, Random Island, eastern Newfoundland.

Timofeevia lancarae (Cramer and Diez de Cramer) Vanguestaine, 1978
Hypotypes 83209-83211
Martin, F. and Dean, W.T., 1988, Geol. Surv. Can., Bull. 381, pl. 12, figs. 12, 15, 16.
Elliott Cove Formation, Upper Cambrian, cliff at Promontory I, coastal section south of Elliotts Cove, northwest Random Island, eastern Newfoundland.

Timofeevia aff. *T. lancarae* (Cramer and Diez de Cramer) Vanguestaine, 1978
Hypotype 57847
Martin, F. and Dean, W.T., 1981, Geol. Surv. Can., Bull. 343, pl. 6, fig. 2.
Elliott Cove Formation, Upper Cambrian, south side of Promontory I, south of Elliott's Cove, Random Island, eastern Newfoundland.

Timofeevia microretis Martin
Holotype 57794; paratypes 57793, 57795-57797
Martin, F. and Dean, W.T., 1981, Geol. Surv. Can., Bull. 343, p. 21, pl. 2, figs. 4, 10, 13, 16, 17.
Martin, F., 1982, Nat. Mus. Wales, Geol. Ser. 3, pl. 1, fig. 15 (59794).
Elliott Cove Formation, Upper Cambrian, cliff immediately south of Promontory II and south side of Promontory I (57795, 57797), south of Elliott's Cove, Random Island, eastern Newfoundland.

Timofeevia microretis Martin *in* Martin and Dean, 1981
Hypotype 83212
Martin, F. and Dean, W.T., 1988, Geol. Surv. Can., Bull. 381, pl. 12, fig. 13.
Elliott Cove Formation, Upper Cambrian, cliff immediately south of Promontory II, coastal section south of Elliotts Cove, northwest Random Island, eastern Newfoundland.

Timofeevia pentagonalis (Vanguestaine) Vanguestaine, 1978
Hypotype 57841
Martin, F. and Dean, W.T., 1981, Geol. Surv. Can., Bull. 343, p. 21, pl. 5, figs. 7, 9.
Elliott Cove Formation, Upper Cambrian, cliff immediately south of Promontory II, south of Elliott's Cove, Random Island, eastern Newfoundland.

Timofeevia pentagonalis (Vanguestaine) Vanguestaine, 1978
Hypotypes 83213, 83214
Martin, F. and Dean, W.T., 1988, Geol. Surv. Can., Bull. 381, pl. 12, figs. 5, 6.
Elliott Cove Formation, Upper Cambrian, cliff at Promontory I, coastal section south of Elliotts Cove, northwest Random Island, eastern Newfoundland.

Timofeevia phosphoritica Vanguestaine, 1978
Hypotypes 57801, 57802, 57830, 57840
Martin, F. and Dean, W.T., 1981, Geol. Surv. Can., Bull. 343, p. 21, pl. 2, figs. 7, 15; pl. 4, fig. 1; pl. 5, figs. 2, 10.
Elliott Cove Formation, Upper Cambrian, cliff immediately south of Promontory II and lower part of cliff approximately 1800m southeast of the southeastern end of Bounds Mead (57840), south of Elliott's Cove, Random Island, eastern Newfoundland.

Timofeevia phosphoritica Vanguestaine, 1978
Hypotypes 83215-83217
Martin, F. and Dean, W.T., 1988, Geol. Surv. Can., Bull. 381, pl. 12, figs. 10, 11, 14.
Elliott Cove Formation, Upper Cambrian, cliff at Promontory I, coastal section south of Elliotts Cove, northwest Random Island, eastern Newfoundland.

Timofeevia sp.
Fig. spec. 72973
Martin, F. and Dean, W.T., 1984, Geol. Surv. Can., Paper 84-1A, pl. 57.3, fig. 10.
Manuels River Formation, Middle Cambrian, west coast of Random Island north-northwest of Weybridge, Newfoundland.

Trachyhystrichosphaera magna Allison
Holotype 81903
Allison, C.W. and Awrimak, S.M., 1989, Precambrian Res., vol. 43, no. 4, p. 288, figs. 11.5-11.8.

Tindir Group, earliest Cambrian or latest Precambrian, headwaters of Tindir Creek between lat. 65°15′N and 65°17′N, long. 141°56′W and long. 141°57′W, Yukon.

Trachyhystrichosphaera vidalii Knoll, 1984
Hypotype 81904
Allison, C.W. and Awrimak, S.M., 1989, Precambrian Res., vol. 43, no. 4, p. 287, figs. 11.9, 11.10.
Tindir Group, earliest Cambrian or latest Precambrian, headwaters of Tindir Creek between lat. 65°15′N and 65°17′N, long. 141°56′W and long. 141°57′W, Yukon.

Transition *Vulcanisphaera turbata V. africana*
Fig. spec. 83248
Martin, F. and Dean, W.T., 1988, Geol. Surv. Can., Bull. 381, p. 43, pl. 18, fig. 2.
Elliott Cove Formation, Upper Cambrian, at and just south of Promontory VII, coastal section south of Elliotts Cove, northwest Random Island, eastern Newfoundland.

Trunculumarium revinium (Vanguestaine) Loeblich and Tappan, 1976
Hypotypes 57798, 57799, 57844, 57845
Martin, F. and Dean, W.T., 1981, Geol. Surv. Can., Bull. 343, p. 22, pl. 2, figs. 5, 18; pl. 5, figs. 4, 6.
Elliott Cove Formation, Upper Cambrian, between Promontory VII and VIII, south of Elliott's Cove, Random Island, eastern Newfoundland.

Trunculumarium revinium (Vanguestaine) Loeblich and Tappan, 1976
Hypotype 65681
Martin, F., 1982, Nat. Mus. Wales, Geol. Ser. 3, pl. 1, fig. 14.
Elliott Cove Formation, Upper Cambrian, between Promontories VII and VIII, south of Elliotts Cove, Random Island, eastern Newfoundland.

Trunculumarium revinium (Vanguestaine) Loeblich and Tappan, 1976
Hypotypes 83218-83225
Martin, F. and Dean, W.T., 1988, Geol. Surv. Can., Bull. 381, p. 41, pl. 15, figs. 7, 12-20.
Elliott Cove Formation, Upper Cambrian, northwest side of Promontory VII, coastal section south of Elliotts Cove, northwest Random Island, eastern Newfoundland.

Veryhachium dumonti Vanguestaine, 1973
Hypotypes 83226-83236
Martin, F. and Dean, W.T., 1988, Geol. Surv. Can., Bull. 381, p. 41, pl. 16, figs. 2, 3, 7, 10-18.
Elliott Cove Formation, Upper Cambrian, at and just south of, and northwest side (83234) of Promontory VII, coastal section south of Elliotts Cove, northwest Random Island, eastern Newfoundland.

Veryhachium cf. *V. dumontii* Vanguestaine, 1973
Hypotypes 83237-83239
Martin, F. and Dean, W.T., 1988, Geol. Surv. Can., Bull. 381, p. 41, pl. 16, figs. 4, 6, 9.
Elliott Cove Formation, Upper Cambrian, northwest side of , and at and just south (83239) of Promontory VII, coastal section south of Elliotts Cove, northwest Random Island, eastern Newfoundland.

Veryhachium sp. A
Fig. specs. 57774, 57833
Martin, F. and Dean, W.T., 1981, Geol. Surv. Can., Bull. 343, p. 22, pl. 1, fig. 12; pl. 4, fig. 5.
Elliott Cove Formation, Upper Cambrian, south of Promontory VIII, south of Elliott's Cove, Random Island, eastern Newfoundland.

Veryhachium sp. A Martin *in* Martin and Dean, 1981
Fig. spec. 83240
Martin, F. and Dean, W.T., 1988, Geol. Surv. Can., Bull. 381, pl. 16, fig. 1.
Elliott Cove Formation, Upper Cambrian, cliff south of Promontory VIII, coastal section south of Elliotts Cove, northwest Random Island, eastern Newfoundland.

Vulcanisphaera africana Deunff, 1961
Hypotypes 57762, 57832
Martin, F. and Dean, W.T., 1981, Geol. Surv. Can., Bull. 343, p. 23, pl. 1, fig. 1; pl. 4, fig. 3; text-fig. 5.
Elliott Cove Formation, Upper Cambrian, south of Promontory VIII, south of Elliott's Cove, Random Island, eastern Newfoundland.

Vulcanisphaera africana Deunff, 1961
Hypotypes 83241, 83242
Martin, F. and Dean, W.T., 1988, Geol. Surv. Can., Bull. 381, pl. 18, figs. 1-4.
Elliott Cove Formation, Upper Cambrian, at and just south of Promontory VII, coastal section south of Elliotts Cove, northwest Random Island, eastern Newfoundland.

Vulcanisphaera lanugo Martin
Holotype 83252; paratypes 83249-83251, 83253-83256
Martin, F. and Dean, W.T., 1988, Geol. Surv. Can., Bull. 381, p. 42, pl. 18, figs. 7, 9-12, 14-17.
Manuels River Formation, Middle Cambrian, west bank of Manuels River, eastern Newfoundland.

Vulcanisphaera turbata Martin
Holotype 57763; paratypes 57764, 57765
Martin, F. and Dean, W.T., 1981, Geol. Surv. Can., Bull. 343, p. 23, pl. 1, figs. 2-4; text-fig. 6.
Martin, F., 1982, Nat. Mus. Wales, Geol. Ser. 3, pl. 1, fig. 19 (57764).
Elliott Cove Formation, Upper Cambrian, Promontory V, south side of Promontory I, and between Promontories IV and, south of Elliott's Cove, Random Island, eastern Newfoundland.

Vulcanisphaera turbata Martin *in* Martin and Dean, 1981
Hypotypes 83243-83247
Martin, F. and Dean, W.T., 1988, Geol. Surv. Can., Bull. 381, p. 42, pl. 18, figs. 3, 5, 6, 8, 13.
Elliott Cove Formation, Upper Cambrian, at and just south of Promontory VII, and cliff immediately south of Promontory II (83245), coastal section south of Elliotts Cove, northwest Random Island, eastern Newfoundland.

Vulcanisphaera aff. *V. turbata* Martin
Hypotype 57766
Martin, F. and Dean, W.T., 1981, Geol. Surv. Can., Bull. 343, p. 24, pl. 1, fig. 5.
Elliott Cove Formation, Upper Cambrian, Promontory III, south of Elliott's Cove, Random Island, eastern Newfoundland.

Ordovician

Acanthodiacrodium achrasi Martin, 1973
Hypotype 48198
Dean, W.T. and Martin, F., 1978, Geol. Surv. Can., Bull. 284, pl. 3, fig. 9.
Bell Island Group, Lower Ordovician, cliff section behind old Scotia Pier and immediately southwest of disused ore conveyor terminal, southeast coast of Bell Island, Newfoundland.

Acanthodiacrodium achrasi Martin, 1973
Hypotype 57825
Martin, F. and Dean, W.T., 1981, Geol. Surv. Can., Bull. 343, pl. 3, fig. 15.
Clarenville Formation, Lower Ordovician, about midway between Promontories IX and X, south of Elliott's Cove, Random Island, eastern Newfoundland.

Acanthodiacrodium costatum Burmann, 1968
Hypotype 46891
Dean, W.T. and Martin, F., 1978, Geol. Surv. Can., Bull. 284, p. 6, pl. 3, fig. 7.
Wabana Group, Lower Ordovician, northeast side of entrance to a disused mine tunnel (Submarine Slope No. 2) between Wabana town centre and northwest coast of Bell Island, Newfoundland.

Acanthodiacrodium tassellii Martin, 1969
Hypotype 48189
Dean, W.T. and Martin, F., 1978, Geol. Surv. Can., Bull. 284, pl. 1, fig. 1.
Wabana Group, Lower Ordovician, Wabana Mine, Bell Island, Newfoundland.

Acanthodiacrodium uniforme Burmann, 1968
Hypotypes 46892, 46893
Dean, W.T. and Martin, F., 1978, Geol. Surv. Can., Bull. 284, p. 7, pl. 2, fig. 14; pl. 3, fig. 17.
Wabana Group, Lower Ordovician, Wabana Mine and disused quarry on northwest coast north of Wabana, Bell Island, Newfoundland.

Acanthodiacrodium cf. *A. uniforme* Burmann, 1968
Hypotype 55926
Martin, F. and Dean, W.T., 1983, Geol. Surv. Can., Bull. 310, p. 21, pl. 8, fig. 2.
Neuville Formation, Upper Ordovician, bay to east of Pointe Delisle, Québec.

Acanthodiacrodium sp.
Fig. spec. 65680
Martin, F., 1982, Nat. Mus. Wales, Geol. Ser. 3, pl. 1, fig. 11.
Bell Island Group, Lower Arenig (Ordovician), 1 m below top of cliff east side of Polls Head, Bell Island, eastern Newfoundland.

Acritarch gen. et sp. nov. C, B, A
Fig. specs. 65661, 65664, 65672
Dean, W.T. and Martin, F., 1982, Nat. Mus. Wales, Geol. Ser. 3, p. 138,139, pl. 1, figs. 1, 4, 7, 13.
Middle member, Survey Peak Formation, Lower Ordovician, northeastern end Wilcox Pass, Jasper Park, Alberta.

Actinotodissus crassus Loeblich and Tappan 1978
Hypotype 80130
Jacobson, S.R. and Achab, A., 1985, Palynology, vol. 9, p. 171, pl. 1, fig. 2.
Vaureal Formation, Upper Ordovician, depth 3200 feet, New Associated Consolidated Paper Anticosti No. 1 well, near 24-mile lodge on Jupiter River, Anticosti Island, Québec.

Actipilion aff. *A. druggii* Loeblich
Hypotype 56667
Martin, F., 1978, Geol. Surv. Can., Paper 78-1B, p. 80, pl. 12.1, fig. 3.
Winterhouse Formation, Long Point Group, Upper Ordovician, quarry immediately southeast of fishing huts near west of Clam Bank Cove, 750 km northwest of Lourdes, western Newfoundland.

Adorfia prolongata Burmann, 1970
Hypotypes 46889, 46890
Dean, W.T. and Martin, F., 1978, Geol. Surv. Can., Bull. 284, p. 7, pl. 2, figs. 6, 9; pl. 3, fig. 27.
Wabana Group, Lower Ordovician, Wabana Mine, Bell Island, Newfoundland.

Arbusculidium destombesii Deunff, 1968
Hypotype 48200
Dean, W.T. and Martin, F., 1978, Geol. Surv. Can., Bull. 284, pl. 3, fig. 12.
Bell Island Group, Lower Ordovician, cliff summit opposite midpoint of shingle beach between its southeastern extremity and northeastern end, The Beach, Bell Island, Newfoundland.

Arbusculidium destombesii Deunff, 1968
Hypotypes 57776, 57777
Martin, F. and Dean, W.T., 1981, Geol. Surv. Can., Bull. 343, pl. 1, figs. 14, 19.
Clarenville Formation, Lower Ordovician, low cliff about 137 m southeast of mouth of stream at Elliott's Cove settlement, and about midway between Promontories IX and X, south of Elliott's Cove, Random Island, eastern Newfoundland.

Arbusculidium filamentosum (Vavrdová, 1965) Vavrdová, 1972
Hypotypes 48190, 48203

Dean, W.T. and Martin, F., 1978, Geol. Surv. Can., Bull. 284, pl. 1, fig. 3; pl. 3, fig. 23.
Wabana Group, Lower Ordovician, disused quarry, northwest coast north of Wabana, Bell Island, Newfoundland.

Arbusculidium stephanum Vavrdová, 1976
Hypotype 46894
Dean, W.T. and Martin, F., 1978, Geol. Surv. Can., Bull. 284, p. 7, pl. 2, fig. 8.
Wabana Group, Lower Ordovician, Wabana Mine, Bell Island, Newfoundland.

Aremoricanium? grootaertii Martin
Holotype 72998; paratypes 72993-72997, 72999-73002
Martin, F., 1984, Geol. Surv. Can., Paper 84-1A, p. 442, pl. 58.1, figs. 1-9; pl. 58.2, figs. 1-4.
Survey Peak Formation, Lower Ordovician, Wilcox Pass 3 km north of Athabasca Glacier, southwestern Alberta.

Aremoricanium squarrosum Loeblich and MacAdam 1971
Hypotype 80132
Jacobson, S.R. and Achab, A., 1985, Palynology, vol. 9, p. 171, pl. 1, fig. 4.
Vaureal Formation, Upper Ordovician, depth 3000 feet, New Associated Consolidated Paper Anticosti No. 1 well, near 24-mile lodge on Jupiter River, Anticosti Island, Québec.

Aremoricanium syringosagis Loeblich et MacAdam, 1971
Hypotype 61548
Martin, F., 1980, Can. J. Earth Sci., vol. 17, no. 1, p. 110, pl. 3, fig. 10.
White Head Formation, Upper Ordovician, Grande Coupe, Gaspé, Québec.

Asketopalla formosula Loeblich et Tappan, 1969
Hypotypes 55927-55930
Martin, F., 1983, Geol. Surv. Can., Bull. 310, p. 21, pl. 9, figs. 1, 8; pl. 10, figs. 9, 13, 15.
Chazy Group, Middle Ordovician, Laval Formation, Bédard quarry, Caughnawaga, Laprairie County, Québec; Rockcliffe Formation, cut on Highway 17, 4.5 km west of Cumberland, lat. 45°31′N, long. 75°27′W, Ontario (55929, 55930).

Athabascaella playfordi Martin
Holotype 73007; paratypes 73003-73006, 73008, 73009
Martin, F., 1984, Geol. Surv. Can., Paper 84-1A, p. 444, pl. 58.3, figs. 1-9.
Survey Peak Formation, Middle Ordovician, Wilcox Pass 3 km north of Athabasca Glacier, southwestern Alberta.

Athabascaella rossii Martin
Holotype 73014; paratypes 73010-73013
Martin, F., 1984, Geol. Surv. Can., Paper 84-1A, p. 446, pl. 58.2, figs. 5-12.
Survey Peak Formation, Middle Ordovician, Wilcox Pass 3 km north of Athabasca Glacier, southwestern Alberta.

Baltisphaeridium accinctum Loeeblich and Tappan 1978
Hypotype 80135
Jacobson, S.R. and Achab, A., 1985, Palynology, vol. 9, p. 172, pl. 2, fig. 1.
Vaureal Formation, Upper Ordovician, depth 3000 feet, New Associated Consolidated Paper Anticosti No. 1 well, near 24-mile lodge on Jupiter River, Anticosti Island, Québec.

Baltisphaeridium annelieae Kjellström, 1976
Hypotypes 55931-55933
Martin, F., 1983, Geol. Surv. Can., Bull. 310, p. 21, pl. 7, figs. 2, 7, 11; pl. 12, fig. 1.
Utica Group, Upper Ordovician, Lotbinière Formation, bay west of Pointe Delisle, lat. 46°41′57″N, long. 71°36′29″W; Neuville Formation, bay to east of Pointe Delisle (55932), Québec.

Baltisphaeridium crinitum Martin
Holotype 46895
Dean, W.T. and Martin, F., 1978, Geol. Surv. Can., Bull. 284, p. 7, pl. 3, fig. 5.
Bell Island Group, Lower Ordovician, cliff summit opposite midpoint of shingle beach between its southeastern extremity and northeastern end, The Beach, Bell Island, Newfoundland.

Baltisphaeridium crinitum Martin, 1978
Hypotype 57804
Martin, F. and Dean, W.T., 1981, Geol. Surv. Can., Bull. 343, pl. 3, fig. 1.
Clarenville Formation, Lower Ordovician, immediately north of Promontory X, south of Elliott's Cove, Random Island, eastern Newfoundland.

Baltisphaeridium crinitum Martin, 1978
Hypotype 65678
Martin, F., 1982, Nat. Mus. Wales, Geol. Ser. 3, pl. 1, fig. 8.
Bell Island Group, Lower Arenig (Ordovician), 1 m below top of cliff east side of Polls Head, Bell Island, eastern Newfoundland.

Baltisphaeridium digitiforme Górka, 1969
Hypotypes 55934-55937
Martin, 1983, Geol. Surv. Can., Bull. 310, p. 21, pl. 9, fig. 6; pl. 12, figs. 4, 5, 7.
Neuville Formation, Utica Group, Upper Ordovician, bay to east of Pointe Delisle (55934); Deschambault Formation, Middle Ordovician, Lafarge Cement quarry, Saint-Constant, Québec.

Baltisphaeridium filosum Kjellström 1971
Hypotype 69396
Legault, J.A., 1982, Can. J. Earth Sci., vol. 19, no. 9, pl. 1, fig. 5.
Middle-Upper Ordovician, Orphan Knoll, lat. 50°33′N, long. 46°6′W, Labrador Sea.

Baltisphaeridium granosum Kjellström, 1971
Hypotype 55938
Martin, F., 1983, Geol. Surv. Can., Bull. 310, p. 22, pl. 6, figs. 1, 2.
Lotbinière Formation, Utica Group, Upper Ordovician, bay to west of Pointe Delisle, lat. 46°41′57″N, long. 71°36′29″W, Québec.

Baltisphaeridium cf. *hirsutoides* (Eisenack) Eisenack 1962
Hypotype 80139
Jacobson, S.R. and Achab, A., 1985, Palynology, vol. 9, p. 172, pl. 2, fig. 5.
Vaureal Formation, Upper Ordovician, depth 2600 feet, New Associated Consolidated Paper Anticosti No. 1 well, near 24-mile lodge on Jupiter River, Anticosti Island, Québec.

Baltisphaeridium cf. *B. multipilosum* (Eisenack) Eisenack, 1959
Hypotypes 55939, 55940
Martin, F., 1983, Geol. Surv. Can., Bull. 310, p. 22, pl. 12, figs. 10, 11.
Deschambault and Montréal formations, Middle Ordovician, Lafarge Cement quarry, Saint-Constant, Québec.

Baltisphaeridium? cf. B. multipilosum (Eisenack) Eisenack, 1959
Hypotype 65662
Dean, W.T. and Martin, F., 1982, Nat. Mus. Wales, Geol. Ser. 3, p. 136, pl. 1, fig. 2.
Middle member, Survey Peak Formation, Lower Ordovician, southeastern end Wilcox Pass, Jasper Park, Alberta.

Baltisphaeridium orbum Legault
Holotype 69392; hypotype 69408
Legault, J.A., 1982, Can. J. Earth Sci., vol. 19, no. 9, p. 1854, pl. 2, figs. 5-8.
Middle-Upper Ordovician, Orphan Knoll, lat. 50°33′N, long. 46°6′W, Labrador Sea.

Baltisphaeridium aff. *B. perclarum* Loeblich et Tappan, 1978
Hypotypes 61549, 61550
Martin, F., 1980, Can. J. Earth Sci., vol. 17, no. 1, p. 110, pl. 3, figs. 2, 14.
White Head Formation, Upper Ordovician, Grande Coupe, Gaspé, Québec.

Baltisphaeridium plicatispinae Gorká
Hypotype 66964
Duffield, S.L. and Legault, J.A., 1981, IUGS Subcommission on Silurian Stratigraphy, Ordovician-Silurian Boundary Working Group. Field Meeting, Anticosti-Gaspé, Québec, vol. II, pl. 1, fig. 3.
Member 6, Ellis Bay Formation, Upper Ordovician, Pointe Laframboise, Anticosti Island, Québec.

Baltisphaeridium plicatispinae Gorká 1969
Hypotype 69412
Legault, J.A., Can. J. Earth Sci., vol. 19, no. 9, pl. 1, fig. 4.
Middle-Upper Ordovician, Orphan Knoll, lat. 50°33′N, long. 46°6′W, Labrador Sea.

Baltisphaeridium plicatispinae Gorká 1969
Hypotype 82883
Martin, F., 1988, Bull British Mus.(Nat. Hist.), Geol. ser., vol. 43, p. 304, fig. 9.
Ellis Bay Formation, Upper Ordovician, west side of Ellis Bay, Anticosti Island, Québec.

Baltisphaeridium ritvae Kjellström, 1971
Hypotypes 61551, 61552
Martin, F., 1980, Can. J. Earth Sci, vol. 17, no. 1, p. 110, pl. 3, figs. 3, 7.
White Head Formation, Upper Ordovician, Grande Coupe, Gaspé, Québec.

Baltisphaeridium trabeculaespinae Gorká 1969
Hypotype 69401
Legault, J.A., 1982, Can. J. Earth Sci., vol. 19, no. 9, pl. 1, fig. 8.
Middle-Upper Ordovician, Orphan Knoll, lat. 50°33′N, long. 46°6′W, Labrador Sea.

Baltisphaeridium trichophorum (Eisenack) Kjellström 1971
Hypotype 69403
Legault, J.A., 1982, Can. J. Earth Sci., vol. 19, no. 9, pl. 1, fig. 9.
Middle-Upper Ordovician, Orphan Knoll, lat. 50°33′N, long. 46°6′W, Labrador Sea.

Baltisphaeridium sp.
Fig spec. 69410
Legault, J.A., 1982, Can. J. Earth Sci., vol. 19, no. 9, pl. 1, fig. 3.
Middle-Upper Ordovician, Orphan Knoll, lat. 50°33′N, long. 46°6′W, Labrador Sea.

?*Baltisphaeridium* sp. I
Fig. spec. 26279
McGregor, D.C. and Cramer, F.H., 1971, Geol. Surv. Can., Bull. 202, pl. 1, figs. 21, 22.
Cat Head Member, Red River Formation, Ordovician, between McBeth Point and Cat Head, Lake Winnipeg, Manitoba.

Baltisphaeridium sp. II, III
Fig. specs. 26284-26286, 26296
McGregor, D.C. and Cramer, F.H., 1971, Geol. Surv. Can., Bull. 202, pl. 1, figs. 29, 30, 32; pl. 2, fig. 7.
Cat Head Member, Red River Formation, Ordovician, between McBeth Point and Cat Head, Lake Winnipeg, Manitoba.

Cheleutochroa gymnobrachiata Loeblich and Tappan 1978
Hypotypes 80133, 80134
Jacobson, S.R. and Achab, A., 1985, Palynology, vol. 9, p. 174, pl. 1, figs. 5, 6.
Vaureal Formation, Upper Ordovician, depth 2800 feet, New Associated Consolidated Paper Anticosti No. 1 well, near 24 mile lodge on Jupiter River, Anticosti Island, Québec.

Comasphaeridium lanugiferum Jacobson and Achab
Holotype 80154; paratype 80155
Jacobson, S.R. and Achab, A., 1985, Palynology, vol. 9, p. 176, pl. 5, figs. 6, 7.
Vaureal Formation, Upper Ordovician, depth 2800 feet, New Associated Consolidated Paper Anticosti No. 1 well, near 24-mile lodge on Jupiter River, Anticosti Island, Québec.

Conochitina poumoti Combaz and Peniguel, 1972
Hypotypes 76622-76624
Landing, E. and Ludvigsen, R., 1984, Can. J. Earth Sci., vol. 21, no. 12, pl. 1, figs. 12-14.
Quebec Group, Lower Ordovician, 1.1 km east of Villa Guay, Québec.

Cornuferifusa bifidipertica Jacobson and Achab
Holotype 80137
Jacobson, S.R. and Achab, A., 1985, Palynology, vol. 9, p. 176, pl. 2, fig. 3.
Vaureal Formation, Upper Ordovician, depth 3000 feet, New Associated Consolidated Paper Anticosti No. 1 well, near 24-mile lodge on Jupiter River, Anticosti Island, Québec.

Corollasphaeridium wilcoxianum Martin
Holotype 65674; paratype 65673
Dean, W.T. and Martin, F., 1982, Nat. Mus. Wales, Geol. Ser. 3, p.136, pl. 1, figs. 15, 16.
Basal silty member, Survey Peak Formation, Lower Ordovician, southeastern end Wilcox Pass, Jasper Park, Alberta.

Coryphidium bohemicum Vavrdová, 1972
Hypotypes 46896-46898
Dean, W.T. and Martin, F., 1978, Geol. Surv. Can., Bull. 284, p. 7, pl. 2, figs. 1, 2, 5; pl. 3, fig. 25.
Wabana Group, Lower Ordovician, Wabana Mine, Bell Island, Newfoundland.

Cymatiogalea bellicosa Deunff, 1961
Hypotypes 57827, 57838
Martin, F. and Dean, W.T., 1981, Geol. Surv. Can., Bull. 343, pl. 3, fig. 18; pl. 4, fig. 8.
Clarenville Formation, Lower Ordovician, low cliff about 137 southeast of mouth of stream at Elliott's Cove settlement, and immeditely north of Promomtory X, south of Elliott's Cove, Random Island, eastern Newfoundland.

Cymatiogalea cf. C. *cristata* (Downie) Deunff et al., 1974
Hypotype 65671
Dean, W.T. and Martin, F., 1982, Nat. Mus. Wales, Geol Ser. 3, pl. 1, fig. 10.
Basal silty member, Survey Peak Formation, Lower Ordovician, southeastern end Wilcox Pass, Jasper Park, Alberta.

Cymatiogalea? membranula Martin
Holotype 46899; paratypes 46900, 46901
Dean, W.T. and Martin, F., 1978, Geol. Surv. Can., Bull. 284, p. 7, pl.1, fig. 8; pl. 3, figs. 29, 31.
Bell Island Group, Lower Ordovician, cliff section behind old Scotia Pier and immediately southwest of the disused ore conveyor terminal, southeast coast (46899) and cliff at east end of Chimmey Cove, near southwest end of Bell Island, Newfoundland.

Cymatiogalea cf. *C. multarea* (Deunff) Deunff et al., 1974
Hypotype 65670
Dean, W.T. and Martin, F., 1982, Nat. Mus. Wales, Geol. Ser. 3, pl. 1, fig. 9.
Basal silty member, Survey Peak Formation, Upper Cambrian, southeastern end Wilcox Pass, Jasper Park, Alberta.

Cymatiosphaera cf. *C. mirabilis* Deunff, 1958
Hypotype 55941
Martin, F., 1983, Geol. Surv. Can., Bull 320, p. 22, pl. 11, fig. 5.
Saint Martin Formation, Chazy Group, Middle Ordovician, about 1.5 km east of Green Creek and Highway 17 crossing, lat. 45°27′N, long. 75°34′W, Carleton County, Ontario.

Cymatiosphaera sp.
Fig. spec. 69414
Legault, J.A., 1982, Can. J. Earth Sci., vol. 19, no. 9, pl. 2, fig. 10.
Middle-Upper Ordovician, Orphan Knoll, lat. 50°33′N, long. 46°6′W, Labrador Sea.

Dactylofusa n. sp.
Fig. spec. 26282
McGregor, D.C. and Cramer, F.H., 1971, Geol. Surv. Can., Bull. 202, pl. 1, fig. 26.
Cat Head Member, Red River Formation, Ordovician, McBeth Point, Lake Winnipeg, Manitoba.

Dasydiacrodium caudatum Vanguestaine, 1973
Hypotypes 57771-57773
Martin, F. and Dean, W.T, 1981, Geol. Surv. Can., Bull. 343, p. 18, pl. 1, figs. 10, 11, 15.
Clarenville Formation, Lower Ordovician, about midway between Promontories IX and X and immediately north of Promontory X (57772), south of Elliott's Cove, Random Island, eastern Newfoundland.

Dicommopalla macadamii Loeblich 1970
Hypotype 69400
Legault, J.A., 1982, Can. J. Earth Sci., vol. 19, no. 9, pl. 1, fig. 13.
Middle-Upper Ordovician, Orphan Knoll, lat. 50°33′N, long. 46°6′W, Labrador Sea.

Dicommopalla macadamii Loeblich, 1970
Hypotype 55942
Martin, F., 1983, Geol. Surv. Can., Bull. 310, p. 22, pl. 12, fig. 18.
Lotbinière Formation, Utica Group, Upper Ordovician, bay to west of Pointe Delisle, lat. 46°41′57″N, long. 71°36′29″W, Québec.

Dicommopalla rissae Bunner and Legault
Holotype 94683; hypotypes 94681, 94682, 94684-94686
Bunner, W.D. and Legault, J.A., 1989, Palynology, vol. 13, p. 60, pl. 1, figs. 1-4, 6, 7.
Middle Ordovician, Simcoe Group, Gull River Formation, depths 1095.1 m (96483) and 1125.86 (94684), Rowe/Ram No. 7 well, Dover tp., 5 km west of Chatham; Bobcaygeon Formation, Little Bob Quarry, 3 km south of Bobcaygem, Ontario.

Diexallophasis remota (Deunff) Playford 1977
Hypotype 82889
Martin, F., 1988, Bull. British Mus. (Nat. Hist.), Geol. ser., vol. 43, p. 306, fig. 15.
Ellis Bay Formation, Upper Ordovician, west side of Ellis Bay, Anticosti Island, Québec.

Dictyotidium? cambriense Slavíková, 1968
Hypotype 48199
Dean, W.T. and Martin, F., 1978, Geol. Surv. Can., Bull. 284, pl. 3, fig. 11.
Bell Island Group, Lower Ordovician, cliff summit opposite midpoint of shingle beach between its southeastern extremity and northeastern end, The Beach, Bell Island, Newfoundland.

Dictyotidium? dentatum (Vavrdová, 1976) comb. nov.
Hypotypes 46902, 46903
Dean, W.T. and Martin, F., 1978, Geol. Surv. Can., Bull. 284, p. 8, pl. 3, figs. 14, 28.
Wabana Group, Lower Ordovician, disused quarry on northwest coast north of Wabana, Bell Island, Newfoundland.

Diexallophasis denticulata (Stockmans and Willière) Loeblich 1970
Hypotype 80138
Jacobson, S.R. and Achab, A., 1985, Palynology, vol. 9, p. 178, pl. 2, fig. 4.
Vaureal Formation, Upper Ordovician, depth 3100 feet, New Associated Consolidated Paper Anticosti No. 1 well, near 24-mile lodge on Jupiter River, Anticosti Island, Québec.

Diexallophasis remota (Deunff, 1955)
Hypotype 61554
Martin, F., 1980, Can. J. Earth Sci., vol. 17, no. 1, pl. 4, fig. 3.
White Head Formation, Upper Ordovician, Grande Coupe, Gaspé, Québec.

Diexallophasis sanpetrensis (Cramer) Dorning 1981
Hypotype 80131
Jacobson, S.R. and Achab, A., 1985, Palynology, vol. 9, p. 178, pl. 1, fig. 3.
Vaureal Formation, Upper Ordovician, depth 3200 feet, New Associated Consolidated Paper Anticosti No. 1 well, near 24-mile lodge on Jupiter River, Anticosti Island, Québec.

Disparifusa borea Jacobson and Achab
Holotype 80140; paratype 80143
Jacobson, S.R. and Achab, A., 1985, Palynology, vol. 9, p. 178, pl. 3, figs. 1,4.
Vaureal Formation, Upper Ordovician, depth 2600 feet, New Associated Consolidated Paper Anticosti No. 1 well, near 24-mile lodge on Jupiter River, Anticosti Island, Québec.

Elektoriskos sp. A
Fig. spec. 80146
Jacobson, S.R. and Achab, A., 1985, Palynology, vol. 9, p. 180, pl. 3, fig. 7.
Vaureal Formation, Upper Ordovician, depth 3000 feet, New Associated Consolidated Paper Anticosti No. 1 well, near 24-mile lodge on Jupiter River, Anticosti Island, Québec.

Estiastra avita Loeblich and Tappan 1978
Hypotype 69397
Legault, J.A, 1982, Can. J. Earth Sci., vol. 19, no. 9, pl. 1, fig. 12.
Middle-Upper Ordovician, Orphan Knoll, lat. 50°33′N, long. 46°6′W, Labrador Sea.

Estiastra magna Eisenack 1959
Hypotype 80147
Jacobson, S.R. and Achab, A., 1985, Palynology, vol. 9, p. 180, pl. 4, fig. 1.
Vaureal Formation, Upper Ordovician, depth 3100 feet, New Associated Consolidated Paper Anticosti No. 1 well, near 24-mile lodge on Jupiter River, Anticosti Island, Québec.

Estiastra aff. *E. magna* Eisenack
Hypotype 66966
Duffield, S.L. and Legault, J.A., 1981, IUGS Subcommission on Silurian Stratigraphy, Ordovician-Silurian Boundary Working Group. Field Meeting, Anticosti-Gaspé, Québec, vol. II, pl. I, fig. 5.
Member 5, Ellis Bay Formation, Upper Ordovician, Cap Henry, Anticosti Island, Québec.

Estiastra cf. *E. magna* Eisenack, 1959
Hypotype 61553
Martin, F., 1980, Can. J. Earth Sci, vol. 17, no. 1, p. 110, pl. 3, fig. 12.
White Head Formation, Upper Ordovician, Grande Coupe, Gaspé, Québec.

cf. *Estiastra oligospinosa* (Eisenack)
Hypotype 80149
Jacobson, S.R. and Achab, A., 1985, Palynology, vol. 9, p. 182, pl. 4, fig. 3.
Vaureal Formation, Upper Ordovician, depth 2800 feet, New Associated Consolidated Paper Anticosti No. 1 well, near 24-mile lodge on Jupiter River, Anticosti Island, Québec.

Eupoikilofusa strita (Staplin, Jansonius, and Pocock) Loeblich and Tappan, 1978
Hypotype 80136
Jacobson, S.R. and Achab, A., 1985, Palynology, vol. 9, p. 182, pl. 2, fig. 2.
Vaureal Formation, Upper Ordovician, depth 2800 feet, New Associated Consolidated Paper Anticosti No. 1 well, near 24-mile lodge on Jupiter River, Anticosti Island, Québec.

Eupoikilofusa striatifera (Cramer) Cramer 1970
Hypotype 50142
Jacobson, S.R. and Achab, A., 1985, Palynology, vol. 9, p. 182, pl. 3, fig. 3.

Vaureal Formation, Upper Ordovician, depth 2600 feet, New Associated Consolidated Paper Anticosti No. 1 well, near 24-mile lodge on Jupiter River, Anticosti Island, Québec.

Eupoikilofusa stratifera typica (Cramer) Cramer and Diez
Hypotype 66974
Duffield, S.L. and Legault, J.A., 1981, IUGS Subcommission on Silurian Stratigraphy, Ordovician-Silurian Boundary Working Group. Field Meeting, Anticosti-Gaspé, Québec, vol. II, pl. I, fig. 13.
Member 6, Ellis Bay Formation, Upper Ordovician, "8 mile pool", Salmon River, Anticosti Island, Québec.

Excultibrachium concinnum Loeblich and Tappan 1978
Hypotype 69398
Legault, J.A., 1982, Can. J. Earth Sci., vol. 19, no. 9, pl.1, fig. 16.
Middle-Upper Ordovician, Orphan Knoll, lat. 50°33′N, long. 46°6′W, Labrador Sea.

Excultibrachium oligocladatum Turner 1984
Hypotype 80152
Jacobson, S.R. and Achab, A., 1985, Palynology, vol. 9, p. 182, pl. 4, fig. 6.
Vaureal Formation, Upper Ordovician, depth 2700 feet, New Associated Consolidated Paper Anticosti No. 1 well, near 24-mile lodge on Jupiter River, Anticosti Island, Québec.

Excultibrachium sp.
Fig. spec. 69394
Legault, J.A., 1982, Can. J. Earth Sci., vol. 19, no. 9, pl. 1, fig. 17.
Middle-Upper Ordovician, Orphan Knoll, lat. 50°33′N, long. 46°6′W, Labrador Sea.

Gen. et sp. nov. cf. *Rhopaliophora* sp.
Fig. spec. 82882
Martin, F., 1988, Bull. British Mus. (Nat. Hist.), Geol. ser., vol. 43, p. 304, fig. 8.
Ellis Bay Formation, Upper Ordovician, west side of Ellis Bay, Anticosti Island, Québec.

Goniosphaeridium conjunctum Kjellström 1976
Hypotype 69404
Legault, J.A., 1982, Can. J. Earth Sci., vol. 19, no. 9, pl. 1, fig. 6.
Middle-Upper Ordovician, Orphan Knoll, lat. 50°33′N, long. 46°6′W, Labrador Sea.

Goniosphaeridium connectum Kjellström
Hypotype 66963
Duffield, S.L. and Legault, J.A., 1981, IUGS Subcommission on Silurian Stratigraphy, Ordovician-Silurian Boundary Working Group. Field Meeting, Anticosti-Gaspé, Québec, vol. II, pl. I, fig. 2.
Member 6, Ellis Bay Formation, Upper Ordovician, Pointe Laframboise, Anticosti Island, Québec.

Goniosphaeridium conjunctum Kjellström, 1971
Hypotype 55943
Martin, F., 1983, Geol. Surv. Can., Bull. 310, p. 22, pl. 12, fig. 8.
Neuville Formation, Upper Ordovician, bay to east of Pointe Delisle, Québec.

Goniosphaeridium polygonale (Eisenack, 1931)
Hypotype 56676
Martin, F., 1978, Geol. Surv. Can., Paper 78-1B, p. 70, pl. 12.1, fig. 12.
Winterhouse Formation, Long Point Group, Upper Ordovician, quarry immediately southeast of fishing huts near west end of Clam Bank Cove, 750 northwest of Lourdes, western, Newfoundland.

Goniosphaeridium polygonale (Eisenack) Eisenack 1969
Hypoype 69407
Legault, J.A., 1982, Can. J. Earth Sci., vol. 19, no. 9, pl. 1, fig. 14.
Middle-Upper Ordovician, Orphan Knoll, lat. 50°33′N, long. 46°6′W, Labrador Sea.

Goniosphaeridium uncinatum (Downie, 1958) Kjellström, 1971
Hypotypes 48191, 48192
Dean, W.T. and Martin, F., 1978, Geol. Surv. Can., Bull. 284, pl. 1, figs. 4, 5.
Wabana Group, Lower Ordovician, disused quarry, northwest coast north of Wabana, and Wabana Mine, Bell Island, Newfoundland.

Goniosphaeridium uncinatum (Downie) Kjellström, 1971
Hypotypes 56023, 56024
Martin, F. 1983, Geol. Surv. Can., Bull. 310, p. 22, pl. 7, fig. 3; pl. 8, fig. 13.
Lobtinière Formation, Utica Group, Upper Ordovician, bay to west of Pointe Delisle, lat. 46°41′57″N, long. 71°36′29″W, Québec.

Gorgonisphaeridium? penna Martin
Holotype 55944; paratype 55945
Martin, F., 1983, Geol. Surv. Can., Bull 310, p. 23, pl. 10, figs. 11, 12, 18.
Rockcliffe Formation, Chazy Group, Middle Ordovician, cut on Highway 17, 4.5 km west of Cumberland, lat. 45°31′N, long. 75°27′W, Ontario.

Goniosphaeridium? sp. A
Fig. specs. 61555, 61556
Martin, F., 1980, Can. J. Earth Sci., vol. 17, no. 1, p. 110, pl. 3, figs. 8, 11.
White Head Formation, Upper Ordovician, Chemin du Falle, southeast of the Amphithéatre, Gaspé, Québec.

Gyalorhethium sp. A
Fig. spec. 80150
Jacobson, S.R. and Achab, A., 1985, Palynology, vol. 9, p. 183, pl. 4, fig. 4.
Vaureal Formation, Upper Ordovician, depth 3000 feet, New Associated Consolidated Paper Anticosti No. 1 well, near 24-mile lodge on Jupiter River, Anticosti Island, Québec.

Hercochitina lindsayensis Melchin and Legault
Holotype 77278; paratypes 77279, 77280
Melchin, M. and Legault, J.A., 1985, Palynology, vol. 9, p. 207, pl. 2, figs. 3-5.
Lindsay Formation, Upper Ordovician, road-cut 6.5 km north of Lindsay, Ontario.

Hercochitina minuta Achab
Hypotype 77290
Melchin, M. and Legault, J.A., 1985, Palynology, vol. 9, pl.1, fig. 6.
Vaureal Formation, Upper Ordovician, English Head, Anticosti Island, Québec.

Hercochitina pinguis Melchin and Legault
Hologype 77281; paratypes 77282, 77283, 77289, 77291
Melchin, M. and Legault, J.A., 1985, Palynology, vol. 9, p. 207, pl.1, figs. 4,5; pl. 2, figs. 1, 2, 7.
Lindsay Formation, Upper Ordovician, road-cut 6.5 km north of Lindsay, Ontario.

Hercochitina spinetum Melchin and Legault
Holotype 77284; paratypes 77285-77288
Melchin, M. and Legault, J.A., 1985, Palynology, vol. 9, p. 208, pl. 2, figs. 6, 8-11.
Lindsay Formation, Upper Ordovician, road-cut 6.5 km north of Lindsay, and Cobourg Hill No. 1 road-cut, lot 15, con. VII, Eldon tp., Victoria co. (77285, 77287, 77288), Ontario.

Hogklintia radicosa (Loeblich) n. comb.
Hypotype 80148
Jacobson, S.R. and Achab, A., 1985, Palynology, vol. 9, p. 183, pl. 4, fig. 2.
Vaureal Formation, Upper Ordovician, depth 3100 feet, New Associated Consolidated Paper Anticosti No. 1 well, near 24-mile lodge on Jupiter River, Anticosti Island, Québec.

Leiofusa asymmetrica Colbath 1979
Hypotype 69416
Legault, J.A., 1982, Can. J. Earth Sci., vol. 19, no. 9, pl. 1, fig. 11.
Middle-Upper Ordovician, Orphan Knoll, lat. 50°33′N, long. 46°6′W, Labrador Sea.

Leiofusa estrecha Cramer
Hypotype 66973
Duffield, S.L. and Legault, J.A., 1981, IUGS Subcommission on Silurian Stratigraphy, Ordovician-Silurian Boundary Working Group. Field Meeting, Anticosti-Gaspé, Québec, vol. II, pl. I, fig. 12.
Member 6, Ellis bay Formation, Upper Ordovician, Pointe Laframboise, Anticosti Island, Québec.

Leiofusa fusiformis (Eisenack) Eisenack 1938
Hypotype 80156
Jacobson, S.R. and Achab, A., 1985, Palynology, vol. 9, p. 183, pl. 5, fig. 10.
Vaureal Formation, Upper Ordovician, depth 2500 feet, New Associated Consolidated Paper Anticosti No. 1 well, near 24-mile lodge on Jupiter River, Anticosti Island, Québec.

Leiofusa aff. *L. striata* Brito et Da Silva Santos, 1963
Hypotype 55946
Martin, F., 1983, Geol. Surv. Can., Bull. 310, p. 23, pl. 12, fig. 14.
Grondines Member, Neuville Formation, Upper Ordovician, west of pier at Neuville, Québec.

Leiofusa aff. *L. tumida* Downie
Hypotype 66972
Duffield, S.L. and Legault, J.A., 1981, IUGS Subcommission on Silurian Stratigraphy, Ordovician-Silurian Boundary Working Group. Field Meeting, Anticosti-Gaspé, Québec, vol. II, pl. I, fig. 11.
Member 6, Ellis Bay Formation, Upper Ordovician, east side Ellis Bay, Anticosti Island, Québec.

Leiofusa sp.
Fig. spec. 26281
McGregor, D.C. and Cramer, F.H., 1971, Geol. Surv. Can., Bull. 202, pl. 1, fig. 25.
Cat Head Member, Red River Formation, Ordovician, McBeth Point, Lake Winnipeg, Manitoba.

Leiofusa sp.
Fig. spec. 46904
Dean, W.T. and Martin, F., 1978, Geol. Surv. Can., Bull. 284, p. 8, pl. 3, fig. 19.
Wabana Group, Lower Ordovician, disused quarry on northwest coast north of Wabana, Bell Island, Newfoundland.

Leiosphaeridia tenuissima Eisenack 1958
Hypotype 80151
Jacobson, S.R. and Achab, A., 1985, Palynology, vol. 9, p. 184, pl. 4, fig. 5.
Vaureal Formation, Upper Ordovician, depth 2900 feet, New Associated Consolidated Paper Anticosti No. 1 well, near 24-mile lodge on Jupiter River, Anticosti Island, Québec.

Leiosphaeridia sp.
Fig. spec. 55947
Martin, F., 1983, Geol. Surv. Can., Bull. 310, p. 23, pl. 11, fig. 10.
Rockcliff Formation, Chazy Group, Middle Ordovician, road-cut on Highway 17, 4.5 km west of Cumberland, lat. 45°31′N, long. 75°27′W, Ontario.

Leiovalia scaberula Loeblich and Tappan 1978
Hypotype 80144
Jacobson, S.R. and Achab, A., 1985, Palynology, vol. 9, p. 183, pl. 3, fig. 5.
Vaureal Formation, Upper Ordovician, depth 2800 feet, New Associated Consolidated Paper Anticosti No. 1 well, near 24-mile lodge on Jupiter River, Anticosti Island, Québec.

Lophosphaeridium cf. *L. citrinum* Downie
Hypotype 26287
McGregor, D.C. and Cramer, F.H., 1971, Geol. Surv. Can., Bull. 202, pl. 1, fig. 31.

Cat Head Member, Red River Formation, Ordovician, between McBeth Point and Cat Head, Lake Winnipeg, Manitoba.

Lophosphaeridium edenense Loeblich and Tappan, 1978
Hypotype 80145
Jacobson, S.R. and Achab, A., 1985, Palynology, vol. 9, p. 184, pl. 3, fig. 6.
Vaureal Formation, Upper Ordovician, depth 2800 feet, New Associated Consolidated Paper Anticosti No. 1 well, near 24-mile lodge on Jupiter River, Anticosti Island, Québec.

Lophosphaeridium papilatum Martin
Holotype 55948; paratypes 55949-55951
Martin, F., 1983, Geol. Surv. Can., Bull 310, p. 23, pl. 9, figs. 9, 10; pl. 10, figs. 17, 19; pl. 11, fig. 13.
Rockcliffe Formation, Chazy Group, Middle Ordovician, south bank Outaouais River, lat. 45°31′N, long. 75°27′W, Russel County, and road-cut on Highway 17, 4.5 km west of Cumberland (55951), Ontario.

Lophosphaeridium sp.
Fig. spec. 69417
Legault, J.A., 1982, Can. J. Earth Sci., vol. 19, no. 9, pl. 1, fig. 10.
Middle-Upper Ordovician, Orphan Knoll, lat. 50°33′N, long. 46°6′W, Labrador Sea.

Lophosphaeridium sp. A
Fig. specs. 55952, 55953
Martin, F., 1983, Geol. Surv. Can., Bull 310, p. 24, pl. 10, fig. 11; pl. 11, fig. 11.
Rockcliffe Formation, Chazy Group, Middle Ordovician, road-cut on Highway 17, 4.5 km west of Cumberland, and south bank Outaouais River, lat. 45°31′N, long. 75°27′W, Russel County, Ontario.

Lusatia aff. *L. dendroidea* Burmann, 1970
Hypotype 46905
Dean, W.T. and Martin, F., 1978, Geol. Surv. Can., Bull. 284, p. 8, pl. 3, fig. 21.
Bell Island Group, Lower Ordovician, cliff section behind old Scotia Pier and immediately southwest of disused ore conveyor terminal, southeast coast of Bell Island, Newfoundland.

Marrocanium simplex Cramer, Kanes et al., 1974
Hypotype 65676
Martin, F., 1982, Nat. Mus. Wales, Geol. Ser. 3, pl. 1, fig. 2.
Bell Island Group, Lower Ordovician, top of cliff in gully immediately southeast of ruin at Gull Island South Head, Bell Island, eastern Newfoundland.

Micrhystridium sp.
Fig. specs. 56031, 56032
Martin, F., 1983, Geol. Surv. Can., Bull 310, pl. 9, fig. 12; pl. 11, fig. 6.
Lotbinière Formation, Utica Group, Upper Ordovician, bay to west of Pointe Delisle, lat. 46°41′57″N, long. 71°36′29″W, Québec; Rockcliffe Formation, Chazy Group, Middle Ordovician, road-cut on Highway 17, 4.5 km west of Cumberland, lat. 45°31′N, long. 75°27′W, Ontario.

Microfossil A
Fig. spec. 80173
Jacobson, S.R. and Achab, A., 1985, Palynology, vol. 9, p. 195, pl. 8, fig. 7.
Vaureal Formation, Upper Ordovician, depth 2800 feet, New Associated Consolidated Paper Anticosti No. 1 well, near 24-mile lodge on Jupiter River, Anticosti Island, Québec.

Multiplicisphaeridium bifurcatum Staplin, Jansonius et Pocock, 1965
Hypotype 61557
Martin, F., 1980, Can. J. Earth Sci., vol. 17, no. 1, pl. 4, fig. 4.
White Head Formation, Upper Ordovician, Mont Joli, Gaspé, Québec.

Multiplicisphaeridium bifurcatum (Staplin, Jansonius and Pocock) Loeblich and Tappan 1978
Hypotype 80157
Jacobson, S.R. and Achab, A., 1985, Palynology, vol. 9, p. 184, pl. 6, fig. 1.
Vaureal Formation, Upper Ordovician, depth 2700 feet, New Associated Consolidated Paper Anticosti No. 1 well, near 24-mile lodge on Jupiter River, Anticosti Island, Québec.

Multiplicisphaeridium birminghamensis (Cramer) Eisenack, Cramer and Diez
Hypotype 66970
Duffield, S.L. and Legault, J.A., 1981, IUGS Subcommission on Silurian Stratigraphy, Ordovician-Silurian Boundary Working Group. Field Meeting, Anticosti-Gaspé, Québec, vol. II, pl. I, fig. 9.
Member 6, Ellis bay Formation, Upper Ordovician, Vaureal RIver, Anticosti Island, Québec.

Multiplicisphaeridium continuatum Kjellström, 1971
Hypotypes 55954, 55955
Martin, F., 1983, Geol. Surv. Can., Bull. 310, p. 24, pl. 12, figs. 15, 16.
Grondines Member, Neuville Formation, Upper Ordovician, west of pier at Neuville, Québec.

Multiplicisphaeridium cf. *continuatum* Kjellström 1971
Hypotypes 80159, 80161
Jacobson, S.R. and Achab, A., 1985, Palynology, vol. 9, p. 186, pl. 6, figs. 3, 5.
Vaureal Formation, Upper Ordovician, depth 2500 feet, New Associated Consolidated Paper Anticosti No. 1 well, near 24-mile lodge on Jupiter River, Anticosti Island, Québec.

Multiplicisphaeridium? cf. M? furcatum (Deunff) Eisenack, Cramer and Diez Rodriguez, 1973
Hypotypes 57828, 57829
Martin, F. and Dean, W.T., 1981, Geol. Surv. Can., Bull. 343, p. 19, pl. 3, figs. 19, 25.

Clarenville Formation, Lower Ordovician, immediately north of Promontory X, south of Elliott's Cove, Random Island, eastern Newfoundland.

Multiplicisphaeridium irregulare Staplin, Jansonius et Pocock, 1965
Hypotype 61558
Martin, F., 1980, Can. J. Earth Sci., vol. 17, no. 1, pl. 4, fig. 6.
White Head Formation, Upper Ordovician, Grande Coupe, Gaspé, Québec.

Multiplicisphaeridium irregulare (Staplin, Jansonius and Pocok) Loeblich and Tappan 1978
Hypotype 80158
Jacobson, S.R. and Achab, A., 1985, Palynology, vol. 9, p. 186, pl. 6, fig. 2.
Vaureal Formation, Upper Ordovician, depth 2600 feet, New Associated Consolidated Paper Anticosti No. 1 well, near 24-mile lodge on Jupiter River, Anticosti Island, Québec.

Multiplicisphaeridium lobeznum (Cramer) Eisenack, Cramer et Diez Rodriguez, 1973
Hypotypes 56025, 56026
Martin, F., 1983, Geol. Surv. Can., Bull. 310, pl. 8, fig. 3; pl. 9, fig. 4.
Grondines Member, Neuville Formation, Upper Ordovician, west of pier at Neuville, and bay to east of Pointe Delisle, Québec.

Multiplicisphaeridium radicosum Loeblich, 1970
Hypotypes 61559, 61560
Martin, F., 1980, Can. J. Earth Sci., vol. 17, no. 1, p. 112, pl. 3, figs. 5, 6.
White Head Formation, Upper Ordovician, Chemin du Falle, southeast of the Amphithéatre, Gaspé, Québec.

Multiplicisphaeridium radicosum Loeblich, 1970
Hypotypes 55956, 55957
Martin, F., 1983, Geol. Surv. Can., Bull. 310, p. 24, pl. 12, figs. 13, 20.
Deschambault Formation, Middle Ordovician, Lafarge Cement quarry, Saint Constant; Neuville Formation, Upper Ordovician, Elzéar Verreault Ltée quarry, Giffard, Québec.

Multiplicisphaeridium cf. *M. varians* (Stockmans et Willière) Eisenack, Cramer et Diez Rodriguez, 1973
Hypotype 55958
Martin, F., 1983, Geol. Surv. Can., Bull. 310, p. 24, pl. 11, fig. 18.
Laval Formation, Chazy Group, Middle Ordovician, Bédard quarry, Caughnawaga, Laprairie County, Québec.

Multiplicisphaeridium wrightii Jacbson and Achab
Holotype 80160
Jacobson, S.R. and Achab, A., 1985, Palynology, vol. 9, p. 186, pl. 6, figs. 4, 7.
Vaureal Formation, Upper Ordovician, depth 2600 feet, New Associated Consolidated Paper Anticosti No. 1 well, near 24-mile lodge on Jupiter River, Anticosti Island, Québec.

Multiplicisphaeridium sp. 1, 2
Fig. specs. 66965, 66968
Duffield, S.L. and Legault, J.A., 1981, IUGS Subcommission on Silurian Stratigraphy, Ordovician-Silurian Boundary Working Group. Field Meeting, Anticosti-Gaspé, Québec, vol. II, pl. I, figs. 4, 7.
Member 6, Ellis Bay Formation, Upper Ordovician, Pointe Laframboise and "8 mile" pool section, Salmon River, Anticosti Island, Québec.

Multiplicisphaeridium sp. I, *sensu* Duffield and Legault, 1981
Fig. spec. 82890
Martin, F., 1988, Bull. British Mus. (Nat. Hist.), Geol. ser., vol. 43, p. 304, fig. 16.
Ellis Bay Formation, Upper Ordovician, west side of Ellis Bay, Anticosti Island, Québec.

Navifusa ancepsipuncta Loeblich 1970
Hypotype 80141
Jacobson, S.R. and Achab, A., 1985, Palynology, vol. 9, p. 188, pl. 3, fig. 2.
Vaureal Formation, Upper Ordovician, depth 3000 feet, New Associated Consolidated Paper Anticosti No. 1 well, near 24-mile lodge on Jupiter River, Anticosti Island, Québec.

Navifusa aff. *N. ancepsipuncta* Loeblich, 1970
Hypotypes 55959-55961
Martin, F., 1983, Geol. Surv. Can., Bull. 310, p. 24, pl. 10, figs. 8, 16; pl. 11, figs. 15, 19.
Laval Formation, Chazy Group, Middle Ordovician, Bédard quarry, Caughawaga, Laprairie County, Québec.

Navifusa brasiliensis lingula (Brito and Santos) Combaz, Lange and Pansart 1967
Hypotype 69409
Legault, J.A., 1982, Can. J. Earth Sci., vol. 19, no. 9, pl. 1, fig. 2.
Middle-Upper Ordovician, Orphan Knoll, lat. 50°33′N, long. 46°6′W, Labrador Sea.

Ordovicidium elegantulum Tappan et Loeblich, 1971
Hypotypes 55962-55965
Martin, F., 1983, Geol. Surv. Can., Bull. 310, p. 24, pl. 8, figs. 8-10; pl. 11, fig. 1; pl. 12, fig. 3.
Upper Ordovician, Grondines Member, Neuville Formation, west of pier at Neuville, and bay to east of Pointe Delisle (55963); Utica Group, southeast of Montmorency Falls (55964), Québec.

Ordovicidium elegantulum Tappan et Loeblich, 1971
Hypotype 61561
Martin, F., 1980, Can. J. Earth Sci., vol. 17, no. 1, pl. 4, fig. 9.
White Head Formation, Upper Ordovician, Grande Coupe, Gaspé, Québec.

Ordovicidium fissum Martin
Holotype 55966; paratypes 55967-55970
Martin, F., 1983, Geol. Surv. Can., Bull. 310, p. 25, pl. 8, fig. 6; pl. 9, figs. 7, 11; pl. 12, figs. 9, 19.

Lotbinière Formation, Utica Group, Upper Ordovician, bay west of Pointe Delisle, lat. 46°41′57″N, long. 71°36′29″W; Grondines Member, Neuville Formation, Upper Ordovician, west of pier at Neuville (55967, 55968); Deschambault Formation, Middle Ordovician, Industries de Neuville Inc. quarry, about 3 km northeast of Neuville (55970), Québec.

Ordovicidium groetlingboense (Kjellström) Loeblich and Tappan 1978
Hypotype 80163
Jacobson, S.R. and Achab, A., 1985, Palynology, vol. 9, p. 188, pl. 7, fig. 1.
Vaureal Formation, Upper Ordovician, depth 3100 feet, New Associated Consolidated Paper Anticosti No. 1 well, near 24-mile lodge on Jupiter River, Anticosti Island, Québec.

Orthosphaeridium gaspesianum Martin
Holotype 61566; paratypes 61564, 61565
Martin, F., 1980, Can. J. Earth Sci., vol. 17, no. 1, p. 112, pl. 4, figs. 1, 2, 7; text-fig. 2.
White Head Formation, Upper Ordovician, Grande Coupe, Gaspé, Québec.

Orthosphaeridium inflatum Loeblich 1970
Hypotype 80162
Jacobson, S.R. and Achab, A., 1985, Palynology, vol. 9, p. 188, pl. 6, fig. 9.
Vaureal Formation, Upper Ordovician, depth 3100 feet, New Associated Consolidated Paper Anticosti No. 1 well, near 24-mile lodge on Jupiter River, Anticosti Island, Québec.

Orthosphaeridium insculptum Loeblich, 1970
Hypotypes 61562, 61563
Martin, F., 1980, Can. J. Earth Sci., vol. 17, no. 1, p. 112, pl. 3, figs. 1, 4.
White Head Formation, Upper Ordovician, Grande Coupe, Québec.

Orthosphaeridium insculptum Loeblich, 1970
Hypotypes 55971-55973
Martin, F., 1983, Geol. Surv. Can., Bull. 310, p. 25, pl. 6, figs. 4, 5; pl. 12, fig. 21.
Upper Ordovician, Grondines Member, Neuville Formation, west of pier at Neuville; Lotbinière Formation, Utica Group, bay west of Pointe Delisle, lat. 46°41′57″N, long. 71°36′29″W (55972), Québec.

Orthosphaeridium insculptum Loebich 1970
Hypotype 80166
Jacobson, S.R. and Achab, A., 1985, Palynology, vol. 9, p. 188, pl. 7, fig. 6.
Vaureal Formation, Upper Ordovician, depth 3200 feet, New Associated Consolidated Paper Anticosti No. 1 well, near 24-mile lodge on Jupiter River, Anticosti Island, Québec.

Orthosphaeridium insculptum Loeblich
Hypotype 82877
Martin, F., 1988, Bull. British Mus. (Nat. Hist.), Geol. ser., vol. 43, p. 304, figs. 3a, b.
Ellis Bay Formation, Upper Ordovician, west side of Ellis Bay, Anticosti Island, Québec.

Orthosphaeridium octospinosum Eisenack, 1968
Hypotypes 55974, 55975
Martin, F., 1983, Geol. Surv. Can., Bull. 310, p. 25, pl. 7, fig. 8; pl. 11, fig. 28.
Upper Ordovician, Gronides Member, Neuville Formation, west of pier at Neuville; Utica Group, southeast of Montmorency Falls, Québec.

Orthosphaeridium rectangulare (Eisenack)
Hypotype 82878
Martin, F., 1988, Bull. British Mus. (Nat. Hist.), Geol. ser., vol. 43, p. 304, figs. 4a, b.
Ellis Bay Formation, Upper Ordovician, west side of Ellis Bay, Anticosti Island, Québec.

Petaloferidium stiggi Jacobson 1978
Hypotype 69411
Legault, J.A., 1982, Can. J. Earth Sci., vol. 19, no. 9, pl. 1, fig. 1.
Middle-Upper Ordovician, Orphan Knoll, lat. 50°33′N, long. 46°6′W, Labrador Sea.

Petaloferidium stigii Jacobson 1978
Hypotype 80167
Jacobson, S.R. and Achab, A., 1985, Palynology, vol. 9, p. 190, pl. 7, fig. 7.
Vaureal Formation, Upper Ordovician, depth 2800 feet, New Associated Consolidated Paper Anticosti No. 1 well, near 24-mile lodge on Jupiter River, Anticosti Island, Québec.

Petaloferidium sp. A
Fig. spec. 80164
Jacobson, S.R. and Achab, A., 1985, Palynology, vol. 9, p. 192, pl. 7, fig. 4.
Vaureal Formation, Upper Ordovician, depth 2500 feet, New Associated Consolidated Paper Anticosti No. 1 well, near 24-mile lodge on Jupiter River, Anticosti Island, Québec.

Peteinosphaeridium breviradiatum (Eisenack) Eisenack, 1969
Hypotypes 58976-55978
Martin, F., 1983, Geol. Surv. Can., Bull. 310, p. 26, pl. 10, figs. 6, 7; pl. 12, fig. 6.
Middle Ordovician, Chazy Group, Rockcliffe Formation, road-cut on Highway 17, 4.5 km west of Cumberland, lat. 45°31′N, long. 75°27′W, Ontario; Laval Formation, Bédard quarry, Caughnawaga, Laprairie County, Québec; Neuville Formation, Upper Ordovician, Charlesbourg Ltée quarry, Charlesbourg-Est, Québec.

Peteinosphaeridium breviradiatum (Eisenack, 1959)
Hypotype 56685
Martin, F., 1978, Geol. Surv. Can., Paper 78-1B, pl. 12.1, fig. 25.

Winterhouse Formation, Long Point Group, Upper Ordovician, quarry immediately southeast of fishing huts near west end Clam Bank Cove, 750 m northwest of Lourdes, western Newfoundland.

Peteinosphaeridium breviradiatum (Eisenack, 1959)
Hypotypes 61567, 61568
Martin, F., 1980, Can. J. Earth Sci., vol. 17, no. 1, pl. 3, fig. 13; pl. 4, fig. 5.
White Head Formation, Upper Ordovician, Grande Coupe and Mont-Joli, Gaspé, Québec.

Peteinosphaeridium breviradiatum (Eisenack) Kjellström 1971
Hypotype 69402
Legault, J.A., 1982, Can. J. Earth Sci., vol. 19, no. 9, pl. 1, fig. 15.
Middle-Upper Ordovician, Orphan Knoll, lat. 50°33′N, long. 46°6′W, Labrador Sea.

Peteinosphaeridium cf. *P. breviradiatum* (Eisenack) Eisenack, 1969
Hypotype 65663
Dean, W.T. and Martin, F., 1982, Nat. Mus. Wales, Geol. Ser. 3, pl. 1, fig. 3.
Outram Formation, Lower Ordovician, southeastern end Wilcox Pass, Jasper Park, Alberta.

Peteinosphaeridium groetlingboensis Kjellström, 1971
Hypotypes 55979-55984
Martin, F., 1983, Geol. Surv. Can., Bull. 310, p. 26, pl. 7, figs. 1, 4, 5; pl. 8, figs. 1, 4, 7, 11; pl. 12, fig. 2.
Upper Ordovician, Lotbinière Formation, Utica Group, bay to west of Pointe Delisle, lat. 46°41′57″N, long. 71°36′29″W, and road extending from bay west of Pointe Delisle to Highway 138(55983); Neuville Formation, bay to east of Pointe Delisle (55982), Québec.

Petinosphaeridium intermedium (Eisenack) Jacobson 1978
Hypotype 80165
Jacobson, S.R. and Achab, A., 1985, Palynology, vol. 9, p. 190, pl. 7, fig. 5.
Vaureal Formation, Upper Ordovician, depth 3000 feet, New Associated Consolidated Paper Anticosti No. 1 well, near 24-mile lodge on Jupiter River, Anticosti Island, Québec.

Peteinosphaeridium micranthum (Eisenack, 1959)
Hypotype 56682
Martin, F., 1978, Geol. Surv. Can., Paper 78-1B, pl. 12.1, fig. 20.
Table Head Formation, Middle Ordovician, north side small bay about 300 m south of Black Point, East Bay, 5.5 km north-northeast of Port au Port village, western Newfoundland.

Peteinosphaeridium cf. *P. nanofurcatum* Kjellström, 1971
Hypotype 56681
Martin, F., 1978, Geol. Surv. Can., Paper 78-1B, pl. 12.1, fig. 16.
Table Head Formation, Middle Ordovician, north side small bay about 300 m south of Black Point, East Bay, 5.5 km north-northeast of Port au Port village, western Newfoundland.

Peteinosphaeridium palmatum Combaz and Peniguel, 1972
Hypotype 46906
Dean, W.T. and Martin, F., 1978, Geol. Surv. Can., Bull. 284, pl. 3, fig. 2.
Wabana Group, Lower Ordovician, northeast side of entrance to a disused mine tunnel (Submarine Slope No. 2), between Wabana town centre and northeast coast of Bell Island, Newfoundland.

Peteinosphaeridium cf. *P. velatum* Kjellström 1971
Hypotype 69413
Legault, J.A., 1982, Can. J. Earth Sci., vol. 19, no. 9, pl. 1, fig. 7.
Middle-Upper Ordovician, Orphan Knoll, lat. 50°33′N, long. 46°6′W, Labrador Sea.

Pheoclosterim fuscinulaegerum Tappan et Loeblich, 1971
Hypotype 55985
Martin, F., 1983, Geol. Surv. Can., Bull. 310, p. 26, pl. 12, fig. 12.
Grondines Member, Neuville Formation, Upper Ordovician, west of pier at Neuville, Québec.

Pheoclosterium sp. A
Fig. spec. 55986
Martin, F., 1983, Geol. Surv. Can., Bull. 310, p. 26, pl. 11, fig. 4.
Deschambault Formation, Middle Ordovician, Lafarge Cement quarry, Saint-Constant, Québec.

Pheoclosterium sp. nov.
Fig. spec. 82881
Martin, F., 1988, Bull British Mus. (Nat. Hist.), Geol. ser., vol. 43, p. 304, figs. 7a, b.
Ellis Bay Formation, Upper Ordovician, west side of Ellis Bay, Anticosti Island, Québec.

Pirea ornata (Burmann) Vavrdová, 1977
Hypotypes 55987, 55988
Martin, F., 1983, Geol. Surv. Can., Bull. 310, p. 27, pl. 11, figs. 14, 25.
Montréal and Deschambault formations, Middle Ordovician, Lafarge Cement quarry, Saint-Constant, Québec.

Pirea aff. *P. ornata* (Burmann, 1970) Vavrdová, 1977
Hypotype 46907
Dean, W.T. and Martin, F., 1978, Geol. Surv. Can., Bull. 284, p. 8, pl. 3, fig. 6.
Wabana Group, Lower Ordovician, disused quarry on northwest coast, north of Wabana, Bell Island, Newfoundland.

Poikilofusa spinata Staplin, Jansonius et Pocock, 1965
Hypotype 61569
Martin, F., 1980, Can. J. Earth Sci, vol. 17, no. 1, p. 114, pl. 4, fig. 8.
White Head Formation, Upper Ordovician, Grande Coupe, Gaspé, Québec.

Poikilofusa spinata (Staplin, Jansonius and Pocock) Loeblich and Tappan 1978
Hypotypes 80169, 80170
Jacobson, S.R. and Achab, A., 1985, Palynology, vol. 9, p. 192, pl. 8, figs. 2, 3.

Vaureal Formation, Upper Ordovician, depths 3000 and 2700 feet, New Associated Consolidated Paper Anticosti No. 1 well, near 24-mile lodge on Jupiter River, Anticosti Island, Québec.

Poikilofusa squama (Deunff) Martin, 1973
Hypotype 57780
Martin, F. and Dean, W.T., 1981, Geol. Surv. Can., Bull. 343, pl. 1, fig. 18.
Clarenville Formation, Lower Ordovician, about midway between Promontories IX and X, south of Elliott's Cove, Random Island, eastern Newfoundland.

Polygonium gracile Vavrdová 1966 emend.
Hypotype 80168
Jacobson, S.R. and Achab, A., 1985, Palynology, vol. 9, p. 192, pl. 7, fig. 8.
Vaureal Formation, Upper Ordovician, depth 3000 feet, New Associated Consolidated Paper Anticosti No. 1 well, near 24-mile lodge on Jupiter River, Anticosti Island, Québec.

Priscogalea cortinula Deunff 1961
Hypotype 48195
Dean, W.T. and Martin, F., 1978, Geol. Surv. Can., Bull. 284, pl. 3, fig. 3.
Bell Island Group, Lower Ordovician, cliff at east end of Chimmey Cove, near southwest end of Bell Island, Newfoundland.

Priscogalea striatula (Vavrdova, 1966) Martin, 1973
Hypotype 48194
Dean, W.T. and Martin, F., 1978, Geol. Surv. Can., Bull. 284, pl. 2, fig. 11.
Wabana Group, Lower Ordovician, Wabana Mine, Bell Island, Newfoundland.

Pseudolunulidia alveolus Martin
Holotype 55990
Martin, F., 1983, Geol. Surv. Can., Bull. 310, p. 27, pl. 11, fig. 3.
Rockcliffe Formation, Chazy Group, Middle Ordovician, road-cut on Highway 17, 4.5 km west of Cumberland, lat. 45°31′N, long. 75°27′W, Ontario.

Pterospermella sp. A
Fig. spec. 55989
Martin, F., 1983, Geol. Surv. Can., Bull. 310, p. 27, pl. 11, fig. 20.
Louville Fromation, Black River Group, Middle Ordovician, Saint-Vincent-de-Paul quarry, Laval, Québec.

Rhopaliophora brevituberculata (Kjellström, 1971) n. comb.
Hypotypes 55991, 55992
Martin, F., 1983, Geol. Surv. Can., Bull. 310, p. 27, pl. 10, figs. 2, 10.
Rockcliffe Formation, Chazy Group, Middle Ordovician, road-cut on Highway 17, 4.5 km west of Cumberland, lat. 45°31′N, long. 75°27′W, Ontario.

Rhopaliophora impexa Tappan and Loeblich 1971
Hypotype 69393
Legault, J.A., 1982, Can. J. Earth Sci., vol. 19, no. 9, pl. 2, fig. 9.
Middle-Upper Ordovician, Orphan Knoll, lat. 50°33′N, long. 46°6′W, Labrador Sea.

Rhopaliophora? sp. nov.
Fig. specs. 65665, 65666
Dean, W.T. and Martin, F., 1982, Nat. Mus. Wales, Geol. Ser. 3, p. 138, pl. 1, figs. 6, 11.
Skoki Formation, Lower Ordovician, southeastern end Wilcox Pass, Jasper Park, Alberta.

Saharidia fragile (Downie, 1958) Combaz, 1967
Hypotype 48204
Dean, W.T. and Martin, F., 1978, Geol. Surv. Can., Bull. 284, pl. 3, fig. 32.
Bell Island Group, Lower Ordovician, cliff at east end of Chimmey Cove, near southwest end of Bell Island, Newfoundland.

Saharidia fragile (Downie) Combaz, 1967
Hypotype 57767
Martin, F. and Dean, W.T. 1981, Geol. Surv. Can., Bull. 343, pl. 1, fig. 6.
Clarenville Formation, Lower Ordovician, about midway between Promontories IX and X, south of Elliott's Cove, Random Island, eastern Newfoundland.

Saharidia fragile (Downie) Combaz, 1967
Hypotype 65677
Martin, F., 1982, Nat. Mus. Wales, Geol. Ser. 3, pl. 1, fig. 7.
Clarenville Formation, Lower Ordovician, about midway between Promontories IX and X, south of Elliott's Cove, Random Island, eastern Newfoundland.

Striatotheca microrugulata (Vavrdová, 1972) Martin, 1977
Hypotypes 48193, 48202
Dean, W.T. and Martin, F., 1978, Geol. Surv. Can., Bull. 284, pl. 2, fig. 10; pl. 3, fig. 16.
Wabana Group, Lower Ordovician, Wabana Mine and disused quarry on northwest coast north of Wabana, Bell Island, Newfoundland.

Striatotheca mutua Burmann, 1970
Hypotypes 46908-46910
Dean, W.T and Martin, F., 1978, Geol. Surv. Can., Bull. 284, p. 9, pl. 2, figs. 12, 13; pl. 3, fig. 18.
Wabana Group, Lower Ordovician, disused quarry on northeast coast, north of Wabana (46908) and Wabana Mine, Bell Island, Newfoundland.

Striatotheca principalis Burmann, 1970
Hypotype 48201
Dean, W.T. and Martin, F., 1978, Geol. Surv. Can., Bull. 284, pl. 3, fig. 13.
Wabana Group, Lower Ordovician, northeast side of entrance to a disused nine tunnel (Submarine Slope No. 2), between Wabana town centre and northwest coast of Bell Island, Newfoundland.

Striatotheca principalis parva Burmann, 1970
Hypotype 48197
Dean, W.T. and Martin, F., 1978, Geol. Surv. Can., Bull. 284, pl. 3, fig. 8.
Wabana Group, Lower Ordovician, Wabana Mine, Bell Island, Newfoundland.

Sylvanidium paucibrachium Loeblich, 1970
Hypotype 61570
Martin, F., 1980, Can. J. Earth Sci., vol. 17, no. 1, p. 114, pl. 3, fig. 9.
White Head Formation, Upper Ordovician, Chemin du Falle, southeast of the Amphithéatre, Gaspé, Québec.

Sylvanidium paucibrachium Loeblich 1970
Hypotype 80177
Jacobson, S.R. and Achab, A., 1985, Palynology, vol. 9, p. 193, pl. 9, fig. 4.
Vaureal Formation, Upper Ordovician, depth 2600 feet, New Associated Consolidated Paper Anticosti No. 1 well, near 24-mile lodge on Jupiter River, Anticosti Island, Québec.

Sylvanidium? sp.
Fig. spec. 55993
Martin, F., 1983, Geol. Surv. Can., Bull. 310, p. 28, pl. 11, fig. 22.
Montréal Formation, Middle Ordovician, Lafarge Cement quarry, Saint-Constant, Québec.

Synsphaeridium sp.
Fig. specs. 55994, 55995
Martin, F., 1983, Geol. Surv. Can., Bull. 310, p. 28, pl. 7, fig. 9; pl. 11, fig. 26.
Laval Formation, Chazy Group, Middle Ordovician, Bédard quarry, Caughnawaga, Laprairie County, Québec.

Tectitheca additionalis Burmann, 1968
Hypotype 46911
Dean, W.T. and Martin, F., 1978, Geol. Surv. Can., Bull. 284, p. 9, pl. 3, fig. 30.
Wabana Group, Lower Ordovician, Wabana Mine, Bell Island, Newfoundland.

Tectitheca filigera Burmann, 1968
Hypotype 46912
Dean, W.T. and Martin, F., 1978, Geol. Surv. Can., Bull. 284, p. 9, pl. 3, fig. 22.
Wabana Group, Lower Ordovician, Wabana Mine, Bell Island, Newfoundland.

Tectitheca spinifera Burmann, 1968
Hypotype 46913
Dean, W.T. and Martin, F., 1978, Geol. Surv. Can., Bull. 284, p. 9, pl. 3, fig. 10.
Wabana Group, Lower Ordovician, northeast side of entrance to a disused mine tunnel (Submarine Slope No. 2), between Wabana town centre and northwest coast of Bell Island, Newfounland.

Tetraniveum arenigum (Vavrdová) Vavrdová, 1976
Hypotype 65675
Martin, F., 1982, Nat. Mus. Wales, Geol. Ser. 3, pl. 1, fig. 1.
Bell Island Group, Lower Ordovician, top of cliff in gully immediately southeast of ruin at Gull Island South Head, Bell Island, eastern Newfoundland.

Tunisphaeridium caudatum Deunff and Evitt 1968
Hypotype 80172
Jacobson, S.R. and Achab, A., 1985, Palynology, vol. 9, p. 193, pl. 8, fig. 6.
Vaureal Formation, Upper Ordovician, depth 2600 feet, New Associated Consolidated Paper Anticosti No. 1 well, near 24-mile lodge on Jupiter River, Anticosti Island, Québec.

Tunisphaeridium eisenackii Loeblich and Tappan 1978
Hypotype 80171
Jacobson, S.R. and Achab, A., 1985, Palynology, vol. 9, p. 193, pl. 8, fig. 5.
Vaureal Formation, Upper Ordovician, depth 3000 feet, New Associated Consolidated Paper Anticosti No. 1 well, near 24-mile lodge on Jupiter River, Anticosti Island, Québec.

Tunisphaeridium? eligmosum Vavrdová, 1973
Hypotype 46914
Dean, W.T. and Martin, F., 1978, Geol. Surv. Can., Bull. 284, p. 9, pl. 3, fig. 1.
Wabana Group, Lower Ordovician, northeast side of entrance to a disused mine tunnel (Submarine Slope No. 2), between Wabana town centre and northwest coast of Bell Island, Newfoundland.

Uranidium semicalvum Jacobson and Achab
Holotype 80153
Jacobson, S.R. and Achab, A., 1985, Palynology, vol. 9, p. 193, pl. 5, fig. 5.
Vaureal Formation, Upper Ordovician, depth 2600 feet, New Associated Consolidated Paper Anticosti No. 1 well, near 24-mile lodge on Jupiter River, Anticosti Island, Québec.

Velatasphaera hudsonii Miller and Williams
Holotype 94671; paratypes 94672-94675; hypotypes 94676-94678
Miller, M.A. and Williams, G.L., 1988, Palynology, vol. 12, p. 124, pl. 1, figs. 1-10; pl. 2, figs. 1-6.
Late Ordovician, CSS Hudson cruise 85-027 core holes Station 90, lat. 61°21.73′N, long. 70°21.27′W, Station 78(94672-94674) and Station 77 (94676-94678), lat. 62°23.09′N, long. 72°44.94′W, Hudson Strait, Northwest Territories.

Veryhachium abditum Martin
Holotype 55996; paratypes 55997, 55998
Martin, F., 1983, Geol. Surv. Can., Bull. 310, p. 28., pl. 9, figs. 2, 3; pl. 12, fig. 22.
Lotbinière Formation, Utica Group, Upper Ordovician, bay to west of Pointe Delisle, lat. 46°41′57″N, long. 71°36′29″W, Québec.

Veryhachium circumornatum Martin
Holotype 55999; paratypes 56000-56002
Martin, F., 1983, Geol. Surv. Can., Bull. 310, p. 28, pl. 7, figs. 6, 10; pl. 11, figs. 8, 9, 23.

Lotbinière Formation, Utica Group, Upper Ordovician, bay to west of Pointe Delisle, lat. 46°41′57″N, long. 71°36′29″W; Neuville Formation, Upper Ordovician, bay to east of Pointe Delisle (56001), Québec.

Veryhachium dumontii Vanguestaine, 1973
Hypotypes 57768, 57769
Martin, F. and Dean, W.T., 1981, Geol. Surv. Can., Bull. 343, p. 22, pl. 1, figs. 7, 8.
Elliott Cove Formation, Upper Cambrian, Promontory V and south of Promontory VIII, south of Elliott's Cove, Random Island, eastern Newfoundland.

Veryhachium dumontii Vanguestaine, 1973
Hypotype 65682
Martin, F., 1982, Nat. Mus. Wales, Geol. Ser. 3, pl. 1, fig. 17.
Elliott Cove Formation, Upper Cambrian, between Promontories VII and VIII south of Elliott's Cove, Random Island, eastern Newfoundland.

Veryhachium hamii Loeblich, 1970
Hypotypes 56003, 56004
Martin, F., 1983, Geol. Surv. Can., Bull. 310, p. 28, pl. 6, figs. 3, 6, 8.
Upper Ordovician, Lotbinière Formation, Utica Group, bay to west of Pointe Delisle, lat. 46°41′57″N, long. 71°36′29″W; Grondines Member, Neuville Formation, west of pier at Neuville, Québec.

Veryhachium hamii Loeblich 1970
Hypotypes 80178, 80179
Jacobson, S.R. and Achab, A., 1985, Palynology, vol. 9, p. 194, pl. 9, figs. 5, 8.
Vaureal Formation, Upper Ordovician, depth 2500 feet, New Associated Consolidated Paper Anticosti No. 1 well, near 24-mile lodge on Jupiter River, Anticosti Island, Québec.

Veryhachium horidum Paris and Deunff, 1970
Hypotype 46915
Dean, W.T. and Martin, F., 1978, Geol. Surv. Can., Bull. 284, p. 3, pl. 3, fig. 15.
Wabana Group, Lower Ordovician, summit of cliff about 125 m west-northwest of western end of disused airfield on Bell Island, Newfoundland.

Veryhachium irroratum Loeblich and Tappan
Hypotype 26280
McGregor, D.C. and Cramer, F.H., 1971, Geol. Surv. Can., Bull. 202, pl. 1, figs. 23, 24.
Cat Head Member, Red River Formation, Ordovician, between McBeth Point and Cat Head, Lake Winnipeg, Manitoba.

Veryhachium irroratum Loeblich and Tappan 1969
Hypotype 69406
Legault, J.A., 1982, Can. J. Earth Sci., vol. 19, no. 9, pl. 2, fig. 1.
Middle-Upper Ordovician, Orphan Knoll, lat. 50°33′N, long. 46°6′W, Labrador Sea.

Veryhachium irroratum Loeblich et Tappan, 1969
Hypotypes 56005, 56006
Martin, F., 1983, Geol. Surv. Can., Bull. 310, p. 29, pl. 6, fig. 7; pl. 8, fig. 12; pl. 4, fig. 24.
Laval Formation, Chazy Group, Middle Ordovician, Bédard quarry, Caughnawaga, Laprairie County, Québec.

Veryhachium lairdi Delfandre ex Deunff 1959
Hypotype 69399
Legault, J.A., 1982, Can. J. Earth Sci., vol. 19, no. 9, pl. 2, fig. 12.
Middle-Upper Ordovician, Orphan Knoll, lat. 50°33′N, long. 46°6′W, Labrador Sea.

Veryhachium lairdii (Delfandre) Deunff ex Downie 1958
Hypotype 80175
Jacobson, S.R. and Achab, A., 1985, Palynology, vol. 9, p. 195, pl. 9, fig. 2.
Vaureal Formation, Upper Ordovician, depth 2600 feet, New Associated Consolidated Paper Anticosti No. 1 well, near 24-mile lodge on Jupiter River, Anticosti Island, Québec.

Veryhachium mameti Martin
Holotype 56007; paratypes 56008-56015
Martin, F., 1983, Geol. Surv. Can., Bull. 310, p. 29, pl. 6, figs. 10-12; pl. 10, figs. 5, 14; pl. 11, figs. 7, 12, 16, 17.
Laval Formation, Chazy Group, Middle Ordovician, Bédard quarry, Caughnawaga, Laprairie County, Québec.

Veryhachium-Micrhystridium complex
Fig. specs. 26283, 26288-26293
McGregor, D.C. and Cramer, F.H., 1971, Geol. Surv. Can., Bull. 202, pl. 1, figs. 27, 28; pl. 2, fig. 1, 2, 4, 6, 8, 9.
Cat Head Member, Red River Formation, Ordovician, between McBeth Point and Cat Head, Lake Winnipeg, Manitoba.

Veryhachium oklahomense Loeblich, 1970
Hypotypes 56016, 56017, 56027
Martin, F., 1983, Geol. Surv. Can., Bull. 310, p. 29, pl. 8, fig. 5; pl. 9, fig. 5; pl. 11, fig. 21.
Grondines Member, Neuville Formation, Upper Ordovician, bay to east of Pointe Delisle, and west of pier at Neuville (56017), Québec.

Veryhachium piliferum Martin, 1966
Hypotype 56030
Martin, F., 1983, Geol. Surv. Can., Bull. 310, p. 29, pl. 11, fig. 27.
Neuville Formation, Upper Ordovician, Elzéar Verrault Ltée. quarry, Giffard, Québec.

Veryhachium setosapelliculum Loeblich, 1970
Hypotypes 56018, 56019
Martin, F., 1983, Geol. Surv. Can., Bull. 310, p. 29, pl. 6, fig. 9; pl. 12, fig. 17.

Grondines Member, Neuville Formation, Upper Ordovician, bay to west of Pointe Delisle; Deschambault Formation, Middle Ordovician, Lafarge Cement quarry, Saint-Constant, Laprairie County, Québec.

Veryhachium valiente Cramer-*V. lairdi* Deunff
Hypotype 66962
Duffield, S.L. and Legault, J.A., 1981, IUGS Subcommission on Silurian Stratigraphy, Ordovician-Silurian Boundary Working Group. Field Meeting, Anticosti-Gaspé, Québec, vol. II, pl. I, fig. 1.
Member 5, Ellis Bay Formation, Upper Ordovician, Cap Henry, Anticosti Island, Québec.

Veryhachium sp.
Fig. spec. 69418
Legault, J.A., 1982, Can. J. Earth Sci., vol. 19, no. 9, pl. 2, fig. 2.
Middle-Upper Ordovician, Orphan Knoll, lat. 50°33′N, long. 46°6′W, Labrador Sea.

Veryhachium sp.
Fig. specs. 56028, 56029
Martin, F., 1983, Geol. Surv. Can., Bull. 310, pl. 11, figs. 2, 29.
Trenton Group, Middle Ordovician, southeast of Montmorency Falls; Neuville Formation, Upper Ordovician, Union Quarries and Paving Ltée quarry, Charlesbourg-Ouest, Québec.

Veryhachium sp. A
Fig. specs. 80174, 80176
Jacobson, S.R. and Achab, A., 1985, Palynology, vol. 9, p. 195, pl. 9, figs. 1, 3.
Vaureal Formation, Upper Ordovician, depths 2700 and 3000 feet, New Associated Consolidated Paper Anticosti No. 1 well, near 24-mile lodge on Jupiter River, Anticosti Island, Québec.

Veryhachium sp. I, II
Fig. specs. 26294, 26295
McGregor, D.C. and Cramer, F.H., 1971, Geol. Surv. Can., Bull. 202, pl. 2, figs. 3, 5.
Cat Head Member, Red River Formation, Ordovician, McBeth Point and between McBeth Point and Cat Head, Lake Winnipeg, Manitoba.

Vogtlandia coalita Martin
Holotype 46916; paratype 48183
Dean, W.T. and Martin, F., 1978, Geol. Surv. Can., Bull. 284, p. 9, pl. 2, figs. 3, 7.
Wabana Group, Lower Ordovician, Wabana Mine, Bell Island, Newfoundland.

Vogtlandia flos Martin
Holotype 48184; paratypes 48185, 48186
Dean, W.T. and Martin, F., 1978, Geol. Surv. Can., Bull. 284, p. 10, pl. 1, figs. 2, 6; pl. 2, figs. 24, 26.
Wabana Group, Lower Ordovician, disused quarry on northwest coast north of Wabana, Bell Island, Newfoundland.

Vogtlandia? perculta Martin
Holotype 56020; paratypes 56021, 56022
Martin, F., 1983, Geol. Surv. Can., Bull. 310, p. 30, pl. 10, figs. 3, 4, 20.
Laval Formation, Chazy Group, Middle Ordovician, Bédard quarry, Caughnawaga, Laprairie County, Québec.

Vogtlandia tenuata Burmann, 1970
Hypotypes 48187, 48188
Dean, W.T. and Martin, F., 1978, Geol. Surv. Can., Bull. 284, p. 10, pl. 1, fig. 7; pl. 3, fig. 20.
Wabana Group, Lower Ordovician, northeast side of entrance to a disused mine tunnel (Submarine Slope No. 2) between Wabana town centre and northwest coast, and disused quarry, northwest coast north of Wabana, Bell Island, Newfoundland.

Vulcanisphaera africana Deunff, 1961
Hypotype 48196
Dean, W.T. and Martin, F., 1978, Geol. Surv. Can., Bull. 284, pl. 3, fig. 4.
Bell Island Group, Lower Ordovician, cliff at east end of Chimney Cove, near southwest end of Bell Island, Newfoundland.

Vulcanisphaera africana Deunff, 1961
Hypotype 65679
Martin, F., 1982, Nat. Mus. Wales, Geol. Ser. 3, pl. 1, fig. 10.
Bell Island Group, Lower Arenig (Ordovician), 1 m below top of cliff east side of Polls Head, Bell Island, eastern Newfoundland.

Vulcanisphaera capillata Jardiné, Combaz, Magloire, Peniguel and Vachey, 1974
Hypotype 57770
Martin, F. and Dean, W.T., 1981, Geol. Surv. Can., Bull. 343, p. 23, pl. 1, fig. 9.
Clarenville Formation, Lower Ordovician, about midway between Promontories IX and X, south of Elliott's Cove, Random Island, eastern Newfoundland.

Vulcanisphaera sp.
Fig. spec. 69415
Legault, J.A., 1982, Can. J. Earth Sci., vol. 19, no. 9, pl. 2, fig. 11.
Middle-Upper Ordovician, Orphan Knoll, lat. 50°33′N, long. 46°6′W, Labrador Sea.

Silurian

Ammonidium cf. *A. microcladum* (Downie)
Hypotype 31625
Thusu, B., 1973, Palaeontology, vol. 16, pt. 4, pl. 105, fig. 5.
Rochester Formation, Middle Silurian, Highway 2 road-cut, Ancaster, Ontario.

Ammonidium sp.
Fig. spec. 55518
McGregor, D.C. and Narbonne, G.M., 1978, Can. J. Earth Sci., vol. 15, no. 8, p. 1296, pl. 1, figs. 23, 24.
Member B 20 m above base, Read Bay Formation, Upper Silurian, south side Goodsir Creek, eastern Cornwallis Island, District of Franklin.

Baltisphaeridium longispinosum Downie
Hypotype 31635
Thusu, B., 1973, Palaeontology, vol. 16, pt. 4, pl. 105, fig. 19.
Rochester Formation, Middle Silurian, access road, Sir Adam Beck-Niagara generating station, Niagara Falls, Ontario.

Baltisphaeridium pilaris Cramer
Hypotypes 31626 a, b
Thusu, B., 1973, Palaeontology, vol. 16, pt. 4, p. 811, pl. 105, figs. 6, 18.
Rochester Formation, Middle Silurian, access road, Robert Moses Niagara Power Station, Niagara Falls, New York, and Clappison Corners road-cut, Ontario.

Cymatiosphaera octoplana Downie
Hypotypes 31644 a, b
Thusu, B., 1973, Palaeontology, vol. 16, pt. 4, pl. 106, figs. 11, 14.
Rochester Formation, Middle Silurian, Highway 20 road-cut, Stoney Creek, Ontario.

Cymatiosphaera wenlockia Downie
Hypotype 31651
Thusu, B., 1973, Palaeontology, vol. 16, pt. 4, pl. 106, fig. 19.
Rochester Formation, Middle Silurian, access road, Sir Adam Beck-Niagara generating station, Niagara Falls, Ontario.

Deunffia furcata Downie
Hypotype 31619
Thusu, B., 1973, Palaeontology, vol. 16, pt. 4, pl. 104, fig. 20.
Rochester Formation, Middle Silurian, access road, Sir Adam Beck-Niagara generating station, Niagara Falls, Ontario.

Deunffia furcata var. *niagarensis* Thusu
Holotype 31616
Thusu, B., 1973, Palaeontology, vol. 16, pt. 4, p. 807, pl. 104, fig. 14.
Rochester Formation, Middle Silurian, Highway 20 road-cut, Stoney Creek, Ontario.

Deunffia monospinosa Downie
Hypotype 31612
Thusu, B., 1973, Palaeontology, vol. 16, pt. 4, pl. 104, fig. 8.
Rochester Formation, Middle Silurian, access road, Sir Adam Beck-Niagara generating station, Niagara Falls, Ontario.

Deunffia monospinosa var. *tonawandensis* Thusu
Holotype 31609; paratype 31609a
Thusu, B., 1973, Palaeontology, vol. 16, pt. 4, p. 807, pl. 104, figs. 4, 24.
Rochester Formation, Middle Silurian, Clappison Corners road-cut and Highway 2 road-cut, Ancaster, Ontario.

Deunffia ramusculosa Downie
Hypotypes 31611a-c
Thusu, B., 1973, Palaeontology, vol. 16, pt. 4, pl. 104, figs. 7, 15, 22.
Rochester Formation, Middle Silurian, Clappison Corners road-cut, Highway 20 road-cut, Stoney Creek, and access road, Sir Adam Beck-Niagara generating station, Niagara Falls, Ontario.

Deunffia ramusculosa var. *rochesterensis* Thusu
Holotype 31618
Thusu, B., 1973, Palaeontology, vol. 16, pt. 4, p. 807, pl. 104, fig. 18.
Rochester Formation, Middle Silurian, access road, Sir Adam Beck-Niagara generating station, Niagara Falls, Ontario.

Dictyotidium dictyotum (Eisenack)
Hypotype 31645
Thusu, B., 1973, Palaeontology, vol. 16, pt. 4, pl. 106, fig. 12.
Rochester Formation, Middle Silurian, access road, Sir Adam Beck-Niagara generating station, Niagara Falls, Ontario.

Diexallophasis denticulata (Stockmans and Willière)
Hypotype 31623
Thusu, B., 1973, Palaeontology, vol. 16, pt. 4, p. 811, pl. 105, fig. 3.
Rochester Formation, Middle Silurian, access road, Sir Adam Beck-Niagara generating station, Niagara Falls, Ontario.

Dilatisphaera williereae (Martin) Lister 1970
Hypotype 82879
Martin, F., 1988, Bull. British Mus. (Nat. His.), Geol. ser., vol. 43, p. 304, fig. 5.
Jupiter Formation, Lower Silurian, 4km southeast of Pointe Sud-Ouest, Anticosti Island, Québec.

Domasia amphora Martin
Hypotypes 31613a, b
Thusu, B., 1973, Palaeontology, vol. 16, pt. 4, pl. 104, figs. 9, 19.

Rochester Formation, Middle Silurian, Highway 2 road-cut, Ancaster and access road, Sir Adam Beck-Niagara generating station, Niagara Falls, Ontario.

Domasia bispinosa Downie
Hypotype 31615
Thusu, B., 1973, Palaeontology, vol. 16, pt. 4, pl. 104, fig. 11.
Rochester Formation, Middle Silurian, Highway 20 road-cut, Stoney Creek, Ontario.

Domasia canadensis Thusu
Holotype 31610; paratypes 31610a, b
Thusu, B., 1973, Palaeontology, vol. 16, pt. 4, p. 808, pl. 104, figs. 6, 12, 21.
Rochester Formation, Middle Silurian, access road, Sir Adam Beck-Niagara generating station, Niagara Falls, Ontario.

Domasia elongata Downie
Hypotypes 31620a, b
Thusu, B., 1973, Palaeontology, vol. 16, pt. 4, pl. 104, figs. 23, 25.
Rochester Formation, Middle Silurian, access road, Sir Adam Beck-Niagara generating station, Niagara Falls, Ontario.

Domasia limaciformis (Stockmans and Willière) Cramer 1970
Hypotype 82880
Martin, F., 1988, Bull. British Mus. (Nat. His.), Geol. ser., vol. 43, fig. 6.
Jupiter Formation, Lower Silurian, 4km southeast of Pointe Sud-Ouest, Anticosti Island, Québec.

Domasia rochesterensis Thusu
Holotype 31607; paratype 31607a
Thusu, B., 1973, Palaeontology, vol. 16, pt. 4, p. 810, pl. 104, figs. 2, 5.
Rochester Formation, Middle Silurian, access road, Sir Adam Beck-Niagara generating station, Niagara Falls, Ontario.

Domasia trispinosa Downie
Hypotype 31606
Thusu, B., 1973, Palaeontology, vol. 16, pt. 4, pl. 104, fig. 1.
Rochester Formation, Middle Silurian, Highway 20 road-cut, Stoney Creek, Ontario.

Duvernaysphaera aranaides (Cramer)
Hypotype 31650
Thusu, B., 1973, Palaeontology, vol. 16, pt. 4, pl. 106, fig. 18.
Rochester Formation, Middle Silurian, access road, Sir Adam Beck-Niagara generating station, Niagara Falls, Ontario.

Elektoriskos pogonius Loeblich
Hypotype 31630
Thusu, B., 1973, Palaeontology, vol. 16, pt. 4, pl. 105, fig. 10.
Rochester Formation, Middle Silurian, access road, Sir Adam Beck-Niagara generating station, Niagara Falls, Ontario.

Elektoriskos simplex Thusu
Holotype 31629
Thusu, B., 1973, Palaeontology, vol. 16, pt. 4, p. 814, pl. 105, fig. 9.
Rochester Formation, Middle Silurian, Highway 2 road-cut, Ancaster, Ontario.

Estiastra magna Eisenack
Hypotype 66977
Duffield, S.L. and Legault, J.A., 1981, IUGS Subcommission on Silurian Stratigraphy, Ordovician-Silurian Boundary Working Group. Field Meeting, Anticosti-Gaspé, Québec, vol. II, pl. II, fig. 3.
Gun River Formation, Middle Silurian, '9-mile pool' section, Jupiter River, Anticosti Island, Québec.

Estiastra sp.
Fig. spec. 69390
Duffield, S.L. and Legault, J.A., 1982, Third North American Palaeontol. Convention, Proc., vol. 1, pl. 1, fig. 14.
Gun River Formation, Middle Silurian, '9-mile pool', Jupiter River, Anticosti Island, Québec.

Eupoikilofusa aff. *E. ampulliformis* Martin
Hypotype 66969
Duffield, S.L. and Legault, J.A., 1981, IUGS Subcommission on Silurian Stratigraphy, Ordovician-Silurian Boundary Working Group. Field Meeting, Anticosti-Gaspé, Québec, vol. II, pl. I, fig. 8.
Member 6, Ellis Bay Formation, Lower Silurian, '8-mile pool' section, Salmon River, Anticosti Island, Québec.

Eupoikilofusa aff. *E. ampulliformis* (Martin)
Hypotypes 69378, 69379
Duffield, S.L. and Legault, J.A., 1982, Third North American Palaeontol. Convention, Proc., vol. 1, p. 138, pl. 1, fugs. 1, 2.
Becscie Formation, Lower Silurian, Jupiter River about 1km upstream from '24-mile lodge', Anticosti Island, Québec.

Eupoikilofusa aff. *E. ampulliformis, sensu* Duffield and Legault, 1981
Martin. F., 1988, Bull. British Mus. (Nat. His.), Geol. ser., vol. 43, p. 304, figs. 14a, b.
Becscie Formation, Lower Silurian, west side of Ellis Bay, Anticosti Island, Québec.

Eupoikilofusa stratifera Cramer
Hypotype 31621
Thusu, B., 1973, Palaeontology, vol. i6, pt. 4, pl. 105, fig. 1.
Rochester Formation, Middle Silurian, Highway 20 road-cut, Stoney Creek, Ontario.

Evittia monterrosa (Cramer)
Hypotypes 31638a, b
Thusu, B., 1973, Palaeontology, vol. 16, pt. 4, p. 815, pl. 106, figs. 2, 7.

Rochester Formation, Middle Silurian, access road, Sir Adam Beck-Niagara generating station, Niagara Falls, and Highway 20 road-cut, Stoney Creek, Ontario.

Evittia remota (Deunff)
Hypotype 31640
Thusu, B., 1973, Palaeontology, vol. 16, pt. 4, pl. 106, fig. 4.
Rochester Formation, Middle Silurian, Highway 2 road-cut, Ancaster, Ontario.

Filisphaeridium bifurcatum Thusu
Holotype 31628; paratype 31628a
Thusu, B., 1973, Palaeontology, vol. 16, pt. 4, p. 814, pl. 105, figs. 8, 12.
Rochester Formation, Middle Silurian, access road, Sir Adam Beck-Niagara generating station, Niagara Falls, and Highway 20 road-cut, Stoney Creek, Ontario.

Goniosphaeridium cylindricum (Cramer) Eisenack, Cramer and Diez
Hypotype 66984
Duffield, S.L. and Legault, J.A., 1981, IUGS Subcommission on Silurian Stratigraphy, Ordovician-Silurian Boundary Working Group. Field Meeting, Anticosti-Québec, vol. II, pl. II, fig. 10.
Jupiter Formation, Middle Silurian, creek section 2.4 km west of mouth of Jupiter River, Anticosti Island, Québec.

Goniosphaeridium oligospinosum (Eisenack) Eisenack
Hypotype 66971
Duffield, S.L. and Legault, J.A., 1981, IUGS Subcommission on Silurian Stratigraphy, Ordovician-Silurian Boundary Working Group. Field Meeting, Anticosti-Gaspé, Québec, vol. II, pl. I, fig. 10.
Duffield, S.L. and Legault, J.A., 1982, Third North American Palentol. Convention, Proc., vol. 1, pl. 1, fig. B.
Member 6, Ellis Bay Formation, Lower Silurian, Pointe Laframboise, Anticosti Island, Québec.

Goniosphaeridium oligospinosum (Eisenack) Eisenack 1969
Hypotype 82887
Martin, F., 1988, Bull. British Mus. (Nat. His.), Geol. ser., vol. 43, p. 306, fig. 13.
Becscie Formation, Lower Silurian, west side of Ellis Bay, Anticosti Island, Québec.

Goniosphaeridium polygonale (Eisenack)
Hypotype 69389
Duffield, S.L. and Legault, J.A., 1982, Third North American Palaeontol. Convention, Proc., vol. 1, pl. 1, fig. 13.
Gun River Formation, Middle Silurian, road-cut 1.8 km south of '24-mile lodge' Jupiter River bridge, Anticosti Island, Québec.

Gorgonisphaeridium wenlockium Thusu
Holotype 31631
Thusu, B., 1973, Palaeontology, vol. 16, pt. 4, p. 814, pl. 105, fig. 11.
Rochester Formation, Middle Silurian, Highway 20 road-cut, Stoney Creek, Ontario.

Gorgonisphaeridium sp.
Fig. spec. 55520
McGregor, D.C. and Narbonne, G.M., 1978, Can. J. Earth Sci., vol. 15, no. 8, p. 1298, pl. 1, fig. 27.
Member B 20m above base, Read Bay Formation, Upper Silurian, south side Goodsir Creek, eastern Cornwallis Island, District of Franklin.

Helosphaeridium latispinosum Lister
Hypotype 31633
Thusu, B., 1973, Palaeontology, vol. 16, pt. 4, pl. 105, fig. 16.
Rochester Formation, Middle Silurian, Clappison Corners road-cut, Ontario.

Hemideunffia trifurcata Thusu
Holotype 31614; paratype 31614a
Thusu, B., 1973, Palaeontology, vol. 16, pt. 4, p. 810, pl. 104, figs. 10, 16.
Rochester Formation, Middle Silurian, access roads, Sir Adam Beck-Niagara generating station and Robert Moses Niagara Power Station, Niagara Falls, Ontario and New York.

'Hogklintia digitata - H. visbyensis' complex
Fig. specs. 82885, 82886
Martin, F., 1988, Bull. British Mus. (Nat. Hist.), Geol. ser., vol. 43, p. 306, figs. 11, 12.
Becscie Formation, Lower Silurian, and Ellis Bay Formation, Upper Ordovician, west side of Ellis Bay, Anticosti Island, Québec.

Leiofusa algerensis Cramer
Hypotype 31617
Thusu, B., 1973, Palaeontology, vol. 16, pt. 4, pl. 104, fig. 17.
Rochester Formation, Middle Silurian, access road, Sir Adam Beck-Niagara generating station, Niagara Falls, Ontario.

Leiofusa fusiformis Eisenack
Hypotypes 31608a, b
Thusu, B., 1973, Palaeontology, vol. 16, pt. 4, pl. 104, figs. 3, 13.
Rochester Formation, Middle Silurian, access road, Sir Adam Beck-Niagara generating station, Niagara Falls, and Clappison Corners road-cut, Ontario.

Leiofusa aff. *L. tumida* Downie
Hypotypes 69380, 69381
Duffield, S.L. and Legault, J.A., 1982, Third North American Palaeontol. Convention, Proc., vol. 1, p. 138, pl. 1, figs. 3-4.
Becscie Formation, Lower Silurian, '9-mile pool', Salmon River, Anticosti Island, Québec.

Lophosphaeridium microgranulosum Thusu
Holotype 31649
Thusu, B., 1973, Palaeontology, vol. 16, pt. 4, p. 818, pl. 106, fig. 17.
Rochester Formation, Middle Silurian, Highway 20 road-cut, Stoney Creek, Ontario.

Lophosphaeridium rugosum Thusu
Holotype 31646
Thusu, B., 1973, Palaeontology, vol. 16, pt. 4, p. 818, pl. 106, fig. 13.
Rochester Formation, Middle Silurian, Highway 2 road-cut, Ancaster, Ontario.

Micrhystridium stellatum Delfandre
Hypotype 31627
Thusu, B., 1973, Palaeontology, vol. 16, pt. 4, pl. 105, fig. 7.
Rochester Formation, Middle Silurian, Highway 20 road-cut, Stoney Creek, Ontario.

Moyeria uticaensis Thusu
Hypotypes 55522-55524
McGregor, D.C and Narbonne, G.M., 1978, Can. J. Earth Sci., vol. 15, no. 8, p. 1298, pl. 1, figs. 29-31.
Member B 20m above base, Read Bay Formation, Upper Silurian, south side Goodsir Creek, eastern Cornwallis Island, District of Franklin.

Moyeria sp.
Fig. spec. 55521
McGregor, D.C and Narbonne, G.M., 1978, Can. J. Earth Sci., vol. 15, no. 8, p. 1298, pl. 1, fig. 28.
Member B 20m above base, Read Bay Formation, Upper Silurian, south side Goodsir Creek, eastern Cornwallis Island, District of Franklin.

Multiplicisphaeridium arbusculiferum Downie
Hypotypes 31624a, b
Thusu, B., 1973, Palaeontology, vol. 16, pt. 4, pl. 105, figs. 4, 15.
Rochester Formation, Middle Silurian, Jolly Cut, Hamilton, Ontario.

Multiplicisphaeridium arbusculiferum (Downie) Lister
Hypotype 66979
Duffield, S.L. and Legault, J.A., 1981, IUGS Subcommission on Silurian Stratigraphy, Ordovician-Silurian Boundary Working Group. Field Meeting, Anticosti-Gaspé, Québec, vol. II, pl. II, fig. 5.
Jupiter Formation, Middle Silurian, coastal section east of mouth of Jupiter River, Anticosti Island, Québec.

Multiplicisphaeridium aff. *M. asturiae* (Cramer)
Hypotype 69388
Duffield, S.L. and Legault, J.A., 1982, Third North American Palaeontol. Convention, Proc., vol. 1, pl. 1, fig. 12.
Gun River Formation, Middle Silurian, road-cut 1.8 km south of '24-mile lodge' Jupiter River bridge, Anticosti Island, Québec.

Multiplicisphaeridium birminghamensis (Cramer)
Hypotype 69386
Duffield, S.L. and Legault, J.A., 1982, Third North American Palaeontol. Convention, Proc., vol. 1, pl. 1, fig. 10.
Gun River Formation, Middle Silurian, road-cut 1.8 km south of '24-mile lodge' Jupiter River bridge, Anticosti Island, Québec.

Multiplicisphaeridium denticulatum gotlandicum (Cramer) Eisenack, Cramer and Diez
Hypotype 66982
Duffield, S.L. and Legault, J.A., 1981, IUGS Subcommission on Silurian Stratigraghy, Ordovician-Silurian Boundary Working Group. Field Meeting, Anticosti-Gaspé, Québec, vol. II, pl. II, fig. 8.
Member 6, Ellis Bay Formation, Lower Silurian, '8-mile pool' section, Salmon River, Anticosti Island, Québec.

Multiplicisphaeridium eoplanktonicum Downie
Hypotypes 31622a, b
Thusu, B., 1973, Palaeontology, vol. 16, pt. 4, pl. 105, figs. 2, 14.
Rochester Formation, Middle Silurian, access road, Sir Adam Beck-Niagara generating station, Niagara Falls and Highway 2 road-cut, Ancaster, Ontario.

Multiplicisphaeridium fermosum (Cramer) Eisenack, Cramer and Diez
Hypotype 66985
Duffield, S.L. and Legault, J.A., 1981, IUGS Subcommission on Silurian Stratigraghy, Ordovician-Silurian Boundary Working Group. Field Meeting, Anticosti-Gaspé, Québec, vol. II, pl. II, fig. 11.
Jupiter Formation, Middle Silurian, The Jumpers, Anticosti Island, Québec.

Multiplicisphaeridium fischerii (Cramer)
Hypotype 31634
Thusu, B., 1973, Palaeontology, vol. 16, pt. 4, pl. 105, fig. 17.
Rochester Formation, Middle Silurian, Highway 20 road-cut, Ancaster, Ontario.

Multiplicisphaeridium lobeznum (Cramer) Eisenack, Cramer and Diez
Hypotype 66976
Duffield, S.L. and Legault, J.A., 1981, IUGS Subcommission on Silurian Stratigraghy, Ordovician-Silurian Boundary Working Group. Field Meeting, Anticosti-Gaspé, Québec, vol. II, pl. II, fig. 2.
Gun River Formation, Middle Silurian, coastal section east of mouth La Loutre River, Anticosti Island, Québec.

Multiplicisphaeridium mergaeferum Loeblich
Hypotypes 69383, 69384, 69391
Duffield, S.L. and Legault, J.A., 1982, Third North American Palaeontol. Convention, Proc., vol. 1, pl. 1, figs. 7, 8, 15.
Jupiter Formation, Middle Silurian, Cap Jupiter north of mouth of Jupiter River, Anticosti, Québec.

Multiplicisphaeridium neahgae (Cramer) Eisenack, Cramer and Diez
Hypotype 66981
Duffield, S.L. and Legault, J.A., 1981, IUGS Subcommission on Silurian Stratigraghy, Ordovician-Silurian Boundary Working Group. Field Meeting, Anticosti-Gaspé, Québec, vol. II, pl. II, fig. 7.
Jupiter Formation, Middle Silurian, Cape Jupiter, west of mouth of Jupiter River, Anticosti Island, Québec.

Multiplicisphaeridium neahgae (Cramer)
Hypotype 69387
Duffield, S.L. and Legault, J.A., 1982, Third North American Palaeontol. Contrib., Proc., vol. 1, pl. 1, fig. 11.
Jupiter Formation, Middle Silurian, Cap Ottawa southeast of mouth of Jupiter River, Anticosti Island, Québec.

Multiplicisphaeridium paraguaferum (Cramer) Eisenack, Cramer and Diez
Hypotype 66978
Duffield, S.L. and Legault, J.A., 1981, IUGS Subcommission on Silurian Stratigraghy, Ordovician-Silurian Boundary Working Group. Field Meeting, Anticosti-Gaspé, Québec, vol. II, pl. II, fig. 4.
Jupiter Formation, Middle Silurian, The Jumpers, Anticosti Island, Québec.

Multiplicisphaeridium aff. *M. ravum* (Downie) Eisenack, Cramer and Diez
Hypotype 66980
Duffield, S.L. and Legault, J.A., 1981, IUGS Subcommission on Silurian Stratigraghy, Ordovician-Silurian Boundary Working Group. Field Meeting, Anticosti-Gaspé, Québec, vol. II, pl. II, fig. 6.
Member 6, Ellis Bay Formation, Lower Silurian, '9-mile pool' section, Salmon River, Anticosti Island, Québec.

Multiplicisphaeridium visbyense (Eisenack)
Hypotype 69385
Duffield, S.L. and Legault, J.A., 1982, Third North American Palaeontol. Convention, Proc., vol. 1, pl. 1, fig. 9.
Becscie Formation, Lower Silurian, Jupiter River about 1km upstream from '24-mile lodge', Anticosti Island, Québec.

Multiplicisphaeridium aff. *M. visbyense* (Eisenack) Eisenack
Hypotype 66975
Duffield, S.L. and Legault, J.A., 1981, IUGS Subcommission on Silurian Stratigraghy, Ordovician-Silurian Boundary Working Group. Field Meeting, Anticosti-Gaspé, Québec, vol. II, pl. II, fig. 1.
Becscie Formation, Lower Silurian, about 1km upstream from '24-mile lodge', north bank Jupiter River, Anticosti Island, Québec.

Multiplicisphaeridium sp.
Fig. spec. 55519
McGregor, D.C and Narbonne, G.M., 1978, Can. J. Earth Sci., vol. 15, no. 8, p. 1298, pl. 1, figs. 25, 26.
Member B 20m above base, Read Bay Formation, Upper Silurian, south side Goodsir Creek, eastern Cornwallis Island, District of Franklin.

Neoveryhachium carminae (Cramer) Cramer, 1970
Hypotype 67799
McGregor, D.C., 1984, Academia Nacional de Ciencias, Misc. No. 69, pl. 4, fig. 15.
Tarabuco Formation, Upper Silurian, Laurani section, Chuquisaca Dept., northwest of Tarabuco, Cordillera Oriental, Bolivia.

Pterospermopsis cf. *P. martinii* Cramer and Cramer
Hypotype 31647
Thusu, B., 1973, Palaeontology, vol. 16, pt. 4, pl. 106, fig. 15.
Rochester Formation, Middle Silurian, access road, Sir Adam Beck-Niagara generating station, Niagara Falls, Ontario.

Pterospermopsis onondagaensis Deunff
Hypotypes 31648a, b
Thusu, B., 1973, Palaeontology, vol. 16, pt. 4, pl. 106, figs. 16, 20.
Rochester Formation, Middle Silurian, Highway 20 road-cut, Stoney Creek, and access road, Sir Adam Beck-Niagara generating station, Niagara Falls, Ontario.

Pulvinosphaeridium sp.
Fig. spec. 69382
Duffield, S.L. and Legault, J.A., 1982, Third North American Paleontol. Convention, Proc., vol. 1, pl. 1, fig. 5.
Becscie Formation, Lower Silurian, '9-mile pool', Salmon River, Anticosti Island, Québec.

Quadraditum fantisticum Cramer
Hypotype 31632
Thusu, B., 1973, Palaeontology, vol. 16, pt. 4, pl. 105, fig. 13.
Rochester Formation, Middle Silurian, Robert Moses Niagara Power Station, Niagara Falls, New York.

Tetrahedraletes sp.
Fig. spec. 66983
Duffield, S.L. and Legault, J.A., 1981, IUGS Subcommission on Silurian Stratigraghy, Ordovician-Silurian Boundary Working Group. Field Meeting, Anticosti-Gaspé, Québec, vol. II, pl. II, fig. 9.
Jupiter Formation, Middle Silurian, Cape Jupiter, north of mouth of Jupiter River, Anticosti Island, Québec.

Tunisphaeridium tentaculiferum (Martin)
Hypotype 31636
Thusu, B., 1973, Palaeontology, vol. 16, pt. 4, pl. 105, fig. 20.
Rochester Formation, Middle Silurian, Clappison Corners road-cut, Ontario.

Tylotopalla sp.
Fig. spec. 82884
Martin, F., 1988, Bull. British Mus. (Nat. Hist), Geol. ser., vol. 43, p. 304, fig. 10.
Jupiter Formation, Lower Silurian, 4km southeast of Pointe Sud-Ouest, Anticosti Island, Québec.

Unidentified acritarch
Fig. specs. 67797, 67798, 67800
McGregor, D.C., 1984, Academia Nacional de Ciencias, Misc. No. 69, pl. 4, figs. 13, 14, 16.
Tarabuco and Santa Rosa (67800) formations, Upper Silurian and Lower Devonian, Laurani section, Chuquisaca Dept., northwest of Tarabuco, Cordillera Oriental, Bolivia.

Veryhachium lairdi (Deflandre)
Hypotypes 31641a, b
Thusu, B., 1973, Palaeontology, vol. 16, pt. 4, p. 815, pl. 106, figs. 5, 6.
Rochester Formation, Middle Silurian, Highway 20 road-cut, Stoney Creek, and access road Sir Adam Beck-Niagara generating station, Niagara Falls, Ontario.

Veryhachium limaciforme Stockmans and Willière
Hypotypes 31642a, b
Thusu, B., 1973, Palaeontology, vol. 16, pt. 4, p. 816, pl. 106, figs. 8, 10.
Rochester Formation, Middle Silurian, Highway 20 road-cut, Stoney Creek, Ontario.

Veryhachium trispinosum Eisenack
Hypotype 31643
Thusu, B., 1973, Palaeontology, vol. 16, pt. 4, pl. 106, fig. 9.
Rochester Formation, Middle Silurian, access road, Sir Adam Beck-Niagara generating station, Niagara Falls, Ontario.

Veryhachium wenlockium (Downie)
Hypotype 31639
Thusu, B., 1973, Palaeontology, vol. 16, pt. 4, pl. 106, fig. 3.
Rochester Formation, Middle Silurian, access road, Sir Adam Beck-Niagara generating station, Niagara Falls, Ontario.

Visbysphaera dilatispinosa (Downie)
Hypotype 31637
Thusu, B., 1973, Palaeontology, vol. 16, pt. 4, pl. 106, fig. 1.
Rochester Formation, Middle Silurian, Highway 20 road-cut, Stoney Creek, Ontario.

Devonian

Apectosphaeridium apoplanium Turner
Holotype 97985; isotypes 97980-97984, 97986, 97987
Turner, R.E, 1991, Can. J. Earth Sci., vol. 28, no. 9, p. 1472, pl. 1, figs. 1-8.
Fort Simpson Formation, Upper Devonian, depth 415.5 m, HW-79-11 core, Mackenzie Highway, lat. 60°48′19″N, long. 116°36′32″W, District of Mackenzie.

Arkonites bilixus Legault
Holotype 30659
Legault, J.A., 1973, Geol. Surv. Can., Bull. 221, p. 52, pl. 11, figs. 22, 23.
Arkona Member, Hamilton Formation, Middle Devonian, Argor Consolidated Morrison, 65-1 well, lot 28, con. II, Moore tp., Lambton co., Ontario.

Asyncosmium isum Turner
Holotype 97988; isotypes 97989-97992
Turner, R.E, 1991, Can. J. Earth Sci., vol. 28, no. 9, p. 1475, pl. 3, figs. 1-3, 5, 6.
Fort Simpson Formation, Upper Devonian, depth 415.5 m, HW-79-11 core, Mackenzie Highway, lat. 60°48′19″N, long. 116°36′32″W, District of Mackenzie.

Baltisphaeridium anfractum Playford
Holotype 48266; paratypes 48267-48272
Playford, G., 1977, Geol. Surv. Can. Bull. 279, p. 11, pl. 1, figs. 1-8.
Stooping River Formation, Lower Devonian, Jaab lake No. 1 well, depths 755-757 (48266), 960-962 (48267, 48268), 725-727 (48269) and 935-937 feet, lat. 51°11′54″N, long. 82°56′W, Ontario.

Baltisphaeridium crebrum Playford
Holotype 48273; paratypes 48274, 48275
Playford, G., 1977, Geol. Surv. Can., Bull. 279, p. 11, pl. 1, figs. 9-12.
Stooping River Formation, Lower Devonian, Jaab Lake No. 1 well, depths 925-927, 785-786 and 898-900 feet, lat. 51°11′54″N, long. 82°56′W, Ontario.

Baltisphaeridium distentum Playford
Holotype 48278; paratypes 48276, 48277, 48279-48281
Playford, G., 1977, Geol. Surv. Can., Bull. 279, p. 12, pl. 1, figs. 13, 14; pl. 2, figs. 1-5.
Stooping River Formation, Lower Devonian, Jaab Lake No. 1 well, depths 960-962 (48278, 48277), 617-619 (48276) and 935-937 feet, lat. 51°11′54″N, long. 82°56′W, Ontario.

Baltisphaeridium juliae Cramer, 1964
Hypotypes 48282-48285
Playford, G., 1977, Geol. Surv. Can., Bull. 279, p. 13, pl. 2, figs. 6-11.
Stooping River Formation, Lower Devonian, Jaab Lake No. 1 well, depth 725-727 feet, lat. 51°11′54″N, long. 82°56′W, Ontario.

Baltisphaeridium microfurcatum (Deunff) Stockmans and Willière, 1962
Hypotypes 48286-48288, 448296, 48297
Playford, G., 1977, Geol. Surv. Can., Bull. 279, p. 14, pl. 2, figs. 12-15; pl. 3, figs. 11, 12.
Stooping River Formation, Lower Devonian, Jaab Lake No. 1 well, depths 765-767, 755-757, 915-917, 925-927 and 1031-1033 feet, lat. 51°11′54″N, long. 82°56′W, Ontario.

Baltisphaeridium stamineum Playford
Holotype 48289; paratypes 48290, 48291
Playford, G., 1977, Geol. Surv. Can., Bull. 279, p. 14, pl. 3, figs. 1-3.
Stooping River Formation, Lower Devonian, Jaab Lake No. 1 well, depths 755-757, 735-737 and 775-777 feet, lat. 51°11′54″N, long. 82°56′W, Ontario.

Baltisphaeridium sp.
Fig. spec. 30650
Legault, J.A., 1973, Geol. Durv. Can., Bull. 221, p. 50, pl. 11, figs. 1, 2.
Bell Member, Hamilton Formation, Middle Devonian, GSC Ipperwash No. 1 well, lot 9, con A, Bosanquet tp., Lambton co., Ontario.

Buedingisphaeridium sp.
Fig. spec. 67801
McGregor, D.C., 1984, Academia Nacional de Ciencias, Misc. No. 69, pl. 4, fig. 17.
Santa Rosa Formation, Lower Devonian, Laurani section, Chuquisaca Dept., northwest of Tarabuco, Cordillera Oriental, Bolivia.

Cerastum pelorum Turner
Holotype 97993; paratypes 97994-97996
Turner, R.E, 1991, Can. J. Earth Sci., vol. 28, no. 9, p. 1477, pl. 4, figs. 1-4.
Ireton Formation, Upper Devonian, depth 2422.6 m, Amoco et al. Bigoray well, lot 16, sec. 4, tp. 52, rge. 8, W.5th mer., depth 6096-6106 feet, Texaco Wizard Lake well, lot 14, sec. 14, tp. 48, rge. 27, W.4th mer., and depths 3093-3098 feet (97995) and 3060-3065 feet (97996), Socony Vegreville No. 1 well, lot 14, sec. 20, tp. 51, rge 15, W.4th mer., Alberta.

Costatilobus undulatus Playford
Holotype 48292; paratypes 48293-48295
Playford, G., 1977, Geol. Surv. Can., Bull. 279, p. 16, pl. 3, figs. 4-9.
Stooping River Formation, Lower Devonian, Jaab Lake No. 1 well, depths 925-927, 960-962 (48293) and 935-937 (48295) feet, lat. 51°11′54″N, long. 82°56′W, Ontario.

Craterisphaeridium sprucegrovense (Staplin)
Topotypes 76252-76258
Turner, R.E, 1986, Can. J. Earth Sci., vol. 23, no. 5, p. 602, pl. 1, figs. 1-6; pl. 2, figs. 1-4.
Duvernay Formation, Upper Devonian, depth 5635-5637 feet, Imperial Labyrinth Lake borehole, 15-14-98-23W4, Alberta.

Cymatiosphaera "canadensis" Deunff
Hypotypes 30651-30653
Legault, J.A, 1973, Geol. Surv. Can., Bull. 221, p. 50, pl. 11, figs. 3-6, 15, 16.
Arkona Member, Hamilton Formation, Middle Devonian, GSC Arkona No. 1 well, Rock Glen Park, lot 4, Con. SB, Bosanquet tp., Lambton co., and GSC Ipperwash No. 1 well, lot 9, con. A, Bosanquet tp., Lambton co. (30652), Ontario.

Cymatiosphaera canadensis Deunff, 1961
Hypotypes 48298, 48299
Playford, G., 1977, Geol. Surv. Can., Bull. 279, p. 16, pl. 4, figs. 1, 2.
Stooping River Formation, Lower Devonian, Jaab Lake No. 1 well, depth 1031-1033 feet, lat. 51°11′54″N, long. 82°56′W, Ontario.

Cymatiosphaera cornifera Deunff, 1955
Hypotypes 48301-48307
Playford, G., 1977, Geol. Surv. Can., Bull. 279, p. 17, pl. 4, figs. 4, 5, 7-13.
Lower Devonian, Stooping River Formation and Middle Devonian Kwataboahegan Formation (48305), Jaab Lake No. 1 well, depths 935-937, 860-862, 649-650, 972-974, 509-511 (48305) and 1031-1033 feet, lat. 51°11′54″N, long. 82°56′W, Ontario.

Cymatiosphaera multisepta Deunff, 1955
Hypotype 48300
Playford, G., 1977, Geol. Surv. Can., Bull. 279, p. 17, pl. 4, fig. 3.
Williams Island Formation, Middle Devonian, Jaab Lake No. 1 well, depth 147-149 feet, lat. 51°11′54″N, long. 82°56′W, Ontario.

Cymatiosphaera sp. cf. *C. pavimenta* (Deflandre) Deflandre, 1954
Fig. specs. 48322-48324
Playford, G., 1977, Geol. Surv. Can., Bull. 279, p. 17, pl. 5, figs. 9-11.
Williams Island Formation, Middle Devonian, Jaab Lake No. 1 well, depth 147-149 feet, lat. 51°11′54″N, long. 82°56′W, Ontario.

Cymatiosphaera winderi Deunff, emend.
Hypotypes 48308-48313
Playford, G., 1977, Geol. Surv. Can., Bull. 279, p. 17, pl. 4, figs. 15-21.
Stooping River Formation, Lower Devonian, Jaab Lake No. 1 well, depths 925-927, 735-737 (48309, 48311), 935-937 (48310) and 785-786 (48313) feet, lat. 51°11′54″N, long. 82°56′W, Ontario.

Cymatiosphaera sp.
Fig. spec. 30654
Legault, J.A., 1973, Geol. Surv. Can., Bull 221, p. 51, pl. 11, figs. 13, 14.
Dundee Formation, Middle Devonian, GSC Arkona No. 1 well, Rock Glen Park, lot 4, Con. SB, Bosanquet tp., Lambton co., Ontario.

Dateriocradus sp. A, B
Fig. specs. 48314, 48325, 48326
Playford, G., 1977, Geol. Surv. Can., Bull. 279, p. 18, pl. 5, figs. 1, 12-14.
Stooping River and Williams Island formations, Lower and Middle Devonian, Jaab Lake No. 1 well, depths 898-900, 147-149 and 755-757 feet, lat. 51°11′54″N, long. 82°56′W, Ontario.

Dictyotidium cavernosulum Playford
Holotype 48318; paratypes 48319-48321
Playford, G., 1977, Geol. Surv. Can., Bull. 279, p. 18, pl. 5, figs. 5-8.
Stooping River Formation, Lower Devonian, Jaab Lake No. 1 well, depths 925-927, 725-727 (48320) and 935-937 (48321) feet, lat. 51°11′54″N, long. 82°56′W, Ontario.

Dictyotidium dictyotum (Eisenack)
Hypotypes 30655-30658
Legault, J.A., 1973, Geol. Surv. Can., Bull. 221, p. 52, pl. 11, figs. 7-10.
Hamilton Formation, Middle Devonian, Bell (30655) and Arkona (306538) members, GSC Ipperwash No. 1 well, lot 9, con. A, Bosanquet tp., Lambton co., and Rockport Quarry (30656) and Arkona members, GSC Arkona No. 1 well, Rock Glen Park, lot 4, Con. SB, Bosanquet tp., Lambton co., Ontario.

Dictyotidium variatum Playford
Holotype 48328; paratypes 48315-48317, 48329, 48330
Playford, G., 1977, Geol. Surv. Can., Bull. 279, p. 18, pl. 5, figs. 2-4; pl. 6, figs. 3, 4, 6.
Stooping River Formation, Lower Devonian, Jaab Lake No. 1 well, depths 925-927, 935-937 (48315, 48329) and 909-911 (48316, 48317) feet, lat. 51°11′54″N, long. 82°56′W, Ontario.

Dictyotidium sp. A
Fig. spec. 48327
Playford, G., 1977, Geol. Surv. Can., Bull. 279, p. 19, pl. 5, fig. 16.
Williams Island Formation, Middle Devonian, Jaab Lake No. 1 well, depth 187.5-189.5 feet, lat. 51°11′54″N, long. 82°56′W, Ontario.

Diexallophasis remota (Deunff), emend.
Hypotypes 48336-48348
Playford, G., 1977, Geol. Surv. Can., Bull. 279, p. 19, pl. 6, figs. 12-14; pl. 7, figs. 1-11.
Stooping River Formation, Lower Devonian, Jaab Lake No. 1 well, depths 871-873 (48336), 915-917 (48337, 48341), 833-834 (48338), 960-962 (48339, 48340), 925-927, and 589-591 (48343) feet, lat. 51°11′54″N, long. 82°56′W, Ontario.

Divetipellis ventricosa Playford
Holotype 48332; paratypes 48331, 48333-48335
Playford, G., 1977, Geol. Surv. Can., Bull. 279, p. 21, pl. 6, figs. 7-11.
Stooping River and Kwataboahegan (48334) formations, Lower and Middle Devonian, Jaab Lake No. 1 well, depths 681-683, 1031-1033, 871-873, 378-380 and 785-786 feet, lat. 51°11′54″N, long. 82°56′W, Ontario.

Duvernaysphaera tenuicingulata Staplin, 1961
Hypotypes 48349-48354
Playford, G., 1977, Geol. Surv. Can., Bull. 279, p. 21, pl. 8, figs. 1-6.
Williams Island and Murray Island formations, Middle Devonian, depths 187.5-189.5, 197-199.6 (48351), 211-213 (48352), and 147-149 (48353, 48354) feet, lat. 51°11′54″N, long. 82°56′W, Ontario.

Goniolopadion prolixum Playford
Holotype 48356; paratype 48355
Playford, G., 1977, Geol. Surv. Can., Bull. 279, p. 22, pl. 8, figs. 7, 10, 11.
Williams Island Formation, Middle Devonian, Jaab Lake No. 1 well, depth 187.5-189.5 feet, lat. 51°11′54″N, long. 82°56′W, Ontario.

Goniosphaeridium rallum Turner
Holotype 98007; paratypes 98008-98010
Turner, R.E, 1991, Can. J. Earth Sci., vol. 28, no. 9, p. 1477, pl. 5, figs. 5-8.
Ireton Formation, Upper Devonian, depth 6096-6106 feet, Texaco Wizard Lake well, lot 14, sec. 14, tp. 48, rge. 27, W.4th mer., and depths 3203-3205 feet, 3177-3182 feet and 3045-3050 feet, Socony Vegreville No. 1 well, lot. 14, sec. 20, tp. 51, rge. 15, W.4th mer., Alberta.

Gorgonisphaeridium cumulatum Playford
Holotype 48360; paratypes 48357-48359
Playford, G., 1977, Geol. Surv. Can., Bull. 279, p. 22, pl. 8, figs. 16-20.
Stooping River Formation, Lower Devonian, Jaab Lake No. 1 well, depths 935-937 and 909-911 (48357) feet, lat. 51°11′54″N, long. 82°56′W, Ontario.

Gorgonisphaeridium disparatum Playford
Holotype 48367; paratypes 48366, 48368-48373
Playford, G., 1977, Geol. Surv. Can., Bull. 279, p. 23, pl. 9, figs. 8-16.
Stooping River Formation, Lower Devonian, Jaab Lake No. 1 well, depths 935-937 (48367, 48372), 925-927, 944-945 (48368, 48370) and 972-974 (48369) feet, lat. 51°11′54″N, long. 82°56′W, Ontario.

Gorgonisphaeridium granatum Playford
Holotype 48362; paratypes 48361, 48363-48365
Playford, G., 1977, Geol. Surv. Can., Bull. 279, p. 23, pl. 9, figs. 1-4, 6, 7.
Stooping River Formation, Lower Devonian, Jaab Lake No. 1 well, depths 589-591, 671-673, 609-610, 745-747, and 925-927 feet, lat. 51°11′54″N, long. 82°56′W, Ontario.

Gorgonisphaeridium sp. cf. *Micrhystridium sericum* Deunff, 1961
Fig. specs. 48374-48377
Playford, G., 1977, Geol. Surv. Can., Bull. 279, p. 24, pl. 9, figs. 17-21.

Stooping River and Murray Island formations, Lower and Middle Devonian, Jaab Lake No. 1 well, depths 909-911 (48374), 211-213 and 197-199.6 (48376) feet, lat. 51°11′54″N, long. 82°56′W, Ontario.

Gorgonisphaeridium sp. A
Fig. spec. 48413
Playford, G., 1977, Geol. Surv. Can., Bull. 279, p. 24, pl. 12, figs. 12.
Stooping River Formation, Lower Devonian, Jaab Lake No. 1 well, depth 599-600 feet, lat. 51°11′54″N, long. 82°56′W, Ontario.

Gorgonisphaeridium sp. B
Fig. specs. 48411, 48412
Playford, G., 1977, Geol. Surv. Can., Bull. 279, p. 24, pl. 12, figs. 10, 11.
Murray Island Formation, Middle Devonian, Jaab Lake No. 1 well, depth 211-213 feet, lat. 51°11′54″N, long. 82°56′W, Ontario.

Hapsidopalla exornata (Deunff), emend.
Hypotypes 48378-48381
Playford, G., 1977, Geol. Surv. Can., Bull. 279, p. 25, pl. 10, figs. 1-6.
Williams Island Formation, Middle Devonian, Jaab Lake No. 1 well, depths 147-149 (48378) and 187.5-189.5 feet, lat. 51°11′54″N, long. 82°56′W, Ontario.

Hapsidopalla sannemannii (Deunff), emend.
Hypotypes 48382-48386
Playford, G., 1977, Geol. Surv. Can., Bull. 279, p. 26, pl. 10, figs. 7-13.
Stooping River Formation, Lower Devonian, Jaab Lake No. 1 well, depths 735-737, 671-673 (48385) and 755-757 (48386) feet, lat. 51°11′54″N, long. 82°56′W, Ontario.

Leiosphaeridia gigantea Turner
Holotype 97997; paratypes 97998-98003
Turner, R.E, 1991, Can. J. Earth Sci., vol. 28, no. 9, p. 1478, pl. 6, figs. 1-8.
Ireton Formation, Upper Devonian, depths 3147-3152 feet, 3177-3182 feet (97998, 98001, 91003) and 3203-3205 feet (98000, 98002), Socony Vegreville No. 1 well, lot. 14, sec. 20, tp. 51, rge. 15, W.4th mer., Alberta.

Leiosphaeridia sp. 1
Fig. specs. 30678, 30679
Legault, J.A., 1973, Geol. Surv. Can., Bull. 221, p. 58, pl. 13, figs. 6, 13.
Dundee Formation and Arkona Member, Hamilton Formation, GSC Arkona No. 1 well, Rock Glen Park, lot. 4, con. SB, Bosanquet tp., Lambton co., Ontario.

Leiosphaeridia sp. 2
Fig. specs. 30680, 30681
Legault, J.A, 1973, Geol. Surv. Can., Bull. 221, p. 59, pl. 13, figs. 9, 11, 12, 15.
Rockport Quarry and Arkona members, Hamilton Formation, Middle Devonian, GSC Ipperwash No. 1 well, lot. 9, con. A, Bosanquet tp., Lambton co., and GSC Arkona No. 1 well, Rock Glen Park, lot 4, con. SB, Bosanquet tp., Lambton co., Ontario.

Lophosphaeridium dumalis Playford
Holotype 48387; paratypes 48388-48391
Playford, G., 1977, Geol. Surv. Can., Bull. 279, p. 26, pl. 11, figs. 1-5.
Kwataboahegan Formation, Middle Devonian, Jaab Lake No. 1 well, depth 378-380 feet, lat. 51°11′54″N, long. 82°56′W, Ontario.

Micrhystridium stellatum Deflandre, 1945
Hypotypes 48392-48395
Playford, G., 1977, Geol. Surv. Can., Bull. 279, p. 27, pl. 11, figs. 6-9.
Stooping River and Murray Island (48394, 48395) formations, Lower and Middle Devonian, Jaab Lake No. 1 well, depths 1031-1033 and 211-213 feet, lat. 51°11′54″N, long. 82°56′W, Ontario.

Multiplicisphaeridium ampliatum Playford
Holotype 48396; paratypes 48397
Playford, G., 1977, Geol. Surv. Can., Bull. 279, p. 28, pl. 11, figs. 11-13.
Stooping River Formation, Lower Devonian, Jaab Lake No. 1 well, depth 925-927 feet, lat. 51°11′54″N, long. 82°56′W, Ontario.

Multiplicisphaeridium ramusculosum (Deflandre) Lister, 1970
Hypotypes 48398-48403
Playford, G., 1977, Geol. Surv. Can., Bull. 279, p. 28, pl. 11, figs. 14-20.
Stooping River and Williams Island (48399) formations, Lower and Middle Devonian, Jaab Lake No. 1 well, depths 898-900 (48398), 735-737 (48400), 755-757 (48401) and 935-937 feet, lat. 51°11′54″N, long. 82°56′W, Ontario.

Navifusa bacillum (Deunff)
Hypotypes 48404-48410
Playford, G., 1977, Geol. Surv. Can., Bull. 279, p. 29, pl. 12, figs. 1-6, 8, 9.
Stooping River and Kwataboahegan (48405, 48406, 48410) formations, Lower and Middle Devonian, Jaab Lake No. 1 well, depths 860-862, 471-473, 390-392, 735-737, 915-917, 925-927, and 439-441 feet, lat. 51°11′54″N, long. 82°56′W, Ontario.

Onondagaella asymmetrica (Deunff) Cramer, 1966 emend.
Hypotypes 48414-48417, 48425, 48426
Playford, G., 1977, Geol. Surv. Can., Bull. 279, p. 30, pl. 12, figs. 13-16; pl. 13, figs. 10, 11.
Stooping River Formation, Lower Devonian, Jaab Lake No. 1 well, depths 765-767, 775-777, 896-898, 915-917 and 960-962 (48425, 48426) feet, lat. 51°11′54″N, long. 82°56′W, Ontario.

Ozotobrachion furcillatus (Deunff)
Hypotypes 48418-48424, 48439-48442
Playford, G., 1977, Geol. Surv. Can., Bull. 279, p. 31, pl. 13, figs. 1-9; pl. 14, figs. 13-16.

Stooping River Formation, Lower Devonian, Jaab Lake No. 1 well, depths 755-757 (48418, 48440), 935-937, 942-974 (48420), 944-945 (48421), 1031-1033 (48422) and 735-737 (48441) feet, lat. 51°11′54″N, long. 82°56′W, Ontario.

Ozotobrachion palidodigitatus (Cramer)
Hypotypes 48437, 48438
Playford, G., 1977, Geol. Surv. Can., Bull. 279, p. 31, pl. 14, figs. 11, 12.
Kwataboahegan Formation, Lower Devonian, Jaab Lake No. 1 well, depths 509-511 and 493-495 feet, lat. 51°11′54″N, long. 82°56′W, Ontario.

Palacanthus ledanoisii (Deunff), emend.
Hypotypes 48427-48436
Playford, G., 1977, Geol. Surv. Can., Bull. 279, p. 32, pl. 14, figs. 1-10.
Stooping River Formation, Lower Devonian, Jaab Lake No. 1 well, depths 649-650 (48427), 699-700 (48428), 765-767 (48429, 48431), 935-937, 755-757 (48432) and 925-927 (48435) feet, lat. 51°11′54″N, long. 82°56′W, Ontario.

Pertusisphaeridium dispar Turner
Holotype 76261; isotypes 76259, 76260, 76262-76268
Turner, R.E, 1986, Can. J. Earth Sci., vol. 23, no. 5, p. 606, pl. 3, figs. 1-6; pl. 4, figs. 1-4; pl. 5, figs. 1-4.
Duvernay Formation, Upper Devonian, depth 3761-3770 feet, Anglo Canadian Beaverhill Lake No. 2 borehole, 11-11-50-17W4, Alberta.

Polyedryxium bathyaster Deunff, 1961
Hypotype 48443
Playford, G., 1977, Geol. Surv. Can., Bull. 279, p. 33, pl. 15, figs. 1, 2.
Kwataboahegan Formation, Lower Devonian, Jaab Lake No. 1 well, depth 493-495 feet, lat. 51°11′54″N, long. 82°56′W, Ontario.

Polyedryxium carnatum Playford
Holotype 48451; paratypes 48448-48450, 48452
Playford, G., 1977, Geol. Surv. Can., Bull. 279, p. 33, pl. 15, figs. 7-10, 12, 13.
Stooping River Formation, Lower Devonian, Jaab Lake No. 1 well, depths 925-927 and 972-974 (48448, 48449) feet, lat. 51°11′54″N, long. 82°56′W, Ontario.

"*Polyedryxium*" *cuboides* Deunff
Hypotypes 30671, 30672
Legault, J.A., 1973, Geol. Surv. Can., Bull. 221, p. 55, pl. 12, figs. 8, 9, 16, 19.
Arkona and Widder members, Hamilton Formation, Middle Devonian, GSC Arkona No. 1 well, Rock Glen Park, lot 4, con. SB, Bosanquet tp., Lambton co., Ontario.

Polyedryxium decorum Deunff, 1955
Hypotypes 48445-48447, 48456, 48457
Playford, G., 1977, Geol. Surv. Can., Bull. 279, p. 33, pl. 15, figs. 4-6; pl. 16, figs. 5-7.
Murray Island and Kwataboahegan (48446) formations, Middle Devonian, Jaab Lake No. 1 well, depths 211-213 and 378-380 feet, lat. 51°11′54″N, long. 82°56′W, Ontario.

Polyedryxium deflandrei Deunff, 1961
Hypotype 48444
Playford, G., 1977, Geol. Surv. Can., Bull. 279, p. 34, pl. 15, fig. 3.
Murray Island Formation, Middle Devonian, Jaab Lake No. 1 well, depth 211-213 feet, lat. 51°11′54″N, long. 82°56′W, Ontario.

Polyedryxium embudum Cramer, 1964
Hypotypes 48453-48455
Playford, G., 1977, Geol. Surv. Can., Bull. 279, p. 34, pl. 16, figs. 1-4.
Stooping River Formation, Lower Devonian, Jaab Lake No. 1 well, depths 571-573, 599-600 and 735-737 feet, lat. 51°11′54″N, long. 82°56′W, Ontario.

Polyedryxium fragosulum Playford
Holotype 48458; paratypes 48459-48461
Playford, G., 1977, Geol. Surv. Can., Bull. 279, p. 34, pl. 16, figs. 8-11.
Stooping River Formation, Lower Devonian, Jaab Lake No. 1 well, depths 925-927 (48458) and 944-945 feet, lat. 51°11′54″N, long. 82°56′W, Ontario.

Polyedryxium leptum Turner
Holotype 98004; paratypes 98005, 98006
Turner, R.E, 1991, Can. J. Earth Sci., vol. 28, no. 9, p. 1482, pl. 5, figs. 1-4.
Ireton Formation, Upper Devonian, depth 3632-3641 feet, Anglo Canadian Beaverhill Lake No. 2 well, lot 11, sec. 11, tp. 50, rge. 17, W.4th mer., depth 1559.6-1561 feet, 79 m, Norcen et al. Evi well, lot 3, sec. 14, tp. 87, rge. 13, W.5 mer., and depth 4688.8-4693.6 feet, Chevron Suptst. Dome Loon well, lot 10, sec. 28, tp. 84, rge. 9, W.5th mer., Alberta.

"*Polyedryxium*"pharaonis Deunff
Hypotypes 30673-30675
Legault, J.A., 1973, Geol. Surv. Can., Bull. 221, p. 56, pl. 12, figs. 10, 12-14, 17, 20.
Hamilton Formation, Middle Devonian, Bell Member, GSC Ipperwash No. 1 well, lot 9, con. A, Bosanquet tp., Lambton co., and Widder Member, GSC Arkona No. 1 well, Rock Glen Park, lot 4, con. SB, Bosanquet tp., Lambton co. (30674), Ontario.

Polyedryxium pharaonis Deunff, 1961
Hypotypes 48462-48470
Playford, G., 1977, Geol. Surv. Can., Bull. 279, p. 35, pl. 17, figs. 1-12.
Stooping River , Williams Island (48468) and Murray Island (48489, 4870) formations, Lower and Middle Devonian, Jaab Lake No. 1 well, depths 609-610, 765-767, 1031-1033, 755-757 (48465, 48467), 944-945 (48466), 187.5-189.5 (48468) and 211-213 (48469, 48470) feet, lat. 51°11′54″N, long. 82°56′W, Ontario.

Pterospermella circumstriata (Jardiné, Combaz, Magloire, Peniguel and Vachey) Eisenack and Cramer, 1973
Hypotypes 48471-48475
Playford, G., 1977, Geol. Surv. Can., Bull. 279, p. 36, pl. 18, figs. 1-6.

Stooping River Formation, Lower Devonian, depths 935-937, 860-862 (48472), 815-817 (48473) and 755-757 (48475) feet, lat. 51°11′54″N, long. 82°56′W, Ontario.

Saharidia perplexa Turner
Holotype 98012; isotype 98011
Turner, R.E, 1991, Can. J. Earth Sci., vol. 28, no. 9, p. 1486, pl. 9, figs. 1-3.
Kakisa Formation, Upper Devonian, Whittaker Falls, Mackenzie Highway, lat. 61°8′23″N, long. 119°50′17″W, District of Mackenzie.

Stellinium micropolygonale (Stockmans and Willière)
Hypotypes 48476-48478
Playford, G., 1977, Geol. Surv. Can., Bull. 279, p. 36, pl. 18, figs. 7-9.
Williams Island and Murray Island (48478) formations, Middle Devonian, Jaab Lake No. 1 well, depths 147-149 and 211-213 feet, lat. 51°11′54″N, long. 82°56′W, Ontario.

Tunisphaeridium concentricum Deunff and Evitt
Hypotypes 30682-30684
Legault, J.A., 1973, Geol. Surv. Can., Bull. 221, p. 59, pl. 13, figs. 1-5, 7.
Arkona Member, Hamilton Formation, Middle Devonian, GSC Arkona No. 1 well, Rock Glen Park, lot. 4, con. SB, Bosanquet tp., Lambton co., Ontario.

Tunisphaeridium tentaculiferum (Martin) Cramer, 1971
Hypotypes 48479-48481
Playford, G., 1977, Geol. Surv. Can., Bull. 279, p. 37, pl. 18, figs. 10-12.
Stooping River Formation, Lower Devonian, Jaab Lake No. 1 well, depths 735-737, 755-757 and 765-767 feet, lat. 51°11′54″N, long. 82°56′W, Ontario.

Tyligmasoma alargadum Cramer)
Hypotypes 48482-48486
Playford, G., 1977, Geol. Surv. Can., Bull. 279, p. 38, pl. 19, figs. 1-6.
Stooping River Formation, Lower Devonian, Jaab Lake No. 1 well, depths 735-737, 815-817, 725-727, 649-650 and 691-693 feet, lat. 51°11′54″N, long. 82°56′W, Ontario.

Variolidium omnium Turner
Holotype 98015; paratype 98014; isotypes 98013, 98016
Turner, R.E, 1991, Can. J. Earth Sci., vol. 28, no. 9, p. 1486, pl. 9, figs. 3-6.
Gilwood Member, Watt Mountain Formation, Middle Devonian, depth 4359.4 feet, H.S. Mobil Arco Carcajou well, lot 11, sec. 1, tp. 100, rge. 20, W.5th mer., and depth 5448.1 feet, Chevron Shell Nina well, lot 6, sec. 5, tp. 98, rge. 21, W.5th mer. (98014), Alberta.

Veryhachium downiei Stockmans and Willière, 1962
Hypotypes 48497-48504
Playford, G., 1977, Geol. Surv. Can., Bull. 279, p. 38, pl. 20, figs. 4-13.
Stooping River and Williams Island (48498, 48500, 48504) formations, Lower and Middle Devonian, depths 925-927, 187.5-189.5 (48498, 48500), 725-727 (48499), 1031-1033 (48501, 48502), 735-737 (48503) and 147-149 (48504) feet, lat. 51°11′54″N, long. 82°56′W, Ontario.

Veryhachium lairdii Deflandre ex Deunff, 1959
Hypotypes 48494-48496
Playford, G., 1977, Geol. Surv. Can., Bull. 279, p. 39, pl. 20, figs. 1-3.
Williams Island (48494) and Stooping River formations, Middle and Lower Devonian, Jaab Lake No. 1 well, depths 147-149 and 960-962 feet, lat. 51°11′54″N, long. 82°56′W, Ontario.

Veryhachium cf. *V. lairdi* (Deflandre)
Hypotypes 30665-30667
Legault, J.A, 1973, Geol. Surv. Can., Bull. 221, p. 53, pl. 12, figs. 3, 6, 7, 11.
Hamilton Formation, Middle Devonian, Bell Member, GSC Ipperwash No. 1 well, lot 9, con. A, Bosanquet tp. Lambton co., and Widder Member, GSC Arkona No. 1 well, Rock Glen Park, lot 4, con. SB, Bosanquet tp., Lambton co. (30666), Ontario.

Veryhachium polyaster Staplin, 1961
Hypotypes 48487-48490
Playford, G., 1977, Geol. Surv. Can., Bull. 279, p. 39, pl. 19, figs. 7-10.
Stooping River Formation, Lower Devonian, Jaab Lake No. 1 well, depths 735-737, 755-757 (48488) and 871-873 (48490) feet, lat. 51°11′54″N, long. 82°56′W, Ontario.

Veryhachium radiosum Playford
Holotype 48493; paratypes 48491, 48492
Playford, G., 1977, Geol. Surv. Can., Bull. 279, p. 39, pl. 19, figs. 11-13.
Stooping River Formation, Lower Devonian, Jaab Lake No. 1 well, depths 1031-1033 and 1027-1029.6 (48492) feet, lat. 51°11′54″N, long. 82°56′W, Ontario.

Veryhachium sp. 1
Fig. specs. 30668, 30669
Legault, J.A., 1973, Geol. Surv. Can., Bull. 221, p. 54, pl. 12, figs. 1, 2, 15, 18.
Arkona and Bell members, Hamilton Formation, Middle Devonian, GSC Arkona No. 1 well, Rock Glen Park, lot 4, con. SB, Bosanquet tp., Lambton co., and GSC Ipperwash No. 1 well, lot 9, con. A, Bosanquet tp., Lambton co., Ontario.

Veryhachium sp. 2
Fig. spec. 30670
Legault, J.A., 1973, Geol. Surv. Can., Bull. 221, p. 54, pl. 12, figs. 4, 5.
Arkona Member, Hamilton Formation, Middle Devonian, GSC Ipperwash No. 1 well, lot 9, con. A, Bosanquet tp., Lambton co., Ontario.

Carboniferous

Actinotodissus sp. A
Fig. spec. 73833
Utting, J., 1987, Geol. Surv. Can., Bull. 374, p. 38, pl. 5, fig. 21.
Windsor Group, Lower Carboniferous, north shore of Port Hood Island, lat. 46°1′N, long. 61°26′W, Cape Breton Island, Nova Scotia.

Diexallophasis remota (Deunff) Playford, 1977
Hypotype 89979
Utting, J., Keppie, J.D. and Giles, P.S., 1989, Geol. Surv. Can., Bull. 396, pl. 5.4, fig. 22.
Wilkie Brook Formation, Horton Group, Lower Carboniferous (reworked?), 380 m north of mouth of Wilkie Brook, western shore of St. George Bay, Antigonish Basin, Nova Scotia.

Elektoriskos sp. A
Fig. spec. 73838
Utting, J., 1987, Geol. Surv. Can., Bull. 374, p. 39, pl. 6, fig. 6.
Mollichignick Member, Robinsons River Formation, Codroy Group, Lower Carboniferous, south bank of Codroy River, 1 km west-southwest of confluence with Mollichignick Brook, lat. 47°53′30″N, long. 59°6′30″W, Newfoundland.

Eisenackidium playfordii Utting
Holotype 73834; hypotypes 73835, 73836
Utting, J., 1987, Geol. Surv. Can., Bull. 374, p. 38, pl. 6, figs. 1-4.
Windsor Group, Lower Carboniferous, north shore of Port Hood Island, lat. 46°1′N, long. 61°26′W, Cape Breton Island, Nova Scotia.

Leiosphaeridia? scotia Utting
Holotype 73839; hypotypes 73840-73842
Utting, J., 1987, Geol. Surv. Can., Bull. 374, p. 39, pl. 6, figs. 7-11.
Windsor Group, Lower Carboniferous, west side of headland, Kneydart Point, lat. 45°42′5″N, long. 62°13′6″W, Nova Scotia.

Leiosphaeridia? sp. A
Fig. spec. 73843
Utting, J., 1987, Geol. Surv. Can., Bull. 374, p. 39, pl. 6, fig. 12.
Windsor Group, Lower Carboniferous, north shore of Port Hood Island, lat. 46°1′N, long. 61°26′W, Cape Breton Island, Nova Scotia.

Michrystridium sp.
Fig. specs. 89980-89983
Utting, J., Keppie, J.D. and Giles, P.S., 1989, Geol. Surv. Can., Bull. 396, pl. 5.4, figs. 15, 16, 18, 19.
Lower Carboniferous, (reworked?), Horton Bluff Formation, Horton Group, west end of Horton Bluff Avon River, and Chevrie Formation, about 200 m north of Chevrie Point (89983), Windsor Basin, Nova Scotia.

Micrhystridium? sp.
Fig. spec. 73837
Utting, J., 1987, Geol. Surv. Can., Bull. 374, pl. 6, fig. 5.
Windsor Group, Lower Carboniferous, northeast of Knoydart Point, lat. 45°42′5″N, long. 62°13′16″W, Nova Scotia.

Pterospermella sp.
Fig. spec. 87187
Utting, J., Keppie, J.D. and Giles, P.S., 1989, Geol. Surv. Can., Bull. 396, pl. 5.4, fig. 20.
Horton Bluff Formation, Horton Group, Lower Carboniferous (reworked?), west end of Horton Bluff, Avon River, Windsor Basin, Nova Scotia.

Veryhachium downiei Stockmans and Willière, 1962
Hypotype 87188
Utting, J., Keppie, J.D. and Giles, P.S., 1989, Geol. Surv. Can., Bull. 396, pl. 5.4, fig. 21.
Horton Bluff Formation, Horton Group, Lower Carboniferous (reworked?), west end of Horton Bluff, Avon River, Windsor Basin, Nova Scotia.

Veryhachium europaeum Stockmans and Willière, 1962
Hypotype 87189
Utting, J., Keppie, J.D. and Giles, P.S., 1989, Geol. Surv. Can., Bull. 396, pl. 5.4, fig. 17.
Horton Bluff Formation, Horton Group, Lower Carboniferous (reworked?), west end of Horton Bluff, Avon River, Windsor Basin, Nova Scotia.

Veryhachium sp.
Fig. spec. 94504
Utting, J., 1991, Geol. Surv. Can., Bull. 412, pl. 4.2, fig. 20.
Hart River Formation, Lower Carboniferous, north bank Peel River, northern Yukon.

Mesozoic

Acritarch Form 2, 1
Fig. specs. 38163, 38164
McIntyre, D.J., 1974, Geol. Surv. Can., Paper 74-14, pl. 13, figs. 8, 19.
Upper Cretaceous, lat. 69°28′N, long. 126°58′W, and lat. 69°27′30″N, long. 126°58′W, Horton River, District of Mackenzie.

Baltisphaeridium debilispinum Wall and Downie
Hypotype 75466
Davies, E.H., 1985, Palynology, vol. 9, pl. 2, fig. 19.
Early Pliensbachian, Liassic, Brehna Road Section, Lusitanian Basin, west-central Portugal.

Baltisphaeridium sp. A
Fig. spec. 75461
Davies, E.H., 1985, Palynology, vol. 9, p. 130, pl. 2, figs. 13, 14.
Early Pliensbachian, Liassic, Zambujal Section, south of Condeixa, Lusitanian Basin, west-central Portugal.

Crassosphaera hexagonalis Wall
Hypotype 75510
Davies, E.H., 1985, Palynology, vol. 9, pl. 4, fig. 22.
Early Pliensbachian, Liassic, Zambujal Section, south of Condeixa, Lusitanian Basin, west-central Portugal.

Cymatosphaera pachytheca Eisenack
Hypotype 75509
Davies, E.H., 1985, Palynology, vol. 9, pl. 4, fig. 21.
Late Pliensbachian, Liassic, Peniche Cliffs, Lusitanian Basin, west-central Portugal.

Lecaniella foveata Singh
Hypotypes 41559, 41560
Brideaux, W.W. and McIntyre, D.J., 1975, Geol. Surv. Can., Bull. 252, p. 39, pl. 14, figs. 10, 11.
Horton River Formation, Lower Cretaceous, lat. 69°19′N, long. 126°50′W, Horton River, District of Mackenzie.

Lophodictyotidium sp. cf. *L. sarjeanti* Pocock
Fig. spec. 80705
Carter, E.S., Cameron, BEB. and Smith, P.L., 1988, Geol. Surv. Can., Bull. 386, p. 63, pl. 15, fig. 10.
Graham Island Formation, Yakoun Group, Middle Jurassic, Branch Road 57, Graham Island, Queen Charlotte Islands, lat. 53°23.65′N, long. 132°16.29′W, British Columbia.

Micrhystridium fragile Deflandre
Hypotype 75465
Davies, E.H., 1985, Palynology, vol. 9, pl. 2, fig. 18.
Early Pliensbachian, Liassic, Brehna Road Section, Lusitanian Basin, west-central Portugal.

Micrhystridium lymensis Wall
Hypotype 75462
Davies, E.H., 1985, Palynology, vol. 9, pl. 2, fig. 15.
Late Pliensbachian, Liassic, Zambujal Section, south of Condeixa, Lusitanian Basin, west-central Portugal.

Micrhystridium rarispinum Sarjeant
Hypotype 75463
Davies, E.H., 1985, Palynology, vol. 9, pl. 2, fig. 16.
Late Pliensbachian, Liassic, Zambujal Section, south of Condeixa, Lusitanian Basin, west-central Portugal.

Micrhystridium sp.
Fig. spec. 53028
van Helden, B.G.T., 1977, Geol. Surv. can., Paper 77-1B, p. 170, pl. 33.1, fig. 17.
Wilkie Point Formation, Jurassic, 1.6 to 8 km north and east of Mould Bay Weather Station, Prince Patrick Island, District of Franklin.

Micrhystridium sp. of Dörhöfer and Davies (1980)
Fig. spec. 75468
Davies, E.H., 1985, Palynology, vol. 9, p. 130, pl. 2, fig. 21.
Toarcian, Liassic, Brehna Road Section, Lusitanian Basin, west-central Portugal.

Micrhystridium sp. A
Fig. specs. 41561, 46063
Brideaux, W.W. and McIntyre, D.J., 1975, Geol. Surv. Can., Bull. 252, p. 39, pl. 14, figs. 12, 13.
Gilmore Lake Member, Langton Bay Formation, Lower Cretaceous, lat. 68°50′N, long. 125°24′W, Horton River, District of Mackenzie.

Micrhystridium sp. D.C
Fig specs. 75464. 75467
Davies, E.H., 1985, Palynology, vol. 9, p. 130, pl. 2, fig. 17, 20.
Late and Early Pliensbachian, Liassic, Zumbujal Section, south of Condeixa, Lusitanian Basin, west-central Portugal.

Pterospermopsis australiensis Deflandre and Cookson
Hypotype 41556
Brideaux, W.W. and McIntyre, D.J., 1975, Geol. Surv. Can., Bull. 252, p. 39, pl. 14, fig. 7.
Gilmore Lake Member, Langton Bay Formation, Lower Cretaceous, lat. 68°50′N, long. 125°24′W, Horton River, District of Mackenzie.

Pterospermopsis helios Sarjeant
Hypotype 75512
Davies, E.H., 1985, Palynology, vol. 9, pl. 4, fig. 24.
Late Pliensbachian, Liassic, Zambujal Section, south of Condeixa, Lusitanian Basin, west-central Portugal.

Pterospermopsis sp.
Fig. spec. 71412
Wilson, M.A., 1978, Palaeontographica, Abt.B, vol. 166, p. 156, pl. 11, fig. 11.
Tent Island Formation, Cretaceous, Little Fish (Cache) Creek, Big Fish River area, District of Mackenzie.

Schizophacus sp. A
Fig. spec. 75508
Davies, E.H., 1985, Palynology, vol. 9, p. 129, pl. 4, fig. 20.
Late Pliensbachian, Liassic, Brehna Road Section, Lusitanian Basin, west-central Portugal.

Veryhachium sp.
Fig. spec. 53043
van Helden, B.G.T., 1977, Geol. Surv. Can., Paper 77-1B, p. 170, pl. 33.2, fig. 16.
Wilkie Point Formation, Jurassic, east side Intrepid Inlet, Prince Patrick Island, District of Franklin.

Visbysphaera sp. A
Fig. spec. 75507
Davies, E.H., 1985, Palynology, vol. 9, p. 130, pl. 4, fig. 19.
Late Pliensbachian, Liassic, Brehna Road Section, Lusitanian Basin, west-central Portugal.
Tertiary-Recent

"Horologinella" sp. A
Fig. specs. 68168, 68169
Norris, G., 1986, Geol. Surv. Can., Bull. 340, p. 46, pl. 13, figs. 8, 9.
Ivik Member, Kugmallit Formation, Middle to Early Oligocene, depths 7682 and 7291 feet, Imperial Nuktak C-22 well, lat. 69°41′07″N, long. 134°51′30″W, District of Mackenzie.

Micrhystridium fragile
Hypotype 68772
Dixon, J. et al., 1984, Geol. Surv. Can., Paper 82-13, pl. 5, fig. 1.
Pliocene to Holocene, 610-637 m interval, Dome Gulf et al. Hunt Kopanoar M-13 well, lat. 70°22′55.1″N, long. 135°05′33.9″W, Beaufort Sea.

Veryhachium sp.
Fig. specs. 47845, 47846
Williams, G.L., 1978, Initial Repts. Deep Sea Drilling Project, vol. 41, pl. 1, figs. 1, 2.
Oligocene, DSDP Site 370, lat. 32°50.2′N, long. 10°46.6′W, northeastern Moroccan continental margin.

Veryhachium sp.
Fig. spec. 68773
Dixon, J. et al., 1984, Geol. Surv. Can., Paper 82-13, pl. 5, fig. 2.
Pliocene to Holocene, 762-789 m interval, Dome Gulf et al. Hunt Kopanoar M-13 well, lat. 70°22′55.1″N, long. 135°05′33.9″W, Beaufort Sea.

CHITINOZOA

Acanthochitina? cancellata Martin
Holotype 55810; paratypes 55811, 55812
Martin, F., 1983, Geol. Surv. Can., Bull. 310, p. 11, pl. 3, fig. 19; pl. 4, figs. 14, 28, 29.
Lotbinière Formation, Utica Group, Upper Ordovician, bay west of Pointe Delisle, lat. 46°41′57″N, long. 71°36′29″W, and road extending from bay west of Pointe Delisle to Highway 138 (55812), Québec.

Acanthochitina cancellata Martin, 1983
Hypotypes 85421-85423
Achab, A., 1987, Can. J. Earth Sci., vol. 24, no. 6, p. 1213, pl. 2, figs. 1-3.
Trenton Group, Middle Ordovician, depth 2250 feet, Bald Mountain Saint-Roch no. 1 borehole, near Sorel, Québec.

Alpenachitina ontariensis Legault
Holotype 30554
Legault, J.A., 1973, Geol. Surv. Can., Bull. 221, p. 17, pl. 1, figs. 1, 3.
Dundee Formation, Middle Devonian, GSC Arkona No. 1 well, 251.1-255.6 interval, Rock Glen Park, lot 4, con. SB, Bosanquet tp., Lambton co., Ontario.

Amphorachitina sp.
Fig. spec. 56684
Martin, F., 1978, Geol. Surv. Can., Paper 78-1B, pl. 12.1, fig. 23.
Cow Head Group, Lower Ordovician, west side of The Ledge, Cow Head Peninsula, western Newfoundland.

Ancyrochitina alaticornis Jenkins, 1967
Hypotype 56675
Martin, F., 1978, Geol. Surv. Can., Paper 78-1B, pl. 12.1, fig. 11.
Winterhouse Formation, Long Point Group, Upper Ordovician, quarry immediately southeast of fishing huts near west end of Clam Bank Cove, 750 m northwest of Lourdes, western Newfoundland.

Ancyrochitina alaticornis Jenkins, 1967
Hypotypes 55813-55815
Martin, F., 1983, Geol. Surv. Can., Bull. 310, p. 11, pl. 3, figs. 11, 17; pl. 4, fig. 15.
Deschambault Formation, Middle Ordovician, Lafarge Cement quarry, Saint-Constant, Québec.

Ancyrochitina arkonensis Legault
Holotype 30579; paratypes 30580-30582
Legault, J.A., 1973, Geol. Surv. Can., Bull. 221, p. 23, pl. 4, figs. 4, 5, 7-9.

Arkona (30579, 30582) and Widder (30581) members, Hamilton Formation, Middle Devonian, Argor Consolidated Morrison 65-1 well, lot 28, con. II, Moore tp., Lambton co., and GSC Arkona No. 1 well, Rock Glen Park, lot 4, con. SB, Bosanquet tp., Lambton co., Ontario.

Ancyrochitina barbescens Martin
Holotype 41214; paratypes 41213, 41223
Martin, F., 1975, Can. J. Earth Sci., vol. 12, no. 6, p. 1016, pl. 3, figs. 9, 10, 20.
Utica Formation, Ordovician, Montmorency Falls, Québec.

Ancyrochitina barbescens Martin, 1975
Hypotype 55816
Martin, F., 1983, Geol. Surv. Can., Bull. 310, p. 11, pl. 5, fig. 18.
Lotbinière Formation, Utica Group, Upper Ordovician, bay to west of Pointe-aux-Trembles-Ouest, lat. 46°40′N, long. 71°39′51″W, Québec.

Ancyrochitina cf. *A. cornigera* Collinson and Scott
Hypotypes 30555-30558
Legault, J.A., 1973, Geol. Surv. Can., Bull. 221, p. 17, pl. 1, figs. 2, 4-12.
Hamilton Formation, Middle Devonian, Arkona Member, Argor Consolidated Morrison 65-1 well, lot 28, con. II, Moore tp., Lambton co. (30555, 30556); Bell Member, GSC Arkona No. 1 well, Rock Glen Park, lot 4, con. SB, Bosanquet tp., Lambton co. (30557); and Hungry Hollow Member, GSC Ipperwash No. 1 well, lot 9, con. A, Bosanquet tp., Lambton co., Ontario.

Ancyrochitina cf. *A. desmea* Eisenack
Hypotype 30559
Legault, J.A., 1973, Geol. Surv. Can., Bull. 221, p. 18, pl. 2, figs. 1-3.
Widder Member, Hamilton Formation, Middle Devonian, GSC Ipperwash No. 1 well, lot 9, con. A, Bosanquet tp., Lambton co., Ontario.

?*Ancyrochitina gordita* Cramer
Hypotypes 30560-30563
Legault, J.A., 1973, Geol. Surv. Can., Bull. 221, p. 19, pl. 2, figs. 4-6, 8.
Hamilton Formation, Middle Devonian, Bell and Arkona members, Argor Consolidated Morrison 65-1 well, lot 28, con. II, Moore tp., Lambton co. (30560, 30562); Bell and Hungry Hollow members, GSC Arkona No. 1 well, Rock Glen Park, lot 4, con. SB, Bosanquet tp., Lambton co., Ontario.

Ancyrochitina hamiltonsensis Legault
Holotype 30578
Legault, J.A., 1973, Geol. Surv. Can., Bull. 221, p. 23, pl. 4, figs. 2, 3.
Arkona Member, Hamilton Formation, Middle Devonian, Argor Consolidated Morrison 65-1 well, lot 28, con. II, Moore tp., Lambton co., Ontario.

Ancyrochitina cf. *A. langei* Sommer and van Boekel
Hypotypes 30564-30571
Legault, J.A., 1973, Geol. Surv. Can., Bull. 221, p. 20, pl. 2, figs. 7, 9-12; pl. 3, figs. 1-6.
Hamilton Formation, Middle Devonian, Rockport Quarry (30564), Bell (30565) and Hungry Hollow (30587, 30570, 30571) members, GSC Arkona No. 1 well, Rock Glen Park, lot 4, con. SB, Bosanquet tp., Lambton co.; Bell Member, GSC Ipperwash No. 1 well, lot. 9, con. A, Bosanquet tp., LAmbton co., Ontario.

Ancyrochitina merga Jenkins, 1970
Hypotypes 55817-55820, 55925
Martin, F., 1983, Geol. Surv. Can, Bull. 310, p. 11, pl. 4, figs. 9, 12, 16; pl. 5, figs. 14, 26.
Billings Formation, Utica Group, Upper Ordovician, Billings Creek west of Billings Bridge, Bank Street, Ottawa, Ontario.

Ancyrochitina merga Jenkins, 1970
Hypotypes 85417, 85418
Achab, A., 1987, Can. J. Earth Sci., vol. 24, no. 6, p. 1213, pl. 1, figs. 6, 7.
Utica Group, Upper Ordovician, depth 1600 feet, Bald Mountain Saint-Rock no. 1 borehole, near Sorel, Québec.

Ancyrochitina cf. *A. multiramosa* Taugourdeau and de Jekhowsy
Hypotype 30572
Legault, J.A., 1973, Geol. Surv. Can., Bull. 221, p. 21, pl. 3, fig. 10.
Hungry Hollow Member, Hamilton Formation, Middle Devonian, GSC Arkona No. 1 well, Rock Glen Park, lot 4, co. SB, Bosanquet tp., Lambton co., Ontario.

Ancyrochitina spongiosa Achab
Holotype 49055; hypotypes 49050-49054, 49056, 49476, 49477, 49480
Achab, A., 1977, Can. J. Earth Sci., vol. 14, no. 10, p. 2195, pl. 1, figs. 1-9, 12.
Vauréal Formation, Upper Ordovician, depths 1500 and 1775 feet, New Associated Consolidated Paper Anticosti No. 1 well, lat. 49°37′20″N, long. 63°26′20″W, Anticosti Island, Québec.

Ancyrochitina spongiosa Achab, 1977
Hypotypes 53590-53592
Achab, A., 1978, Rev. Palaeobot. Palynol., vol. 25, nos. 3/4, p. 299, pl. I, figs. 1-3.
Vauréal Formation, Upper Ordovician, depths 3850 and 3800 feet, New Associated Consolidated Paper Anticosti No. 1 well, lat. 49°37′20″N, long. 63°26′20″W, Anticosti Island, Québec.

Ancyrochitina spongiosa Achab, 1977
Hypotypes 85412-85414
Achab, A., 1987, Can. J. Earth Sci., vol. 24, no. 6, p. 1214, pl. 1, figs. 1-3.
Utica Group, Upper Ordovician, depth 1500 feet, Bald Mountain Saint-Rock No. 1 borehole, near Sorel, Québec.

Ancyrochitina tomentosa Taugourdeau and de Jekhowsky
Hypotype 30573
Legault, J.A., 1973, Geol. Surv. Can., Bull. 221, p. 22, pl. 3, fig. 11.

Widder Member, Hamilton Formation, Middle Devonian, GSC Arkona No. 1 well, Rock Glen Park, lot 4, con. SB, Bosanquet tp., Lambton co., Ontario.

Ancyrochitina cf. *A. tomentosa* Taugourdeau et de Jekhowsky, 1960
Hypotype 96832
Asselin, E., Achab, A. and Bourque, P.A, 1989, Can. J. Earth Sci., vol. 26, no. 12, p. 2446, pl. 5, fig. 23.
Anse à Pierre-Loiselle Formation, Lower Silurian, Anse à Pierre-Loiselle, Gaspé, Québec.

Ancyrochitina cf. *A. tumida* Taugourdeau and de Jekhowsky
Hypotypes 30574-30577
Legault, J.A., 1973, Geol. Surv. Can., Bull. 221, p. 22, pl. 3, figs. 7-9, 12; pl. 4, fig. 1.
Bell Member, Hamilton Formation, Middle Devonian, GSC Arkona No. 1 well, Rock Glen Park, lot 4, con. SB, Bosanquet tp., Lambton co., Ontario.

Ancyrochitina sp.
Fig. specs. 85415, 85418
Achab, A., 1977, Can. J. Earth Sci., vol. 24, no. 6, pl. 1, figs. 4, 9.
Lorraine Group, Upper Ordovician, depth 750 feet, Bald Mountain Saint-Roch no. 1 borehole, near Sorel, Québec.

Ancyrochitina sp. 1
Fig. specs. 96801-96803, 96820
Asselin, E., Achab, A. and Bourque, P.A., 1989, Can. J. Earth Sci., vol. 26, no. 12, p. 2442, pl. 2, figs. 6, 11, 12; pl. 3, fig. 14.
Clemville and Weir (96820) formations, Lower Silurian, Little Port-Daniel River, Clemville, Gaspé, Québec.

Ancyrochitina spp. 1-8
Fig. specs. 30583-30590
Legault, J.A., 1973, Geol. Surv. Can., Bull. 221, p. 24-27, pl. 4, figs. 6, 10-13; pl. 5, figs. 1, 2, 4.
Hamilton Formation, Middle Devonian, Arkona Member, Argor Consolidated Morrison 65-1 well, lot 28, con. II, Moore tp., Lambton co. (30583, 30584, 30587); Widder (30585), Hungry Hollow (30586), Bell (30588) and Arkona (30590) members, GSC Arkona No. 1 well, Rock Glen Park, lot 4, con. SB, Bosanquet tp., Lambton co.; and Bell Member, GSC Ipperwash No. 1 well, lot 9, con. A, Bosanquet tp., Lambton co., Ontario.

Ancyrochitina sp. 3
Fig. specs. 96831, 96839
Asselin, E. Achab, A. and Bourque, P.A., 1989, Can. J. Earth Sci., vol. 26, no. 12, p. 2446, pl. 5, fig. 20; pl. 6, fig. 23.
Anse à Pierre-Loiselle Formation, Lower Silurian, Anse à Pierre-Loiselle, Gaspé, Québec.

?*Ancyrochitina* sp. 9
Fig. spec. 30591
Legault, J.A., 1973, Geol. Surv. Can., Bull. 221, p. 28, pl. 5, fig. 3.
Arkona Member, Hamilton Formation, Middle Devonian, GSC Arkona No. 1 well, Rock Glen Park, lot 4, con. SB, Bosanquet tp., Lambton co., Ontario.

Ancyrochitina sp. A
Fig. spec. 55821
Martin, F., 1983, Geol. Surv. Can., Bull. 310, p. 12, pl. 5, fig. 24.
Sherman Fall Formation, Ottawa Group, Middle Ordovician, Val-Tétreau, Hull, Québec.

Angochitina calcarata Legault
Holotype 30606; paratype 30607
Legault, J.A., 1973, Geol. Surv. Can., Bull. 221, p. 33, pl. 10, figs. 8, 9.
Arkona Member, Hamilton Formation, Middle Devonian, GSC Ipperwash No. 1 well, lot 9, con. A, Bosanquet tp., Lambton co., Ontario.

Angochitina cf. *A.? collinsoni* Taugourdeau and de Jekhowsky
Hypotypes 30592, 30593
Legault, J.A., 1973, Geol. Surv. Can., Bull. 221, p. 29, pl. 5, figs. 5, 10.
Widder Member, Hamilton Formation, Middle Devonian, GSC Ipperwash No. 1 well, lot 9, con. A, Bosanquet tp., Lambton co., and GSC Arkona No. 1 well, Rock Glen Park, lot 4, con. SB, Bosanquet tp., Lambton co., Ontario.

Angochitina devonica Eisenack
Hypotypes 30594-30596
Legault, J.A., 1973, Geol. Surv. Can., Bull. 221, p. 29, pl. 5, figs. 9, 11, 13.
Widder Member, Hamilton Formation, Middle Devonian, GSC Ipperwash No. 1 well, lot 9, con. A, Bosanquet tp., Lambton co., and GSC Arkona No. 1 well, Rock Glen Park, lot 4, con. SB, Bosanquet tp. Lambton co., Ontario.

Angochitina huronensis Legault
Holotype 30605
Legault, J.A., 1973, Geol. Surv. Can., Bull. 221, p. 32, pl. 6, fig. 4.
Widder Member, Hamilton Formation, Middle Devonian, GSC Ipperwash No. 1 well, lot 9, con. A, Bosanquet tp., Lambton co., Ontario.

Angochitina aff. *A. longicollis* Eisenack, 1959
Hypotype 96837
Asselin, E., Achab, A. and Bourque, P.A., 1989, Can. J. Earth Sci., vol. 26, no. 12, p. 2446, pl. 6, fig. 16.
La Vieille Formation, Lower Silurian, Anse à Pierre-Loiselle, Gaspé, Québec.

Angochitina milanensis Collinson and Scott
Hypotypes 30597, 30598
Legault, J.A., 1973, Geol. Surv. Can., Bull. 221, p. 30, pl. 5, figs. 6, 7, 12.
Arkona and Widder members, Hamilton Formation, Middle Devonian, GSC Arkona No. 1 well, Rock Glen Park, lot 4, con. SB, Bosanquet tp., Lambton co., Ontario.

Angochitina cf. *A. ramusculosa* Cramer
Hypotypes 30599-30601
Legault, J.A., 1973, Geol. Surv. Can., Bull. 221, p. 31, pl. 5, fig. 8; pl. 6, figs. 7, 9.

Arkona and Widder members, Hamilton Formation, Middle Devonian, GSC Arkona No. 1 well, Rock Glen Park, lot 4, con. SB, Bosanquet tp., Lambton co., Ontario.

Angochitina toyetae Cramer
Hypotypes 30602-30604
Legault, J.A., 1973, Geol. Surv. Can., Bull. 221, p. 32, pl. 6, fig. 10; pl. 7, figs. 1, 2.
Widder Member, Hamilton Formation, Middle Devonian, GSC Arkona No. 1 well, Rock Glen Park, lot 4, con. SB, Bosanquet tp., Lambton co., Ontario.

Angochitina sp. 1
Fig. specs. 96833, 96836
Asselin, E., Achab, A. and Bourque, P.A., 1989, Can. J. Earth Sci., vol. 26, no. 12, p. 2446, pl. 6, figs. 7, 8, 15.
Anse à Pierre-Loiselle and La Vieille formations, Lower Silurian, Anse à Pierre-Loiselle, Gaspé, Québec.

Angochitina spp. 1, 2
Fig. specs. 30608, 30609
Legault, J.A., 1973, Geol. Surv. Can., Bull. 221, p. 33, 34, pl. 6, figs. 3, 5, 6.
Widder and Bell members, Hamilton Formation, Middle Devonian, GSC Ipperwash No. 1 well, lot 9, con. A, Bosanquet tp., Lambton co., and GSC Arkona No. 1 well, Rock Glen Park, lot 4, con. SB, Bosanquet tp., Lambton co., Ontario.

Angochitina sp. 2
Fig. spec. 96838
Asselin, E., Achab, A. and Bourque, P.A., 1989, Can. J., Earth Sci., vol. 26, no. 12, p. 2446, pl. 6, fig. 19.
Anse à Pierre-Loiselle Formation, Lower Silurian, Anse à Pierre-loiselle, Gaspé, Québec.

Aremoricanium? matapediensis Achab
Holotype 43879
Achab, A., 1977, Can. J. Earth Sci., vol. 13, no. 9, p. 1314, pl. 3, fig. 28.
Awantjish Formation, Middle Silurian, depth 401 feet, Val Brillant No. 1 borehole, lat. 48°32′30″N, long. 67°34′W, Matapedia Valley, Québec.

Calpichitina megastrophica Achab
Holotype 77238; paratypes 77235-77237
Achab, A., 1984, Review of Palaeobotony and Palynology, vol. 43, no. 1-3, p. 126, pl. I, figs. 1-4.
Middle Ordovician, depths 4350 and 4390 feet, New Associated Consolidated Paper Anticosti No. 1 borehole, Anticosti Island, Québec.

Calpichitina sp.
Fig. spec. 30612
Legault, J.A., 1973, Geol. Surv. Can., Bull. 221, p. 36, pl. 7, fig. 5.
Rockport Quarry Member, Hamilton Formation, Middle Devonian, GSC Ipperwash No. 1 well, lot 9, con. A, Bosanquet tp., Lambton co., Ontario.

Chitinozoa
Fig. specs. 55533-55535
McGregor, D.C. and Narbonne, G.M., 1978, Can. J. Earth Sci., vol. 15, no. 8, p. 1301, pl. 2, figs. 9-11.
Member B 20 m above base, Read Bay Formation, Upper Silurian, south side Goodsir Creek, eastern Cornwallis Island, District of Franklin.

Chitinozoa (?)
Fig. specs. 53554, 53555, a-c
Bolton, T.E., 1980, Geol. Surv. Can., Paper 80-1C, pl. 2.5, figs. 5, 8.
White Head Formation, Upper Ordovician, Grande Coupe, Percé, Gaspé Peninsula, Québec.

Chitinozoa n. gen., n. sp.
Fig. spec. 30611
Legault, J.A., 1973, Geol. Surv. Can., Bull. 221, p. 35, pl. 6, fig. 8.
Bell Member, Hamilton Formation, Middle Devonian, GSC Arkona No. 1 well, Rock Glen Park, lot 4, con. SB, Bosanquet tp. Lambton co., Ontario.

Clathrochitina sylvanica Jenkins
Hypotypes 61520-61524
Martin, F., 1980, Can. J. Earth Sci., vol. 17, no. 1, p. 107, pl. 1, figs. 1-3, 6, 7.
White Head Formation, Upper Ordovician, Grande Coupe, Gaspé, Québec.

Clathrochitina sylvanica Jenkins, 1970
Hypotypes 55822, 55823
Martin, F., 1983, Geol. Surv. Can., Bull. 310, p. 12, pl. 5, figs. 20, 23.
Lotbinière Formation, Utica Group, Upper Ordovician, bay to west of Pointe-aux-Trembles-Ouest, lat. 46°40′N, long. 71°39′51″W, Québec.

Conochitina accuminata Eisenack, 1959
Hypotype 96834
Asselin, E., Achab, A. and Bourque, P.A., 1989, Can. J. Earth Sci., vol. 26, no. 12, p. 2446, pl. 6, fig. 9.
La Vieille Formation, Lower Silurian, Anse à Pierre-Loiselle, Gaspé, Québec.

Conochitina armifera Achab
Holotype 49483; hypotypes 49484-49488
Achab, A., 1977, Can. J. Earth Sci., vol. 14, no. 10, p. 2198, pl. 2, figs. 3-8.
Vauréal Formation, Upper Ordovician, depths 1700 and 1600 feet, New Associated Consolidated Paper Anticosti No. 1 well, lat. 49°37′20″N, long. 63°26′20″W, Anticosti Island, Québec.

Conochitina baculata Achab
Holotype 49515; hypotypes 49512-49514
Achab, A., 1977, Can. J. Earth Sci., vol. 14, no. 10, p. 2197, pl. 5, figs. 2, 6-8.
Vauréal Formation, Upper Ordovician, depths 1800, 1600 and 1750 feet, New Associated Consolidated Paper Anticosti No. 1 well, lat. 49°37′20″N, long. 63°26′20″W, Anticosti Island, Québec.

Conochitina brevis Taugourdeau et de Jekhowsky, 1960
Hypotypes 77851-77856, 77866
Achab, A., 1977, Can. J. Earth Sci., vol. 23, no. 5, p. 687, pl. 1, figs. 13-18; pl. 2, fig. 10.

Lower Ordovician, Levis Formation, Côte Bégin, Lévis, Québec; Cow Head Group, Cow Head, Newfoundland (77866).

Conochitina bulmani Jansonius
Hypotypes 49481, 49482
Achab, A., 1977, Can. J. Earth Sci., vol. 14, no. 10, p. 2197, pl. 2, figs. 1, 2.
Vauréal Formation, Upper Ordovician, depths 1700 and 1775 feet, New Associated Consolidated Paper Anticosti No. 1 well, lat. 49°37′20″N, long. 63°26′20″W, Anticosti Island, Québec.

Conochitina bulmani Jansonius, 1964
Hypotypes 53601-53603
Achab, A., 1978, Rev. Palaeobot. Palynol., vol. 25, nos. 3/4, p. 300, pl. I, figs. 12, 14.
Vauréal Formation, Upper Ordovician, depths 3875 and 3700 feet, New Associated Consolidated Paper Anticosti No. 1 well, lat. 49°37′20″N, long. 63°26′20″W, Anticosti Island, Québec.

Conochitina bulmani Jansonius
Hypotypes 61525-61528
Martin, F., 1980, Can. J. Earth Sci., vol. 17, no. 1, p. 107, pl. 2, figs. 8-10, 13.
White Head Formation, Upper Ordovician, Grande Coupe, Gaspé, Québec.

Conochitina cf. *bulmani* Jansonius 1964
Hypotype 41221
Martin, F., 1975, Can. J. Earth Sci., vol. 12, no. 6, p. 1008, pl. 2, fig. 16.
Utica Formation, Ordovician, Montmorency Falls, Québec.

Conochitina cf. *C. bulmani* Jansonius, 1964
Hypotypes 69648-69652
Achab, A., 1983, Can. J. Earth Sci., vol. 20, no. 6, p. 922, pl. 1, figs. 1-5.
Table Head Formation, Middle Ordovician, quarry, Highway 49, West Bay Centre, western Newfoundland.

Conochitina aff. *cactacea* Eisenack
Hypotypes 41211, 41219
Martin, F., 1975, Can. J. Earth Sci., vol. 12, no. 6, p. 1008, pl. 2, figs. 4, 6, 17; pl. 3, fig. 6.
Utica Formation, Ordovician, Montmorency Falls, Québec.

Conochitina chydea Jenkins, 1967
Hypotype 56680
Martin, F., 1978, Geol. Surv. Can., Paper 78-1B, pl. 12.1, fig. 14.
Table Head Formation, Middle Ordovician, gully on shore at north end of "Beach Gravels", Table Point aera, western Newfoundland.

Conochitina cf. *C. clava-herculi* Eisenack, 1959
Hypotypes 77844-77847
Achab, A., 1986, Can. J. Earth Sci., vol. 23, no. 5, p. 689, pl. 1, figs. 6-9.
Levis Formation, Lower Ordovician, Côte Bégin, Lévis, Québec.

Conochitina decipiens Taugourdeau et de Jekhlowsky, 1960
Hypotypes 77857-77859
Achab, A., 1986, Can. J. Earth Sci., vol. 23, no. 5, p. 689, pl. 2, figs. 1-3.
Cow Head Group, Lower Ordovician, Cow Head, Newfoundland.

Conochitina aff. *dolosa* Laufeld
Hypotypes 49516-49519
Achab, A., 1977, Can. J. Earth Sci., vol. 14, no. 10, p. 2198, pl. 5, figs. 3, 4, 9, 10.
Vauréal Formation, Upper Ordovician, depths 2000, 1850 and 1800 feet, New Associated Consolidated Paper Anticosti No. 1 well, lat. 49°37′20″N, long. 63°26′20″W, Anticosti Island, Québec.

Conochitina edjelensis Taugourdeau
Hypotype 30635
Legault, J.A., 1973, Geol. Surv. Can., Bull. 221, p. 43, pl. 8, fig. 7.
Arkona Member, Hamilton Formation, Middle Devonian, GSC Arkona No. 1 well, Rock Glen Park, lot 4, con. SB, Bosanquet tp., Lambton co., Ontario.

Conochitina cf. *C. edjelensis* Taugourdeau, 1963
Hypotype 96805
Asselin, E., Achab, A. and Bourque, P.A, 1989, Can. J. Earth Sci., vol. 26, no. 12, p. 2442, pl. 2, fig. 16.
Clemville Formation, Lower Silurian, Little Port-Daniel River, Clemville, Gaspé, Québec.

Conochitina edjelensis alargada Cramer, 1966
Hypotype 96826
Asselin, E., Achab, A. and Bourque, P.A., 1989, Can. J. Earth Sci., vol. 26, no. 12, p. 2442, pl. 4, figs. 15, 16.
Anse Cascons Formation, Lower Silurian, Little Port-Daniel River, Clemville, Gaspé, Québec.

Conochitina cf.*C. edjelensis elongata* Taugourdeau, 1963
Hypotype 96804
Asselin, E., Achab, A. and Bourque, P.A., 1989, Can. J. Earth Sci., vol. 26, no. 12, p. 2442, pl. 2, fig. 13.
Clemville Formation, Lower Silurian, Little Port-Daniel River, Clemville, Gaspé, Québec.

Conochitina cf. *C. electa* Nestor, 1980
Hypotype 96828
Asselin, E., Achab, A. and Bourque, P.A., 1989, Can. J. Earth Sci., vol. 26, no. 12, p. 2446, pl. 5, fig. 4.
Anse à Pierre-Loiselle Formation, Lower Silurian, Anse à Pierre-Loiselle, Gaspé, Québec.

Conochitina aff. *elegans* Eisenack 1931
Hypotype 41215
Martin, F., 1975, Can. J. Earth Sci., vol. 12, no. 6, p. 1008, pl. 2, fig. 13.
Utica Formation, Ordovician, Montmorency Falls, Québec.

Conochitina flamma Laufeld, 1974
Hypotype 96835
Asselin, E., Achab, A. and Bourque, P.A., 1989, Can. J. Earth Sci., vol. 26, no. 12, p. 2446, pl. 6, fig. 10.
Anse à Pierre-Loiselle Formation, Lower Silurian, Anse à Pierre-Loiselle, Gaspé, Québec.

Conochitina hirsuta Laufeld, 1967
Hypotypes 77216-77221
Achab, A, 1984, Review of Palaeobotany and Palynology, vol. 43, no. 1-3, p. 126, pl. II, figs. 2-5, 8, 9.
Middle Ordovician, depth 4474 feet, New Associated Consolidated Paper Anticosti No. 1 borehole, Anticosti Island, Québec.

Conochitina cf. *C. insueta* Umnova, 1969
Hypotypes 69675, 69676
Achab, A., 1983, Can. J. Earth Sci., vol. 20, no. 6, p. 928, pl. 2, figs. 13, 14.
Table Head Formation, Middle Ordovician, quarry, Highway 49, West Bay Centre, western Newfoundland.

Conochitina kjellstromi Martin
Holotype 41214; paratypes 41208, 41215, 41216
Martin, F., 1975, Can. J. Earth Sci., vol. 12, no. 6, p. 1009, pl. 2, figs. 1, 3; pl. 3, figs. 15, 19, 21.
Utica Formation and Trenton Group (41208), Ordovician, Montmorency Falls, Québec.

Conochitina kjellstromi Martin, 1975
Hypotypes 55824-55827, 55924
Martin, F., 1983, Geol. Surv. Can., Bull. 310, p. 12, pl. 3, fig. 18; pl. 5, figs. 1-3, 10, 12.
Upper Ordovician, Cobourg Formation, Geological Survey of Canada, borehole COB-1-1, Ottawa, Ontario; Lotbinière Formation, Utica Group, bay to west of Pointe Delisle, lat. 46°41′57″N, long. 71°36′29″W, Québec (55825, 55924).

Conochitina kjellstromi Martin, 1975
Hypotypes 85444-85455
Achab, A., 1987, Can. J. Earth Sci., vol. 24, no. 6, p. 1216, pl. 5, figs. 1-4, 9-16.
Trenton Group, Middle Ordovician, depths 2300 and 2400 (85448) feet; Upper Ordovician, Utica Group, depths 1650 (85449), 1750 (85451, 85452), 1900 (85453), 1500 (85455) feet, and Lorraine Group, depths 1350 (85450) and 1250 (85454) feet, Bald Mountain Saint-Roch no. 1 borehole, near Sorel, Québec.

Conochitina kryos Bockelie, 1980
Hypotypes 77817-77820
Achab, A., 1986, Can. J. Earth Sci., vol. 23, no. 5, p. 689, pl. 3, figs. 5-8.
Levis Formation, Lower Ordovician, Côte Bégin, Levis, Québec.

Conochitina cf. *C. kryos* Bockelie, 1980
Hypotypes 63995-63998
Achab, A., 1982, Can. J. Earth Sci., vol. 19, no. 6, p. 1302, pl. 2, figs. 4-7.
Levis Formation, Lower Ordovician, Côte Fréchette (Davidson Street), Lévis, Québec.

Conochitina cf. *C. kryos* Bockelie, 1980
Hypotypes 77842, 77843
Achab, A., 1986, Can. J. Earth Sci., vol. 23, no. 5, p. 689, pl. 1, figs. 4, 5.
Levis Formation, Lower Ordovician, Côte Bégin, Levis, Québec.

Conochitina micracantha Eisenack, 1931
Hypotypes 55828, 55829
Martin, F., 1983, Geol. Surv. Can., Bull. 310, p. 12, pl. 5, figs. 21, 38.
Billings Formation, Utica Group, Upper Ordovician, Billings Creek west of Billings Bridge, Bank Stret, Ottawa, Ontario.

Conochitina micracantha Eisenack, 1931
Hypotypes 85456-85463
Achab, A., 1987, Can. J. Earth Sci., vol. 24, no. 6, p. 1216, pl. 6, figs. 1-8.
Upper Ordovician, Lorraine Group, depths 850, 1050 (85457), 950 (85458-85460), and 1150 (85461) feet; Utica Group, depth 1950 feet (85462, 85463), Bald Mountain Saint-Roch no. 1 borehole, near Sorel, Québec.

Conochitina cf. *C. micracantha* Eisenack, 1931
Hypotypes 69655, 69656
Achab, A., 1983, Can. J. Earth Sci., vol. 20, no. 6, p. 924, pl. 1, figs. 7, 8.
Table Head Formation, Middle Ordovician, quarry, Highway 49, West Bay Centre, western Newfoundland.

Conochitina minnesotensis (Stauffer, 1933)
Hypotypes 55830, 55831
Martin, F., 1983, Geol. Surv. Can., Bull. 310, p. 13, pl. 3, figs. 16, 20.
Neuville Formation, Trenton Group, Middle-Upper Ordovician, Charlesbourg Ltée. quarry, Charlesbourg-Est, and Saint-Laurent cement quarry, Villeneuve, Québec.

Conochitina minnesotensis (Stauffer, 1933)
Hypotypes 69688-69691
Achab, A., 1983, Can. J. Earth Sci., vol. 20, no. 6, p. 930, pl. 4, figs. 1-4.
Table Head Formation, Middle Ordovician, quarry, Highway 49, West Bay Centre, western Newfoundland.

Conochitina minnesotensis (Stauffer, 1933)
Hypotypes 85424-85427
Achab, A., 1987, Can. J. Earth Sci., vol. 24, no. 6, p. 1216, pl. 2, figs. 5-8.
Utica Group, Upper Ordovician, depth 1500 feet, Bald Mountain Saint-Roch no. 1 borehole, near Sorel, Québec.

Conochitina aff. *minnesotensis* Staufer 1933
Hypotype 41209
Martin, F., 1975, Can. J. Earth Sci., vol. 12, no. 6, p. 1009, pl. 2, fig. 9.
Utica Formation, Ordovician, Montmorency Falls, Québec.

Conochitina naevia Martin
Holotype 41208; paratype 41208
Martin, F., 1975, Can. J., Earth Sci., vol. 12, no. 6, p. 1009, pl. 2, figs. 8, 15; pl. 3, fig. 1.
Trenton Group, Middle Ordovician, Montmorency Falls, Québec.

Conochitina naevia Martin, 1975
Hypotypes 55832-55834
Martin, F., 1983, Geol. Surv. Can., Bull. 310, p. 13, pl. 1, figs. 7, 8; pl. 5, fig. 32.
Neuville Formation, Trenton Group, Middle Ordovician, about 710 m east of pier at Neuville, and Chateau Richer (55834), Québec.

Conochitina oklahomensis (Taugourdeau, 1965)
Hypotype 41220
Martin, F., 1975, Can. J. Earth Sci., vol. 12, no. 6, p. 1009, pl. 3, fig. 4.
Utica Formation, Ordovician, Montmorency Falls, Québec.

Conochitina oklahomensis (Taugourdeau, 1965)
Hypotypes 55535, 55836
Martin, F., 1983, Geol. Surv. Can., Bull. 310, p. 13, pl. 1, fig. 6; pl. 5, fig. 29.
Neuville Formation, Trenton Group, Middle-Upper Ordovician, about 710 m east of pier at Neuville, and bay to east of Pointe Delisle, Québec.

Conochitina cf. *C. ordinaria* Achab, 1980
Hypotypes 64009-64011
Achab, A., 1982, Can. J. Earth Sci., vol. 19, no. 6, p. 1304, pl. 4, figs. 1, 5, 6.
Levis Formation, Lower Ordovician, Côte Fréchette (Davidson Street), Lévis, Québec.

Conochitina pirum Achab
Holotype 64000; hypotypes 63999, 64001-64008
Achab, A., 1982, Can. J. Earth Sci., vol. 19, no. 6, p. 1298, pl. 3, figs. 1-12.
Levis Formation, Lower Ordovician, Côte Fréchette (Davidson Street), Lévis, Québec.

Conochitina pirum Achab, 1982
Hypotypes 69677-69687
Achab, A., 1983, Can. J. Earth Sci., vol. 20, no. 6, p. 928, pl. 3, figs. 1-12.
Table Head Formation, Middle Ordovician, quarry, Highway 49, West Bay Centre, western Newfoundland.

Conochitina poumoti Combaz et Péniguel, 1972
Hypotypes 63985-63990
Achab, A., 1982, Can. J. Earth Sci., vol. 19, no. 6, p. 1302, pl. 1, figs. 6-11.
Levis Formation, Lower Ordovician, Côte Fréchette (Davidson Street), and falaise face à la gare de Lévis (63989), Québec.

Conochitina poumoti Combaz et Péniguel, 1972
Hypotypes 69692-69699
Achab, A., 1983, Can. J. Earth Sci., vol. 20, no. 6, p. 930, pl. 4, figs. 5-13.
Table Head Formation, Middle Ordovician, quarry, Highway 49, West Bay Centre, western Newfoundland.

Conochitina poumoti Combaz et Péniguel, 1972
Hypotypes 77839-77841
Achab, A., 1986, Can. J. Earth Sci., vol. 23, no. 5, p. 689, pl. 1, figs. 1-3.
Levis Formation, Lower Ordovician, Côte Bégin, Lévis, Québec.

Conochitina primitiva Eisenack, 1939
Hypotypes 55837-55841
Martin, F., 1983, Geol. Surv. Can., Bull 310, p. 13, pl. 2, figs. 1, 3; pl. 3, fig. 5; pl. 5, fig. 5, 6.
Deschambault Formation, Middle Ordovician, Neuville Industries Inc. quarry, about 3 km northeast of pier at Neuville, Québec.

Conochitina cf. *C. proboscifera* Eisenack, 1937
Hypotype 96827
Asselin, E., Achab, A. and Bourque, P.A., 1989, Can. J. Earth Sci., vol. 26, no. 12, p. 2446, pl. 5, figs. 1-3.
Anse à Pierre-Loiselle Formation, Lower Silurian, Anse à Pierre-Loiselle, Gaspé, Québec.

Conochitina pygmaea Achab
Holotype 85479; paratypes 85480-85485
Achab, A., 1987, Can. J. Earth Sci., vol. 24, no. 6, p. 1216, pl. 9, figs. 1-5, 9, 10.
Utica Formation, Ordovician, depths 1700, 1500 (85480, 85481) and 1650 (85482-85485) feet, Bald Mountain Saint-Roch no. 1 borehole, near Sorel, Québec.

Conochitina raymondi Achab, 1980
Hypotypes 77860-77862, 77821-77824
Achab, A., 1986, Can. J. Earth Sci., vol. 23, no. 5, p. 691, pl. 2, figs. 4-6; pl. 3, figs. 9-12.
Lower Ordovician, Cow Head Group, Cow Head, Newfoundland; Levis Formation, Côte Bégin, Levis, Quebec (77821-77842).

Conochitina robusta Eisenack 1959
Hypotypes 41201, 41211
Martin, F., 1975, Can. J. Earth Sci., vol. 12, no. 6, p. 1011, p. 1, figs. 1, 8; pl. 3, fig. 7.
Utica Formation, Ordovician, Montmorency Falls, Québec.

Conochitina robusta Eisenack, 1959
Hypotype 55842
Martin, F., 1983, Geol. Surv. Can., Bull 310, p. 14, pl. 3, fig. 3.
Lotbinière Formation, Utica Group, Upper Ordovician, bay to west of Pointe-aux-Trembles-Ouest, lat. 46°40′N, long. 71°39′51″W, Québec.

Conochitina robusta Eisenack 1959
Hypotypes 777248-77251
Achab, A., 1984, Review of Palaeobotany and Palynology, vol. 43, no. 1-3, p. 130, pl. III, figs. 2-7.
Middle Ordovician, depths 4000, 4350, 4390 and 4600 feet, New Associated Consolidated Paper Anticosti no. 1 borehole, Anticosti Island, Québec.

Conochitina schopfi ssp. *americana* Taugourdeau
Hypotypes 49489-49492
Achab, A., 1977, Can. J. Earth Sci., vol. 14, no. 10, p. 2198, pl. 2, figs. 9-12.

Vauréal Formation, Upper Ordovician, depths 2000 and 2250 feet, New Associated Consolidated Paper Anticosti No. 1 well, lat. 49°37′20″N, long. 63°26′20″W, Anticosti Island, Québec.

Conochitina senta Achab
Holotype 53595; hypotypes 53593, 53594, 53596
Achab, A., 1978, Rev. Palaebot. Palynol., vol. 25, nos. 3/4, p. 302, pl. I, figs. 4-6, 9.
Vauréal Formation, Upper Ordovician, depths 3850, 3300 and 3500 feet, New Associated Consolidated Paper Anticosti No. 1 well, lat. 49°37′20″N, long. 63°26′20″W, Anticosti Island, Québec.

Conochitina seriespinosa Jenkins, 1969
Hypotype 69405
Legault, J.A., 1982, Can. J. Earth Sci., vol. 19, no. 9, pl. 2, fig. 3.
Middle-Upper Ordovician, Orphan Knoll, lat. 50°33′N, long. 46°6′W, Labrador Sea.

Conochitina aff. *simplex* Eisenack 1931
Hypotypes 41204, 41219
Martin, F., 1975, Can. J. Earth Sci., vol. 12, no. 6, p. 1011, pl. 1, fig. 9; pl. 3, fig. 3.
Utica Formation, Ordovician, Montmorency Falls, Québec.

Conochitina aff. *C. simplex* Eisenack, 1931
Hypotype 55843
Martin, F., 1983, Geol. Surv. Can., Bull. 310, p. 14, pl. 5, fig. 27.
Neuville Formation, Trenton Group, Upper Ordovician, Union Quarries and Paving Ltée quarry, Charlesbourg-Ouest, Québec.

Conochitina subcylindrica Combaz et Péniguel, 1972
Hypotypes 69653, 69654
Achab, A., 1983, Can. J. Earth Sci., vol. 20, no. 6, p. 924, pl. 1, figs. 6, a.
Table Head Formation, Middle Ordovician, quarry, Highway 49, West Bay Centre, western Newfoundland.

Conochitina symmetrica Taugourdeau et de Jekhowsky, 1960
Hypotypes 77813-77816
Achab, A., 1986, Can. J. Earth Sci., vol. 23, no. 5, p. 691, pl. 3, figs. 1-4.
Levis Formation, Lower Ordovician, Côte Bégin, Lévis, Québec.

Conochitina tribulosa Jenkins 1969
Hypotypes 41206, 41216
Martin, F., 1975, Can. J. Earth Sci., vol. 12, no. 6, p. 1011, pl. 1, fig. 2; pl. 2, fig. 5.
Utica Formation, Ordovician, Montmorency Falls, Québec.

Conochitina tribulosa Jenkins, 1969
Hypotype 55844
Martin, F., 1983, Geol. Surv. Can., Bull 310, p. 14, pl. 5, fig. 39.
Lotbinière Formation, Trenton Group, Upper Ordovician, road extending from bay west of Pointe Delisle to Highway 138, Québec.

Conochitina aff. *C. turris* Taugourdeau, 1961
Hypotypes 96821, 96825
Asselin, E., Achab, A. and Bourque, P.A., 1989, Can. J. Earth Sci., vol. 26, no. 12, p. 2446, pl. 4, figs. 9, 14.
Anse Cascons Formation, Lower Silurian, Little Port-Daniel River, Clemville, Gaspé, Québec.

Conochitina wesenbergensis Eisenack, 1959
Hypotypes 55845-55854
Martin, F., 1983, Geol. Surv. Can., Bull. 310, p. 14, pl. 2, figs. 4, 7; pl. 3, figs. 13-15; pl. 4, figs. 20, 21; pl. 5, figs. 4, 33, 34.
Upper Ordovician, Neuville Formation, west of pier at Neuville; Lotbinière Formation, bay west of Pointe Delisle, lat. 46°41′57″N, long. 71°36′29″W (55846) and road extending from bay west of Pointe Delisle to Highway 138 (55853), Québec; Billings Formation, Utica Group, Billings Creek east of Billings Bridge, Bank Street, Ottawa, Ontario (55852); Montreal and Deschambault formations, Lafarge Cement quarry, Saint-Constant (55847-55850), and Neuville Industries Inc. quarry, about 3 km northeast of pier at Neuville (55851, 55854), Québec.

Conochitina sp.
Fig. specs. 69455, 69456
Achab, A., 1982, Can. J. Earth Sci., vol. 19, no. 6, p. 1304, pl. 4, figs. 9, 10.
Levis Formation, Lower Ordovician, falaise face à la gare de Lévis, Québec.

Conochitina sp.
Fig. specs. 77848-77850
Achab, A., 1986, Can. J. Earth Sci., vol. 23, no. 5, p. 691, pl. 1, figs. 10-12.
Levis Formation, Lower Ordovician, Côte Bégin, Lévis, Québec.

Conochitina sp.
Fig. specs. 85464, 85465
Achab, A., 1987, Can. J. Earth Sci., vol. 24, no. 6, pl. 6, figs. 9, 10.
Trenton Group, Middle Ordovician, depth 2300 feet, Bald Mountain Saint-Roch no. 1 borehole, near Sorel, Quebec.

Conochitina sp. 1
Fig. specs. 85435, 85436
Achab, A., 1987, Can. J. Earth Sci., vol. 24, no. 6, p. 1218, pl. 4, figs. 3, 4.
Lorraine Group, Upper Ordovician, depth 400 feet, Bald Mountain Saint-Roch no. 1 borehole, near Sorel, Quebec.

Conochitina sp. 1
Fig. specs. 96793, 96794, 96807
Asselin, E., Achab, A. and Bourque, P.A., 1989, Can. J. Earth Sci., vol. 26, no. 12, p. 2442, pl. 1, figs. 12, 13; pl. 2, fig. 18.
Clemville Formation, Lower Silurian, Little Port-Daniel River, Clemville, Gaspé, Québec.

Conochitina sp. 2
Fig. specs. 85433, 85434
Achab, A., 1987, Can. J. Earth Sci., vol. 24, no. 6, p. 1218, pl. 3, figs. 5, 6.
Lorraine Group, Upper Ordovician, depth 850 feet, Bald Mountain Saint-Roch no. 1 borehole, near Sorel, Québec.

Conochitina sp. 2
Fig. specs. 96795-96800
Asselin, E., Achab, A. and Bourque, P.A., 1989, Can. J. Earth Sci., vol. 26, no. 12, p. 2442, pl. 2, figs. 1-4, 8, 9.
Clemville Formation, Lower Silurian, Little Port-Daniel River, Gaspé, Québec.

Conochitina sp. 3
Fig. specs. 96806, 96818
Asselin, E., Achab, A. and Bourque, P.A., 1989, Can. J. Earth Sci., vol. 26, no. 12, p. 2442, pl. 2, fig. 17; pl. 3, fig. 12.
Clemville and Weir formations, Lower Silurian, Little Port-Daniel River, Clemville, Gaspé, Québec.

Conochitina sp. 5
Fig. specs. 96811-96813, 96815, 98616
Asselin, E., Achab, A. and Bourque, P.A., 1989, Can. J. Earth Sci., vol. 26, no. 12, p. 2442, pl. 3, fig. 4-6, 8, 9.
Wier and Clemville (98616) formations, Lower Silurian, Little Port-Daniel River, Clemville, Gaspé, Québec.

Conochitina sp. 6, 7
Fig. specs. 96822-96824
Asselin, E., Achab, A. and Bourque, P.A., 1989, Can. J. Earth Sci., vol. 26, no. 12, p. 2442, 2446, pl. 4, figs. 6, 12, 13.
Anse Cascons Formation, Lower Silurian, Little Port-Daniel River, Clemville, Gaspé, Québec.

Conochitina sp. 9
Fig. specs. 96829, 96830
Asselin, E., Achab, A. and Bourque, P.A., 1989, Can. J. Earth Sci., vol. 26, no. 12, p. 2446, pl. 5, figs. 6, 7.
Anse à Pierre-Loiselle Formation, Lower Silurian, Anse à Pierre-Loiselle, Gaspé, Québec.

Conochitina sp. A
Fig. specs. 53597-53600
Achab, A., 1978, Rev. Palaeobot. Palynol., vol. 25, nos. 3/4, p. 302, pl. I, figs. 7, 8, 10, 11.
Vauréal Formation, Upper Ordovician, depth 3875 feet, New Associated Consolidated Paper Anticosti No. 1 well, lat. 49°37′20″N, long. 63°26′20″W, Anticosti Island, Québec.

Conochitina sp. A
Fig. specs. 64012-64014
Achab, A., 1982, Can. J. Earth Sci., vol. 19, no. 6, p. 1302, pl. 4, figs. 2, 3, 7.
Levis Formation, Lower Ordovician, Côte Fréchette (Davidson Street), Lévis, Québec.

Cyathochitina anticostiensis Achab
Holotype 47965
Achab, A., 1977, Can. J. Earth Sci., vol. 14, no. 3, p. 422, pl. 4, figs. 1, 2, 5.
Vauréal Formation, Upper Ordovician, depth 2700 feet, New Associated Consolidated Paper Anticosti No. 1 well, near '24-mile lodge' on Jupiter River, Anticosti Island, Québec.

Cyathochitina cf. *C. campanulaeformis* (Eisenack) 1931
Hypotypes 77233, 77234
Achab, A, 1984, Review of Palaebotany and Palynology, vol. 43, no. 1-3, p. 133, pl. IV, figs. 3, 4.
Middle Ordovician, depth 4600 feet, New Associated Consolidated Paper Anticosti no. 1 borehole, Anticosti Island, Québec.

Cyathochitina hyalophrys Eisenack, 1959
Hypotypes 85428-85431
Achab, A., 1987, Can. J. Earth Sci., vol. 24, no. 6, p. 1218, pl. 2, figs. 9-12.
Lorraine Group, Upper Ordovician, depth 400 feet, Bald Mountain Saint-Roch no. 1 borehole, near Sorel, Québec.

Cyathochitina cf. *C. jenkinsi* Neville 1974
Hypotypes 77222-77226
Achab, A, 1984, Review of Palaebotany and Palynology, vol. 43, no. 1-3, p. 134, pl. V, figs. 1-3, 6, 9.
Middle Ordovician, depth 4450 feet, New Associated Consolidated Paper Anticosti no. 1 borehole, Anticosti Island, Québec.

Cyathochitina kuckersiana (Eisenack)
Hypotypes 49526-49528
Achab, A., 1977, Can. J. Earth Sci., vol. 14, no. 10, p. 2200, pl. 6, figs. 5, 8, 9.
Vauréal Formation, Upper Ordovician, depths 1150 and 1000 feet, New Associated Consolidated Paper Anticosti No. 1 well, lat. 49°37′20″N, long. 63°26′20″W, Anticosti Island, Québec.

Cyathochitina kuckersiana (Eisenack, 1934)
Hypotypes 85488-85490
Achab, A., 1987, Can. J. Earth Sci., vol. 24, no. 6, p. 1218, pl. 9, figs. 13-15.
Upper Ordovician, Utica Group, depth 1850 feet; Lorraine Group, depths 1200 (85489) and 1250 (85490) feet, Bald Mountain Saint-Roch no. 1 borehole, near Sorel, Quebec.

Cyathochitina kuckersiana brevis Eisenack 1962
Hypotypes 41208, 41221
Martin, F., 1975, Can. J. Earth Sci., vol. 12, no. 6, p. 1015, pl. 2, figs. 11, 20, 21; pl. 3, fig. 12.
Trenton Group and Utica Formation (41221), Ordovician, Montmorency Falls, Québec.

Cyathochitina kuckersiana brevis Eisenack, Martin
Hypotypes 61529-61531
Martin, F., 1979, Can. J. Earth Sci., vol. 17, no. 1, p. 108, pl. 1, figs. 12, 14, 16.
White Head Formation, Upper Ordovician, Grande Coupe, Gaspé, Québec.

Cyathochitina kuckersiana brevis Martin ex. Eisenack, 1962
Holotype 55855; paratypes 55856, 55857
Martin, F., 1983, Geol. Surv. can., Bull 310, p. 15, pl. 2, fig. 2 ; pl. 3, figs. 8, 22.
Upper Ordovician, Lotbinière Formation, Utica Group, bay west of Pointe Deslisle, lat. 46°41'57"N, long. 71°36'29"W; Neuville Formation, Trenton Group, Upper Ordovician, Charlesbourg Ltée. quarry, Charlesbourg-Est (55857), Québec.

aff. *Cyathochitina kuckersiana kuckersiana* (Eisenack)
Hypotype 30644
Legault, J.A., 1973, Geol. Surv. Can., Bull. 221, p. 47, pl. 10, fig. 2.
Ipperwash Member, Hamilton Formation, Middle Devonian, GSC No. 1 well, lot 9, con. A, Bosanquet tp., Lambton co., Ontario.

Cyathochitina kuckersiana latipatagium Jenkins, 1969
Hypotypes 53606-53608
Achab, A, 1978, Rev. Palaeobot. Palynol, vol. 25, nos. 3/4, p. 303, pl. II, figs. 3, 4, 9.
Vauréal Formation, Upper Ordovician, depths 3200 and 3500 feet, New Associated Consolidated Paper Anticosti No. 1 well, lat. 49°37'20"N, long. 63°26'20"W, Anticosti Island, Québec.

Cyathochitina kuckersiana latipatagium Jenkins, 1969
Hypotypes 55858-55861
Martin, F., 1983, Geol. Surv. Can., Bull 310, p. 15, pl. 2, fig. 10; pl. 3, figs. 1, 21, 23.
Neuville Formation, Trenton Group, Middle Ordovician, about 710 m east of pier at Neuville and Charlesbourg Ltée. quarry, Charlesbourg-Est (55861), Québec.

Cyathochitina kuckersiana patagiata Jenkins, 1969
Hypotype 55862
Martin, F., 1983, Geol. Surv. Can., Bull 310, p. 15, pl. 3, fig. 10.
Tetreauville Formation, Upper Ordovician, Rouge River, south bank, northwest of Crabtree, Québec.

Cyathochitina kuckersiana patigiata Jenkins 1969
Hypotypes 41206, 41207, 41209
Martin, F., 1975, Can. J. Earth Sci., vol. 12, no. 6, p. 1015, pl. 1, figs. 3, 7; pl. 2, fig. 2.
Utica Formation, Ordovician, Montmorency Falls, Québec.

Cyathochitina macastyensis Achab
Holotype 53609; hypotypes 53610, 53611
Achab, A., 1978, Rev. Palaeobot. Palynol., vol. 25, nos. 3/4, p. 305, pl. II, figs. 5, 6, 10.
Vauréal Formation, Upper Ordovician, depth 3950 feet, New Associated Consolidated Paper Anticosti No. 1 well, lat. 49°37'20"N, long. 63°26'20"W, Anticosti Island, Québec.

Cyathochitina cf. *C. macastyensis* Achab, 1978
Hypotype 85491
Achab, A., 1987, Can. J. Earth Sci., vol. 24, no. 6, p. 1220, pl. 9, fig. 17.
Utica Formation, Ordovician, depth 1950 feet, Bald Mountain Saint-Roch no. 1 borehole, near Sorel, Québec.

Cyathochitina vaurealensis Achab
Holotype 47964
Achab, A., 1977, Can. J. Earth Sci., vol. 14, no. 3, p. 420, pl. 4, fig 8.
Vauréal Formation, Upper Ordovician, depth 2500 feet, New Associated Consolidated Paper Anticosti No. 1 well, near '24-mile lodge' on Jupiter River, Anticosti Island, Québec.

Cyathochitina vaurealensis Achab
Hypotypes 49523-49525
Achab, A., 1977, Can. J. Earth Sci., vol. 14, no. 10, p. 2200, pl. 6, figs. 3, 4, 6.
Vauréal Formation, Upper Ordovician, depths 1950 and 2000 feet, New Associated Consolidated Paper Anticosti No. 1 well, lat. 49°37'20"N, long. 63°26'20"W, Anticosti Island, Québec.

Cyathochitina vaurealensis Achab, 1977
Hypotypes 53604, 53605
Achab, A., 1978, Rev. Palaebot. Palynol., vol. 25, nos. 3/4, p. 303, pl. II, figs. 1, 2.
Vauréal Formation, Upper Ordovician, depth 3200 feet, New Associated Consolidated Paper Anticosti No. 1 well, lat. 49°37'20"N, long. 63°26'20"W, Anticosti Island, Québec.

Cyathochitina vaurealensis Achab
Hypotypes 61532-61536
Martin, F., 1980, Can. J. Earth Sci., vol. 17, no. 1, p. 108, pl. 1, figs. 4, 5, 8, 9, 15.
White Head Formation, Upper Ordovician, Grande Coupe, Gaspé, Québec.

Cyathochitina spp.
Fig. spec. 96819
Asselin, E., Achab, A. and Bourque, P.A., 1989, Can. J. Earth Sci., vol. 26, no. 12, p. 2442, pl. 3, fig. 13.
Weir Formation, Lower Silurian, Little Port-Daniel River, Clemville, Gaspé, Québec.

Cyathochitina sp. 1
Fig. specs. 85486, 85487
Achab, A., 1987, Can. J. Earth Sci., vol. 24, no. 6, p. 1220, pl. 9, figs. 11, 12.
Utica Formation, Ordovician, depth 1450 feet, Bald Mountain Saint-Roch no. 1 borehole, near Sorel, Québec.

?*Desmochitina bulla* Taugourdeau et de Jekhowsky, 1960
Hypotype 77828
Achab, A., 1986, Can. J. Earth Sci., vol. 23, no. 5, p. 691, pl. 4, fig. 4.
Levis Formation, Lower Ordovician, Côte Bégin, Lévis, Québec.

Desmochitina bursa Taugourdeau and de Jekhowsky
Hypotypes 30613, 30614
Legault, J.A., 1973, Geol. Surv. Can., Bull. 221, p. 37, pl. 7, figs. 4, 7.
Arkona Member, Hamilton Formation, Middle Devonian, GSC Arkona No. 1 well, Rock Glen Park, lot 4, con. SB, Bosanquet tp., lambton co., Ontario.

Desmochitina complanata Eisenack, 1932
Hypotypes 53612, 53613
Achab, A., 1978, Rev. Palaeobot, Palynol., vol. 25, nos. 3/4, p. 306, pl. II, figs. 7, 8.
Vauréal Formation, Upper Ordovician, depths 3850 and 3875 feet, New Associated Consolidated Paper Anticosti No. 1 well, lat. 49°37′20″N, long. 63°26′20″W, Anticosti Island, Québec.

Desmochitina complanata Eisenack
Hypotypes 61537, 61538
Martin, F., 1980, Can. J. Earth Sci., vol. 17, no. 1, pl. 1, figs. 10, 11.
White Head Formation, Upper Ordovician, Grande Coupe, Gaspé, Québec.

Desmochitina lata Schallreuter, 1963
Hypotype 85499
Achab, A., 1987, Can. J. Earth Sci., vol. 24, no. 6, p. 1220, pl. 10, fig. 10.
Lorraine Group, Upper Ordovician, depth 750 feet, Bald Mountain Saint-Roch no. 1 borehole, near Sorel, Quebec.

Desmochitina cf. *D. lata* sensu Neville, 1974
Hypotypes 56669, 56670
Martin, F., 1978, Geol. Surv. Can., Paper 78-1B, pl. 12.1, figs. 6, 8.
Table Head Formation, Middle Ordovician, Table Point and excavation west of road No. 59 leading to Fox Island River 3 km north-northeast of Port au Port village, western Newfoundland.

Desmochitina cf. *D. lata* Schallreuter, 1963
Hypotypes 69662-69665
Achab, A., 1983, Can. J. Earth Sci., vol. 20, no. 6, p. 924, pl. 2, figs. 1-4.
Table Head Formation, Middle Ordovician, quarry, Highway 49, West Bay Centre, western Newfoundland.

Desmochitina minor Eisenack, 1931
Hypotype 69395
Legault, J.A., 1982, Can. J. Earth Sci., vol. 19, no. 9, pl. 2, fig. 4.
Middle-Upper Ordovician, Orphan Knoll, lat. 50°33′N, long. 46°6′W, Labrador Sea.

Desmochitina minor Eisenack 1931
Hypotypes 77239-77244
Achab, A, 1984, Review of Palaebotany and Palynology, vol. 43, no. 1-3, p. 136, pl. I, figs. 8-13.
Middle Ordovician, depths 4390, 4600, 4492 and 4550 feet, New Associated Consolidated Paper Anticosti no. 1 borehole, Anticosti Island, Québec.

Desmochitina minor Eisenack, 1958
Hypotypes 85492-85495
Achab, A., 1987, Can. J. Earth Sci., vol. 24, no. 6, p. 1220, pl. 10, figs. 1-3, 5.
Upper Ordovician, Utica Group, depth 1650 feet, and Lorraine Group, depth 750 feet (85495), Bald Mountain Saint-Roch no. 1 borehole, near Sorel, Québec.

Desmochitina minor typica Eisenack 1958
Hypotypes 41202, 41204, 41213, 41214, 41220
Martin, F., 1975, Can. J. Earth Sci., vol. 12, no. 6, p. 1016, pl. 1, figs. 5, 11, 12; pl. 2, fig. 10; pl. 3, figs 11, 13, 18.
Utica Formation, Ordovician, Montmorency Falls, Québec.

Desmochitina minor f. *typica* Eisenack, 1958
Hypotypes 55863, 55864
Martin, F., 1983, Geol. Surv. can., Bull 310, p. 16, pl. 5, figs. 22, 30.
Upper Ordovician, Neuville Formation, Trenton Group, Saint-Laurent Ltée. Cement Company quarry, Villeneuve; Lotbinière Formation, Utica Group, road extending from bay west of Pointe Delisle to Highway 138, Québec.

Desmochitina aff. *D. nodosa* Eisenack, 1931
Hypotype 55865
Martin, F., 1983, Geol. Surv. Can., Bull 310, p. 16, pl. 4, fig. 3.
Lotbinière Formation, Utica Group, Upper Ordovician, road extending from bay west of Pointe Delisle to Highway 138, Québec.

Desmochitina sp.
Fig. specs. 30615, 30616
Legault, J.A., 1973, Geol. Surv. Can., Bull. 221, p. 37, pl. 7, figs. 3, 6.
Arkona Member, Hamilton Formation, Middle Devonian, GSC Ipperwash No. 1 well, lot 9, con. A, Bosanquet tp. Lambton co., and GSC Arkona No. 1 well, Rock Glen Park, lot 4, con. SB, Bosanquet tp., Lambton co., Ontario.

Desmochitna sp.
Fig. spec. 85432
Achab, A., 1987, Can. J. Earth Sci., vol. 24, no. 6, pl. 2, fig. 13.
Utica Group, Upper Ordovician, depth 1650 feet, Bald Mountain Saint-Roch no. 1 borehole, near Sorel, Québec.

Eisenackitina canadensis Legault
Holotype 30621; paratypes 30622, 30623
Legault, J.A., 1973, Geol. Surv. Can., Bull. 221, p. 39, pl. 7, figs. 8, 11; pl. 8, fig. 1.
Hamilton Formation, Middle Devonian, Rockport Quarry and Bell members, GSC Ipperwash No. 1 well, lot 9, con. A, Bosanquet tp., Lambton co., and Arkona Member, GSC Arkona No. 1 well, Rock Glen Park, lot 4, con. SB, Bosanquet tp., Lambton co. (30622), Ontario.

Eisenackitina castor Jansonius
Hypotypes 30617-30620
Legault, J.A., 1973, Geol. Surv. Can., Bull. 221, p. 38, pl. 7, figs. 9, 10, 12; pl. 8, fig. 2.
Bell and Arkona members, Hamilton Formation, Middle Devonian, GSC Arkona No. 1 well, Rock Glen Park, lot 4, con. SB, Bosanquet tp., Lambton co. (30617), and GSC Ipperwash No. 1 well, lot. 9, con. A, Bosanquet tp., Lambton co., Ontario.

Eisenackitina obsoleta Martin
Holotype 55866; paratype 55867
Martin, F., 1983, Geol. Surv. can., Bull 310, p. 16, pl. 4, figs. 7, 8.
Deschambault Formation, MIddle Ordovician, Lafarge Cement quarry, Saint-Constant, Québec.

Eisenackitina uter Martin
Holotype 55868; paratype 55869
Martin, F., 1983, Geol. Surv. Can., Bull 310, p. 17, pl. 4, figs. 6, 13.
Pamelia Formation, Black River Group, Middle Ordovician, Saint-Vincent-de-Paul quarry, Laval, Québec.

Eisenackitina sp.
Fig. specs. 30624, 30625
Legault, J.A., 1973, Geol. Surv. Can., Bull. 221, p. 40, pl. 8, figs. 3, 9.
Rockport Quarry Member, Hamilton Formation, Middle Devonian, GSC Ipperwash No. 1 well, lot 9, con. A, Bosanquet tp., Lambton co., and GSC Arkona No. 1 well, Rock Glen Park, lot 4, con. SB, Bosanquet tp., Lambton co., Ontario.

Fustichitina langei (Combaz et Péniguel, 1972)
Hypotypes 63980-63984
Achab, A., 1982, Can. J. Earth Sci., vol. 19, no. 6, p. 1298, pl. 1, figs. 1-5.
Levis Formation, Lower Ordovician, Côte Fréchette (Davidson Street), Lévis, Québec.

Fustichitina langei (Combaz et Péniguel, 1972)
Hypotypes 69666-69672
Achab, A., 1983, Can. J. Earth Sci., vol. 20, no. 6, p. 928, pl. 2, figs. 5-10, 15.
Table Head Formation, Middle Ordovician, quarry, Highway 49, West Bay Centre, western Newfoundland.

Fustichitina langei (Combaz et Péniguel, 1972)
Hypotypes 77835-77838
Achab, A., 1986, Can. J. Earth Sci., vol. 23, no. 5, p. 693, pl. 4, figs. 11-14.
Levis Formation, Lower Ordovician, Côte Bégin, Levis, Québec.

Fustichitina cf. *F. pervulgata* (Umnova, 1969)
Hypotypes 64015-64017, 69454
Achab, A., 1982, Can. J. Earth Sci., vol. 19, no. 6, p. 1304, pl. 4, figs. 4, 8, 11, 12.
Levis Formation, Lower Ordovician, Côte Fréchette (Davidson Street), and falaise face à la gare de Lévis (64016), Québec.

Fustichitina cf. *F. pervulgata* (Umnova, 1969)
Hypotypes 77831, 77832
Achab, A., 1986, Can. J. Earth Sci., vol. 23, no. 5, p. 693, pl. 4, figs. 7, 8.
Levis Formation, Lower Ordovician, Côte Bégin, Lévis, Québec.

Fustichitina sp.
Fig. specs. 77870, 77871
Achab, A., 1986, Can. J. Earth Sci., vol. 23, no. 5, p. 694, pl. 2, figs. 14, 15.
Cow Head Group, Lower Ordovician, Cow Head, Newfoundland.

Fustichitina? sp.
Fig. specs. 69673, 69674
Achab, A., 1983, Can. J. Earth Sci., vol. 20, no. 6, p. 928, pl. 2, figs. 11, 12.
Table Head Formation, Middle Ordovician, quarry, Highway 49, West Bay Centre, western Newfoundland.

Hercochitina crickmayi Jansonius
Hypotypes 49493-49495
Achab, A., 1977, Can. J. Earth Sci., vol. 14, no. 10, p. 2202, pl. 3, figs. 1-3.
Vauréal Formation, Upper Ordovician, depths 1250 and 1300 feet, New Associated Consolidated Paper Anticosti No. 1 well, lat. 49°37′20″N, long. 63°26′20″W, Anticosti Island, Québec.

Hercochitina crickmayi Jansonius
Hypotypes 61539-61545
Martin, F., 1980, Can. J. Earth Sci., vol. 17, no. 1, p. 108, pl. 2, figs. 1-7.
White Head Formation, Upper Ordovician, Chemin du Falle, southeast of the Amphithéatre, Gaspé, Québec.

Hercochitina cf. *H. crickmayi* Jansonius, 1964
Hypotypes 55870-55873
Martin, F., 1983, Geol. Surv. Can., Bull 310, p. 17, pl. 1, fig. 14; pl. 2, figs. 5, 6, 9; pl. 4, fig. 32.
Middle and Upper Ordovician, Neuville Formation, Trenton Group, Charlesbourg Ltée. quarry, Charlesbourg-Est; Lotbinière Formation, Utica Group, bay west of Point Delisle, lat. 46°41′57″N, long. 71°36′29″W (55871), and road extending from bay west of Pointe Delisle to Highway 138 (55872), Québec.

Hercochitina cristata Achab
Holotype 85471; paratypes 85470, 85472, 85473
Achab, A., 1987, Can. J. Earth Sci., vol. 24, no. 6, p. 1220, pl. 7, figs. 3, 6, 9, 10.
Utica Formation, Ordovician, depths 1550 and 1650 (85472, 85473) feet, Bald Mountain Saint-Roch no. 1 borehole, near Sorel, Québec.

Hercochitina aff. *H. downiei* Jenkins, 1967
Hypotypes 56671-56673
Martin, F., 1978, Geol. Surv. Can., Paper 78-1B, p. 80, pl. 12.1, figs. 7, 17, 24.
Winterhouse Formation, Long Point Group, Upper Ordovician, quarry immediately southeast of fishing huts near west end of Clam Bank Cove 750 m northwest of Lourdes, western Newfoundland.

Hercochitina? duplicitas Martin
Holotype 55874; paratypes 55875-55881
Martin, F., 1983, Geol. Surv. Can., Bull 310, p. 17, pl. 1, fig. 11,; pl. 2, figs. 8, 11, 12; pl. 3, fig. 2; pl. 4, figs. 5, 33; pl. 5, figs, 15, 17.

Middle and Upper Ordovician, Neuville Formation, Trenton Group, Charlesbourg Ltée. quarry, Charlesbourg-Est, about 710 m east of (55875) and west of (55880) pier at Neuville, and bay east of Pointe Delisle (55876); Lotbinière Formation, Utica Group, bay west of Pointe Delisle, lat. 46°41′57″N, long. 71°36′29″W (55877), Québec.

Hercochitina? duplicitas Martin, 1983
Hypotypes 85474, 85475
Achab, 1987, Can. J. Earth Sci., vol. 24, no. 6, p. 1224, pl. 8, fig. 1, 2.
Trenton Group, Middle Ordovician, depths 2150 and 2300 feet, Bald Mountain Saint-Roch no. 1 borehole, near Sorel, Québec.

Hercochitina filamentosa Achab
Holotype 49506; hypotypes 49504, 49505, 49507
Achab, A., 1977, Can. J. Earth Sci., vol. 14, no. 10, p. 2208, pl. 4, figs. 1-5.
Vauréal Formation, Upper Ordovician, depth 1800 feet, New Associated Consolidated Paper Anticosti No. 1 well, lat. 49°37′20″N, long. 63°26′20″W, Anticosti Island, Québec.

Hercochitina grandispina Achab
Holotype 49509; hypotypes 49508, 49510, 49511
Achab, A., 1977, Can. J. Earth Sci., vol. 14, no. 10, p. 2209, pl. 4, figs. 6, 7; pl. 5, figs. 1, 5.
Vauréal Formation, Upper Ordovician, depth 1950 feet, New Associated Consolidated Paper Anticosti No. 1 well, lat. 49°37′20″N, long. 63°26′20″W, Anticosti Island, Québec.

Hercochitina lindsayensis Melchin and Legault, 1985
Hypotype 85478
Achab, A., 1987, Can. J. Earth Sci., vol. 24, no. 6, p. 1222, pl. 8, fig. 8.
Trenton Group, Middle Ordovician, depth 2400 feet, Bald Mountain Saint-Roch no. 1 borehole, near Sorel, Québec.

Hercochitina lineola Martin
Holotype 55882; paratypes 55883-55888
Martin, F., 1983, Geol. Surv. Can., Bull 310, p. 18, pl. 1, figs. 2, 3; pl. 3, fig. 12; pl. 4, figs. 1, 27; pl. 5, figs. 7, 19.
Lotbinière Formation, Utica Group, Upper Ordovician, bay to west of Pointe-aux-Trembles-Ouest, lat. 46°40′N, long. 71°39′51″W, and bay to west of Pointe Delisle, lat. 46°41′57″N, long. 71°36′29″W (55887); Sherman Fall Formation, Ottawa Group, Middle Ordovician, Val-Tétreau, Hull, Québec.

Hercochitina lineola Martin, 1983
Hypotypes 85466-85469
Achab, A., 1987, Can. J. Earth Sci., vol. 24, no. 6, p. 1222, pl. 6, figs. 11-14.
Utica Formation, Upper Ordovician, depth 1850 feet, Bald Mountain Saint-Roch no. 1 borehole, near Sorel, Québec.

Hercochitina minuta Achab
Holotype 49499; hypotypes 49497, 49498, 49500
Achab, A., 1977, Can. J. Earth Sci., vol. 14, no. 10, p. 2209, pl. 3, figs. 4-8.
Vauréal Formation, Upper Ordovician, depths 2349 and 2375 feet, New Associated Consolidated Paper Anticosti No. 1 well, lat. 49°37′20″N, long. 63°26′20″W, Anticosti Island, Québec.

Hercochitina aff. *H. minuta* Achab
Hypotype 61546
Martin, F., 1980, Can. J. Earth Sci., vol. 17, no. 1, p. 110, pl. 2, figs. 11, 12.
White Head Formation, Upper Ordovician, Grande Coupe, Gaspé, Québec.

Hercochitina modesta Achab
Holotype 53614; hypotypes 53615, 53616
Achab, A., 1978, Rev. Palaeobot. Palynol, vol. 25, nos. 3/4, p. 306, pl. III, figs. 1, 6, 7.
Vauréal Formation, Upper Ordovician, depths 3875 and 3850 feet, New Associated Consolidated Paper Anticosti No. 1 well, lat. 49°37′20″N, long. 63°26′20″W, Anticosti Island, Québec.

Hercochitina normalis Achab
Holotype 49501; hypotypes 49502, 49503
Achab, A., 1977, Can. J. Earth Sci., vol. 14, no. 10, p. 2208, pl. 3, figs. 9-12.
Vauréal Formation, Upper Ordovician, depths 1850, 1600 and 1300 feet, New Associated Consolidated Paper Anticosti No. 1 well, lat. 49°37′20″N, long. 63°26′20″W, Anticosti Island, Québec.

Hercochitina pinguis Melchin and Legault, 1985
Hypotypes 85476, 85477
Achab, A., 1987, Can. J. Earth Sci., vol. 24, no. 6, p. 1222, pl. 8, figs. 5, 6.
Trenton Group, Middle Ordovician, depth 2300 feet, Bald Mountain Saint-Roch no. 1 borehole, near Sorel, Quebec.

Hercochitina aff. *H. turnbulli* Jenkins
Hypotypes 30639, 30640
Legault, J.A., 1973, Geol. Surv. Can., Bull. 221, p. 45, pl. 9, figs. 13, 14.
Arkona Member, Hamilton Formation, Middle Devonian, GSC Arkona No. 1 well, Rock Glen Park, lot 4, con. SB, Bosanquet tp. Lambton co., Ontario.

Hercochitina sp.
Fig. spec. 49496
Achab, A., 1977, Can. J. Earth Sci., vol. 14, no. 10, p. 2202, pl. 3, fig. 4.
Vauréal Formation, Upper Ordovician, New Associated Consolidated Paper Anticosti No. 1 well, lat. 49°37′20″N, long. 63°26′20″W, Anticosti Island, Québec.

Hoegisphaera cf. *H. glabra* Staplin
Hypotypes 30626-30630, 32590, 32591
Legault, J.A., 1973, Can. J. Earth Sci., vol. 10, no. 5, figs. 2-8.
Legault, J.A., 1973, Geol. Surv. Can., Bull. 221, p. 40, pl. 8, figs. 4-6, 8, 10.

Rockport Quarry and Bell (30627) members, Hamilton Formation, Middle Devonian, GSC No. 1 Arkona No. 1 well, Rock Glen Park, lot 4, con. SB, Bosanquet tp., Lambton co., Ontario.

Hoegisphaera scabiosa (Wilson and Hedlund, 1964)
Hypotypes 41201, 41209, 41212, 41217, 41222
Martin, F., 1975, Can. J. Earth Sci., vol. 12, no. 6, p. 1016, pl. 1, fig. 4; pl. 2, figs. 7, 12, 18, 19; pl. 3, fig. 5.
Ordovician, Utica Formation, Montmorency Falls, Québec; Eastview Formation, behind Steinberg's Store, Montreal Road southeast St. Laurent Boulevard, Ottawa, Ontario (41222).

Hoegisphaera scabiosa (Wilson et Hedlund, 1964)
Hypotype 55890
Martin, F., 1983, Geol. Surv. Can., Bull. 310, p. 18, pl. 4, fig. 26.
Neuville Formation, Middle Ordovician, bay to east of Pointe Delisle, Québec.

?*Hoegisphaera scabiosa* (Wilson et Hedlund, 1964)
Hypotype 55889
Martin, F., 1983, Geol. Surv. Can., Bull 310, p. 18, pl. 1, fig. 4.
Lotbinière Formation, Utica Group, Upper Ordovician, bay to west of Point Delisle, lat. 46°41′57″N, long. 71°36′29″W, Québec.

?*Illichitina* sp.
Fig. spec. 30641
Legault, J.A., 1973, Geol. Surv. Can., Bull. 221, p. 46, pl. 9, fig. 9.
Bell Member, Hamilton Formation, Middle Devonian, GSC Ipperwash No. 1 well, lot 9, con A., Bosanquet tp., Lambton co., Ontario.

Kalochitina hirsuta (Laufeld, 1967) n. comb.
Hypotypes 55891-55894
Martin, F., 1983, Geol. Surv. Can., Bull 310, p. 18, pl. 4, fig. 19; pl. 5, figs. 11, 16, 28.
Upper Ordovician, Cobourg Formation, Ottawa Group, Geological Survey of Canada borehole COB-1-1, Ottawa, Ontario; Lotbinière Formation, Utica Group, bay to west of Pointe-aux-Trembles-Ouest, lat. 46°40′N, long. 71°39′51″W (55893), and road extending from bay west of Pointe Delisle to Highway 138 (55894), Québec; Neuville Formation, Trenton Group, Middle Ordovician, about 170 m east of pier at Neuville, Québec (55892).

Kalochitina multispinata Jansonius 1964
Hypotypes 41203, 41205, 41210, 41214, 41216, 41217
Martin, F., 1975, Can. J. Earth Sci., vol. 12, no. 6, p. 1013, pl. 1, figs. 6, 10, 13; pl. 2, fig. 14; pl. 3, figs. 2, 14, 16, 17.
Utica Formation, Ordovician, Montmorency Falls, Québec.

Kalochitina multispinata Jansonius, 1964
Hypotypes 56677-56679
Martin, F., 1978, Geol. Surv. Can., Paper 78-1B, pl. 12.1, figs. 13, 19, 21.
Winterhouse Formation, Long Point Group, Upper Ordovician, quarry immediately southeast of fishing huts mear west end of Clam Bank Cove, 750 m northwest of Lourdes, western Newfoundland.

Kalochitina multispinata Jansonius, 1964
Hypotypes 53617-53620
Achab, A., 1978, Rev. Palaeobot. Palynol., vol. 25, nos. 3/4, p. 308, pl. III, figs. 2-5.
Vauréal Formation, Upper Ordovician, depths 3850, 3875 and 3200 feet, New Associated Consolidated Paper Anticosti No. 1 well, lat. 49°37′20″N, long. 63°26′20″W, Anticosti Island, Québec.

Kalochitina multispinata Jansonius, 1964
Hypotypes 55895-55899
Martin, F., 1983, Geol. Surv. Can., Bull 310, p. 19, pl. 1, fig. 10; pl. 4, figs. 10; pl. 5, figs. 8, 9, 31.
Upper Ordovician, Lotbinière Formation, Utica Group, bay to west of Pointe Delisle, lat. 46°41′57″N, long. 71°36′29″W, and Neuville Formation, abandoned quarry at Saint-Joachim (55896), bay to east of Pointe Delisle (55897), Château-Richer (55898); Neuville Formation, Middle Ordovician, Charlesbourg Ltée. quarry, Charlesbourg-Est (55899), Québec.

Kalochitina multispinata Jansonius, 1964
Hypotypes 85500-85502
Achab, A., 1987, Can. J. Earth Sci., vol. 24, no. 6, p. 1224, pl. 10, figs. 11, 13, 14.
Upper Ordovician, Lorraine Group, depth 750 feet; Utica Group, depths 1500 (85501) and 1600 (85502) feet, Bald Mountain Saint-Roch no. 1 borehole, near Sorel, Québec.

Kalochitina? sp.
Fig. specs. 30642, 30643
Legault, J.A., 1973, Geol. Surv. Can., Bull. 221, p. 46, pl. 10, figs. 1, 10.
Arkona Member, Hamilton Formation, Middle Devonian, GSC Arkona No. 1 well, Rock Glen Park, lot 4, con. SB, Bosanquet tp., Lambton co., Ontario.

Lagenochitina cf. *L. amottensis* Grignani and Mantovani
Hypotype 30645
Legault, J.A., 1973, Geol. Surv. Can., Bull. 221, p. 47, pl. 10, fig. 3.
Bell Member, Hamilton Formation, Middle Devonian, GSC Ipperwash No. 1 well, lot 9, con. A, Bosanquet tp., Lambton co., Ontario.

Lagenochitina cf. *L. baltica* Eisenack, 1931
Hypotype 56683
Martin, F., 1978, Geol. Surv. Can., Paper 78-1B, pl. 12.1, fig. 22.
Table Head Formation, Middle Ordovician, south side Table Point, western Newfoundland.

Lagenochitina cf. *L. brevicollis* Taugourdeau and de Jekhowsky
Hypotypes 30646, 30647
Legault, J.A., 1973, Geol. Surv. Can., Bull. 221, p. 48, pl. 10, figs. 6, 7.

Widder Member, Hamilton Formation, Middle Devonian, GSC Arkona No. 1 well, Rock Glen Park, lot 4, con. SB, Bosanquet tp., Lambton co., Ontario.

Lagenochitina cf. *L. combazi* Finger, 1982
Hypotypes 77863-77865, 77833, 77834
Achab, A., 1986, Can. J. Earth Sci., vol. 23, no. 5, p. 691, pl. 2, figs. 7-9, pl. 4, figs. 9, 10.
Lower Ordovician, Cow Head Group, Cow Head, Newfoundland; Levis Formation, Côte Bégin, Lévis, Québec (77833, 77834).

Lagenochitina esthonica Eisenack, 1955
Hypotypes 77872-77874, 77825-77827
Achab, A., 1986, Can. J. Earth Sci., vol. 23, no. 5, p. 693, pl. 2, figs. 16-18; pl. 4, figs. 1-3.
Lower Ordoviciasn, Cow Head Group, Cow Head, Newfoundland; Levis Formation, Côte Bégin, Lévis, Québec (77825-77827).

Lagenochitina cf. *L. magnifica* Umnova, 1969
Hypotypes 77829, 77830
Achab, A., 1986, Can. J. Earth Sci., vol. 23, no. 5, p. 693, pl. 4, figs. 5, 6.
Levis Formation, Lower Ordovician, Côte Bégin, Lévis, Québec.

Lagenochitina sp.
Fig. specs. 30648, 30649
Legault, J.A., 1973, Geol. Surv. Can., Bull. 221, p. 48, pl. 10, figs. 4, 5.
Arkona and Widder members, Hamilton Formation, Middle Devonian, GSC Arkona No. 1 well, Rock Glen Park, lot 4, con. SB, Bosanquet tp., Lambton co., Ontario.

Lagenochitina sp.
Fig. specs. 63991-63994
Achab, A., 1982, Can. J. Earth Sci., vol. 19, no. 6, p. 1298, pl. 2, figs. 1-3, 8.
Levis Fromation, Lower Ordovician, Côte Fréchette (Davidson Street), Lévis, Québec.

Lagenochitina? sp.
Fig. spec. 66170
Fortey, R.A, Landing, E. and Skevington, D., 1982, National Mus. Wales, Geol. Ser. No. 3, text-fig. 7B.
Cow Head Group, Early Ordovician, Broom Point North, western Newfoundland.

?*Lagenochinita* sp. A
Fig. specs. 77227-77230
Achab, A, 1984, Review of Palaebotany and Palynology, vol. 43, no. 1-3, p. 136, pl. VI, figs. 4-7.
Middle Ordovician, depths 4492 and 4450 feet, New Associated Consolidated Paper Anticosti no. 1 borehole, Anticosti Island, Québec.

Parachitina curvata Eisenack
Hypotypes 49478, 49479
Achab, A., 1977, Can. J. Earth Sci., vol. 14, no. 10, p. 2210, pl. 1, figs. 10, 11.
Vauréal Formation, Upper Ordovician, depth 1150 feet, New Associated Consolidated Paper Anticosti No. 1 well, lat. 49°37′20″N, long. 63°26′20″W, Anticosti Island, Québec.

Poteriochitina briarca Legault
Holotype 30633; paratypes 30631, 30632, 30634
Legault, J.A., 1973, Geol. Surv. Can., Bull. 221, p. 42, pl. 9, figs. 1-8.
Hamilton Formation, Middle Devonian, Bell Member, GSC Arkona No. 1 well, Rock Glen Park, lot 4, con. SB, Bosanquet tp., Lambton co. (30633, 30634), and Arkona and Bell members, GSC Ipperwash No. 1 well, lot 9, con. A, Bosanquet tp., Lambton co. Ontario.

Pterochitina hymenelytrum Jenkins, 1969
Hypotypes 55900-55902
Martin, F., 1983, Geol. Surv. Can., Bull 310, p. 19, pl. 3, figs. 6, 7, 9.
Lotbinière Formation, Utica Group, Upper Ordovician, road extending from bay west of Pointe Delisle to Highway 138, Québec.

Pterochitina hymenelytrum Jenkins, 1969
Hypotypes 85496-85498
Achab, A., 1987, Can. J. Earth Sci., vol. 24, no. 6, p. 1228, pl. 10, figs. 6-8.
Lorraine Group, Upper Ordovician, depths 800 and 750 (85498) feet, Bald Mountain Saint-Roch no. 1 borehole, near Sorel, Quebec.

Rhabdochitina discriminata Martin
Holotype 55903; paratypes 55904-55907, 559222, 55923
Martin F., 1983, Geol. Surv. Can, Bull. 310, p. 19, pl. 1, figs. 1, 12, 13; pl. 3, fig. 4; pl. 4, figs. 2, 4, 30, 31.
Chazy Group, Middle Ordvician, Laval Formation, Bédard quarry Caughnawaga, Laprairie County, Québec; Rockcliffe Formation, road-cut on Highway 17, 4.5 km west of Cumberland, lat. 45°31′N, long. 75°27′W, Russell County, Ontario (55904, 55906, 55922, 55923).

Rhabdochitina magna Eisenack, 1931
Hypotypes 53621-53623
Achab, A., 1978, Rev. Palaebot. Palynol., vol. 25, nos. 3/4, p. 310, pl. III, figs. 8-10.
Vauréal Formation, Upper Ordovician, depths 3500 and 3700 feet, New Associated Consolidated Paper Anticosti No. 1 well, lat. 49°37′20″N, long. 63°26′20″W, Anticosti Island, Québec.

Rhabdochitina tubularis Umnova, 1976
Hypotypes 77867-77869
Achab, A., 1986, Can. J. Earth Sci., vol. 23, no. 5, p. 694, pl. 2, figs. 11-13.
Cow Head Group, Lower Ordovician, Cow Head, Newfoundland.

Rhabdochitina turgida Jenkins, 1967
Hypotype 56674
Achab, A., 1978, Geol. Surv. Can., Paper 78-1B, pl. 12.1, fig. 10.
Table Head Formation, Middle Ordovician, excavation west of road No. 59 leading to Fox Island River 3 km north-northeast of Port au Port village, western Newfoundland.

Rhabdochitina turgida Jenkins, 1967
Hypotype 55908
Martin, F., 1983, Geol. Surv. Can., Bull. 310, p. 19, pl. 5, fig. 36.
Neuville Formation, Trenton Group, Middle Ordovician, about 710 m east of pier at Neuville, Québec.

Rhabdochitina? turgida Jenkins, 1967
Hypotypes 69657-69661
Achab, A., 1983, Can. J. Earth Sci., vol. 20, no. 6, p. 924, pl. 1, figs. 9-13.
Table Head Formation, Middle Ordovician, quarry, Highway 49, West Bay Centre, western Newfoundland.

Rhabdochitina usitata Jenkins, 1967
Hypotypes 56661-56666
Martin, F., 1978, Geol. Surv. Can., Paper 78-1B, p. 80, pl. 12.1, figs. 1, 2, 4, 9, 15, 18.
Cow Head Group, Lower Ordovician, east side of The Ledge (56661) and west side of The Head, Cow Head Peninsula, western Newfoundland.

Rhabdochitina usitata Jenkins 1967
Hypotypes 77231, 77232
Achab, A, 1984, Review of Palaebotany and Palynology, vol. 43, no. 1-3, p. 138, pl. IV, figs. 1, 2.
Middle Ordovician, depth 4474 feet, New Associated Consolidated Paper Anticosti no. 1 borehole, Anticosti Island, Québec.

Rhabdochitina sp. 1
Fig. spec. 96792
Asselin, E., Achab, A. and Bourque, P.A., 1989, Can. J. Earth Sci., vol. 26, no. 12, p. 2438, pl. 1, fig. 1.
Clemville Formation, Lower Silurian, Little Port-Daniel River, Gaspé, Québec.

Rhabdochitina spp. 1-3
Fig. specs. 30636-30638
Legault, J.A., 1973, Geol. Surv. Can., Bull. 221, p. 43, 44, pl. 9, figs. 10-12.
Ipperwash and Rockport Quarry members, Hamilton Formation, Middle Devonian, Argor Consolidated 65-1 well, lot 28, con. II, Moore tp., Lambton co. (30636), and GSC Ipperwash No. 1 well, lot 9, con. A, Bosanquet tp., Lambton co., Ontario.

Sphaerochitina compactilis Jenkins 1969
Hypotype 41223
Martin, F., 1975, Can. J. Earth Sci., vol. 12, no. 6, p. 1013, pl. 3, fig. 8.
Utica Formation, Ordovician, Montmorency Falls, Québec.

Sphaerochitina compactilis Jenkins
Hypotype 61547
Martin, F., 1980, Can. J. Earth Sci., vol. 17, no. 1, pl. 1, fig. 13.
White Head Formation, Upper Ordovician, Grande Coupe, Gaspé, Québec.

Sphaerochitina compactilis Jenkins, 1969
Hypotypes 55905-55912
Martin, F., 1983, Geol. Surv. Can., Bull. 310, p. 20, pl. 1, fig. 5; pl. 4, fig. 25; pl. 5, figs. 17, 25.
Upper and Middle Ordovician, Neuville Formation, Trenton Group, west of pier at Neuville, Charlesbourg Ltée quarry, Charlesbourg-Est (55910), and bay west of Pointe Delisle, lat. 46°41′57″N, long. 71°36′29″W (55911); Lotbinière Formation, Utica Group, road extending from bay west of Pointe Delisle to Highway 138 (55912), Québec.

Sphaerochitina gracqui Martin
Holotype 55913; paratypes 55914-55920
Martin, F., 1983, Geol. Surv. Can., Bull. 310, p. 20, pl. 1, fig. 9; pl. 4, figs. 11, 17, 18, 23, 24; pl. 5, figs. 13, 35.
Upper Ordovician, Neuville Formation, Trenton Group, Charlesbourg Ltée quarry, Charlesbourg-Est, Elzéar Verreault Ltée quarry, Gifford (55916, 55917, 55919), and Château-Richer (55918), Québec; Billings Formation, Utica Group, Billings Creek east of (55914) and west of (55920) Billings Bridge, Bank Street, Ottawa, Ontario.

Sphaerochitina gracqui Martin, 1983
Hypotypes 85419, 85420
Achab, A., 1987, Can. J. Earth Sci., vol. 24, no. 6, p. 1228, pl. 1, figs. 13, 14.
Trenton Group, Middle Ordovician, depth 2450 feet, Bald Mountain Saint-Roch no. 1 borehole, near Sorel, Quebec.

Sphaerochitina aff. *S. mundana Taugourdeau,* 1961
Hypotype 55921
Martin, F., 1983, Geol. Surv. Can., Bull 310, p. 20, pl. 4, fig. 22.
Laval Formation, Chazy Group, Middle Ordovician, Bédard quarry, Caughnawaga, Laprairie County, Québec.

Sphaerochitina rockglenensis Legault
Hypotype 30610
Legault, J.A., 1973, Geol. Surv. Can., Bull. 221, p. 34, pl. 6, figs. 1, 2.
Arkona Member, Hamilton Formation, Middle Devonian, GSC Arkona No. 1 well, Rock Glen Park, lot 4, con. SB, Bosanquet tp., Lambton co., Ontario.

Spinachitina bulmani (Jansonius, 1964)
Hypotypes 85437-85443
Achab, A., 1987, Can. J. Earth Sci., vol. 24, no. 6, p. 1228, pl. 4, figs. 5-14.
Upper Ordovician, Lorraine Group, depths 1350, 1250, 950 feet, and Utica Group, depth 1850 feet, Bald Mountain Saint-Rock no. 1 borehole, near Sorel, Quebec.

Spinachitina sp. A
Fig. specs. 77245-77247
Achab, A, 1984, Review of Palaebotany and Palynology, vol. 43, no. 1-3, p. 138, pl. III, figs. 1-3.
Middle Ordovician, depth 4000 feet, New Associated Consolidated Paper Anticosti no. 1 borehole, Anticosti Island, Québec.

Tanuchitina anticostiensis (Achab)
Hypotypes 49520-49522
Achab, A., 1977, Can. J. Earth Sci., vol. 14, no. 10, p. 2200, pl. 6, figs. 1, 2, 7.
Vauréal Formation, Upper Ordovician, depths 1400 and 1500 feet, New Associated Consolidated Paper Anticosti No. 1 well, lat. 49°37′20″N, long. 63°26′20″W, Anticosti Island, Québec.

Tornacia sp.
Fig. spec. 30676
Legault, J.A., 1973, Geol. Surv. Can., Bull. 221, p. 57, pl. 13, fig. 14.
Dundee Formation, Middle Devonian, GSC Arkona No. 1 well, Rock Glen Park, lot 4, con. SB, Bosanquet tp., Lambton co., Ontario.

Triangulina cf. *T. alargada* Cramer
Hypotype 30677
Legault, J.A., 1973, Geol. Surv. Can., Bull. 221, p. 58, pl. 11, figs. 11, 12.
Arkona Member, Hamilton Formation, Middle Devonian, GSC Ipperwash No. 1 well, lot 9, con. A, Bosanquet tp., Lambton co., Ontario.

BACTERIA

Archaeoellipsoides obesus Zhang, 1985
Hypotype 98883
Hofmann, H.J. and Jackson, G.D, 1991, J. Paleontol., vol. 65, no. 3, p. 372, fig. 7.17.
Victor Bay Formation, Uluksan Group, Proterozoic, Eclipse Sound, lat. 72.33°N, long. 79.78°W, northwestern Baffin Island, District of Franklin.

Archaeotrichion? sp.
Fig. spec. 98861
Hofmann, H.J. and Jackson, G.D., 1991, J. Paleontol., vol. 65, no. 3, p. 367, fig. 5.5.
Society Cliffs Formation, Uluskan Group, Proterozoic, Eclipse Sound, lat. 72°24′N, long. 79.49°W, northwestern Baffin Island, District of Franklin.

Brachypleganon sp.
Fig. specs. 98904, 98905
Hofmann, H.J. and Jackson, G.D, 1991, J. Paleontol., vol. 65, no. 3, p. 368, figs. 8.16, 8.17.
Victor Bay Formation, Uluksan Group, Proterozoic, Eclipse Sound, lat. 72.33°N, long. 79.78°W, northwestern Baffin Island, District of Franklin.

Cephalophytarion majesticum Allison
Holotype 81877
Allison, C.W. and Awramik, S.M., 1989, Precambrian Res., vol. 43, no. 4, p. 273, figs. 8.7, 8.8.
Tindir Group, earliest Cambrian or latest Precambrian, headwaters of Tindir Creek between lat. 65°15′N and lat. 65°17′N, long. 141°56′W and long. 141°57′W, Yukon.

Eogloeocapsa bella Golovenok and Belova, 1984
Hypotype 98922
Hofmann, H.J. and Jackson, G.D, 1991, J. Paleontol., vol. 65, no. 3, p. 377, fig. 10.17.
Victor Bay Formation, Uluksan Group, Proterozoic, Eclipse Sound, lat. 72.33°N, long. 79.78°W, northwestern Baffin Island, District of Franklin.

Eogloecapsa bella Galovenok and Belova, 1984?
Hypotype 98923
Hofmann, H.J. and Jackson, G.D, 1991, J. Paleontol., vol. 65, no. 3, p. 377, fig. 10.18.
Victor Bay Formation, Uluksan Group, Proterozoic, Eclipse Sound, lat. 72.33°N, long. 79.78°W, northwestern Baffin Island, District of Franklin.

Eomicrocoleus sp.
Fig. specs. 98873, 98875, 98876, 98903
Hofmann, H.J. and Jackson, G.D, 1991, J. Paleontol., vol. 65, no. 3, p. 368, figs. 7.7, 7.9, 7.10, 8.15.
Victor Bay and Society Cliffs (98903) formations, Uluksan Group, Proterozoic, Eclipse Sound, lat. 72.33°N, long. 79.78°W and lat. 72.24°N, long. 79.49°W, northwestern Baffin Island, District of Franklin.

Eomycetopsis robusta Schopf, 1968 em. Knoll and Golubic, 1979
Hypotype 81881
Allison, C.W. and Awramik, S.M., 1989, Precambrian Res., vol. 43, no. 4, p. 274, fig. 8.9.
Tindir Group, earliest Cambrian or latest Precambrian, headwaters of Tindir Creek between lat. 65°15′N and lat. 65°17′N, long. 141°56′W and long. 141°57′W, Yukon.

Eomycetopsis robusta Schopf, 1968, emend. Knoll and Golubic, 1979
Hypotypes 98858, 98859, 98864, 98869, 98870
Hofmann, H.J. and Jackson, G.D, 1991, J. Paleontol., vol. 65, no. 3, p. 367, figs. 5.1-5.3, 5.8, 7.3, 7.4.
Uluksan Group, Proterozoic, Society Cliffs Formation, Eclipse Sound, lat. 72.28°N, long. 79.82°W and lat. 72°25′N, long. 79.97°W (98864) and Victor Bay Formation, lat. 72.33°N, long. 79.78°W (98869, 98870), northwestern Baffin Island, District of Franklin.

Eosynechococcus grandis Hofmann, 1976
Hypotypes 98884, 98885
Hofmann, H.J. and Jackson, G.D, 1991, J. Paleontol., vol. 65, no. 3, p. 372, figs. 7.18, 7.19.
Victor Bay Formation, Uluksan Group, Proterozoic, Eclipse Sound, lat. 72.33°N, long. 79.78°W, northwestern Baffin Island, District of Franklin.

Eosynechococcus medius Hofmann, 1976
Hypotypes 98880, 98881
Hofmann, H.J. and Jackson, G.D, 1991, J. Paleontol., vol. 65, no. 3, p. 371, figs. 7.14, 7.15.

Victor Bay Formation, Uluksan Group, Proterozoic, Eclipse Sound, lat. 72.33°N, long. 79.78°W, northwestern Baffin Island, District of Franklin.

Eosynechococcus sp. Knoll, 1982
Fig. spec. 81873
Allison, C.W. and Awramik, S.M., 1989, Precambrian Res., vol. 43, no. 4, p. 271, fig. 8.2.
Tindir Group, earliest Cambrian or latest Precambrian, headwaters of Tindir Creek between lat. 65°15′N and lat. 65°17′N, long. 141°56′W and long. 141°57′W, Yukon.

Filiconstrictus diminutus Schopf and Blacic, 1971
Hypotype 81878
Allison, C.W. and Awramik, S.M., 1989, Precambrian Res., vol. 43, no. 4, p. 274, fig. 8.4.
Tindir Group, earliest Cambrian or latest Precambrian, headwaters of Tindir Creek between lat. 65°15′N and lat. 65°17′N, long. 141°56′W and long. 141°57′W, Yukon.

Glenobotrydion aenigmatis Schopf, 1968 em. Nyberg and Schopf, 1984
Hypotype 81869
Allison, C.W. and Awramik, S.M., 1989, Precambrian Res., vol. 43, no. 4, p. 276, fig. 7.3.
Tindir Group, earliest Cambrian or latest Precambrian, headwaters of Tindir Creek between lat. 65°15′N and lat. 65°17′N, long. 141°56′W and long. 141°57′W, Yukon.

Gloeodiniopsis lamellosa Schopf, 1968 em. Knoll and Golubic, 1979
Hypotype 81872
Allison, C.W. and Awramik, S.M., 1989, Precambrian Res., vol. 43, no. 4, p. 269, fig. 8.1.
Tindir Group, earliest Cambrian or latest Precambrian, headwaters of Tindir Creek between lat. 65°15′N and lat. 65°17′N, long. 141°56′W and long. 141°57′W, Yukon.

Gloeodiniopsis magna Nyberg and Schopf, 1984
Hypotypes 98931-98937, 98941-98944
Hofmann, H.J. and Jackson, G.D, 1991, J. Paleontol., vol. 65, no. 3, p. 377, figs. 13.1-13.7, 13.11-13.14.
Victor Bay Formation, Uluksan Group, Proterozoic, Eclipse Sound, lat. 72.33°N, long. 79.78°W, northwestern Baffin Island, District of Franklin.

Gloeodiniopsis mikros Knoll, 1982
Hypotype 98882
Hofmann, H.J. and Jackson, G.D, 1991, J. Paleontol., vol. 65, no. 3, p. 378, fig. 7.16.
Victor Bay Formation, Uluksan Group, Proterozoic, Eclipse Sound, lat. 72.33°N, long. 79.78°W, northwestern Baffin Island, District of Franklin.

Gloeodiniopsis sp. 1
Fig. specs. 98938-98940
Hofmann, H.J. and Jackson, G.D, 1991, J. Paleontol., vol. 65, no. 3, p. 377, figs. 13.8-13.10.
Victor Bay Formation, Uluksan Group, Proterozoic, Eclipse Sound, lat. 72.33°N, long. 79.78°W, northwestern Baffin Island, District of Franklin.

Heliconema australiensis Schopf, 1968
Hypotype 81879
Allison, C.W. and Awramik, S.M., 1989, Precambrian Res., vol. 43, no. 4, p. 274, fig. 8.5.
Tindir Group, earliest Cambrian or latest Precambrian, headwaters of Tindir Creek between lat. 65°15′N and lat. 65°17′N, long. 141°56′W and long. 141°57′W, Yukon.

Heliconema bulbosa Allison
Holotype 81880
Allison, C.W. and Awramik, S.M., 1989, Precambrian Res., vol. 43, no. 4, p. 274, fig. 8.6.
Tindir Group, earliest Cambrian or latest Precambrian, headwaters of Tindir Creek between lat. 65°15′N and lat. 65°17′N, long. 141°56′W and long. 141°57′W, Yukon.

Melasmatosphaera sp.
Fig. spec. 98920 (partim)
Hofmann, H.J. and Jackson, G.D, 1991, J. Paleontol., vol. 65, no. 3, p. 375, figs. 10.15.
Victor Bay Formation, Uluksan Group, Proterozoic, Eclipse Sound, lat. 72.33°N, long. 79.78°W, northwestern Baffin Island, District of Franklin.

Microagglomeratus borealis Awramik
Holotype 81868
Allison, C.W. and Awramik, S.M., 1989, Precambrian Res., vol. 43, no. 4, p. 268, fig. 7.2.
Tindir Group, earliest Cambrian or latest Precambrian, headwaters of Tindir Creek between lat. 65°15′N and lat. 65°17′N, long. 141°56′W and long. 141°57′W, Yukon.

Myxococcoides cantabrigiensis Knoll, 1982
Hypotype 81875
Allison, C.W. and Awramik, S.M., 1989, Precambrian Res., vol. 43, no. 4, p. 272, fig. 8.3.
Tindir Group, earliest Cambrian or latest Precambrian, headwaters of Tindir Creek between lat. 65°15′N and lat. 65°17′N, long. 141°56′W and long. 141°57′W, Yukon.

Myxococcoides grandis Horodyski and Donaldson, 1980
Hypotypes 98894-98896, 98913, 98921, 98924, 98925, 98928
Hofmann, H.J. and Jackson, G.D, 1991, J. Paleontol., vol. 65, no. 3, p. 374, figs. 8.6-8.8, 10.10, 10.16, 12.1-12.5, 12.8.
Uluksan Group, Proterozoic, Society Cliffs Formation, Navy Board Inlet, lat. 73.48°N, long. 80.76°W and lat. 73.38°N, long. 80.39°W (98925), Bylot Island; Victor Bay Formation, lat. 72.33°N, long. 79.78°W (98913, 98921), northwestern Baffin Island, District of Franklin.

Myxococcoides grandis? Horodyski and Donaldson, 1980
Hypotypes 98926, 98927
Hofmann, H.J. and Jackson, G.D, 1991, J. Paleontol., vol. 65, no. 3, p. 374, figs. 12.6, 12.7.
Society Cliffs Formation, Uluksan Group, Proterozoic, Eclipse Sound, lat. 72.25°N, long. 79.97°W and lat. 72.24°N, long. 79.81°W, northwestern Baffin Island, District of Franklin.

Myxococcoides minor Schopf, 1968
Hypotypes 98889-98893
Hofmann, H.J. and Jackson, G.D, 1991, J. Paleontol., vol. 65, no. 3, p. 374, figs. 8.1-8.5.
Society Cliffs Formation, Uluksan Group, Proterozoic, Eclipse Sound, lat. 72.28°N, long. 79.82°W, northwestern Baffin Island, District of Franklin.

Myxococcoides or Melasmatosphaera
Fig. spec. 98919
Hofmann, H.J. and Jackson, G.D, 1991, J. Paleontol., vol. 65, no. 3, p. 374, fig. 10.14.
Victor Bay Formation, Uluksan Group, Proterozoic, Eclipse Sound, lat. 72.33°N, long. 79.78°W, northwestern Baffin Island, District of Franklin.

Palaeoanacystis magna Awramik
Holotype 81871
Allison, C.W. and Awramik, S.M., 1989, Precambrian Res., vol. 43, no. 4, p. 269, fig. 7.6.
Tindir Group, earliest Cambrian or latest Precambrian, headwaters of Tindir Creek between lat. 65°15′N and lat. 65°17′N, long. 141°56′W and long. 141°57′W, Yukon.

Palaeoanacystis sp.
Fig. specs. 98900, 98901, 98911
Hofmann, H.J. and Jackson, G.D, 1991, J. Paleontol., vol. 65, no. 3, p. 375, figs. 8.12, 8.13, 10.6.
Society Cliffs and Victor Bay (98911) formations, Uluksan Group, Proterozoic, Eclipse Sound, lat. 72.13°N, long. 79.40°W and lat. 72.33°N, long. 79.78°W, northwestern Baffin Island, District of Franklin.

Phacelogeminus lineatus Awramik
Holotype 81870
Allison, C.W. and Awramik, S.M., 1989, Precambrian Res., vol. 43, no. 4, p. 273, fig. 7.4.
Tindir Group, earliest Cambrian or latest Precambrian, headwaters of Tindir Creek between lat. 65°15′N and lat. 65°17′N, long. 141°56′W and long. 141°57′W, Yukon.

Phanerosphaerops capitaneus Schopf and Blacic, 1971
Hypotype 98929
Hofmann, H.J. and Jackson, G.D, 1991, J. Paleontol., vol. 65, no. 3, p. 375, fig. 12.9.
Society Cliffs Formation, Uluksan Group, Proterozoic, Navy Board Inlet, lat. 73.38°N, long. 80.39°W, Bylot Island, District of Franklin.

Polybessurus bipartitus Fairchild, 1975, ex Green et al., 1987
Hypotype 98874
Hofmann, H.J. and Jackson, G.D, 1991, J. Paleontol., vol. 65, no. 3, p. 378, fig. 7.8.
Victor Bay Formation, Uluksan Group, Proterozoic, Eclipse Sound, lat. 72.33°N, long. 79.78°W, northwestern Baffin Island, District of Franklin.

Rhicnonema antiquum Hofmann, 1976
Hypotypes 98863 (partim), 98877 (partim)
Hofmann, H.J. and Jackson, G.D, 1991, J. Paleontol., vol. 65, no. 3, p. 368, figs. 5.7, 7.11.
Society Cliffs and Victor Bay formations, Uluksan Group, Proterozoic, Eclipse Sound, lat. 72.75°N, long. 79.97°W and lat. 72.33°N, long. 79.78°W, northwestern Baffin Island, District of Franklin.

Siphonophycus inornatum Y. Zhang, 1981
Hypotypes 98860, 98862, 98871
Hofmann, H.J. and Jackson, G.D, 1991, J. Paleontol., vol. 65, no. 3, p. 368, figs. 5.4, 5.6, 7.5.
Society Cliffs and Victor Bay (98871) formations, Uluksan Group, Proterozoic, Eclipse Sound, lat. 72.24°N, long. 79.49°W and lat. 72.33°N, long. 79.78°W, northwestern Baffin Island, District of Franklin.

Siphonophycus kestron Schopf, 1968
Hypotypes 98865, 98866
Hofmann, H.J. and Jackson, G.D, 1991, J. Paleontol., vol. 65, no. 3, p. 368, figs. 5.9, 5.10.
Society Cliffs Formation, Uluksan Group, Proterozoic, Navy Board Inlet, lat. 73.38°N, long. 80.39°W, Bylot Island, District of Franklin.

Sphaerophycus medium Horodyski and Donaldson, 1980
Hypotypes 98863 (partim), 98909, 98910, 98920 (partim)
Hofmann, H.J. and Jackson, G.D, 1991, J. Paleontol., vol. 65, no. 3, p. 374, figs. 5.7, 10.4, 10.5, 10.15.
Society Cliffs (98863) and Victor Bay formations, Uluksan Group, Proterozoic, Eclipse Sound, lat. 72.25°N, long. 79.97°W and lat. 72.33°N, long. 79.78°W, northwestern Baffin Island, District of Franklin.

Sphaerophycus parvum Schopf, 1968
Hypotype 81874
Allison, C.W. and Awramik, S.M., 1989, Precambrian Res., vol. 43, no. 4, p. 271, fig. 7.5.
Tindir Group, earliest Cambrian or latest Precambrian, headwaters of Tindir Creek between lat. 65°15′N and lat. 65°17′N, long. 141°56′W and long. 141°57′W, Yukon.

Sphaerophycus parvum Schopf, 1968
Hypotypes 98879 (partim), 98897, 98898, 98906-98908
Hofmann, H.J. and Jackson, G.D, 1991, J. Paleontol., vol. 65, no. 3, p. 374, figs. 7.13, 8.9, 8.10, 10.1-10.3.
Society Cliffs and Victor Bay (98879, 98906-98908) formations, Uluksan Group, Proterozoic, Eclipse Sound, lat. 72.28°N, long. 79.82°W (98897), lat. 72.13°N, long. 79.4°W (98898) and lat. 72.33°N, long. 79.78°W, northwestern Baffin Island, District of Franklin.

Sphaerophycus? sp.
Fig. spec. 98899
Hofmann, H.J. and Jackson, G.D, 1991, J. Paleontol., vol. 65, no. 3, p. 374, fig. 8.11.
Society Cliffs Formation, Uluksan Group, Proterozoic, Eclipse Sound, lat. 72.13°N, long. 79.40°W, northwestern Baffin Island, District of Franklin.

Talakania? sp.
Fig. spec. 98872
Hofmann, H.J. and Jackson, G.D, 1991, J. Paleontol., vol. 65, no. 3, p. 371, fig. 7.6.

Victor Bay Formation, Uluksan Group, Proterozoic, Eclipse Sound, lat. 72.33°N, long. 79.78°W, northwestern Baffin Island, District of Franklin.

Tenuofilum septatum Schopf, 1968
Hypotypes 98867, 98868 (partim)
Hofmann, H.J. and Jackson, G.D, 1991, J. Paleontol., vol. 65, no. 3, p. 367, figs. 7.1, 7.2.
Victor Bay Formation, Uluksan Group, Proterozoic, Eclipse Sound, lat. 72.33°N, long. 79.78°W, northwestern Baffin Island, District of Franklin.

Tetraphycus hebeiensis Liu, 1982
Hypotypes 98912, 98914-98918
Hofmann, H.J. and Jackson, G.D, 1991, J. Paleontol., vol. 65, no. 3, p. 375, figs. 10.7, 10.9-10.13.
Victor Bay Formation, Uluksan Group, Proterozoic, Eclipse Sound, lat. 72.33°N, long. 79.78°W, northwestern Baffin Island, District of Franklin.

Uluksanella baffinensis Hofmann and Jackson
Holotype 98888; paratypes 98886, 98887
Hofmann, H.J. and Jackson, G.D, 1991, J. Paleontol., vol. 65, no. 3, p. 371, fig. 7.20-7.22.
Victor Bay Formation, Uluksan Group, Proterozoic, Eclipse Sound, lat. 72.33°N, long. 79.78°W, northwestern Baffin Island, District of Franklin.

Yukonosphaeridiom interior Awramik
Holotype 81867
Allison, C.W. and Awramik, S.M., 1989, Precambrian Res., vol. 43, no. 4, p. 266, fig. 7.1.
Tindir Group, earliest Cambrian or latest Precambrian, headwaters of Tindir Creek between lat. 65°15′N and lat. 65°17′N, long. 141°56′W and long. 141°57′W, Yukon.

DIATOMS

Achnanthes flexella (Kützing) Brun
Hypotypes 61769-61772
Lichti-Federovich, S., 1979, Geol. Surv. Can., Paper 79-1B, pl. 11.1, figs. 1-5.
Recent, shore of largest low-level labe in northwestern part of Cape Herschel Peninsula, lat. 78°37.6′N, long. 74°43′W (61769) and mud-water interface in a pond on raised beaches north of head of Herschel Bay, lat. 78°36.3′N, long. 74°45′W, Ellesmere Island, District of Franklin.

Amphora veneta Var. *capitata* Haworth
Hypotypes 61773-61775
Lichti-Federovich, S., 1979, Geol. Surv. Can, Paper 79-1B, pl. 11.2, figs. 1-5.
Recent, snowlake along north side of river valley, ca. 6.5 km north-northwest of Cape Storm, lat. 76°23.9′N, long. 87°32′W (61773), and mud-water interface, pond on raised beaches north of head of Herschel Bay, lat. 78°36.3′N, long. 74°45′W, Ellesmere island, District of Franklin.

Ceratoneis arcus (Ehrenberg) Kützing
Hypotypes 61776-61780
Lichti-Federovich, S., 1979, Geol. Surv. Can, Paper 79-1B, pl. 11.3, figs. 1-5.
Recent, stream water on north side of the northernmost tidal glacier on west side of Bentham Fiord, Makinson Inlet, lat. 77°10.5′N, long. 80°10′W, Ellesmere Island, District of Franklin.

'Coscinodiscus' sp. 1
Fig. spec. 64264
Gradstein, F.M. and Agterberg, F.P., 1982, Quantitative Stratigraphic Correlation, pl. 7, fig. 21.
Pliocene-Pleistocene, depth 2860 feet, Eastcan et al. Bjarni H-81 well, Labrador Shelf, East Coast Canada.

Coscinodiscus? sp. 2400
Fig. specs. 89516, 89518
McNeil, D.H., 1990, Bull. Can. Petrol. Geol., vol. 38, no. 1, p. 21, pl. 1, fig. 1a, b, 3.
Richards Formation, Middle to Late Eocene, depths 2875-2885 and 4290-4305 m, Dome Gulf Tarsiult A-25 well, lat. 69°54′10″N, long. 136°20′20″W, Beaufort Sea.

Cyclotella antiqua Smith
Hypotypes 61781-61785
Lichti-Federovich, S., 1979, Geol. Surv. Can, Paper 79-1B, pl. 11.4, figs. 1-5.
Recent, pond adjacent to RCMP post at Alexandra Fiord, lat. 78°52.9′N, long. 75°47′W (61781), and mud-water interface, pond on raised beaches north of head of Herschel Bay, lat. 78°36.3′N, long. 74°45′W, Ellesmere Island, District of Franklin.

Diatom Type A, B, C, D, E
Fig. specs. 84430-84438
Given, M.M. and Wall, J.H., 1971, Bull. Can. Petrol. Geol., vol. 19, no. 1, p. 540-542, pl. 3, figs. 16-27.
Bearpaw Formation, Upper Cretaceous, RCA Castor well, depths 477-480.5, 533-536.5 (84434-84436) and 484-487.5 (84437) feet, l.s.d. 13, sec. 34, tp. 37, rge. 13, W.4th mer., Alberta.

Eunotia praerupta Ehr.
Hypotype 80489
Burden, E. and Holloway, D., 1985, Can. J. Earth Sci., vol. 22, no. 10, pl. 1, fig. 8.
Quatermary, Scott Inlet area, Baffin Island, District of Franklin.

Eunotia suecica A.Cl.
Hypotype 80490
Burden, E. and Holloway, D., 1985, Can. J. Earth Sci., vol. 22, no. 10, pl. 1, fig. 9.
Quatermary, Scott Inlet area, Baffin Island, District of Franklin.

Genus indeterminate
Fig. spec. 89519
McNeil, D.H., 1990, Bull. Can. Petrol Geol., vol. 38, no. 1, p. 21, pl. 1, figs. 4a, b.
Richards Formation, Middle to Late Eocene, depth 2725-2735m, Dome Gulf Tarsiut A-25 well, lat. 69°54′10″N, long. 136°20′20″W, Beaufort Sea.

Morphotype A-4, A-5, A-2, A-3
Fig. specs. 84442, 84443, 84446, 84447
Wall, J.H., 1975, Geol. Assoc. Can., Sp. Paper 13, p. 396, pl. 1, figs. 4, 5, 8, 9.
Lower Cretaceous, Clearwater Formation, Birch River, N.W. portion tp. 103, rge. 24, W.4th mer.; Joli Fou Formation, Athabasca River, sec. 1, tp. 86, rge. 18, W.4th mer (84443); Smoky Group, Upper Cretaceous, Alice Creek, N.W. 1/4 sec. 6, tp. 104, rge. 17, W.4th mer., Alberta.

Morphotype B
Fig. specs. 84444, 84445
Wall, J.H., 1975, Geol Assoc. Can., Sp. Paper 13, p. 396, pl. 1, figs. 6, 7.
Smoky Group, Upper Cretaceous, Alice Creek, N.W. 1/4 sec. 6, tp. 104, rge. 7, W.4th mer., Alberta.

Morphotype C
Fig. spec. 84456
Wall, J.H., 1975, Geol. Assoc. Can., Sp. Paper 13, p. 396, pl. 1, fig. 18.
Bearfaw Formation, Upper Cretaceous, RCA Castor test hole, depth 355-362 feet, l.s.d. 13, sec. 34, tp. 37, rge. 13, W.4th mer., Alberta.

Morphotype D-1, D-2
Fig. specs. 84439, 84440
Wall, J.H., 1975, Geol. Assoc. Can., Sp. Paper 13, p. 396, pl. 1, figs. 1, 2.
Clearwater Formation, Lower Cretaceous, Birch River, N.W. portion tp. 103, rge. 24, W.4th mer., Alberta.

Morphotype E-2, E-1
Fig. specs. 84441, 84454
Wall, J.H., 1975, Geol. Assoc. Can., Sp. Paper 13, p. 396, pl. 1, figs. 3, 16.
Clearwater Formation, Lower Cretaceous, Birch River, N.W. portion tp. 103, rge. 24, W.4th mer.; Bearpaw Formation, Upper Cretaceous, RCA Thelma test hole, depth 921-925 feet, l.s.d. 15, sec. 31, tp. 6, rge. 2, W.4th mer., Alberta.

Morphotype F
Fig. spec. 84448
Wall, J.H., 1975, Geol. Assoc. Can., Sp. Paper 13, p. 396, pl. 1, fig. 10.
Smoky Group, Upper Cretaceous, Alice Creek, N.W. 1/4 sec. 6, tp. 104, rge. 17, W.4th mer., Alberta.

Morphotype G
Fig. spec. 84449
Wall, J.H., 1975, Geol. Assoc. Can., Sp. Paper 13, p. 396, pl. 1, fig. 11.
Smoky Group, Upper Cretaceous, Alice Creek, N.W. 1/4 sec. 6, tp. 104, rge. 17, W.4th mer., Alberta.

Morphotype H-2, H-1, H-3
Fig. specs. 84450-84452
Wall, J.H., 1975, Geol. Assoc. Can., Sp. Paper 13, p. 396, pl. 1, figs. 12-14.
Bearpaw Formation, Upper Cretaceous, St. Mary River near Lethbridge, NE. 1/4 sec. 33, tp. 6, rge. 22, W.4th mer., Alberta.

Morphotype J
Fig. spec. 84453
Wall, J.H., 1975, Geol. Assoc. Can., Sp. Paper 13, p. 396, pl. 1, fig. 15.
Bearpaw Formation, Upper Cretaceous, St. Mary River near Lethbridge, NE. 1/4 sec. 33, tp. 6, rge. 22, W.4th mer., Alberta.

Morphotype K
Fig. spec. 84455
Wall, J.H., 1975, Geol. Assoc. Can., Sp. Paper 13, p. 396, pl. 1, fig. 17.
Bearpaw Formation, Upper Cretaceous, St. Mary River near Lethbridge, NE. 1/4 sec. 32, tp. 6, rge. 22, W.4th mer., Alberta.

Navicula tuscula Ehrenberg
Hypotypes 61786-61789
Lichti-Federovich, S., 1979, Geol. Surv. Can, Paper 79-1B, pl. 11.5, figs. 1-5.
Recent, mud-water interface, pond on raised beaches north of head of Herschel Bay, lat. 78°36.3′N, long. 74°45′W, Ellesmere Island, District of Franklin.

Navicula sp.
Fig. spec. 80492
Burden, E. and Holloway, D., 1985, Can. J. Earth Sci., vol. 22, no. 10, pl. 1, fig. 11.
Quatermary, Scott Inlet area, Baffin Island, District of Franklin.

Pinnularia sp.
Fig. spec. 80491
Burden, E. and Holloway, D., 1985, Can. J. Earth Sci., vol. 22, no. 10, pl. 1, fig. 10.
Quatermary, Scott Inlet area, Baffin Island, District of Franklin.

Stellarima? spp. 2405, 2414
Fig. specs. 89517, 89520
McNeil, D.H., 1990, Bull. Can. Petrol Geol., vol. 38, no. 1, p. 21, pl. 1, figs. 2a-c, 5a-c.
Richards Formation, Middle to Late Eocene, depths 2930-2945 m and 2875-2885 m, Dome Gulf Tassiut A-25 well, lat. 69°54′10″N, long. 136°20′20″W, Beaufort Sea.

Triceratium? spp. 2460, 2465
Fig. specs. 89521, 89522
McNeil, D.H., 1990, Bull. Can. Petrol Geol., vol. 38, no. 1, p. 21, pl. 1, figs. 6a, b, 7a-c.
Richards Formation, Middle to Late Eocene, depths 2865-2870 m and 2875-2885 m, Dome Gulf Tassiut A-25 well, lat. 69°54′10″N, long. 136°20′20″W, Beaufort Sea.

PLANT INCERTAE SEDIS

Altarmilla multistriata Allison and Hilgert
Holotype 81029
Allison, C.W. and Hilgert, J.W., 1986, J. Paleontol., vol. 60, no. 5, p. 1009, figs. 12.1, 12.2.
Tindir Group, Early Cambrian, east of headwaters of upper Tindir Creek a few kilometers east of U.S.-Canada boundary, west-central Yukon.

Aqualisquama centritubula Allison and Hilgert
Holotype 81024
Allison, C.W. and Hilgert, J.W., 1986, J. Paleontol., vol. 60, no. 5, p. 1003, fig. 10.3.
Tindir Group, Early Cambrian, east of headwaters of upper Tindir Creek a few kilometers east of U.S.-Canada boundary, west-central Yukon.

Archeoxybaphon amydrum Allison and Hilgert
Holotype 81002; paratype 81034
Allison, C.W. and Hilgert, J.W., 1986, J. Paleontol., vol. 60, no. 5, p. 987, figs. 5.3, 5.4.
Tindir Group, Early Cambrian, east of headwaters of upper Tindir Creek a few kilometers east of U.S.-Canada boundary, west-central Yukon.

Archeoxybaphon diminutum Allison and Hilgert
Holotype 81003
Allison, C.W. and Hilgert, J.W., 1986, J. Paleontol., vol. 60, no. 5, p. 987, figs. 5.5.
Tindir Group, Early Cambrian, east of headwaters of upper Tindir Creek a few kilometers east of U.S.-Canada boundary, west-central Yukon.

Archeoxybaphon polykeramoides Allison and Hilgert
Holotype 81001; paratype 81033
Allison, C.W. and Hilgert, J.W., 1986, J. Paleontol., vol. 60, no. 5, p. 987, figs. 5.1, 5.2.
Tindir Group and Funnel Creek Limestone, Early Cambrian, east of headwaters of upper Tindir Creek a few kilometers east of U.S.-Canada boundary, west-central Yukon.

Balbariniella praestans Oehler, 1978
Hypotype 81876
Allison, C.W. and Awramik, S.M., 1989, Precambrian Res., vol. 43, no. 4, p. 276, fig. 9.13.
Tindir Group, earliest Cambrian or latest Precambrian, headwaters of Tindir Creek between lat. 65°15′N and lat. 65°17′N, and long. 141°56′W, and long. 141°57′W, Yukon.

Bicorniculum brochum Allison and Hilgert
Holotype 81023
Allison, C.W. and Hilgert, J.W., 1986, J. Paleontol., vol. 60, no. 5, p. 1002, figs. 10.1, 10.2.
Tindir Group, Early Cambrian, east of headwaters of upper Tindir Creek a few kilometers east of U.S.-Canada boundary, west-central Yukon.

cell cluster
Fig. spec. 71115
Duffield, S.L., 1985, J. Palentol, vol. 59, no. 4, p. 1007, fig. 3, 10.
Jupiter Formation, member 3, Early Silurian, cliff north of mouth of Jupiter River, Anticosti Island, Québec.

Characodictyon diskolopium Allison and Hilgert
Holotype 81013
Allison, C.W. and Hilgert, J.W., 1986, J. Paleontol., vol. 60, no. 5, p. 993, figs. 7.1-7.3.
Tindir Group, Early Cambrian, east of headwaters of upper Tindir Creek a few kilometers east of U.S.-Canada boundary, west-central Yukon.

Characodictyon diskolopium var. *circulare* Allison and Hilgert
Holotype 81014
Allison, C.W. and Hilgert, J.W., 1986, J. Paleontol., vol. 60, no. 5, p. 995, figs. 7.4, 7.5.
Tindir Group, Early Cambrian, east of headwaters of upper Tindir Creek a few kilometers east of U.S.-Canada boundary, west-central Yukon.

Characodictyon skolopium Allison and Hilgert
Holotype 81010
Allison, C.W. and Hilgert, J.W., 1986, J. Paleontol., vol. 60, no. 5, p. 993, figs. 6.7, 6.8.
Tindir Group, Early Cambrian, east of headwaters of upper Tindir Creek a few kilometers east of U.S.-Canada boundary, west-central Yukon.

Characodictyon skolopium var. *soleniscum* Allison and Hilgert
Holotype 81012
Allison, C.W. and Hilgert, J.W., 1986, J. Paleontol., vol. 60, no. 5, p. 993, fig. 6.11.
Tindir Group, Early Cambrian, east of headwaters of upper Tindir Creek a few kilometers east of U.S.-Canada boundary, west-central Yukon.

Characodictyon skolopium var. *tetragonum* Allison and Hilgert
Holotype 81011
Allison, C.W. and Hilgert, J.W., 1986, J. Paleontol., vol. 60, no. 5, p. 993, figs. 6.9, 6.10.
Tindir Group, Early Cambrian, east of headwaters of upper Tindir Creek a few kilometers east of U.S.-Canada boundary, west-central Yukon.

Chilodictyon caliporum Allison and Hilgert
Holotype 81015
Allison, C.W. and Hilgert, J.W., 1986, J. Paleontol., vol. 60, no. 5, p. 996, figs. 7.6, 7.7.
Tindir Group, Early Cambrian, east of headwaters of upper Tindir Creek a few kilometers east of U.S.-Canada boundary, west-central Yukon.

Chilodictyon caliporum var. *striatimarginatum* Allison and Hilgert
Holotype 81016
Allison, C.W. and Hilgert, J.W., 1986, J. Paleontol., vol. 60, no. 5, p. 996, figs. 7.8, 7.9.
Tindir Group, Early Cambrian, east of headwaters of upper Tindir Creek a few kilometers east of U.S.-Canada boundary, west-central Yukon.

Chilodictyon myriocanthum Allison and Hilgert
Holotype 81017; paratypes 81036, 81037
Allison, C.W. and Hilgert, J.W., 1986, J. Paleontol., vol. 60, no. 5, p. 996, figs. 8.1- 8.3.
Tindir Group, Early Cambrian, east of headwaters of upper Tindir Creek a few kilometers east of U.S.-Canada boundary, west-central Yukon.

Chuaria-Tawuia assemblage
Fig. specs. 66185, 66186
Hofmann, H.J., 1985 Paleoalgology: Contemporary Research and Applications, eds. D.F. Toomey and M.H. Nitecki, p.26, figs. 4A,B.
Little Dal Group, Precambrian, lat. 64°48′N, long. 128°52.5′W, Mackenzie Mountains, District of Mackenzie.

Confinisquama fimbriata Allison and Hilgert
Holotype 81025; paratype 81043
Allison, C.W. and Hilgert, J.W., 1986, J. Paleontol., vol. 60, no. 5, p. 1003, figs. 10.4-10.6.
Tindir Group, Early Cambrian, east of headwaters of upper Tindir Creek a few kilometers east of U.S.-Canada boundary, west-central Yukon.

Fusilabellum brevistriatum Awramik
Holotype 81882
Allison, C.W. and Awramik, S.M., 1989, Precambrian Res., vol. 43, no. 4, p. 277, fig. 9.2.
Tindir Group, earliest Cambrian or latest Precambrian, headwaters of Tindir Creek between lat. 65°15′N and lat. 65°17′N, long. 141°56′W and long. 141°57′W, Yukon.

Hyalocyrillium clardyi Allison
Holotype 81900; hypotypes 81898, 81899
Allison, C.W. and Awramik, S.M., 1989, Precambrian Res., vol. 43, no. 4, p. 285, figs. 10.9-10.11.
Tindir Group, earliest Cambrian or latest Precambrian, headwaters of Tindir Creek between lat. 65°15′N and lat. 65°17′N, long. 141°56′W and long. 141°57′W, Yukon.

Hyaloxybaphon monokeramoides Allison and Hilgert
Holotype 81004; paratype 81035
Allison, C.W. and Hilgert, J.W., 1986, J. Paleontol., vol. 60, no. 5, p. 989, figs. 5.6, 5.7.
Tindir Group, Early Cambrian, east of headwaters of upper Tindir Creek a few kilometers east of U.S.-Canada boundary, west-central Yukon.

Invaginatibalteus depressus Allison and Hilgert
Holotype 81027
Allison, C.W. and Hilgert, J.W., 1986, J. Paleontol., vol. 60, no. 5, p. 1007, figs. 11.4, 11.5.
Tindir Group, Early Cambrian, east of headwaters of upper Tindir Creek a few kilometers east of U.S.-Canada boundary, west-central Yukon.

Invaginatibalteus dimorphus Allison and Hilgert
Holotype 81026
Allison, C.W. and Hilgert, J.W., 1986, J. Paleontol., vol. 60, no. 5, p. 1005, figs. 11.1-11.3.
Tindir Group, Early Cambrian, east of headwaters of upper Tindir Creek a few kilometers east of U.S.-Canada boundary, west-central Yukon.

Obruchevella minuta Allison
Holotype 81889
Allison, C.W. and Awramik, S.M., 1989, Precambrian Res., vol. 43, no. 4, p. 275, fig. 9.10.
Tindir Group, earliest Cambrian or latest Precambrian, headwaters of Tindir Creek between lat. 65°15′N and lat. 65°17′N, long. 141°56′W and long. 141°57′W, Yukon.

?*Obruchevella* sp.
Fig. spec. 81890
Allison, C.W. and Awramik, S.M., 1989, Precambrian Res., vol. 43, no. 4, p. 275, fig. 9.11.
Tindir Group, earliest Cambrian or latest Precambrian, headwaters of Tindir Creek between lat. 65°15′N and lat. 65°17′N, long. 141°56′W and long. 141°57′W, Yukon.

Organic spiral tube
Fig. spec. 71116
Duffield, S.L., 1985, J. Palentol, vol. 59, no. 4, p. 1007, fig. 3.11.
Jupiter Formation, Member 3, Early Silurian, cliff north of mouth of Jupiter River, Anticosti Island, Québec.

Paleocrassilimbus spinosus Allison and Hilgert
Holotype 81031
Allison, C.W. and Hilgert, J.W., 1986, J. Paleontol., vol. 60, no. 5, p. 1010, figs. 12.6, 12.7.
Tindir Group, Early Cambrian, east of headwaters of upper Tindir Creek a few kilometers east of U.S.-Canada boundary, west-central Yukon.

Paleohexadictyon coroniforme Allison and Hilgert
Holotype 81007
Allison, C.W. and Hilgert, J.W., 1986, J. Paleontol., vol. 60, no. 5, p. 992, fig. 6.5.
Tindir Group, Early Cambrian, east of headwaters of upper Tindir Creek a few kilometers east of U.S.-Canada boundary, west-central Yukon.

Paleohexadictyon coroniforme var. *delicatum* Allison and Hilgert
Holotype 81009
Allison, C.W. and Hilgert, J.W., 1986, J. Paleontol., vol. 60, no. 5, p. 992, fig. 6.6.
Tindir Group, Early Cambrian, east of headwaters of upper Tindir Creek a few kilometers east of U.S.-Canada boundary, west-central Yukon.

Paleohexadictyon coroniforme var. *tetragonum* Allison and Hilgert
Holotype 81008

Allison, C.W. and Hilgert, J.W., 1986, J. Paleontol., vol. 60, no. 5, p. 992, fig. 6.4.
Tindir Group, Early Cambrian, east of headwaters of upper Tindir Creek a few kilometers east of U.S.-Canada boundary, west-central Yukon.

Paleohexadictyon litosum Allison and Hilgert
Holotype 81005
Allison, C.W. and Hilgert, J.W., 1986, J. Paleontol., vol. 60, no. 5, p. 991, figs. 6.1, 6.2.
Tindir Group, Early Cambrian, east of headwaters of upper Tindir Creek a few kilometers east of U.S.-Canada boundary, west-central Yukon.

Paleohexadictyon myriotrematum Allison and Hilgert
Holotype 81006
Allison, C.W. and Hilgert, J.W., 1986, J. Paleontol., vol. 60, no. 5, p. 991, fig. 6.3.
Tindir Group, Early Cambrian, east of headwaters of upper Tindir Creek a few kilometers east of U.S.-Canada boundary, west-central Yukon.

Paleomegasquama coccolithoides Allison and Hilgert
Holotype 81032; paratypes 81045, 81046
Allison, C.W. and Hilgert, J.W., 1986, J. Paleontol., vol. 60, no. 5, p. 1011, figs. 13.1-13.7.
Funnel Creek Limestone, Early Cambrian, Tindir Creek area a few kilometers east of U.S.-Canada boundary, west-central Yukon.

Paterisquama crassa Allison and Hilgert
Holotype 81028
Allison, C.W. and Hilgert, J.W., 1986, J. Paleontol., vol. 60, no. 5, p. 1007, figs. 11.6, 11.7.
Tindir Group, Early Cambrian, east of headwaters of upper Tindir Creek a few kilometers east of U.S.-Canada boundary, west-central Yukon.

Patiniquama cirratomarginata Allison and Hilgert
Holotype 81030; paratype 81044
Allison, C.W. and Hilgert, J.W., 1986, J. Paleontol., vol. 60, no. 5, p. 1010, figs. 12.3-12.5.
Tindir Group, Early Cambrian, east of headwaters of upper Tindir Creek a few kilometers east of U.S.-Canada boundary, west-central Yukon.

Petasisquama alta Allison and Hilgert
Holotype 81020
Allison, C.W. and Hilgert, J.W., 1986, J. Paleontol., vol. 60, no. 5, p. 1001, figs. 9.1, 9.2.
Tindir Group, Early Cambrian, east of headwaters of upper Tindir Creek a few kilometers east of U.S.-Canada boundary, west-central Yukon.

Petasisquama laciniata Allison and Hilgert
Holotype 81022
Allison, C.W. and Hilgert, J.W., 1986, J. Paleontol., vol. 60, no. 5, p. 1002, fig. 9.6.
Tindir Group, Early Cambrian, east of headwaters of upper Tindir Creek a few kilometers east of U.S.-Canada boundary, west-central Yukon.

Petasisquama versicorona Allison and Hilgert
Holotype 81021; paratype 81041, 81042
Allison, C.W. and Hilgert, J.W., 1986, J. Paleontol., vol. 60, no. 5, p. 1002, figs. 9.3-9.5.
Tindir Group, Early Cambrian, east of headwaters of upper Tindir Creek a few kilometers east of U.S.-Canada boundary, west-central Yukon.

Radiocerniculum inornatum Allison and Hilgert
Holotype 81018; paratypes 81038, 81039
Allison, C.W. and Hilgert, J.W., 1986, J. Paleontol., vol. 60, no. 5, p. 997, figs. 8.4-8.6.
Tindir Group, Early Cambrian, east of headwaters of upper Tindir Creek a few kilometers east of U.S.-Canada boundary, west-central Yukon.

Sheets of cells
Fig. specs. 55527, 55528
McGregor, D.C. and Narbonne, G.M., 1978, Can. J. Earth Sci., vol. 15, no. 8, p. 1300, pl. 2, figs. 2, 3.
Member B 20 m above base, Read Bay Formation, Upper Silurian, south side Goodsir Creek, eastern Cornwallis Island, District of Franklin.

Sphaeranasillos irregularis Allison
Holotype 81883
Allison, C.W. and Awramik, S.M., 1989, Precambrian Res., vol. 43, no. 4, p. 279, figs. 9.3, 9.4.
Tindir Group, earliest Cambrian or latest Precambrian, headwaters of Tindir Creek between lat. 65°15′N and lat. 65°17′N, long. 141°56′W and long. 141°57′W, Yukon.

Spinicerniculum tribulosum Allison and Hilgert
Holotype 81019; paratype 81040
Allison, C.W. and Hilgert, J.W., 1986, J. Paleontol., vol. 60, no. 5, p. 999, figs. 8.7, 8.8.
Tindir Group, Early Cambrian, east of headwaters of upper Tindir Creek a few kilometers east of U.S.-Canada boundary, west-central Yukon.

Spirally banded tube-like structure
Fig. spec. 67796
McGregor, D.C., 1984, Academia Nacional de Ciencias, Misc. No. 69, pl. 4, fig. 12.
Santa Rosa Formation, Lower Devonian, Laurani section, Chuquisaca Dept., northwest of Tarabuco, Cordillera Oriental, Bolivia.

Spirally thickened tubes
Fig. specs. 55529, 55530
McGregor, D.C. and Narbonne, G.M., 1978, Can. J. Earth Sci., vol. 15, no. 8, p. 1298, pl. 2, figs. 4-6.
Member B 20 m above base, Read Bay Formation, Upper Silurian, south side Goodsir Creek, eastern Cornwallis Island, District of Franklin.

Tetraporina sp.
Fig. spec. 94482
Utting, J., 1991, Geol. Surv. Can., Bull. 412, pl. 4.1, fig. 25.
Hart River Formation, Lower Carboniferous, north bank Peel River, northern Yukon.

Thecamoebian(?) A
Fig. specs. 64771, 64772
McLean, J.R. and Wall, J.H., 1982, Bull. Can. Petrol. Geol., vol. 29, no. 3, 1981, pl. 6, figs. 18, 19.
Gladstone Formation, 25 m below top, Lower Cretaceous, Malcom Creek, sec. 6, tp. 57, rge. 8, W.6th mer., Alberta.

Thecamoebian (?) B
Fig. specs. 64773, 64774
McLean, J.R. and Wall, J.H., 1982, Bull. Can. Petrol. Geol., vol. 29, no. 3, 1981, pl. 6, figs. 20, 21.
Moosebar Member, 20 to 22 m above base, Malcom Creek Formation, Lower Cretaceous, Ruby Creek, sec. 11, tp. 45, rge. 21, W.5th mer., Alberta.

Tindiromacula hofmanni Allison
Holotype 81897
Allison, C.W. and Awramik, S.M., 1989, Precambrian Res., vol. 43, no. 4, p. 281, figs. 10.7, 10.8.
Tindir Group, earliest Cambrian or latest Precambrian, headwaters of Tindir Creek between lat. 65°15′N and lat. 65°17′N, long. 141°56′W and long. 141°57′W, Yukon.

Tube-like structure
Fig. spec. 55531
McGregor, D.C. and Narbonne, G.M., 1978, Can. J. Earth Sci., vol. 15, no. 8, p. 1298, pl. 2, fig. 7.
Member B 20 m above base, Read Bay Formation, Upper Silurian, south side Goodsir Creek, eastern Cornwallis Island, District of Franklin.

Tubomorphophyton nubillum (Korde) Korde (1973)
Hypotype 90277
Voronova, L.G. et al., 1987, Acad. Nauk. SSSR, Trans. Palaeontol. Instit., vol. 224, p. 60, pl. 26, fig. 4; pl. 27, fig. 2.
Sekwi Formation, Lower Cambrian, lat. 63°26 1/3-2/3′N, long. 129°26′-27′W, Mackenzie Mountains, District of Mackenzie.

Unnamed Bulbous Form
Fig. spec. 81895
Allison, C.W. and Awramik, S.M., 1989, Precambrian Res., vol. 43, no. 4, p. 284, fig. 10.5.
Tindir Group, earliest Cambrian or latest Precambrian, headwaters of Tindir Creek between lat. 65°15′N and lat. 65°17′N, long. 141°56′W and long. 141°57′W, Yukon.

Unnamed Coccoid Form A, B, C, D
Fig. specs. 81885-81888
Allison, C.W. and Awramik, S.M., 1989, Precambrian Res., vol. 43, no. 4, p. 280, figs. 9.6-9.9.
Tindir Group, earliest Cambrian or latest Precambrian, headwaters of Tindir Creek between lat. 65°15′N and lat. 65°17′N, long. 141°56′W and long. 141°57′W, Yukon.

Unnamed Filament Form A, B, C
Fig. specs. 81891-81894
Allison, C.W. and Awramik, S.M., 1989, Precambrian Res., vol. 43, no. 4, p. 281, 283, 284, figs. 9.12, 10.1-10.4.
Tindir Group, earliest Cambrian or latest Precambrian, headwaters of Tindir Creek between lat. 65°15′N and lat. 65°17′N, long. 141°56′W and long. 141°57′W, Yukon.

Unnamed Form D Knoll, 1984
Fig. spec. 81884
Allison, C.W. and Awramik, S.M., 1989, Precambrian Res., vol. 43, no. 4, p. 279, fig. 9.5.
Tindir Group, earliest Cambrian or latest Precambrian, headwaters of Tindir Creek between lat. 65°15′N and lat. 65°17′N, long. 141°56′W and long. 141°57′W, Yukon.

Unnamed Spiral Grooved Form
Fig. spec. 81896
Allison, C.W. and Awramik, S.M., 1989, Precambrian Res., vol. 43, no. 4, p. 284, fig. 10.6.
Tindir Group, earliest Cambrian or latest Precambrian, headwaters of Tindir Creek between lat. 65°15′N and lat. 65°17′N, long. 141°56′W and long. 141°57′W, Yukon.

Unnamed Tringular Form
Fig. spec. 81905
Allison, C.W. and Awramik, S.M., 1989, Precambrian Res., vol. 43, no. 4, p. 288, fig. 11.11.
Tindir Group, earliest Cambrian or latest Precambrian, headwaters of Tindir Creek between lat. 65°15′N and lat. 65°17′N, long. 141°56′W and long. 141°57′W, Yukon.